THE UNIVERSITY OF ARIZONA SPACE SCIENCE SERIES

RICHARD P. BINZEL, GENERAL EDITOR

Planetary Astrobiology
V. S. Meadows, G. N. Arney, B. E. Schmidt, and
D. J. Des Marais, editors, 2020, 534 pages

Enceladus and the Icy Moons of Saturn
P. M. Schenk, R. N. Clark, C. J. A. Howett, A. J. Verbiscer, and
J. H. Waite, editors, 2018, 475 pages

Asteroids IV
P. Michel, F. E. DeMeo, and W. F. Bottke, editors, 2015, 895 pages

Protostars and Planets VI
Henrik Beuther, Ralf S. Klessen, Cornelis P. Dullemond, and
Thomas Henning, editors, 2014, 914 pages

Comparative Climatology of Terrestrial Planets
Stephen J. Mackwell, Amy A. Simon-Miller, Jerald W. Harder,
and Mark A. Bullock, editors, 2013, 610 pages

Exoplanets
S. Seager, editor, 2010, 526 pages

Europa
Robert T. Pappalardo, William B. McKinnon,
and Krishan K. Khurana, editors, 2009, 727 pages

The Solar System Beyond Neptune
M. Antonietta Barucci, Hermann Boehnhardt, Dale P. Cruikshank,
and Alessandro Morbidelli, editors, 2008, 592 pages

Protostars and Planets V
Bo Reipurth, David Jewitt, and Klaus Keil, editors, 2007, 951 pages

Meteorites and the Early Solar System II
D. S. Lauretta and H. Y. McSween, editors, 2006, 943 pages

Comets II
M. C. Festou, H. U. Keller,
and H. A. Weaver, editors, 2004, 745 pages

Asteroids III
William F. Bottke Jr., Alberto Cellino, Paolo Paolicchi,
and Richard P. Binzel, editors, 2002, 785 pages

TOM GEHRELS, GENERAL EDITOR

Origin of the Earth and Moon
R. M. Canup and K. Righter, editors, 2000, 555 pages

Protostars and Planets IV
Vincent Mannings, Alan P. Boss,
and Sara S. Russell, editors, 2000, 1422 pages

Pluto and Charon
S. Alan Stern and David J. Tholen, editors, 1997, 728 pages

**Venus II—Geology, Geophysics, Atmosphere,
and Solar Wind Environment**
S. W. Bougher, D. M. Hunten,
and R. J. Phillips, editors, 1997, 1376 pages

Cosmic Winds and the Heliosphere
J. R. Jokipii, C. P. Sonett,
and M. S. Giampapa, editors, 1997, 1013 pages

Neptune and Triton
Dale P. Cruikshank, editor, 1995, 1249 pages

Hazards Due to Comets and Asteroids
Tom Gehrels, editor, 1994, 1300 pages

Resources of Near-Earth Space
John S. Lewis, Mildred S. Matthews,
and Mary L. Guerrieri, editors, 1993, 977 pages

Protostars and Planets III
Eugene H. Levy and Jonathan I. Lunine, editors, 1993, 1596 pages

Mars
Hugh H. Kieffer, Bruce M. Jakosky, Conway W. Snyder,
and Mildred S. Matthews, editors, 1992, 1498 pages

Solar Interior and Atmosphere
A. N. Cox, W. C. Livingston,
and M. S. Matthews, editors, 1991, 1416 pages

The Sun in Time
C. P. Sonett, M. S. Giampapa,
and M. S. Matthews, editors, 1991, 990 pages

Uranus
Jay T. Bergstralh, Ellis D. Miner,
and Mildred S. Matthews, editors, 1991, 1076 pages

Asteroids II
Richard P. Binzel, Tom Gehrels,
and Mildred S. Matthews, editors, 1989, 1258 pages

Origin and Evolution of Planetary and Satellite Atmospheres
S. K. Atreya, J. B. Pollack,
and Mildred S. Matthews, editors, 1989, 1269 pages

Planetary Astrobiology

Planetary Astrobiology

Edited by

V. S. Meadows, G. N. Arney, B. E. Schmidt, and D. J. Des Marais

With the assistance of

Renée Dotson

With 50 collaborating authors

THE UNIVERSITY OF ARIZONA PRESS
Tucson

in collaboration with

LUNAR AND PLANETARY INSTITUTE
Houston

About the front cover:

This 2004 painting depicts an hypothesized extrasolar super-Jupiter system in which a Ganymede-like moon could be a habitable Earth-like world with oceans of liquid water. Warming could come not only from the the system's star, but possibly from the host planet's reflected and thermal infrared radiation, and tidal interaction with the moon. Painting copyright William K. Hartmann.

About the back cover:

This HiRISE image shows aeolian drift deposits and bedrock in Aram Chaos near the outlet to Ares Valles on Mars. Aram Chaos is a 1300-km-diameter depression where large volumes of groundwater emerged long ago and flowed northward across the southern highlands, helping to carve the ~2000-km Ares Valles channel system. Credit: NASA/JPL/University of Arizona.

The introductory statement by Dr. Lori Glaze is a work of the U.S. Government.

The Lunar and Planetary Institute is operated by the Universities Space Research Association under Award No. NNX08AC28A with the Science Mission Directorate of the National Aeronautics and Space Administration. Any opinions, findings, and conclusions or recommendations expressed in this volume are those of the author(s) and do not necessarily reflect the views of the National Aeronautics and Space Administration.

The University of Arizona Press
in collaboration with the Lunar and Planetary Institute
© 2020 The Arizona Board of Regents
All rights reserved
∞ This book is printed on acid-free, archival-quality paper.
Manufactured in the United States of America

25 24 23 22 21 20 6 5 4 3 2 1

Library of Congress Cataloging-in-Publication Data
Names: Meadows, Victoria S., editor. | Arney, Giada Nicole, editor. |
   Schmidt, Britney E., editor. | Marais, David J., editor.
Title: Planetary astrobiology / edited by Victoria S. Meadows, Giada N.
   Arney, Britney E. Schmidt, David J. Marais ; with the assistance of
   Renee Dotson.
Other titles: University of Arizona space science series.
Description: Tucson : The University of Arizona Press in collaboration with
   Lunar and Planetary Institute, 2019. | Series: The University of Arizona
   space science series | Includes bibliographical references and index. |
   Summary: "Planetary Astrobiology provides an accessible,
   interdisciplinary gateway to the frontiers of knowledge in astrobiology
   via results from the exploration of our own solar system and
   exoplanetary systems"-- Provided by publisher.
Identifiers: LCCN 2019018468 | ISBN 9780816540068 (cloth)
Subjects: LCSH: Exobiology. | Planetary science.
Classification: LCC QH326 .P48 2019 | DDC 576.8/39--dc23
LC record available at https://lccn.loc.gov/2019018468

Contents

PART 1: EARTH — LESSONS LEARNED FROM AN INHABITED TERRESTRIAL PLANET

PART 2: ASTROBIOLOGY IN OUR SOLAR SYSTEM

List of Contributing Authors

Amador E. S.   153

Anderson R. E.   71

Arney G. N.   355, 505

Bains W.   37

Baross J. A.   71

Brain D.   419

Cable M. L.   217, 247

Carlson R. W.   3

Cronin L.   477

Davila A.   37, 169

Del Genio A. D.   419

Des Marais D. J.   121, 169, 505

Domagal-Goldman S.   477

Drew A.   477

Ehlmann B. L.   153

Eyster A.   93

Fisher T.   477

Fournier G. P.   93

Graham H. V.   121

Hazen R. M.   121

Hoehler T. M.   37, 217

Hörst S. M.   247

Hsu H.-W.   217

Izidoro A.   287

Johnson S. S.   121

Kahre M. A.   169

Kane S.   355

Kopparapu R. K.   449

Line M.   477

Lunine J. I.   247

Meadows V. S.   449, 505

Meech K.   325

Millsaps C.   477

Morbidelli A.   287

Neveu M.   217

Noack L.   419

Parenteau M. N.   37

Pohorille A.   37

Quinn R.   169

Rahm M.   247

Raymond S. N.   287, 325

Reinhard C. T.   379

Robinson T. D.   379

Rummel J. D.   267

Schaefer L.   419

Schmidt B. E.   185, 505

Stüeken E. E.   71, 93

Walker S. I.   477

Wolf E. T.   449

Zahnle K. J.   3

Scientific Organizing Committee

Giada Arney (co-editor)

Vikki Meadows (co-editor)

David Des Marais (co-editor)

Sean Raymond

Shawn Domagal-Goldman

Britney Schmidt (co-editor)

Aaron Goldman

Mary Voytek

Timothy Lyons

Acknowledgment of Reviewers

The editors gratefully acknowledge the following individuals, as well as several anonymous reviewers, for their time and effort in reviewing chapters in this volume:

Dorian S. Abbot

Conel M. O'D. Alexander

Laura M. Barge

John A. Baross

Zachory K. Berta-Thompson

David C. Catling

John Chambers

Catharine Conley

Nicolas B. Cowan

David Crisp

Steve Desch

Alberto G. Fairén

Hidenori Genda

Christopher Glein

Tori M. Hoehler

Carly J. A. Howett

Bo Barker Jørgensen

Helmut Lammer

Ralph Lorenz

Timothy W. Lyons

Noah Planavsky

Olga Prieto-Ballesteros

Edward Schwieterman

Christophe Sotin

Roger E. Summons

Kevin Walsh

Robin Wordsworth

Introductory Statement

This is the golden age of astrobiology as we advance the study of the origin, evolution and distribution of life in the universe. In the last several decades, scientists have plumbed the depths of understanding of how life came to be and how it has evolved on our own planet. The range of different conditions that were present during the early stages of Earth's history would appear as extreme alien worlds compared to the Earth we know today, and yet somehow life managed to take hold and persist in those environments. Scientists have found organisms in every possible water-bearing environment of modern Earth, demonstrating the adaptability and pervasiveness of life. In our solar system, environments analogous to some of the most extreme environments on Earth, with the possibility of hosting life as we know it, have been found. The recent discovery of likely terrestrial planets orbiting within their star's habitable zone has provided enticing targets that may even allow us to extend the search for life beyond the solar system.

Astrobiology is fundamentally a systems science. It is not possible to recognize habitability nor life's impact on a world without understanding the environmental context of that world. Consequently, threads from multiple disciplines must be interwoven to determine the characteristics and processes that influence those environments. We have learned that life itself is a planetary process, embedded in the geochemical cycles that impact the interior, surface, and atmospheric systems. Planetary scientists have been a part of astrobiology since its inception and as space agencies identify the search for life as a top priority, planetary scientists will continue to play a crucial role in planning and implementing missions to investigate astrobiological destinations within our solar system and beyond. As we look to the future, it is time to synthesize what we have learned so far about astrobiology and work together to better understand how and where to look for life!

Dr. Lori Glaze
Director, NASA Planetary Science Division
Washington, DC
February 2020

Preface

Are we alone in the universe? This profound question has echoed down through the millennia, yet only now is the answer within our scientific and technical grasp. The new science of astrobiology enables and drives our search for life beyond Earth, which will place our glittering jewel of a home world in its broader cosmic context. As an emerging interdisciplinary field, astrobiology seeks to synthesize humanity's knowledge in multiple disciplines to inform both the current and future robotic exploration of the solar system, as well as telescopic observations of worlds orbiting other stars. Exciting initiatives at the frontiers of space exploration are planned to return samples from Mars, explore the icy moons of Jupiter and Saturn, and build next-generation groundbased and spacebased telescopes to investigate the atmospheres and surfaces of exoplanets for signs of habitability and life. Astrobiology inspires and enables these endeavors by providing the scientific foundation to identify where and how we should search for life elsewhere.

Astrobiologists recognize that life itself is a planetary process, and Earth's biosphere is the only one currently known. Thus astrobiology necessarily builds outward from Earth. Astrobiology broadens our understanding of the diversity of environments that could harbor life and helps us identify its processes and signatures, including possible attributes of life as we don't know it. Astrobiologists seek to understand the environment and mechanisms that led to the origin(s) of life and to examine its fundamental physics and chemistry, which informs our view of the requirements and limits of life. By identifying life's fundamental requirements we can recognize environments that potentially satisfy those requirements and therefore are more likely to support life. Characterizing the factors and processes whereby a planet acquires, maintains — and could lose — habitable environments will enhance our ability to identify the most promising targets for further study. The co-evolution of life with its environment is now known to have impacted the nature and evolution of Earth's atmosphere, surface, and interior. Life's modification of its planetary environment may also provide signs of life — biosignatures — that could be sought in martian soils, alien oceans of icy worlds, or atmospheres and surfaces of exoplanets. The efforts of many scientists in multiple disciplines have culminated in one of astrobiology's clearest lessons: Life cannot be divorced from the planetary environments in which it originated and evolved, and therefore the search for life in the universe must be informed by its planetary, stellar, and planetary system contexts.

Two of the key characteristics of astrobiology are its fundamentally interdisciplinary nature and how it focuses investigations of a very wide range of processes and environments on their implications for life. Astrobiology provides a lens through which to view planetary science with a fresh and complementary perspective. For example, exploring the martian surface reveals the fascinating geology and history of the planet, but these studies can also be used to understand Mars' current and past capacity to support habitable environments and life. Astrobiology is also a critical means to connect solar system and exoplanetary science; for example, understanding the processes that govern solar wind interaction and atmospheric loss from Mars can inform our understanding of atmospheric retention and the likelihood of habitability for terrestrial exoplanets. Venus' environment and its evolutionary path illuminate the processes by which a terrestrial planet can lose its ocean and severely modify its atmosphere. The icy moons of the giant planets are alternative ocean worlds that complement our Earth, and the processes that shape them may have analogs in the population of volatile-rich exoplanets found orbiting in the habitable zones of red-dwarf stars.

Astrobiology provides a framework to address scientific questions that are simply too large to be addressed by a single traditional discipline. The field of astrobiology pioneers a new way of doing science by engaging complementary expertise to cross disciplinary boundaries. For example, understanding how Earth, an inhabited world, formed and evolved requires an understanding of processes within the purview of planetary science, such as planet formation and volatile delivery, interior processes and outgassing, and solar-planetary atmosphere interactions, as well as the rise, spread, and co-evolution of life with our changing environment.

As such, this example draws upon and synthesizes information and scientific techniques from the fields of planet formation, small-body science, heliophysics, geophysics, geochemistry, oceanography, atmospheric science, chemistry, and biology.

Roadmap to *Planetary Astrobiology*

Planetary Astrobiology provides a gateway for entering this growing field by illuminating the many ways that the interdisciplinary sciences of astrobiology and planetary science complement each other, while highlighting the synergies between solar system and exoplanet science that inform our search for life in the universe. The collective expertise of the book's Scientific Organizing Committee spans planet formation to evolutionary biology, and many of the authors first met to discuss the book at the Nexus for Exoplanet System Science workshop on Habitable Worlds in Laramie, Wyoming, in 2017. Our authors were selected first and foremost for their scientific expertise, but also for their ability and interest in spanning at least two of the fields of astrobiology — planetary science and exoplanets — and for their skill in writing clearly for an interdisciplinary audience. The book is divided into four major sections: (1) our understanding of life in the context of its environment, using Earth and its history as an exemplar; (2) the astrobiological significance of a suite of solar system bodies; (3) the synergies between, and insights provided by, studies of the solar system and exoplanets; and (4) a synthesis of an even broader overview of key concepts in planetary astrobiology that incorporates strong aspects of comparative planetology, and life in the context of its environment. We conclude *Planetary Astrobiology* with a short summary of the key themes presented in the book and the identification of frontiers for future research. The majority of the chapters are strongly interdisciplinary and the authors take an immersive approach to presenting the different fields that comprise astrobiology by introducing concepts in context and defining jargon as they go.

The first section, "Earth: Lessons Learned from an Inhabited Terrestrial Planet," uses the formation, evolution, and current characteristics of the only known inhabited planet to introduce the reader to relevant concepts in biology and biosignature science, to emphasize the importance of planetary environmental context, and to illustrate life as a planetary process over time. The introductory chapter by Zahnle and Carlson provides a sweeping review of the formation and early evolution of Earth, interweaving planet formation, meteoritics, planetary interiors, impactors, and atmospheric science to paint a vivid picture of Earth's origin and the ancient global environments in which life may have gotten its first toehold. The chapter by Hoehler et al. delves into biochemistry and geochemistry to understand how life's requirements are met on Earth, and how these requirements may be reduced to their fundamentals to provide a means of assessing whether alternative biochemistries are also possible. In Baross et al., the early Earth is explored in more detail to synthesize and distill the critical environmental, geochemical, and biochemical factors that may have led to the origin of life. Expanding on the theme of life as a planetary process, the chapter by Stüeken et al. describes how life on Earth has coevolved with its environment over eons, and is strongly integrated into Earth's geochemical cycles. In the final chapter in this section, Johnson et al. describe how searching for life in the rock record of ancient Earth has helped us identify processes and biosignatures that can inform our search for life elsewhere in the solar system.

In the section "Astrobiology In Our Solar System," we review the astrobiological significance of several solar system worlds to better understand their potential for habitability and their significance in the search for life. We start with Mars, and assess its habitability both in the past and today. Amador and Ehlmann explore what can be learned of Mars' past habitability from its rock record and atmosphere, and discuss the likelihood of preservation of potential biosignatures on Mars. Davila et al. review the environmental aspects of present-day Mars that are most relevant to its habitability and biological potential, emphasizing the near-surface environment that will be sampled and investigated by future robotic missions. Moving outward, Schmidt reviews the prospects for habitability and life for Europa and the jovian system, with an emphasis on the processes that sculpt these alien ocean worlds. Cable et al. survey Saturn's moon Enceladus, which shows evidence for a global subsurface ocean that sustains a plume of material ejected from fissures in the icy surface. Lunine et al. introduce us to the many environments that could harbor exotic life on Saturn's moon Titan, including its atmosphere, surface, and subsurface ocean, and describe how planetary processes, including impacts, could drive pre-biotic chemistry on this intriguing world. Rounding out this section, Rummel reviews the importance to planetary exploration of guarding against both forward contamination of other planetary environments and the possibility of bringing alien microbes back to Earth.

In the third section, "The Solar System — Exoplanet Synergy," we compare and explore key links between solar system and exoplanetary science, with a focus on the terrestrial worlds that may harbor or lose surface oceans. This focus is a byproduct of the restriction that exoplanets can only be studied using astronomical remote-sensing observations, and so their exploration is limited to their atmospheres and surfaces, including surface oceans. Raymond et al. build on the overview of Earth's formation provided in Zahnle and Carlson to provide a comprehensive review of why the solar system is so unusual when compared to known planetary systems. They review global models of our solar system's formation and migration, including the role of giant planets, in the context of what is known about these processes from studying extrasolar planetary systems. After touching upon water content in terrestrial planet formation in Raymond et al., Meech and Raymond survey in detail the many mechanisms that may deliver water to nascent habitable planets, with a focus on the origin of Earth's water. And while acquiring water early in a terrestrial planet's evolution is one key aspect of habitability, retaining that water over geologic timescales is also required. Arney and Kane look at Venus as a nearby exemplar for highly irradiated, close-in terrestrial exoplanets. Venus' history of runaway greenhouse and ocean and atmospheric loss processes possibly exemplifies a common outcome of terrestrial planet evolution, and Venus can provide key lessons on the loss of habitability and abiotic generation of biosignatures. This section concludes by examining Earth as if it were an exoplanet, and Robinson and Reinhard use this comparison to understand a distant Earth's potentially observable signs of habitability and life today, as well as over 4 G.y. of evolution. By reaching back into Earth's past we can explore many examples of inhabited, but alien, environments that are constrained by what is known of the geology and biology at those times.

In the final section, "Synthesis," we take an even broader view to understand the frameworks that help us synthesize bigger-picture questions in planetary astrobiology. Del Genio et al. survey and compare what is known of the divergent evolution and key processes responsible for the maintenance or loss of habitability for Venus, Earth, and Mars, and show how these solar system terrestrials provide an invaluable context for thinking about the habitability of exoplanets. Broadening this comparative planetology theme, Kopparapu et al. provide a comprehensive survey of how we might characterize habitable exoplanets, first by understanding the characteristics of and interactions between the planet, its star, and its planetary system that support liquid water on its surface, and then by identifying which of these characteristics or interactions may be modeled or observed. Walker et al. then address the frontiers of biosignature science by describing the ongoing development of a comprehensive framework for biosignature assessment. This framework acknowledges that separating the impact of life on an environment from abiotic processes acting in that environment requires the probabilistic combination of multiple types of information. This framework also allows for searches for agnostic biosignatures, namely attributes of environmental complexity that may indicate the actions of life as we do not know it. Finally, Meadows et al. provide a brief summary of key themes and future avenues for exploration at the forefront of planetary astrobiology.

This book is the result of selfless work by many members of the community. The editors are deeply grateful and forever indebted to the many authors and referees who contributed their considerable efforts and expertise, and gave an unwavering commitment to making these chapters accessible to the broadest readership possible. The editors would also like to thank General Editor of the Space Science Series Richard Binzel for his excellent advice on numerous occasions, and for helping to keep this complicated book on track through the editing and review process. We would especially like to thank Renée Dotson at the Lunar and Planetary Institute for her tireless and cheerful efforts in the compilation and production of the book, which, like all good scientific efforts, produced an exponentially larger workload the closer we got to the deadline. Other invaluable LPI editorial and graphic assistance was provided by Heidi Lavelle, Kevin Portillo, Linda Chappell, and Jamie Shumbera. Special thanks are also given to William Hartmann for permission to use his stunning painting on the book's cover. Finally, we wish to thank our colleagues from our many communities: Earth sciences, heliophysics, solar system, and exoplanets, in fields spanning biology, geology, chemistry, and astronomy, who are so open to learning from, working with, and training others. You make this grand interdisciplinary experiment that is astrobiology such a fun and exhilarating experience.

— *V. S. Meadows, G. N. Arney, B. E. Schmidt, D. J. Des Marais*
November 2019

Part 1:

Earth —
Lessons Learned from an
Inhabited Terrestrial Planet

Zahnle K. and Carlson R. (2020) Creation of a habitable planet. In *Planetary Astrobiology* (V. Meadows et al., eds.), pp. 3–36. Univ. of Arizona, Tucson, DOI: 10.2458/azu_uapress_9780816540068-ch001.

Creation of a Habitable Planet

Kevin J. Zahnle
NASA Ames Research Center

Richard W. Carlson
Carnegie Institution for Science

Earth is the type example of a habitable planet. This chapter begins in the cloud of gas and dust that gave birth to the solar system, and ends on Earth before the dawn of life. Solid grains condensed or evaporated in the solar nebula according to temperature. How grains were gathered into planetesimals is not yet known; the leading hypothesis is that gravito-aerodynamic mechanisms concentrated grains into denser clumps that collapsed directly into full-grown planetesimals. Radioactive ^{26}Al melted the first generation of planetesimals. Isotopic trends in several elements show that solid materials in the solar nebula were early separated into an inner solar system reservoir, denoted "NC," and an outer solar system reservoir, denoted "CC," divided by the creation of the giant planets. Earth and Moon are genetically related to enstatite chondrites (ECs), and not to NC materials in general, nor to the outer solar system CC materials. Earth and the ECs appear to sample complementary reservoirs descended from a common ancestor. A collision between two planets formed the Moon and Earth some 50–150 m.y. after the Sun formed. An excess of lightly fractionated highly siderophile elements in Earth's mantle implies that Earth accreted about 0.5% of its mass as a "late veneer" after the Moon-forming impact. Isotopes group the late veneer with dry EC-like inner solar system material, hence Earth's water apparently predates the late veneer. The origin of life on Earth implies a reduced early atmosphere, rich in CH_4 and H_2, but geological evidence suggests that Earth's mantle has always been relatively oxidized and volcanic gases dominated by CO_2 and H_2O. Transient CH_4-rich, H_2-rich atmospheres created by iron-rich EC-like impacts big enough to vaporize the oceans can resolve the paradox. The near absence of solar neon on Earth suggests that Earth's building blocks either did not attract significant nebular atmospheres, or that any nebular atmospheres were lost after the nebula dispersed. The heavier noble gases provide other evidence constraining histories of atmospheric escape. A complete picture of Earth requires a better understanding of Venus, because Venus contains nearly half the mass of the inner solar system, and therefore holds secret the other half of the story.

For hot, cold, moist, and dry, four Champions fierce
Strive here for Maistrie, and to Battel bring
Thir embryon Atoms; Chaos Umpire sits,
And by decision more imbroiles the fray
By which he Reigns: next him high Arbiter
Chance governs all . . .
— Milton

1. INTRODUCTION — THE MEANING OF EARTH

Astrobiology is intended to be the science of life in the cosmos. When fully developed it will encompass the origins of life and all of life's manifestations in the full variety of astrophysical environments in which it will be found. It is a big picture, but with one exception it is all future conditional. This is why one begins a book on planetary astrobiology with Earth: Earth is our only known role model for how to succeed in the cosmos. This chapter explores the processes involved in planet formation and which of them may play important roles in establishing the characteristics of Earth that led it to develop a habitable surface.

1.1. Rare Earth

One question that gets asked is, how rare is Earth? The question gets asked in many forms, such as "Are we alone?" or "Where are they?" or posed in the symbolic form of a Drake equation. The question gets asked of Earth even before we define what we mean by Earth, and it was certainly asked long before humans had developed anything resembling a realistic concept of what Earth is (*Wooten*, 2015), or what the facts of Earth are that make it special if it is special.

One way to bracket our thoughts is to compare and contrast two modern memes of cosmology with impressive names. The "Copernican Principle" is a name applied to the postulate that there is nothing special about our point of view (*Bondi*, 1952). It is more precept than truth, but it has to date served cosmology well. The Copernican Principle warns us to be wary of making Earth the center of the universe. The counterpoint to the Copernican principle is the "Anthropic Principle," which is the name we use to acknowledge that what we see is profoundly biased by the fact that we are the observers (*Livio and Rees*, 2005). Putting people first does not necessarily move Earth back to the center of the universe — for all we know, we could be so commonplace that no one has bothered to tell us (*Adams*, 1979). Distinguishing the "Anthropic Principle" from a simple tautology can be hard, but we think there is something more to it, and in any event it is undeniably true.

When the Copernican Principle is applied to a discussion of the origin and evolution of Earth, it suggests that the solar system should be typical of planetary systems, that planets like Earth should be typical of planets in typical planetary systems, and that life evolving on Earth should be typical of what typically evolves on typical Earths. To date there is no evidence that any of this typecasting is true. Indeed, observations of exoplanet systems show hot Jupiters and an abundance of planets 2–3× Earth's mass, neither one of which appear in our solar system (*Lissauer et al.*, 2014; *Petigura et al.*, 2018). In practice the Copernican Principle encourages theorists to devise blow-by-blow deterministic narratives of how our solar system had to be made and how our Earth came to be in it. The theories then feed back into an expectation that many or most solar systems should generally resemble our own and that Earths and life emerge almost inevitably.

The Anthropic Principle can lead to a very different perspective. There are in the known universe roughly 100 billion galaxies each containing roughly 100 billion stars. If most stars have planetary systems — a presumption that exoplanet science is close to verifying (*Petigura et al.*, 2018) — we might expect on the order of 10^{22} planetary systems (roughly the number of carbon atoms in a sugar cube). If Earth as we know it is a near miracle that occurs once in say 10^{19} chances, there would still be 1000 Earths in the cosmos, and 1000 sets of Anthropic Principles contending. To put this another way, there is nothing unlikely about Earth as we know it being a 10^{-19} chance outcome in a universe as big as ours. Indeed, it is a statistical near certainty, which means that we cannot reject a story of Earth and its inhabitants simply because it seems extremely unlikely. Unlikely Earth was the ruling paradigm a century ago, when the solar system was imagined as congealing from a stream of matter pulled out of the Sun during a near-collision with another star (see *Oparin*, 1938). We will not overemphasize unlikely Earth here — modern planetary scientists are too habituated

to seeking plausible explanations to abandon the practice — but we think it important to point out at the beginning of a book entitled *Planetary Astrobiology* that, given what we know now, our universe could be a vast unfriendly desert that's as big as it is because it has to be.

1.2. Light on Water

The important astrobiological fact about Earth is simple: It is at the right distance from the Sun to provide a platform for liquid water to interact with sunlight. This happy place is usually called the "habitable zone" (*Kasting et al.*, 1993). The other necessary ingredients — carbon and nitrogen and the molecules that they make — are abundant enough and widespread enough in our solar system that they may seem a given.

Why water? The late Yutaka Abe argued that liquid is the best medium for life because it combines mobility with the potential for high concentrations; solids are less suitable because mobility is limited, while gases allow mobility but at low concentrations. Of liquids, water is best because it is abundant (H_2O is probably the most abundant condensible molecule in the cosmos under ordinary conditions), and it is liquid at temperatures warm enough for complex organic chemistry to take place but not so hot that complex organic molecules denature or dissolve in a sea of entropy. The coincidence of life-as-we-know-it's preferred temperature range with liquid water's temperature range may seem as unremarkable as the Anthropic Principle, but it emerges from the properties of hydrogen bonds. Both liquid water and the kind of complex organic chemistry that we see practiced by life depend on making and breaking hydrogen bonds, which become too strong to quicken when too cold, and too weak to hold things together when too hot. As for sunlight — by far the biggest source of free energy available to life on Earth — our solar system provides a small constellation of ice-covered ocean worlds to test whether sunlight is truly necessary for life as we know it. (See the chapter by Hoehler et al. in this volume for a more detailed description of the physics and chemistry of water and life.)

Our presence here is evidence that Earth has an astrobiologically auspicious amount of water. In assessing the rareness of Earth, one can ask how much water is too much water, how much is too little, and how much is just right. We — who descend from organisms that crawled out of the seas — might even imagine Earth's oceans optimal, because they give a roughly even division between land and sea. Would a planet with oceans tenfold deeper than Earth's be as suitable for life as Earth is (*Kite and Ford*, 2018)? Would a planet with just 1% as much water be wholly unsuitable? If the oceans are too vast, the dissolved solutes might be too dilute; if too deep, the bottom may freeze to an unfamiliar high-pressure form of ice. A deep ocean may suppress traffic between the liquid water and the rocky mantle, and the planet may lack any solid surface other than ice, which may be unsuitable. Too little water is problematic in more obvious ways: When present, liquid water is likely to be an uninhabitable concentrated brine, and continuous habitability would be frustrated by the infrequency and inaccessibility of nontoxic liquid water. To some extent Europa and Mars provide local examples of these extremes, but neither world is Earth-like enough in other particulars to properly test the hypothesis. It may take centuries before this question can be addressed empirically by telescopes far more capable than any currently planned.

2. ORIGIN OF EARTH IN THE SOLAR SYSTEM

The story of Earth could be traced back to the primordial asymmetry that physicists tell us gave us matter, or to the old stars that built the periodic table of the elements from hydrogen and helium, or to the pooling of stars and their effluents into galaxies, or to the pooling of the effluents into molecular clouds. The prehistory would have to be told in generalities, albeit each atom has its own particular history. Here we will begin with the particular story of our Sun and our solar system.

Our Sun formed at the eye of a swirling disk of gas and dust we call the solar nebula, or often just the nebula. Throughout its existence the gas, dust, boulders, and planets of the solar nebula were continuously falling inward to the Sun. At the end the remains of the nebula may also have been herded into the Sun by migrating giant planets, or its gases photoevaporated from its surface by ultraviolet (UV) radiation from the young Sun itself or from nearby blue giant stars.

The solar nebula appears to have been nearly but not exactly elementally and isotopically homogenous (*Carlson et al.*, 2018; *Burkhardt et al.*, 2019). This is consistent with the solar nebula sampling a cumulative reservoir of elements made in many stars over the 9-b.y. history of the universe prior to the formation of the Sun. It is difficult to separate chemical processing from elemental heterogeneities, but the isotopic heterogeneities — which trace back to nucleosynthesis in different stars (*Burbidge et al.*, 1957; *Truran and Heger*, 2003) — are more conserved and have proved rather telling, and will be discussed in more detail later in this chapter.

The Sun itself probably took significantly less than a million years to grow to its present size (the free-fall time is 0.1 m.y.), while meteoritic radiometric ages tell us that solid planetesimals grew large enough to undergo melting and internal differentiation into core, mantle, and crust within 0.2 to 4 m.y. (*Nyquist et al.*, 2009; *Kruijer et al.*, 2014a), which is a typical lifespan for the kind of nebula that gives birth to Sun-like stars (e.g., *Haisch et al.*, 2001). Planet formation in the solar system took as little as 1 m.y. or as long as 150 m.y., depending on the planet and on what one defines as the end of planet formation. Here our focus is on Earth, which according to theory (*Wetherill*, 1990; *Chambers*, 2004) is composed of bodies that may themselves have grown in less than 3 m.y., but did not resolve into the Moon and Earth as we know them until

much later. Radiometric ages for Earth and the Moon range widely and can be interpreted to date Earth formation from 30 m.y. (*Kleine et al.*, 2009; *Barboni et al.*, 2017) to as late as 100–200 m.y. (*Patterson*, 1956; *Wetherill*, 1975; *Pepin*, 2006; *Touboul et al.*, 2007; *Borg et al.*, 2015). What these longer durations actually date is a good question, one that is better addressed after we review some basic concepts of planet formation theory.

In its heyday the solar nebula would have been hottest and most vertically compressed near the Sun (where the gravity is strongest), becoming generally colder and more vertically extended at ever greater distances from the central vortex. The thermal structure of the nebula derives mostly from the gravitational potential energy released as matter flows downhill to the Sun (Fig. 1). The temperature profile maps to a condensation (or evaporation) sequence of solids (*Grossman*, 1972). Near the Sun only the most refractory (least volatile) elements can exist in a condensed state. As one moves outward and the temperature drops more elements can exist in condensates, until at the edge of space the whole periodic table other than H, He, and Ne is in solids. The most important condensation/evaporation fronts are those associated with rock and iron, and the "snow line" associated with water ice, as these are the most abundant types of solids by mass. Other condensation/evaporation fronts are possible. Ices more volatile than water might define more distant condensation fronts — CO_2, CH_3OH, and NH_3 comprise one possibility, CO and allied ices (N_2, CH_4, Ar) another. Condensation/evaporation fronts between the rock line and the snow line are more speculative. The high sensitivity of condensation to temperature predicts that an element is either present or not, but abundances in actual meteorites are mixtures in the middle (*Cassen*, 1996). Organic molecules remain an open question (the "tar line") (*Lodders*, 2004), because organic molecules can condense at almost any temperature. In principle tars can be abundant. If there is a particular temperature to associate with a tar line, it would probably be in the range of 600–1100 K, in the same range as the "moderately volatile" inorganic elements (*Palme et al.*, 2014).

The theoretical condensation sequence maps directly to the general structure of our solar system, with its inner planets made of rock and iron and its outer planets wrapped in ice and gas. The water ice snow line in particular has played an outsized role in theories of how Jupiter in our solar system formed, with the conventional view targeted on the snow line as a region where the density of solids is particularly high, which makes the snow line peculiarly well suited to rapid growth of planets (*Stevenson and Lunine*, 1988). A minority view correlates Jupiter with the tar line, as organic molecules in principle could have been very abundant and it is easy to imagine them sticky (*Lodders*, 2004). Overall, the temperature-condensation sequence and the special importance of water ice are so obvious that both seemed fundamental to planet formation, which is why the discovery of "hot Jupiters" as the first known exoplanets came as a major surprise (*Mayor and Queloz*, 1995).

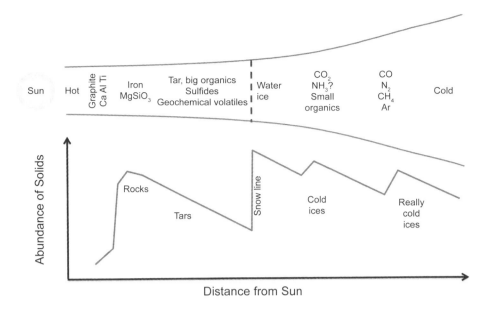

Fig. 1. Schematic cartoon of a solar nebula with condensation (evaporation) fronts at a particular time. The fronts move inward or outward depending mostly on the rate that mass flows toward the Sun, and to a lesser degree on the brightness of the Sun. H_2, He, and Ne do not condense.

2.1. Accretion

Planetary accretion is usually imagined as hierarchical, beginning with small particles that condense from a hot gas that is cooling (*Grossman*, 1972). The condensates ("grains") are small enough to couple strongly to the gas. Little grains coalesce into bigger grains by low-speed electromagnetic or chemical (sticky) interactions until they grow to the size of "pebbles" or "cobbles" or "snowballs," a size category that can be functionally defined as solid bodies that respond significantly to both hydrodynamic forces and gravity (*Weidenschilling and Cuzzi*, 1993; *Levison et al.*, 2015). Note that rocky words like "pebble" or "cobble" may overstate the strength and hardness of the particles. Because they are blown about by the winds as they fall under the influence of gravity, the pebbles and cobbles and snowballs tend to move at velocities of tens of meters per second with respect to grains and with respect to each other, which makes it challenging for them to merge by mutual collisions; rather, they are doomed to bounce, chip, or shatter, a barrier to accretion known as the bouncing barrier (*Brauer et al.*, 2008a,b; *Ormel*, 2017). In classic accretionary models, provided that particles grow in size past this barrier, they continue to accumulate into objects of increasing size from planetesimals to planetary embryos and eventually to planets (*Weidenschilling and Cuzzi*, 1993). (Planetary accretion is described more fully in the chapter by Raymond et al. in this volume.)

"Planetesimals" describe a class of bigger bodies held together by self-gravity. Planetesimals are by definition big enough to be virtually unaffected by gas drag yet numerous enough that their mutual interactions are important. How grains, dust bunnies, pebbles, snowballs, boulders, etc. gather into planetesimals is one of the outstanding problems in the theory of planet formation (*Youdin and Goodman*, 2005; *Johansen et al.*, 2006; *Levison et al.*, 2015; *Ormel*, 2017; *Hartlep et al.*, 2019). Two general families of notional solutions are (1) sticky interactions and (2) gravito-aerodynamic concentration. Sticky interactions are currently out of favor on scales larger than about a centimeter. Stickiness, if effective on larger scales, is most likely at particular locations, such as condensation fronts (cf. *Brauer et al.*, 2008b) or tar lines, where welding can occur.

Gravito-aerodynamic mechanisms envision solids first settling to the midplane of the disk, and then being further concentrated aerodynamically until they become collectively self-gravitating (*Youdin and Goodman*, 2005; *Johansen et al.*, 2006; *Hartlep et al.*, 2017, 2019). The leading idea is the "streaming instability" (*Youdin and Goodman*, 2005; *Johansen et al.*, 2006, 2015). In a disk, particles orbit a little faster than the gas (because the gas is partially pressure supported), and hence tend to spiral in toward the Sun by gas drag. In the streaming instability, a local concentration of particles provides some self-shielding of the swarm against aerodynamic drag, and hence the swarm spirals in a little more slowly than non-swarm particles. The swarm grows as non-swarm particles fall into it. The streaming instability is predicted to be highly effective when turbulence in the nebula is negligible (*Johansen et al.*, 2015). If nebular turbulence is too great the rapid growth of the streaming instability can be frustrated and other effects, such as turbulent concentration, can become competitive (*Estrada et al.*, 2016). Turbulence in the solar nebula is debated; here we simply note that the formation of planetesimals remains unresolved.

"Planetary embryo" is a theoretical concept that refers to a body in the nebula that is substantially larger than other bodies, and relatively isolated from the other embryos.

Embryos are often imagined as the alpha predators of their feeding zones. In the classical model, collisional mergers of smaller planetesimals form larger embryos (*Wetherill*, 1980; *Kokubo and Ida*, 1998; *Chambers*, 2004) that then grow substantially larger than competing lesser bodies by runaway gravitational accretion of planetesimals (*Kokubo and Ida*, 1998).

"Pebble accretion" enhances the role of pebbles and reduces the role of planetesimals as feedstock of embryos. In pebble accretion, gas drag causes pebbly particles to drift toward a seed planetesimal or embedded embryo around which the gas flows (*Hayashi et al.*, 1986; *Levison et al.*, 2015; *Ormel*, 2017; *Johansen and Lambrechts*, 2017). Pebble accretion is thought to be fast. It can resolve some puzzles of accretion (e.g., it can build big embryos very quickly, yet because the particles are small it can also explain how a large world like Callisto can accrete quickly from snowballs without differentiating), and it can provide for very rapid growth of favored planets from seed planetesimals, but it does not explain how seed planetesimals arise in the first place; i.e., it does not entirely resolve the bouncing problem (*Ormel*, 2017). A quality of pebble accretion is that the number and sizes of embryos can be limited by the number of seeds, rather than by the width of feeding zones as in conventional runaway accretion. A corollary quality of pebble accretion is that the seeds could be created under local conditions that are peculiarly favorable to planetesimal formation.

If an embryo grows big enough, it can pull in nebular gas and thus grow much greater still. If the big embryo grows big enough, it can carve a gap out of the nebula, after which growth is metered to the rate that material flows into the gap. Gas accretion is how giant planets can grow quickly (*Mizuno et al.*, 1978), but to trigger it requires a sufficiently massive rocky/icy core that itself must grow quickly, which in turn may depend on rapid operation of the planetesimal accumulation methods mentioned previously (*Youdin and Goodman*, 2005; *Chambers*, 2014; *Levison et al.*, 2015). Something like this took place in the outer solar system, but in the inner solar system the feeding zones may have been too closely spaced to grow embryos big enough to effectively capture nebular gas (*Lissauer*, 1987), or the absence of water ice left inner solar system embryos hopelessly slow-growing compared to their colder outer solar system counterparts (*Chambers*, 2014). In any event, inner solar system accretion is usually imagined to have stalled until the nebula dissipated, at which point the embryos began to stray.

In standard N-body terrestrial planet accretion models the regular swarm of embryos at the moment the gas goes away is treated as an initial condition, as is the presence of an already-formed Jupiter (*Wetherill*, 1980; *Kokubo and Ida*, 1998; *Chambers*, 2004). In gas-free N-body models the embryos and/or "protoplanets" (the latter comprising merged embryos, or super-embryos) wander, collide, and merge over millions or tens of millions of years, with the eventual emergence of a small number of long-lasting "planets." In this way of looking at things, a planet would be any embryo or protoplanet that is no longer likely to collide with another planet in the lifetime of the solar system. For Earth, planetary status begins with the Moon-forming impact, while Mars can be both an embryo and a planet (*Dauphas and Pourmand*, 2011), which is more diplomatic than calling it a big asteroid.

Obtaining a final configuration resembling the terrestrial planets of our solar system in a gas-free N-body simulation requires restricting the initial swarm of embryos to an annulus between 0.7 and 1.1 AU (*Wetherill*, 1990; *Hansen*, 2009). Theoretical interest has focused on the outer boundary of the annulus and the small mass of Mars. One way to stunt the growth of Mars is for Jupiter to migrate, first inward (pushbrooming inner solar system embryos inward as well), then outward to where it is now (*Walsh et al.*, 2012; *Raymond and Morbidelli*, 2014). This scenario is called the "Grand Tack." Outward migration in the Grand Tack requires a second giant planet in 2:1 resonance with Jupiter — here Saturn — which may be pertinent to generalizing the story of our solar system to planetary systems in general. By stirring up the embryos of the inner solar system into chaotic combat against each other (*Goldblatt et al.*, 2010), the Grand Tack might lead to the collisional disintegration of a first generation of inner solar system planets (*Batygin and Laughlin*, 2015), which for our purposes might be the Grand Tack's most important consequence. Like other models positing outer planet migration, the Grand Tack also throws volatile-rich outer solar system material into the inner solar system, thereby contributing to the volatile inventories of the terrestrial planets (*Walsh et al.*, 2012; *Raymond and Morbidelli*, 2014).

A different explanation of small Mars assumes that inefficient pebble accretion beyond 1.5 AU frustrated growth of a large planet at Mars' distance, while allowing Jupiter to form quickly (*Levison et al.*, 2015). Once Jupiter formed, its malign influence (which could include a Grand Tack) on the outer terrestrial planet region would have further decreased the likelihood of a large planet forming at Mars' distance.

2.2. Radioactivity and Aluminum

The previous section described planetary accretion from the top down, but for the most part the data have been acquired from the bottom up, beginning with the elements and their isotopes. Stellar nucleosynthesis produces all the isotopes of all the elements heavier than boron. Not all are stable. Many are radioactive with half-lives that range from seconds to hundreds of billions of years. The longest-lived help keep the Earth warm and tectonically active, while serving scientists as chronometers and tracers of planetary differentiation. Extinct isotopes with half-lives in the range of 10^5–10^8 yr provide both high-precision chronometers and tracers for chemical differentiation events occurring in the early solar system (*Nyquist et al.*, 2009; *Davis and McKeegan*, 2014). As a tool, the decay of ^{182}Hf (half-life 8.9 m.y.) to ^{182}W has been exceptionally useful. Hafnium is a lithophile element (i.e., an element with a strong tendency to partition into silicates), while tungsten is a siderophile element (i.e., an element with a strong tendency to partition into metals). Hence the ^{182}Hf-^{182}W system is well-suited for dating the separation of iron cores from silicate mantles. Another of the short-lived radionuclides, ^{26}Al (0.717 m.y.), was abundant enough to be a major source of radioactive heating in early-formed planetesimals. A third, ^{60}Fe (2.6 m.y.), was briefly thought to be important, but recent measurements are pushing its initial abundance in the solar system down to levels of relative insignificance (*Tang and Dauphas*, 2015).

Aluminum was by far the most important radioactive element in the early solar system. ^{26}Al can melt planetesimals as small as 5 km at t = 0, and it remained capable of melting planetesimals in the 50–100-km-diameter range at times as late as 3–4 m.y. (*Moskovitz and Gaidos*, 2011). Evidence for the formation of iron meteorites, presumably the segregated cores of small planetesimals, in the range 0.5 to a few million years after the 4.567 Ga age of the solar system comes from ^{182}Hf-^{182}W (*Kruijer et al.*, 2014b).

The short half-life of ^{26}Al indicates that it was made shortly before the solar nebula formed, presumably in a star in the Sun's birth cluster (*Dauphas and Chaussidon*, 2011; *Gounelle and Meynet*, 2012). From the perspective of astrobiology, an interesting feature of ^{26}Al is that its abundance likely varies significantly from solar system to solar system, given its short half-life and the likelihood that different stars make and eject vastly different amounts of it (*Gounelle and Meynet*, 2012). It may be pertinent that the initial ^{26}Al/^{27}Al ratio in the solar system was 6× bigger than the average galactic ^{26}Al/^{27}Al ratio determined from γ-rays (*Diehl et al.*, 2006). The widespread ^{26}Al-induced melting inferred in our solar system may be unusual (*Ciesla et al.*, 2015; *Young*, 2016; *Fujimoto et al.*, 2018). Melted planetesimals may accrete by different processes than unmelted bodies, collisions between them may have different outcomes, and they have different means of acquiring, storing, or shedding volatile elements than planetesimals composed only of collections of solids. These differences in turn could lead to different kinds of planets in different kinds of planetary systems.

3. WHAT IS EARTH MADE OF?

To first approximation, Earth comprises an iron core enveloped by a magnesium silicate mantle (*Palme and O'Neill*, 2014). This general elemental composition is determined by the relatively high cosmic abundances of O, Mg, Si, and Fe, and the clear geochemical and physical distinctions (iron is dense) between lithophile and siderophile elements. Impurities are present at the 15% level. Many of the more incompatible impurities are concentrated into the crust (*Rudnick and Gao*, 2003). Earth is depleted in moderately volatile elements (those with nebular condensation temperatures between 600° and 1250°C) and extremely depleted in atmophile elements compared to solar abundances (*Palme and O'Neill*, 2014).

Because Earth's composition derives directly from typical cosmic abundances, other planetary systems may well share compositional similarities. However, our solar system provides several examples of alternative compositions that might also be scaled up to host habitable environments. Mercury is an example of a planet made mostly of iron. The Moon is made mostly of refractory lithophile elements. Several outer solar system satellites have deep mantles of water, which in our solar system present as icy surfaces, but in a different solar system might present as liquid water oceans. At least one icy satellite, Tethys, is almost pure water ice. Normal cosmic abundances allow the contemplation of a planet rich in sulfur, perhaps presenting as a layer between the silicates and iron, or perhaps presenting as a surface volatile as on Io, or as a sulfide on a highly reduced surface such as Mercury's (*Nittler et al.*, 2011). Finally, different cosmic abundances may allow for other less familiar compositions, say iron-diamond planets condensed from tar in places where the C/O ratio exceeds one (*Kuchner and Seager*, 2005), or carbon monoxide planets, or other oddities (*Lewis*, 1999) if there are planets orbiting the oldest stars that have a paucity of heavy elements.

3.1. Note About Some Geochemical Jargon

Goldschmidt (1937) classified the elements of the periodic table into five categories: "siderophile" (elements that prefer to sit in metallic phases); "lithophile" (elements that prefer to sit in silicates or other refractory oxides, hence they can also be called "oxyphile"); "chalcophile" (elements that readily form sulfides); "atmophile" (elements that partition into the atmosphere); and "biophile" (the elements of life). For the most part this classification scheme remains in general use. Lithophile elements are subdivided between "compatible" and "incompatible" depending on how well they fit into the crystal structures of the dominantly magnesium silicate mantle. Compatible elements tend to stay in solid phases when the mantle partially melts, while the incompatible elements are preferentially expelled into the melt to accumulate in the crust. Another classification axis is set by the condensation temperature. Substances that evaporate easily are "volatile," and substances that do not evaporate easily are "refractory." Refractory and volatile are usually measured with respect to magnesium silicates and metallic iron, which are abundant and have accidently similar condensation temperatures under putative solar nebular conditions (*Lodders*, 2003). Each of these classifications can be further modified by "moderately" or "highly" or "extremely." Thus uranium can be described as a highly incompatible refractory lithophile element, arsenic a moderately volatile chalcophile element, and iridium an extremely siderophile element. The adjectives are not necessarily exclusive, as few elements are simple enough to do only one thing. Iron, for example, is a moderately siderophile element that is also compatible. Finally, "atmophile" and "biophile" seem not to be in general use. It is more usual to refer to atmophile elements as volatiles.

This dual usage of volatile lumps "highly volatile" rock-forming elements like cadmium and indium together with nitrogen. Goldschmidt's original idea is better, and in this chapter we will refer to nitrogen, water, carbon dioxide, and the noble gases as atmophile elements. We do not attempt to define biophile elements.

3.2. Note About Meteorite Classes

Modern Earth is still accreting a variety of meteorites. Chondrites are rocky meteorites usually composed of cemented chondrules, which are submillimeter rock spherules of uncertain but highly contentious origin that are ubiquitous in solar system materials dating from 2–4 m.y. after time zero. Ordinary chondrites, like Earth, are rich in the silicate mineral olivine. Enstatite chondrites have low Mg/Si ratios and feature the silicate mineral enstatite instead of olivine. Carbonaceous chondrites comprise a wide variety of meteorites that are generally more oxidized than ordinary chondrites and often carry organic material and hydrated minerals. Achondrites are rocky fragments from the crusts or mantles of melted bodies, often older than chondrites. Irons — the composition is self-explanatory — are samples of the cores of disintegrated melted bodies. These too are often older than the chondrites.

3.3. Earth Apparently Accreted After the Disappearance of the Nebula

If Earth accreted to planetary size in the nebula, it should have captured an atmosphere of nebular gases (*Hayashi et al., 1979; Wetherill, 1980; Walker, 1982; Sasaki and Nakazawa, 1988; Pepin, 1991; Ikoma and Genda, 2006*). There are four such atmospheres in our own solar system. Kepler has revealed that gassy planets near the central stars of exoplanetary systems are common (*Lissauer et al., 2013, 2014; López and Fortney, 2014; Petigura et al., 2018*). Many of these exoplanets are not much more massive than Earth, and most of those observed to date are much hotter, and thus harder put to hold on to a hydrogen-helium atmosphere. Given that Hayashi's hypothesis is both reasonable and often seemingly verified, it must be taken seriously. In light of this, the striking near-absence of solar nebular atmophile elements in Earth (*Aston, 1924; Russell and Menzel, 1933; Brown, 1949*) places an important boundary condition on the accretion of Earth.

The most useful tracer of the solar nebula is neon, because it does not condense under solar system conditions and it is heavy enough to not escape easily from planet-sized bodies. The solar N/Ne ratio is roughly unity, while Earth's atmospheric ratio is 8×10^4. But the ratio of terrestrial N to the amount of terrestrial Ne that came directly from the solar nebula is even higher than this, because air's Ne/^{36}Ar ratio and Ne isotopes betray atmospheric Ne as chondritic (from meteorites), not solar. There is a small amount of solar Ne in the mantle (*Moreira, 2013; Tucker and Mukhopadhyay, 2014*). We know that about 99% of Earth's ^{36}Ar (a nonradiogenic isotope) is in the atmosphere, an estimated based on radiogenic ^{40}Ar being ~50% degassed (as the abundance of 40 K, the parent of ^{40}Ar, is known) and the ratios of ^{40}Ar/^{36}Ar in air (300) and in the mantle (3×10^4) (*Moreira, 2013*). Neon is probably less degassed than ^{36}Ar, but not all the Ne in the mantle is solar (*Moreira, 2013*); for specificity we will apply the same factor of 100 to neon. We therefore estimate that the ratio of N to *nebular* Ne on Earth is at least 10^7; i.e., we conclude that less than 10^{-7} of Earth's N came directly from the solar nebula. The rest of Earth's nitrogen must have accreted in solids. This is the classic argument of *Aston* (1924), *Russell and Menzel* (1933), and *Brown* (1949), using modern data.

For Earth to end up with such a tiny contribution from solar nebular gases is remarkable. It's not zero — there is a tiny amount of solar He and Ne in Earth; moreover it has more than one source (*Tucker and Mukhopadhyay, 2014; Williams and Mukhopadhyay, 2019*) — but the quantity is so small that almost any story might do [e.g., solar wind implants in dust grains later swept up by Earth and subducted (*Moreira, 2013*)]. For our part we agree with *Tucker and Mukhopadhyay* (2014) and *Williams and Mukhopadhyay* (2019) that Earth's solar Ne and He are probably the last tiny remnants of solar nebular atmospheres of one or more of Earth's contributing embryos. It is expected that embryos that accreted in the nebula experienced some ingassing of nebular gas, especially those that were melted by ^{26}Al (*Walker, 1982*). As these gases were mostly H_2 they were profoundly reducing and the effect on the embryos must have been reducing as well. Such H_2-dominated nebular atmospheres are easily lost after the nebula disperses if the embryo is Mars-sized or smaller: A giant impact leaves the gas too hot to be gravitationally bound; failing that, solar EUV would drive it off (cf. *Walker, 1982*). Nonetheless, it seems inescapable that the construction of Earth and its bigger building blocks postdate the solar nebula. This is the philosophical basis behind gas-free N-body simulations of terrestrial planet accretion (*Wetherill, 1980; Mizuno and Wetherill, 1984*).

The message in neon is that Earth was not directly assembled from large bodies that accreted in the nebula, but rather was assembled from bodies that had lost their nebular gas. This is best accomplished during post-nebular evolution. While the nebula was present, its gravity tamped the orbits of the embryos, and it shielded the embryos from solar EUV radiation. Removing the nebula was like releasing the jack in the box (cf. *Juric and Tremaine, 2008*): Orbits sprang free and the embryos hammered into each other, with many breaking up and the others being badly damaged by hit and run collisions. All were exposed to the Sun. Earth accreted from the debris. Kepler saw many solar systems whose planets may have been built to large sizes directly from embryos in the presence of gas (*Lissauer et al., 2013, 2014; López and Fortney, 2014; Petigura et al., 2018*). At least in the realm of the terrestrial planets, our solar system does not look like one of these.

3.4. Geochemical Models of Earth's Accretion

Traditional geochemical models of Earth's accretion are informed by the nebular condensation sequence and by the assumption that on average Earth accreted nearby materials before it accreted more distant materials. First-generation models divided the accreting material into two classes that differ by redox state and volatile content, usually called A and B, and divided accretion into two episodes, an early phase dominated by a highly reduced component A that loosely resembled ordinary chondrites (but more reduced) and a later phase mixing A with a more oxidized and more volatile-rich component B that loosely resembled carbonaceous chondrites (*Ringwood*, 1979; *Dreibus and Wänke*, 1989).

Components A and B are supplemented by a late-accreting component C — often called the late veneer — that contributes about 0.5% of Earth's mass. In this hypothesis, the late veneer is the source of highly siderophile elements (HSEs) that are found stranded in Earth's mantle in chondritic (i.e., solar nebular) relative abundances at much higher absolute abundances than expected from mantle-core equilibrium. The HSEs (also known as platinum group elements, PGEs, or as the noble metals) are seven heavy elements (Ru, Rh, Pd, Os, Ir, Pt, and Au) that are chemically rather inert and show very strong tendencies to partition into metals and, on a planetary scale, into cores (*Day et al.*, 2016).

If Earth's mantle and core were fully equilibrated almost all the HSEs would be in the core. The tiny amount remaining in the mantle would be highly chemically and isotopically fractionated (*Rubie et al.*, 2016). But this is not what is seen. Rather, the mantle contains a modest cohort of excess HSEs that, to first approximation, are present in the same relative proportions as they are in chondritic meteorites. A good deal of experimental effort has been expended to evaluate the idea that siderophile elements become less so under the high temperatures and pressures expected for core formation on a planet the size of Earth (*Murthy*, 1991; *Righter et al.*, 2014). This certainly seems to be the case, but no experiments have yet been able to produce sufficiently low, and similar, metal-silicate partition coefficients for the HSEs to explain both their abundance and their near chondritic relative abundances in Earth's mantle. The straightforward explanation is that the excess HSEs were dropped into the mantle and left stranded there after core formation was complete (*Day et al.*, 2016).

The ABC models satisfy both the general chemical composition of Earth and the theoretical expectations of the order and nature of terrestrial planet accretion. Component A represents local inner solar system high-temperature condensates, while later-accreting material on average fell from ever further away, tapping into colder outer solar system condensates that were richer in volatile elements and more strongly influenced by oxidation by H_2O. Grand Tack models assign the transition from A to B to scattering of outer solar system planetesimals (B) into the inner solar system when Jupiter and Saturn reversed their inward migration (*Raymond and Izadoro*, 2017). In the ABC models, Component C is expected to have a distant outer solar system source, beyond Jupiter and perhaps beyond all the giant planets. Component C has therefore drawn considerable attention as a possible source of Earth's volatile and atmophile elements (*Albarède*, 2009; *Marty*, 2012; *Halliday*, 2013; *Marty et al.*, 2017).

Modern counterparts to the early ABC models integrate the geochemistry of a range of differently-behaving elements with modern N-body codes that simulate terrestrial planet accretion (e.g., *Rubie et al.*, 2015, 2016). The N-body codes label the different embryos by their birthplaces as proxies for their volatile compositions. The N-body model then tracks the scattering and merging of embryos into planets while the geochemical model tracks core-mantle partitioning of several geochemically disparate elements in the growing planets. Best results are obtained by restricting chemical interaction between the cores of the accreting planetesimals and the larger planet's mantle; i.e., most of the bigger mantle remains unaffected by the core of the smaller. A recent ABC model (*Rubie et al.*, 2016) introduces a late sulfidation stage that resets the mantle abundances of moderately siderophile elements. Some of the models incorporate aspects of the Grand Tack scenario (*Rubie et al.*, 2015). A subset of the models create an Earth-like planet in the context of a solar-system-like inner solar system, picked from a diverse set of model outcomes that should in principle look like solar systems still to be discovered.

3.5. Isotopically Redefining Earth's Building Blocks

An important recent development in meteoritics has been the genetic sorting of meteorites by isotopes rather than by phenotypic classes determined by mineralogy and alteration (*Warren*, 2011; *Scott et al.*, 2018; *Carlson et al.*, 2018). The program began with oxygen, in which the isotopic signal is large. It was realized quite early that Earth and Moon were linked by a common mass fractionation relation but that most meteorites, Vesta, and Mars did not plot on this line, and hence were in some sense genetically unrelated to Earth (*Clayton et al.*, 1973; *Clayton*, 2003). The oxygen isotopes were initially interpreted as a nucleosynthetic signal (*Clayton et al.*, 1973), but are now usually interpreted as a consequence of photochemistry taking place in the UV-irradiated nebular gas (*Thiemens and Heidenreich*, 1983), either before (*Yurimoto and Kuramoto*, 2004) or after (*Lyons and Young*, 2005) collapse of the molecular cloud to a nebular disk. Whatever their origin, the variable oxygen isotope composition seen in the terrestrial planets and their meteoritic building blocks provides the basis for evaluating the contributions of inner and outer solar system materials (*Young*, 2007; *McKeegan et al.*, 2011).

The ability to use isotopic differences between different planetary objects to track their contribution to Earth formation was advanced by the detection of nucleosynthetic variability in a wide variety of elements in meteorites and the terrestrial planets (e.g., *Carlson et al.*, 2018). These differences reflect different contributions from different stars. The terms of art are to distinguish between r-process nuclei (rapid nucleosynthesis, creating neutron-rich isotopes) and s-process nuclei (slow nucleosynthesis, favoring isotopes with fewer neutrons). Differences in the relative amounts of r and s nuclei are now being used as tracers to revise our understanding of the link between the inner and outer solar system and the nature of the materials that make up Earth (*Warren*, 2011; *Fischer-Gödde et al.*, 2015; *Kruijer et al.*, 2017; *Dauphas*, 2017; *Carlson et al.*, 2018; *Fischer et al.*, 2018; *Scott et al.*, 2018; *Budde et al.*, 2018).

To first approximation the isotopes map rather well to traditional classes of meteorites, which is comforting. But an unexpected feature of the new isotopic maps is that they consistently fall into two well-separated clumps (*Warren*, 2011). Figure 2, adapted from *Scott et al.* (2018) for O isotopes and Cr isotopes, provides a nice illustration. One of the clumps contains the carbonaceous chondrites and is designated "CC". The other clump does not include carbonaceous chondrites and has been denoted "not-CC" or "NC." Both the CC and the NC clumps contain iron meteorites

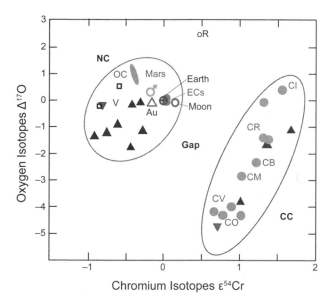

Fig. 2. Solar system solids fall into two sprawling yet impressively well-separated clumps when sorted by isotopes (*Warren*, 2011; *Scott et al.*, 2018). This particular figure — redrafted from *Scott et al.* (2018) — uses oxygen isotopes and chromium isotopes, but similar patterns are seen in other elements (e.g., Ti, Mo). The "CC" clump includes the carbonaceous chondrites (disks; several varieties are indicated) and also some achondrites (triangles), pallasites (inverted triangle), and irons (although none are plotted here). The "NC" clump (denoting "not-CC") includes the ordinary chondrites (OC), the enstatite chondrites (EC), some achondrites (triangles), pallasites (inverted triangle), aubrites (aka enstatite achondrites, open triangle), and irons (black squares). The NC outlier is Rumuruti (R). The NC clump includes the terrestrial planets for which we have data and the HED parent body (here indicated by the "V" for Vesta). The gap between the clumps may map to the gap in the solar nebula opened by Jupiter.

and achondrites derived from differentiated ~100-km-sized worlds that were heated by ^{26}Al before the heating element decayed away (*Kruijer et al.*, 2017). Both clumps sprawl over a considerable area in isotope space, but they are well-separated by a gap that is easily imagined to map the gap in the solar nebula carved out by Jupiter as it formed (*Kruijer et al.*, 2017). In this picture mixing between the two parts of the solar nebula was frustrated by the gap, while presumably the material within the gap itself was ingested by Jupiter. Moreover, the NC and CC clouds have been separated since 1 m.y., and remained separated until 4 m.y., after which primary chondrites stopped forming (*Kruijer et al.*, 2017). This leads to the hypothesis that the NC bodies formed between Jupiter and the Sun, while the CC bodies formed beyond Jupiter. The origin of the difference between the NC and the CC isotopic reservoirs is a topic of active research (*Burkhardt et al.*, 2019).

In Fig. 2, Earth, Moon, and the enstatite meteorites (EC; enstatite chondrites and aubrites) plot in a tight little clump on the edge of the NC cloud. Mars is not all that far away. The EC subclump is far from the center of the NC distribution. Since almost all the mass in this plot is in Earth, this plot highlights the singular relevance of the ECs to the nature of the material that made Earth, or put another way, it highlights the apparent irrelevance of all other meteorite classes to the nature of Earth (*Jacobsen et al.*, 2013; *Dauphas et al.*, 2014b; *Fischer-Gödde and Kleine*, 2017; *Carlson et al.*, 2018).

An important advance provided by the number of elements now with measurable nucleosynthetic anomalies is that the elements have different chemistries, in particular their propensity to separate into core or silicate mantle. For an element that is entirely concentrated in the mantle, the isotopic composition of that element in the mantle averages the isotopic composition of all material accreted to form Earth. In contrast, elements that are siderophile will be segregated from the mantle with each core-forming event, leading the isotopic composition of these elements in the mantle to be weighted toward later additions to Earth. *Dauphas* (2017) used this observation to construct a new kind of ABC model of Earth's accretion constrained almost entirely by isotopes, in which geochemistry and accretion are treated as highly idealized analytic functions. As in traditional ABC models, Earth's accretion is divided into three epochs, with A′ comprising the first 60% of Earth, B′ the middle 39.5%, and C′ the last 0.5%.

A source of uncertainty in the A′B′C′ model stems from whether Nd (a highly lithophile element that is obsessively measured because it has two radiogenic isotopes) is used as one of the constraints. Nucleosynthetic variability in Nd is most clearly expressed in ^{142}Nd (*Carlson et al.*, 2007), but that isotope is affected by radioactive decay from extinct ^{146}Sm (103-m.y. half-life), complicating whether the measured relative variability in ^{142}Nd abundances is nucleosynthetic or the result of the chemical changes driven by early planetary differentiation (*Boyet and Carlson*, 2005; *Burkhardt et al.*, 2016). If Nd is not used as a constraint,

Dauphas (2017) finds room for significant contributions by ordinary chondrites (OC, ~25% of Earth's mass) and carbonaceous chondrites (CC, ~5% of Earth's mass) in A′. There is relatively little scope for non-EC material in the later accreting B′, and none at all in C′ (the late veneer). In total, Earth would be about 70% isotopically EC-like material. But if Nd is included as an additional constraint, the best fit to Earth is, isotopically, very nearly all EC all the time.

Of particular interest in this approach is the new light shed on the last 0.5% of accreted material — Component C — which has historically borne a heavy burden as the source of the highly siderophile (*Walker,* 2009), the volatile (*Dreibus and Wänke,* 1989), and the atmophile (*Halliday,* 2013) elements. Two elements with strong siderophile tendencies, molybdenum and ruthenium, display nucleosynthetic isotopic anomalies (*Dauphas et al.,* 2002; *Fischer-Gödde et al.,* 2015; *Fischer-Gödde and Kleine,* 2017; *Kruijer et al.,* 2017; *Budde et al.,* 2018, 2019). Earth's Ru isotopes are consistent with either an enstatite chondrite or type IAB iron heritage (*Fischer-Gödde and Kleine,* 2017). There are no CC isotopes in them. This suggests that Component C encompassed pure inner solar system material that was probably chemically very reducing and water-poor.

Molybdenum is not as siderophile as Ru. Nonetheless, its retention in the mantle (2%) (*Lodders and Fegley,* 1998) suggests that it was delivered late in Earth's accretion (*Dauphas,* 2017). *Budde et al.* (2019) and *Bermingham et al.* (2018) show that Mo isotopes in meteorites sort neatly into CC and NC flavors. The terrestrial Mo standard is consistent with an NC Earth, but *Budde et al.* (2018, 2019) show that the terrestrial Mo standard is not representative of bulk silicate Earth, and they argue that Earth's actual Mo has composition intermediate between NC and CC. This would imply that Earth's Mo is at least 30% CC, but the actual nature of this material is unknown, as there are no known CC materials of the required composition. *Budde et al.* (2019) suggest that these data imply that Theia — the Moon-forming impactor — was itself the CC carrier. However, given present understanding of the details of the Moon-forming impact, Earth's CC-Mo signature could just as easily have been acquired by Earth before the Moon-forming impact. This is clearly an unsettled topic of active research.

3.6. E is for Enstatite

Despite the excellent isotopic match, Earth is almost certainly not made from enstatite chondrites. Earth's upper mantle has a much higher Mg/Si ratio (1.09) compared to any type of chondrite (C ~ 0.9; O ~ 0.8) and particularly E (0.68). One model (*Javoy et al.,* 2010) for a bulk Earth of E-chondrite composition accounts for the excess Si in E chondrites by assuming a core Si concentration of nearly 8% and a lower mantle with 51.6% SiO_2 compared to typical bulk-mantle estimates of 47.5% (*Palme and O'Neill,* 2014).

Enstatite chondrites have two other salient properties: They are highly reduced (so much so that some of their Si has melded with metallic iron), and they are volatile- and atmophile-rich compared to Earth (i.e., they are rich in moderately volatile elements like K on one hand, and they have a great deal of C, N, S, and the noble gases on the other; see Fig. 3). These qualities unite in C, N, and S, which are present in ECs not as volatile compounds but rather as refractory minerals that form when oxygen is busy elsewhere. Carbon is chiefly in graphite, N in unfamiliar ceramics like sinoite (Si_2N_2O) and osbornite (TiN, commonly used as a coating for drill bits), and S in refractory sulfides like oldhamite (CaS). Yet the heavier noble gases — Ar, Kr, and Xe — are also very abundant in ECs compared to Earth (*Pepin,* 1991).

The missing volatile is water. Water in ECs is variable, but never abundant. Some ECs are remarkably dry, even drier than bulk Earth. The little water in ECs is found as hydroxide minerals in the matrix material that fills the spaces between the chondrules. Dessication may be a consequence of reduction rather than a primary property of the EC material. Sulfidation (as opposed to oxidation) has been proposed as an explanation of how enstatite chondrites evolved from more ordinary material (*Lehner et al.,* 2013).

Aubrites are the E-type achondrites, which as achondrites may be preferred as analogs to Earth's building blocks (*Greenwood et al.,* 2018); however, like any enstatite meteorites, they have the wrong elemental composition to comprise Earth. Like ECs, aubrites are dry, but some aubrites are highly endowed with apparently nebular noble gases (*Pepin,* 1991).

The nature of the familial relationship between Earth and ECs is hinted at in Fig. 4. Here we plot the abundances of

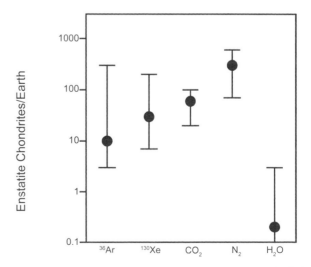

Fig. 3. Atmophile elements in enstatite chondrites (ECs) compared to Earth. In ECs, N and C are present as refractory ceramics or graphite; abundances are converted to N_2 and CO_2 for purposes of the comparison. ECs are H_2O-poor. EC data (for Indarch, type EH, meaning high iron) are from *Schaefer and Fegley* (2017). Bulk Earth data are from *Lodders and Fegley* (1998).

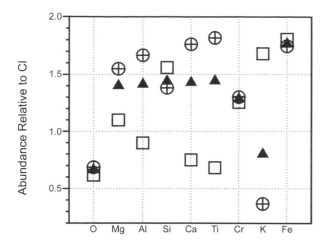

Fig. 4. Abundances of selected elements in Earth (circle-pluses) and an EC (squares) relative to CI chondrites. Earth and the ECs show opposite and complementary abundance patterns in the major refractory oxyphile elements Mg, Si, Al, Ti, and Ca. The primary CI abundances for these five elements (triangles) are recovered by mixing EC and bulk Earth in a 1:2 ratio.

the most abundant refractory oxyphile elements in Earth and in a representative EH chondrite [Indarch, as compiled by *Schaefer and Fegley* (2017)]. Both are normalized with respect to CI chondrite abundances. For Mg, Al, Si, Ca, and Ti, Earth and Indarch are mirror images of each other. This detailed elemental complementarity among the refractory oxyphile elements suggests that Earth materials and EH materials evolved from a common source material with standard solar nebular elemental composition. A possible explanation for this complementarity notes that olivine and clinopyroxene have slightly higher condensation temperatures than enstatite (*Lodders,* 2003), therefore ECs may have formed from nebular leftovers after the more olivine-rich materials that make up Earth condensed and accreted into the terrestrial planets. Exact complementarity does not extend to K or Cr, which are more volatile and, in the case of Cr, more siderophile.

If Earth and ECs did indeed partition their refractory lithophiles in a 2:1 ratio from a common source material, one might ask what happened to the EC complement to Earth, which would if gathered together have comprised a planet half the mass of Earth. Other questions that arise are, what was the form of the common source material, and how did it come to be partitioned? Put another way, how did the two pools of material (isotopically almost identical but highly fractionated chemically) diverge to create Earth on one hand and leave only a tiny remnant population of meteorite parent bodies on the other?

4. THE MOON-FORMING IMPACT AND ITS IMMEDIATE AFTERMATH

The leading explanation for the origin of the Moon-Earth system is that it formed when two planets collided

(*Benz et al.,* 1986; *Stevenson,* 1987; *Canup and Righter,* 2000; *Canup and Asphaug,* 2001; *Canup,* 2004). The event is widely known as the Moon-forming impact, but for all practical purposes it was also the Earth-forming impact, in the sense that it obliterated most, but not all, of the characteristics of Earth that predate the impact.

Earth's Moon is a rocky world made overwhelming of refractory lithophile elements. It has only a tiny iron core, is greatly depleted in volatile elements compared to Earth, and is nearly devoid of atmophile elements. In general the Moon looks a lot like the refractory mantle of a differentiated body, but without a core to match. The paradox was resolved when early smoothed particle hydrodynamics (SPH) simulations showed that a suitable giant impact between two differentiated planets could leave the bigger planet orbited by an iron-poor debris disk of mantle-like material (*Benz et al.,* 1986). A compositionally plausible Moon could then be made from the debris disk.

Standard accretion theory suggests that the Moon was made very quickly and thus was made very hot (*Stevenson,* 1987). It was also made close to Earth, and hence was certain to experience significant tidal heating. The hot young Moon predicted by the giant impact meshed beautifully with the Apollo-era recognition that the Moon had experienced a global melting event, the so-called lunar magma ocean, early in its history (*Smith et al.,* 1970; *Wood et al.,* 1970; *Elkins-Tanton,* 2012).

For the next several decades, the Moon-forming impact seemed a settled matter, in which the particulars of the so-called "canonical" model — a low-speed off-center impact of a Mars-sized planet (call it "Theia") with a planet 90% the mass of Earth (call it "Tellus") — were determined by the mass of the Moon and conservation of angular momentum of the Earth-Moon system (*Canup and Righter,* 2000; *Canup and Asphaug,* 2001; *Canup,* 2004). The picture was filled out with a debris disk rapidly cooling and accreting over some 10–1000 yr into one to several moonlets at or beyond the Roche limit, which in turn aggregated into our one true Moon (*Salmon and Canup,* 2012).

There remained, however, the puzzle of the isotopes: Earth and Moon are very nearly identical isotopically, at a level of identity that seems to go beyond good luck (*Zhang et al.,* 2012; *Dauphas et al.,* 2014b; *Kleine and Walker,* 2017; *Kleine and Kruijer,* 2017). The problem was first recognized as such in the oxygen isotopes (*Wiechert et al.,* 2001). But with increasing instrumental precision the same isotopic unity is seen in many other elements (exceptions are Zn and Si, each interesting in its own way, but more likely tracking chemical processes associated with Moon formation than the parentage of lunar materials). In the canonical impact the Moon is mostly made from the mantle of Theia. *Meier et al.* (2014) outline the characteristics of possible Theias, proto-Earths, and impact parameters that could satisfy the isotopic similarity of Earth and Moon. Although they could not wholly rule out Mars-like or CI-like compositions, the most likely case is that Theia had an isotopically Earth-like composition.

Possible explanations fall into three broad categories that need not be exclusive:

1. The first retains the low-energy canonical impact parameters while hypothesizing that a mechanism existed for intimately mixing the mantles of the Moon and Earth, presumably in the protolunar disk before the Moon accretes (*Pahlevan and Stevenson,* 2007). This works best if Earth's mantle is itself circulated through the disk. Such models are difficult to make work because material mixing in the disk is closely linked to the transport of angular momentum in the disk, and hence to the evolution and dissipation of the disk.

2. The second suggestion drops angular momentum conservation as a constraint on the evolution of the Earth-Moon system (Ćuk and Stewart, 2012; *Canup,* 2012; *Reufer et al.,* 2012). Tidal interactions taking place when the precession period of the Moon's orbit was resonant with the year would have been able to transfer considerable angular momentum from the Earth-Moon system to Earth's orbit around the Sun (Ćuk and Stewart, 2012). This happened when the Moon was still quite near Earth. The idea met considerable resistance at first, but presently it is in danger of becoming the standard model.

Dropping conservation of angular momentum as a constraint greatly expands the phase space of possible solutions to include a galaxy of "high-energy" Moon-forming impacts. One scenario posits an extremely rapidly-rotating Tellus whose rotational angular momentum supplies the mass of the disk (Ćuk and Stewart, 2012); a second posits two almost identical impactors for which a well-mixed outcome is required by symmetry (*Canup,* 2012). Current interest in high-energy impacts is in their potential for effective mixing by creating a transient, very fast-rotating "synestia" with properties intermediate between a rigidly-rotating planet and a Keplerian disk (*Lock and Stewart,* 2017; *Lock et al.,* 2018). Synestias are a topic of active research.

3. The third way (*Jacobsen et al.,* 2013; *Dauphas et al.,* 2014b) drops the presumption that Theia and Tellus were isotopically different. The presumption instead is that Tellus, Theia, and the enstatite chondrites and aubrites were all made from the same isotopic pool, which was by far the largest isotopic reservoir in the inner solar system. The logic here is that, if enstatite chondrites are isotopically representative of inner solar system material, it is inevitable that Theia and Tellus should share that same isotopic composition (*Jacobsen et al.,* 2013).

The chief objection raised against the third scenario is that the Moon and Earth also have very similar radiogenic tungsten isotopic compositions. As Hf is strongly lithophile while W is moderately siderophile, metal-silicate separation causes large changes in the mantle's Hf/W ratios that are expressed as variations in $^{182}W/^{184}W$ if core formation occurs while ^{182}Hf is extant (*Halliday and Kleine,* 2006; *Kleine et al.,* 2009). Recent measurements (*Touboul et al.,* 2007, 2015) show lunar materials to have a $^{182}W/^{184}W$ ratio only marginally different from early Earth materials (*Willbold et al.,* 2011; *Touboul et al.,* 2015). Similar tungsten isotope compositions require Theia and Tellus to have experienced similar histories of core formation, which adds a layer of improbability to Theia being compositionally identical to Tellus (*Kleine and Walker,* 2017; *Kleine and Kruijer,* 2017).

A way to have our cake and eat it too is to insert earlier giant impacts into the chronology to ensure that the Hf-W isotopic pools are also the same (*Dauphas et al.,* 2014b). *Tucker and Mukhopadhyay* (2014) drew a parallel inference from ^{3}He-^{22}Ne systematics. For example, a previous hit-and-run giant impact (*Asphaug et al.,* 2006) between progenitors of Tellus and Theia might stamp the two planets with the same Hf-W age. Such an impact could be highly disruptive of both bodies, and is not wholly unexpected given the proximity of the two planets that eventually do collide to form the Moon. The debris annulus would quickly accrete to form new planets in adjacent orbits.

4.1. The Ages of the Moon and Earth

The first evidence for a lunar magma ocean (LMO) dates to the discovery that the lunar highlands crust is dominated by anorthosite, a rock type composed of essentially a single mineral: anorthite plagioclase. *Wood et al.* (1970) proposed that the way to explain this extreme mineralogical segregation was to assume that at least the outer portion of the Moon had at one time been molten. The LMO crystallized as it cooled. The first phases to crystallize would be dense Mg silicates such as olivine and pyroxene that sank to the bottom of the LMO; much later, when remelted, these served as the source of mare basalt volcanism (*Walker et al.,* 1975; *Snyder et al.,* 1992). Plagioclase, next to crystallize, is buoyant compared to the magma and hence floated to the surface to form what is now the highlands crust. The residual magma trapped between the crust and the solidified mantle was enriched in incompatible elements such as K, REE, and P, which inspired lunar scientists to adopt the acronym KREEP for this readily identified component of the Moon (*Warren and Wasson,* 1979; *Jolliff et al.,* 2000). (Because it concentrates radioactive K, U, and Th, KREEP emits γ rays that can be mapped from orbit.) In the absence of a lunar atmosphere, it takes much less than 1 m.y. for the LMO to develop an insulating crust (*Solomatov,* 2009), but once a crust forms cooling slows and final crystallization can be delayed another $\sim 10^{8}$ yr (*Solomon and Longhi,* 1977).

The LMO theory predicts that the mare basalt sources formed first, then the highland crust, and finally KREEP, with the expectation that these events took place over a time interval that is not resolvable by radioactive chronometers appropriate for determining the age of the samples. Inferred radiometric ages for lunar crustal rocks span more than 200 m.y. (*Carlson et al.,* 2014); more recent work has narrowed this range, at least for a few samples, to a surprisingly youthful range of 4.34 to 4.37 Ga (*Borg et al.,* 2015). Age scatter may reflect the true ages of crystallization or the consequences of shock metamorphism experienced by essentially every sample of ancient lunar crust. The oldest U-Pb age obtained for a lunar zircon is 4.42 Ga (*Nemchin et al.,* 2009), which, for the time being, provides the most convincing oldest age for a lunar material. Support for the ca. 4.35–4.45-Ga LMO age comes from Sm-Nd and Lu-Hf model ages for both the initial differentiation of the mare basalt sources (*Nyquist et al.,* 1996; *Boyet and Carlson,* 2007; *Brandon et al.,* 2009) and KREEP (*Carlson and Lugmair,* 1979; *Sprung et al.,* 2013; *Gaffney and Borg,* 2014). Tungsten isotopic measurements of lunar samples resolve no variability in $^{182}W/^{184}W$ that can be attributed to decay of ^{182}Hf, despite the wide range in Hf/W ratios of lunar materials. The Hf-W data thus provide only an upper limit of 4.51 Ga for the LMO (*Touboul et al.,* 2007). The chronological significance of the Hf-W data, however, has been called into question due to the sensitivity of W to late input of chondritic impactors that have W isotopic compositions dramatically different from those seen in terrestrial and lunar materials (*Touboul et al.,* 2015). Initial Hf isotope data for lunar zircons provide model ages in the 4.38–4.48-Ga range (*Taylor et al.,* 2009) with newer data extending this model age to 4.51 Ga (*Barboni et al.,* 2017). A complication with the older zircon Hf model ages is that they require an extremely low Lu/Hf ratio in the parental materials of the melt that the zircons crystallized from. Using typical KREEP Lu/Hf ratios for the sources of these zircons leads to model ages older than the solar system (*Barboni et al.,* 2017). The zircon data also conflict with whole-rock Lu-Hf data that give model ages in the 4.35–4.43-Ga range (*Sprung et al.,* 2013; *Gaffney and Borg,* 2014), although the zircon and whole-rock ages need not date the same event.

Dating the giant impact using Earth rocks is complicated by the fact that only about a part per million of Earth's surface area consists of rocks 3.8 Ga or older. The Hf-W system for terrestrial rocks requires that the greater part of core segregation took place within 30–50 m.y. of solar system formation, but does not constrain the timing of the Moon-forming impact (*Kleine et al.,* 2009; *Kleine and Walker,* 2017). The I-Xe system shows that most of Earth's Xe outgassed before 50–100 m.y. (*Wetherill,* 1975; *Ozima and Podosek,* 2002; *Avice et al.,* 2017). Given their short half-lives, however, these systems would be incapable of recording events occurring much later than that because the radioactive parent was extinct by that time. What these systems do provide is evidence that core-forming, and mantle-outgassing, events were occurring on the early Earth.

Hydrocode simulations of giant impacts of the scale needed to form the Moon show only limited, if any, ability to remix core and mantle (*Canup,* 2004; *Nimmo and Agnor,* 2006) and may not even be capable of fully melting and mixing Earth's mantle (*Melosh,* 1990; *Nakajima and Stevenson,* 2015); on the other hand, any such conclusions must be treated as provisional at best, because all such simulations are extremely underresolved numerically, which hobbles their ability to model mixing.

If the atmosphere was not expelled by the Moon-forming impact (cf. *Genda and Abe,* 2005), both the terrestrial Hf-W and I-Xe records could predate it. Longer-lived radiometric systems on Earth show a preponderance of 4.4 ± 0.1 Ga ages from U-Pb systematics of terrestrial rocks (*Allègre*

et al., 2008) to ^{146}Sm-^{142}Nd model ages for ancient rocks from Greenland (*Caro et al.,* 2006; *Bennett et al.,* 2007) and Quebec (*Roth et al.,* 2013) or the oldest ages for terrestrial zircons (*Froude et al.,* 1983; *Harrison,* 2009). Whether any of these ages date the giant impact is likely to remain a topic of debate for some time, but the preponderance of both lunar and terrestrial "ages" near 4.4 Ga (*Carlson et al.,* 2014) is at least suggestive that the Moon-forming impact was a relatively late phenomenon in solar system history, occurring perhaps 100–150 m.y. after the start of solar system formation.

4.2. A Recap On the Show So Far

Figure 5 is a pictorial summary of a mostly conventional story of terrestrial planet formation in the solar system. The Sun forms as the central condensation of an accretion disk. Planet formation models typically begin with a disk of gas and disk in orbit about the young Sun. The zero age is defined by the condensation of calcium-aluminum-rich inclusions (CAIs), small jagged particles made of highly refractory minerals with very ancient radiometric ages.

Fully differentiated 50–100-km-sized planetesimals, heated and melted by ^{26}Al, formed in the NC isotopic region almost immediately. Planetesimals may have formed just as quickly at the snow line, and been just as quickly melted and differentiated. It has been suggested that water vapor from much of the nebula could have frozen out at one particular distance and time, thus promoting the rapid assembly of snowballs into proto-Jupiter (*Stevenson and Lunine,* 1988). In any event, the observed separation of NC from CC isotopes implies that planetesimals in the general vicinity of the snow line grew rapidly into a planet big enough to create a gap in the nebula. It is possible that this planet was Jupiter, although the mass that had been in the gap may have spawned several giant planets. However it was accomplished, there seems little doubt that Jupiter and Saturn formed quickly compared to the current terrestrial planets.

In the inner solar system the nebula lasted some 3–4 m.y., by which point ^{26}Al was no longer the force that it had been. Large bodies composed of chondrules formed and survived without fully melting in the NC region. It is not clear whether these chondrite parent bodies made up a significant amount of mass compared to the melted bodies that had formed earlier. The planetesimals and embryos of the inner solar system may have been shepherded by the nebula such that they did not interact, or they may have sometimes merged to make bigger protoplanets. In the Grand Tack hypothesis, Jupiter migrated inward in response to gravitational interactions with the nebula and then outward in response to Saturn capturing it in a 2:1 resonance (*Walsh et al.,* 2012). Planetesimals, embryos, and protoplanets in the inner solar system were either scattered away or bulldozed inward; the region of Mars and other asteroids is mostly evacuated at this time (*Batygin and Laughlin,* 2015). After the nebular gas dissipates, collisions between remaining embryos and protoplanets from different regions create the terrestrial planets on a 3–150-m.y. timescale.

4.3. Earth After the Moon-Forming Impact

If there were no atmosphere, Earth's cooling after the Moon-forming impact would have been governed by the physics of silicate convection, and the surface would have been very hot, on the order of 2000 K (*Solomatov*, 2009). An atmosphere changes this fundamentally, because the atmosphere controls thermal radiation to space (*Zahnle et al.*, 2007, 2010; *Lebrun et al.*, 2013; *Sleep et al.*, 2014; *Lupu et al.*, 2014; *Zahnle et al.*, 2015). At first, when the atmosphere was dominated by silicate vapors (*Lupu et al.*, 2014) or hydrated silicate vapors (*Fegley et al.*, 2016), cooling was fast. But once Earth had cooled enough for the hot silicates to sink from sight, control over cooling shifted to greenhouse gases such as H_2O, CO_2, H_2, and CH_4.

Thereafter thermal blanketing by the greenhouse gases was very effective and Earth cooled slowly (*Lupu et al.*, 2014).

Water vapor was the most important of the greenhouse gases. For a water vapor atmosphere in equilibrium with condensed water (oceans or clouds), the upper limit on radiative cooling is on the order of 280 W m^{-2} (*Goldblatt et al.*, 2013). Earth's cooling is limited to the difference between the runaway greenhouse cooling rate of 280 W m^{-2} and the amount of sunlight absorbed (*Zahnle et al.*, 2007). With a cloud-dominated albedo in the range of 50%, the net insolation supplied by the faint young Sun would have been 120 W m^{-2}. Thus Earth's interior blanketed by a greenhouse water vapor atmosphere would have cooled at a net rate of 160 W m^{-2}.

Heat flow of 160 W m^{-2} is three orders of magnitude higher than radioactive heating at that time. Such high heat flow implies vigorous mantle convection, which is only possible in silicates when near their melting points. The heat flow remained high until the silicates cooled to a

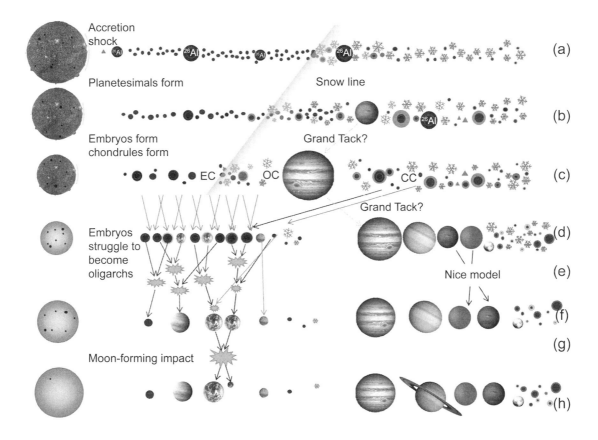

Fig. 5. See Plate 1 for color version. A pictorial synopsis of standard solar system formation. **(a)** 0–1 m.y.: 100-km-sized NC planetesimals quickly assemble and melt from heating by ^{26}Al. **(b)** 0–2 m.y.: Snowline planetesimals/boulders/pebbles assemble into a proto-Jupiter the size of Neptune, which balloons into something more Jupiter-like with the gravitational capture of nebular gas. CC planetesimals assemble and melt from heating by ^{26}Al. Jupiter separates NC (inside Jupiter's orbit) from CC planetesimals (outside); the two isotopic reservoirs remain separate while the nebula endures. **(c)** 0–4 m.y.: Planetesimals and embryos merge to make proto-planets. The Grand Tack is optional. The snowline moves inward as the nebula thins and the Sun fades. **(d),(e)** 4–60 m.y.: Embryos and proto-planets mix and merge to create planets. Mars survives as a fossil protoplanet or embryo by practicing nonintervention. (e),(f) Hypothesized Nice model rearrangement (timing uncertain). **(g)** 60–150 m.y.: The Moon and Earth form as products of the collision of two planets. **(h)** Later. Saturn acquires rings. Throughout, the size of the Sun suggests how bright it is (*D'Antona and Mazzitelli*, 1994), and the indications of activity (spots, prominences) on it suggest its EUV luminosity.

point where they become too viscous to convect easily, after which the heat flow decayed quickly (*Abe,* 1997; *Solomatov,* 2009). Depending on the energy released by the impact and the composition of the atmosphere, it would have taken 1–10 m.y. for Earth to cool to the point where geothermal heat flow was no longer big enough to significantly affect the climate (*Lebrun et al.,* 2013; *Zahnle et al.,* 2015).

Tidal heating (the conversion of Earth's rotational energy into heat by Moon-raised tides) can prolong the runaway greenhouse state by several million years (*Zahnle et al.,* 2015). Tidal heating stems from friction within the Earth as it flexes to follow the Moon. Frictional heating is concentrated in mantle materials that are just beginning to freeze; neither hard rock nor low-viscosity liquids are as easily heated by flexing. Because mantle cooling is fixed at the rate the atmosphere radiates, a higher rate of tidal heating would lead to more melting, thus reducing tidal heating. This is a stable negative feedback. While the negative feedback holds, tidal evolution of the Earth-Moon system is locked to the modest net radiative cooling rate of Earth's runaway greenhouse atmosphere. The feedback breaks when the Moon has spun out far enough that tidal heating can no longer supply the 160-W m^{-2} heat flow needed to sustain the runaway greenhouse state. At this point the mantle hardens and tidal heating dwindles to insignificance compared to insolation, and surface cooling resuming. The "magma ocean" (under a solid crust) may last another 10^8 yr, as on the Moon.

As the mantle solidifies, water, which is soluble in magma but fails to make stable low-pressure minerals that could persist in hot solid silicates, is mostly squeezed out of the mantle and accumulates at the surface (*Matsui and Abe,* 1986; *Abe and Matsui,* 1988; *Zahnle et al.,* 1988, 2007; *Elkins-Tanton,* 2008). Other less-magma-soluble gases such as CO_2 and H_2 will also partition into the atmosphere (*Holland,* 1984). For specificity we presume that 100 bars of CO_2 accumulate in the atmosphere, which is about half Earth's total CO_2 inventory. A 100-bar CO_2-H_2O atmosphere supports a surface temperature on the order of ~500 K (*Abe,* 1997). The hot, high-pressure ocean would be a fiercely effective solvent of seafloor basalts, presumably able to extract and dissolve much of the periodic table, which among other things could hasten the fixing of CO_2 into carbonate. The actual chemical composition of the carbon- and hydrogen-containing species depends on the oxidation state of the mantle at the time. If relatively oxidized, the chief gases will be H_2O and CO_2; if relatively reduced, the gases will be a mix of H_2O, H_2, and CH_4 (*Hirschmann,* 2012).

Further cooling exsolved what had been dissolved in the oceans. The precipitates fell to the surface or plated the seafloor or encrusted hydrothermal systems. Salts, NaCl in particular, would have been prominent among these. The pace of events is governed by how quickly greenhouses gases can be removed from the atmosphere. Methane is removed by photolysis (either polymerized to tars or oxidized to CO_2 consequent to photolysis of H_2O), and hydrogen escapes. Carbon dioxide must be removed by fixing it as carbonate rock and then subducting it into the mantle. Mass balance demands that the seafloor must be recycled several times to supply cations enough to remove 100 bars of CO_2 as carbonates (*Sleep et al.,* 2001). Titanium thermometry of Hadean zircons demonstrates that, at least locally, subducting material was cool enough even at 4.4 Ga for carbonate subduction to take place (*Hopkins et al.,* 2008; *Harrison,* 2009). How long it actually took to tuck the CO_2 into the mantle is an open question.

Figure 6 illustrates broad aspects of Earth's cooling after a minimum energy (canonical) Moon-forming impact (*Zahnle et al.,* 2015). Two cases are shown that illustrate the influence of the atmosphere. The "100-bar" atmosphere represents a case in which most of the H_2O is in the mantle and other greenhouse gases have limited effect. The "1000-bar" case is a proxy for the very strong greenhouse effect that we actually expect given the wealth of gases that can be made from moderately volatile elements present as vapors in a 1000–1500-K atmosphere.

Figure 7 shows the evolution of Earth's surface environment after a canonical (low-energy) Moon-forming impact. It is assumed that the equivalent of 100 bars of CO_2 are released from the magma ocean. We assume that the composition of the post-impact atmosphere is controlled by interaction with the mantle while the surface magma ocean endures, and we have arbitrarily assumed that the redox chemistry is governed by a weakly reduced QFM-1.5 mineral buffer that approximates that of carbonaceous chondrites (*Schaefer and Fegley,* 2017). The abbreviation QFM refers to the minerals quartz (SiO_2), fayalite (Fe_2SiO_4, the Fe-rich end-member of olivine), and magnetite (Fe_3O_4). The redox power — aka oxygen fugacity, denoted f_{O_2}, which has units of pressure — is determined by the interplay between Fe^{+2} and Fe^{+3} in the minerals and the O_2 gas that would be evolved (or consumed) by reducing (or oxidizing) the iron. An oxygen fugacity described as QFM-1.5 means that the oxygen fugacity is smaller than that of the QFM buffer by a factor of $10^{-1.5}$. For comparison, the modern mantle is usually approximated by the QFM buffer.

After the magma ocean freezes and geothermal heating no longer supports a runaway greenhouse effect, the atmosphere cools and after a few thousand years the water condenses and rains out. In Fig. 7, most of the CO and H_2 in the atmosphere combine to make methane in the thousands of years it takes to cool from ~1500 K to ~800 K. The methane is later destroyed photolytically on a timescale determined by solar Lyman α irradiation to make organic hazes, CO, and CO_2; some of the organic hazes precipitate and much of the evolved hydrogen escapes to space. For the illustration, Fig. 7 assumes that CO_2 is fixed in carbonates that are subducted by the mantle on an arbitrary timescale on the order of 100 m.y. Then winter comes.

5. THE HADEAN AND THE ORIGIN OF LIFE

The Hadean Earth can be broadly thought of as the period of interpolation between the Moon-forming impact and the gradual emergence of a recognizable rock record between 3.8 and 4.0 Ga. Its importance here is that the Hadean Earth is currently our best guess for the cradle of life.

5.1. Origin of Life

The modern science of the origin of life on Earth began with a short essay by J. B. S. Haldane (*Haldane,* 1929) and a fully realized monograph by A. I. Oparin (*Oparin,* 1938). In perspective the two works are strikingly similar. Haldane was a great biologist and a gifted writer; his essay is readily downloaded as a PDF file from his Wikipedia biography and it is highly recommended reading. Oparin, or at least his translator, also writes well, and he provides a highly readable history of the subject (including when possible pithy quotes from Friedrich Engels). A key point is that the origin of life was not recognized to be a scientific problem until the seventeenth century, as before that life was well known to emerge spontaneously from Earth. The full extent of the problem was made clear in the mid-nineteenth century by Pasteur, who proved that only life could beget life. Thus, at the end of the nineteenth century a modern scientist (e.g., Pasteur) could divide the world into living and nonliving streams of descent, with the burden of proof seeming to lie equally on those who would argue for a material origin of life and those who would argue for a divine origin.

Oparin was a materialist. He emphasized a point made by Darwin: Life cannot spontaneously originate on Earth now, because any chemical building blocks that might spontaneously appear would be eaten; but before the origin of life, such chemical building blocks would persist, accumulate, ramify, and evolve. Oparin originated the metaphor of ancient ponds, lakes, and seas as soups. Oparin and Haldane fully appreciated the improbability of developing a rich organic chemistry in highly oxidized conditions like those at the surface of modern Earth. Only a highly reduced early terrestrial environment would do. In Oparin's picture the reducing power came from the hydrogen of the Sun, which he knew to be the greatest part of the stream of matter pulled out of the Sun by the passing star, and from which the solar system formed. *Oparin* (1938) takes chemical evolution forward in considerable detail in the second half of his book and we will not follow him here — we simply note the emphasis he placed on methane, ammonia, formaldehyde, and especially hydrogen cyanide as primordial materials suitable for further development, a recurring theme in origin of life studies that is as relevant today as it was in 1938 (*Oró and Kamat,* 1961; *Ferris et al.,* 1978; *Stribling and Miller,*

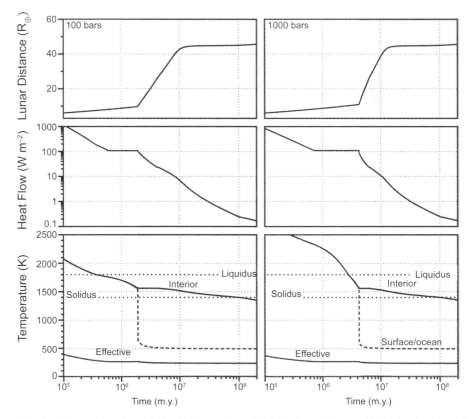

Fig. 6. Post-Moon-forming impact cooling of Earth for modest (100 bar) and strong (1000 bar) atmospheric greenhouses (after *Zahnle et al.,* 2015). Shading indicates runaway greenhouse regimes. The top panels show the Earth-Moon distance, the middle panels the geothermal heat flow, and the bottom panels the effective radiating temperature of the atmosphere, the surface (or ocean) temperature, and the mantle's internal temperature with respect to the liquidus and solidus. The near discontinuity in the surface temperature at 2 m.y. (5 m.y. at 1000 bars) is triggered by the phase transition in the mantle's viscosity that corresponds to freezing, which is accompanied by a rapid decay of geothermal heat flow.

1987; *Oró et al.,* 1990; *Ricardo et al.,* 2004; *Powner et al.,* 2009; *Benner et al.,* 2019).

The link between the origin of life and a more modern understanding of the origin of Earth and the solar system was developed by *Urey* (1952). Urey's overall perspective could be our chapter summary: "In order to estimate the early conditions of the Earth, it is necessary to ask and answer the questions of how the Earth originated, and how the primitive Earth developed into the present Earth." To Urey, the presence of life on Earth meant that early Earth must have been conducive to the origin of life, and therefore conducive to abiotic synthesis of complex organic molecules from simple monomers. This means that conditions on early Earth were reducing, and that the atmosphere was rich in H_2, CH_4, and either NH_3 or HCN (see the chapter by Baross et al. in this volume).

The hypothesized reduced atmosphere was the setting for the famous Miller-Urey experiments, in which sparked or UV-irradiated gas mixtures spontaneously generated interesting organic molecules (*Miller,* 1953, 1955; *Miller and Urey,* 1959; *Cleaves et al.,* 2008; *Johnson et al.,* 2008). A modern observer might note that simple and complicated organic molecules of every conceivable kind are found in abundance in many meteorites, and therefore might conclude that the spontaneous formation of organic

molecules of considerable complexity is actually rather easy under commonplace conditions in the solar nebula or on the surfaces or interiors of meteorite parent bodies, and probably a great many other places. There is no particular reason for not including the surface or near-surface of Earth among these places other than the assumption that the earliest Earth's crust and mantle were as oxidized as they are today. The difficulty for life is not in making organic molecules; the difficulty lies in making only certain ones in large numbers on demand.

There are alternative visions to a reduced atmosphere on early Earth. One that has at times been fairly popular is the hypothesis that the important abiotic materials for the origin of life were delivered directly by comets and asteroids (*Anders,* 1989; *Chyba et al.,* 1990; *Chyba and Sagan,* 1992; *Whittet,* 1997; *Pierazzo and Chyba,* 1999; *Pasek and Lauretta,* 2008). The strength of this hypothesis is that comets and asteroids are often well-endowed with complex organic molecules. An exogenous source has the advantage of being able to deliver concentrated chunks of organic matter to particular places at particular times (*Clark,* 1988; *Whittet,* 1997), but it has the disadvantage that at the typically high impact velocities of stray bodies with Earth the surviving fraction of fragile organic molecules upon impact is expected to be rather low. Of course

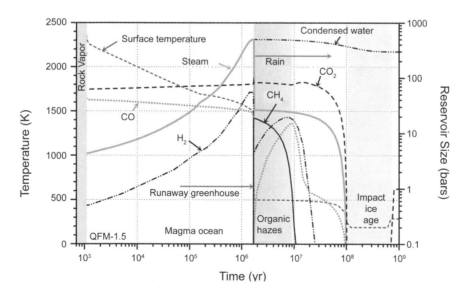

Fig. 7. Schematic evolution of Earth's surface environment after a Moon-forming impact (low-energy canonical impact, 100-bar CO_2 atmosphere). The first phase is rock vapor. Thereafter the atmosphere remains in thermochemical equilibrium with the mantle while the magma ocean endures. Water degasses from the mantle as the magma ocean solidifies. Cooling is controlled by water's runaway greenhouse effect. The surface magma ocean stage ends relatively abruptly on a timescale on the order of 10^3–10^4 yr. Most of the CO and H_2 are combined to make methane in the thousands of years it takes for the surface and atmosphere to cool from ~1500 K to ~800 K. Methane is later destroyed photolytically to make organic hazes, CO and CO_2; some of the organic hazes precipitate and much of the evolved hydrogen escapes to space. CO_2 is eventually fixed in carbonates that are subducted by the mantle. Reservoirs are shown in "bars" to facilitate comparison between them. Speciation is consistent with the QFM-1.5 mineral buffer, which itself is an arbitrary choice.

a mound of organic molecules is not the same thing as life (*Davies*, 2000), and it may seem to the reader that an active chemosynthetic pathway stands a better chance of evolving life (see the chapter by Hoehler et al. in this volume).

Another alternative is synthesis of organic molecules in hydrothermal systems. In this scenario, organic molecules form spontaneously from dissolved CO_2 and N_2 in hot waters circulating through ferrous silicates (*Shock*, 1990; *Shock et al.*, 1995; *Shock and Schulte*, 1998). The relatively low temperatures and high pressures of hydrothermal systems favor methane and ammonia, even for mantle materials as oxidized as today. These sorts of conditions may have played a part in the synthesis of organic molecules in asteroid parent bodies. However, actual evidence of de novo synthesis of organic molecules in real hydrothermal conditions from difficult molecules like CO_2 and N_2 has been scant; a recent report from *Ménez et al.* (2018) seems the best case yet. Having said this, it is quite clear that the hydrothermal origin of life is a testable hypothesis in our solar system, as there are many icy worlds in the outer solar system that should be alive according to the hypothesis (e.g., Triton) (*Shock and McKinnon*, 1993), and at least one of which is freely offering up samples (see the chapter by Cable et al. in this volume).

The geological argument against a reduced early atmosphere has held sway for more than half a century (*Poole*, 1951; *Holland*, 1964, 1984; *Abelson*, 1966; *Walker*, 1977; *Frost and McCammon*, 2008; *Armstrong et al.*, 2019). Modern volcanic gases are consistent with the QFM mineral buffer. An atmosphere dominated by mantle degassing at QFM is only weakly reducing, with percent levels of H_2 and CO compared to H_2O and CO_2 [with plausibly somewhat more reducing power in sulfur gases SO_2, S_2, or H_2S (*Clark et al.*, 1998)]. There is little evidence that the early Archean was different (*Delano*, 2001; *Canil*, 2002), and the only Hadean evidence — rare earth elements in Hadean zircons (*Trail et al.*, 2012) — is also consistent with QFM. There is now increasing experimental evidence suggesting that ferrous silicates were disproportionate to ferric and metallic iron phases at the high pressures encountered in Earth, with the latter likely to make their way to the core, leaving the upper mantle in a QFM-like state of oxidation from the time our planet first became big enough to be called Earth (*Frost and McCammon*, 2008; *Armstrong et al.*, 2019).

Given the hostility of a QFM mantle to a reduced atmosphere, workers have turned to gases equilibrating with the impactors themselves, which are usually much more reduced and often better endowed in atmophile elements than Earth (*Urey*, 1952; *Tyburczy et al.*, 1986; *Ahrens et al.*, 1989; *Schaefer and Fegley*, 2007; *Hashimoto et al.*, 2007; *Sugita and Schultz*, 2009; *Schaefer and Fegley*, 2010; *Kuwahara and Sugita*, 2015; *Schaefer and Fegley*, 2017). *Urey* (1952) put it this way: "That such objects fell on the Moon and Earth at the terminal stage of their formation I regard as certain, and it is difficult in this subject to be certain about anything. The objects contained metallic iron-nickel alloy, silicates, graphite, iron carbide, water or water of crystal-lization, ammonium salts and nitrides, that is substances which would supply the volatile and nonvolatile constituents of the Earth. [Moreover, these] materials would have fallen through the atmosphere in the form of iron and silicate rains and would have reacted with the atmosphere [and hydrosphere] in the process." The overall picture is that the biggest impacts would have distributed hot, highly-reduced material on a global scale creating, at least transiently, reduced conditions suitable to abiotic synthesis of hopeful molecules. To address Urey's hypothesis we need first to address the nature and quantity of the impacts.

5.2. Good Impacts and Bad Impacts

The mantle's excess HSEs set an upper bound on late impacts that might be relevant to origin of life. The late veneer measured in this way is very big. Viewed as a wrapper, 0.5% of Earth's mass corresponds to a shell 20 km thick. Gathered into a sphere, this corresponds to a rocky world 2300 km in diameter — as big as Pluto or Triton, but denser and more massive. If the veneer were sourced from fragments of differentiated worlds, the veneer mass could be a little smaller or very much bigger. A stray iron core would be efficient, while a mantle-like composition like the Moon might have to be as big as Earth itself to do the job.

Essentially all types of chondrites have similar HSE abundances (*Horan et al.*, 2003), but only a few carbonaceous chondrites have enough water to be significant if restricted to 0.5% of Earth's mass. If only the wettest carbonaceous chondrites are used, 0.5% of Earth's mass is enough to deliver 2–3 oceans. This rough coincidence between HSEs and water led directly to the once-popular hypothesis (*Wänke and Dreibus*, 1988; *Dreibus and Wänke*, 1989; *Albarède et al.*, 2013) that most or all of Earth's atmophile elements were delivered by a carbonaceous-chondrite-like late veneer (see also the chapter by Meech and Raymond in this volume).

However, the late veneer now appears constrained by Ru isotopes to resemble either E-type material (enstatite chondrites or aubrites) or IAB-iron-type material, and thus maps to a deep inner solar system reservoir (*Dauphas*, 2017; *Fischer-Gödde et al.*, 2015; *Fischer-Gödde and Kleine*, 2017; *Bermingham et al.*, 2018; *Hopp and Kleine*, 2018). This apparently excludes the late veneer as the source of water on Earth (*Fischer-Gödde and Kleine*, 2017).

To illustrate, assume that the late veneer was no more than 10% CC material. (Recall that the CC class is defined by isotopes, and does not necessarily imply water- or organic-rich.) The most water-rich CC materials, the CI chondrites, have on average 2% H (*Palme et al.*, 2014), so 10% CI in a 0.5% of Earth's mass late veneer delivers 5×10^{20} kg of H_2O, or about 30% of Earth's ocean. Even ignoring the fact that most known comets have decidedly different isotopic compositions of many of the key atmophile elements such as N and H (*Alexander*, 2017) and Xe (*Marty et al.*, 2017), the best that comets can do is supply 50% of an ocean, and in all likelihood they contribute much less than this (*Dauphas and Marty*, 2002; *Marty et al.*, 2017). The impotence of the late veneer as a source of atmophile elements is unique to H_2O, because C, N, S, and even noble gases are all abundant in E-type chondrites.

An important, and often overlooked aspect, of the late veneer is that the HSEs provide no time constraints on when the veneer was added. The chemistry tells us only that it has to be added after chemical communication between core and mantle stopped. Evidence from HSE abundance variations in komatiites (high-temperature mantle melts) (*Maier et al.*, 2009) and of W isotopic composition of early Earth rocks (*Willbold et al.*, 2011) have been interpreted to suggest that the late veneer took at least a billion years to be mixed into Earth's mantle. The komatiite HSE evidence has not been supported by more recent work (*Puchtel et al.*, 2014) and a variety of explanations have been proposed for the W isotope results (*Touboul et al.*, 2012, 2014; *Rizo et al.*, 2016).

The size-number distributions of solar system bodies suggest that packing much of the late veneer into a single large body is likely (*Sleep et al.*, 1989; *Tremaine and Dones*, 1993; *Bottke et al.*, 2010; *Raymond et al.*, 2013; *Brasser et al.*, 2016; *Genda et al.*, 2017a). With that said, it does not follow that the late veneer was added to Earth in a single moment. There is a considerable likelihood, estimated by *Agnor and Asphaug* (2004) to be on the order of 50%, that the final interaction of Earth with a smaller body is the disintegration of the latter. The debris would be swept up by Earth and Moon over hundreds of thousands of years or more (*Genda et al.*, 2017a,b). This kind of distributed event is more likely to strand all its HSEs in the mantle than the direct impact of a Pluto-sized body, which might be expected to drive the impactor's core directly into our own. Stranding all the accreted HSEs in the mantle fits a second requirement imposed by the mantle's Ru isotopes, which were not mass-fractionated by partitioning between the mantle and core (*Fischer-Gödde and Kleine*, 2017; *Hopp and Kleine*, 2018). Distributed delivery would also be more effective at chemically reducing Earth's atmosphere (*Genda et al.*, 2017b), using the process envisioned by *Urey* (1952).

There is a caveat to this neat picture: It is possible that the HSEs came from Theia itself (*Newsom and Taylor*, 1989; *Sleep*, 2016; *Brasser et al.*, 2016; *Morbidelli et al.*, 2018). We know that some of Theia's core was incorporated into the Moon-forming disk, as the Moon has a small core of its own that must have accreted from that disk. Presumably there would have been more Theian core material in the inner parts of the Moon-forming disk that fell to Earth.

If the HSEs came from Theia, the lower bound on late impacts must be extrapolated from the lunar cratering record (*Sleep et al.*, 1989; *Zahnle and Sleep*, 1997, 2006). From the observed numbers and sizes of lunar impact basins, we estimate that the total mass hitting the Moon was on the order of $2–6 \times 10^{19}$ kg (*Zahnle and Sleep*, 2006), comparable to the 10^{20} kg estimated by *Chyba* (1991). Taking into account Earth's bigger cross section and the top-heavy size-number distribution of impactors, the corresponding total mass hit-

ting Earth would have been in the range of $1–10 \times 10^{21}$ kg (*Zahnle and Sleep*, 2006). This corresponds to 3–30% of the maximum late veneer permitted by the excess HSEs. In the upper half of this range, most of the mass must be delivered by one or two E-type or IAB-iron-type bodies to remain consistent with the Ru isotopes. If on the other hand the true veneer were as thin as 3% of the maximum veneer mass, its composition is not constrained.

5.2.1. Good impacts. Impacts, especially the big impacts of the late veneer, were sources of reducing power to the atmosphere and ocean (*Benner et al.*, 2019). Reduced gases can be directly released from the impactor itself (impact degassing) (*Hashimoto et al.*, 2007; *Schaefer and Fegley*, 2007, 2010; *Kuwahara and Sugita*, 2015; *Marchi et al.*, 2016; *Schaefer and Fegley*, 2017), or they can be reduced from the atmosphere and ocean by the iron metal delivered by the impactor (*Kasting*, 1990; *Benner et al.*, 2019).

Most impact degassing studies focus on equilibrium gas compositions with respect to a highly reduced mineral buffer established by the impactor; e.g., the iron-wüstite (IW) buffer (established by the chemical reaction $Fe + \frac{1}{2}O_2 \leftrightarrow FeO$) or the quartz-fayalite-iron (QFI) buffer, the latter appropriate to enstatite and ordinary chondrites (*Schaefer and Fegley*, 2017). The metallic iron extracts oxygen from the gas to make iron oxides. Gas compositions also depend on the atmophile elements in the impactors and the quench conditions — the temperature and pressure — at which chemical reactions stop happening in a cooling gas. Impact degassing effectively presumes that each impact injects a plume of reduced gases into a preexisting atmosphere. Unless catalysts are available to lower the quench temperature (*Kress and McKay*, 2004), gases released by smaller impacts cool quickly enough to quench to CO, CO_2, H_2, and N_2 — the composition favored at high temperatures — even if CH_4 and NH_3 are thermodynamically favored at ordinary temperatures. Moreover, the IW and QFI mineral buffers fail when the available iron is used up. *Schaefer and Fegley* (2017) take this into account for a mixture of impacting bodies; the result is a less-reduced gas plume.

Great impacts acting on the atmosphere and ocean work in a different regime than impact degassing. Availability of metallic iron becomes limiting. To illustrate the order of magnitude, the maximum late veneer corresponds to 3×10^{22} kg of chondritic material containing 1×10^{22} kg of metallic iron, or 1.8×10^{23} moles of Fe. This is enough Fe to reduce 2.3 oceans of H_2O to H_2, or reduce 1500 bars of CO_2 to CO. Therefore only the biggest impact has a realistic chance of fully reducing the atmosphere and ocean (*Benner et al.*, 2019). Lesser impacts can reduce only as much H_2O and CO_2 as their supplies of metallic iron permit. Thus to first approximation, the reducing power of big impacts scales linearly with their mass.

Great impacts — especially those that evaporate the oceans — differ from smaller impacts in two other important ways. First, the high H_2O vapor pressures in a steam atmosphere strongly favor CH_4 and NH_3 over CO or N_2, a preference that goes as the square of the pressure. Second, cooling is slow because the thermal inertia of a hot atmosphere containing hundreds of bars of gas is large. It takes more than a thousand years to cool 270 bars (an ocean) of steam at the runaway greenhouse cooling rate. Quench temperatures for gas phase reactions in the CH_4-CO-CO_2-H_2O-H_2 system drop to 700–800 K, and quench temperatures for the NH_3-N_2-H_2O-H_2 system drop to 1100 K. (Quench temperatures are determined for gas phase chemistry using fits to calculations obtained by time-stepping thermochemical kinetics simulations of brown dwarf atmospheres (*Zahnle and Marley,* 2014).) High pressures and low quench temperatures put these atmospheres deep into the CH_4 stability field, and even NH_3 can become abundant. How much methane actually forms depends on how much CO_2 was in the atmosphere before the impact and on how much reducing power was delivered by the impact. If for example we presume that a maximum-late-veneer scale impact took place on an Earth with a 100-bar CO_2 atmosphere [perhaps left over from the Moon-forming impact (*Zahnle et al.,* 2007)] and 1.85 oceans of water at the surface, essentially all the carbon is converted to methane, diluted in several tens of bars of H_2. The resulting atmosphere is rather Neptune-like.

From the point of view of the origin of life and medieval metaphors, the great impacts are two-edged swords. In their aftermath they leave Earth primed and ready to start life under a classic Urey-Miller H_2-rich, CH_4-rich, possibly even NH_3-rich, atmosphere that origin of life theorists have long favored. Unfortunately, their first act is an attempt to wipe out everything that had been accomplished earlier. What this suggests is that the transient reduced atmosphere may give a planet exactly one highly favorable roll of the dice. Smaller later impacts can inject significant amounts of H_2 or CO into the system, but they require unidentified catalysts to generate large amounts of CH_4 or NH_3.

Powers of reduction, extinction, and creation are estimated in Table 1. Quantities in Table 1 assume the impact of reduced iron, EC- or OC-like bodies into a terrestrial surface environment characterized by 500 bars (1.85 oceans) of water, 1 bar of N_2, and 5 bars of CO_2. (These initial conditions are arbitrary.) Impact-generated gases are given

in partial pressures, measured in bars. Note that nearly all of the 5 bars of CO_2 are converted to CH_4 in the largest impacts. The low-seeming CH_4 and NH_3 partial pressures are consequent to these being H_2-rich atmospheres. The impactors are assigned a density of 3.3×10^3 kg m^{-3} and and impact velocity of 17 km s^{-1}. It is assumed that 25% of the impact energy is used to evaporate oceans (*Zahnle and Sleep,* 1997, 2006). Quotes indicate that we are considering Vesta- and Ceres-sized bodies, not Vesta or Ceres themselves. The South Pole-Aitken basin ("S.P.-Aitken," or "SPA") refers to the Moon's biggest verified crater. Impacts of the SPA scale are certain to have taken place on the Earth. Such impacts are big enough to sterilize the photic zone, but would not be expected to annihilate life.

5.2.2. Photochemistry of the earliest atmosphere. *Kasting* (1990) found using a one-dimensional photochemical code that the reducing power of meteors during the late bombardment can make CO more stable photochemically than CO_2 before 4.0 Ga; however, this calculation assumes that the reducing power of the late bombardment acted on the atmosphere continuously. Because impacts are stochastic and, when viewed individually, are focused events concentrated in time and space, a steady-state description of the Hadean atmosphere driven by impacts may be suspect. It is difficult to reduce CO much further by atmospheric photochemistry alone, but photochemistry in the presence of hydrogen or water vapor readily generates formaldehyde (HCHO) (*Pinto et al.,* 1980; *Wen et al.,* 1989). There is some interest in CO as a constituent of the earliest mantle in the form of Fe carbonyl (*Hirschmann,* 2013).

Transient CH_4-rich atmospheres are more exciting. Methane photochemistry generates hydrocarbons that form organic hazes and tars while CH_4 is abundant (*Lasaga et al.,* 1971; *Yung and Pinto,* 1978; *Zahnle,* 1986; *Trainer et al.,* 2006; *Hörst et al.,* 2018). When CO_2 is preponderant, methane is more often oxidized to formaldehyde (HCHO) or CO. Water photolysis oxidizes CO and CH_4 to CO_2 with the collateral production of H_2. In all cases (hazes, CO, CO_2) a great deal of H_2 is formed that escapes to space.

Nitrogen (N_2) photolysis produces N atoms (*Huebner et al.,* 1992) that have a strong tendency to react with

TABLE 1.   Representative Hadean impacts into 5 bars CO_2 and 5 km global ocean.

Category	Mass (kg)	Vap*	Red†	N‡	Products, Dry Atmosphere, Partial Pressure (bars)				
					CO_2	H_2	CO	CH_4	NH_3
Max HSE	2×10^{22}	130	2	0–1	2(–7)	50	2(–7)	0.24	0.04
Pretty big	2.5×10^{21}	14	0.2	1–2	4(–4)	7.4	6(–6)	0.34	0.012
"Ceres"	1×10^{21}	5	0.08	1–4	0.06	3.9	3(–4)	0.52	0.006
"Vesta"	2.5×10^{20}	1.5	0.02	2–8	1.6	4.0	0.004	0.14	0.002
Sub-Vesta	1×10^{20}	0.7	0.008	5–15	2.6	2.7	0.013	0.005	6(–4)
S.P.-Aitken	1×10^{19}	0.1	8(–4)	20–50	4.2	0.38	0.06	3(–7)	2(–5)

*Oceans vaporized by the impact, using 25% of impact energy.
†Reducing power of the impact, expressed as oceans reduced to H_2.
‡This is the total number in each size class.

hydrocarbon radicals from CH_4 photolysis to make HCN and other nitriles (*Zahnle*, 1986; *Tian et al.*, 2011). Photochemical nitriles and other hydrocarbons have been hypothesized as a source of reduced nitrogen and reduced carbon that can be subducted by the mantle (*Wordsworth*, 2016). Photolytic production of HCN from CH_4 can be efficient in a CO_2 atmosphere (*Zahnle*, 1986; *Tian et al.*, 2011). Ammonia is more problematic because (1) it is extremely soluble in cool waters, and (2) it is highly susceptible to UV photolysis and thus needs a very effective UV shield, possibly supplied by organic hazes, to endure for long (*Kuhn and Atreya*, 1979; *Sagan and Chyba*, 1997; *Pavlov et al.*, 2001; *Wolf and Toon*, 2010). Although photolysis is very fast, rainout may be faster, which would leave impact-generated ammonium in play for the purposes of origin of life.

Figure 8 presents a highly approximated quantitative illustration of possible atmospheric evolution after a maximum late veneer impact ("Max HSE" in Table 1). We assume that, before the impact, Earth had 500 bars of water and 100 bars of CO_2 at the surface, a relatively oxidized initial condition that could have been created in the aftermath of the Moon-forming impact (*Zahnle et al.*, 2007). The composition of the gases for T > 1200 K is governed by thermochemical equilibrium with the IW buffer. Thereafter total atmospheric oxygen ($CO + 2CO_2 + H_2O$) is conserved. The prominent phase transition — lasting less than 10,000 yr in this example — is when water condenses and rains out (note that the water is supercritical, so an abrupt phase transition is a crude approximation to what would actually be more drawn out). Thereafter the atmosphere evolves photochemically on a timescale set by solar UV radiation and hydrogen escape.

At first, methane photolysis leads mostly to polymerization of organic molecules that form hazes that fall to the surface. As methane is depleted and hydrogen escapes, oxidation of methane to make oxidized organics, then CO, and

finally CO_2 becomes monotonically more important. Nitriles from photolysis of CH_4 and N_2 can continue to form and precipitate even after the hydrocarbon hazes have cleared. What eventually happens to the tars and nitriles is left to the imagination. Overall, this view of early Earth is reminiscent enough of modern Titan that the analogy seems appropriate.

By the end of this particular maximum HSE event, hydrogen corresponding to 200 bars of H_2O escapes, leaving 300 bars of H_2O (1.1 oceans) behind. Fifty bars of "tars" and a few tenths of a bar of nitriles fall from the atmosphere. These particular numbers are determined mostly by the initial conditions of 500 bars of H_2O and 100 bars of CO_2.

Figure 9 provides a highly schematic overview of good impacts through the Hadean. It should be noted that there are very few events in total in Table 1, and they need not have occurred in any particular order, so that there is a considerable likelihood, perhaps 10%, that the largest ocean-vaporizing impact was also the last one.

5.2.3. Bad impacts. Since the day 66 m.y. ago when the golden age of dinosaurs ended on Earth, it has been known that the occasional stray body striking an inhabited planet can profoundly impact its biology. For Earth today the impact hazard remains but events are infrequent. The frequency and magnitude of cosmic impacts were both much higher in the Archean, and higher still in the Hadean (*Sleep et al.*, 1989; *Zahnle and Sleep*, 1997).

There have been many studies of impact-generated assaults on Earth's habitability. Most have tended to emphasize the dangers inherent in great impacts that evaporate the oceans and heat the underlying sediments to temperatures that life cannot survive (*Sleep et al.*, 1989; *Zahnle and Sleep*, 1997; *Abramov et al.*, 2013; *Marchi et al.*, 2014; *Shibaike et al.*, 2016; *Abramov and Mojzsis*, 2018), although some have concluded that the magnitude of the impact bombardment has been overstated (*Boehnke and Harrison*, 2016; *Brasser et al.*, 2016).

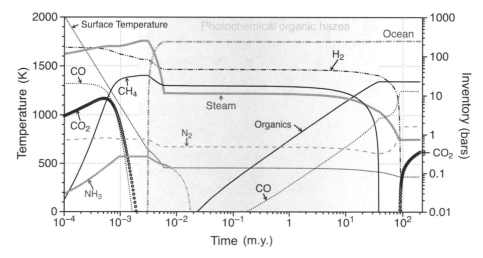

Fig. 8. Schematic atmospheric evolution after a maximum late veneer impact into an Earth with 500 bars of water and 100 bars of CO_2 at the surface. Inventories are given in bars (= partial pressure for gases). Nearly all the CO_2 is thermochemically converted to CH_4. In this particular example half the carbon that was in CH_4 is polymerized into organic hazes that eventually precipitate to the surface. Hydrogen corresponding to 170 bars of H_2O escapes over tens of millions of years. The timescale shown here is semi-quantitative.

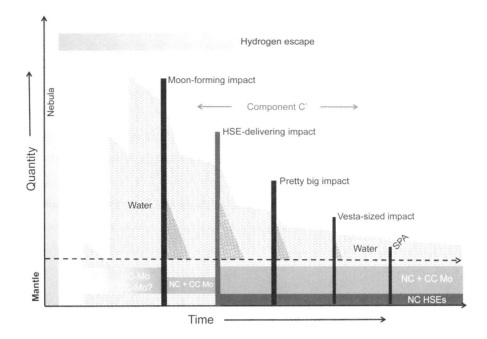

Fig. 9. A notional schematic history of water, methane, and highly siderophile elements on Earth in response to late great impacts. Water on Earth probably accreted before the Moon-forming impact, and at least partly lived in the mantle. Transient methane-rich atmospheres are generated by ocean-vaporizing impacts. Here we presume that the mantle's excess HSEs were delivered mostly by a single impact, and that the CC component of the mantle's Mo predates the Moon-forming impact. "SPA" refers to an impact on the scale of the one that excavated the lunar South Pole-Aitken (SPA) basin — too small to evaporate the oceans — which is implied here by submerging SPA's pole under the waves. Major impacts are shown decreasing in magnitude as time passed, but the true order leaves much to chance; it is possible that the largest of the ocean-vaporizers was also the last.

Figure 10 presents a graphical summary of some first-order thermal consequences of large impacts on Earth as functions of impact energy, following *Segura et al.* (2013). In these calculations Earth is presumed to have continents and the ocean where present is 4 km deep. The top panel addresses temperatures in the atmosphere, on dry land, and at the surface of the ocean. The atmosphere heats quickly and efficiently when global impact ejecta reenter the atmosphere. Exposed land surfaces also heat quickly because thermal conduction severely limits transport of heat into the interior of the continent. Oceans evaporate more slowly, and until they have fully evaporated, they remain relatively cool.

The bottom panel addresses the depth that the thermal wave penetrates into the continents, the amount of water evaporated, and the depth of hot brine left behind by evaporation that initially is left floating on the cold deep ocean. Unevaporated sea waters become saltier. Mixing between the hot salty surface waters and the cold fresher deep waters is governed by the competition between the higher density of salt water vs. the lower density of hot water. It is only after the ocean has evaporated that the exposed seafloor gets hot. As with the continents, subsequent heating of the seafloor is rate-limited by thermal conduction (*Zahnle and Sleep,* 1997). Although we regard it as reasonably likely that a small number of ocean-vaporizing impacts took place, we suspect

that only the greatest of them could have exterminated life on Earth. We expect rather that the largest impacts severely pruned the bush of life — in particular, annihilating any photosynthetic ecosystems that may once have existed — leaving only thermophiles as survivors (*Sleep et al.,* 1989).

5.3. Hadean Climates

An important boundary condition on the Hadean climate is that the Sun was only about 70% as bright as it is now. At first, in the aftermath of the Moon-forming impact or even a maximum HSE impact, we expect deep ~100-bar CO_2-H_2-CH_4 atmospheres with correspondingly potent greenhouse effects (cf. *Wordsworth and Pierrehumbert,* 2013a) that keep the surface and ocean very warm, perhaps 500 K or more. Cooling is drawn out over tens of millions of years as CH_4 is photochemically destroyed and CO_2 is removed from the atmosphere into the mantle as carbonate rock. As the Earth cools, the climate must pass through a temperate period suitable to warm little ponds, but unless a strong greenhouse effect is provided by gases other than CO_2 (e.g., H_2) (*Wordsworth and Pierrehumbert,* 2013a), there is little reason to expect the climate of a lifeless Earth to settle into a temperate steady state, because seafloor weathering is unlikely to provide a buffer to sustain the needed ~1-bar CO_2 atmosphere (*Sleep et al.,* 2001) (Fig. 11).

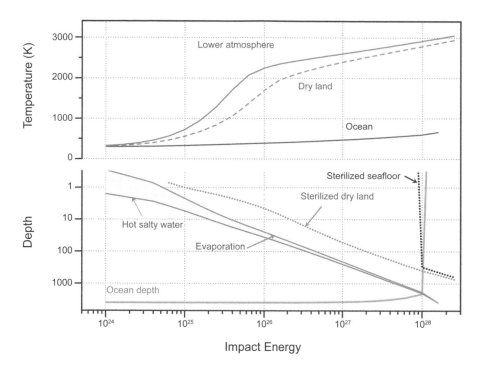

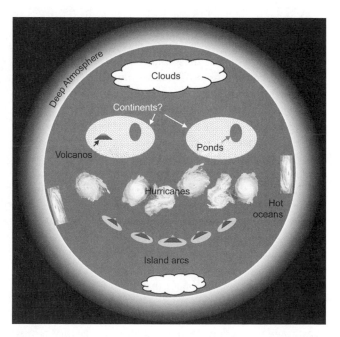

Fig. 10. Global warming after very large impact events (after *Zahnle and Sleep,* 1997, 2006; *Segura et al.,* 2013). The figure shows various first-order cumulative consequences as functions of the energy of the impact. See Table 1 for context.

Smaller ocean-vaporizing impacts have somewhat similar consequences to those produced by the maximum HSE event, but they make less CH_4, greenhouse effects are smaller, and recovery is quicker. These events are more likely than the maximum HSE impact to leave Earth's surface in a temperate state. Much depends on exactly how much CO_2 was in the atmosphere at the time of the impact.

Eventually, in the absence of great impacts — after the CH_4 is gone, the H_2 lost, and the CO_2 subducted into the mantle — a purely abiotic Earth would have grown cold under the faint Sun and its oceans would have frozen over (Fig. 12). In regions of low heat flow (say, over subduction zones) the ice may have been kilometers thick, while in the then-extensive regions of high heat flow (spreading zones and hotspot volcanos), the ice would have been less than 100 m thick and subject to cracking and opening leads. Active subaerial volcanos would likely have surrounded themselves with moats of open water. Other subaerial land would likely have been exposed because the generally cold conditions (~230 K) of a white Earth would have severely limited precipitation. A stable cold steady state can be imagined in which volcanic CO_2 emitted into the atmosphere is balanced against transport of CO_2 through windows in the ice. Ponds of fresh meltwater may have formed on the summer surfaces of the ice. Soluble airborne pollutants like HCN and formaldehyde — two simple photochemical products of a methane-spiked Hadean atmosphere (*Pinto et al.,* 1980; *Zahnle and Kasting,* 1986; *Tian et al.,* 2011) that have both figured prominently in the origin of life literature (*Oparin,*

1938; *Oró and Kamat,* 1961; *Abelson,* 1966; *Ricardo et al.,* 2004; *Powner et al.,* 2009; *Benner et al.,* 2019) — would have concentrated in the ponds as they froze. The frozen road to the origin of life has been at times popular (*Bronowski,* 1973); we are not aware of a good reason why it should not still be in play.

Fig. 11. A thick CO_2 atmosphere permits a hot Hadean with surface temperature as warm as 500 K.

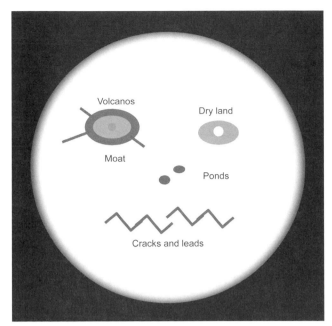

Fig. 12. The faint young Sun suggests an iceball Hadean in the absence of a strong greenhouse effect.

5.4. Hadean Surface Environments

There are very few rocks or fragments of rocks surviving on Earth from before 4.0 Ga. Detrital Hadean zircons are found in a small number of much younger Archean sediments, mostly in Western Australia (*Harrison*, 2009). There are also some 4.02-Ga rocks from the Slave Craton in the Northwest Territories of Canada that just make the 4.0-Ga cut (*Reimink et al.,* 2018). A very ancient age of 4.28 Ga has been reported for metabasalts from the Nuvvuagittuq complex on shores of Hudson's Bay, Canada (*O'Neil et al.,* 2013), but whether the lavas themselves are that old is in debate, as they may have been contaminated with an older hypothetical early Hadean component (*Roth et al.,* 2013). None of these materials are ideal recorders of surface conditions on Earth in the Hadean, but some broad conclusions can be drawn from the data for these ancient materials.

Zircons are crystals of zirconium silicate ($ZrSiO_4$) that crystallize from evolved magmas such as those that make up the continental crust. Uranium substitutes for Zr in the zircon crystal structure but Pb, the decay product of U, does not fit, so zircons crystallize with substantial U but little or no Pb. Hence any Pb present is a decay product. The U-Pb dating scheme has the advantage that two isotopes of U decay to two different isotopes of Pb, thus providing an independent check of the accuracy of the age, which should agree in the two decay schemes. In addition, quite high temporal resolution (occasionally better than plus or minus a million years) is possible due to the relatively rapid (703.8×10^6 yr half-life) decay of ^{235}U. The other advantage of zircon for geochronology is that the mineral structure is very robust, limiting the elemental diffusion that causes resetting of

isotope systematics during metamorphism. Zircon is thus becoming the gold standard for geochronology for old rocks. Unfortunately, zircon does not appear in all rock types, and is particularly rare in rocks derived by melting of the mantle, such as the basalts that make up the oceanic crust.

Besides its excellence as a geochronometer, other impurities in zircon can be used to estimate the temperature at which the zircon crystallized (*Watson and Harrison,* 2005) or the redox state of the magma (*Trail et al.,* 2012) from which they crystallized. Zircons incorporate small amounts of titanium; how much depends on the temperature and pressure of formation (*Watson and Harrison,* 2005; *Harrison,* 2009). Titanium thermometry suggests that many Hadean zircons formed in magmas that were, by magma standards, quite cool (*Watson and Harrison,* 2005), which could be accounted for by water in the magma, as would be expected if the rock that melted had at some earlier time incorporated water during weathering. Oxygen isotopic compositions of some zircons support this scenario *(Cavosie et al.,* 2005; *Harrison,* 2009). Zircon also can grow around, and therefore preserve, mineral inclusions from its source rocks. Such inclusions can be used to gain information about the conditions of melting and the composition of the source rocks of the magma from which the zircon crystallizes (*Hopkins et al.,* 2008). A very few Hadean zircons have incorporated carbon, now graphite, that exhibits the $\delta^{13}C$ isotopic signature that is often regarded as residual to biological carbon fractionation (*Bell et al.,* 2015), although there is some question when the carbon was introduced into the zircon.

Rocks falling under the broad term "granite" (more accurately, tonalite, trondhjemite, and granodiorite: TTG) that make up the majority of the ancient continental crust cannot be generated directly by melting the Mg-rich rocks of the mantle, but instead most likely reflect multiple stages of differentiation. The first step is the production of basaltic rocks by melting of the mantle. The next step is hydration of those basalts in the surface environment followed by melting when temperatures are raised either by burial or by subduction of the basaltic crust into the mantle. The mineral inclusion suite in some Hadean zircons has been interpreted to suggest a petrogenetic setting similar to a modern subduction zone (*Hopkins et al.,* 2008). Subduction does not necessarily imply plate tectonics, and it says nothing about the presence or absence of continents. Subduction merely implies that cold dense crustal blocks sank into the mantle, regardless of whether they pulled plates down with them. When plate tectonics began on Earth is an open question, perhaps related to the history of continental growth, which is also an open question (*Korenaga,* 2013).

What Earth's crust was composed of prior to the formation of these most ancient preserved rocks/minerals is less clear. Some insight is being gained on this topic from the initial Hf and Nd isotopic compositions of ancient rocks and zircons. Zircon has proven particularly useful for this purpose because Hf is a minor element in zircon whereas Lu, which decays very slowly to Hf (^{176}Lu-^{176}Hf, $\tau_{1/2} =$

37 G.y.), is present in extremely low concentrations. Consequently, zircon freezes in the Hf isotopic composition at the time of its formation. Hafnium isotope studies of the ancient zircons from Western Australia point to formation of the source of the zircon-bearing rocks (not the rocks themselves) being ca. 4.4 Ga and having Lu/Hf ratios within the range of basalts (*Kemp et al.*, 2010). Similar results indicating mafic composition progenitors of 4.2 to 4.4 Ga age for the oldest crustal rocks are seen in data for both the Slave Craton (*Bauer et al.*, 2017) and Nuvvuagittuq (*O'Neil et al.*, 2013; *Guitreau et al.*, 2014). The initial Hf evidence for a broadly basaltic precursor for the first preserved crust on Earth is supported by studies of the short-lived (103-m.y. half-life) decay of ^{146}Sm to ^{142}Nd in both of these areas (*O'Neil et al.*, 2013; *Roth et al.*, 2013) and also the ancient rocks from Greenland (*Caro et al.*, 2006; *Bennett et al.*, 2007; *Rizo et al.*, 2013; *O'Neil et al.*, 2016). How this early mafic crust was generated is not known, but basalt is the typical product of low-degree (5–15%) partial melting of Earth's mantle, so a compositionally distinct early crust produced by magma ocean differentiation, similar to that seen on the Moon, is not apparent in the terrestrial data.

5.5. Noble Gases as Witnesses

In the absence of true atmospheric fossils, the noble gases are often treated as tracers of other atmophile elements, much as the HSEs are treated as tracers of siderophile elements (*Anders and Owen*, 1977; *Pollack and Black*,

1979; *Pepin*, 1991; *Marty*, 2012; *Halliday*, 2013; *Alexander*, 2017). Elemental abundances of the nonradiogenic noble gases can be classified as "planetary" or "solar." In the solar pattern, the heavier ones — Ar, Kr, Xe — are present in the same relative abundances as in the Sun. The solar wind and Jupiter are examples. In the planetary pattern, the relative elemental abundances decline precipitously in order of volatility. The planetary pattern is not unique, neither to abundances nor to isotopes, but it has certain universal characteristics, in particular a low Ne/Ar ratio on the order of 0.3% of solar. Noble gas patterns are shown in Fig. 13.

We give four examples of how planetary noble gases relate to each and other: (1) Earth and carbonaceous chondrites have similar isotopic compositions and elemental abundances, but Earth has less than 1% as much of them. However, the Xe is wrong: Earth's Xe is some twenty-fold depleted, extremely mass-fractionated, and uniquely deficient in the two heaviest Xe isotopes (*Pepin*, 1991; *Zahnle et al.*, 2019). (2) Earth and Mars show very similar relative abundances of noble gases in their atmospheres — Mars is also depleted in Xe, and its Xe is also strongly mass-fractionated — yet in detail the Kr and Xe isotopes of the two planets are profoundly different. The martian gases were initially isotopically solar, while Earth's, although isotopically unique, are kin to chondrites. (3) Enstatite chondrites and carbonaceous chondrites have similar quantities of noble gases, but sometimes the EC isotopes are solar. (4) With Venus we have only the abundances of Ar and Ne. The reported ^{20}Ne/^{36}Ar ratio is planetary. But what is

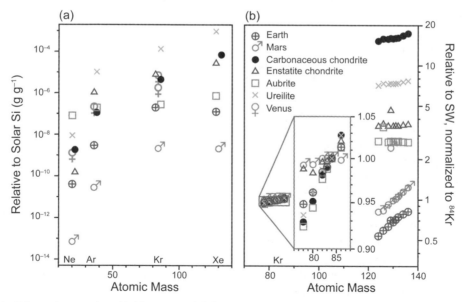

Fig. 13. See Plate 2 for color version. Noble gases. **(a)** Elemental abundances normalized by mass with respect to solar abundances. The planetary pattern — monotonically increasing relative abundances of the heavier elements — is evident in the meteorites and the planets, although offset by orders of magnitude. Note that Venus' bulk Ne and Ar abundances are similar to those in carbonaceous chondrites and enstatite chondrites. Both reported measurements of Kr on Venus are shown. Earth is depleted in all noble gases compared to Venus and the chondrites, and Mars is extremely depleted. **(b)** Comparison of isotopic structures of Xe and Kr; Kr has been magnified to make the differences between chondritic Kr and solar Kr visible. Xenon in Earth and Mars is depleted and very strongly mass fractionated, observations that together imply that Xe has escaped from both planets. No data from Venus are plotted. The ureilite is Havero, the aubrite is Pesyanoe, the enstatite chondrite is South Oman.

striking is that the bulk abundance of nonradiogenic Ar in Venus (gram-per-gram) is similar to what it is in enstatite chondrites or carbonaceous chondrites, and 70 greater than it is on Earth. Other data for Venus have been reported but are either in severe conflict with each other (Kr) or have errors that exceed any useful signal (^{20}Ne/^{22}Ne, ^{36}Ar/^{38}Ar, all things xenon) (cf. *Von Zahn et al., 1983*).

The ubiquity but nonuniqueness of the planetary pattern suggests parallel evolution. A plausible hypothesis is that the noble gases are captured from the nebula in the pores of planetesimals, in which they linger in proportion to their adsorptivity, and are trapped inside when pores are shut (*Ozima and Podosek, 2002*). A reasonable time and place for trapping is during water condensation, which fixes the temperature and so fixes the relative abundances of noble gases. Differential escape of gravitationally captured nebular atmospheres seems extremely unlikely to produce the same planetary pattern again and again, while direct condensation of noble gas as ices is more likely related to the compositions of giant planets.

In contrast to nonradiogenic Ar, Venus' atmosphere is four-fold impoverished in radiogenic ^{40}Ar (from ^{40}K, T $\tau_{1/2}$ = 1.25 G.y.) compared to Earth (*Donahue and Pollack, 1983; Von Zahn et al., 1983*). This could mean that Venus was born much depleted in K compared to Earth; however, the evidence from crustal rocks investigated by Russian space missions is that it was not (*Fegley, 2014*). Evidently there has been little outgassing of Venus compared to Earth in the past 4 G.y. The implication is that Venus' degassing style or convective style have differed from Earth's (*Watson et al., 2007*). The latter options imply that Venus' mantle is dry (to inhibit melting and plate tectonics), or CO_2-depleted (to inhibit degassing), or both.

5.5.1. Atmospheric escape in general.

There is little doubt that the escape of atmophiles to space has played a significant role in the development of the terrestrial planets. The leading hypotheses are thermal hydrodynamic escape driven by extreme ultraviolet radiation (EUV) from the young Sun (*Sekiya et al., 1980, 1981; Watson et al., 1981; Zahnle and Kasting, 1986; Pepin, 1991*), and the expulsion of atmospheric gases to space by cosmic impacts (*Walker, 1986; Zahnle et al., 1988; Melosh and Vickery, 1989; Zahnle, 1993; Chen and Ahrens, 1997; Brain and Jakosky, 1998; Genda and Abe, 2003, 2005; de Niem et al., 2012; Schlichting et al., 2015; Schlichting and Mukhopadhyay, 2018*). Evaluation of the latter is hampered by the absence of an agreed-upon model, yet the combination of enormous across-the-board depletions of martian atmophiles, coupled with relatively modest elemental and isotopic fractionations — easily seen in Fig. 13 — suggests that an efficient nonfractionating escape process is called for. Impact erosion can be such a process. In our solar system, the comparison between Venus and Earth reveals two clear signals of escape, one from each planet: (1) Venus has only $10^{-5}\times$ as much water as Earth, and (2) Venus has 70× more nonradiogenic ^{36}Ar than Earth. The former has been much discussed. The latter could be explained by preferential impact erosion of

Earth's atmosphere over a condensed water ocean (*Genda and Abe, 2003, 2005*); or seen as a specific consequence of the Moon-forming impact (*Pepin, 1997, 2006*); or it might date to escape from protoplanets (*Walker, 1982; Tucker and Mukhopadhyay, 2014*), especially if the biggest of these ended up in Venus; or it might have nothing to do with escape (e.g., *Wetherill, 1981*).

5.5.2. Xenon escape.

Xenon is the heaviest noble gas, yet evidence that Xe has escaped from Earth is strong. First, the isotopes of nonradiogenic Xe are strongly mass fractionated; the fractionation is easily seen on Fig. 13. Second, Xe is depleted in air by a factor of 4–20 compared to Kr, also obvious on Fig. 13. Third, radiogenic Xe is much less abundant than it should be compared to solar system abundances of the extinct parents. There are two of these. About 7% of air's ^{129}Xe (*Pepin, 2006*) derives from decay of ^{129}I (15.7-m.y. half-life), which is only 1% of what a planet of Earth's mass with Earth's iodine content would have, had none escaped (*Tolstikhin et al., 2014*). About 4.5% of air's ^{136}Xe derives from spontaneous fission of ^{244}Pu (half-life 80 m.y.) (*Pepin, 2000, 2006*). About 20% of the expected amount of plutogenic Xe is in the atmosphere, and much less remains in the mantle (*Tolstikhin et al., 2014*). It therefore seems that Xe escape was still taking place more than 200 m.y. into the life of the solar system.

Xenon recovered from Archean and early Proterozoic (2–3.5 Ga) rocks (*Srinivasan, 1976; Pujol et al., 2009, 2011, 2013; Holland et al., 2013; Avice et al., 2018*) is found to resemble Xe in air, but is less strongly mass fractionated, and least fractionated in the oldest samples. When the rock record begins, Xe is about half as fractionated as it finally becomes. These data suggest that xenon's mass fractionation evolved over the first half of Earth's history, with much of the evolution taking place in the Hadean.

Heavy gases like Xe can escape as passengers during episodes of vigorous hydrodynamic hydrogen escape (*Sekiya et al., 1980; Zahnle and Kasting, 1986; Hunten et al., 1987; Sasaki and Nakazawa, 1988; Pepin, 1991, 2006; Tolstikhin and O'Nions, 1994; Dauphas, 2003*). The puzzle is why this process would affect Xe alone among the noble gases. A way to solve the puzzle is for Xe to escape as an ion, a process that can be unique to Xe because Xe is the only noble gas more easily ionized than hydrogen, which is the major constituent of the escaping planetary wind (*Zahnle et al., 2019*). In the model, Xe$^+$ ions are dragged to space by H$^+$ ions along polar magnetic field lines that open into interplanetary space, with hydrogen escape fueled by EUV radiation from the young Sun. The model requires that the atmosphere be at least 1% H_2 by volume (equivalent to 0.5% CH_4 in the lower atmosphere) for the escape flux to be big enough to carry off Xe$^+$. *Zahnle et al.* (2019) therefore concluded that H_2 was at times an abundant gas in the Hadean and Archean atmospheres. To account for the observed Xe loss and mass fractionation, Earth must also have lost hydrogen equivalent to a significant fraction of an ocean of water during the Hadean, with a consequent net oxidation of the crust and mantle. Similar fractionations and depletions of Xe were established on Mars before 4.1 Ga (*Cassata,* 2017) and could be attributed to hydrogen escape when Mars still had a planetary magnetic field.

5.5.3. Hydrogen escape. Hydrogen escape offers a plausible explanation for the oxygenation of Earth's atmosphere (*Catling et al.,* 2001; *Zahnle et al.,* 2013). The mechanism is straightforward: Hydrogen escape irreversibly oxidizes the Earth, beginning with the atmosphere, and then working its way down. The importance of hydrogen in Earth's early atmosphere in promoting the formation of molecules suitable to the origin of life was recognized before the dawn of the space age (*Urey,* 1952) and continues to be recognized (*Tian et al.,* 2005). The subsequent importance of hydrogen escape in pushing chemical evolution in a direction suitable to creating life was also recognized by *Urey* (1952). The potential importance of hydrogen escape in driving biological evolution toward oxygen-using aerobic ecologies has not been as fully appreciated, but it is quite clear that the bias is there. There is considerable evidence in the mass fractionation of Xe isotopes on Earth that hydrogen escape was a major term in shaping Earth's surface environments during the first half of Earth's history, from the Hadean and through the Archean (*Zahnle et al.,* 2019). However, as this topic is more relevant to the origin and development of life on Earth, we will not address it further here (see the chapters by Baross et al., Stüeken et al., and Robinson and Reinhard in this volume).

6. WATER AGAIN

The attentive reader will note that we have offered no specific guidance to where Earth's water comes from. As an issue of quantity there is, as of yet, nothing to get flustered about. Earth in bulk is extremely dry. It is roughly as dry as the enstatite chondrites and it is drier than the ordinary chondrites and much drier than the carbonaceous chondrites. On the other hand, it is probably wetter than the enstatite achondrites that the intersection of conventional notions of hierarchical accretion, ^{26}Al heating, and the isotopic evidence suggest should comprise the bulk of the planet. The origin of Earth's water is central to our status as a habitable planet. It is clearly an important issue for future work.

The usual suspects are (1) nebular hydrogen captured gravitationally, with oxygen scavenged from silicates; (2) hydrous silicates formed as primary or secondary condensates in the inner solar nebula; (3) outer solar system water-rich materials injected into the realm of the inner planets before their completion; and (4) outer solar system water-rich materials added after the inner planets were completed. We have argued here against the first option on the grounds that Earth retains only 10^{-7} of the solar Ne it would have had, had it captured its hydrogen directly from the nebula; nevertheless, scenario (1) has valiant supporters (*Ikoma and Genda,* 2006; *Sharp,* 2016, 2017; *Sharp and Olson,* 2018, 2019; *Wu et al.,* 2018; *Williams and Mukhopadhyay,* 2019). The second option was ascendent in the 1970s when everything was explained by condensation temperatures, but

has not been as popular since, because hydrous silicates are now thought to be slow to form directly from rocks and water vapor (cf. *Ciesla et al.,* 2003). The fourth scenario has been popular for several decades but has run afoul of the Ru isotopic evidence that rather firmly aligns any late accretion to Earth with inner solar system materials — there is no longer much room in the late veneer for comets or carbonaceous chondrites to deliver an ocean. This leaves the third scenario, which indeed seems the leading contender at present (*Raymond and Izadoro,* 2017; *Morbidelli et al.,* 2018; see also the chapter by Meech and Raymond in this volume). Grand Tack models do predict this. Very recently, *Budde et al.* (2019) proposed that Theia itself was the water-bearer. As *Budde et al.* (2019) also shows, there are no known solar system materials resembling the Theia they require, which may or may not seem an attractive feature of their hypothesis.

To first approximation, Earth's water carries the same D/H signature as meteorites (*Alexander et al.,* 2012). This meteoritical D/H signature is widespread in today's inner solar system; it is plausibly imagined as being determined by chemical or photochemical fractionation between H_2 and H_2O in the general vicinity of the snow line, the moving front where water condenses and evaporates (*Alexander,* 2017). A common D/H is inconsistent with scenarios (1) and (2), consistent with (3) and (4).

Snow line D/H inspires speculation of a fifth scenario for the origin of Earth's water. It is likely that the snow line moved Sunward as the nebula cooled and the pre-main-sequence Sun faded to half its current luminosity (*Oka et al.,* 2011; *Desch et al.,* 2018). As the snow line swept through the inner solar system, previously dry embryos and planetesimals would have accumulated water as snow. In this picture water is added to E-type materials by the dying nebula and held onto because the Sun was faint, and through these bodies accreted to Earth. This picture separates water from the other atmophile elements, all of which are abundant in ECs and hence none of which pose any difficulties of supply. Indeed, the question that surrounds the other atmophile elements is how Earth has arranged to show so few of them.

In contrast to Earth, Venus is extremely dry: Its atmosphere contains about 0.003 bars of H_2O, the equivalent of a global puddle no deeper than the width of a finger. Venus' high D/H ratio — 100× higher than Earth's — indicates that more than 99% of Venus' original hydrogen was lost (*Donahue et al.,* 1982; *Donahue and Pollack,* 1983), but this falls far short of the full factor of 2×10^5 that separates Venus and Earth; the D/H ratio could be satisfied by an initial water inventory as small as 1% that of Earth (*Kasting and Pollack,* 1983; *Zahnle and Kasting,* 1986).

One reason to suppose that Venus was born wet is that Venus is well-endowed in other atmophile elements, although in detail these do not tell a consistent story: Nonradiogenic ^{36}Ar is 70× more abundant on Venus than on Earth, there is 3× more N_2 in Venus' atmosphere, and there is 50% more CO_2 in Venus' atmosphere than in the carbonate rocks and black shales of Earth's crust. If one insists that Earth and

Venus should have been born alike, Earth must have lost almost all of its ^{36}Ar and much of its nitrogen, perhaps by impact erosion (*Genda and Abe*, 2005). The other reason to suppose that Venus was born wet is that we have developed theories that can account for Venus gaining and losing an ocean of water (*Ingersoll*, 1969). The runaway greenhouse theory in particular plays a central role in how we currently view water and habitability of planets (*Kasting et al.*, 1993).

When measured against the enstatite chondrites, it is Venus, not Earth, that looks as it ought to look, at least insofar as the atmophile elements are concerned. We can say such a thing because we know so little about Venus, a consequence of Venus being little explored compared to other nearby planets (see chapter by Arney and Kane in this volume). It may seem odd to end an essay on Earth as an astrobiological planet with a sentence addressing Earth's evil twin, but Venus is where the other half of the inner solar system's mass is, and until we learn the truth of what Venus has to tell us, we have only seen half the story.

REFERENCES

Abe Y. (1997) Thermal and chemical evolution of the terrestrial magma ocean. *Phys. Earth Planet. Inter., 100,* 27–39.

Abe Y. and Matsui T. (1988) Evolution of an impact-generated H_2O-CO_2 atmosphere and formation of a hot proto-ocean on Earth. *J. Atmos. Sci., 45,* 3081–3101.

Abelson P. H. (1966) Chemical events on the primitive Earth. *Proc. Natl. Acad. Sci. U.S.A., 55,* 1365–1372.

Abramov O. and Mojzsis S. J. (2018) Impact bombardment on terrestrial planets during late accretion. *Lunar Planet. Sci. XLIX,* Abstract #2962. Lunar and Planetary Institute, Houston.

Abramov O., Kring D. A., and Mojzsis S. J. (2013) The impact environment of the Hadean Earth. *Chem. Erde Geochem., 73,* 227–248.

Adams D. (1979) *The Hitchhikers Guide to the Galaxy.* Pan Books, UK. 240 pp.

Agnor C. B. and Asphaug E. (2004) Accretion efficiency during planetary collisions. *Astrophys. J. Lett., 613,* L157–L160.

Ahrens T. J., O'Keefe J. D., and Lange M. A. (1989) Formation of atmospheres during accretion of the terrestrial planets. In *Origin and Evolution of Planetary and Satellite Atmospheres* (S. K. Ateeya et al., eds.), pp. 328–385. Univ. of Arizona, Tucson.

Albarède F. (2009) Volatile accretion history of the terrestrial planets and dynamic implications. *Nature, 461,* 1227–1233.

Albarède F., Ballhaus C., Blichert-Toft J., Lee C.-T., Marty B., Moynier F., and Yin Q.-Z. (2013) Asteroidal impacts and the origin of terrestrial and lunar volatiles. *Icarus, 222,* 44–52.

Allègre C. J., Manhès G, and Göpel C. (2008) The major differentiation of the Earth at 4.45 Ga. *Earth Planet. Sci. Lett., 267,* 386–398.

Alexander C. M. O. D. (2017) The origin of inner solar system water. *Philos. Trans. R. Soc. London, Ser. A, 375,* 20150384.

Alexander C. M. O. D., Bowden R., Fogel M. L., Howard K. T., Herd C. D. K., and Nittler L. R. (2012) The provenances of asteroids, and their contributions to the volatile inventories of the terrestrial planets. *Science, 337,* 721–723.

Anders E. (1989) Pre-biotic organic matter from comets and asteroids. *Nature, 342,* 255–257.

Anders E. and Owen T. (1977) Mars and Earth – Origin and abundance of volatiles. *Science, 198,* 453–465.

Armstrong K., Frost D. J., McCammon C. A., Rubie D. C., and Ballaran T. B. (2019) Deep magma ocean formation set the oxidation state of Earth's mantle. *Science, 365,* 903–906.

Asphaug E., Agnor C., and Williams Q. (2006) Hit-and-run planetary collisions. *Nature, 439,* 155–160.

Aston F. W. (1924) The rarity of the inert gases on Earth. *Nature, 114,* 786.

Avice G., Marty B., and Burgess R. (2017) The origin and degassing history of the Earth's atmosphere revealed by Archean xenon. *Nature Commun., 8,* 1–9.

Avice G., Marty B., Hofmann A., Zahnle K. J., Philippot P., and Zakharov D. (2018) Evolution of atmospheric xenon and other noble gases inferred from the study of Archean to Paleoproterozoic rocks. *Geochim. Cosmochim. Acta, 232,* 82–100.

Barboni M., Boehnke P., Keller C. B., Kohl I. E., Schoene B., Young E. D., and McKeegan K. D. (2017) The age of the Moon. *Lunar Planet. Sci. XLIX,* Abstract #1900. Lunar and Planetary Institute, Houston.

Batygin K. and Laughlin G. (2015) Jupiter's decisive role in the inner solar system's early evolution. *Proc. Natl. Acad. Sci. U.S.A., 112,* 4214–4217.

Bauer A. M., Fisher C. M., Vervoort J. D., and Bowring S. A. (2017) Coupled zircon Lu-Hf and U-Pb isotopic analyses of the oldest terrestrial crust, the >4.03 Ga Acasta Gneiss Complex. *Earth Planet. Sci. Lett., 458,* 37–48.

Bell E. A., Boehnke P., Harrison T. M., and Mao W. L. (2015) Potentially biogenic carbon preserved in a 4.1 billion-year-old zircon. *Proc. Natl. Acad. Sci. U.S.A., 112,* 14518–14521.

Benner S. A., Bell E. A., Biondi E., Brasser R., Carell T., Kim H.-J., Mojzsis S. J., Omran A., Pasek M. A., and Trail D. (2019) When did life likely emerge on Earth in an RNA-first process? *ChemSystChem,* DOI: 10.1002/syst.201900035.

Bennett V. C., Brandon A. D., and Nutman A. P. (2007) Coupled ^{142}Nd-^{143}Nd isotopic evidence for Hadean mantle dynamics. *Science, 318,* 1907–1910.

Benz W., Slattery W. L., and Cameron A. G. W. (1986) The origin of the moon and the single-impact hypothesis. *Icarus, 66,* 515–535.

Bermingham K. R., Worsham E. A., and Walker R. J. (2018) New insights into Mo and Ru isotope variation in the nebula and terrestrial planet accretionary genetics. *Earth Planet. Sci. Lett., 487,* 221–229.

Boehnke P. and Harrison T. M. (2016) Illusory late heavy bombardments. *Proc. Natl. Acad. Sci. U.S.A., 113,* 10802–10806.

Bondi H. (1952) *Cosmology.* Cambridge Univ., Cambridge. 179 pp.

Borg L. E., Gaffney A. M., and Shearer C. K. (2015) A review of lunar chronology revealing a preponderance of 4.34–4.37 Ga ages. *Meteoritics & Planet. Sci., 50,* 715–732.

Bottke W. F., Walker R. J., Day J. M. D., Nesvorny D., and Elkins-Tanton L. (2010) Stochastic late accretion to Earth, the Moon, and Mars. *Science, 330,* 1527–1530.

Boyet M. and Carlson R. W. (2005) ^{142}Nd evidence for early (>4.53 Ga) global differentiation of the silicate Earth. *Science, 309,* 576–581.

Boyet M. and Carlson R. W. (2007) A highly depleted Moon or a non-magma ocean origin for the lunar crust? *Earth Planet. Sci. Lett., 262,* 505–516.

Brain D. and Jakosky B. (1998) Atmospheric loss since the onset of the martian geologic record: Combined role of impact erosion and sputtering. *J. Geophys. Res., 103,* 22689–22694.

Brandon A. D., Lapen T. J., Debaille V., Beard B. L., Rankenburg K., and Neal C. (2009) Re-evaluating ^{142}Nd/^{142}Nd in lunar mare basalts with implications for the early evolution and bulk Sm/Nd of the Moon. *Geochim. Cosmochim. Acta, 73,* 6421–6445.

Brasser R., Mojzsis S. J., Werner S. C., Matsumura S., and Ida S. (2016) Late veneer and late accretion to the terrestrial planets. *Earth Planet. Sci. Lett., 455,* 85–93.

Brauer F., Dullemond C. P., and Henning T. (2008a) Coagulation, fragmentation, and radial motion of solid particles in protoplanetary disks. *Astron. Astrophys., 480,* 859–877.

Brauer F., Henning T., and Dullemond C. P. (2008b) Planetesimal formation near the snow line in MRI-driven turbulent protoplanetary disks. *Astron. Astrophys., 487,* L1–L4.

Bronowski J. (1973) *The Ascent of Man.* Little Brown and Co., New York. 448 pp.

Brown H. (1949) Rare gases and the formation of the Earth's atmosphere. In *The Atmosphere of the Earth and Planets* (G. Kuiper, ed.), pp. 258–266. Univ. of Chicago, Chicago.

Budde G., Burkhardt C., and Kleine T. (2018) Early solar system dynamics inferred from molybdenum isotope anomalies in meteorites. *Lunar Planet. Sci. XLIX,* Abstract #2353. Lunar and Planetary Institute, Houston.

Budde G., Burkhardt C., and Kleine T. (2019) Molybdenum isotopic evidence for the late accretion of outer solar system material to Earth. *Nature Astron., 3,* 736–741.

Burbidge E. M., Burbidge G. R., Fowler W. A., and Hoyle F. (1957) Synthesis of the elements in stars. *Rev. Mod. Phys., 29,* 547–650.

Burkhardt C., Borg L. E., Brennecka G. A., Shollenberger Q. R., Dauphas N., and Kleine T. (2016) A nucleosynthetic origin for the Earth's anomalous ^{142}Nd composition. *Nature, 537,* 394–398.

Burkhardt C., Dauphas N., Hans U., Bourdon B., and Kleine T. (2019) Elemental and isotopic variability in solar system materials by mixing and processing of primordial disk reservoirs. *Geochim. Cosmochim. Acta, 261,* 145–170.

Canil D. (2002) Vanadium in peridotites, mantle redox, and tectonic environments: Archean to present. *Earth Planet. Sci. Lett., 195,* 75–90.

Canup R. M. (2004) Dynamics of lunar formation. *Annu. Rev. Earth Planet. Sci., 42,* 441–475.

Canup R. M. (2012) Forming a moon with an Earth-like composition via a giant impact. *Science, 338,* 1052–1055.

Canup R. M. and Asphaug E. (2001) Origin of the Moon and the single impact hypothesis. *Nature, 421,* 708–712.

Canup R. M. and Righter K. (2000) *Origin of the Earth and Moon.* Univ. of Arizona, Tucson. 555 pp.

Carlson R. W. and Lugmair G. W. (1979) Sm-Nd constraints on early lunar differentiation and the evolution of KREEP. *Earth Planet. Sci. Lett., 45,* 123–132.

Carlson R. W., Boyet M., and Horan M. (2007) Chondrite barium, neodymium, and samarium isotopic heterogeneity and early Earth differentiation. *Science, 316,* 1175–1178.

Carlson R. W., Garnero E., Harrison T. M., Li J., Manga M., McDonough W. F., Mukhopadhyay S., et al. (2014) How did early Earth become our modern world? *Annu. Rev. Earth Planet. Sci., 42,* 51–178.

Carlson R. W., Brasser R., Yin Q.-Z., Fischer-Gödde M., and Qin L. (2018) Feedstocks of the terrestrial planets. *Space Sci. Rev., 214(121),* DOI: 10.1007/s11214-018-0554-x.

Caro G., Bourdon B., Birck J.-L., and Moorbath S. (2006) High-precision ^{142}Nd/^{144}Nd measurements in terrestrial rocks: Constraints on the early differentiation of the Earth's mantle. *Geochim. Cosmochim. Acta, 70,* 164–191.

Cassata W. S. (2017) Meteorite constraints on martian atmospheric loss and paleoclimate. *Earth Planet. Sci. Lett., 479,* 322–329.

Cassen P. (1996) Models for the fractionation of moderately volatile elements in the solar nebula. *Meteoritics & Planet. Sci., 31,* 793–806.

Catling D. C., Zahnle K. J., and McKay C. P. (2001) Biogenic methane, hydrogen escape, and the irreversible oxidation of early Earth. *Science, 293,* 839–843.

Cavosie A. J., Valley J. W., and Wilde S. A. (2005) Magmatic δ^{18}O in 4400–3900 Ma detrital zircons: A record of the alteration and recycling of crust in the Early Archean. *Earth Planet. Sci. Lett., 235,* 663–681.

Chambers J. E. (2004) Planetary accretion in the inner solar system. *Earth Planet. Sci. Lett., 223,* 241–252.

Chambers J. E. (2014) Giant planet formation with pebble accretion. *Icarus, 233,* 83–100.

Chen G. Q. and Ahrens T. J. (1997) Erosion of terrestrial planet atmosphere by surface motion after a large impact. *Phys. Earth Planet. Inter., 100,* 21–26.

Chyba C. F. (1991) Terrestrial mantle siderophiles and the lunar impact record. *Icarus, 92,* 217–233.

Chyba C. F. and Sagan C. (1992) Endogenous production, exogenous delivery, and impact-shock synthesis of organic molecules: An inventory for the origins of life. *Nature, 355,* 125–132.

Chyba C. F., Thomas P. J., Brookshaw L., and Sagan C. (1990) Cometary delivery of organic molecules to the early Earth. *Science, 249,* 366–373.

Ciesla F. J., Lauretta D. S., Cohen B. A., and Hood L. L. (2003) A nebular origin for chondritic fine-grained phyllosilicates. *Science, 299,* 549–552.

Ciesla F. J., Mulders G. D., Pascucci I., and Apai D. (2015) Volatile delivery to planets from water-rich planetesimals around low mass stars. *Astrophys. J., 804(9),* DOI: 10.1088/0004-637X/804/1/9.

Clark B. C. (1988) Primeval procreative comet pond. *Origins Life Evol. Biospheres, 18,* 209–238.

Clark P. D., Dowling N. I., and Huang M. (1998) Comments on the role of H$_2$S in the chemistry of Earth's early atmosphere and in prebiotic synthesis. *J. Mol. Evol., 47,* 127–132.

Clayton R. N., Grossman L., and Mayeda T. K. (1973) A component of primitive nuclear composition in carbonaceous meteorites. *Science, 182,* 485–488.

Clayton R. N. (2003) Oxygen isotopes in the solar system. *Space Sci. Rev., 106,* 19–32.

Cleaves H. J., Chalmers J. H., Lazcano A., Miller S. L., and Bada J. L. (2008) A reassessment of prebiotic organic synthesis in neutral planetary atmospheres. *Origins Life Evol. Biospheres, 38,* 105–1156.

Ćuk M. and Stewart S. (2012) Making the Moon from a fast-spinning Earth: A giant impact followed by resonant despinning. *Science, 338,* 1047–1051.

D'Antona F. and Mazzitelli I. (1994) New pre-main-sequence tracks for M less than or equal to 2.5 solar mass as tests of opacities and convection model. *Astrophys. J. Suppl., 90,* 467–500.

Dauphas N. (2003) The dual origin of the terrestrial atmosphere. *Icarus, 165,* 326–339.

Dauphas N. (2017) The isotopic nature of the Earth's accreting material through time. *Nature, 541,* 521–524.

Dauphas N. and Chaussidon M. (2011) A perspective from extinct radionuclides on a young stellar object: The Sun and its accretion disk. *Annu. Rev. Earth Planet. Sci., 39,* 351–386.

Dauphas N. and Marty B. (2002) Inference on the nature and the mass of Earth's late veneer from noble metals and gases. *J. Geophys. Res., 107(E12),* 12-1 to 12-7.

Dauphas N. and Pourmand A. (2011) Hf-W-Th evidence for rapid growth of Mars and its status as a planetary embryo. *Nature, 473,* 489–492.

Dauphas N., Marty B, and Reisberg L. (2002) Molybdenum evidence for inherited planetary scale isotope heterogeneity of the protosolar nebula. *Astrophys. J., 565,* 640–644.

Dauphas N., Davis A. M., Marty B., and Reisberg L. (2014a) The cosmic molybdenum-ruthenium isotope correlation. *Earth Planet. Sci. Lett., 226,* 465–475.

Dauphas N., Burkhardt C., Warren P., and Teng F.-Z. (2014b) Geochemical arguments for an Earth-like Moon-forming impactor. *Philos. Trans. R. Soc. London, Ser. A, 372,* 2013.0244.

Day J. M. D., Brandon A. D., and Walker R. J. (2016) Highly siderophile elements in Earth, Mars, the Moon, and asteroids. *Rev. Mineral. Geochem., 81,* 161–238.

Davies P. (2000) *The Fifth Miracle.* Simon and Schuster, New York. 304 pp.

Davis A. M. and McKeegan K. D. (2014) Short-lived radionuclides and early solar system chronology. In *Treatise on Geochemistry, Vol. 1: Meteorites and Cosmochemical Processes* (A. M. Davis, ed.), pp. 361–395. Elsevier-Pergamon, Oxford.

Delano J. W. (2001) Redox history of the Earth's interior since approximately 3900 Ma: Implications for prebiotic molecules. *Origins Life Evol. Biospheres, 31,* 311–341.

de Niem D., Kührt E., Morbidelli A., and Motschmann U. (2012) Atmospheric erosion and replenishment induced by impacts upon the Earth and Mars during a heavy bombardment. *Icarus, 221,* 495–507.

Desch S. J., Kalyaan A., and Alexander C. M. O. D. (2018) The effect of Jupiter's formation on the distribution of refractory elements and inclusions in meteorites. *Astrophys. J. Suppl., 238,* 11.

Diehl R. and 15 colleagues (2006) Radioactive ^{26}Al from massive stars in the galaxy. *Nature, 439,* 45–47.

Donahue T. M. and Pollack J. B. (1983) Origin and evolution of the atmosphere of Venus. In *Venus* (D. M. Hunten et al., eds.), pp. 1003–1036. Univ. of Arizona, Tucson.

Donahue T. M., Hoffman J. H., Hodges R. R., and Watson A. J. (1982) Venus was wet? A measurement of the ratio of deuterium to hydrogen. *Science, 216,* 630–633.

Dreibus G. and Wänke H. (1989) Supply and loss of volatile constituents during the accretion of the terrestrial planets. In *Origin and Evolution of Planetary and Satellite Atmospheres* (S. K. Atreya et al., eds.), pp. 268–288. Univ. of Arizona, Tucson.

Elkins-Tanton L. (2008) Linked magma ocean solidification and atmospheric growth for Earth and Mars. *Earth Planet. Sci. Lett., 271,* 181–191.

Elkins-Tanton L. (2012) Magma oceans in the inner solar system. *Annu. Rev. Earth Planet. Sci., 40,* 113–139.

Estrada P. R., Cuzzi J. N., and Morgan D. A. (2016) Global modeling of nebulae with particle growth, drift, and evaporation fronts. I. Methodology and typical results. *Astrophys. J., 818,* 200.

Fegley B. Jr. (2014) Venus. In *Treatise on Geochemistry, Vol. 1:*

Planets, Asteroids, and the Solar System (A. M. Davis, ed.), pp. 127–148. Elsevier-Pergamon, Oxford.

Fegley B. Jr., Jacobson N. S., Williams K. B., Plane J. M. C., Schaefer L., and Lodders K. (2016) Solubility of rock in steam atmospheres of planets. *Astrophys. J., 824,* 103.

Ferris J. F., Joshi P. C., Edelson E. H., and Lawless J. G. (1978) HCN: A plausible source of purines, pyrimidines, and amino acids on the primitive Earth. *J. Mol. Evol., 11,* 293–311.

Fischer-Gödde M. and Kleine T. (2017) Ruthenium isotopic evidence for an inner solar system origin of the late veneer. *Nature, 541,* 525–527.

Fischer-Gödde M., Burkhardt C., Kruijer T. S., and Kleine T. (2015) Ru isotope heterogeneity in the solar protoplanetary disk. *Geochim. Cosmochim. Acta, 168,* 151–171.

Fischer R., Nimmo F., and O'Brien D. P. (2018) Radial mixing and Ru-Mo isotope systematics under different accretion scenarios. *Earth Planet. Sci. Lett., 482,* 105–114.

Frost D. J. and McCammon C. A. (2008) The redox state of Earth's mantle. *Annu. Rev. Earth Planet. Sci., 36,* 389–420.

Froude D. O., Ireland T. R., Kinny P. D., Williams I. S., Compston W., Williams I. R., and Myers J. S. (1983) Ion microprobe identification of 4,100–4,200 Myr-old terrestrial zircons *Nature, 304,* 616–618.

Fujimoto Y., Krumholz M. R., and Tachibana S. (2018) Short-lived radioisotopes in meteorites from galactic-scale correlated star formation. *Mon. Not. R. Astron. Soc., 480,* 4025–4039.

Gaffney A. M. and Borg L. E. (2014) A young solidification age for the lunar magma ocean. *Geochim. Cosmochim. Acta, 140,* 227–240.

Genda H. and Abe Y. (2003) Survival of a proto-atmosphere through the stage of giant impacts: The mechanical aspects. *Icarus, 164,* 149–162.

Genda H. and Abe Y. (2005) Enhanced atmospheric loss on protoplanets at the giant impact phase in the presence of oceans. *Nature, 433,* 842–844.

Genda H., Brasser R., and Mojzsis S. J. (2017a) The terrestrial late veneer from core disruption of a lunar-sized impactor. *Earth Planet. Sci. Lett., 480,* 25–32.

Genda H., Iizuka T., Sasaki T., Ueno Y., and Ikoma M.(2017b) Ejection of iron-bearing giant-impact fragments and the dynamical and geochemical influence of the fragment re-accretion *Earth Planet. Sci. Lett., 470,* 87–95.

Goldblatt C., Nisbet E. G., Sleep N. H., and Zahnle K. J. (2010) The eons of Chaos and Hades. *Solid Earth, 1,* 1–3.

Goldblatt C. Z., Robinson T. D., Zahnle K. J., and Crisp D. (2013) Low simulated radiation limit runaway greenhouse climates. *Nature Geosci., 6,* 661–667.

Goldschmidt V. M. (1937) The principles of distribution of chemical elements in minerals and rocks. *J. Chem. Soc., 1937,* 655–673.

Gounelle M. and Meynet G. (2012) Solar system genealogy revealed by extinct short-lived radionuclides in meteorites. *Astron. Astrophys., 554,* A4.

Greenwood R. C., Barrat J.-A., Miller M. F., Anand M., Dauphas N., Franchi I. A., Sillard P., and Starkey N. A. (2018) Oxygen isotopic evidence for accretion of Earth's water before a high-energy Moon-forming giant impact. *Sci. Adv., 4,* eaao5928.

Grossman L. (1972) Condensation in the primitive solar nebula. *Geochim. Cosmochim. Acta, 36,* 597–619.

Guitreau M., Blichert-Toft J., Mojzsis S. J. K., Roth A. S. G., Bourdon B., Cates N. L., and Bleeker W. (2014) Lu-Hf isotope systematics of the Hadean-Eoarchean Acasta Gneiss Complex (Northwest Territories, Canada). *Geochim. Cosmochim. Acta, 135,* 251–269.

Haisch K. E., Lada E. A., and Lada C. J. (2001) Circumstellar disks in the IC 348 cluster. *Astron. J., 121,* 2065–2074.

Haldane J. B. S. (1929) The origin of life. *Rationalist Annual, 148,* 3–10.

Halliday A. (2013) The origins of volatiles in the terrestrial planets. *Geochim. Cosmochim. Acta, 105,* 46–171.

Halliday A. and Kleine T. S. (2006) Meteorites and the timing, mechanisms, and conditions of terrestrial planet accretion and early differentiation. In *Meteorites and the Early Solar System II* (D. S. Lauretta and H. Y. McSween, eds.), pp. 775–801. Univ. of Arizona, Tucson.

Hansen B. (2009) Formation of the terrestrial planets from a narrow annulus. *Astrophys. J., 703,* 1131–1140.

Harrison T. M. (2009) The Hadean crust: Evidence from >4 Ga zircons. *Annu. Rev. Earth Planet. Sci., 37,* 479–505.

Hartlep T., Cuzzi J., and Weston B. (2017) Scale dependence of multiplier distributions for particle concentration, enstrophy, and dissipation in the inertial range of homogeneous turbulence. *Phys. Rev. E, 95,* 033115.

Hartlep T., Cuzzi J. N., and Umurhan O. M. (2019) Planetesimal formation in the outer nebula in the presence of turbulence. *Lunar Planet. Sci. L,* Abstract #3044. Lunar and Planetary Institute, Houston.

Hashimoto G. L., Abe Y., and Sugita S. (2007) The chemical composition of the early terrestrial atmosphere: Formation of a reducing atmosphere from CI-like material. *J. Geophys. Res., 112,* E05010.

Hayashi C., Nakazawa K., and Mizuno H. (1979) Earth's melting due to the blanketing effect of the primordial dense atmosphere. *Earth Planet. Sci. Lett., 43,* 22–28.

Hayashi C., Nakazawa K., and Nakagawa Y. (1986) Formation of the solar system. In *Protostars and Planets II* (D. C. Black and M. S Mathews, eds.), pp. 1100–1154. Univ. of Arizona, Tucson.

Hirschmann M. M. (2012) Magma ocean influence on early atmosphere mass and composition. *Earth Planet. Sci. Lett., 341–344,* 48–57.

Hirschmann M. M. (2013) Fe-carbonyl is a key player in planetary magmas. *Proc. Natl. Acad. Sci. U.S.A., 110,* 7967–7968.

Holland H. D. (1964) On the chemical evolution of the terrestrial and cytherian atmospheres. In *The Origin and Evolution of Atmospheres and Oceans.* (P. J. Brancazio and A. G. W. Cameron, eds.), pp. 86–101. Wiley, New York.

Holland H. D. (1984) *The Chemical Evolution of the Atmosphere and Oceans.* Princeton Univ., Princeton. 598 pp.

Holland G., Sherwood-Lollar B., Li L., Lacrampe-Couloume G., Slater G. F., and Ballentine C. J. (2013) Deep fracture fluids isolated in the crust since the Precambrian era. *Nature, 497,* 357–360.

Hopkins M., Harrison T. M., and Manning C. E. (2008) Low heat flow inferred from >4 Gyr zircons suggests Hadean plate boundary interactions. *Nature, 456,* 493–496.

Hopp T. and Kleine T. (2018) Nature of late accretion to Earth inferred from mass-dependent Ru isotopic compositions of chondrites and mantle peridotites. *Earth Planet. Sci. Lett., 494,* 50–59.

Horan M. F., Walker R. J., Morgan J. W., Grossman J. N., and Rubin A. (2003) Highly siderophile elements in chondrites. *Chem. Geol., 196,* 5–20.

Hörst S. M., He C., Ugelow M. S., Jellinek A. M., Pierrehumbert R. T., and Tolbert M. A. (2018) Exploring the atmosphere of neoproterozoic Earth: The effect of O_2 on haze formation and composition. *Astrophys. J., 858,* 119.

Huebner W. F., Keady J. J., and Lyon S. P. (1992) Solar photo rates for planetary atmospheres and atmospheric pollutants. *Astrophys. Space Sci., 195,* 1–294.

Hunten D. M., Pepin R. O., and Walker J. C. G. (1987) Mass fractionation in hydrodynamic escape. *Icarus, 69,* 532–549.

Ikoma M. and Genda H. (2006) Constraints on the mass of a habitable planet with water of nebular origin. *Astrophys. J., 648,* 696–706.

Ingersoll A. P. (1969) The runaway greenhouse: A history of water on Venus. *J. Atmos. Sci., 26,* 1191–1198.

Jacobsen S. B., Petaev M. I., Huang S., and Sasselov D. (2013) An isotopically homogeneous region of the inner terrestrial planet region (Mercury to Earth): Evidence from E chondrites and implications for giant Moon-forming impact scenarios. *Lunar Planet. Sci. XLIV,* Abstract #2344. Lunar and Planetary Institute, Houston.

Javoy M., Kaminski E., Guyot F., Andrault D., Sanloup C., Moreira M., Labrosse S., et al. (2010) The chemical composition of the Earth: Enstatite chondrite models. *Earth Planet. Sci. Lett., 293,* 259–268.

Johansen A. and Lambrechts M. (2017) Forming planets via pebble accretion. *Annu. Rev. Earth Planet. Sci., 45,* 359–387.

Johansen A., Klahr H., and Henning T. (2006) Gravoturbulent formation of planetesimals. *Astrophys. J., 636,* 1121–1134.

Johansen A., Jacquet E., Cuzzi J. N., Morbidelli A., and Gounelle M. (2015) New paradigms for asteroid formation. In *Asteroids IV* (P. Michel et al., eds.), pp. 471–492. Univ. of Arizona, Tucson.

Johnson A. P., Cleaves H. J., Dworkin J. P., Glavin D. P., Lazcano A., and J. L. Bada (2008) The Miller volcanic spark discharge experiment. *Science, 322,* 404.

Joliff B. L., Gillis J. J., Haskin L. A., Korotev R. L., Wieczorek M. A. (2000) Major lunar crustal terranes: Surface expressions and crust-mantle origins. *J. Geophys. Res., 105,* 4197–4216.

Juric M. and Tremaine S (2008) Dynamical origin of extrasolar planet eccentricity distribution. *Astrophys. J., 686,* 603–620.

Kasting J. F. (1990) Bolide impacts and the oxidation state of carbon in the Earth's early atmosphere. *Origins Life Evol. Biospheres, 20,* 199–231.

Kasting J. F. and Pollack J. B. (1983) Loss of water from Venus. I. Hydrodynamic escape of hydrogen. *Icarus, 53,* 479–508.

Kasting J. F., Whitmire D. P., and Reynolds R. T. (1993) Habitable zones around main sequence stars. *Icarus, 101,* 108–128.

Kemp A. I. S., Wilde S. A., Hawkesworth C. J., Coath C. D., Nemchin A., Pidgeon R. T., Vervoort J. D., and DuFrane S. A. (2010) Hadean crustal evolution revisited: New constraints from Pb-Hf isotope systematics of the Jack Hills zircons. *Earth Planet. Sci. Lett., 296,* 45–56.

Kite E. S. and Ford E. B. (2018) Habitability of exoplanet waterworlds. *Astrophys. J., 864,* 75.

Kleine T. and Kruijer T. S. (2017) Tungsten isotopes and the origin of the Moon. *Lunar Planet. Sci. XLVIII,* Abstract #2987. Lunar and Planetary Institute, Houston.

Kleine T. and Walker R. J. (2017) Tungsten isotopes in planets. *Annu. Rev. Earth Planet. Sci., 45,* 389–417.

Kleine T., Touboul M., Bourdon B., Nimmo F., Mezger K., Palme H., Jacobsen S. B., Yin Q.-Z., and Halliday A. N. (2009) Hf-W chronology of the accretion and early evolution of asteroids and terrestrial planets. *Geochim. Cosmochim. Acta, 73,* 5150–5188.

Kokubo E. and Ida S. (1998) Oligarchic growth of protoplanets. *Icarus, 131,* 171–178.

Korenaga J. (2013) Initiation and evolution of plate tectonics on Earth: Theories and observations. *Annu. Rev. Earth Planet. Sci., 41,* 117–151.

Kress M. E. and McKay C. P. (2004) Formation of methane in comet impacts: Implications for Earth, Mars, and Titan. *Icarus, 168,* 475–483.

Kruijer T. S., Kleine T., Fischer-Gödde M., Burkhardt C., and Wieler R. (2014a) Nucleosynthetic W isotope anomalies and the Hf-W chronometry of Ca-Al-rich inclusions. *Earth Planet. Sci. Lett., 403,* 317–327.

Kruijer T. S., Touboul M., Fischer-Gödde M., Bermingham K. R., Walker R. J., and Kleine T. (2014b) Protracted core formation and rapid accretion of protoplanets. *Science, 344,* 1150-1154.

Kruijer T. S., Burkhardt C., Budde G., and Kleine T. (2017) Age of Jupiter inferred from the distinct genetics and formation times of meteorites. *Proc. Natl. Acad. Sci. U.S.A., 114,* 6712–6716.

Kuchner M. J. and Seager S. (2005) Extrasolar carbon planets. *ArXiv e-prints,* arXiv:astro-ph/0504214.

Kuhn W. R and Atreya S. K. (1979) Ammonia photolysis and the greenhouse effect in the primordial atmosphere of the Earth. *Icarus, 37,* 207–213.

Kuwahara H. and Sugita S. (2015) The molecular composition of impact-generated atmospheres on terrestrial planets during the post-accretion stage. *Icarus, 257,* 290–301.

Lasaga A. C., Holland H. D., and Dwyer M. J. (1971) Primordial oil slick. *Science, 174,* 53–55.

Lehner S. W., Petaev M. I., Zolotov M. Y., and Buseck P. R. (2013) Formation of niningerite by silicate sulfidation in EH3 enstatite chondrites. *Geochim. Cosmochin. Acta, 101,* 34–56.

Lebrun T., Massol H., Chassefiere E., Davaille A., Marcq E., Sarda P., Leblanc F., and Brandeis G. (2013) Thermal evolution of an early magma ocean in interaction with the atmosphere: Conditions for the condensation of a water ocean. *J. Geophys. Res.–Planets, 118,* 1155–1176.

Levison H. F., Kretke K. A., Walsh K. J., and Bottke W. F. (2015) Growing the terrestrial planets from the gradual accumulation of sub-meter sized objects. *Proc. Natl. Acad. Sci. U.S.A., 112,* 14180–14185.

Lewis J. S. (1999) *Worlds Without End.* Helix, Reading. 256 pp.

Lissauer J. J. (1987) Timescales for planetary accretion and the structure of the protoplanetary disk. *Icarus, 69,* 249–265.

Lissauer J. J. and 16 colleagues (2013) All six planets known to orbit Kepler-11 have low densities. *Astrophys. J., 770,* 131.

Lissauer J. J., Dawson R. I., and Tremaine S. (2014) Advances in exoplanet science from Kepler. *Nature, 513,* 336–344.

Livio M. and Rees M. J. (2005) Anthropic reasoning. *Science, 309,* 1022–1023.

Lock S. J. and Stewart S. T. (2017) The structure of terrestrial bodies: Impact heating, corotation limits and synestias. *J. Geophys. Res.–*

Planets, 122, 950–982.

Lock S. J., Stewart S. T., Petaeva M. I., Leinhardt Z. M., Mace M. T., Jacobsen S. B., and Cúk M. (2018) The origin of the Moon within a terrestrial synestia. *J. Geophys. Res.–Planets, 123,* 910–951.

Lodders K. (2003) Solar system abundances and condensation temperatures of the elements. *Astrophys. J., 591,* 1220–1247.

Lodders K. (2004) Jupiter formed with more tar than ice. *Astrophys. J., 611,* 587–597.

Lodders K. and Fegley B. Jr. (1998) *The Planetary Scientist's Companion.* Oxford Univ., Oxford. 400 pp.

López E. and Fortney J. J. (2014) Understanding the mass-radius relation for sub-Neptunes: Radius as a proxy for composition. *Astrophys. J., 792,* 1.

Lupu R., Marley M., Schaefer L., Fegley B. Jr., Morley C., Cahoy K., Freedman R., Fortney J. J., and Zahnle K. J. (2014) The atmospheres of Earth-like planets after giant impact events. *Astrophys. J., 784,* 27.

Lyons J. R. and Young E. D. (2005) CO self-shielding as the origin of oxygen isotope anomalies in the early solar nebula. *Nature, 435,* 317–320.

Maier W. D., Barnes S. J., Campbell I. H., Fiorentini M. L., Peltonen P., Barnes S.-J., and Smithies R. H. (2009) Progressive mixing of meteoritic veneer into the early Earth's deep mantle. *Nature, 460,* 620-623.

Marchi S., Bottke W. F., Elkins-Tanton L., Bierhaus M., Wuennemann K., Morbidelli A., and Kring D. A. (2014) Widespread mixing and burial of Earth's Hadean crust by asteroid impacts. *Nature, 511,* 578–582.

Marchi S., Black B. A., Elkins-Tanton L. T., and Bottke W. F. (2016) Massive impact-induced release of carbon and sulfur gases in the early Earth's atmosphere. *Earth Planet. Sci. Lett., 449,* 96–104.

Marty B. (2012) The origins and concentrations of water, carbon, nitrogen, and noble gases on Earth. *Earth Planet. Sci. Lett., 313,* 56–66.

Marty B. and 29 colleagues (2017) Xenon isotopes in 67P/Churyumov-Gerasimenko show that comets contributed to Earth's atmosphere. *Science, 356,* 1069–1072.

Matsui T. and Abe Y. (1986) Impact-induced atmosphere and oceans on Earth and Venus. *Nature, 322,* 526–528.

Mayor M. and Queloz D. (1995) A Jupiter-mass companion to a solar-type star. *Nature, 378,* 355–359.

McKeegan K. D. and 12 colleagues (2011) The oxygen isotopic composition of the Sun inferred from captured solar wind. *Science, 332,* 1528–1531.

Meier M. M. M., Reufer A., and Wieler R. (2014) On the origin and composition of Theia: Constraints from new models of the giant impact. *Icarus, 242,* 316–328.

Melosh H. J. (1990) Giant impacts and the thermal state of the Earth. In *Origin of the Earth* (H. E. Newsom and J. H. Jones, eds.), pp. 69–84. Oxford Univ., Oxford.

Melosh H. J. and Vickery A. M. (1989) Impact erosion of the primordial atmosphere of Mars. *Nature, 338,* 487–489.

Ménez B., Pisapia C., Andreani M., Jamme F., Vanbellingen Q. P., Brunelle A., Richard L., Dumas P., and Réfrégiers M. (2018) Abiotic synthesis of amino acids in the recesses of the oceanic lithosphere. *Nature, 564,* 59–63.

Miller S. L (1953) A production of amino acids under possible primitive Earth conditions. *Science, 117,* 528–529.

Miller S. L. (1955) Production of some organic compounds under possible primitive Earth conditions. *J. Am. Chem. Soc., 77,* 2351–2361.

Miller S. L. and Urey H. C. (1959) Organic compound synthesis on the primitive Earth. *Science, 130,* 245–251.

Mizuno H. and Wetherill G. (1984) Grain abundance in the primordial atmosphere of the Earth. *Icarus, 59,* 74–86.

Mizuno H., Nakazawa K., and Hayashi C. (1978) Instability of gaseous envelope surrounding planetary core and formation of giant planets. *Prog. Theor. Phys., 60,* 699–710.

Moreira M. (2013) Noble gas constraints on the origin and evolution of Earth's volatiles. *Geochem. Perspect., 2,* 229–403.

Morbidelli A., Nesvorny D., Laurenz V., Marchi S., Rubie D. C., Elkins-Tanton L., Wieczorek W., and Jacobson S. (2018) The timeline of the lunar bombardment: Revisited. *Icarus, 305,* 262–276.

Moskovitz N. and Gaidos E. (2011) Differentiation of planetesimals and the thermal consequences of melt migration. *Meteoritics & Planet. Sci., 46,* 903–918.

Murthy V. R. (1991) Early differentiation of the Earth and the problem of mantle siderophile elements — A new approach. *Science, 253,* 303–306.

Nakajima M. and Stevenson D. J. (2015) Melting and mixing states of the Earth's mantle after the Moon-forming impact. *Earth Planet. Sci. Lett., 427,* 286–295.

Nemchin A., Timms N., Pidgeon R., Geisler T., Reddy S., and Meyer C. (2009) Timing of crystallization of the lunar magma ocean constrained by the oldest zircon. *Nature Geosci., 2,* 133–136.

Newsom H. E. and Taylor S. R. (1989) Geochemical implications of the formation of the Moon by a single giant impact. *Nature, 338,* 360.

Nimmo F. and Agnor C. B. (2006) Isotopic outcomes of N-body accretion simulations: Constraints on equilibration processes during large impacts from Hf/W observations. *Earth Planet. Sci. Lett., 243,* 26–43.

Nittler L. R. and 14 colleagues (2011) The major-element composition of Mercury's surface from MESSENGER X-ray spectrometry. *Science, 333,* 1847–1850.

Nyquist L. E., Wiesmann H., Shih C.-Y., and Dasch J. (1996) Lunar meteorites and the lunar crustal Sr and Nd isotopic compositions. *Lunar Planet. Sci. XXVII,* Abstract #1486. Lunar and Planetary Institute, Houston.

Nyquist L. E., Kleine T., Shih C.-Y., and Reese Y. D. (2009) The distribution of short-lived radioisotopes in the early solar system and the chronology of asteroid accretion, differentiation, and secondary mineralization. *Geochim. Cosmochim. Acta, 73,* 5115–5136.

Oka A., Nakamoto T., and Ida S. (2011) Evolution of snow line in optically thick protoplanetary disks: Effects of water ice opacity and dust grain size. *Astrophys. J., 738,* 141.

O'Neil J., Boyet M., Carlson R. W., and Paquette J.-L. (2013) Half a billion years of reworking of Hadean mafic crust to produce the Nuvvuagittuq Eoarchean felsic crust. *Earth Planet. Sci. Lett., 379,* 13–25.

O'Neil J., Rizo H., Boyet M., Carlson R. W., and Rosing M. T. (2016) Geochemistry and Nd isotopic characteristics of Earth's Hadean mantle and primitive crust. *Earth Planet. Sci. Lett., 442,* 194–205.

Oparin A. I. (1938) *Origin of Life.* Dover, Mineola. 270 pp.

Ormel C. (2017) The emerging paradigm of pebble accretion. In *Formation, Evolution, and Dynamics of Young Solar Systems* (M. Pessah and O. Gressel, eds.), pp. 197–228. Springer, Berlin.

Oró J. and Kamat S. (1961) Amino-acid synthesis from hydrogen cyanide under possible primitive Earth conditions. *Nature, 190,* 442–443.

Oró J., Miller S. L., and Lazcano A. (1990) The origin and early evolution of life on Earth. *Annu. Rev. Earth Planet. Sci., 18,* 317–356.

Ozima M. and Podosek F. A. (2002) *Noble Gas Geochemistry, 2nd edition.* Cambridge Univ., Cambridge. 286 pp.

Pahlevan K. and Stevenson D. J. (2007) Equilibration in the aftermath of the lunar-forming giant impact. *Earth Planet. Sci. Lett., 262,* 438–449.

Palme H. and O'Neill H. S. C. (2014) Cosmochemical estimates of mantle composition. In *Treatise on Geochemistry, Vol. 1: Planets, Asteroids, Comets and The Solar System* (A. M. Davis, ed.), pp. 1–39. Elsevier-Pergamon, Oxford.

Palme H., Lodders K., and Jones A. (2014) Solar system abundances of the elements. In *Treatise on Geochemistry, Vol. 1: Planets, Asteroids, Comets and The Solar System* (A. M. Davis, ed.), pp. 15–36. Elsevier-Pergamon, Oxford.

Pasek M. and Lauretta D. (2008) Extraterrestrial flux of potentially prebiotic C, N, and P to the early Earth. *Origins Life Evol. Biospheres, 38,* 5–21.

Patterson C. (1956) Age of meteorites and the Earth. *Geochim. Cosmochim. Acta, 10,* 230–237.

Pavlov A. A., Brown L. L., and Kasting J. F. (2002) UV shielding of NH_3 and O_2 by organic hazes in the Archean atmosphere. *J. Geophys. Res., 106,* 1–21.

Peitgura E., Marcy G. W., Winn J. N., Weiss L. M., Fulton B. J., Howard A. W., Sinukoff E., Isaacson H., Morton T. D., and Johnson J. A. (2018) The California-Kepler Survey. IV. Metal-rich stars host a greater diversity of planets. *Astron. J. ,155,* 89.

Pepin R. O. (1991) On the origin and early evolution of terrestrial planet atmospheres and meteoritic volatiles. *Icarus, 92,* 2–79.

Pepin R. O. (1997) Evolution of Earth's noble gases: Consequences of assuming hydrodynamic loss driven by giant impact. *Icarus, 126,* 148–156.

Pepin R. O. (2000) On the isotopic composition of primordial xenon in terrestrial planet atmospheres. *Space Sci. Rev., 92,* 371–395.

Pepin R. O. (2006) Atmospheres on the terrestrial planets: Clues to origin and evolution. *Earth Planet. Sci. Lett., 252,* 1–14.

Pierazzo E. and Chyba C. F. (1999) Amino acid survival in large cometary impacts. *Meteoritics & Planet. Sci., 34,* 909–918.

Pinto J. P., Gladstone G. R., and Yung Y. L. (1980) Photochemical production of formaldehyde in Earth's primitive atmosphere. *Science, 210,* 183–185.

Pollack J. B. and Black D. (1979) Implications of the gas compositional measurements of Pioneer Venus for the origin of planetary atmospheres. *Science, 205,* 56–59.

Poole J. H. J. (1951) The evolution of the Earth's atmosphere. *Sci. Proc. R. Dublin Soc., 25,* 201–224.

Powner M. W., Gerland B., and Sutherland J. D. (2009) Synthesis of activated pyrimidine ribonucleotides in prebiotically plausible conditions. *Nature, 459,* 239–242.

Puchtel I. S., Walker R. J., Touboul M., Nisbet E. G., and Byerly G. R. (2014) Insights into early Earth from the Pt-Re-Os isotope and highly siderophile element abundance systematics of Barberton komatiites. *Geochim. Cosmochim. Acta, 125,* 394–413.

Pujol M., Marty B., Burnard P., and Phillipot P. (2009) Xenon in Archean barite: Weak decay of ^{130}Ba, mass-dependent isotopic fractionation and implication for barite formation. *Geochim. Cosmochim. Acta, 73,* 6834–6846.

Pujol M., Marty B., and Burgess R. (2011) Chondritic-like xenon trapped in Archean rocks: A possible signature of the ancient atmosphere. *Earth Planet. Sci. Lett., 308,* 298–306.

Pujol M., Marty B., and Burgess R. (2013) Reply to comment on "Chondritic-like xenon trapped in Archean rocks: A possible signature of the ancient atmosphere by Pujol M., Marty B., and Burgess R., Earth and Planetary Science Letters 308 (2011) 298–306 by Pepin R.O." *Earth Planet. Sci. Lett., 371–372,* 296–298.

Raymond S. and Izadoro A. (2017) Origin of water in the inner solar system: Planetesimals scattered inward during Jupiter and Saturn's rapid gas accretion. *Icarus, 297,* 134–148.

Raymond S. and Morbidelli A. (2014) The Grand Tack model: A critical review. In *Complex Planetary Systems* (Z. Knežević and A. Lemaitre, eds.), pp. 194–203. IAU Symp. 310, Cambridge Univ., Cambridge.

Raymond S., Schlichting H. E, Hersant F., and Selsis F. (2013) Dynamical and collisional constraints on a stochastic late veneer on the terrestrial planets. *Icarus, 226,* 671–681.

Reimink J. R., Chacko T., Carlson R. W., Shirey S. B., Liu J., Stern R. A., Bauer A. M., Pearson D. G., and Heaman L. M. (2018) Petrogenesis and tectonics of the Acasta Gneiss Complex derived from integrated petrology and ^{142}Nd and ^{182}W extinct nuclide-geochemistry. *Earth Planet. Sci. Lett., 494,* 12–22.

Ricardo A., Carrigan M. A., Olcott A. N., and Benner S. A. (2004) Borate minerals stabilize ribose. *Science, 303,* 196–199.

Righter K., Pando K. A., Danielson L. R., and Nickodem K. A. (2014) Core-mantle partitioning of volatile elements and the origin of volatile elements in Earth and Moon. *Lunar Planet. Sci. XLV,* Abstract #2130. Lunar and Planetary Institute, Houston.

Ringwood A. E. (1979) *Origin of the Earth and Moon.* Springer, New York. 295 pp.

Rizo H., Boyet M., Blichert-Toft J., and Rosing M. T. (2013) Early mantle dynamics inferred from ^{142}Nd variations in Archean rocks from southwest Greenland. *Earth Planet. Sci. Lett., 377,* 324–335.

Rizo H., Walker R. J., Carlson R. W., Touboul M., Horan M. F., Puchtel I. S., Boyet M., and Rosing M. T. (2016) Early Earth differentiation investigated through ^{142}Nd, ^{182}W, and highly siderophile element abundances in samples from Isua, Greenland. *Geochim. Cosmochim. Acta, 175,* 319–336.

Roth A. S. G., Bourdon B., Mojzsis S. J., Touboul M., Sprung P., Guitreau M., and Blichert-Toft J. (2013) Inherited ^{142}Nd anomalies in Eoarchean protoliths. *Earth Planet. Sci. Lett., 361,* 50–57.

Rubie D. C., Jacobson S. A., Morbidelli A., O'Brien D. P., Young E. D., de Vries J., Nimmo F., Palme H., and Frost D. J. (2015) Accretion and differentiation of the terrestrial planets with implications for the compositions of early-formed solar system bodies and accretion of water. *Icarus, 248,* 89–108.

Rubie D. C., Laurenz V., Jacobson S. A., Morbidelli A., Palme H., Vogel A. K., and Frost D. J. (2016) Highly siderophile elements were stripped from Earth's mantle by iron sulfide segregation. *Science, 353,* 1141–1144.

Rudnick R. L. and Gao S (2003) Composition of the continental crust. In *Treatise on Geochemistry, Vol. 3: The Crust* (R. L. Rudnick, ed.), pp. 1–64. Elsevier, Oxford.

Russell H. N. and Menzel D. H. (1933) The terrestrial abundance of the permanent gases. *Proc. Natl. Acad. Sci. U.S.A., 19,* 997–1001.

Sagan C. and Chyba C. F. (1997) The early faint Sun paradox: Organic shielding of ultraviolet-labile greenhouse gases. *Science, 276,* 1217–1221.

Salmon J. and Canup R. M. (2012) Lunar accretion from a Roche-interior fluid disk. *Astrophys. J., 760,* 83.

Sasaki S. and Nakazawa K. (1988) Origin of isotopic fractionation of terrestrial Xe: Hydrodynamic fractionation during escape of the primordial H_2-He atmosphere. *Earth Planet. Sci. Lett., 89,* 323–334.

Schaefer L. and Fegley B. Jr. (2007) Outgassing of ordinary chondritic material and some of its implications for the chemistry of asteroids, planets, and satellites. *Icarus, 186,* 462–483.

Schaefer L. and Fegley B. Jr. (2010) Chemistry of atmospheres formed during accretion of the Earth and other terrestrial planets. *Icarus, 208,* 438–448.

Schaefer L. and Fegley B. Jr. (2017) Redox states of initial atmospheres outgassed on rocky planets and planetesimals. *Astrophys. J., 843,* 120.

Schlichting H. E. and Mukhopadhyay S. (2018) Atmosphere impact losses. *Space Sci. Rev., 214,* 34.

Schlichting H. E., Sari R., and Yalinewich A. (2015) Atmospheric mass loss during planet formation: The importance of planetesimal impacts. *Icarus, 247,* 81–94.

Scott E. R. D., Krot A. N., and Sanders I. S. (2018) Isotopic dichotomy among meteorites and its bearing on the protoplanetary disk. *Astrophys. J., 854,* 164.

Segura T. L., Zahnle K. J., Toon O. B., and McKay C. P. (2013) The effects of impacts on the climates of terrestrial planets. In *Comparative Climatology of Terrestrial Planets* (S. J. Mackwell et al., eds.), pp. 417–437. Univ. of Arizona, Tucson.

Sekiya M., Nakazawa K., and Hayashi C. (1980) Dissipation of the rare gases contained in the primordial Earth's atmosphere. *Earth Planet. Sci. Lett., 50,* 197–201.

Sekiya M., Hayashi C., and Kanazawa K. (1981) Dissipation of the primordial terrestrial atmosphere due to irradiation of the solar FUV during T-Tauri stage. *Prog. Theor. Phys., 66,* 1301–1316.

Sharp Z. D. (2016) Evidence for a nebular contribution to the Earth's water inventory. *Lunar Planet. Sci. XLVII,* Abstract #3021. Lunar and Planetary Institute, Houston.

Sharp Z. D. (2017) Nebular ingassing as a source of volatiles to the inner planets. *Lunar Planet. Sci. XLVIII,* Abstract #1307. Lunar and Planetary Institute, Houston.

Sharp Z. D. and Olson P. L. (2018) Atmospheric ingassing and outgassing during terrestrial planet accretion: Implications for water, helium-3, and mantle oxidation. *Lunar Planet. Sci. XLIX,* Abstract #1365. Lunar and Planetary Institute, Houston.

Sharp Z. D. and Olson P. L. (2019) Nebular ingassing of water and noble gases. *Lunar Planet. Sci. L,* Abstract #2555. Lunar and Planetary Institute, Houston.

Shibaike Y., Sasaki T., and Ida S. (2016) Excavation and melting of the Hadean continental crust by late heavy bombardment. *Icarus, 266,* 189–203.

Shock E. L. (1990) Geochemical constraints on the origin of organic compounds in hydrothermal systems. *Origins Life Evol. Biospheres, 20,* 331–367.

Shock E. L. and McKinnon W. B. (1993) Hydrothermal processing of cometary volatiles — Applications to Triton. *Icarus, 106,* 464–477.

Shock E. L. and Schulte M. (1998) Organic synthesis during fluid mixing in hydrothermal systems. *J. Geophys. Res., 103,* 28513–28528.

Shock E. L., McCollom T., and Schulte M. (1995) Geochemical constraints on chemolithoautotrophic reactions in hydrothermal systems. *Origins Life Evol. Biospheres, 25,* 141–159.

Sleep N. H. (2016) Asteroid bombardment and the core of Theia as possible sources for the Earth's late veneer component. *Geochem. Geophys. Geosyst., 17,* DOI: 10.1002/2016GC006305.

Sleep N. H., Zahnle K. J., Kasting J. F., and Morrowitz H. J. (1989) Annihilation of ecosystems by large asteroid impacts on the early Earth. *Nature, 342,* 139–142.

Sleep N. H., Zahnle K. J., and Neuhoff P. S. (2001) Initiation of clement surface conditions on the earliest Earth. *Proc. Natl. Acad. Sci. U.S.A., 98,* 3666–3672.

Sleep N. H., Zahnle K. J., Lupu R. S. (2014) Terrestrial aftermath of the Moon-forming impact. *Philos. Trans. R. Soc. London, Ser. A, 372,* 20130172.

Smith J. V., Anderson A. T., Newton R. C., Olsen E. J., Crewe A. V., Isaacson M. S., Johnson D., and Wyllie P. J. (1970) Petrologic history of the Moon inferred from petrography, mineralogy and petrogenesis of Apollo 11 rocks. *Proc. Apollo 11 Lunar Sci. Conf.,* p. 897. Pergamon, New York.

Snyder G. A., Taylor L. A., and Neal C. R. (1992) A chemical model for generating the sources of mare basalts — Combined equilibrium and fractional crystallization of the lunar magmasphere. *Geochim. Cosmochim. Acta, 56,* 3809–3823.

Solomatov V. S. (2009) Magma oceans and primordial mantle differentiation. In *Treatise on Geophysics, Vol. 9: Evolution of the Earth* (D. J. Stevenson, ed.), pp. 91–120. Elsevier-Pergamon, Oxford.

Solomon S. and Longhi J. (1977) Magma oceanography. I — Thermal evolution. *Proc. Lunar Planet. Sci. Conf. 8th,* pp. 583–599.

Sprung P., Kleine T., and Scherer E. E. (2013) Isotopic evidence for chondritic Lu/Hf and Sm/Nd of the Moon. *Earth Planet. Sci. Lett., 380,* 77–87.

Srinivasan B. (1976) Barites: Anomalous xenon from spallation and neutron-induced reactions. *Earth Planet. Sci. Lett., 31,* 129–141.

Stevenson D. J. (1987) Origin of the Moon — The collision hypothesis. *Annu. Rev. Earth Planet. Sci., 15,* 271–315.

Stevenson D. J. and Lunine J. I. (1988) Rapid formation of Jupiter by diffuse redistribution of water vapor in the solar nebula. *Icarus, 75,* 146–155.

Stribling R. and Miller S. L. (1987) Energy yields for hydrogen cyanide and formaldehyde syntheses: The HCN and amino acid concentrations in the primitive ocean. *Orig. Life, 17,* 261–273.

Sugita S. and Schultz P. (2009) Efficient cyanide formation due to impacts of carbonaceous bodies on a planet with a nitrogen-rich atmosphere. *Geophys. Res. Lett., 36,* L20204.

Tang H. and Dauphas N. (2015) Low ^{60}Fe abundance in Semarkona and Sahara 99555. *Astrophys. J., 802,* 22.

Taylor D. J., McKeegan K. D., and Harrison T. M. (2009) Lu-Hf zircon evidence for rapid lunar differentiation. *Earth Planet. Sci. Lett., 279,* 157–164.

Thiemens M. H. and Heidenreich J. E. (1983) The mass-independent fractionation of oxygen — A novel isotope effect and its possible cosmochemical implications. *Science, 219,* 1073–1075.

Tian F., Toon O. B., Pavlov A. A., and De Sterck H. (2005) A hydrogen-rich early Earth atmosphere. *Science, 308,* 1014–1017.

Tian F., Kasting J. F., and Zahnle K. J. (2011) Revisiting HCN formation in Earth's early atmosphere. *Earth Planet. Sci. Lett., 308,* 417–423.

Tolstikhin I. N. and O'Nions R. K. (1994) The Earth's missing xenon: A combination of early degassing and of rare gas loss from the atmosphere. *Chem. Geol., 115,* 1–6.

Tolstikhin I., Marty B., Porcelli D., and Hofmann A. (2014) Evolution of volatile species in the Earth's mantle: A view from xenology. *Geochim. Cosmochim. Acta, 136,* 229–246.

Touboul M., Kleine T., Bourdon B., and Palme H. (2007) The duration of magma ocean crystallization on the Moon — Evidence from new W isotope data for metals from high-Ti and low-Ti mare basalts. *Lunar Planet. Sci. XXXVIII,* Abstract #2385. Lunar and Planetary Institute, Houston.

Touboul M., Liu J. G., O'Neil J., Puchtel I. S., and Walker R. J. (2012) Time constraints on late accretion to the Earth and Moon, and new evidence for early mantle differentiation derived from coupled investigations of W and Os isotope compositions. *Lunar Planet. Sci. XLIII,* Abstract #1923. Lunar and Planetary Institute, Houston.

Touboul M., Walker R. J., and Puchtel I. S. (2014) High-precision W isotope composition of the moon for constraining late accretion and lunar formation. *Lunar Planet. Sci. XLV,* Abstract #1851. Lunar and Planetary Institute, Houston.

Touboul M., Puchtel I. S., and Walker R. J. (2015) Tungsten isotopic evidence for disproportional late accretion to the Earth and Moon. *Nature, 520,* 530–533.

Trail D., Watson E. B., and Tailby N. D. (2012) Ce and Eu anomalies in zircon as proxies for the oxidation state of magmas. *Geochim. Cosmochim. Acta, 97,* 70–87.

Trainer M. G., Pavlov A. A., DeWitt H. L., Jimenez J. L., McKay C. P., Toon O. B., and Tolbert M. A. (2006) Organic haze on Titan and the early Earth. *Proc. Natl. Acad. Sci. U.S.A., 103,* 18035–18042.

Tremaine S. and Dones L. (1993) On the statistical distribution of massive impactors. *Icarus, 106,* 335–341.

Truran J. W. and Heger A. (2003) Origin of the elements. In *Treatise on Geochemistry, Vol. 1: Meteorites and Cosmochemical Processes* (A. M. Davis, ed.), pp. 1–15. Elsevier-Pergamon, Oxford.

Tucker J. M. and Mukhopadhyay S. (2014) Evidence for multiple magma ocean outgassing and atmospheric loss episodes from mantle noble gases. *Earth Planet. Sci. Lett., 393,* 254–265.

Tyburczy J. A., Frisch B., and Ahrens T. J. (1986) Shock-induced volatile loss from a carbonaceous chondrite: Implications for planetary accretion. *Earth Planet. Sci. Lett., 80,* 201–207.

Urey H. C. (1952) On the early chemical history of the earth and the origin of life. *Proc. Natl. Acad. Sci., 38,* 351–363.

Von Zahn U., Komer S., Wiemann H., and Prinn R. (1983) Composition of the Venus atmosphere. In *Venus* (D. M Hunten et al., eds.), pp. 297–430. Univ. of Arizona, Tucson.

Walker D., Longhi J., Stolper E. M., Grove T. L., and Hays J. F. (1975) Origin of titaniferous lunar basalts. *Geochim. Cosmochim. Acta, 39,* 1219–1235.

Walker J. C. G. (1977) *Evolution of the Atmosphere.* Macmillan, New York. 318 pp.

Walker J. C. G. (1982) The earliest atmosphere of the Earth. *Precambrian Res., 17,* 147–171.

Walker J. C. G. (1986) Impact erosion of planetary atmospheres. *Icarus, 68,* 87–98.

Walker R. J. (2009) Highly siderophile elements in the Earth, Moon, and Mars: Update and implications for planetary accretion and differentiation. *Chem. Erde–Geochem., 69,* 101–125.

Walsh K. J., Morbidelli A., Raymond S. N., O'Brien D. P., and Mandell A. M. (2012) Populating the asteroid belt from two parent source regions due to the migration of giant planets: The Grand Tack. *Meteoritics & Planet. Sci., 47,* 1941–1947.

Wänke H. and Dreibus G. (1988) Chemical composition and accretion history of terrestrial planets. *Philos. Trans. R. Soc. London, Ser. A, 325,* 545–557.

Warren P. H. (2011) Stable-isotopic anomalies and the accretionary assemblage of the Earth and Mars: A subordinate role for carbonaceous chondrites. *Earth Planet. Sci. Lett., 311,* 93–100.

Warren P. H. and Wasson J. T. (1979) The origin of KREEP. *Rev. Geophys. Space Phys., 17,* 73–88.

Watson A., Donahue T. M., and Walker J. C. G. (1981) The dynamics of a rapidly escaping atmosphere: Applications to the evolution of Earth and Venus. *Icarus, 48,* 150–166.

Watson E. B. and Harrison T. M. (2005) Zircon thermometer reveals minimum melting conditions on earliest Earth. *Science, 308,* 841–844.

Watson E. B., Thomas J. B., and Cherniak D. J. (2007) ^{40}Ar retention in the terrestrial planets. *Nature, 449,* 299–304.

Weidenschilling S. J. and Cuzzi J. N. (1993) Formation of planetesimals in the solar nebula. In *Protostars and Planets III* (E. H. Levy and J. I. Lunine, eds.), pp. 1031–1060. Univ. of Arizona, Tucson.

Wen J. S., Pinto J. P., and Yung Y. L. (1989) Photochemistry of CO and H_2O: Analysis of laboratory experiments and applications to the prebiotic Earth's atmosphere. *J. Geophys. Res., 94,* 14957–14970.

Wetherill G. W. (1975) Radiometric chronology of the early solar system. *Annu. Rev. Nucl. Sci., 25,* 283–328.

Wetherill G. W. (1980) Formation of the terrestrial planets. *Annu. Rev. Astron. Astrophys., 18,* 77–113.

Wetherill G. W. (1981) Solar wind origin of ^{36}Ar on Venus. *Icarus, 46,* 70–80.

Wetherill G. W. (1990) Formation of the Earth. *Annu. Rev. Earth Planet. Sci., 18,* 205–256.

Whittet D. C. (1997) Is extraterrestrial organic matter relevant to the origin of life on Earth? *Origins Life Evol. Biospheres, 27,* 249–62.

Wiechert U., Halliday A. N., Lee D.-C., Snyder G. A., Taylor L. A., and Rumble D. (2001) Oxygen isotopes and the Moon-forming giant impact. *Science, 294,* 345–348.

Willbold M., Elliott T., and Moorbath S. (2011) The tungsten isotopic composition of the Earth's mantle before the terminal bombardment. *Nature, 477,* 195–198.

Williams C. D. and Mukhopadhyay S. (2019) Capture of nebular gases during Earth's accretion is preserved in deep-mantle neon. *Nature, 565,* 78–81.

Wolf E. T. and Toon O. B. (2010) Fractal organic hazes provided an ultraviolet shield for early Earth. *Science, 128,* 1266–1268.

Wood J. A., Dickey J. S., Marvin U. B., and Powell B. N. (1970) Lunar anorthosites and a geophysical model of the moon. *Proc. Apollo 11 Lunar Sci. Conf.,* p. 965. Pergamon, New York.

Wooten D. (2015) *The Invention of Science: A New History of the Scientific Revolution.* Penguin Random House, UK. 769 pp.

Wordsworth R. (2016) Atmospheric nitrogen evolution on Earth and Venus. *Earth Planet. Sci. Lett., 447,* 103–111.

Wordsworth R. and Pierrehumbert R. T. (2013a) Hydrogen-nitrogen greenhouse warming in Earth's early atmosphere. *Science, 339,* 64–67.

Wu J., Desch S. J., Schaefer L., Elkins-Tanton L. T., Pahlevan K., and Buseck P. R. (2018) Origin of Earth's water: Chondritic inheritance plus nebular ingassing and storage of hydrogen in the core. *J. Geophys. Res.–Planets, 123,* 2691–2712.

Youdin A. N. and Goodman J. (2005) Streaming instabilities in protoplanetary disks. *Astrophys. J., 620,* 459–469.

Young E. D. (2007) Time-dependent oxygen isotopic effects of CO self shielding across the solar protoplanetary disk. *Earth Planet. Sci. Lett., 262,* 468–483.

Young E. D. (2016) Bayes' theorem and early solar short-lived radionuclides: The case for an unexceptional origin for the solar system. *Astrophys. J., 826,* 129.

Yung Y. L. and Pinto J. P. (1978) Primitive atmosphere and implications for the formation of channels on Mars. *Nature, 273,* 730–732.

Yurimoto H. and Kuramoto K. (2004) Molecular cloud origin for the oxygen isotope heterogeneity in the solar system. *Science, 305,* 1763–1766.

Zahnle K. J. (1986) Photochemistry of methane and the formation of hydrocyanic acid (HCN) in the Earth's early atmosphere. *J. Geophys. Res., 91,* 2819–2834.

Zahnle K. J. (1993) Xenological constraints on the impact erosion of the early martian atmosphere. *J. Geophys. Res., 98,* 10899–10913.

Zahnle K. J. and Kasting J. F. (1986) Mass fractionation during transonic escape and implications for loss of water from Mars and Venus. *Icarus, 68,* 462–480.

Zahnle K. J. and Marley M. S. (2014) Methane, carbon monoxide, and ammonia in brown dwarfs and self-luminous giant planets. *Astrophys. J., 797,* 41.

Zahnle K. and Sleep N. (1997) Impacts and the early evolution of life. In *Comets and the Origin and Evolution of Life* (P. Thomas et al., eds.), pp. 175–208. Springer-Verlag, Berlin.

Zahnle K. and Sleep N. (2006) Impacts and the early evolution of life. In *Comets and the Origin and Evolution of Life, 2nd edition* (P. Thomas et al., eds.), pp. 207–251. Springer-Verlag, Berlin.

Zahnle K. J., Kasting J. F., and Pollack J. B. (1988) Evolution of a steam atmosphere during Earth's accretion. *Icarus, 74,* 62–97.

Zahnle K. J., Arndt N., Cockell C., Halliday A. N., Nisbet E., Selsis F., and Sleep N. H. (2007) Emergence of a habitable planet. *Space Sci. Rev., 129,* 35–78.

Zahnle K. J., Schaefer L., and Fegley B. Jr. (2010) Earth's earliest atmospheres. In *The Origin of Cellular Life* (D. Deamer and J. W. Szostak, eds.), pp. 49–65. Cold Spring Harbor Laboratory, New York.

Zahnle K. J., Catling D. C., and Claire M. W. (2013) The rise of oxygen and the hydrogen hourglass. *Chem. Geol., 362,* 26–34.

Zahnle K. J., Dobrovolskis A. R., Lupu R., and Sleep N. H. (2015) The tethered Moon. *Earth Planet. Sci. Lett., 427,* 74–82.

Zahnle K. J., Catling D. C., and Gacesa M. (2019) Strange messenger: A new history of hydrogen on Earth, as told by Xenon. *Geochium. Cosmochim. Acta, 244,* 56–85.

Zhang J., Dauphas N., Davis A. M., Leya I., and Fedkin A. (2012) The proto-Earth as a significant source of lunar material. *Nature Geosci., 5,* 251–255.

Hoehler T. M., Bains W., Davila A., Parenteau M. N., and Pohorille A. (2020) Life's requirements, habitability, and biological potential.
In *Planetary Astrobiology* (V. Meadows et al., eds.), pp. 37–69. Univ. of Arizona, Tucson, DOI: 10.2458/azu_uapress_9780816540068-ch002.

Life's Requirements, Habitability, and Biological Potential

T. M. Hoehler
NASA Ames Research Center

W. Bains
Massachusetts Institute of Technology

A. Davila, M. N. Parenteau, and A. Pohorille
NASA Ames Research Center

A vast diversity of environments exists within and beyond Earth. The life-hosting potential of these environments — not just whether they support life, but how abundantly and robustly they could support it — is expectedly as diverse as the conditions and processes that prevail there. Assessing life's requirements offers a basis for constraining this potential. Generically, life requires energy; a supply of elements from which to construct molecules with a diversity of shapes and properties; a solvent that can support the synthesis, maintenance, and interaction of those molecules; and physicochemical conditions that support the full range of molecules and molecular interactions upon which biochemistry depends. For life on Earth, these requirements are respectively satisfied by light and redox chemical energy, the biogenic elements (carbon, hydrogen, nitrogen, oxygen, phosphorus, and sulfur) and a variety of metals, and specific ranges in environmental parameters. Understanding why and how these specific solutions satisfy life's requirements provides a means of assessing whether alternative solutions might also be viable.

1. INTRODUCTION

A vast diversity of environments exists within and beyond Earth. To the extent life depends on and interacts materially with its environment, it can be expected that life-hosting potential among these environments varies as much as the physical and chemical properties that define them. To understand this potential demands that we first understand life's requirements in some detail. Discussion of these requirements is frequently undertaken in considering "habitability." *Cockell et al.* (2016) argue that the term "habitability" is inherently binary: An environment either can or cannot sustain life. This is an important starting point in evaluating the life-hosting potential of the universe beyond Earth, as it serves to calibrate our full sense of possibility: Where could life be possible? However, Earth's biosphere vividly illustrates that inhabited environments vary widely in the abundance, diversity, productivity, and robustness of life within them. We propose the collective term "biological potential" to capture the observation that environments vary in their ability to support this broader set of measures, above and beyond the yes or no question of "habitability." Understanding how biological potential varies as a function of environmental conditions offers the possibility to evaluate not only where life could be possible, but also which environments may have the greatest probability of hosting life or the greatest potential to

express the existence of a biosphere through observable features. The latter question will be important to consider if we are faced with the need to prioritize search-for-life targets among a range of nominally habitable environments.

A variety of previous work has considered life's requirements, with the report of the Committee on the Limits of Organic Life in Planetary Systems (*Baross et al.,* 2007) providing a useful working example. The report concluded that life requires thermodynamic disequilibrium; an environment capable of maintaining covalent bonds, especially between carbon, hydrogen, and other atoms; a liquid environment; and a molecular system that can support Darwinian evolution. This list is often abbreviated to energy, specific materials (elements), solvent, and clement environmental conditions. But what is the origin of these requirements? Are they universal? And how diverse is the range of possible solutions within these broad categories?

Our understanding of life's requirements necessarily derives from the study of "life as we know it" — extant Earth organisms and their physical or molecular fossils — and is thus unavoidably incomplete in regard to both life's origins and the potential for biochemistry very different from that on Earth. The present understanding of the origin of life is limited, with competing theories differing significantly in the environments and conditions envisioned to foster life's origins (see detailed discussion in the chapter in this volume by Baross et al.). The focus of the present discussion is thus on evolved, cellular-type life, but it must be stressed that a full evaluation of biological potential should ultimately also encompass life's origin. This remains as a critical area to pursue in supporting the search for life beyond Earth. The possibility also exists that life elsewhere could be based on a much different form of biochemistry than we observe in Earthly organisms. To embrace this possibility, it is helpful to think of Earthly life as a specific expression of a more general set of properties that can be expected in all biological systems. Below, we attempt to identify some of these general properties, discuss how they are expressed in Earthly biology, and and discuss how this, in turn, defines the requirements of Earthly life. With this basis, we ask whether or to what extent the requirements of life elsewhere might differ.

Are there attributes, and therefore requirements, that are universal to biological systems? An often-cited definition holds that "life is a self-sustaining chemical system capable of Darwinian evolution" (*Joyce et al.,* 1994); this implies that life must utilize chemical means to process (and allow for modification of) information that describes itself. In turn, this requires a flow of energy and biochemistry capable of harnessing that energy, encoding and decoding information, and catalyzing the construction and assembly of the chemical components of the organism. The laws of physics and principles of chemistry, which presumably apply throughout the universe, constrain the possible solutions to this problem, and thus the range of life's requirements and limitations. The following discussion evaluates life's requirements in this frame of reference.

General attributes of living systems. Earth organisms, even the simplest cells, are formidable machines. They regulate the exchange of materials with the environment; catalyze chemical reactions with great specificity and millions-fold enhancement of rates; process information with error rates as low as 10^{-10} [one mistake for every 10 billion "bits" of information processed (*Drake et al.,* 1998)]; harvest, store, and expend energy to do work; and self-regulate these functions in response to changing environmental conditions. Each of these capabilities would seem necessary, at least qualitatively, to support Darwinian evolution. If so, every biochemistry should be minimally capable of expressing the same behavior.

These remarkable capabilities of Earthly cells arise from systems of molecules that interact extensively with one another, often in extremely specific fashion. More specifically, many of these properties arise from the extensive use of polymers and/or compartmentalization. Earthly life establishes molecular boundaries both within and around itself. Such compartmentalization protects against the dilution or loss of biomolecules and serves as the basis for both energy harvesting and regulated exchange of material with the environment. Moreover, compartmentalization establishes discrete entities whose contents can vary and thereby confer differential "fitness," which forms the basis for evolution (*Szostak et al.,* 2001). Earth life uses polymers for information processing, catalysis of chemical reactions, capture and

management of energy, and regulated exchange of materials with the environment. Not only are these polymers sufficiently large and complex to provide the efficiency and specificity required in these processes, but they provide an elegant and possibly unique solution to an important challenge in the construction of living systems. Specifically, to impart the requisite speed and specificity to the very large number of chemical transformations that life mediates requires a comparably large number of very specific catalysts, which themselves must be constructed by other very specific catalysts, and so on. The use of polymers provides a ready solution, because a vast number of different catalysts can be constructed by stringing together a small set of monomers (e.g., 22 amino acids, in our proteins). This requires only the small number of catalysts required to produce each monomer, plus a small set of catalysts capable of connecting them via a recurring bond type (*Bains,* 2004). Similarly, the genetic instruction set is made of only a few types of chemical units (4 "nucleotides," in the case of both DNA and RNA) that can be strung together in a vast number of combinations. While the specific polymers used in Earth's biochemistry — proteins, RNA, and DNA — need not be universal across biological systems, it seems likely that the utilization of some form of polymer system is.

Significant constraints are imposed on biochemistry, and in turn on life's requirements, by assuming only that these three basic attributes of terrestrial life — its basis in molecular interactions, use of compartmentalization, and use of polymers — are universal. In fact, compartmentalization and the functionality of Earth's catalytic and information-storing polymers are just specific manifestations among a broader set of molecular interactions. These interactions are governed not only by the properties of the molecules themselves, but to a great extent also by the physical and chemical properties of their medium (solvent) and broader environment. We utilize this structure — molecules, solvent, environment — to explore the biochemical basis of life's requirements.

2. MOLECULAR AND ELEMENTAL REQUIREMENTS

Life needs elements to form the skeleton of molecules, and to endow those molecules with chemical function. Multiple studies have surveyed the elements used by life (*Silver and Phung,* 2005; *Chopra and Lineweaver,* 2010), and all emphasize the central role of carbon, hydrogen, nitrogen, oxygen, sulfur, and phosphorus (hereafter referred to as "SPONCH"). A number of metals are also used. Why are these specific elements essential in the function of Earth's biology, and could other elements take their place in different models of biochemistry?

2.1. Basic Chemical Requirements

Life-like behavior is rooted in many specific interactions among a diversity of molecules. For example, the human genome encodes >20,000 different proteins and it is estimated that more than half a million unique binary interactions occur among these proteins (*Stumpf et al.,* 2008). To support such interactions requires molecules that are large and complex enough to have a diversity of shape and chemical function, at least slightly soluble in their solvent, and stable over the timescales required to serve their biochemical purpose. Those timescales range from centuries, in the case of the molecules that form the genetic code of trees or tortoises, to milliseconds, for example, in the processes underlying nerve impulse conduction. The need for both fleeting and lasting associations among atoms and molecules demands a variety of strengths of interaction. Because the types of bonds and associations that can form between atoms and molecules depend on the specific elements involved, it is helpful to briefly review the types of bonds possible between atoms as a basis for assessing life's elemental requirements. For a more extensive introduction to chemical bonding see, for example, *Greenwood and Earnshaw* (2005).

2.1.1. Strong associations: Covalent bonds. *Covalent bonds* are formed when electrons are shared by two atoms — for example, between the carbon and hydrogen atoms in methane — and provide the basis for construction of molecules from atoms. The electrons need not be shared equally. For example, the atoms of different elements attract electrons to themselves with different strengths (a property called "electronegativity"), so that electrons are shared unequally in bonds between such atoms. In the case of methane, carbon has a slightly greater affinity for electrons (is more electronegative) than hydrogen, and so the electrons are pulled toward the carbon, resulting in a slight negative charge on the carbon and a slight positive one on the hydrogen. Such bonds, and the molecules formed from them, are thus "polarized." Polarity in molecules provides an important basis for both reactivity and weak ("non-covalent") interactions with other molecules.

2.1.2. Weak associations: Non-covalent interactions. In addition to the covalent bonds that form molecules, life requires a variety of transient, yet specific, *non-covalent* associations among molecules — for example, the interaction of a hormone with its unique receptor or the recognition of specific nucleotides in the assembly of DNA. As *Stryer* (1988) notes, "reversible molecular interactions are at the heart of the dance of life." Three main types of non-covalent interaction are important in Earthly biochemistry.

Coulomb interactions are the forces of attraction or repulsion between atoms or molecules having uneven distribution of electron density (polarity). Like charges repel; unlike charges attract. Earthly biochemistry includes multiple examples of both attractive and repulsive forces playing important roles, but attractive forces play an overall larger part in molecular interaction and reactivity. One of the most important of the attractive Coulomb interactions is the hydrogen bond (*Jeffrey,* 1997). When hydrogen is linked to a much more electronegative element — especially nitrogen or oxygen — it is left with a partial positive charge that is strongly attracted to the partial negative charges borne by oxygen or nitrogen atoms, either within the same molecule or in other molecules (Fig. 1). Hydrogen bonding involving O, N, and (to a lesser extent) S is pervasive and essential as a mode of non-covalent interactions in Earth's biochemistry. Hydrogen bonds can also form between hydrogen atoms and almost any other atom that is more electronegative, including carbon (*Desiraju and Steiner,* 2001), but these are usually too weak to have any effect at the temperatures of Earthly life.

London dispersion forces (a component of *van der Waal's forces*) are weakly, but always, attractive at distances of interest in this discussion. These forces increase in strength with increasing molecular or atomic packing (decreasing distance) and are non-specific (everything is attracted to everything else).

The hydrophobic effect, as the name implies, causes non-polar molecules to associate with one another in order to minimize association with water. The tendency of oil to form discrete droplets in water is a well-known example. In this case, the association is not based on attractive forces between the non-polar molecules themselves, but rather derives from the force required to "create space" in the water to accommodate those molecules. In this regard, the nature and strength of the hydrophobic effect derives entirely from the properties of the solvent, water; as such, its discussion is deferred to section 3.

Some non-covalent associations can be directional — that is, create preferential geometric orientations between or among molecules — and this directionality plays a very important role in the specificity of molecular interactions. Hydrogen bonds are a key example. These bonds are directional because the electrons on, for example, oxygen or nitrogen atoms are held in "orbitals" — concentrations of electron density that point in a specific direction (Fig. 1). It is these electrons that attract the hydrogen atom, and the strength of attraction varies with the relative geometric orientation of the two atoms. Other forms of non-covalent interaction do not have a preferred direction, which means that they cannot on their own cause a preferential orientation of molecules, only make them attract or repel. This is generally the case for London dispersion forces, but important exceptions exist. For example, "aromatic" ring systems — rings of atoms joined by alternating single and double bonds, in which the component atoms all lie within the same plane (Fig. 1) — have concentrations of electron density on their flat faces. London dispersion forces between such molecules are strongest when these areas of electron density associate, such that aromatic ring systems may "stack" as a result. Such stacking plays an important organizing role in the structure of the DNA double helix (Fig. 1).

2.1.3. Molecular shape. Many molecular interactions must occur with a high degree of specificity in order to serve their biochemical purpose. For example, the overall error rate in human DNA replication — which requires that nucleobases be assembled in a specific sequence — approaches 10^{-10} (*Drake et al.,* 1998). Directional non-covalent interactions

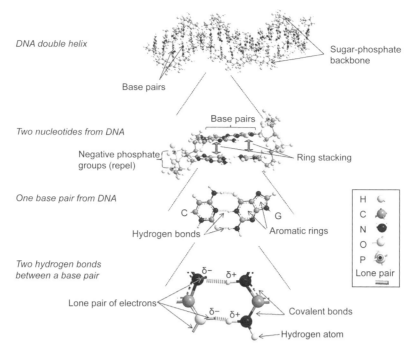

DNA double helix

Base pairs

Sugar-phosphate backbone

Two nucleotides from DNA

Base pairs

Negative phosphate groups (repel)

Ring stacking

One base pair from DNA

C   G

Hydrogen bonds

Aromatic rings

H
C
N
O
P
Lone pair

Two hydrogen bonds between a base pair

δ–   δ+

Lone pair of electrons

δ–   δ+

Covalent bonds

Hydrogen atom

Fig. 1. A portion of the DNA double helix, examined at successive levels of magnification and detail. The uppermost view illustrates the recurring negative phosphate charge along the external surface. It is hypothesized that this feature dominates the interaction of the molecule with its solvent, water, providing consistent properties even as the specific base pair composition changes. Moving incrementally closer (second view from the top), the attractive "stacking" of planar aromatic rings in the molecule's interior provides directional non-covalent stabilization that supports the paired strand structure. Closer still (second view from the bottom), molecular recognition, the cornerstone of information processing, is provided by Watson-Crick base pairing: the formation of three directional hydrogen bonds based on complementary molecular shapes. The rigidity of the planar aromatic systems (denoted by dashed circles) and the unique role of nitrogen in H-bonding within such rings supports this puzzle-piece effect. The highest level of resolution (bottom view) serves to distinguish covalent (shared electron) and non-covalent (electrostatic H) bonds. H bonds form between the partial positive (δ+) charges on H and partial negative (δ–) charges on N and O. Directionality in these bonds is supported by the directionality of the "lone pair" electrons of N and O.

provide a basis for such specificity, but a single hydrogen bond, for example, may be insufficient to promote highly specific interactions. Hydrogen bonds can potentially form between H and *any* nearby N or O, including those in the solvent itself, resulting in the potential for many interactions of comparable strength. Increased specificity is provided by molecular shape. Two molecules that have complementary shapes have the potential for forming multiple, complementary weak bonds whose attractive forces add up to form an overall stronger intermolecular link. A classic example is the complementary base pairs in DNA, which interact on the basis of two or three complementary hydrogen bonds plus the "stacking" of aromatic rings in the bases (Fig. 1). The very many specific interactions inherent in life-like chemistry imply the need for very many specific molecular shapes. Such diversity is possible only in molecules made up of tens, or even hundreds, of atoms, and we can see this as a basic requirement: No matter the specific chemical basis for a given form of life, it will be made of molecules with many atoms that can assemble into a diversity of shapes and support multiple non-covalent interactions.

2.1.4. Redox chemistry. At its core, the chemistry of life, like all chemistry, is rooted in the distribution, sharing, and movement of electrons within and between atoms and molecules. The movement of electrons from one atom or molecule to another is called oxidation-reduction or "redox" chemistry. The need for redox chemistry defines requirements for the elemental composition of life's molecules.

Redox chemistry changes the "oxidation state" of an atom — its complement of electrons, relative to the elemental state, which is defined as zero. Electrons in covalent bonds are shared between the bonded atoms but, as noted above, they are often not shared equally. An atom or molecule is "oxidized" when it loses electrons to another atom or molecule and "reduced" when it gains them. The oxidation state of carbon when bonded to itself is zero, because the two carbons share the electrons with equal strength. The oxidation state increases (representing "loss" of electrons) through bonding to more electronegative elements like nitrogen and oxygen, and decreases to negative values (representing "gain" of electrons) through bonding to less electronegative elements like hydrogen.

Redox chemistry is central to life for two reasons. First, along with light, redox chemistry is the only environmental energy source known to be utilized by life on Earth. This is described extensively in section 4. Redox chemistry is central to life in another, more subtle way. Life requires many molecules comprising a diversity of shapes and forms, and such diversity is not possible if the carbon is in fully oxidized or fully reduced form. The only compound comprised of the SPONCH elements in which carbon is completely reduced is methane, and while there are a few dozen compounds in which carbon is completely oxidized, this is far short of the number or the diversity needed to make life (*Bains and Seager,* 2012). On the other hand, the average oxidation state of carbon in most of our biomolecules is approximately (but slightly less than) zero, because of the prevalence of C-C bonds (C has an oxidation state of zero when bonded to itself) and because the number of instances in which C bonds to H outweighs the number of instances in which it bonds to O or N. On Earth, however, most available carbon in the environment is in the completely oxidized (+4) form of carbon dioxide. Oil and sedimentary organic matter are exceptions, but these are ultimately derived from living organisms. To use this carbon in producing biomolecules, life must *partly* reduce it from +4 to slightly less than zero. The oxygen in Earth's atmosphere is a direct result of life's acquiring electrons (from water) in order to carry out this reduction of carbon. Note that the opposite would be required in environments such as Titan, where the carbon is present as methane. There, life would have to take the methane and *partly* oxidize it (from –4 to about zero).

2.1.6. Summary of basic requirements. Life needs a set of diverse molecules that interact in specific ways and with a variety of strengths, which in turn requires a variety of molecular shapes and distribution of charge. They must be capable of reliably reacting with each other, *not* reacting when stability is needed, and promoting electron transfer. The properties of any such molecules are conferred by the properties of the specific atoms that comprise them. Therefore, life requires elements that:

- Yield large molecules (minimally tens to hundreds of atoms) having a diversity of shapes and stability over timescales consistent with their function in biology (minutes to decades or more)
- Provide the potential for non-covalent interactions, particularly (but not exclusively) directional bonds, within and among molecules
- Provide routes for chemical reactivity, including electron transfer, thus enabling molecular synthesis and redox chemistry

The elements used in Earth's biochemistry confer these properties in different ways. Understanding in detail how they do so provides a basis for asking whether a reduced or alternative set of elements could also suitably support a life-like chemistry.

2.2. Scaffolding: Elements to Create Molecular Structure and Shape

Formation of large, stable molecules having a diversity of possible shapes requires an element (or elements) whose atoms can covalently bond to one another in a diversity of patterns. Only five elements — boron, carbon, silicon, sulfur, and germanium — routinely form covalent compounds in which many atoms of the same type are joined together to form large molecules.

Carbon, the "scaffolding" element used in Earth's biochemistry, is by far the most versatile in supporting a diversity of stable molecular shapes. Within a single molecule, carbon-carbon bonds can produce linear sequences and branches in both planar and tetrahedral geometries, which collectively yield a staggering diversity of possible molecular shapes. Importantly, carbon can also form aromatic molecules composed either exclusively of carbon or with incorporation of oxygen, sulfur, or multiple nitrogen atoms. The properties of these molecules not only support directional non-covalent interactions (e.g.,"ring-stacking") but also confer functionality in the molecules that serve to transfer electrons [e.g., flavin adenine dinucleotide (FAD) and nicotinamide adenine dinucleotide (NAD)] and capture light energy in photosynthesis (e.g., *Stryer*, 1988; *Blankenship*, 2010).

The possible alternatives to carbon — boron, silicon, sulfur, and germanium — provide much less flexibility and functionality, especially in the presence of water. Among these, silicon is the most popularly considered alternative. Silicon (and sulfur) can form chains with itself, and also in alternation with carbon, nitrogen, or oxygen. Should water prove to be a uniquely suitable solvent for life, however, silicon chemistry is non-viable: While silicon-carbon bonds are stable to water, silicon-silicon, silicon-hydrogen, silicon-nitrogen, and most other silicon-atom bonds are very rapidly hydrolyzed. The bonding pattern of sulfur supports the formation of chains but very limited branched structures, which severely limits the diversity of possible shapes in sulfur-based systems. Boron chemistry has the reverse problem, tending to form clusters of atoms rather than smaller, isolated molecules. Thus, for life in water, carbon is uniquely suited to form the skeleton. Even in different solvents, the possible alternatives to carbon probably provide considerably less geometric and chemical diversity, as many others have noted (*Baross et al.,* 2007).

2.3. Heteroatoms: Elements to Enable Reactivity and Directional Bonding

In order to provide function, molecular skeletons need to be "decorated" with groups that can confer chemical reactivity and provide a basis for directional non-covalent interactions. Carbon-carbon and carbon-hydrogen bonds alone provide very little of this functionality. Life-like function thus requires the incorporation of other elements that can form covalent bonds to carbon (or other scaffolding element) but are more electropositive or electronegative. In organic chemistry, the atoms of any element other than carbon or hydrogen are termed "heteroatoms." For Earthly life, nitrogen, oxygen, phosphorus, and sulfur — the N, O, P, and S of "SPONCH" — are the critical heteroatoms. As *Benner* (2010) notes, "The interesting parts of terran organic biomolecules, the parts that permit metabolism at terran temperatures, are the parts that include non-carbon atoms." By considering the way in which these heteroatoms confer life-like behavior in Earth's biology, we can then ask whether suitable alternatives might exist.

2.3.1. Nitrogen. Earthly life uses nitrogen extensively and in many roles, so much so that it is difficult to conceive of a biochemistry that does not rely on nitrogen. The bonding pattern of nitrogen — three covalent bonds, plus a "free" pair of electrons (a directional concentration of negative charge) — supports a wide range of covalent and non-covalent chemistry (*Greenwood and Earnshaw,* 2005). In non-covalent chemistry, nitrogen can be a hydrogen bond acceptor, and N-H groups are hydrogen bond donors. Nitrogen is unique in forming both stable covalent bonds to carbon (and other elements) and, through acquisition of a proton, positively charged forms that are stable in water. Thus, N-containing molecules can have overall or localized positive charge. This, in turn, can significantly impact their solubility, reactivity with respect to other molecules, and potential for non-covalent interactions within the molecule and with other molecules. Chemistry involving positively charged molecules also plays an important role in redox chemistry of NAD and other closely related molecules, which act as redox intermediates in the cell, coupling reduction of one metabolite to oxidation of another (*Stryer,* 1988). Sulfur can also form a stable positive charge, but such molecules are more likely to react, and so are probably less well suited to being stable, long-term components of structures such as DNA (*Shapiro and Schlenk,* 2006). Overall, nitrogen is unique in this biochemical role.

The covalent chemistry of nitrogen supports single-, double-, and triple-bonding to other atoms. Among other behaviors, this allows it to form aromatic ring systems that have planar geometries and, through nitrogen, the potential for both protonation (formation of positive species) and non-covalent interactions via free electron pairs. The value of this is beautifully illustrated by the bases in DNA, where both ring-stacking and hydrogen bonding enabled by nitrogen atoms contribute to the base pair stability. Nitrogen-carbon bonds, in the form of (–CO-NH—) "amide bonds" provide the basic unit by which individual amino acids are polymerized into proteins. The relative rigidity of this bond allows proteins to form regular repeating structures such as the alpha helix. Replacing C-N (amide) with C-O (ester) bonds would result in greater freedom of rotation around the bond, with less constraint within the polymer chain and correspondingly greater difficulty in forming regular structures (*Fischer,* 2000; *Choudhary and Raines,* 2011).

2.3.2. Oxygen. Oxygen is, on average, the third most abundant atom in biomolecules, after carbon and hydrogen. Chemically, oxygen is less flexible than nitrogen; instead, oxygen's extensive role in biochemistry is based principally on three features:

1. O-H groups form strong hydrogen bonds, which contribute to the solubility and directional non-covalent interactions of the molecules that contain them.

2. The carboxyl group, COOH, easily loses a proton to form the negatively charged carboxylate ionic group, COO⁻. This serves in similar but complementary fashion to the potential for acquisition of positive charge by nitrogen-containing molecules. Among other things, this has the potential to make carboxylate-containing compounds soluble in both water (when deprotonated and charged) and non-polar media like lipid membranes (when protonated and uncharged). OH groups bonded to other heteroatoms (e.g., P and S) can also lose a proton to form a negative ion. The formation of negatively charged P-O bonds in phosphate groups has important consequences for the stability of DNA (*Benner et al.,* 2004) and the energetics of adenosine triphosphate (ATP), the primary molecular carrier of energy in Earthly life (*Stryer,* 1988). Almost all the negative charges on Earthly biomolecules are due to oxygen-based anions.

3. The carbonyl group (carbon joined by a double bond to oxygen) is a very flexible one for enabling organic chemistry. The electronegativity of oxygen polarizes the carbon-oxygen bond, leaving carbon with a partial positive charge that renders it more reactive (*Benner et al.,* 2004).

Is oxygen an essential element for life? Conceivably the hydrogen-bonding function of oxygen could be carried out by nitrogen and sulfur, and sulfur can also substitute for oxygen in forming negative charges. But given the ubiquity of oxygen in water, it would be surprising for water-based life not to utilize oxygen chemistry. It is not clear, however, that O is an irreplaceable requirement for all life.

2.3.3. Phosphorus. Life uses phosphorus extensively, but mostly as phosphate (P bonded to four oxygen atoms) or its close analogs (one O replaced by N, S or C) (*Seto and Kuzuyama,* 1999; *Wang et al.,* 2007); there is only one report of phosphorus being used by life in a non-phosphate-like form (*Bains et al.,* 2019). Thus, the biochemistry of phosphorus is essentially the chemistry of phosphate.

Phosphate is unique in its ability to form two P-O-X bonds ("ester" bonds, where X is a carbon-based group of atoms) that are stable in water and still retain a negative charge. Other multivalent anions, such as sulfate and silicate, do not form charged esters. This property of phosphate plays a key role in the nucleic acid polymers DNA and RNA. In nucleic acids, phosphate bonds to two sugar molecules but still retains a negative charge, so that the backbones of DNA and RNA contain a repeating negative charge — one per base unit, where individual DNA or RNA polymers may be thousands or even millions of base units long. *Benner and Hutter* (2002) argued that this repeating charge is critical for the function of any genetic system because it is the dominant control on both the conformation and solubility of such molecules, rendering these properties unchanged even as the chemical composition of the molecules varies by virtue of differing base pairs (or other coding molecules). The same ability of phosphate groups to form adjacent negative charges is at work in the molecules involved in energy storage. Phosphate-anhydride (P-O-P) bonds are easily made and broken, and their hydrolysis releases significant energy, because both phosphates in a diphosphate group are charged. The electrostatic repulsion between these adjacent negative charges makes it highly favorable for one group to "leave," with considerable energy released in the process. The paradigmatic example is the use of ATP as an "energy carrier," in which the energetic P-O-P bond is easily formed and broken (*Westheimer*, 1987). Many other phosphates are used in life for the same overall purpose.

2.3.4. Sulfur. Life's use of sulfur is surprisingly limited, given the versatility of this element's chemistry. Sulfur is the seventh most abundant element in the combined crust, ocean, and atmosphere on the modern Earth (Fig. 2), and can be readily converted by life from all its mineral forms to forms that can be utilized in biochemistry. It can form compounds in all oxidation states between –2 and +6, can form stable positively and negatively charged compounds, can act as a redox intermediate, can form chains and rings, and is excellent at coordinating metals (see below). Life uses all these features of sulfur chemistry, but overall to a much smaller degree than it utilizes nitrogen or oxygen.

Sulfur is a component of the protein-forming amino acids cysteine and methionine, where it plays a role in several aspects of protein structure and function. Sulfur forms relatively weak covalent bonds with itself, such that sulfur-containing amino acids within different regions of a protein may bond to one another. The resulting "disulfide bridge" serves as a structural element that, by virtue of its effect on protein folding, can directly impact the catalytic potential of the protein (*Sevier and Kaiser*, 2002). Sulfur (especially from cysteine) also often participates in the chemistry occurring at enzyme active sites, as a hydrogen bond donor, acid or base catalytic group, or through binding of metal ions, such as in the formation of the widespread "iron-sulfur clusters" (*Johnson et al.*, 2005).

The thioester group — a sulfur bonded to a carbon that is also doubly bonded to oxygen — plays a key role in the metabolism of fatty acids including acetic acid in the cofactor Coenezyme-A. Some ideas about the origin of life suggest that thioesters were key to energy transfer in early metabolism, before ATP took over this role (*de Duve*, 1995). Sulfur as sulfate provides negative charges on a number of biopolymers, and sulfur as a positively charged atom in the molecule S-adenosylmethionine plays the role of a carrier of methyl groups in central metabolism. Thus life does utilize much of sulfur's chemical flexibility, but in a relatively limited *number* of molecular types. This may simply be because some of these roles can just as readily be carried out by other elements.

2.3.5. Alternative heteroatoms. Potential alternatives to N, O, P, and S must at the very least establish stable covalent bonds to, and differ in electronegativity from, the scaffolding element, and elements meeting these basic criteria are explored below. A variety of additional factors, such as the bonding patterns, redox, and acid-base properties described above, ultimately determine whether such elements could replace or add to the SPONCH elements.

Boron. Boron can form stable, polarized bonds with oxygen and carbon and B-OH groups are good hydrogen bond donors (*Baker et al.*, 2009; *Kim et al.*, 2011). However, boron is both rare in the crust and energetically difficult to extract from is primary (borate) form. Possibly for these reasons, Earthly life almost never uses boron: There is only one known case of boron being used as a heteroatom (i.e., bonded directly to a carbon atom) in biochemistry, in a chemical isolated from cranberries (*Murphy et al.*, 2003). Why cranberries (alone among the Earth's life as far as we know) use boron in this way is unknown.

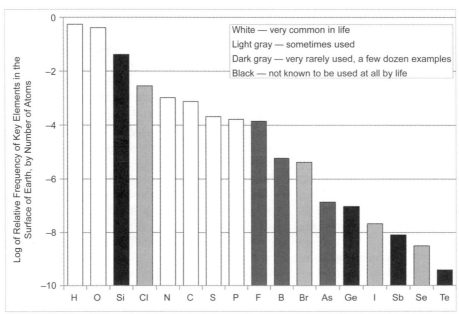

Fig. 2. Abundance of the non-metallic elements at Earth's surface (atmosphere, oceans, and upper 3 km of crust) vs. their use by life as heteroatoms on biomolecules. In general, life uses the elements available to it in greatest abundance. Note that life does use silicon and fluorine, but in mineral structures (teeth, shells, etc.) and not as heteroatoms.

Halogens. Halogens bonded to carbon can serve as hydrogen bond acceptors with a strength that decreases monotonically from fluorine to iodine. Fluorine is very rarely used by life (*O'Hagan and Harper,* 1999), probably because it is so reactive that incorporating it into biomolecules involves highly reactive, dangerous intermediates. Life does utilize chlorine, bromine, and iodine in biochemical roles, but their bonds to carbon are often quite labile (i.e., easily broken down) and reactive — a fact that makes some organohalogen compounds toxic. Indeed, some species make and release compounds containing chlorine and bromine, probably to poison other, competing organisms (*Gribble,* 2015). Halomethanes (compounds with one carbon and one or more halogens) have been studied as potential signs of life on other worlds (*Segura et al.,* 2005; *Seager and Bains,* 2015).

Silicon. Silicon seems implausible as a scaffolding element in the presence of water, and possibly even other solvents, but the stability of carbon-silicon and oxygen-silicon bonds in water makes its use as a heteroatom in carbon-based biochemistry at least theoretically feasible. Indeed, enzymes have been created synthetically that can make carbon-silicon bonds (*Kan et al.,* 2016). However, as noted above, bonds between silicon and most other atoms are not stable in water and, as with boron, it also takes a lot of energy to reduce silicon from silicate (the primary anion in Earth's crustal and mantle rocks). Perhaps for these reasons, life's use of silicon is exclusively as silicate or silica in, for example, the shells of diatoms, in the stems and leaves of grasses, and as defensive spikes in sponges (SiO_2) (*Tacke,* 1999). There is no known cases of silicon's use in a heteroatom role.

Arsenic. Arsenate-arsenite [arsenic (V) and (III)] transitions are exploited by life as sources of redox energy and electrons, and the stability of As-C bonds allows the use of arsenic as a heteroatom in biomolecules (*Reimer et al.,* 2010), but only a few instances of such use are known. Arsenate ester bonds — the chemical equivalent of the phosphate ester bonds that play a role in nucleic acid chemistry and energy metabolism — are extremely unstable in water, implying that As could not substitute for P in these critical biochemical roles. [Claims of arsenate replacing phosphate in DNA have been widely refuted (e.g., *Benner et al.,* 2013).] Arsenic could be a useful redox reagent, but Earthly life has sulfur and selenium to play these roles; this is an example of where life seems to have evolved to use one chemistry when it could have "chosen" either.

Selenium. Selenium is chemically similar to sulfur, but much rarer in the crust. Life uses selenium in a few specific enzymes (*Burk and Hill,* 2015), where its subtly different redox chemistry is advantageous. However, some species have been found that need no selenium (*Romero et al.,* 2005), so selenium must be considered useful but not essential in a biochemical role.

Metalloids. For completeness, it should be mentioned that germanium, antimony, tellurium, mercury, silver, cadmium, tin, lead, and gold can form covalent bonds to C that are stable in water. However, the first three are extremely rare, none add unique chemistry, and several are extremely toxic. Perhaps for some or all of these reasons, life on Earth is not known to use any of these elements in a biochemical role.

2.4. Metals

While most compilations of life's elemental requirements focus on the "biogenic elements" (SPONCH), metals have a ubiquitous and often underappreciated role in biology. Over a third of the proteins in bacterial and mammalian cells are metalloproteins, in which tightly associated metals are pivotal to function (*Shi and Chance,* 2011). The oxygen-carrying protein hemoglobin (iron bound by porphyrin, a large molecule comprising several aromatic rings, which can bind metal atoms at its center) and the light-harvesting pigments, such as chlorophyll (magnesium bound by porphyrin), serve as prime examples of metals functioning in key biochemical roles. About fifteen metals are commonly used by a wide range of organisms (Table 1). There are patterns (*Williams,* 2002) but not inviolable rules in life's use of metals, primarily as counter ions, coordinating groups, redox active (electron transfer) agents, and basic charge units in the electrochemical gradients that some organisms employ in energy metabolism:

Counter ions. In this role, metals provide a positive charge that can balance negative charges either at the molecular or, in some cases, cellular level. Many biological molecules are charged; they would repel each other if that charge was not balanced by an equal and opposite charge. While the requisite positive charge could, in principle, be supplied by any metal ion, sodium and potassium serve frequently in this role, perhaps because their single positive charge allows for more easily reversible interactions with negatively charged molecules.

Coordinating groups. Several metals serve to "coordinate" or organize the structure and activity of associated molecules through electrostatic interaction. For example, many enzymes involved in hydrolysis, such as proteases, have a metal atom (often zinc) in their active site. The strong electric field around the metal atom polarizes nearby bonds, weakening them for subsequent attack (*Lipscomb and Sträter,* 1996). These metals may also form temporary bonds that stabilize chemical intermediates (chemical states midway between reactants and products), and thereby facilitate and catalyze reactions. Transition metals, especially Mn, Fe, Co, Ni, Cu and Zn, serve commonly in such roles. In coordination chemistry, specific metals have affinities for different elements, and this provides a basis for selective interactions between metals and non-metals in biochemistry. It would be reasonable to expect life of any sort to require a selection of metal ions, to provide selectivity for these different functions.

Electron transfer. Some metals play a key role in electron transfer because they can interconvert between oxidized and reduced forms with relative ease. Iron is a key player in this role, with Fe^{2+} and Fe^{3+} states being used as electron donors and acceptors respectively. Copper (Cu^{1+} and Cu^{2+})

TABLE 1. Metals commonly used by life.

Metal	Common functions[*]	Universal?[†]	Species that can do without[‡]	Replacement[§]
Na	Charge balancing, maintaining membrane potentials	Yes		Li in some circumstances, although with significant physiological effect; K in some
K	Charge balancing, protein stabilization, maintaining membrane potential	Yes		Can be replaced by Rb in many processes, Na in some
Mg	All ATP-using reactions; other protein active sites; component of chlorophyll	Yes		Mn can substitute in many enzymes (*Bock et al.,* 1999)
Ca	Signalling, protein stabilization, shells and bones	Yes		By Sr in bone (apatite), Sr, Ba, and Cd in some signalling proteins (*Habermann et al.,* 1983)
Fe	Many redox enzymes, haeme proteins, iron-sulfur proteins; O_2 transport in many species	Yes		By Zn in some enzymes (*Nguyen et al.,* 2007), Cu (*Peers and Price,* 2006), Mn (*Peers and Price,* 2004; *Aguirre et al.,* 2013), Co (*Hoffman and Petering,* 1970), of replaced by non-metal redox systems (*Roche et al.,* 1996)
Zn	Structural role in the ribosome, catalytic centre of many enzymes, including enzymes making RNA	Yes		By Fe (*Ferrer et al.,* 2007), Co (*Holmquist and Vallee,* 1974; *Yee and Morel,* 1996), Mn in many enzymes (*Bock et al.,* 1999)
Mn	A number of enzymes, especially hydrolases; also urease and SOD in some species (replace Fe enzymes)	Yes		Some classes of enzyme can use alternative metals (e.g., *Abreu and Cabelli,* 2010)
Cu	Many redox proteins, oxidases; O_2 transport in some species	No	Around 30% of prokaryotes need no copper (*Ridge et al.,* 2008)	Many Cu enzymes are oxygen-activating, a function also carried out by Fe (*Decker and Solomon,* 2005)
Co	Mostly used as cobalamine (vitamin B12); also some enzymes	No	Chlamydomonas does not need Co (*Merchant et al.,* 2006); plants and fungi do not need vitamin B12 (*Rodionov et al.,* 2003)	Metabolism of odd-C fatty acids and reduction of ribonucleotides to deoxyribonucleotides performed by different pathways
Mo	A range of enzymes, especially nitrogenase; nitrogenase (FeMo enzyme)	No	5% or microorganisms (25% of archaea) do not need Mo (*Zhang and Gladyshev,* 2010)	Non-Mo nitrogenases widespread (*Eady,* 1996; *MacKay and Fryzuk,* 2004); W used by some species (*Hille,* 2002)
Ni	Range of redox reactions in methanogenesis, hydrogenesis; urease is the only Ni-dependent enzyme in many organisms	No	Chlamydononas does not need Ni (*Merchant and Helmann,* 2012)	Ni-independent ureases known (*Merchant and Helmann,* 2012), Ni-independent hydrogenases (*Thauer et al.,* 2010)
W	Redox enzymes in a wide range of species	No	Only a few organisms, mostly archaea, require W (*Stiefel,* 2002)	Often find a V enzyme in one organism is replaced by a W enzyme in another (Kisker et al., 1998; Hille, 2002)
V	Redox enzymes in a wide range of species	No	V not needed by higher animals (*Nielsen and Uthus,* 1990)	

[*] Examples of principal functions of the metal in metabolism. This is an illustrative list, not exhaustive.

[†] Is this metal known to be essential for all free-living organisms tested?

[‡] Notes and references for species that can live without the metal.

[§] Evidence that the metal can be replaced in at least some biological processes in at least one species.

is also widely used. A number of other metals serve key redox roles — notably, molybdenum in some nitrogenases (enzymes that convert nitrogen to ammonia) and manganese, which plays the key role in generating oxygen from water during photosynthesis. Iron and manganese are unusual in serving both coordinating and redox-active functions, although the significance of this — whether it has a special advantage for Earthly life — is unknown. The tendency of many metals to change oxidation state through one-electron reactions (whereas C, N, P, and S reactions typically involve pairs of electrons) may underlie, perhaps even uniquely, their pivotal role in some biochemical redox processes.

Thus, metals play a diverse and essential role in the biochemistry of all life on Earth, but deciding *which* metals are absolutely essential is difficult. On Earth, all known organisms require iron, zinc, sodium, potassium, and magnesium (see Table 1). However, even these may not be truly essential. For example, the role that iron plays in many processes can be replaced by other metals, and zinc can be replaced with other metals in many enzymes and they still function, albeit much less efficiently. Over evolutionary time, life's use of metals has changed from being predominantly metals that are soluble in ancient, anoxic oceans such as Fe, Mn, Ni, Co to ones soluble in modern, oxygenated oceans (Cu, Mo, Zn). This suggests substantial flexibility.

Could life exist without metals entirely? This is a very speculative subject. For Earthly biochemistry the answer is definitely "no," but the broad classes of functions summarized above can all be carried out by non-metal compounds. Charge balancing can be done with ammonium compounds (NH_3^+ groups) or carboxylic acids. The many enzymes that work without a metal ion in their active site demonstrate that metals are not universally essential for catalysis. And redox chemistry, even single electron redox chemistry, can be done with organic compounds, especially sulfur compounds and derivatives of nitrogen oxides. Thus, we cannot logically conclude that metals are absolutely essential for life, only that on Earth, where metals are available in the crust, they are widely and universally used. We also note that many scenarios for the origin of life depend on metals and minerals, so metals might be essential for life to arise in the first place.

2.5. Summary of Molecular and Elemental Requirements

Life needs an element to form the skeleton of molecules, in which role carbon is most plausible. Heteroatoms are needed to provide bond polarization and hence non-covalent interactions and for chemical function; in these roles life uses S, P, O, and N, and to a lesser extent Se, the halogens and very occasionally As. Specifically, strong hydrogen bonds are created using N and O; negative charges using combinations of O, S, and P; positive charges using N; polarized carbon atoms using O; and redox chemistry using S and Se. A number of metals are also used. In many cases other elements could be used, but are less abundant (Fig. 2) or are extremely hard to incorporate into organic chemicals (e.g., Si and F). So we conclude that life on Earth needs hydrogen carbon, oxygen, nitrogen, a selection of other "volatile" heteroatoms, and a few metals. Within our own solar system, at least, the composition of chondritic and cometary materials suggests that the absolute inventory of

SPONCH elements likely does not preclude habitability or require the use of alternatives. Rather, their availability in soluble and biochemically accessible forms may serve to constrain the abundance and productivity of life, as it does on Earth, and therefore potentially our ability to detect it. Moreover, constraints on the elements that could serve in biochemical roles provides criteria by which we might draw a long but not infinite list of chemicals from which life might be constructed (*Seager et al.,* 2016), and hence by which it might be detected on other worlds.

3. SOLVENT REQUIREMENTS

The construction and interaction of molecules is deeply dependent on the medium in which such chemistry occurs. If life's molecules are guests at a party, then solvent plays the role of host, creating context that fosters desired interactions. In the case of life-like chemistry, the environment, and in particular the solvent, should (1) assist in self-organization, (2) help in establishing a balance between synthetic and degradative processes, and (3) bring the strength of molecular interactions into a range that is both robust and responsive to changes in environmental conditions (*Pohorille and Pratt,* 2012). Chemical synthesis and self-organization are both heavily dependent on the medium but are mechanistically quite distinct and, for this reason, warrant separate treatment. Understanding why these features are necessary in living systems, and how they are met by Earth's use of water as the solvent, provide a basis for understanding whether alternatives to water may be viable.

3.1. Synthesis and Degradation

In general, complex chemical synthesis takes place in solution. It is also possible to a lesser extent in the gas phase or on natural solid surfaces, and can be driven by light. From the point of view of organic chemists, water is not a unique, or even the best solvent. As has been pointed out in the context of the origin of life, rich synthetic chemistry is also possible in a host of other organic solvents (*Bains,* 2004; *Benner et al.,* 2004). Some of them, such as formamide, might be astrobiologically relevant by virtue of their possible abundance in planetary environments (*Benner et al.,* 2004; *Adam et al.,* 2018).

Synthesis of Earthly life's biopolymers in water is, in fact, not simple. For example, the formation of the peptide bond in proteins is associated with the release of a water molecule to bulk aqueous solution. Because of this, the presence of water as a solvent shifts the chemical equilibrium away from the products and the polymerization reaction is bound to be thermodynamically unfavorable. From this standpoint, there is no imperative, and perhaps a seeming detriment, for water to be the medium for life.

This perspective, however, might not be appropriate for biology. Excessive chemical diversity, as might result if a solvent was widely and highly favorable to organic synthesis, is not necessarily an advantage. Even the large numbers of molecules used in Earthly biochemistry represent a tiny fraction of the overall diversity possible within the same chemical "space," suggesting that such diversity must ultimately be reduced and selectively channeled to a relatively small number of biochemical molecules and pathways. Supporting much larger arrays of molecules and pathways would impose considerable burden on a living system in terms of additional catalysts and regulatory networks. Perhaps more importantly, in biological systems, it is essential to achieve a balance between synthetic and degradative processes. Solvent-mediated degradation serves a central role in the regulation of cell content and functions, preventing molecular overcrowding and allowing for recycling of molecular resources into new synthetic products. Most polymers and metabolites are synthesized in response to environmental signals or at different stages of an organism's life cycle, and many of them are needed only for a short period of time. Once their biochemical function is complete, they must be inactivated and disposed of in a regulated fashion, usually with the aid of enzymes. The presence of water influences both thermodynamic equilibria and kinetic barriers for many of these degradation reactions, making them more facile. If the thermodynamic equilibrium in an alternative solvent for life were strongly shifted toward the polymer formation, degradation processes would be inefficiently slow or would require potent, energy-utilizing catalysts. An example from Earthly biology is mRNA, a molecule that serves as an intermediate in translation of genes into protein. Once a protein corresponding to a given mRNA molecule has been synthesized, the corresponding mRNA transcript, which is labile in water, is rapidly degraded enzymatically and its building blocks become available for synthesis of new transcripts. Most proteins meet the same fate; eventually they are degraded to amino acid monomers by specialized enzymes, proteases, to be reused in other proteins. This process is thermodynamically assisted by the availability of water. Thus, from this perspective, the so-called "water paradox" — according to which water appears to be both necessary to life and toxic to synthesis of biopolymers (*Benner*, 2014) — is not as paradoxical as it might appear. The problem of synthesizing biopolymers in water at the origin of life remains, but their relative instability might be actually a desirable trait.

3.2. Self-Organization

The existence of a rich pool of organic molecules, or even polymers, is necessary but insufficient to yield life-like properties. This is just molecular hardware that has to self-assemble into a functional entity. Both self-assembly and functionality are largely driven by non-covalent interactions. These interactions are profoundly affected by the chemical properties of the solvent. Water-mediated, non-covalent interactions abound in Earthly biology. For example, assembly of membranous cell walls from lipid monomers, stability of the DNA double helix, and folding of proteins to their active, three-dimensional structures are all driven by such interactions. Water influences non-covalent interactions mainly by

two mechanisms: modulation of the Coulomb interactions (electrostatic attraction or repulsion; section 2.1.2) between molecules, chemical groups or atoms that bear electric charges or dipoles, and the so-called "hydrophobic effect."

3.2.1. Modulation of electrostatic interactions. In classical electrostatics, it is common to represent the medium as a featureless, isotropic continuum. In this approximation, which has yielded many important results, the influence of the solvent on Coulomb interactions can be conveniently characterized by the dielectric constant, which measures to what extent the solvent molecules undergo polarization. Somewhat loosely, polarization can be described as the tendency of electron densities in the solvent molecules to undergo deformation in response to an external electric field or charges in dissolved species. As the dielectric constant of the solvent increases, the absolute value of Coulomb interactions between dissolved species is reduced. This means that both attractive and repulsive interactions are weaker in a high dielectric constant (polar) medium than in a low dielectric constant (non-polar) medium.

With a dielectric constant of 80, water is one of the most polar common liquids, whereas hydrocarbons, such as ethane, hexane, or decane, have dielectric constants of approximately 2. The dielectric constant of formamide is 111, even higher than the dielectric constant of water. A familiar illustration of the role of solvent in mediating electrostatic interactions is the differing capacities of polar and non-polar solvents, like water and hexane (respectively), to dissolve ionic materials like sodium chloride (salt). This difference can be quantified in thermodynamic terms as the Gibbs energy change (ΔG) associated with transferring sodium ions from water to hexane. The Gibbs energy change is a measure of the differing amounts of "useful energy" (energy that can do work) in the final and initial states of a system. A positive ΔG indicates that the final state of the system contains more useful energy, and thus that work must be done to move the system from initial to final state. For the transfer of sodium ions from water to hexane, ΔG is about $+188$ kJ mol^{-1}, meaning that, at room temperature, sodium ions are more readily dissolved in water than in hexane by a factor of 10^{33}! On the other hand, there is hardly any Gibbs energy change associated with the same transfer from water to formamide. A qualitatively similar effect applies to the transfer of any polar molecules between any media having different dielectric constants. Thus, dissolving charged and polar species or chemical groups in polar (high dielectric) liquids is strongly favored, whereas non-polar (low dielectric) environments form a very effective barrier to charged or even highly polar species. This simple, universal principle is very frequently operational in living systems. In particular, it is responsible for the ability to maintain steep ion gradients across cell membranes, the key step in Earthly bioenergetics.

A second important impact of electrostatic interactions arises when one considers that liquids are not featureless isotropic continua, but rather consist of molecules that have chemical structure. A particularly important structural feature of water is its ability to act as both a donor and an acceptor of hydrogen bonds. As we have already seen, hydrogen bonds are pervasive in biochemistry, where they play an important structural and functional role. As with other electrostatic interactions, solvent properties strongly modulate hydrogen bonding, although through a somewhat different mechanism in which the atomic structure of solvent molecules is essential. For example, interactions between the complementary nucleic acid bases, adenine-thymine (A-T), and guanine-cytosine (G-C), which form two and three hydrogen bonds respectively, are quite strong in the gas phase. The corresponding energies of association, which measure the strength of interactions between the bases, are equal to 50.6 and 87.9 kJ mol^{-1} (*Yanson et al.,* 1979). At room temperature, this means that the paired form of the bases is favored by factors of approximately 10^9 and 10^{15} to 1 respectively over the unpaired state. In chloroform, this energy for the G-C pair is reduced to 24.3 kJ mol^{-1} (*Williams et al.,* 1987), and in water, individual nucleic acid bases do not form hydrogen bonds at all, but instead tend to stack on top of each other, as illustrated in Fig. 1. This occurs because water, with its two hydrogen atoms attached to an oxygen atom, can serve as both donor (via the two lone pairs of electrons on oxygen) and acceptor (via the two hydrogen atoms) in hydrogen bonding. That is, it can compete with the pairing of A-T and G-C by forming hydrogen bonds with each of the hydrogen bonding sites in both molecules. This is critical to the function of the double helix. If base pairing were as strong in water as it is in vacuum, the replication function of DNA would be difficult to accomplish.

A similar dependence on solvent is observed in proteins, the catalytic properties of which often depend on "folding" of the protein chain into specific three-dimensional structures. The most common structural motifs in proteins, α-helices and β-strands, involve hydrogen bonds formed along the protein chain. However, if the fragments representing these motifs are excised from the protein and placed in aqueous solution, they remain disordered. Even in fully intact proteins, when hydrogen bonds within the polymer may be partially shielded from competition with water, the contribution of H-bonds to the overall energy of folding is only in the range of 2.1 and 7.5 kJ mol^{-1} — far less that it would be in a non-polar environment (*Fersht,* 1987; *Pace et al.,* 2004). In the gas phase or in non-polar solvents (such as hydrocarbons), *each* hydrogen bond in these α-helices and β-strands would contribute nearly 33.5 kJ mol^{-1}, rendering these structural elements very rigid and stable, and thus lacking in the flexibility and potential for denaturation (destruction) that may be required for proper biochemical function.

3.2.2. *The hydrophobic effect.* Besides modulating electrostatic interactions, water influences biological systems through the hydrophobic effect (*Tanford,* 1978): the tendency of non-polar (hydrophobic) molecules and groups to minimize direct contact with aqueous solution. This effect causes non-polar groups or molecules to aggregate and interact with each other, effectively creating the equivalent of an attractive force among them. Perhaps the most familiar

example of the hydrophobic effect is the spontaneous separation of water and oil. In organic chemistry, hydrophobic species are usually characterized by high content of carbon atoms, in particular C-H bonds, and the absence or near absence of polar groups containing, for example, nitrogen and oxygen atoms. In living systems, the hydrophobic effect is largely responsible for self-organization of molecules into more complex structures. An example that underlies life's critical ability to compartmentalize — e.g., form cells or subcellular structures — is the formation of biological membranes, which are aggregates of lipid molecules that contain hydrophilic (water-loving) polar "head groups" and non-polar (hydrophobic) hydrocarbon tails. In water, these molecules spontaneously arrange themselves into bilayers, such that their head groups are exposed to water on both sides of the bilayer while hydrophobic chains aggregate to form a membrane interior that is isolated from the bulk aqueous medium (Fig. 3a). Remarkably, lipid bilayers are very tight, robust structures, even though they are held together entirely by water-mediated non-covalent interactions. Other molecules that also have hydrophilic head groups and hydrophobic chains can form different aggregates, such as micelles, in which the tails aggregate into the interior of a spherical structure, leaving the head groups in contact with water (Fig. 3b). Soaps and detergents are examples of molecules that form micelles in water, and are effective because their non-polar interiors effectively dissolve and concentrate (non-polar) oils and fats.

Hydrophobic interactions also contribute importantly to the folding and stability of proteins. In most cases, adopting a unique, folded state is required for efficient and selective protein functions. In particular, folding of an enzyme ensures recognition of its substrates and significantly reduces the free energy barrier to the chemical reaction that it catalyzes, compared to unfolded or partially folded states. In an unfolded, disordered form, the protein chain is quite flexible and can adopt a huge number of possible conformational states. In a folded form, the number of possible states is greatly reduced, to just a small number. The result is a large, unfavorable change in entropy (a measure of how "disordered" a system is, or of the number of possible states available to it). This, in turn, makes a significant, unfavorable contribution to the overall Gibbs energy of protein folding, which must be counterbalanced by interactions that preferentially stabilize the folded structure relative to unfolded states. Some of the needed stabilizing interactions come from electrostatic and van der Waals interactions, as described above, but this is not enough. The critical stabilizing contribution is provided by hydrophobic interactions (*Dill,* 1990). Although water-soluble proteins have a great diversity of shapes, they share a common structural characteristic: The protein chain arranges itself such that hydrophobic amino acids are removed from contact with water and, instead, become buried inside a shell of hydrophilic or charged amino acids, as shown in Fig. 3c. In fact, the rapid adoption of such conformation, called "hydrophobic collapse," is the first step in protein folding and dramatically decreases the time required for this process (*Dill and Chan,* 1997; *Karplus,* 1997). The importance of this effect is illustrated by Levinthal's paradox of protein folding (*Levinthal,* 1969). A protein is a linear chain of amino acids, in which consecutive units are joined by chemical bonds capable of rotating with respect to each other. If each of these connections could adopt just three possible angles of rotation, the number of possible structures becomes astronomically large, and finding the "correct" three-dimensional structure of even a small protein through random search would require longer than the age of the universe. Yet, most proteins reliably fold to their native, biologically active structure in a matter of seconds or, at most, minutes. Protein folding can thus not be a random search process, but must instead be guided by specific forces. Hydrophobic collapse radically reduces the number of degrees of freedom in the protein chain and, by doing so, greatly accelerates its further folding. Thus, water's induction of the hydrophobic effect influences not only the stability of biological structures, but also the kinetics of biological processes such as protein folding.

Finally, the hydrophobic effect of water promotes spontaneous aggregation of individual proteins into larger, multiunit structures. A striking, although unwanted, demonstration of such interactions is sickle-cell anemia: A single amino acid mutation in hemoglobin, from charged glutamic acid to hydrophobic valine, induces aggregation of hemoglobin molecules into disease-causing fibrous precipitates (*Clancy,* 2008). But such aggregations also confer many positive functions, for example, the assembly of membrane-spanning proteins into structures that have a central, hydrophilic channel (Fig. 3d). These channels serve in the transport of ions and polar molecules across an otherwise forbidding non-polar membrane environment, and are essential in both energy metabolism and regulated exchange of materials with the environment (*Hille,* 2001).

Hydrophobic interactions are entirely due to the solvent. The presence of a solute causes highly unfavorable disruptions in the structure of liquid water due to the high energetic cost of creating "space" in the water to accommodate a solute molecule. If the solute is polar, this is compensated by favorable, largely electrostatic solute-water interactions, but for non-polar solutes no such compensation occurs. The unfavorable effect of disrupting the structure of liquid water was traditionally ascribed to an accompanying disruption of the hydrogen-bonding network between water molecules (*Kauzmann,* 1959). However, this disruption is not directly responsible for the hydrophobic effect. In fact, other, hydrogen bonded liquids — for example, ammonia — do not exhibit an equivalent of the hydrophobic effect. Instead, the effect follows from the small size of water molecules and the unusual thermodynamic properties of liquid water (*Pratt and Pohorille,* 1992; *Ashbaugh and Pratt,* 2006; *Pohorille and Pratt,* 2012). In particular, water is a "stiff" liquid, which means that, in contrast to most other liquids, its thermodynamic parameters change only slightly with physical parameters such as temperature or pressure. For example, the compressibility of water — a measure of

volume change in response to pressure change — is small and nearly independent of temperature. This reflects the difficulty (high free-energy cost) of disrupting the structure of liquid water — including making space for solute molecules. It is this "stiffness," rather than the breaking of hydrogen bonds between water molecules, that is largely responsible for the hydrophobic effect.

3.2.3. Summary: The effect of water on non-covalent interactions. Generally, in order to survive in constantly changing surroundings, living systems have to be both robust and nimble. In other words, molecular structures that carry out biological functions must be stable but also have flexibility to respond quickly and reliably to signals from the environment. This means that molecular interactions cannot be too strong. Otherwise, response to signals would be

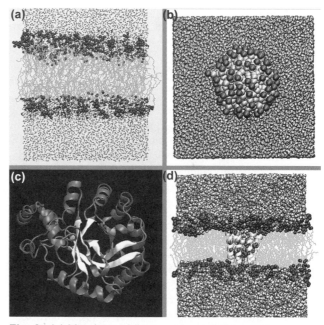

Fig. 3. (a) Membrane bilayer surrounded by water on both sides. Polar head groups (black) are exposed to water, whereas nonpolar, hydrocarbon chains (gray) are in the middle of the bilayer, sequestered away from the solvent. (b) Spherical micelle in water. Polar head groups are at the surface of the sphere, exposed to water. Nonpolar chains (white) are buried inside. (c) Protein chain in water represented as black background. In the native structure, polar segments of the protein (in gray) are exposed to water whereas nonpolar segments (white) are buried inside the protein away from water. (d) A protein ion channel traversing a membrane bilayer surrounded by water. As in (a), polar head groups (black) of the membrane-forming molecules are in contact with water and nonpolar chains (gray) form the membrane core removed from water. The nonpolar segments of the channel (white) are in contact with the nonpolar core of the membrane. The polar segments (black) are directed toward a water-filled pore (also in black) in the middle of the channel that creates a favorable environment for the transfer on ions. An ion in the pore is represented as a white ball. All figures are snapshots from molecular dynamics simulations of the corresponding systems.

sluggish and energetically costly. For these reasons, a system built almost completely as a chemically bonded structure, taking little advantage of non-covalent interactions, would be incapable of adjusting to environmental demands. On the other hand, if molecular interactions were too weak, random fluctuations in physical or chemical parameters could cause havoc in the system. For many processes and structures, the best balance is achieved by energies of association in the range of tens of kilojoules per mole — only a few-fold higher than the average thermal energy at room temperature. For example, even in molecules as large and complex as proteins, the combined effect of electrostatic, van der Waals, and hydrophobic interactions gives proteins a net stability in water of only about 20–40 kJ mol^{-1}, which means that biologically active protein structures are near the edge of conformational stability (*Pace,* 1990). The importance of water in this context is to simultaneously reduce the strength of electrostatic interactions and increase the strength of interaction between non-polar species, bringing all into this range of "just right" stability.

3.3. Alternative Solvents

The purpose of the discussion above is to exemplify the numerous roles that a solvent must play in supporting a life-like chemistry: helping to balance synthetic and degradative reactions, promoting self-organization, and modulating molecular interactions to ensure that a living system is both robust and responsive to environmental changes. In addition, it is an obvious advantage if, like water, a potential solvent for life is a common compound in the universe, exists in the liquid state over a wide range of temperatures, and has properties that change little with changing physical parameters, such as temperature, pressure, or radiation (*Pohorille and Pratt,* 2012). Could other solvents be suitable for life? Earthly biochemistry would not work in a solvent other than water: Most polymers and molecular assemblies lose their structure and function in non-aqueous solvents. But this is hardly surprising, given that Earthly life evolved in an aqueous medium, and is not the metric for whether alternative solvents could support life. Rather, the question is whether an alternative solvent could play an analogous set of roles in promoting self-organization and modulating molecular interactions and reactivity.

Are there other liquids that have these characteristics? Consider first "life as we know it," i.e., assume that the general blueprint of a living system remains the same, but the molecular hardware might be somewhat different. For such forms of life, alternatives to water are hard to find. In particular, no known solvent exhibits properties analogous to the hydrophobic effect, although several non-aqueous liquids, such as formamide, glycerol, diethylene glycol, ethylene glycol, 2-aminoethanol, 1–4 butanediol, and 1–5 pentanediol, exhibit some degree of "stiffness" (*Marcus and Hefter,* 1997). Formamide is of particular interest, as it can be formed from H_2O and HCN, which are both present in planetary environments. It exists in a liquid state between

2° and 210°C. Several other stiff liquids can be synthesized through oxidation of small hydrocarbon molecules known to exist, for example, on the surface of Titan. Furthermore, glycerol and formamide show rudimentary capabilities to mediate the formation of lipid bi-layers (*McDaniel et al.,* 1983; *McIntosh et al.,* 1989). Thus, it has been argued that formamide might be a plausible alternative to water (*Benner et al.,* 2004), despite its being photochemically unstable. Formamide is indeed a good solvent for organic synthesis relevant to the origin of life (*Schoffstall,* 1976; *Philipp and Seliger,* 1977) and may even be a precursor for synthesis of RNA (*Pino et al.,* 2015), but it is not nearly as good in supporting self-organization because it exhibits an equivalent of the hydrophobic effect only to a very limited extent. This once again stresses the uniqueness of water in supporting both covalent and non-covalent chemistry.

A number of other liquids have been also proposed as alternatives for water (*Schultze-Makuch and Irvin,* 2006), such as ammonia or hydrocarbons. The latter possibility has been used to justify considering Titan (see the chapter by Lunine et al. in this volume) as the prime target in a search for extraterrestrial life (*McKay and Smith,* 2005; *Shapiro and Schultze-Makuch,* 2009). It has even been suggested that a "hydrocarbon solvent may actually improve the chances for the origin of life" (*Schultze-Makuch and Irvin,* 2006). While this may be true of the capability of a solvent to assist in synthetic organic chemistry, it is only one piece of the puzzle of life. Other, perhaps more important, contributions to the formation and maintenance of living systems must also be addressed in considering alternative solvents. Similar questions apply to hypothetical life on Venus (see the chapter by Arney and Kane in this volume), where it is suggested that life might originate in mixtures of water and sulfuric acid that are presumably present in lower and middle cloud layers (*Cockell,* 1999; *Limaye et al.,* 2018). However, the physical and chemical properties of such mixtures that are most relevant to life, especially their bearing on molecular self-organization, have not been determined. They must be, if the suggested habitability of Venus (*Schultze-Makuch et al.,* 2004) is to be rigorously considered.

The absence of an equivalent to the hydrophobic effect, which is critical for the formation of lipid bilayers, is another key challenge in considering alternative solvents. If compartmentalization is an essential feature of living systems, this requires that different structures be available to bound living systems. For example, in environments of non-polar liquids that have at least some water content, it might be possible to form cell-like compartments from "inverse micelles" (*Schultze-Makuch and Irvin,* 2006). These are spherical structures in which hydrophobic tails are exposed to the non-polar solvent, while hydrophilic head groups, pointing toward the inside of the micelle, surround a small water droplet. Such droplets might support water-based life similar to ours. Although inverse micelles are known to be good chemical reactors, questions remain about their robustness, sufficient droplet size, and the ability to modulate regulated exchange of materials. The opposite problem faces the proposal that cell walls do not have to be built of organic molecules at all, but instead could be, for example, mineral structures (*Barge et al.*, 2015). Such systems may face challenges in supporting self-replication and in regulating, vs. simply retaining, chemical and physical gradients across cell walls.

The modulation of Coulomb interactions is also a challenge to life in non-polar solvents, in that it affects the ability to include both ions and hydrogen bonding as part of biochemistry. Ions are almost completely insoluble in non-polar solvents, and the extensive use of metals in our biochemistry depends on their ionic forms. Thus, life in non-polar solvents would either have to use metals in uncharged complexes or do without them. As noted in section 2.1.4, the possibility of life without metals is highly speculative. Moreover, all biochemistry that makes use of charged forms, such as the use of polyionic species as energy carriers (like ATP), information carriers (like DNA), or redox intermediates (like NADH), would not work in non-polar solvents, and alternative solutions would be required. Hydrogen bonding could conceivably still be utilized in supporting molecular interactions, but would have to be much weaker, as provided by a very different chemistry. In particular, OH groups would form hydrogen bonds that, in non-polar solvents, would be completely rigid, inflexible, and far too strong to support a role in transient molecular interactions. While alternative chemistry could in principle overcome this, it remains to be demonstrated whether such chemistry could be viable in all the respects necessary for life.

Finally, in the question of solvents for life, it is worth remarking on abundance. Water is made from the first and third most abundant elements in the cosmos, and is quite stable against a wide range of processes and conditions. Moreover, water exists in the liquid state over a wide range of temperatures, and its properties — including the properties that make it a suitable solvent for life — change little with changing physical parameters, such as temperature, pressure, or radiation. Water ice is abundant in the outer solar system, and at least five moons there are thought to have global-scale liquid water oceans (see chapters in this volume by Schmidt and by Cable et al.). In the rocky inner solar system, Earth, Mars (see chapters in this volume by Davila et al. and by Amador and Ehlmann), and possibly Venus (see chapter by Arney and Kane) have had (or still have) globally significant inventories of surface liquid water during their histories. Alternative solvents are considerably more rare, at least in our own solar system. Not including the molten materials of planetary interiors, only Titan's frigid lakes of ethane and methane represent remotely comparable volumes of non-aqueous liquid (*Cordier et al.*, 2009) (see the chapter by Lunine et al. in this volume). Indeed, formamide, ammonia, and the hydrocarbons are all photolabile: Stellar ultraviolet (UV) and X-rays break them down in the atmosphere, as is happening today to the methane on Titan. Thus, while future work can consider whether alternative solvents might meet all the needs of life-like chemistry, it appears that a solvent known to fill this role, water, is pervasive.

3.4. Summary: Solvent Requirements

We do not know of a viable alternative to water as the solvent for life. Several liquids exhibit some desired characteristics, but are deficient in others that appear quite essential for life, at least life that we can readily imagine. This does not mean that a search for alternatives to water is futile. A number of potentially interesting solvents, including some mixtures and supercritical fluids, have not yet been examined in sufficient detail. And we cannot dismiss the possibility of life so different from Earthly life that it imposes significantly different demands on the role of solvent. We do not know enough about these possibilities to exclude them with full confidence. But to elevate possible alternative solvents from "speculative" to "scientifically viable" will require a comprehensive set of observations that demonstrate the full suite of properties required by a solvent to support life-like chemistry.

4. ENERGY REQUIREMENTS

Energy is required to construct and maintain the complex, orderly, non-equilibrium states found in living organisms. But what type of energy, and how much, is needed to serve this purpose?

4.1. Type and Quantity of Energy Requirements

Life on Earth harnesses energy from only a subset of the many processes that liberate it. Specifically, life is only known to utilize the Gibbs energy liberated in redox reactions, and delivered by light in the visible to near-infrared portion of the electromagnetic spectrum. Both forms of energy can drive a flow of electrons, either by the chemically spontaneous transfer of electrons from one atom to another (redox chemistry) or via the excitation of electrons out of the ground state by absorption of a photon (light energy), followed by transfer to another atom or molecule. Our energy metabolism — the capture, storage, and deployment of energy — is constructed around and fundamentally depends on this flow. Chemical processes that release energy but do not drive a flow of electrons (for example, dissolution, precipitation, or hydration reactions) are apparently not used by life. Likewise, life is not known to drive light-based energy capture using wavelengths that do not also initiate electron flow. The use of energy sources beyond those known in Earth life would thus require either a mechanism of coupling to electron (or, more generally, charge) flow, or a means of energy capture and metabolism that does not require such flow at all.

In order to be useful, energy must also be delivered at levels compatible with biochemical mechanisms of energy capture. This is manifested in minimum requirements for both the amount of energy delivered per unit of energy carrier (e.g., energy per photon or per mole of chemical reacted), and energy delivered per unit of time.

The first measure is reflected in the Gibbs energy change (ΔG) in the case of chemical reactions and the wavelength or frequency of light, either of which must be sufficient to allow the production of ATP, the primary energy-storing compound in Earth's biochemistry (*Schink,* 1988). Under physiological conditions, the production of ATP from ADP (the formation of the high-energy phosphate-phosphate bond) typically requires about 50–60 kJ mol^{-1}, but the specific mechanism of formation allows the needed energy to be collected in smaller "bits" and then used collectively to form ATP. Specifically, the energy yield of metabolism may be temporarily captured in transmembrane ion (usually proton) gradients. This temporary store of energy is tapped for ATP production by releasing the ions back across the membrane; the down-gradient flow of ions can be tapped to do work, just as electricity (electrons flowing from high to low potential) can. Synthesis of a single ATP is typically coupled to the down-gradient release of several (usually 3–4) ions at once. Because the energy yield of metabolism can be used to "store" ions in varying numbers, the amount of ATP produced per metabolic reaction can range from about one-fourth to large quantities — for example, 36 ATP are produced in the oxidation of a single glucose molecule by oxygen. The low end of the range sets the minimum amount of energy that a metabolic reaction can yield and still be usefully captured by biology. A combination of theoretical considerations and observations suggest that this energy is in the range of 10–20 kJ mol^{-1} (*Schink,* 1988; *Hoehler et al.,* 2001; *Hoehler,* 2004). Below this level, the energy yield of redox processes cannot be usefully captured by biology. Although the same requirement exists whether the ultimate source of energy is light or redox chemistry, it is more tangible and restrictive in the latter case. A variety of redox processes may occur with Gibbs energy yields that, under prevailing environmental conditions, fail to satisfy life's minimum requirement. In contrast, visible to near-infrared wavelengths represent per-photon energies that significantly exceed the minimum required for ATP synthesis.

The second dimension of energy requirement has units of energy per time (or power, e.g., watts = joules per second). This requirement derives from life's need to expend energy to maintain a metabolically viable state (*Hoehler and Jørgensen,* 2013). For example, life must expend energy to repair or newly synthesize damaged biomolecules. This requirement has been termed "maintenance energy" (*Pirt,* 1965), and is, in essence, a determinant of how much life a given system could potentially support. More power can potentially support a larger standing biosphere. It can be expected that maintenance energy depends not only on the physiological specifics of the organism in question, but also on the physicochemical conditions of its surroundings. For example, increases in temperature or radiation may induce higher rates of molecular damage that require higher rates of (energy-expending) repair, while extremes of pH or salinity may obligate an organism to "work harder" (expend more energy) to maintain an internally clement environment. While the Gibbs energy requirement is quite well constrained, measured values of the maintenance energy

requirement range over 3 or more orders of magnitude. Understanding this variability within Earthly organisms and environments will be an important first step in understanding how much maintenance energy requirements may vary for different forms of life living in different environmental conditions. Doing so is important, however, because it will help to inform our expectation about how robustly inhabited, and thus how detectable, a given system could be.

For completeness, it must be noted that both "voltage" and power can be delivered at levels that overwhelm the biochemical machinery that serves to capture energy, for example, when sunlight is focused through a magnifying glass onto a leaf. In such a case, otherwise useful energy sources can become destructive to the very biochemistry that is designed to harness them. Thus, the utility of energy sources to biology is constrained not only to specific types, but also to defined ranges of power and voltage (*Hoehler,* 2007).

4.2. Chemotrophy

Chemotrophy ("eating of chemicals") is the capture and utilization of the Gibbs energy released during chemically spontaneous redox reactions. By their very nature, redox reactions (section 2.1.4) encompass a transfer of electrons between atoms or molecules, and are thus readily coupled to energy metabolism — which, as described above, has its basis in flowing charge. Earth's diverse environments offer a range of both oxidants (or electron acceptors) and reductants (or electron donors), and thus many possible redox combinations to potentially serve as sources of energy for life. Example electron acceptors include O_2, NO_3^-, many transition metals in their higher oxidation states (e.g., MnO_2, Fe^{3+}, AsO_4^{2-}), SO_4^{2-}, and CO_2. Example electron donors include H_2, H_2S, CH_4, Fe^{2+}, and myriad organic compounds. These lists are by no means exhaustive; many other compounds can serve as donors or acceptors of electrons. In some cases, electron transfer can occur *within* a molecule, such that a portion becomes more oxidized and a portion becomes more reduced. These "disproportionation" reactions, sometimes also called "fermentation" reactions, can occur in both organic (e.g., $CH_3COOH \rightarrow CH_4 + CO_2$) and inorganic (e.g., $S_2O_3^{2-} + H_2O \rightarrow SO_4^{2-} + H_2S$) compounds, and life can often capitalize on the Gibbs energy released in such reactions. For example, we are indebted to yeasts for catalyzing the alcohol-yielding fermentation of sugars. Much of the metabolic diversity to be found on Earth is represented in evolutionary solutions that allow the myriad possible redox combinations to be tapped for energy.

Some donor-acceptor combinations yield more energy than others. In some cases, the energy yield may be insufficient to support ATP synthesis (as described above), and in still others, electron transfer may not be chemically spontaneous (yield energy) at all. For example, the oxidation of methane (CH_4) with O_2, SO_4^{2-}, and CO_2 occurs with standard Gibbs energy changes ($\Delta G°$) that are, respectively, highly favorable, barely enough to support ATP synthesis, and thermodynamically unfavorable

$$CH_4 + O_2 \rightarrow CO_2 + 2H_2O \qquad \Delta G° = -859 \text{ kJ/mol}$$

$$CH_4 + SO_4^{2-} \rightarrow HCO_{3-} + HS^- + H_2O$$
$$\Delta G° = -16.6 \text{ kJ/mol}$$

$$CH_4 + CO_2 \rightarrow CH_3COOH \qquad \Delta G° = +23.9 \text{ kJ/mol}$$

The term "standard" Gibbs energy change refers to yield of the reaction under specific conditions of temperature (298°K), pressure (1 bar), and with all reactants and products having unit activity (e.g., 1 M for dissolved substances and 1 bar for gases). Of course, temperature, pressure, and abundance of reactants and products can all vary widely in nature, and the Gibbs energy change varies with them. The Gibbs energy change for reactions that occur under non-standard conditions is given by

$$\Delta G_{rxn} = \Delta G°_{(T,P)} + RT \ln Q$$

where $\Delta G°_{(T,P)}$ is the standard Gibbs energy change adjusted for ambient temperature and pressure, R is the universal gas constant, T is the temperature in degrees Kelvin, and the reaction quotient, Q, consists of the mathematical product of the activities of the reaction products, divided by the mathematical product of the activities of the reactants, with the activity of each reactant and product raised to the power of its stoichiometric coefficient. For example, in the reaction $2A + B \rightarrow 3C + 4D$, $Q = (\{C\}^3 \cdot \{D\}^4)/(\{A\}^2 \cdot \{B\}^4)$. The sensitivity of Gibbs energy change to these factors means that some reactions (e.g., those having ΔG_{rxn} or $\Delta G°$ close to zero, such as the oxidation of methane with sulfate, above) can be quite sensitive to even modest variations in temperature or abundance of products or reactants. These reactions may move in and out of thermodynamic and bioenergetic favorability as local conditions vary. For metabolisms based on such reactions, the minimum Gibbs energy requirement may be an important constraint on energy availability. For reactions that have large positive or negative Gibbs energy changes, such as the oxidation of methane with O_2, above, very large deviations from standard conditions in temperature (hundreds to thousands of degrees Kelvin) and/or reactant/product concentrations (tens or orders of magnitude) are required to change the sign of ΔG. The minimum Gibbs energy requirement is generally not a tangible consideration for metabolisms based on these reactions.

Redox-based metabolisms are often most productive where large fluxes of oxidant and reductant combine. For example, at deep-sea hydrothermal vents, fluids laden with the reductant H_2S jet into an ocean filled with the powerful oxidant O_2. Microorganisms capable of reacting H_2S with O_2 and capturing the resultant energy yield thrive at such sites, often forming the basis of a food chain that includes dense communities of animals. A less dramatic, but globally more important example is the diffusion of the oxidant SO_4^{2-} from seawater into organic matter-containing (reductant-containing) sediments on the sea bed. Diffusive fluxes deliver materials at much slower rates than do jetting hydrothermal vents, so the density of microbial biomass that is supported

in any given site is much lower than at hydrothermal vents. Lower fluxes simply support smaller communities. However, because sediments are pervasive across most of the sea bed, while vents are scarce, the energy released (and biomass supported) by the combination of sulfate and organic matter in sediments is globally much larger (*Bowles et al.*, 2014) than that released by hydrothermal venting of reducing fluids into oxygenated seawater (*Elderfield and Schultz*, 1996).

In global terms, the largest biologically-accessible "redox" flux, by far, is the production of organic matter by photosynthetic organisms at the O_2-filled surface of the Earth. Global net primary productivity (NPP) (the amount of CO_2 converted into organic matter) due to photosynthesis on land and in the oceans is estimated at 105 Pg (8.8×10^{15} mol) organic carbon per year. By contrast, the combined global flux of reductants such as H_2, H_2S, and CH_4 from hydrothermal vents is estimated at 10^{11}–10^{12} mol per year — a 5–6-order-of-magnitude smaller flux. The Earth's chemotrophic biosphere can thus be (and is) many orders of magnitude more abundant and productive by virtue of photosynthetic production of organic matter and O_2. This vast difference in productivity becomes important when considering the *detectability* of biospheres.

All animals and almost all fungi, as well as many microorganisms, harness the energy offered in the combination of O_2 with organic matter. Here, "organic matter" refers to any of a wide range of organic molecules produced in living systems — for example, fats, sugars, and proteins. Plants (along with other photosynthetic organisms) distinguish themselves in the ability to harness light as their primary energy source (see below), but also fuel their energy metabolism during dark periods by combining O_2 with the sugars they produce and store via photosynthesis. Thus, virtually the entire "macroscopic" biosphere (essentially all organisms that can be seen with the naked eye), along with many microscopic organisms, is fueled by the photosynthetic production of organic matter and O_2, as described below.

4.3. Photosynthesis

Photosynthesis, the conversion of light energy to chemical energy, is an elegant solution to life's requirement for energy. Indeed, the flux of sunlight to Earth provides such an abundant energy source that the productivity of our biosphere overall is not energy limited. However, there are limits on the wavelengths and intensities of light that can drive photosynthesis, which bear directly on the detectability of photosynthetic biospheres on rocky habitable planets.

4.3.1. Chemical requirements of photosynthesis. As described in section 2.1, our biochemistry is comprised of molecules with an average oxidation state slightly less than zero, while the carbon available in the environment is largely in the most oxidized form, CO_2. Life must reduce (add electrons to) this carbon in order to produce the molecules of biochemistry — a process called autotrophy. A generalized scheme for *photo*autotrophy — the use of light energy to carry out this fixation — is shown in equation (1). Here, "reductant" is an electron-donating species, "oxidized product" is its oxidized equivalent, and "CH_2O" is a generic representation of organic matter [e.g., the sugar glucose, $C_6H_{12}O_6$, is $(CH_2O)_6$]. Many phototrophs can also use organic carbon as their carbon source, a process called photoheterotrophy, and in this case do not need an external source of reductants. The conversion of carbon into cellular material is called the "dark reaction" of photosynthesis.

$$Reductant + CO_2 + 2H^+ \rightarrow Oxidized\ product + CH_2O \quad (1)$$

In addition to light energy and CO_2, the photosynthetic process requires a source of electrons, or reductant. In the case of oxygen-evolving ("oxygenic") photosynthesis in plants, algae, and cyanobacteria, the reductant is H_2O, which is converted to O_2 via the removal of electrons and protons. Other reductants such as H_2, H_2S, and reduced iron or Fe^{2+} can be utilized in non-oxygen evolving ("anoxygenic") photosynthesis. Anoxygenic photosynthesis is biochemically less complex and evolutionarily more ancient than oxygenic photosynthesis (*Blankenship*, 2010).

On the early Earth, primitive anoxygenic phototrophs likely used strong reductants, such as H_2 or H_2S, from sources such as volcanism and water-rock reactions (e.g., serpentinization). These compounds have easily extractable electrons (they have a low "reduction potential"), but were probably in limited supply on the early Earth, and thus rapidly depleted. Limitation of reductant contributed to selection pressure for the development of more complicated photosynthetic machinery that could support the extraction of electrons from weaker but more abundant reductants such as water (*Blankenship*, 2010). Due to the wide availability of water as an electron donor compared to the limited availability of the volcanogenic reductants, oxygenic photosynthesis spread to dominate aquatic and terrestrial habitats, and is by far the dominant mode of photosynthesis, and the overall most dominant metabolism, on the modern Earth.

4.3.2. Details of the photosynthetic process.

Understanding the wavelength requirements of photosynthesis is key to assessing the detectability of photosynthetic biospheres on exoplanets. To build this understanding, it is necessary to first understand the four key events that underlie photosynthesis in all photosynthetic organisms: light absorption, charge separation, electron transport, and energy storage. The wavelength requirements of photosynthesis are rooted in the specific mechanisms associated with these processes.

Light absorption is achieved by pigments associated with protein complexes that are embedded in membranes within the cell. The wavelengths of light absorbed by the pigment molecules are determined by their chemical composition and structure. In general, pigments are based on a molecular structure called a macrocycle — a large central ring created by joining together four separate five-atom rings (Fig. 4). The three main types of macrocycles — porphyrin, chlorin, and bacteriochlorin — give rise to the major types

of chlorophyll and bacteriochlorophyll pigments (Fig. 4). The pattern of alternating double and single ("aromatic") bonds in the macrocycle, the ability of the nitrogen atoms to coordinate metal cations (Mg), and the conformational flexibility of the macrocycle and side chains allow for the unique biological function of pigments.

The wavelengths of light absorbed by pigments are controlled by the oxidation state of the macrocycle, the chemical groups arrayed around its periphery, and interactions with surrounding proteins. The central macrocycle contains a series of "conjugated" double bonds (alternating double and single bonds), which leads to a broad area of "delocalized" π electrons across the body of the structure (Fig. 4). (Delocalized means, in essence, that the electrons that form the double bond can be equivalently associated with, and move among, all the conjugated atoms, rather than existing as part of one discrete bond. The effects of an event occurring at any point in the conjugated network — for example, the impact of a photon — can thus be "felt by" and distributed across the entire network; this phenomenon often serves to stabilize molecules and intermediate states to a much greater extent than is possible in unconjugated systems.) The delocalized π electrons can be excited to higher energy levels by absorption of light; the specific excitation wavelength shifts as the peripheral substituents are varied. For example, adding more polar functional groups around the perimeter of the molecule shifts the excitation wavelength toward the red end of the spectrum, as occurs with chlorophylls *d* and *f* (Table 2; Fig. 4). Chlorophyll and bacteriochlorophyll pigments as a collective set absorb across a range of wavelengths in the spectrum, not just at the absorption maximum we typically associate with chlorophyll *a* (670–680 nm). There are two pairs of absorption bands in the blue and red area of the spectrum (Table 2).

Photosynthetic pigments can be broadly grouped into two functional categories: accessory light-harvesting pigments and reaction center pigments. Accessory pigments are those that harvest light and transfer the energy to the reaction center pigments. Accessory pigments allow for absorption of light across a broader range of wavelengths, enabling an organism to harvest more of the spectrum to power photosynthesis. Phycobilins, such as phycoerythrin and phycocyanin, are water-soluble pigment-protein complexes that absorb in the green, yellow, orange, and red areas of the spectrum. Carotenoids, which absorb in the blue, can function in light gathering as the energy decays to the reaction center, but in the cell, they primarily function in photoprotection and preventing oxidative cellular damage.

The reaction center contains a chlorophyll or bacteriochlorophyll molecule in a special environment. The absorption of light by a reaction center chlorophyll causes an electron to be excited to a higher energy level, from which it is easily donated to an acceptor molecule. This step is called *charge separation*, and represents the first chemical change in the conversion of light energy to chemical energy.

Transfer of an excited electron from chlorophyll to an acceptor molecule initiates a process of *electron transport*, in which electrons are rapidly separated from chlorophyll via a series of transfers to molecules having successively higher reduction potentials (the electrons "flow downhill" to successively more powerful oxidants). Ultimately, this flow serves to capture the electron in a biologically useful form, nicotinamide adenine dinucleotide phosphate (NADPH), and to generate a proton motive force that allows for the synthesis of ATP (*Ort, 1996*). NADPH and ATP are, in turn, used to reduce CO_2 into cellular material via the Calvin cycle, thus securing the *storage of light energy as chemical energy*. Meanwhile, the loss of an electron from the reaction center pigment changes its redox potential so that it becomes a strong oxidant, which promotes the extraction of an electron from an external reductant, such as H_2O, H_2, H_2S, or Fe^{2+}. Upon extracting an electron, the

reaction center pigment is "reset" to its original electronic state, and the cycle can begin again.

Collectively, the special chlorophyll or bacteriochlorophyll pigments and electron-transporting molecules are called a "reaction center." Anoxygenic photosynthesis utilizes a single reaction center that is driven by absorption of one photon per cycle. Oxygenic photosynthesis utilizes *two* reaction centers or photosystems linked together, and couples their activities on nanosecond timescales (*Rappaport and Diner*, 2008). Because each photosystem is driven by absorption of a photon, the conduct of oxygenic photosynthesis requires the absorption of two photons in rapid succession.

4.3.3. Light intensity limits for photosynthesis. The maintenance energy requirement in phototrophic organisms manifests as a requirement for a minimal flux of photons

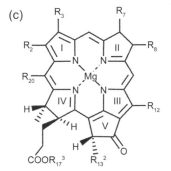

Porphyrin			Chlorin								
Chl c	R_7	R_8	Chlorophylls	R_2	R_3	R_7	R_8	R_{12}	R_{13}^2	R_{20}	R_{17}^3
Chl c1	CH₃	CH₂-CH₃	Chl a	CH₃	CH=CH₂	CH₃	CH₂-CH₃	CH₃	COOCH₃	H	P
Chl c2	CH₃	CH=CH₂	8-vinyl Chl a	CH₃	CH=CH₂	CH₃	CH=CH₂	CH₃	COOCH₃	H	P
Chl c3	COOCH₃	CH=CH₂	Chl b	CH₃	CH=CH₂	CHO	CH₂-CH₃	CH₃	COOCH₃	H	P
			8-vinyl Chl b	CH₃	CH=CH₂	CHO	CH=CH₂	CH₃	COOCH₃	H	P
			Chl d	CH₃	CHO	CH₃	CH₂-CH₃	CH₃	COOCH₃	H	P
			Chl f	CHO	CH=CH₂	CH₃	CH₂-CH₃	CH₃	COOCH₃	H	P
			Bacterio-chlorophylls								
			Bchl c	CH₃	CHOCH₃	CH₃	various¹	various²	H	CH₃	F, S
			Bchl d	CH₃	CHOCH₃	CH₃	various¹	various²	H	H	F
			Bchl e	CH₃	CHOCH₃	CHO	various¹	various²	H	CH₃	F

[1]Various isomers such as CH₃, C₂H₅, C₃H₇, etc.; [2]various isomers such as CH₃ or C₂H₅; [3]P = phytyl ester; F = farnesyl ester; S = stearyl alcohol

Bacteriochlorin				
Bacterio-chlorophylls	R_2	R_3	R_7	R_8
Bchl a	CH₃	COCH₃	CH₃	C₂H₅
Bchl b	CH₃	COCH₃	CH₃	=CH-CH3
Bchl g	CH₃	CH=CH₂	CH₃	C₂H₅
	R_{12}	R_{13}^2	R_{20}	R_{17}^3
Bchl a	CH₃	COOCH₃	H	P, Gg
Bchl b	CH₃	COOCH₃	H	P
Bchl g	CH₃	COOCH₃	H	F

P = phytyl ester; Gg = geranylgeraniol ester; F = farnesyl ester

Fig. 4. Chemical structures of chlorophyll (Chl) with IUPAC/IUBMB numbering of the carbon atoms. Chemical structure of **(a)** porphyrin-type chlorophyll (Chl *c*), **(b)** chlorin-type chlorophylls (Chls *a, b, d,* and *f*) and bacteriochlorophylls (Bchls *c, d, e*), and **(c)** bacteriochlorin-type bacteriochlorophylls (Bchls *a, b, g*). The dashed lines bisecting the macrocycle in **(b)** are the x and y axes of the compound, which give rise to the Q_x and Q_y absorption bands (after *Chen and Sheer, 2013*).

TABLE 2. *In vivo* absorption maxima of light-harvesting (bacterio) chlorophylls in living cells or photosynthetic membranes (after *Pierson et al.,* 1992; *Mielke et al.,* 2013).

Pigment	*In vivo* absorption maxima (nm)
Bacteriochlorophyll *a*	375, 590, 790–810, 830–920
Bacteriochlorophyll *b*	400, 600–610, 835–850, 1015–1040
Bacteriochlorophyll *c*	325, 450–460, 740–755
Bacteriochlorophyll *d*	325, 450, 725–745
Bacteriochlorophyll *e*	345, 450–460, 710–725
Bacteriochlorophyll *g*	420, 575, 670?, 788
Chlorophyll *a*	435, 670–680 in PS II, 700 in PS I
Chlorophyll *d*	710–720 in PS II, 740 in PS I

(intensity) in the needed wavelength range. Conceptually, this flux should be sufficient to support a viable metabolic state. Mechanistically, it must also be capable of powering the primary photochemical events in photosynthesis, leading to the synthesis of ATP. The mechanistic constraint is important in the case of oxygenic photosynthesis in that it requires the absorption of two photons within nanoseconds of each other (*Rappaport and Diner,* 2008). Theoretically, this could impose a more stringent constraint on photon flux than the maintenance energy requirement.

The lowest photon fluxes that appear to support a metabolically viable state in several anoxygenic phototrophs and a species of red algae that grow deep underwater are 0.00075–0.01 μmol photons m^{-2} s^{-1} (5×10^{14}–6×10^{15} photons m^{-2} s^{-1}) in the 400–700-nm wavelength range (*Littler et al.,* 1986; *Manske et al.,* 2005; *Marschall et al.,* 2010) — about 5–6 orders of magnitude lower than the average flux at Earth's surface. To receive such a flux at its surface, Earth would have to move 500–1000 times farther out in its orbit, where its surface water would be completely frozen. The water worlds of our solar system are frozen at the surface from Jupiter's orbit (5 times farther out than Earth) outward. Thus, it seems unlikely that photon flux would be limiting at the organism (maintenance energy) level for any world on which the radiation budget allows liquid surface water. Rather, low light intensities may be more likely to affect whether a photosynthetic biosphere can be sufficiently productive to yield a detectable signal. There have also been reports of anoxygenic phototrophs (a green sulfur bacterium) subsisting on blackbody radiation emitted from a deep sea hydrothermal vent (*Beatty et al.,* 2005), which would raise the possibility of photosynthetic metabolism on completely dark (e.g., deeply ice-covered) but hydrothermally active worlds. However, the phototroph identified in these studies has not been unequivocally shown to survive on such extremely low light fluxes *in situ,* or in the laboratory.

4.3.4. Light availability and wavelength requirements.
The intensity and quality of light available to drive photosynthesis depends on both stellar and planetary properties. Stellar temperature governs the relative intensity of emission at different wavelengths, with hotter stars emitting more light at shorter wavelengths than cooler ones. For example, the

peak emission of our Sun, a G-type star, is in the green-blue region (Fig. 5), while cooler M-type stars emit primarily at longer wavelengths. Moreover, the intensity and quality of light that is ultimately available to phototrophs may differ from that reaching the top of a planet's atmosphere, as shown in Fig. 5 for Earth. Specifically, light may be attenuated in certain wavelength ranges through scattering and absorption by atmospheric gases such as CO_2 and H_2O, and by the specific environment in which the phototroph lives. Phototrophs can respond to the wavelengths of light (light quality) in their local environment by synthesizing more of a particular type of accessory pigment, or tuning the absorption maxima of their pigments to be able to exploit the available light niches. For example, as sunlight penetrates the water column of lakes and oceans, it's rapidly attenuated in the blue and red regions. The only light available at depth is in the green region, and phototrophs that live deeper in the water column synthesize more pigments to absorb in this region (*Overmann et al.,* 1992).

Given the seeming ability to tune photosynthetic pigments to absorb over a wide range of wavelengths, is there a long wavelength limit (a minimum energy per photon) beyond which photons are no longer useful in photosynthesis? This question is particularly important in regard to M stars, which are by far the most abundant stars in the universe, and which emit mostly at longer wavelengths. The light-harvesting (bacterio) chlorophylls exhibit absorption maxima at wavelengths as short as 325 nm and as long as 1040 nm. What sets these limits, and are they absolute?

Recalling the mechanism of photosynthesis, a photon must have sufficient energy to (1) excite pigment electrons into the electron transport chain and (2) support the capture of energy as ATP. Even the longest wavelengths thus far known to be used in photosynthesis equate to much more energy than the 10–20 kJ mol^{-1} needed to make ATP [1040 nm ≈ 115 kJ (mol photons)$^{-1}$], so the energy storage requirement does not account for the observed limit. Rather, it is the need for photons having sufficient energy to excite electrons in the pigment molecule from ground state to excited state, where they can be readily lost to an acceptor molecule. Here, it is important to note that absorptions at very long wavelengths are attributable to accessory pigments that collect and transfer photon energy uphill to the reaction center, but do not, themselves, participate in the primary photochemical events of photosynthesis. The reaction center pigments, where these events do occur, absorb at wavelengths in the range of about 800–960 nm in anoxygenic photosynthesis and 680–727 nm in oxygenic photosynthesis (*Nürnberg et al.,* 2018). The shorter wavelength requirements for oxygenic photosynthesis, as well as the use of two coupled photosystems, is ultimately attributable to the need not only to excite electrons in the reaction center chlorophyll *a*, but also to initiate chemistry that leads to the extraction of tightly held water electrons.

The long wavelength limits of photosynthesis thus depend on both the ability to promote electronic transitions in pigments and to facilitate the extraction of electrons from external donors like H_2, H_2S, or H_2O. The activity of antenna pigments at wavelengths as long as 1040 nm indicates that electronic transitions are possible to wavelengths at least this long, and *Kiang et al.* (2007a,b) suggest that there is no clear theoretical limit to what biological molecules could allow. Among Earthly organisms, the function of extracting electrons from external donors may thus set the most stringent wavelength limits, and this is particularly true of oxygenic photosynthesis. It has been proposed that even this requirement could theoretically be relaxed to wavelengths of ~1000 nm by coupling the energies of three or more photon absorptions in three linked photosystems, in the same fashion that phototrophs on Earth couple two-photon absorptions in two linked photosystems (*Wolstencroft and Raven,* 2002). Such a system would, however, require even greater molecular coordination and biochemical complexity than is represented in our two-photon system.

What are the shortest wavelengths of light that can power photosynthesis? On Earth, wavelengths as short as about

400 nm are absorbed by antenna pigments, but this energy is partially decayed as it is transferred to the reaction center pigments, and for oxygenic phototrophs, still drives the 680-nm excitation. Shorter, UV wavelengths are destructive or inhibitory to photosynthetic organisms on Earth, and can damage DNA and generate reactive oxygen species that inhibit photosynthesis (*Caldwell,* 1979; *White and Jahnke,* 2002). Phototrophs combat these effects through the production of UV-screening pigments and anti-oxidants (*White and Jahnke,* 2002; *McCree and Keener,* 1974), but these mechanisms limit damage by UV radiation, rather than enable it to be used directly in photosynthesis. The direct use of UV and shorter wavelengths to power photosynthesis would require both "antenna" pigments capable of capturing and transferring photon energy, and a biochemistry capable of withstanding damage from exposure to such radiation. Such behavior is not yet known among organisms on Earth.

4.3.5. Photosynthesis and the detectability of biospheres. On rocky planets in the habitable zone (the region around a star in which liquid water could be stable at a planet's surface), energy considerations suggest that photosynthetic biospheres may be more detectable to space- or groundbased telescopes than non-photosynthetic biospheres, for two reasons.

First, photosynthesis has the potential to yield and sustain a globally larger and more productive biosphere. Non-photosynthetic life is reliant on planetary chemistry (including atmospheric chemistry resulting from stellar radiation) to meet its energy needs, and there can be significant limitations on the availability of this energy. In contrast, photosynthesis eliminates the reliance on planetary chemistry to provide energy through provision of an external source (light). Oxy-

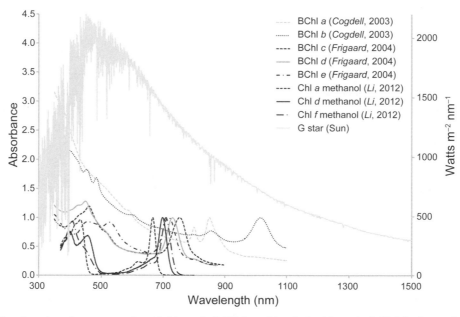

Fig. 5. In vivo and in vitro absorbance spectra of chlorophyll (Chl) and bacteriochlorophyll (Bchl) pigments in photosynthetic microbes. Note that the Bchl pigments absorb in the near-infrared area of the spectrum. Pigment spectra and references are hosted in the NAI VPL spectral databases and tools (http://vplapps.astro.washington.edu/pigments). The Sun's incident radiation (W m^{-2} nm^{-1}) to the top of the atmosphere is plotted to show the peak photon flux in the visible (gray line) (after Kiang, 2007b). Other stellar types have peak photon fluxes at other areas of the spectrum.

genic photosynthesis in particular effectively alleviates the potential for chemical reductants to be limiting because it uses water, which is widely available in the habitable zone, as an electron donor for photosynthesis. By virtue of this capability, the modern (oxygenic photosynthetic) biosphere is thought to be more productive by 3 or more orders of magnitude than was the biosphere before the emergence of photosynthesis (*Des Marais*, 2000; *Canfield et al.*, 2006).

Could light energy be limiting to the global productivity of a photosynthetic biosphere, and thereby diminish the detectable signal of life? As discussed above, it seems unlikely that low light flux could be prohibitive to photosynthetic life in absolute terms. If the planet receives sufficient light to support liquid water at the surface, the flux would likely be enough to power photosynthesis based on the low-light-adapted phototrophs described above. Moreover, Earth's biosphere utilizes only about 14% of the light energy incident in the 400–700-nm range (*Field et al.*, 1998), such that photon fluxes would have to fall well below the levels required to sustain liquid water before they proved limiting to photosynthesis. However, even if the light flux were enough to sustain oxygenic photosynthesis, one must keep in mind that oxygen may never build up to detectable levels in the atmosphere (a false negative). Studies of the "Great Oxidation Event" on the early Earth have revealed that the process of oxygen accumulation was likely a complex set of feedback mechanisms between oxygenic photosynthesis and the fixation of organic carbon (see the chapter by Stüeken et al. in this volume), aerobic respiration, oxidation of other reductants such as H_2 and H_2S, oxidation of mineral species, and burial of organic carbon (*Holland*, 2006). All these sinks must be saturated before oxygen levels can begin to rise.

For cooler stars that emit a much smaller fraction of their light in visible wavelengths than does our Sun, it may be possible for the stellar flux to prove limiting to global photosynthetic productivity, particularly for oxygenic photosynthesis (*Lehmer et al.*, 2018). This could be true particularly for planets near the outer edge of the habitable zone and/or those in which atmospheric absorption (e.g., by photochemical hazes) significantly diminishes the flux of visible photons to the surface. While these lower photon fluxes would not necessarily challenge the maintenance energy requirements of individual photosynthetic organisms, they could nonetheless decrease overall productivity. Should diminished productivity translate to less or no buildup of oxygen (*Lehmer et al.*, 2018), life on such worlds might be more difficult to detect. In this context, and because oxygen is an important target in the search for exoplanet life (*Meadows et al.*, 2018), it is particularly important to understand the long wavelength limits for oxygenic photosynthesis.

Second, photosynthetic life is capable of creating *gaseous* products that are far from equilibrium relative to the contextual chemistry, as reflected in large ΔG values for the production of such species. For example, O_2 is created during photosynthesis when electrons from water are extracted and ultimately used to reduce CO_2 to organic compounds, such as glucose: $CO_2 + H_2O \rightarrow O_2 + 1/6\ C_6H_{12}O_6$. The large disequilibrium represented in the gaseous product O_2 is reflected in the large positive Gibbs energy change for this reaction, $\Delta G = +478$ kJ · (mol O_2)$^{-1}$. This closely approximates the Gibbs energy change associated with the hypothetical extraction of electrons from water: $2H_2O \rightarrow O_2 + 4H^+ + 4$ electrons, $\Delta G = +475$ kJ · (mol O_2)$^{-1}$ and +19 kJ · (mol electrons)$^{-1}$. It is the significant energy content of visible photons that enables the establishment of such large Gibbs energy changes: The primary excitation wavelength in oxygenic photosynthesis, 680 nm, is equivalent to a photon energy of 176 kJ mol^{-1}.

It is important to note that non-photosynthetic life can also invest energy that results in the creation of disequilibrium products, for example, when (energy-yielding) ATP hydrolysis is coupled to (energy-requiring) biosynthesis. However, the energy represented in ATP under physiological conditions is approximately 50 kJ mol^{-1}, so that multiple instances of ATP hydrolysis would have to b+e coupled to the synthesis of a single product in order to achieve disequilibria such as that represented in O_2. Although this occurs routinely in biochemistry — for example, in the synthesis of proteins and nucleic acid polymers — the products of such reactions are typically large, complex molecules that are not volatile.

4.4. Alternative Energy Sources

A wide range of energy sources can be used to do work, as demonstrated by humanity's use of thermal gradient, mechanical, nuclear, electromagnetic, and chemical energy. Yet only the last two are used at a metabolic level, and in both cases only a subset thereof. Several energy sources on Earth would seem, superficially, to be sufficiently large and long lived as to represent an attractive niche for life to fill, if that were possible to do.

About one-third of the total solar irradiance falls within a wavelength range that is usable in photosynthesis, meaning that two-thirds of that irradiance, or about 100,000 TW of energy, is apparently not usable by life. Factors that might limit the extension of the photosynthetically useful range to shorter or longer wavelengths have been discussed above. Moreover, Earth's biosphere overall is not light-limited, so relatively limited impetus may exist to push into biochemically challenging wavelength ranges. However, as mentioned above, light-limited biospheres are possible on planets within the classical habitable zone (planets on which liquid water is stable at the surface). There, utilization of longer wavelengths could significantly increase overall biospheric productivity.

Earth's total heat flux, driven primarily by radioactive decay within the crust and mantle, is on the order of 40 TW. However, according to classical heat engine theory, such flux only represents usable energy in cases where advective transport (for example, fluid flow) delivers material into surroundings of significantly lower temperature. Perhaps a quarter of Earth's heat flux, represented in high-temperature fluid venting into a cold ocean, fits this criterion. At typical hydrothermal temperatures, 50–60% of the delivered heat flux could theoretically be extracted to do useful work, so the usable worldwide thermal gradient energy is on the order of 5–6 TW. Although a metabolism that uses thermal gradients to create flowing charge (which then supports energy capture) can be envisioned, to date no organisms are known that take advantage of such energy.

The reaction of crustal rocks with water is chemically spontaneous at temperatures below a few hundred degrees, corresponding to roughly the upper 10 km of Earth's crust. Thus, about 5 billion cubic kilometers of rock is hosted in conditions where its reaction with water would release energy, although the rates of such release could be limited by infiltration of water and the relatively slow kinetics of rock dissolution. The majority of energy released in such reactions comes from hydration reactions, rather than redox chemistry, and no organisms are known to harness the former into metabolism. This may result in part from Earth life's requirement that energy sources must support a flow of charge if they are to be usable. Nevertheless, the chemical energy represented in rock hydration constitutes a major source of disequilibrium on any world where rocks and liquid water are both abundant, and this would be particularly important for systems that lack habitable surface conditions, such as Mars or the icy moons of the outer solar system. To tap such a source would seemingly require an energy metabolism that does not depend mechanistically on flowing charge.

5. THE PHYSICOCHEMICAL ENVIRONMENT

Empirical observations and theoretical considerations show that biochemistry can only occur within a limited range of physical and chemical parameters — including salinity, temperature, pressure, pH, and ionizing radiation — that affect the stability and interactions of biomolecules (Table 3) (e.g., *Rothschild and Mancinelli,* 2001; *McKay,* 2014). Together these parameters define a multidimensional space that represents the habitability envelope of life on Earth.

Most Earthly organisms comfortably inhabit the range of physical and chemical conditions that we consider "normal" — that is, the range that is most prevalent in Earth's surface environments, and to which we, ourselves, are adapted. But some specialized organisms have evolved strategies that allow them to occupy the fringes of this multidimensional space, where competition for resources and predatory pressures are less severe, but where physical and/or chemical conditions fall dangerously close to crippling and potentially deadly levels. These are the so-called *extreme environments,* and their inhabitants are the *extremophiles.* Organisms that face more than one extreme are called *polyextremophiles* (*Rothschild and Mancinelli,* 2001).

In the following we summarize the most important physical and chemical parameters that define the habitability envelope of life on Earth, and the known threshold values for each parameter that are still compatible with basic biological processes such as metabolism and growth. Then, we place these laboratory observations in a natural context by looking at specific regions on Earth where the effects of physical and chemical parameters combine to define the actual limits of habitability on our planet.

5.1. Water Availability

As we have just seen, the myriad interactions that underlie life-like behavior depend critically on solvent. On Earth, few organisms can tolerate any loss of cellular water, mak-

TABLE 3. Biological limits for different physical and chemical parameters (updated from McKay, 2014).

Parameter	Limit	Note
Low/High temperature	−15°C/122°C	Growth demonstrated
Low/High pH	0–12.5	"
Water activity	0.6	"
Salinity	Saturated NaCl	"
Low pressure	10–700 Pa	"
High pressure	130 MPa	"
Low pressure	10^{-6} Pa	Survival of spores in space
High pressure	1200–1500 MPa	Survival of spores
UV radiation*	~9000 ergs mm^{-2} (265 nm)	<1% survival
Ionizing radiation	50 G.y. h^{-1}	Growth under chronic irradiation
	12,000 G.y.	Limit of survival
Pressure	1100 atm	

*From *Setlow and Duggan* (1964).

ing water availability one of the most important ecological factors and evolutionary pressures on life (*Alpert,* 2005; *Noy-Meir,* 1973; *Potts,* 1994). Cells can lose their water content when exposed to an environment with low water activity (A_w), a measure of how "accessible" water is when present as part of a solution or mixed medium, relative to its accessibility in pure form. Water activity is the ratio of the partial vapor pressure of water (p) in a substance to the partial vapor pressure of pure water (p_0) at that temperature, and can be empirically estimated based on relative humidity (RH) measurements ($A_w = \frac{p}{p_0} = \frac{RH}{100}$). A difference in water activity between intracellular and external environments constitutes an *osmotic potential* — a driving force for transfer of water from the region of higher to lower water activity.

The water activity of a typical cell is close to unity. A cell in a liquid solution with a small osmotic potential (e.g., seawater) will tend to lose its water content, so that active regulation of osmotic potential is required. Drying to equilibrium with air even at high relative humidity is lethal to most species of animals and plants. They tolerate low RH only by having structures that keep their internal environment at effectively 100% RH — skin, bark, lungs, and the special structures on leaves called stomata. Drought-tolerant organisms have evolved even more complex structures for water retention that extend the period required to reach that equilibrium, which nonetheless remains lethal. For that reason, water stress is different than other stressors such as extremes of temperature, pH, or salinity discussed below, in that increasing water deficit *always* causes a decrease in metabolic function.

Small organisms (e.g., bacteria) lack structures for water retention and readily lose their cellular water when exposed to even moderately dry air. Microorganisms that are resistant to desiccation often secrete conspicuous amounts of exopolysaccharides (EPS), which contain high-viscosity polymeric substances that tend to be hygroscopic (water-retaining), thus helping to decrease the rate of water loss. Colony structure can also be a tool to mitigate water loss and many bacterial colonies that grow exposed to air tend to be spherical, which exposes the minimal surface area to the vapor phase, thus minimizing the net rate of evaporation (*Potts,* 1994). But even desert-adapted cyanobacteria cease to grow or perform photosynthesis when dried to equilibrium with air at T ~30°C and RH <98%, equivalent to A_w ~0.98 (*Potts and Friedmann,* 1981; *Potts,* 1994). Some lichens are exceptional in that they can supplement their water requirements with atmospheric water vapor, and in some species RH ~80% is sufficient to activate photosynthesis (*Lange et al.,* 2001).

A few groups of small organisms, which include taxa in all three domains of life, have evolved the unique ability to dry up without dying, and resume metabolic activity quickly after rehydration. These are the so-called desiccation-resistant organisms, and they can survive even after drying to 10% water content. This is equivalent to air-dryness at 50% RH and 20°C (*Alpert,* 2005). The threshold of 10% water content appears to correspond to the lowest amount of cellular water that can still form a monolayer around macromolecules (*Billi and Potts,* 2002). Desiccation-resistant organisms require a number of adaptations to survive in the dry state (*Potts,* 1994), such as the synthesis and accumulation of disaccharides (trehalose and sucrose) that protect enzymes during air drying and can also stabilize membranes. These features preserve molecular integrity and thus allow for longer periods of stasis, but rehydration is required for the resumption of metabolic activity. Desiccation-resistant organisms can also efficiently repair DNA damage that accumulates during dehydration, and this ability has a common basis with radiation resistance (see section 5.5); for this reason some of the most desiccation-resistant microorganisms can also tolerate high levels of UV and ionizing radiation (*Billi et al.,* 2000).

5.2. Salinity

Dissolved salts can have two effects on biology. Generally, dissolved salts lower water activity in a manner proportional to their concentration. A solution saturated in NaCl has a water activity of 0.75 at room temperature, and such low water activity can cause an efflux of water molecules from the cell until an equilibrium is reached between the water activities within and outside the cell. *Halophilic* organisms can still grow optimally in such low water activity solutions by creating internally low water activity using solutes that are biochemically tolerated, and thereby reducing or eliminating the osmotic potential. Commonly, this is accomplished by synthesizing so-called "compatible" organic solutes, such as glutamate, glutamine, proline, quaternary amines, and the sugars trehalose and glucosylglycerol, and less frequently by accumulating K^+ within the cell (*Potts,* 1994). This requires additional adaptations to allow proteins and other biomolecules to function at high solute concentrations (*Oren,* 1994, 2002; *Potts,* 1994). The lowest water activity at which growth has been recorded is 0.61. In general, filamentous fungi appear to be better adapted to survive such low water activity levels (*Barbosa-Canovas et al.,* 2008), although it has been suggested that microbial tolerance to high osmotic stress might ultimately converge on a common value of A_w ~0.6 for all three domains of life (*Stevenson et al.,* 2015).

In addition to lowering water activity, a group of soluble compounds that includes common salts such as $MgCl_2$, $CaCl_2$ has chaotropic activity, which disrupts the regular hydrogen bond structure in water, and destabilizes biological macromolecules such as DNA, RNA, and proteins. At sufficiently high concentrations, chaotropic compounds are thought to weaken the hydrophobic effect, and thereby interfere with the role of water in stabilizing the shape and structure of macromolecule solutes. As a result, most chaotropic solutes are powerful inhibitors of enzyme activity (*Hallsworth et al.,* 2007). Chaotropic effects can actually limit metabolism at water activity levels that are otherwise compatible with biological processes. For example, *in vitro* enzyme activity can be totally inhibited at $MgCl_2$ concentrations below 1 M, and concentrations around 2.3 M inhibit microbial growth, even though the water activity of such solutions is compatible with halophilic bacteria ($A_w = 0.75$).

5.3. Temperature

Temperature affects the thermodynamics and kinetics of biochemical reactions, as well as the relative stability of both covalent and non-covalent bonds, and this carries a variety of implications for biological function. As thermal energy increases, the overall strength of association in, for example, folded proteins and the DNA double helix must increase in concert if stability is to be maintained. For example, the strength of bonding between the complementary DNA bases cytosine and guanine (three H bonds) is greater than that between adenosine and thymine (two H bonds), and the relative proportion of G-C pairs in DNA is generally higher in organisms that live at higher temperatures. Organisms that grow at temperatures greater than 90°C also appear to rely on specialized enzymes that help stabilize DNA and help protect it from degradation (*Kampmann and Stock,* 2004). In the case of proteins, molecule stabilization at high temperatures can be achieved by increasing the number of non-covalent bonds (i.e., hydrogen bonds, salt bridges), modifying protein packing, or using shorter oligomers to form larger protein complexes (*Razvi and Scholtz,* 2006; *Petsko,* 2001). All these adaptations increase the number of intramolecular interactions throughout the protein.

Changes in ambient temperature also affect the fluidity of lipid membranes, and organisms must compensate these changes to maintain an optimal membrane fluidity, which is of critical importance for cell integrity, the stability of membrane proteins, diffusion and transport across the membrane, and cell division. Cold temperatures make membranes more rigid, and therefore organisms in cold environments *increase* their membrane fluidity by producing a higher content of unsaturated, polyunsaturated and methyl-branched fatty acids, and/or a shorter acyl-chain length (*Chintalapati et al.,* 2004; *Russell,* 1997). High temperatures make membranes more fluid, and therefore organisms in hot environments *decrease* their membrane fluidity. Microorganisms can accomplish this by modifying the molecular structure and size of their membrane lipids (*Ray et al.,* 1971; *Nordstrom,* 1993).

The highest temperatures yet observed to support growth are 122°C for life overall (*Takai et al.,* 2008) and 73°C and 66°–69°C for oxygenic phototrophs (*Peary and Castenholz,* 1964) and anoxygenic phototrophs (*Pierson and Castenholz,* 1995), respectively. *Corkrey et al.* (2018) suggest that, in Earthly life, increasing temperatures will correspond to a "failure hierarchy" that progresses through ribosome conformational stability, DNA stability, lipid membrane stability, cell wall integrity, and ultimately the stability of small molecules such as ATP. If the latter is seen as an ultimate limit, it is instructive that the half-life of ATP at 122°C [the highest temperature thus far shown to support microbial growth (*Takai et al.,* 2008)] is, at several minutes, still considerably longer than the average turnover time of ATP in

living organisms (*Daniel et al.,* 2004). The currently known temperature tolerance of life would thus be significantly below this perceived upper limit. At the other end of the temperature spectrum, the actual low temperature limit of growth is less well constrained. *Planococcus halocryophilus* strain Or1, isolated from high Arctic permafrost, is able to grow and divide at –15°C, and remains metabolically active (without growth) at –25°C (*Mykytczuk et al.,* 2013). While high temperatures can be lethal, very low temperatures appear to be compatible with extremely weak metabolism of immobile, probably dormant communities (*Price and Sowers,* 2004), but the actual low-temperature limit for this type of survival metabolism is difficult to determine, primarily due to technological constraints of detecting extremely low rates of metabolism and cell division (*Rummel et al.,* 2014).

5.4. Hydrogen Ion Concentration (pH)

Changes in hydrogen ion concentration can affect both covalent and non-covalent bonding in biomolecules. The three-dimensional conformation of proteins necessary for their chemical reactivity can be compromised by pH changes that affect the ionization of acidic or basic amino acids, and therefore change their capacity for hydrogen bonding. Changes in pH can also alter the shape or charge of substrates, preventing them from binding to protein active sites. Different proteins can have different pH optima, so rather than the pH value itself, it is often the *change* in pH that can affect the structure of proteins. Even with the potential to optimize to different pH values, most proteins denature at very low (<1) and very high (>11) pH. In the case of DNA, both low (<1) and high (>11) pH affect the integrity of the backbone structure, which splits the DNA molecule into smaller fragments. Changes in pH also affect hydrogen bonding in the lipid membrane, which can alter its viscosity and therefore its integrity, and the stability of membrane proteins.

Organisms in all three domains of life have developed a range of strategies for coping with extremes of pH (e.g., *Krulwich et al.,* 2011). Commonly, organisms maintain the concentration of hydrogen ions in the cytoplasm in a narrow (near-neutral) range by expending energy to actively pump protons (H^+) into or out of the cell, as required by the external pH (e.g., *Gross,* 2000; *Krulwich et al.,* 2011).

5.5. Ultraviolet and Ionizing Radiation

The stability and interactions of biomolecules can also be impaired by ionizing and electromagnetic radiation, both of which are sufficiently energetic to break covalent bonds. Ultraviolet wavelengths of light are inhibitory or destructive to all organisms on Earth. Ultraviolet-B (280–315 nm) and shorter wavelengths cause damage to DNA (*Caldwell,* 1979; *Cockell,* 1998), while both UV-B and UV-A (315–400 nm) cause cellular damage due to generation of reactive oxygen species (ROS) (*White and Jahnke,* 2002). Some organisms combat these effects through production of UV screening pigments and anti-oxidants, but these are mechanisms that limit damage by short wavelength radiation, rather than enable it to be used directly in metabolic processes. Spores of organisms are rapidly inactivated within a few minutes to a few hours after exposure to UV, but a thin layer (<1 mm) of dust or soil can be an effective shield for life. Radiation-resistant microorganisms, such as *Deinococcus radiodurans*, are as susceptible to damage by ionizing radiation as radiation-sensitive organisms; however, they can survive exposure to higher doses of radiation because they activate several cellular mechanisms to scavenge or diminish the amount of ROS in the cytoplasm, thereby protecting oxidant-sensitive proteins that are needed for DNA repair and structural maintenance (*Ghosal et al.,* 2005; *Battista,* 1997; *Daly et al.,* 2007).

The ionizing radiation field produced by galactic cosmic rays (GCRs) and solar energetic particles (SEPs) is also harmful to life (*Hutchinson,* 1966). Direct cellular damage occurs when energy deposited by ionizing radiation excites electrons in biomolecules, leading to ionization and radiolysis. In addition, ionizing radiation interacting with cellular water causes radiolysis and the production of hydrated free electrons (e^-_{aq}) and ROS (e.g., OH^-), which are probably the most damaging agents derived from the interaction of ionizing radiation and biomass (*Zirkle,* 1954; *Hutchinson,* 1966). In the case of Earthly organisms, cell death from irradiation is believed to be primarily due to protein and DNA damage (*Blok and Loman,* 1973; *Daly et al.,* 2007; *Daly,* 2012; *Krisko and Radman,* 2010), and under γ irradiation, roughly 80% of DNA damage is due to irradiation-induced ROS (*Ghosal et al.,* 2005).

On Earth most of the energy deposited by GCRc and SEPs is absorbed in the upper layers of the atmosphere, and their impact on life on the surface of the planet is negligible. But GCRs and SEPs can be a significant habitability constraint in planetary bodies with thin or negligible atmospheres such as Mars, because they directly impinge on the surface where they interact with crustal materials to produce energetic secondary particles: mesons, γ-ray photons, electrons/positrons, neutrons, and highly-charged/highly-ionized ions (HZE) (*Teodoro et al.,* 2018). While the flux of primary and secondary particles on the surface of a planet such as Mars would not be sufficient to kill extant forms of life (*Hassler et al.,* 2014), it does impose a strict upper boundary on the amount of time that a cell can be dormant and remain viable near the surface.

5.6. Pressure

High and low pressures (measured with respect to the standard pressure at sea level) have an effect on the reproduction and survival of organism. Based on laboratory experiments it appears that most bacteria cease to grow and divide at pressures below 2500 Pa (*Schuerger and Nicholson,* 2006). However, some bacterial isolates have been found to proliferate at pressures below 700 Pa (*Nicholson et al.,* 2013), and even as low as 1–10 Pa (*Pavlov et al.,*

2010). Viable spores of *B. subtilis* have been recovered after exposure to space vacuum (10^{-6} Pa), which suggest that low pressure does not represent a primary limit for life (*Horneck et al.,* 1994).

Pressures between 10 and 150 MPa also effect the growth and viability of organisms (*Bartlett,* 2002; *Picard and Daniel,* 2013). Pressure-sensitive processes in *E. Coli* include motility (abolished at 10 MPa); cell division (impaired at 20–50 MPa); or growth (ceases at 50 MPa) (*Bartlett,* 2002), although pressure effects appear to be modulated by other physicochemical factors such as temperature and pH, which together can destabilize protein structure and can also alter the phase state of cellular membranes (*Bartlett et al.,* 1995; *Bartlett,* 2002). There are *piezophilic* microorganisms that have a higher growth rate at high hydrostatic pressure (*Yayanos,* 1995), and *hyperpiezophiles* possess optimal growth rates at pressures >60 MPa (*Bartlett,* 2002). The most *piezophilic* microorganism (*Moritella yayanosii* MT-41) was isolated from the Mariana Trench and is able to grow up to 130 MPa (*Yayanos et al.,* 1981). Molecular and physiological responses to high pressure include modulating membrane fluidity, viscosity and composition; increases in DNA replication; and changes in cellular morphology toward filamentous shapes (*Bartlett et al.,* 1995; *Bartlett,* 2002; *Picard and Daniel,* 2013). The high pressure limit of survivability in metabolically active microorganisms is in the range of 200–600 Mpa, while inactivation of spores requires higher pressures in the range of 1200–1500 MPa (*Picard and Daniel,* 2013). Hence, while pressure might not be a limiting factor for life in the deep sea, it might be a limit for microbial survival in the deep continental subsurface.

5.7. The Environmental Limits of Habitability

Laboratory investigations have provided valuable information regarding the physical and chemical limits of survival and growth of Earthly organisms. But it is important to recognize that such investigations are generally conducted under conditions that are very different from the natural environment, both in terms of the range of parameters that organisms must face in their habitats, and the temporal variability of those parameters in timescales ranging from seconds to seasons. In addition to the physical and chemical environment, a natural community is also a complex web of intra- and interspecies interactions, and this biological complexity is typically lost in laboratory experiments that focus on one, or few, microbial isolates. The dynamic multidimensional space that organisms must face in their natural habitats defines the true habitability envelope of life on Earth, and it is worth contrasting the physical and chemical limits of life as measured in the confines of the laboratory, with observations in the more complex natural world.

With very few exceptions, the limits of biological activity and growth in natural environments are sharply defined with regard to temperature, pH, and salinity, and they seem to coincide with the limits measured in the laboratory. Bacteria sulfate reduction has been measured at hydrothermal vents of the Guaymas Basin, in the Gulf of California at temperatures up to 110°C, with an optimum rate at 103° to 106°C (*Jorgensen et al., 1992*). It has been suggested that the upper temperature limit for life might be closer to 150°C (*Holden and Daniel, 2004*), but field measurements of biological processes at such high temperatures are difficult and remain inconclusive. At the low end of the temperature spectrum, Deep Lake, a hypersaline (32% salt) Antarctic lake that remains ice-free year-round at temperatures down to –20°C, is dominated by halophilic Archaea, which under those conditions divide ~6 times per year (*DeMaere et al., 2013*). Also in Antarctica, the unfrozen, anoxic, NaCl brine of Lake Vida (~20% salinity, –13°C) contains active bacteria that live under very high levels of reduced metals, ammonia, molecular hydrogen, and dissolved organic carbon (*Murray et al., 2012*).

With respect to pH, culture-independent molecular methods in the Richmond Mine at Iron Mountain have revealed active pyrite-associated microbial communities in pH 0.5–1.4 at temperatures ranging between 27° and 50 C (*Baker and Banfield, 2003*). On the other end of the spectrum, slow metabolic activity and growth have been reported for microorganisms in hyperalkaline springs (pH ~12) (*Pedersen et al., 2004*), but this limit might be slightly lower (*Sorokin, 2005*). In terms of salinity, halophilic organisms belonging to the three domains of life are found in saturated NaCl brines (A_w ~0.75) (*Stevenson et al., 2015; Oren, 2002*), and while active biological processes have also been reported in the $MgCl_2$ saturated brine in the Discovery Basin and the Kyros Basin (A_w ~0.4), located on the Mediterranean Sea floor (*van der Wielen et al., 2005; Steinle et al., 2018*), and the $CaCl_2$ saturated brine of Don Juan Pond in the McMurdo Dry Valleys of Antarctica (*Siegel et al., 1979*), these results remain controversial (*Hallsworth et al., 2007; Horowitz et al., 1972*).

Both *piezosensitive* and *piezophilic* microorganisms have been isolated at the oceans' deepest locations. Carbon turnover rates suggest that microbial cells in the Mariana Trench are alive and active at a pressure of ~110 MPa (*Yayanos et al., 1981; Glud et al., 2013*), and live microbial cells have been detected at pressure as high as ~78 MPa in the deep subseafloor sediments of the Newfoundland margin, and a depth of 1613 m below the seafloor under a water column of 4570 m (*Roussel et al., 2008*). Given the much higher survival pressures of spores, in the gigapascal range, it appears that the high pressure limit of life has not yet been crossed in nature, at least within the current reach of deep drilling technology.

Environments that are very dry are different from other extreme environments in that increasing water stress (decreasing water potential) always causes a decrease in metabolic function. Community-level activity in mineral soils ceases on average at Aw ~0.95, which is comparable to the water potential at which soil diffusion becomes impaired (*Manzoni et al., 2012*). Physiological adaptations to water stress largely focus on survival and dormancy, rather than on optimizing or even maintaining active metabolism or growth (see section 2.3.1). Remarkably, despite the acute sensitivity of most forms of life to water stress and the lack of physiological adaptations to remain metabolically active even after small amounts of water loss, microorganisms can be found in virtually all desert environments, including the driest regions of the hyperarid Atacama Desert in northern Chile and the McMurdo Dry Valleys of Antarctica (*Pointing and Belnap, 2012; Wierzchos et al., 2012a*). The type of adaptation that has enabled microorganisms to colonize even the driest regions on Earth is not physiological but ecological, and consists of inhabiting the ventral surfaces, fissures, and internal fabrics of translucent rocks (e.g., *Wierzchos et al., 2012a; Friedmann, 1980, 1982; Warren-Rhodes et al., 2006; Cary et al., 2010; Pointing and Belnap, 2012; Davila and Schulze-Makuch, 2016*). These lithic substrates maximize water collection during wetting events and retain that water subsequently by surface tension (*Wierzchos et al., 2012b; Davila et al., 2008; Friedmann, 1982; Pointing and Belnap, 2012; Walker and Pace, 2007*). In addition, the rock habitat shields UV radiation and ameliorates extreme temperature fluctuations, which are additional stressors commonly encountered in dry environments.

The most extreme example of lithic colonization that has enabled microorganisms to survive water stress is found in the hyperarid core of the Atacama Desert. Here, the mean annual temperature is around 20°C and the mean RH is about 40% (*McKay et al., 2003*), although soil temperature can reach 65°C and RH often falls below 5%. The average soil water activity in this hyperarid region is below 0.4, and under these conditions lithic substrates that are typically colonized in deserts, such as the ventral sides of translucent quartz rocks or the interior of gypsum crusts, are devoid of life (*Warren-Rhodes et al., 2006; Wierzchos et al., 2011*). However, active microbial communities can still be found inside porous hygroscopic salt (NaCl) nodules (*Wierzchos et al., 2006*) that cover the surface of ancient salt-encrusted playas (*Artieda et al., 2015*). Microorganisms inside the salt nodules use liquid brines that condense from the vapor phase via deliquescence (*Davila et al., 2013, 2008; Wierzchos et al., 2012b*), which typically occurs at night but is sufficient to keep the interior of the salt nodules moist and the communities active for several days despite the surrounding extremely dry conditions (*Davila et al., 2013, 2015*).

5.8. The Limits of Habitability and Life Elsewhere

A deeper understanding of the limits of habitability on Earth — in particular, the tolerance to physicochemical "extremes" — has significantly advanced our understanding of the potential for elsewhere. All locations where life could exist, or have existed, in the solar system can be considered extreme environments. Mars appears to have been a habitable and relatively benign planet early in its history, but for the most part it has remained a cold hyperarid desert with a surface bathed in UV and ionizing radiation. Based on our understanding of the environmental limits of Earthly life, the probability of extant life on Mars, either in vegetative or dormant state, appears to be low at least near the surface (see the chapter by Davila et al. in this volume). The subsurface oceans of Europa and Enceladus, which are considered potential abodes for life, are overall cold, dark, probably energy-poor, and, in the case of Enceladus, mildly to quite alkaline (*Glein et al.*, 2015). They may also host sites of hydrothermal activity at the ocean-sediment boundary where water reaches temperatures in excess of 100°C (*Sekine et al.*, 2015; *Hsu et al.*, 2015), and hydrostatic pressure ranges between 130 and 260 MPa (Europa seafloor) and 5–10 MPa (Enceladus seafloor). In both moons, salinity (equivalent to Earth's oceans) would be compatible with mesophilic microorganisms, whereas temperature, pH, and pressure would require specialized adaptations within a range already realized on Earth. Other, more exotic locations that have been considered possible abodes for life include the acidic clouds of Venus and the hydrocarbon lakes of Titan (*Schulze-Makuch and Irwin*, 2006), but lacking specific examples of possible biochemistries in these worlds makes it difficult to assess their habitability potential with respect to Earth's environments, even extreme ones.

The ability of life on Earth to fill a wide range of extreme niches speaks to the potential for these alien environments to support life, but with an important caveat. In particular, the range of conditions tolerated by extant Earthly organisms could be significantly broader than the range of conditions conducive to the origin of life. Modern Earthly organisms employ a variety of sometimes complex and energy-intensive mechanisms for adapting to the broad range of physical and chemical parameters reflected in Table 2, and such mechanisms would likely not be available to prebiotic chemistry or early organisms. The origin of life on Earth (see the chapter by Baross et al. in this volume) is not understood sufficiently to predict what reduced portion of the physicochemical space could give rise to Earth-like life, much less to alternative biochemistries. Absent that understanding, it should simply be borne in mind that environments capable of supporting extant life are not necessarily equivalent to environments where life can emerge. The diversity of environmental niches on Earth offers the possibility for life to have emerged within a benign physicochemical envelope and radiated into more extreme niches by evolving adaptive mechanisms. Whether or not the same possibility exists in other planetary environments — whether conditions exist that favor life's *emergence*, and not just its ongoing maintenance – remains as a critical factor in understanding the habitability of worlds beyond Earth.

6. SUMMARY

Understanding life's requirements in qualitative and quantitative terms provides a basis for assessing not only the habitability of other worlds, but also the detectability of any biospheres that may inhabit them. Identifying even

a few core attributes of life — for example, information processing — defines broad requirements for a source of energy and a system of complex, interacting molecules. In turn, this levies specific requirements on the properties of the elements that comprise the molecules, the solvent in which those molecules interact, and the envelope of physiochemical conditions in which such chemistry can proceed.

Among a broad range of processes that liberate energy, Earthly life is only known to utilize light and the energy of redox chemical reactions. This limited range may result from mechanistic constraints (the dependence of our energy metabolism on flowing charge), because some alternative energy sources may be inherently destructive toward our biochemistry, or because some may fail to meet minimum requirements. As a result, several quantitatively important energy fluxes on Earth — about two-thirds of the incident solar energy, Earth's thermal gradient energy, the mechanical energy inherent in winds and tides, and the chemical energy associated with phase changes and the potential hydration of crustal rocks — go untapped by life.

All organisms on Earth require C, H, N, O, P, S, and at least some metals, all of which confer specific properties that support the function and interaction of the molecules they comprise. All are relatively abundant in the crust/ocean/atmosphere, but not all abundant elements (e.g, chlorine) are required or widely used. Among these elements, carbon would appear to be unique in its potential to serve as a molecular scaffolding element, particularly in the presence of water, and it is difficult to envision a remotely Earth-like biochemistry without nitrogen. The roles played by O, P, and S might conceivably be replaced, so that a smaller or alternative list of essential elements might be possible; however, the viable alternatives might, in many cases, be less abundant or more difficult to obtain than the elements they replace.

All life on Earth depends on liquid water as a solvent to support the wide range of molecular interactions that confer life-like behavior. Water's role as a mediator of molecular interactions and self-assembly derives from a collection of properties that are perhaps unique to water. In particular, water serves both to decrease the strength of electrostatic interactions and induce a "hydrophobic" effect that leads to several critical instances of molecular self-organization. Potential alternatives to liquid water must be evaluated on their potential to support not only synthetic (covalent) chemistry, but also the diverse range of interactions that occur in living systems.

The specifics of our biochemistry — both covalent and non-covalent — define an envelope of physical and chemical conditions in which the complete range of chemical functionality is supported. The presently known limits of life extend to water activity as low as about 0.6, from tens of degrees below zero to 122°C, to salinities beyond halite saturation, and across a pH range of about 0 to 13. Importantly, the survival of life at some of these extremes, pH in particular, results from its ability to use compartmentalization and energy-expending solutions to maintain clement internal conditions. It is likely that the envelope of conditions that supports life's origin is considerably narrower than that tolerated by modern life.

An understanding of the requirements of Earthly life — in particular, of the basis of those requirements — is a starting point for understanding how similar or different alien biology might be. In turn, this can inform our sense of both possibility and priority as we begin to search for life beyond Earth.

REFERENCES

Abreu I. A. and Cabelli D. E. (2010) Superoxide dismutases — A review of the metal-associated mechanistic variations. *Biochim. Biophys. Acta, Proteins Proteomics, 1804(2),* 263–274.

Adam Z. R., Hongo Y., Cleaves H. J., Yi R., Fahrenbach A. C., Yoda I., and Aono M. (2018) Estimating the capacity for production of formamide by radioactive minerals on the prebiotic Earth. *Sci. Rept., 8,* 265–273.

Aguirre J. D., Clark H. M., McIlvin M., Vazquez C., Palmere S. L., Grab D. J., and Culotta V. C. (2013) A manganese-rich environment supports superoxide dismutase activity in a Lyme disease pathogen, *Borrelia burgdorferi. J. Biol. Chem, 288(12),* 8468–8478.

Alpert P. (2005) The limits and frontiers of desiccation-tolerant life. *Integr. Comp. Biol., 45,* 685–695.

Artieda O., Davila A., Wierzchos J., et al. (2015) Surface evolution of salt-encrusted playas under extreme and continued dryness. *Earth Surf. Processes Landforms, 40,* 1939–1950.

Ashbaugh H. S. and Pratt L. R. (2006) Colloquium: Scaled particle theory and the length scales of hydrophobicity. *Rev. Mod. Phys., 78,* 159.

Bains W. (2004) Many chemistries could be used to build living systems. *Astrobiology, 4(2),* 137–167.

Bains W. and Seager S. (2012) A combinatorial approach to biochemical space: Description and application to the redox distribution of metabolism. *Astrobiology, 12,* 271–281.

Bains W., Petkowski J. P., and Seager S. (2019) Trivalent phosphorus and phosphines as components of biochemistry in anoxic environments. *Astrobiology, 19(7),* DOI: 10.1089/ast.2018.1958.

Baker B. J. and Banfield J. F. (2003) Microbial communities in acid mine drainage. *FEMS Microbiol. Ecol., 44,* 139–152.

Baker S. J., Ding C. Z., Akama T., Zhang Y.-K., Hernandez V., and Xia Y. (2009) Therapeutic potential of boron-containing compounds. *Future Med. Chem., 1(7),* 1275–1288.

Barbosa-Canovas G. V., Fontana A., Schmidt S. J., and Labuza T. (2008) *Water Activity in Foods: Fundamentals and Applications.* Blackwell, Hoboken. 440 pp.

Barge L. M., Abedian Y., Russell M. J., Doloboff I. J., Cartwright J. H., Kidd R. D., and Kanik I. (2015) From chemical gardens to fuel cells: Generation of electrical potential and current across self-assembling iron mineral membranes. *Angew. Chem., Intl. Ed., 54,* 8184–8187.

Baross J. and the Committee on the Limits of Organic Life in Planetary Systems (2007) *The Limits of Organic Life in Planetary Systems.* National Academies, Washington DC. 116 pp.

Bartlett D. H. (2002) Pressure effects on *in vivo* microbial processes. *Biochim. Biophys. Acta, Protein Struct. Mol. Enzymol., 1595 (1–2),* 367–381.

Bartlett D. H., Kato C., and Horokoshi K. (1995) High pressure influences on gene and protein expression. *Res. Microbiol., 146(8),* 697–706.

Battista J. R. (1997) Against all odds: The survival strategies of *Deinococcus radiodurans. Annu. Rev. Microbiol., 51,* 203–224.

Beatty J. T., Overmann J., Lince M. T., Manske A. K., Lang A. S., Blankenship R. L., van Dover C. L., Martinson T. A., and Plumley G. F. (2005) An obligately photosynthetic bacterial anaerobe from a deep-sea hydrothermal vent. *Proc. Natl. Acad. Sci. USA, 102(26),* 9306–9310, DOI: 10.1073/pnas.0503674102.

Benner S. A. (2010) Defining life. *Astrobiology, 10 (10),* 1021–1030, DOI: 10.1089/ast.2010.0524.

Benner S. A. (2014) Paradoxes in the origin of life. *Origins Life Evol. Biospheres, 44,* 339–343.

Benner S. A. and Hutter D. (2002) Phosphates, DNA, and the search for nonterrean life: A second generation model for genetic molecules. *Bioorg. Chem., 30(1),* 62–80.

Benner S. A., Ricardo A., and Carrigan M. A. (2004) Is there a common chemical model for life in the universe? *Curr. Opin. Chem. Biol., 8,* 672–689.

Benner S. A., Bains W., and Seager S. (2013) Models and standards of proof in cross-disciplinary science: The case of arsenic DNA. *Astrobiology, 13,* 510–513.

Billi D. and Potts M. (2002) Life and death of dried prokaryotes. *Res. Microbiol., 153,* 7–12.

Billi D., Friedmann E. I., Hofer K. G., Caiola M. G., and Ocampo-Friedmann R. (2000) Ionizing-radiation resistance in the desiccation-tolerant cyanobacterium *Chroococcidiopsis. Microb Ecol., 66(4),* 1489–1492.

Blankenship R. E. (2010) Early evolution of photosynthesis. *Plant Physiol., 154,* 434–438.

Blok J. and Loman H. (1973) The effects of gamma-radiation in DNA. *Curr. Top. Radiat. Res. Q., 9,* 165–245.

Bock C. W., Katz A. K., Markham G. D., and Glusker J. P. (1999) Manganese as a replacement for magnesium and zinc: Functional comparison of the divalent ions. *J. Am. Chem. Soc., 121(32),* 7360–7372.

Bowles M. W. W., Mogollòn J. M., Kasten S., Zabel M., and Hinrichs K.-U. (2014) Global rates of marine sulfate reduction and implications for sub-seafloor metabolic activities. *Science, 344,* 889–891.

Burk R. F. and Hill K. E. (2015) Regulation of selenium metabolism and transport. *Annu. Rev. Nutr., 35,* 109–134.

Caldwell M. M. (1979) Plant life and ultraviolet radiation: Some perspective in the history of the Earth's UV climate. *BioScience, 29,* 520–525.

Canfield D. E., Rosing M. T., and Bjerrum C. (2006) Early anaerobic metabolisms. *Philos. Trans. R. Soc. B, 361,* 1819–1836.

Cary S. C., McDonald I. R., Barrett J. E., and Cowan D. A. (2010) On the rocks: The microbiology of Antarctic Dry Valley soils. *Nature Rev. Microbiol., 8(2),* 129–138. DOI: 10.1038/nrmicro2281.

Chen M. and Scheer H. (2013) Extending the limits of natural photosynthesis and implications for technical light harvesting. *J. Porphyrins Phthalocyanines, 17,* 1–15.

Chintalapati S., Kiran M. D., and Shivaji S. (2004) Role of membrane lipid fatty acids in cold adaptation. *Cell. Mol. Biol., 50,* 631–642.

Chopra A. and Lineweaver C. (2010) The stoichiometry of the essential elements of life. In *Astrobiology Science Conference, 2010,* Abstract #1538. Lunar and Planetary Institute, Texas.

Choudhary A. and Raines R. T. (2011) An evaluation of peptide-bond isosteres. *ChemBioChem, 12(12),* 1801–1807.

Clancy S. (2008) Genetic mutation. *Nature Education, 1(1),* 187.

Cockell C. S. (1998) Biological effects of high ultraviolet radiation on early Earth — a theoretical evaluation. *J. Theor. Biol., 193,* 717–729.

Cockell C. S. (1999) Life on Venus. *Planet. Space Sci., 47,* 1487–1501.

Cockell C. S., Bush T., Bryce C., Direito S., Fox-Powell M., Harrison J. P., Lammer H., Landenmark H., Martin-Torres J., Nicholson N., Noack L., O'Malley-James J., Payler S. J., Rushby A., Samuels T., Schwendner P., Wadsworth J., and Zorzano M. P. (2016) Habitability: A review. *Astrobiology, 16,* 89–117.

Cordier D., Mousis O., Lunine J. I., Lavvas P., and Vuitton V. (2009) An estimate of the chemical composition of Titan's lakes. *Astrophys. J., Lett.,707(2),* L128.

Corkrey R., McMeekin T. A., Bowman J. P., et al. (2018) The maximum growth rate of life on Earth. *Intl. J. Astrobiol., 17,* 17–33.

Daly M. J. (2012) Death by protein damage in irradiated cells. *DNA Repair, 11,* 12–21.

Daly M. J., Gaidamakova E. K., Matrosova V. Y., et al. (2007) Protein oxidation implicated as the primary determinant of bacterial radioresistance. *PLoS Biol., 5,* 769–779.

Daniel R. M., van Eckert R., Holden J. F., et al. (2004) The stability of biomolecules and the implications for life at high temperatures. *Geophys. Monogr., 144,* 25–39.

Dartnell L. R., Desorgher L., Ward J. M., and Coates A. J. (2007) Modelling the surface and subsurface martian radiation environment: Implications for astrobiology. *Geophys. Res. Lett., 34,* 4–9.

Davila A. F. and Schulze-Makuch D. (2016) The last possible outposts for life on Mars. *Astrobiology, 16,* 159–168.

Davila A. F., Gómez-Silva B., de los Rios A., et al. (2008) Facilitation of endolithic microbial survival in the hyperarid core of the Atacama Desert by mineral deliquescence. *J. Geophys. Res., 113,* G01028.

Davila A. F., Hawes I., Ascaso C., and Wierzchos J. (2013) Salt deliquescence drives photosynthesis in the hyperarid Atacama Desert. *Environ. Microbiol. Rep., 5,* 583–587.

Davila A. F., Hawes I., Araya J. G., et al. (2015) *In situ* metabolism in halite endolithic microbial communities of the hyperarid Atacama Desert. *Front. Microbiol., 6,* DOI: 10.3389/fmicb.2015.01035.

Decker A. and Solomon E. I. (2005) Dioxygen activation by copper, heme, and non-heme iron enzymes: Comparison of electronic structures and reactivities. *Curr. Opin. Chem. Biol., 9(2),* 152–163.

de Duve C. (1995) The beginnings of life on Earth. *Am. Sci., 83,* 428–437.

DeMaere M. Z., Williams T. J., Allen M. A., et al. (2013) High level of intergenera gene exchange shapes the evolution of haloarchaea in an isolated Antarctic lake. *Proc. Natl. Acad. Sci. USA, 110,* 16939–16944.

Des Marais D. J. (2000) When did photosynthesis emerge on Earth? *Science, 289,* 1703–1705.

Desiraju G. R. and Steiner T. (2001) *The Weak Hydrogen Bond: In Structural Chemistry and Biology.* Oxford Univ., Oxford. 526 pp.

Dill K. A. (1990) Dominant forces in protein folding. *Biochemistry, 29(31),* 7133–7155.

Dill K. A. and Chan H. S. (1997) From Levinthal to pathways to funnels. *Nature Struct. Mol. Biol., 4,* 10–19.

Drake J. W., Charlesworth B., Charlesworth D., and Crow J. F. (1998) Rates of spontaneous mutation. *Genetics, 148,* 1667–1686.

Eady R. R. (1996) Structure-function relationships of alternative nitrogenases. *Chem. Rev., 96(7),* 3013–3030.

Elderfield H. and Schultz A. (1996) Hydrothermal fluxes and the chemical composition of the ocean. *Annu. Rev. Earth Planet. Sci., 24,* 191–224.

Ferrer M., Golyshina O. V., Beloqui A., Golyshin P. N., and Timmis K. N. (2007) The cellular machinery of *Ferroplasma acidiphilum* is iron-protein-dominated. *Nature, 445(7123),* 91.

Fersht A. R. (1987) The hydrogen bond in molecular recognition. *Trends Biochem. Sci., 12,* 301–304.

Field C. B., Behrenfeld M. J., Randerson J. T., and Falkowski P. (1998) Primary production of the biosphere: Integrating terrestrial and oceanic components. *Science, 281,* 237–240.

Fischer G. (2000) Chemical aspects of peptide bond isomerisation. *Chem. Soc. Rev., 29,* 119–127, DOI: 10.1039/A803742F.

Friedmann E. I. (1980) Endolithic microbial life in hot and cold deserts. *Orig. Life, 10,* 223–235.

Friedmann E. I. (1982) Endolithic microorganisms in the Antarctic cold desert. *Science, 215,* 1045–1053.

Ghosal D., Omelchenko M. V., Gaidamakova E. K., et al. (2005) How radiation kills cells: Survival of *Deinococcus radiodurans* and *Shewanella oneidensis* under oxidative stress. *FEMS Microbiol. Rev., 29,* 361–375.

Glein C. R., Baross J. A., and Waite J. H. Jr. (2015) The pH of Enceladus' ocean. *Geochim. Cosmochim. Acta, 162,* 202–219, DOI: 10.1016/j.gca.2015.04.017.

Glud R. N., Wenzhöfer F., Middleboe M., Oguri K., Turnewitsch R., Canfield D. E., and Kitazato H. (2013) High rates of microbial carbon turnover in sediments in the deepest ocean trench on Earth. *Nature Geosci., 6,* 284–288.

Greenwood N. N. and Earnshaw A. (2005) *Chemistry of the Elements, 2nd edition.* Butterworth-Heinemann, Oxford. 1600 pp.

Gribble G. W. (2015) A recent survey of naturally occurring organohalogen compounds. *Environ. Chem., 12,* 396–405.

Gross W. (2000) Ecophysiology of algae living in highly acidic environments. *Hydrobiologia, 433,* 31–37.

Habermann E., Crowell K., and Janicki P. (1983) Lead and other metals can substitute for Ca^{2+} in calmodulin. *Arch. Toxicol., 54(1),* 61–70.

Hallsworth J. E., Yakimov M. M., Golyshin P. N., et al. (2007) Limits of life in $MgCl_2$-containing environments: Chaotropicity defines the window. *Environ. Microbiol., 9,* 801–813.

Hassler D. M., Zeitlin C., Wimmer-Schweingruber R. F., et al. (2014) Mars' surface radiation environment measured with the Mars Science Laboratory's Curiosity Rover. *Science, 343(6169),* 1244797.

Hille B. (2001) *Ion Channels of Excitable Membranes, 3rd edition.* Sinauer Associates, Inc., Sunderland. 814 pp.

Hille R. (2002) Molybdenum and tungsten in biology. *Trends Biochem. Sci., 27(7),* 360–367.

Hoehler T. M. (2004) Biological energy requirements as quantitative boundary conditions for life in the subsurface. *Geobiology, 2,* 205–215.

Hoehler T. M. (2007) An energy balance concept of habitability. *Astrobiology, 7,* 824–838.

Hoehler T. M. and Jorgensen B. B. (2013) Microbial life under extreme energy limitation. *Nature Rev. Microbiol., 11,* 83–94.

Hoehler T. M., Alperin M. J., Albert D. B., and Martens C. S. (2001) Apparent minimum free energy requirements for methanogenic archaea and sulfate-reducing bacteria in an anoxic marine sediment. *FEMS Microbiol. Ecol., 38,* 33–41.

Hoffman B. M. and Petering D. H. (1970) Coboglobins: Oxygen-carrying cobalt-reconstituted hemoglobin and myoglobin. *Proc. Natl. Acad. Sci. USA, 67(2),* 637–643.

Holden J. F. and Daniel R. M. (2004) The upper temperature limit for life based on hyperthermophile culture experiments and field observations. *Geophys. Monogr., 144,* 13–24.

Holland H. D. (2006) The oxygenation of the atmosphere and oceans. *Philos. Trans. R. Soc. B, 361,* 903–916.

Holmquist B. and Vallee B. L. (1974) Metal substitutions and inhibition of thermolysin: Spectra of the cobalt enzyme. *J. Biol. Chem., 249(14),* 4601–4607.

Horneck G., Bücker H., and Reitz G. (1994) Long-term survival of bacterial spores in space. *Adv. Space Res., 14(10),* 41–45.

Horowitz N. H., Cameron R. E., and Hubbard J. S. (1972) Microbiology of the dry valleys of Antarctica. *Science, 176,* 242–245.

Hsu H.-W., Postberg F., Sekine Y., et al. (2015) Ongoing hydrothermal activities within Enceladus. *Nature, 519,* 207–210.

Hutchinson F. (1966) The molecular basis for radiation effects on cells. *Cancer Res., 26,* 2045–2052.

Jeffrey G. A. (1997) *An Introduction to Hydrogen Bonding.* Oxford Univ., New York. 320 pp.

Johnson D. C., Dean D. R., Smith A. D., and Johnson M. K. (2005) Structure, function, and formation of biological iron-sulfur clusters. *Annu. Rev. Biochem., 74,* 247–281.

Jorgensen B. B., Isaksen M. F., and Jannasch H. W. (1992) Bacterial sulfate reduction above 100°C in deep-sea hydrothermal vent sediments. *Science, 258,* 1756–1757, DOI: 10.1126/science.258.5089.1756.

Joyce G. F., Deamer D. W., and Fleischaker G. (1994) *Origins of Life: The Central Concepts.* Jones and Bartlett, Boston. 431 pp.

Kampmann M. and Stock D. (2004) Reverse gyrase has heat-protective DNA chaperone activity independent of supercoiling. *Nucleic Acids Res., 32,* 3537–3545.

Kan S. B. J., Lewis R. D., Chen K., and Arnold F. H. (2016) Directed evolution of cytochrome c for carbon-silicon bond formation: Bringing silicon to life. *Science, 354,* 1048–1051.

Karplus M. (1997) The Levinthal paradox: Yesterday and today. *Folding Des., 2,* S69–S75.

Kauzmann W. (1959) Some factors in the interpretation of protein denaturation. *Adv. Protein Chem., 14,* 1–63.

Kiang N. Y., Segura A., Tinetti G., Govindjee, Blankenship R. E., Cohen M., Siefert J., Crisp D., and Meadows V. S. (2007a) Spectral signatures of photosynthesis II: Coevolution with other stars and the atmosphere on extrasolar worlds. *Astrobiology, Special Issue on M Stars, 7,* 252–274.

Kiang N. Y., Siefert J., Govindjee, and Blankenship R. E. (2007b) Spectral signatures of photosynthesis I: Review of Earth organisms. *Astrobiology, Special Issue on M Stars, 7,* 222–251.

Kim H.-J., Ricardo A., Illangkoon H. I., Kim M. J., Carrigan M. A., Frye F., and Benner S. A. (2011) Synthesis of carbohydrates in mineral-guided prebiotic cycles. *J. Am. Chem. Soc., 133,* 9457–9468.

Kisker C., Schindelin H., Baas D., Rétey J., Meckenstock R. U., and Kroneck P. M. (1998) A structural comparison of molybdenum cofactor-containing enzymes. *FEMS Microbiol. Rev., 22(5),* 503–521.

Krisko A. and Radman M. (2010) Protein damage and death by radiation in *Escherichia coli* and *Deinococcus radiodurans. Proc. Natl. Acad. Sci. USA, 107,* 14373–14377.

Krulwich T. A., Sachs G., and Padan E. (2011) Molecular aspects of bacterial pH sensing and homeostasis. *Nature Rev. Microbiol., 9(5),* 330–343, DOI: 10.1038/nrmicro2549.

Lange O. L., Green T. G., and Heber U. (2001) Hydration-dependent photosynthetic production of lichens: What do laboratory studies tell us about field performance? *J. Exp. Bot., 52,* 2033–2042.

La Roche J., Boyd P. W., McKay R. M. L., and Geider R. J. (1996) Flavodoxin as an *in situ* marker for iron stress in phytoplankton. *Nature, 382(6594),* 802.

Lehmer O. R., Catling D. C., Parenteau M. N., and Hoehler T. M. (2018) The productivity of oxygenic photosynthesis around cool M dwarf stars. *Astrophys. J., 859,* 171.

Leibrock E., Bayer P., and Lüdemann H. D. (1995) Nonenzymatic hydrolysis of adenosine triphosphate (ATP) at high temperatures and high pressures. *Biophys. Chem., 54,* 175–180.

Levinthal C. (1969) How to fold graciously. In *Mossbauer Spectroscopy in Biological Systems: Proceedings of a Meeting held at Allerton House, Monticello, Illinois* (P. Debrunner et al., eds.), pp. 22–24. Univ. of Illinois, Urbana.

Limaye S. S., Mogul R., Smith D. J., Ansari A. H., Słowik G. P., and Vaishampayan P. (2018) Venus' spectral signatures and the potential for life in the clouds. *Astrobiology, 18(9),* 1181-1198.

Lipscomb W. N. and Sträter N. (1996) Recent advances in zinc enzymology. *Chem. Rev., 96,* 2375–2434.

Littler M. M., Littler D. S., Blair S. M., and Norris J. M. (1986) Deep-water plant communities from an uncharted seamount off San Salvador Island, Bahamas — Distribution, abundance, and primary productivity. *Deep Sea Res., Part A, 33,* 881–892.

MacKay B. A. and Fryzuk M. D. (2004) Dinitrogen coordination chemistry: On the biomimetic borderlands. *Chem. Rev., 104,* 385–401.

Manske A. K., Glaeser J., Kuypers M. M. M., and Overmann J. (2005) Physiology and phylogeny of green sulfur bacteria forming a monospecific phototrophic assemblage at a depth of 100 meters in the Black Sea. *Appl. Environ. Microbiol., 71,* 8049–8060.

Manzoni S., Schimel J. P., and Porporato A. (2012) Responses of soil microbial communities to water stress: Results from a meta-analysis. *Ecology, 93(4),* 930–938.

Marcus Y. and Hefter G. T. (1997) The compressibility of liquids at ambient temperature and pressure. *J. Mol. Liq., 73,* 61–74.

Marschall E., Jogler M., Henssge U., and Overmann J. (2010) Large-scale distribution and activity patterns of an extremely low-light-adapted population of green sulfur bacteria in the Black Sea. *Environ. Microbiol., 12,* 1348–1362.

McCree K. J. and Keener M. E. (1974) Effect of atmospheric turbidity on the photosynthetic rates of leaves. *Agr. Meteorol., 13,* 349–357.

McDaniel R. V., McIntosh T. J., and Simon S. A. (1983) Nonelectrolyte substitution for water in phosphatidylcholine bilayers. *Biochim. Biophys. Acta, Biomembr., 731,* 97–108.

McIntosh T. J., Magid A. D., and Simon S. A. (1989) Range of the solvation pressure between lipid membranes: Dependence on the packing density of solvent molecules. *Biochemistry, 28,* 7904–7912.

McKay C. P. (2014) Requirements and limits for life in the context of exoplanets. *Proc. Natl. Acad. Sci. USA, 111,* 12628–12633.

McKay C. P. and Smith H. D. (2005) Possibilities for methanogenic life in liquid methane on the surface of Titan. *Icarus, 178,* 274–276.

McKay C. P., Friedmann E. I., Gómez-Silva B., et al. (2003) Temperature and moisture conditions for life in the extreme arid region of the Atacama Desert: Four years of observations including the El Niño of 1997–1998. *Astrobiology, 3,* 393–406.

Meadows V. S., Reinhard C. T., Arney G. N., Parenteau M. N., Schwieterman E. W., Domagal-Goldman S. D., Lincowski A. P., Stapelfeldt K. R., Rauer H., DasSarma S., Hegde S., Narita N., Deitrick R., Lustig-Yaeger J., Lyons T. W., Siegler N., and Grenfell J. L. (2018) Exoplanet biosignatures: Understanding oxygen as a biosignature in the context of its environment. *Astrobiology, 18,* 630–662.

Merchant S. and Helmann J. D. (2012) Elemental economy: Microbial strategies for optimizing growth in the face of nutrient limitation. *Adv. Microb. Physiol., 60,* 91–210.

Merchant S. S., Allen M. D., Kropat J., Moseley J. L., Long J. C., Tottey S., and Terauchi A. M. (2006) Between a rock and a hard place: Trace element nutrition in *Chlamydomonas. Biochim. Biophys. Acta, Mol. Cell Res., 1763(7),* 578–594.

Murphy B. T., Mackinnon S. L., Yan X., Hammond G. B., Vaisberg A. J., and Neto C. C. (2003) Identification of triterpene hydroxycinnamates with in vitro antitumor activity from whole cranberry fruit (*Vaccinium macrocarpon*). *J. Agric. Food. Chem., 51,* 3541–3545.

Murray A. E., Kenig F., Fritsen C. H., et al. (2012) Microbial life at −13°C in the brine of an ice-sealed Antarctic lake. *Proc. Natl. Acad. Sci. USA, 109,* 20626–20631.

Mykytczuk N. C. S., Foote S. J., Omelon C. R., et al. (2013) Bacterial growth at −15°C; molecular insights from the permafrost bacterium

Planococcus halocryophilus Or1. *ISME J., 7,* 1211–1226.

Nguyen K. T., Wu J. C., Boylan J. A., Gherardini F. C., and Pei D. (2007) Zinc is the metal cofactor of *Borrelia burgdorferi* peptide deformylase. *Arch. Biochem. Biophys., 468(2),* 217–225.

Nicholson W. L., Krivushin K., Gilichinsky D., et al. (2013) Growth of *Carnobacterium* spp. from permafrost under low pressure, temperature, and anoxic atmosphere has implications for Earth microbes on Mars. *Proc. Natl. Acad. Sci. USA, 110,* 666–671.

Nielsen F. H. and Uthus E. O. (1990) The essentiality and metabolism of vanadium. In *Vanadium in Biological Systems: Physiology and Biochemistry* (N. D. Chasteen, ed.), pp. 51–62. Springer, Dordrecht.

Nordstrom K. M. (1993) Effect of temperature on fatty acid composition of a white thermus strain. *Appl. Environ. Microbiol., 59,* 1975–1976.

Noy-Meir I. (1973) Desert ecosystems: Environment and producers. *Annu. Rev. Ecol. Syst., 4,* 25–51.

Nürnberg D. J., Morton J., Santabarbara S., Telfe A., Joliot P., Antonaru L. A., Ruban A. V., Cardona T., Krausz E., Boussach A., and Rutherford A. W. (2018) Photochemistry beyond the red limit in chlorophyll f-containing photosystems. *Science, 360,* 1210–1213.

O'Hagan D. and Harper D. B. (1999) Fluorine-containing natural products. *J. Fluorine Chem., 100,* 127–133.

Oren A. (1994) The ecology of the extremely halophilic archaea. *FEMS Microbiol. Rev., 13,* 415–439.

Oren A. (2002) Diversity of halophilic microorganisms: Environments, phylogeny, physiology, and applications. *J. Ind. Microbiol. Biotechnol., 28,* 56–63.

Ort D. R. and Yocum C. F. (1996) Electron transfer and energy transduction in photosynthesis: An overview. In *Oxygenic Photosynthesis: The Light Reactions* (D. R. Ort et al. eds.), pp. 1–9. Springer, Dordrecht.

Overmann J., Cypionka H., and Pfennig N. (1992) An extremely low-light-adapted phototrophic sulfur bacterium from the Black Sea. *Limnol. Oceanogr., 37,* 150–155.

Pace C. N. (1990) Measuring and increasing protein stability. *Trends Biotechnol., 8,* 93–98.

Pace C. N., Trevino S., Prabhakaran E., and Scholtz J. M. (2004) Protein structure, stability and solubility in water and other solvents. *Philos. Trans. R. Soc. B, 359,* 1225–1235.

Pace C. N., Fu H., Fryar K., Landua J., Trevino S. R., Schell D., and Sevcik J. (2014) Contribution of hydrogen bonds to protein stability. *Protein Sci., 23(5),* 652–661.

Pavlov A. K., Shelegedin V. N., Vdovina M. A., et al. (2010) Growth of microorganisms in martian-like shallow subsurface conditions: Laboratory modelling. *Intl. J. Astrobiol., 9,* 51–58.

Peary J. A. and Castenholz R. W. (1964) Temperature strains of a thermophilic blue-green alga. *Nature, 202,* 720–721.

Pedersen K., Nilsson E., Arlinger J., et al. (2004) Distribution, diversity and activity of microorganisms in the hyper-alkaline spring waters of Maqarin in Jordan. *Extremophiles, 8,* 151–164.

Peers G. and Price N. M. (2004) A role for mananganese in superoxide dismutases and growth of iron-deficient diatoms. *Limnol. Oceanogr., 49(5),* 1774–1783.

Peers G. and Price N. M. (2006) Copper-containing plastocyanin used for electron transport by an oceanic diatom. *Nature, 441,* 341–344.

Petsko G. A. (2001) Structural basis of thermostability in hyperthermophilic proteins, or "There's more than one way to skin a cat." *Methods Enzymol., 344,* 469–478.

Philipp M. and Seliger H. (1977) Spontaneous phosphorylation of nucleosides in formamide-ammonium phosphate mixtures. *Naturwissenschaften, 64,* 273–273.

Picard A. and Daniel I. (2013) Pressure as an environmental parameter for microbial life — A review. *Biophys. Chem., 183,* 30–41.

Pierson B. K. and Castenholz R. W. (1995) Taxonomy and physiology of filamentous anoxygenic phototrophs. In *Anoxygenic Photosynthetic Bacteria* (R. E Blankenship et al., eds.), pp. 31–47. Kluwer, Dordrecht.

Pierson B., Bauld J., Castenholz R., D'Amelio E., Marais D., Farmer J., and Ward D. (1992) Modern mat-building microbial communities: A key to the interpretation of Proterozoic stromatolitic communities. In *The Proterozoic Biosphere: A Multidisciplinary Study* (J. Schopf and C. Klein, eds.), pp. 245–342. Cambridge Univ., Cambridge, DOI: 10.1017/CBO9780511601064.008.

Pino S., Sponer J. E., Costanzo G., Saladino R., and Mauro E. D. (2015) From formamide to RNA, the path is tenuous but continuous. *Life, 5,* 372–384.

Pirt S. J. (1965) The maintenance energy of bacteria in growing cultures. *Proc. R. Soc. London, Ser. B, 163,* 224–231.

Pohorille A. and Pratt L. R. (2012) Is water the universal solvent for life? *Origins Life Evol. Biospheres, 42,* 405–409.

Pointing S. B. and Belnap J. (2012) Microbial colonization and controls in dryland systems. *Nature Rev. Microbiol., 10,* 654–654.

Potts M. (1994) Desiccation tolerance of prokaryotes. *Microbiol. Rev., 58,* 755–805.

Potts M. and Friedmann E. I. (1981) Effects of water stress on cryptoendolithic cyanobacteria from hot desert rocks. *Arch. Microbiol., 130,* 267–271.

Pratt L. R. and Pohorille A. (1992) Theory of hydrophobicity: Transient cavities in molecular liquids. *Proc. Natl. Acad. Sci. USA, 89,* 2995–2999.

Price P. B. and Sowers T. (2004) Temperature dependence of metabolic rates for microbial growth, maintenance, and survival. *Proc. Natl. Acad. Sci. USA, 101,* 4631–4636.

Rappaport F. and Diner B. A. (2008) Primary photochemistry and energetics leading to the oxidation of the $(Mn)_4Ca$ cluster and to the evolution of molecular oxygen in Photosystem II. *Coord. Chem. Rev., 252,* 259–272.

Ray P. H., White D. C., and Brock T. D. (1971) Effect of temperature on the fatty acid composition of *Thermus aquaticus. J. Bacteriol., 106,* 25–30.

Razvi A. and Scholtz J. M. (2006) Lessons in stability from thermophilic proteins. *Protein Sci., 15,* 1569–1578.

Reimer K. J., Koch I., and Cullen W. R. (2010) Organoarsenicals. Distribution and transformation in the environment. In *Organometallics in Environment and Toxicology: Metal Ions in Life Sciences* (A. Sigel et al., eds.), pp. 165–229. Royal Society of Chemistry, Cambridge.

Ridge P. G. et al. (2008) Comparative genomic analyses of copper transporters and cuproproteomes reveal evolutionary dynamics of copper utilization and its link to oxygen. *PLoS One, 3(1),* e1378.

Rodionov D. A., Vitreschak A. G., Mironov A. A., and Gelfand M. S. (2003) Comparative genomics of the vitamin B12 metabolism and regulation in prokaryotes. *J. Biol. Chem., 278(42),* 41148–41159.

Romero H., Zhang Y., Gladyshev V., and Salinas G. (2005) Evolution of selenium utilization traits. *Genome Biol., 6,* R66.

Rothschild L. J. and Mancinelli R. L. (2001) Life in extreme environments. *Nature, 409,* 1092–1101.

Roussel E. G., Bonavita M.-A. C., Querellou J., Cragg B. A., Webster G., Prieur D., and Parkes J. (2008) Extending the sub-seafloor biosphere. *Science, 320,* 1046.

Rummel J. D., Beaty D. W., Jones M. A., et al. (2014) A new analysis of Mars "special regions": Findings of the second MEPAG Special Regions Science Analysis Group (SR-SAG2). *Astrobiology, 14,* 887–968.

Russell N. J. (1997) Psychrophilic bacteria-molecular adaptations of membrane lipids. *Comp. Biochem. Physiol., Part A: Mol. Integr. Physiol., 118,* 489–493.

Schink B. (1988) Principles and limits of anaerobic degradation: Environmental and technological aspects. In *Biology of Anaerobic Microorganisms* (J. B. A. Zehnder, ed.), 872 pp. Wiley and Sons, New York.

Schoffstall A. M. (1976) Prebiotic phosphorylation of nucleosides in formamide. *Orig. Life, 7,* 399–412.

Schuerger A. and Nicholson W (2006) Interactive effects of hypobaria, low temperature, and CO_2 atmospheres inhibit the growth of mesophilic *Bacillus* spp. under simulated martian conditions. *Icarus, 185,* 143–152.

Schulze-Makuch D. and Irwin L. N. (2006) The prospect of alien life in exotic forms on other worlds. *Naturwissenschaften, 93,* 155–172.

Schulze-Makuch D., Grinspoon D. H., Abbas O., Irwin L. N., and Bullock M. A. (2004) A sulfur-based survival strategy for putative phototrophic life in the venusian atmosphere. *Astrobiology, 4,* 11–18.

Seager S. and Bains W. (2015) The search for signs of life on exoplanets at the interface of chemistry and planetary science. *Sci. Adv., 1(2),* e1500047.

Seager S., Bains W., and Petkowski J. J. (2016) Toward a list of molecules as potential biosignature gases for the search for life on exoplanets and applications to terrestrial biochemistry. *Astrobiology, 16,* 465–485.

Segura A., Kasting J. F., Meadows V., Cohen M., Scalo J., Crisp D., Butler R. A. H., and Tinetti G. (2005) Biosignatures from Earth-like planets around M dwarfs. *Astrobiology, 5,* 706–725.

Sekine Y., Shibuya T., Postberg F., et al. (2015) High-temperature water-rock interactions and hydrothermal environments in the chondrite-like core of Enceladus. *Nature Commun., 6,* 1–8. DOI: 10.1038/ncomms9604.

Setlow J. K. and Duggan D. E. (1964) The resistance of *Micrococcus radiodurans* to ultraviolet radiation. *Biochim. Biophys. Acta, Spec. Sect. Nucleic Acids Relat. Subj., 87,* 664–668.

Seto H. and Kuzuyama T. (1999) Bioactive natural products with carbon-phosphorus bonds and their biosynthesis. *Nat. Prod. Rep., 16,* 589–596.

Sevier C. S. and Kaiser C. A. (2002) Formation and transfer of disulphide bonds in living cells. *Nature Rev. Mol. Cell Biol., 3,* 836–847.

Shapiro S. K. and Schlenk F. (2006) The biochemistry of sulfonium compounds. In *Advances in Enzymology and Related Areas of Molecular Biology, Vol. 22* (F. F. Nord, ed.), pp. 237–276. Interscience, Hoboken.

Shapiro R. and Schulze-Makuch D. (2009) The search for alien life in our solar system: Strategies and priorities. *Astrobiology, 9,* 335–343.

Shi W. and Chance M. R. (2011) Metalloproteomics: Forward and reverse approaches in metalloprotein structural and functional characterization. *Curr. Opin. Chem. Biol., 15,* 144–148.

Siegel B. Z., McMurty G., Siegel S. M., et al. (1979) Life in the calcium chloride environment of Don Juan Pond, Antarctica. *Nature, 280,* 828–829.

Silver S. and Phung L. T. (2005) A bacterial view of the periodic table: Genes and proteins for toxic inorganic ions. *J. Ind. Microbiol. Biotechnol., 32,* 587–605.

Sorokin D. Y. (2005) Is there a limit for high-pH life? *Intl. J. Syst. Evol. Microbiol., 55,* 1405–1406.

Steinle L., Knittel K., Felber N., et al. (2018) Life on the edge: Active microbial communities in the Kryos $MgCl_2$-brine basin at very low water activity. *ISME J., 12,* 1414–1426.

Stevenson A., Cray J. A., Williams J. P., et al. (2015) Is there a common water-activity limit for the three domains of life? *ISME J., 9,* 1333–1351.

Stiefel E. I. (2002) The biogeochemistry of molybdenum and tungsten. *Met. Ions. Biol. Syst., 39,* 1–29.

Stryer L. (1988) *Biochemistry.* Freeman and Company, New York. 1089 pp.

Stumpf M. P. H., Thorne T., de Silva E., Stewart R., Hyeong J. A., Lappe M., and Wiuf C. (2008) Estimating the size of the human interactome. *Proc. Natl. Acad. Sci. USA, 105,* 6959–6964.

Szostak J. W., Bartel D. P., and Luisi P. L. (2001) Synthesizing life. *Nature, 409,* 387–390.

Tacke R. (1999) Milestones in the biochemistry of silicon: From basic research to biotechnological applications. *Angew. Chem., Intl. Ed., 38,* 3015–3018.

Takai K., Nakamura K., Toki T., et al. (2008) Cell proliferation at 122 C and isotopically heavy CH_4 production by a hyperthermophilic methanogen under high-pressure cultivation. *Proc. Natl. Acad. Sci. USA, 105,* 10949–10954.

Tanford C. (1978) The hydrophobic effect and the organization of living matter. *Science, 200,* 1012–1018.

Teodoro L., Davila A., Elphic R. C., Hamilton D., McKay C., and Quinn R. (2018) Habitability and biomarker preservation in the martian near-surface radiation environment. In *From Habitability to Life on Mars* (N. Cabrol and E. Grin, eds.), pp. 211–231. Elsevier, Netherlands, DOI: 10.1016/B978-0-12-809935-3.00012-8.

Thauer R. K., Kaster A. K., Goenrich M., Schick M., Hiromoto T., and Shima S. (2010) Hydrogenases from methanogenic archaea, nickel, a novel cofactor, and H_2 storage. *Annu. Rev. Biochem., 79,* 507–536.

van der Wielen P. W., Bolhuis H., Borin S., et al. (2005) The enigma of prokaryotic life in deep hypersaline anoxic basins. *Science, 307(5706),* 121-123.

Walker J. J. and Pace N. R. (2007) Endolithic microbial ecosystems. *Annu. Rev. Microbiol., 61,* 331–347.

Wang L., Chen S., Xu T., Taghizadeh K., Wishnok J. S., Zhou X., You D., Deng Z., and Dedon P. C. (2007) Phosphorothioation of DNA in bacteria by *dnd* genes. *Nature Chem. Biol., 3,* 709.

Warren-Rhodes K. A., Rhodes K. L., Pointing S. B., et al. (2006) Hypolithic cyanobacteria, dry limit of photosynthesis, and microbial ecology in the hyperarid Atacama Desert. *Microb. Ecol., 52,* 389–398.

Westheimer F. H. (1987) Why nature chose phosphates. *Science, 235,* 1173–1178.

White A. L. and Jahnke L. S. (2002) Contrasting effects of UV-A and UV-B on photosynthesis and photoprotection of beta-carotene in two *Dunaliella* spp. *Plant Cell Physiol., 43,* 877–884.

Wierzchos J., Ascaso C., and McKay C. P. (2006) Endolithic cyanobacteria in halite rocks from the hyperarid core of the Atacama Desert. *Astrobiology, 6,* 415–422.

Wierzchos J., Cámara B., de los Ríos A., et al. (2011) Microbial colonization of Ca-sulfate crusts in the hyperarid core of the Atacama Desert: Implications for the search for life on Mars. *Geobiology, 9,* 44–60.

Wierzchos J., de los Ríos A., and Ascaso C. (2012a) Microorganisms in desert rocks: The edge of life on Earth. *Intl. Microbiol., 15,* 173–183.

Wierzchos J., Davila A. F., Sánchez-Almazo I. M., et al. (2012b) Novel water source for endolithic life in the hyperarid core of the Atacama Desert. *Biogeosciences, 9,* 3071–3098.

Williams L. D., Chawla B., and Shaw B. R. (1987) The hydrogen bonding of cytosine with guanine: Calorimetric and 1H-NMR analysis of the molecular interactions of nucleic acid bases. *Biopolymers, 26(4),* 591–603.

Williams R. J. P. (2002) The fundamental nature of life as a chemical system: The part played by inorganic elements. *J. Inorg. Biochem., 88,* 241–250.

Wolstencroft R. D. and Raven J. A. (2002) Photosynthesis: Likelihood of occurrence and possibility of detection on Earth-like planets. *Icarus, 157,* 535–548.

Yanson I. K., Teplitsky A. B., and Sukhodub L. F. (1979) Experimental studies of molecular interactions between nitrogen bases of nucleic acids. *Biopolymers, 18(5),* 1149–1170.

Yayanos A A. (1995) Microbiology to 10,500 meters in the deep sea. *Annu. Rev. Microbiol., 49,* 777–805.

Yayanos A. A., Dietz A. S., and van Boxtel R. (1981) Obligately barophilic bacterium from the Mariana trench. *Proc. Natl. Acad. Sci. USA, 78,* 5212–5215.

Yee D. and Morel F. M. M. (1996) *In vivo* substitution of zinc by cobalt in carbonic anhydrase of a marine diatom. *Limnol. Oceanogr., 41(3),* 573–577.

Zhang Y. and Gladyshev V. N. (2010) General trends in trace element utilization revealed by comparative genomic analyses of Co, Cu, Mo, Ni, and Se. *J. Biol. Chem., 285(5),* 3393–3405.

Zirkle R. E. (1954) Biological effects of external X and gamma radiation, part I. *Phys. Today, 7,* 20–22.

Baross J. A., Anderson R. E., and Stüeken E. E. (2020) The environmental roots of the origin of life. In *Planetary Astrobiology* (V. Meadows et al., eds.), pp. 71–92. Univ. of Arizona, Tucson, DOI: 10.2458/azu_uapress_9780816540068-ch003.

The Environmental Roots of the Origin of Life

J. A. Baross
University of Washington

R. E. Anderson
Carleton College

E. E. Stüeken
University of St Andrews

The ongoing quest to understand how life emerged and evolved on Earth converges on four interdependent lines of inquiry. These include the earliest time in Earth history when there is evidence of life, the geological and geochemical properties of the environmental setting where life appeared, and the characteristics of the earliest life forms based on molecular and biochemical data of both extant organisms and models that infer ancient genetic and protein lineages. The fourth line of inquiry is how life got started. This involves the complex set of organic chemical and biochemical reactions that resulted in metabolism, genetics, and an evolving organism. Here, we discuss these four lines of inquiry and point out that the "how" question may not be linked to the inferred environmental setting of the earliest microbial communities, or the last common ancestor of life, and may require multiple settings that might better represent a dynamic and interactive Hadean global Earth. Recent discoveries and models of characteristics of the Hadean Earth indicate a more dynamic mafic crust dominated by plume volcanism. This scenario would include multiple physical and chemical gradients that could be analogous to a chemical reactor resulting in the synthesis of organic compounds and production of energy sources including hydrogen. These environmental conditions are consistent with the inferred metabolic characteristics of the earliest microbes and the geological settings where there is evidence of life detected earlier than 3.7 b.y. ago. One of the goals of this review is to identify the interrelationship of the Hadean environmental and geochemical conditions with critical biochemical determinants involved in the origin of life that could be helpful in our search for Earth-like life elsewhere.

Let me learn the lesson that you have hidden in every leaf and rock.
— Native American prayer, attributed to Chief Yellow Hawk of the Lakota Sioux

1. INTRODUCTION

The origin of life is perhaps the most perplexing unanswered question in science. The main reason for this perplexity is the overlap of four critical components to life's origin: when and where life emerged, how it formed, and what the earliest life forms looked like. Each of these components has their own set of questions that cross multiple disciplines. For example, insights into the question of *when* and *where* life began are predicated on the early Archaean rock record, which is limited due to metamorphism and erosion. With time, continued study of the rock record will reveal more about the earliest appearance of discernible life and perhaps constrain the environmental conditions while providing clues to early metabolisms. However, the rock record is limited in that it may not necessarily shed much light on our understanding of the multiple steps leading to a replicating and

evolving "living" system. Here, we will review the current research results and models related to these critical components and attempt to present a synthesis. One of our primary goals is to distill the critical environmental, geochemical, and biochemical determinants leading to the origin of life that could be helpful in our search for Earth-like life elsewhere.

Fundamentally, life is composed of the major elements carbon, hydrogen, nitrogen, oxygen, phosphorus, and sulfur (CHNOPS), which are arranged into amino acids, sugars, fatty acids, and nucleotide bases (see the chapter by Hoehler et al. in this volume). These are the monomers that form the macromolecular building blocks of living cells: proteins for execution of functions, polysaccharides for energy storage, lipids for the formation of cell membranes, and nucleic acids for the storage of hereditary information. These building blocks are linked through metabolism and replication. Metabolism comprises the processes by which a cell catalyzes reactions between metabolites (catabolism) to channel energy into the construction of new organic molecules (anabolism). Enzymes, which are frequently made of proteins, act as catalysts and often contain transition metals in their catalytic center. During the process of replication, a strand of deoxyribonucleic acid (DNA) is first transcribed into ribonucleic acid (RNA), which is then translated into proteins that aid in the replication of DNA. This is known as the "central dogma" in biology. RNA takes on the prominent role of being able to take on hereditary information from DNA while acting as an enzyme itself. The translation of nucleotide sequences into amino acid sequences follows the genetic code, which translates a specific amino acid from sets of three nucleotides referred to as codons. A plausible model for the origin of life on Earth must explain not only the sources of the major elements and catalytic metals and the formation of monomers and polymers, but also the origin of the genetic code. The advent of heritable information in the form of the genetic code was essential for the onset of Darwinian evolution.

We only know about one form of life, and it is a generally held view that specific properties of Earth life will persist on other planetary bodies that share geological, chemical, and physical characteristics of Earth. These specific properties considered to be universal for life on an Earth-like planet include carbon-based macromolecules, metal and trace-element dependence, the use of chemical or light energy for metabolism, and the need for water as the solvent to carry out oxidation and reduction reactions (*Dass et al.,* 2016; *Stüeken et al.,* 2013; *Westall and Brack,* 2018) (see also the chapter by Hoehler et al. in this volume). Carbon-based life is emphasized not only because carbon is one of the most abundant elements in the universe, but also because many of the organic compounds used by life are synthesized abiotically under a range of Earth-like conditions. Organic compounds are also recorded in comets, in meteorites, in interplanetary dust, and as a result of bolide impacts (*Westall et al.,* 2018).

Furthermore, a strong argument can be made that Darwinian evolution is essential for all life since it is the primary mechanism for creating diversity and complexity (*Baross,* 2007). This is the rationale for including evolution as one of the essential characteristics listed in multiple definitions of life (*Joyce,* 2005). However, it leaves open the possibility that organisms could have evolved different genetic codes, different mechanisms for translating the code, and structural differences in membranes and proteins to allow growth under different environmental conditions. Is it possible that the origin of Earth life involved multiple alternative organic chemical frameworks with distinct classes of polymers and energy sources before the selection of the most fit lineage that later became the so-called last universal common ancestor (LUCA)? The concept of LUCA acknowledges that all extant organisms share common biochemical and genetic characteristics referred to as the "unity of biochemistry" (see section 4) (*Baross,* 1998; *Hud et al.,* 2013; *Kitadai and Maruyama,* 2017; *Koonen and Novozhilov,* 2017). An argu-

ment against alternative biochemistries is the strong link between specific nucleotide bases and specific amino acids, which points out that there are "rules of organic chemistry" that either favor or presuppose specific chemical reactions and macromolecular structures (*Copley et al.,* 2005, 2007; *Koonen and Novozhilov,* 2017). A second genesis on Earth or elsewhere on an Earth-like planet, given the contingency rule of Darwinian evolution, could result in the selection of organisms and ecosystems that are significantly different. Yet it is likely, compared to present-day organisms, that they would share similar biochemical characteristics and evolve many of the same phenotypes (i.e., the observable characteristics of an organism resulting from its genetic makeup), albeit possibly with different genotypes (i.e., the genetic makeup of an organism), which are adapted to multiple environmental conditions.

Given our working hypothesis that specific properties of Earth life would persist on other planetary bodies where a second origin of life had occurred, our goal in this chapter is to outline the general requirements for an origin of life so that we may conduct a more informed search for habitable or inhabited bodies beyond our planet.

2. A BRIEF HISTORY OF ORIGIN OF LIFE STUDIES AND PARADIGMS

Among origin-of-life scholars, there have been two contrasting views regarding the first essential step leading to a living organism: (1) the synthesis of a replicating molecule (*Oparin,* 1952), believed at that time to be a protein, and (2) a metabolic network that would have provided the energy to drive and sustain a living system (*Haldane,* 1929). The central idea of the "metabolism first" scenario is that energy-driven networks of small molecules are more likely to have constituted the first replicating entities, prior to the advent of a self-replicating molecule similar to the RNA and DNA that forms the informational basis of life today (*Shapiro,* 2007). At some point, early life gained the capability to encode heritable information — the basis of Darwinian evolution. Life's informational molecule is DNA, which encodes its functional molecule, proteins, via RNA. Walter Gilbert first suggested that RNA may have acted as both informational and functional molecule in the earliest stages of life's evolution in a theory now called the "RNA world hypothesis" (*Gilbert,* 1986), which posits that early RNA molecules created self-replicating networks. This idea has profoundly influenced the focus of origin of life research by adding environmental and biochemical complexity to the prebiotic synthesis of the first self-replicating nucleic acid. This complexity is best discussed by *de Duve* (1995), who hypothesizes that the merging of both the "replicator first" and "metabolism first" approaches are necessary for a *de novo* origin of life. De Duve argued that an RNA-protein world could not have been "supported or initiated" without a source of chemical energy that de Duve proposed was the high-energy thioester bond synthesized by a protometabolic network. Subsequent research has upheld and updated the view that RNA molecules were key players in early self-replicating networks. This RNA world model is now supported both through *in vitro* chemical evolution studies in the laboratory (i.e., *Robertson and Joyce,* 1990; *Beaudry and Joyce,* 1992; *Ellington and Szostak,* 1990; *Higgs and Lehman,* 2015) and through studies that examine the evolutionary relationships between ribozyme-like structures that are very similar across all known organisms today (in other words, they are universally conserved) (i.e., *Hsiao et al.,* 2009). Life arising elsewhere may arrive at higher complexity through other means. However, if RNA were one of the crucial molecules in early replicating networks on Earth and possibly elsewhere, then the environments in which it could be synthesized and polymerized should be of key interest for early Earth scientists and planetary geologists.

The early investigations into the origin of life also focused on the assumption that a prebiotic organic soup, one that was exposed to wet and dry cycles like an evaporitic lake, was the spawning ground for all chemical and biochemical stages leading to a self-replicating entity. The "organic soup" model was supported by Stanley Miller's "lightning discharge" experiments, demonstrating that organic molecules such as amino acids could be synthesized by a mixture of hydrogen, methane, and ammonia, which were believed at the time to be the dominant gases present in Earth's earliest atmosphere (*Miller,* 1953). These extraordinary results sparked the new disciplines of prebiotic chemistry and exobiology. While the lacustrine (lake) "organic soup" can plausibly explain some aspects of the origin of life, other environments including hydrothermal vents, geothermal environments, and ice brine pockets have also emerged as possible settings (*Baross and Hoffman,* 1985; *Stüeken et al.,* 2013; *Mulkidjanian et al.,* 2012; *Adam,* 2007; *Westall et al.,* 2018). All these environmental settings have their strengths and weaknesses, and combining their strengths is a challenge. These will be discussed further in section 5.

These early studies highlight key issues that researchers continue to explore to better understand the origin of life. These include a number of important questions: How were prebiotic organic compounds synthesized and polymerized into macromolecules? What environmental settings were involved in these early chemical stages and how early in Earth's history did they start? What were the sources and nature of the catalysts that drove these chemical reactions? And finally, what factors were involved in producing the increasing chemical complexity that would eventually lead to life? [For a more comprehensive discussion of the early history of origin of life research, see *Hazen* (2005), *Lazcano* (2010), and *Oro et al.* (1990).]

Origin of life research can be broadly categorized into "top-down" and "bottom-up" studies, according to their approach (Fig. 1). The "top-down" approach uses paleo-genetic information from extant organisms to reconstruct the earliest genes, genomes, and proteins, and uses that information to infer other important physiological characteristics. This approach relies mostly on bioinformatics techniques based on relics that remain in the genomes and proteomes of modern organisms, providing clues that trace back to the most deeply rooted organisms on the tree of life. In contrast, the "bottom-up" approach relies on fundamental principles of organic chemistry and molecular assembly. This includes recapitulating the most likely steps for the formation of prebiotic molecules, polymers, and early metabolic networks. Crucially, both the "top-down" and "bottom-up" approaches

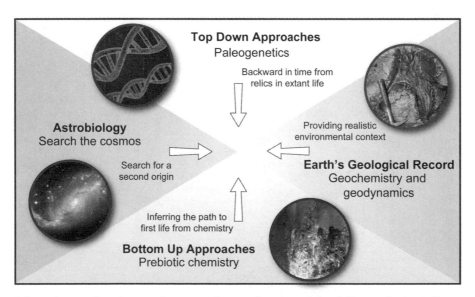

Fig. 1. See Plate 3 for color version. Approaches to understanding the origin of life can be broadly categorized into "top-down" and "bottom-up" approaches, with top-down approaches focusing on bioinformatics methods that can reveal clues about the nature of our common ancestors by examining the genomes of extant organisms, whereas bottom-up approaches focus on principles of synthetic and experimental chemistry to recreate the most likely steps leading to early life. Earth's geological record, by providing information about the Hadean Earth, contextualizes the results from both top-down and bottom-up approaches in a realistic environment and can validate results from both methods. Astrobiology could yield further information by identifying a second origin of life.

must be constrained and informed by Earth's geological record in order to provide a realistic environmental context for processes occurring during the Hadean and Archaean eras, when these processes were taking place. Bottom-up experiments can reveal something about likely settings and processes for the origin of life; these approaches can be merged with top-down approaches based on modern biology, which inform us about the earliest stages of life's evolution. Below, we review the major results gained from top-down (section 3) and bottom-up studies (section 4). To further advance our understanding of the origin of life, both approaches need to be combined within a realistic environmental framework. To this end, new geochemical data from the rock record provide important insights for the environmental context of prebiotic chemistry and early evolution (section 5).

3. TOP-DOWN PERSPECTIVES

The top-down approach is based on the tree of life (Fig. 2), which is predicated on the recognition that the ribosomal RNA genes in all extant organisms share conserved sequences, thus making it possible to group organisms based on their genetic relatedness (*Woese,* 1987, 1998; *Woese et al.,* 1990; *Iwabe et al.,* 1989). The tree of life determines the evolutionary relationship of all organisms, revealing that there are three groups, or domains, of life. The domain of life that includes multicellular organisms, Eukarya, is more closely related to a group of single-celled organisms called the Archaea than to the Bacteria. Currently, the top-down approach makes inferences about the nature of early life based on great advances in comparative genomics and

metagenomics approaches. These techniques have allowed us to track gene and genome evolution over time, revealed novel lineages in the cellular and viral trees of life, revealed ancient relics of the RNA world within molecular mechanisms of modern cells, and have allowed us to reconstruct early organisms and protein sequences. Comparative genomics analyses can identify genes that are widely shared across the tree of life and thus provide insights into the genes that were most likely present in the LUCA, allowing us to deduce likely metabolisms and physiologies that evolved very early in life's history. Although comparative genomics approaches will miss genes that have been lost since LUCA, various attempts at reconstructing LUCA's genomic repertoire have suggested that it possessed a relatively small genome with between 140 and 600 genes, yet this genome encoded most of the essential functions present in modern organisms (*Koonin,* 2003; *Mirkin et al.,* 2003; *Ranea et al.,* 2006). Both the universal common ancestor and the last archaeal common ancestor have been inferred to have small genomes that expanded later through gene duplication and horizontal gene transfer (or the process by which foreign genes are introduced into new genomes through uptake of free DNA, introduction by viruses, or direct sharing of genetic material between microbes) (*David and Alm,* 2011; *Williams et al.,* 2017). More recent work has argued that LUCA was a thermophile (i.e., a heat-loving organism; see the chapter by Hoehler et al. in this volume), used hydrogen as an energy source, fixed atmospheric N_2 into ammonium, and possessed many proteins that contained iron-sulfur clusters (*Weiss et al.,* 2016; *Sousa et al.,* 2016). This work has been used to argue that LUCA lived near hydrothermal vents, supporting previous hypotheses (*Baross and Hoffman,* 1985; *Martin and*

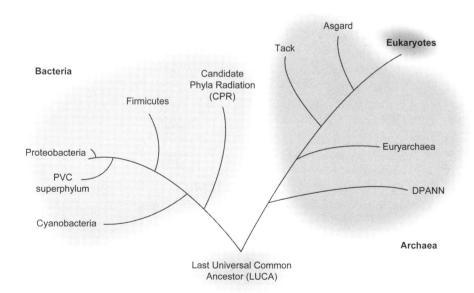

Fig. 2. See Plate 4 for color version. Schematic representation of our current understanding of the tree of life. Discovery of novel lineages through genomes assembled from environmental metagenomes has revealed archaeal superphyla that may provide the closest link yet discovered between the archaeal domain and the eukaryotic ancestors. While debate remains ongoing, if true these results would indicate that the tree of life is a two-domain tree of life rather than a three-domain tree of life, with the eukaryotes falling within the archaeal clade. The last universal common ancestor (LUCA) is at the root of the tree, representing the ancestor of all extant cellular life.

Russell, 2007; *Martin et al.*, 2008; *Gaucher et al.*, 2003, 2008; *di Giulio*, 2003), although some have argued that these methods produced a skewed gene set, and therefore the LUCA was not necessarily a vent-dwelling organism (*Gogarten and Deamer*, 2016). Moreover, others have examined the most likely genomic characteristics of LUCA, such as the relative abundance of guanine and cytosine relative to adenine and thymine (also called "GC content"), and inferred that LUCA inhabited a more temperate environment (*Galtier et al.*, 1999; *Penny and Poole*, 1999; *Boussau et al.*, 2008; *Cantine and Fournier*, 2018).

Genomic analyses are necessarily restricted to inferring the ancestor of the lineages that survived, and not those that later went extinct. It is thus possible that even if LUCA was mesophilic, we may only see the descendants of a "hyperthermophilic Noah" (*Nisbet and Sleep*, 2001). Moreover, such analyses can only go as far back as LUCA and can make only limited inferences about prebiotic chemistry. Comparative genomics methods are made more difficult by the fact that the universal gene set is quite small (*Becerra et al.*, 2007), and the signal is clouded by the loss of extinct lineages, gene loss, and gene gain through horizontal gene transfer. Nevertheless, work such as this can generate hypotheses about life's earliest habitat and physiology that can be used for further hypotheses development and testing, in particular when combined with plausible environmental reconstructions from the rock record.

Top-down approaches can also teach us about the energy sources and metabolic pathways used by the earliest organisms, which ties directly into geochemical and photochemical processes that generated these metabolites (*Braakman and Smith*, 2012) (section 5). Phylogenomic analysis of ancient anaerobic metabolic pathways points to the CO_2-fixing, H_2-dependent Wood-Ljungdahl pathway used by methanogens as a potentially ancient pathway (*Fuchs*, 2011). This model is supported by conserved protein sequences indicating that the most ancient lineage of the archaea consisted of methanogens or acetogens (*Raymond et al.*, 2004; *Weiss et al.*, 2016) and it is consistent with the detection of biogenic methane in fluid inclusions in early Archaean rocks (~3.5 b.y. ago) (*Ueno et al.*, 2006). More recently, *Williams et al.* (2017) developed a new model for the root of the archaeal tree based on evidence from more than 30,000 archaeal gene families, including gene duplications, horizontal transfers, and gene losses. Their metabolic reconstruction indicates an early archaeal common ancestor that was anaerobic and reduced CO_2 to acetate via the Wood-Ljungdahl pathway and grew at temperatures greater than 70°C. However, the placement of the root of this early archaea was not within the euryarchaea, where the methanogens cluster, but between other groups of newly discovered archaeal lineages.

Our understanding of the overall structure of the tree of life, and thus our ability to infer characteristics of early lineages, has been revolutionized by new bioinformatics methods that allow researchers to access the genomes of uncultivated microorganisms, which constitute the majority of microbes on Earth (*Castelle and Banfield*, 2018; *Eme et al.*, 2017; *Hug et al.*, 2016; *Koonin*, 2018; *Spang and Ettema*, 2016; *Spang and Offre*, 2019; *Zaremba-Niedzwiedzka et al.*, 2017). Through metagenomics methods, researchers are able to collect short DNA sequences directly from the environment, assemble those shorter sequences into longer sequences, and then cluster those longer sequences into metagenome-assembled genomes (MAGs) based on sequence similarity and other characteristics (i.e., *Imelfort et al.*, 2014; *Wu et al.*, 2014; *Eren et al.*, 2015; *Kang et al.*, 2015; *Graham et al.*, 2017). Perhaps the most profound discovery using these techniques was that of the Lokiarchaeota at a hydrothermal vent environment in the Arctic (*Spang et al.*, 2015). The Lokiarchaeota comprise a lineage within the newly discovered "Asgard" superphylum within the Archaea. Phylogenetic analyses indicate that this phylum represents the closest living archaeal lineage to the eukaryotes (*Zaremba-Niedzsiedzka et al.*, 2017; *Raymann et al.*, 2015; *Zhou et al.*, 2018). The genomes of Asgard archaea contain more eukaryotic-like genes than any other known Archaea. These findings imply that some of the defining features of eukaryotes, including membrane remodeling, ubiquitin, cytoskeleton, and endocytosis and/or phagocytosis, might have already evolved in the last common ancestor of eukaryotes and archaea (*Spang and Ettema*, 2016). There remains some question as to whether the Asgard superphylum bridges the gap between eukaryotes and prokaryotes, or whether the archaea are a sister group to the Asgards (*Da Cunha et al.*, 2017, 2018). The Asgard archaea either acquired the eukaryotic genes from lateral gene transfer or they are in fact the source of these genes to the early lineage of eukaryotes (*Nasir et al.*, 2016). The archaeal ancestor to the first eukaryote may have acquired genes from bacteria as part of a symbiotic association (*Dey et al.*, 2016). Nevertheless, discoveries like these are challenging our notions of who our earliest eukaryotic ancestors were and expanding our understanding of the evolution of complex life on Earth. It is conceivable that some traits of complexity were developed very early in the history of biochemistry.

The rock record can play an important role in informing these top-down approaches. For example, a recent study revealed putative microfossils associated with a hydrothermal vent that is at least 3.77 b.y. old (*Dodd et al.*, 2017). If correct, this would be the oldest evidence of life on Earth and support the view that thermophiles arose early. Rocks of similar age (3.95 b.y.) show carbon isotope fractionation consistent with biological CO_2 fixation (*Tashiro et al.*, 2017) and microbial structures resembling stromatolites (3.7 b.y.) that were growing in shallow marine environments (*Allwood*, 2016; *Nutman et al.*, 2016). While these results are controversial given the deformation and alteration of these rocks, they open up the possibility that the origin of life occurred more than 4 b.y. ago, i.e., predating the oldest preserved sedimentary rocks (*Pearce et al.*, 2018). Silicon and oxygen isotope data from Hadean zircons confirm that the hydrological and geological cycles that would have been necessary for an origin of life date back to at least 4.1 b.y. (*Trail et al.*, 2018). Other important discoveries from the rock record indicate that microbial sulfide oxidation and sulfate reduction occurred at 3.47 b.y. based on isotopic evidence and barite deposits (*Shen et al.,* 2001; *Wacey et al.,* 2011). The source of the electron acceptors involved in the oxidation of sulfides is inferred to be CO_2, possibly indicating anoxygenic photosynthesis involving sulfides as the electron donor (*Farquhar et al.,* 2000). The time of the origin of oxygenic photosynthesis is thought to be between 3.1 and 2.7 b.y., based on geochemical evidence for manganese and ammonium oxidation (*Koehler et al.,* 2018; *Planavsky et al.,* 2014).

4. BOTTOM-UP APPROACHES

The goal of the "bottom-up" approach is to understand the steps leading to the "unity of biochemistry" through laboratory experiments and theoretical thermodynamics. Much of this work has focused on principles of organic chemistry and chemical synthesis. For example, experiments focused on the synthesis of prebiotic molecules have led to continued discussion regarding whether the steps leading to early prebiotic molecules were targeted through specific catalysis and sparsity, yielding specific reaction products in high concentrations (*Copley et al.,* 2007); or whether early life was characterized by "messy chemistry" from which complex chemical networks emerged (*Guttenberg et al.,* 2017). Similarly, recent work from experimental organic chemistry has provided key insights into the likely reaction steps for the synthesis of ribonucleotide bases, which are thought to have formed the basis of the RNA world (*Powner et al.,* 2009; *Hud et al.,* 2013).

After nucleotide synthesis, those nucleotides must have been polymerized in order to form the informational molecules that later formed the backbone of unity of biochemistry. Some experimental work suggests that mineral catalysis on clays may have played a key role (*Ferris,* 2006; *Hazen and Sverjensky,* 2010). Later, as prebiotic chemistry merged into early living systems, natural selection most likely began to operate on early networks of prebiotic molecules, possibly on self-replicating molecules or networks that eventually included RNA molecules. Bottom-up work focused on modeling early prebiotic replication networks has revealed insights into possible mechanisms for early replication and has emphasized the importance of heredity through encapsulation or localization of networks in physical space (*Takeuchi et al.,* 2017; *Bansho et al.,* 2016). Much of this work has focused on experimental organic chemistry and modeling approaches, but to mature and advance the field, these experiments should be reconciled with evidence from the rock record, models of the early Earth, and investigation of modern analog environments to determine whether the reactions proposed in the laboratory and computer can take place in feasible prebiotic conditions. Below we outline the most likely sources of prebiotic compounds, nutrients, and catalysts on the early Earth to constrain the most likely conditions under which the origin and early evolution of life may have occurred.

5. THE ENVIRONMENTAL CONTEXT OF THE HADEAN EARTH

Reconstructing environmental conditions on the early Earth is essential for understanding which settings were present that could have hosted important prebiotic reactions. One key factor is the geological activity, which controls the extent of land masses and the degree and distribution of volcanism (section 5.1.). Abiotic cycles of essential elements (CHNOPS and metals) and their incorporation into biomolecules build upon these planetary parameters (section 5.2).

5.1. The Geological Setting of the Hadean Earth

The origin of life on the Hadean Earth has often been viewed within the broad tectonic framework of the modern Earth with its characteristic horizontal motion of large oceanic and continental plates, bounded by mid-ocean ridges and subduction zones. In this setup, exposed landmasses of buoyant felsic (Si-rich) crust offer extensive stable platforms for non-marine settings, hydrothermal activity in the ocean is largely restricted to plate boundaries, and subduction of oceanic crust recycles essential nutrient elements to the ocean and atmosphere. However, multiple lines of evidence are beginning to paint a more dynamic picture of the early Earth. It has long been argued that Earth's mantle was several hundred degrees hotter at a time when radioactive elements were still more abundant and residual heat from accretion and core formation was higher (e.g., *Herzberg et al.,* 2010). Under these conditions heat loss by modern-style plate tectonics would likely have been inefficient. A growing number of geodynamic models and field observations therefore suggest that for more than a billion years into Earth's history, heat loss was largely accommodated by plume volcanisms that led to voluminous outpourings of lava similar to oceanic plateaus and flood basalts (*Debaille et al.,* 2013; *Moore and Webb,* 2013). The crust would have been overall more mafic (Fe-Mg-rich) in composition (*Lee et al.,* 2016; *Smit and Mezger,* 2017). Some felsic continental crust existed as evidenced by the geochemical zircon record (*Boehnke et al.,* 2018), but time series analyses suggest that the volume of felsic continents that characterize our planet today was probably significantly smaller (*Dhuime et al.,* 2012). Rather than being fragmented into distinct rigid plates, Earth's surface may have been largely covered by a single stagnant lid, perhaps comparable to Venus (*Solomatov and Moresi,* 1997; *Lenardic and Crowley,* 2012; *Moore and Webb,* 2013; *O'Neill and Debaille,* 2014; *Rozel et al.,* 2017). Plate subduction possibly occurred locally or intermittently (*O'Neill and Debaille,* 2014), but modern-style plate tectonics probably did not start until 3.0–2.5 b.y., when the mantle had cooled sufficiently (*Dhuime et al.,* 2012, 2015; *Laurent et al.,* 2014; *Condie,* 2016; *Tang et al.,* 2016; *Nebel-Jacobsen et al.,* 2018). In other words, the origin and earliest evolution of life most likely occurred in a different tectonic regime than what we experience today. This milieu evidently created environmental conditions that were favorable for abiotic organic chemical reactions, likely in part because it resulted in a diverse suite of catalytic minerals that may have formed the backbone of numerous protometabolic reactions (*Hazen,* 2013, 2017). A summary of these processes is depicted in Fig. 3.

5.1.1. Widespread hydrothermal activity on the Hadean seafloor.
In the modern ocean, hydrothermal convection is focused along mid-ocean ridges, where it is driven by heat from magma that rises as a consequence of mantle convection. Such rift-related hydrothermal fields can extend over kilometers as shown by examples from the more recent rock record (e.g., *Martin-Izard et al.,* 2016). In contrast, abundant plume volcanism on the Hadean Earth driven by a hotter mantle would have caused extrusions of large volumes of hot lava over extensive areas within a relatively short time. For example, the Cretaceous Ontong-Java Plateau (~120 m.y.) — a large igneous province in the southern Pacific and the most recent analog of a voluminous submarine plume eruption — covered approximately 1% of Earth's surface with a thickness of more than 25 km (*Coffin and Eldholm,* 1994). Magmatic production rates likely exceeded those of the entire mid-ocean ridge system. Calculations suggest that hydrothermal fluids circulating between individual basalt flows of the Ontong-Java Plateau may have leached such large quantities of iron from the rock that algal productivity spiked globally and subsequent respiration drew down dissolved oxygen levels in seawater (*Sinton and Duncan,* 1997; *Kerr,* 1998). Additional oxygen was likely consumed by direct reaction with ferrous minerals and sulfides, leading to widespread anoxia. These estimates illustrate the strong impact of flood volcanism on global ocean chemistry. If a large number of such plume events occurred on the Hadean Earth, large parts of the ocean floor would at any given time have been covered by fresh extrusive rocks that were reactive to liquid water and driving vigorous hydrothermal convection. Indeed, geochemical data indicate a strong hydrothermal signature in early Archean seawater (*Viehmann et al.,* 2015) (section 5.2.3).

Extensive circulation of seawater through fresh piles of volcanic rocks would have created convection cells encompassing strong physicochemical gradients in temperature, pressure, and chemical composition. Gradients generate thermodynamic disequilibria, which are a fundamental requirement for chemical synthesis reactions (e.g., *Shock and Schulte,* 1998). Much of the hydrothermal activity in flood basalts was perhaps comparable to diffuse off-axis flow in modern oceanic crust with temperatures on the order of several tens of degrees Celsius (*Johnson and Pruis,* 2003), i.e., not as destructive to prebiotic macromolecules as axial black smoker fluids where temperatures can reach 400°C. Furthermore, these warm conditions may have favored the production of H_2 — an important reductant in (proto-) biochemistry (*Mayhew et al.,* 2013) (section 3), especially if oceanic crust was relatively more ultramafic in composition than it is today. Ultramafic rock reacts exothermically with water during the process of serpentinization, where Fe-rich olivine is altered to magnetite, serpentine, and brucite; H_2O

is reduced to H_2; and H^+ ions are consumed, leading to highly alkaline pH values. These conditions are relatively rare on the modern Earth (*Kelley et al.,* 2002, 2005), but in the Hadean, large parts of the ocean floor may have been undergoing serpentinization at any given time — if resurfacing events were more frequent and extensive and more ultramafic in composition. The resulting high pH leads to extensive deposition of calcium carbonate, which may have played an important role in chiral selection of organic building blocks (*Hazen et al.,* 2001; *Jiang et al.,* 2017). H_2 is a strong reductant that can react with Fe^{2+} and Ni^{2+} in basalts to form metallic FeNi alloys (*Klein and Bach,* 2009), which have been shown to be efficient catalysts for the reduction of CO_2 and N_2 to hydrocarbons, organic acids, and ammonium, respectively (*Brandes et al.,* 1998; *Lang et al.,* 2010; *Proskurowski et al.,* 2008; *Sousa et al.,* 2018; *Varma et al.,* 2018). Reactions catalyzed by these FeNi alloys may thus have played a central role in providing some of the essential monomers leading to more complex macromolecules. Another potentially important product of H_2-rich settings on the seafloor is iron carbide (*Preiner et al.,* 2018). It was recently shown that iron carbide also forms the core of the modern N_2 fixing enzyme nitrogenase, which could conceivably be a relic of ancient geochemical processes (*Preiner et al.,* 2018; *Lancaster et al.,* 2011).

Besides serpentinizing settings, hot magma-driven hydrothermal vents were also likely abundant along volcanic fissures on the seafloor. Examples of these are preserved from the Archean rock record (e.g., *Vearncombe et al.,* 1995). These settings are characterized by acidic pH and extensive deposition of sulfide minerals, which, like metal alloys, can serve as catalysts of CO_2 reduction in Fischer-Tropsch-type reactions (*Cody,* 2005; *McCollom and Seewald,* 2007; *Holm and Neubeck,* 2009). Metallic particles from these vents may have been transported for many kilometers into the open ocean (*Gartman et al.,* 2014) and served as catalysts even under low-temperature conditions (*Roldan et al.,* 2015).

Another key aspect of hydrothermal convection cells and ocean circulation in general is that they would not have existed in isolation. Like today, mixing would have occurred between atmospheric and hydrothermal reaction products. The Archean rock record shows evidence of photochemically-induced mass-independent sulfur isotope fractionation preserved in hydrothermal deposits (*Jamieson et al.,* 2013; *Siedenberg et al.,* 2016; *Aoyama and Ueno,* 2018), which is direct evidence of such mixing between atmospheric and magmatic compounds. Similarly, the abundance of iron formation on the continental shelf in the Archean rock record is evidence for the reverse transport pathway from the deep ocean to the surface, because the iron contained in these rocks (the major source of iron ore to modern economy) contains chemical signatures indicative of a hydrothermal source (*Isley and Abbott,* 1999). Hence it is also likely that other potential products of atmospheric reactions, including amino acids (*Miller,* 1953) and nitrogen oxides from lightning reactions (*Navarro-González et al.,* 1998), were carried down into the deep ocean and interacted with hydrothermal fluids while hydrothermal reaction products became entrained in surface waters. Such interactions between different environmental settings may have been essential for the formation of complex organic macromolecules with diverse functional groups (*Stüeken et al.,* 2013).

5.1.2. The early atmosphere. The early atmosphere has long been regarded as an important factor in the origin of life, because it may have hosted numerous key reactions induced by lightning and photolysis. Furthermore, the amount of greenhouse gases would have constrained global climate and hydrological cycles. Although the young Sun was significantly fainter than it is today (*Sagan and Mullen,* 1972), state-of-the-art three-dimensional climate models suggest that ice-free conditions could have been maintained with millibar quantities of CO_2 and CH_4 (*Charnay et al.,* 2013). CH_4 could have been generated in serpentinization reactions (*Proskurowski et al.,* 2008), and CO_2 was continuously degassed from volcanos. Reactions between dissolved CO_2 and mafic crust would have drawn down CO_2 as carbonate, but stacking of flood basalt provinces followed by delamination of the lowest stack and absorption into the mantle would have recycled this CO_2 efficiently even in the absence of modern plate tectonics (*Foley and Smye,* 2018). While some geochemical data support previous arguments for a hot climate on the early Earth (*Tartèse et al.,* 2017), new sedimentological observations suggests temporary episodes of glacial conditions at as far back as 3.5 b.y. ago (*de Wit and Furnes,* 2016). It is conceivable that climate fluctuated between extremes, similar to more recent times in Earth's history. Hence both warm and frigid conditions should be considered in origin of life scenarios.

In the presence of CO_2 and CH_4, photochemical reactions could have produced moderate amounts of HCN and formaldehyde (HCHO) in the upper atmosphere (*Tian et al.,* 2011; *Cleaves,* 2008), which can serve as an important reactant in the formation of amino acids and nucleotides (*Deamer,* 2007). Although the efficiency of this reaction has so far only been appraised in numerical models, sulfur isotopes indicate a strong impact of UV photolysis on the sulfur cycle back to at least 3.8 b.y. ago and likely earlier (*Thomassot et al.,* 2015; *Siedenberg et al.,* 2016), which highlights the potential importance of photochemistry on early element cycles.

The most widely-known atmospheric reaction in the origin of life is the synthesis of amino acids during lightning strikes (*Miller and Urey,* 1959). This discovery was undermined by studies showing that the yields of the reaction are much lower than previously thought, if the atmosphere had a relatively more neutral redox state (*Schlesinger and Miller,* 1983). However, conditions conducive to the production of amino acids could have been created in volcanic eruption clouds (*Johnson et al.,* 2008). Furthermore, recent models suggest that the early atmosphere may in fact have been richer in H_2 than commonly thought (*Wordsworth and Pierrehumbert,* 2013), which would make the original data directly applicable and maintain the idea that light-

ning reactions played a critical role in delivering organic monomers to Earth's surface.

In addition to gas-phase reactions, the atmosphere likely hosted myriad aerosols, in particular water droplets ejected from the surface ocean (*Dobson et al.,* 2000; *Tuck,* 2002). Several recent studies have revealed that molecular crowding, paired with changes in pH, temperature, and radiation could have induced chemical reactions inside such droplets, including the phosphorylation of organic compounds, which is an essential step toward the formation of RNA and DNA (*Vaida,* 2017). The organic precursors for these reactions may have been picked up from the sea surface microlayer, where hydrophobic organic molecules produced in hydrothermal vents or elsewhere may have become concentrated similar to an oil slick (*Lasaga et al.,* 1971). If so, then these aerosol reactors would represent another example of interactions between multiple environmental settings in the origin of life, including deep-marine hydrothermal circulation, the surface ocean, and the atmosphere.

5.1.3. Land surfaces. Land masses exposed to the atmosphere are sometimes invoked as important settings in the origin of life, because wet/dry cycles in evaporitic environments could have favored polymerization reactions of organic monomers (*Mamajanov et al.,* 2014; *Ross and Deamer,* 2016). Furthermore, protocells composed of fatty acids have been shown to form spontaneously in freshwater (*Segré et al.,* 2001; *Black et al.,* 2013). On the modern Earth, felsic continental crust provides a buoyant platform above sea level that covers roughly 30% of the globe. In contrast, multiple geochemical datasets suggest that the volume of felsic crust was less than half the modern value prior to 3 b.y. ago (reviewed by *Hawkesworth et al.,* 2016). Furthermore, the ocean volume may have been 25–100% larger (*Pope et al.,* 2012; *Korenaga et al.,* 2017), and a hotter mantle may not have been able to support tall mountain ranges (*Lee et al.,* 2018). Collectively, this could mean that the earliest continents were largely under water. However, this does not rule out the presence of dry land. In the absence of horizontal plate motion, it is conceivable that some volcanic islands grew to much higher altitude than today, perhaps akin to the martian volcano Olympus Mons, which is over 21 km in elevation. Such landmasses would have been largely mafic to ultramafic in composition, which would have favored widespread serpentinization in terrestrial settings (*Smit and Mezger,* 2017). Importantly, unlike today where H_2 production is restricted to subsurface settings that are devoid of O_2, H_2 production could have been widespread at the surface of the anoxic Hadean Earth (*Smit and Mezger,* 2017). Olivine-rich green sand beaches such as those found around modern Hawaii may thus have been significant sources of H_2, conceivably in direct contact with reagents raining out from atmospheric reactions. Some of the oldest well-preserved packages of sedimentary rocks on Earth from 3.5 b.y. ago have been interpreted as lacustrine hot spring deposits (*Djokic et al.,* 2017). Such terrestrial hot springs may have combined some of the features of deep marine hydrothermal vents and exposed wet/dry cycles on land (*Mulkidjanian et al.*, 2012).

5.2. Abiotic Nutrient Cycles

The geological framework of the early Earth sets the conditions for geochemical cycles of essential nutrients that eventually assembled into organic macromolecules: CHNOPS and various minor elements, including transition metals (see the chapter by Hoehler et al. in this volume). Carbon would have been abundantly available as volcanic CO_2, which could be converted into simple organic compounds in Fischer-Tropsch-type reactions under hydrothermal conditions (*McCollom and Seewald*, 2007; *McDermott et al.*, 2015) and in Miller-Urey-type reactions in the atmosphere (*Miller and Urey*, 1959). Additional organic compounds may have been delivered by comets and meteorites (*Chyba and Sagan*, 1992). Hydrogen and oxygen would have been supplied by liquid water; reactive forms of sulfur could be sourced from volcanogenic SO_2 and H_2S gases (*Farquhar et al.*, 2000), which may have become incorporated into organic compounds under hydrothermal conditions (*Martin and Russell* 2007). More uncertain are the availability of reactive phosphorus and nitrogen species as well as transition metals. Nitrogen is an integral component of amino acids and nucleobases, while phosphorus is contained in some lipids and forms the backbone of DNA and RNA. Both elements are often limiting nutrients on the modern Earth.

5.2.1. Phosphorus. Phosphorus has an average crustal abundance of 654 ppm (*Rudnick and Gao*, 2014) but is mostly contained in apatite minerals, which are only sparingly soluble in water and thus limit the flux of phosphorus to the ocean (*Filippelli*, 2008). Although models suggest that this flux could have been up to 10 times higher on the early Earth where weathering occurred under more acidic conditions (*Hao et al.*, 2017a), dissolved phosphate levels were probably still lower in Hadean seawater than today (2 μM) for several reasons. Most importantly, freely dissolved phosphate would likely have been scavenged from anoxic water columns by adsorption to iron oxides and carbonate minerals (*Jones et al.*, 2015; *Reinhard et al.*, 2017). Furthermore, aerobic recycling of biomass, which sustains the marine phosphate reservoir today and greatly extends the residence time of phosphate in seawater, would have been strongly suppressed prior to the emergence of respiratory metabolisms (*Kipp and Stüeken*, 2017). Life's dependence on phosphorus for essential macromolecules therefore poses a geochemical conundrum that has led to the suggestion that the earliest macromolecules may have relied on thioesters instead of phosphate bonds for energy storage (*Goldford et al.*, 2017). Another possible solution may have been the formation of dissolved phosphite (PIII) from phosphate (Pv) under hydrothermal conditions (*Herschy et al.*, 2018). Phosphite is markedly more soluble and reactive than phosphate and may thus have participated more readily in prebiotic reaction networks that ultimately led to

the phosphorylation of key organic compounds (*Pasek et al.*, 2017). Importantly, however, the formation of phosphite by this mechanism would have been limited by the relatively small Hadean phosphate reservoir circulating through hydrothermal systems. Hence dissolved phosphite levels were probably still lower than modern marine phosphate levels.

5.2.2. Nitrogen. The largest reservoir of nitrogen at Earth's surface is atmospheric N_2 gas, which is unreactive under standard conditions. While it was once thought that the Hadean atmosphere might have been rich in ammonia (NH_3) (*Sagan and Mullen*, 1972), this model has fallen out of favor, because NH_3 is unstable under strong UV radiation (*Kuhn and Atreya*, 1979). It might be protected under a UV-shielding organic haze in a CH_4-rich atmosphere (*Wolf and Toon*, 2010), but it is uncertain if such a haze could develop prior to the origin of life. In the absence of an NH_3-rich atmosphere, incorporation of nitrogen into prebiotic organic molecules would thus have depended on abiotic N_2 fixation pathways that continuously generated dissolved ammonium (NH_4^+). Several such pathways have been proposed, including mineral-catalyzed reduction of N_2 to NH_4^+ under hydrothermal conditions (*Brandes et al.*, 1998; *Schoonen and Xu*, 2001; *Smirnov et al.*, 2008), including serpentinization conditions (*Ménez et al.*, 2018), photochemical production of HCN by UV radiation (*Zahnle*, 1986; *Tian et al.*, 2011), and volcanic-, lightning- or impact-induced production of NO_x species from the reaction of N_2 and CO_2 (*Kasting*, 1990; *Navarro-González et al.*, 1998; *Mather et al.*, 2004), followed by reduction to NH_4^+ in hydrothermal settings or on green rust (*Summers and Chang*, 1993; *Brandes et al.*, 1998; *Hansen et al.*, 2001). The major sinks of nitrogen from the ocean in the absence of biological consumption would have been NO_x reduction to N_2, dissociation of NH_4^+ into NH_3 and adsorption of NH_4^+ onto clay minerals (*Stüeken*, 2016). Recent steady-state models suggest that the standing reservoir of dissolved NH_4^+ and NO_x may have been up to 10 μM and 1 μM, respectively (*Stüeken*, 2016; *Wong et al.*, 2017; *Laneuville et al.*, 2018). These values are within an order of magnitude of the 30 μM of NO_3^- that sustains a large biosphere in the modern ocean and could thus conceivably have been sufficient to maintain a constant production of ammoniated organic molecules. Importantly, NH_4^+ can react with reduced carbon and form HCN — the potential building block of nucleotides and amino acids — under hydrothermal conditions (250°–325°C) catalyzed by zeolite minerals, which are common alteration products of basaltic glass (*Holm and Neubeck*, 2009). Nitrogen may only have become a truly limiting resource once life emerged and actively scavenged it from seawater.

5.2.3. Transition metals. Besides the major elements (CHNOPS), many first-row transition metals fulfill key functions in modern biochemistry, often as catalytic centers in proteins. Phylogeny suggests that many of these metalloenzymes are deeply rooted in the tree of life and may have played an essential role in prebiotic chemistry (*Dupont et al.*, 2010; *Schoepp-Cothenet et al.*, 2012). Sources of metals on the Hadean Earth may thus have placed strong constraints on the origin of metabolism. Today, most of the metals entering the ocean are liberated due to oxidative weathering, and they are relatively stable in oxic seawater in the absence of dissolved sulfide. In contrast, thermodynamic models of the anoxic early Earth once predicted markedly lower concentrations of transition metals in seawater, based on the assumption that metals would have been efficiently scavenged into sulfide minerals throughout the early ocean (*Saito et al.,* 2003). However, recent compilations of metal concentrations in Archean sediments revealed relatively high levels of Ni, Cr, Co, and Zn in the marine realm (*Konhauser et al.,* 2009; *Scott et al.,* 2012; *Robbins et al.,* 2013, 2016; *Swanner et al.,* 2014). Isotopically fractionated Mo in 2.95-b.y.-old sedimentary rocks indicate non-trivial levels of dissolved Mo in seawater at that time (*Planavsky et al.,* 2014). A plausible explanation for these patterns is a high influx of metals from hydrothermal sources (*Robbins et al.,* 2016). Indeed, large-scale leaching of transition metals has been documented from a hydrothermally altered Archean seafloor (~3.24 b.y.), and mass balance calculations suggest that a large proportion of the leached metals was vented into the open ocean (*Brauhart et al.,* 2001; *Huston et al.,* 2001). Today, most transition metals emanating from hydrothermal vents precipitate in sulfide minerals near the vent site, or are adsorbed to newly formed iron-manganese oxides. However, prior to the onset of oxidative sulfate delivery to the ocean around 2.75 b.y. ago (*Stüeken et al.,* 2012), sulfate concentrations in seawater would have been markedly lower, which likely decreased the concentrations of sulfur in hydrothermal fluids and led to relatively smaller sulfide deposits around hydrothermal vents, allowing leached metals to mix with seawater. Even on the modern Earth, hydrothermal metal emissions can be traced for hundreds of kilometers (*Wu et al.,* 2011; *Fitzsimmons et al.,* 2014; *Resing et al.,* 2015), and complexation by organic matter produced in Fischer-Tropsch-type reactions may have further stabilized the dissolved metal load, similar to today (*Sander and Koschinsky,* 2011).

New independent evidence for a strong hydrothermal imprint on the early ocean comes from europium anomalies in rare Earth element (REE) patterns from ancient sediments (*Viehmann et al.,* 2015). Unlike other REEs, Eu can be reduced from a 3+ to a 2+ oxidation state under hydrothermal conditions where the temperature exceeds 200°C (*Michard and Albarede,* 1986). This reduction step alters the chemical behavior of Eu and causes an anomaly in REE concentrations archived in chemical sedimentary rocks. These Eu anomalies are largest in 3.5–3.8-b.y.-old samples, and they show another peak in the Neoarchean (2.7–2.5 b.y.), concurrent with the appearance of abundant flood basalts (*Viehmann et al.,* 2015; *Condie,* 2016). These data suggest that plume volcanism may have imparted a strong hydrothermal signature on the composition of Archean and Hadean seawater. If ancient oceanic crust was more ultramafic than it is today, this may explain particularly high abundances of Ni, Cr, and Co, which are relatively more abundant in ultramafic rocks (*Nna-Mvondo and Martinez-Frias,* 2007).

In addition to this hydrothermal source, some metals would also have been liberated during anoxic, acidic weathering of land-masses. Under high atmospheric CO_2 levels, rainwater would have been more acidic, thus accelerating chemical weathering rates (*Hao et al.,* 2017a,b). Models and paleosol data suggest elevated riverine fluxes of Mn, Co, Ni, Zn, and W (*Hessler and Lowe,* 2006; *Murakami et al.,* 2016; *Hao et al.,* 2017a). However, if land masses were overall significantly smaller, then this flux was probably less important for dissolved metal concentrations in the open ocean.

6. STEPS LEADING TO THE FIRST "ORGANISMS"

Understanding the settings for the origin of life is not only important for providing a realistic hypothesis for the origin of life on Earth, but is also important for astrobiological studies aiming to determine the likelihood that life could have originated on other planetary bodies. Given what we know of the Hadean Earth, robust models for an independent origin of life must consider the environmental context for each of the five key steps in the bottom-up scenario (see also *Kitadai and Maruyama,* 2017). These steps include (1) the synthesis of nucleotide bases, amino acids, and lipids; (2) the polymerization of these compounds into complex molecules, culminating in RNA and DNA, proteins, and fatty acids; (3) harnessing a redox or pH gradient to power early metabolism; (4) compartmentalization; and (5) the subsequent evolution of the first cells into LUCA (Table 1). Although some variants of these steps may have built up on each other and thus be protobiological rather than geochemical inventions, at the most fundamental level each of these processes required suitable host environments and physicochemical linkages, which poses some constraints on origin of life models. Many models attempt to combine most or all of these steps into a single environmental setting such as a "warm pond" with wet/dry cycles, hydrothermal vents, or terrestrial volcanic environments. Alternatively, the entire Hadean Earth could be considered as a "global chemical reactor," in which separate environmental settings were linked through physical exchange of reactants and products (Fig. 4) (*Stüeken et al.,* 2013). Such transport processes include ocean and atmospheric circulation, rainout into the surface ocean and onto land surfaces, river runoff, and hydrothermal convection through the crust. As noted above (section 5.1.1), there is direct geochemical evidence from the Archean rock record that sulfur compounds were transported from the upper atmosphere into deep-sea hydrothermal vents, while ferrous iron and other metals traveled from deep-sea vents to the continental shelf. Hence these mixing processes are inescapable for any realistic origin of life scenario. Transportable molecules can include dissolved species as well as adsorbed compounds on mineral particles. Organic matter has a strong affinity to adsorb to clay minerals, which constitute a major shuttle of biomass into deep-sea sediments even today (*Playter et al.,* 2017; *Mayer,*

TABLE 1. Key steps leading to the origin of life on Earth.

	Likely Sources/ Necessary Conditions	Possible Settings	References
Synthesis of nucleotide bases	UV radiation, dry-wet cycles, pH ~3–6	Lagoons, ponds on land	*Powner et al.* (2009), *Stairs et al.* (2017)
Synthesis of lipids	Fischer-Tropsch-type reactions; delivery by meteorites	Hydrothermal vents; surface settings	*McCollom et al.* (1999), *Deamer* (1997)
Synthesis of amino acids	Spark discharge (Miller-Urey synthesis, requiring lightning), synthesis in hydrothermal solutions	Hydrothermal vents; surface settings	*Hennet et al.* (1992), *Amend and Shock* (1998), *Johnson et al.* (2008), *Parker et al.* (2011)
Polymerization	Clays and other mineral surfaces of nucleotides	Ocean floor, lake beds	*Ferris et al.* (1996), *Ferris* (2006), *Baaske et al.* (2007)
Harnessing a proton gradient (metabolism)	Redox gradient	Hydrothermal vents	*Martin and Russell* (2007)
Compartmentalization	Lipid micelles/amphiphiles OR FeS "bubbles"	Lakes/ponds OR hydrothermal vents	*Deamer* (1997), *Koonin and Martin* (2005)
Last Universal Common Ancestor (LUCA)	Hydrogen for metabolism, FeS for protein cofactors, CO_2 for C fixation	Hydrothermal vents?	*Weiss et al.* (2016)

1994). Lacustrine settings would have received significant input from the atmosphere (section 5.1.2) and some input from marine incursions. These reactants may have undergone further reactions during wet/dry cycles. Lacustrine products that were relatively more resistant to UV damage or protected by protocells or adsorption to mineral surfaces would eventually have been washed into the ocean and accumulated in seawater. Deep marine settings likely received a relatively "filtered" selection of transportable atmospheric and lacustrine compounds from the seawater reservoir. High-temperature vents contributed metals, reactive phosphorus, ammonium, and Fischer-Tropsch-type (FTT) organics to the marine reservoir (section 5.1.1). Lower-temperature crustal settings could selectively accumulate organics from various sources and perhaps facilitated polymerization reactions in dehydrated rock fractures (section 6.2).

6.1. High-Energy Organic Synthesis Reactions

The most important processes for the production of simple organic carbon molecules on the Hadean Earth were probably photochemical and lightning reactions in the atmosphere and mineral-catalyzed CO_2 reduction under hydrothermal conditions. Together, these sources would have produced a diverse suite of molecular building blocks that would have ultimately led to key molecules for life. While atmospheric reactions have been noted for the production of HCN, formaldehyde, amino acids, and nucleobases (*Miller and Urey,* 1959; *Johnson et al.,* 2008; *Deamer,* 2007; *Cleaves,* 2008; *Ferus et al.,* 2017), deep-marine vents likely contributed alkanes, metal-complexes, and diverse functional groups that may have acted as protometabolic intermediates (*Muchowska et al.,* 2019). As noted above, ammonium,

which is produced from NO_x and N_2 in hydrothermal settings, can react with reduced carbon at high temperature to form organic amines (*Holm and Neubeck,* 2009; *Barge et al.,* 2019). A recent study demonstrated the abiotic production of the amino acid tryptophan in the marine serpentinization environment known as Lost City (*Ménez et al.,* 2018). The synthesis of tryptophan is dependent on two chemical reactions that have important origin of life implications. These include the findings that (1) nitrogen was reduced to ammonia by the iron-clay mineral saponite, previously not found at Lost City, and (2) indole, a heterocyclic nitrogen compound that is an intermediate in the synthesis of tryptophan, was abiotically synthesized. The abiotic synthesis of heterocyclic amines such as indole opens the possibility that purines and pyrimidines, components of RNA and DNA, might also be synthesized abiotically in marine serpentinization environments (*Baross,* 2018), especially when considering that hydrothermally generated phosphite would have been reactive toward organic carbon (*Pasek et al.,* 2017). The production of reduced organic carbon from CO_2 by Fischer-Tropsch-type reactions as well as subsequent additions of metals, ammonification, phosphorylation, and sulfurization reactions have all been demonstrated to be thermodynamically feasible under hydrothermal conditions, aided by mineral catalysts (*Martin and Russell,* 2007; *Holm and Neubeck,* 2009; *Pasek et al.,* 2017; *Preiner et al.,* 2018; *Sousa et al.,* 2018; *Varma et al.,* 2018). It is likely that the Hadean ocean was enriched in small organic carbon molecules that were continuously dispersed from hydrothermal sources, given the abundance of hydrothermal vents on the Hadean seafloor inferred from the europium record in ancient sediments (*Viehmann et al.,* 2015) and the distinct signatures of these vents on seawater chemistry. These latter signatures can be traced for hundreds of kilometers even in the modern ocean (*Wu et al.*, 2011; *Fitzsimmons et al.*, 2014; *Resing et al.*, 2015). Mixing would have occurred with compounds sourced from the atmosphere, as evidenced by sulfur isotope data that indicate transport of photochemical products from the Archean atmosphere into the deep ocean (section 5.1.1). Collectively, this organic carbon inventory — present as dissolved or mineral-sorbed molecules — would have cycled through low-temperature diffuse hydrothermal vents and washed up onto beaches around exposed land masses (Figs. 3 and 4). It is important to note that some of these hydrothermal synthesis reactions require temperatures above 100°C, which is not conducive to the stabilization of polymeric organic compounds. Hence sites of organic monomer production were probably spatially separated from sites of polymerization.

6.2. Polymerization Reactions in Dehydrating Settings

While the production of monomers could have occurred in high-energy environments in deep-marine hydrothermal vents and in the atmosphere, settings for the polymerization of organic monomers are more tightly constrained by the requirement of dehydration, i.e., the removal of hydroxyl groups for molecular bond formation. Furthermore, polymerization of a chain of amino acids or nucleotides, for example, requires a process that selectively enriches these compounds within a small volume from a diverse mixture of organic monomers. A plausible mechanism is adsorption of monomers onto mineral templates, clay minerals in particular (*Ferris*, 2005; *Deamer et al.*, 2006; *Hazen and Sverjensky*, 2010). Wet/dry cycles on land have been invoked as a way to concentrate organics within sediments and induce dehydration and polymerization reactions (*Mamajanov et al.*, 2014; *Ross and Deamer*, 2016). In the marine realm, under hydrothermal conditions, only short oligomers have so far been synthesized (*Burcar et al.*, 2015). However, it is conceivable that more extensive polymerization can occur deeper within warm crust where serpentinization reactions effectively consume water as it penetrates into the rock along fractures (*Kelemen and Hirth*, 2012). This reaction leads to a volume expansion of the rock that cuts off the resupply of water and results in water-limited reaction halos within the rock. Multiple phases of cracking and sealing has been documented from the ultramafic rock record (*Cathelineau et al.*, 2017). It is an intriguing yet unexplored possibility that this setting may have facilitated the dehydration and polymerization of entrained organic compounds that were then later re-released through new fractures.

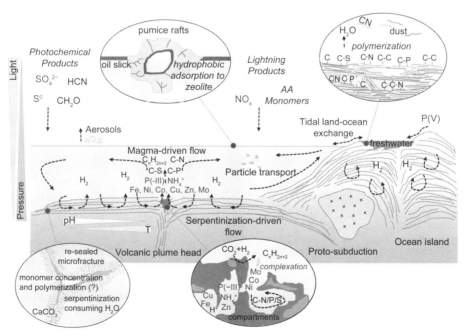

Fig. 3. See Plate 5 for color version. Schematic cross section of Hadean surface environments. Extensive plume volcanism as inferred from the early rock record (*O'Neill and Debaille*, 2014) would have led to widespread hydrothermal circulation in the crust (black arrows). Magma-driven hydrothermal vents may have formed proximal to volcanic fissures whereas distal circulation may have been driven by serpentinization of fresh (ultra-)mafic lava flows. Also, land masses were likely more mafic in composition and supporting H_2 production through serpentinization (*Smit and Mezger*, 2017). Major localities for the production of simple organic molecules were probably lightning and photochemistry in the atmosphere (*Tian et al.*, 2011; *Miller and Urey*, 1959) and Fischer-Tropsch-type reactions in high-temperature hydrothermal vents (*McCollom and Seewald*, 2007). The latter would also have released reactive forms of nitrogen, phosphorus, and transition metals. Polymerization into larger more complex molecules likely occurred in lower-energy environments such as evaporitic lakes and perhaps in dehydrated fractures in serpentinized crust. It is unlikely that monomers and polymers were produced in the same environmental setting, but the respective settings were likely linked by numerous transport processes, including adsorption onto suspended minerals with diverse surface properties.

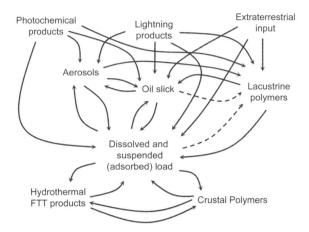

Fig. 4. Illustration of physical linkages between important sources and reservoirs of organic compounds. Transport processes include water circulation, river runoff, and rainout, and affects dissolved as well as adsorbed molecules (section 6).

In any case, polymerization would probably not have been possible under the same conditions as organic monomer synthesis, which occurred near high-temperature hydrothermal vents favoring abiotic CO_2 reduction, sulfurization, ammonification, and phosphorylation (section 5.1). Instead, the formation of stable polymers may have required relatively cooler temperatures and a more quiescent environment. The most limiting factor in the formation of RNA in particular is the provision of ribose sugar, which is unstable in water. The most promising solution to this problem is the stabilization of sugar molecules by borate ions, which are relatively more abundant in alkaline streams and serpentinizing vents (*Ricardo et al.*, 2004; *Grew et al.*, 2011). This requirement may further narrow down the number of possible settings for the formation of hereditary polymers.

6.3. Electrochemical Energy Sources

All modern biochemistry is ultimately powered by proton gradients that are established through metabolic redox reactions (*Schoepp-Cothenet et al.*, 2013). The universality of this strategy attests to its antiquity, suggesting that it may date back to the roots of metabolism itself. Multiple sources of electrochemical energy probably existed on the Hadean Earth. In the deep ocean, the H_2/CO_2 redox couple may have been most important, as H_2 would have been abundantly produced in serpentinizing hydrothermal vents that were likely widespread on the Hadean seafloor (section 5.1.1). The H_2/CO_2 redox couple provides a relatively small amount of energy, but it is sufficient for sustaining autotrophic methanogens and acetogens through the Wood-Ljungdahl pathway and may have been the root to autotrophic carbon-fixation (section 3) (*Fuchs*, 2011; *Martin et al.*, 2014; *Preiner et al.*, 2018; *Raymond et al.*, 2004; *Weiss et al.*, 2016). Recent studies have further highlighted the potential importance of Fe^0/CO_2 coupling in hydrothermal

settings that are often enriched in metallic iron and other metals (*Sousa et al.*, 2018; *Varma et al.*, 2018).

In contrast, shallow marine and terrestrial settings would perhaps have hosted relatively higher abundances of photochemical and lightning reaction products, including sulfate and nitrogen oxides that rained out from the atmosphere (*Farquhar et al.*, 2000; *Navarro-Gonzalez et al.*, 1998). Their reduction could have been coupled to the oxidation of H_2 generated in terrestrial serpentinizing settings (*Smit and Mezger*, 2017). As the Archean record of sulfur isotopes and iron formations provides direction evidence of transport from the atmosphere to the deep ocean and vice versa (section 5.1.1), it is likely that also many other solutes, gases, and adsorbed compounds were cycled between environmental settings (Fig. 4). In surface environments, the reduction of atmospheric nitrogen oxides coupled to the oxidation of hydrothermally-sourced ferrous iron could have provided significant amounts of electrochemical energy (*Schoepp-Cothenet et al.*, 2013). In fact, nitrogen oxide reduction has been proposed as another important metabolic pathway in early life that may date back to protometabolic networks (*Ducluzeau et al.*, 2009; *Nitschke and Russell*, 2013). If shallow-water settings received greater contributions of atmospheric redox couples while the deep ocean was relatively more dominated by hydrothermal products, then it is conceivable that the early ocean displayed a chemocline with a relatively more oxidized surface cap and highly reducing bottom waters. One may speculate that the diversity of electrochemical energy sources on the Hadean Earth may have triggered an early diversification of metabolic strategies that would ultimately have led to increased organismal diversity and increased ecological stability for the young biosphere.

6.4. Compartmentalization and the First Cells

While much of the discussion in origin of life research focuses on the synthesis and polymerization of early prebiotic compounds and how this might have occurred within a realistic environmental context, less emphasis is generally placed on the environmental settings of the subsequent prebiotic reactions, in which self-replicating entities arose that obeyed the rules of natural selection. At some point, a self-replicating entity that was contained within a compartment must have formed. Central to the origin of the first replicator is the RNA world hypothesis, in which a simple RNA strand arose spontaneously and increased in complexity over time (*Gilbert*, 1986). However, these early self-replicating entities needed to maintain heritability and avoid parasitism, or the takeover of replicating networks by selfish molecular entities. Most researchers have proposed that early replicating cycles could have maintained heritability and limited the impact of parasites through limited diffusion or compartmentalization *(Eigen and Schuster,* 1979; *Szathmáry and Demeter,* 1987; *Szabó et al.,* 2002; *Bansho et al.,* 2016)*, which may place some environmental constraints on the requirements for the RNA world.

As a result of the need to maintain heritability and avoid parasitism, many researchers have focused on understanding how early life would have become compartmentalized. Some have suggested that membranes were crucial early on and therefore that the first self-replicating networks must have arisen in a hydrating/dehydrating freshwater pool where such membranes are stable (*Deamer,* 2017; *Segré et al.,* 2001). Others have argued that diffusion could have been limited through association with clays or other minerals (*Koonin and Martin,* 2005; *Martin and Russell,* 2007; *Baross and Martin,* 2015). In any case, the first genomes may have formed from these associated self-replicating networks, eventually leading to the first independent "organisms."

Importantly, in its early stages and prior to the origin of enzymatic CO_2 fixation, this "organism" would have depended on a continuous supply of organic substrates, including specific nucleotides and perhaps amino acids, but excluding potential chemical interferents. This process may have required a network of protometabolic cycles and mineral templates that "selected" specific reactants over others (*Copley et al.,* 2007). Perhaps the most productive form of such a protometabolic network was one in which the replicator provided progressively more positive feedback in the form of a molecular template, similar to modern RNA and DNA, which could be copied and then form the basis for new copies. Such a feedback mechanism may, for example, have occurred in a seafloor circulation system where reagents were cycled along gradients and selectively enhanced and suppressed through a combination of physicochemical factors, minerals, and organic templates. Once interactions between organic compounds took increasing control of these selection processes, a feedback loop would have been established that could perhaps have been the first step toward a co-evolution of replication and metabolism. Until this stage in life's evolution, we can only infer likely settings and processes through bottom-up experiments; it is at this point, when life approached the LUCA, that top-down approaches can begin to inform our hypotheses about life's earliest stages.

6.5. Linking Metabolism and Replication

While protometabolic networks were necessary to generate a replicating RNA capable of evolving, it remains unclear how and under what environmental conditions metabolism and RNA merged. An important sequence of unknown events that stems from the RNA-world hypothesis is how to initiate the evolutionary sequence of events that would lead to the ribosome, the universal genetic code, and the "unity of biochemistry." Exchange of genetic "ideas" via horizontal gene transfer may have been necessary to select the best-fit genes and build a genome with a sufficient number of genes to eventually result in the emergence of free-living organisms (*Baross and Hoffman,* 1985; *Woese,* 1998). Perhaps an alternative scenario is the emergence of a community of RNA pre-cells co-evolving with small RNAs capable of transferring genes in a way similar to transposable elements. This scenario would require an interacting community of RNAs embedded in an environment with protometabolic catalysts for continuous RNA synthesis and selection of base sequences that could result in a genetic code, catalytic replication of RNA, RNA-protein interaction, and eventually the origin of the ribosome and self-replicating RNAs with increasing size and complexity. *Baross and Martin* (2015) have postulated the concept of the "ribofilm" as a potential early stage to the evolution of LUCA and eventually to "free-living" organisms. The idea is based on a known archaeal biofilm where horizontal gene transfer is a dominant factor in physiological adaptation (*Brazelton et al.,* 2011). Archaeal biofilms might serve as a model system to better understand the selection systems that led to the "unity of biochemistry" and the genome of LUCA. This model would also postpone the emergence of "free-living" encapsulated organisms and potentially imply a spatially vast film of pre-cellular RNA with the potential to perhaps lead to DNA genomes in which the community of nucleic acid chromosomes would be mutually dependent for survival, replication, and increasing evolutionary complexity. The LUCA may have emerged from such a nucleic acid film.

7. A GLOBAL CHEMICAL REACTOR

Increasing knowledge about the likely environmental conditions on the Hadean Earth, as described above, make it increasingly likely that no single setting provided sufficient prebiotic ingredients or conditions to satisfy all of the five steps that eventually led to the development of the first living organism. *Stüeken et al.* (2013) originally proposed four hypotheses as a means to synthesize the various chemical, geological, and environmental parameters required for a *de novo* origin of life:

- Multiple environments contributed to the building blocks of life, and these niches were not necessarily inhabitable by the first organisms
- Mineral catalysts were the backbone of prebiotic reaction networks that led to modern metabolism
- Multiple local and global transport processes were essential for linking reactions occurring in separate locations
- Global diversity and local selection of reactants and products provided mechanisms for the generation of most of the diverse building blocks necessary for life

Updated information about our deep phylogenetic history and Hadean geology as reviewed above support these hypotheses and strengthen the conclusion that a global chemical reactor with diverse environmental settings was not only an inescapable reality but may have been a necessary requirement for life to emerge on Earth. The global context for the origin of life greatly expands the likelihood of productive mineral-catalyzed reactions (*Hazen,* 2017), and it opens up the possibility that all key building blocks could be generated over a relatively short time span through contributions from numerous environmental settings (*Kitadai and Murayama,* 2017). If life is envisioned as a natural outgrowth of geochemical cycles rather than the outcome of a long sequence of unlikely events, then one secret to the success of the earliest biosphere was perhaps its inherent adaptation to the great diversity of environmental conditions from which it arose. In the same vein, an independent origin of life may depend on, and reflect, the diverse environmental settings present on another planetary body.

The traditional "bottom-up" approach is based on the assumption that all the necessary precursor organic and inorganic chemicals for the formation of nucleic acids and nucleotides would have accumulated in an environment in which wet/dry, temperature gradients, and minerals would catalyze the formation of long-chained polynucleotides including complementary strands of which one, upon melting, would yield a single-stranded, self-replicating ribozyme leading to the emergence of a living organism. In reality, all the above steps would have required a wide-range of environmental conditions that included multiple protometabolic catalytic reactions to produce the precursors, form polymers, and eventually progress to increasing biochemical complexity and an evolving organism (Fig. 4). Protometabolism and a living organism both have the same requirements for a sustained source of energy driven by chemical disequilibrium and initially driven by geological processes. Furthermore, both depend on the presence of key elements including transition metals to provide catalytic potential for synthesis of organic material (*Cody and Scott,* 2007). Initially, would the protometabolic network be focused in one environmental setting, or dispersed in multiple environmental settings? If transition-metal minerals were the early source of catalysis even in protometabolic networks, then it is likely that multiple environmental settings were involved or that precursor components of the network formed in one or more locations and was transported to other locations. An implication is that the reaction network would span tens to thousands of kilometers (*Stüeken et al.,* 2013). As discussed in sections 5.1.1 and 6, such dispersal mechanisms and global mixing processes are widely recognized in the modern ocean and can be directly inferred from the isotopic record.

The broad range of environmental conditions that led to life were most likely decoupled from the narrow subset of the conditions of the first habitat for life. This is consistent with models indicating that early life and LUCA became established and evolved in hydrothermal environments that provided both the chemical energy such as hydrogen, thioesters, and ferrodoxin, and a pH-dependent proton gradient for generating organic energy compounds (*Martin,* 2012; *Martin and Russell,* 2007; *Martin and Thauer,* 2017; *Martin et al.,* 2014; *Nitschke and Russell,* 2009; *Preiner et al.,* 2018; *Russell et al.,* 2010; *Goldford et al.,* 2017; *Goldford and Segré,* 2018). The dynamic, multiple-environmental-conditions model also overlaps somewhat with the flow chemistry model for the synthesis of nucleotides from precursors like cyanide (*Ritson et al.,* 2018), and the multiple physical and geochemical conditions involved in the synthesis of amino acids (*Hennet et al.,* 1992), sugars (*Benner et al.,* 2010; *Ricardo et al.,* 2004), and organic acids (*McCollum and Seewald,* 2013; *Proskurowski et al.,* 2008; *Schrenk et al.,* 2013). Finally, one may speculate that a global chemi-

cal reactor during the origin of life may favor a "unity of biochemistry" on all planetary bodies where life arises. If multiple environmental settings are required to account for the specific organic and inorganic components of life, then it is more likely that the most fit lineage leading to life will dominate by selection at each setting, ultimately leading to a "unity of biochemistry" via natural selection both within and between environments. Some have argued that it is possible that more than one evolutionary lineage developed independently in different environmental settings that could be out of the physiological bounds of its sister lineage, resulting in a "shadow biosphere" (*Davies et al.,* 2009). Earth is currently our only model for life's origin and evolution. There are important reasons for the biochemistry that we are familiar with, some we know and some we don't understand, but Earth provides our knowledge base for how to conduct our search — even our search for "weird life" with different biochemistries and different energy sources (*National Research Council,* 2007).

8. INDEPENDENT ORIGINS OF LIFE ELSEWHERE

There are major differences between the multiple unknowns involved in the complex synthesis of life and the environmental conditions that could support Earth-like organisms. It is very likely that there are millions of planetary bodies outside our solar system that have liquid water and the geophysical processes to extract key elements, gases, and organic compounds from rock that can support Earth-like organisms, but do not have all the properties for a *de novo* synthesis of life. Icy moons in our solar systems appear to be such examples of planetary bodies that have the prerequisite conditions to support life but may not have had a *de novo* origin. Enceladus exhibits many of the chemical properties that have been observed in serpentinizing environments on Earth that do support life (*Waite et al.,* 2006, 2009, 2017; *Glein et al.,* 2015). It is of interest that a methanogen has been demonstrated to grow and produce methane under Enceladus-like conditions, including with the low levels of hydrogen recently detected by Cassini (*Taubner et al.,* 2018; *Waite et al.,* 2017). These environments have also been implicated as a possible setting for the origin of life, and particularly as support for the metabolism-first scenario for the origin of life (*Martin and Russell,* 2007; *Russell et al.,* 2010, 2014, 2017). However, these bodies lack many characteristics of the early Earth, including an ocean-atmosphere interface and exposed land masses. If environmental diversity was required for the origin of life on Earth, then icy moons may not be suitable for an independent origin. However, while Enceladus and possibly Europa may not be "living moons," they could provide insight into early chemical stages leading to life and add support to the hypothesis that hydrothermal systems played an essential role in the origin of life. It has also been suggested that Saturn's moon Titan could host extant life or could have been a site for the origin of life due to the abundance of organic compounds found there (*Fortes,* 2000; *McKay and Smith,* 2005; *Schulze-Makuch and Grinspoon,* 2005), although there is no liquid water on Titan and thus any Titan biology would represent a very different origin of life than seen here on Earth (see the chapter by Lunine et al. in this volume).

9. SUMMARY AND FUTURE PROSPECTS

What can be concluded about the Hadean Earth at the time of life's origin? There is evidence, based on rare Earth metals, for a strong hydrothermal imprint at the time of life's origin. It is also likely that during the first billion years of Earth's history, there was too much heat from the mantle for plate tectonics to occur and that consequently heat was primarily lost through plume volcanism. As a result, the crust would have been more mafic, allowing for a high incidence of serpentinization and resulting in the production of hydrogen and organic compounds. Extensive circulation of seawater through volcanic rock would have created convective cells that resulted in gradients of temperature, pressure, and chemical composition that would have increased the probability for synthesis of organic compounds. Unexplored is the possibility of intracrustal wet/dry cycles associated with the serpentinization reaction that consumes water and could potentially lead to dehydration of rock pores, perhaps inducing organic dehydration reactions. Subsequent cracking and fluid circulation could release these reaction products back into the environment.

A combination of atmospheric and seafloor processes was probably critical for the provision of nutrient elements to the origin of life. Nitrogen as N_2 was abundant in the early atmosphere, and although the abiotic source reactions of NH_4^+ are not well understood, it is likely that mineral-catalyzed reduction of N_2 or atmospheric NO_x species in hydrothermal settings played a pivotal role. Most recent estimates suggest that there was sufficient NH_4^+ for the origin of life, but this flux may have limited the expansion of the early biosphere. There is considerable evidence that biological nitrogen fixation is ancient and may have evolved in early methanogens and in LUCA, possibly in response to nitrogen limitation (*Preiner et al.,* 2018; *Raymond et al.,* 2004; *Weiss et al.,* 2016). Future research should focus on possible mineral catalysis in the reduction of N_2 that may have made its way into the earliest nitrogenase enzymes. Similarly, the availability of phosphorus at the time of life's origin and the source reactions involved in removal of phosphorus from apatite is mostly unknown, as is the early biochemistry in the synthesis of organic phosphate compounds, including adenosine mono-, di-, and triphosphate and phospholipids. However, new data suggesting the hydrothermal production of phosphite — a more reactive form of phosphorus — may pose a solution to the conundrum of prebiotic phosphorylation. Future research might focus on the antiquity and distribution of enzymes known to transport and metabolize phosphorus in order to constrain the likely sources and limitations of phosphorus on the early Earth. In addition to the canonical elements of life, carbon, hydrogen, oxygen, phosphorus, and sulfur, organisms utilize numerous transition metals as cores to metabolic and catabolic proteins, and hydrogen is a common electron donor. Mineral catalysts

may have been the backbone of the prebiotic reaction networks that led to modern metabolism, and future research can better constrain which enzymes were ancient, and which metal cofactors were most crucial for the enzymes of early organisms. For example, there is increasing evidence that the earliest organisms used the Wood-Ljungdahl pathway, which uses hydrogen as an electron donor. Hydrogen was likely very abundant in the Hadean with multiple sources and was likely important in numerous chemical reactions as well as a source of energy for early organisms. The Wood-Ljungdahl pathway also has requirements for Ni, Zn, Co, Fe, and W, and in fact the active site for of some of the enzymes involved in this pathway occurs at the metal-sulfur core (*Cody and Scott,* 2007; *Martin,* 2012). Hydrothermal water/rock interactions are a major source of these and other transition metals. Future bioinformatics research could better constrain which metal-containing enzymes are most ancient, which could be combined with geochemical modeling efforts to better constrain the source and distribution of those transition metals on the early Earth.

The "top-down" approach to reconstructing early life not only helps to understand the earliest metabolisms, the origin of catalytic proteins and protein diversity (*Goldman et al.,* 2010, 2012), the environmental settings of the earliest microbial communities, and the evolution of eukaryotes, but it has also verified key elements of the "bottom-up" approach, including the importance of RNA as a predecessor to DNA-based life. Top-down approaches have also helped to better constrain the environmental parameters, including geochemical and energy cycles and physical gradients, involved in the origin of life. However, as discussed above, limitations inherent to the "top-down" approach require a new synthesis of these approaches, as well as an awareness of both top-down and bottom-up results in order to constrain results to a realistic environmental context. One promising avenue forward is molecular reconstruction, including the incidence of ancient protein folds (e.g., *Goldman et al.,* 2010), of early proteins involved in important catalysis reactions involving geochemical nitrogen, sulfur, and metal cycles. Combined with paleogeological reconstructions these could point to environmental settings that perhaps are sources of key catalytic minerals that predated the formation of specific enzymes. Other promising avenues include reconciling comparative genomics approaches with updated geochemical models and evidence from the rock record in order to better understand the metabolic and physiological potential of early life and to better constrain where and how life's earliest stages began (e.g., *Gaucher et al.,* 2003). Future research efforts would benefit from synthesizing top-down and bottom-up results to better constrain the likely requirements, metabolism, and habitats for early life.

The Hadean Earth was likely an open system with continuous exchange between the deep-subsurface mantle, crust, water column, terrain environments, and atmosphere. Geochemical archives contain direct evidence of these mixing processes and need to be considered in any realistic origin of life model. Such extensive spatial and temporal circulation patterns on the early Earth would allow all the necessary ingredients for life generated from different global sources a greater chance to accumulate, mix, and combine. This framework allows for the possibility that different components of life arose in distinct settings. For example, simple organic building blocks were probably generated primarily in high-energy settings such as lightning, photolysis, and hydrothermal systems, whereas polymerization was likely restricted to localities that experienced dehydration and were less dynamic for at least some period of time. This concept may be extrapolated to the notion that the synthesis of organic compounds and polymers essential for life, metabolism, and replication began in different settings and were only combined at a later stage. Catabolism and anabolism may have distinct environmental roots, as does perhaps the origin of lipid membranes. Our rapidly advancing technological abilities will provide ever more access to the secrets of life's origins wrapped up in the genetic code and locked in the rock record. To unlock those secrets, we need only be perceptive enough to read the clues hidden there.

REFERENCES

Adam Z. (2007) Actinides and life's origins. *Astrobiology, 7(6),* 852–872.

Allwood A. C. (2016) Evidence of life in Earth's oldest rocks. *Nature, 537,* 500–501.

Amend J. P. and Shock E. (1998) Energetics of amino acid synthesis in hydrothermal ecosystems. *Science, 281,* 1659–1662.

Aoyama S. and Ueno Y. (2018) Multiple sulfur isotope constraints on microbial sulfate reduction below an Archean seafloor hydrothermal system. *Geobiology, 16(2),* 107–120.

Baaske P., Weinert F. M., Duhr S., Lemke K. H., Russell M. J., and Braun D. (2007) Extreme accumulation of nucleotides in simulated hydrothermal pore systems. *Proc. Natl. Acad. Sci. U.S.A., 104,* 9346–9351.

Bansho Y., Furubayashi T., Ichihashi N., and Yomo T. (2016) Host-parasite oscillation dynamics and evolution in a compartmentalized RNA replication system. *Proc. Natl. Acad. Sci. U.S.A., 113,* 4045–4050.

Barge L. M., Flores E., Baum M. M., VanderVelde D. G., and Russell M. J. (2019) Redox and pH gradients drive amino acid synthesis in iron oxyhydroxide mineral systems. *Proc. Natl. Acad. Sci. U.S.A., 116(11),* 4828–4833.

Baross J. A. (1998) Do the geological and geochemical records of the early Earth support the prediction from global phylogenetic models of a thermophilic cenancestor? In *Thermophiles: The Keys to Molecular Evolution and the Origin of Life* (J. Wiegel and M. Adams, eds.), pp. 3–18. Taylor and Francis, London.

Baross J. A. (2007) Evolution: An essential feature of life. In *Planets and Life: The Emerging Science of Astrobiology* (W. T. Sullivan and J. Baross, eds.), pp. 213–221. Cambridge Univ., Cambridge.

Baross J. A. (2018) The rocky road to biomolecules. *Nature, 564,* 42–43.

Baross J. A. and Hoffman S. E. (1985) Submarine hydrothermal vents and associated gradient environments as sites for the origin and evolution of life. *Origins Life Evol. Biosph., 15,* 327–345.

Baross J. A. and Martin W. F. (2015) The ribofilm as a concept for life's origins. *Cell, 162,* 13–15.

Beaudry A. A. and Joyce G. F. (1992) Directed evolution of an RNA enzyme. *Science, 257,* 635–641.

Becerra A., Delaye L., Islas S., and Lazcano A. (2007) The very early stages of biological evolution and the nature of the last common ancestor of the three major cell domains. *Annu. Rev. Ecol. Evol. Syst., 38,* 361–379.

Benner S. A., Kim H.-J., Kim M.-J., and Ricardo A. (2010) Planetary organic chemistry and the origin of biomolecules. *Cold Spring Harbor Perspect. Biol.,* 2a003467.

Black R. A., Blosser M. C., Stottrup B. L., Tavakley R., Deamer D. W.,

and Keller S. L. (2013) Nucleobases bind to and stabilize aggregates of a prebiotic amphiphile, providing a viable mechanism for the emergence of protocells. *Proc. Natl. Acad. Sci. U.S.A., 110(33),* 13272–13276.

Boehnke P., Bell E. A., Stephan T., Trappitsch R., Keller C. B., Pardo O. S., Davis A. M., Harrison T. M., and Pellin M. J. (2018) Potassic, high-silica Hadean crust. *Proc. Natl. Acad. Sci. U. S. A., 115,* 6353–6356.

Boussau B., Blanquart S., Necsulea A., Lartillot N., and Gouy M. (2008) Parallel adaptations to high temperatures in the Archaean eon. *Nature, 456,* 942–945.

Braakman R. and Smith E. (2012) The emergence and early evolution of biological carbon-fixation. *PLoS Comput. Biol., 8(4),* e1002455.

Brandes J. A., Boctor N. Z., Cody G. D., Cooper B. A., Haze R. M., and Yoder H. S. (1998) Abiotic nitrogen reduction on the early Earth. *Nature, 395,* 365–367.

Brauhart C. W., Huston D. L., Groves D. I., Mikucki E. J., and Gardoll S. J. (2001) Geochemical mass-transfer patterns as indicators of the architecture of a complete volcanic-hosted massive sulfide hydrothermal alteration system, Panorama district, Pilbara, Western Australia. *Econ. Geol., 96(5),* 1263–1278.

Brazelton W. J., Mehta M. P., Kelley D. S., and Baross J. A. (2011) Multicellular characteristics of a single-species biofilm fueled by serpentinization. *mBio, 2(4),* e00127-11.

Burcar B. T., Barge L. M., Trail D., Watson E. B., Russell M. J., and McGown L. B. (2015) RNA oligomerization in laboratory analogues of alkaline hydrothermal vent systems. *Astrobiology, 15(7),* 509–522.

Cantine M. D. and Fournier G. P. (2018) Environmental adaptation from the origin of life to the last universal common ancestor. *Origins Life Evol. Biosph., 48,* 35–54.

Castelle C. J. and Banfield J. F. (2018) Major new microbial groups expand diversity and alter our understanding of the tree of life. *Cell, 172,* 1181–1197.

Cathelineau M., Myagkiy A., Quesnel B., Boiron M. C., Gautier P., Boulvais P., Ulrich M., Truche L., Golfier F., and Drouillet M. (2017) Multistage crack seal vein and hydrothermal Ni enrichment in serpentinized ultramafic rocks (Koniambo massif, New Caledonia). *Miner. Deposita, 52(7),* 945–960.

Charnay B., Forget F., Wordsworth R., Leconte J., Millour E., Codron F., and Spiga A. (2013) Exploring the faint young Sun problem and the possible climates of the Archean Earth with a 3-D GCM. *J. Geophys. Res.–Atmos., 118,* 10414–10431.

Chyba C. and Sagan C. (1992) Endogenous production, exogenous delivery and impact-shock synthesis of organic molecules: An inventory for the origins of life. *Nature, 355,* 125–132.

Cleaves H. J. II (2008) The prebiotic geochemistry of formaldehyde. *Precambrian Res., 164(3–4),* 111–118.

Cody G. D. (2005) Geochemical connections to primitive metabolism. *Elements, 1(3),* 139–143.

Cody G. D. and Scott J. H. (2007) The roots of metabolism. In *The Emerging Science of Astrobiology* (W. T. Sullivan and J. Baross, eds), pp. 174–186. Cambridge Univ., Cambridge.

Coffin M. F. and Eldholm O. (1994) Large igneous provinces: Crustal structure, dimensions, and external consequences. *Rev. Geophys., 32(1),* 1–36.

Condie K. C. (2016) A planet in transition: The onset of plate tectonics on Earth between 3 and 2 Ga? *Geosci. Front., 9(1),* 51–60.

Copley S. D., Smith D. E., and Morowitz H. J. (2005) A mechanism for the association of amino acids with their codons and the origin of the genetic code. *Proc. Natl. Acad. Sci. U.S.A., 102,* 4442–4447.

Copley S. D., Smith E., and Morowitz H. J. (2007) The origin of the RNA world: Co-evolution of genes and metabolism. *Bioorg. Chem., 35,* 430–443.

Da Cunha V., Gaia M., Gadelle D., Nasir A., and Forterre P. (2017) Lokiarchaea are close relatives of Euryarchaeota, not bridging the gap between prokaryotes and eukaryotes. *PLoS Genet., 13,* e1006810.

Da Cunha V., Gaia M., Nasir A., and Forterre P. (2018) Asgard archaea do not close the debate about the universal tree of life topology. *PLOS Genet., 14,* e1007215.

Dass A. V., Hickman-Lewis K., Brack A., Kee T. P., and Westall F. (2016) Stochastic prebiotic chemistry within realistic geological systems. *Chemistry Select, 1,* 4906–4926.

David L. A. and Alm E. J. (2011) Rapid evolutionary innovation during an Archaean genetic expansion. *Nature, 469,* 93–96.

Davies P. C. W., Benner S. A., Cleland C. E., Lineweaver C. H., McKay

C. P., and Wolfe-Simon F. (2009) Signatures of a shadow biosphere. *Astrobiology, 9,* 241–249.

Deamer D. W. (1997) The first living systems: A bioenergetic perspective. *Microbiol. Mol. Biol. Rev., 61,* 239–261.

Deamer D. W. (2007) The origin of cellular life. In *Planets and Life: The Emerging Science of Astrobiology* (W. T. Sullivan and J. Baross, eds.), pp. 187–209. Cambridge Univ., Cambridge.

Deamer D. (2017) The role of lipid membranes in life's origin. *Life, 7(1),* 5.

Deamer D., Singaram S., Rajamani S., Kompanichenko V., and Guggenheim S. (2006) Self-assembly processes in the prebiotic environment. *Philos. Trans. R. Soc., B, 361(1474),* 1809–1818.

Debaille V., O'Neill C., Brandon A. D., Haenecour P., Yin Q. Z., Mattielli N., and Treiman A. H. (2013) Stagnant-lid tectonics in early Earth revealed by ^{142}Nd variations in late Archean rocks. *Earth Planet. Sci. Lett., 373,* 83–92.

de Duve C. (1995) The beginning of life on Earth. *Am. Sci., 83,* 428–437.

Dey G., Thattai M., and Baum B. (2016) On the archaeal origins of eukaryotes and the challenges of inferring phenotype from genotype. *Trends Cell. Biol., 26,* 476–448.

de Wit M. J. and Furnes H. (2016) 3.5-Ga hydrothermal fields and diamictites in the Barberton Greenstone Belt — Paleoarchean crust in cold environments. *Sci. Adv., 2(2),* e1500368, DOI: 10.1126/sciadv.1500368.

Dhuime B., Hawkesworth C. J., Cawood P. A., and Storey C. D. (2012) A change in the geodynamics of continental growth 3 billion years ago. *Science, 335,* 1334–1336.

Dhuime B., Wuestefeld A., and Hawkesworth C. J. (2015) Emergence of modern continental crust about 3 billion years ago. *Nature Geosci., 8(7),* 552–555.

di Giulio R. M. (2003) The universal ancestor was a thermophile or a hyperthermophile: Tests and further evidence. *J. Theor. Biol., 221,* 425–436.

Djokic T., Van Kranendonk M. J., Campbell K. A., Walter M. R., and Ward C. R. (2017) Earliest signs of life on land preserved in ca. 3.5 Ga hot spring deposits. *Nature Commun., 8,* 15263, DOI: 10.1038/ncomms15263.

Dobson C. M., Ellison G. B., Tuck A. F., and Vaida V. (2000) Atmospheric aerosols as prebiotic chemical reactors. *Proc. Natl. Acad. Sci. U.S.A., 97(22),* 11864–11868.

Dodd M. S., Papineau D., Grenne T., Slack J. F., Rittner M., Pirajno F., O'Neil J., and Little C. T. (2017) Evidence for early life in Earth's oldest hydrothermal vent precipitates. *Nature, 543(7643),* 60–64.

Ducluzeau A. L., Van Lis R., Duval S., Schoepp-Cothenet B., Russell M. J., and Nitschke W. (2009) Was nitric oxide the first deep electron sink? *Trends Biochem. Sci., 34(1),* 9–15.

Dupont C. L., Butcher A., Vala R. E., Bourne P. E., and Caetano-Anollés G. (2010) History of biological metal utilization inferred through phylogenomic analysis of protein structures. *Proc. Natl. Acad. Sci. U.S.A., 107(23),* 10567–10572.

Eigen M. and Schuster P. (1979) *The Hypercycle: A Principle of Natural Self-Organization.* Springer-Verlag Berlin, Heidelberg. 92 pp.

Ellington A. D. and Szostak J. W. (1990) *In vitro* selection of RNA molecules that bind specific ligands. *Nature, 346,* 818–822.

Eme L., Spang A., Lombard J., Stairs C. W., and Ettema T. J. G. (2017). Archaea and the origin of eukaryotes. *Nature Rev. Microbiol., 15,* 711–723.

Eren A. M., Esen Ö. C., Quince C., Vineis J. H., Morrison H. G., Sogin M. L., and Delmont T. O. (2015) Anvi'o: An advanced analysis and visualization platform for 'omics data. *PeerJ,* 3:e1319, DOI: 10.7717/peerj.1319.

Farquhar J., Bao H., and Thiemens M. H. (2000) Atmospheric influence of Earth's earliest sulfur cycle. *Science, 289,* 756–758.

Ferris J. P. (2005) Mineral catalysis and prebiotic synthesis: Montmorillonite-catalyzed formation of RNA. *Elements, 1(3),* 145–149.

Ferris J. P. (2006) Montmorillonite-catalysed formation of RNA oligomers: The possible role of catalysis in the origins of life. *Philos. Trans. R. Soc., B, 361(1474),* 1777–1786.

Ferris J. P., Hill A. R., Liu R., and Orgel L. E. (1996) Synthesis of long prebiotic oligomers on mineral surfaces. *Nature, 381,* 59–61.

Ferus M., Pietrucci F., Saitta A. M., Knížek A., Kubelík P., Ivanek O., Shestivska V., and Civiš S. (2017) Formation of nucleobases in a Miller-Urey reducing atmosphere. *Proc. Natl. Acad. Sci. U.S.A., 114(17),* 4306–4311.

Filippelli G. M. (2008) The global phosphorus cycle: Past, present, and future. *Elements, 4(1),* 89–95.

Fitzsimmons J. N., Boyle E. A., and Jenkins W. J. (2014) Distal transport of dissolved hydrothermal iron in the deep South Pacific Ocean. *Proc. Natl. Acad. Sci. U.S.A., 111(47),* 16654–16661.

Foley B. J. and Smye A. J. (2018) Carbon cycling and habitability of Earth-size stagnant lid planets. *Astrobiology, 18(7),* DOI: 10.1089/ast.2017.1695.

Fortes A. D. (2000) Exobiological implications of a possible ammonia-water ocean inside Titan. *Icarus, 146,* 444–452.

Fuchs G. (2011) Alternative pathways of carbon dioxide fixation: Insights into the early evolution of life. *Annu. Rev. Microbiol., 65,* 631–658.

Galtier N., Tourasse N., and Gouy M. (1999) A nonhyperthermophilic common ancestor to extant life forms. *Science, 283,* 220–221.

Gartman A., Findlay A. J., and Luther G. W. (2014) Nanoparticulate pyrite and other nanoparticles are a widespread component of hydrothermal vent black smoker emissions. *Chem. Geol., 366,* 32–41.

Gaucher E. A., Thomson J. M., Burgan M. F., and Benner S. A. (2003) Inferring the palaeoenvironment of ancient bacteria on the basis of resurrected proteins. *Nature, 425(6955),* 285–288.

Gaucher E. A., Govindarajan S., and Ganesh O. K. (2008) Palaeotemperature trend for Precambrian life inferred from resurrected proteins. *Nature, 451,* 704–707.

Gilbert W. (1986) Origin of life: The RNA world. *Nature, 319,* 618.

Glein C. R., Baross J. A., and Waite J. H. (2015) The pH of Enceladus' ocean. *Geochim. Cosmochim. Acta, 163,* 302–319.

Gogarten J. P. and Deamer D. (2016) Is LUCA a thermophilic progenote? *Nature Microbiol., 1,* 16229.

Goldford J. E., Hartman H., Smith T. F., and Segrè D. (2017) Remnants of an ancient metabolism without phosphate. *Cell, 168(6),* 1126–1134.

Goldford J. E. and Segrè D. (2018) Modern views of ancient metabolic networks. *Curr. Opin. Syst. Biol., 8,* 117–124.

Goldman A., Samudraia D. R., and Baross J. A. (2010) The evolution and functional repertoire of translation proteins during the origin of life. *Biol. Direct, 5,* 15.

Goldman A., Baross J. A., and Samurraia D. R. (2012) The enzymatic and metabolic capabilities of early life. *PLoS One, 7(9),* e39912.

Graham E. D., Heidelberg J. F., and Tully B. J. (2017) BinSanity: Unsupervised clustering of environmental microbial assemblies using coverage and affinity propagation. *PeerJ,* 5:e3035.

Guttenberg N., Virgo N., Chandru K., Scharf C., and Mamajanov I. (2017) Bulk measurements of messy chemistries are needed for a theory of the origins of life. *Philos. Trans. R. Soc., A, 375(2109).*

Grew E. S., Bada J. L., and Hazen R. M. (2011) Borate minerals and origin of the RNA world. *Origins Life Evol. Biospheres, 41(4),* 307–316.

Haldane J. B. S. (1929) The origin of life. *Rationalist Annual, 148,* 3–10.

Hansen H. C. B., Guldberg S., Erbs M., and Koch C. B. (2001) Kinetics of nitrate reduction by green rusts — Effects of interlayer anion and Fe(II):Fe(III) ratio. *Appl. Clay Sci., 18,* 81–91.

Hao J., Sverjensky D. A., and Hazen R. M. (2017a) Mobility of nutrients and trace metals during weathering in the late Archean. *Earth Planet. Sci. Lett., 471,* 148–159.

Hao J., Sverjensky D. A., and Hazen R. M. (2017b) A model for late Archean chemical weathering and world average river water. *Earth Planet. Sci. Lett., 457,* 191–203.

Hazen R. M. (2005) *Genesis: The Scientific Quest for Life's Origin.* Joseph Henry, Washington, DC. 368 pp.

Hazen R. M. (2013) Paleomineralogy of the Hadean eon: A preliminary species list. *Am. J. Sci., 313(9),* 807–843.

Hazen R. M. (2017) Chance, necessity, and the origins of life: A physical sciences perspective. *Philos. Trans. R. Soc., A, 375(2109),* DOI: 10.1098/rsta.2016.0353.

Hazen R. M. and Sverjensky D. A. (2010) Mineral surfaces, geochemical complexities, and the origin of life. *Cold Spring Harb. Perspect. Biol,* DOI: 10.1101/cshperspect.a002162.

Hazen R. M., Filley T. R., and Goodfriend G. A. (2001) Selective adsorption of L-and D-amino acids on calcite: Implications for biochemical homochirality. *Proc. Natl. Acad. Sci. U.S.A., 98(10),*

5487–5490.

Hawkesworth C. J., Cawood P. A., and Dhuime B. (2016) Tectonics and crustal evolution. *GSA Today, 26(9)*, 4–11.

Hennet R. J.-C., Holm N. G., and Engel M. H. (1992) Abiotic synthesis of amino acids under hydrothermal conditions and the origin of life: A perpetual phenomenon? *Naturwissenschaften, 79*, 361–365.

Herschy B., Chang S. J., Blake R., Lepland A., Abbott-Lyon H., Sampson J., Atlas Z., Kee T. P., and Pasek M. A. (2018) Archean phosphorus liberation induced by iron redox geochemistry. *Nature Commun., 9*, 1346, DOI: 10.1038/s41467-018-03835-3.

Hessler A. M. and Lowe D. R. (2006) Weathering and sediment generation in the Archean: An integrated study of the evolution of siliciclastic sedimentary rocks of the 3.2 Ga Moodies Group, Barberton Greenstone Belt, South Africa. *Precambrian Res.,151(3)*, 185–210.

Herzberg C., Condie K., and Korenaga J. (2010) Thermal history of the Earth and its petrological expression. *Earth Planet. Sci. Lett., 292(1–2)*, 79–88.

Higgs P. G. and Lehman N. (2015) The RNA world: Molecular cooperation at the origins of life. *Nature Rev. Genet., 16*, 7–17.

Holm N. G. and Neubeck A. (2009) Reduction of nitrogen compounds in oceanic basement and its implications for HCN formation and abiotic organic synthesis. *Geochem. Trans., 10(1)*, DOI: 10.1186/1467-4866-10-9.

Hsiao C., Mohan S., Kalahar B. K., and Williams L. D. (2009) Peeling the onion: Ribosomes are ancient molecular fossils. *Mol. Biol. Evol., 26(11)*, 2415–2425.

Hud N. V., Cafferty B. J., Krishnamurthy R., and Williams L. D. (2013) The origin of RNA and "my grandfather's axe." *Chem. Biol., 20*, 466–474.

Hug L. A., Baker B. J., Anantharaman K., Brown C. T., Probst A. J., Casatelle C. J., Butterfield C. N., Hernsdorf A. W., Amano Y., Ise K., Suzuko Y., Dudek N., Relman D. A., Finstad K. M., Amundson R., Thomas B. C., and Banfield J. F. (2016) A new view of the tree of life. *Nature Microbiol., 1*, 16408.

Huston D. L., Brauhart C. W., Drieberg S. L., Davidson G. J., and Groves D. I. (2001) Metal leaching and inorganic sulfate reduction in volcanic-hosted massive sulfide mineral systems: Evidence from the paleo-Archean Panorama district, Western Australia. *Geology, 29(8)*, 687–690.

Imelfort M., Parks D., Woodcroft B. J., Dennis P., Hugenholtz P., and. Tyson G. W. (2014) GroopM: An automated tool for the recovery of population genomes from related metagenomes. *PeerJ, 2:e603.*

Isley A. E. and Abbott D. H. (1999) Plume-related mafic volcanism and the deposition of banded iron formation. *J. Geophys. Res.–Solid Earth, 104(B7)*, 15461–15477.

Iwabe N., Kuma K., Hasegawa M., Osawa S., and Miyata T. (1989) Evolutionary relationship of archaebacteria, eubacteria, and eukaryotes inferred phylogenetic trees of duplicated genes. *Proc. Natl. Acad. Sci. U.S.A., 86*, 9355–9359.

Jiang W., Pacella M. S., Athanasiadou D., Nelea V., Vali H., Hazen R. M., Gray J. J., and McKee M. D. (2017) Chiral acidic amino acids induce chiral hierarchical structure in calcium carbonate. *Nature Commun., 8*, 15066, DOI: 10.1038/ncomms15066.

Jamieson J. W., Wing B. A., Farquhar J., and Hannington M. (2013) Neoarchaean seawater sulphate concentrations from sulphur isotopes in massive sulphide ore. *Nature Geosci., 6*, 61–64.

Johnson A. P., Cleaves H. J., Dworkin J. P., Glavin D. P., Lazcano A., and Bada J. L. (2008) The Miller volcanic spark discharge experiment. *Science, 322*, 404.

Johnson H. P. and Pruis M. J. (2003) Fluxes of fluid and heat from the oceanic crustal reservoir. *Earth Planet. Sci. Lett., 216(4)*, 565–574.

Jones C., Nomosatryo S., Crowe S. A., Bjerrum C. J., and Canfield D. E. (2015) Iron oxides, divalent cations, silica, and the early Earth phosphorus crisis. *Geology, 43(2)*, 135–138.

Joyce G. F. (2005) The RNA world: Life before DNA and protein. In *Extraterrestrials: Where Are They?* (B. Zuckerman and M. H. Hart, eds.), pp. 139–151. Cambridge Univ., Cambridge.

Kang D. D., Froula J., Egan R., and Wang Z. (2015) MetaBAT, an efficient tool for accurately reconstructing single genomes from complex microbial communities. *PeerJ, 3:e1165.*

Kasting J. F. (1990) Bolide impacts and the oxidation state of carbon in the Earth's early atmosphere. *Origins Life Evol. Biospheres, 20(3–4)*, 199–231.

Kelemen P. B. and Hirth G. (2012) Reaction-driven cracking during

retrograde metamorphism: Olivine hydration and carbonation. *Earth Planet. Sci. Lett., 345*, 81–89.

Kelley D. S., Baross J. A., and Delaney J. R. (2002) Volcanoes, fluid, and life at mid-ocean ridge spreading centers. *Annu. Rev. Earth Planet. Sci., 30*, 385–491.

Kelley D. S., Karson J. A., Früh-Green G. L., Yoerger D. R., Shank T. M., Butterfield D. A., Hayes J. M., Schrenk M. O., Olson E. J., Proskurowski G., Jakuba M, Bradley A., Larson B., Ludwig L., Glickson D., Buckman K, Bradley A. S. Brazelton W. J., Roe K., Elend M. J., Delacour A., Bernasconi S. M., Lilley M. D., Baross J. A., Summons R. E., and Silva S. P. (2005) A serpentinite-hosted ecosystem: The Lost City hydrothermal field. *Science, 307,*1428–1434.

Kerr A. C. (1998) Oceanic plateau formation: A cause of mass extinction and black shale deposition around the Cenomanian-Turonian boundary? *J. Geol. Soc. London, 155(4)*, 619–626.

Kipp M. A. and Stüeken E. E. (2017) Biomass recycling and Earth's early phosphorus cycle. *Sci. Adv., 3(11)*, eaao4795, DOI: 10.1126/sciadv.aao4795.

Kitadai N. and Maruyama S. (2018) Origins of building blocks of life: A review. *Geosci. Front., 9(4)*, 1117–1153, DOI: 10.1016/j.gsf.2017.07.007.

Klein F. and Bach W. (2009) Fe-Ni-Co-O-S phase relations in peridotite-seawater interactions. *J. Petrol., 50(1)*, 37–59.

Koehler M. C., Buick R., Kipp M. A., Stüeken E. E., and Zaloumis J. (2018) Transient surface ocean oxygenation recorded in the ~2.66-Ga Jeerinah Formation, Australia. *Proc. Natl. Acad. Sci. U.S.A., 115(30)*, 7711–7716.

Konhauser K. O., Pecoits E., Lalonde S. V., Papineau D., Nisbet E. G., Barley M. E., Arndt N. T., Zahnle K., and Kamber B. S. (2009) Oceanic nickel depletion and a methanogen famine before the Great Oxidation Event. *Nature, 458(7239)*, 750–753.

Koonin E. V. (2003) Comparative genomics, minimal gene-sets, and the last universal common ancestor. *Nature Rev. Microbiol., 1*, 127–136.

Koonin E. V. (2018) Environmental microbiology and metagenomics: The Brave New World is here, what's next? *Environ. Microbiol., 20*, 4210-4212, DOI: 10.1111/1462- 2920.14403.

Koonin E. V. and Martin W. (2005) On the origin of genomes and cells within inorganic compartments. *Trends Genet., 21*, 647–654.

Koonin E. V. and Novozhilov A. S. (2017) Origin and evolution of the universal code. *Annu. Rev. Genet., 51*, 45–62.

Korenaga J., Planavsky N. J., Evans D. A. (2017) Global water cycle and the coevolution of the Earth's interior and surface environment. *Philos. Trans. R. Soc., A, 375(294)*, DOI: 10.1098/rsta.2015.0393.

Kuhn W. R. and Atreya S. K. (1979) Ammonia photolysis and the greenhouse effect in the primordial atmosphere of the Earth. *Icarus, 37(1)*, 207–213.

Lancaster K. M., Roemelt M., Ettenhuber P., Hu Y., Ribbe M. W., Neese F., Bergmann U., and DeBeer S. (2011) X-ray emission spectroscopy evidences a central carbon in the nitrogenase iron-molybdenum cofactor. *Science, 334(6058)*, 974–977.

Laneuville M., Kameya M., and Cleaves H. J. (2018) Earth without life: A systems model of a global abiotic nitrogen cycle. *Astrobiology,18(7)*, 897–914, DOI: 10.1089/ast.2017.1700.

Lang S. Q., Butterfield D. A., Schulte M., Kelley D. S., and Lilley M. D. (2010) Elevated concentrations of formate, acetate, and dissolved organic carbon found at the Lost City hydrothermal field. *Geochim. Cosmochim. Acta, 74*, 941–952.

Lasaga A. C., Holland H. D., and Dwyer M. J. (1971) Primordial oil slick. *Science, 174(4004)*, 53–55.

Laurent O., Martin H., Moyen J. F. and Doucelance R. (2014) The diversity and evolution of late-Archean granitoids: Evidence for the onset of "modern-style" plate tectonics between 3.0 and 2.5 Ga. *Lithos, 205*, 208–235.

Lazcano A. (2010) Which way to life? *Origins Life Evol. Biospheres, 40(2)*, 161–167.

Lee C. T. A., Caves J., Jiang H., Cao W., Lenardic A., McKenzie N. R., Shorttle O., Yin Q. Z., and Dyer B. (2018) Deep mantle roots and continental emergence: Implications for whole-Earth elemental cycling, long-term climate, and the Cambrian explosion. *Intl. Geol. Rev., 60(4)*, 431–448.

Lee C. T. A., Yeung L. Y., McKenzie N. R., Yokoyama Y., Ozaki K., and Lenardic A. (2016) Two-step rise of atmospheric oxygen linked to the growth of continents. *Nature Geosci., 9(6)*, 417–424.

Lenardic A. and Crowley J. W. (2012) On the notion of well-defined tectonic regimes for terrestrial planets in this solar system and others. *Astrophys. J., 755(2)*, DOI: 10.1088/0004-637X/755/2/132.

Mamajanov I., MacDonald P. J., Ying J., Duncanson D. M., Dowdy G. R., Walker C. A., Engelhart A. E., Fernández F. M., Grover M. A., Hud N. V., and Schork F. J. (2014) Ester formation and hydrolysis during wet-dry cycles: Generation of far-from-equilibrium polymers in a model prebiotic reaction. *Macromolecules, 47(4)*, 1334–1343.

Martin W. F. (2012) Hydrogen, metals, bifurcating electrons, and proton gradients: The early evolution of biological energy conservation. *FEBS Lett., 586*, 485–493.

Martin W. and Russell M. J. (2007) On the origin of biochemistry at an alkaline hydrothermal vent. *Philos. Trans. R. Soc., B, 362*, 1887–1925.

Martin W. F. and Thauer R. K. (2017) Energy in ancient metabolism. *Cell, 168*, 953–955.

Martin W., Baross J., Kelley D., and Russell M. J. (2008) Hydrothermal vents and the origin of life. *Nature Rev. Microbiol., 6*, 805–814.

Martin W. F., Sousa F. L., and Lane N. (2014) Energy at life's origin. *Science, 344*, 1092–1093.

Martin-Izard A., Arias D., Arias M., Gumiel P., Sanderson D. J., Castañon C., and Sánchez J. (2016) Ore deposit types and tectonic evolution of the Iberian Pyrite Belt: From transtensional basins and magmatism to transpression and inversion tectonics. *Ore Geol. Rev., 79*, 254–267.

Mather T. A., Pyle D. M., and Allen A. G. (2004) Volcanic source for fixed nitrogen in the early Earth's atmosphere. *Geology, 32(10)*, 905–908.

Mayer L. M. (1994) Relationships between mineral surfaces and organic carbon concentrations in soils and sediments. *Chem. Geol., 114(3–4)*, 347–363.

Mayhew L. E., Ellison E. T., McCollom T. M., Trainor T. P., and Templeton A. S. (2013) Hydrogen generation from low-temperature water-rock reactions. *Nature Geosci., 6(6)*, 478–484.

McCollom T. M. and Seewald J. S. (2007) Abiotic synthesis of organic compounds in deep-sea hydrothermal environments. *Chem. Rev., 107*, 382–401.

McCollom T. M. and Seewald J. S. (2013) Serpentinites, hydrogen, and life. *Elements, 9*, 129–134.

McCollom T. M., Ritter G., and Simoneit B. R. T. (1999) Lipid synthesis under hydrothermal conditions by Fischer-Tropsch-type reactions. *Origins Life Evol. Biosph., 29*, 153–166.

McDermott J. M., Seewald J. S., German C. R., and Sylva S. P. (2015) Pathways for abiotic organic synthesis at submarine hydrothermal fields. *Proc. Natl. Acad. Sci. U.S.A., 112*, 7668–7672.

McKay C. P. and Smith H. D. (2005) Possibilities for methanogenic life in liquid methane on the surface of Titan. *Icarus, 178*, 274–276.

Ménez B., Pisipia C., Andreami M., Jamme F., Vanbellingen Q. P., Brunelle A., Richard L., Dumas P., and Réfrégiers M. (2018) Abiotic synthesis of amino acids in the recesses of the oceanic lithosphere *Nature, 564*, 59–63, DOI: 10.1038/s41586-018-0684-z.

Michard A. and Albarede F. (1986) The REE content of some hydrothermal fluids. *Chem. Geol., 55(1–2)*, 51–60.

Miller S. L. (1953) A production of amino acids under possible primitive Earth conditions. *Science, 117*, 528–529.

Miller S. L. and Urey H. C. (1959) Organic compound synthesis on the primitive Earth. *Science, 130(3370)*, 245–251.

Mirkin B. G., Fenner T. I., Galperin M. Y., and Koonin E. V. (2003) Algorithms for computing parsimonious evolutionary scenarios for genome evolution, the last universal common ancestor and dominance of horizontal gene transfer in the evolution of prokaryotes. *BMC Evol. Biol., 3*, 2.

Moore W. B. and Webb A. A. G. (2013) Heat-pipe earth. *Nature, 501(7468)*, 501–505.

Muchowska K. B., Varma S. J., and Moran J. (2019) Synthesis and breakdown of universal metabolic precursors promoted by iron. *Nature, 569*, 104–107.

Mulkidjanian A. Y., Bychkov A. Y., Dibrova D. Λ., Galperin M. Y., and Koonin E. V. (2012) Origin of the first cells at terrestrial, anoxic geothermal fields. *Proc. Natl. Acad. Sci. U.S.A., 109*, 821–840.

Murakami T., Matsuura K., and Kanzaki Y. (2016) Behaviors of trace elements in Neoarchean and Paleoproterozoic paleosols: Implications for atmospheric oxygen evolution and continental oxidative weathering. *Geochim. Cosmochim. Acta, 192*, 203–219.

Nasir A., Kim K. M., and Caetano-Anolles G. (2016) Lokiarchaeota: Eukaryote-like missing links from microbial dark matter? *Trends Microbiol., 23*, 448–450.

National Research Council (2007) *The Limits of Organic Life in Planetary Systems*. National Academies, Washington, DC. 116 pp.

Navarro-González R., Molina M. J., and Molina L. T. (1998) Nitrogen fixation by volcanic lightning in the early Earth. *Geophys. Res. Lett., 25(16)*, 3123–3126.

Nebel-Jacobsen Y., Nebel O., Wille M., and Cawood P. A. (2018) A non-zircon Hf isotope record in Archean black shales from the Pilbara craton confirms changing crustal dynamics ca. 3 Ga ago. *Sci. Rept., 8(1)*, DOI: 10.1038/s41598-018-19397-9.

Nisbet E. G. and Sleep N. H. (2001) The habitat and nature of early life. *Nature, 409*, 1083–1091.

Nitschtke W. and Russell M. J. (2009) Hydrothermal focusing of chemical and chemiosmotic energy, supported by delivery of catalytic Fe, Ni, Mo/W, Co, S, and Se, forced life to emerge. *J. Mol. Evol., 69(5)*, 481–496.

Nitschke W. and Russell M. J. (2013) Beating the acetyl coenzyme A-pathway to the origin of life. *Philos. Trans. R. Soc., B, 368(1622)*, DOI: 10.1098/rstb.2012.0258.

Nna-Mvondo D. and Martinez-Frias J. (2007) Review komatiites: From Earth's geological settings to planetary and astrobiological contexts. *Earth Moon Planets, 100(3–4)*, 157–179.

Nutman A. P., Bennett V. C., Friend C. L., Van Kranendonk M. J., and Chivas A. R. (2016) Rapid emergence of life shown by discovery of 3700-million-year-old microbial structures. *Nature, 537*, 535–538.

O'Neill C. and Debaille V. (2014) The evolution of Hadean-Eoarchaean geodynamics. *Earth Planet. Sci. Lett., 406*, 49–58.

Oparin A. I. (1952) *The Origin of Life, 2nd edition.* Dover, New York. 270 pp.

Oro J., Miller S. L., and Lazcano A. (1990) The origin and early evolution of life on Earth. *Annu. Rev. Earth Planet. Sci.,18*, 317–356.

Parker E. T., Cleaves H. J., Dworkin J. P., Glavin D. P., Callahan M., Aubrey A., Lazcano A., and Bada J. L. (2011) Primordial synthesis of amines and amino acids in a 1958 Miller H_2S-rich spark discharge experiment. *Proc. Natl. Acad. Sci. U.S.A., 108*, 5526–5531.

Pasek M. A., Gull M., and Herschy B. (2017) Phosphorylation on the early earth. *Chem. Geol., 475*, 149–170.

Pearce B. K. D., Tupper A. S., Pudritz R. F., and Higgs P. G. (2018) Constraining the time interval for the origin of life on Earth. *Astrobiology, 18*, 1–22.

Penny D. and Poole A. (1999) The nature of the last universal common ancestor. *Curr. Opin. Genet. Dev., 9*, 672–677.

Planavsky N. J., Asael D., Hofmann A., Reinhard C. T., Lalonde S. V., Knudsen A., Wang X., Ossa F. O., Pecoits E., Smith A. J., and Beukes N. J. (2014) Evidence for oxygenic photosynthesis half a billion years before the Great Oxidation Event. *Nature Geosci., 7(4)*, 283–286.

Playter T., Konhauser K., Owttrim G., Hodgson C., Warchola T., Mloszewska A. M., Sutherland B., Bekker A., Zonneveld J. P., Pemberton S. G., and Gingras M. (2017) Microbe-clay interactions as a mechanism for the preservation of organic matter and trace metal biosignatures in black shales. *Chem. Geol., 459*, 75–90.

Powner M. W., Gerland B., and Sutherland J. D. (2009) Synthesis of activated pyrimidine ribonucleotides in prebiotically plausible conditions. *Nature, 459*, 239–242.

Pope E. C., Bird D. K., and Rosing M. T. (2012) Isotope composition and volume of Earth's early oceans. *Proc. Natl. Acad. Sci. U.S.A., 109(12)*, 4371–4376.

Preiner M., Xavier J. C., Sousa F. L., Zimorski V., Neubeck A., Lang S. Q., Greenwell H. C., Leinermanns K., Tuysuz H., McCollum T. M., Holm N. G., and Martin W. F. (2018) Serpentinization: Connecting geochemistry, ancient metabolism, and industrial hydrogenation. *Life, 8(41)*, DOI: 10.3390/life8040041.

Proskurowski G., Lilley M. D., Seewald J. S., Früh-Green G. L., Olson E. J., Lupton J. E., Sylva S. P., and Kelley D. S. (2008) Abiogenic hydrocarbon production at Lost City hydrothermal field. *Science, 319(5863)*, 604–607.

Raymond J., Siefert J. L., Staples C. R., and Blankenship R. K. (2004) The natural history of nitrogen fixation. *Mol. Biol. Evol., 21*, 541–554.

Ricardo A., Carrigan M. A., Olcott A. N., and Benner S. A. (2004) Borate minerals stabilize ribose. *Science, 303(5655)*, 196.

Ritson D. J., Battilocchio C., Ley S. V., and Sutherland J. D. (2018)

Mimicking the surface and prebiotic chemistry of early Earth using flow chemistry. *Nature Commun., 9,* 1821, DOI: 10.1038/s41467-018-04147-2.

Roldan A., Hollingsworth N., Roffey A., Islam H. U., Goodall J. B. M., Catlow C. R. A., Darr J. A., Bras W., Sankar G., Holt K. B., and Hogarth G. (2015) Bio-inspired CO_2 conversion by iron sulfide catalysts under sustainable conditions. *Chem. Commun., 51(35),* 7501–7504.

Ross D. S. and Deamer D. (2016) Dry/wet cycling and the thermodynamics and kinetics of prebiotic polymer synthesis. *Life, 6(3),* DOI: 10.3390/life6030028.

Rozel A. B., Golabek G. J., Jain C., Tackley P. J., and Gerya T. (2017) Continental crust formation on early Earth controlled by intrusive magmatism. *Nature, 545(7654),* 332–335.

Ranea J. A. G., Sillero A., Thornton J. M., and Orengo C. A. (2006) Protein superfamily evolution and the Last Universal Common Ancestor (LUCA). *J. Mol. Evol., 63,* 513–525.

Raymann K., Brochier-Armanet C., and Gribaldo S. (2015) The two-domain tree of life is linked to a new root for the Archaea. *Proc. Natl. Acad. Sci. U.S.A., 112(21),* 6670–6675, DOI: 10.1073/pnas.1420858112.

Reinhard C. T., Planavsky N. J., Gill B. C., Ozaki K., Robbins L. J., Lyons T. W., Fischer W. W., Wang C., Cole D. B., and Konhauser K. O. (2017) Evolution of the global phosphorus cycle. *Nature, 541(7637),* 386–389.

Resing J. A., Sedwick P. N., German C. R., Jenkins W. J., Moffett J. W., Sohst B. M., and Tagliabue A. (2015) Basin-scale transport of hydrothermal dissolved metals across the South Pacific Ocean. *Nature, 523(7559),* 200–203.

Robbins L. J., Lalonde S. V., Saito M. A., Planavsky N. J., Mloszewska A. M., Pecoits E., Scott C., Dupont C. L., Kappler A., and Konhauser K. O. (2013) Authigenic iron oxide proxies for marine zinc over geological time and implications for eukaryotic metallome evolution. *Geobiology, 11,* 295–306.

Robbins L. J., Lalonde S. V., Planavsky N. J., Partin C. A., Reinhard C. T., Kendall B., Scott C., Hardisty D. S., Gill B. C., Alessi D. S., and Dupont C. L. (2016) Trace elements at the intersection of marine biological and geochemical evolution. *Earth-Sci. Rev., 163,* 323–348.

Robertson D. L. and Joyce G. F. (1990) Selection *in vitro* of an RNA enzyme that specifically cleaves single-stranded DNA. *Nature, 344,* 467–468.

Rudnick R. L. and Gao S. (2014) Composition of the continental crust. *Treatise Geochem., 4,* 1–51.

Russell M. J., Hall A. J., and Martin W. (2010) Serpentinization as a source of energy at the origin of life. *Geobiology, 8,* 355–371.

Russell M. J., Barge L. M., Bhartia R., Bocanegra D., Bracher P. J., Branscomb E., Kidd R., McGlynn S., Meier D. H., Nitschke W., Shibuya T., Vance S., White L., and Kanik I. (2014) The drive to life on wet and icy worlds. *Astrobiology, 14,* 308–343.

Russell M. J., Murray A. E., and Hand K. P. (2017) The possible emergence of life and differentiation of a shallow biosphere on irradiated icy worlds: The example of Europa. *Astrobiology, 17,* 1265–1273.

Sagan C. and Mullen G. (1972) Earth and Mars: Evolution of atmospheres and surface temperatures. *Science, 177(4043),* 52–56.

Saito M. A., Sigman D. M., and Morel F. M. M. (2003) The bioinorganic chemistry of the ancient ocean: The co-evolution of cyanobacterial metal requirements and biogeochemical cycles at the Archean-Proterozoic boundary? *Inorg. Chim. Acta, 356,* 308–318.

Sander S. G. and Koschinsky A. (2011) Metal flux from hydrothermal vents increased by organic complexation. *Nature Geosci., 4,* 145–150.

Schlesinger G. and Miller S. L. (1983) Prebiotic synthesis in atmospheres containing CH_4, CO, and CO_2. *J. Mol. Evol., 19(5),* 383–390.

Schoepp-Cothenet B., van Lis R., Philippot P., Magalon A., Russell M. J., and Nitschke W. (2012) The ineluctable requirement for the trans-iron elements molybdenum and/or tungsten in the origin of life. *Sci. Rept., 2,* 263.

Schoepp-Cothenet B., van Lis R., Atteia A., Baymann F., Capowiez L., Ducluzeau A. L., Duval S., ten Brink F., Russell M. J., and Nitschke W. (2013) On the universal core of bioenergetics. *Biochim. Biophys. Acta, Bioenerg., 1827(2),* 79–93.

Schoonen M. A. and Xu Y. (2001) Nitrogen reduction under hydrothermal vent conditions: Implications for the prebiotic

synthesis of CHON compounds. *Astrobiology, 1(2),* 133–142.

Schrenk M. O., Brazelton W. J., and Lang S. Q. (2013) Serpentinization, carbon, and deep life. *Rev. Mineral. Geochem., 75,* 575–606.

Schulze-Makuch D. and Grinspoon D. H. (2005) Biologically enhanced energy and carbon cycling on Titan? *Astrobiology, 5,* 560–567.

Scott C., Planavsky N. J., Dupont C. L., Kendall B., Gill B. C., Robbins L. J., Husband K. F., Arnold G. L., Wing B. A., Poulton S. W., and Bekker A. (2012) Bioavailability of zinc in marine systems through time. *Nature Geosci., 6,* 125–128.

Segré D., Ben-Eli D., Deamer D. W., and Lancet D. (2001) The lipid world. *Origins Life Evol. Biospheres, 31(1–2),* 119–145.

Shapiro R. (2007) A simpler origin for life. *Sci. Am., 296(6),* 46–53.

Shen Y., Buick R., and Canfield D. E. (2001) Isotopic evidence for microbial sulfate reduction in the early Archaean era. *Nature, 410,* 77–81.

Shock E. L. and Schulte M. D. (1998) Organic synthesis during fluid mixing in hydrothermal systems. *J. Geophys. Res.–Planets, 103(E12),* 28513–28527.

Siedenberg K., Strauss H., and Hoffmann E. J. (2016) Multiple sulfur isotope signature of early Archean oceanic crust, Isua (SW-Greenland). *Precambrian Res., 283,* 1–12.

Sinton C. W. and Duncan R. A. (1997) Potential links between ocean plateau volcanism and global ocean anoxia at the Cenomanian-Turonian boundary. *Econ. Geol., 92(7–8),* 836–842.

Smirnov A., Hausner D., Laffers R., Strongin D. R., and Schoonen M. A. (2008) Abiotic ammonium formation in the presence of Ni-Fe metals and alloys and its implications for the Hadean nitrogen cycle. *Geochem. Trans., 9(5),* DOI: 10.1186/1467-4866-9-5.

Smit M. A. and Mezger K. (2017) Earth's early O_2 cycle suppressed by primitive continents. *Nature Geosci., 10,* 788–792.

Solomatov V. S. and Moresi L. N. (1997) Three regimes of mantle convection with non-Newtonian viscosity and stagnant lid convection on the terrestrial planets. *Geophys. Res. Lett., 24(15),* 1907–1910.

Sousa F. L., Nelson-Sathi S., and Martin W. F. (2016) One step beyond a ribosome: The ancient anaerobic core. *Biochim. Biophys. Acta, 1857,* 1027–1038.

Sousa F. L., Preiner M., and Martin W. F. (2018) Native metals, electron bifurcation, and CO_2 reduction in early biochemical evolution. *Curr. Opin. Microbiol., 43,* 77–83.

Spang A., Saw J. H., Jørgensen S. L., Zaremba-Niedzwiedzka K., Martijn J., Lind A. E., van Eijk R., Schleper C., Guy L., and Ettema T. J. G. (2015) Complex archaea that bridge the gap between prokaryotes and eukaryotes. *Nature, 521,* 173–178.

Spang A. and Ettema T. J. G. (2016) The tree of life comes of age. *Nature Microbiol., 1,* 1–2.

Spang A. and Offre P. (2019) Towards a systematic understanding of differences between archaeal and bacterial diversity. *Environ. Microbiol. Rept., 11(1),* 9–12.

Stairs S., Nikmal A., Bučar D.-K., Zheng S.-L., Szostak J. W., and Powner M. W. (2017) Divergent prebiotic synthesis of pyrimidine and 8-oxo-purine ribonucleotides. *Nature Commun., 8,* 15270, DOI: 10.1038/ncomms15270.

Stüeken E. E. (2016) Nitrogen in ancient mud: A biosignature? *Astrobiology, 16(9),* 730–735.

Stüeken E. E., Anderson R. E., Bowman J. S., Brazelton W. J., Colangelo-Lillis J., Goldman A. D., Som S. M., and Baross J. A. (2013) Did life originate from a global chemical reactor? *Geobiology, 11(2),* 101–126.

Stüeken E. E., Catling D. C., and Buick R. (2012) Contributions to late Archaean sulphur cycling by life on land. *Nature Geosci., 5,* 722–725.

Summers D. P. and Chang S. (1993) Prebiotic ammonia from reduction of nitrite by iron (II) on the early Earth. *Nature, 365,* 630–633.

Swanner E. D., Planavsky N. J., Lalonde S. V., Robbins L. J., Bekker A., Rouxel O. J., Saito M. A., Kappler A., Mojzsis S. J., and Konhauser K. O. (2014) Cobalt and marine redox evolution. *Earth Planet. Sci. Lett., 390,* 253–263.

Szabó P., Scheuring I., Czárán T., and Szathmáry E. (2002) *In silico* simulations reveal that replicators with limited dispersal evolve towards higher efficiency and fidelity. *Nature, 420,* 340–343.

Szathmáry E. and Demeter L. (1987) Group selection of early replicators and the origin of life. *J. Theor. Biol., 128,* 463–486.

Takeuchi N., Hogeweg P., and Kaneko K. (2017) Conceptualizing the origin of life in terms of evolution. *Philos. Trans. R. Soc., A, 372 (2109),* DOI: 10.1098/rsta.2016.0346.

Tang M., Chen K., and Rudnick R. L. (2016) Archean upper crust transition from mafic to felsic marks the onset of plate tectonics. *Science, 351(6271),* 372–375.

Tartèse R., Chaussidon M., Gurenko A., Delarue F., and Robert F. (2017) Warm Archean oceans reconstructed from oxygen isotope composition of early-life remnants. *Geochem. Perspect. Lett., 3,* 55–65.

Tashiro T., Ishida A., Hori M., Igisu M., Koike M., Méjean P., Takahata N., Sano Y., and Komiya T. (2017) Early trace of life from 3.95 Ga sedimentary rocks in Labrador, Canada. *Nature 549(7673),* 516–518.

Taubner R.-S., Pappenreiter P., Zwicker J., Smrzka D., Pruckner C., Kolar P., Bernacchi S., Seifert A. H., Krajete A., Bach W., and Peckmann J. (2018) Biological methane production under putative Enceladus-like conditions. *Nature Commun., 9,* 748, DOI: 10.1038/s41467-018-02876-y.

Thomassot E., O'Neil J., Francis D., Cartigny P., and Wing B. A. (2015) Atmospheric record in the Hadean Eon from multiple sulfur isotope measurements in Nuvvuagittuq Greenstone Belt (Nunavik, Quebec). *Proc. Natl. Acad. Sci. U.S.A., 112(3),* 707–712.

Tian F., Kasting J. F., and Zahnle K. (2011) Revisiting HCN formation in Earth's early atmosphere. *Earth Planet. Sci. Lett., 308(3),* 417–423.

Trail D., Boehnke P., Savage P. S., Liu M.-C., Miller M. L., and Bindeman R. (2018) Origin and significance of Si and O isotope heterogeneities in Phanerozoic, Archean, and Hadean zircon. *Proc. Natl. Acad. Sci. U.S.A., 115,* 10287–10392.

Tuck A. (2002) The role of atmospheric aerosols in the origin of life. *Surv. Geophys., 23(5),* 379–409.

Ueno Y., Yamada K., Yoshida N., Maruyama S., and Isozaki Y. (2006) Evidence from fluid inclusions for microbial methanogenesis in the early Archaean era. *Nature, 440,* 516–519.

Vaida V. (2017) Prebiotic phosphorylation enabled by microdroplets. *Proc. Natl. Acad. Sci. U.S.A., 114(47),* 12359–12361.

Varma S. J., Muchowska K. B., Chatelain P., and Moran J. (2018) Native iron reduces CO_2 to intermediates and end-products of the acetyl-CoA pathway. *Nature Ecol. Evol., 2,* 1019–1024, DOI: 10.1038/s41559-018-0542-2.

Vearncombe S., Barley M. E., Groves D. I., McNaughton N. J., Mikucki E. J. and Vearncombe J. R. (1995) 3.26 Ga black smoker-type mineralization in the Strelley belt, Pilbara craton, Western Australia. *J. Geol. Soc. (London, U.K.), 152(4),* 587–590.

Viehmann S., Bau M., Hoffmann J. E., and Münker C. (2015) Geochemistry of the Krivoy Rog Banded Iron Formation, Ukraine, and the impact of peak episodes of increased global magmatic activity on the trace element composition of Precambrian seawater. *Precambrian Res., 270,* 165–180.

Waite J. H., Combi M. R., Ip W. H., Cravens T. E., McNutt R. L., Kasprzak W., Yelle R., Luhmann J., Niemann H., Gell D., and Magee B. (2006) Cassini Ion and Neutral Mass Spectrometer: Enceladus plume composition and structure. *Science, 311,* 1419–1423.

Waite J. H., Lewis, W. S., Magee B. A., Lunine J. I., McKinnon W. B., Glein C. R., Mousis O., Young D. T., Brockwell T., Westlake J., and Nguyen M. J. (2009) Liquid water on Enceladus from observations of ammonia and ^{40}Ar in the plume. *Nature, 460,* 487–490.

Waite J. H., Glein C. R., Perryman R. S., Teolis B. D., Magee B. A., Miller G., Grimes J., Perry M. E., Miller K. E., Bouquet A., Lunine J. I., Brockwell T., and Bolton S. J. (2017) Cassini finds molecular hydrogen in the Enceladus plume: Evidence for hydrothermal processes. *Science, 356,* 155–159.

Wacey D., Saunders M., Brasier M. D., and Kilburn M. R. (2011) Earliest microbially mediated pyrite oxidation in ~3.4 billion-year-old sediments. *Earth Planet. Sci. Lett., 301(1–2),* 393–402.

Weiss M. C., Sousa F. L., Mrnjavac N., Neukirchen S., Roettger M., Nelson-Sathi S., and Martin W. F. (2016) The physiology and habitat of the last universal common ancestor. *Nature Microbiol., 1,* 16116.

Westall F. and Brack A. (2018) The importance of water for life. *Space Sci. Rev., 214,* 50.

Westall F., Hickman-Lewis K. K., Hinman N., Gautret P., Campbess K. A., Breheret J. G., Foucher F., Hubert A., Sorieul S., Dass A. V., Kee T. P., Georgelin T., and Brack A. (2018) A hydrothermal-sedimentary context for the origin of life. *Astrobiology, 18,* 259–293.

Williams T. A., Szöllősi G. J., Spang A., Foster P. G., Heaps S. E., Boussau B., Ettema T. J., and Embley T. M. (2017) Integrative modeling of gene and genome evolution roots the archaeal tree of life. *Proc. Natl. Acad. Sci. U.S.A., 114(23),* E4602–E4611.

Woese C. (1987) Bacterial evolution. *Microbiol. Rev., 51(2),* 221–271.

Woese C. (1998) The universal ancestor. *Proc. Natl. Acad. Sci. U.S.A., 95(12),* 6854–6859.

Woese C., Kandler O., and Wheelis M. L. (1990) Towards a natural system of organisms: Proposal for the domains Archaea, Bacteria, and Eucarya. *Proc. Natl. Acad. Sci. U.S.A., 87,* 4576–4579.

Wolf E. T. and Toon O. B. (2010) Fractal organic hazes provided an ultraviolet shield for early Earth. *Science, 328(5983),* 1266–1268.

Wong M. L., Charnay B. D., Gao P., Yung Y. L., and Russell M. J. (2017) Nitrogen oxides in early Earth's atmosphere as electron acceptors for life's emergence. *Astrobiology, 17(10),* 975–983.

Wordsworth R. and Pierrehumbert R. (2013) Hydrogen-nitrogen greenhouse warming in Earth's early atmosphere. *Science, 339(6115),* 64–67.

Wu H., Wells M. L., and Rember R. (2011) Dissolved iron anomaly in the deep tropical-subtropical Pacific: Evidence for long-range transport of hydrothermal iron. *Geochim. Cosmochim. Acta, 75(2),* 460–468.

Wu Y.-W., Tang Y.-H., Tringe S. G., Simmons B. A., and Singer S. W. (2014) MaxBin: An automated binning method to recover individual genomes from metagenomes using an expectation-maximization algorithm. *Microbiome, 2,* 26.

Zahnle K. J. (1986) Photochemistry of methane and the formation of hydrocyanic acid (HCN) in the Earth's early atmosphere. *J. Geophys. Res.: Atmos., 91(D2),* 2819–2834.

Zaremba-Niedzwiedzka K., Caceres E. F., Saw J. H., Bäckström D., Juzokaite L., Vancaester E., Seitz K. W., Anantharaman K., Starnawski P., Kjeldsen K. U., Stott M. B., Nunoura T., Banfield J. F., Schramm A., Baker B. J., Spang A., and Ettema T. J. G. (2017) Asgard archaea illuminate the origin of eukaryotic cellular complexity. *Nature, 541,* 353–358.

Zhou Z., Liu Y., Li M., and Gu J-D. (2018) Two or three domains: A new view of tree of life in the genomics era. *Appl. Microbiol. Biotechnol., 102,* 3049–3058.

Stüeken E. E., Fournier G., and Eyster A. (2020) Life as a planetary process. In *Planetary Astrobiology* (V. Meadows et al., eds.), pp. 93–120. Univ. of Arizona, Tucson, DOI: 10.2458/azu_uapress_9780816540068-ch004.

Life as a Planetary Process

E. E. Stüeken
University of St. Andrews

G. P. Fournier and A. Eyster
Massachusetts Institute of Technology

Life on Earth is strongly integrated into geochemical cycles. Just as biological metabolic processes are dependent on environmental sources of redox couples, so too can life alter the redox state of surface environments. This relationship between Earth and its biosphere has evolved through distinct stages over the last 4 b.y., each of which can be thought of as a potential analog for other inhabited worlds. As such, the geobiological evolution of our planet reveals several key principles that may guide our search for life elsewhere: (1) On a completely anoxic planet with an anaerobic biosphere, the sustenance of life is contingent upon fundamental geological properties of the planet, including mantle differentiation, volcanism, hydrothermal activity, and photochemical reactions in the atmosphere. (2) The transitions toward crustal oxidation and atmospheric oxygenation may require feedbacks between global tectonic transformations and biological innovations. (3) It is conceivable that extended intervals of redox stratification are common on both inhabited and uninhabited worlds as planets are likely to become more oxic over their lifetime via multiple biotic and abiotic processes. (4) An evolving and diversifying terrestrial (land-based) biosphere likely played an important role for several major planetary changes, including atmospheric and ocean oxygenation. Conversely, an environmentally diverse planet with oceans and land masses may be favorable for the development of a diverse and resilient biosphere. Understanding Earth through time provides a process-based, universally relevant framework that should illuminate the fundamental controls on planetary surface environments elsewhere. We focus here on the earlier parts of Earth's history, 3.5 to 0.5 b.y. ago. The microbially-dominated biosphere during that time is part of a very different Earth system than we observe today, broadening our perspective on how life can persist across a wide range of planetary states. We end with a discussion of a more oxidizing younger Earth.

"Nature is ever at work building and pulling down, creating and destroying, keeping everything whirling and flowing, allowing no rest but in rhythmical motion, chasing everything in endless song out of one beautiful form into another." — John Muir

1. INTRODUCTION

Earth is the only planet in our solar system with active geological and hydrological cycles. It is also the only planet we know of that hosts a flourishing biosphere at its surface. All living organisms on Earth are descendant from a single origin and share one last universal common ancestor (LUCA) or ancestral community. Since then, life has adapted to a wide range of environments, from the deep crust to atmospheric aerosols, from hydrothermal vents to sea ice brine pockets. This diverse biosphere that we see on Earth today must be a product of both biological evolution and planetary processes. The term "planetary" is used here for geologic, atmospheric, and environmental processes that do not require life.

The biosphere has modified the planet through metabolic and behavioral traits. Fundamentally, biological metabolism and replication require that complex organic structures are generated from inorganic substrates using chemical energy obtained from the environment. Thermodynamically speaking, life decreases entropy within itself, and doing so requires a constant supply of chemical energy to maintain net exergonic (i.e., energy-releasing) reactions (e.g., *Schoepp-Cothenet et al.,* 2012). All life as we know it is dependent on electron transfer, i.e., energy released in exothermic (i.e., heat releasing) redox reactions. Cells harvest this energy to produce phosphorylated compounds such as nicotinamide adenine dinucleotide phosphate (NADP), NADPH (a reduced form of NADP), and adenosine triphosphate (ATP), the universal energy currency of life. These compounds fuel biosynthesis and many other cellular processes, including nutrient uptake and CO_2 fixation in autotrophs (organisms that can derive energy by forming organic compounds from simple substances). This basic thermodynamic concept has two major implications for the linkage between Earth and life: (1) The sustenance of the biosphere requires environmental sources of nutrients and redox couples that can maintain the supply of chemical energy, and (2) biological activity catalyzes reactions between oxidized and reduced substrates that are otherwise unreactive and thereby produces a new suite of gases, solutes, and minerals with implications for the evolution of the crust, ocean, and atmosphere (e.g., *Hazen and Ferry,* 2010). The expansion of the biosphere has thus converted geochemical cycles into biogeochemical cycles and modified those planetary processes that initially created a habitable platform.

The co-evolutionary trajectory of Earth and life can be subdivided by the appearance of what is arguably the most influential metabolic characteristic of our biosphere: the ability to produce free O_2 gas from water and sunlight through oxygenic photosynthesis. With the invention of this metabolism, life became capable of accessing a source of energy that is orders of magnitude larger than that of hydrothermal reductants and volcanism alone (*Canfield et al.,* 2006). Because O_2 is a powerful metabolic substrate for respiration and chemotrophic reactions (which obtain energy by the oxidation of electron donors in their environments) (*Schoepp-Cothenet et al.,* 2012), many other non-photosynthetic organisms also thrived as a consequence. With the oxygenation of surface environments, the size and productivity of the biosphere has thus become progressively more independent from internal abiotic planetary processes.

It is uncertain if oxygenic photosynthesis is likely to evolve on other inhabited worlds, because it was an evolutionary singularity on Earth. When using Earth as an analog for life elsewhere, the first important stage to consider is therefore the time before the appearance or radiation (i.e., diversification and environmental expansion) of oxygenic photosynthesis, which may be representative of a biosphere that is strongly shaped by planetary sources of energy and nutrients through volcanism, hydrothermal fluids and photochemical reactions (section 2). On Earth, this probably encompasses the early Archean before 3 b.y. ago (3 Ga) (*Lyons et al.,* 2014). The Archean spans from about 3.8 to 2.5 Ga and is preceded by the Hadean (4.5–3.8 Ga), of which no significant rock record is preserved. The second major interval in Earth's redox evolution is that of a redox-stratified world, where O_2 accumulates in surface settings proximal to its biogenic source but is quantitatively consumed in the deep ocean by high abundances of geochemical reductants (section 3). This scenario likely characterized the Proterozoic Earth (2.5–0.54 Ga) and it may be common on other inhabited planets where the biological production of O_2 in the photic zones competes with geological fluxes of reducing agents from depth. Such a stratified world is defined by strong zonation of habitability. The most recent stage observed on Earth is that of complete oxygenation of the

deep ocean and rise of oxygen in the atmosphere to levels stabilized at or near current levels. This allowed complex aerobic organisms to thrive in all surface settings and perhaps leads to the progressive oxidation of the upper mantle over billion-year timescales (*Evans,* 2012). On Earth, this transition happened between the Neoproterozoic and early Phanerozoic (0.54 Ga to modern) (*Lyons et al.,* 2014), and it was perhaps a necessary requirement for the evolution of large animals and intelligent life (section 4). Importantly, the transitions between these three stages of oxygenation coincided with large-scale geological transformations of our planet that may have played a critical role in the history of the ocean, atmosphere, and biosphere. These feedbacks may be what ultimately constrains the evolution of habitability on Earth and other Earth-like planets.

2. LIFE BEFORE OXYGEN: THE EARLY ARCHEAN

From the origin of life to the evolution of oxygenic photosynthesis, the planetary biosphere persisted and diversified in a world without molecular oxygen. In such an anaerobic biosphere, not only was oxygen unavailable for aerobic respiration, but it was also unavailable for geochemical and atmospheric processes that produce a whole series of secondary oxidants important to life today, such as sulfate and nitrate. With only very small volcanic and photochemical sources of these substrates (section 2.2.2), the earliest biosphere would thus have been challenged by a low supply of metabolic energy sources with high redox coupling potentials. At the same time, geological processes may have been fundamentally distinct, with slowly decreasing rates of asteroid impacts (*Boehnke and Harrison,* 2016; *Zellner,* 2017), smaller continental land masses, and perhaps a different style of tectonics (*Condie,* 2016). Earth before oxygen is thus a truly alien world that may serve as an important analog for other planets with life where oxygenic photosynthesis has not evolved.

2.1. Setting the Stage: Conditions on the Early Earth

Life emerged on a geologically and hydrologically active Earth and can thus perhaps be considered an outgrowth of global geochemical cycles (*Stüeken et al.,* 2013). If so, the early biosphere needs to be considered in a global planetary context, shaped by fundamental properties of Earth's mantle, crust, hydrosphere, and atmosphere.

2.1.1. Mantle differentiation and temperatures. Earth's largest geological component is the mantle, which acts as a major control of surface processes, because it regulates the style of plate tectonics and volcanism and determines initial conditions for the subsequent evolution of the planet. Two parameters of a planet's mantle are particularly relevant for surface processes: (1) temperature, which sets the pace for convection and plate tectonics, and (2) composition and differentiation, which influence core formation as well as the redox state of volcanic gases. Multiple studies of ancient volcanic rocks suggest that Earth's early mantle was at least a hundred degrees hotter than it is today (*Grove and Parman,* 2004; *Herzberg et al.,* 2010), in part due to the abundance of radioactive elements (*Arevalo et al.,* 2009) and residual heat from core crystallization and planetary accretion (*Labrosse et al.,* 1997; *Herzberg et al.,* 2010; *Jaupart et al.,* 2016). One major consequence of a hotter mantle may have been an overall higher degree of mantle melting with larger volumes of magma production, a thicker oceanic lithosphere (*Sleep and Windley,* 1982), and possibly high rates of outgassing (*Richter,* 1985; *Martin et al.,* 2007). Alternative interpretations suggest similar to slower plate velocities for the early Earth coupled with enhanced activity of arc magmatism (*Korenaga,* 2006, 2013). Life on the young Earth may have been exposed to higher rates of volcanic and hydrothermal activity.

The composition of volcanic gases is partially determined by planetary differentiation. The mantle, particularly the upper mantle, is oxidized relative to chondrites that are thought to represent Earth's building blocks (*Frost and McCammon,* 2008; *Rubie et al.,* 2011; *Gu et al.,* 2016). Importantly, the term "oxidized" here is relative; although the upper mantle is oxidized relative to chondrites, it is reduced relative to Earth's surface today. The earliest state of the mantle is determined by accreting material composition and core-formation dynamics (*McCammon,* 2005; *Frost and McCammon,* 2008). Measurements generally suggest that the upper mantle redox state has not changed significantly since at least 3.5 Ga (*Nicklas et al.,* 2018). Thus the upper mantle became oxidized relatively early in its history, possibly via removal of metallic iron during core formation (*McCammon,* 2005). The resultant mantle redox state regulates the fate and transfer of metals during magmatism and the inputs of volcanic gases to the atmosphere. Relative to the modern atmosphere, the mantle thus provides a major source of reductants to Earth's surface with varying fluxes through geologic time (e.g., *Frost and McCammon,* 2008, and references therein; *Nicklas et al.,* 2018). If so, one could postulate that the initial composition and core formation of a terrestrial planet is an important determinant for the oxygenation timescales of its surface habitats (discussed in more detail in section 3.3.3).

Another linkage between the deep Earth and the atmosphere is the establishment of the magnetic field after the formation of a metallic core, which is important for the protection of the atmosphere from solar wind stripping (*Lammer et al.,* 2014). There is evidence for an active dynamo and for geomagnetic fields with intensity 50–70% of the present day at 3.45 Ga (*Biggin et al.,* 2011). However, the nature of the magnetic field before that time — and thus its importance for the origin of life — has yet to be established (*Labrosse,* 2003; *Tarduno et al.,* 2015; *Weiss et al.,* 2015).

2.1.2. Tectonics and the growth of continental crust. Plate tectonics may be one of the most important geodynamic features that determine Earth's habitability, as it imposes the tempo of large-scale geochemical fluxes and the growth of continental crust. Continental crust may in turn have been important for early life, because its formation

modified the composition of the mantle and the atmosphere. Additionally, felsic (silica-rich, magnesium-poor) continental crust provided key nutrients to the biosphere through erosion, offered shallow water habitats, and contributed to modulating atmospheric pCO_2 through weathering. Today, felsic magmas are formed when hydrated oceanic crust is subducted and partially molten. The existence of continental crust itself has therefore been interpreted as evidence of subduction and modern-style plate tectonics (*Condie,* 2016; *Greber et al.,* 2017). There are two end-member models of continental crust growth over Earth's history. The first one suggests early rapid growth prior to 4.0 Ga followed by little increase in volume thereafter (*Armstrong,* 1968; *Campbell,* 2003; *Harrison et al.,* 2005; *Rosas and Korenaga,* 2018), whereas the other posits a continuous growth of continental crust over time (*Veizer and Jansen,* 1979; *Allègre and Othman,* 1980; *McLennan and Taylor,* 1982). Several recent studies suggest an inflection point in crustal growth rates at ca. 3.0–2.5 Ga with a transition from a predominantly mafic crust to andesitic or felsic compositions (*Lee et al.,* 2016; *Smit and Mezger,* 2017) (Fig. 1), possibly coinciding with the onset of modern-style plate tectonics (*Dhuime et al.,* 2012, 2015; *Laurent et al.,* 2014; *Condie,* 2016; *Tang et al.,* 2016; *Nebel-Jacobsen et al.,* 2018) (reviewed by *Hawkesworth et al.,* 2016). The geodynamic regime of the early Earth is still being explored, with mobile lid plate tectonics or stagnant lid convection with frequent large-scale magmatic events suggested (*Solomatov and Moresi,* 1997; *Lenardic and Crowley,* 2012; *Moore and Webb,* 2013; *O'Neill and Debaille,* 2014; *Rozel et al.,* 2017). It is also possible that subduction could have been intermittent, which may reconcile the stagnant lid model with geochemical evidence of early plate tectonics (*O'Neill and Debaille,* 2014). This emerging view would imply that Earth's earliest biosphere may have lived on a planet with a different tectonic style than what we have today, and the same may be true for life elsewhere.

In addition, the early Earth may have been a water world. It was suggested that the accumulation of heat-producing radioactive isotopes in the early continental crust could have led to hotter, more ductile, and weaker continents, perhaps unable to support high topography and delaying continental emergence until the Archean-Proterozoic transition (*Rey and Houseman,* 2006; *Rey and Coltice,* 2008). More recent work continues to explore the emergence of continents. Some results suggest that although Hadean oceans may have been twice as voluminous, continental freeboard has been roughly constant since at least the early Proterozoic (*Korenaga et al.,* 2017). Alternatively, others have proposed that large-scale emergence of the continents may not have occurred until the Neoproterozoic (*Lee et al.,* 2018).

2.1.3. Surface temperature and atmospheric composition. The composition of a planetary atmosphere is critical for global climate and habitability. On the early Earth, billions of years ago, the relatively weak solar luminosity would have required a strong greenhouse effect to prevent permanent glaciations (*Sagan and Mullen,* 1972). Very warm surface temperatures (50°–70°C) were once inferred from

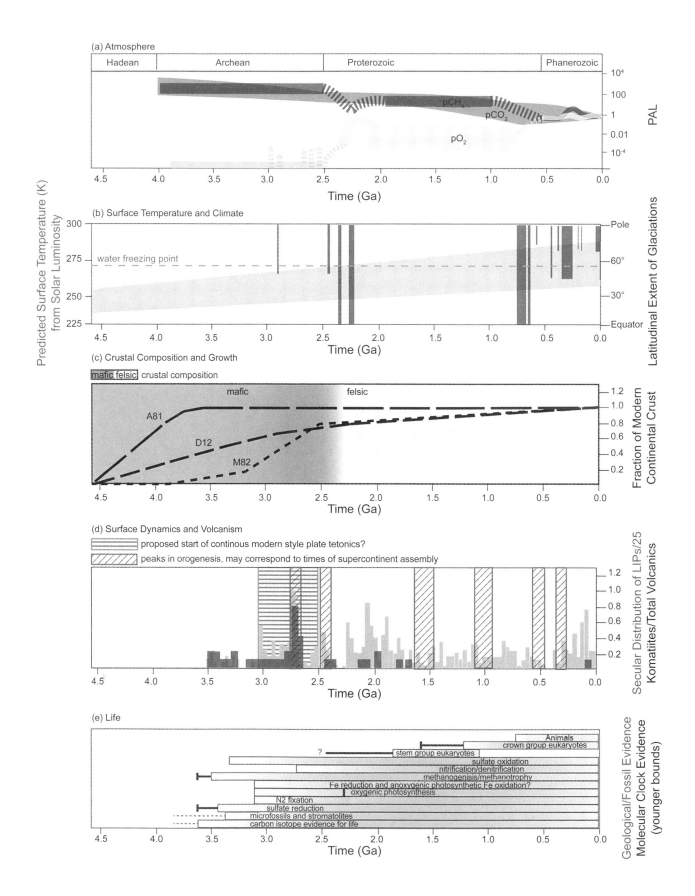

(a) Atmosphere

(b) Surface Temperature and Climate

(c) Crustal Composition and Growth

(d) Surface Dynamics and Volcanism

(e) Life

the oxygen isotope record of marine Archean cherts (*Knauth,* 2005). However, oxygen isotopes in phosphates (*Blake et al.,* 2010) and combined oxygen-deuterium isotopes in cherts (*Hren et al.,* 2009) point to more moderate estimates of 26°–35°C for the Archean surface ocean. Sedimentological evidence of at least regional glaciations at 3.5 Ga and 2.9 Ga support relatively cool ocean temperatures (*Young et al.,* 1998; *de Wit and Furnes,* 2016). A lower albedo resulting from smaller continental land masses and fewer clouds could have provided some warming under the faint young Sun (*Rosing et al.,* 2010), but the major burden likely rested on greenhouse gases such as CO_2 and CH_4 (*Goldblatt and Zahnle,* 2011), with possible enhancement of the greenhouse effect by higher N_2 pressures and potentially higher H_2 pressures early in Earth's history (*Goldblatt et al.,* 2009; *Wordsworth and Pierrehumbert,* 2013).

The early Archean atmosphere was likely rich in CO_2 as a result of rapid volcanic outgassing (*Walker et al.,* 1981), although its evolution to Phanerozoic levels is difficult to constrain from geochemical proxies (*Sheldon,* 2006). Methanogenic organisms and water-rock reactions (serpentinization) could have sustained high levels of atmospheric CH_4 during the Archean (*Pavlov et al.,* 2000; *Kasting,* 2005; *Smit and Mezger,* 2017), perhaps to the point of hydrocarbon haze formation (*Domagal-Goldman et al.,* 2008; *Haqq-Misra et al.,* 2008; *Zerkle et al.,* 2012; *Izon et al.,* 2015), which may have, in turn, provided a net cooling effect on global climate. Overall, the vast uncertainty about Archean surface temperature is a clear impediment to the recognition of general co-evolutionary trends between early life and global climate. But it is possible that life already played a critical role in the Archean carbon cycle, in particular for the generation of CH_4 (*Kasting et al.,* 2001; *Haqq-Misra et al.,* 2008; *Wolfe and Fournier,* 2018).

Another key characteristic of the Archean atmosphere was the lack of free O_2 gas and ozone shielding. The most compelling evidence for a largely anoxic Archean atmosphere is the record of mass-independent sulfur isotope fractionation (S-MIF) in sedimentary rocks up until 2.3 Ga (*Farquhar et al.,* 2000). This isotopic signature is caused by photolysis and/or photoexcitation of volcanogenic SO_2 via ultraviolet light penetrating an anoxic atmosphere without an ozone shield (*Farquhar et al.,* 2000, 2001). Although the fractionation mechanisms are still under investigation (*Ono et al.,* 2013; *Claire et al.,* 2014), results of one-dimensional photochemical modelling show that less than 10^{-5}–$10^{-7}\times$ the present atmospheric level (PAL) of O_2 are required for the production and preservation of MIF signals (*Pavlov and Kasting,* 2002). Under these conditions the ocean would also have been devoid of oxygen and likely dominated by dissolved Fe^{2+} (*Poulton and Canfield,* 2011), consistent with the frequent and voluminous occurrence of banded iron formations.

2.2. What Sustained the Earliest Biosphere?

A non-photosynthetic biosphere would be entirely dependent on and limited by environmental sources of chemical disequilibrium. The disequilibrium is needed as a source of electrochemical energy by redox coupling through chemotrophy. Chemotrophy is in contrast to phototrophy, which generates high-energy electrons from low-energy electron donors and photons, by far the dominant source of biological energy on Earth today. The evolution of phototrophy occurred very early in Earth's history, likely before 3.5 Ga (the oldest evidence of phototrophy; see section 2.3). However, before the later invention of oxygenic photosynthesis, even these early phototrophs were in their own way likely limited by geochemical fluxes, as they used electron donors other than water, e.g., Fe^{2+} or sulfide. As a consequence, geological and atmospheric processes likely had greater influence in shaping and sustaining the earliest biosphere than at other times in Earth's history.

2.2.1. Sources of electron donors. On the early Archean Earth, prior to the onset of photosynthesis, electron donors were abundant and continuously supplied by hydrothermal and volcanic processes as well as by anoxic weathering of continental crust. Volcanic outgassing would likely have created a major flux of CO and H_2 into the early atmosphere (reviewed by *Catling,* 2014). Additional H_2 would have been supplied by hydrothermal hydration reactions of olivine — known as serpentinization (e.g., *Neubeck et al.,* 2014). Hydrothermal vents would also have supplied large amounts of Fe^{2+} and Mn^{2+} into the early ocean, as evidenced by massive banded iron formations in the rock record of this time (*Lyle,* 1976; *Morris and Horwitz,* 1983; *Isley and Abbott,* 1999). An additional Fe^{2+} source was probably enhanced continental weathering of mafic crust under a CO_2-rich atmosphere that created relatively more acidic rainwater in which Fe^{2+} was soluble (*Hao et al.,* 2017b). Dissolved Fe^{2+} levels in the ocean may have been on the order of 0.1–1 μM (*Holland,* 2003; *Tosca et al.,* 2016). Dissolved sulfide levels were probably low in Archean seawater,

Fig. 1. (facing page) **(a)** Atmospheric pO_2 after *Lyons et al.* (2014), CO_2 after *Kah and Riding* (2007), Archean CH_4 levels after *Claire et al.* (2006), Proterozoic CH_4 after *Olson et al.* (2016b), and Phanerozoic CH_4 after *Bartdorff et al.* (2008). **(b)** Surface temperature predicted from solar luminosity after *Rampino and Caldeira* (1994) and latitudinal extent of glaciations after *Hoffman and Schrag* (2002) with updated Neoproterozoic age constraints from *Rooney et al.* (2014). **(c)** Change from mafic to felsic crust after *Lee et al.* (2016) and *Smit and Mezger* (2017) in addition to others (see text). Proposed onset of continuous plate tectonics after *Condie* (2016) and *O'Neill and Debaille* (2014) in addition to others (see text). Models for crustal growth after *Armstrong* (1981) (A81), *McLennan and Taylor* (1982) (M82), and *Dhuime et al.* (2012) (D12). Peaks in global orogenies after *Bradley* (2011) and *Campbell and Allen* (2008). Mantle LIPs after *Condie et al.* (2015) and Komatiites after *Isley and Abbot* (1999) and *Dostal* (2008). **(d)** Record of biological metabolisms as inferred from geochemical and paleontological proxies after *Lyons et al.* (2015), modified after references in the text.

as sulfide sourced from hydrothermal processes would have been titrated by Fe^{2+} to form insoluble mineral deposits, thus maintaining Fe^{2+}-dominated (ferruginous) conditions in the ocean (*Poulton and Canfield, 2011*). These hydrothermal sulfide deposits themselves could have acted as electron donors in microbial systems.

Another important source of electron donors would have been reduced carbon. Under anoxic conditions, organic matter and CH_4 — either biogenic or produced in Fischer-Tropsch-type reactions between CO or CO_2 and H_2 in hydrothermal settings (*McCollom and Seewald, 2006; Lazar et al., 2012; McCollom, 2013*) — would have been able to accumulate in surface environments. Hence, overall, reductants would have been bioavailable in all habitats on the Archean Earth, from land surfaces to the bottom of the ocean. The major limitation of the chemotrophic biosphere was the supply of oxidizing agents (*Lyons et al., 2014*).

2.2.2. Sources of electron acceptors. Prior to the rise of oxygenic photosynthesis, the only significant sources of free O_2 gas were photochemical reactions in the upper atmosphere with modeled mixing ratios of around 10^{-12} (*Haqq-Misra et al., 2011; Kurzweil et al., 2013*). Only vertical and lateral transport of this abiotically produced O_2 gas may have locally sustained low levels of aerobic respiration (*Haqq-Misra et al., 2011*). Oxygen alone could therefore not have supported a significant chemotrophic biosphere. However, another consequence of low atmospheric O_2 and the absence of an ozone shield was a strong UV flux into the lower atmosphere, where it could induce a number of photochemical reactions that generated alternative oxidants (Fig. 2). Sulfur isotopes in Archean sedimentary rocks

provide evidence of photochemical disproportionation of volcanogenic SO_2 into elemental sulfur and sulfate (*Farquhar et al., 2000*). Although the exact mechanism for this reaction is still debated, it is widely agreed that it is a relic of UV photochemistry that was shut down with the rise of atmospheric O_2 and ozone, which limited the production and preservation of this photochemical signature (*Pavlov and Kasting, 2002; Domagal-Goldman et al., 2008; Halevy, 2013; Claire et al., 2014*). All three constituents in this reaction — the volcanic SO_2 gas, which dissolves in water to form SO_3^{2-}, and the two major products S_8 and SO_4^{2-} — can serve as electron acceptors in biological metabolism.

Besides sulfur, iron and manganese may also have been affected by UV radiation. Experimental data and models suggest that under Archean conditions, dissolved Fe^{2+} and Mn^{2+} could have been oxidized to Fe^{3+} and Mn^{4+}, respectively (*Braterman et al., 1983; François, 1986; Anbar and Holland, 1992*). It remains uncertain how important these reactions were on the Archean Earth, but given the impact of photochemistry on the sulfur cycle, it is conceivable that at least minor amounts of Fe^{3+} and Mn^{4+} were produced by this mechanism even before microbial iron-oxidizers evolved and took control over iron redox chemistry (*Konhauser et al., 2002; Konhauser et al., 2007a*). Another source of atmospheric oxidants may have been lightning, which can catalyze the reaction between CO_2 and N_2 gas to form various nitrogen oxides (NO, NO_2, HNO_2, and HNO_3) (*Kasting and Walker, 1981; Navarro-González et al., 1998*). Similar reactions have also been proposed for volcanic eruptions that could have generated HNO_3 (*Mather et al., 2004*). There is so far no evidence that the resulting NO_x fluxes were large

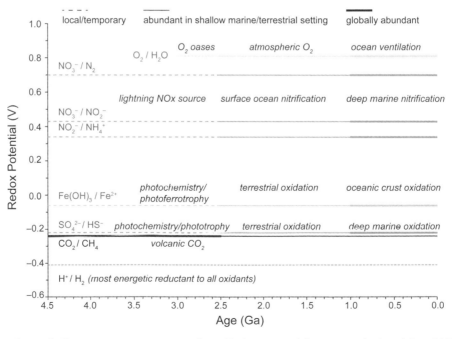

Fig. 2. Availability of metabolic energy sources over time. Redox potentials were calculated for pH 7, using thermodynamic data from *Stumm and Morgan* (1996). Oxidants with high redox potentials can be combined with reductants of lower redox potentials. For example, O_2 can oxidize all the reduced compounds shown in the figure, but NO_3^- can only oxidize Fe^{2+}, HS^-, CH_4, and H_2.

enough to sustain early ecosystems over long timescales, but this process likely contributed to the overall pool of oxidants on the Archean Earth.

One of the most abundant oxidants on the Archean Earth would probably have been volcanogenic CO_2, which can be used as an electron acceptor by autotrophic methanogens and acetogens, coupled to H_2 as an electron donor. However, these metabolisms are energetically less favorable than sulfate reduction, iron reduction, denitrification, or aerobic respiration (*Schoepp-Cothenet et al., 2012*). Hence a large proportion of the pre-photosynthetic biosphere on the early Earth was probably sustained through oxidants supplied by atmospheric reactions and triggered by sunlight and lightning discharge (*Kipp and Stüeken, 2017*). Biogeochemical models of the pre-photosynthetic Archean biosphere point to relatively low levels of biological productivity compared to today (*Kharecha et al., 2005; Canfield et al., 2006*). This result is conceivably transferrable to other non-photosynthetic biospheres.

2.2.3. Sources of other nutrients. Besides sources of electrochemical energy, life also requires a constant supply of major and minor nutrients that form the building blocks of biological macromolecules. The most bio-limiting nutrients are typically nitrogen, phosphorus, or various transition metals (*Frausto da Silva and Williams, 2001; Wackett et al., 2004*).

On the Archean Earth — as today — the largest reservoir of nitrogen would have been atmospheric N_2 gas. Abiotic fixation of N_2 into bioavailable ammonium and nitrogen oxides via hydrothermal reactions and lightning discharge may have been sufficient to sustain a small biosphere (*Kasting and Walker, 1981; Brandes et al., 1998; Navarro-González et al., 1998; Schoonen and Xu, 2001; Mather et al., 2004; Tian et al., 2011*). However, a spatially and temporally heterogeneous distribution of these sources could have favored an early origin of biological N_2 fixation (*Kasting and Siefert, 2001; Stüeken et al., 2015b, 2016a; Weiss et al., 2016*). This metabolic innovation would have made the early biosphere slightly less dependent on abiotic planetary processes and was perhaps similar in importance to the emergence of autotrophic CO_2 fixation. Furthermore, an early onset of biological N_2 drawdown without an oxidative return flux of organic ammonium to N_2 gas may have resulted in a relatively low atmospheric N_2 reservoir (*Som et al., 2012, 2016; Stüeken et al., 2016b*).

If N_2 fixation evolved early, phosphorus would likely have been the more bio-limiting macronutrient. Geochemical data and thermodynamic models both suggest relatively low concentrations of dissolved phosphorus in Precambrian seawater (*Bjerrum and Canfield, 2002; Konhauser et al., 2007b; Planavsky et al., 2010; Reinhard et al., 2017*). Although the input of phosphorus via continental weathering may have been enhanced by more acidic rainwater (*Hao et al., 2017b*), recycling of phosphorus within the ocean would have been suppressed under anoxic conditions (*Kipp and Stüeken, 2017; Laakso and Schrag, 2017*). Furthermore, a large fraction of any liberated phosphorus was likely scav-

enged by iron minerals and diagenetic carbonate fluorapatites (*Laakso and Schrag, 2014; Derry, 2015; Jones et al., 2015; Reinhard et al., 2017*).

With regard to transition metals, their availability was once thought to have been orders of magnitude lower under anoxic conditions than it is today (*Saito et al., 2003*) because (1) oxidative weathering fluxes would have been insignificant before the Great Oxidation Event (GOE) at around 2.3 Ga (see section 3) and (2) metals were sequestered rapidly into authigenic (i.e., formed in place) sulfide minerals in marine sediments, resulting in small marine inventories (*Anbar and Knoll, 2002; Scott et al., 2008*). However, new data indicate at least moderate levels of certain metals in Archean seawater. First, acidic continental weathering may have created significant riverine fluxes of metals, such as Mn, Co, Ni, Zn, and W (*Hessler and Lowe, 2006; Murakami et al., 2016; Hao et al., 2017a*). Second, seafloor weathering of (ultra-)mafic crust and enhanced hydrothermal activity were probably major sources of Ni (*Konhauser et al., 2009*) and Co (*Swanner et al., 2014*) and possibly minor sources of Mo (*McManus et al., 2002; Wheat et al., 2002, 2017; Greaney et al., 2016*) and Zn (*Scott et al., 2012; Robbins et al., 2013*) and perhaps other elements. Geochemical data suggest that hydrothermal processes likely imparted a stronger control on Archean ocean chemistry than they do today in the form of higher fluxes (*Viehmann et al., 2015; Robbins et al., 2016*). The biological nutrient demand probably adapted to differing supplies (*Zerkle et al., 2005; Dupont et al., 2006, 2010*), meaning that the relative proportions of elements in marine biomass — known as Redfield ratios — may have been markedly different in the Archean ocean compared to today (*Planavsky, 2014; Konhauser et al., 2017*). Such adaptability in Archean biochemistry could serve as an analog for life on other planets with different inventories of major and minor nutrients.

2.3. Relics of an Anaerobic Biosphere

Despite erosion, subduction, and alteration of most Archean sedimentary rocks, some direct evidence of Earth's early habitability is preserved in the form of body fossils, trace fossils, isotopic fractionations and inherited genetic traits in living organisms (Fig. 1). These direct and indirect remains inform us about metabolic activities and behavioral traits in deep time.

2.3.1. Geochemical evidence. The most widely used tool to track biological metabolism is isotope fractionation. Enzymatic processes within living cells often show a slight preference for substrates containing the relatively lighter isotopes of the constituent atoms. Therefore, when a substrate is not limiting (that is, not all of the available source is consumed), the biologically processed product will tend to be isotopically lighter than the background material. Perhaps the most ancient and basic proposed evidence of a biosphere on the early Earth is in the form of the fractionation of carbon isotopes, specifically $\delta^{13}C$ {the difference from the expected background ratio of

^{13}C and ^{12}C, relative to an international standard: $\delta^{13}C = [(^{13}C/^{12}C)_{sample}/(^{13}C/^{12}C)_{standard}-1] \times 1000\}$. When organisms die, this isotopic signature will be preserved in their residual biomass. The oldest proposed isotopic evidence of early life comes from detrital zircon grains that formed ~4.1 Ga but were subsequently reworked into younger rock (*Bell et al.,* 2015). Isotopically light organic material incorporated during their crystallization may represent a small sample of the earliest biosphere. The oldest sedimentary rocks with a possibly biogenic carbon isotopic signal are from 3.7–3.9-Ga rocks in Greenland (*Mojzsis et al.,* 1996; *Rosing,* 1999; *Tashiro et al.,* 2017). Similarly, relatively light nitrogen isotope data from 3.2 Ga point to an early appearance of biological N$_2$ fixation to ammonium and subsequent incorporation into organic molecules (*Stüeken et al.,* 2015b). If these inferences are correct (cf. *van Zuilen et al.,* 2002), they would imply that life became independent from abiotic sources of organic carbon and fixed nitrogen very early in its history, making it less dependent on (although not fully independent from) high-energy processes such as lightning and hydrothermal circulation. These processes may still have been needed as sources of oxidants for chemotrophic organisms, but they were no longer required for the provision of key biomolecules.

Isotopic fractionations are produced by other biological processes as well, specifically dissimilatory processes (metabolic reactions performed for the purpose of generating energy) where organisms preferentially consume the isotopically lighter versions of substrates because they are kinetically favorable. In these cases, the fractionation is apparent in the waste products released into the environment, rather than incorporated into the body of the organism. Dissimilatory sulfate and sulfur reduction produce fractionations of multiple sulfur isotopes as microbes generate energy from a redox couple of sulfur species to a reductant, such as hydrogen (*Schidlowski,* 1979; *Canfield and Thamdrup,* 1994; *Habicht and Canfield,* 1997; *Philippot et al.,* 2007; *Sim et al.,* 2011). Such fractionations are seen as early as 3.49 Ga (*Shen et al.,* 2001; *Aoyama and Ueno,* 2018) and may be evidence of sulfate reduction, elemental sulfur reduction, or elemental sulfur disproportionation — if abiotic mechanisms can be ruled out (*Bao et al.,* 2008; *Muller et al.,* 2016). All these substrates could have been provided by volcanism and photochemical reactions, and hence these data may reflect an ecosystem sustained by planetary (i.e., abiotic geological and atmospheric) energy sources (section 2.2.2).

Dissimilatory energy metabolisms can also generate carbon isotope fractionation, as is the case with methanogenesis, the reduction of inorganic single-carbon substrates to methane. Methane dissolved in fluid inclusions in 3.46 Ga hydrothermal chert veins from the Pilbara craton contains isotopically light carbon with $\delta^{13}C$ values similar to those produced by microbial methanogenesis (*Ueno et al.,* 2006). Additional evidence for this metabolic process has been reported from light carbon isotopic signatures in similarly aged microfossils (*Schopf et al.,* 2018). There is also evidence proposed for Archean biogenic fractionation of iron isotopes

through dissimilatory iron reduction and/or photoferrotrophy (*Crosby et al.,* 2005; *Tangalos et al.,* 2010; *Percak-Dennett et al.,* 2011; *Czaja et al.,* 2013; *McCoy et al.,* 2017). Abiotic fractionation pathways for iron isotopes are difficult to rule out (e.g., *Dauphas et al.,* 2004; *Guilbaud et al.,* 2011), but a biological contribution is supported by co-occurrences of iron and biogenic carbon isotope fractionations in the same sample (e.g., *Heimann et al.,* 2010). Broadly speaking, the geochemical evidence for chemotrophic metabolisms on the Archean Earth matches the expected repertoire of an anoxic world with volcanic and hydrothermal supplies of CO$_2$, H$_2$, SO$_2$, and Fe^{2+}.

2.3.2. Fossil evidence. Perhaps the most compelling argument for an Archaean biosphere is direct fossil evidence. This can take the form of body fossils of individual cells or colonies of cells, trace fossils that show direct physical evidence of biological activity, or abiotic structures that form in processes mediated by biological structures. The oldest proposed microfossils are filamentous and cell-like structures from the Apex cherts of the Warrawoona Group in Western Australia, ~3.46 Ga, which have been proposed to correspond to several groups of extant microbes, including Cyanobacteria and methanogens, albeit with distinct features that may represent novel morphologies (e.g., *Schopf and Packer,* 1987; *Schopf,* 1993; *Brasier et al.,* 2015). Although these fossils often resemble cellular morphologies, they remain contentious, and abiogenic origins have been proposed (*Brasier et al.,* 2004, 2006). Recent improvements in technology permitting precise isotopic measurements that are spatially correlated with individual microfossils may help to further constrain the possible biogenicity and even physiology of these objects (*Schopf et al.,* 2018).

Perhaps the most striking fossil evidence for Archaean life are the preserved record of stromatolites, which are highly distinct laminar sediment structures produced during layered growth of microbial communities. Such features are still observed in some microbial ecosystems today. Records of stromatolites go back to 3.49 Ga (*Walter et al.,* 1980) and possibly earlier (*Nutman et al.,* 2016). An important aspect of morphology that distinguishes these structures from abiotic laminar structures is the specific spacing of the conical features, which is best explained by biophysical processes (*Petroff et al.,* 2010). For the most part, stromatolites preserve insufficient information to identify the specific microbial group(s) that may have produced them or to fingerprint their metabolisms; however, phototrophy, specifically, anoxygenic phototrophy, is the most consistent interpretation for stromatolites from the 3.43-Ga Strelley Pool chert in Western Australia (*Allwood et al.,* 2006; *Wacey et al.,* 2011; *Duda et al.,* 2016). These rocks also show isotopic evidence of microbial S-cycling (*Bontognali et al.,* 2012; *Allen,* 2016).

Collectively, these findings of microfossils and stromatolites — if correct — suggest that early microbes resembled modern descendants in terms of morphology and colonial behavior. Cellular organization and specifically formation of microbial mats are evidently very ancient traits that have persisted over billions of years of planetary and biological evolution. Another key aspect of early paleontological and geochemical archives of Archean life is the wide range of habitats that they span — including deep-marine hydrothermal vents (*Rasmussen*, 2000; *Kiyokawa et al.*, 2006; *Dodd et al.*, 2017), mud volcanos (*Pons et al.*, 2011), terrestrial hot springs (*Djokic et al.*, 2017), offshore marine settings (e.g., *Guy et al.*, 2012; *Stüeken et al.*, 2015b), intertidal zones (e.g., *Gamper et al.*, 2012; *House et al.*, 2013; *Flannery et al.*, 2018), soils, rivers, and lakes (*Guy et al.*, 2012; *Nabhan et al.*, 2016; *Stüeken and Buick*, 2018). Earth's early biosphere apparently adapted to a wide range of environmental conditions relatively quickly despite the suppressed supply of metabolic substrates and harsh UV radiation. This observation may bode well for the likelihood of finding life on other Archean-like worlds.

2.3.3. Genomic evidence.

Other important evidence for reconstructing the earliest biosphere is preserved within genes and genomes of modern organisms. This genetic information is an indirect record of the evolving surface conditions of the Archean Earth as they affected the evolutionary trajectory of subsequent generations through organismal lineages. The history of this genetic information can be traced back in time through phylogenetic and phylogenomic studies, representing the evolutionary histories of species as well as particular sets of genes. Comparative genomics studies support the extreme antiquity of some microbial metabolisms, which is consistent with a biosphere fueled by planetary processes. For example, methanogenesis is likely extremely old, certainly evolving early within the diversification of the archeal superkingdom Euryarchaeota (*Borrel et al.*, 2013). Some phylogenomic interpretations place methanogenesis even earlier, evolving in the ancestor lineage of Archaea itself (*Martin and Sousa*, 2016). Molecular clock studies also support the antiquity of methanogenesis, suggesting the last common ancestor of extant methanogens existed by 3.6 Ga (*Wolfe and Fournier*, 2018). Similarly, phylogenetic studies of the reaction center proteins common to both oxygenic and anoxygenic photosynthetic machineries support a very early origin of photosystem-based phototrophy within microbes (*Cardona et al.*, 2015). However, the limited conservation of sequence between these highly divergent protein families prevents conclusive molecular clock estimates for their origin (*Magnabosco et al.*, 2018). Several molecular clock studies have estimated the ages of extant phototrophic lineages (e.g., *Cardona et al.*, 2018; *Sánchez-Baracaldo et al.*, 2017; *Shih et al.*, 2017; *Magnabosco et al.*, 2018), including Cyanobacteria. While these studies vary in their assumptions, calibrations, and resulting age estimates, they generally agree that extant lineages of phototrophic bacteria are too young to be represented by the earliest proposed paleontological evidence of phototrophic communities, such as the 3.5-Ga stromatolites (section 2.3.2). The distribution of photosystem types across extant microbial lineages also shows a pattern consistent with extensive horizontal gene transfer (HGT), which is the lateral movement of genetic material between organisms rather than from parent to offspring (*Hohmann-Marriott and Blankenship*, 2011). Ancient patterns of lineage speciation, extinction, and HGT could therefore obscure the identities of the first phototrophic lineages, which may have been members of now-extinct groups. Specific phototrophic metabolisms may also have arisen multiple times independently, albeit using the same photosynthetic machinery. For example, while photoferrotrophy is proposed to be an important microbial metabolism in the early Archean, and possibly responsible for iron isotope fractionation and banded iron formations at this time (*Camacho et al.*, 2017), extant photoferrotrophic organisms appear to be members of much younger lineages that diversified in the Paleoproterozoic (*Magnabosco et al.*, 2018).

Sulfur metabolism genes also have an ancient history. Genes involved in dissimilatory sulfate reduction show a broad phylogenetic distribution, originating before the ancestors of the major extant groups of sulfate reducing microbes, including Deltaproteobacteria, Firmicutes, Chloroflexi, and Archaeoglobales. These genes were likely shared among these groups, and possibly other lineages, via HGT (*Klein et al.*, 2001). Preliminary molecular clock studies suggest that sulfate reduction was likely present by ~3.5 Ga (Momper et al., 2018, personal communication), supporting hypotheses for the antiquity of this metabolism based on geochemical evidence — specifically isotopic fractionations. The inferred low levels of sulfate in the Archaean ocean (*Habicht et al.*, 2002; *Crowe et al.*, 2014) suggest that microbial sulfate reducers were perhaps restricted to local, sulfate-rich environments, unlike in the post-Archean when microbial sulfate reduction became more widespread (*Canfield*, 1998). This would imply that the evolution of microbial metabolic pathways may, in general, predate their ecological abundance and their widespread geochemical evidence in the rock record.

Some ancient biological processes that undoubtedly date back to the early Archean, such as anoxygenic photosynthesis and methanogenesis, are staggeringly complex, requiring dozens of genes and multiple cofactors and coenzymes with their own complex biosynthetic pathways and, in the case of photosynthesis, massive reaction centers for finely tuned photochemistry and electron transfer processes (*Hohmann-Marriott and Blankenship*, 2011). Therefore, even the most ancient microbial metabolisms are sophisticated and highly derived, i.e., having evolved from earlier, simpler energy-generating processes. These primordial metabolisms may not have any surviving analogs, limiting what we can infer via comparative genomics studies. This observation has two important implications. First, much of the genetic information about the first and simplest microbial communities has likely been lost in the intervening billions of years, especially with the rise of oxygen and an entirely new biogeochemical regime. This loss hinders accurate reconstructions of the earliest biogeochemical feedbacks on Earth. Second, a diverse and biochemically complex biosphere capable of a wide range of energy metabolisms — including sulfate and/or sulfur reduction, methanogenesis, methanotrophy, and anoxygenic phototrophy, as well as heterotrophic (i.e., consumption of organic carbon) anaerobic metabolisms like fermentation — evidently evolved relatively quickly on the early Earth despite a limited supply of electrochemical energy sources. Metabolic and ecological sophistication may thus be achievable relatively soon after life's origin.

3. LIFE IN A REDOX-STRATIFIED WORLD: THE NEOARCHEAN TO NEOPROTEROZOIC

The biological invention of oxygenic photosynthesis — the ability to generate large amounts of chemical energy from electromagnetic radiation — is probably unrivaled in its profound effects on global biogeochemical cycles and long-term evolutionary trends, because it allowed life to tap into sources of metabolic energy (through diverse electron acceptors) that were collectively orders of magnitude greater than what could be supplied by geological and atmospheric reactions alone (*Rosing,* 2005; *Canfield et al.,* 2006). Once free O_2 began to accumulate in the atmosphere around 2.4–2.3 Ga (*Lyons et al.,* 2014; *Luo et al.,* 2016), aerobic life began to dominate ecosystems, while anaerobic organisms had to recede to anoxic refugia. The "Great Oxidation Event" (GOE), as it is called, was arguably a turning point for global climate and habitability without which the rise of complex life would perhaps never have been possible. As of yet, the amount of time that elapsed between the origin of oxygenic photosynthesis and oxygen's rise in the atmosphere at 2.4–2.3 Ga is unknown, as is the specific cause of this rise (*Sessions et al.,* 2009) (section 3.3). Some estimates place this gap at hundreds of millions of years (*Lyons et al.,* 2014). In other words, during the latest stages of the anaerobic biosphere in the Archean, oxidation was likely occurring even in the absence of atmospheric accumulation of oxygen, with significant consequences for both living systems and the preserved record of their activities.

3.1. The Innovation of Oxygenic Photosynthesis

Phototrophy is a metabolism using multiple photochemical and biochemical innovations for capturing energy from light and converting it into biologically usable forms. It is, in essence, the use of photons to generate high-energy electrons to do chemical work in biological systems (Fig. 3). The process also produces abundant energy for ecologically critical processes such as carbon and nitrogen fixation. Phototrophy, especially oxygenic photosynthesis, so greatly increases the available energy to a biosphere that it is tempting to assume its discovery by evolution is inevitable, and that this process would be a key feature on any planet with a biosphere. However, the phototrophic machinery is extremely complex, as is the evolutionary history of these systems across microbial lineages. As such, we should avoid making any astrobiological generalizations solely based on utility, without careful consideration

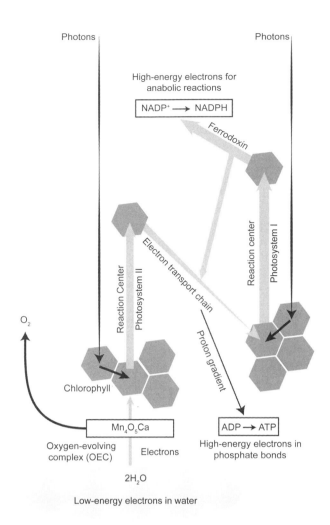

Fig. 3. Schematic of the photosystem in oxygenic photosynthesis.

of the highly contingent evolutionary processes that gave rise to the diversity and complexity we observe in organisms today.

The complex biological machines that carry out phototrophic reactions are photosystems, collections of proteins that — together with antenna molecules (chlorophyll or bacteriochlorophyll) — channel the energy from photons to reaction centers where low-energy electrons are excited to higher-energy states. This energy can then be harnessed by the cell (*Nelson and Junge,* 2015). These low-energy electrons are provided by an electron donor, ideally, a small, soluble molecule in abundance, with an oxidized product that is easily removed from the system. In anoxygenic phototrophs, the most common electron donors are sulfide and ferrous iron, although other substrates may be used (*Griffin et al.,* 2007; *Frigaard and Dahl,* 2008; *Camacho et al.,* 2017). In light-abundant environments, these microbes are limited by the availability of their electron donors, which may become scarce if consumed, or may always be at very low concentrations or absent in some habitats. One of the most important evolutionary events in the history of Earth was the first discovery of a new electron donor for phototrophy, one in inexhaustible abundance. That electron donor was water itself, triggering the evolution of oxygenic photosynthesis.

It is difficult to oxidize water, requiring an electrical potential greater than that produced using bacteriochlorophyll. The higher-energy photons captured by a different antenna molecule, chlorophyll, permit the traversing of larger spans of redox potential, bringing the very-low-energy electrons in water to the higher potential energy of the excited state of the reaction center (*Blankenship and Hartman,* 1998; *Saer and Blankenship,* 2017). This process also requires an oxygen evolving complex (OEC) — i.e., an enzymatic site for "water splitting" at the photosystem II (PSII) reaction center, wherein four manganese ions, aided by a calcium ion, remove each of the four electrons from the outer orbitals of the water molecule. The electrons are transferred to chlorophyll, releasing a fully oxidized product, molecular oxygen (*Barber,* 2016; *Saer and Blankenship,* 2017). In fact, the specific dependence on manganese oxidation in the system has led some researchers to propose that the ancestral electron donor of oxygenic phototrophs was, in fact, manganese (*Fischer et al.,* 2015; *Deshmukh et al.,* 2018). The electrons flowing from this process generate a proton gradient, which the cell can use for ATP synthesis. The precise electrochemistry involved is still poorly understood, as is the evolutionary origin of this remarkable enzymatic process (*Vinyard and Brudvig,* 2017). In anoxygenic phototrophy, the activated reaction center passes electrons through an electron transport chain. In the case of PSII, this process is cyclical and continues to generate a proton motive force. For photosystem I (PSI), the process is acyclical, reducing NAD+ and NADP+ to NADH and NADPH, respectively, which can then do biochemical work. In oxygenic phototrophy, both photosystems are used in series. An electron transport chain transfers electrons from the activated PSII reaction center to the PSI reaction center, where another photon excites the electron to energy levels high enough to eventually generate NADPH (*Saer and Blankenship,* 2017). In this two-step manner, both proton motive force and reducing agents for biosynthetic processes are generated from very-low-energy electrons of an unlimited supply.

Oxygenic photosynthesis likely evolved in the ancestor of Cyanobacteria at some point after their divergence from their closest non-photosynthetic relatives, Melainabacteria (*Shih et al.,* 2017). This evolutionary innovation may have been the most important in the history of life on Earth, both dramatically increasing the potential primary productivity of ecosystems and, incidentally, producing vast quantities of molecular oxygen. Accumulation of atmospheric oxygen fundamentally altered every geochemical system on Earth and provided a ubiquitous oxidant for high-energy redox couplings, including those tied to the degradation of organic matter. How did this metabolism evolve? While the question is complex, the key appears to have been the integration of two photosystems by the cyanobacterial ancestor lineage.

While highly divergent at the gene and protein sequence level, both PSI and PSII are clearly homologous in their structures and functions and share a common ancestry (*Schubert et al.,* 1998; *Cardona et al.,* 2015). Three hypotheses have been proposed to explain the origin of the two-photosystem arrangement. The "duplication" hypothesis states that the photosystems duplicated and diverged within the ancestor lineage of Cyanobacteria, which originally was an anoxygenic phototroph with a single ancestral photosystem (*Olson and Pierson,* 1987; *Mulkidjanian et al.,* 2006). The "fusion" hypothesis states that the two photosystems were horizontally transferred into the ancestor lineage of Cyanobacteria (*Mathis,* 1990; *Hohmann-Marriott and Blankenship,* 2011). Finally, the "selective loss" hypothesis states that both photosystems diverged even earlier in bacterial evolution and were present before the cyanobacterial ancestor lineage diverged (*Cardona et al.,* 2017).

The phylogenetic distribution of single-photosystem anoxygenic photosynthesis is important to each hypothesis. In the duplication hypothesis, single photosystems were horizontally transferred to anoxygenic recipient lineages from within the cyanobacterial ancestor lineage after the duplication occurred. Some of these transfers may be from even older lineages containing the ancestral PS type (either the PSI or PSII ancestor), if a single PS ancestor was also originally acquired by the cyanobacterial ancestor lineage. In the fusion hypothesis, PSI and PSII diverged across other anoxygenic lineages, which may also have transferred them to other bacterial groups, including extant anoxygenic phototrophs. Eventually, through horizontal gene transfer both photosystems were acquired by the cyanobacterial ancestor lineage where they evolved to operate in tandem. Finally, in the selective loss hypothesis, both PS diverged very early in bacterial evolution, with the last common ancestor of all phototrophic bacteria having both PSI and PSII, and possibly also performing oxygenic photosynthesis itself. Many lineages subsequently lost either PSI or PSII, or both, resulting in the observed phylogenetic distribution.

Each of these hypotheses is both informed and complicated by the possibility of extinct phototrophic lineages and the observed additional reaction center protein duplications and diversifications seen across different phototroph lineages (*Cardona et al.,* 2015; *Magnabosco et al.,* 2018). Additional clues can be surmised from the evolutionary history of chlorophyll biosynthesis genes (*Sousa et al.,* 2012), although this history is also complicated by extensive HGT and our inability to trace gene tree histories through extinct species tree lineages. Future discovery of additional anoxygenic phototrophic lineages may be able to resolve these debates. The discovery of an anoxygenic phototrophic lineage with a single photosystem more closely related to that of Cyanobacteria than their non-photosynthetic outgroups, for example, would be able to resolve the entire question, because it would provide strong support for the idea that oxygenic photosynthesis arose from anoxygenic photosynthesis. However, such a lineage, if it ever existed, may be extinct.

Each of these hypotheses has consequences for the relative timing of the evolution of oxygenic photosynthesis on

Earth, as well as our understanding of the Archaean biosphere and the likelihood of oxygenesis elsewhere in the universe. The selective loss hypothesis is consistent with a very early origin of oxygenic photosynthesis, predating even the cyanobacterial stem lineage, regardless of the age estimates for those groups. The duplication hypotheses presupposes a more recent origin of oxygenic photosynthesis within the cyanobacterial stem lineage and is most consistent with a long phototrophic history within this group, giving time for both divergence and diversification of photosystems and their subsequent HGT to other clades. The fusion hypothesis permits the most recent origin of oxygenic photosynthesis in the ancestor of cyanobacteria, with all of PSI and PSII evolution occurring outside of the cyanobacterial lineage, although not necessarily before it diverged. However, even under this hypothesis extensive photosystem evolution must have occurred in the cyanobacterial lineage, as there are many cyanobacterial-specific evolutionary events within photosystems, including additional reaction center protein duplications (*Cardona et al.*, 2015), as well as the OEC and other derived traits.

The complexity of oxygenic photosynthesis and its convoluted evolutionary history are important to bear in mind when using Earth's early biosphere as an analog for life elsewhere. It is conceivable that life on other planets never evolves the capacity to produce large quantities of free O_2. Even if phototrophy does evolve, there is no certainty that the proper chlorophyll-like molecules or reaction center configurations for oxygenic phototrophy are evolutionarily accessible from pre-existing anoxygenic systems. At the same time, the utilization of H_2O as an electron donor should be strongly selected for in any water-rich setting. It is a singular solution that may be converged upon across a wide diversity of suitably adaptable alien biochemistries. If oxygenic photosynthesis does evolve on another world, the oxygenation of the atmosphere (and therefore the rise of remotely detectable oxygen as a biosignature) is not necessarily an immediate process; there may be a significant lag between the start of biogenic oxygen production and atmospheric oxygenation as described in the sections below.

3.2. Oxidation: Traces of O_2 Production in an Anoxic World

The biological invention of oxygenic photosynthesis is the fundamental reason why Earth's atmosphere and ocean contain substantial amounts of free and dissolved O_2 gas. However, the transition from the first appearance of this metabolism to the full oxygenation of Earth's surface environments was protracted, because numerous planetary reservoirs of reductants first needed to be saturated before oxygen could build up (*Lyons et al.*, 2014). The timescales of complete planetary oxygenation are therefore strongly controlled by the timescales of oxidation of reduced minerals and gases, i.e., by crustal composition, metamorphic and volcanic gas production, and atmospheric processes.

Geochemical records of oxidation can provide evidence of O_2 production long before the Paleoproterozoic oxygenation of the atmosphere at 2.3–2.4 Ga. The major tool for tracking oxidation is the mobility or immobilization of certain elements, inferred from accumulations or isotopic fractionations in the rock record. The rationale behind many of these isotopic systems, including Mo, Cr, Se, U, N, and others is that fractionations — regardless of mechanism — can only occur if the element is present in dissolved form, which may only be the case under oxic conditions. The oldest proposed evidence of oxidation is based on U enrichments in >3.7-Ga metamorphosed sedimentary rocks from Greenland (*Rosing and Frei*, 2004). Uranium is only mobile in its oxidized form, and hence this enrichment could represent U(IV)– to –U(VI) oxidation by free biogenic O_2. However, recent Cr isotope data from similarly aged rocks were interpreted as abiotic Cr oxidation (*Frei et al.*, 2016). From the 2.9-Ga Pongola Supergroup, large sulfur isotope fractionations indicative of locally abundant sulfate (*Eickmann et al.*, 2018) and Mo isotopes correlated with Mn/Fe ratios in marine banded iron formations indicative of Mn(II)– to –Mn(IV) oxidation (2014a) have been interpreted as evidence of a local "oxygen oasis" in the surface ocean (*Riding et al.*, 2014).

Evidence of oxidation becomes more common in Neoarchean rocks (2.8–2.5 Ga). Rising sulfur abundances in marine sediments from 2.75 Ga onward are most plausibly interpreted as increasing rates of oxidative sulfide weathering on land, followed by drawdown into marine sediments (*Stüeken et al.*, 2012). Starting around 2.7 Ga, multiple proxies, including Mo, Fe, S, and N isotopes from several localities indicate a mildly oxygenated surface ocean overlying anoxic deep waters (*Wille et al.*, 2007; *Godfrey and Falkowski*, 2009; *Kendall et al.*, 2010; *Koehler et al.*, 2018). Importantly, none of these proxies for local oxidation require atmospheric oxygenation above 10^{-5} PAL, which would otherwise contradict the S-MIF evidence of less than 1 ppm O_2 throughout the Archean (*Farquhar et al.*, 2000; *Pavlov and Kasting*, 2002). Most likely, oxidative weathering occurred locally under phototrophic microbial mats on land (*Lalonde and Konhauser*, 2015; *Sumner et al.*, 2015). These mats may have mobilized redox sensitive elements like Mo or S, which could then be washed into the ocean, despite very low atmospheric O_2 levels. Evidence of such phototrophic mats may be preserved by 2.7-Ga lacustrine stromatolites in the Tumbiana Formation, Western Australia, with morphologies, isotopic signatures, and fenestral (i.e., window-like) structures consistent with O_2 production (*Buick*, 1992; *Bosak et al.*, 2009; *Flannery and Walter*, 2012; *Stüeken et al.*, 2017b). Furthermore, active oxygenic photosynthesis in shallow marine waters could have resulted in oxygen oases with up to a few micromolars of dissolved O_2 in disequilibrium with the anoxic deep ocean and atmosphere (*Olson et al.*, 2013). Such a heterogeneous redox landscape may be characteristic of planets where biological O_2 production evolves but is trumped for long periods by planetary supplies of reductants.

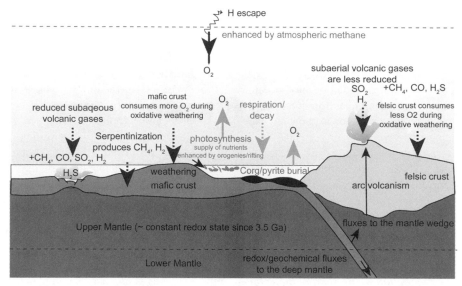

Fig. 4. Schematic of O_2 sources and sinks. The large black arrow with a solid line highlights O_2 sourced by reductant loss from the upper atmosphere (top down). The large black arrows with dotted lines indicate sinks of O_2 via volcanic or crustal reductants (bottom up). The large gray arrows with solid/dotted lines indicate biologically influenced sources/sinks (oxygenation from the center). Finally, the small black arrows indicate the directions of certain redox/geochemical fluxes. See text for further discussion.

3.3. Oxygenation: The Buildup of Free O_2 in a Planetary Atmosphere

Free O_2 can build up in a planet's atmosphere once a tipping point is reached where the major sinks have been oxidized and reductants are separated from O_2 gas (*Catling,* 2014). In principle, three pathways can be envisioned (Fig. 4): (1) a loss of reductants from the upper atmosphere (top down), (2) a removal of reductants to the mantle and/or a decline of reductant input from the deep Earth (bottom up), and (3) a biological innovation and associated organic burial that greatly enhanced the rate of biological O_2 production. While the governing mechanisms behind the Paleoproterozoic GOE are widely debated, all these processes could conceivably play an important role on other inhabited worlds — as they likely did on Earth.

3.3.1. Proxies for the oxygenation of the atmosphere.
The oldest temporary rise in atmospheric O_2 may have occurred at 2.5 Ga as evidenced by the Mount McRae Shale in Western Australia. Here, datasets of molybdenum, rhenium (*Anbar et al.,* 2007), sulfur (*Kaufman et al.,* 2007), iron (*Reinhard et al.,* 2009), selenium (*Stüeken et al.,* 2015a), and nitrogen (*Garvin et al.,* 2009) are all interpreted as recording a transient "whiff" of oxygen before the Paleoproterozoic GOE. Unlike the older Archean records of reductant oxidation, this end-Archean "whiff" of oxygen in the Mt. McRae Shales shows evidence for oxidative weathering on land, in this case recorded in osmium, which is difficult to explain without an increase in atmospheric O_2 levels to about 0.1% modern levels (*Kendall et al.,* 2015). This event can be reconciled with concurrent S-MIF in the same rocks, if those S-MIF signatures were inherited by recycling of older sediments (*Reinhard et al.,* 2013a). More recent work suggests that similar pulses of oxygenation may have occurred in the

Paleoproterozoic, leading up to the GOE at 2.33 Ga, when S-MIF essentially vanished from the rock record (*Bekker et al.,* 2004; *Luo et al.,* 2016; *Gumsley et al.,* 2017). From this time onward, the sulfur isotope record is dominated by mass-dependent fractionation (MDF), induced by biological metabolisms acting on a growing sulfate reservoir sourced from oxidative weathering on land.

3.3.2. Top-down oxygenation: Hydrogen escape and climatic perturbations.
One possible cause of atmospheric oxygenation that may have operated on Earth — as it does on other planetary bodies in the solar system — is atmospheric hydrogen escape to space (*Catling,* 2014). Today hydrogen-escape rates are relatively minor, but in the past the abiotic photo-dissociation of abundant biogenic CH_4, followed by the escape of hydrogen, could have constituted a major reductant sink from Earth's system (*Catling et al.,* 2001; *Claire et al.,* 2006; *Zahnle et al.,* 2013). Unlike H_2O, which is cold-trapped to the lower atmosphere, CH_4 could have risen to high altitudes in the Archean atmosphere, where it was readily dissociated by UV radiation, allowing hydrogen escape into space. This process leads to the irreversible oxidation and ultimately oxygenation of Earth's surface, because oxygen and oxidized equivalents stay behind.

A possible complication in this model is that CH_4 itself is a reducing gas and thus a sink for O_2, meaning that a decrease in CH_4 production may have been required as the ultimate tipping point for O_2 accumulation (*Zahnle et al.,* 2006). One possible reason for decreasing rates of methanogenesis is an increase in the supply of sulfate to the ocean via an expansion of oxidative terrestrial ecosystems. This in turn could have stimulated microbial sulfate reducers in seawater where they outcompeted methanogens in H_2- and organic-limited systems (*Zahnle et al.,* 2006; *Stüeken et al.,* 2012; *Izon et al.,* 2017; *Fakhraee et al.,* 2018). Furthermore, an initial rise in O_2 and sulfate may have led to microbial CH_4 oxidation via an anaerobic oxidation pathway, thus suppressing the CH_4 flux into the atmosphere and creating a positive feedback on atmospheric oxygenation (*Olson et al.,* 2016b; *Gumsley et al.,* 2017). Alternatively, it is possible that methane production rates waned because oceanic crust became less ultramafic, leading to a drop in the supply of nickel, which is an essential co-factor in the enzymatic machinery of methanogens (*Konhauser et al.,* 2009, 2015) (section 3.3.3). In any case, the predicted decline of the potent greenhouse gas CH_4 around the time of the GOE is supported by occurrences of widespread low-latitude glaciations in the Paleoproterozoic (*Kasting,* 2005; *Bekker and Kaufman,* 2007; *Gumsley et al.,* 2017). If this climatic perturbation was indeed the result of oxygenation, then it represents perhaps the oldest dramatic example of biological impacts on global climate.

3.3.3. Bottom-up oxygenation: Differentiation, tectonics and volcanism.
Although hydrogen escape is a plausible mechanism for progressive planetary oxidation and oxygenation, it is possible that the GOE itself was triggered or aided by transformations deep within Earth. If so, then global redox transitions may be strongly tied to geological properties. These potential linkages can be broadly classified into three categories: (1) tectonically enhanced burial of reductants or stimulation of O_2 production, (2) declining volcanic supplies of reductants, and (3) declining crustal supplies of reducing minerals.

Regarding the first option (burial of reductants), on a tectonically active planet, continental rifting can lead to increased erosion, accelerating nutrient and sediment fluxes into the ocean. Furthermore, continental rifting can enhance growth of continental shelves as loci for organic burial. Collectively, these processes may have enhanced photosynthetic O_2 production rates (*Squire et al.,* 2006; *Campbell and Allen,* 2008) as well as the burial of organic carbon and pyrite in sediments, which leads to a net increase in O_2 because both organic carbon and pyrite are reductants and thus O_2 sinks (*Lindsay and Brasier,* 202; *Berner,* 1989; *Godderis and Veizer,* 2000; *Campbell and Allen,* 2008). An even stronger and more permanent effect may have been achieved through subduction of carbon into the mantle (*Duncan and Dasgupta,* 2017). Traditionally, organic carbon burial has been tracked with paired organic and inorganic carbon isotopes because of the strong isotope effect associated with photosynthesis (*Des Marais,* 2001; *Bjerrum and Canfield,* 2004; *Krissansen-Totton et al.,* 2015). However, recent studies question the underlying assumptions about the CO_2 source composition (*Mason et al.,* 2017) and the sensitivity of the record to organic carbon burial (*Daines et al.,* 2017). Alternatively, it has been suggested that reductant burial may be linked to mountain-building events (orogenies). On the early Earth, global orogenies — possibly related to supercontinent formation — have been identified based on distinct peaks in zircon age spectra at 2.75–2.6 Ga and 2.40–2.5 Ga (*Campbell and Allen,* 2008). However, these mountain-building events are not coincident with the GOE, suggesting that this mechanism

may not have been dominant or that the dynamics of these feedbacks are not straightforward. More recent analysis of the record suggests that in fact rapid continental erosion and weathering, rather than collisional activities, coincide with oxidation events, further complicating this relationship (*Planavsky,* 2018). Tectonics are likely to play an important role, but more analyses are needed before the governing principles can be applied to other planets.

For option two (declining volcanic supply of reductants), it has been proposed that the composition of volcanic gases could have changed over time in two ways: first, through progressive oxidation of the upper mantle (*Kump et al.,* 2001; *Gu et al.,* 2016), and second, by a switch from largely subaqueous to largely subaerial volcanism, which may change the redox state of gases because the removal of water pressure can affect the thermodynamics of magmatic reactions (*Kump and Barley,* 2007). Regarding the former, most evidence is inconsistent with a change in the mantle's redox state shortly before the GOE (*Gaillard et al.,* 2015; *Gu et al.,* 2016; *Nicklas et al.,* 2018), but this question is still subject to debate (*Aulbach and Stagno,* 2016). Regarding the latter, large subaerial flood basalts are indeed preserved in the Neoarchean and Paleoproterozoic rock record (e.g., *Ciborowski et al.,* 2015), and a strong effect of the confining pressure of subaqueous eruptions on the redox state of volcanic gases is supported by models, which suggest a switch to more oxidized gases in subaerial volcanic eruptions (*Gaillard et al.,* 2011). This is because the reaction $H_2S + 2H_2O = SO_2 + 3H_2$ shifts to the right with increasing pressure. However, this pressure effect on redox state has recently been questioned based on field observations (*Brounce et al.,* 2017), meaning that a change in the style of volcanism may not have been as important as previously proposed (*Kasting,* 1993; *Kump et al.,* 2001; *Holland,* 2002).

For option three, which requires a declining supply of crustal reductants, crustal differentiation of continental crust can contribute to oxygenation rates, because felsic iron-poor minerals consume less O_2 during weathering and metamorphism than do mafic (iron-rich) minerals. Isotopic studies have identified a transition in the continental crust composition from mafic to intermediate felsic crust around 2.5 Ga (*Gaschnig et al.,* 2016; *Lee et al.,* 2016; *Tang et al.,* 2016; *Smit and Mezger,* 2017). This transition could thus have allowed O_2 to accumulate in the atmosphere (*Lee et al.,* 2016). New data, however, suggest abundant felsic lithologies back to 3.5 Ga (*Greber et al.,* 2017), perhaps calling into question the effect of crustal evolution. Alternatively, the changing phosphorous content of igneous rocks could have increased the efficiency of oxygenic photosynthesis (*Cox et al.,* 2018), if the additional phosphorus, a key nutrient, was indeed soluble and bioavailable.

Similarly, changes in the composition of oceanic crust also could have impacted the size of the atmospheric O_2 reservoir. A hot early Earth generated relatively more ultramafic oceanic crust (komatiite) due to a higher percentage of partial melting (*Konhauser et al.,* 2009; *Kasting and Canfield,* 2012; *Greber et al.,* 2017; *Smit and Mezger,* 2017).

Ultramafic rocks are significant O_2 sinks not only because they are rich in ferrous iron minerals that consume O_2 but also because serpentinization — the hydration of ferrous iron silicates — produces large amounts of reduced gases, in particular H_2, which act as additional O_2 sinks (*Sleep et al.*, 2004; *Smit and Mezger*, 2017). As the crust transitioned from ultramafic to mafic, this O_2 sink would have decreased in relative importance. Importantly, crustal differentiation on the early Earth may be linked to the proposed onset of modern-style plate tectonics at 3.0–2.5 Ga, i.e., coincident with the first appearances of oxygen oases and leading up to the GOE (*Dhuime et al.*, 2012, 2015; *Laurent et al.*, 2014; *Condie*, 2016; *Tang et al.*, 2016; *Nebel-Jacobsen et al.*, 2018). This temporal coincidence may point to a linkage between the deep Earth and the redox state of the atmosphere. If so, this connection would have important implications for the oxygenation timescales of planets with different compositions and degrees of mantle differentiation. These in turn can impact the redox properties of crustal lithologies and mantle sourced volcanic gases.

3.3.4. Oxygenation from the center: Possible roles of biological innovations.

Finally, it is possible that biology itself controlled the timing and tempo of atmospheric oxygenation against a background of geological and atmospheric processes. Some phylogenetic data suggest an expansion of oxygenic photosynthesizers and possibly the evolution of multicellular cyanobacteria with greater metabolic fitness just prior to the GOE (*Kopp et al.*, 2005; *Schirrmeister et al.*, 2013; *Magnabosco et al.*, 2018). An increase in O_2 production could be correlated with several changes to the photosynthetic apparatus (*Magnabosco et al.*, 2018). Additionally, diversification of Cyanobacteria and expansion into a wider array of environments may have greatly increased their ecological role. Evolutionary innovations in detoxification of reactive oxygen species may also have removed a negative feedback on the growth of cyanobacterial populations. Finally, higher rates of O_2 production from increased cyanobacterial activity likely increased the recycling of phosphorus and other nutrients via respiration, leading to a positive feedback on biological productivity in the photic zone (*Kipp and Stüeken*, 2017).

3.3.5. Universality of stratification and oxygenation?

On Earth, several hundred million years may have elapsed between the first appearance of O_2-producing bacteria and the first major rise of free O_2 gas to >1 ppm in the global atmosphere (*Lyons et al.*, 2014). Even then, the deep ocean remained anoxic and dominated by Fe^{2+} throughout most of the Proterozoic (2.5–0.54 Ga) and possibly extending into the early Paleozoic (0.55–0.25 Ga) (*Berry and Wilde*, 1978; *Shen et al.*, 2002, 2003; *Kah et al.*, 2004; *Dahl et al.*, 2010; *Planavsky et al.*, 2011; *Poulton and Canfield*, 2011; *Gilleaudeau and Kah*, 2013; *Sperling et al.*, 2015). Fe^{2+}-buffered redox stratification was recently inferred from a martian crater lake, which shows that redox stratification is not unique to Earth (*Hurowitz et al.*, 2017). Since O_2 and other oxidizing equivalents are produced in surface environments through photosynthesis or photochemical reactions,

while reductants in the form of reduced minerals and gases are primarily supplied from the planetary interior, it may be a general rule that surface habitats, including terrestrial settings and shallow seawater, become oxygenated long before the deep ocean. Models suggest that deep-water oxygenation through downwelling of atmospheric O_2 only became dominant above ~2.5% present atmospheric O_2 levels, which may have been attained relatively late in Earth's history, perhaps in the Neoproterozoic (*Reinhard et al.*, 2016).

The oxygenation history of Earth may thus reveal how difficult it is to saturate planetary O_2 sinks, which could indicate that oxygenation is likely to be a protracted process. On the other hand, some processes, in particular hydrogen escape to space (*Catling*, 2014) and planetary and crustal differentiation resulting in distinct redox properties of the crust and volcanic gases (*Frost and McCammon*, 2008), are probably widespread beyond our own solar system. Therefore, even in the absence of life, planetary surfaces may thus evolve to relatively more oxidized states over the course of their lifetime. By contrast, it is difficult to envision a scenario where a planetary surface becomes progressively more reduced. Whether or not free O_2 can accumulate would then depend on the initial endowment of reductants to a planetary body and the vigor with which the reductant reservoir at the planet's surface is replenished from the interior through the establishment of different tectonic regimes. Given this balance between reductants from depth and oxidants from the planet's surface, an episode of stratification — akin to the Proterozoic Earth — may not be uncommon on planets where oxygenic photosynthesis takes progressively stronger control over biogeochemical cycles.

3.4. Consequences of Redox Stratification

3.4.1. Transformation of terrestrial habitats.

On Earth, the GOE has traditionally been recognized by the appearance of red beds (i.e., mudstones rich in iron oxides) in surface environments, the retention of Fe^{3+} in soils, and the disappearance of detrital pyrite and uraninite from river deposits (*Holland*, 1984, 2006; *Rye and Holland*, 1998; *Johnson et al.*, 2014). These proxies are evidence that O_2 was widely available on land at Earth's surface from the mid-Paleoproterozoic onward. With the build-up of O_2, Earth's atmosphere would have developed an ozone shield that protected land surfaces from harmful UV radiation and made them more habitable for a greater variety of organisms (*Cockell*, 2000; *Cockell and Horneck*, 2011).

The extent of Proterozoic land habitation has been difficult to test with the geological record, because soils and freshwater sediments have a relatively low preservation potential. Reported lines of evidence for life on land — including biogenic weathering profiles, carbonaceous laminae, and stromatolites — go back to the Archean (*Rye and Holland*, 2000; *Neaman et al.*, 2005; *Heinrich*, 2015; *Flannery et al.*, 2016; *Nabhan et al.*, 2016; *Djokic et al.*, 2017; *Stüeken et al.*, 2017b; *Stüeken and Buick*, 2018), and this style of preservation continues into the mid-Proterozoic

(*Cheadle*, 1986; *Horodyski and Knauth*, 1994; *Gutzmer and Beukes*, 1998; *Eriksson et al.*, 2000; *Prave*, 2002; *Hahn et al.*, 2015). Although an increase in terrestrial bioproductivity across the GOE is not evident from the rock record, it is likely that such an increase occurred for the reasons listed above. Higher O_2 availability, enhanced rock weathering of nutrients (see below), and lower UV radiation should collectively have allowed land-based ecosystems to thrive. A transition does appear to occur in the late Proterozoic, as evidenced by several studies of eukaryotic biomarkers and microfossils from proposed lacustrine basins (*Pratt et al.*, 1991; *Strother et al.*, 2011; *Battison and Brasier*, 2012; *Fedorchuk et al.*, 2016). It is conceivable that lakes were favorable settings for eukaryotic life, because they may have been relatively more oxygenated and nutrient-rich than deep marine habitats (*Parnell et al.*, 2010, 2015). However, some of the studied sections have been interpreted as non-lacustrine (*Stüeken et al.*, 2017a), suggesting that careful scrutiny is required in interrogating these archives.

The importance of identifying bona fide freshwater settings in the Proterozoic rock record is highlighted by phylogenetic indications that Cyanobacteria and algal eukaryotes may have originated in freshwater in the Neoarchean and early Proterozoic, respectively (*Battistuzzi et al.*, 2004; *Battistuzzi and Hedges*, 2008; *Blank and Sanchez-Baracaldo*, 2010; *Sánchez-Baracaldo et al.*, 2017). If so, the scarcity of eukaryotic biosignatures from mid-Proterozoic marine sediments (*Knoll et al.*, 2006; *Brocks et al.*, 2017) may record habitat restrictions. One such challenge may have been frequent upwelling of O_2-poor, sulfide-rich waters into the shallow marine realm (*Reinhard et al.*, 2016; *Hardisty et al.*, 2017). Habitability on a redox-stratified planet such as the Proterozoic Earth would thus have been strongly zoned into anoxic and oxic settings.

3.4.2. Transformation of marine habitats. While the deep ocean remained anoxic and buffered by hydrothermal sources of ferrous iron throughout most of the Proterozoic (*Planavsky et al.*, 2011; *Poulton and Canfield*, 2011), the initial effects of higher atmospheric O_2 levels would have been felt in shallow and mid-depth marine habitats. First, increasing oxidative weathering rates on land, perhaps paired with the generation of sulfuric acid, likely increased the delivery of redox-sensitive micronutrients, such as vanadium, chromium, molybdenum, copper, selenium, and uranium to the ocean (*Konhauser et al.*, 2011; *Murakami et al.*, 2016). The seawater reservoir of these and other elements increased as evidenced by abundances and isotopic signatures in marine sediments (*Scott et al.*, 2008; *Partin et al.*, 2013; *Kipp et al.*, 2017; *Sheen et al.*, 2018). More acidic weathering probably also raised the riverine delivery of phosphorus (*Bekker and Holland*, 2012), culminating in widespread phosphorite deposition at the marine chemocline, i.e., the interface between the oxic surface ocean and the anoxic deep ocean (*Papineau*, 2010; *Lepland et al.*, 2014). Collectively, the increasing supply of O_2, phosphorus, and micronutrients to shallow marine habitats probably led to a dramatic rise in biological productivity, O_2 production, and episodes of

intense carbon burial expressed in positive carbon isotope values in carbonate rocks (*Martin et al.*, 2015). In fact, the most prolonged and pronounced carbon isotope excursion in the sedimentary rock record is recorded in mid-Paleoproterozoic strata and referred to as the Lomagundi-Jatuli Event and initially thought to be related to an "oxygen overshoot" where atmospheric O_2 may have exceeded modern levels (*Schidlowski et al.*, 1976; *Karhu and Holland*, 1996; *Bekker and Holland*, 2012; *Martin et al.*, 2013).

While plausible models exist for the onset of this event, its magnitude and ending are poorly understood, which hinders inferences that may be applied to general concepts of planetary habitability. Oxidative weathering of the continental crust and specifically organic-rich shales deposited during the Lomagundi-Jatuli Event may eventually have pulled atmospheric O_2 levels back down to intermediate levels (0.1–1%) until the end of the Proterozoic (*Kump et al.*, 2011). Alternatively or in addition, submarine volcanism may have released abundant reductants that upwelled into the surface ocean and consumed O_2 (*Ossa Ossa et al.*, 2018). Furthermore, the nearly complete oxidation of the Archean siderite reservoir — a source of CO_2 and a sink of O_2 — may eventually have limited the supply of carbon and lowered primary productivity (*Bachan and Kump*, 2015). It is also possible that atmospheric O_2 did not change significantly from the Paleoproterozoic GOE to the Neoproterozoic oxygenation of the deep ocean (section 4), although mid-Proterozoic O_2 levels are still widely debated (*Planavsky et al.*, 2014b; *Zhang et al.*, 2016; *Daines et al.*, 2017).

A hallmark of Proterozoic ocean chemistry was widespread sulfidic conditions along some continental margins as a result of microbial sulfate reduction in productive water columns, perhaps akin to modern oxygen-minimum zones (*Jackson and Raiswell*, 1991; *Poulton et al.*, 2010; *Gilleaudeau and Kah*, 2015; *Partin et al.*, 2015). Under these conditions, some micronutrients, including vanadium, copper, zinc, and molybdenum, may have been much less soluble in seawater (*Saito et al.*, 2003). Models suggest more than 10× lower molybdenum concentrations in the mid-Proterozoic relative to today as a consequence of euxinia (anoxic and sulfidic conditions) that spanned roughly 1–10% of the seafloor (*Reinhard et al.*, 2013b). This state may have had important implications for the marine nitrogen cycle, because molybdenum is a key constituent of the most efficient N_2-fixing enzyme nitrogenase, as well as for nitrate reductase (*Anbar and Knoll*, 2002; *Glass et al.*, 2009). Although nitrogen isotopes suggest widespread N_2 fixation in offshore settings (*Stüeken*, 2013; *Koehler et al.*, 2017), perhaps driven by anaerobes as in modern anoxic basins (*Fernandez et al.*, 2011; *Jayakumar et al.*, 2012; *Farnelid et al.*, 2013), absolute fixation rates and biological productivity may have been suppressed, especially if cyanobacteria were less active than today (*Sánchez-Baracaldo et al.*, 2014; *Olson et al.*, 2016a).

In addition to suppressed N_2 fixation, the supply of nitrate (NO_3^-) would have been lower in a redox-stratified world. Nitrate production likely began with the oxygenation of surface waters during the end of the Archean (*Garvin et al.,* 2009; *Godfrey and Falkowski,* 2009), but nitrate bioavailability was probably limited to shallow oxic waters, which may have placed an additional constraint on the radiation of eukaryotic life that requires a source of nitrate (*Anbar and Knoll,* 2002; *Stüeken,* 2013; *Hardisty et al.,* 2017; *Koehler et al.,* 2017). Some models suggest that loss of nitrate from the water column via extensive denitrification at the oxic-anoxic interface could have induced a "nitrogen crisis" (*Falkowski and Godfrey,* 2008) in the Proterozoic. However, such as a crisis could perhaps have been avoided by dissimilatory nitrate reduction to ammonium (DNRA) instead of N_2, which seems to be favored in iron-rich settings and has the advantage over regular denitrification in that it keeps fixed nitrogen in the system (*Michiels et al.,* 2017). Geochemical evidence for DNRA is yet to be found. If true, however, phosphorus limitation may have imparted a stronger control than nitrogen on biological productivity, because it is rapidly scavenged by authigenic iron minerals (i.e., iron minerals that formed *in situ*, such as iron carbonates and phosphate) under ferruginous conditions, where Fe^{2+} is freely dissolved in the water column (*Reinhard et al.,* 2017).

Generally speaking, the degree of secondary productivity — driven by the increasing supply of oxidants from the surface — was likely much higher in the redox-stratified Proterozoic world than in the largely anoxic Archean (*Kipp and Stüeken,* 2017). However, redox stratification, nutrient limitation, and local accumulation of sulfide in the water column likely imposed stronger spatial constraints on habitability than during any other time in Earth's history. The Fe^{2+}-buffered deep ocean could have represented a large refugium for the anaerobic Archean biota, while complex aerobic organisms were perhaps restricted to the more rapidly oxygenated shallow waters, lakes, and rivers.

4. A FULLY OXYGENATED WORLD: THE NEOPROTEROZOIC AND ONWARD

Habitats for obligate aerobic organisms (i.e., organisms that strictly depend on O_2) expanded with the second major rise in atmospheric O_2 levels during the Neoproterozoic Oxygenation Event (NOE) at ca. 800–550 Ma, when O_2 was at least above 1% of modern levels (*Och and Shields-Zhou,* 2012; *Lyons et al.,* 2014; *Planavsky et al.,* 2014b; *Cole et al.,* 2016). This event may have resulted in pervasive and prolonged "oxic events," where dissolved O_2 penetrated into the deep ocean for perhaps the first time in Earth's history (*Sahoo et al.,* 2016). The radiation of land plants around 470–400 Ma may have been the final trigger for the rise of atmospheric O_2 to modern levels and the complete oxygenation of the deep ocean (*Dahl et al.,* 2010; *Sperling et al.,* 2015; *Lenton et al.,* 2016; *Wallace et al.,* 2017). In addition to these redox changes, the Neoproterozoic Era is also marked by extreme excursions in the global carbon isotope record (*Halverson et al.,* 2010), two low-latitude glaciations (*Hoffman et al.,* 2017), widespread rifting of the putative supercontinent Rodinia and amalgamation of

Gondwanaland (*Li et al.,* 2008), and the first appearance of macroscopic metazoan biota during the Ediacaran, followed soon after by the "Cambrian explosion" of extant metazoan phyla in the early Phanerozoic. This time interval thus marks the transition from a largely microbially dominated planet to one that was habitable for complex organisms and intelligent life. The cause of this transition is widely debated, and it is probably a complex interplay between geological transformations, biological innovations, and bio-geochemical feedbacks (*Och and Shields-Zhou,* 2012).

4.1. How Can Stratification Be Overcome?

4.1.1. Possible planetary drivers. Similar to the GOE, geodynamic processes may have provided a mechanism for the NOE through (1) tectonically enhanced organic carbon burial and/or (2) increasing weathering fluxes of nutrients, in particular phosphorus, that spurred biological activity and O_2 production. Widespread rifting at low latitudes near the terminal Proterozoic has been tied to breakup of the supercontinent Rodinia (*Li et al.,* 2008). Along with this tectonic activity, this time was marked by eruptions of large igneous provinces (*Ernst et al.,* 2008), which likely weathered relatively rapidly, releasing significant amounts of phosphorus into the ocean (*Horton,* 2015; *Cox et al.,* 2016). Overall, this configuration in a low-latitudinal climatic regime probably enhanced biological productivity, while creating accommodation space for sediment storage in rift basins and along passive continental margins (*Bradley,* 2008) — leading to a dramatic increase in carbon burial. Indeed, records show an increase in sediment accumulation on continental crust at this time, lending credence to the idea that O_2 was able to build up because large amounts of organic carbon — a reductant and O_2 sink — became locked up in sedimentary reservoirs (*Husson and Peters,* 2017). In addition, atmospheric CO_2 levels likely dropped as a result of biological consumption and rapid chemical weathering, causing global cooling and widespread glaciations between 720 Ma and 630 Ma (*Hoffman et al.,* 2017). Glacial erosion may have further enhanced phosphorus delivery to the ocean (*Planavsky et al.,* 2010). Starting ca. 650 Ma, phosphorus delivery may have increased even further due to weathering of the newly formed mountain ranges resulting from the amalgamation of Gondwanaland (*Campbell and Squire,* 2010; *Goddéris et al.,* 2017). Expanding land biota at this time may have contributed to higher weathering fluxes (*Mills et al.,* 2014a). Overall, such a massive flux of a critical nutrient to the ocean would have led to an increase in primary production as well as carbon burial, which could explain the Neoproterozoic rise of oxygen. If so, then the NOE would ultimately have been driven by tectonics with biology playing an important but relatively passive role.

4.1.2. Possible biological drivers. Alternatively, or in addition to geological factors, the NOE may also have been driven by biological innovations, in particular by the increasing complexity of eukaryotic life relative to their simpler mid-Proterozoic ancestors. The first "modern"

eukaryotes capable of aerobic respiration via mitochondria likely evolved in the early Proterozoic against the backdrop of redox-stratification that strongly limited their habitat. According to molecular clocks, diversification into the major extant groups probably occurred around ~2.0–1.6 Ga (*Hedges et al.,* 2004; *Parfrey et al.,* 2011). Shortly after this initial diversification, one lineage of eukaryotes acquired cyanobacterial endosymbionts, a population of oxygen-producing cyanobacteria living within their cells. This was the ancestor of the plastid (also known as chloroplast) lineage within eukaryotic cells. The descendants of this group — the Archaeplastida — underwent dramatic diversification, beginning with the ancestors of the three major primary plastid lineages ~1.6–1.4 Ga (*Hedges et al.,* 2004; *Yoon et al.,* 2004; *Parfrey et al.,* 2011): the green algae (Viridiplantae), red algae (Rhodophyta), and Glaucophyta. The oldest widely accepted fossil evidence for members of an extant group of eukaryotes is Bangiomorpha, a type of red alga, which dates back to ~1.05 Ga (*Butterfield,* 2007; *Gibson et al.,* 2017), constraining the divergence of red and green algae to older than the age of this fossil. These groups diversified throughout the Mesoproterozoic and Neoproterozoic, likely including both freshwater and marine lineages, and increasingly dominating primary production, shifting the source of both organic carbon and O_2 away from cyanobacterial systems (*Brocks et al.,* 2017; *Isson et al.,* 2018).

Later, likely during the late Neoproterozoic (*Parfrey et al.,* 2011; *Jackson et al.,* 2018), multiple independent secondary endosymbioses occurred involving other groups of heterotrophic eukaryotes that enveloped red and green algae. These secondary algae continued to evolve, giving rise to major constituents of the Phanaerozoic planktonic system, including diatoms, dinoflagellates, haptophytes, cryptomonads, chlorarachniophytes, euglenids, and others (reviewed by *Keeling,* 2010). By the Mesoproterozoic, and certainly by the Neoproterozoic, unambiguous, diagnostic microfossil and lipid biomarker evidence of eukaryotic algae is abundant in the geological record (*Moldowan et al.,* 1990; *Knoll,* 1994; *Javaux,* 2007; *Kodner et al.,* 2008; *Hoshino et al.,* 2017). Although the interpretation of this record requires careful consideration of context, including potential preservational bias overprinting evolutionary trends (*Knoll and Sergeev,* 1995; *Brocks et al.,* 2017), it is conceivable that these biological innovations triggered environmental perturbations in the Neoproterozoic.

Phytoplankton are an important part of the "biological pump" that delivers nutrients and organic material from the productive areas of the ocean to deep ocean waters and sediments (*Meyer et al.,* 2016). The larger bodies of eukaryotic algae contain more organic carbon and nitrogen than smaller cyanobacterial cells and thus promote more efficient delivery and possibly burial of organic material (*Butterfield,* 1997; *Rothman et al.,* 2003). The diversification and expansion of the eukaryotic biota could therefore have enhanced the removal of a major reductant and contributed significantly to the buildup of O_2 in the atmosphere (*Lenton et al.,* 2014). The onset of filter feeding by benthic siliceous sponges in the late Neoproterozoic and the appearance of eukaryotic plankton with rapidly-sinking carbonate skeletons in the early Phanerozoic may have further accelerated organic carbon burial and O_2 accumulation (*Bartley and Kah,* 2004; *Tatzel et al.,* 2017). These eukaryotic innovations were perhaps influenced by the rise of picocyanobacteria (small cyanobacteria with cell sizes ranging 0.2–2.0 μm and important constituents of phytoplankton populations), and possibly nitrogen-fixing cyanobacteria in the Neoproterozoic (*Sánchez-Baracaldo et al.,* 2014), which may have led to a greater supply of fixed nitrogen in the open ocean, thus expanding the "habitable zone" of eukaryotic life.

These evolutionary events are important to our understanding of habitability in a broader context. Increasingly efficient and successful phototrophic organismal systems evolved via multiple, serial, and parallel endosymbiotic events — independently between different lineages during a relatively narrow interval of Earth's history. This scenario argues for endosymbiotic events being a general feature of the evolution of complexity on a habitable world in response to certain environmental conditions, although such events may still be highly contingent upon the evolution of cells with the necessary cellular machinery for predation and vesicle/protein transport systems.

4.2. Consequences of a Fully Oxygenated World

4.2.1. *Energy supply and biological complexification.*
While the initial rise of atmospheric O_2 during the GOE was a profound geochemical and biological event, the resulting atmospheric concentration throughout most of the Proterozoic was much lower than levels consistently maintained during the Phanerozoic (*Lyons et al.,* 2014). The second rise of O_2 during the NOE was not only quantitatively different but was also qualitatively different in its impact on the biosphere and organismal evolution.

The low levels of O_2 (<1–10% PAL) of the stratified Proterozoic world after the GOE were likely sufficient for an expansion of certain cytological functions in aerobic organisms and have explanatory power for the early evolution of single-celled eukaryotes, which first appear in the fossil record around 1.6–1.8 Ga (*Blackstone,* 2016; *Knoll and Nowak,* 2017). For example, aerobic respiration in mitochondria appears to have been established before the diversification of extant Eukarya, and likely arose substantially before 1.6 Ga — well before the NOE (e.g., *Hedges et al.,* 2004; *Parfrey et al.,* 2011; *Eme et al.,* 2014). Also, the O_2-dependent biosynthesis of eukaryote-specific sterols (a specific type of lipid molecule with an unsaturated, fused four-ring structure found in cell membranes) is possible at dissolved O_2 concentrations as low as ~7 nM compared to 280 μM in modern saturated surface waters (*Waldbauer et al.,* 2011). Phylogenetic and molecular clock analyses of sterol biosynthesis genes show that these genes diversified around the time of the GOE, and possibly earlier, when oxygen levels were extremely low (*Gold et al.,* 2017). It is even conceivable that simple sponges evolved under Proterozoic conditions with dissolved O_2 levels of less than 1% modern (*Mills et al.*, 2014b). Hence, a substantial degree of eukaryal evolution, including the initial diversification of algae (*Butterfield*, 2007), may have taken place under microaerophilic (i.e., low-oxygen) conditions. In fact, the early evolution of some major eukaryal features may not have been contingent on the availability of oxygen at all, because some eukaryal lineages have evolved to be entirely anaerobic, suggesting that maintenance of eukaryal physiology is possible under these conditions (*Takishita et al.*, 2012).

There is so far, however, no convincing evidence for macroscopic organisms, specifically metazoans, prior to the appearance of demosponges in the Cryogenian (~635 Ma) (*Love et al.*, 2009). The next level of complexification of eukaryal life beyond algae appears to only have been reached around the NOE (*Erwin*, 2015). While it is possible that biological complexification simply required time and even had a causal effect on the NOE (*Lenton et al.*, 2014), it is likely that tissue-grade organization and diversification of complex life was in turn dependent on high concentrations of O_2 for energy-intensive physiological processes (*Shen et al.*, 2008; *Chen et al.*, 2015; *Edwards et al.*, 2017). Although the specific causal mechanisms of these evolutionary processes remain uncertain, it seems clear that habitability for Phanerozoic-type complex animal life, including intelligent organisms, is contingent upon levels of O_2 higher than those sustained during most of the Proterozoic and closer to those at surface habitats during the NOE.

4.2.2. Oxidation of oceanic lithosphere and sediments.
The oxygenation of the deep oceans during and after the NOE also could have had substantial implications for the solid Earth, because it resulted in oxidation of upper seafloor sediments and basalts that may have ultimately influenced arc volcanism and the chemical redox evolution of the mantle. When exposed to seawater, oceanic crust undergoes hydration and oxidative alteration and accumulates oxidized sediments and precipitates such as iron oxide particles and sulfates (*Kelley and Cottrell*, 2009). Furthermore, dissolved O_2 and other oxidized compounds like nitrate and sulfate can penetrate tens of meters into marine sediment packages in areas of low biological productivity (*D'Hondt et al.*, 2015). When this material is subducted, it transfers oxidizing potential into Earth's mantle and has the capacity to influence the redox state of magmas and volcanic gases over hundred-million-year timescales (*Kelley and Cottrell*, 2009; *Evans*, 2012). The exact fate of the oxidants is complex. While some may be lost from the subducted slab during dehydration and melting reactions (*Hayes and Waldbauer*, 2006), a considerable fraction may be transferred to the overlying mantle wedge (*Kelley and Cottrell*, 2009), and another portion may be carried to the upper mantle and possibly even to the lower mantle (*Lécuyer and Ricard*, 1999). Prior to the NOE, the flux of oxidants into the mantle was probably small, with the most important source being iron oxides such as in banded iron formations. This relationship may explain why the redox state of the mantle shows no detectable change over most of geologic history (*Rhodes and Vollinger*, 2005; *Nicklas et al.*, 2018). In fact, subduction of organic carbon throughout the Precambrian may have maintained a low mantle redox state (*Sleep et al.*, 2012; *Duncan and Dasgupta*, 2017). From the NOE onward, the oxidant flux likely became more important because deep marine waters became more oxidizing. However, this process may not express itself in the redox budget of the mantle until a billion years after the event, 500 m.y. in the future (*Evans*, 2012). The retention of oxidizing potential in the mantle may be further facilitated by progressive mantle cooling, which increases the compatibility of oxidized species with mantle minerals (*Frost and McCammon*, 2008). Although this mechanism may not yet have had major consequences on Earth, e.g., in the form of magmas that are highly depleted in Fe^{2+} or H_2S, such a scenario is conceivable for other planets with an oxygenic biosphere. The future Earth could thus be thought of as another "alternative" example of an inhabited planet.

5. CONCLUSIONS AND IMPLICATIONS

Earth and its biosphere have co-evolved from a fully anoxic and anaerobic state in the early Archean, through a redox-stratified interval with strongly zoned habitability in the Proterozoic, to a fully oxic and aerobic state conducive to the evolution of complex life and intelligence in the Phanerozoic. The transitions between these intervals were likely driven by a combination of geological and biological factors. When using the early Earth as an analog for other inhabited worlds, several aspects should be noted.

1. Earth's earliest biosphere was probably strongly dependent on planetary sources of nutrients via volcanism, hydrothermal circulation, and photochemical reactions. The rock record suggests that life was ubiquitous during this interval, but total primary productivity was likely much lower before the invention of oxygenic photosynthesis, i.e., a metabolism that could exploit solar radiation as an energy source and couple it to the oxidation of liquid water — the most abundant metabolite on the planet (*Canfield et al.*, 2006; *Lyons et al.*, 2014). Characterizing and quantifying hydrothermal and volcanic fluxes and atmospheric reaction pathways on the early Archean Earth is therefore important for our ability to constrain habitability of other analog worlds as well as to recognize biosignatures of anoxygenic life elsewhere.

2. Through atmospheric hydrogen escape and mantle differentiation, most planets probably evolve toward relatively more oxidized states over the course of their lifetime (*Frost and McCammon*, 2008; *Catling*, 2014), and biological metabolisms can greatly accelerate this process through oxygenesis. Generally speaking, oxidized products are produced at a planet's surface, while reductants are supplied from depth through magma generation and volcanic outgassing. An interval of redox stratification, such as on the Proterozoic Earth, may therefore be a common occurrence on other inhabited and perhaps even uninhabited planets. Furthermore, progressive subduction of oxidants subsequent

to an interval of stratification may lead to a state where a planet's upper mantle becomes oxidized with possible implications for crustal mineralogies, nutrient supplies, and volcanic gas compositions (*Evans,* 2012). This state has not yet been reached on Earth, but it may exist on other worlds.

3. Several lines of evidence suggest that oxidation (the first appearance of oxidized compounds) and oxygenation (the first rise of free O_2 gas in Earth's atmosphere) before and during the GOE, respectively (*Lyons et al.,* 2014), roughly coincided with the proposed onset of modern-style plate tectonics around 3.0–2.5 Ga (e.g., *Condie,* 2016). Similarly, the second rise of oxygen to high levels in the Neoproterozoic probably followed continental breakup and extrusions of flood basalts at low latitudes (*Och and Shields-Zhou,* 2012). Although biological innovations alone have been invoked to explain these redox transitions (*Kirschvink and Kopp,* 2008; *Butterfield,* 2018), their co-occurrence with large-scale geological transformations may not be coincidental but rather evidence for a strong linkage between the deep Earth and its biosphere. It has been proposed that the onset of biologically enhanced weathering of land masses in the Archean may have led to increased sedimentation rates on the seafloor, which in turn accelerated the subduction of water into the mantle, enhanced the production of granitic rocks, and stabilized continental crust (*Rosing et al.,* 2006; *Höning and Spohn,* 2016). In addition, biologically enhanced silicate weathering may have contributed to global cooling and prevented a return to a stagnant lid regime (*Foley,* 2015). Although these hypotheses are untested, any such feedbacks between the solid Earth and its surface processes would imply that the duration and timing of redox intervals on other planets are contingent upon a planet's fundamental geological properties. If so, then it is imperative for future studies of exoplanetary biosignatures to identify mechanisms that govern the initial endowment and exchange of oxidants and reductants between Earth's geological reservoirs.

4. Involvement of a terrestrial (land-based) biosphere has been invoked for several important events in Earth's history, including (1) a contribution to global warming and hydrocarbon haze formation due to enhanced methanogenesis in lakes and rivers from at least the Mesoarchean onward (*Zhao et al.,* 2017; *Stüeken and Buick,* 2018), possibly leading to a positive feedback on land colonization through haze-induced UV shielding (*Arney et al.,* 2016); (2) biotic enhancement of weathering rates, affecting topography and uplift (*Schwartzman and Volk,* 1989; *Dietrich and Perron,* 2006) and potentially contributing to continental stabilization (*Rosing et al.,* 2006); (3) the origin of Cyanobacteria in freshwater habitats in the Archean (*Blank and Sanchez-Baracaldo,* 2010); (4) production of sulfate by weathering of Neoarchean continents, leading to a decline of methanogenesis, increasing biomass remineralization and higher rates of oxygen production in seawater prior to the GOE (*Zahnle et al.,* 2006; *Stüeken et al.,* 2012; *Lalonde and Konhauser,* 2015; *Kipp and Stüeken,* 2017); (5) the origin of eukaryotes (*Sánchez-Baracaldo et al.,* 2017); (6) enhanced liberation of phosphate through organic acids from the Archean onward (*Neaman et al.,* 2005) and possibly a major acceleration of this process in the Neoproterozoic, contributing to ocean ventilation (*Mills et al.,* 2014a); (7) additional production of O_2 gas with the rise of land plants in the early Paleozoic, leading to modern levels of atmospheric O_2 (*Lenton et al.,* 2016); and finally (8) the provision of a diversity of habitats, which perhaps allowed for a higher degree of biological diversification and the evolution of intelligent life (*Pennisi,* 2017). Examples abound, and this list is by no means exhaustive. Although most of these propositions are still subject to debate, these linkages may point to the importance of a heterogeneous planetary surface with a combination of large water bodies and extensive dry land for the evolution of a large, diverse, and resilient biosphere.

In conclusion, biological evolution on Earth has been strongly shaped by geological and atmospheric parameters that are likely to play a major role on any terrestrial planet. Life is truly a planetary process, which should be reflected in our search for biosignatures on other worlds.

Acknowledgments. We thank T. Lyons and N. Planavsky for constructive comments that improved the manuscript. V. Meadows and G. Arney are thanked for editorial comments and for making this contribution possible.

REFERENCES

Ader M., Sansjofre P., Halverson G. P., Busigny V., Trindade R. I., Kunzmann M. and Nogueira A. C. (2014) Ocean redox structure across the Late Neoproterozoic Oxygenation Event: A nitrogen isotope perspective. *Earth Planet. Sci. Lett., 396,* 1–13.

Allègre C. J. and Othman D. B. (1980) Nd-Sr isotopic relationship in granitoid rocks and continental crust development: a chemical approach to orogenesis. *Nature, 286(5771),* 335–342.

Allen J. F. (2016) A proposal for formation of Archaean stromatolites before the advent of oxygenic photosynthesis. *Front Microbiol., 7,* DOI: 10.3389/fmicb.2016.01784.

Allwood A. C., Walter M. R., Kamber B. S., Marshall C. P., and Burch I. W. (2006) Stromatolite reef from the Early Archaean era of Australia. *Nature, 441(7094),* 714–718.

Anbar A. D. and Holland H. D. (1992) The photochemistry of manganese and the origin of banded iron formations. *Geochim. Cosmochim. Acta, 56(7),* 2595–2603.

Anbar A. D. and Knoll A. H. (2002) Proterozoic ocean chemistry and evolution: A bioinorganic bridge? *Science, 297,* 1137–1142.

Anbar A. et al. (2007) A whiff of oxygen before the Great Oxidation Event? *Science, 317(5846),* 1903–1906.

Aoyama S. and Ueno Y. (2018) Multiple sulfur isotope constraints on microbial sulfate reduction below an Archean seafloor hydrothermal system. *Geobiology, 16(2),* 107–120.

Arevalo R., McDonough W. F., and Luong M. (2009) The K/U ratio of the silicate Earth: Insights into mantle composition, structure, and thermal evolution. *Earth Planet. Sci. Lett., 278(3),* 361–369.

Armstrong R. L. (1968) A model for the evolution of strontium and lead isotopes in a dynamic Earth. *Rev. Geophys., 6(2),* 175–199.

Armstrong R. (1981) Radiogenic isotopes: The case for crustal recycling on a near-steady-state no-continental-growth Earth. *Philos. Trans. R. Soc., A, 301(1461),* 443–472.

Arney G. et al. (2016) The pale orange dot: The spectrum and climate of hazy Archean Earth. *Astrobiology, 16(11),* 873–899.

Aulbach S. and Stagno V. (2016) Evidence for a reducing Archean ambient mantle and its effects on the carbon cycle. *Geology, 44(9),* 751–754.

Bachan A. and Kump L. R. (2015) The rise of oxygen and siderite oxidation during the Lomagundi Event. *Proc. Natl. Acad. Sci. U.S.A., 112(21),* 6562–6567.

Bao H., Sun T., Kohl I., and Peng Y. (2008) Comment on "Early Archaean microorganisms preferred elemental sulfur, not sulfate." *Science, 319(5868)*, 1336.

Barber J. (2016) Photosystem II: The water splitting enzyme of photosynthesis and the origin of oxygen in our atmosphere. *Q. Rev. Biophys., 49*, DOI: 10.1017/S0033583516000093.

Bartdorff O., Wallmann K., Latif M., and Semenov V. (2008) Phanerozoic evolution of atmospheric methane. *Global Biogeochem. Cycles, 22(1)*, DOI: 10.1029/2007GB002985.

Bartley J. K. and Kah L. C. (2004) Marine carbon reservoir, C_{org}-C_{carb} coupling, and the evolution of the Proterozoic carbon cycle. *Geology, 32(2)*, 129–132.

Battison L. and Brasier M. D. (2012) Remarkably preserved prokaryote and eukaryote microfossils within 1 Ga-old lake phosphates of the Torridon Group, NW Scotland. *Precambrian Res., 196*, 204–217.

Battistuzzi F. U. and Hedges S. B. (2008) A major clade of prokaryotes with ancient adaptations to life on land. *Mol. Biol. Evol., 26(2)*, 335–343.

Battistuzzi F. U., Feijao A., and Hedges S. B. (2004) A genomic timescale of prokaryote evolution: Insights into the origin of methanogenesis, phototrophy, and the colonization of land. *BMC Evol. Biol., 4(1)*, DOI: 10.1186/1471-2148-4-44.

Bekker A. and Holland H. D. (2012) Oxygen overshoot and recovery during the early Paleoproterozoic. *Earth Planet. Sci. Lett., 317*, 295–304.

Bekker A. and Kaufman A. J. (2007) Oxidative forcing of global climate change: A biogeochemical record across the oldest Paleoproterozoic ice age in North America. *Earth Planet. Sci. Lett., 258(3–4)*, 486–499.

Bekker A. et al. (2004) Dating the rise of atmospheric oxygen. *Nature, 427*, 117–120.

Bell E. A., Boehnke P., Harrison T. M., and Mao W. L. (2015) Potentially biogenic carbon preserved in a 4.1 billion-year-old zircon. *Proc. Natl. Acad. Sci. U.S.A., 112(47)*, 14518–14521.

Berner R. A. (1989) Biogeochemical cycles of carbon and sulfur and their effect on atmospheric oxygen over Phanerozoic time. *Global Planet. Change, 1(1–2)*, 97–122.

Berry W. B. and Wilde P. (1978) Progressive ventilation of the oceans; an explanation for the distribution of the lower Paleozoic black shales. *Am. J. Sci., 278(3)*, 257–275.

Biggin A. et al. (2011) Palaeomagnetism of Archaean rocks of the Onverwacht Group, Barberton Greenstone Belt (southern Africa): Evidence for a stable and potentially reversing geomagnetic field at ca. 3.5 Ga. *Earth Planet. Sci. Lett., 302*, 314–328.

Bjerrum C. J. and Canfield D. E. (2002) Ocean productivity before about 1.9 Gyr ago limited by phosphorus adsorption onto iron oxides. *Nature, 417(6885)*, 159–162.

Bjerrum C. J. and Canfield D. E. (2004) New insights into the burial history of organic carbon on the early Earth. *Geochem. Geophys. Geosyst., 5(8)*, Q08001, DOI: 10.1029/2004GC000713.

Blackstone N. W. (2016) An evolutionary framework for understanding the origin of eukaryotes. *Biology, 5(2)*, DOI: 10.3390/biology5020018.

Blake R. E., Chang S. J., and Lepland A., (2010) Phosphate oxygen isotopic evidence for a temperate and biologically active Archaean ocean. *Nature, 464(7291)*, 1029–1032.

Blank C. E. and Sanchez-Baracaldo P. (2010) Timing of morphological and ecological innovations in the cyanobacteria — A key to understanding the rise in atmospheric oxygen. *Geobiology, 8*, 1–23.

Blankenship R. E. and Hartman H. (1998) The origin and evolution of oxygenic photosynthesis. *Trends Biochem. Sci., 23(3)*, 94–97.

Boehnke P. and Harrison T. M. (2016) Illusory late heavy bombardments. *Proc. Natl. Acad. Sci. U.S.A., 113(39)*, 10802–10806.

Bontognali T. R. R. et al. (2012) Sulfur isotopes of organic matter preserved in 3.45-billion-year-old stromatolites reveal microbial metabolism. *Proc. Natl. Acad. Sci. U.S.A., 109(38)*, 15146–15151.

Borrel G. et al. (2013) Phylogenomic data support a seventh order of methylotrophic methanogens and provide insights into the evolution of methanogenesis. *Genome Biol. Evol., 5(10)*, 1769–1780.

Bosak T., Liang B., Sim M. S., and Petroff A. P. (2009) Morphological record of oxygenic photosynthesis in conical stromatolites. *Proc. Natl. Acad. Sci. U.S.A., 106(27)*, 10939–10943.

Bradley D. C. (2008) Passive margins through Earth history. *Earth-Sci. Rev., 91(1–4)*, 1–26.

Bradley D. C. (2011) Secular trends in the geologic record and the supercontinent cycle. *Earth-Sci. Rev., 108(1–2)*, 16–33.

Brandes J. A. et al. (1998) Abiotic nitrogen reduction on the early Earth. *Nature, 395*, 365–367.

Brasier M., Green O., Lindsay J., and Steele A. (2004) Earth's oldest (approximately 3.5 Ga) fossils and the 'Early Eden hypothesis': Questioning the evidence. *Origins Life Evol. Biospheres, 34(1–2)*, 257–269.

Brasier M., McLoughlin N., Green O., and Wacey D. (2006) A fresh look at the fossil evidence for early Archaean cellular life. *Philos. Trans. R. Soc., B, 361(1470)*, 887–902.

Brasier M. D., Antcliffe J., Saunders M., and Wacey D. (2015) Changing the picture of Earth's earliest fossils (3.5–1.9 Ga) with new approaches and new discoveries. *Proc. Natl. Acad. Sci. U.S.A., 112(16)*, 4859–4864.

Braterman P. S., Cairns-Smith A. G., and Sloper R. W. (1983) Photo-oxidation of hydrated Fe^{2+} — Significance for banded iron formations. *Nature, 303*, 163–164.

Brocks J. J. et al. (2017) The rise of algae in Cryogenian oceans and the emergence of animals. *Nature, 548(7669)*, 578–581.

Brounce M., Stolper E., and Eiler J. (2017) Redox variations in Mauna Kea lavas, the oxygen fugacity of the Hawaiian plume, and the role of volcanic gases in Earth's oxygenation. *Proc. Natl. Acad. Sci. U.S.A., 114(34)*, 8997–9002.

Buick R. (1992) The antiquity of oxygenic photosynthesis: Evidence from stromatolites in sulphate-deficient Archaean lakes. *Science, 255*, 74–77.

Butterfield N. J. (1997) Plankton ecology and the Proterozoic-Phanerozoic transition. *Paleobiology, 23(2)*, 247–262.

Butterfield N. J. (2007) Macroevolution and macroecology through deep time. *Palaeontology, 50(1)*, 41–55.

Butterfield N. J. (2018) Oxygen, animals, and aquatic bioturbation: An updated account. *Geobiology, 16(1)*, 3–16.

Camacho A., Walter X. A., Picazo A., and Zopfi J. (2017) Photoferrotrophy: Remains of an ancient photosynthesis in modern environments. *Front Microbiol., 8*, DOI: 10.3389/fmicb.2017.00323.

Campbell I. H. (2003) Constraints on continental growth models from Nb/U ratios in the 3.5 Ga Barberton and other Archaean basalt-komatiite suites. *Am. J. Sci., 303(4)*, 319–351.

Campbell I. H. and Allen C. M. (2008) Formation of supercontinents linked to increases in atmospheric oxygen. *Nature Geosci., 1(8)*, 554–558.

Campbell I. H. and Squire R. J. (2010) The mountains that triggered the Late Neoproterozoic increase in oxygen: The Second Great Oxidation Event. *Geochim. Cosmochim. Acta, 74(15)*, 4187–4206.

Canfield D. E. (1998) A new model for Proterozoic ocean chemistry. *Nature, 396(6710)*, 450.

Canfield D. E. and Thamdrup B. (1994) The production of ^{34}S-depleted sulfide during bacterial disproportionation of elemental sulfur. *Science, 266*, 1973–1975.

Canfield D. E., Rosing M. T., and Bjerrum C. (2006) Early anaerobic metabolisms. *Philos. Trans. R. Soc., B, 361*, 1819–1836.

Cardona T., Murray J. W., and Rutherford A. W. (2015) Origin and evolution of water oxidation before the last common ancestor of the cyanobacteria. *Mol. Biol. Evol., 32(5)*, 1310–1328.

Cardona T., Sanchez-Baracaldo P., Rutherford A. W., and Larkum A. (2018) Early Archean origin of Photosystem II. *Geobiology, 17(2)*, 127–150, DOI: 10.1111/gbi.12322.

Catling D. (2014) The Great Oxidation Event transition. In *Treatise on Geochemistry, 2nd edition, Volume 6: The Atmosphere — History* (H. Elderfield, ed.), pp. 177–195. Elsevier, Amsterdam.

Catling D. C., Zahnle K. J. and McKay C. P. (2001) Biogenic methane, hydrogen escape, and the irreversible oxidation of early Earth. *Science, 293*, 839–843.

Cheadle B. A. (1986) Alluvial-playa sedimentation in the lower Keweenawan Sibley Group, Thunder Bay District, Ontario. *Can. J. Earth Sci., 23(4)*, 527–542.

Chen X. et al. (2015) Rise to modern levels of ocean oxygenation coincided with the Cambrian radiation of animals. *Nature Commun., 6*, 7142.

Ciborowski T. J. R. et al. (2015) The Early Proterozoic Matachewan large igneous province: Geochemistry, petrogenesis, and implications for Earth evolution. *J. Petrol., 56(8)*, 1459–1494.

Claire M. W., Catling D. C., and Zahnle K. J. (2006) Biogeochemical modelling of the rise in atmospheric oxygen. *Geobiology, 4(4)*, 239–269.

Claire M. W. et al. (2014) Modeling the signature of sulfur mass-independent fractionation produced in the Archean atmosphere. *Geochim. Cosmochim. Acta, 141,* 365–380.

Cockell C. S. (2000) The ultraviolet history of the terrestrial planets — Implications for biological evolution. *Planet. Space Sci., 48,* 203–214.

Cockell C. S. and Horneck G. (2011) The history of the UV radiation climate of the Earth — Theoretical and space-based observations. *Photochem. Photobiol., 73(4),* 447–451.

Cole D. B. et al. (2016) A shale-hosted Cr isotope record of low atmospheric oxygen during the Proterozoic. *Geology, 44(7),* 555–558.

Condie K. C. (2016) A planet in transition: The onset of plate tectonics on Earth between 3 and 2 Ga? *Geosci. Front., 9(1),* 51–60.

Condie K. C., Davaille A., Aster R. C., and Arndt N. (2015) Upstairs-downstairs: Supercontinents and large igneous provinces, are they related? *Int. Geol. Rev., 57(11–12),* 1341–1348.

Cox G. M. et al. (2016) Continental flood basalt weathering as a trigger for Neoproterozoic Snowball Earth. *Earth Planet. Sci. Lett., 446,* 89–99.

Cox G. M., Lyons T. W., Mitchell R. N., Hasterok D., and Gard M. (2018) Linking the rise of atmospheric oxygen to growth in the continental phosphorus inventory. *Earth Planet. Sci. Lett., 489,* 28–36.

Crosby H. A., Johnson C. M., Roden E. E., and Beard B. L. (2005) Coupled Fe(II)-Fe(III) electron and atom exchange as a mechanism for Fe isotope fractionation during dissimilatory iron oxide reduction. *Environ. Sci. Tech., 39(17),* 6698–6704.

Crowe S. A. et al. (2014) Sulfate was a trace constituent of Archean seawater. *Science, 346(6210),* 735–739.

Czaja A. D. et al. (2013) Biological Fe oxidation controlled deposition of banded iron formation in the ca. 3770 Ma Isua Supracrustal Belt (West Greenland). *Earth Planet. Sci. Lett., 363,* 192–203.

Dahl T. W. et al. (2010) Devonian rise in atmospheric oxygen correlated to the radiations of terrestrial plants and large predatory fish. *Proc. Natl. Acad. Sci. U.S.A., 107(42),* 17911–17915.

Daines S. J., Mills B. J., and Lenton T. M. (2017) Atmospheric oxygen regulation at low Proterozoic levels by incomplete oxidative weathering of sedimentary organic carbon. *Nature Commun., 8,* 14379, DOI: 10.1038/ncomms14379.

Dauphas N. et al. (2004) Clues from Fe isotope variations on the origin of early Archean BIFs from Greenland. *Science, 306(5704),* 2077–2080.

de Wit M. J. and Furnes H. (2016) 3.5-Ga hydrothermal fields and diamictites in the Barberton Greenstone Belt — Paleoarchean crust in cold environments. *Sci. Adv., 2(2),* DOI: 10.1126/sciadv.1500368.

Derry L. A. (2015) Causes and consequences of mid-Proterozoic anoxia. *Geophys. Res. Lett., 42(20),* 8538–8546.

Des Marais D. J. (2001) Isotopic evolution of the biogeochemical carbon cycle during the Precambrian. *Rev. Mineral. Geochem., 43,* 555–578.

Deshmukh S. S., Protheroe C., Ivanescu M. A., Lag S., and Kálmán L. (2018) Low potential manganese ions as efficient electron donors in native anoxygenic bacteria. *Biochim. Biophys. Acta, Bioenerg., 1859(4),* 227–233.

D'Hondt S. et al. (2015) Presence of oxygen and aerobic communities from sea floor to basement in deep-sea sediments. *Nature Geosci., 8(4),* 299–304.

Dhuime B., Hawkesworth C. J., Cawood P. A., and Storey C. D. (2012) A change in the geodynamics of continental growth 3 billion years ago. *Science, 335,* 1334–1336.

Dhuime B., Wuestefeld A., and Hawkesworth C. J. (2015) Emergence of modern continental crust about 3 billion years ago. *Nature Geosci., 8(7),* 552–555.

Dietrich W. E. and Perron J. T. (2006) The search for a topographic signature of life. *Nature, 439(7075),* 411–418.

Djokic T., Van Kranendonk M. J., Campbell K. A., Walter M. R., and Ward C. R. (2017) Earliest signs of life on land preserved in ca. 3.5 Ga hot spring deposits. *Nature Commun., 8,* 15263, DOI: 10.1038/ncomms15263.

Dodd M. S. et al. (2017) Evidence for early life in Earth's oldest hydrothermal vent precipitates. *Nature, 543(7643),* 60–64.

Domagal-Goldman S. D., Kasting J. F., Johnston D. T., and Farquhar J. (2008) Organic haze, glaciations, and multiple sulfur isotopes in the Mid-Archean Era. *Earth Planet. Sci. Lett., 269,* 29–40.

Dostal J. (2008) Igneous rock associations 10. Komatiites. *Geosci. Can., 35(1).*

Duda J. P. et al. (2016) A rare glimpse of Paleoarchean life: Geobiology of an exceptionally preserved microbial mat facies from the 3.4 Ga Strelley Pool Formation, Western Australia. *PLoS One, 11(1),* DOI: 10.1371/journal.pone.0147629.

Duncan M. S. and Dasgupta R. (2017) Rise of Earth's atmospheric oxygen controlled by efficient subduction of organic carbon. *Nature Geosci., 10(5),* 387–392.

Dupont C. L., Yang S., Palenik B., and Bourne P. E. (2006) Modern proteomes contain putative imprints of ancient shifts in trace metal geochemistry. *Proc. Natl. Acad. Sci. U.S.A., 103(47),* 17822–17827.

Dupont C. L., Butcher A., Valas R. E., Bourne P. E., and Caetano-Anollés G. (2010) History of biological metal utilization inferred through phylogenomic analysis of protein structures. *Proc. Natl. Acad. Sci. U.S.A., 107(23),* 10567–10572.

Edwards C. T., Saltzman M. R., Royer D. L., and Fike D. A. (2017) Oxygenation as a driver of the Great Ordovician Biodiversification Event. *Nature Geosci., 10(12),* 925–929.

Eickmann B. et al. (2018) Isotopic evidence for oxygenated Mesoarchaean shallow oceans. *Nature Geosci., 11,* 133–138.

Eme L., Sharpe S. C., Brown M. W., and Roger A. J. (2014) On the age of eukaryotes: Evaluating evidence from fossils and molecular clocks. *Cold Spring Harb. Perspect. Biol., 6(8),* DOI: 10.1101/cshperspect.a016139.

Eriksson P. G. et al. (2000) Muddy roll-up structures in siliclastic interdune beds of the c. 1.8 Ga Waterberg Group, South Africa. *Palaios, 15,* 177–183.

Ernst R. E., Wingate M. T. D., Buchan K. L., and Li Z. X. (2008) Global record of 1600–700 Ma large igneous provinces (LIPs): Implications for the reconstruction of the proposed Nuna (Columbia) and Rodinia supercontinents. *Precambrian Res., 160(1–2),* 159–178.

Erwin D. H. (2015) Early metazoan life: Divergence, environment and ecology. *Philos. Trans. R. Soc., B, 370(1684),* DOI: 10.1098/rstb.2015.0036.

Evans K. A. (2012) The redox budget of subduction zones. *Earth-Sci. Rev., 113(1–2),* 11–32.

Fakhraee M., Crowe S. A., and Katsev S. (2018) Sedimentary sulfur isotopes and Neoarchean ocean oxygenation. *Sci. Adv., 4(1),* DOI: 10.1126/sciadv.1701835.

Falkowski P. G. and Godfrey L. V. (2008) Electrons, life, and the evolution of Earth's oxygen cycle. *Philos. Trans. R. Soc., B, 363,* 2705–2716.

Farnelid H. et al. (2013) Active nitrogen-fixing heterotrophic bacteria at and below the chemocline of the central Baltic Sea. *ISME J, 7(7),* DOI: 10.1038/ismej.2013.26.

Farquhar J., Bao H., and Thiemens M. H. (2000) Atmospheric influence of Earth's earliest sulfur cycle. *Science, 289,* 756–758.

Farquhar J., Savarino J., Airieau S., and Thiemens M. H. (2001) Observation of wavelength-sensitive mass-independent sulfur isotope effects during SO_2 photolysis: Implications for the early atmosphere. *J. Geophys. Res., 106(E12),* 32829–32839.

Fedorchuk N. D. et al. (2016) Early non-marine life: Evaluating the biogenicity of Mesoproterozoic fluvial-lacustrine stromatolites. *Precambrian Res., 275,* 105–118.

Fernandez C., Farías L., and Ulloa O. (2011) Nitrogen fixation in denitrified marine waters. *PloS One, 6(6),* DOI: 10.1371/journal.pone.0020539.

Fischer W. W., Hemp J., and Johnson J. E. (2015) Manganese and the evolution of photosynthesis. *Origins Life Evol. Biospheres, 45(3),* 351–357.

Flannery D. T. and Walter M. R. (2012) Archean tufted microbial mats and the Great Oxidation Event: New insights into an ancient problem. *Aust. J. Earth Sci., 59(1),* 1–11.

Flannery D. T., Allwood A. C., and van Kranendonk M. J. (2016) Lacustrine facies dependence of highly ^{13}C-depleted organic matter during the global age of methanotrophy. *Precambrian Res., 285,* 216–241.

Flannery D. T. et al. (2018) Spatially-resolved isotopic study of carbon trapped in ~3.43 Ga Strelley Pool Formation stromatolites. *Geochim. Cosmochim. Acta, 223,* 21–35.

Foley B. J. (2015) The role of plate tectonic-climate coupling and exposed land area in the development of habitable climates on rocky planets. *Astrophys. J., 812(1),* DOI: 10.1088/0004-637X/812/1/36.

François L. M. (1986) Extensive deposition of banded iron formations was possible without photosynthesis. *Nature, 320,* 352–354.

Frausto da Silva J. J. R. and Williams R. J. (2001) *The Biological Chemistry of the Elements: The Inorganic Chemistry of Life.* Oxford Univ., Oxford. 575 pp.

Frei R. et al. (2016) Oxidative elemental cycling under the low O_2 Eoarchean atmosphere. *Sci. Rept., 6,* DOI: 10.1038/srep21058.

Frigaard N. U. and Dahl C. (2008) Sulfur metabolism in phototrophic sulfur bacteria. *Adv. Microb. Physiol., 54,* 103–200.

Frost D. J. and McCammon C. A. (2008) The redox state of Earth's mantle. *Annu. Rev. Earth Planet. Sci., 36,* 389–420.

Gaillard F., Scaillet B., and Arndt N. T. (2011) Atmospheric oxygenation caused by a change in volcanic degassing pressure. *Nature, 478,* 229–232.

Gaillard F., Scaillet B., Pichavant M., and Iacono-Marziano G. (2015) The redox geodynamics linking basalts and their mantle sources through space and time. *Chem. Geol., 418,* 217–233.

Gamper A., Heubeck C., Demske D., and Hoehse M. (2012) Composition and microfacies of Archean microbial mats (Moodies Group, ca. 3.22 Ga, South Africa). In *Microbial Mats in Siliciclastic Depositional Systems Through Time* (N. Noffke and H. Chafetz, eds.), pp. 65–74. Society for Sedimentary Geology, Tulsa.

Garvin J., Buick R., Anbar A. D., Arnold G. L., and Kaufman A. J. (2009) Isotopic evidence for an aerobic nitrogen cycle in the latest Archean. *Science, 323,* 1045–1048.

Gaschnig R. M. et al. (2016) Compositional evolution of the upper continental crust through time, as constrained by ancient glacial diamictites. *Geochim. Cosmochim. Acta, 186,* 316–343.

Gibson T. M. et al. (2017) Precise age of Bangiomorpha pubescens dates the origin of eukaryotic photosynthesis. *Geology, 46(2),* 135–138.

Gilleaudeau G. J. and Kah L. C. (2013) Carbon isotope records in a Mesoproterozoic epicratonic sea: carbon cycling in a low-oxygen world. *Precambrian Res., 228,* 85–101.

Gilleaudeau G. J. and Kah L. C. (2015) Heterogeneous redox conditions and a shallow chemocline in the Mesoproterozoic ocean: Evidence from carbon-sulfur-iron relationships. *Precambrian Res., 257,* 94–108.

Glass J. B., Wolfe-Simon F., and Anbar A. D. (2009) Coevolution of metal availability and nitrogen assimilation in cyanobacteria and algae. *Geobiology, 7,* 100–123.

Goddéris Y. et al. (2017) Paleogeographic forcing of the strontium isotopic cycle in the Neoproterozoic. *Gondwana Res., 42,* 151–162.

Godderis Y. and Veizer J. (2000) Tectonic control of chemical and isotopic composition of ancient oceans; the impact of continental growth. *Am. J. Sci., 300(5),* 434–461.

Godfrey L. V. and Falkowski P. G. (2009) The cycling and redox state of nitrogen in the Archaean ocean. *Nature Geosci., 2,* 725–729.

Gold D. A., Caron A., Fournier G. P., and Summons R. E. (2017) Paleoproterozoic sterol biosynthesis and the rise of oxygen. *Nature, 543(7645),* 420–423.

Goldblatt C. and Zahnle K. J. (2011) Faint young Sun paradox remains. *Nature, 474(7349),* DOI: 10.1038/nature09961.

Goldblatt C. et al. (2009) Nitrogen-enhanced greenhouse warming on early Earth. *Nature Geosci., 2(12),* 891–896.

Greaney A., Rudnick R. L., and Gaschnig R. (2016) Crustal sources of molybdenum. Goldschmidt Conference, Abstract #982, Yokohama, Japan.

Greber N. D. et al. (2017) Titanium isotopic evidence for felsic crust and plate tectonics 3.5 billion years ago. *Science, 357(6357),* 1271–1274.

Griffin B. M., Schott J., and Schink B. (2007) Nitrite, an electron donor for anoxygenic photosynthesis. *Science, 316(5833),* 1870–1870.

Grove T. L. and Parman S. W. (2004) Thermal evolution of the Earth as recorded by komatiites. *Earth Planet. Sci. Lett., 219(3–4),* 173–187.

Gu T., Li M., McCammon C., and Lee K. K. (2016) Redox-induced lower mantle density contrast and effect on mantle structure and primitive oxygen. *Nature Geosci., 9(9),* 723–727.

Guilbaud R., Butler I. B., and Ellam R. M. (2011) Abiotic pyrite formation produces a large Fe isotope fractionation. *Science, 332(6037),* 1548–1551.

Gumsley A. P. et al. (2017) Timing and tempo of the Great Oxidation Event. *Proc. Natl. Acad. Sci. U.S.A., 114(8),* 1811–1816.

Gutzmer J. and Beukes N. J. (1998) Earliest laterites and possible evidence for terrestrial vegetation in the Early Proterozoic. *Geology, 26(3),* 263–266.

Guy B. M. et al. (2012) A multiple sulfur and organic carbon isotope record from non-conglomeratic sedimentary rocks of the Mesoarchean Witwatersrand Supergroup, South Africa. *Precambrian Res., 216–219,* 208–231.

Habicht K. S. and Canfield D. E. (1997) Sulfur isotope fractionation during bacterial sulfate reduction in organic-rich sediments. *Geochim. Cosmochim. Acta, 61(24),* 5351–5361.

Habicht K. S., Gade M., Thamdrup B., Berg P., and Canfield D. E. (2002) Calibration of sulfate levels in the Archean Ocean. *Science, 298,* 2372–2374.

Hahn K. E., Turner E. C., Babechuk M. G., and Kamber B. S. (2015) Deep-water seep-related carbonate mounds in a Mesoproterozoic alkaline lake, Borden Basin (Nunavut, Canada). *Precambrian Res., 271,* 173–197.

Halevy I. (2013) Production, preservation, and biological processing of mass-independent sulfur isotope fractionation in the Archean surface environment. *Proc. Natl. Acad. Sci. U.S.A., 110,* 17644–17649.

Halverson G. P., Wade B. P., Hurtgen M. T., and Barovich K. M. (2010) Neoproterozoic chemostratigraphy. *Precambrian Res., 182(4),* 337–350.

Hao J., Sverjensky D. A., and Hazen R. M. (2017a) Mobility of nutrients and trace metals during weathering in the late Archean. *Earth Planet. Sci. Lett., 471,* 148–159.

Hao J., Sverjensky D. A., and Hazen R. M. (2017b) A model for late Archean chemical weathering and world average river water. *Earth Planet. Sci. Lett., 457,* 191–203.

Haqq-Misra J. D., Domagal-Goldman S. D., Kasting P.J., and Kasting J. F. (2008) A revised, hazy methane greenhouse for the Archean Earth. *Astrobiology, 8(6),* 1127–1137.

Haqq-Misra J., Kasting J. F., and Lee S. (2011) Availability of O_2 and H_2O_2 on pre-photosynthetic Earth. *Astrobiology, 11(4),* 293–302.

Hardisty D. S. et al. (2017) Perspectives on Proterozoic surface ocean redox from iodine contents in ancient and recent carbonate. *Earth Planet. Sci. Lett., 463,* 159–170.

Harrison T. M. et al. (2005) Heterogeneous Hadean hafnium: Evidence of continental crust at 4.4 to 4.5 Ga. *Science, 310,* 1947–1950.

Hawkesworth C. J., Cawood P. A., and Dhuime B. (2016) Tectonics and crustal evolution. *GSA Today, 26(9),* 4–11.

Hayes J. M. and Waldbauer J. R. (2006) The carbon cycle and associated redox processes through time. *Philos. Trans. R. Soc., B, 361(1470),* 931–950.

Hazen R. M. and Ferry J. M. (2010) Mineral evolution: Mineralogy in the fourth dimension. *Elements, 6(1),* 9–12.

Hedges S. B., Blair J. E., Venturi M. L., and Shoe J. L. (2004) A molecular timescale of eukaryote evolution and the rise of complex multicellular life. *BMC Evol. Biol., 4(1),* DOI: 10.1186/1471-2148-4-2.

Heimann A. et al. (2010) Fe, C, and O isotope compositions of banded iron formation carbonates demonstrate a major role for dissimilatory iron reduction in ~2.5 Ga marine environments. *Earth Planet. Sci. Lett., 294(1–2),* 8–18.

Heinrich C. A. (2015) Witwatersrand gold deposits formed by volcanic rain, anoxic rivers and Archaean life. *Nature Geosci., 8(3),* 206–209.

Herzberg C., Condie K., and Korenaga J. (2010) Thermal history of the Earth and its petrological expression. *Earth Planet. Sci. Lett., 292(1–2),* 79–88.

Hessler A. M. and Lowe D. R. (2006) Weathering and sediment generation in the Archean: An integrated study of the evolution of siliciclastic sedimentary rocks of the 3.2 Ga Moodies Group, Barberton Greenstone Belt, South Africa. *Precambrian Res., 151(3),* 185–210.

Hoffman P. F. and Schrag D. P. (2002) The snowball Earth hypothesis: Testing the limits of global change. *Terra Nova, 14(3),* 129–155.

Hoffman P. F. et al. (2017) Snowball Earth climate dynamics and Cryogenian geology-geobiology. *Sci. Adv., 3(11),* e1600983.

Hohmann-Marriott M. F. and Blankenship R. E. (2011) Evolution of photosynthesis. *Annu. Rev. Plant Biol., 62,* 515–548.

Holland H. D. (1984) *The Chemical Evolution of the Atmosphere and Oceans.* Princeton Univ., Princeton. 598 pp.

Holland H. D. (2002) Volcanic gases, black smokers, and the Great Oxidation Event. *Geochim. Cosmochim. Acta, 66(21),* 3811–3826.

Holland H. D. (2003) The geologic history of seawater. In *Treatise on Geochemistry, Volume 6: The Atmosphere — History* (H. Elderfield, ed.), pp. 583–625. Elsevier, Amsterdam.

Holland H. D. (2006) The oxygenation of the atmosphere and oceans. *Philos. Trans. R. Soc., B, 361(1470),* 903–915.

Höning D. and Spohn T. (2016) Continental growth and mantle

hydration as intertwined feedback cycles in the thermal evolution of Earth. *Phys. Earth Planet. Inter., 255,* 27–49.

Horodyski R. J. and Knauth L. P. (1994) Life on land in the Precambrian. *Science, 263,* 494–498.

Horton F. (2015) Did phosphorus derived from the weathering of large igneous provinces fertilize the Neoproterozoic ocean? *Geochem. Geophys. Geosyst., 16(6),* 1723–1738.

Hoshino Y. et al. (2017) Cryogenian evolution of stigmasteroid biosynthesis. *Sci. Adv., 3(9),* DOI: 10.1126/sciadv.1700887.

House C. H., Oehler D. Z., Sugitani K., and Mimura K. (2013) Carbon isotopic analyses of ca. 3.0 Ga microstructures imply planktonic autotrophs inhabited Earth's early oceans. *Geology, 41(6),* 651–654.

Hren M. T., Tice M. M., and Chamberlain C. P. (2009) Oxygen and hydrogen isotope evidence for a temperate climate 3.42 billion years ago. *Nature, 462,* 205–208.

Hurowitz J. A. et al. (2017) Redox stratification of an ancient lake in Gale crater, Mars. *Science, 356,* DOI: 10.1126/science.aah6849.

Husson J. M. and Peters S. E. (2017) Atmospheric oxygenation driven by unsteady growth of the continental sedimentary reservoir. *Earth Planet. Sci. Lett., 460,* 68–75.

Isley A. E. and Abbott D. H. (1999) Plume-related mafic volcanism and the deposition of banded iron formation. *J. Geophys. Res.–Solid Earth, 104(B7),* 15461–15477.

Isson T. T. et al. (2018) Tracking the rise of eukaryotes to ecological dominance with zinc isotopes. *Geobiology, 16(4),* 341–352.

Izon G. et al. (2015) Multiple oscillations in Neoarchaean atmospheric chemistry. *Earth Planet. Sci. Lett., 431,* 264–273.

Izon G. et al. (2017) Biological regulation of atmospheric chemistry en route to planetary oxygenation. *Proc. Natl. Acad. Sci. U.S.A., 114(13),* 2571–2579.

Jackson M. J. and Raiswell R. (1991) Sedimentoogy and carbon-sulphur geochemistry of the Velkerri Formation, a mid-Proterozoic potential oil source in northern Australia. *Precambrian Res., 54,* 81–108.

Jackson C., Knoll A. H., Chan C. X., and Verbruggen H. (2018) Plastid phylogenomics with broad taxon sampling further elucidates the distinct evolutionary origins and timing of secondary green plastids. *Sci. Rept., 8(1),* DOI: 10.1038/s41598-017-18805-w.

Jaupart C., Mareschal J. C., and Iarotsky L. (2016) Radiogenic heat production in the continental crust. *Lithos, 262,* 398–427.

Javaux E. J. (2007) The early eukaryotic fossil record. In *Eukaryotic Membranes and Cytoskeleton* (G. Jékely, ed.), pp. 1–19. Springer, New York.

Jayakumar A., Al-Rshaidat M. M., Ward B. B., and Mulholland M. R. (2012) Diversity, distribution, and expression of diazotroph nifH genes in oxygen-deficient waters of the Arabian Sea. *FEMS Microbiol. Ecol., 82(3),* 597–606.

Johnson J. E., Gerpheide A., Lamb M. P., and Fischer W. W. (2014) O_2 constraints from Paleoproterozoic detrital pyrite and uraninite. *Geol. Soc. Am. Bull., 126 (5-6),* 813-830, DOI: 10.1130/B30949.1.

Jones C., Nomosatryo S., Crowe S. A., Bjerrum C. J., and Canfield D. E. (2015) Iron oxides, divalent cations, silica, and the early earth phosphorus crisis. *Geology, 43(2),* 135–138.

Kah L. C. and Riding R. (2007) Mesoproterozoic carbon dioxide levels inferred from calcified cyanobacteria. *Geology, 35(9),* 799–802.

Kah L. C., Lyons T. W., and Frank T. D. (2004) Low marine sulphate and protracted oxygenation of the Proterozoic biosphere. *Nature, 431,* 834–838.

Karhu J. A. and Holland H. D. (1996) Carbon isotopes and the rise of atmospheric oxygen. *Geology, 24(10),* 867–870.

Kasting J. F. (1993) Earth's early atmosphere. *Science, 259(5097),* 920–926.

Kasting J. F. (2005) Methane and climate during the Precambrian era. *Precambrian Res., 137,* 119–129.

Kasting J. F. and Canfield D. E. (2012) The global oxygen cycle. In *Fundamentals of Geobiology* (A. H. Knoll et al., eds.), pp. 93–104. Blackwell, Chichester.

Kasting J. F. and Siefert J. L. (2001) Biogeochemistry: The nitrogen fix. *Nature, 412(6842),* 26–27.

Kasting J. F. and Walker J. C. (1981) Limits on oxygen concentration in the prebiological atmosphere and the rate of abiotic fixation of nitrogen. *J. Geophys. Res.–Oceans, 86(C2),* 1147–1158.

Kasting J. F., Pavlov A. A., and Siefert J. L. (2001) A coupled ecosystem-climate model for predicting the methane concentration in the Archean atmosphere. *Origins Life Evol. Biospheres, 31(3),* 271–285.

Kaufman A. J. et al. (2007) Late Archean biospheric oxygenation and atmospheric evolution. *Science, 317,* 1900–1903.

Keeling P. J. (2010) The endosymbiotic origin, diversification and fate of plastids. *Philos. Trans. R. Soc., B, 365(1541),* 729–748.

Kelley K. A. and Cottrell E. (2009) Water and the oxidation state of subduction zone magmas. *Science, 325,* 605–607.

Kendall B. et al. (2010) Pervasive oxygenation along late Archaean ocean margins. *Nature Geosci., 3,* 647–652.

Kendall B., Creaser R. A., Reinhard C. T., Lyons T. W., and Anbar A. D. (2015) Transient episodes of mild environmental oxygenation and oxidative continental weathering during the late Archean. *Sci. Adv., 1(10),* DOI: 10.1126/sciadv.1500777.

Kharecha P., Kasting J., and Siefert J. (2005) A coupled atmosphere-ecosystem model of the early Archean Earth. *Geobiology, 3(2),* 53–76.

Kipp M. A. and Stüeken E. E. (2017) Biomass recycling and Earth's early phosphorus cycle. *Sci. Adv., 3(11),* DOI: 10.1126/sciadv. aao4795.

Kipp M. A., Stüeken E. E., Bekker A., and Buick R. (2017) Selenium isotopes record extensive marine suboxia during the Great Oxidation Event. *Proc. Natl. Acad. Sci. U.S.A., 114(5),* 875–880.

Kirschvink J. L. and Kopp R. E. (2008) Palaeoproterozoic ice houses and the evolution of oxygen-mediating enzymes: The case for a late origin of photosystem II. *Philos. Trans. R. Soc., B, 363(1504),* 2755–2765.

Kiyokawa S., Ito T., Ikehara M., and Kitajima F. (2006) Middle Archean volcano-hydrothermal sequence: Bacterial microfossil-bearing 3.2 Ga Dixon Island Formation, coastal Pilbara terrane, Australia. *Geol. Soc. Am. Bull., 118(1–2),* 3–22.

Klein M. et al. (2001) Multiple lateral transfers of dissimilatory sulfite reductase genes between major lineages of sulfate-reducing prokaryotes. *J. Bacteriol., 183(20),* 6028–6035.

Knauth L. P. (2005) Temperature and salinity history of the Precambrian ocean: Implications for the course of microbial evolution. *Palaeogeogr. Palaeoclimatol. Palaeoecol., 219(1),* 53–69.

Knoll A. H. (1994) Proterozoic and Early Cambrian protists: Evidence for accelerating evolutionary tempo. *Proc. Natl. Acad. Sci. U.S.A., 91(15),* 6743–6750.

Knoll A. H. and Nowak M. A. (2017) The timetable of evolution. *Sci. Adv., 3(5),* DOI: 10.1126/sciadv.1603076.

Knoll A. H. and Sergeev V. N. (1995) Taphonomic and evolutionary changes across the Mesoproterozoic-Neoproterozoic transition. *Neues Jahrb. Geol. P-A, 195(1–3),* 289–302.

Knoll A. H., Javaux E. J., Hewitt D., and Cohen P. (2006) Eukaryotic organisms in Proterozoic oceans. *Philos. Trans. R. Soc., B, 361,* 1023–1038.

Kodner R. B., Pearson A., Summons R. E., and Knoll A. H. (2008) Sterols in red and green algae: Quantification, phylogeny, and relevance for the interpretation of geologic steranes. *Geobiology, 6(4),* 411–420.

Koehler M. C., Stüeken E. E., Kipp M. A., Buick R., and Knoll A. H. (2017) Spatial and temporal trends in Precambrian nitrogen cycling: A Mesoproterozoic offshore nitrate minimum. *Geochim. Cosmochim. Acta, 198,* 315–337.

Koehler M. C., Buick R., Kipp M. A., Stüeken E. E., and Zaloumis J. (2018) An Archean transient oxygen oasis recorded in the ~2.66 Ga Jeerinah Formation, Australia. *Proc. Natl. Acad. Sci. U.S.A., 115(30),* 7711–7716, DOI: 10.1073/pnas.1720820115.

Konhauser K. O. et al. (2002) Could bacteria have formed the Precambrian banded iron formations? *Geology, 30(12),* 1079–1082.

Konhauser K. O. et al. (2007a) Decoupling photochemical Fe(II) oxidation from shallow-water BIF deposition. *Earth Planet. Sci. Lett., 258(1–2),* 87–100.

Konhauser K. O., Lalonde S. V., Amskold L., and Holland H. D. (2007b) Was there really an Archean phosphate crisis? *Science, 315(5816),* 1234.

Konhauser K. O. et al. (2009) Oceanic nickel depletion and a methanogen famine before the Great Oxidation Event. *Nature, 458(7239),* 750–753.

Konhauser K. O. et al. (2011) Aerobic bacterial pyrite oxidation and acid rock drainage during the Great Oxidation Event. *Nature, 478,* 369–373.

Konhauser K. O. et al. (2015) The Archean nickel famine revisited. *Astrobiology, 15(10),* 804–815.

Konhauser K. O. et al. (2017) Phytoplankton contributions to the trace-element composition of Precambrian banded iron formations. *Geol. Soc. Am. Bull., 130(5–6)*, 941–951, DOI: 10.1130/B31648.1.

Kopp R. E., Kirschvink J. L., Hilburn I. A., and Nash C. Z. (2005) The Paleoproterozoic snowball Earth: A climate disaster triggered by the evolution of oxygenic photosynthesis. *Proc. Natl. Acad. Sci. U.S.A., 102(32)*, 11131–11136.

Korenaga J. (2006) Archean geodynamics and the thermal evolution of Earth. In *Archean Geodynamics and Environments* (K. Benn et al., eds.), pp. 7–32. American Geophysical Union, Washington, DC.

Korenaga J. (2013) Initiation and evolution of plate tectonics on Earth: Theories and observations. *Annu. Rev. Earth Planet. Sci., 41*, 117–151.

Korenaga J., Planavsky N., and Evans D. A. (2017) Global water cycle and the coevolution of the Earth's interior and surface environment. *Philos. Trans. R. Soc., A, 375(2094)*, 20150393.

Krissansen-Totton J., Buick R., and Catling D. C. (2015) A statistical analysis of the carbon isotope record from the Archean to Phanerozoic and implications for the rise of oxygen. *Am. J. Sci., 315(4)*, 275–316.

Kump L. R. and Barley M. E. (2007) Increased subaerial volcanism and the rise of atmospheric oxygen 2.5 billion years ago. *Nature, 448(7157)*, 1033–1036.

Kump L. R., Kasting J. F., and Barley M. E. (2001) Rise of atmospheric oxygen and the "upside-down" Archean mantle. *Geochem. Geophys. Geosyst., 2(1)*, DOI: 10.1029/2000GC000114.

Kump L. R. et al. (2011) Isotopic evidence for massive oxidation of organic matter following the Great Oxidation Event. *Science, 334(6063)*, 1694–1696.

Kurzweil F. et al. (2013) Atmospheric sulfur rearrangement 2.7 billion years ago: Evidence for oxygenic photosynthesis. *Earth Planet. Sci. Lett., 366*, 17–26.

Laakso T. A. and Schrag D. P. (2014) Regulation of atmospheric oxygen during the Proterozoic. *Earth Planet. Sci. Lett., 388*, 81–91.

Laakso T. A. and Schrag D. P. (2017) A theory of atmospheric oxygen. *Geobiology, 15(3)*, 366–384.

Labrosse S. (2003) Thermal and magnetic evolution of the Earth's core. *Phys. Earth Planet. Inter., 140*, 127–143.

Labrosse S., Poirier J. P., and Le Mouël J. L. (1997) On cooling of the Earth's core. *Phys. Earth Planet. Inter., 99(1–2)*, 1–17.

Lalonde S. V. and Konhauser K. O. (2015) Benthic perspective on Earth's oldest evidence for oxygenic photosynthesis. *Proc. Natl. Acad. Sci. U.S.A., 112(4)*, 995–1000.

Lammer H. et al. (2014) Origin and loss of nebula-captured hydrogen envelopes from 'sub'-to 'super-Earths' in the habitable zone of Sun-like stars. *Mon. Not. R. Astron. Soc., 439(4)*, 3225–3238.

Laurent O., Martin H., Moyen J. F., and Doucelance R. (2014) The diversity and evolution of late-Archean granitoids: Evidence for the onset of "modern-style" plate tectonics between 3.0 and 2.5 Ga. *Lithos, 205*, 208–235.

Lazar C., McCollom T. M., and Manning C. E. (2012) Abiogenic methanogenesis during experimental komatiite serpentinization: Implications for the evolution of the early Precambrian atmosphere. *Chem. Geol., 326*, 102–112.

Lécuyer C. and Ricard Y. (1999) Long-term fluxes and budget of ferric iron: Implication for the redox states of the Earth's mantle and atmosphere. *Earth Planet. Sci. Lett., 165(2)*, 197–211.

Lee C. T. A. et al. (2016) Two-step rise of atmospheric oxygen linked to the growth of continents. *Nature Geosci., 9(6)*, 417–424.

Lee C. T. A. et al. (2018) Deep mantle roots and continental emergence: Implications for whole-Earth elemental cycling, long-term climate, and the Cambrian explosion. *Intl. Geol. Rev., 60(4)*, 431–448.

Lenardic A. and Crowley J. W. (2012) On the notion of well-defined tectonic regimes for terrestrial planets in this solar system and others. *Astrophys. J., 755(2)*, DOI: 10.1088/0004-637X/755/2/132.

Lenton T. M., Boyle R. A., Poulton S. W., Shields-Zhou G. A., and Butterfield N. J. (2014) Co-evolution of eukaryotes and ocean oxygenation in the Neoproterozoic era. *Nature Geosci., 7(4)*, 257–265.

Lenton T. M. et al. (2016) Earliest land plants created modern levels of atmospheric oxygen. *Proc. Natl. Acad. Sci. U.S.A., 113(35)*, 9704–9709.

Lepland A. et al. (2014) Potential influence of sulphur bacteria on Palaeoproterozoic phosphogenesis. *Nature Geosci., 7*, 20–24.

Li Z. X. et al. (2008) Assembly, configuration, and break-up history of Rodinia: A synthesis. *Precambrian Res., 160(1–2)*, 179–210.

Lindsay J. F. and Brasier M. D. (2002) Did global tectonics drive early biosphere evolution? Carbon isotope record from 2.6 to 1.9 Ga carbonates of Western Australian basins. *Precambrian Res., 114(1–2)*, 1–34.

Love G. D. et al. (2009) Fossil steroids record the appearance of Demospongiae during the Cryogenian period. *Nature, 457(7230)*, 718–721.

Luo G. et al. (2016) Rapid oxygenation of Earth's atmosphere 2.33 billion years ago. *Sci. Adv., 2(5)*, DOI: 10.1125/sciadv.1600134.

Lyle M. (1976) Estimation of hydrothermal manganese input to the oceans. *Geology, 4(12)*, 733–736.

Lyons T. W., Reinhard C. T., and Planavsky N. J. (2014) The rise of oxygen in Earth's early ocean and atmosphere. *Nature, 506*, 307–315.

Lyons T. W., Fike D. A., and Zerkle A. (2015) Emerging biogeochemical views of Earth's ancient microbial worlds. *Elements, 11(6)*, 415–421.

Magnabosco C., Moore K. R., Wolfe J. M., and Fournier G. P. (2018) Dating phototropic microbial lineages with reticulate gene histories. *Geobiology, 16(2)*, 179–189.

Martin A. P., Condon D. J., Prave A. R., and Lepland A. (2013) A review of temporal constraints for the Palaeoproterozoic large, positive carbonate carbon isotope excursion (the Lomagundi-Jatuli Event). *Earth-Sci. Rev., 127*, 242–261.

Martin A. P. et al. (2015) Multiple Palaeoproterozoic carbon burial episodes and excursions. *Earth Planet. Sci. Lett., 424*, 226–236.

Martin R. S., Mather T. A., and Pyle D. M. (2007) Volcanic emissions and the early Earth atmosphere. *Geochim. Cosmochim. Acta, 71*, 3673–3685.

Martin W. F. and Sousa F. L. (2016) Early microbial evolution: The age of anaerobes. *Cold Spring Harbor Perspect. Biol., 8(2)*, DOI: 10.1101/cshperspect.a018127.

Mason E., Edmonds M., and Turchyn A. V. (2017) Remobilization of crustal carbon may dominate volcanic arc emissions. *Science, 357(6348)*, 290–294.

Mather T. A., Pyle D. M., and Allen A. G. (2004) Volcanic source for fixed nitrogen in the early Earth's atmosphere. *Geology, 32(10)*, 905–908.

Mathis P. (1990) Compared structure of plant and bacterial photosynthetic reaction centers. Evolutionary implications. *Biochim. Biophys. Acta Bioenerg., 1018(2–3)*, 163–167.

McCammon C. (2005) The paradox of mantle redox. *Science, 308(5723)*, 807–808.

McCollom T. M. (2013) Laboratory simulations of abiotic hydrocarbon formation in Earth's deep subsurface. *Rev. Mineral. Geochem., 75*, 467–494.

McCollom T. M. and Seewald J. S. (2006) Carbon isotopic composition of organic compounds produced by abiotic synthesis under hyhdrothermal conditions. *Earth Planet. Sci. Lett., 243(1)*, 74–84.

McCoy V. E., Asael D., and Planavsky N. (2017) Benthic iron cycling in a high-oxygen environment: Implications for interpreting the Archean sedimentary iron isotope record. *Geobiology, 15(6)*, 619–627.

McLennan S. M. and Taylor S. R. (1982) Geochemical constraints on the growth of the continental crust. *J. Geol., 90(4)*, 347–361.

McManus J., Nägler T. F., Siebert C., Wheat C. G., and Hammond D. E. (2002) Oceanic molybdenum isotope fractionation: Diagenesis and hydrothermal ridge-flank alteration. *Geochem. Geophys. Geosyst., 3(12)*, 1–9.

Meyer K. M., Ridgwell A., and Payne J. L. (2016) The influence of the biological pump on ocean chemistry: Implications for long-term trends in marine redox chemistry, the global carbon cycle, and marine animal ecosystems. *Geobiology, 14(3)*, 207–219.

Michiels C. C. et al. (2017) Iron-dependent nitrogen cycling in a ferruginous lake and the nutrient status of Proterozoic oceans. *Nature Geosci., 10(3)*, 217–221.

Mills B., Lenton T. M., and Watson A. J. (2014a) Proterozoic oxygen rise linked to shifting balance between seafloor and terrestrial weathering. *Proc. Natl. Acad. Sci. U.S.A., 111(25)*, 9073–9078.

Mills D. B. et al. (2014b) Oxygen requirements of the earliest animals. *Proc. Natl. Acad. Sci. U.S.A., 111(11)*, 4168–4172.

Mojzsis S. J. et al. (1996) Evidence for life on Earth before 3,800 million years ago. *Nature, 384*, 55–59.

Moldowan J. M. et al. (1990) Sedimentary 24-n-propylcholestanes, molecular fossils diagnostic of marine algae. *Science, 247*, 309–312.

Moore W. B. and Webb A. A. G. (2013) Heat-pipe earth. *Nature,*

501(7468), 501–505.

Morris R. C. and Horwitz R. C. (1983) The origin of the iron-formation-rich Hamersley Group of Western Australia — Deposition on a platform. *Precambrian Res., 21(3–4)*, 273–297.

Mulkidjanian A. Y. et al. (2006) The cyanobacterial genome core and the origin of photosynthesis. *Proc. Natl. Acad. Sci. U.S.A., 103(35)*, 13126–13131.

Muller É., Philippot P., Rollion-Bard C., and Cartigny P. (2016) Multiple sulfur-isotope signatures in Archean sulfates and their implications for the chemistry and dynamics of the early atmosphere. *Proc. Natl. Acad. Sci. U.S.A., 113(27)*, 7432–7437.

Murakami T., Matsuura K., and Kanzaki Y. (2016) Behaviors of trace elements in Neoarchean and Paleoproterozoic paleosols: Implications for atmospheric oxygen evolution and continental oxidative weathering. *Geochim. Cosmochim. Acta, 192*, 203–219.

Nabhan S., Wiedenbeck M., Milke R., and Heubeck C. (2016) Biogenic overgrowth on detrital pyrite in ca. 3.2 Ga Archean paleosols. *Geology, 44(9)*, 763–766.

Navarro-González R., Molina M. J., and Molina L. T. (1998) Nitrogen fixation by volcanic lightning in the early Earth. *Geophys. Res. Lett., 25(16)*, 3123–3126.

Neaman A., Chorover J., and Brantley S. L. (2005) Element mobility patterns record organic ligands in soils on early Earth. *Geology, 33(2)*, 117–120.

Nebel-Jacobsen Y., Nebel O., Wille M., and Cawood P. A. (2018) A non-zircon Hf isotope record in Archean black shales from the Pilbara craton confirms changing crustal dynamics ca. 3 Ga ago. *Sci. Rep., 8(1)*, DOI: 10.1038/s41598-018-19397-9.

Nelson N. and Junge W. (2015) Structure and energy transfer in photosystems of oxygenic photosynthesis. *Annu. Rev. Biochem., 84*, 659–683.

Neubeck A. et al. (2014) Olivine alteration and H_2 production in carbonate-rich, low temperature aqueous environments. *Planet. Space Sci., 96*, 51–61.

Nicklas R. W., Puchtel I. S., and Ash R. D. (2018) Redox state of the Archean mantle: Evidence from V partitioning in 3.5–2.4 Ga komatiites. *Geochim. Cosmochim. Acta, 222*, 447–466.

Nutman A. P., Bennett V. C., Friend C. R., van Kranendonk M. J., and Chivas A. R. (2016) Rapid emergence of life shown by discovery of 3,700-million-year-old microbial structures. *Nature, 537(7621)*, 535–538.

O'Neill C. and Debaille V. (2014) The evolution of Hadean-Eoarchean geodynamics. *Earth Planet. Sci. Lett., 406*, 49–58.

Och L. M. and Shields-Zhou G. A. (2012) The Neoproterozoic oxygenation event: Environmental perturbations and biogeochemical cycling. *Earth-Sci. Rev., 110(1)*, 26–57.

Olson J. M. and Pierson B. K. (1987) Evolution of reaction centers in photosynthetic prokaryotes. *Intl. Rev. Cytol., 108*, 209–248.

Olson S. L., Kump L. R., and Kasting J. F. (2013) Quantifying the areal extent and dissolved oxygen concentrations of Archean oxygen oases. *Chem. Geol., 362*, 35–43.

Olson S. L., Reinhard C. T., and Lyons T. W. (2016a) Cyanobacterial diazotrophy and Earth's delayed oxygenation. *Front. Microbiol., 7*, DOI: 10.3389/fmicb.2016.01526.

Olson S. L., Reinhard C. T., and Lyons T. W. (2016b) Limited role for methane in the mid-Proterozoic greenhouse. *Proc. Natl. Acad. Sci. U.S.A., 113(41)*, 11447–11452.

Ono S., Whitehill A. R., and Lyons J. R. (2013) Contribution of isotopologue self-shielding to sulfur mass-independent fractionation during sulfur dioxide photolysis. *J. Geophys. Res.–Atmospheres, 118(5)*, 2444–2454.

Ossa Ossa F. et al. (2018) Two-step deoxygenation at the end of the Paleoproterozoic Lomagundi Event. *Earth Planet. Sci. Lett., 486*, 70–83.

Papineau D. (2010) Global biogeochemical changes at both ends of the Proterozoic: Insights from phosphorites. *Astrobiology, 10(2)*, 165–181.

Parfrey L. W., Lahr D. J., Knoll A. H., and Katz L. A. (2011) Estimating the timing of early eukaryotic diversification with multigene molecular clocks. *Proc. Natl. Acad. Sci. U.S.A., 108(33)*, 13624–13629.

Parnell J., Boyce A. J., Mark D., Bowden S., and Spinks S. (2010) Early oxygenation of the terrestrial environment during the Mesoproterozoic. *Nature, 468*, 290–293.

Parnell J., Spinks S., Andrews S., Thayalan W., and Bowden S. (2015) High molybdenum availability for evolution in a Mesoproterozoic lacustrine environment. *Nature Commun., 6*, DOI: 10.1038/ncomms7996.

Partin C. A. et al. (2013) Uranium in iron formations and the rise of atmospheric oxygen. *Chem. Geol., 362*, 82–90.

Partin C. A., Bekker A., Planavsky N. J., and Lyons T. W. (2015) Euxinic conditions recorded in the ca. 1.93 Ga Bravo Lake Formation, Nunavut (Canada): Implications for oceanic redox evolution. *Chem. Geol., 417*, 148–162.

Pavlov A. A. and Kasting J. F. (2002) Mass-independent fractionation of sulfur isotopes in Archean sediments: Strong evidence for an anoxic Archean atmosphere. *Astrobiology, 2(1)*, 27–41.

Pavlov A. A., Kasting J. F., Brown L. L., Rages K. A., and Freedman R. (2000) Greenhouse warming by CH_4 in the atmosphere of early Earth. *J. Geophys. Res.–Planets, 105(E5)*, 11981–11990.

Pennisi E. (2017) Evolution accelerated when life set foot on land. *Science, 358(6360)*, 158–159.

Percak-Dennett E. M. et al. (2011) Iron isotope fractionation during microbial dissimilatory iron oxide reduction in simulated Archaean seawater. *Geobiology, 9(3)*, 205–220.

Petroff A. P. et al. (2010) Biophysical basis for the geometry of conical stromatolites. *Proc. Natl. Acad. Sci. U.S.A., 107(22)*, 9956–9961.

Philippot P. et al. (2007) Early Archaean microorganisms preferred elemental sulfur, not sulfate. *Science, 317*, 1534–1537.

Planavsky N. J. (2014) The elements of marine life. *Nature Geosci., 7(12)*, 855–856.

Planavsky N. J. (2018) From orogenies to oxygen. *Nature Geosci., 11(1)*, 9.

Planavsky N. J. et al. (2010) The evolution of the marine phosphate reservoir. *Nature, 467*, 1088–1090.

Planavsky N. J. et al. (2011) Widespread iron-rich conditions in the mid-Proterozoic ocean. *Nature, 477*, 448–451.

Planavsky N. J. et al. (2014a) Evidence for oxygenic photosynthesis half a billion years before the Great Oxidation Event. *Nature Geosci., 7*, 283–286.

Planavsky N. J. et al. (2014b) Low Mid-Proterozoic atmospheric oxygen levels and the delayed rise of animals. *Science, 346(6209)*, 635–638.

Pons M.-L. et al. (2011) Early Archean serpentine mud volcanoes at Isua, Greenland, as a niche for early life. *Proc. Natl. Acad. Sci. U.S.A., 108(43)*, 17639–17643.

Poulton S. W. and Canfield D. E. (2011) Ferruginous conditions: A dominant feature of the ocean through Earth's history. *Elements, 7*, 107–112.

Poulton S. W., Fralick P. W., and Canfield D. E. (2010) Spatial variability in oceanic redox structure 1.8 billion years ago. *Nature Geosci., 3*, 486–490.

Pratt L. M., Summons R. E., and Hieshima G. B. (1991) Sterane and triterpane biomarkers in the Precambrian Nonesuch Formation, North American Midcontinent Rift. *Geochim. Cosmochim. Acta, 55(3)*, 911–916.

Prave A. R. (2002) Life on land in the Proterozoic: Evidence from the Torridonian rocks of northwest Scotland. *Geology, 30(9)*, 811–814.

Rampino M. R. and Caldeira K. (1994) The Goldilocks problem: Climatic evolution and long-term habitability of terrestrial planets. *Annu. Rev. Astron. Astrophys., 32(1)*, 83–114.

Rasmussen B. (2000) Filamentous microfossils in a 3,235-million-year-old volcanogenic massive sulphide deposit. *Nature, 405(6787)*, 676–679.

Reinhard C. T., Raiswell R., Scott C. T., Anbar A., and Lyons T. W. (2009) A late Archean sulfidic sea stimulated by early oxidative weathering of the continents. *Science, 326*, 713–716.

Reinhard C. T., Planavsky N. J., and Lyons T. W. (2013a) Long-term sedimentary recycling of rare sulphur isotope anomalies. *Nature, 497(7447)*, 100–103.

Reinhard C. T. et al. (2013b) Proterozoic ocean redox and biogeochemical stasis. *Proc. Natl. Acad. Sci. U.S.A., 110(14)*, 5357–5362.

Reinhard C. T., Planavsky N., Olson S. L., Lyons T. W., and Erwin D. H. (2016) Earth's oxygen cycle and the evolution of animal life. *Proc. Natl. Acad. Sci. U.S.A., 113(32)*, 8933-8938, DOI: 10.1073/pnas.1521544113.

Reinhard C. T. et al. (2017) Evolution of the global phosphorus cycle. *Nature, 541(7637)*, 386–389.

Rey P. F. and Coltice N. (2008) Neoarchean lithospheric strengthening and the coupling of Earth's geochemical reservoirs. *Geology, 36(8)*,

635–638.

Rey P. F. and Houseman G. (2006) Lithospheric scale gravitational flow: The impact of body forces on orogenic processes from Archaean to Phanerozoic. *Geol. Soc. Spec. Publ., 253(1)*, 153–167.

Rhodes J. M. and Vollinger M. J. (2005) Ferric/ferrous ratios in 1984 Mauna Loa lavas: A contribution to understanding the oxidation state of Hawaiian magmas. *Contrib. Mineral. Petrol., 149*, 666–674.

Richter F. M. (1985) Models for the Archean thermal regime. *Earth Planet. Sci. Lett., 73(2–4)*, 350–360.

Riding R., Fralick P., and Liang L. (2014) Identification of an Archean marine oxygen oasis. *Precambrian Res., 251*, 232–237.

Robbins L. J. et al. (2013) Authigenic iron oxide proxies for marine zinc over geological time and implications for eukaryotic metallome evolution. *Geobiology, 11*, 295–306.

Robbins L. J. et al. (2016) Trace elements at the intersection of marine biological and geochemical evolution. *Earth-Sci. Rev., 163*, 323–348.

Rooney A. D. et al. (2014) Re-Os geochronology and coupled Os-Sr isotope constraints on the Sturtian snowball Earth. *Proc. Natl. Acad. Sci. U.S.A., 111(1)*, 51–56.

Rosas J. C. and Korenaga J. (2018) Rapid crustal growth and efficient crustal recycling in the early Earth: Implications for Hadean and Archean geodynamics. *Earth Planet. Sci. Lett., 494*, 42–49.

Rosing M. T. (1999) ^{13}C-depleted carbon microparticles in >3700-Ma sea-floor sedimentary rocks from West Greenland. *Science, 283(5402)*, 674–676.

Rosing M. T. (2005) Thermodynamics of life on the planetary scale. *Intl. J. Astrobiol., 4(1)*, 9–11.

Rosing M. T. and Frei R. (2004) U-rich Archaean sea-floor sediments from Greenland — indications of >3700 Ma oxygenic photosynthesis. *Earth Planet. Sci. Lett., 217(3)*, 237–244.

Rosing M. T., Bird D. K., Sleep N. H., Glassley W., and Albarede F. (2006) The rise of continents — An essay on the geologic consequences of photosynthesis. *Palaeogeogr. Palaeoclimatol. Palaeoecol., 232(2–4)*, 99–113.

Rosing M. T., Bird D. K., Sleep N. H., and Bjerrum C. J. (2010) No climate paradox under the faint early Sun. *Nature, 464(7289)*, 744–747.

Rothman D. H., Hayes J. M., and Summons R. E. (2003) Dynamics of the Neoproterozoic carbon cycle. *Proc. Natl. Acad. Sci. U.S.A., 100(14)*, 8124–8129.

Rozel A. B., Golabek G. J., Jain C., Tackley P. J., and Gerya T. (2017) Continental crust formation on early Earth controlled by intrusive magmatism. *Nature, 545(7654)*, 332–335.

Rubie D. C. et al. (2011) Heterogeneous accretion, composition and core-mantle differentiation of the Earth. *Earth Planet. Sci. Lett., 301(1–2)*, 31–42.

Rye R. and Holland H. D. (1998) Paleosols and the evolution of atmospheric oxygen: A critical review. *Am. J. Sci., 298(8)*, 621–672.

Rye R. and Holland H. D. (2000) Life associated with a 2.76 Ga ephemeral pond?: Evidence from Mount Roe #2 paleosol. *Geology, 28(6)*, 483–486.

Saer R. G. and Blankenship R. E. (2017) Light harvesting in phototrophic bacteria: Structure and function. *Biochem. J., 474(13)*, 2107–2131.

Sagan C. and Mullen G. (1972) Earth and Mars: Evolution of atmospheres and surface temperatures. *Science, 177(4043)*, 52–56.

Sahoo S. K. et al. (2016) Oceanic oxygenation events in the anoxic Ediacaran ocean. *Geobiology, 14(5)*, 457–468.

Saito M. A., Sigman D. M., and Morel F. M. M. (2003) The bioinorganic chemistry of the ancient ocean: The co-evolution of cyanobacterial metal requirements and biogeochemical cycles at the Archean-Proterozoic boundary? *Inorg. Chim. Acta, 356*, 308–318.

Sánchez-Baracaldo P., Ridgwell A., and Raven J. A. (2014) A Neoproterozoic transition in the marine nitrogen cycle. *Curr. Biol., 24(6)*, 652–657.

Sánchez-Baracaldo P., Raven J. A., Pisani D., and Knoll A. H. (2017) Early photosynthetic eukaryotes inhabited low-salinity habitats. *Proc. Natl. Acad. Sci. U.S.A., 114(37)*, DOI: 10.1073/pnas.1620089114.

Schidlowski M. (1979) Antiquity and evolutionary status of bacterial sulfate reduction: Sulfur isotope evidence. *Orig. Life, 9(4)*, 299–311.

Schidlowski M., Eichmann R., and Junge C. E. (1976) Carbon isotope geochemistry of the Precambrian Lomagundi carbonate province, Rhodesia. *Geochim. Cosmochim. Acta, 40(4)*, 449–455.

Schirrmeister B. E., de Vos J. M., Antonelli A., and Bagheri H. C. (2013) Evolution of multicellularity coincided with increased diversification of cyanobacteria and the Great Oxidation Event. *Proc. Natl. Acad. Sci. U.S.A., 110(5)*, 1791–1796.

Schoepp-Cothenet B. et al. (2012) On the universal core of bioenergetics. *Biochim. Biophys. Acta, Bioenerg., 1827*, 79–93.

Schoonen M. A. A. and Xu Y. (2001) Nitrogen reduction under hydrothermal vent conditions: Implications for the prebiotic synthesis of C-H-O-N compounds. *Astrobiology, 1(2)*, 133–142.

Schopf J. W. (1993) Microfossils of the Early Archean Apex chert: New evidence of the antiquity of life. *Science, 260(5108)*, 640–646.

Schopf J. W. and Packer B. M. (1987) Early Archean (3.3-billion to 3.5-billion-year-old) microfossils from Warrawoona Group, Australia. *Science, 237(4810)*, 70–73.

Schopf J. W., Kitajima K., Spicuzza M. J., Kudryavtsev A. B., and Valley J. W. (2018) SIMS analyses of the oldest known assemblage of microfossils document their taxon-correlated carbon isotope compositions. *Proc. Natl. Acad. Sci. U.S.A., 115(1)*, 53–58.

Schubert W. D. et al. (1998) A common ancestor for oxygenic and anoxygenic photosynthetic systems: A comparison based on the structural model of photosystem II. *J. Mol. Biol., 280(2)*, 297–314.

Schwartzman D. W. and Volk T. (1989) Biotic enhancement of weathering and the habitability of Earth. *Nature, 340*, 457–460.

Scott C. et al. (2008) Tracing the stepwise oxygenation of the Proterozoic ocean. *Nature, 452*, 456–459.

Scott C. et al. (2012) Bioavailability of zinc in marine systems through time. *Nature Geosci., 6*, 125–128.

Sessions A. L., Doughty D. M., Welander P. V., Summons R. E., and Newman D. K. (2009) The continuing puzzle of the great oxidation event. *Curr. Biol., 19(14)*, 567–574.

Sheen A. I. et al. (2018) A model for the oceanic mass balance of rhenium and implications for the extent of proterozoic ocean anoxia. *Geochim. Cosmochim. Acta, 227*, 75–95, DOI: 10.1016/j.gca.2018.01.036.

Sheldon N. D. (2006) Precambrian paleosols and atmospheric CO_2 levels. *Precambrian Res., 147(1)*, 148–155.

Shen Y., Buick R., and Canfield D. E. (2001) Isotopic evidence for microbial sulphate reduction in the early Archaean era. *Nature, 410*, 77–81.

Shen Y., Canfield D. E., and Knoll A. H. (2002) Middle Proterozoic ocean chemistry: Evidence from the McArthur Basin, Northern Australia. *Am. J. Sci., 302*, 81–109.

Shen Y., Knoll A. H., and Walter M. R. (2003) Evidence for low sulphate and anoxia in a mid-Proterozoic marine basin. *Nature, 423*, 632–635.

Shen Y., Zhang T., and Hoffman P. F. (2008) On the coevolution of Ediacaran oceans and animals. *Proc. Natl. Acad. Sci. U.S.A., 105(21)*, 7376–7381.

Shih P. M., Hemp J., Ward L. M., Matzke N. J., and Fischer W. W. (2017) Crown group Oxyphotobacteria postdate the rise of oxygen. *Geobiology, 15(1)*, 19–29.

Sim M. S., Bosak T., and Ono S. (2011) Large sulfur isotope fractionation does not require disproportionation. *Science, 333(6038)*, 74–77.

Sleep N. H. and Windley B. F. (1982) Archean plate tectonics: Constraints and inferences. *J. Geol., 90(4)*, 363–379.

Sleep N. H., Meibom A., Fridriksson T., Coleman R. G., and Bird D. K. (2004) H_2-rich fluids from serpentinization: Geochemical and biotic implications. *Proc. Natl. Acad. Sci. U.S.A., 101(35)*, 12818–12823.

Sleep N. H., Bird D. K., and Pope E. (2012) Paleontology of Earth's mantle. *Annu. Rev. Earth Planet. Sci., 40*, 277–300.

Smit M. A. and Mezger K. (2017) Earth's early O_2 cycle suppressed by primitive continents. *Nature Geosci., 10(10)*, 788–792.

Solomatov V. S. and Moresi L. N. (1997) Three regimes of mantle convection with non-Newtonian viscosity and stagnant lid convection on the terrestrial planets. *Geophys. Res. Lett., 24(15)*, 1907–1910.

Som S. M., Catling D. C., Harnmeijer J. P., Polivka P. M., and Buick R. (2012) Air density 2.7 billion years ago limited to less than twice modern levels by fossil raindrop imprints. *Nature, 484(7394)*, 359–362.

Som S. M. et al. (2016) Earth's air pressure 2.7 billion years ago constrained to less than half of modern levels. *Nature Geosci., 9*, 448–451.

Sousa F. L., Shavit-Grievink L., Allen J. F., and Martin W. F. (2012) Chlorophyll biosynthesis gene evolution indicates photosystem gene duplication, not photosystem merger, at the origin of oxygenic photosynthesis. *Genome Biol. Evol., 5(1)*, 200–216.

Sperling E. A. et al. (2015) Statistical analysis of iron geochemical data suggests limited late Proterozoic oxygenation. *Nature, 523,* 451–454.

Squire R. J., Campbell I. H., Allen C. M., and Wilson C. J. (2006) Did the Transgondwanan Supermountain trigger the explosive radiation of animals on Earth? *Earth Planet. Sci. Lett., 250(1–2),* 116–133.

Strother P. K., Battison L., Brasier M. D., and Wellman C. H. (2011) Earth's earliest non-marine eukaryotes. *Nature, 473,* 505–509.

Stüeken E. E. (2013) A test of the nitrogen-limitation hypothesis for retarded eukaryote radiation: Nitrogen isotopes across a Mesoproterozoic basinal profile. *Geochim. Cosmochim. Acta, 120,* 121–139.

Stüeken E. E. and Buick R. (2018) Environmental control on microbial diversification and methane production in the Mesoarchean. *Precambrian Res., 304,* 64–72.

Stüeken E. E., Catling D. C., and Buick R. (2012) Contributions to late Archaean sulphur cycling by life on land. *Nature Geosci., 5,* 722–725.

Stüeken E. E. et al. (2013) Did life originate from a global chemical reactor? *Geobiology, 11(2),* 101–126.

Stüeken E. E., Buick R., and Anbar A. D. (2015a) Selenium isotopes support free O_2 in the latest Archean. *Geology, 43(3),* 259–262.

Stüeken E. E., Buick R., Guy B. M., and Koehler M. C. (2015b) Isotopic evidence for biological nitrogen fixation by Mo-nitrogenase at 3.2 Gyr. *Nature, 520,* 666–669.

Stüeken E. E. et al. (2017a) Not so non-marine? Revisiting the Torridonian Stoer Group and the Mesoproterozoic biosphere. *Geochem. Perspect. Lett., 3(2),* DOI: 10.7185/geochemlet.1725.

Stüeken E. E. et al. (2017b) Environmental niches and biodiversity in Neoarchean lakes. *Geobiology, 15(6),* 767–783.

Stüeken E. E., Kipp M. A., Koehler M. C., and Buick R. (2016a) The evolution of Earth's biogeochemical nitrogen cycle. *Earth-Sci. Rev., 160,* 220–239.

Stüeken E. E. et al. (2016b) Modeling pN_2 through geologic time: Implications planetary climates and atmospheric biosignatures. *Astrobiology, 16(12),* 949–963.

Stumm W. and Morgan J. J. (1996) *Aquatic Chemistry.* Wiley, New York. 1040 pp.

Sumner D. Y., Hawes I., Mackey T. J., Jungblut A. D., and Doran P. T. (2015) Antarctic microbial mats: A modern analog for Archean lacustrine oxygen oases. *Geology, 43(10),* 887–890.

Swanner E. D. et al. (2014) Cobalt and marine redox evolution. *Earth Planet. Sci. Lett., 390,* 253–263.

Takishita K. et al. (2012) Lateral transfer of tetrahymanol-synthesizing genes has allowed multiple diverse eukaryote lineages to independently adapt to environments without oxygen. *Biol. Direct, 7(1),* DOI: 10.1186/1745-6150-7-5.

Tang M., Chen K., and Rudnick R. L. (2016) Archean upper crust transition from mafic to felsic marks the onset of plate tectonics. *Science, 351(6271),* 372–375.

Tangalos G. E. et al. (2010) Microbial production of isotopically light iron(II) in a modern chemically precipitated sediment and implications for isotopic variations in ancient rocks. *Geobiology, 8(3),* 197–208.

Tarduno J. A., Cottrell R. D., Davis W. J., Nimmo F., and Bono R. K. (2015) A Hadean to Paleoarchean geodynamo recorded by single zircon crystals. *Science, 349(6247),* 521–524.

Tashiro T. et al. (2017) Early trace of life from 3.95 Ga sedimentary rocks in Labrador, Canada. *Nature, 549(7673),* 516–518.

Tatzel M., Blanckenburg F., Oelze M., Bouchez J., and Hippler D. (2017) Late Neoproterozoic seawater oxygenation by siliceous sponges. *Nature Commun., 8(1),* DOI: 10.1038/s41467-017-00586-5.

Tian F., Kasting J. F., and Zahnle K. (2011) Revisiting HCN formation in Earth's early atmosphere. *Earth Planet. Sci. Lett., 308(3),* 417–423.

Tosca N. J., Guggenheim S., and Pufahl P. K. (2016) An authigenic origin for Precambrian greenalite: Implications for iron formation and the chemistry of ancient seawater. *Geol. Soc. Am. Bull., 128(3–4),* 511–530.

Ueno Y., Yamada K., Yoshida N., Maruyama S., and Isozaki Y. (2006) Evidence from fluid inclusions for microbial methanogenesis in the early Archaean era. *Nature, 440,* 516–519.

van Zuilen M. A., Lepland A., and Arrhenius G. (2002) Reassessing the evidence for the earliest traces of life. *Nature, 418(6898),* 627–630.

Veizer J. and Jansen S. L. (1979) Basement and sedimentary recycling and continental evolution. *J. Geol., 87(4),* 341–370.

Viehmann S., Bau M., Hoffmann J. E., and Münker C. (2015) Geochemistry of the Krivoy Rog Banded Iron Formation, Ukraine, and the impact of peak episodes of increased global magmatic activity on the trace element composition of Precambrian seawater. *Precambrian Res., 270,* 165–180.

Vinyard D. J. and Brudvig G. W. (2017) Progress toward a molecular mechanism of water oxidation in photosystem II. *Annu. Rev. Phys. Chem., 68,* 101–116.

Wacey D., Kilburn M. R., Saunders M., Cliff J., and Brasier M. D. (2011) Microfossils of sulphur-metabolizing cells in 3.4-billion-year-old rocks of Western Australia. *Nature Geosci., 4(10),* 698–702.

Wackett L. P., Dodge A. G., and Ellis L. B. M. (2004) Microbial genomics and the periodic table. *Appl. Environ. Microbiol., 70(2),* 647–655.

Waldbauer J. R., Newman D. K., and Summons R. E. (2011) Microaerobic steroid biosynthesis and the molecular fossil record of Archean life. *Proc. Natl. Acad. Sci. U.S.A., 108(33),* 13409–13414.

Walker J. C., Hays P. B., and Kasting J. F. (1981) A negative feedback mechanism for the long-term stabilization of Earth's surface temperature. *J. Geophys. Res.–Oceans, 86(C10),* 9776–9782.

Wallace M. W., Shuster A., Greig A., Planavsky N. J., and Reed C. P. (2017) Oxygenation history of the Neoproterozoic to early Phanerozoic and the rise of land plants. *Earth Planet. Sci. Lett., 466,* 12–19.

Walter M. R., Buick R., and Dunlop J. S. R. (1980) Stromatolites 3,400–3,500 Myr old from the North Pole area, Western Australia. *Nature, 284,* 443–445.

Weiss B. P. et al. (2015) Pervasive remagnetization of detrital zircon host rocks in the Jack Hills, Western Australia and implications for records of the early geodynamo. *Earth Planet. Sci. Lett., 430,* 115–128.

Weiss M. C. et al. (2016) The physiology and habitat of the last universal common ancestor. *Nature Microbiol., 1,* 16116, DOI: 10.1038/nmicrobiol.2016.116.

Wheat C. G., Mottl M. J., and Rudnicki M. (2002) Trace element and REE composition of a low-temperature ridge-flank hydrothermal spring. *Geochim. Cosmochim. Acta, 66(21),* 3693–3705.

Wheat C. G., Fisher A. T., McManus J., Hulme S. M., and Orcutt B. N. (2017) Cool seafloor hydrothermal springs reveal global geochemical fluxes. *Earth Planet. Sci. Lett., 476,* 179–188.

Wille M. et al. (2007) Evidence for a gradual rise of oxygen between 2.6 and 2.5 Ga from Mo isotopes and Re-PGE signatures in shales. *Geochim. Cosmochim. Acta, 71,* 2417–2435.

Wolfe J. M. and Fournier G. P. (2018) Horizontal gene transfer constrains the timing of methanogen evolution. *Nature Ecol. Evol., 2,* 897–903, DOI:10.1038/s41559-018-0513-7.

Wordsworth R. and Pierrehumbert R. (2013) Hydrogen-nitrogen greenhouse warming in Earth's early atmosphere. *Science, 339(6115),* 64–67.

Yoon H. S., Hackett J. D., Ciniglia C., Pinto G., and Bhattacharya D. (2004) A molecular timeline for the origin of photosynthetic eukaryotes. *Mol. Biol. Evol., 21(5),* 809–818.

Young G. M., Brunn V. V., Gold D. J., and Minter W. E. L. (1998) Earth's oldest reported glaciation: Physical and chemical evidence from the Archean Mozaan Group (~2.9 Ga) of South Africa. *J. Geol., 106(5),* 523–538.

Zahnle K. J., Claire M. W., and Catling D. C. (2006) The loss of mass-independent fractionation in sulfur due to a Palaeoproterozoic collapse of atmospheric methane. *Geobiology, 4(4),* 271–283.

Zahnle K. J., Catling D. C., and Claire M. W. (2013) The rise of oxygen and the hydrogen hourglass. *Chem. Geol., 362,* 26–34.

Zellner N. E. (2017) Cataclysm no more: New views on the timing and delivery of lunar impactors. *Origins Life Evol. Biospheres, 47(3),* 261–280.

Zerkle A., House C. H., and Brantley S. L. (2005) Biogeochemical signatures through time as inferred from whole microbial genomes. *Am. J. Sci., 305,* 467–502.

Zerkle A. L., Claire M. W., Domagal-Goldman S. D., Farquhar J., and Poulton S. W. (2012) A bistable organic-rich atmosphere on the Neoarchaean Earth. *Nature Geosci., 5,* 359–363.

Zhang S. et al. (2016) Sufficient oxygen for animal respiration 1,400 million years ago. *Proc. Natl. Acad. Sci. U.S.A., 113(7),* 1731–1736.

Zhao M., Reinhard C. T., and Planavsky N. (2017) Terrestrial methane fluxes and Proterozoic climate. *Geology, 46(2),* 139–142.

Johnson S., Graham H., Des Marais D. J., and Hazen R. (2020) Detecting life on Earth and the limits of analogy. In *Planetary Astrobiology*
(V. Meadows et al., eds.), pp. 121–150. Univ. of Arizona, Tucson, DOI: 10.2458/azu_uapress_9780816540068-ch005.

Detecting Life on Earth and the Limits of Analogy

S. S. Johnson
Georgetown University

H. V. Graham
NASA Goddard Space Flight Center

D. J. Des Marais
NASA Ames Research Center

R. M. Hazen
Carnegie Institution for Science

This chapter covers the types of evidence used for detecting life on Earth, both in the past and the present. As a whole this evidence constitutes what we call "biosignatures" — physical indications of present or past life forms. These indications can be the organisms themselves, compounds left behind by the organisms or consequent to their degradation, or an imprint in the physical environment. We consider multiple lines of molecular and chemical evidence as well as structures, textures, and mineral assemblages. We also discuss the important roles of taphonomy and diagenesis. While no biosignatures are impervious to time, some are more resilient to certain environmental pressures than others. We conclude that it is important for astrobiologists to not only rely on the approaches to life detection that work readily here on Earth but also strive to develop ways to identify unknowable, unfamiliar features and chemistries that may represent processes of life as yet unrecognized.

1. INTRODUCTION

Detecting life on Earth may seem trivial in the present. After all, life permeates our planet, from airborne bacteria living in our clouds (*DeLeon-Rodriguez et al.*, 2013) to carpets of plants covering nearly every possible surface (*World Wildlife Fund*, 2016). Microscopic organisms infuse the ocean, joining animals on every scale, from enormous to tiny, at every depth (*McIntyre*, 2010). Vast communities of invertebrates, fungi, and prokaryotes abound in soils, and *Bacteria* and *Archaea* pervade marine sediments even in the deepest parts of the seas (*Whitman et al.,* 1998*; Fry et al.*, 2008; *Kallmeyer et al.*, 2012). The past record of life, however, can be much more enigmatic.

By current standards, Earth was a far less hospitable place early in its history (*Lunine*, 2006). The atmosphere and oceans had little or none of the oxygen that multicellular life requires, and euxinic (sulfidic) seawater was more widespread (*Berner and Canfield*, 1989; *Canfield*, 1998; *Farquhar et al.,* 2001; *Holland*, 2006; *Planavsky et al.,* 2011). The earliest hints of life are not recognizable fossil creatures but accumulations of minerals or chemicals, the byproducts of metabolism or molecules that can be linked to extant organisms. Our active planet has buried, heated, flooded, eroded, or altered much of the evidence of life and has left few pristine places for us to search for its nascent forms. Similarly, as we plumb more inaccessible parts of our biosphere where life is far less charismatic and the habitats less hospitable, we often need to use increasingly novel and rigorous methods to detect life.

It would seem that there are few environmental limits to life — our continued exploration of Earth has revealed it in many habitats previously considered sterile (*Kieft*, 2016). But our detection methods must be careful to distinguish between organisms native to these regions vs. immigrants from more hospitable habitats. The paleo-environmental research community has long been aware of the possibility that extant ecosystems can overprint evidence of past ecosystems. The methods used to explore prospective living systems on Earth have obvious application to our search for life elsewhere in the solar system, as well as our planetary protection efforts to prevent contamination of non-terran ecosystems on other worlds.

On Earth, we are not only interested in detecting life, but also understanding the role of life in the context of its environment, and using biosignatures to identify and reconstruct environmental characteristics. Like life, biosignatures occupy a variety of scales. For the purposes of simplification, we arrange the biosignatures in this chapter along a continuum of scale from molecular and chemical to structural and textural biosignatures to mineral biosignatures. This ordering of biosignatures is also along a continuum of the survivability of each biosignature. Increased preservation potential is often associated with a decrease in organismal specificity [see the Ladder of Life (*Neveau et al.*, 2018)]. For example, nucleic acids are the most ephemeral of the biomolecules but remarkably diagnostic for particular organisms, while less polar molecules and patterns of molecule fragments preserve on longer timescales but with less organismal specificity. Carbohydrates and proteins are also excellent indicators of life, but they are recycled or utilized on short timescales (*Jørgensen et al.*, 1993).

In this chapter, we also describe how the general classes of biosignatures arise from within a living system and how time and environment can change these signals. We also consider the abiotic features that can mimic biosignatures. This is an especially important consideration for defining the threshold where certain Earthly environments are inappropriate as relevant analogs for other solar system destinations.

2. NUCLEIC ACIDS, PROTEINS, AND METABOLITES

One of the key ways that life is recognized and analyzed on Earth is through the detection of particular classes of molecules, particularly nucleic acids. Decades ago Francis Crick proposed a central dogma to describe storage and flow of information through biological systems (*Watson and Crick*, 1953; *Crick*, 1970). Still the prevailing paradigm today, in its simplest form the central dogma posits that DNA is the primary reservoir of biological complexity, and DNA molecules are transcribed to RNA molecules and then translated to amino acids (*Crick*, 1970). By 1972, scientists began to characterize the exact order of nucleotide bases in DNA molecules, otherwise known as DNA sequencing (*Padmanab and Wu*, 1972; *Jay et al.*, 1974). Within 10 years Carl Woese and Norman Pace were able to detect and study a single gene known as the small subunit ribosomal RNA (SSU rRNA) and use it to determine what kinds of organisms were present in the environment (*Woese*, 1987; *Lane et al.*, 1985; *Woese*, 2002).

Before the advent of these molecular methods, microbial communities were primarily studied using culturing and microscopy. In 1660, the first bacterium was observed (*Dobell*, 1932). In the nineteenth century germ theory began to be developed (*Smith et al.*, 2012), and by 1860, Louis Pasteur and Robert Koch had begun cultivating microorganisms on growth media (*Falkow*, 1988; *Dahm*, 2005). By these measures some of the most Mars-like environments on Earth were thought to be sterile. For example, a *Science* article published in 1969 concluded that that while Antarctic soils contained significant amounts of organic carbon, the building blocks of life, parts of the cold and barren Dry Valleys were incapable of hosting life, as no microbes from these soils grew as part of an array of laboratory culturing experiments (*Horowitz et al.*, 1969). In time, however, this assumption was overturned, as growth was detected on glass slides left in the Antarctic soils. This discovery was followed by low-temperature laboratory culturing experiments (*Vishniac and Hempfling*, 1979).

The widespread utilization of genomic sequencing techniques has now revealed that as many as 99% of prokaryotes in environmental samples are unculturable in a laboratory setting (*Amann et al.*, 1995; *Schloss and Handelsman*, 2005). This key fact was unknown in the 1970s, when NASA first sent life-detection experiments that relied in part on the idea of growth and culturing to probe for life in martian soils as part of the Viking 1 and 2 missions. Efforts to coax more microorganisms into culture have improved in recent decades, with a recognition of the importance of microbial consortia (e.g., *Dekas et al.*, 2009) and the roles of dormancy and slow growth (e.g., *Kirchman*, 2016). While culturing still plays an important role in understanding the microbial world, particularly with regard to the diagnosis of infectious diseases, genomic methods enable a fuller range of microorganisms — both culturable and unculturable — to be studied simply by analyzing their nucleic acid contents.

To not only detect life but also characterize it and understand its interactions with the surrounding environment, there are several methods that allow for access to the information stored within molecules, including targeted approaches for specific genes of interest, whole genome sequencing for cultured or isolated cells, and metagenomic sequencing wherein a whole microbial community can be simultaneously studied using molecular information. These approaches can now determine not only what kinds of organisms are present in an environment, but also the metabolic processes and ecotypes (i.e., genetically distinct geographically separated populations) present (see Fig. 1).

2.1. DNA: 16S and 18S rRNA Genes

Because genes coding for ribosomal RNA — the RNA component of the ribosome, which is essential for protein synthesis in all known organisms — are ubiquitous and highly conserved, species can be rapidly identified via comparison to databases. These genes include the 16S small ribosomal subunit RNA (SSU rRNA) gene in *Bacteria* and *Archaea* and the 18S rRNA gene in eukaryotes. These small segments of DNA can be amplified, generating millions of copies, and then sequenced to enable a deep understanding of microbial community structure. Following sequencing, the reads are typically screened to control for quality and compared to one another. Unique sequences are then classified as operational taxonomic units (OTUs), also known as phylotypes. These terms are used in lieu of species, as the definition of species as individuals that can interbreed is of little use within the realm of microbiology. The abundance of each OTU or

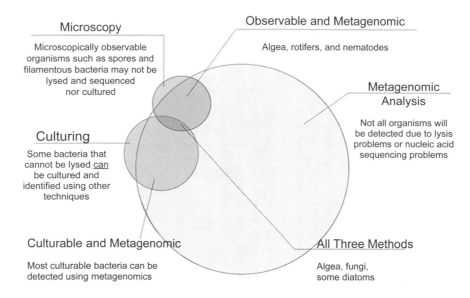

Microscopy

Microscopically observable organisms such as spores and filamentous bacteria may not be lysed and sequenced nor cultured

Observable and Metagenomic

Algea, rotifers, and nematodes

Metagenomic Analysis

Not all organisms will be detected due to lysis problems or nucleic acid sequencing problems

Culturing

Some bacteria that cannot be lysed <u>can</u> be cultured and identified using other techniques

Culturable and Metagenomic

Most culturable bacteria can be detected using metagenomics

All Three Methods

Algea, fungi, some diatoms

Fig. 1. This diagram shows which organisms are detectable via the three major microbiology techniques of microscopy, culturing, and metagenomics. As shown in areas of overlap, some organisms are accessible via multiple methods. While the more traditional techniques of microscopy and culturing can reveal many organisms, much of modern life detection on Earth focuses on characterization of nucleic acids via metagenomic analysis, the study of genetic material recovered directly from environmental samples. Figure adapted from *Tighe et al.* (2017).

phylotope can then be reported as the number or count of reads of their respective subunit genes.

DNA provides evidence that life has been present and provides a snapshot of its metabolic potential, whereas RNA can be used to identify active microbial communities and the genes they are expressing. SSU rRNA molecules are not only captured in DNA form, as once they are transcribed, it is possible to capture and sequence rRNA transcripts. However, the detection of DNA is only evidence that life has been present; it does not necessarily indicate live cells. In contrast, rRNA transcripts, while more difficult to isolate and sequence, strongly indicate cellular activity and thus are often used as indicators for active microbial communities (see section 7.2 on ancient DNA and nucleic acid diagenesis). Not all individuals have the same number of SSU rRNA gene copies within their genomes, so the number of copies recovered is not always related to the abundances of the species present. While this often complicates specific analyses of microbial communities, when used in tandem, rRNA:rDNA ratios can help determine the relative activity of different microbial groups (*DeAngelis et al.*, 2010).

SSU rRNA molecules can also be used to identify and describe microbial community structure and trophic relationships (i.e., relationships at various levels of the food chain) within microbial communities. This is typically done by modeling the mathematical relationships between the genes recovered (*Chaffron et al.*, 2010). Often, 16S and 18S rRNA data are analyzed with phylogenetic trees like "the tree of life," branching diagrams depicting the evolutionary relationships among organisms based on similarities in genetic traits (see Fig. 2). Some examples in soil and water include *Freilich et al.* (2010), *Lupatini et*

al. (2014), *Ruan et al.* (2006), *Lima-Mendez et al.* (2015), *Williams et al.* (2014), and *Ma et al.* (2016).

Although massively parallel analyses of SSU rRNA genes have enabled a deeper understanding of microbial communities, complications due to error rates, primer bias, and amplification complications can present challenges (*Koboldt et al.*, 2013). Some limitations of rRNA analysis are that the primers required, the short nucleic acid sequences that provide a starting point for DNA synthesis, are known to mismatch certain targets (e.g., *Isenbarger et al.*, 2008). Also, common DNA extraction methods, including many commercially available DNA extraction kits, may not achieve full and nonbiased lysis (cell rupturing needed to extract DNA) of all cells (*Tighe et al.*, 2017).

2.2. DNA: Whole Genome

The next generation sequencing, also known as massively parallel sequencing, began to take hold in the 1990s (*Xu*, 2014), allowing high throughput characterization of whole genomes of single organisms. The first complete bacterial genome was sequenced in 1995 (*Fleischmann et al.*, 1995), followed by the first draft human genome in 2001 (*International Human Genome Sequencing Consortium*, 2001; *Venter et al.*, 2001). In 2004, shotgun sequencing of diverse ocean microbial communities in the Sargasso Sea was completed by simultaneously sequencing fragments of DNA (*Venter et al.*, 2004). Metagenomics, formally defined as the analysis of a mixture of microbial genomes by sequencing (*Handelsman et al.*, 1998), thus came into vogue. By sequencing the entirety of the pool of genes within the sample, metagenomics enables scientists to uncover not only which types of organisms are present, but also the metabolic capabilities of those microbes.

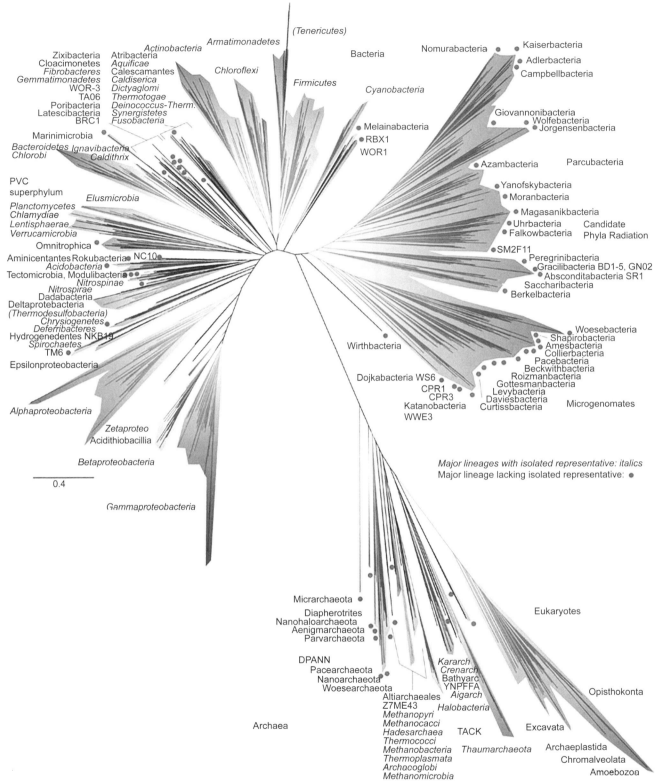

Fig. 2. See Plate 6 for color version. Using new genomic data from over 1000 uncultivated and little-known organisms, together with published sequences, *Hug et al.* (2016) recently published this expanded version of the tree of life.

Given the limitations associated with DNA extraction, amplification, and sequencing, it remains challenging to sequence to completion each genome within a microbial community. Even sequencing a single microbial genome requires dozens of individual copies of each base pair (bp)

in order to create a reliable reference. Using "short reads" (often >300 bp), however, a researcher can assemble, or overlap, the ends of short sequences to create longer pieces of DNA. "Long read" technologies, developed more recently (thousands to millions of bp in length) (e.g. *Jain et al.*, 2016;

Koren and Phillipy, 2015; *Payne et al.*, 2018), may also be used directly to study functional genes of a single organism and uncover cell lineage relationships (*Shapiro et al.*, 2013).

After the longest pieces of DNA possible are captured, segments of nucleotide sequences delimited by a start and stop codon, called open reading frames (ORFs) until their function is verified, are often predicted. ORFs can be predicted based on complete genomes or DNA fragments. These genes are then compared to databases of genes with known function. This process, called a homology search, is based on the identification of shared ancestry between the query (the new ORF) and the references (genes in the database). Homology searches are performed through the alignment of query nucleotide or amino acid sequences to carefully curated databases of known reference genes and proteins. This often represents one of the most computationally intensive tasks in molecular biology. If there is close alignment between the query and the reference, the ORF is putatively assigned the same function as the reference gene. Sometimes the putative function and the abundance of single genes can be used in combination to investigate the metabolic potential of an organism or a community.

Genomic and metagenomic reads also contain phylogenetic information, thus algorithms can be used to assign even single genes to taxonomic groups (e.g., *Huson et al.*, 2007). It is also often possible to "bin" pieces of DNA into phylogenetically-related groups based on genomic properties such as the frequency of certain amino acids and the ratios between guanine/cytosine and adenine/thymine nucleotides. The canonical approaches of predicting and assigning functions to ORFs can then be applied to the bins, and function and taxonomy can be related.

2.3. RNA

The genomes of microbes are rich in information, and databases of comparative sequence data continue to expand, with implications for understanding even the most remote parts of our biosphere. But actual gene expression within a microbial community depends on a number of factors, from oxidative stress and nutrient availability to external pH and temperature. The analysis of RNA molecules can shed light on which metabolic pathways are in use under the conditions at hand. While DNA reflects functional and metabolic potential, RNA reflects what the cell is attempting to do. RNA molecules come in many flavors, such as SSU rRNA (section 4.1), and it is possible to target specific rRNA transcripts, such as the small subunit. It is also possible to target some specific types of RNA, such as transfer RNA (tRNA), which can help identify genes transferred between organisms, small RNA (sRNA), which can reveal viral infection and environmental stresses, and, most commonly, messenger RNA (mRNA), which tracks the activity of specific genes and can reveal how organisms cope with environmental stressors. As with DNA, RNA can be collected from single cells or from an entire community simultaneously with metatranscriptomics. In the broadest

sense, RNA can be used to study the attempted synthesis of individual proteins.

Transfer RNA (tRNA) molecules are typically 75–90 base pairs long and serve as molecular "go betweens" for messenger RNA and amino acid sequences, thus playing an essential role in translation. More recently, they have also been implicated in antibiotic resistance and synthesis as well as processes involved in cell membrane maintenance (*Shepherd and Ibba*, 2013, 2015). tRNAs are also widely studied due to their role in the identification of "genomic islands" or genes that have been horizontally transferred (*Hudson et al.*, 2015). Horizontal gene transfer, the process by which an individual acquires genes via a process other than parent-offspring transmission (for example, through the uptake of free DNA, or plasmid- or virus-mediated transfer), has important implications for evolution (*Juhas et al.*, 2009). It can occur among different species and essentially provides new functional potential to a cell, such as antibiotic resistance and virulence genes. Genomic islands typically appear at tRNA sites along prokaryotic chromosomes and thus identifying tRNAs can often lead to the discovery of horizontally transferred genes (*Hudson et al.*, 2015).

Small RNA (sRNA) are found in both eukaryotes and prokaryotes. There are many types of sRNAs and they are known to play roles in regulating protein synthesis. For example, antisense RNA, the opposite or complementary strand of a transcript, may bind onto mRNA within a cell and inhibit or down-regulate protein activity (*Gottesman and Storz*, 2011). sRNAs have also been found to play important roles in prokaryotic responses to viral infections and environmental stresses (e.g., *Bhaya et al.*, 2011).

Messenger RNA (mRNA) encodes the sequence of amino acids to be translated within the ribosome. Analyses of mRNA enable studies of the activity of specific genes. mRNA can provide a snapshot of the metabolic activity of an individual or community to discover proteins needed to cope with a particular environmental stress.

2.4. Proteins and Metabolites

Proteomics or metaproteomics — the characterization of the entire protein complement of environmental microbiota at a given point in time — not only allows for a comprehensive look at the modern community but also the proteins produced or modified by the microbiome (*Wilmes and Bond*, 2004; *Tanca et al.*, 2014). While analyzing RNA can be highly informative of which genes are being actively expressed, there are additional levels of cellular localization and regulation at the protein level, such as post-translational modifications and controlled proteolysis/protein turnover, that aren't captured by analyzing RNA alone (*Hettich et al.*, 2013). Implementing proteomics via mass spectroscopy can expose active pathways and functional gene products, including the actual molecular machinery produced and utilized by organisms (see Fig. 9). Protein or peptide data can be compared with data predicted from genomic information for complex samples in natural environments, such as biofilms, soils, or seawater microbial communities (e.g., *Ram et al.,* 2005; *Singer et al.,* 2008; *Dong et al.,* 2014; *Urich et al.,* 2014; *Bergauer et al.,* 2017), or under different growth conditions for microbial isolates (*Thompson et al.,* 2007; *Hettich et al.,* 2013), to provide an unprecedented view of microbial functionality. Analytical protocols typically include sample preparation including protein extraction, purification, and concentration; protein denaturation and reduction; protein (or peptide) separation, enzymatic digestion, and analysis; and protein identification (*Schneider and Riedel,* 2010).

Metabolomics is the study of metabolites — the intermediates and products of metabolism. Unlike the genome and transcriptome, which are based on different nucleotides, or the proteome, which is based on different amino acids, the metabolome consists of a huge variation of diverse chemical compounds, ranging from ionic inorganic species to hydrophilic carbohydrates, volatile alcohols and ketones, amino and non-amino organic acids, hydrophobic lipids, and complex natural products (*Villas-Boas et al.,* 2007). Because a single metabolite can participate in multiple pathways of cell metabolism, it can be difficult to establish a direct link between a gene and particular metabolite. Still, metabolomics enables a comprehensive study of the small molecule profiles of life-sustaining chemical transformations in the cells of living organisms (*Ahn et al.,* 2010). Together, these "-omics" approaches can be used to characterize microbial communities across multiple levels of specificity (see Fig. 3).

3. ORGANIC BIOMARKERS

3.1. Biomarker Definition

Organic biomarkers are molecules that are synthesized by an organism and can be traced back to a habitat, community, or specific taxa (*Van Bergen et al.,* 1995; *Brocks and Summons,* 2003; *Summons et al.,* 2007). For clarity, biomarkers are a type of biosignature and the term biomarker refers to specific molecules. These molecules and their alteration products can include structural components (lignin, cellulose, membrane lipids), pigments (chlorophylls,

carotenoids), or metabolic regulators and defense compounds (sterols, isoprenoids) (*Eglinton and Logan,* 1991). Organic biomarkers have a well-established utility for understanding the history of life on Earth (e.g., *Bada et al.,* 1973; *Hayes,* 1983; *Buick,* 1992; *Koopmans et al.,* 1996; *Parrish,* 1998; *Brocks et al.,* 1999, 2003; *Rosing,* 1999; *Schouten et al.,* 2000, 2007a; *Xiong et al.,* 2000; *Shen et al.,* 2001; *Holba et al.,* 2003; *Pancost and Boot,* 2004; *Turich et al.,* 2007; *Brocks and Banfield,* 2009; *Turich and Freeman,* 2011; *Uno et al.,* 2016). Organic biomarkers are frequently used to search the geologic past as well as the few environments on Earth where proof of the presence of life is elusive. Our search for chemical evidence of life on Earth can be applied across the scales of time and place: How far back can we find this record of life and how can we identify life in a variety of environments?

3.1.1. Functionalization vs. diagenesis. Age is an important consideration of organic biomarker preservation potential and/or diagenesis (the biotic/abiotic alteration of organic matter and minerals that occurs in sediments after deposition). In this review, we have organized biomarkers on a stability continuum with nucleic acids as the most ephemeral and mineral signatures as the most robust. Biomarker structures must be considered in concert with fossilization processes, or "taphonomy" (see section 7.2). As molecules fossilize, predictable structural changes occur that must be accounted for when trying to trace the precursor molecule. Defunctionalization is the process where reactive functional groups (i.e., oxygen, nitrogen, sulfur-bearing groups) are lost and most double bonds are reduced, resulting in fewer polar hydrocarbon alteration products.

3.1.2. Source vs. reservoir. Biomarker interpretation requires consideration of the source environment of the original organism as well as the preservation potential of their chemical constituents (*Briggs et al.,* 2000). The sedimentary archive of biomarkers — whether they arise on land or in the seas — is valuable for understanding recent as well as ancient life, whereas aqueous organic biomarkers are valuable primarily for modern organisms. In some cases, abiotic sources such as geochemical activity and extrater-

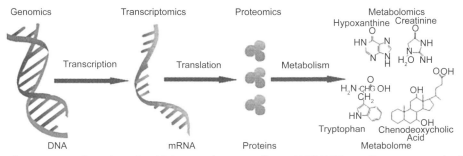

Fig. 3. Genomics allows researchers to ask which organisms are there (16S/18S amplicon sequencing) and what is the metabolic potential (metagenomics). Transcriptomics (via transcription), proteomics (via translation), and metabolomics (via metabolism) enable researchers to ask what organisms are actually doing. Figure from *Zhao and Lin* (2014).

restrial materials must also be considered when assessing the fidelity of a biomarker signature (*Summons et al.*, 2007). Specific molecular features can be used to distinguish these sources from biotic signals.

3.2. Biomarkers and Their Interpretation

The four major categories of biomolecules include carbohydrates, lipids, proteins, and nucleic acids (see Figs. 4a,b for examples). Given the short recycle time of carbohydrates, proteins, and nucleic acids, the presence of these compounds in a sedimentary reservoir are often an indicator of a modern source rather than a signature of an ancient community (*Jørgensen et al.,* 1993; *Bada et al.,* 1999; *Torti et al.,* 2015). Fossilized alteration products derived from the complex carbohydrate lignin (and occasionally cellulose) can be preserved on geologic timescales in peats, lignites (coals), and fossilized plant cuticle (*Orem and Hatcher*, 1987; *Briggs and Eglinton*, 1994; *Van Bergen et al.*, 1995; *Collinson et al.*, 1998). Very few of examples of potentially fossilized protein exist and these (controversial) examples represent specific collagen proteins dried in a very specific preservation environment (*Asara et al.*, 2007; *Schweitzer et al.*, 2009) where the protein precipitation process may have altered their structure (*San Antonio et al.*, 2011). Likewise, nucleic acids do not fossilize well in the rock record but have been found preserved in evaporites, ices, and permafrost cores that date to a few million years old (e.g., *Willerslev et al.*, 2007*; Bidle et al.,* 2007*; Briggs et al.*, 2007*; Brocks and Banfield*, 2009; *Boere et al.*, 2011; *Corinaldesi et al.*, 2008). While these biomolecules may not preserve well over geologic timescales, these compounds are very useful for understanding extant life and life in the recent past. Studies detailing the value of these biomolecules are discussed in section 2.4.

By contrast, structural hydrocarbon biopolymers and lipids have excellent preservation potential on much longer timescales (*Briggs et al.*, 2000). Unlike the other major biomolecules (carbohydrates, proteins, and nucleic acids), the greater danger for preservation of hydrocarbons is alteration during diagenesis rather than loss (*Cranwell*, 1981). Structural hydrocarbons from vascular land plants (waxes, lignin, and cellulose) and aquatic plants and algae (algenans) are common components of peats and coals and individual recalcitrant wax components and membrane lipids are replete in the sedimentary record (*Eglinton and Hamilton*, 1967; *Kolattukudy*, 1976). Heat and pressure induced by long periods of burial rearrange carbon-carbon bonds. The resulting product (kerogen) is a solvent insoluble macromolecular aggregate of organic matter. Given even more time and heat, kerogen will degrade to form smaller hydrocarbons. These lighter hydrocarbons make up fossil fuels (oil and gas) (*Tissot and Welte, 1984; de Leeuw et al.,* 1991; *Collinson et al.*, 1994). While some chemical characteristics preserved in these products can indicate the broad origin of the organic matter in alteration products, thermal degradation destroys much of the specific molecular structure information that makes them particularly valuable for paleobiological interpretation.

3.3. Lipids by Class

Those sedimentary lipids that escape alteration can be highly diagnostic for specific taxa and are commonly used to reconstruct communities, ecosystems, and environmental conditions, both on land and in aquatic settings. These compounds are preserved on billion-year timescales and the possible range of combinations of these lipids offers significant interpretive value. The simplest of these hydrocarbon biomarkers are *n*-alkanes, straight hydrocarbon chains that derive from bacterial and archaeal membranes (phospholipids and sphingolipids) as well as from the leaf waxes of land plants (*Eglinton and Hamilton*, 1967; *Tegelaar et al.*, 1989). The origination of these compounds can be distinguished from one another by molecular weight (see section 4.3.6). Similarly, very lightweight ($<n$-C_{15}) alkanes are produced during thermal degradation of organic matter, and an abundance of these compounds is an indicator of this thermal "maturation" process (*Brocks and Summons*, 2003).

3.3.1. Alkanes. *n*-Alkanes (see Fig. 4a) are useful proxies for ecosystem reconstruction since patterns of chain length can indicate the organism of origin. In fact, many biomarkers, including *n*-alkanes, can be traced back to specific biosynthetic pathways still expressed in extant organisms. It is generally then assumed that these same taxa or their close ancestors were active in the ecosystem and that the environment reflected current growth requirements and trophic associations. This assumption could be faulty given how few representatives of the major microbial domains have been studied either in culture or in the field. Furthermore, some molecules can be both a biosynthetic product and the product of a diagenetic process. For example, branched alkanes in sediments can be matched to compounds produced by freshwater algae, cyanobacterial mats in hypersaline and hydrothermal settings, and even insect cuticles (*Metzger et al*, 1985; *Shiea et al.*, 1990; *Kenig et al.*, 1995; *Gupta et al.*, 2007; *Brocks et al.*, 2018; *Graham et al.*, 2019). However, there is also evidence suggesting that these compounds are the products of abiotic reactions involving bacterial fatty acids, which changes the interpretation of a biomarker's origin ecosystem (*Perry et al.*, 1979).

3.3.2. Alkyl lipids. In addition to alkanes there are many other alkyl lipids (see Fig. 4a) that can be preserved on geologic timescales and can be used to reconstruct communities and ecosystems. Long-chain fatty acids, alkenols, alkenoates, and alkenones, as well as diols and diacids, are common organic biomarkers found in sediments (*Eglinton and Hamilton*, 1967; *Schouten et al.*, 1998; *Diefendorf et al.*, 2011). Like *n*-alkanes, long-chain linear fatty acids and alkenols can be linked to terrestrial sources, whereas long-chain alkenones and alkenoates and phytosterols are associated with phytoplankton (*Volkman et al.*, 1980; *Cranwell*, 1985; *Freeman and Wakeham*, 1992; *Conte et al.*, 1998; *Schouten et al.*, 1998) (more details are given on polymerization in section 4.4).

3.3.3. Ether membrane lipids. Membrane lipids are comprised of the geologically well-preserved core lipids as well as the many variations of bound polar head groups. The intact polar lipids (IPLs) and polar headgroups are rapidly lost upon cell senescence and rare in the geologic record. Membrane lipids with much greater taxonomic specificity include the many forms of the glycerol dialkyl glycerol tetraethers (GDGTs) (see Fig. 4c). These lipids exist either as pairs of head-to-head isoprenoid chains with variable numbers of five- and six-sided rings (isoprenoidal GDGTs) or as head-to-head alkyl chains substituted with methyl groups (branched GDGTs). The core isoprenoidal and branched lipid structures can include many taxon-specific ether-bound head groups (the -R position of the GDGT molecule in Fig. 4c). These membrane-spanning lipids have been found in both *Archaea* and *Bacteria* and while these lipids have been linked to lacustrine (lake-dwelling), marine, and terrestrial (soil-dwelling) representatives of these groups, some structures are more diagnostic of particular physical and chemical conditions. For example, the GDGT crenarchaeol has been linked almost exclusively with the Thaumarchaota (*Sinninghe Damsté et al.*, 2002a; *Koga and Nakano, 2008; Schouten et al., 2013*), while calditol-GDGT is associated with pH tolerant thermoacidophiles (*Zeng et al., 2018*). The core lipid archaeol is found in halophilic archaea where GDGTs are absent (*Bauersachs et al.*, 2015). Specific ratios of archaeol with other GDGTs can be associated with methanogenic archaea, lending greater insight into the redox conditions of their origin environment (*Koga et al.*, 1993). In fact, given the incredible ubiquity of these biomarkers across taxa and in the sedimentary record, paleoenvironmental reconstructions rely heavily on ratios of structures to determine sourcing. A few of the common ratio proxies based on GDGTs include an annual mean sea surface temperature proxy based on 86-carbon GDGTs (*Schouten et al.*, 2002), an alkenone proxy to reconstruct water temperature during growth (*Brassell et al.*, 1986), an alkenone-CO_2 proxy to reconstruct atmospheric composition (*Pagani*, 2002), a proxy that uses branched GDGTs to estimate terrestrial input in a marine sediment (*Hopmans et al.*, 2004), and a proxy that uses both distributions of methylation and cyclization of branched tetraethers to reconstruct mean annual soil temperature and pH (*Weijers et al.*, 2007).

3.3.4. Hopanes.

Hopanes are another very abundant membrane component found in sedimentary archives (*Ourisson and Albrecht*, 1992). Hopanes are the diagenetically altered form of membrane hopanoids or bacteriohopanepolyols (BHPs), isoprenoid lipids found in *Bacteria* (see Fig. 4c for an example) (*Summons et al.*, 1999). The ability to synthesize BHPs is scattered throughout the bacterial domain and occurs in a wide variety of aerobic metabolic pathways including methylotrophy, methanotrophy, oxygenic and anoxygenic photosynthesis, purple non-sulfur bacteria, and acetic acid fermenters (*Ourisson and Albrecht*, 1992; *Summons et al.*, 1999; *Talbot et al.*, 2003, 2008). The basic BHP structure is a pentacyclic moiety of four six-sided rings and one five-sided ring bonded to a six-carbon side chain. This basic structure can also feature up to 63 possible functional groups that include alcohols, sugars, fatty acids, and amino

groups as well as methylated sites on the rings (*Talbot et al.*, 2001). Structural specificities in BHP molecules can be linked to specific metabolic processes (photosynthesis, iron, sulfur and nitrogen metabolism) and to environmental conditions such as salinity, pH, and temperature. While much of the hopanoid molecule's structural information is lost during diagenetic transformation, the stereochemistry, methylation sites, and truncation patterns of the hydrocarbon tail on a hopane can still be used to discern the metabolism of the originating organism and estimate the thermal history of the sediment. In particular, methylation patterns are used to distinguish oxygenic photosynthesizers from methanotrophs and fermenters, and carbon tail truncation can indicate redox conditions as well as sulfur presence in the deposition site (*Summons et al.*, 1994; *Welander et al.*, 2010; *Welander and Summons*, 2012).

3.3.5. Steranes.

Steranes, like hopanes, are membrane components that can be preserved on geologic timescales and are diagenetically altered forms of steroids. These molecules are linked to a much wider range of taxa and serve metabolic as well as structural purposes (see Fig. 4c for an example) (*Volkman*, 1986; *Hannich et al.*, 2011). Steroids are produced by eukaryotic organisms, including plants, animals, and fungi (*Summons and Walter*, 1990; *Ourisson et al.*, 1987), and also a limited number of anaerobic bacteria. Steroids are involved in many cellular functions besides membrane fluidity, including development regulation and the synthesis of hormones and signaling molecules. All steranes derive from the tetracyclic triterpenoid steroid with side-chain substitutions particular to specific broad taxanomic groups. Notable examples include abundant dinosterane derived from dinoflagellates and, importantly, 650–540-Ma isopropylcholestane potentially linked to sponges that could signal the dawn of advanced life even before the Cambrian explosion (*Love et al.*, 2009; *Zumerge et al.*, 2018).

3.3.6. Isoprenoids.

Hopanes and steranes are just a few of the geologically significant molecules that are based on isoprenoid lipids. Isoprene, a five-carbon molecule, is produced by all domains of life and polymerized to form a variety of isoprenoid lipids (*Lange et al.*, 2000) (see Fig. 4b). For example, carotenoids are based on isoprenoid chains that include either cyclic or linear end groups and a wide variety of functional groups in these end groups. These combinations can yield hundreds of individual compounds that can serve as photosynthetic and/or photoprotective pigments and for cell-signaling molecules. Thermal degradation of functional groups does erase much of the diagnostic value, although important exceptions exist. For example, the presence of certain aromatic carotenoids (notably okenane, chlorobactane, and isorenieratane) can be linked almost exclusively to aquatic green and purple sulfur bacteria (*Chlorobiaceae* and *Chromatiaceae*) (*Brocks and Schaeffer*, 2008).

Chlorophyll pigments are constructed from a large heterocyclic aromatic (tetrapyrrole) ring surrounding four nitrogen atoms and a magnesium atom and have a long isoprenoid-derived hydrocarbon tail (see Fig. 4d). Slight differences in substitution and stereochemistry distinguish chlorophylls found in various members of the domain *Bacteria* from the chlorophyll found in land plants. Thermal alteration removes the isoprenoid tail from this structure and defunctionalizes groups on the ring structure, resulting in a wide array of generic porphyrin structures. Only rarely are ring structures that are diagnostic for either plants or bacteria preserved in the fossil record. The hydrocarbon tail is reduced to phytol that transforms to either pristane or phytane. The ratio of these two compounds is commonly used as an indicator of redox state in depositional environments (*Koopmans et al.*, 1999).

Cyclization of isoprene units yield terpenes that oxidize to form an enormous variety of terpenoid compounds that serve primarily as defense compounds, but also as growth factors in plants and some *Bacteria* (see beta-amyrin in Fig. 4c). Substitution of these rings can result in thousands of individual structures with precise taxonomic specificity, and the number of rings in these structures can indicate particular groups of vascular land plants distinguished by their reproductive style. These compounds are well-preserved in sediments and can give very specific taxonomic information on paleoecosystems and, by extension, detailed insight into environmental conditions (*Noble et al.,* 1985; *Otto and Simoneit,* 2001; *Bouvier et al.,* 2005; *Schouten et al.*, 2007b). Notable examples include lupane, oleanane, and ursane from angiosperms (flowering plants) (*Ten Haven and Rullkötter,* 1988) and pimarane and abietane from gymnosperms (conifers) (*Otto and Simoneit,* 2001).

3.4. Polymerization

Biochemical macromolecules are generally expressed as polymers of smaller repeating units with many features of functional convergence across the domains of life (see Fig. 4a, a carbohydrate and nucleic acid polymer). In the absence of highly specific structural features based on functional groups lost during diagenesis, these molecular distribution patterns can serve as proxies for organisms or paleoenvironments. Lipid molecules are built on only two basic molecules — the acetyl and the isopentenyl pyrophosphate (IPP) precursor — so these patterns are particularly noticeable. A typical example is the odd-over-even (OEP) preference in the length of alkane chains synthesized by land plants (*Eglinton and Hamilton,* 1967). Shorter alkane chains ($C_{15}–C_{19}$) are an indication of marine (phytoplankton and macroalgae) inputs (*Meyers and Ishiwatari,* 1993). Even number alkane chains can indicate bacterial, fungal, or algal inputs (*Nishimura and Baker,* 1986). Ratios of the odd and even alkanes can be used to source organic matter in a coastal depositional system (*Chevalier et al.,* 2015). A sedimentary suite of long-chain ($>n–C_{27}$) alkanes without OEP can be an indication of diagenesis (*Tegelaar et al.,* 1989). Likewise, the addition of acetyl units to form a fatty acid results in only even-carbon-numbered chains with carbon numbers between $C_{14}–C_{20}$.

Another example of polymerization is the addition patterns of isoprene units that leads to characteristic structural patterns in functional biochemicals (see Fig. 4b). Further tail-to-tail addition of these chains will cyclize to squalene, the precursor structure of steroids and hopanoids, functional analogs in cell membranes of *Eukarya* and *Bacteria*, respectively (*Ourisson et al.,* 1987; *Flesch and Rohmer,* 1988; *Saenz et al.,* 2015). Hence, the myriad possible lipid structures can be constrained to only a few precursor molecules that combine with regularity to form base unit structures: acyclic isoprenoids that are C_{15}, C_{20}, or C_{25}; cyclic terpenoids and steroids that are C_{20} or C_{30}; BHPs that are C_{35}; and C_{40} carotenoids. An understanding of synthetic pathways that use isoprene units also helps our interpretation of their alteration products as well. For example, pristane and phytane are recognized as the alteration product of phytol tails from chlorophyll degradation and can thus be used to identify the presence of photosynthetic organisms (*Koopmans et al.,* 1999) (see Fig. 4d, chlorophyll degradation).

4. CHEMISTRY

4.1. Molecular Patterns as a Biosignature

While time and diagenesis may erase the diagnostic specificity of many biomarkers, there are recalcitrant molecular features that can still be used to indicate the presence of life and even provide general paleoenvironmental clues. Biosynthetic pathways of organic molecules follow distinctive patterns when compared with abiotic pathways. These patterns are driven by preferred reaction pathways based on a conserved number of precursor materials as well as the functions and thermodynamic stability of these compounds.

4.2. Stereochemistry — The Geometric Orientation of Atoms in a Molecule

Often the same atoms in organic molecules could be arranged with a variety of geometries. Biological systems, however, generally synthesize preferred spatial configurations of compounds. These geometric characteristics of molecules, when well preserved, are also used for deriving paleoecological information from biomarkers. Amino acids in nearly all living systems are manufactured with the lefthanded orientation of the functional groups around the central carbon atom: the chiral center. Deviations from these expressions indicate either racemization (inclusion of left- and righthanded orientations) due to an extraterrestrial source (*Bada et al.,* 1973) or during diagenesis: the biotic and abiotic mechanisms by which compounds are transformed in the rock. Likewise, preferred configurations of organic acids and sugars with multiple chiral centers are produced by biologic systems. This preference means that only a small subset of the number of chemically possible combinations is produced by life and thus stereochemistry can be used as a metric of preservation. Thermal alteration reduces the stereochemical specificity of hopanes (section 4.3.4) such that the relative abundance of 20S– and

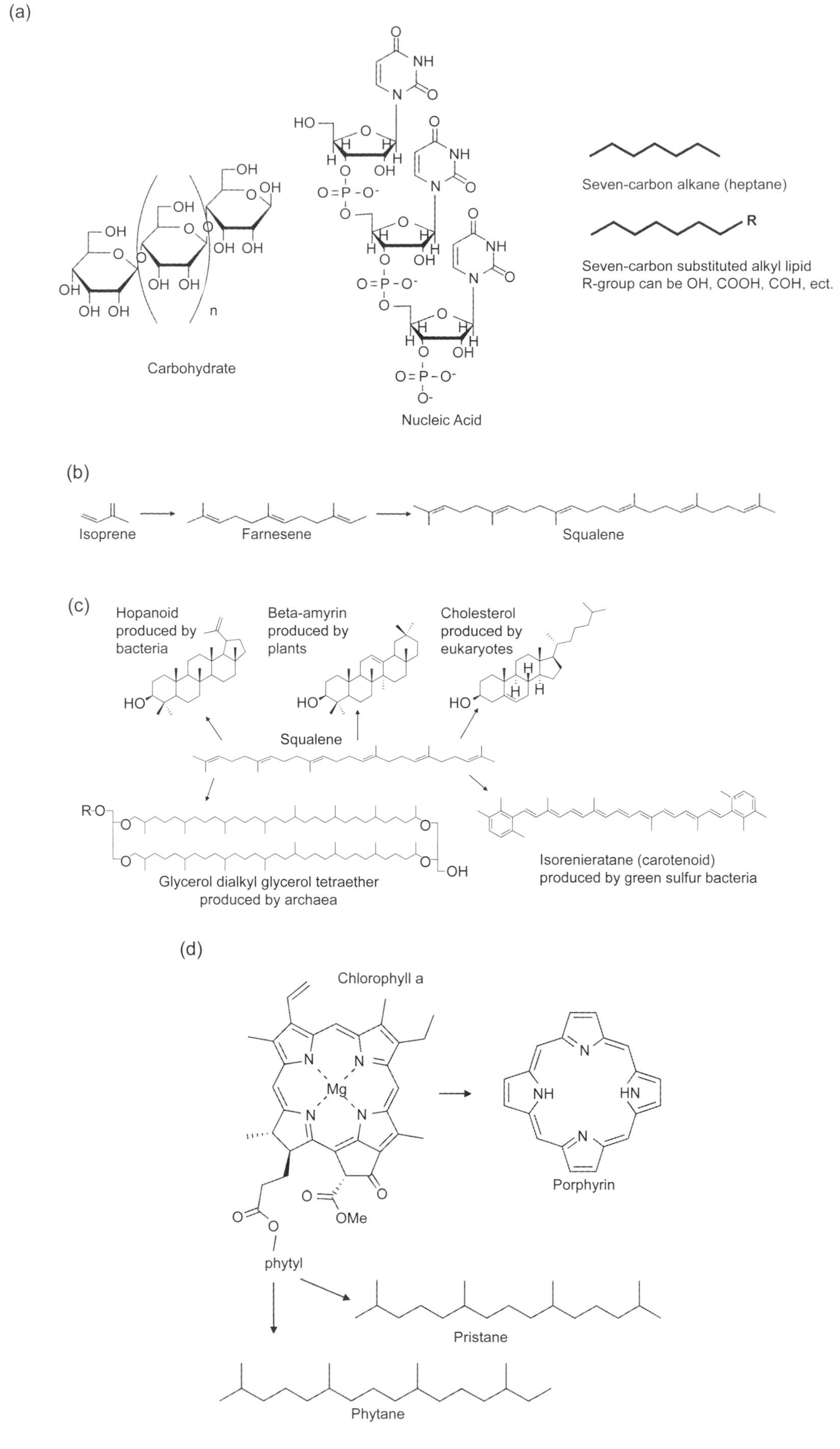

(a)

Carbohydrate

Nucleic Acid

Seven-carbon alkane (heptane)

Seven-carbon substituted alkyl lipid
R-group can be OH, COOH, COH, ect.

(b)

Isoprene

Farnesene

Squalene

(c)

Hopanoid
produced by
bacteria

Beta-amyrin
produced by
plants

Cholesterol
produced by
eukaryotes

Squalene

Glycerol dialkyl glycerol tetraether
produced by archaea

Isorenieratane (carotenoid)
produced by green sulfur bacteria

(d)

Chlorophyll a

Porphyrin

phytyl

Pristane

Phytane

20R– isomers can be used as an indicator of the degree of thermal alteration that may have been applied to preserved biomarkers (*Mackenzie and Maxwell*, 1981; *Innes et al.*, 1997; *Sinnghe Damsté et al.*, 1995).

4.3. Isotope Patterns

Just as stereoisomers and ratios of biomarkers can express patterns, the isotopic composition of these molecules (especially the carbon isotope signature) can be diagnostic of specific metabolisms or ecosystems (*Hayes*, 2001). The carbon isotopic composition ($\delta^{13}C$) of alkanes in higher plants can be used to deduce which photosynthetic carbon assimilation pathway (C_3 or C_4) was the most common in the past, and, by extension, the broad paleoenvironmental conditions (water availability, mean annual temperature) that were prevalent (*Schouten et al.*, 2007b; *Smith et al.*, 2007). Carbon isotopic patterns can distinguish pigments called carotenoids that arise from oxygenic phototrophs vs. those from phototrophic sulfur bacteria. Isotopic patterns in bacterial and archaeal lipids can be used to identify methanotrophy and methanogenesis (*Summons et al.*, 1994). The stability and molecular size of core membrane lipids tend to minimize the magnitude of any fractionation during diagenesis and storage in geologic deposits, thereby favoring excellent fidelity of their original carbon isotope compositions. Broadly, the carbon isotopic composition of preserved biomass can be used to constrain concentrations of inorganic carbon species in the atmosphere and in seawater in the past. This informs our understanding of the carbon cycle over long timescales and can also be linked to ocean temperatures, given the reduced solubility of gases in warmer waters (*Schidlowski, 1995; Pagani et al., 1999; Pancost and Boot, 2004*).

Fig. 4. (facing page) **(a)** Biomolecules are polymers of smaller monomer subunits. Simple sugars polymerize to form carbohydrates. Nucleic acids polymerize to form DNA and RNA macromolecules. Long-chain hydrocarbons, like n-alkanes, form simple lipids and can accommodate many functional group substitutions, creating a variety of alkyl lipids. Proteins (not shown) are formed from individual amino acids. Binding between monomers can result in complex tertiary structures, like the familiar DNA double helix. **(b)** Many lipids found in the geologic record are polymers of the relatively simple isoprene unit. Synthesis by this precursor molecule is universal to all domains of life. **(c)** An example of how one universal molecule (squalene) is used by different organisms to synthesize distinctive, diagnostic lipids ranging from archaeal and bacterial membrane components and pigments to eukaryotic sterols and terpenoids. **(d)** Diagenetic alteration of chlorophyll molecules results in many geomolecules, such as porphyrins, from the central ring as well as pristane and phytane formed from the cleaved phytyl tail.

4.4. Enantiomeric Excesses and Abiotic Organics

Together, the steroisomeric preferences and isotopic patterns observed in biochemicals can be used to distinguish them from the diverse organic complements of meteoritic infall. Meteorites carry many carboxylic, sulfonic, and hydroxyl acids; aliphatic and polar hydrocarbons; purines; pyrimidines; amines; and diverse amino acids (*Cronin et al.*, 1988; *Schmitt-Kopplin et al.*, 2010). Carbonaceous chondrites can contain up to 2 wt% carbon and represent 4% of all meteorite falls (*Sephton*, 2002). With roughly 30,000 tons of cosmic dust falling on Earth yearly, this organic complement is significant (*Pizzarello et al.*, 2006). These compounds are also potential false positives for identification of biomolecules on other planets as well. Three metrics can be used to determine if a biomarker is of biological or extraterrestrial origin: structural diversity, stereochemistry, and isotopic composition (*Epstein et al.*, 1987; *Cronin and Chang*, 1993).

As mentioned above, biosynthetic pathways produce a limited number of molecules compared with the possible structural diversity. Organic molecules produced by abiotic pathways produce more diverse (although generally smaller) compounds. For example, organisms rely heavily on 20 amino acids but well over 80 individual amino acids have been identified in meteorites (*Glavin et al.*, 2006; *Schmitt-Kopplin et al.*, 2010). Organisms use almost exclusively only one enantiomer of amino acids, but meteoritic amino acids contain either racemic mixtures or moderate enantiomeric excesses.

Biosynthesis has greatly narrowed the ranges of carbon and hydrogen isotopic compositions observed in organic molecules on Earth. The same organic molecules from extraterrestrial sources will express the enormous isotopic ranges typical of interstellar elements. $\delta^{13}C$ values of biologic organics on Earth range from approximately –80‰ to –10‰, whereas meteoritic organics range from –20‰ to +50‰. Hydrogen isotopes are an even more diagnostic indicator. Similarly, the deuterium isotopic composition (δD) in biological organics on Earth range from approximately –250‰ to +250‰, whereas δD values in meteoritic organics range from +200‰ to +2500‰. Sources of ambiguous molecules can sometimes be traced using isotopic ratios alone (*Epstein et al.*, 1987; *Cronin and Chang*, 1993; *Elsila et al.*, 2012; *Aponte et al.*, 2014).

5. STRUCTURES AND TEXTURES

5.1. Introduction to Structures and Textures

The order and function that are intrinsic to living systems can be expressed as structures characteristic of the organisms themselves and as structures and fabrics that they create in their surrounding environments. A microbial cell is the most obvious example of such a feature, but additional important examples include organized cell clusters, biofilms, bioherms (carbonate mounds built by microbes), and bodies of multicellular life. Accordingly, the array of structural and textural biosignatures spans broad spatial scales, from 10^{-6} m (cellular structures), through 10^{-3} to 10^{-2} m (e.g., cell clusters and biofilms), to 10^{-1} to 10^2 m or larger (e.g., bioherms and reefs). Many larger structures and textures can survive burial and storage in geologic deposits more robustly than the microbial populations that created them.

Several examples of structures and textures are briefly discussed below and represent a range of spatial scales. The scope of this section is limited to features created by microorganisms and their communities. Microbial communities situated at the bottom of water bodies (i.e., the "benthos") receive particular emphasis because the array of biosignatures that they create is particularly well represented in the early geologic record (e.g., *Walter*, 1983). The antiquity and widespread distribution of biogenic structures and fabrics qualify them as highly promising potential exploration targets on Mars and elsewhere.

5.2. Modern Microbial Communities

Microbial biofilms are microbial communities that are typically embedded in macroscopic, structurally coherent organic layers. Microbial mats are accumulations of biofilms that can achieve thicknesses of several millimeters or more. They occur in a surprisingly wide variety of environments from equatorial zones to both polar regions and can vary in extent from terrestrial and hypersaline marine mats that cover hundreds of square kilometers to tiny mats covering only a few square centimeters in small thermal springs or in coral reef crevices (e.g., *Seckbach*, 2010) to meter-scale domes in freshwater systems (e.g., *Laval et al.*, 2000; *Gischler et al.*,

2008). Mats have diverse community structures, physiological characteristics, and morphologies.

Although photosynthetic microbial mats are remarkably complete self-sustaining ecosystems that flourish at the millimeter scale (e.g., *Des Marais*, 2010), they can also substantially affect sedimentation and other environmental processes at much larger scales. These mats very likely are direct descendants of the most ancient biological communities known, perhaps even those in which oxygenic photosynthesis developed (*Des Marais*, 2003).

Microbial mats are excellent natural laboratories that can help us learn how microbial communities created sedimentary structures and textures that have been preserved in the geologic record (e.g., Fig. 5). Biofilms and mats permeate and stabilize accreting sediment surfaces and can create distinctive structures and fabrics. Stabilized sediments called "microbialites" (*Burne and Moore*, 1987) are organo-sedimentary structures built by some bacteria and algae. Microbialite fabrics reflect the attributes of benthic microbial populations. Filamentous cyanobacteria and other phototrophs can create finely laminated microbialites (e.g., *Des Marais*, 2003) commonly called stromatolites. Microbialites having internal clotted fabrics are called "thrombolites." The origins of thrombolitic fabrics have been attributed to communities dominated by unicellular cyanobacteria or eukaryotic algae, or to grazing of microbial communities by meiofauna (*Burne and Moore*, 1987; *Moore and Burne*, 1994).

Stromatolites in modern subtidal marine environments are rare compared to the widespread elaborate forms documented in Precambrian carbonates. The abundance and diversity of stromatolites, most notably the deeper water forms, declined after about 1 b.y. ago due to the rise of

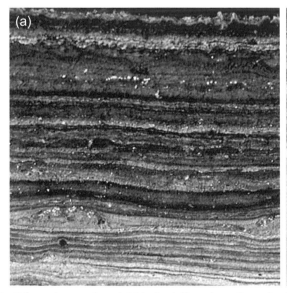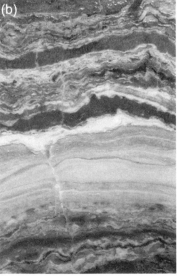

Fig. 5. See Plate 7 for color version. **(a)** Vertical (3 cm) section of a hypersaline mat from Guerrero Negro, Mexico. The crenulated fabric at the surface is lost and the mat layers become thinner with depth due to the degradation of the organic matter. **(b)** Vertical (3 cm) stromatolite section from an evaporite deposit, 1.35-Ga Sibley Formation, Ontario, Canada. The crenulated mat fabric was retained at depth, indicating that degradation rates were attenuated likely because mineralization had rapidly entombed the organic matter of the mat.

grazing predators and competing flora (*Walter and Heys,* 1985). Shark Bay, Western Australia, and the Bahamian platform provide informative modern examples.

Studies of modern marine stromatolites in the Bahamas have shown that the development of their characteristic laminated textures arises from a dynamic balance between sedimentation and intermittent lithification of cyanobacterial mats (*Reid et al., 2000*). During periods of rapid sediment accretion, a pioneer community of gliding filamentous cyanobacteria migrates upward as it traps and binds sediment. These episodes alternate with periods when sedimentation is minimal and surface biofilms that are bound by exopolymers accumulate and also sustain decomposition by heterotrophic bacteria, forming thin crusts of microcrystalline carbonate. During prolonged periods of low sedimentation rates, coccoid (spherical) cyanobacteria further modify the sediment and form thicker lithified laminae. As a stromatolite continues to grow or is buried, these lithified layers are preserved, creating millimeter-scale laminations. This scheme for the development of marine stromatolites may be highly relevant to ancient stromatolites, and it illustrates that their structures and textures arise from interactions between microbial communities and environmental conditions, notably water chemistry and sedimentation rates.

5.3. Stromatolite and Microbialite Structures and Fabrics in the Geologic Record

Tice and Lowe (2006) reported some of the earliest known evidence of benthic microbial communities in 3416-Ma sediments deposited in normal open shallow to deep marine environments. They documented curved and/ or contorted sheet-like structures indicating the former presence of flexible organic biofilms. They concluded from these features and their associated sedimentary structures that photosynthetic mats within the euphotic zone formed most of the carbonaceous matter that subsequently was distributed as detrital debris by waves and currents to surrounding environments.

Stromatolites constitute the most ancient, widespread evidence of life on early Earth (*Walter*, 1983). Stacks of laminations formed and stabilized by microbial biofilms can accumulate in aqueous sedimentary environments to form stromatolites.

Allwood et al. (2006, 2009) described a reef-like assemblage of stromatolites in the 3450-Ma Strelley Pool Formation in Western Australia. They found relict fabrics and organic layers that varied in concert with stromatolite morphology. Stromatolite morphologies included wavy/ rippled laminae, small crested/conical laminites, large complex cones, encrusting domical laminites, and iron-rich laminites. Their observations related morphologic diversity to changes in sedimentation, seafloor mineral precipitation, and the development of microbial mats.

Duda et al. (2016) described a microbial mat facies in the Strelley Pool Formation that had been deposited in a shallow marine environment with associated hydrothermal activity. This facies featured laminated black cherts and small domical stromatolites. Geochemical analyses of carbonates, reduced carbon, and framboidal pyrite within the black cherts provided additional evidence that this deposit preserved the remnants of an ancient microbial ecosystem.

Assessing the origins of stromatolite-like structures requires caution because non-biological processes can mimic their laminated fabrics and their compositions. This concern is particularly acute where postdepositional deformation and thermal alteration of the deposits have been extensive, as is the case for many rocks of Archean age. For example, *Nutman et al.* (2016) reported 1–4-cm-high macroscopically layered structures in 3700-m.y.-old metacarbonates in the Isua supracrustal belt, Greenland. Observations of storm-wave-generated breccias and seawater-like signatures in rare Earth and other trace elements indicated that the carbonates were deposited in a shallow marine environment. However, *Allwood et al.* (2018) contended that these layered structures "...are more plausibly interpreted as part of an assemblage of deformation structures formed in carbonate-altered metasediments long after burial." Allwood et al. highlight "...the importance of three-dimensional, integrated analysis of morphology, rock fabrics, and geochemistry at appropriate scales."

The Proterozoic Eon hosted the most diverse, globally extensive assemblages of stromatolites in the geologic record. In fact, there are very few Precambrian carbonate sequences without them. Figure 6 illustrates examples of common stromatolite morphologies. The variety of morphologies apparently indicates the many ways that benthic microbial communities interacted with physical and chemical environmental processes, e.g., sedimentation rates and the degree of carbonate saturation of seawater. Stromatolite assemblages from the Paleoproterozoic, Mesoproterozoic, and Neoproterozoic time intervals differ from each other. Many stromatolitic carbonates older than 1900 Ma resemble tufa-like precipitates, whereas those younger than ca. 1000 Ma consist principally of trapped and bound micritic carbonate. These secular changes might indicate changes in both physical and biological factors that influenced the growth of microbial mats and bioherms (e.g., *Grotzinger*, 1990; *Knoll and Swett,* 1990), consistent with observations of modern benthic ecosystems.

5.4. Microbially Induced Sedimentary Structures (MISS)

Microbially Induced Sedimentary Structures (MISS) are primary structures (e.g., Fig. 7) formed as microbial communities interact with sediments and the processes of transportation, deposition, and erosion (*Noffke et al.,* 2001). *Noffke* (2010) has described 17 main types of MISS whose features span a range of spatial scales. Examples include microbial mat chips, wrinkle structures, sinusoidal structures, multidirectional ripples, polygonal oscillation cracks, and pockets (gas domes). The most ancient examples of these structures might be billions of years old.

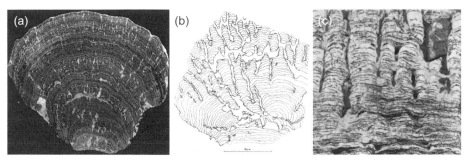

Fig. 6. Vertical cross-sections of stromatolites as examples of morphologies. **(a)** See Plate 8 for color version. Domal stromatolite (16 cm wide), Morrison Formation, Triassic, New Mexico, USA. **(b)** Branching columnar stromatolite (35 cm wide), Bitter Springs Formation, Neoproterozoic Eon, Australia (*Walter*, 1972). **(c)** Stratiform and columnar stromatolites (8 cm wide), Karelian Series, Paleoproterozoic, Russia. Credit: D. Des Marais.

As with all biosignatures, it is important to devise criteria for discriminating MISS from similar structures that can arise non-biologically. Their depositional environments should be characterized to establish their compatibility with microbial mat development. The potential roles of diagenesis and metamorphism should also be determined.

5.5. Microfossils and Microtextures

Opportunities to observed ancient fossilized microorganisms directly can provide definitive evidence of life as well as provide clues about their attributes. But challenges arise, both from the small size and relatively simple morphologies of many microfossils, and from processes that can compromise the potential for their preservation in the geologic record (*Knoll*, 1996). This section reviews briefly the knowledge gained from the microfossil record and the challenges associated with its interpretation.

In modern microbial mats, microorganisms typically disappear visually without any morphological trace over relatively short timescales (e.g., *Aizenshtat et al.*, 1984). Very early lithification seems required in order to preserve most microfossils and their associated mat microtextures (e.g., Fig. 5). However, extracellular sheaths of cyanobacteria and perhaps other bacteria persist to greater depths in mats and as such are more robust candidates for preservation. Because some organic constituents are more resistant than others and

because not all microorganisms necessarily produce these resistant constituents, the fossil record typically represents a biased subset of the original biological community. On the related topic of taphonomy, *Oehler* (1976) documented the structural and chemical changes that accompanied the simulated fossilization of cyanobacteria in synthetic chert.

In order to confirm microscopic structures as fossils and to identify specific microbial taxa, one must identify particular features of these structures. Microorganisms that produce more robust chemical constituents and that exhibited particularly diagnostic features can be recognized more readily in the fossil record. Cyanobacteria (e.g., Fig. 8a) exemplify a bacterial group whose features are relatively well preserved and whose morphological variety forms the basis of their classification systematics. Planktonic protists (e.g., Fig. 8b) can produce chemically recalcitrant cell walls and cysts, and these have ornamentation that facilitates their classification systematics.

Environmental conditions that prevail during the growth, burial, and sequestration of biological components in rocks are critical for the survival of microfossils. Evaporites, carbonates, and mudstones deposited in quiet, shallow marine environments are more favorable for preservation than other environments. To the extent that microbial populations differ in different environments, the fossil record will be biased accordingly. Metamorphism of the host rock also degrades fossils, particularly in the early geologic record. Progressive alteration and recrystallization first obliterates those delicate features used in fossil taxonomy; ultimately it obscures any evidence of biogenicity. Accordingly, interpretations of microfossils depend critically on characterizing the original depositional environments and quantifying the extent of postdepositional alteration.

5.5.1. Archean record. While at least some potential Archean (>2500 Ma) microfossils have been confirmed as authentic biosignatures (e.g., *Schopf et al.*, 2007), the morphological simplicity of many such features and the effects of metamorphism illustrate the challenges associated with the Archean record (e.g., *Garcia-Ruiz et al.*, 2002). Geochemical lines of evidence have been developed to support the biological origin of some of these earliest fossils.

Fig. 7. Example of a microbially induced sedimentary structure (MISS) — a crenulated fabric on the bedding plane of quartz sandstone (view is 20 cm wide).

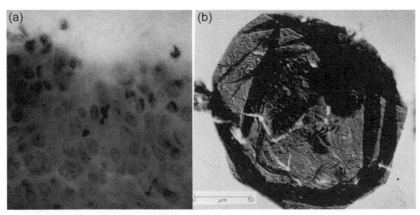

Fig. 8. Images of Proterozoic microfossils. **(a)** Eoentophysalis cyanobacteria, Paleoproteroic (image is 40 µm wide), Canada. **(b)** Kildinosphaera acritarch, planktonic protist, 850 Ma, Arizona.

Schopf et al. (2007) used Raman imaging to characterize carbonaceous matter in ~3465 Ma cherts.

For example, *Schopf and Barghoorn* (1967) and *Knoll and Barghoorn* (1977) observed microscopic objects having size distributions consistent with those of microbial cells, but their definitive interpretation as fossils has been difficult due to their simple morphologies. Silica precipitates can assume diverse morphologies and some of these resemble microbial fossils (e.g., *Garcia-Ruiz et al.,* 2003). The discovery of filamentous structures in Archean chert grainstones in the Apex Basalt, Western Australia (*Schopf,* 1993), led to a vigorous debate (e.g., *Brasier,* 2004). The ultimate determination of these filaments and other cell-like structures as microfossils required the combination of the following lines of evidence: a habitable paleoenvironment, a carbonaceous composition, carbon isotopic composition, three-dimensional structures preserved by permineralization, biological morphologies, and the presence of multiple taxa (*Schopf et al.,* 2007).

Postdepositional alteration by metamorphism has consequences even in cases where biogenicity can be established. *Knoll et al.* (1988) observed in a Proterozoic deposit that the apparent diversity of microfossil assemblages decreased along a gradient of increasing metamorphism. Metamorphism can alter a deposit that once had diverse microfossils to an extent where it resembles an Archean deposit exhibiting low diversity. Very likely the surviving Archean fossil record vastly underrepresents the diversity of microfossil assemblages that once existed.

5.5.2. Proterozoic record. Paleoproterozic (2500 to 1600 Ma) sediments have yielded relatively well-preserved diverse microfossils. The Paleoproterozoic Gunflint Iron Formation provides an excellent early example of these (*Barghoorn and Tyler*, 1965). Stratiform to digitate stromatolites that had been preserved in chert exhibit abundant populations of microfossils. These include filaments, clusters of cocci, rods, and apparent iron bacteria. Non-stromatolitic cherts include similar morphologies as well as umbrella-shaped fossils. Although these populations probably include cyanobacteria, no definitive taxonomic assignments have yet been made.

Some Paleoproterozoic microfossils have definitive taxonomic affinities. Microfossils (Fig. 8a) that are morphologically and developmentally identical to the modern cyanobacteria *Entophysalis* were discovered in the ca. 2-Ga Belcher Supergroup, Canada (*Hofman,* 1976). These are accompanied by numerous additional taxa, including cyanobacteria that strongly resemble younger microfossils with known taxonomic affinities.

The known record of eukaryotic algae is relatively more recent than that of the *Bacteria* (*Perasso et al.,* 1989). Perhaps the oldest known examples are large populations of *Grypania* preserved in the ca. 2.1-Ga Negaunee Iron Formation. *Han and Runnegar* (1992) proposed that these large coiled compressions were photosynthetic algae. Acritarchs are interpreted to be planktonic algae (see Fig. 8b for a Neoproterozoic example). Large spherical acritarchs having simple morphologies (some >100 µm in diameter) were documented in the Paleoproterozoic Chuanlingguo Formation, China (*Zhang,* 1986).

In stark contrast with the older rock record, the Mesoproterozoic (1600 to 1000 Ma) record exhibits abundant microfossil assemblages, including some with known taxonomic affinities. Observations of stratiform stromatolitic carbonates yielded oscillatorian, entophysalid, and nostocalean fossil cyanobacteria, similar to cyanobacteria still found today (e.g., *Horodyski and Donaldson,* 1983, *Zhang,* 1981, *Sergeev et al.,* 1995). Shales in the Roper Group in northern Australia yielded a variety of abundant sphaeromorphic acritarchs (*Peat et al.,* 1978). When compared to the Paleoproterozoic record, Mesoproterozoic microfossil assemblages are much more abundant but morphologically they are generally similar (*Knoll,* 1996). However, *Knoll* (1992) presented paleontological evidence to demonstrate that the rapid diversification of *Eukarya* indicated by analyses of molecular phylogeny (e.g., *Sogin,* 1994) occurred ca. 1200 to 1000 Ma.

The abundance, diversity, and preservation of the Neoproterozoic (1000 to 544 Ma) fossil record far exceed that found in earlier records. This includes the morphological diversity of prokaryotes, acritarchs (e.g., Fig. 6b), and multicellular algae (*Knoll,* 1996). *Knoll* (1991) examined silicified carbonates and shales deposited in a tidal flat-lagoonal complex and documented changes in microfossil assemblages across a sedimentologically-defined paleoenvironmental gradient. They identified prokaryote-dominated low diversity assemblages in upper intertidal to supratidal microbial mats. The diversity of microfossils in the mats increased across the intertidal to subtidal zones, and lagoonal shales even exhibited a modest diversity of protists. Multiple examples of well-preserved, diverse Neoproterozoic microfossil assemblages have since been described in the contexts of their environments.

Microscopic tunnels or "microborings" in basaltic glasses have been proposed as "trace fossil" evidence of past life (see *Staudigel et al.,* 2008, for a review). These microtextures have attracted considerable attention because they resemble microbial borings in carbonates and other sedimentary rocks and they can be well preserved in basalts. But many if not all such features can have non-biological origins. *French and Blake* (2016) have reinterpreted "these enigmatic microtextures as end products of the preferential corrosion/ dissolution of radiation damage (alpha-recoil tracks and fission tracks) in the glass by seawater, possibly combined with pressure solution etch-tunneling." The uncertainties surrounding the origins of these features illustrates the challenges associated with discriminating between potentially biogenic microtextures vs. features that are very similar in appearance but can be created by non-biological processes.

5.6. Synergy with Other Biosignature Categories

Clearly the most important challenge in paleontological studies is to prove the biogenic origins of potential microfossils through detailed studies of their morphologies, population distributions, and other evidence. But attempts to assess biogenic origins, taxonomic affinities, and diversity of microfossil assemblages must consider the physical and chemical attributes of paleoenvironments and the quality of their preservation. Interpreting the morphological microfossil record is an inherently interdisciplinary effort.

The combination of confocal laser scanning microscopy and Raman imagery of potential biosignatures has helped to confirm the biological origins of microfossils (e.g., *Schopf and Kudryavtsev,* 2009). These non-destructive techniques can assay in three dimensions "the molecular-structural composition of the kerogenous components of fossils and their surrounding matrices" and also quantify the fidelity of their preservation. The combination of cathodoluminescence and light microscopy, Raman spectroscopy, time of flight-secondary ion mass spectroscopy, and stable isotopic analyses enhanced the interpretations of fossilized microbial mats in the Strelley Pool Formation (*Duda et al.,* 2016).

Organic geochemical analyses can greatly enhance paleontological research (e.g., *Brocks,* 2018). For example, eukaryotes are dominant sources of sterols (*Ourisson et al.,* 1987), which are reduced to steranes upon burial and preservation (see also section 4.3.5). However, caution is

in order because geologic deposits can be compromised by organic contamination from groundwater, drilling fluids, and other sources. New techniques and protocols are addressing this challenge (e.g., *Jarrett et al.,* 2013; *Flannery and George,* 2014).

These investigations and others exemplify how modern analytical techniques can work in concert to characterize multiple types of biosignatures in individual samples, thereby greatly enhancing our understanding of our early biosphere.

6. MINERALOGICAL BIOSIGNATURES

Rocks and minerals represent the most robust, long-lived features in many planetary systems — objects that can preserve clues to ancient environments over billions of years. Furthermore, minerals are information-rich objects, with trace and minor elements, isotopes, solid and fluid inclusions, and a host of physical attributes that preserve insights to the environments of the origins and subsequent alteration. Accordingly, mineralogical evidence is increasingly cited as worthy of astrobiological consideration at local, regional, and planetary scales.

At least three attributes of mineralogical systems might contribute to the search for life on other worlds: (1) the presence of specific mineral species or suites of species that are known to be biologically produced, or form as a result of biological modifications to the environment; (2) morphological characteristics of minerals and their aggregates, especially the precipitation of nano-crystalline structures due to rapid biological redox reactions; and (3) global-scale diversity and distributions of minerals in space and time, notably the distinctive occurrence of numerous rare minerals in near-surface biologically-influenced environments. The following sections consider each of these potential mineral biosignatures in more detail.

6.1. Mineral Species as Biosignatures

Minerals form as a consequence of local chemical and physical conditions. More than 5400 mineral species, each of which possesses a unique combination of chemical composition and crystal structure, have been approved by the International Mineralogical Association (IMA) (see *http://rruff.info/ima* for a complete up-to-date list). Several thousand additional species (*Hazen et al.,* 2015b; *Hystad et al.,* 2015, 2019) are predicted to occur on Earth but have yet to be discovered and described. These diverse natural crystalline compounds reveal a wide range of processes of formation — the consequence of varied physical, chemical, and/or biological environments. Two groups of minerals in particular — organic minerals and oxidized minerals — deserve special attention as potential biosignatures.

6.1.1. Organic minerals. It is likely that some carbon-bearing mineral species occur exclusively as a consequence of biological activity, and thus they represent promising biosignatures both on Earth and possibly on other worlds. The most unambiguous mineral biosignatures are several dozen organic species that arise from the decay of plant material or alteration of animal excrement (*Hazen et al.*, 2013a). For example, some molecular crystals are unique to fossilized roots and wood [e.g., refikite ($C_{20}H_{32}O_2$) and flagstaffite ($C_{10}H_{22}O_3$)], or bat guano deposits in caves, e.g., the purines uricite ($C_5H_4N_4O_3$) and guanine ($C_5H_5N_5O$), along with urea [$CO(NH_2)_2$]. Coal hosts several hydrocarbon minerals, including both aromatic (e.g., carpathite, $C_{24}H_{12}$; equivalent to coronene) and aliphatic (dinite; $C_{20}H_{36}$) compounds, as well as acetamide (CH_3CONH_2) and the nickel porphyrin abelsonite [$Ni(C_{31}H_{32}N_4)$; *Hummer et al.* (2017)]. Burning coal mines also produce molecular crystals through sublimation, including kladnoite [$C_6H_4(CO)_2NH$] and hoelite ($C_{14}H_8O_2$). An unusual example is tinnunculite ($C_5H_4N_4O_3 \cdot 2H_2O$) (*Pekov et al.*, 2016), which arises when bird excrement is baked in a coal fire.

Two provisos limit the utility of organic minerals as universal biosignatures. First, on Earth today they are rare and localized in occurrence, and thus are unlikely to be detected on other planets or moons. Second, and more importantly, certain organic minerals will be abundant on worlds without obvious Earth-like biota. Hydrocarbon-rich Titan, for example, likely hosts a range of organic crystals that are stable at its near-constant ambient temperature of 92 K (*Maynard-Caseley et al.*, 2018). Prebiotic Earth, which boasted significant quantities of endogenous and exogenous organic compounds (*Chyba and Sagan*, 1992), may also once have featured a wide range of organic crystals, including reactive carbohydrates, ferrocyanide, and amino acids, most of which were subsequently consumed as food by cellular life (*Benner et al.*, 2010; *Hazen*, 2018). In a sense, the absence of these edible organic minerals — crystalline phases that should be present in small amounts in organic-rich solutions — may represent a mineralogical biosignature.

6.1.2. Oxidized minerals. Suites of oxidized weathering minerals that arose on Earth following the Great Oxidation Event may represent a much more widespread potential biosignature. Hazen and colleagues (*Hazen*, 2013; *Hazen et al.*, 2008) speculated that two-thirds of mineral species on Earth are the consequence of oxygenic photosynthesis and thus would not have occurred prior to 2.5 b.y. ago. In subsequent studies (e.g., *Golden et al.*, 2013; *Hazen et al.*, 2014, 2015b; *Liu et al.*, 2016, 2017; *Hummer et al.*, 2020) they hypothesize that the global distribution of oxidized species, including Fe^{3+}-bearing clay minerals, secondary ore minerals, and weathering products of metals such as Cr and Mn, are a consequence of the co-evolving geosphere and biosphere.

The hypothesis that the occurrence of thousands of oxidized mineral species represents a global-scale mineralogical biosignature might be challenged on the grounds that planets can display purely physico-chemical processes that create significant redox gradients and oxidizing conditions at various scales. At the largest scale of planetary differentiation, the immiscibility of silicate- and metal-rich melts, and the consequent gravitational segregation of a metal-rich core, creates a significant redox gradient at the core-mantle boundary and a permanent relative oxidation of the outer layers as electrons are sequestered in the native iron-nickel alloys of the deepest interior. Non-living planets with an oxygen-to-iron ratio significantly greater than Earth's might thus display a surface mineralogy similar to Earth today. A similar planetary-scale development of redox gradients can occur when upper atmosphere photo-dissociation of water and subsequent hydrogen escape to space leads to a gradual increase in atmospheric oxygen fugacity.

Other non-biological processes, including lightning strikes, the formation and separation of immiscible fluids in the crust or upper mantle, reactions with phases in extremely reduced iron or enstatite chondrite meteorites, and photo-oxidation, may result in local environments that are exceptionally oxidized or reduced (*Morrison et al.*, 2018a). Consequently, additional research on these planetary processes is necessary to assess potential abiotic mineral diversity.

6.2. Morphological Characteristics of Minerals and Their Aggregates

The sizes and shapes of minerals reflect a wide variety of physical, chemical, and biological processes. Apart from the obvious biologically directed forms of biomineralized shells, bones, and teeth, mineral morphologies may provide evidence for cellular life. The examples of nanocrystals of magnetite and uraninite — minerals that in abiotic environments typically nucleate relatively few macroscopic crystals, but form nanocrystallites when nucleated by numerous individual cells — exemplify this approach.

6.2.1. Magnetite. Significant focus has been placed on nanocrystals of magnetite (Fe_3O_4) (see Fig. 9), which some microbes grow in chain-like structures called magnetosomes (*Bazylizinki et al.*, 1995; *Pósfai et al.*, 2013). These microbial organelles act as compass needles that sense "up" vs. "down" from Earth's dipping magnetic

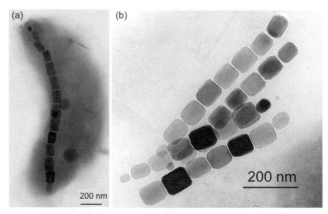

Fig. 9. Biomagnetite can occur in chain-like structures, as shown in these elongated magnetosomes from magnetotactic Alphaproteobacteria (*Pósfai et al*, 2013).

field — an ability that enables microbes to move to optimal levels within a vertical redox gradient (*Dusenbery*, 2009). The magnetite nanocrystals of magnetosomes have been described as distinctive, perhaps in ways that are unique to biology.

The focus on possible biological origins of magnetic nanocrystals was heightened by the discovery of nano-magnetite in martian meteorite Allan Hills (ALH) 84001. *Thomas-Keprta et al.* (2000, 2001, 2002) describe characteristics of these crystals, including (1) a narrow size range (on the order of 100 nm), (2) unusual chemical purity, (3) unusually perfect single-domain structure, (4) distinctive truncated hexa-octahedral crystal form, (5) an elongated crystal morphology, and (6) occurrence in chains of nanocrystals. They conclude that this combination of features, all of which are present in a subset of ALH 84001 magnetite grains, represents a "presumptive biosignature." While the claims of *Thomas-Keprta et al.* (2002) have come under scrutiny (*Barber and Scott*, 2002; *Golden et al.*, 2004; *Weiss et al.*, 2004; see also *Jimenez-Lopez et al.*, 2010, and references therein), the occurrence of magnetite nanocrystals remains an intriguing topic of debate.

6.2.2. Uraninite. *Hazen et al.* (2009) estimated that as many as 90% of uranium minerals — those incorporating the uranyl (U^{6+}) ion — are a consequence of atmospheric oxygenation on Earth and thus an indirect consequence of the biosphere. Some microbial strains also mediate uranium mineral formation directly through metabolic processes that lead to potentially distinctive nano-crystalline morphology of the reduced uranium ore mineral, uraninite (UO_2). For ex-

ample, strains of *Desulfovibrio, Geobacter*, and *Shewanella* are able to produce uraninite concentrations by coupling enzymatic metabolic acetate oxidation to the reduction of aqueous uranyl cations $(U^{6+}O_2)^{2+}$ (*Lovely et al.*, 1991; *Sharp et al.*, 2008). The resulting uraninite ore concentrations in "roll-front" deposits, all of which formed during the Phanerozoic Eon when atmospheric oxygen was abundant, constitute a major source of uranium ore.

The nano-crystalline morphology of these uraninite deposits contrasts with the coarser crystals of igneous, metamorphic, and hydrothermal ore-forming process, and thus may uniquely reflect processes occurring at the scale of cell membranes (*Fayek et al.*, 2005). It is possible, therefore, that uraninite morphology in the appropriate geological context could provide a biosignature. Note, however, that other mechanisms might lead to localized nano-uraninite. For example, *Utsunomiya et al.* (2005) demonstrated that ionizing radiation from uranium decay can induced uranyl minerals to decompose and form nano-crystals of UO_2 (*Utsunomiya et al.*, 2005).

6.2.3. Other nanophases. In addition to magnetite and uraninite, other nanocrystalline materials are known to be due to microbial mineralization, which is of special note because their relatively common occurrences are oxide-hydroxides of iron, including ferrihydrite [$Fe_{10}O_{14}(OH)_2$] and goethite [$FeO(OH)$], which form rapidly in near-surface environments. Subsequently, as most of the Fe has been precipitated and Mn concentration increases, several manganese oxide/hydrates, including birnessite [$(Na,Ca,K)_{0.6}(Mn^{4+},Mn^{3+})_2O_4 \cdot 1.5H_2O$], hollandite [$Ba(Mn^{4+}_6Mn^{3+}_2)O_{16}$],

Fig. 10. See Plate 9 for color version. Iron and manganese biominerals. The distinctive morphologies of many oxide-hydroxide iron and manganese mineral deposits reflect their nanocrystalline constituents, which are rapidly precipitated at numerous nucleation sites through microbial redox reactions. Massive ferric iron oxide-hydroxide deposits [principally goethite, $FeO(OH)$] form with sufficient rapidity to preserve localized fluid flow patterns [**(a)** field of view ~1 m; photo by Robert Hazen] and to encase recently lost objects such as this watchband [**(b)** photo courtesy of Stephen Godfrey]. Nodules and rosettes of nanophase Mn^{3+}-Mn^{4+} minerals, including birnessite [$(Na,Ca,K)_{0.6}(Mn^{4+},Mn^{3+})_2O_4 \cdot 1.5H_2O$], hollandite [$Ba(Mn^{4+}_6Mn^{3+}_2)O_{16}$], romanechite [$(Ba,H_2O)_2(Mn^{4+},Mn^{3+})_5O_{10}$], and todorokite [$(Na,Ca,K,Ba,Sr)_{1-x}(Mn,Mg,Al)_6O_{12} \cdot 3-4H_2O$], may form at the end of this process after most of the iron has been precipitated [**(c)** (height ~3 cm), **(d)** height ~3 cm), and **(e)** courtesy of Rob Lavinsky]. All specimens from Driftwood Beach, Calvert County, Maryland.

romanechite $[(Ba,H_2O)_2(Mn^{4+},Mn^{3+})_5O_{10}]$, and todorokite $[(Na,Ca,K,Ba,Sr)_{1-x}(Mn,Mg,Al)_6O_{12}\cdot3-4H_2O]$, may form. Biomineralization occurs rapidly as reduced subsurface waters rich in Fe^{2+} and Mn^{2+} are quickly oxidized when exposed to the air (*Nealson, 2006; Andrews et al., 2013*). The resulting Fe^{3+} and $Mn^{3+/4+}$ oxidized minerals invariably form nano-crystalline deposits of distinctive morphologies, some of which preserve transient fluid flow features (*Greene and Madgwick,* 1991; *Tebo et al.,* 2005; *Nealson,* 2006; *Druschel et al.,* 2008; *Posth et al.,* 2014; *Yang et al.,* 2013) (see Fig. 10).

While the microbiology of such deposits has received considerable attention, less focus has been paid to the range of macro-scale structures that can arise from the rapid nano-scale crystallization of these phases. Future studies on Fe-Mn biomineral morphologies are thus warranted in the context of the search for reliable biosignatures.

6.3. Global-Scale Diversity and Distributions of Minerals in Space and Time

Large and growing data resources on Earth's mineral species, localities, and ages are revealing patterns of diversity and distribution that may reflect the influences of life at a planetary scale. Hystad and colleagues (*Hystad et al.,* 2015; *Hazen et al.,* 2015a) demonstrated that the diversity and distribution of Earth's minerals conform to a "large number of rare events" (LNRE) distribution, by which a few mineral species are common but the major-ity of species are rare, documented from five or fewer localities (see Fig 11). LNRE distributions are found for the more than 5400 mineral species on Earth, as well as for many subsets of these minerals (*Hazen et al.,* 2015a; *Liu et al.,* 2017; *Hystad et al.,* 2017, 2019).

The distinctive LNRE frequency spectrum for minerals on Earth contrasts with preliminary analyses of lunar and martian minerals (*Hazen et al.,* 2015a; *Morrison et al.,* 2018b), which do not appear to show this pattern. While much more extensive surveys of varied environments on these terrestrial worlds will be required, the relatively low mineral diversity and uniform distribution of a few dominant mineral species on the Moon and Mars contrast sharply with Earth's more heterogeneous near-surface mineralogy. Hazen and colleagues have thus postulated that the LNRE mineral distribution may, itself, be a global biosignature that arises from the numerous varied microenvironments caused by biological activity. The influence of life on mineral evolution and ecology will continue to be a central focus in the search for unambiguous mineral biosignatures.

6.4. Future Directions for Mineral Biosignatures

We now know that minerals are physically and chemically complex, information-rich objects that can preserve details of their formation environments for billions of years. This brief review of mineral biosignatures focused on three emerging aspects of mineral diversity and distribution that

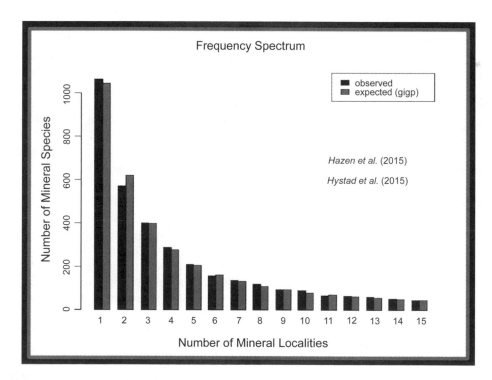

Fig. 11. Minerals conform to a large number of rare events (LNRE) distribution. Observed (black) and modeled (gray) generalized inverse Gaussian-Poisson frequency distribution for rare minerals on Earth. Most of Earth's >5400 mineral species are rare, occurring at 5 or fewer localities. Based on preliminary observations, this distribution has not been observed for minerals on the Moon or Mars.

might point unambiguously to life: diagnostic mineral species, distinctive morphologies, and global patterns of mineral distribution. This list is by no means complete. For example, it is plausible that certain characteristic patterns of trace elements and isotopic fractionations in minerals arise from biological processes.

One promising approach for future research is the development of an "evolutionary system of mineralogy" that will improve current classification schemes by incorporating the impact of biology on mineral formation. The official IMA definition of a mineral species is based on a unique combination of fictive end-member composition and idealized crystal structure, and it is ill-suited to understanding roles of minerals in the co-evolution of the geosphere and biosphere. To address this problem, a new evolutionary system of mineralogy, based on natural kind clustering, has been proposed (*Hazen*, 2019). This system will focus on idiosyncratic chemical and physical attributes of minerals that may more clearly point to biological origins. The more we understand those biotic formation processes, the better able we will be to recognize mineral biosignatures.

7. TAPHONOMY

7.1. Overview

While no biosignature is impervious to time, some are more resilient to certain environmental pressures than others. In most cases, biomarker persistence depends greatly on the preservation environment (*Derenne and Largeau*, 2001). Taphonomy is the study of how organisms decay and become fossilized. Processes of biomarker degradation include decomposition, transport, burial, compaction, and any other chemical, biological, or physical phenomena that can alter biological remains. In many cases these processes are well enough understood that identification and interpretation of altered biomolecules is still possible.

Upon death, any organism — from the simplest prokaryote to large plants and animals — will begin the process of decomposition. Many of the immediate changes in necromass are due to bacterial metabolism. However, even before this process has started, enzymatic autolysis will expose cellular contents to the environment. During this initial phase of decomposition biomolecules that are highly functionalized or easily metabolized — such as amino acids, nucleobases, and carbohydrates — will be consumed and remineralized, primarily by bacterial organisms, prior to burial (*Henrichs and Farrington*, 1987; *Mayer and Rice*, 1992; *Poinar et al.*, 1996; *Bada et al.*, 1999). The extent to which this decomposition occurs depends on the availability of water and redox state of the site.

Terrestrial remains preserved in the geologic record are generally transported into sedimentary basins where burial occurs (allochthonous burial) (*Spicer*, 1981; *Hedges and Keil*, 1995). Only rarely are remains entombed *in situ*

(autochthonous burial) (*Swift et al.*, 1979). Microbial mats remain one of the few examples of organisms preserved in their growth positions. Paleosols can preserve molecular remnants in stratigraphic order but are very often reworked by bioturbation or tectonic forces. Organismal and molecular remains are generally transported by water and occasionally by wind to topographic low points where they can be preserved on geologic timescales (*Rieley et al.*, 1991; *Simoneit*, 1997; *Schlünz and Schneider*, 2000). The most common depositional sites are marine basins and some lacustrine sites (*Gagosian and Peltzer*, 1986; *Goñi et al.*, 1997). In either site there are many physical stressors that can degrade the physical remains and thereby the biomarker signal fidelity (*Hedges et al.*, 1997). The quality of preservation depends heavily on the nature of the transport mechanism and the speed of the burial process (*Bada et al.*, 1999). Anoxic conditions promote the best preservation and this can be provided by prompt burial of remains (*Sinninghe Damsté et al.*, 2002b). Similarly, efficient compaction of buried sediments also promotes preservation, hence the observed relationship between well-preserved organics and small grain size of the burial substrate (*Mayer*, 1994; *Eigenbrode and Freeman*, 2006). Thus, tight shales promote better organic preservation than sands. The type and specific surface area of minerals in a depositional site can also promote organic preservation. Adsorption to mineral surfaces stabilizes organic matter, preventing degradation reactions. For example, salt precipitation is a uniquely rapid depositional mode, resulting in the entrapment of biological material within the mineral structure and fluid inclusions (*Jaakkola et al.*, 2016), preserving not only cellular material, but in some cases also viable cells (*Mormile et al.*, 2003; *Lowenstein*, 2012). An important consideration when assessing any assemblage of fossils and/or biomarkers is to remember that the expressed components represent an accumulation of preservation biases and may, for many reasons, exclude important compounds that would alter the interpretation of a sedimentary deposit (*Van Bergen et al.*, 1995; *Briggs*, 1999).

Regardless of the efficiency of the transport to a marine depositional environment or the large amounts of primary production that occurs in the photic zone of the oceans, only ~1% of all of this organic matter is preserved on the seafloor (*Burdige*, 2007; *Schlünz and Schneider*, 2000; *Schmidt et al.*, 2010). Most is remineralized in the water column by fauna and microbial communities. Organismal remains may travel great distances through the water column before settling on the floor of the body of water. The depth of the reservoir and the settling velocity of these sinking particles can affect the fidelity of their biomarker signal since molecular photo-oxidative degradation and remineralization due to grazing can continue during transport. If remains are metabolized, waste products can continue settling in fecal pellets where intact organic biomarkers will still be preserved in artificial concentrations (*Goutx et al.*, 2007). Provided the area is tectonically quiescent, the ocean floor can be an excellent preservation environment for macroscopic and molecular remains as long as the bottom waters remain hypoxic or anoxic.

7.2. Diagenesis

Even after burial, diagenesis — the alteration of the biomarker structures — can occur due to biological, chemical, or physical processes. Some biomolecules, like nucleic acids, are extremely fragile and not typically preserved over lengthy time periods (hundred-thousand-year to possibly million-year timescales) (e.g., *Willerslev et al.,* 2004; *Schubert et al.,* 2012). RNA molecules, for example, have very short half-lives, typically ranging from seconds to minutes (*Deutscher,* 2006). The half-life of RNA is further complicated as stability varies among genes (*Bernstein et al.,* 2002; *Hambraeus et al.,* 2003; *Selinger et al.,* 2003; *Redon et al.,* 2005), species (*Bernstein et al.,* 2002; *Redon et al.,* 2005), and resource availability (*Carvalhais et al.,* 2012), adding further complexity to transcriptomic and metatranscriptomic studies. Due to the short half-life of RNA, the presence of RNA is considered a better indicator for live organisms than DNA. For the same reason, ancient RNA studies are very rare, although there have been some strides forward in the realm of ancient seed transcriptomics (*Fordyce et al.,* 2013).

In situ DNA decay can be described by first-order kinetics, and even in bone, most bonds between nucleotides in the backbone of a sample have been estimated to break over timescales of hundreds of years (*Allentoft et al.,* 2012). Improvements in our ability to analyze ancient DNA molecules — overcoming challenges from post-mortem damage and mitigating the risk of contamination — have made the study of ancient DNA feasible over lengthy time periods (hundreds of thousands of years) (e.g., *Schubert et al.,* 2012*)*. Scores of ancient genomes have been sequenced, enabling a deep look into rapid evolutionary changes and the genetics of domestication and climatic adaptation (*Hofreiter et al.,* 2015). While the short fragment lengths associated with most preserved DNA molecules can create alignment ambiguities, protocol optimizations can overcome some of these limitations (*Kircher et al.,* 2011).

Peptides persist exponentially longer than DNA but have lifetimes far shorter than lipids. Because peptide bonds are subject to hydrolysis, proteins are easily depolymerized and degraded, particularly when not enclosed within mineralized tissues of larger macrofossils (*Toporski and Steele,* 2002). Multi-subunit proteins or aggregates may be able to persist longer in the geologic record because of non-covalent interactions with the detrital organic matrix, and covalent cross-linking could also make proteinaceous material more refractory (*Nguyen and Harvey,* 2001). Acidic proteins may also degrade more rapidly than basic or neutral proteins, although this remains to be thoroughly explored.

For long-term preservation, water-insoluble structural polymers are the most likely to withstand degradation. Depolymerization will continue as microbiota in the soils and sediment metabolize organic remains (*Maduireira et*

al., 1995). Exoenzymes can further reduce polymers to monomers and labile biomolecules such as carbohydrates, amino acids, and nucleobases will either be used by microorganisms to synthesize new cellular material or be used as a source of energy and subsequently mineralized. Examples of water-insoluble structural polymers include the proteins keratin, collagen, and elastin (*Briggs,* 1999).

Any amino acids or sugar monomers that survive biological degradation will be subjected to increased pressure and heat in the burial setting. These conditions promote random repolymerization and condensation reactions with preserved hydrocarbons to produce fulvic and humic acids (*Peters et al.,* 1981). Over time these compounds will fuse into increasingly complex refractory structures. This maturation process leads to the formation of kerogen, an insoluble matrix of organic compounds. The relatively non-reactive lipids and free hydrocarbons are spared much of this alteration, hence they constitute most of the readily interpretable geochemical fossil biomarkers.

Lipids may not necessarily be spared any alteration by time, heat, and pressure but the alteration that occurs can be used to gauge burial conditions. Molecular alteration reactions can include changes in stereochemistry and isotopic changes. Diagenesis will preserve the more thermodynamically stable configurations of molecules and force the less-stable molecules into the more stable configuration. Thus, ratios of these conformers are often used to understand the extent to which a source rock has been heated or the depth to which it was buried. These stereochemical indices are widely applied to preserved hopanes and sterols (*Mackenzie and Maxwell,* 1981).

Aside from stereochemical changes, diagenesis can result in the systematic cleavage of reactive molecular moieties, functional groups, or diagnostic double bonds. A canonical example of this process is the formation of various porphyrin rings and tail components from chlorophyll degradation (see Fig. 4e) (*Koopmans et al.,* 1999). The ratios of these compounds can be used to reconstruct the redox conditions of a burial environment. Similarly, physical maturation can remove indications of even or odd chain-length preferences in preserved alkanes, and biodegradation can cause an overall decrease in the abundance of low molecular weight alkane homologs.

7.3. Taphonomy in Icy Environments

Rapid freezing and entombment in ice represent unique preservation conditions. Freezing temperatures can retard the biological and chemical processes associated with degradation of necromass and organic biomarkers. Remains with excellent protein, DNA, and lipid preservation have been discovered encased in ice. The preservation and reconstructive potential of ice is limited, however, by its ephemeral and dynamic nature. While the poles have been covered in ice for at least 30 m.y. (*Zachos et al.,* 1996), this ice is constantly in motion and lost to the ocean. The oldest ice reserves on

Earth are only ~2–3 Ma (*Yan et al.*, 2017), and ice itself imposes considerable physical taphonomic forces on preserved remains. While cold temperatures prevent biologic degradation, tectonic and chemical diagenetic changes are still possible. As ice deforms, remains can be disarticulated and subjected to increased pressure, and small-scale thawing events can facilitate oxidative degradation.

As with sediments, biosignatures entombed in ice generally represent transported material. Wind and water can deposit remains and organic biomarkers in ice or freezing reservoirs. Stratigraphy in ice reservoirs can be affected by stratinomic sorting during firnification (the recrystallization of snow due to partial melting, refreezing, and compaction) as well as ice ploughing events that rework deeper sedimentary materials into surface ice. Ice floes can also break loose and deposit surface materials into seafloor sediments that present as abnormalities in the biomarker record. Despite these depositional constraints and the relatively brief record of ice on Earth, these reservoirs do represent regions of exceptional preservation and excellent sites of consideration on other worlds, especially if they are protected from radiation, induced oxidation (a concern on icy moons), and thawing.

8. DETECTING LIFE AND THE LIMITS OF ANALOGY

As outlined in the preceding sections, biosignatures on Earth can take many forms ranging from nucleic acids, proteins, and metabolites; chemistry; organics; structures and textures; and mineral assemblages. Our ability to detect life on our planet, both extant and extinct, is continually improving, as is our understanding of how traces of life persist over ancient timescales against the forces of entropy. Within the context of astrobiology, we can draw upon the wealth of information that these decades of collective research have provided as we consider the possibilities of detecting life beyond Earth.

Yet it is important to remember that methods for detecting life on our own planet rely primarily on identification of well-established and widely accepted features associated with terrestrial life and signatures of biologic processes. These include particular classes of molecules and isotopic signatures, enantiomeric excesses, and patterns within the molecular weights of fatty acids or other lipids. However, Earth is a very particular and dynamic environment. Most of our ancient crust has disappeared. What remains is heavily metamorphosed, making it difficult to untangle the record of early life even when we have highly sophisticated tools at our disposal. The earliest biomarker and isotope record of life on our planet has also been heavily overprinted by the widespread dominance and immense biomass of photosynthetic life — a metabolism that may never have developed on other planets (*Michalski et al.*, 2018). While amino acids and nucleobases are present in space and abundant in extraterrestrial materials (*Kvenvolden et al.*, 1970; *Ehrenfreund et al.*, 2001; *Kuan et al., 2003; Callahan et al.*, 2011), it is

unclear whether they would form the basis for life elsewhere.

As we have explored the limits of life in extreme environments on Earth and delved into the rock record to reconstruct our past, we have developed a robust set of biosignatures associated with life on Earth, which should be drawn upon by astrobiologists. We can use these established approaches that work readily on Earth to inform our search for life on less familiar worlds while not relying specifically on terran biosignatures. While non-Earth life may exhibit different chemistries, the techniques perfected for detecting biosignatures on Earth and our knowledge of molecular preservation and taphonomy may inform our search for life without restricting our definitions. That said, the farther we travel from Earth, the likelihood that life shares a common prebiotic pathway with Earth diminishes. Life on other moons and planets may have developed with different biochemistries and different structural components. For instance, life based on unfamiliar characteristics may be present in our solar system, particularly on non-water worlds like Titan.

Thus, alternative approaches — looking for patterns of complexity, disequilibrium, or evidence of energy transfer — may enable us to look for life in a more inclusive manner (e.g., *Conrad and Nealson*, 2001; *Marshall et al.*, 2017; *Johnson et al.*, 2018). Although these approaches may be less definitive than uncovering a hopane or DNA sequence, they may allow us to identify unknowable, unfamiliar features and chemistries that represent processes of life as-yet unrecognized. The most effective approach to life detection for future space missions will require a synthesis of techniques based on a thoughtful evaluation of the scope and quality of each line of evidence.

REFERENCES

Ahn Y. Y., Bagrow J. P., and Lehmann S. (2010) Link communities reveal multiscale complexity in networks. *Nature, 466,* 761–764.

Aizenshtat L., Lipiner G., and Cohen Y. (1984) Biogeochemistry of the carbon and sulfur cycle in the microbial mats of the Solar Lake (Sinai). In *Microbial Mats, Stromatolites* (Y. Cohen et al., eds.), pp. 281–312. Liss, New York.

Allentoft M. E., Collins M., Harker D., Haile J., Oskam C. L., Hale M. L., Campos P. F., Samaniego J. A., Gilbert M. T. P., Willerslev E., and Zhang G. (2012) The half-life of DNA in bone: Measuring decay kinetics in 158 dated fossils. *Proc. R. Soc. B, 279(1748),* 4724–4733.

Allwood A. C., Walter M. R., Kamber B. S., Marshall C. P., and Burch I. W. (2006) Stromatolite reef from the Early Archaean era of Australia. *Nature, 441,* 714–718.

Allwood A. C., Grotzinger J. P., Knoll A. H., Burch I. W., Anderson M. S., Coleman M. L., and Kanik I. (2009) Controls on development and diversity of Early Archean stromatolites. *Proc. Natl. Acad. Sci. USA, 106,* 9458–9555, DOI: 10.1073/pnas.0903323106.

Allwood A. C., Rosing M. T., Flannery D. T., Hurowitz J. A., and Heirwegh C. M. (2018) Reassessing evidence of life in 3700-million-year-old rocks of Greenland. *Nature, 563,* 241–244.

Amann R. I., Ludwig W., and Schleifer K. H. (1995) Phylogenetic identification and in situ detection of individual microbial cells without cultivation. *Microbiol. Rev., 59,* 143–169.

Andrews S., Norton I., Salunkhe A. S., Goodluck H., Aly W. S. M., Mourad-Agha H., and Cornelis P. (2013) Control of iron metabolism in bacteria. In *Metallomics and the Cell.* (L. Banci, ed.), pp. 203–239. Metal Ions in Life Sciences, Vol. 12, Springer, Dordrecht.

Aponte J. C., Dworkin J. P. and Elsila J. E. (2014) Assessing the origins of aliphatic amines in the Murchison meteorite from their compound-specific carbon isotopic ratios and enantiomeric composition.

Geochim. Cosmochim. Acta, 141, 331–345.

Asara J. M., Schweitzer M. H., Freimark L. M., Phillips M., and Cantley L. C. (2007) Protein sequences from mastodon and Tyrannosaurus rex revealed by mass spectrometry. *Science, 316,* 280–285.

Bada J. L., Kvenvolden K. A., and Peterson E. (1973) Racemization of amino acids in bones. *Nature, 245,* 308–310.

Bada J. L., Wang X. S., and Hamilton H. (1999) Preservation of key biomolecules in the fossil record: Current knowledge and future challenges. *Philos. Trans. R. Soc., B, 354,* 77–87.

Barber D. J. and Scott E. R. D. (2002) Origin of supposedly biogenic magnetite in the martian meteorite Allan Hills 84001. *Proc. Natl. Acad. Sci. USA, 99(10),* 6556–6561.

Barghoorn E. S. and Tyler S. M. (1965) Microorganisms from the Gunflint chert. *Science, 147,* 563–577.

Bauersachs T., Weidenbach K., Schmitz R. A., and Schwark L. (2015) Distribution of glycerol ether lipids in halophilic, methanogenic, and hyperthermophilic archaea. *Org. Geochem., 83–84,* 101–108.

Bazylinski D. A., Frankel R. B., Heywood B. R., Mann S., King J. W., Donaghay P. L., and Hanson A. K. (1995) Controlled biomineralization of magnetite (Fe_3O_4) and griegite (Fe_3S_4) in a magnetotactic bacterium. *Appl. Environ. Microbiol., 61(9),* 3232–3239.

Benner S. A., Kim H.-J., Kim M.-J., and Ricardo A. (2010) Planetary organic chemistry and the origins of biomolecules. In *The Origin of Life, 4th edition* (D. Deamer and J. W. Szostak, eds.), pp. 67–87. Cold Spring Harbor, New York.

Bergauer K., Fernandez-Guerra A., Garcia J. A., Sprenger R. R., Stepanauskas R., Pachiadaki M. G., Jensen O. N., and Herndl G. J. (2017) Organic matter processing by microbial communities throughout the Atlantic water column as revealed by metaproteomics. *Proc. Natl. Acad. Sci. USA, 115 (3),* E400-E408, DOI: 10.1073/pnas.1708779115.

Berner R. A. and Canfield D. E. (1989) A new model for atmospheric oxygen over Phanerozoic time. *Am. J. Sci., 289,* 333–361.

Bernstein J. A., Khodursky A. B., Lin P. H., Lin-Chao S., and Cohen S. N. (2002) Global analysis of mRNA decay and abundance in *Escherichia coli* at single-gene resolution using two-color fluorescent DNA microarrays. *Proc. Natl. Acad. Sci. USA, 99(15),* 9697–9702.

Bhaya D., Davison M., and Barrangou R. (2011) CRISPR-Cas systems in bacteria and archaea: Versatile small RNAs for adaptive defense and regulation. *Annu. Rev. Genet., 45,* 273–297.

Bidle K. D. et al. (2007) Fossil genes and microbes in the oldest ice on Earth. *Proc. Natl. Acad. Sci. USA, 104 (33),* 13455–13460.

Boere A., Rijpstra W., De Lange G., Sinninghe Damsté J., and Coolen M. (2011) Preservation potential of ancient plankton DNA in Pleistocene marine sediments. *Geobiology, 9,* 377–393.

Bouvier F., Rahier A., and Camara B. (2005) Biogenesis, molecular regulation, and function of plant isoprenoids. *Prog. Lipid Res., 44(6),* 347–429.

Brasier M. D., Green O. R., and McLoughlin N. (2004) Characterization and critical testing of potential microfossils from the early Earth: The Apex 'microfossil debate' and its lessons for Mars sample return. *Intl. J. Astrobiol., 3,* 139–150.

Brassell S. C., Eglinton G., Marlowe I. T., Pflaumann U., Sarnthein M. (1986) Molecular stratigraphy: A new tool for climatic assessment. *Nature, 320,* 129–133.

Briggs D. E. G. (1999) Molecular taphonomy of animal and plant cuticles: Selective preservation and diagenesis. *Philos. Trans. R. Soc., B , 354,* 7–17.

Briggs D. E. G. and Eglinton G. (1994) Chemical traces of ancient life. *Chem. Br., 31,* 907–912.

Briggs D. E. G., Evershed R. P., and Lockheart M. J. (2000) The biomolecular paleontology of continental fossils. *Paleobiology, 26,* 169–193.

Briggs A. W., Stenzel U., Johnson P. L., Green R. E., Kelso J., Prüfer K., Meyer M., Krause J., Ronan M. T., and Lachmann M. (2007) Patterns of damage in genomic DNA sequences from a Neanderthal. *Proc. Natl. Acad. Sci. USA, 104,* 14616–14621.

Brocks J. J. (2018) The transition from a cyanobacterial to algal world and the emergence of animals. *Emerging Top. Life Sci.,* ETLS20180039, DOI: 10.1042/ETLS20180039.

Brocks J. J. and Banfield J. (2009) Unravelling ancient microbial history with community proteogenomics and lipid geochemistry. *Nature Rev. Microbiol., 7,* 601–609.

Brocks J. and Schaeffer P. (2008) Okenane, a biomarker for purple sulfur bacteria (Chromatiaceae), and other new carotenoid derivatives from the 1640 Ma Barney Creek Formation. *Geochim. Cosmochim. Acta, 72,* 1396–1414.

Brocks J. J and Summons R. E. (2003) Sedimentary hydrocarbons, biomarkers for early life. In *Treatise in Geochemistry, Vol. 10: Biogeochemistry* (D. M. Karl and W. H. Schlesinger, eds.), pp. 63–115, Elsevier-Pergamon, Oxford, DOI: 10.1016/B978-0-08-095975-7.00803-2.

Brocks J. J., Logan G. A., Buick R., and Summons R. E. (1999) Archean molecular fossils and the early rise of eukaryotes. *Science, 285(5430),* 1033–1036.

Brocks J. J., Buick R., Summons R. E., and Logan G. A. (2003) A reconstruction of Archean biological diversity based on molecular fossils from the 2.78 to 2.45 billion-year-old Mount Bruce Supergroup, Hamersley Basin, Western Australia. *Geochim. Cosmochim. Acta, 67,* 4321–4335.

Buick R. (1992) The antiquity of oxygenic photosynthesis: Evidence from stromatolites in sulphate-deficient Archaean lakes. *Science, 255,* 74–77.

Burdige D. J. (2007) Preservation of organic matter in marine sediments: Controls, mechanisms, and an imbalance in sediment organic carbon budgets? *Chem. Rev., 107,* 467–485.

Burne R. V. and L. S. Moore (1987) Microbialites: Organosedimentary deposits of benthic microbial communities. *Palaios, 2,* 241–254.

Callahan M. P., Smith K. E., Cleaves H. J., Ruzicka J., Stern J. C., Glavin D. P., House C. H., and Dworkin J. P. (2011) Carbonaceous meteorites contain a wide range of extraterrestrial nucleobases. *Proc. Natl. Acad. Sci. USA, 108(34),* 13995–13998.

Canfield D. E. (1998) A new model for Proterozoic ocean chemistry. *Nature, 396,* 450–453.

Carvalhais L. C., Dennis P. G., Tyson G. W., and Schenk P. M. (2012) Application of metatranscriptomics to soil environments. *J. Microbiol. Methods, 91(2),* 246–251.

Chaffron S., Rehrauer H., Pernthaler J., and von Mering C. (2010) A global network of coexisting microbes from environmental and whole-genome sequence data. *Genome Res., 20(7),* 947–959.

Chevalier N., Savoye N., Dubois S., Lama Lama M., David V., Lecroart P., le Ménach K., and Budzinski H. (2015) Precise indices based on *n*-alkane distribution for quantifying sources of sedimentary organic matter in coastal systems. *Org. Geochem., 88,* 69–77.

Chyba C. F. and Sagan C. (1992) Endogenous production, exogenous delivery, and impact-shock synthesis of organic molecules: An inventory for the origins of life. *Nature, 355,* 125–132.

Collinson M. E., van Bergen P. M., Scott A. C., and de Leeuw J. W. (1994) The oil-generating potential of plants from coal and coal-bearing strata through time: A review with new evidence from carboniferous plants. In *Coal and Coal-Bearing Strata as Oil-Prone Source Rocks* (A. C. Scott and A. J. Fleet, eds.), pp. 31–70. Geol. Soc. London Special Publ. 77.

Collinson M. E. B., Mösle P., Finch A., Scott C., and Wilson R. (1998) The preservation of plant cuticle in the fossil record: A chemical and microscopical investigation. *Ancient Biomol., 2,* 251–265.

Conrad P. G. and Nealson K. H. (2001) A non-Earthcentric approach to life detection. *Astrobiology, 1(1),* 15–24.

Conte M. H., Thompson A., Lesley D., and Harris R. P. (1998) Genetic and physiological influences on the alkenone/alkenoate versus growth temperature relationship in *Emiliania huxleyi* and *Gephyrocapsa oceanica. Geochim. Cosmochim. Acta, 62,* 51–68.

Corinaldesi C., Dell'Anno A., and Danovaro R. (2008) Early diagenesis and trophic role of extracellular DNA in different benthic ecosystems. *Limnol. Oceanogr., 52,* 1710–1717.

Cranwell P. A. (1981) Diagenesis of free and bound lipids in terrestrial detritus deposited in a lacustrine sediment. *Org. Geochem., 3(3),* 79–89.

Cranwell P. A. (1985) Long-chain unsaturated ketones in recent lacustrine sediments. *Geochim. Cosmochim. Acta, 49,* 1545–1551.

Crick F. H. (1970) Central dogma of molecular biology. *Nature, 227(5258),* 561–563.

Cronin J. R. and Chang S. (1993) Organic matter in meteorites: Molecular and isotopic analyses of the Murchison meteorite. In *The Chemistry of Life's Origins* (J. M. Greenberg and V. Pirronello, eds.), pp. 209–258. Kluwer, Dordrecht.

Cronin J. R., Pizzarello S., and Cruikshank D. P. (1988) Organic matter in carbonaceous chondrites, planetary satellites, asteroids, and comets. In *Meteorites and the Early Solar System* (J. M. Greenberg et al., eds.), pp. 819–857. Univ. of Arizona, Tucson.

Dahm R. (2005) Friedrich Miescher and the discovery of DNA. *Dev.*

Biol., 278(2), 274–288.

DeAngelis K. M., Silver W. L., Thompson A. W., and Firestone M. K. (2010) Microbial communities acclimate to recurring changes in soil redox potential status. *Environ. Microbiol., 12,* 3137–3149.

Dekas A. E., Poretsky R. S., and Orphan V. J., (2009) Deep-sea archaea fix and share nitrogen in methane-consuming microbial consortia. *Science, 326(5951),* 422–426.

de Leeuw J. W., van Bergen P. F., van Aarssen B. G. K., Gatellier J.-P. L. A., Sinninghe Damsté J. S., and Collinson M. E. (1991) Resistant biomacromolecules as major contributors to kerogen. Philos. Trans. R. Soc., B, 333, 329–333.

DeLeon-Rodriguez N., Lathem T. L., Rodriguez-R L. M., Barazesh J. M., Anderson B. E., Beyersdorf A. J., Ziemba L. D., Bergin M., Nenes A. and Konstantinidis K. T. (2013) Microbiome of the upper troposphere: Species composition and prevalence, effects of tropical storms, and atmospheric implications. *Proc. Natl. Acad. Sci. USA, 110(7),* 2575–2580.

Derenne S. and Largeau C. (2001) A review of some important families of refractory macromolecules: Composition, origin, and fate in soils and sediments. *Soil Sci., 166,* 833–847.

Des Marais D. J. (2003) The biogeochemistry of hypersaline microbial mats illustrates the dynamics of modern microbial ecosystems and the early evolution of the biosphere. *Biol. Bull., 204(2),* 160–167.

Des Marais D. J. (2010) Marine hypersaline *Microcoleus*-dominated cyanobacterial mats in the saltern at Guerrero Negro, Baja California Sur, Mexico. In *Cellular Origin, Life in Extreme Habitats and Astrobiology, Vol. 14: Microbial Mats* (J. Seckbach, ed.), pp. 401–420. Springer, Dordrecht.

Deutscher M. P (2006) Degradation of RNA in bacteria: Comparison of mRNA and stable RNA. *Nucleic Acids Res., 34(2),* 659–666.

Diefendorf A. F., Freeman K. H., Wing S. L., and Graham H. V. (2011) Production of *n*-alkyl lipids in living plants and implications for the geologic past. *Geochim. Cosmochim. Acta, 75,* 7472–7485.

Dobell C. (1932) *Antony van Leeuwenhoek and His "Little Animals."* Harcourt Brace, New York. 512 pp.

Dong H.-P., Hong Y.-G., Lu S., and Xie L.-Y. (2014) Metaproteomics reveals the major microbial players and their biogeochemical functions in a productive coastal system in the northern South China Sea. *Environ. Microbiol. Rept., 6,* 683–695.

Druschel G. K., Emerson D., Sutka R., Suchecki P., and Luther G. W. (2008) Low-oxygen and chemical kinetic constraints on the geochemical niche of neutrophilic iron(II) oxidizing microorganisms. *Geochim. Cosmochim. Acta, 72,* 3358–3370.

Duda J.-P., Van Kranendonk M. J., Thiel V., Ionescu D., Strauss H., Schafer N., and Reitner J. (2016) A rare glimpse of Paleoarchean life: Geobiology of an exceptionally preserved microbial mat facies from the 3.4 Strelley Pool Formation, Western Australia. *PLoS One, 11(1),* e0147629, DOI: 10.1371/journal.pone.0147629.

Dusenbery D. B. (2009) *Living at Micro Scale: The Unexpected Physics of Being Small.* Harvard Univ., Cambridge. 416 pp.

Eglinton G. and Hamilton R. J. (1967) Leaf epicuticular waxes. *Science, 156,* 1322–1335.

Eglinton G. and Logan G. A. (1991) Molecular preservation. *Philos. Trans. R. Soc., B, 333,* 315–328.

Ehrenfreund P., Glavin D. P., Botta O., Cooper G., and Bada J. L. (2001) Extraterrestrial amino acids in Orgueil and Ivuna: Tracing the parent body of CI type carbonaceous chondrites. *Proc. Natl Acad. Sci. USA, 98,* 2138–2141.

Eigenbrode J. L. and Freeman K. H. (2006) Late Archean rise of aerobic microbial ecosystems. *Proc. Natl. Acad. Sci. USA, 103,* 15759–19764.

Elsila J. E., Charnley S. B., Burton A. S., Glavin D. P., and Dworkin J. P. (2012) Compound-specific carbon, nitrogen, and hydrogen isotopic ratios for amino acids in CM and CR chondrites and their use in evaluating potential formation pathways. *Meteoritics & Planet. Sci., 47,* 1517–1536.

Epstein S., Krishnamurthy R. V., Cronin J. R., Pizzarello S., and Yuen G. U. (1987) Unusual stable isotope ratios in amino acid and carboxylic acid extracts from the Murchison meteorite. *Nature, 326,* 477–479.

Falkow S. (1988) Molecular Koch's postulates applied to microbial pathogenicity. *Rev. Infect. Dis., 10(2),* S274–276.

Farquhar J., Bao H., and Thiemens M. (2000) Atmospheric influence of Earth's earliest sulfur cycle. *Science, 289(5480),* 756–758.

Fayek M. J., Utsunomiya S., Pfiffner S. M., Anovitz L. M., White D. C.,

Riciputi L. R., Ewing R. C., and Stadermann F. J. (2005) Nanoscale chemical and isotopic characterization of *Geobacter sulfurreducens* surfaces and bio-precipitated uranium minerals. *Can. Mineral., 43,* 1631–1641.

Flannery E. N. and George S. C. (2014) Assessing the syngeneity and indigeneity of hydrocarbons in the ~1.4 Ga Velkerri Formation, McArthur Basin, using slice experiments. *Org. Geochem., 77,* 115–125.

Fleischmann R. D., Adams M. D., White O., Clayton R. A., Kirkness E. F., Kerlavage A. R., Bult C. J., Tomb J. F., Dougherty B. A., and Merrick J. M. (1995) Whole-genome random sequencing and assembly of *Haemophilus influenzae* Rd. *Science, 269(5223),* 496–512.

Flesch G. and Rohmer M. (1988) Prokaryotic hopanoids: The biosynthesis of the bacteriohopane skeleton. *FEBS J., 175(2),* 405–411.

Fordyce S. L., Avila-Arcos M. C., Rasmussen M., Cappellini E., Romero-Navarro J. A., Wales N., Alquezar-Planas D. E., Penfield S., Brown T. A., Vielle-Calzada J. P., and Montiel R. (2013) Deep sequencing of RNA from ancient maize kernels. *PLoS One, 8(1),* e50961.

Freeman K. H. and Wakeham S. G. (1992) Variations in the distributions and isotopic compositions of alkenones in Black Sea particles and sediments. *Org. Geochem., 19,* 277–285.

Freilich S., Kreimer A., Meilijson I., et al. (2010) The large-scale organization of the bacterial network of ecological co-occurrence interactions. *Nucleic Acids Res., 38(12),* 3857–3868.

French J. E. and Blake D. F. (2016) Discovery of naturally etched fission tracks and alpha-recoil tracks in submarine glasses: Reevaluation of a putative biosignature for Earth and Mars. *Intl. J. Geophys., 2016,* 2410573, DOI: 10.1155/2016/2410573.

Fry J. C., Parkes R. J., Cragg B. A., Weightman A. J., and Webster G. (2008) Prokaryotic biodiversity and activity in the deep subseafloor biosphere. *FEMS Microbiol. Ecol., 66,* 181–196.

Gagosian R. B. and Peltzer E. T. (1986) The importance of atmospheric input of terrestrial organic material to deep-sea sediments. *Org. Geochem., 10,* 661–669.

García Ruiz J. M., Carnerup A., Christy A. G., Welham N. J., and Hyde S. T. (2002) Morphology: An ambiguous indicator of biogenicity. *Astrobiology, 2,* 353–369.

Garcia-Ruiz J. M., Hyde S. T., Carnerup A. M., Christy A. G., Van Kranendonk M. J., and Welham N. J. (2003) Self-assembled silica-carbonate structures and detection of ancient microfossils. *Science, 302,* 1194–1197.

Gischler E., Gibson M. A., and Oschmann W. (2008) Giant Holocene freshwater microbialites, Laguna Bacalar, Quintana Roo, Mexico. *Sedimentology, 55,* 1293–1309.

Glavin D. P., Dworkin J. P., Aubrey A., Botta O., Doty J. H. III, Martins Z., and Bada J. L. (2006) Amino acid analyses of Antarctic CM2 meteorites using liquid chromatograph-time of flight-mass spectrometry. *Meteoritics & Planet. Sci., 41,* 889–902.

Golden D. C., Ming D. W., Morris R. V., Brearley A., Lauer H. V., Treiman A. H., Zolensky M. E., Schwandt C. S., Lofgren G. E., and McKay G. A. (2004) Evidence for exclusively inorganic formation of magnetite in martian meteorite ALH84001. *Am. Mineral., 89,* 681–695.

Golden J., McMillan M., Downs R. T., Hystad G., Stein H. J., Zimmerman A., Sverjensky D. A., Armstrong J., and Hazen R. M. (2013) Rhenium variations in molybdenite (MoS$_2$): Evidence for progressive subsurface oxidation. *Earth Planet. Sci. Lett., 366,* 1–5.

Goñi M. A., Ruttenberg K. C., and Eglinton T. I. (1997) Source and contribution of terrigenous organic carbon to surface sediments in the Gulf of Mexico. *Nature, 389,* 275–278.

Gottesman S. and Storz G. (2011) Bacterial small RNA regulators: Versatile roles and rapidly evolving variations. *Cold Spring Harbor Perspect. Biol., 3(12),* a003798.

Goutx M., Wakcham S. G., Lee C., Duflos M., Guigue C., Liu Z., Moriceau B., Sempére R., Tedetti M., and Xue. J. (2007) Composition and degradation of marine particles with different settling velocities in the northwestern Mediterranean Sea. *Limnol. Oceanogr., 52(4),* 1645–1664.

Graham H-P., Tegner C., and Lesher C. E. (2019) Strontium isotope systematics for plagioclase of the Skaergaard intrusion (East Greenland): A window to crustal assimilation, differentiation, and magma dynamics. *Geology, 47(4),* 313–316.

Greene A. and Madgwick J. C. (1991) Microbial formation of manganese oxides. *Appl. Environ. Microbiol., 57,* 1114–1120.

Grotzinger J. P. (1990) Geochemical model for Proterozoic stromatolite decline. *Am. J. Sci., 290,* 80–103.

Gupta N. S., Briggs D. E. G., Collinson M. E., Evershed R. P., Michels R., and Pancost R. D. (2007) Molecular preservation of plant and insect cuticles from the Oligocene Enspel Formation, Germany: Evidence against derivation of aliphatic polymer from sediment. *Org. Geochem, 38,* 404–418.

Hambraeus G., von Wachenfeldt C., and Hederstedt L. (2003) Genome-wide survey of mRNA half-lives in *Bacillus subtilis* identifies extremely stable mRNAs. *Mol. Genet. Genomics, 269(5),* 706–714.

Han T. M. and Runnegar B. (1992) Megascopic eukaryotic algae from the 2.1-billion-year-old Negaunee Iron-Formation, Michigan. *Science, 257,* 232–235.

Handelsman J., Rondon M. R., Brady S. F., Clardy J., and Goodman R. M. (1998) Molecular biological access to the chemistry of unknown soil microbes: A new frontier for natural products. *Chem. Biol., 5(10),* R245–R249.

Hannich J. T., Umebayashi K., and Riezman H. (2011) Distribution and functions of sterols and sphingolipids. *Cold Spring Harbor Perspect. Biol., 3(5).*

Hayes J. M. (1983) Geochemical evidence bearing on the origin of aerobiosis, a speculative hypothesis. In *Earth's Earliest Biosphere, Its Origin and Evolution* (J. W. Schopf, ed.), pp. 291–301. Princeton Univ., Princeton, New Jersey.

Hayes J. M. (2001) Fractionation of carbon and hydrogen isotopes in biosynthetic processes. *Rev. Mineral. Geochem., 43,* 225–277.

Hazen R. M. (2013) Paleomineralogy of the Hadean Eon: A preliminary list. *Am. J. Sci., 313,* 807–843.

Hazen R. M. (2018) Titan mineralogy: A window on organic mineral evolution. *Am. Mineral., 103,* 341–342.

Hazen R. M. (2019) An evolutionary system of mineralogy: Proposal for a classification based on natural kind clustering. *Am. Mineral., 104,* 810–816.

Hazen R. M., Papineau D., Bleeker W., Downs R. T., Ferry J., McCoy T., Sverjensky D. A., and Yang H. (2008) Mineral evolution. *Am. Mineral., 93,* 1693–1720.

Hazen R. M., Ewing R. C., and Sverjensky D. A. (2009) Evolution of uranium and thorium minerals. *Am. Mineral., 94 (10),* 1293–1311, DOI: 10.2138/am.2009.3208.

Hazen R. M., Downs R. T., Jones A. P., and Kah L. (2013a) The mineralogy and crystal chemistry of carbon. In *Carbon in Earth* (R. M. Hazen et al., eds), pp. 7–46. Mineralogical Society of America, Washington, DC.

Hazen R. M., Sverjensky D. A., Azzolini D., Bish D. L., Elmore S., Hinnov L., and Milliken R. E. (2013b) Clay mineral evolution. *Am. Mineral., 98,* 2007–2029.

Hazen R. M., Liu X.-M., Downs R. T., Golden J. J., Pires A. J., Grew E. S., Hystad G., Estrada C., and Sverjensky D. A. (2014) Mineral evolution: Episodic metallogenesis, the supercontinent cycle, and the coevolving geosphere and biosphere. In *Building Exploration Capability for the 21st Century.* (K. D. Kelley and H. C. Golden, eds.) pp. 1–15. Society of Economic Geologists Special Publication Vol. 18, Society of Economic Geologists, Littleton, Colorado.

Hazen R. M., Grew E. S., Downs R. T., Golden J., and Hystad G. (2015a) Mineral ecology: Chance and necessity in the mineral diversity of terrestrial planets. *Can. Mineral., 53,* 295–323.

Hazen R. M., Hystad G., Downs R. T., Golden J., Pires A., and Grew E. S. (2015b) Earth's "missing" minerals. *Am. Mineral., 100,* 2344–2347.

Hedges J. I. and Keil R. G. (1995) Sedimentary organic matter preservation: An assessment and speculative synthesis. *Mar. Chem., 49,* 81–115.

Hedges J. I., Keil R. G., and Benner R. (1997) What happens to terrestrial organic matter in the ocean? *Org. Geochem., 27,* 195–212.

Henrichs S. M. and Farrington J. W. (1987) Early diagenesis of amino acids and organic matter in two coastal marine sediments. *Geochim. Cosmochim. Acta, 51,* 1–15.

Hettich R. L., Pan C., Chourey K., and Giannone R. J. (2013) Metaproteomics: Harnessing the power of high performance mass spectrometry to identify the suite of proteins that control metabolic activities in microbial communities. *Anal. Chem., 85(9),* 4203–4214.

Hofmann H. J. (1976) Precambrian microflora, Belcher Islands, Canada: Significance and systematics. *J. Paleontol., 50,* 1040–1073.

Hofreiter M., Paijmans J. L., Goodchild H., Speller C. F., Barlow A., Fortes G. G., Thomas J. A., Ludwig A., and Collins M. J. (2015) The future of ancient DNA: Technical advances and conceptual shifts. *BioEssays, 37(3),* 284–293.

Holba A. G., Dzou L. I. P., Wood G. D., Ellis L., Adam P., Schaeffer P., Albrecht P., Green T., and Hughes W. B. (2003) Application of tetracyclic polyprenoids as indicators of input from fresh-brackish water environments. *Org. Geochem., 34,* 441–469.

Holland H. D. (2006) The oxygenation of the atmosphere and oceans. *Philos. Trans. R. Soc., B, 361(1470),* 903–915.

Hopmans E. C., Weijers J. W. H., Schefuß E., Herfort L., and Sinninghe Damsté J. S. (2004) A novel proxy for terrestrial organic matter in sediments based on branched and isoprenoid tetraether lipids. *Earth Planet. Sci. Lett., 24,* 107–116.

Horodyski R. J. and Donaldson J. A. (1983) Distribution and significance of microfossils in cherts of the middle Proterozoic Dismal Lakes Group, District of Mackenzie, Northwest Territories. *J. Paleontol., 57,* 271–288.

Horowitz N. H., Bauman A. J., Cameron R. E., Geiger P. J., Hubbard J. S., Shulman G. P., Simmonds P. G., and Westberg K. (1969) Sterile soil from Antarctica: Organic analysis. *Science, 164(3883),* 1054–1056.

Hudson C. M., Lau B. Y., and Williams K. P. (2015) Islander: A database of precisely mapped genomic islands in tRNA and tmRNA genes. *Nucleic Acids Res., 43,* D48–D53.

Hug L. A. and 16 colleagues (2016) A new view of the tree of life. *Nature Microbiol., 1, 16048.*

Hummer D. R., Noll B. C., Hazen R. M., and Downs R. T. (2017) Crystal chemistry of abelsonite, the only known crystalline geoporphyrin. *Am. Mineral., 102,* 1129–1132.

Hummer D. R., Hazen R. M., Golden J. J., Hystad G., Liu C., and Downs R. T. (2020) The oxidation of Earth's crust: Evidence from the evolution of manganese minerals. *Nature Commun.,* in press.

Huson D. H., Auch A. F., Qi J., and Schuster S. C. (2007) MEGAN analysis of metagenomic data. *Genome Res., 17(3),* 377–386, DOI: 10.1101/gr.5969107.

Hystad G., Downs R. T., and Hazen R. M. (2015) Mineral frequency distribution data conform to a LNRE model: Prediction of Earth's "missing" minerals. *Math. Geosci., 47,* 647–661.

Hystad G., Downs R. T., Hazen R. M., and Golden J. J. (2017) Relative abundances for the mineral species on Earth: A statistical measure to characterize Earth-like planets based on Earth's mineralogy. *Math. Geosci., 49(2),* 179–194.

Hystad G., Eleish A., Downs R. T., Morrison S. M., and Hazen R. M. (2019) Bayesian estimation of the number of distinct mineral species in Earth's crust using noninformative priors. *Math. Geosci., 51(4),* 401–417.

Innes H. E., Bishop A. N., Head I. M., and Farrimond P. (1997) Preservation and diagenesis of hopanoids in recent lacustrine sediments of Priest Pot, England. *Org. Geochem., 26,* 565–576.

International Human Genome Sequencing Consortium (2001) Initial sequencing and analysis of the human genome. *Nature, 409(6822),* 860.

Isenbarger T. A., Finney M., Ríos-Velázquez C., Handelsman J., and Ruvkun G. (2008) Miniprimer PCR, a new lens for viewing the microbial world. *Appl. Environ. Microbiol., 74(3),* 840–849.

Jaakola S. T., Ravantti J. J., Oksanen H. M., and Bamford D. H. (2016) Buried alive: Microbe from ancient halite. *Trends Microbiol., 24,* 148–160.

Jain M., Olsen H. E., Paten B., and Akeson M. (2016) The Oxford Nanopore MinION: Delivery of nanopore sequencing to the genomics community. *Genome Boil., 17(1),* 239.

Jarrett A. J. M., Schinteie R., Hope J. M., and Brocks J. J. (2013) Micro-ablation, a new technique to remove drilling fluids and other contaminants from fragmented and fissile rock material. *Org. Geochem., 61,* 57–65.

Jay E., Bambara R., Padmanabhan R., and Wu R. (1974) DNA sequence analysis: A general, simple and rapid method for sequencing large oligodeoxyribonucleotide fragments by mapping. *Nucleic Acids Res., 1(3),* 331–353.

Jimenez-Lopez C., Romanek C. S., and Bazylinski D. A. (2010) Magnetite as a prokaryotic biomarker: A review. *J. Geophys. Res., 115,* G00G03.

Johnson S. S., Anslyn E. V., Graham H. V., Mahaffy P. R., and Ellington A. D. (2018) Fingerprinting non-terran biosignatures. *Astrobiology,*

18(7), 915–922.

Jørgensen N., Kroer N., Coffin R., Yang X.-H., and Lee C. (1993) Dissolved free amino acids, combined amino acids, and DNA as sources of carbon and nitrogen to marine bacteria. *Mar. Ecol. Prog. Ser., 98(1–2),* 135–148.

Juhas M., Roelof van der Meer J., Gaillard M., Harding R. M., Hood D. W., Crook D. W. (2009) Genomic islands: Tools of bacterial horizontal gene transfer and evolution. *FEMS Microbiol. Rev., 33(2),* 376–393.

Kallmeyer J., Pockalny R., Adhikari R. R., Smith D. C., and D'Hondt S. D. (2012) Global distribution of microbial abundance and biomass in subseafloor sediment. *Proc. Natl. Acad. Sci. USA, 109(40),* 16213–16216.

Kenig F., Sinninghe Damsté J. S., Kock-van Dalen A. C., Rijpstra W. I. C., Huc A. Y., and de Leeuw J. W. (1995) Occurrence and origin of mono-, di-, and trimethylalkanes in modern and Holocene cyanobacterial mats from Abu Dhabi, United Arab Emirates. *Geochim. Cosmochim. Acta, 59,* 2999–3015.

Kieft T. L. (2016) Microbiology of the deep continental biosphere. In *Advances in Environmental Microbiology, Vol. 1: Their World: A Diversity of Microbial Environments* (C. J. Hurst, ed.), pp. 225–249. Springer, Switzerland.

Kircher M., Heyn P., and Kelso J. (2011) Addressing challenges in the production and analysis of illumina sequencing data. *BMC Genomics, 12(1),* 382.

Kirchman D. L. (2016) Growth rates of microbes in the oceans. *Annu. Rev. Marine Sci., 8,* 285–309.

Knoll A. H. (1991) Paleobiology of a Neoproterozoic tidal flat/lagoonal complex: The Draken Conglomerate Formation, Spitsbergen. *J. Paleontol., 65,* 531–570.

Knoll A. H. (1992) The early evolution of eukaryotes: A geological perspective. *Science, 256,* 622–627.

Knoll A. H. (1996) Archean and Proterozoic paleontology. In *Principles and Applications Vol. 1: Palynology* (J. Jasonius and D. C. MacGregor, eds.), pp. 51–80. American Association of Stratigraphic Palynologists Foundation, Tulsa.

Knoll A. H. and Barghoorn E. S. (1977) Archean microfossils showing cell division from the Swaziland System of South Africa. *Science, 198,* 396–398.

Knoll A. H. and Swett K. (1990) Carbonate deposition during the Late Proterozoic Era: An example from Spitsbergen. *Am. J. Sci., 290A,* 104–132.

Knoll A. H., Strother P. K., and Rossi S. (1988) Distribution and diagenesis of microfossils from the Lower Proterozoic Duck Creek Dolomite, Western Australia. *Precambrian Res., 38,* 257–279.

Koboldt D. C., Steinberg K. M., Larson D. E., Wilson R. K., and Mardis E. R. (2013) The next-generation sequencing revolution and its impact on genomics. *Cell, 155(1),* 27–38.

Koga Y. and Nakano M. (2008) A dendrogram of archaea based on lipid component parts composition and its relationship to rRNA phylogeny. *Syst. Appl. Microbiol., 31,* 169–182.

Koga Y., Nishihara M., Morii H., and Akagawa-Matsushita M. (1993) Ether polar lipids of methanogenic bacteria: Structures, comparative aspects, and biosynthesis. *Microbiol. Rev., 57(1),* 164–182.

Kolattukudy P. E. (1976) *The Chemistry and Biochemistry of Natural Waxes.* Elsevier, Amsterdam. 459 pp.

Koopmans M. P., Köster J., van Kaam-Peters H. M. E., Kenig F., Schouten S., Hartgers W. A., de Leeuw J. W., and Sinninghe Damste J. S. (1996) Diagenetic and catagenetic products of isorenieratene: Molecular indicators for photic zone anoxia. *Geochim. Cosmochim. Acta, 60,* 4467–4496.

Koopmans M. P., Rijpstra W. I. C., Klapwijk M. M., de Leeuw J. W., Lewan M. D., and Sinninghe Damsté J. S. (1999) A thermal and chemical degradation approach to decipher pristane and phytane precursors in sedimentary organic matter. *Org. Geochem., 30,* 1089–1104.

Koren S. and Phillippy A. M. (2015) One chromosome, one contig: Complete microbial genomes from long-read sequencing and assembly. *Curr. Opin. Microbiol., 23,* 110–120.

Kuan Y.-J., Charnley S. B., Huang H.-C., Tseng W. L., and Kisiel Z. (2003) Interstellar glycine. *Astrophys. J., 593,* 848.

Kvenvolden K., Lawless J., Pering K., Peterson E., Flores J., Ponnamperuma C., Kaplan I. R., and Moore C. (1970) Evidence for extraterrestrial amino-acids and hydrocarbons in the Murchison meteorite. *Nature, 228(5275),* 923.

Lane D. J., Pace B., Olsen G. J., Stahl D. A., Sogin M. L., and Pace N. R. (1985) Rapid determination of 16S ribosomal RNA sequences for phylogenetic analyses. *Proc. Natl. Acad. Sci. USA, 82(20),* 6955–6959.

Lange B. M., Rujan T., Martin W., and Croteau R. (2000) Isoprenoid biosynthesis: The evolution of two ancient and distinct pathways across genomes. *Proc. Natl. Acad. Sci. USA, 97(24),* 13172-13177.

Laval B., Cady S. L., Pollack J. C., McKay C. P., Bird J. S., Grotzinger J. P., Ford D. C., and Bohm H. R. (2000) Modern freshwater microbialite analogues for ancient dendritic reef structures. *Nature, 407,* 626–629.

Lima-Mendez G., Faust K., Henry N., et al. (2015) Determinants of community structure in the global plankton interactome. *Science, 348,* 6237.

Liu C., Hystad G., Golden J. J., Hummer D. R., Downs R. T., Morrison S. M., Grew E. S., and Hazen R. M. (2017) Chromium mineral ecology. *Am. Mineral., 102,* 612–619.

Liu X.-M., Kah L. C., Knoll A. H., Cui H., Kaufman A. J., Shahar A., and Hazen R. M. (2016) Tracing Earth's CO_2 evolution using Zn/Fe ratios in marine carbonate. *Geochem. Perspect. Lett., 2,* 24–34.

Love G. D., Grosjean E., Stalvies C., Fike D. A., Grotzinger J. P., Bradley A. S., Kelly A. E., Bhatia M., Meredith W., Snape C. E., Bowring S. A., Condon D. J., and Summons R. E. (2009) Fossil steroids record the appearance of Demospongiae during the Cryogenian period. *Nature, 457,* 718–721.

Lovely D. R., Phillips E. J. P., Gorby Y. A., and Landa E. R. (1991) Microbial reduction of uranium. *Nature, 350,* 413–416.

Lowenstein T. K. (2012) Microorganisms in evaporites: Review of modern geomicrobiology. In *Advances in Understanding the Biology of Halophilic Microorganisms* (R. H. Vreeland, ed.), pp. 117–139. Springer, Dordrecht.

Lunine J. I. (2006) Physical conditions on the early Earth. *Philos. Trans. R. Soc., B, 361(1474),* 1721–1731.

Lupatini M., Suleiman A. K. A., Jacques R. J. S., et al. (2014) Network topology reveals high connectance levels and few key microbial genera within soils. *Front. Environ. Sci., 2,* DOI: 10.3389/fenvs.2014.00010.

Ma B., Wang H., Dsouza M., et al. (2016) Geographic patterns of co-occurrence network topological features for soil microbiota at continental scale in eastern China. *ISME J., 10,* 1891–1901.

Mackenzie A. S. and Maxwell J. R. (1981) Assessment of thermal maturation in sedimentary rocks by molecular measurements. In *Organic Maturation Studies and Fossil Fuel Exploration* (J. Brooks, ed.), pp. 239–254. Academic, San Diego.

Maduireira L. A. S., Conte M. H., and Eglinton G. (1995) Early diagenesis of lipid biomarker compounds in North Atlantic sediments. *Paleoceanography, 10(3),* 627–642, DOI: 10.1029/94PA03299.

Marshall S. M., Murray A. R., and Cronin L. (2017) A probabilistic framework for identifying biosignatures using Pathway Complexity. *Philos. Trans. R. Soc., A, 375,* 20160342.

Mayer L. M. (1994) Surface area control of organic carbon accumulation in continental shelf sediments *Geochim. Cosmochim. Acta, 58,* 1271–1284.

Mayer L. M. and Rice D. L. (1992) Early diagenesis of proteins: A seasonal study. *Limnol. Oceanogr., 37,* 280–295.

Maynard-Casely H. E., Cable M. L., Malaska M. J., Vu T. H., Choukroun M., and Hodyss R. (2018) Prospects for mineralogy on Titan. *Am. Mineral., 103(3),* 343-349.

McIntyre A. D., ed. (2010) *Life in the World's Oceans: Diversity, Distribution, and Abundance.* Blackwell, Oxford. 384 pp.

Metzger P., Casadevall E., Pouet M. J., and Pouet Y. (1985) Structures of some botryococcenes: Branched hydrocarbons from the b-race of the green alga *Botryococcus braunii. Phytochemistry, 24,* 2995–3002.

Meyers P. A. and Ishiwatari R. (1993) Lacustrine organic geochemistry — An overview of indicators of organic matter sources and diagenesis in lake sediments. *Org. Geochem., 20,* 867–900.

Michalski J. R., Onstott T. C., Mojzsis S. J., Mustard J., Chan Q. H., Niles P. B., and Johnson S. S. (2018) The martian subsurface as a potential window into the origin of life. *Nature Geosci., 11,* 21–26.

Mormile M. R., Biesen M. A., Gutierrez M. C., Ventosa A., Pavlovich J. B., Onstott T. C., and Frederickson J. K. (2003) Isolation of Halobacterium salinarum retrieved directly from halite brine inclusions. *Environ. Microbiol., 5,* 1094–1102.

Moore L. S. and Burne R. V. (1994) The modern thrombolites of Lake

Clifton, Australia. In *Phanerozoic Stromatolites II* (J. Bertrand-Sarfati and C. Monty, eds.), pp. 3–29. Kluwer, Dordrecht.

Morrison S. M., Runyon S. E., and Hazen R. M. (2018a) The paleomineralogy of the Hadean Eon revisited. *Life, 8(4),* 64.

Morrison S. M., Downs R. T., Blake D. F., Vaniman D. T., Ming D. W., Rampe E. B., Bristow T. F., Achilles C. N., Chipera S. J., Yen A. S., Morris R. V., Treiman A. H., Hazen R. M., Sarrazin P. C., Fendrich K. V., Morookian J. M., Farmer J. D., Des Marais D. J., and Craig P. I. (2018b) Crystal chemistry of martian minerals from Bradbury Landing through Naukluft Plateau, Gale crater, Mars. *Am. Mineral., 103 (6),* 857-871.

Nealson K. H. (2006) The manganese-oxidizing bacteria. In *The Prokaryotes* (M. Dworkin et al., eds.), pp. 222–231. Springer, New York.

Neveu M. et al. (2018) The ladder of life detection. *Astrobiology, 18(11),* 1375–1402.

Nguyen R. T. and Harvey H. R. (2001) Preservation of protein in marine systems: Hydrophobic and other noncovalent associations as major stabilizing forces. *Geochim. Cosmochim. Acta, 65(9),* 1467–1480.

Nishimura M. and Baker E. W. (1986) Possible origin of *n*-alkanes with a remarkable even to odd predominants in recent marine sediments. *Geochim. Cosmochim. Acta, 50,* 299–305.

Noble R. A., Alexander R., Kagi R. I., and Knox J. (1985) Tetracyclic diterpenoid hydrocarbons in some Australian coals, sediments and crude oils. *Geochim. Cosmochim. Acta, 49(10),* 2141–2147.

Noffke N. (2010) *Microbial Mats in Sandy Deposits from the Archean Era to Today.* Springer-Verlag, Heidelberg. 193 pp.

Noffke N., Gerdes G., Klenke T., and Krumbein W. E. (2001) microbially induced sedimentary structures: A new category within the classification of primary sedimentary structures. *J. Sediment. Res., 71(5),* 649.

Nutman A. P., Bennett V. C., Friend C. R. L., Van Kranendonk M. J., and Chivas A. R. (2016) Rapid emergence of life shown by discovery of 3,700-million-year-old microbial structures. *Nature, 537,* 535–538.

Oehler J. H. (1976) Experimental studies in Precambrian paleobiology: Structural and chemical changes in blue-green algae during simulated fossilization in synthetic chert. *Geol. Soc. Am. Bull., 87,* 117–129.

Orem W. H. and Hatcher P. G. (1987) Early diagenesis of organic matter in a sawgrass peat from the Everglades, Florida. *Intl. J. Coal Geol., 8,* 33–54.

Otto A. and Simoneit B. R. T. (2001) Chemosystematics and diagenesis of terpenoids in fossil conifer species and sediment from the Eocene Zeitz formation, Saxony, Germany. *Geochim. Cosmochim. Acta, 65(20),* 3505–3527.

Ourisson G. and Albrecht P. (1992) Hopanoids. 1 Geohopanoids: The most abundant natural products on Earth? *Acc. Chem. Res., 25,* 398–402.

Ourisson G., Rohmer M., and Poralla K. (1987) Prokaryotic hopanoids and other polyterpenoid sterol surrogates. *Annu. Rev. Microbiol., 41,* 301–334.

Padmanab R. and Wu R. (1972) Nucleotide sequence analysis of DNA use of oligonucleotides of defined sequence as primers in DNA sequence analysis. *Biochem. Biophys. Res. Commun., 48(5),* 1295.

Pagani M. (2002) The alkenone-CO$_2$ proxy and ancient atmospheric carbon dioxide. *Philos. Trans. R. Soc., A, 360,* 609–632.

Pagani M., Arthur M. A., and Freeman K. H. (1999) The Miocene evolution of atmospheric carbon dioxide. *Paleoceanography, 14,* 273–292.

Pancost R. D. and Boot C. S. (2004) The palaeoclimatic utility of terrestrial biomarkers in marine sediments. *Mar. Chem., 92,* 239–261.

Parrish J. T. (1998) *Interpreting Pre-Quaternary Climate from the Geologic Record.* Columbia Univ., New York. 348 pp.

Payne A., Holmes N., Rakyan V., and Loose M. (2018) Whale watching with BulkVis: A graphical viewer for Oxford Nanopore bulk FAST5 files. *bioRxiv, Genomics,* 312256.

Peat C. R., Muir M. B., Plumb K. A., McKirdy D. M., and Norvick M. S. (1978) Proterozoic microfossils from the Roper Group, Northern Territory, Australia. *BMR J. Aust. Geol. Geophys., 3,* 1–17.

Pekov I. V., Chukanov N. V., Belakovskiy D. I., Lykova I. S., Yapaskurt V. O., Zubkova N. V., Shcherbakova E. P., and Britvin S. N. (2016) Tinnunculite IMA 2015-021a. *Mineral. Mag., 80(1),* 199–205.

Perasso R., Baroin A., Liang Huqu J., Bachellerie P., and Adoutte A. (1989) Origin of the algae. *Nature, 339,* 142–144.

Perry G. J., Volkman J. K., Johns R. B., and Bevor H. J. Jr. (1979) Fatty acids of bacterial origin in contemporary marine sediments. *Geochim. Cosmochim. Acta, 43,* 1715–1725.

Peters K. E., Rohrback B. G., and Kaplan I. R. (1981) Geochemistry of artificially heated humic and sapropelic sediments — I: Protokerogen. *AAPG Bull., 65,* 688–705.

Pizzarello S., Cronin J. R., and Flynn G. (2006) The nature and distribution of the organic material in carbonaceous chondrites and interplanetary dust particles. In *Meteorites and the Early Solar System II* (D. S. Lauretta and H. Y. McSween, ed.), pp. 625–651. Univ. of Arizona, Tucson.

Planavsky N. J., McGoldrick P., Scott C. T., Li C., Reinhard C. T., Kelley A. E., Chu X., Bekker A., Love G. D., and Lyons T. W. (2011) Widespread iron-rich conditions in the mid-Proterozoic ocean. *Nature, 477,* 448–451.

Poinar H., Hoss M., Bada J. L., and Pääbo S. (1996) Amino acid racemization and the preservation of ancient DNA. *Science, 272,* 864–866.

Pósfai M., Lefèvre C. T., Trubitsyn D., Bazylinski D. A., and Frankel R. B. (2013) Phylogenetic significance of composition and crystal morphology of magnetosome minerals. *Front. Microbiol., 4,* 344.

Posth N. R., Canfield D. E., and Kappler A. 2014. Biogenic Fe(III) minerals: From formation to diagenesis and preservation in the rock record. *Earth-Sci. Rev., 135,* 103–121.

Ram R. J., Verberkmoes N. C., Thelen M. P., Tyson G. W., Baker B. J., Blake R. C. II, Shah M., Hettich R. L., and Banfield J. F. (2005) Community proteomics of a natural microbial biofilm. *Science, 308,* 1915–1920.

Redon E., Loubiere P., and Cocaign-Bousquet M. (2005) Role of mRNA stability during genome-wide adaptation of *Lactococcus lactis* to carbon starvation. *J. Biol. Chem., 280(43),* 36380–36385.

Reid R. P., Visscher P. T., Decho A. W., Stolz J., Bebout B. M., MacIntyre I. G., Paerl H. W., Pinckney J. L., Prufert-Bebout L., Steppe T. F., and Des Marais D. J. (2000) The role of microbes in accretion, lamination, and early lithification of modern marine stromatolites. *Nature, 406,* 989–992.

Rieley G., Collier R. J., Jones D. M., and Eglinton G. (1991) The biogeochemistry of Ellesmere Lake, UK: I. Source correlation of leaf wax inputs to the sedimentary lipid record. *Org. Geochem., 17,* 901–912.

Rosing M. T. (1999) ^{13}C-depleted carbon microparticles in 3700-Ma sea-floor sedimentary rocks from West Greenland. *Science, 283,* 674–676.

Ruan Q. S., Dutta D., Schwalbach M. S., et al. (2006) Local similarity analysis reveals unique associations among marine bacterioplankton species and environmental factors. *Bioinformatics, 22(20),* 2532–2538.

Saenz J. P., Grosser D., Bradley A. S., Lagny T. J., Lavrynenko O., Broda M. and Simons K. (2015) Hopanoids as functional analogues of cholesterol in bacterial membranes. *Proc. Natl. Acad. Sci. USA, 112(38),* 11971–11976.

San Antonio J. D. S., Schweitzer M. H., Jensen S. T., Kallur R., Buckley M., and Orgel J. P. (2011) Dinosaur peptides suggest mechanisms of protein survival. *PLoS One, 6(6),* e20381.

Schidlowski M. (1995) Isotope fractionations in the terrestrial carbon cycle: A brief overview. *Adv. Space Res., 15(3),* 441–449.

Schmidt F., Hinrichs K. U., and Elvert M. (2010) Sources, transport, and partitioning of organic matter at a highly dynamic continental margin. *Mar. Chem., 118,* 37–55.

Schloss P. D. and Handelsman J. (2005) Introducing DOTUR, a computer program for defining operational taxonomic units and estimating species richness. *Appl. Environ. Microbiol., 71(3),* 1501–1506.

Schlünz B. and Schneider R. R. (2000) Transport of terrestrial organic carbon to the oceans by rivers: Re-estimating flux- and burial-rates. *Intl. J. Earth Sci., 88,* 599–606.

Schmitt-Kopplin P., Gablica Z., Gougeon R. D., Fekete A., Kanawati B., Harir M., Gebefuegi I., Eckel G., and Hertkorn N. (2010) High molecular diversity of extraterrestrial organic matter in Murchison meteorite revealed 40 years after its fall. *Proc. Natl. Acad. Sci. USA, 107,* 2763–2768.

Schneider T. and Riedel K. (2010) Environmental proteomics: Analysis of structure and function of microbial communities. *Proteomics, 10(4),* 785–798.

Schopf J. W. (1993) Microfossils of the early Archean Apex Chert: New evidence for the antiquity of life. *Science, 260,* 640–646.

Schopf J. W. and Barghoorn E. S. (1967) Algae-like fossils from the Early Precambrian of South Africa. *Science, 156,* 508–512.

Schopf J. W. and Kudryavtsev A. B. (2009) Confocal laser scanning microscopy and Raman imagery of ancient microbial fossils. *Precambrian Res., 173,* 39–49.

Schopf J. W., Chang S., Ernst W. G., Holland H. D., Kasting J. F., and Lowe D. R. (1992) Geology and paleobiology of the Archean Earth. In *The Proterozoic Biosphere* (J. W. Schopf and C. Klein, eds.), pp. 5–42. Cambridge Univ., Cambridge.

Schopf J. W., Kudryavtsev A. B., Czaja A. D., and Tripathi A. B. (2007) Evidence of Archean life: Stromatolites and microfossils. *Precambrian Res., 158,* 141–155.

Schouten S., Hoefs M. J. L., Koopmans M. P., Bosch H.-J., and Sinninghe Damsté J. S. (1998) Structural identification, occurrence, and fate of archaeal ether-bound acyclic and cyclic biphytanes and corresponding diols in sediments. *Org. Geochem., 29,* 1305–1319.

Schouten S., Hopmans E. C., Pancost R. D., and Sinninghe Damsté J. S. (2000) Widespread occurrence of structurally diverse tetraether membrane lipids: Evidence for the ubiquitous presence of low-temperature relatives of hyperthermophiles. *Proc. Natl. Acad. Sci. USA, 97,* 14421–14426.

Schouten S., Hopmans E. C., Schefuss E., and Sinninghe Damsté J. S. (2002) Distributional variations in marine crenarchaeotal membrane lipids: A new tool for reconstructing ancient sea water temperatures? *Earth Planet. Sci. Lett., 204,* 265–274.

Schouten S., Huguet C., Hopmans E. C., Kjienhuis M. V. M., and Sinninghe Damsté J. S. (2007a) Analytical methodology for TEX86 paleothermometry by high pressure liquid chromatography/atmospheric pressure chemical ionization-mass spectrometry. *Anal. Chem., 79,* 2940–2944.

Schouten S., Woltering M., Rijpstra W. I. C., Slujis A., Brinkhuis H., and Sinninghe Damsté J. S. (2007b) The Paleocene-Eocene carbon isotope excursion in higher plant organic matter: Differential fractionation of angiosperms and conifers in the Arctic. *Earth Planet. Sci. Lett. 258(3–4),* 581–592.

Schouten S., Hopmans E. C., and Sinninghe Damsté J. S. (2013) The organic geochemistry of glycerol dialkyl glycerol tetraether lipids: A review. *Org. Geochem., 54,* 19–61.

Schubert M., Ginolhac A., Lindgreen S., Thompson J. F., Al-Rasheid K. A., Willerslev E., Krogh A., and Orlando L. (2012) Improving ancient DNA read mapping against modern reference genomes. *BMC Genomics, 13(1),* 178.

Schweitzer M. H., Zheng W., Organ C. L., Avci R., Suo Z., Freimark L. M., and Lebleu V. S. (2009) Biomolecular characterization and protein sequences of the Campanian hadrosaur *B. canadensis. Science, 324(5927),* 626–631.

Seckbach J. and Oren A. (2010) *Microbial Mats: Ancient Microorganisms in Stratified Systems.* Springer, Dordrecht. 606 pp.

Selinger D. W., Saxena R. M., Cheung K. J., Church G. M., and Rosenow C. (2003) Global RNA half-life analysis in *Escherichia coli* reveals positional patterns of transcript degradation. *Genome Res., 13(2),* 216–223.

Sephton M. A. (2002) Organic compounds in carbonaceous meteorites. *Nat. Prod. Rept., 19,* 292–311.

Sergeev V. N., Knoll A. H., and Grotzinger J. P. (1995) Paleobiology of the Mesoproterozoic Billyakh Group, Anbar Uplift, northern Siberia. *J. Paleontol., 69(S39),* 1-37, DOI: 10.1017/S0022336000062375.

Shapiro E., Biezuner T. and Linnarsson S. (2013) Single-cell sequencing-based technologies will revolutionize whole-organism science. *Nature Rev. Genet., 14,* 618–630.

Sharp J. O., Schofield E., Junier P., Veeramani H., Suvorova E., Bargar J. R., and Bernier-Latmani R. (2008) Systematic investigation of the product of microbial U(VI) reduction by different bacteria. *Geochim. Cosmochim. Acta, 72,* A852.

Shen Y., Buick R., and Canfield D. E. (2001) Isotopic evidence for microbial sulphate reduction in the early Archaean era. *Nature, 410,* 77–81.

Shepherd J. and Ibba M. (2013) Lipid II-independent trans editing of mischarged tRNAs by the penicillin resistance factor MurM. *J. Biol. Chem., 288(36),* 25915–25923.

Shepherd J. and Ibba M. (2015) Bacterial transfer RNAs. *FEMS Microbiol. Rev., 39(3),* 280–300.

Shiea J., Brassell S. C., and Ward D. M. (1990) Mid-chain branched mono- and dimethyl alkanes in hot spring cyanobacterial mats: A direct biogenic source for branched alkanes in ancient sediments? *Org. Geochem., 15(3),* 223–231.

Simoneit B. R. T. (1997) Compound-specific carbon isotope analyses of individual long-chain alkanes and alkanoic acids in Harmatten aerosols. *Atmos. Environ., 31,* 2225–2233.

Singer S. W., Chan C. S., Zemla A.; VerBerkmoes N. C., Hwang M., Hettich R. L., Banfield J. F., and Thelen M. P. (2008) *Appl. Environ. Microbiol., 74,* 4454–4462.

Sinninghe Damsté J. S., van Duin A. C. T., Hollander D., Kohnen M. E. L., and de Leeuw J. W. (1995) Early diagenesis of bacteriohopanepolyol derivatives: Formation of fossil homohopanoids. *Geochim. Cosmochim. Acta, 59(24),* 5141–5147.

Sinninghe Damsté J. S., Hopmans E. C., Schouten S., van Duin A. C. T., and Geenevasen J. A. J. (2002a) Crenarchaeol: The characteristic core glycerol dibiphytanyl glycerol tetraether membrane lipid of cosmopolitan pelagic crenarchaeota. *J. Lipid Res., 43,* 1641–1651.

Sinninghe Damsté J. S., Rijpstra W. I. C., and Reichart G.-J. (2002b) The influence of oxic degradation on the sedimentary biomarker record II. Evidence from Arabian Sea sediments. *Geochim. Cosmochim. Acta, 66,* 2737–2754.

Sinninghe Damsté J. S., Rijpstra W. I. C., Hopmans E. C., Jung M.-Y., Kim J.-G., Rhee S.-K., Stieglmeier M., and Schleper C. (2012) Intact polar and core glycerol dibipytanyl glycerol tetraether lipids of Group I.1a and I.1b Thaumarchaeota in soil. *Appl. Environ. Microbiol. 78(19),* 6866–6874.

Smith F. A., Wing S. L., and Freeman K. H. (2007) Magnitude of the carbon isotope excursion at the Paleocene-Eocene thermal maximum: The role of plant community change. *Earth Planet. Sci. Lett. 262,* 50–65.

Smith P. W., Watkins K., and Hewlett A. (2012) Infection control through the ages. *Am. J. Infect. Control, 40(1),* 35–42.

Sogin M. (1994) The origin of eukaryotes and evolution into major kingdoms. In *Early Life on Earth* (S. Bengtson, ed.), pp. 181–192. Nobel Symposium No. 84, Columbia Univ., New York.

Spalding R. F. et al. (2011) Long-term groundwater monitoring results at large, sudden denatured ethanol releases. *Groundwater Monit. Rem., 31,* 69–81.

Spicer R. A. (1981) *The Sorting and Deposition of Allochthonous Plant Material in a Modern Environment at Sillwood Lake, Sillwood Park, Berkshire, England.* U.S. Geological Survey Professional Paper 1143, DOI: 10.3133/pp1143.

Staudigel H., Furnes H., McLoughlin N., Banerjee N. R., Connell L. B., and Templeton A. (2008) 3.5 billion years of glass bioalteration: Volcanic rocks as a basis for microbial life? *Earth-Sci. Rev., 89 (3–4),* 156–176.

Summons R. E. and Walter M. R. (1990) Molecular fossils and microfossils of prokaryotes and protists from Proterozoic sediments. *Am. J. Sci., 290-A,* 212–244.

Summons R. E., Jahnke L. L., and Roksandic Z. (1994) Carbon isotopic fractionation in lipids from methanotrophic bacteria: Evidence for interpretation of the geochemical record of biomarkers. *Geochim. Cosmochim. Acta, 58,* 2853–2863.

Summons R. E., Jahnke L. L., Hope J. M., and Logan G. A. (1999) 2-Methylhopanoids as biomarkers for cyanobacterial oxygenic photosynthesis. *Nature, 400,* 554–557.

Summons R. E., Albrecht P., McDonald G., and Moldowan J. M. (2007) Molecular biosignatures: Generic qualities of organic compounds that betray biological origins. *Space Sci. Rev., 133,* 133–157.

Swift M. J., Heal O. W., and Anderson J. M. (1979) *Decomposition in Terrestrial Ecosystems.* Univ. of California, Berkeley.

Talbot H. M., Watson D. F., Murrell J. C., Carter J. F., and Farrimond P. (2001) Analysis of intact bacteriohopanepolyols from methanotrophic bacteria by reversed-phase high-performance liquid chromatography-atmospheric pressure chemical ionisation mass spectrometry. *J. Chromatogr. A, 921(2),* 175–185.

Talbot H. M., Watson D. F., Pearson E. J., and Farrimond P. (2003) Diverse biohopanoid compositions of non-marine sediments. *Org. Geochem, 34,* 1353–1371.

Talbot H. M., Summons R. E., Jahnke L. L., Cockell C. S., Rohmer M., and Farrimond P. (2008) Cyanobacterial bacteriohopanepolyol signatures from cultures and natural environmental settings. *Org. Geochem, 39,* 232–263.

Tanca A., Palomba A., Pisanu S., Deligios M., Fraumene C., Manghina

V., Pagnozzi D., Addis M. F., and Uzzau S. (2014) A straightforward and efficient analytical pipeline for metaproteome characterization. *Microbiome, 2(1),* 49.

Tebo B. M., Johnson H. A., McCarthy J. K., and Templeton A. S. (2005) Geomicrobiology of manganese(II) oxidation. *Trends Microbiol., 13,* 421–428.

Tegelaar E. W., de Leeuw J. W., Derenne S., and Largeau C. (1989) A reappraisal of kerogen formation. *Geochim. Cosmochim. Acta, 53,* 3103–3106.

Ten Haven H. L. and Rullkötter J. (1988) The diagenetic fate of taraxer-14-ene and oleanane isomers. *Geochim. Cosmochim. Acta, 52(10),* 2543–2548.

Thomas-Keprta K., Bazylinski D. A., Kirschvink J. L., Clemett S. J., McKay D. S., Wentworth S. J., Vali H., Gibson E. K. Jr., and Romanek C. S. (2000) Elongated prismatic magnetite crystals in ALH84001 carbonate globules: Potential martian magnetofossils. *Geochim. Cosmochim. Acta, 64,* 4049–4081.

Thomas-Keprta K., Clemett S. J., Bazylinski D. A., Kirschvink J. L., McKay D. S., Wentworth S. J., Vali H., Gibson E. K. Jr., McKay M. F., and Romanek C. S. (2001) Truncated hexa-octahedral magnetite crystals in ALH84001: Presumptive biosignatures. *Proc. Natl. Acad. Sci. USA, 98,* 2164–2169.

Thomas-Keprta K., Clemett S. J., Bazylinski D. A., Kirschvink J. L., McKay D. S., Wentworth S. J., Vali H., Gibson E. K. Jr., and Romanek C. S. (2002) Magnetofossils from ancient Mars: A robust biosognature in the martian meteorite ALH84001. *Appl. Environ. Microbiol., 68,* 3663–3672.

Thompson M. R., VerBerkmoes N. C., Chourey K., Shah M., Thompson D. K., and Hettich R. L. (2007) *J. Proteome Res., 6,* 1745–1757.

Tice M. M. and Lowe D. R. (2006) The origin of carbonaceous matter in pre-3.0 Ga greenstone terrains: A review and new evidence from the 3.42 Ga Buck Reef Chert. *Earth-Sci. Rev., 76,* 259–300.

Tighe S., Afshinnekoo E., Rock T. M., McGrath K., Alexander N., McIntyre A., Ahsanuddin S., Bezdan D., Green S. J., Joye S., Johnson S. S., Baldwin D. A., Bivens N., Ajami N., Carmical J. R., Herriott I. C., Colwell R., Donia M., Foox J., Greenfield N., Hunter T., Hoffman J., Hyman J., Jorgensen E., Krawczyk D., Lee J., Levy S., Garcia-Reyero N., Settles M., Thomas K., Gómez F., Schriml L., Kyrpides N., Zaikova E., Penterman J., and Mason C. E. (2017) Genomic methods and microbiological technologies for profiling novel and extreme environments for the Extreme Microbiome Project (XMP). *J. Biomol. Tech., 28(1),* 31–39, DOI: 10.7171/jbt.17-2801-004.

Tissot B. P. and Welte D. H. (1984) *Petroleum Formation and Occurrence: A New Approach to Oil and Gas Exploration.* Springer-Verlag, Berlin.

Toporski J. and Steele A. (2002) The relevance of bacterial biomarkers in astrobiological research. In *Proceedings of the Second European Workshop on Exo/Astrobiology* (H. Sawaya-Lacoste, ed.), pp. 239-242. ESA Special Publication SP-518, European Space Agency, Noordwijk, Netherlands.

Torti A., Lever M. A., and Jørgensen B. B. (2015) Origin, dynamics, and implications of extracellular DNA pools in marine sediments. *Mar. Genomics, 24,* 185–196.

Turich C. and Freeman K. H. (2011) Archaeal lipids record paleosalinity in hypersaline systems. *Org. Geochem., 42,* 1147–1157.

Turich C., Freeman K. H., Bruns M. A., Conte M., Jones A. D., and Wakeham S. G. (2007) Lipids of marine Archaea: Patterns and provenance in the water-column and sediments. *Geochim. Cosmochim. Acta, 71,* 3272–3291.

Uno K. T., Polissar P. J., Kahle E., Feibel C., Harmand S., Roche H., and deMenocal P. B. (2016) A Pleistocene palaeovegation record from plant wax biomarkers from the Nachukui Fm., West Turkana, Kenya. *Philos. Trans. R. Soc., B, 371(1698),* 315–328, DOI: 10.1098/rstb.2015.0235.

Urich T., Lanzén A., Stokke R., Pedersen R. B., Bayer C., Thorseth I. H. Schleper C., Steen I. H., and Øvreas L. (2014) Microbial community structure and functioning in marine sediments associated with diffuse hydrothermal venting assessed by integrated meta-omics. *Environ. Microbiol., 16,* 2699–2710.

Utsunomiya S., Ewing R. C., and Wang L. M. (2005) Radiation-induced decomposition of U(VI) phases to nanocrystals of UO_2. *Earth Planet. Sci. Lett., 240,* 521–528.

Van Bergen P. F., Collinson M. E., and Briggs D. E. G. (1995) Resistant biomacromolecules in the fossil record. *Acta Bot. Neerl., 44(4),* 319–342.

Venter J. C., Adams M. D., Myers E. W., Li P. W., Mural R. J., Sutton G. G., Smith H. O., Yandell M., Evans C. A., Holt R. A., and Gocayne J. D. (2001) The sequence of the human genome. *Science, 291(5507),* 1304–1351.

Venter J. C., Remington K., Heidelberg J. F., Halpern A. L., Rusch D., Eisen J. A., Wu D., Paulsen I., Nelson K. E., Nelson W., and Fouts D. E. (2004) Environmental genome shotgun sequencing of the Sargasso Sea. *Science, 304(5667),* 66–74.

Villas-Boas S. G., Nielsen J., Smedsgaard J., Hansen M. A., and Roessner-Tunali U. (2007) *Metabolome Analysis: An Introduction.* Wiley Series on Mass Spectrometry Book 24, Wiley, Hoboken. 320 pp.

Vishniac H. S. and Hempfling W. P. (1979) Evidence of an indigenous microbiota (yeast) in the dry valleys of Antarctica. *Microbiology, 112(2),* 301–314.

Volkman J. K. (1986) A review of sterol markers for marine and terrigenous organic matter. *Org. Geochem., 9,* 83–99.

Volkman J. K., Eglinton G., Corner E. D. S., and Forsberg T. E. V. (1980) Long-chain alkenes and alkenones in the marine coccolithophorid *Emiliania huxleyi. Phytochemistry, 19,* 2619–2622.

Walter M. R. (1972) Stromatolites and the biostratigraphy of the Australian Precambrian and Cambrian. *Special Papers in Palaeontology, 11,* 1–190.

Walter M. R. (1983) Archean stromatolites: Evidence of the Earth's earliest benthos. In *Earth's Earliest Biosphere, Its Origin and Evolution* (J. W. Schopf, ed.), pp. 187–213. Princeton Univ., Princeton.

Walter M. R. and Heys G. R. (1985) Links between the rise of the metazoa and the decline of stromatolites. *Precambrian Res., 29,* 149–174.

Watson J. D. and Crick F. H. (1953) Molecular structure of nucleic acids; a structure for deoxyribose nucleic acid. *Nature, 171(4356),* 737–738.

Weijers J. W. H., Schouten S., van Den Donker J. C., Hopmans E. C., and Sinninghe Damsté J. S. (2007) Environmental controls on bacterial tetraether membrane lipid distribution in soils. *Geochim. Cosmochim. Acta, 71,* 703–713.

Weiss B. P., Kim S. S., Kirschvink J. L., Kopp R. E., Sankaran M., Kobayashi A., and Komelli A. (2004) Magnetic tests for magnetosome chains in martian meteorite ALH84001. *Proc. Natl. Acad. Sci. USA, 101,* 8281–8284.

Welander P. V. and Summons R. E. (2012) Discovery, taxonomic distribution, and phenotypic characterization of a gene required for 3-methylhopanoid production. *Proc. Natl. Acad. Sci. USA, 109,* 12905–12910.

Welander P. V., Coleman M., Sessions A. L., Summons R. E., and Newman D. K. (2010) Identification of a methylase required for 2-methylhopanoid production and implications for the interpretation of sedimentary hopanes. *Proc. Natl. Acad. Sci. USA, 107,* 8537–8542.

Whitman W. B., Coleman D. C., and Wiebe W. J. (1998) Prokaryotes: The unseen majority. *Proc. Natl. Acad. Sci. USA, 95,* 6578–6583.

Willerslev E., Hansen E. J., Rønn R., Brand T. B., Barnes I., Wiuf C., Gilinchinsky D., Mitchell D., and Cooper A. (2004) Long-term persistence of bacterial DNA. *Curr. Biol., 6,* R9–R10.

Willerslev E., Cappellini E., Boomsma W., Nielsen R., Hebsgaard M. B., et al. (2007) Ancient biomolecules from deep ice cores reveal a forested southern Greenland. *Science, 317,* 111–114.

Williams R. J., Howe A., and Hofmockel K. S. (2014) Demonstrating microbial co-occurrence pattern analyses within and between ecosystems. *Front. Microbiol., 5,* 358, DOI: 10.3389/fmicb.2014.00358.

Wilmes P. and Bond P. L. (2004) The application of two-dimensional polyacrylamide gel electrophoresis and downstream analyses to a mixed community of prokaryotic microorganisms. *Environ. Microbiol., 6,* 911–920.

Woese C. R. (1987) Bacterial evolution. *Microbiol. Rev., 51(2),* 221–271.

Woese C. R. (2002) On the evolution of cells. *Proc. Natl. Acad. Sci. USA, 99(13),* 8742–8747.

World Wide Fund for Nature (2016) *Living Planet Report 2016: Risk and Resilience in a New Era.* WWF International, Gland,

Switzerland.

Xiong J., Fischer W. M., Inoue K., Nakahara M., and Bauer C. E. (2000) Molecular evidence for the early evolution of photosynthesis. *Science, 289,* 1724–1730.

Xu J., ed. (2014) *Next-Generation Sequencing: Current Technologies and Applications.* Caister, Poole. 160 pp.

Yan Y., Ng J., Higgins J., Kurbatov A., Clifford H., Spaulding N., Severinghaus J., Brook E., Mayewski P., and Bender M. (2017) 2.7-million-year-old ice from Allan Hills blue ice areas, East Antarctica reveals climate snapshots since early Pleistocene. *Goldschmidt Abstracts, 2017,* 4359.

Yang W., Zhang Z., Zhang Z., Chen H., Liu J., Ali M., Liu F., and Li L. (2013) Population structure of manganese-oxidizing bacteria on stratified soils and properties of manganese oxide aggregates under manganese-complex medium enrichment. *PLoS One, 8(9),* e73778.

Zachos J. C., Quinn T. M., and Salamy K. A. (1996) *Paleoceanogr., 11,* 251–266.

Zeng Z., Liu X.-L., Wei J. H., Summons R. E., and Welander P. V. (2018) Calditol-linked membrane lipids are required for acid tolerance in Sulfolobus acidocaldarius. *Proc. Natl. Acad. Sci., 115(51),* 12932–12937, DOI: 10.1073/pnas.1814048115.

Zhang Y. (1981) Proterozoic stromatolite microfloras of the Gaoyuzhuang Formation (early Sinian Riphean), Hebei, China. *J. Paleontol., 55,* 485–506.

Zhang Z. (1986) Clastic facies microfossils from the Chuanlinggou Formation (1800 Ma) near Jixian, North China. *J. Micropalaeontol., 5,* 9–16.

Zhao Y. Y. and Lin R. C. (2014) UPLC-MSE application in disease biomarker discovery: The discoveries in proteomics to metabolomics. *Chem.-Biol. Interact., 215,* 7–16.

Zumberge J. A., Love G. D., Cárdenas P., Sperling, E. A., Gunasekera S., Rohrssen M., Grosjean E., Grotzinger J. P., and Summons R. E. (2018) Demosponge steroid biomarker 26-methylstigmastane provides evidence for Neoproterozoic animals. *Nature Ecol. Evol., 2,* 1709–1714, DOI: 10.1038/s41559-018-0676-2.

Part 2:

*Astrobiology in
Our Solar System*

Amador E. S. and Ehlmann B. L. (2020) Planetary astrobiology: Early Mars. In *Planetary Astrobiology* (V. Meadows et al., eds.), pp. 153–167. Univ. of Arizona, Tucson, DOI: 10.2458/azu_uapress_9780816540068-ch006.

Planetary Astrobiology: Early Mars

Elena S. Amador
Division of Geological and Planetary Sciences, California Institute of Technology

Bethany L. Ehlmann
Division of Geological and Planetary Sciences, California Institute of Technology, and
Jet Propulsion Laboratory, California Institute of Technology

Mars' surface preserves the rich history of a habitable planet. Nearly half of the rocks exposed on the surface today were formed over 3.7 b.y. ago, providing a compelling record for understanding the early history of rocky planets in our solar system. Continued studies ranging from the global to the microscopic scale have demonstrated that habitats on Mars were widespread, diverse, and dynamic with surface and subsurface habitats that ebbed and flowed, depending on global and local conditions. This chapter reviews the diversity of habitats that existed during Mars' most geologically active period and describes key questions that remain outstanding regarding Mars' evolution and its implications for available habitats over time. Current and future studies of the planet will be guided by several key questions, including determining the absolute chronology of martian surfaces in order to constrain the timing of key events that impact planetary habitability and understanding the evolution and fate of the early martian atmosphere. Lastly, given the evidence that early Mars was indeed once habitable, an emphasis has been placed on understanding the biosignatures potential life may have left behind, and how those signatures would be preserved in the evolving martian subsurface and surface environment.

"Studying whether there's life on Mars or studying how the universe began, there's something magical about pushing back the frontiers of knowledge. That's something that is almost part of being human, and I'm certain [it] will continue."
— Sally Ride, first American woman in space

1. INTRODUCTION

1.1. Background

Mars has been of interest to planetary scientists and the lay public alike for well over a century. Early telescopic observations showed global linear features trending from the poles through the equator and changes in surface albedo that stirred the imagination, leading to speculation about "unnatural" irrigation canals and seasonal vegetation cycles on Mars (*Lowell*, 1906). However, with the dawn of space exploration in the mid-twentieth century, the first images from the Mariner 4 spacecraft in 1965 did not show evidence for surface life, but rather a planet with a tenuous atmosphere, craters, large valleys, volcanos, and surface features that showed a once geologically active planet whose habitability in the past and at present was uncertain (*Leighton et al.,* 1965). Over the next decade, several subsequent flybys and orbiters sent by NASA and the Soviet Union continued to explore Mars with measurements of the surface and atmosphere (e.g., *Masurksy,* 1973).

Although the surface conditions on Mars appeared superficially to be inhospitable to life, NASA landed two planetary science missions that included life detection experiments. The Viking 1 and 2 missions each consisted of an orbiter and an associated lander. Among their planetary science objectives were three biology experiments intended to test for the presence of extant life (*Klein,* 1978). Although some of the results from the individual life detection experiments have been interpreted as consistent with active biota, the totality of results from all three experiments was more consistent with abiotic geochemical reactions in the oxidizing surface environment (*Horowitz et al.,* 1976), thus leaving the question of extant life on the surface of Mars largely inconclusive. After the Viking missions, the search for martian life, and the exploration of Mars in general, went on a hiatus for nearly 20 years. In the interim, study of life in extreme environments progressed on Earth, finding active or viable microbial life wherever there was water: the bottom of the oceans, within sea ice, deep in the subsurface, in the stratosphere, in hot springs, and arid deserts — e.g., *Rothschild and Mancinelli,* 2001.

Given what is now known about early terrestrial conditions, the antiquity of Earth's fossil record, and the vast array of environments where life can survive (see the chapter by Johnson et al. in this volume), planetary scientists have prioritized study of early Mars history through the lens of habitability and, in particular, searching for locales that may have once been habitable. Habitability is defined as more than merely access to liquid water, which is indeed a good solvent for biochemical reactions. A habitable environment, or "habitat" (*Cockell,* 2011), must also meet geochemical constraints with respect to temperature, pH, and water activity to maintain biochemical reactions (*Rothschild and Mancenelli,* 2001; *Hoehler,* 2007; see also the chapter by Hoehler et al. in this volume). There must be a supply of key elements (e.g., the elements S, P, O, N, C, and H) to make up biomass and an energy source for metabolisms (e.g., energy via photosynthesis or environmental redox gradients) (*Hoehler,* 2007). An environment is therefore habitable for a particular form of life if it meets all these criteria for that organism at any one time. For example, Gale Crater, where the Curiosity rover is currently exploring, has been established as a lake, once habitable to chemolithoautotrophic life given the evidence for aqueous conditions with neutral pH, low salinity, and variable redox states, as well as the observed, and in some cases inferred, availability of biogenic elements (*Grotzinger et al.,* 2014).

Similarly, many criteria for habitability can now be assessed from orbital data, in particular by including observed surface mineralogy that identifies the aqueous chemistry that once existed there (e.g., *Murchie et al.,* 2009; *Ehlmann et al.,* 2011). This has led to the discovery of many more locales on Mars that were also likely habitable in the past (e.g., Jezero Crater, where the Mars 2020 rover will search for evidence of biosignatures). It is now clear that early Mars hosted multiple habitable environments, varying in space and time. Consequently, the question that many planetary scientists ask has transitioned in the last decade from "Was an ancient Mars habitable?" to "Was an ancient Mars inhabited?" Additional questions concern whether the observed habitats were intermittent or long-lived and the processes that sustained them. It is possible that some (or all) martian habitats were "uninhabited habitats" (*Cockell,* 2011; *Cockell et al.,* 2012), i.e., vacant of any life. Searching for life on Mars, and understanding the processes of habitat generation there, will inform our understanding of how rare or how common life is in the universe and what planetary-scale geologic and geochemical controls are needed for life to exist on a planet with important implications for the search of life on terrestrial exoplanets. Therefore, in this chapter, we will explore the history of early Mars through the lens of astrobiology: what its geologic history implies for the habitability of the planet, and which habitats we can explore to determine if Mars was ever inhabited.

1.2. Mars Basics

Mars is a small rocky planet, approximately half the diameter and a tenth of the mass of Earth. At its orbital distance of ~1.5 AU from the Sun, Mars receives ~600 W m^{-2} of solar irradiance, less than half that of the Earth. From geophysical measurements, its average density is ~3.9 g cm^{-3}, implying a differentiated interior with an Fe or Fe-S core and a silicate-rich mantle and crust (*Helffrich,* 2017). The present-day climate is cold and dry with an average surface temperature of 220 K (–53°C), temperature extremes ranging from 145 K to 300 K, and an average surface pressure of 6 mbar (~0.6% of Earth's average surface pressure). Mars' thin atmosphere is dominated by CO_2 gas (~96%). In this modern-day cold and arid climate, most of the surface is below the triple point of water and surface liquid water is unstable to evaporation and freezing (*Haberle et al.,* 2001); however, water is commonly present in the form of vapor and ice clouds and as ice in the surface and subsurface at high to mid-latitudes (*Feldman et al.,* 2004; *Hecht,* 2002). Additionally, high relative humidity occurs even in equatorial regions at night, which can lead to condensation including fog observed in Valles Marineris (*Mohlmann et al.,* 2009). Whether any water exists at depth in the modern subsurface is debated (e.g., *Clifford et al.,* 2010). The thin atmosphere means that the surface is highly irradiated with ultraviolet (UV) and charged particle radiation (*Guo et al.,* 2017). In all, the surface of Mars is hardly habitable today (see the chapter by Avila et al. in this volume), but years of observations have led scientists to understand that many surface and subsurface locales on ancient Mars were indeed once habitable.

Mars has striking physiography when viewed from orbital data, including a global topographic boundary that separates the northern and southern hemispheres, and large shield volcanos, impact basins, and outflow channels (*Carr and Head,* 2010) (Fig. 1). The southern hemisphere is higher in elevation and dominated by heavily cratered ancient terrains, while the low-lying northern hemisphere is dominated by volcanic and sedimentary plains that have few craters due to resurfacing and erosion (e.g., *Aharonson et al.,* 2001; *Smith and Zuber,* 1996; *Smith et al.,* 1999). Mars' crustal dichotomy likely formed due to a major impact event ~100 m.y. after accretion, around the time of the Earth–Moon-forming impact (e.g., *Nimmo et al.,* 2008); as such, it is one of the oldest surface features on Mars. This global-scale impact event likely influenced internal mantle/core interactions, including potentially influencing the magnetic dynamo (e.g., *Grott et al.,* 2013; *Roberts et al.,* 2009). Also notable are the large shield volcanos surrounded by large fault structures. Unlike Earth, there is little evidence for active plate tectonics on early Mars (*Golombek and Phillips,* 2010); therefore, as the planet cooled after accretion and heat was transported to the surface, volcanic centers were fixed, allowing volcanos to grow over long periods of time. The volcanos are thought to have been active as recently as 20–200 m.y. ago

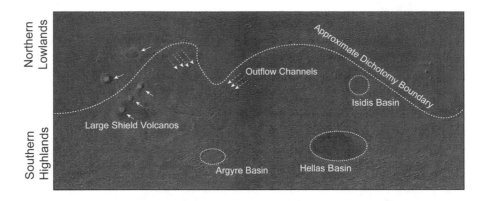

Fig. 1. See Plate 10 for color version. MOLA colorized topography of Mars with major landforms referenced. Dashed white arrows mark the large outflow channels, solid white arrows indicate large shield volcanos, dashed circles indicate large impact basins, and dashed white line indicates the approximate boundary of the hemispheric dichotomy (*Andrews-Hanna et al.,* 2008).

(*Berman and Hartmann,* 2002; *McSween,* 2002; *Werner,* 2009). Large impact basins indicate well-preserved ancient surfaces, in some cases with layering exposed by tectonics and erosion. Thousands of outflow channels, many of which are kilometers wide, imply high volumes of fluid transported from the southern highlands into the northern lowlands. At smaller spatial scales, dendritic valley networks and fan and delta deposits are reminiscent of river systems on Earth (e.g., *Seybold et al.,* 2018).

Mars' early history is well preserved for study within the oldest martian rocks, which, unlike on Earth, make up a significant percentage of the exposed surface today. Terrains of Noachian age (those >~3.7 G.y.; Fig. 2) make up ~45% of the modern exposed surface (*Tanaka et al.,* 2014), while Hesperian-aged surfaces (~3.7—~3.0 G.y.; Fig. 2) (*Hartmann and Neukam,* 2001) make up another ~30% of the modern martian surface. As such, nearly three-quarters of the exposed martian surface today is older than ~3 G.y. These terrains provide a unique view of the early geologic evolution of terrestrial planets in our solar system when the Sun was faint (*Ulrich,* 1975; *Feulner,* 2012), impacts were bombarding bodies in the inner solar system (*Fassett and Minton,* 2013), life was first emerging on Earth (*Allwood et al.,* 2006; *Mojzsis et al.,* 1996), and Mars was particularly dynamic and sustained an active hydrologic system and chemical weathering (e.g., *Carr and Head,* 2010; *Ehlmann and Edwards,* 2014) (Fig. 2). By studying rocks from these early martian surfaces we may begin to unpack planetary processes that led to the very different planetary fates of Mars and Earth.

2. BUILDING MARS

How did Mars become a habitable planet? We briefly summarize some of Mars' key initial conditions, all of which would have contributed to its habitability. Geochemical observations of excess ^{182}W and ^{142}Nd in martian meteorites indicate that Mars likely accreted and differentiated earlier than Earth (*Lee and Halliday,* 1997). In fact, it may be a planetary embryo rather than the product of the accretion of several planetary embryos with most accretion and differentiation completed by 5 m.y. after the initiation of solar system formation (*Dauphas and Pourmand,* 2011). Mars likely formed near its current orbital position of ~1.52 AU from the Sun; given that at the time the Sun was fainter (*Ulrich,* 1975), this placed Mars just outside the circumstellar habitable zone, conventionally defined as the region within which a terrestrial planet can sustain surface liquid water (*Kasting et al.,* 1993; *Kopparapu et al.,* 2013) (for further discussion, see section 4).

A magnetic field shielded an early Mars from solar charged particles. This dynamo-driven field is recorded by strong remnant magnetic anomalies detected in very old martian terrains, namely in Terra Cimmeria and Terra Sirenum, which are dated to be >4.1 Ga (*Acuña et al.,* 1999; *Lillis et al.,* 2013; *Tanaka et al.,* 2014) and in the remnant magnetism found in carbonate-associated magnetite in the Mars meteorite ALH 84001, dated at 3.9 Ga (e.g., *Weiss et al.,* 2008). This magnetic field likely had global implications for planetary habitability. First, the field would have protected the surface from radiation that could affect biological processes; this protection would have prevented some genetic mutations but could have also slowed down certain prebiotic reactions (*Ranjan and Sasselov,* 2017). Additionally, magnetic fields prevent equatorial and mid-latitude atmospheric loss but may also promote loss from high latitudes (e.g., *Engwall et al.,* 2008). Therefore, the extent to which the early martian magnetic field was a benefit or a detriment to the planet's overall habitability is still debated. Its cessation appears to have been early

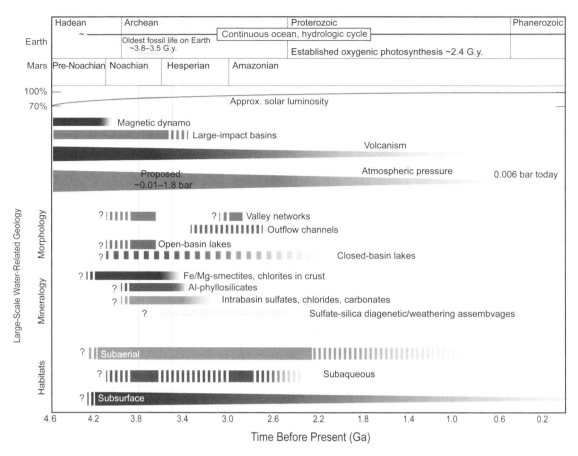

Fig. 2. Timeline of major events and habitats in Mars history with Earth's geologic eons displayed above for reference. Approximate ages for major martian transitions are from *Hartmann and Neukam* (2001). Absolute ages contain some uncertainty but relative sequences for events are robust. Question marks indicate where events, processes, and habitats could have started earlier, but that record does not exist. Dashed boxes indicate an episodic occurrence and faded boxes indicate the gradual observed decline of the process.

(~4 Ga), as no preserved magnetic anomalies are found on the large basin floors.

The martian atmosphere has evolved several times since its formation. Mars' initial atmosphere was formed by the impact devolatilization of accreting planetesimals (e.g., *Hashimoto et al.*, 2007) and degassing from an early magma ocean (*Elkins-Tanton et al.*, 2005). It was likely initially composed principally of H_2 and other hydrides (e.g., H_2O, CH_4, NH_3) with some noble gas constituents (e.g., *Pepin*, 2006). This primordial atmosphere was heated by the young Sun's intense extreme UV flux, causing hydrogen to be rapidly lost on timescales of <100 m.y. via hydrodynamic escape (*Haberle et al.*, 2017). This outward flow of hydrogen also exerted upward drag on heavier atmospheric components (e.g., *Hunten et al.*, 1987). Because lighter components are more easily lost to space, a mass-dependent fractionation signature of the residual atmosphere is now observed in water vapor deuterium to hydrogen ratios (D/H) and noble gases in both martian meteorites (SNC) and the modern martian atmosphere (e.g., *Hunten et al.*, 1987; *Jakosky and Jones*, 1997; *Pepin*, 2006; *Webster et al.*, 2013). Loss of this

primary atmosphere may have also been supplemented by impact erosion (*Lammer et al.*, 2008), although whether asteroid and cometary impacts cause net addition or loss of atmosphere is debated (for a review, see *Ehlmann et al.*, 2016a).

The isotopically-fractionated early atmosphere that remained after these early loss processes was then supplemented by CO_2-rich volatiles released by volcanic degassing as basalts were erupted following partial melting of the mantle (e.g., *Elkins-Tanton*, 2008; *Gaillard and Scaillet*, 2014). Formed by volcanism extending over most of Mars' history (*Grott et al.*, 2011), this secondary atmosphere was lost more slowly given a decrease in extreme UV flux from the Sun (*Zahnle and Walker*, 1982). The composition, evolution, and subsequent fate of this secondary atmosphere within the context of observed surface geomorphology and mineralogy will be discussed in detail in section 4.

In contrast to Earth where the spin axis orientation (obliquity) is ~23 ± 1°, stabilized in the Earth-Moon system, Mars' orbital eccentricity and obliquity are modeled to have undergone significant variations throughout its history, ranging from 0 to 0.125 for orbital eccentricity and

$10°-60°$ in obliquity (*Laskar et al.*, 2004). This would have had profound effects on the martian climate throughout its history with dramatic changes in solar insolation across latitudes (e.g., *Jakosky et al.*, 1995; *Head et al.*, 2003; *Forget et al.*, 2006), thus creating dynamic surface conditions on timescales of $\sim 10^6$ years.

3. THE HABITATS OF EARLY MARS

The geologic record of Mars has been assessed over the past several decades via remote sensing from orbital spacecraft and *in situ* via landers and rovers. Geomorphological features such as river channels, lake basins, and perhaps also ocean basins point to conditions that were conducive to the stability of surface liquid water (see summary in *Carr and Head*, 2010). Surface mineralogy also indicates variable degrees of aqueous alteration of primary igneous rocks and the precipitation of secondary minerals (*Ehlmann and Edwards*, 2014). These landforms and hydrated minerals mostly occur in terrains $>\sim 3$ Ga — Noachian — and Hesperian-aged terrains. Valley network systems are found typically in Noachian terrains (*Fassett and Head*, 2008a), while shorter-lived but massive outflow channels are found in Hesperian and younger terrains (*Tanaka*, 1997). Hydrated silicates are the dominant water-formed minerals found in most Noachian surfaces and include Fe/Mg phyllosilicates (smectites, chlorites), Al phyllosilicates (kaolinite, montmorillonite), opaline silica, prehnite, and zeolites. Hesperian units less commonly have phyllosilicates and more often have sulfates and opaline silica (*Bibring et al.*, 2006; *Murchie et al.*, 2009). Chloride deposits of ancient but uncertain age are also found in several locations in the southern highlands (*Osterloo et al.*, 2010) and carbonate-enriched rocks are found more rarely (*Niles et al.*, 2013).

Given this ample evidence for surface liquid water shaping the landscape, its role in chemical alteration, and our notional understanding of habitability and its requirements (see the chapter by Hoehler et al. in this volume), data from rovers and orbiters have now shown in detail that multiple types of habitable environments existed during the Noachian and Hesperian eras. Current research now focuses on better constraining the characteristics of specific martian habitats, explaining what processes contributed to their habitability, how they evolved over time, and ultimately, if they were ever inhabited. This includes assessing the potential of candidate locales to both provide and sustain key biologically relevant resources (e.g., energy, nutrients) and conditions over time. Here we broadly separate the assessed habitats into three endmembers: subaerial, subaqueous, and subsurface (Fig. 3). In reality it is rare for any one habitat to be in isolation and free from the influence of another; therefore, it is important to keep in mind how one environment might interact with and influence another over time and space.

3.1. Subaerial Environments

Subaerial environments are rock or soil surfaces that interact with liquid water but are not covered by standing bodies of water. Rather, dew, frost, precipitation, snow/ice-melt, or discharged groundwater chemically alter soils and rocks to create deposits that range from weakly altered rinds or rock coatings that are millimeter-thick (e.g., *Kraft et al.*, 2003; *Salvatore et al.*, 2013) to meter-scale weathering sequences (e.g., *Velbel*, 1988). These types of environments ultimately promote the formation of soils; water percolates through the rocks or sediments to cause chemical weathering and gradients in geochemical conditions, leading to the formation and accumulation of iron and aluminum

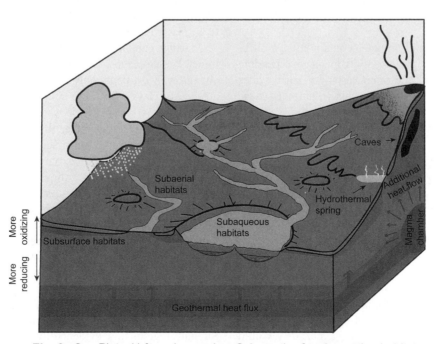

Fig. 3. See Plate 11 for color version. Schematic of early martian habitats.

oxides, smectites, kaolins, other phyllosilicates, silica, and sometimes carbonate (*Jackson et al.,* 1957). Depending on whether these areas are well or poorly drained, the remnant soils can form highly oxidizing or highly reducing conditions (e.g., *Jackson et al.,* 1957; *Chamley,* 1989). Paleosols (preserved ancient soil profiles) typically contain distinct horizons with mineralogical stratification that can be identified using remote sensing techniques.

On Mars there is mineralogical evidence for long-lived water-rock interactions in the form of Al phyllosilicates. These either formed from aqueous alteration of felsic materials or from water percolating from the surface downward, leading to the leaching of Fe, Mg, and Ca cations and the formation of Al-rich soil horizons (e.g., *Wray et al.,* 2008; *Bishop et al.,* 2008; *Carter et al.,* 2015;). At Mawrth Vallis, the detailed examination of mineral signatures has revealed characteristic stratigraphy that implies extended aqueous activity in an open system (i.e., one in which chemical constituents are free to move between reservoirs), namely an Al phyllosilicate-bearing horizon formed above an Fe/Mg smectite-bearing unit (*Bishop et al.,* 2008; *Wray et al.,* 2008; *McKeown et al.,* 2009). Similar stratigraphy is seen in Nili Fossae with kaolin minerals overlying Fe/Mg smectites (*Ehlmann et al.,* 2009) and elsewhere globally (*Ehlmann et al.,* 2011; *Carter et al.,* 2013, 2015).

Still lesser degrees of water-rock reactions can create ephemeral potentially habitable conditions within the pore space of rocks and soils at the surface. At the Mars Exploration Rover (MER) Spirit landing site, the presence of near-surface water (upper centimeters) in pore spaces in relatively recent history (Amazonian) is indicated by hydrated salts (sulfates, chlorides, bromides) uncovered by the rover (*Wang et al.,* 2006). Thin film water-rock reactions also formed coatings on the surfaces of igneous rocks (*Hurowitz et al.,* 2006), which for more porous rocks might provide sufficient water for habitable conditions as in Antarctica (*Friedmann,* 1982).

On Earth, subaerial environments contain a range of redox, temperature, pH, and water activity gradients depending on the local environment and often varying at small scales (see the chapter by Hoehler et al. in this volume). Because these environments are on or near the surface, they can support both photo- and chemotrophic microbial communities, such as microbial mats or endoliths (e.g., *Hays et al.,* 2017). These types of habitats may have been more extensive on an early Mars but discontinuous later in its history (e.g., see the chapter by Davila et al. in this volume). For example, during early wetter periods, there would likely be more environments similar to that indicated at Mawrth Vallis. During drier times subaerial environments may have been limited to periodic frosts and dews in most locations. Some locations, near long-lived snowpack (*Christensen,* 2003) or ice melt that survives obliquity cycles (*Bramson et al.,* 2017), may have provided refugia for surface life (*Hays et al.,* 2017; *Cousins,* 2011). But subaerial habitats can be limited by episodic aridity and freezing, i.e., they would have been intermittent and affected by any climate changes

that occurred over time. Furthermore, the immediate surface (upper ~2 m) would have been affected by galactic and solar radiation (*Pavlov et al.,* 2012).

3.2. Subaqueous Environments

Subaqueous habitats occur within a water column and therefore can include shallow playa systems, rivers, open and closed lake basins, and deep oceans. Aqueous environments can range in pH, temperature, and redox state depending on the source of water and whether they are open or closed systems (see the chapter by Hoehler et al. in this volume). Additionally, a single subaqueous environment can contain multiple potential habitats with varying geochemical characteristics according to, for example, depth — as in a stratified lake system with both oxic and anoxic waters. Sediment deposition in aqueous environments commonly creates characteristic landforms that are observable from orbiting spacecraft (e.g., *Malin and Edgett,* 2000) and, if water were saline or sufficiently long-lived to promote chemical weathering, can leave behind distinctive mineral deposits.

For much of its history Mars has hosted a range of subaqueous habitats that were first identified and characterized by their geomorphology (e.g., *Carr and Head,* 2010). An early discovery during the Mariner 9 mission was well-developed valley-network systems found across the southern highlands (e.g., *Marsusky,* 1973). These dendritic valley networks have relatively high stream-order and likely formed from precipitation and surface runoff, implying a once active hydrologic cycle on Mars (e.g., *Irwin et al.,* 2002; *Howard et al.,* 2005; *Hynek et al.,* 2010). It is still debated whether these networks were fed by rain or snow. They have been dated as being active during the Noachian through the early Hesperian (*Fassett and Head,* 2008a) and are often associated with hundreds of open- and closed-basin paleolake basins, some of which have fluviodeltaic deposits (e.g., *Fassett and Head,* 2008b; *Goudge et al.,* 2012).

Martian paleolake deposits show a range of aqueous mineral assemblages, indicating diverse aqueous paleogeochemistry. Most commonly associated with the paleolakes are Fe/Mg clays, which are compositionally similar to the surrounding terrains in their watersheds and therefore are likely detrital in origin rather than having formed *in situ*. Gale Crater, where the Mars Science Laboratory's (MSL) Curiosity rover is presently exploring, once hosted a closed-basin lake that persisted for thousands to millions of years (*Grotzinger et al.,* 2015). The suite of sedimentary rocks exposed along the rover's traverse includes clay-bearing fine-grained sediments, comprising detrital smectites both from the Gale walls and those formed *in situ*. These deposits are cross-cut by later mineral deposits, particularly Ca sulfate veins (*Rapin et al.,* 2016). Other irregular depressions in the southern highlands show evidence for clays in association with salts like chloride (*Osterloo et al.,* 2008; 2010), implying an evaporative setting. Still other craters in the Terra Sirenum region show evidence for more sulfur-rich waters with clay in association with polyhydrated Mg-sulfates and acid sulfates like alunite and jarosite in Cross and Columbus craters (*Wray et al.,* 2011; *Ehlmann et al.,* 2016b). These likely formed in a closed basin fed by acidic groundwater. Similar deposits are found in the Valles Marineris system, Aram Chaos, and Meridiani Planum (*Murchie et al.,* 2009). In contrast, Jezero Crater and McLaughlin Crater show evidence for a more alkaline groundwater-fed system with sedimentary clays and Mg carbonate (*Ehlmann et al.,* 2008, 2009; *Michalski et al.,* 2013; *Goudge et al.,* 2015).

Unlike the preserved valley network systems, the large Hesperian-aged outflow channels have a low stream order, implying less evolved systems that formed quickly and flowed directly into the northern lowlands (Fig. 1) (*Tanaka,* 1997). These landforms and observations of associated equipotential contacts (those that follow similar elevations as in terrestrial coastlines) along the hemispheric dichotomy and associated fan-shaped deposits have been interpreted as shorelines and deltas, respectively. This has led some to suggest that there was once a large body of water, an ocean, present in the northern hemisphere (e.g., *Baker et al.,* 1991; *Di Achille and Hynek,* 2010; *DiBiase et al.,* 2013). Much of the initial evidence for such an ocean came from early Viking Orbiter observations of landforms interpreted to be paleoshorelines (e.g., *Parker et al.,* 1989, 2010; *Clifford and Parker,* 2001). However, higher-resolution imagery from subsequent missions show many of the proposed shorelines undulate from the proposed equipotential, at times by kilometers, and others may be more consistent with other landforms like volcanic or glacial contacts (e.g., *Carr and Head,* 2003; *Malin and Edgett,* 2000; *Sholes et al.,* 2019). Advocates for the paleoshorelines have argued that these putative shorelines could have been modified over time by wind, isostatic rebound, true polar wander, or faulting leading to the more ambiguous modern features (*Parker et al.,* 2010). Additionally, geochemical evidence for a past ocean is also absent (e.g., evaporites) (*Ehlmann and Edwards,* 2014; *Bibring et al.,* 2006; *Pan et al.,* 2017), but some have explained this by invoking short-lived oceans, which received little to no sediment load (*Ghatan and Zimbelman,* 2006). Work by *Carr and Head* (2015) shows that it can be difficult to reconcile the volumes of water necessary to meet the proposed shorelines and instead suggest that the observed geomorphic landforms are consistent with short-lived fluvial and lacustrine features. In all, the observation of large outflow channels into the northern lowlands imply that large volumes of water flowed into the area at least episodically. Whether those waters were stable for extended periods in the form of an ocean or were short-lived is still an open debate.

Subaqueous settings on Earth have highly diverse biological communities that are dominated by phototrophs and microbes that can live both within the water column or in the lake (or ocean) sediments. In terrestrial lacustrine settings, near-shore, shallow waters that typically see a steady input of nutrients and are well within the photic zone are richly inhabited. The overlying water can also protect against UV radiation (*Fleischmann,* 1989). These near-shore regions can host microbial mats that trap mineral detritus (e.g., *Reid et al.,* 2000). On Earth, biosignatures are well preserved within deltas as sediment is quickly deposited and buried in near-shore shelves as precipitating minerals entomb organisms. These areas tend to be reducing and protect organic material from oxidative decomposition. On Mars, a challenge to lacustrine environments as long-term abodes of life is the surface radiation and regular, high-frequency climate change that is driven by obliquity and eccentricity cycles (see section 2) that have no Earth equivalent but would have caused cyclical aridity and subfreezing temperatures on <1-m.y. timescales.

3.3. Subsurface Environments

Subsurface habitats include any space with liquid water, sufficiently available nutrients, and an energy source with low enough temperatures (<129°C limit for active biota on Earth) and large enough pore space for life to exist (*Hays et al.,* 2017; *Onstott et al.,* 2019, and references therein). This includes shallow subsurface aquifers and caves, low-temperature groundwater and hydrothermal systems at depths from deep to shallow, and ice deposits with liquid water in the pore space. The potential for subsurface habitability is highly controlled by access to biochemically available energy (*Hoehler et al.,* 2004). Unlike surface environments with abundant solar energy, subsurface environments are typically heterogeneous with redox boundaries and fractures or zones of higher permittivity that support larger concentrations of biomass because there is greater energy from chemical reactions, and reactants can be replenished and waste products removed (*Onstott et al.,* 2019, and references therein). As such, subsurface habitats are typically dominated by chemolithotrophic processes. *Michalski et al.* (2013) proposed that martian life during the Noachian could exist as deep as ~6 km in depth (the modeled 120°C isotherm) before elevated temperatures or reduced pore space would become a limiting factor, implying that a quite large volume of the subsurface was accessible to putative life. On Earth, subsurface habitats are recognized primarily by specific mineral assemblages, which indicate subsurface fluid temperatures and chemistries, and by surface fluid chemistry and biotic activity detected at discharge sites.

Minerals and mineral suites that indicate long-lived subsurface aqueous alteration, as well as regions where subsurface fluids have emerged at the martian surface, have been detected on Mars. Many hydrated silicate minerals detected on the surface today are found within eroded Noachian terrains and are most consistent with a subsurface origin via diagenesis, hydrothermalism, or low-grade metamorphism, formed at temperatures that range from ambient to ~400°C in relatively closed systems (*Ehlmann et al.,* 2011). This includes spectroscopic evidence for serpentinization processes (e.g, *Ehlmann et al.,* 2010; *Amador et al.,* 2017, 2018). Specific minerals include Fe/Mg smectites, serpentine, chlorite, prehnite, illite, silica phases, and zeolites in distinct terrains. Some of the Fe/Mg phyllosilicates are associated with fracture systems (*Saper and Mustard*, 2013). Evidence for sulfates and Fe oxides formed by groundwater are also widespread (e.g., *Bibring et al.*, 2006; *Murchie et al.*, 2009).

Zones of discharge from hydrothermal systems are also detected from orbit (*Skok et al.*, 2010; *Thollot et al.*, 2012), as evidenced by silica, sulfate, and phyllosilicate-bearing deposits. The Spirit rover encountered volcaniclastic rocks and regolith that were highly enriched in opaline silica (*Squyres et al.*, 2008) and are interpreted to have formed either via fumarole-related acid-sulfate leaching or from the precipitation of silica-rich fluids from hot springs (*Squyres et al.*, 2008; *Ruff and Farmer*, 2016). As discussed in section 3.2, at both Gale Crater and in Meridiani Planum, *in situ* evidence indicates that multiple generations of groundwater circulated within the lake deposits well after deposition. These observations indicate that a vast array of potentially habitable subsurface environments existed during Mars' early history, varying in chemistry.

On Earth, chemoautotrophic microorganisms in the subsurface have metabolisms that exploit redox gradients that arise from the aqueous chemical interactions with the surrounding rock or sediments. For example, where fluids interact with olivine- and pyroxene-enriched primary rocks, serpentinization and other reactions from iron oxidation release bioaccessible H_2 (e.g., *Schulte et al.*, 2006). This H_2 can either be directly consumed by microbial methanogens or can react abiotically to form CH_4, which can also be used by methanotrophs. Fluids dissolve and transport nutrients and biologically active elements (e.g., C, N, P, S, K, Mb, Mg, Fe, etc.) from the surrounding bedrock or from the surface in circulating groundwaters. Redox fronts or zones of mixing — e.g., fracture surfaces, near-surface pore space where deep waters encounter diffusing atmospheric gases, or discharge areas at the surface — are particularly favorable because of the energy available from chemical reactions. These include interfaces between subsurface and surface (section 3.1) and subaqueous (section 3.2) environments. Protected from ionizing radiation and climate instability, subsurface environments may have been the most extensive habitat both volumetrically and throughout time on early Mars and perhaps extending to present. On Earth, concentrations of organic material in the subsurface are variable and range over orders of magnitude at small spatial scales (*Magnabosco et al.*, 2018; *Onstott et al.*, 2019) but are elevated in zones of enhanced permeability, in mineralized fractures and pore spaces, and in mineral deposits at springs.

4. A MARTIAN HABITABILITY PARADOX: RECONCILING EVIDENCE FOR SURFACE LIQUID WATER ON AN EARLY MARS WITH CLIMATE MODELS

The early Mars habitats discussed thus far all require liquid water and several require this water to be present at the surface. However, it is difficult to model an early Mars climate that would have sustained surface liquid water. Surface liquid water depends upon the surface temperature of the planet, which in turn is strongly controlled by solar radiation. However, early Mars was limited in the amount of solar radiation it received, not only due to its distance from the Sun but also due to the "faint young Sun" problem (*Ulrich*, 1975; *Feulner*, 2012). Initially, young stars are less luminous because as the star burns hydrogen, its core contracts and heats up, increasing the star's luminosity over time (*Hoyle and Schwarzchild*, 1955; *Schwarzchild*, 1958). This profoundly affects the amount of solar insolation a planetary body receives, and this in turn can have large climatic and geologic consequences, as with our own Earth (*Sagan and Mullen*, 1972; *Ulrich*, 1975). Early Mars only received ~75% of the solar insolation it does today. Assuming a planetary albedo of zero (i.e., all incoming radiation is absorbed by the planet), Mars' equilibrium surface temperature would be ≤210 K (*Wordsworth*, 2016). Even with this unrealistically low albedo where all light energy is absorbed, this temperature is far below the requisite temperature for stable liquid water (>273 K), yet we have discussed above the unequivocal evidence for surface liquid water during this time on Mars. Therefore, in order to reconcile this faint young Sun with the empirical evidence for surface liquid water during this time, the atmosphere needed to be a minimum of 63 K warmer, likely due to atmospheric greenhouse processes. This warming is 9× that which occurs on Mars today (for further review, see *Wordsworth*, 2016; *Haberle et al.*, 2017).

4.1. How to Warm a Cold Planet?

For the martian surface to have been sufficiently warm to sustain liquid water in the Noachian and Hesperian, there must have been a decrease in the amount of outgoing long-wavelength radiation (OLR) passing through the atmosphere to space. This can be achieved by introducing greenhouse gases (e.g., CO_2, H_2O, CH_4, H_2) that increase the atmospheric opacity in the infrared. There are three main questions that are still not fully understood: (1) What was the composition of this atmosphere? (2) How thick must it have been? (3) Can we properly account for the fate of this atmosphere?

The atmosphere's ability to retain infrared radiation is controlled in large part by its greenhouse gas constituents. This in turn is almost entirely determined by volcanic outgassing with minor contributions from impactors (*Wordsworth*, 2016). Carbon dioxide and water are typically considered to be the major constituents of the early martian atmosphere (*Grott et al.*, 2011), but the exact concentration is dependent on the oxygen fugacity of the mantle (*Gaillard and Scaillet*, 2014). Studies of martian meteorites indicate that the mantle was more reducing than Earth's mantle (*Wadhwa*, 2001), so there were likely to be contributions from H_2, CO, CH_4, and H_2S as well — all of which are efficient greenhouse gases. More recent work by *Wordsworth et al.* (2017) showed that collision-induced absorption (CIA) between CO_2-H_2 and CO_2-CH_4 can be promising mechanisms for retaining OLR, even when H_2 and CH_4 are at low atmospheric abundance levels (<10%), because these CIA absorptions occur within critical spectral windows that are otherwise transparent to the OLR. Such H_2 and CH_4 could result from iron oxidation or serpentinization in the surface and subsurface. Another proposed way to decrease the OLR is via CO_2 clouds in the high atmosphere; scattered infrared (IR) light from these clouds returns IR back down to the surface (*Forget and Pierrehumbert*, 1997). However, current models indicate early Mars would need to have had nearly 100% cloud coverage to significantly raise temperatures (*Forget et al.,* 2013; *Wordsworth et al.,* 2013), which can be difficult to realistically achieve. Still other workers have proposed outgassed sulfurous gases for raising the greenhouse temperature, given the ample observations of sulfur-bearing surface constituents (*Johnson et al.,* 2009; *Halevy et al.,* 2007).

Estimates for how thick the atmosphere must have been to maintain surface liquid water range between 2 and 10 bar of pCO_2 in the ancient atmosphere (*Ramirez and Craddock,* 2018; *Pollack et al.,* 1987). However, it has been difficult for modelers to realistically recreate these levels of atmospheric CO_2 (e.g., *Wordsworth,* 2016), and some suggest that Mars' early outgassed atmosphere was likely to contain less than 1 bar of CO_2 (*Lammer et al.,* 2013). Empirically derived estimates of paleoatmospheric pressures come from the size distributions of impact craters in Noachian terrains, given that the density of atmosphere will control the size of the impactor that can survive atmospheric entry. Estimates from impact craters in Dorsa Aeolis indicate paleopressures between 0.9 and 3.8 bar (*Kite et al,* 2014). Constraints from coupled modeling of $^{13}C/^{12}C$ values, volcanism, and loss to space (*Hu et al.,* 2015) suggest that the most likely maximum atmospheric pressure is between 1 and 1.8 bar. Furthermore, estimates of how much CO_2 could have been released from an early mantle range between 40 mbar to 1.4 bar of pCO_2, based on magma composition and estimates from observed lava flows (*Hirschmann and Withers,* 2008; *Craddock and Greeley,* 2009).

This is all in contrast to the modern 6-mbar average atmosphere of Mars. If the early martian atmosphere was indeed greater than at a minimum 1 bar, we must consider the loss mechanisms for much of this original atmosphere. Atmospheric CO_2 has one of two fates: It can either be lost to space, or it can be integrated into the martian surface as ice or carbonate. Loss mechanisms include impact ejection, thermal processes, and non-thermal processes. Impact ejection can remove either the entire atmosphere, or small amounts depending on impactor size and velocity (*Schlichting and Mukhopadhyay,* 2018). Thermal escape (i.e., Jean's escape) occurs when a particle reaches escape velocity and is able to escape Mars' gravitational well. Non-thermal escape includes processes such as sputtering (*Luhmann et al.,* 1992), photodissociation (*Lillis et al.,* 2017), and solar-wind pickup (*Hunten,* 1987; *Brain et al.,* 2015). Recent results informed by the Mars Atmosphere and Volatile Evolution (MAVEN) mission suggest that the process producing the largest loss of the early martian atmosphere was atmospheric escape, which

could have removed 1–2 bar [for a recent review of the processes governing the martian CO_2 inventory, see *Jakosky* (2019)]. Presently, the non-atmospheric carbon available for exchange on short timescales, the free-carbon reservoir (*Hu et al.,* 2015), amounts to ~47 mbar; this includes CO_2 ice at the martian poles (*Smith et al.,* 2009), in the subsurface at high latitudes (*Phillips et al.,* 2011), and CO_2 adsorbed in the martian regolith (*Zent and Quinn,* 1995). Carbonate minerals should also be a significant sink for atmospheric CO_2 (e.g., *Kahn,* 1985; *Pollack et al.,* 1987; *Catling,*1999); however, observations of carbonate have been limited to minor concentrations within dust (*Bandfield et al.,* 2003), martian meteorites (*Bridges et al.,* 2001), and *in situ* observations at both the Phoenix landing site (*Boynton et al.,* 2009) and at Columbia Hills (*Morris et al.,* 2010). Even accounting for the most extensive carbonate occurrences within an olivine-enriched unit in the Nili Fossae region (*Ehlmann et al.,* 2008; *Edwards and Ehlmann,* 2015), the observed mineralized CO_2 only accounts for a fraction of that necessary to warm the surface of Mars. Estimates of the global carbonate reservoir from current outcrops observed from orbit, as well as *in situ* carbonate soil detections, are ~300 mbar (*Hu et al.,* 2015). There could be up to several more bars of pCO_2 stored in the subsurface as carbonate that has yet to be exposed (*Wray et al.,* 2016), and glimpses of this possible subsurface reservoir are observed where large impact craters have exposed deeper rocks. As such, work continues to comb the martian surface for indicators of this potential deep carbon reservoir.

4.2. An Episodically Habitable Martian Surface?

Although atmospheric loss and the formation of a significant subsurface carbonate reservoir could perhaps have removed an early atmosphere that was sufficient for greenhouse warming (*Jakosky,* 2019), others have proposed that Mars' early history was characterized by a thin atmosphere with intermittent episodes of warmer conditions that would allow for surface liquid water and the persistence of deep lakes for thousands to few millions of years (i.e., the amount of time that Gale Crater hosted an active lake). For example, the "icy highlands" hypothesis (*Wordsworth et al.,* 2016) suggests that a cold early martian atmosphere may have still transported water in the form of ice and snow to the southern highlands, the source region for the valley network systems. Subsequent transient warming episodes could have melted these cold traps and produced the observed landforms. Perturbations from, for example, obliquity changes, release of reducing gases from water-rock reactions, volcanic activity, or impact events, may have supported transient warm climates. Periodic obliquity fluctuations change the distribution of water ice and the latitudes at which temperatures may exceed the melting point of ice (*Palumbo et al.,* 2018). Episodic release of reduced greenhouse gases, like CH_4 and H_2, when, for example, permafrost thaws, could also episodically elevate temperatures (*Wordsworth et al.,* 2017). Serpentinization is a geochemical process that can produce these reduced molecules as well, and there is evidence for this process early in Mars' history (*Chassefiere et al.,* 2013; *Amador et al.,* 2018). Methane and H_2 may also be outgassed during volcanism or produced during large impact events (*Segura et al.,* 2012; *Wordsworth et al.,* 2017). Thus, variability would have been a constant feature affecting early martian habitability, determining which habitats would have been more or less favorable for life over time. It is likely that subsurface habitats would have been the most protected from these dramatic climate variations.

5. PRESERVATION OF BIOSIGNATURES FROM ANCIENT MARS

Key to detecting evidence for any past martian life is the long-term preservation of biosignatures. Any early Mars biosignatures, whether from a surficial or subsurface environment, would need to have survived 3 b.y. of radiation, diagenesis, weathering, and erosion.

Mars is in some ways a challenging environment for the preservation of biosignatures, with several mechanisms that can lead to their destruction in the martian rock record. For example, ionizing radiation can break down organic molecules. Cosmic rays and highly energetic solar particles interact with the top 3 m of the surface, and models predict that in the top 4–6 cm of the surface, up to 99% of organic material would be altered or degraded after only 650 m.y. of exposure (*Hassler et al.,* 2014). Galactic cosmic rays and solar energetic particles interacting with strong oxidants can also create reactive oxygen species that then degrade organic molecules, including DNA (*Hassler et al.,* 2014; *Wadsworth and Cockell,* 2017). This is particularly troublesome when considering both the habitability and the preservation potential of the martian surface throughout its history. Ionizing radiation has been bombarding the surface for several billion years with little attenuation due to (1) the lack of a magnetic field to shield near-surface environments since ~4 Ga and (2) a thin atmosphere since ~3 Ga and possibly earlier. Thus habitats that are either protected from this surface radiation (e.g., subsurface habitats) or ones in which there may be quick burial of biomaterials (e.g., lacustrine with rapid burial) have higher potential for protecting biosignatures from radiative or oxidative breakdown.

The long-term preservation of putative biosignatures is also controlled by chemical alteration and physical weathering. Some Noachian-aged terrains have been fractured, heated, and chemically altered by impacts, volcanism, and groundwater interactions. These processes can also lead to the alteration and destruction of biosignatures. However, compared to Earth, these processes have been relatively mild. Without plate tectonics, crustal materials on Mars remain near the surface and are at most subject to low-grade metamorphism due to impact-induced geothermal processes, for example, as evidenced by the mineral phases observed to date across Mars (*Ehlmann et al.,* 2011). Volcanism, although active through the mid-Amazonian, was limited to specific regions. Although groundwater alteration was active through the mid-Hesperian, alteration assemblages observed on the surface today are generally mineralogically juvenile (*Tosca and Knoll,* 2009), implying limited duration of alteration. Thus, a cold, arid late Mars is favorable for the preservation of >3-Ga biosignatures, especially in comparison to similarly aged rocks on Earth.

The most favorable regions for the preservation of biosignatures are those where mineralizing fluids entomb and protect putative biomarkers or where rapid sediment accumulation buries biomarkers so that they are protected from radiation and oxidation. Given the modern-day high-radiation surface environment, the most promising exposures to search for preserved organics will be those that are more recently exposed, for example, along eroded valley walls or scarps.

6. KEY OPEN QUESTIONS

There is still much to explore and understand about the early astrobiological history of Mars. For a detailed overview of key questions and measurements for future Mars exploration, see *Ehlmann et al.* (2016a). Here, we highlight three important questions that are relevant to the discussions in this chapter, and the key measurements that will help shed light on them.

1. *What is the absolute chronology for Mars's surface and interior processes?* Martian chronology is currently established by combining radiometrically-derived ages from returned lunar samples from the Apollo missions and models of the rate of inner solar system impact flux for the Earth-Moon system, extrapolated to Mars (e.g., *Tanaka et al.,* 1992). The method is based on the assumptions that (1) older terrains have relatively more craters than younger terrains, because they have had more time to be impacted, and (2) Mars was impacted at the same relative rate as the Moon, accounting for scaling due to location and body properties. Although broadly these may be fair first-order assumptions, the evolution of each planetary surface is unique. The absolute age of key martian rock units is of the utmost importance for understanding the context for the habitable environments that have been discussed in this chapter, i.e., when they formed and the rate at which they changed. Additionally, by dating significant units, we will gain an important understanding of the interior evolution of the planet; for example, how the thermal and chemical evolution of the planet changed with time and when exactly the magnetic dynamo ceased to provide a magnetic field. Ideal samples will be volcanic lava flows, ash deposits, or impact melts, and the measurement of radioactive parent-radiogenic daughter isotopes. Some measurements are possible *in situ* with landers or rovers (*Farley et al.,* 2013; *Cohen et al.,* 2014; *Anderson et al.,* 2015), but the most precise radiometric measurements may necessitate returning samples to Earth.

2. *What was the composition and fate of the early martian atmosphere?* As discussed in section 4, the com-

position and fate of Mars' early atmosphere, during its most habitable time, is poorly understood (*Wordsworth,* 2016). Although progress has been made through modeling and now with measurements of Mars' modern atmosphere by the MAVEN spacecraft (*Jakosky et al.,* 2015), much more is to be gained by detailed study of rocks and minerals that formed in contact with, or with waters derived from, this early, elusive atmosphere. Detailed petrology, and isotopic analyses of these rocks at a hand-sample scale — including the possibility of sampling gas inclusions — will provide insight into the atmospheric composition, redox state, and pressure at the time of formation, as well as enhance our understanding of surface vs. subsurface volatile reservoirs. These are important considerations when trying to understand the factors that control habitability; constraining them will inform our understanding of which habitats might have been the longest lived or most conducive to life. Key measurements of these rock samples should include the isotopic composition of H, C, S, O, and N in order to understand fractionation between reservoirs and their petrologic contexts, as well as measurements of trapped gases in fluid inclusions in lavas and impact melts.

3. *Was early (or modern) Mars inhabited?* Ultimately, a key question that drives much of the Mars exploration program today is whether or not life ever took root on early Mars (*National Research Countil,* 2011). This question is key, not only to expand our understanding of the planet Mars, but also to better understand the context in which life on Earth exists within our solar system and indeed the universe. Answering this question is complicated, because a null result from one location does not necessarily mean that Mars was always uninhabited everywhere. As such, selecting sites and rocks to be sampled for biosignatures should be done diligently with great consideration for locales that would have the highest potential for having long-lived habitats on an early Mars and that may be representative of the multiple types of habitats that were widespread on the planet. Multiple sites across the planet must be examined in the search for ancient life, particularly given the extraordinary diversity of habitat types (see section 3). Biosignatures might exist as structures (e.g., microbialites, including stromatolites) like the oldest fossils on Earth at ~3.5 Ga. We should be prepared, however, for a search where microscopic biosignatures are of equal or greater importance than macrostructures, given the greater age and perhaps shorter cumulative duration of habitability in some martian habitats as well as the relative importance of groundwaters through time. The search for organic compounds and/or compositional patterns (e.g., in relative abundances of stable carbon or sulfur isotopes, specific mineral-organic assemblages and their morphologies) is key for finding many types of microbial life and can be larger-scale signatures that are pointers to microfossils. Consequently, an array of observational techniques applied *in situ* and on samples at a variety of sites on Mars will be required in the pursuit of the question of life on Mars (see the chapter by Johnson et al. in this volume).

7. SUMMARY

Mars has long been of interest to scientists working toward understanding life's place on Earth, our solar system, and beyond. The martian surface is hardly habitable today, with low atmospheric pressure and frigid surface temperatures. However, the planet's well-preserved ancient rocks provide abundant evidence for a once-active hydrologic cycle that maintained spatially widespread and chemically diverse habitats — surface, subaqueous, and subsurface — early in its history. The evidence is recorded in both geomorphic features and mineralogical signatures that imply flowing water on the surface and long-lived waters in the subsurface. There is still much to learn about how Mars was able to maintain these habitats for hundreds of millions of years, if not longer, given its orbital configuration and the "faint young Sun," which sustained equilibrium temperatures tens of degrees below the freezing point of water. With future detailed measurements from orbital and landed missions, as well as hopefully one day samples returned to Earth, we will be able to piece together the complicated and important history of early Mars, understand its changing habitability, and determine whether Mars was ever inhabited.

REFERENCES

Acuña M. H., Connerney J. E. P., Ness N. F., Lin R. P., Mitchell D., Carlson C. W., McFadden J., Anderson K. A., Rème H., Mazelle C., Vignes D., Wasilewski P., and Cloutier P. (1999) Global distribution of crustal magnetization discovered by the Mars Global Surveyor MAG/ER experiment. *Science, 284(5415),* 790–793.

Aharonson O., Zuber M. T., and Rothman D. H. (2001) Statistics of Mars' topography from the Mars Orbiter Laser Altimeter: Slopes, correlations, and physical models. *J. Geophys. Res.–Planets, 106(E10),* 23723–23735, DOI: 10.1029/2000JE001403.

Allwood A. C., Walter M. R., Kamber B. S., Marshall C. P., and Burch I. W. (2006) Stromatolite reef from the Early Archaean era of Australia. *Nature, 441,* 714.

Amador E. S., Bandfield J. L., Brazelton W. J., and Kelley D. (2017) The Lost City hydrothermal field: A spectroscopic and astrobiological analogue for Nili Fossae, Mars. *Astrobiology, 17(11),* 1138–1160, DOI: 10.1089/ast.2016.1606.

Amador E. S., Bandfield J. L., and Thomas N. H. (2018) A search for minerals associated with serpentinization across Mars using CRISM spectral data. *Icarus, 311,* 113–134, DOI: 10.1016/j.icarus.2018.03.021.

Anderson F. S., Levine J., and Whitaker T. J. (2015) Rb-Sr resonance ionization geochronology of the Duluth Gabbro: A proof of concept for *in situ* dating on the Moon. *Rapid Commun. Mass Spectrom., 29(16),* 1457–1464, DOI: 10.1002/rcm.7253.

Baker V. R., Strom R. G., Gulick V., Kargel J. S., Komatsu G., and Kale V. S. (1991) Ancient oceans, ice sheets and the hydrological cycle on Mars. *Nature, 352,* 589–594.

Bandfield J. L., Glotch T. D., and Christensen P. R. (2003) Spectroscopic Identification of carbonate minerals in the martian dust. *Science, 301(5636),* 1084–1087, DOI: 10.1126/science.1088054.

Berman D. C. and Hartmann W. K. (2002) Recent fluvial, volcanic, and tectonic activity on the Cerberus Plains of Mars. *Icarus, 159(1),* 1–17, DOI: 10.1006/icar.2002.6920.

Bibring J.-P., Langevin Y., Mustard J. F., Poulet F., Arvidson R. E., Gendrin A., Gondet B., et al. (2006) Global mineralogical and aqueous mars history derived from OMEGA/Mars Express data. *Science, 312(5772),* 400–404, DOI:10.1126/science.1122659.

Bishop J. L., Dobrea E. Z. N., McKeown N. K., Parente M., Ehlmann B. L., Michalski J. R., Milliken R. E., et al. (2008) Phyllosilicate diversity and past aqueous activity revealed at Mawrth Vallis, Mars. *Science, 321(5890),* 830–833.

Boynton W. V, Ming D. W., Kounaves S. P., Young S. M. M., Arvidson R. E., Hecht M. H., Hoffman J., et al. (2009) Evidence for calcium carbonate at the Mars Phoenix landing site. *Science, 325(5936),* 61–64, DOI: 10.1126/science.1172768.

Brain D. A., McFadden J. P., Halekas J. S., Connerney J. E. P., Bougher S. W., Curry S., Dong C. F., et al. (2015) The spatial distribution of planetary ion fluxes near Mars observed by MAVEN. *Geophys. Res. Lett. 42 (21),* 9142–9148.

Bramson A. M., Byrne S., and Bapst J. (2017) Preservation of midlatitude ice sheets on Mars. *J. Geophys. Res.–Planets, 122(11),* 2250–2266, DOI: 10.1002/2017JE005357.

Bridges J. C., Catling D. C., Saxton J. M., Swindle T. D., Lyon I. C., and Grady M. M. (2001) Alteration assemblages in martian meteorites: Implications for near-surface processes. *Space Sci. Rev., 96(1),* 365–392, DOI: 10.1023/A:1011965826553.

Carr M. H. and Head J. W. (2003) Oceans on Mars: An assessment of the observational evidence and possible fate. *J.Geophys. Res., 108(E5),* 5042, DOI: 10.1029/2002JE001963.

Carr M. H. and Head J. W. (2010) Geologic history of Mars. *Earth Planet. Sci. Lett., 294(3–4),* 185–203, DOI: 10.1016/j. epsl.2009.06.042.

Carr M. H. and Head J. W. (2015) Martian surface/near-surface water inventory: Sources, sinks, and changes with time. *Geophys. Res. Lett., 42(3),* 726–732, DOI: 10.1002/2014GL062464.

Carter J., Poulet F., Bibring J.-P., Mangold N., and Murchie S. (2013) Hydrous minerals on Mars as seen by the CRISM and OMEGA imaging spectrometers: Updated global view. *J. Geophys. Res.– Planets, 118(4),* 831–858, DOI: 10.1029/2012JE004145.

Carter J., Loizeau D., Mangold N., Poulet F., and Bibring J.-P. (2015) Widespread surface weathering on early Mars: A case for a warmer and wetter climate. *Icarus, 248,* 373–382, DOI: 10.1016/j. icarus.2014.11.011.

Catling D. C. (1999) A chemical model for evaporites on early Mars: Possible sedimentary tracers of the early climate and implications for exploration. *J. Geophys. Res.–Planets, 104(E7),* 16453–16469, DOI: 10.1029/1998JE001020.

Chamley H. (1989) *Clay Sedimentology.* Springer-Verlag, Berlin. 623 pp.

Chassefière E., Langlais B., Quesnel Y., and Leblanc F. (2013) The fate of early Mars' lost water: The role of serpentinization. *J. Geophys. Res.–Planets, 118(5),* 1123–1134, DOI: 10.1002/jgre.20089.

Christensen P. R. (2003) Formation of recent martian gullies through melting of extensive water-rich snow deposits. *Nature, 422(6927),* 45–48, DOI: 10.1038/nature01436.

Clifford S. M. and Parker T. J. (2001) The evolution of the martian hydrosphere: Implications for the fate of a primordial ocean and the current state of the northern plains. *Icarus, 154(1),* 40–79, DOI: 10.1006/icar.2001.6671.

Clifford S. M., Lasue J., Heggy E., Boisson J., McGovern P., and Max M. D. (2010) Depth of the martian cryosphere: Revised estimates and implications for the existence and detection of subpermafrost groundwater. *J. Geophys. Res., 115(E07001),* DOI: 10.1029/2009JE003462.

Cockell C. S. (2011) Vacant habitats in the universe. *Trends Ecol. Evol., 26(2),* 73–80, DOI: 10.1016/j.tree.2010.11.004.

Cockell C. S., Balme M., Bridges J. C., Davila A., and Schwenzer S. P. (2012) Uninhabited habitats on Mars. *Icarus, 217(1),* 184–193, DOI: 10.1016/j.icarus.2011.10.025.

Cohen B. A., Miller J. S., Li Z.-H., Swindle T. D., and French R. A. (2014) The Potassium-Argon Laser Experiment (KArLE): *In situ* geochronology for planetary robotic missions. *Geostand. Geoanal. Res., 38(4),* 421–439, DOI: 10.1111/j.1751-908X.2014.00319.x.

Cousins C. R. (2011) Volcano-ice interaction: A haven for life on Mars? *Astron. Geophys, 52(1),* 1.36–1.38.

Craddock R. A. and Greeley R. (2009) Minimum estimates of the amount and timing of gases released into the martian atmosphere from volcanic eruptions. *Icarus, 204(2),* 512–526, DOI: 10.1016/j. icarus.2009.07.026.

Dauphas N. and Pourmand A. (2011) Hf-W-Th evidence for rapid growth of Mars and its status as a planetary embryo. *Nature, 473,* 489.

Di Achille G. and Hynek B. M. (2010) Deltas and valley networks on Mars: Implications for a global hydrosphere. In *Lakes on Mars* (N. A. Cabrol and E. A. Grin, eds.), pp. 223–248. Elsevier, Amsterdam.

DiBiase R. A., Limaye A. B., Scheingross J. S., Fischer W. W., and

Lamb M. P. (2013) Deltaic deposits at Aeolis Dorsa: Sedimentary evidence for a standing body of water on the northern plains of Mars. *J. Geophys. Res.–Planets, 118(6),* 1285–1302, DOI: 10.1002/ jgre.20100.

Edwards C. S. and Ehlmann B. L. (2015) Carbon sequestration on Mars. *Geology, 43(10),* 863–866, DOI: 10.1130/G36983.1.

Ehlmann B. L. and Edwards C. S. (2014) Mineralogy of the martian surface. *Annu. Rev. Earth Planet. Sci., 42(1),* 291–315, DOI: 10.1146/annurev-earth-060313-055024.

Ehlmann B. L., Mustard J. F., Murchie S. L., Poulet F., Bishop J. L., Brown A. J., Calvin W. M., et al. (2008) Orbital identification of carbonate-bearing rocks on Mars. *Science, 322(5009),* 1828–1832.

Ehlmann B. L., Mustard J. F., Swayze G. A., Clark R. N., Bishop J. L., Poulet F., Des Marais D. J., et al. (2009) Identification of hydrated silicate minerals on Mars using MRO-CRISM: Geologic context near Nili Fossae and implications for aqueous alteration. *J. Geophys. Res.–Planets, 114(10),* 1–33, DOI: 10.1029/2009JE003339.

Ehlmann B. L., Mustard J. F., and Murchie S. L. (2010) Geologic setting of serpentine deposits on Mars. *Geophys. Res. Lett., 37(6),* 1–5, DOI: 10.1029/2010GL042596.

Ehlmann B. L., Mustard J. F., Murchie S. L., Bibring J.-P., Meunier A., Fraeman A. A., and Langevin Y. (2011) Subsurface water and clay mineral formation during the early history of Mars. *Nature, 479(7371),* 53–60, DOI: 10.1038/nature10582.

Ehlmann B. L., Anderson F. S., Andrews-Hanna J., Catling D. C., Christensen P. R., Cohen B. A., Dressing C. D., et al. (2016a) The sustainability of habitability on terrestrial planets: Insights, questions, and needed measurements from Mars for understanding the evolution of Earth-like worlds. *J. Geophys. Res.–Planets, 121(10),* 1927–1961, DOI: 10.1002/2016JE005134.

Ehlmann B. L., Swayze G. A., Milliken R. E., Mustard J. F., Clark R. N., Murchie S. L., Breit G. N., et al. (2016b) Discovery of alunite in Cross crater, Terra Sirenum, Mars: Evidence for acidic, sulfurous waters. *Am. Mineral., 101(7),* 1527–1542, DOI: 10.2138/am-2016-5574.

Elkins-Tanton L. T. (2008) Linked magma ocean solidification and atmospheric growth for Earth and Mars. *Earth Planet. Sci. Lett., 271(1),* 181–191, DOI: 10.1016/j.epsl.2008.03.062.

Elkins-Tanton L. T., Zaranek S. E., Parmentier E. M., and Hess P. C. (2005) Early magnetic field and magmatic activity on Mars from magma ocean cumulate overturn. *Earth Planet. Sci. Lett., 236(1–2),* 1–12, DOI: 10.1016/j.epsl.2005.04.044.

Engwall E., Eriksson A. I., Cully C. M., André M., Torbert R., and Vaith H. (2008) Earth's ionospheric outflow dominated by hidden cold plasma. *Nature Geosci., 2,* 24, DOI: 10.1038/ngeo387.

Farley K. A., Hurowitz J. A., Asimow P. D., Jacobson N. S., and Cartwright J. A. (2013) A double-spike method for K-Ar measurement: A technique for high precision *in situ* dating on Mars and other planetary surfaces. *Geochim. Cosmochim. Acta, 110,* 1–12, DOI: 10.1016/j.gca.2013.02.010.

Fassett C. I. and Head J. W. (2008a) The timing of martian valley network activity: Constraints from buffered crater counting. *Icarus, 195(1),* 61–89, DOI: 10.1016/j.icarus.2007.12.009.

Fassett C. I. and Head J. W. (2008b) Valley network-fed, open-basin lakes on Mars: Distribution and implications for Noachian surface and subsurface hydrology. *Icarus, 198(1),* 37–56, DOI: 10.1016/j. icarus.2008.06.016.

Fassett C. I. and Minton D. A. (2013) Impact bombardment of the terrestrial planets and the early history of the solar system. *Nature Geosci., 6,* 520, DOI: 10.1038/ngeo1841.

Feldman W. C., Prettyman T. H., Maurice S., Plaut J. J., Bish D. L., Vaniman D. T., Mellon M. T., et al. (2004) Global distribution of near-surface hydrogen on Mars. *J. Geophys. Res.–Planets, 109(E9),* DOI: 10.1029/2003JE002160.

Feulner G. (2012) The faint young Sun problem. *Rev. Geophys., 50(2),* 1–30, DOI: 10.1029/2011RG000375.

Fleischmann E. M. (1989) The measurement and penetration of ultraviolet radiation into tropical marine water. *Limnol. Oceanogr., 34(8),* 1623–1629, DOI: 10.4319/lo.1989.34.8.1623.

Forget F. and Pierrehumbert R. T. (1997) Warming early Mars with carbon dioxide clouds that scatter infrared radiation. *Science, 278(5341),* 1273–1276, DOI: 10.1126/science.278.5341.1273.

Forget F., Haberle R. M., Montmessin F., Levrard B., and Head J. W. (2006) Formation of glaciers on Mars by atmospheric precipitation at high obliquity. *Science, 311(5759),* 368–371, DOI: 10.1126/

science.1120335.

Forget F., Wordsworth R., Millour E., Madeleine J.-B., Kerber L., Leconte J., Marcq E., and Haberle R. M. (2013) 3D modeling of the early martian climate under a denser CO_2 atmosphere: Temperatures and CO_2 ice clouds. *Icarus, 222,* 81–99, DOI: 10.1016/j.icarus.2012.10.019.

Friedmann E. I. (1982) Endolithic microorganisms in the Antarctic cold desert. *Science, 215(4536),* 1045–1053, DOI: 10.1126/science.215.4536.1045.

Gaillard F. and Scaillet B. (2014) A theoretical framework for volcanic degassing chemistry in a comparative planetology perspective and implications for planetary atmospheres. *Earth Planet. Sci. Lett., 403,* 307–316, DOI: 10.1016/j.epsl.2014.07.009.

Ghatan G. J. and Zimbelman J. R. (2006) Paucity of candidate coastal constructional landforms along proposed shorelines on Mars: Implications for a northern lowlands-filling ocean. *Icarus, 185(1),* 171–196, DOI: 10.1016/j.icarus.2006.06.007.

Golombek M. P. and Phillips R. J. (2010) Mars tectonics. In *Planetary Tectonics* (T. R. Watters and R. A. Schultz, eds.), pp. 183–232. Cambridge Univ., Cambridge.

Goudge T. A., Head J. W., Mustard J. F., and Fassett C. I. (2012) An analysis of open-basin lake deposits on Mars: Evidence for the nature of associated lacustrine deposits and post-lacustrine modification processes. *Icarus, 219(1),* 211–229, DOI: 10.1016/j.icarus.2012.02.027.

Goudge T. A., Aureli K. L., Head J. W., Fassett C. I., and Mustard J. F. (2015) Classification and analysis of candidate impact crater-hosted closed-basin lakes on Mars. *Icarus, 260,* 346–367, DOI: 10.1016/j.icarus.2015.07.026.

Grott M., Morschhauser A., Breuer D., and Hauber E. (2011) Volcanic outgassing of CO_2 and H_2O on Mars. *Earth Planet. Sci. Lett., 308(3–4),* 391–400, DOI: 10.1016/j.epsl.2011.06.014.

Grott M., Baratoux D., Hauber E., Sautter V., Mustard J., Gasnault O., Ruff S. W., et al. (2013) Long-term evolution of the martian crust-mantle system. *Space Sci. Rev., 174(1),* 49–111, DOI: 10.1007/s11214-012-9948-3.

Grotzinger J. P., Sumner D. Y., Kah L. C., Stack K., Gupta S., Edgar L., Rubin D., et al. (2014) A habitable fluvio-lacustrine environment at Yellowknife Bay, Gale Crater, Mars. *Science, 343(6169),* 1242777.

Grotzinger J. P., Gupta S., Malin M. C., Rubin D. M., Schieber J., Siebach K., Sumner D. Y., et al. (2015) Deposition, exhumation, and paleoclimate of an ancient lake deposit, Gale Crater, Mars. *Science, 350(6257),* aac7575, DOI: 10.1126/science.aac7575.

Guo J., Zeitlin C., Wimmer-Schweingruber R., Hassler D. M., Köhler J., Ehresmann B., Böttcher S., Böhm E., and Brinza D. E. (2017) Measurements of the neutral particle spectra on Mars by MSL/RAD from 2015-11-15 to 2016-01-15. *Life Sci. Space Res., 14,* 12–17, DOI: 10.1016/J.LSSR.2017.06.001.

Haberle R. M. et al. (2001) On the possibility of liquid water on present-day Mars. *J. Geophys. Res., 106,* E10.

Haberle R. M., Catling D. C., Carr M. H., and Zahnle K. J. (2017) The early Mars climate system. In *The Atmosphere and Climate of Mars* (R. M. Haberle et al., eds.), pp. 526–568. Cambridge Univ., Cambridge.

Halevy I., Zuber M. T., and Schrag D. P. (2007) A sulfur dioxide climate feedback on early Mars. *Science, 318(5858),* 1903–1907, DOI: 10.1126/science.1147039.

Hartmann W. and Neukam G. (2001) Cratering chronology and the evolution of Mars. *Space Sci. Rev., 96,* 165–194.

Hashimoto G. L., Abe Y., and Sugita S. (2007) The chemical composition of the early terrestrial atmosphere: Formation of a reducing atmosphere from CI-like material. *J. Geophys. Res.–Planets, 112(5),* 1–12, DOI: 10.1029/2006JE002844.

Hassler D. M., Zeitlin C., Wimmer-Schweingruber R. F., Ehresmann B., Rafkin S., Eigenbrode J. L., Brinza D. E., et al. (2014) Mars' surface radiation environment measured with the Mars Science Laboratory's Curiosity Rover. *Science, 343(6169),* 124497, DOI: 10.1126/science.1244797.

Hays L. E., Graham H. V., Des Marais D. J., Hausrath E. M., Horgan B., McCollom T. M., Parenteau M. N., et al. (2017) Biosignature preservation and detection in Mars analog environments. *Astrobiology, 17(4),* 363–400, DOI: 10.1089/ast.2016.1627.

Head J. W., Mustard J. F., Kreslavsky M. A., Milliken R. E., and Marchant D. R. (2003) Geological evidence for recent ice ages on Mars. In *AGU Fall Meeting Abstracts,* Abstract #P32B-02. American Geophysical Union, Washington, DC.

Hecht M. H. (2002) Metastability of liquid water on Mars. *Icarus, 156(2),* 373–386, DOI: 10.1006/icar.2001.6794.

Helffrich G. (2017) Mars core structure — Concise review and anticipated insights from InSight. *Progr. Earth Planet. Sci., 4(1),* 24, DOI: 10.1186/s40645-017-0139-4.

Hirschmann M. M. and Withers A. C. (2008) Ventilation of CO_2 from a reduced mantle and consequences for the early martian greenhouse. *Earth Planet. Sci. Lett., 270(1),* 147–155, DOI: 10.1016/j.epsl.2008.03.034.

Hoehler T. M. (2004) Biological energy requirements as quantitative boundary conditions for life in the subsurface. *Geobiology, 2,* 205–215.

Hoehler T. M. (2007) An energy balance concept for habitability. *Astrobiology, 7(6),* 824–838, DOI: 10.1089/ast.2006.0095.

Horowitz N. H., Hobby G. L., and Hubbard J. S. (1976) The Viking carbon assimilation experiments: Interim report. *Science, 194(4271),* 1321–1322, DOI: 10.1126/science.194.4271.1321.

Howard A. D., Moore J. M., and Irwin R. P. (2005) An intense terminal epoch of widespread fluvial activity on early Mars: 1. Valley network incision and associated deposits. *J. Geophys. Res.–Planets, 110(E12),* DOI: 10.1029/2005JE002459.

Hoyle F. and Schwarzschild M. (1955) On the evolution of type II stars. *Astrophys. J. Suppl., 2,* 1, DOI: 10.1086/190015.

Hu R., Kass D. M., Ehlmann B. L., and Yung Y. L. (2015) Tracing the fate of carbon and the atmospheric evolution of Mars. *Nature Commun., 6,* 10003, DOI: 10.1038/ncomms10003.

Hunten D. M., Pepin R. O., and Walker J. C. G. (1987) Mass fractionation in hydrodynamic escape. *Icarus, 69,* 532–549.

Hurowitz J. A., McLennan S. M., Tosca N. J., Arvidson R. E., Michalski J. R., Ming D. W., Schröder C., and Squyres S. W. (2006) *In situ* and experimental evidence for acidic weathering of rocks and soils on Mars. *J. Geophys. Res.–Planets, 111(E2),* DOI: 10.1029/2005JE002515.

Hynek B. M., Beach M., and Hoke M. R. T. (2010) Updated global map of martian valley networks and implications for climate and hydrological processes. *J. Geophys. Res., 115,* E09008, DOI: 10.1029/2009JE003548.

Hynek B. M., Osterloo M. K., and Kierein-Young K. S. (2015) Late-stage formation of martian chloride salts through ponding and evaporation. *Geology, 43(9),* 787–790, DOI: 10.1130/G36895.1.

Irwin R. P., Maxwell T. A., Howard A. D., Craddock R. A., and Leverington D. W. (2002) A large paleolake basin at the head of Ma'adim Vallis, Mars. *Science, 296(5576),* 2209–2212, DOI: 10.1126/science.1071143.

Jackson M. L. (1957) Frequency distribution of clay minerals in major great soil groups as related to the factors of soil formation. *Clays Clay Miner., 6(1),* 133–143, DOI: 10.1346/CCMN.1957.0060111.

Jakosky B. M. (2019) The CO_2 inventory on Mars. *Planet. Space Sci., 175,* 52–59, DOI: 10.1016/j.pss.2019.06.002.

Jakosky B. M. and Jones J. (1997) The history of martian volatiles. *Rev. Geophys., 35(1),* 1–16.

Jakosky B. M., Henderson B. G., and Mellon M. T. (1995) Chaotic obliquity and the nature of the martian climate. *J. Geophys. Res.–Planets, 100(E1),* 1579–1584, DOI: 10.1029/94JE02801.

Jakosky B. M., Lin R. P., Grebowsky J. M., Luhmann J. G., Mitchell D. F., Beutelschies, G., Priser T., et al. (2015) The Mars Atmosphere and Volatile Evolution (MAVEN) mission. *Space Sci. Rev., 195(1),* 3–48, DOI: 10.1007/s11214-015-0139-x.

Johnson S. S., Pavlov A. A., and Mischna M. A. (2009) Fate of SO_2 in the ancient martian atmosphere: Implications for transient greenhouse warming. *J. Geophys. Res.–Planets, 114(11),* 1–16, DOI: 10.1029/2008JE003313.

Kahn R. (1985) The evolution of CO_2 on Mars. *Icarus, 62(2),* 175–190, DOI: 10.1016/0019-103(85)90116-2.

Kasting J. F., Whitmire D. P., and Reynolds R. T. (1993) Habitable zones around main sequence stars. *Icarus, 101(1),* 108–128, DOI: 10.1006/icar.1993.1010.

Kite E. S., Williams J.-P., Lucas A., and Aharonson O. (2014) Low palaeopressure of the martian atmosphere estimated from the size distribution of ancient craters. *Nature Geosci., 7,* 335, DOI: 10.1038/ngeo2137.

Kite E. S. (2019) Geologic constraints on early Mars climate. *Space Sci. Rev., 215(1),* 10, DOI: 10.1007/s11214-018-0575-5.

Klein H. P. (1978) The Viking biological experiments on Mars. *Icarus,*

34(3), 666–674, DOI: 10.1016/0019-1035(78)90053-2.

Kopparapu R. K., Ramirez R., Kasting J. F., Eymet V., Robinson T. D., Mahadevan S., Terrien R. C., Domagal-Goldman S., Meadows V., and Deshpande R. (2013) Habitable zones around main-sequence stars: New estimates. *Astrophys. J., 765(2)*, DOI: 10.1088/0004-637X/765/2/131.

Kraft M. D. (2003) Effects of pure silica coatings on thermal emission spectra of basaltic rocks: Considerations for martian surface mineralogy. *Geophys. Res. Lett., 30(24)*, 2288, DOI: 10.1029/2003GL018848.

Lammer H., Kasting J. F., Chassefière E., Johnson R. E., Kulikov Y. N., and Tian F. (2008) Atmospheric escape and evolution of terrestrial planets and satellites. *Space Sci. Rev., 139(1–4)*, 399–436, DOI: 10.1007/s11214-008-9413-5.

Lammer H., Chassefière E., Karatekin Ö., Morschhauser A., Niles P. B., Mousis O., Odert P., et al. (2013) Outgassing history and escape of the martian atmosphere and water inventory. *Space Sci. Rev., 174(1–4)*, 113–154, DOI: 10.1007/s11214-012-9943-8.

Laskar J., Correia A. C. M., Gastineau M., Joutel F., Levrard B., and Robutel P. (2004) Long term evolution and chaotic diffusion of the insolation quantities of Mars. *Icarus, 170(2)*, 343–364, DOI: 10.1016/j.icarus.2004.04.005.

Lee D.-C. and Halliday A. N. (1997) Core formation on Mars and differentiated asteroids. *Nature, 388*, 854, DOI: 10.1038/42206.

Leighton R. B., Murray B. C., Sharp R. P., Allen J. D., and Sloan R. K. (1965) Mariner IV photography of Mars: Initial results. *Science, 149(3684)*, 627–630, DOI: 10.1126/science.149.3684.627.

Lillis R. J., Robbins S., Manga M., Halekas J. S., and Frey H. V. (2013) Time history of the martian dynamo from crater magnetic field analysis. *J. Geophys. Res.–Planets, 118(7)*, 1488–1511, DOI: 10.1002/jgre.20105.

Lillis R. J., Deighan J., Fox J. L., Bougher S. W., Lee Y., Combi M. R., Cravens T. E., et al. (2017) Photochemical escape of oxygen from Mars: First results from MAVEN *in situ* data. *J. Geophys. Res.—Space Phys., 122(3)*, 3815–3836, DOI: 10.1002/2016JA023525.

Lowell P. (1906) *Mars and Its Canals*. MacMillan, New York.

Luhmann J. G., Johnson R. E., and Zhang M. H. G. (1992) Evolutionary impact of sputtering of the martian atmosphere by OfI pickup ions. *Geophys. Res. Lett., 19*, 2151–2154.

Magnabosco C., Moore K. R., Wolfe J. M., and Fournier G. P. (2018) Dating phototrophic microbial lineages with reticulate gene histories. *Geobiology, 16*, 179–189, DOI: 10.1111/gbi.12273.

Malin M. C. and Edgett K. S. (2000) Sedimentary rocks of early Mars. *Science, 290(5498)*, 1927–1937, DOI: 10.1126/science.290.5498.1927.

Masursky H. (1973) An overview of geological results from Mariner 9. *J. Geophys. Res., 78(20)*, 4009–4030, DOI: 10.1029/JB078i020p04009.

McKeown N. K., Bishop J. L., Noe Dobrea E. Z., Ehlmann B. L., Parente M., Mustard J. F., Murchie S. L., et al. (2009) Characterization of phyllosilicates observed in the central Mawrth Vallis region, Mars, their potential formational processes, and implications for past climate. *J. Geophys. Res.–Planets, 114(E2)*, DOI: 10.1029/2008JE003301.

McSween H. Y. (2002) The rocks of Mars, from far and near. *Meteoritics & Planet. Sci., 37*, 7–25.

Michalski J. R., Cuadros J., Niles P. B., Parnell J., Rogers A. D., and Wright S. P. (2013) Groundwater activity on Mars and implications for a deep biosphere. *Nature Geosci., 6(2)*, 133–138, DOI: 10.1038/ngeo1706.

Möhlmann D., Niemand M., Formisano V., Savijärvi H., and Wolkenberg P. (2009) Fog phenomena on Mars. *Planet. Space Sci., 57*, 14–15, DOI: 10.1016/j.pss.2009.08.003.

Mojzsis S. J., Arrhenius G., McKeegan K. D., Harrison T. M., Nutman A. P., and Friend C. R. L. (1996) Evidence for life on Earth before 3,800 million years ago. *Nature, 384(6604)*, 55–59, DOI: 10.1038/384055a0.

Morris R. V, Ruff S. W., Gellert R., Ming D. W., Arvidson R. E., Clark B. C., Golden D. C., et al. (2010) Identification of carbonate-rich outcrops on Mars by the Spirit Rover. *Science, 329(5990)*, 421–424, DOI: 10.1126/science.1189667.

Murchie S. L., Mustard J. F., Ehlmann B. L., Milliken R. E., Bishop J. L., McKeown N. K., Noe Dobrea E. Z., et al. (2009) A synthesis of martian aqueous mineralogy after 1 Mars year of observations from the Mars Reconnaissance Orbiter. *J. Geophys. Res.–Planets,*

114(E2), DOI: 10.1029/2009JE003342.

Niles P. B., Catling D. C., Berger G., Chassefière E., Ehlmann B. L., Michalski J. R., Morris R., Ruff S. W., and Sutter B. (2013) Geochemistry of carbonates on Mars: Implications for climate history and nature of aqueous environments. *Space Sci. Rev., 174(1–4)*, 301–328, DOI: 10.1007/s11214-012-9940-y.

Nimmo F., Hart S. D., Korycansky D. G., and Agnor C. B. (2008) Implications of an impact origin for the martian hemispheric dichotomy. *Nature, 453(7199)*, 1220–1223, DOI: 10.1038/nature07025.

National Research Council (2011) *Vision and Voyages for Planetary Science in the Decade 2013–2022*. National Academies, Washington, DC. 398 pp.

Onstott T. C., Ehlmann B. L., Sapers H., Colemna M., Ivarsson M., Marlow J. J., Neubeck A., and Niles P. (2019) Paleo-rock-hosted life on Earth and the search on Mars: A review and strategy for exploration. *Astrobiology, 19(10)*, 1230–1262, DOI: 10.1089/ast.2018.1960.

Osterloo M. M., Hamilton V. E., Bandfield J. L., Glotch T. D., Baldridge A. M., Christensen P. R., Tornabene L. L., and Anderson F. S. (2008) Chloride-bearing materials in the southern highlands of Mars. *Science, 319(5870)*, 1651–1654, DOI: 10.1126/science.1150690.

Osterloo M. M., Anderson F. S., Hamilton V. E., and Hynek B. M. (2010) Geologic context of proposed chloride-bearing materials on Mars. *J. Geophys. Res.–Planets, 115(E10)*, DOI: 10.1029/2010JE003613.

Palumbo A., Head J., and Wordsworth R. (2018) Late Noachian icy highlands climate model: Exploring the possibility of transient melting and fluvial/lacustrine activity through peak temperatures. *Icarus, 300*, 261–286.

Pan L., Ehlmann B. L., Carter J., and Ernst C. M. (2017) The stratigraphy and history of Mars' northern lowlands through mineralogy of impact craters: A comprehensive survey. *J. Geophys. Res.–Planets, 122(9)*, 1824–1854, DOI: 10.1002/2017JE005276.

Parker T. J., Saunders R. S., and Schneeberger D. M. (1989) Transitional morphology in West Deuteronilus Mensae, Mars: Implications for modification of the lowland/upland boundary. *Icarus, 82(1)*, 111–145, DOI: 10.1016/0019-1035(89)90027-4.

Parker T. J., Grant J. A., and Franklin B. J. (2010) The northern plains: A martian oceanic basin? In *Lakes on Mars* (N. A. Cabrol and E. A. Grin, eds.), pp. 249–273. Elsevier, Amsterdam.

Pavlov A. A., Vasilyev G., Ostryakov V. M., Pavlov A. K., and Mahaffy P. (2012) Degradation of the organic molecules in the shallow subsurface of Mars due to irradiation by cosmic rays. *Geophys. Res. Lett., 39*, L13202, DOI: 10.1029/2012GL052166.

Pepin R. O. (2006) Atmospheres on the terrestrial planets: Clues to origin and evolution. *Earth Planet. Sci. Lett., 252(1–2)*, 1–14, DOI: 10.1016/j.epsl.2006.09.014.

Phillips R. J., Davis B. J., Tanaka K. L., Byrne S., Mellon M. T., Putzig N. E., Haberle R. M., et al. (2011) Massive CO_2 ice deposits sequestered in the south polar layered deposits of Mars. *Science, 332(6031)*, 838–841, DOI: 10.1126/science.1203091.

Pollack J. B., Kasting J. F., Richardson S. M., and Poliakoff K. (1987) The case for a wet, warm climate on early Mars. *Icarus, 71(2)*, 203–224, DOI: 10.1016/0019-1035(87)90147-3.

Ramirez R. M. and Craddock R. A. (2018) The geological and climatological case for a warmer and wetter early Mars. *Nature Geosci., 11(4)*, 230–237, DOI: 10.1038/s41561-018-0093-9.

Ranjan S. and Sasselov D. D. (2017) Constraints on the early terrestrial surface UV environment relevant to prebiotic chemistry. *Astrobiology, 17(3)*, 169–204, DOI: 10.1089/ast.2016.1519.

Rapin W., Meslin P.-Y., Maurice S., Vaniman D., Nachon M., Mangold N., Schröder S., et al. (2016) Hydration state of calcium sulfates in Gale Crater, Mars: Identification of bassanite veins. *Earth Planet. Sci. Lett., 452*, 197–205, DOI: 10.1016/j.epsl.2016.07.045.

Reid R. P., Visscher P. T., Decho A. W., Stolz J. F., Bebout B. M., Dupraz C., Macintyre I. G., et al. (2000) The role of microbes in accretion, lamination and early lithification of modern marine stromatolites. *Nature, 406(6799)*, 989–992, DOI: 10.1038/35023158.

Roberts J. H., Lillis R. J., and Manga M. (2009) Giant impacts on early Mars and the cessation of the martian dynamo. *J. Geophys. Res., 114(E040009)*, DOI: 10.1029/2008JE003287.

Rothschild L. J. and Mancinelli R. L. (2001) Life in extreme environments. *Nature, 409(6823)*, 1092–1101, DOI: 10.1038/35059215.

Ruff S. W. and Farmer J. D. (2016) Silica deposits on Mars with features resembling hot spring biosignatures at El Tatio in Chile. *Nature Commun., 7,* 13554, DOI: 10.1038/ncomms13554.

Sagan C. and Mullen G. (1972) Earth and Mars: Evolution of atmospheres and surface temperatures. *Science, 177(4043),* 52–56, DOI: 10.1126/science.177.4043.52.

Salvatore M. R., Mustard J. F., Head J. W., Cooper R. F., Marchant D. R., and Wyatt M. B. (2013) Development of alteration rinds by oxidative weathering processes in Beacon Valley, Antarctica, and implications for Mars. *Geochim. Cosmochim. Acta, 115,* 137–161, DOI: 10.1016/j.gca.2013.04.002.

Saper L. and Mustard J. F. (2013) Extensive linear ridge networks in Nili Fossae and Nilosyrtis, Mars: Implications for fluid flow in the ancient crust. *Geophys. Res. Lett., 40,* 245–249, DOI: 10.1002/grl.50106.

Schulte M., Blake D., Hoehler T., and McCollom T. (2006) Serpentinization and its implications for life on the early Earth and Mars. *Astrobiology, 6(2),* 364-376.

Schwarzschild M. (1958) *Structure and Evolution of the Stars.* Princeton Univ., Princeton. 360 pp.

Segura T. L., McKay C. P., and Toon O. B. (2012) An impact-induced, stable, runaway climate on Mars. *Icarus, 220(1),* 144–148, DOI: 10.1016/j.icarus.2012.04.013.

Seybold H. J., Kite E., and Kirchner J. W. (2018) Branching geometry of valley networks on Mars and Earth and its implications for early martian climate. *Sci. Adv., 4(6),* eaar6692, DOI: 10.1126/sciadv.aar6692.

Schlichting H. E. and Mukhopadhyay S. (2018) Atmosphere impact losses. *Space Sci. Rev., 214,* DOI: 10.1007/s11214-018-0471-z.

Sholes S. F., Montgomery D. R., and Catling D. C. (2019) Quantitative high-resolution reexamination of a hypothesized ocean shoreline in Cydonia Mensae on Mars. *J. Geophys. Res.–Planets, 124(2),* 316–336, DOI: 10.1029/2018JE005837.

Skok J. R., Mustard J. F., Ehlmann B. L., Milliken R. E., and Murchie S. L. (2010) Silica deposits in the Nili Patera caldera on the Syrtis Major volcanic complex on Mars. *Nature Geosci., 3(12),* 838–841, DOI: 10.1038/ngeo990.

Smith D. E. and Zuber M. T. (1996) The shape of Mars and the topographic signature of the hemispheric dichotomy. *Science, 271(5246),* 184–188, DOI: 10.1126/science.271.5246.184.

Smith D. E., Zuber M. T., Solomon S. C., Phillips R. J., Head J. W., Garvin J. B., Banerdt W. B., et al. (1999) The global topography of Mars and implications for surface evolution. *Science, 284(5419),* 1495–1503.

Smith M. D., Wolff M. J., Clancy R. T., and Murchie S. L. (2009) Compact Reconnaissance Imaging Spectrometer observations of water vapor and carbon monoxide. *J. Geophys. Res.–Planets, 114(E2),* DOI: 10.1029/2008JE003288.

Squyres S. W., Arvidson R. E., Ruff S., Gellert R., Morris R. V., Ming D. W., Crumpler L., et al. (2008) Detection of silica-rich deposits on mars. *Science, 320(5879),* 1063–1067, DOI: 10.1126/science.1155429.

Tanaka K. L. (1997) Sedimentary history and mass flow structures of Chryse and Acidalia Planitiae, Mars. *J. Geophys. Res.–Planets, 102(E2),* 4131–4149, DOI: 10.1029/96JE02862.

Tanaka K. L., Scott D. H., and Greeley R. (1992) Global stratigraphy. In *Mars* (H. H. Kieffer et al., eds.), pp. 354–382. Univ. of Arizona, Tucson.

Tanaka K. L., Skinner J. A., Dohm J. M., Irwin R. P., Kolb E. J., Fortezzo C. M., Platz T., et al. (2014) *Geologic Map of Mars.* USGS Sci. Inv. Map I-3292, DOI: 10.3133/sim3292. USGS Astrogeology Science Center, Flagstaff.

Thollot P., Mangold N., Ansan V., Mouélic S., Milliken R., Bishop J.,

Weitz C., Roach L., Mustard J. F., and Murchie S. (2012) Most Mars minerals in a nutshell: Various alteration phases formed in a single environment in Noctis Labyrinthus. *J. Geophys. Res., 117,* E00J06, DOI: 10.1029/2011JE004028.

Tosca N. J. and Knoll A. H. (2009) Juvenile chemical sediments and the long term persistence of water at the surface of Mars. *Earth Planet. Sci. Lett., 286(3–4),* 379–386, DOI: 10.1016/j.epsl.2009.07.004.

Ulrich R. K. (1975) Solar neutrinos and variations in the solar luminosity. *Science, 190(4215),* 619–624.

Velbel M. A. (1988) Weathering and soil-forming processes. In *Forest Hydrology and Ecology at Coweeta. Ecological Studies (Analysis and Synthesis), Vol. 66* (W. T. Swank and D. A. Crossley, eds.), pp. 93–102. Springer, New York.

Wadhwa M. (2001) Redox state of Mars's upper mantle and crust from Eu anomalies in shergottite pyroxene. *Science, 291(5508),* 1527–1530.

Wadsworth J. and Cockell C. S. (2017) The Janus face of iron on anoxic worlds: Iron oxides are both protective and destructive to life on the early Earth and present-day Mars. *FEMS Microbiol. Ecol., 93(5),* 1–9, DOI: 10.1093/femsec/fix056.

Wang A., Haskin L. A., Squyres S. W., Jolliff B. L., Crumpler L., Gellert R., Shröder C., et al. (2006) Sulfate deposition in subsurface regolith in Gusev crater, Mars. *J. Geophys. Res.–Planets, 111(E2),* DOI: 10.1029/2005JE002513.

Webster C. R., Mahaffy P. R., Flesch G. J., Niles P. B., Jones J. H., Leshin L. A., Atreya S. K., et al. (2013) Isotope ratios of H, C, and O in CO_2 and H_2O of the martian atmosphere. *Science, 341(6143),* 260–263, DOI: 10.1126/science.1237961.

Weiss B. P., Fong L. E., Vali H., Lima E. A., and Baudenbacher F. J. (2008) Paleointensity of the ancient martian magnetic field. *Geophys. Res. Lett., 35(23),* 1–5, DOI: 10.1029/2008GL035585.

Werner S. C. (2009) The global martian volcanic evolutionary history. *Icarus, 201(1),* 44–68, DOI: 10.1016/j.icarus.2008.12.019.

Wordsworth R. (2016) The climate of early Mars. *Annu. Rev. Earth Planet. Sci., 44,* 381–408, DOI: 10.1146/annurev-earth-060115-012355.

Wordsworth R. D. and Pierrehumbert R. T. (2013) Water loss from terrestrial planets with CO_2-rich atmospheres. *Astrophys. J., 778(2),* DOI: 10.1088/0004-637X/778/2/154.

Wordsworth R., Kalugina Y., Lokshtanov S., Vigasin A., Ehlmann B., Head J., Sanders C., and Wang H. (2017) Transient reducing greenhouse warming on early Mars. *Geophys. Res. Lett., 44(2),* 665–671, DOI: 10.1002/2016GL071766.

Wray J. J., Ehlmann B. L., Squyres S. W., Mustard J. F., and Kirk R. L. (2008) Compositional stratigraphy of clay-bearing layered deposits at Mawrth Vallis, Mars. *Geophys. Res. Lett., 35(12),* DOI: 10.1029/2008GL034385.

Wray J. J., Milliken R. E., Dundas C. M., Swayze G. A., Andrews-Hanna J. C., Baldridge A. M., Chojnacki M., et al. (2011) Columbus crater and other possible groundwater-fed paleolakes of Terra Sirenum, Mars. *J. Geophys. Res.–Planets, 116(1),* 1–41, DOI: 10.1029/2010JE003694.

Wray J. J., Murchie S. L., Bishop J. L., Ehlmann B. L., Milliken R. E., Wilhelm M. B., Seelos, K. D., and Chojnacki M. (2016) Orbital evidence for more widespread carbonate-bearing rocks on Mars. *J. Geophys. Res.–Planets, 121,* 4, DOI: 10.1002/2015JE004972.

Zahnle K. J. and Walker J. C. G. (1982) The evolution of solar ultraviolet luminosity. *Rev. Geophys., 20(2),* 280–292, DOI: 10.1029/RG020i002p00280.

Zent A. P. and Quinn R. C. (1995) Simultaneous adsorption of CO_2 and H_2O under Mars-like conditions and application to the evolution of the martian climate. *J. Geophys. Res., 100(E3),* 5341–5349, DOI: 10.1029/94JE01899.

Davila A., Kahre M. A., Quinn R., and Des Marais D. J. (2020) The biological potential of present-day Mars. In *Planetary Astrobiology*
(V. Meadows et al., eds.), pp. 169–184. Univ. of Arizona, Tucson, DOI: 10.2458/azu_uapress_9780816540068-ch007.

The Biological Potential of Present-Day Mars

A. Davila, M. A. Kahre, R. Quinn, and D. J. Des Marais
NASA Ames Research Center

In the mid-1970s the twin Viking Landers conducted three biological experiments in two different locations on the surface of Mars. Results from these experiments did not provide conclusive evidence of biological activity, and instead they were consistent with abiotic chemical reactions in the regolith. The paradigm established by the Viking missions is that the near-surface environment on Mars is too cold, dry, and hostile for life as we know it. Our understanding of the limits of metabolic activity and growth in terrestrial organisms, together with ecological studies in terrestrial analog environments, some of which approximate the temperature and dryness conditions on Mars, seem to support this view. But data gathered by surface and orbital assets in the past two decades suggests that liquid water could still be an important agent of landscape evolution over timescales ranging from seasons to millions of years. In addition, the tantalizing discovery of both liquid water buried deep beneath the southern polar deposits and fluctuating levels of methane in the atmosphere make the existence of a subsurface biosphere both possible and realistic. These discoveries have sparked a renewed interest on the biological potential of present-day Mars, and the possibility of extant life. This chapter reviews the environmental aspects most relevant to the biological potential of present-day Mars, emphasizing the near-surface environment that will be sampled and investigated by future robotic missions.

1. INTRODUCTION

In 1975 the Viking 1 and Viking 2 spacecraft each placed an orbiter around Mars and a lander on the surface (*Klein et al.,* 1972). The orbiters mapped the surface of the planet for several years and offered the first close look of an alien and yet familiar world, providing photographic evidence of ancient lakes, rivers, and perhaps even oceans. Equipped with a large instrument payload, each Viking lander conducted atmospheric and chemical measurements as well as biological experiments on opposite sides of the planet. Measurements of the atmosphere revealed an absolute desert drier and colder than any place on Earth, with a thin atmosphere composed mainly of CO_2. Three biological experiments sought evidence of life by incubating martian soil under conditions seemingly favorable for biological activity and growth. None of the experiments yielded evidence of viable forms of life (*Klein,* 1979). The results of the biology experiments combined with the failure by both landers to detect organic matter even at trace levels (*Biemann,* 1979) led the broader scientific community to conclude that the martian surface might be too extreme for life as we know it.

Results from the Viking missions shaped the exploration of Mars for decades. The thousands of photographs collected from orbit painted a canvas of a dynamic planet with a complex geologic history that included large meteorite impacts with global effects, cataclysmic volcanism, gigantic eruptions of water from the subsurface, substantial loss of the primordial atmosphere, and global climate change that caused temperatures to plummet and the surface to freeze and dry up (*Carr and Head,* 2010). During the decades that followed, the exploration of Mars focused on unraveling the details of this complex geologic history, and in the process we have learned that early Mars has some similarities to early Earth and that conditions conducive to the origin of life might have coexisted in both planets. The Mars Exploration Rover (MER) Opportunity was the first surface asset to provide direct evidence of an ancient aqueous environment on Mars, based on visual and geochemical characterizations of rocks exposed along the walls of a small impact crater at Meridiani Planum (*Squyres et al.,* 2004). Almost a decade later, the Mars Science Laboratory (MSL) rover Curiosity discovered evidence that a habitable lake environment existed in Gale Crater approximately 3.5 b.y. ago (*Grotzinger et al.,* 2014). Collectively, data gathered by orbital and surface assets shows that the physical, chemical, and energetic requirements necessary to establish a habitable environment were realized on Mars early in the history of the planet (see the chapter in this volume by Amador and Ehlmann).

Following up on these recent scientific discoveries, current Mars exploration goals primarily focus on the search for evidence of life that might have existed during an earlier, more habitable period (*Mustard et al.,* 2013). But the possibility of extant martian organisms still eking out a cryptic existence in the most inhospitable conditions has not entirely left the collective scientific imagination (*McKay et al.*, 2013; *Schulze-Makuch et al.*, 2013; *Levin and Straat,* 2016), and scenarios are still proposed for recent or even extant life on the planet (*McKay et al., 2013*). Such scenarios need to be considered critically, taking into account the current physical and chemical environment on Mars and emphasizing those factors most relevant to habitability (Table 1). In the following sections we review what is known regarding the current physical and chemical environment near the surface of Mars, and how it impacts habitability. We acknowledge that the environmental boundaries of a putative martian biosphere might not coincide with the environmental boundaries of Earth's biosphere. But because the former cannot be elucidated, our account must be necessarily Earth-centric.

2. WATER, WATER EVERYWHERE / NOR ANY DROP TO DRINK

Liquid water, essential elements and energy define the basic requirements for life, as we know it (see the chapter by Hoehler et al. in this volume). Organisms can only grow and reproduce in liquid water; no terrestrial organisms are known that can subsist on frozen water or water vapor or use other liquids. Essential elements are needed to build cells and to provide function to biomolecules. Terrestrial organisms rely on six essential elements [carbon, hydrogen, nitrogen, oxygen, phosphorous, and sulfur (CHNOPS)] and other minor elements such as sodium (Na), potassium (K), magnesium (Mg), iron (Fe), and calcium (Ca). A source of energy must be present to fuel metabolism. Some terrestrial organisms use sunlight while others can oxidize organic matter with inorganic electron acceptors (e.g., nitrates, perchlorates). There are also organisms that can subsist on simple molecules such as hydrogen gas (H_2), carbon dioxide (CO_2), and methane (CH_4). Arguably, of these basic requirements for life, the limiting ingredient on Mars is liquid water.

The surface of Mars is perhaps not the absolute frozen desert most people envision, but it is close. The mean annual air temperature in equatorial regions is ~210 K (*Martínez et al., 2017*), almost 90° colder than in equatorial regions on Earth, and the amount of water vapor in the atmosphere is 10–100× lower than in the driest terrestrial regions, such as the Atacama Desert and the Antarctic Plateau (*Smith, 2004, 2008; Davila and Schulze-Makuch, 2016*). Yet, despite the extreme cold and dryness, the landscape is still dynamic and there is now photographic evidence of surface processes and landscape evolution over multiple timescales, from seasons to millions of years. One of the agents of landscape change could be liquid water.

TABLE 1. Physicochemical factors relevant to present-day habitability and the potential for extant life on Mars.

Parameter	Mars Range	Notes	Ref.
Air temperature (K) Annual Max/Min/Ave	Equator: 280/185/210 Pole: 245/150/185	Lowest limit for cell division = 258 K Low limit for metabolism unconstrained	*Pla-Garcia et al.* (2016); *Haberle et al.* (2001); *Davy et al.* (2010)
Ground temperature (K) Annual Max/Min/Ave	Equator: 295/170/215 Pole: 255/145/160	Lowest limit for cell division = 258 K Low limit for metabolism unconstrained	*Martínez et al.* (2017)
Relative Humidity (%) Annual Max/Min	Equator: 70/10 Pole: 100/1		*Hudson et al.* (2009); *Martínez et al.* (2017)
Atmospheric H_2O content (pr µm) Global average	10–20	Driest regions on Earth: Antarctic Plateau: 200–670 Atacama Desert: 680–3700	*Jakosky and Farmer* (1982); *Smith* (2004, 2008); *Chamberlin et al.* (1997)
Pressure (mbar)	Annual range: 6–12	Triple point of water = 6.1 *Vibrio sp.* proliferates under 0.1	*Martínez et al.* (2017); *Pavlov et al.* (2010)
GCR (mGy day^{-1})	0.18–0.22	*D. radiodurans*: GCR lethal dose ~12 kGy	*Hassler et al.* (2014); *Cockell et al.* (2000a); *Martínez et al.* (2014); *Diaz and Schulze-Makuch* (2006)
Daily UV flux (W m^{-2})	8–67	>50% viability loss at 37 W m^{-2}	*Hassler et al.* (2014); *Cockell et al.* (2000a); *Martínez et al.* (2014); *Diaz and Schulze-Makuch* (2006)
Soil conductivity (mS cm^{-1})	Phoenix Lander: 1.5	Saturated NaCl brine: 65	*Kounaves et al.* (2010); *Hecht et al.* (2009)
Soil pH	Phoenix Lander: 7–8	Habitable range on Earth = 0–12.5	*Kounaves et al.* (2010); *Hecht et al.* (2009)

2.1. Contemporary Liquid Water

No direct measurements yet exist of liquid water activity on the surface of Mars, but global climate models predict that pure water could be seasonally stable against boiling or freezing in regions such as the Hellas impact basin and the northern lowland terrains, where the atmospheric pressure is permanently above the triple point (*Haberle et al.,* 2001). The stability range of liquid water increases in the presence of salts that form brines with low freezing temperature. For instance, perchlorate salts, which appear to be common regolith components (*Hecht et al.,* 2009; *Ming et al.,* 2014; *Glavin et al.,* 2013), can form stable brines at temperature as low as 205 K (*Chevrier et al.,* 2009). Thin films of perchlorate brine could theoretically form in the shallow regolith from equatorial to polar regions during warm summer months, when relative humidity in the air and the temperature on the ground approximate the stability boundary of saturated perchlorate solutions (*Martín-Torres et al.,* 2015; *Chevrier et al.,* 2009; *Smith et al.,* 2009; *Renno et al.,* 2009). In equatorial regions this is most likely to happen at nighttime, resulting in thin-films of liquid brines that then evaporate at sunrise (*Martín-Torres et al.,* 2015). The mechanism behind the formation of these brines could be deliquescence, a phase-change that some salts such as perchlorates undergo when exposed to high humidity (*Tang,* 1980; *Nuding et al.,* 2014; *Zorzano et al.,* 2009; *Renno et al.,* 2009; *Davila et al.,* 2010; *Toner et al.,* 2014), and which leads to liquid brines of high salinity and very low freezing points.

Deliquescence has been invoked to explain the formation of recurring slope lineae (RSL), imaged by the High Resolution Imaging Science Experiment (HiRISE) onboard the Mars Reconnaissance Orbiter (MRO) (*McEwen et al.,* 2011). RSL are relatively dark markings on certain steep slopes (25°–40°) that grow and wane with the seasons in a manner that is consistent with cycles of liquid water wicking and evaporating (*McEwen et al.,* 2011) (Fig. 1a). These features are only a few meters across but tens of meters long, and their possible link to liquid water has been strengthened by the detection in several RSL of perchlorate salts (*Ojha et al.,* 2015), because brines of perchlorate with a freezing temperature of ~223 K could explain the observed seasonality in RSL activity (*Chevrier and Rivera-Valentin,* 2012; *Kossacki and Markiewicz,* 2014). However, the spectral detection of perchlorates on RSL has been disputed (*Leask et al.,* 2018) and the thermophysical properties of RSL constrain the water content along the tracks to 0.5–3 wt%, an observation more consistent with (dry) granular flows (*Edwards and Piqueux,* 2016; *Dundas et al.,* 2017). As such, the possible role of liquid water in the formation of RSL remains uncertain.

2.2. Liquid Water in Geologically Recent Times

There is also photographic evidence of more widespread and intense liquid water in geologically recent times. Images acquired by the Mars Global Surveyor (MGS) Mars Orbiter Camera (MOC) since March 1999 revealed hundreds of locales with morphological features that resemble terrestrial gullies (Fig. 1b) (*Malin and Edgett,* 2000; *Heldmann et al.,* 2007; *Balme et al.,* 2006; *Kneissl et al.,* 2010; *Dickson and Head,* 2009). Martian gullies and their associated deposits are not cratered and they are stratigraphically imposed upon aeolian and periglacial landforms. Both observations suggest that relatively little time has passed since their formation, perhaps only a few million years, or that these landforms could still be evolving. In fact, fresh terminal deposits were imaged by the MOC in two martian gullies (*Malin et al.,* 2006), and continued monitoring using HiRISE has provided more evidence of seasonal gully activity, including substantial channel incision, large-scale mass movements, and the formation of new terminal deposits (*Dundas et al.,* 2010, 2012).

Gully formation and recent gully activity have been interpreted as the result of liquid water (*Malin and Edgett,* 2000; *Heldmann et al.,* 2005; *Head et al.,* 2008; *Christensen,* 2003; *Heldmann et al.,* 2007), although the source of the liquid has been much debated. Initial explanations involved seepage of groundwater water, perhaps up to present times (*Malin and Edgett,* 2000; *Mellon and Phillips,* 2001; *Gaidos,* 2001), or melting of ice or snow under past climate conditions (*Costard et al.,* 2002; *Christensen,* 2003), followed by dry degradation of the older water-formed features (*Kolb et al.,* 2010; *Schon and Head,* 2011). But recent gully activity correlates strongly with seasonal deposition and defrosting of CO_2 frost (*Dundas et al.,* 2012), which argues instead that CO_2-driven processes could be responsible for the formation of gullies and their recent activity with no involvement of liquid water (*Dundas et al.,* 2015). Considering all plausible explanations, different processes, with and without liquid water, might be responsible for the formation and evolution of martian gullies.

Another potential source of liquid water in geologically recent times could be the shallow icy regolith found only a few centimeters deep at mid- and high latitudes. The Neutron and Gamma-ray Spectrometers (NS and GRS, respectively) on the Mars Odyssey orbiter first hinted at the presence of water ice deposits ranging between 20% and 100% water-equivalent hydrogen (WEH) by mass, within the top 1 m of the regolith at latitudes poleward of 55° (*Feldman et al.,* 2002; *Boynton et al.,* 2002). The widespread presence of shallow icy regolith across mid- and high latitudes has now been confirmed with images of ground ice excavated by recent impact craters obtained by the HiRISE camera (*Byrne et al.,* 2009; *Viola et al.,* 2015; *Dundas et al.,* 2014) (Fig. 1c), as well as images of scarps in mantle terrain also at mid-latitudes revealing deposits of water ice that can be >100 m thick buried beneath 1–2 m of regolith (*Dundas et al.,* 2018).

Numerical models predict that the shallow ground ice could partially melt due to changes in Mars' obliquity, which have recurred approximately every 125,000 years (k.y.) during the past 20 m.y. or longer (*Laskar et al.,* 2002, 2004). The current obliquity of Mars is 25.2°, which results in temperatures at high northern latitudes too cold for liquid water. But when obliquity exceeds 35° the associated increase in insolation at high latitudes could partially melt shallow ground ice (*Richardson and Mischna,* 2005; *Costard et al.,* 2002; *Zent,* 2008; *Stoker et al.,* 2010; *McKay et al.,* 2013), particularly in the northern lowlands, where the atmospheric pressure is always above the thermodynamic triple point of water (Fig. 2). If present, microorganisms in the icy regolith could theoretically remain frozen in a dormant state during periods of low obliquity and become metabolically active during periods of higher obliquity, provided they survive while in stasis (see section 3). Indeed, terrestrial microorganisms have been resuscitated from dry ecosystems after a resting period of >10 k.y. (*Zhu et al.,* 2018) from Antarctic ground ice that is between 100 k.y. and 8 m.y. old (*Bidle et al.,* 2007) and from amber that is 25–40 m.y. old (*Cano and Borucki,* 1995), which underscores the capacity of some terrestrial organisms to remain viable in a dormant state for significantly long periods of time.

The potential habitability of icy regolith was one of the main motivations behind the 2008 Mars Phoenix mission. The Phoenix spacecraft landed at latitude 68°N in a region with the highest WEH signal outside the polar caps. Digging through the top dry regolith layer, the Phoenix lander uncovered water ice only a few centimeters deep (*Smith et al.,* 2009) (Fig. 1d). The heterogeneous aspect and distribution of the ice appeared most consistent with two different mechanisms of ice emplacement, one involving the direct freezing of atmospheric water vapor in the pore space of the regolith (*Fisher,* 2005), and another involving the formation of ice lenses, needle ice, or similar ice structures (*Mellon et al.,* 2009). Phoenix's Wet Chemistry Laboratory (WCL) analyzed three samples of the dry regolith layer. There were no significant differences in soluble ion concentrations, pH, or conductivity between the samples. The leached portion of the regolith was slightly alkaline (pH 7–8) and dominated by carbonate and perchlorate, with Ca^{2+}, Mg^{2+}, Na^+, and K^+ as main cations (*Hecht et al.,* 2009; *Kounaves et al.,* 2010). Considering all available data collected by the Phoenix lander, the chemistry of the regolith in the high northern latitudes appears compatible with life, and a recurrence of transient habitable conditions during high obliquity periods is still a valid hypothesis (*Stoker et al.,* 2010; *McKay et al.,* 2013; *Heldmann et al.,* 2014). However, the Phoenix lander ended mission operations short of testing whether organic compounds might be present in the icy regolith, and therefore the possibility of life in these icy soils remains to be conclusively addressed.

2.3. Liquid Water and the Deep Subsurface

Considerations of habitability on Mars are not constrained to near-surface environments. The existence of a deep, dark energy biosphere on Earth (*Edwards et al.,* 2012; *Lovley and Chapelle,* 1995) implies that microbial communities could also exist on Mars deep beneath the surface, completely independent from sunlight (*Boston et al.,* 1992). Water-rock

interactions in deep seated aquifers could generate fluids with high concentrations of H_2 and CO_2 (*Neal and Stanger,* 1983; *Abrajano et al.,* 1990), which on Earth are used by certain microorganisms to generate CH_4 and in the process obtain energy and fix carbon for growth (*Madigan et al.,* 2012). This could also be the metabolic foundation for a deep subsurface biosphere on Mars (*Schulte et al.,* 2006). Numerical models predict that the modern geothermal gradient on Mars could stabilize liquid water beneath a layer of regolith ranging from a few kilometers at the equator to ~20 km near the poles (*Clifford et al.,* 2010). More recently, radar soundings provided evidence of liquid water trapped 1.5 km below the ice of the south polar layered deposits (*Orosei et al.,* 2018) (Fig. 1e), much shallower that predicted by numerical models. Such relatively shallow bodies of liquid water might be stabilized by high concentrations of salts with low eutectic temperature, such as perchlorates. Independent confirmation of deep liquid water aquifers and a better understanding of their distribution could lay the foundation for future missions that explore the deep subsurface and search for evidence of life (*Stamenković et al.,* 2019). However, significant technological developments beyond current capabilities, or a human presence, might be needed to implement such missions.

A CH_4-based subsurface biosphere would be a tempting explanation for the discovery of low but seasonally-fluctuating levels of CH_4 measured in the atmosphere (*Mumma et al.,* 2009; *Webster et al.,* 2015; *Yung et al.,* 2018). The detection of CH_4 has been reported from both remote observations and *in situ* measurements (e.g., *Mumma et al.,* 2009; *Webster et al.,* 2015). On Earth, methanogenic archaea are a source. Metamorphism of remnant organic matter from any past or present martian life could yield CH_4. Because the half-life of CH_4 in the martian atmosphere (~300 years) (*Yung et al.,* 2018) is short relative to geological timescales, its detection indicates the presence of a recently active source that very likely is located somewhere beneath the surface. Abiotic processes involving water can also produce CH_4 (*Sherwood-Lollar et al.,* 2014), and abiotic CH_4 would still be relevant as an indicator of potentially habitable environments. The presence of CH_4 in the overwhelmingly oxidizing surface environment of Mars could provide a biologically useful source of redox energy. Organic matter from meteorites could be a biologically useful source of reduced carbon, and its metamorphism could produce CH_4 (e.g., *Oehler and Etiope,* 2017).

However, details regarding CH_4 sources and their locations have proved elusive. For example, the episodicity of the CH_4 detections has posed major uncertainties. Persistent CH_4 concentrations in the range 0.2–0.8 parts per billion by volume (ppbv) have been punctuated by occasional brief "spikes" in the range 6–10 ppbv (*Webster et al.,* 2015, 2018). But as of this writing, multiple observations by the ExoMars Trace Gas Orbiter have not detected CH_4 (*Korablev et al.,* 2019). The reported rapid changes in CH_4 concentrations imply that its lifetime in the atmosphere is shorter than 1 year (*Lefèvre and Forget,*

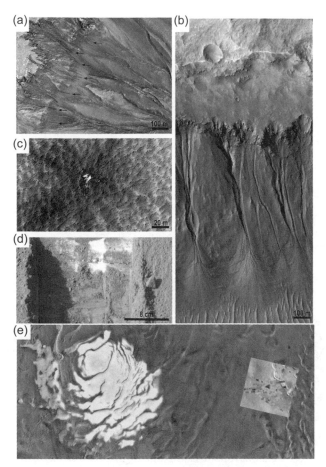

Fig. 1. See Plate 12 for color version. **(a)** Recurring slope lineae on the equator-facing wall of Palikir Crater. Credit: NASA/JPL-Caltech/Univ. of Arizona. **(b)** Gullies with clearly defined alcoves, channels, and terminal aprons (HiRISE image ESP_011727_1490). Credit: NASA/JPL/University of Arizona. **(c)** Ground ice (white) uncovered by NASA's Phoenix Lander. Credit: NASA/JPL-Caltech/University of Arizona/Texas A&M University. **(d)** Newly formed impact crater that has excavated down to buried ice (HiRISE image ESP_016954_2245). Credit: NASA/JPL/University of Arizona. **(e)** Mosaic image of the south polar region on Mars where a body of liquid water might exist 1.5 km below the surface (blue area in the center of the color tracks). Adapted from USGS Astrogeology Science Center, Arizona State University, INAF.

2009), which is inconsistent with numerical models of atmospheric chemistry (e.g., *Yung et al.*, 2018). Conceivably the episodic elevated CH_4 concentrations might arise from nearby plumes emitted from the subsurface. Changes in wind direction could influence the detection of nearby plumes by the Curiosity rover. And the plumes might be diluted to concentrations too low for detection by the Exo-Mars orbiter before they reach the altitudes interrogated by the orbiter's limb scans. Also, these scans might not yet have targeted regions adjacent to the plume sources. Accordingly, additional CH_4 measurements by the Curiosity rover should be correlated with wind directions, and the ExoMars orbiter should continue to target additional regions on Mars.

Current technology can only probe the top few meters of the martian surface and there is no conclusive evidence of springs, playas, or standing bodies of water that could link the modern surface to those deep-seated bodies of water. As such, the hypothesis of a martian deep-subsurface biosphere is currently not testable through direct sample analysis, but could be indirectly tested with a more comprehensive study of atmospheric CH_4, including the measurement of stable isotopic signatures in the CH_4 itself and its potential carbon sources and of additional products of hydrocarbon synthesis (e.g., ethane, propane) (*Yung et al.*, 2018).

In summary, observations obtained since the Viking missions both from orbit and on the surface of Mars, as well as numerical data from global climate models, portray an extreme and yet dynamic planet where liquid water activity could be very rare but possible under the current climate, but likely under different climate regimes in the past few million years. This has rekindled theories about the possibility of extant life on the planet. However, liquid water activity is a condition necessary but not sufficient for life; nutrients and energy are also required, as well as physical and chemical conditions compatible with biology (see the chapter by Hoehler et al. in this volume). Taking terrestrial life as a guide, the nutrients necessary to sustain biological processes ought to include compounds that contain bioavailable forms of CHNOPS. Of these, N-bearing compounds have been the most elusive, but the Curiosity rover finally detected a small source of fixed nitrogen, probably in the form of nitrate (NO_3^-) (*Stern et al.*, 2015). Therefore, nutrients do not appear to be a limiting factor for life on Mars. Energy ought not to be a life-limiting factor either. Forms of life inside translucent rocks near the surface could use sunlight while avoiding lethal levels of UV exposure, and several other sources of chemical energy have been identified, including potential electron donors such as organic compounds (*Freissinet et al.*, 2015; *Sutter et al.*, 2017; *Eigenbrode et al.*, 2018) and electron acceptors such as perchlorates (*Hecht et al.*, 2009), sulfates (*Bibring et al.*, 2006; *Mustard et al.*, 2008; *Squyres et al.*, 2004), and nitrates (*Stern et al.*, 2015). But as discussed in the next section, the main challenge for life on Mars near the surface is the physical and chemical environment and particularly the synergistic, lethal effects of extremely low temperature, radiation, and dryness.

3. DEATH WAS A FRIEND / AND SLEEP WAS DEATH'S BROTHER

3.1. The Physical and Chemical Environment

Direct exposure to the martian surface environment would kill any terrestrial microorganism within seconds (*Cockell et al.*, 2000; *Wadsworth and Cockell*, 2017; *Rummel et al.*, 2014). Even the most resistant terrestrial microorganisms that can survive dormant without nutrients and under extreme cold and dryness would perish due to UV radiation exposure, particularly when combined with strong

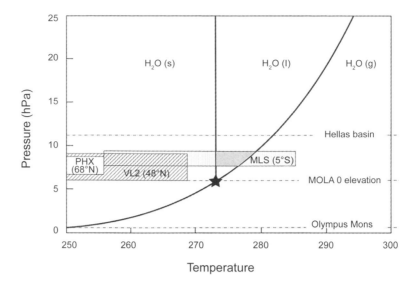

Fig. 2. Pure phase water diagram in relation to different regions on Mars and several landing sites. The average surface pressure on Mars ranges from 1 mbar at the top of Olympus Mons to 12 mbar at the bottom of the Hellas impact basin. The 0 MOLA elevation is defined as the altitude where the average surface pressure equals the triple point (6.1 mbar). Also shown is the range of surface pressure at three landing sites (Viking 2, Phoenix, and MSL), which define a transect from equatorial to polar regions. The only landing site where pure water could be liquid under current conditions is Gale Crater (MSL), but the liquid phase would only be stable in a narrow window of temperature and pressure, and prone to boiling, freezing, or rapid evaporation.

oxidative chemistry (*Wadsworth and Cockell*, 2017). The total integrated UV flux over 200–400 nm on the surface of Mars is comparable to the Earth's flux (~55 W m^{-2}). But the thinner martian atmosphere is more transparent to the most biocidal UV wavelengths (UVC, 200–280 nm; UVB, 280–315 nm) (13 W m^{-2} on Mars vs. 2 W m^{-2} on Earth) and as a result, biologically weighted UV irradiances on Mars are approximately 3 orders of magnitude higher than typical terrestrial values (*Cockell et al.*, 2000).

Beyond the sterilizing levels of UV radiation, it is the synergistic effect of radiation, low pressure, lack — or meager amounts — of liquid water, the chemistry of martian regolith, and the extremely low temperatures that truly make the surface of Mars inhospitable (*Cockell*, 1998; *Cockell et al.*, 2000; *Rummel et al.*, 2014; *Schuerger et al.*, 2013). Atmospheric pressure, as measured by multiple surface observatories over an entire martian year, oscillates seasonally between 6 and 11 mbar, with a global average surface atmospheric pressure close to 6 mbar (Haberle et al., 2008), near the vapor pressure of water at the triple point (Fig. 2) (*Hess et al.*, 1972, 1980; *Zurek et al.*, 1992; *Smith*, 2008; *Martínez et al.*, 2017). Pressures below ~25 mbar inhibit growth in most terrestrial bacteria both in Earth-like and Mars-like air (*Schuerger and Nicholson*, 2006; *Schuerger et al.*, 2013), although some studies claimed bacterial growth between 0.01 and 0.1 mbar (*Pavlov et al.*, 2010), and also after 28 days of exposure to a simulated martian atmosphere (7 mbar, 273 K, and mostly CO$_2$ air) (*Nicholson et al.*, 2013; *Schuerger et al.*, 2013). However, the ability to grow in a simulated martian

atmosphere is rare — observed in only 1 out of 26 strains tested — and it requires assumptions that are not easily met on current-day Mars, including shielding from biocidal UV irradiation, continuous exposure to stable hydrated substrates rich in carbon and nitrogen sources, and protection from rapid evaporation (*Schuerger et al.*, 2013). In addition to its direct impact on biological activity, low atmospheric pressure also places a strict boundary on the stability of liquid water by allowing only a modest 10° warming above freezing before liquid water starts to boil (*Haberle et al.*, 2001).

The thin martian atmosphere is also extremely dry, and water vapor is only present at trace levels representing ~0.02% of its composition (*Mahaffy et al.*, 2013), but highly variable depending on time of day, season, and latitude. The column vapor abundance measured by the Viking Orbiter Mars Atmospheric Water Detectors (MAWD) from June 1976 through April 1979 ranged between 0 and 75 precipitable micrometers (pr µm) and averaged global values around 10–20 pr µm (*Jakosky and Farmer*, 1982). The MGS Thermal Emission Spectrometer (TES) provided the current best description of seasonal and spatial dependence of water vapor on Mars, with somewhat lower overall values (*Smith*, 2004, 2008). For comparison, the lowest column vapor abundance on Earth has been measured near the South Pole and ranges between 200 pr µm and 670 pr µm (*Chamberlin et al.*, 1997), and in the driest parts of the Atacama Desert, the driest temperate region on Earth, it ranges between 680 and 3700 pr µm. Hence, the column vapor abundance in the driest places on Earth is on average 20–100× higher than the highest values measured on Mars. If the total atmospheric water on Mars were to precipitate in the liquid form, that liquid layer would still be unstable and evaporate within a few minutes at a rate of 1 mm hr^{-1} (*Sears and Moore,* 2005), because the vapor pressure of water in the atmosphere at temperatures above freezing is much less than the saturation vapor pressure of liquid on the surface. Water on Mars is physically driven to be in the frozen or vapor state, but rarely the liquid state.

As discussed earlier, hygroscopic salts such as perchlorates could force the condensation of water at temperatures below freezing, and generate transient briny films in the pore space of the regolith (*Zorzano et al.,* 2009; *Martín-Torres et al.,* 2015). However, periods conducive to the formation of briny films would be few and far between, and microorganisms in those briny films would still have to endure rapid wet/dry cycles, with wet conditions only lasting minutes to a few hours during each cycle (*Rivera-Valentín et al.,* 2018). Rapid wet/dry cycles are an efficient way to kill microorganisms, because the removal of water from cells, the period of cell dormancy in the dry state, and the rewetting of dry cells can inflict physiological damage that few organisms tolerate (*Potts,* 1994). Desiccation and rehydration impose large changes in the volume of cells, and therefore cause strain on their cellular membranes. The dissolution and precipitation of salts during these cycles also leads to large osmotic changes in the briny films, for which microorganisms must compensate by changing the concentration of compatible solutes in their cells. To make matters worse, wet/dry cycles on Mars would most likely occur along the liquid/ice equilibrium of briny solutions (i.e., freeze drying), rather than along the liquid/vapor equilibrium (i.e., air drying) (*Chevrier et al.,* 2009), and eutectic crystallization could cause irreparable physical damage to membranes and cell walls (*Han and Bischof,* 2004). $Mg(ClO_4)_2$ and $Ca(ClO_4)_2$ solutions could be an exception, because these brines form glasses (i.e., non-crystalline amorphous solids) at very low temperatures (*Toner et al.,* 2014), and vitrification could help preserve cell structures (*MacKenzie et al.,* 1977).

Any hypothetical form of life protected from UV radiation under a layer of dust or regolith that could still survive on meager and transient amounts of briny solution would also need to deal with the chemistry of the resulting liquid environment. The amount of salt needed to stabilize a thin film solution at subfreezing temperatures under martian conditions would translate into an impossibly low water activity (a_w), defined as the partial vapor pressure of the solution divided by the partial vapor pressure of pure water at the same temperature — water activity is expressed on a scale of 0 to 1 where 1 is for pure water. The vast majority of terrestrial microorganisms are metabolically active within the range of 1–0.90 a_w, and the current low limit of cell division and metabolism is 0.60 a_w (*Stevenson et al.,* 2015). These physiologically-relevant values must be contrasted with the a_w and temperature of a thin film of perchlorate brine in martian regolith. Such thin films would have ~0.55 a_w at

a temperature ~205 K (*Rivera-Valentín et al.,* 2018), more than 50° colder than the lowest temperature at which growth has been recorded in terrestrial cold-adapted microorganisms (*Mykytczuk et al.,* 2013). At higher temperatures, a_w values would rapidly lower to <0.1 a_w (e.g., *Chevrier et al.,* 2009).

In addition to very low a_w, perchlorate brines, as well as brines of $MgCl_2$ or $CaCl_2$ have strong chaotropic activity. Chaotropic compounds weaken the hydrophobic effect of water and at sufficiently high concentrations they disrupt the structure of biological macromolecules such as DNA, RNA, and proteins. As a result, *in vitro* enzyme activity can be totally inhibited at $MgCl_2$ concentrations below 1 mol l^{-1}, and concentrations around 2.3 mol l^{-1} inhibit microbial growth, even though a_w of such solutions is compatible with halophilic bacteria (0.75 a_w) (*Hallsworth et al.,* 2007). As such, chaotropic compounds can limit metabolism at a_w levels that are otherwise compatible with biological activity.

The biology experiments conducted by the Viking landers also showed that reactive chemical species accumulate in the surface regolith, and recent work suggests that these chemical species could be highly detrimental to both vegetative and dormant cells. In the Viking Gas Exchange (GEx) and Labeled Release (LR) biology experiments, when solutions of water containing dissolved organic compounds were added to regolith, CO_2 was immediately released. Additionally, in the Viking GEx experiment O_2 was released when regolith samples came in contact with either water vapor or liquid water (*Klein et al.,* 1976). The release of CO_2 was interpreted by some as a possible indication of metabolism (*Levin and Straat,* 1976, 2016). However, the observed chemical reactivity can also be explained by the reaction of the added organic compounds with reactive inorganic species present in the dry martian regolith (*Oyama et al.,* 1977).

Earlier abiotic hypotheses to explain the release of CO_2 and O_2 invoked the presence of peroxide and superoxide, respectively (cf. *Zent and McKay,* 1994). However, the discovery of perchlorate salts by the WCL on the Phoenix lander (*Hecht et al.,* 2009) provides a more plausible explanation. Perchlorate salts are chemically stable at Mars surface temperatures and are not reactive under the conditions of the Viking biology experiments. Therefore, perchlorate by itself cannot explain the observed chemical reactivity of the regolith. However, when perchlorates are exposed to UV radiation or to ionizing radiation they are reduced and produce other oxychlorine species (e.g., hypochlorite, the active ingredient in chlorine bleach) and reactive oxygen (*Quinn et al.,* 2013). The presence of hypochlorite in martian regolith explains the reactivity of the regolith observed in the Viking LR experiment, while the presence of reactive oxygen explains the oxygen release observed in the Viking GEx experiment (*Quinn et al.,* 2013). Hypochlorite and other reactive compounds produced during perchlorate irradiation can act as antimicrobial agents (*Deborde and von Gunten,* 2008). Ultraviolet-irradiated perchlorate solutions have biocidal properties, killing cells twice faster than UV alone, and populations of cells protected from UV in a simulated rock habitat suffered a ninefold drop in viability after only 60 seconds of exposure to a solution treated with perchlorate and UV radiation under Mars-like conditions (*Wadsworth and Cockell*, 2017). Because perchlorate salts appear to be ubiquitous on Mars, it must be assumed that reactive chlorine oxyanions might also be globally distributed near the surface and could therefore be a pervasive biocidal agent when martian regolith is wetted.

Finally, microorganisms on Mars would need to endure extremely low average temperatures and large daily temperature fluctuations (*Martínez et al.*, 2017). Global circulation models of Mars accurately predict a mean annual air temperature of 210 K near the equator and 185 K near the poles (*Haberle et al.*, 2001). For comparison, the mean annual temperature near the Earth's South Pole is 223 K. Both Viking landers measured daily changes in air temperature between 250 K and 190 K, with temperatures peaking around local noon, and plummeting rapidly after sunset (*Hess et al.*, 1976). Higher air temperatures have been measured by the Curiosity rover near the equator, with a steady pattern of daily highs at about 273 K and lows at about 203 K (*Gómez-Elvira et al.*, 2014). Data compiled from five surface assets, some spanning one martian year or more shows daily mean air temperature values between 160 K (high latitudes) and 230 K (low latitudes), with extreme diurnal amplitude values as high as 70 K (*Martínez et al.*, 2017).

It has long been recognized that some terrestrial microorganisms possess adaptations that allow them to function and grow at low temperatures. The standing bacterial record of growth at low temperatures is 258 K (*Mykytczuk et al.*, 2013), and there are reports of cell division of yeast at 255 K (*Collins and Buick*, 1989). Some microorganisms can metabolize without growing at temperatures as low as 240 K (*Junge et al.*, 2006; *Panikov and Sizova*, 2007; *De Vera et al.*, 2014), and some biophysical processes may be functional at lower temperatures (*Rummel et al.*, 2014). Terrestrial microorganisms must modify the chemistry of their cell membranes in response to large temperature fluctuations because temperature affects membrane fluidity and therefore its capability to exchange molecules with the surrounding environment. After a downward shift in temperature of 10°–15°, cyanobacterial cells cease to grow and modify (desaturate) the fatty acid composition of their membrane. When the degree of unsaturation reaches a certain level the cells begin to grow again and synthesize fatty acids. This process can last for about 10 hours (*Sato and Murata*, 1980). However, microorganisms on Mars would need to change the composition of cell membranes in response to daily shifts in temperature in excess of 60°–70° in much shorter timescales. Considering the additional challenges imposed by the cold lifestyle, including reduced enzyme activity, altered transport of nutrients and waste products, decreased rates of transcription, translation and cell division, protein cold-denaturation, inappropriate protein folding, and intracellular ice formation (*D'Amico et al.*, 2006), surviving the cold temperatures on Mars would require physiological adaptations beyond those observed in the most psychrophilic microorganisms on Earth.

3.2. Dormancy and Ionizing Radiation

Rather than facing the insurmountable challenges of surviving in the current martian environment, microbial cells could enter a protective state of stasis or dormancy until more clement conditions ensue. Some terrestrial organisms can enter a state of dormancy in response to environmental stresses such as lack of water or nutrients, or extreme temperatures, and quickly become active when conditions are more favorable. As explained earlier, more favorable conditions for biology could have happened in the polar regions of Mars due to changes in the planet's obliquity every ~125 k.y. for the past 20 m.y. or more. Cells active during those periods would now be dormant in the regolith, awaiting the next favorable obliquity cycle (*McKay et al.*, 2013). Such dormant cells might have been undetectable to the Viking landers if the incubation experiments performed by both landers failed to provide the adequate conditions for metabolic activity and growth. Indeed, only a small fraction of the microbial population on Earth can be stimulated to grow in the laboratory (*Stewart*, 2012). However, even dormant cells in the martian regolith would be under threat by the slow but deadly effects of ionizing radiation.

The ionizing radiation field produced by galactic cosmic rays (GCR) is harmful to life (*Hutchinson*, 1966). This is a significant habitability constraint in planetary bodies with thin or negligible atmospheres such as Mars, because it imposes a strict upper boundary on the amount of time that a cell can be dormant and remain viable near the surface [a maximum dormancy limit (MDL)]. Unlike UV radiation, which is quickly absorbed within a skin depth of the regolith, GCR affect the top few meters of regolith (*Dartnell et al.*, 2007a; *Kminek and Bada*, 2006; *Pavlov et al.*, 2002; *Teodoro et al.*, 2018). As GCR impinge on the surface of Mars, interactions with the shallow materials produce energetic secondary particles and this leads to a highly structured shower of ionizing radiation that propagates with depth, and in the case of Mars, reaches a maximum flux just below the surface of the planet (*Dartnell et al.*, 2007a; *Teodoro et al.*, 2018).

Ionizing radiation could, in principle, cause cellular damage to dormant martian microorganisms directly by exciting electrons in biomolecules, leading to molecular ionization and radiolysis (*Dartnell*, 2011; *Hutchinson*, 1966; *Dartnell et al.*, 2007a; *Pavlov et al.*, 2002), and indirectly through the production of hydrated free electrons and reactive oxygen species (ROS) (*Krisko and Radman*, 2010; *Zirkle*, 1954; *Hutchinson*, 1966; *Dartnell*, 2011; *Rao et al.*, 1998). Some terrestrial organisms have evolved adaptations that enhance their resistance to direct and indirect effects of ionizing radiation. On the non-resistant end of the spectrum a typical bacterium like *E. coli* is capable of surviving exposures up to 0.6 kiloGray (kGy). On the resistant end of the spectrum the terrestrial champion is *D. radiodurans*, capable of surviving exposure to ~12 kGy (*Krisko and Radman*, 2010). While we cannot discount the possibility of an organism on Mars evolving radiation resistance capabilities superior to that of *D. radiodurans*, we can also not speculate on what those capabilities might be.

The Radiation Assessment Detector (RAD) on the Curiosity rover obtained direct measurements of GCR at Gale Crater (*Hassler et al., 2014*). The biologically weighted absorbed dose measured by the RAD was 232 mGy yr^{-1} at the surface. At this absorbed dose, the MDL for *D. radiodurans* in the top few centimeters of the surface regolith would be ~50 k.y., much shorter than the pace of obliquity changes. Microorganisms in the shallow subsurface would be somewhat protected by the overlying regolith, but the effectiveness of this natural shielding depends on the composition of the regolith, with dry regolith providing better shielding than icy regolith or pure ice (*Dartnell et al., 2007a; Teodoro et al., 2018*). This is due to the fact that the neutron flux is particularly sensitive to hydrogen content in water and hydrated minerals — this is the basis of neutron spectroscopy employed to map subsurface water (*Feldman et al., 1993, 2004, 2011*). Depending on the water content, and assuming the same absorbed dose measured at Gale Crater, MDL values at 1 m depth could oscillate from hundreds of thousands (pure ice) to several millions of years (dry regolith). Under these conditions, microorganisms could in principle survive long stretches of dormancy, and become metabolically active during periods of high obliquity.

Ionizing radiation is an important and often overlooked long-term habitability parameter on Mars. The need to survive cumulative radiation effects during long stretches of dormancy in between punctuated periods of habitable conditions might be one of the most important handicaps for extant life near the surface. In addition, long-term exposure to ionizing radiation destroys organic matter (*Kminek and Bada, 2006; Dartnell et al., 2007b*), and regions of high ionizing radiation flux could damage or destroy molecular biosignatures near the surface in timescales of tens of millions of years (*Blanco et al., 2018*). Therefore, a better understanding of the ionizing radiation environmental near the surface is important both for habitability models and for life detection strategies.

4. MARS ON EARTH

The odds of microorganisms eking out an existence near the surface of Mars appear slim when considering multiple environmental stressors such as extreme dryness, subfreezing temperature, radiation, and oxidizing chemistry acting in unison, and often synergistically (*Rummel et al., 2014*). The combined biological effects of multiple environmental stressors are difficult to investigate in the laboratory. For that, it is better to turn to environments on Earth where microorganisms are exposed to a similar cocktail of extreme conditions, although the magnitude of those extremes might not be quite as severe as on Mars (*Fairén et al., 2010*). Two relevant environments are the upper reaches of the McMurdo Dry Valleys of Antarctica and the hyperarid core of the Atacama Desert in northern Chile (Fig. 3).

These so-called analog environments are Mars-like, but

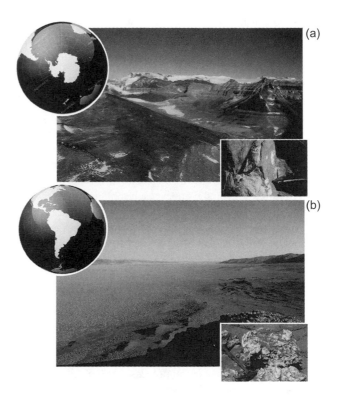

Fig. 3. See Plate 13 for color version. **(a)** The upper reaches of the McMurdo Dry Valleys in Antarctica. This is the largest ice-free region in Antarctica. Here, primary productivity almost exclusively occurs a few millimeters under the surface of sandstone cliffs and boulders (inset). **(b)** Salt-encrusted playa in the hyperarid core of the Atacama Desert in northern Chile. Here, primary productivity almost exclusively occurs inside hygroscopic halite nodules (inset).

they are not Mars. The McMurdo Dry Valleys are one of the coldest regions on Earth, but their mean annual temperature is 40°–80° higher than on Mars. The Atacama Desert is the driest temperate region on our planet, but its average atmospheric water vapor content is 50–100× higher than on Mars. Despite these differences, important lessons can be learned from these extreme terrestrial environments regarding the survival strategies of microorganisms when faced with multiple environmental stressors. Furthermore, because these terrestrial environments lie at the edge of habitability, they represent a baseline to assess the biological potential of Mars today and in the past.

4.1. The McMurdo Dry Valleys

Collectively, the McMurdo Dry Valleys are considered a polar desert environment, but the coldest and driest regions are located at the highest elevations (~1000–2500 m above sea level) (*Marchant and Head, 2007*). This is the only place on Earth that contains Mars-like permafrost — i.e., a layer of dry regolith overlying ground ice. Some of the ground ice was vapor-deposited and never transitioned through the liquid phase (*Lacelle et al., 2013*), a phenomenon that is rare on Earth but common on Mars (*Fisher,* 2005). The mean annual air temperature in the upper reaches of the McMurdo Dry Valleys is ~250 K and the mean annual soil surface temperature ranges between 247–250 K, with mean daily values always below 273 K (*Goordial et al.,* 2016; *Lacelle et al.,* 2016). Mean annual precipitation, always in the form of snow, ranges between 10 and 20 mm (*Marchant and Head,* 2007). Further to the south, near the pole, conditions are even more extreme, with a mean annual air temperature of 231 K and a mean annual atmospheric water abundance <1 mm (*Burton,* 2010; *Chamberlin et al.,* 1997).

Permafrost soils in the upper reaches of the McMurdo Dry Valleys appear to be almost uninhabitable. The soils are nutrient poor (0.01–0.05% total carbon; undetectable to 0.09% total nitrogen) with near neutral pH (7.5–8), and they have an extremely low microbial biomass, around 1000 cells g^{-1} soil (*Goordial et al.,* 2016). Respiration assays and molecular biology analyses suggest that microbial communities in these permafrost soils are rarely capable of metabolic activity under the current climate conditions, due to the combination of extremely cold temperatures, hyperaridity, and nutrient deficiency (*Goordial et al.,* 2016). Carbon:nitrogen ratios suggest that physicochemical processes in the soils dominate over biological ones (*Faucher et al.,* 2017).

Life can still exist under these frigid conditions but only inside specialized rock habitats (*Friedmann and Ocampo,* 1976) (Fig. 3a). Lithobiontic (i.e., rock inhabiting) microbial communities occupy the interior of sandstone boulders and cliffs where they survive and function as the cornerstone of primary productivity (*Friedmann,* 1982; *Friedmann and Ocampo,* 1976). The main habitat for life is a narrow zone (~0.5 mm) under the sandstone rock surface (*Friedmann,* 1982). The rock substrate provides key advantages to its inhabitants, such as maximizing water collection during wetting events; efficiently retaining liquid water by surface tension; absorbing harmful UV radiation; and still granting access to light for photosynthetic reactions, which can sustain relatively large and complex communities. Not surprisingly, the lithobiontic lifestyle is the most successful survival strategy in arid and hyperarid regions on Earth (*Goordial et al.,* 2017; *Davila and Schulze-Makuch,* 2016; *Pointing et al.,* 2009; *Wierzchos et al.,* 2012; *Pointing and Belnap,* 2012; *Friedmann,* 1982; *Warren-Rhodes et al.,* 2006).

However, lithobiontic communities in the upper reaches of the McMurdo Dry Valleys also appear to be at the brink of extinction (*Friedmann et al.,* 1994). The net ecosystem productivity (yearly accretion of organic matter) is about 3 mg C m^{-2} yr^{-1}, which represents 0.1% of the net primary production. The rest of the net photosynthetic gain is used to survive near the lower temperature limit of life (*Friedmann et al.,* 1993). In some of the coldest regions the microbial communities inside sandstone boulders and cliffs are largely dead and fossilized (*Wierzchos and Ascaso,* 2001, 2002). Here, the mean annual temperature is ~245 K and the maximum temperature rarely, if ever, reaches 273 K (*Friedmann et al.,* 1994). These conditions appear to define a limit of habitability on our planet.

4.2. The Atacama Desert

Until recently, the hyperarid core of the Atacama Desert was considered the only place on Earth where not even lithobiontic communities could survive (*Navarro-González et al.,* 2003; *Warren-Rhodes et al.,* 2006). Long-term, mean annual precipitation rates in this region average <2 mm and surface soil biomass is comparable to permafrost soils in the upper reaches of the McMurdo Dry Valleys (*Crits-Christoph et al.,* 2013; *Schulze-Makuch et al.,* 2018), although the distribution of soil bacteria is very heterogeneous with some areas that have no recoverable microorganisms or DNA (*Navarro-González et al.,* 2003). Like in the McMurdo Dry Valleys, low biomass levels in Atacama hyperarid soils are likely due to a combination of extreme water stress (*McKay et al.,* 2003; *Azua-Bustos et al.,* 2015), oligotrophy; the accumulation of ROS such as superoxides and hydroxyl radicals (*Georgiou et al.,* 2015); and high levels of UV radiation (*Cockell et al.,* 2008). As a result of multiple environmental stressors, microbial activity in these soils appears to be non-existent except during extremely rare rainfall events (*Schulze-Makuch et al.,* 2018).

Lithobiontic communities typically found in arid deserts are virtually absent in the driest parts of the Atacama (*Warren-Rhodes et al.,* 2006; *Wierzchos et al.,* 2011), which has been taken to indicate a fundamental threshold in habitability (*Navarro-González et al.,* 2003; *Warren-Rhodes et al.,* 2006). But this paradigm changed with the discovery of active microbial communities inside halite nodules that are part of salt-encrusted playas (*Wierzchos et al.,* 2006) (Fig. 3b). Cell densities inside halite nodules range between 10^5 and 10^7 cell g^{-1}, 2–4 orders of magnitude higher than in the surrounding soils (*Robinson et al.,* 2015). The communities are dominated by archaea, but also include cyanobacteria and heterotrophic bacteria (*Rios et al.,* 2010; *Robinson et al.,* 2015). The presence of a relatively abundant and diverse community in an environment that excludes practically every other form of life points to a novel survival strategy. Long-term, *in situ* monitoring of the environment inside halite nodules revealed that the salt substrate provides important ecological advantages compared to soils and other lithic substrates. Chief among them are the deliquescence of the salt substrate and its capacity to retain liquid water against evaporation (*Davila et al.,* 2008; *Wierzchos et al.,* 2012). Deliquescence represents a source of liquid water that is inaccessible to microorganisms in soils and in other non-hygroscopic substrates. In the driest parts of the desert this can result in periods suitable for photosynthesis inside halite nodules ~3× longer than in soils (*Davila et al.,* 2008).

4.3. The Martian Habitability Gap

The hyperarid core of the Atacama Desert and the upper reaches of the McMurdo Dry Valleys define the habitability threshold of photosynthesis-based life on Earth. We can use these environments to define a semi-quantitative threshold of habitability based on measurable parameters such as mean annual temperature and atmospheric water abundance (as an indicator of water availability), and use this threshold as a baseline to assess the present-day habitability on Mars (Fig. 4). Upper temperature limits in equatorial regions on Mars approximate lower temperature limits in the upper reaches of the McMurdo Dry Valleys, but dryness on Mars is several orders of magnitude more severe. On the other hand, water availability in the martian polar regions is "only" 1 order of magnitude less than in the driest regions on Earth, but here mean annual temperatures are 20°–30° lower. Hence, there is a significant habitability gap between the coldest/driest regions on Earth and the surface of Mars. When temperature and dryness are considered together with other environmental stressors, such as low pressure, UV and ionizing radiation, water chemistry, and oxidative stress, the resulting habitability gap between the surface of Mars and the coldest and driest regions on Earth is likely to widen.

4.4. The Biogeographic Problem

Microbial ecology studies in the hyperarid core of the Atacama Desert and in the upper reaches of the McMurdo Dry Valleys also point to a particular sequence of ecological successions as the environment becomes increasingly dry, whereby widespread soil communities gradually retreat to localized lithic habitats and finally to hygroscopic habitats, reflecting the need for microorganisms to maximize exposure to liquid water during rare wet events (*Davila and Schulze-Makuch*, 2016; *Pointing and Belnap*, 2012; *Wierzchos et al.,* 2012). This sequence of ecological successions results in a distribution of life with a strong biogeographic component

that is rarely recognized in martian habitability models, despite the fact that it could have an important impact on strategies to search for evidence of recent or extant life.

For instance, the twin Viking landers investigated soils in regions chosen primarily based on EDL entry, descent, and landing (EDL) risk parameters — the original landing coordinates for both landers were modified after inspection of images acquired by Viking Orbiter 1. Implicit in this approach was the assumption that if extant life existed on Mars, it would be distributed rather homogenously over the planet's surface, and therefore selection of a particular landing site should not impact the likelihood of finding evidence of life. As a result, the biological experiments on both landers were conducted on materials that largely consisted of aeolian sand and dust (*Moore et al.,* 1977).

In retrospect, and taking the hyperarid core of the Atacama Desert and the upper reaches of the McMurdo Dry Valleys as an ecological analog, the assumption that extant life on Mars would be homogeneously distributed might have been flawed. Life can be found almost anywhere in these extreme terrestrial environments, but native microbial communities are almost exclusively confined to the inside of lithic habitats. The cell abundance, taxonomy, and function of the lithic microbial communities reflect adaptations for active metabolism and growth (*Goordial et al.,* 2017; *DiRuggiero et al.,* 2013). However, cell abundance in microbial assemblages found in adjacent, non-lithic substrates is significantly lower, and their taxonomy and function are skewed toward dormancy and survival (*Crits-Christoph et al.,* 2013; *Goordial et al.,* 2016; *Schulze-Makuch et al.,* 2018), which likely reflects

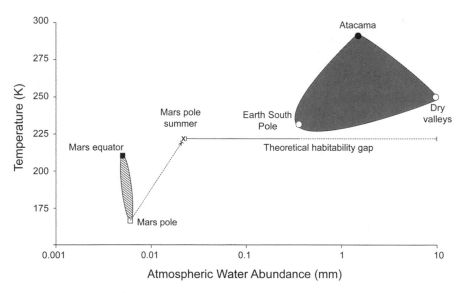

Fig. 4. Semi-quantitative assessment of martian habitability based on comparisons with terrestrial analog environments. The plot shows the mean annual temperature and atmospheric water abundance in the driest places on Earth (circles) and on Mars (squares). Temperature and atmospheric water abundance in the Atacama Desert, the McMurdo Dry Valleys, and the South Pole define the dry limit of habitability on Earth (dark gray area). Current conditions near the surface of Mars from the equator to the poles (ellipse with diagonal lines) fall outside this limit, even during the most favorable conditions (summer months in polar regions). The habitability gap is particularly wide in terms of atmospheric water abundance.

biomass transport and accumulation from more biologically rich regions (*Goordial et al.,* 2016, 2017; *Connon et al.,* 2007; *Wilhelm et al.,* 2018).

5. MARS — NEXT STEPS

The distribution of life in an environment as extreme as the surface of Mars would likely have a much stronger geographical component, with "biological hotspots" associated with certain substrates surrounded by a vast expanse of sterile ground — a biological oasis in an abiotic desert. In this scenario, intelligent landing site selection could be critical to the success of a life detection mission. In that respect, it is worth mentioning the large chloride-bearing deposits are found scattered through the martian southern highlands (*Osterloo et al.,* 2008, 2010), which bear morphological and composition similarities to deliquescent salt crusts in the Atacama Desert (*Davila et al.,* 2010). Like in the Atacama, these martian salt deposits might have been one of the last substrates capable of supporting biological activity, and one of the best locations to search for evidence of life near the surface (*Davila and Schulze-Makuch,* 2016).

The recent selection of Jezero Crater as the destination for NASA's Mars 2020 rover mission illustrates how assessments of past habitability and preservation can inform the site selection process. This crater once hosted an ancient (Noachian) lake, as evidenced by delta deposits situated at the mouths of the fluvial valleys that enter the crater from the west and north (*Fassett and Head,* 2005). The delta deposits include phyllosilicates (*Ehlmann et al.,* 2008) and silica (*Tarnas et al.,* 2019) that, on Earth, are significant repositories for preserving organic matter and fossils. Carbonates and phyllosilicates have also been detected in fan deposits within the crater (*Goudge et al.,* 2015). On Earth many carbonates are also important fossil repositories. Should the rover remain viable beyond its nominal mission period, ultimately it could exit Jezero Crater and drive into the Northeast Syrtis region (*Mustard et al.,* 2009), where diverse aqueous mineral assemblages and exhumed deposits might reveal evidence of past subsurface habitable environments.

Attributes such as those in Jezero Crater and the Northeast Syrtis region that have made them compelling sites for astrobiology should be sought in other regions where habitable environments, and potentially life, might have persisted much later in Mars history. Further research in the most extreme terrestrial deserts can help identify other potential "hot spots" of biological activity — present, recent, or ancient — and guide landing site selection in future life detection missions.

Strategies for detecting evidence of life on Mars depend on the geologic age of the materials that are sampled and the expected state, or preservation of potential, of biosignatures therein. Geologically ancient materials (i.e., billions of years old) call for strategies that emphasize time-resistant biosignatures, resembling those used to identify Earth's early biosphere (see the chapter by Johnson et al. in this volume). Geologically younger materials (i.e., hundreds of millions of years old) call for strategies that emphasize information-rich biosignatures, resembling those used to identify extant biospheres (chapter by Johnson et al.). In both instances, the most robust strategies simultaneously seek multiple types of biosignatures, including diagnostic organic compounds, inorganic chemical patterns, physical structures, sedimentary textures, and mineral assemblages.

Accordingly, a definitive detection of martian life very likely will require analyses of samples returned to Earth (e.g., *iMOST,* 2019). Mars sample return offers multiple advantages. For example, sophisticated observation-guided sample preparation and dissection are required to explore microscale spatial relationships between fabrics, minerals, and any organic matter. A single prepared sample often must be shared between multiple state-of-the-art laboratories. Sophisticated analytical techniques will likely be required that will never be flight-compatible due to restrictions in weight, size, and sample preparation needs. Preliminary analyses of returned samples might call for analytical methods that had been unanticipated. Because curatorial practices will ensure that returned samples are never exhausted, they will enable future analytical innovations to be applied.

6. CONCLUDING REMARKS

There is mounting evidence that the martian surface is dynamic, and evolves locally over seasonal periods and globally due to changes in orbital parameters. Some changes in landscape could be due to liquid water activity in the near-surface regolith, and this has triggered a renewed interest in the possibility of extant life on the planet.

Liquid water is a requirement for life, but it is not the only one. In order to survive, grow, and reproduce in the current climate, extant life in martian regolith would have to endure the combined and often synergistic effects of physical and chemical conditions that are deadly to the most resistant terrestrial microorganisms. Dormant microorganisms could still exist several meters beneath the surface of the planet, protected from hazardous ionizing radiation by natural rock shielding, and awaiting more clement conditions. It is to be noted, however, that such clement conditions would still represent an extreme environment comparable to or worse than the driest and coldest regions on Earth, and resuscitated microorganisms would still have to face a battle for survival. A deeper subsurface biosphere, completely cut off from the surface, is both possible and realistic, particularly in light of the tentative discovery of liquid water buried deep beneath the southern polar deposits. However, the deep subsurface is beyond the reach of current and planned robotic missions. Habitability models of present-day Mars must also consider the possibility that extant or dormant life might only exist in localized "hotspots," as suggested by biogeographic studies in extreme deserts on Earth. Intelligent landing site selection could be a decisive factor for the success of future life detection missions.

Acknowledgments. Writing of this chapter was supported with funding from the NASA Astrobiology Institute to A.D. and R.Q. (NAI grant to the SETI Institute) and with funding from the NASA Internal Scientist Funding Model (ISFM) to all authors. We would like to acknowledge the comments and suggestions provided by A. Fairén and D. Catling, which greatly improved the chapter.

REFERENCES

Abrajano T. A., Sturchio N. C., Kennedy B. M., et al. (1990) Geochemistry of reduced gas related to serpentinization of the Zambales Ophiolite, Philippines. *Appl. Geochem., 5*, 625–630.

Azua-Bustos A., Caro-Lara L., and Vicuña R. (2015) Discovery and microbial content of the driest site of the hyperarid Atacama Desert, Chile. *Environ. Microbiol. Rep., 7*, 388–394.

Balme M., Mangold N., Baratoux D., et al. (2006) Orientation and distribution of recent gullies in the southern hemisphere of Mars: Observations from High Resolution Stereo Camera/Mars Express (HRSC/MEX) and Mars Orbiter Camera/Mars Global Surveyor (MOC/MGS) data. *J. Geophys. Res., 111*, E05001.

Bibring J.-P., Langevin Y., Mustard J. F., et al. (2006) Global mineralogical and aqueous Mars history derived from OMEGA/Mars Express data. *Science, 312*, 400–404.

Bidle K. D., Lee S., Marchant D. R., and Falkowski P. G. (2007) Fossil genes and microbes in the oldest ice on Earth. *Proc. Natl. Acad. Sci. U.S.A., 104*, 13455–13460.

Biemann K. (1979) The implications and limitations of the findings of the Viking organic analysis experiment. *J. Mol. Evol., 14*, 65–70.

Blanco Y., de Diego-Castilla G., Viúdez-Moreiras D., et al. (2018) Effects of gamma and electron radiations on the structural integrity of organic molecules and macromolecular biomarkers measured by microarray immunoassays and their astrobiological implications. *Astrobiology, 18*, 1497–1516.

Boston P. J., Ivanov M. V., and McKay C. P. (1992) On the possibility of chemosynthetic ecosystems in subsurface habitats on Mars. *Icarus, 95*, 300–308.

Boynton W. V., Feldman W. C., Squyres S. W., et al. (2002) Distribution of hydrogen in the near surface of Mars: Evidence for subsurface ice deposits. *Science, 297*, 81–85.

Burton M. G. (2010) Astronomy in Antarctica. *Astronomy Astrophys. Rev., 18*, 417–469.

Byrne S., Dundas C. M., Kennedy M. R., et al. (2009) Distribution of mid-latitude ground ice on Mars from new impact craters. *Science, 325*, 1674–1676.

Cano R. J. and Borucki M. K. (1995) Revival and identification of bacterial spores in 25- to 40-million-year-old Dominican amber. *Science, 268(5213)*, 1060–1064, DOI: 10.1126/science.7538699.

Carr M. H. and Head J. W. (2010) Geologic history of Mars. *Earth Planet. Sci. Lett., 294*, 185–203.

Chamberlin R. A., Lane A. P., and Stark A. A. (1997) The 492 GHz atmospheric opacity at the geographic South Pole. *Astrophys. J., 476*, 428–433.

Chevrier V. F. and Rivera-Valentin E. G. (2012) Formation of recurring slope lineae by liquid brines on present-day Mars. *Geophys. Res. Lett., 39*, L21202, DOI: 10.1029/2012GL054119.

Chevrier V. F., Hanley J., and Altheide T. S. (2009) Stability of perchlorate hydrates and their liquid solutions at the Phoenix landing site, Mars. *Geophys. Res. Lett., 36*, L10202.

Christensen P. R. (2003) Formation of recent Martian gullies through melting of extensive water-rich snow deposits. *Nature, 422*, 45–48.

Clifford S. M., Lasue J., Heggy E., et al. (2010) Depth of the Martian cryosphere: Revised estimates and implications for the existence and detection of subpermafrost groundwater. *J. Geophys. Res., 115*, E07001.

Cockell C. S. (1998) Biological effects of high ultraviolet radiation on early Earth — A theoretical evaluation. *J. Theor. Biol., 193*, 717–729.

Cockell C. S., Catling D. C., Davis W. L., et al. (2000) The ultraviolet environment of Mars: Biological implications past, present, and future. *Icarus, 146*, 343–359.

Cockell C. S., McKay C. P., Warren-Rhodes K., and Horneck G. (2008) Ultraviolet radiation-induced limitation to epilithic microbial growth in arid deserts — Dosimetric experiments in the hyperarid core of the Atacama Desert. *J. Photochem. Photobiol., B, 90*, 79–87.

Collins M. A. and Buick R. K. (1989) Effect of temperature on the spoilage of stored peas by *Rhodotorula Glutinis. Food Microbiol., 6*, 135–141.

Connon S. a., Lester E. D., Shafaat H. S., et al. (2007) Bacterial Diversity in Hyperarid Atacama Desert Soils. *J. Geophys. Res. Biogeosciences 112*, 1–9.

Costard F., Forget F., Mangold N., and Peulvast J. P. (2002) Formation of recent Martian debris flows by melting of near-surface ground ice at high obliquity. *Science, 295*, 110–113.

Crits-Christoph A., Robinson C. K., Barnum T., Fricke W. F., Davila A. F., Jedynak B., McKay C. P., and DiRuggiero J. (2013) Colonization patterns of soil microbial communities in the Atacama Desert. *Microbiome, 1*, 28, DOI: 10.1186/2049-2618-1-28.

D'Amico S., Collins T., Marx J.-C., et al. (2006) Psychrophilic microorganisms: Challenges for life. *EMBO Rep., 7*, 385–89.

Dartnell L. R. (2011) Ionizing radiation and life. *Astrobiology, 11*, 551–582.

Dartnell L. R., Desorgher L., Ward J. M., and Coates A. J. (2007a) Modelling the surface and subsurface Martian radiation environment: Implications for astrobiology. *Geophys. Res. Lett., 34*, 4–9.

Dartnell L. R., Desorgher L., Ward J., and Coates A. (2007b) Martian sub-surface ionizing radiation: Biosignatures and geology. *Biogeosciences, 4*, 545–558.

Davila A. F. and Schulze-Makuch D. (2016) The last possible outposts for life on Mars. *Astrobiology, 16*, 159–168.

Davila A. F., Gómez-Silva B., de los Rios A., et al. (2008) Facilitation of endolithic microbial survival in the hyperarid core of the Atacama Desert by mineral deliquescence. *J. Geophys. Res., 113*, G01028.

Davila A. F., Duport L. G., Melchiorri R., et al. (2010) Hygroscopic salts and the potential for life on Mars. *Astrobiology, 10*, 617–628.

Davy R., Davis J. A., Taylor P. A., et al. (2010) Initial analysis of air temperature and related data from the Phoenix MET station and their use in estimating turbulent heat fluxes. *J. Geophys. Res., 115*, E00E13.

Deborde M. and von Gunten U. (2008) Reactions of chlorine with inorganic and organic compounds during water treatment-kinetics and mechanisms: A critical review. *Water Res., 42*, 13–51.

De Vera J. P., Schulze-Makuch D., Khan A., et al. (2014) Adaptation of an Antarctic lichen to Martian niche conditions can occur within 34 days. *Planet. Space Sci., 98*, 182–190.

Diaz B. and Schulze-Makuch D. (2006) Microbial survival rates of *Escherichia Coli* and *Deinococcus Radiodurans* under low temperature, low pressure, and UV-irradiation conditions, and their relevance to possible Martian life. *Astrobiology, 6*, 332–347.

Dickson J. L. and Head J. W. (2009) The formation and evolution of youthful gullies on Mars: gullies as the late-stage phase of Mars' most recent ice age. *Icarus, 204*, 63–86.

DiRuggiero J., Wierzchos J., Robinson C. K., et al. (2013) Microbial colonization of chasmoendolithic habitats in the hyper-arid zone of the Atacama Desert. *Biogeosciences, 10*, 2439–2450.

Dundas C. M., McEwen A. S., Diniega S., et al. (2010) New and recent gully activity on Mars as seen by HiRISE. *Geophys. Res. Lett., 37(7)*, DOI: 10.1029/2009GL041351.

Dundas C. M., Diniega S., Hansen C. J., et al. (2012) Seasonal activity and morphological changes in Martian gullies. *Icarus, 220*, 124–143.

Dundas C. M., Byrne S., McEwen A. S., et al. (2014) HiRISE observations of new impact craters exposing Martian ground ice. *J. Geophys. Res.–Planets, 119*, 109–127.

Dundas C. M., Diniega S., and McEwen A. S. (2015) Long-term monitoring of Martian gully formation and evolution with MRO/HiRISE. *Icarus, 251*, 244–263.

Dundas C. M., McEwen A. S., Chojnacki M., et al. (2017) Granular flows at recurring slope lineae on Mars indicate a limited role for liquid water. *Nat. Geosci., 10*, 903–907.

Dundas C. M., Bramson A. M., Ojha L., et al. (2018) Exposed subsurface ice sheets in the Martian mid-latitudes. *Science, 359*, 199–201.

Edwards C. S. and Piqueux S. (2016) The water content of recurring slope lineae on Mars. *Geophys. Res. Lett., 43*, 8912–8919.

Edwards K. J., Becker K., and Colwell F. (2012) The deep, dark energy biosphere: Intraterrestrial life on Earth. *Annu. Rev. Earth Planet. Sci., 40*, 551–568.

Ehlmann B. L., Mustard J. F., Fassett C. I., Schon S. C., Head III J. W., Des Marais D. J., et al. (2008) Clay minerals in delta deposits and organic preservation potential on Mars. *Nature Geosci., 1(6)*, 355–358, DOI: 10.1038/ngeo207.

Eigenbrode J. L., Summons R. E., Steele A., Freissinet C., Millan M., Navarro-González R., Sutter B., et al. (2018) Organic matter preserved in 3-billion-year-old mudstones at Gale Crater, Mars. *Science, 360(6393),* 1096–1101, DOI: 10.1126/science.aas9185.

Fairén A. G., Davila A. F., Lim D., et al. (2010) Astrobiology through the ages of Mars: The study of terrestrial analogues to understand the habitability of Mars. *Astrobiology, 10,* 821–843.

Fassett C. I. and Head J. W. (2005) Fluvial sedimentary deposits on Mars: Ancient deltas in a crater lake in the Nili Fossae region. *Geophys. Res. Lett., 32,* L14201.

Faucher B., Lacelle D., Davila A., et al. (2017) Physicochemical and biological controls on carbon and nitrogen in permafrost from an ultraxerous environment, McMurdo Dry Valleys of Antarctica. *J. Geophys. Res.–Biogeosci., 122,* 2593–2604.

Feldman W. C., Boynton W. V., Jakosky B. M., and Mellon M. T. (1993) Redistribution of subsurface neutrons caused by ground ice on Mars. *J. Geophys. Res., 98,* 20855.

Feldman W. C., Boynton W. V, Tokar R. L., et al. (2002) Global distribution of neutrons from Mars: Results from Mars Odyssey. *Science, 297,* 75–78.

Feldman W. C., Prettyman T. H., Maurice S., et al. (2004) Global distribution of near-surface hydrogen on Mars. *J. Geophys. Res., 109,* E09006.

Feldman W. C., Pathare A., Maurice S., et al. (2011) Mars Odyssey neutron data: 2. Search for buried excess water ice deposits at nonpolar latitudes on Mars. *J. Geophys. Res., 116,* E11009.

Fisher D. (2005) A process to make massive ice in the martian regolith using long-term diffusion and thermal cracking. *Icarus, 179,* 387–397.

Freissinet C., Glavin D. P., Mahaffy P. R., et al. (2015) Organic molecules in the Sheepbed Mudstone, Gale Crater, Mars. *J. Geophys. Res.–Planets, 120,* 495–514.

Friedmann E. I. (1982) Endolithic microorganisms in the Antarctic cold desert. *Science, 215,* 1045–1053.

Friedmann E. I. and Ocampo R. (1976) Endolithic blue-green algae in the dry valleys: Primary producers in the Antarctic desert ecosystem. *Science, 193,* 1247–1249.

Friedmann E. I., Kappen L., Meyer M. A., and Nienow J. A. (1993) Long-term productivity in the cryptoendolithic microbial community of the Ross Desert, Antarctica. *Microb. Ecol., 25,* 51–69.

Friedmann E., Druk A., and McKay C. P. (1994) Limits of life and microbial extinction in the Antarctic desert. *Antarct. J., 29,* 176–179.

Gaidos E. J. (2001) Cryovolcanism and the recent flow of liquid water on Mars. *Icarus, 153(1),* 218–223.

Georgiou C. D., Sun H. J., McKay C. P., et al. (2015) Evidence for photochemical production of reactive oxygen species in desert soils. *Nature Commun., 6,* 7100, DOI: 10.1038/ncomms8100.

Glavin D. P., Freissinet C., Miller K. E., et al. (2013) Evidence for perchlorates and the origin of chlorinated hydrocarbons detected by SAM at the Rocknest aeolian deposit in Gale Crater. *J. Geophys. Res.–Planets, 118,* 1955–1973.

Gómez-Elvira J., Armiens C., Carrasco I., et al. (2014) Curiosity's Rover Environmental Monitoring Station: Overview of the first 100 sols. *J. Geophys. Res.–Planets, 119,* 1680–1688.

Goordial J., Davila A., Lacelle D., et al. (2016) Nearing the cold-arid limits of microbial life in permafrost of an Upper Dry Valley, Antarctica. *ISME J., 10,* 1613–1624.

Goordial J., Davila A., Greer C. W., et al. (2017) Comparative activity and functional ecology of permafrost soils and lithic niches in a hyper-arid polar desert. *Environ. Microbiol., 19(2),* 443–458.

Goudge T. A., Mustard J. F., Head J. W., Fassett C. I., and Wiseman S. M. (2015) Assessing the mineralogy of the watershed and fan deposits of the Jezero crater paleolake system, Mars. *J. Geophys. Res.–Planets, 120(4),* 775-808, DOI: 10.1002/2014JE004782.

Grotzinger J. P., Sumner D. Y., Kah L. C., et al. (2014) A habitable fluvio-lacustrine environment at Yellowknife Bay, Gale Crater, Mars. *Science, 343(6169),* 1242777, doi: 10.1126/science.1242777.

Haberle R. M., McKay C. P., Schaeffer J., et al. (2001) On the possibility of liquid water on present-day Mars. *J. Geophys. Res., 106,* DOI: 10.1029/2000JE001360.

Haberle R. M., Forget F., Colaprete A., et al. (2008) The effect of ground ice on the martian seasonal CO_2 cycle. *Planet. Space Sci., 56,* 251–255.

Hallsworth J. E., Yakimov M. M., Golyshin P. N., et al. (2007) Limits of life in $MgCl_2$-containing environments: Chaotropicity defines the window. *Environ. Microbiol., 9,* 801–813.

Han B. and Bischof J. C. (2004) Direct cell injury associated with eutectic crystallization during freezing. *Cryobiology, 48,* 8–21.

Hassler D. M., Zeitlin C., Wimmer-Schweingruber R. F., et al. (2014) Mars' surface radiation environment measured with the Mars Science Laboratory's Curiosity Rover. *Science, 343,* 1244797.

Head J. W., Marchant D. R., and Kreslavsky M. A. (2008) Formation of gullies on Mars: Link to recent climate history and insolation microenvironments implicate surface water flow origin. *Proc. Natl. Acad. Sci. U.S.A., 105,* 13258–13263.

Hecht M. H., Kounaves S. P., Quinn R. C., et al. (2009) Detection of perchlorate and the soluble chemistry of Martian soil at the Phoenix lander site. *Science, 325,* 64–67.

Heldmann J. L., Toon O. B., Pollard W. H., et al. (2005) Formation of martian gullies by the action of liquid water flowing under current Martian environmental conditions. *J. Geophys. Res., 110,* E05004.

Heldmann J. L., Carlsson E., Johansson H., et al. (2007) Observations of martian gullies and constraints on potential formation mechanisms. *Icarus, 188,* 324–344.

Heldmann J. L. L., Schurmeier L., McKay C. P., et al. (2014) Midlatitude ice-rich ground on Mars as a target in the search for evidence of life and for *in situ* resource utilization on human missions. *Astrobiology, 14,* 102–118.

Hess S. L., Henry R. M., Kuettner J., et al. (1972) Meteorology experiments: The Viking Mars Lander. *Icarus, 16,* 196–204.

Hess S. L., Henry R. M., Leovy C. B., et al. (1976) Preliminary meteorological results on Mars from the Viking 1 Lander. *Science, 193,* 788–791.

Hess S. L., Ryan J. A., Tillman J. E., et al. (1980) The annual cycle of pressure on Mars measured by Viking Landers 1 and 2. *Geophys. Res. Lett., 7,* 197–200.

Hudson T. L., Zent A., Hecht M. H., et al. (2009) Near-surface humidity at the Phoenix landing site as measured by the Thermal and Electrical Conductivity Probe (TECP). In *Lunar Planet. Sci. XL,* Abstract #1804. Lunar and Planetary Institute, Houston.

Hutchinson F. (1966) The molecular basis for radiation effects on cells. *Cancer Res., 26,* 2045–2052.

International MSR Objectives and Samples Team (iMOST) (2019) The potential science and engineering value of samples delivered to Earth by Mars sample return. *Meteorit. Planet. Sci., 54 (3),* 667–671, DOI: 10.1111/maps.13232.

Jakosky B. M. and Farmer C. B. (1982) The seasonal and global behavior of water vapor in the Mars atmosphere — Complete global results of the Viking Atmospheric Water Detector Experiment. *J. Geophys. Res., 87,* 2999–3019.

Junge K., Eicken H., Swanson B. D., and Deming J. W. (2006) Bacterial incorporation of leucine into protein down to −20° C with evidence for potential activity in sub-eutectic saline ice formations. *Cryobiology, 52,* 417–429.

Klein H. P. (1979) The Viking mission and the search for life on Mars. *Rev. Geophys., 17,* 1655.

Klein H. P., Lederberg J., and Rich A. (1972) Biological experiments: The Viking Mars lander. *Icarus, 16,* 139–146.

Klein H. P., Horowitz N. H., Levin G. V, et al. (1976) The Viking biological investigation: Preliminary results. *Science, 194,* 99–105.

Kminek G. and Bada J. (2006) The effect of ionizing radiation on the preservation of amino acids on Mars. *Earth Planet. Sci. Lett., 245,* 1–5.

Kneissl T., Reiss D., van Gasselt S., and Neukum G. (2010) Distribution and orientation of northern-hemisphere gullies on Mars from the evaluation of HRSC and MOC-NA data. *Earth Planet. Sci. Lett., 294,* 357–367.

Kolb K. J., McEwen A. S., and Pelletier J. D. (2010) Investigating gully flow emplacement mechanisms using apex slopes. *Icarus, 208,* 132–142.

Korablev O., Vandaele A. C., Montmessin F., et al. (2019) No detection of methane on Mars from early ExoMars Trace Gas Orbiter observations. *Nature, 568,* 517–520.

Kossacki K. J. and Markiewicz W. J. (2014) Seasonal flows on dark Martian slopes, thermal condition for liquescence of salts. *Icarus, 233,* 126–130.

Kounaves S. P., Hecht M. H., Kapit J., et al. (2010) Soluble sulfate in the Martian soil at the Phoenix landing site. *Geophys. Res. Lett., 37,* DOI: 10.1029/2010GL042613.

Krisko A. and Radman M. (2010) Protein damage and death by radiation in *Escherichia Coli* and *Deinococcus Radiodurans. Proc. Natl. Acad. Sci. U.S.A., 107,* 14373–14377.

Lacelle D., Davila A. F., Fisher D., et al. (2013) Excess ground ice of condensation-diffusion origin in University Valley, Dry Valleys of Antarctica: Evidence from isotope geochemistry and numerical modeling. *Geochim. Cosmochim. Acta, 120,* 280–297.

Lacelle D., Lapalme C., Davila A. F., et al. (2016) Solar radiation and air and ground temperature relations in the cold and hyper-arid Quartermain Mountains, McMurdo Dry Valleys of Antarctica. *Permafrost Periglac., 27,* 163–173.

Laskar J., Levrard B., and Mustard J. F. (2002) Orbital forcing of the martian polar layered deposits. *Nature, 419,* 375–377.

Laskar J., Correia A. C. M., Gastineau M., et al. (2004) Long term evolution and chaotic diffusion of the insolation quantities of Mars. *Icarus, 170,* 343–364.

Leask E. K., Ehlmann B. L., Dundar M. M., et al. (2018) Challenges in the search for perchlorate and other hydrated minerals with 2.1-μm absorptions on Mars. *Geophys. Res. Lett., 45,* 12180–12189.

Lefévre F. and Forget F. (2009) Observed variations of methane on Mars unexplained by known atmospheric chemistry and physics. *Nature, 460,* 720–723.

Levin G. V. and Straat P. A. (1976) Viking labeled release biology experiment: Interim results. *Science, 194,* 1322–1329.

Levin G. V. and Straat P. A. (2016) The case for extant life on Mars and its possible detection by the Viking Labeled Release experiment. *Astrobiology, 16,* 798–810.

Lovley D. R. and Chapelle F. H. (1995) Deep subsurface microbial processes. *Rev. Geophys., 33(3),* 365–381.

MacKenzie A. P., Derbyshire W., and Reid D. S. (1977) Non-equilibrium freezing behaviour of aqueous systems. *Philos. Trans. R. Soc., B, 278,* 167–189.

Madigan M. T., Martinko J. M., Stahl D. A., and Clark D. P. (2012) *Brock Biology of Microorganisms, 13th Edition.* Pearson, New York. 1043 pp.

Mahaffy P. R., Webster C. R., Atreya S. K., et al. (2013) Abundance and isotopic composition of gases in the Martian atmosphere from the Curiosity Rover. *Science, 341,* 263–266.

Malin M. C. and Edgett K. S. (2000) Evidence for recent groundwater seepage and surface runoff on Mars. *Science, 288,* 2330–2335.

Malin M. C., Edgett K. S., Posiolova L. V., et al. (2006) Present-day impact cratering rate and contemporary gully activity on Mars. *Science, 314,* 1573–1577.

Marchant D. R. and Head J. W. (2007) Antarctic Dry Valleys: Microclimate zonation, variable geomorphic processes, and implications for assessing climate change on Mars. *Icarus, 192,* 187–222.

Martín-Torres F. J., Zorzano M.-P., Valentín-Serrano P., et al. (2015) Transient liquid water and water activity at Gale Crater on Mars. *Nature Geosci., 8,* 357–361.

Martínez G. M., Rennó N., Fischer E., et al. (2014) Surface energy budget and thermal inertia at Gale Crater: Calculations from ground-based measurements. *J. Geophys. Res.–Planets, 119,* 1822–1838.

Martínez G. M., Newman C. N., De Vicente-Retortillo A., et al. (2017) The modern near-surface martian climate: A review of *in-situ* meteorological data from Viking to Curiosity. *Space Sci. Rev., 212,* 295–338.

McEwen A. S., Ojha L., Dundas C. M., et al. (2011) Seasonal flows on warm Martian slopes. *Science, 333,* 740–743.

McKay C. P., Friedmann E. I., Gómez-Silva B., et al. (2003) Temperature and moisture conditions for life in the extreme arid region of the Atacama Desert: Four years of observations including the El Niño of 1997–1998. *Astrobiology, 3,* 393–406.

McKay C. P., Stoker C. R., Glass B. J., et al. (2013) The Icebreaker Life mission to Mars: A search for biomolecular evidence for Life. *Astrobiology, 13,* 334–354.

Mellon M. T. and Phillips R. J. (2001) Recent gullies on Mars and the source of liquid water. *J. Geophys. Res.–Planets, 106,* 23165–23179.

Mellon M. T., Arvidson R. E., Sizemore H. G., et al. (2009) Ground ice at the Phoenix Landing Site: Stability state and origin. *J. Geophys. Res., 114,* E00E07.

Ming D. W., Archer P. D., Glavin D. P., et al. (2014) Volatile and organic compositions of sedimentary rocks in Yellowknife Bay, Gale Crater, Mars. *Science, 343,* 1245267.

Moore H. J., Hutton R. E., Scott R. F., et al. (1977) Surface materials of the Viking Landing sites. *J. Geophys. Res.–Planets, 82,* 4497–4523.

Mumma M. J., Villanueva G. L., Novak R. E., et al. (2009) Strong release of methane on Mars in northern summer 2003. *Science, 323,* 1041–1045.

Mustard J. F., Adler M., Allwood A., et al. (2013) *Report of the Mars 2020 Science Definition Team.* Mars Exploration Program Analysis Group (MEPAG), 154 pp. Available online at *https://mepag.jpl.nasa. gov/reports/MEP/Mars_2020_SDT_Report_Final.pdf.*

Mustard J. F., Murchie S. L., Pelkey S. M., et al. (2008) Hydrated silicate minerals on Mars observed by the Mars Reconnaissance Orbiter CRISM instrument. *Nature, 454,* 305–309.

Mustard J. F., Ehlmann B. L., Murchie S. L., Poulet F., Mangold N., Head J. W., Bibring J.-P., and Roach L. H. (2009) Composition, morphology, and stratigraphy of Noachian crust around the Isidis basin. *J. Geophys. Res., 114,* DOI: 10.1029/2009JE003349.

Mykytczuk N. C. S., Foote S. J., Omelon C. R., et al. (2013) Bacterial growth at −15°C; molecular insights from the permafrost bacterium *Planococcus Halocryophilus* Or1. *ISME J., 7,* 1211–1226.

Navarro-González R., Rainey F. A., Molina P., et al. (2003) Mars-like soils in the Atacama Desert, Chile, and the dry limit of microbial life. *Science, 302,* 1018–1021.

Neal C. and Stanger G. (1983) Hydrogen generation from mantle source rocks in Oman. *Earth Planet. Sci. Lett., 66,* 315–320.

Nicholson W. L., Krivushin K., Gilichinsky D., and Schuerger A. C. (2013) Growth of *Carnobacterium* spp. from permafrost under low pressure, temperature, and anoxic atmosphere has implications for Earth microbes on Mars. *Proc. Natl. Acad. Sci. U.S.A., 110,* 666–671.

Nuding D. L., Rivera-Valentin E. G., Davis R. D., et al. (2014) Deliquescence and efflorescence of calcium perchlorate: An investigation of stable aqueous solutions relevant to Mars. *Icarus, 243,* 420–428.

Oehler D. Z. and Etiope G. (2017) Methane seepage on Mars: Where to look and why. *Astrobiology, 17,* 1233–1264.

Ojha L., Wilhelm M. B., Murchie S. L., et al. (2015) Spectral evidence for hydrated salts in recurring slope lineae on Mars. *Nature Geosci., 8,* 829–832.

Orosei R., Lauro S. E., Pettinelli E., et al. (2018) Radar evidence of subglacial liquid water on Mars. *Science, 6401,* 490–493.

Osterloo M. M., Anderson F. S., Hamilton V. E., and Hynek B. M. (2010) Geologic context of proposed chloride-bearing materials on Mars. *J. Geophys. Res., 115,* 1–29.

Osterloo M. M., Hamilton V. E., Bandfield J. L., et al. (2008) Chloride-bearing materials in the southern highlands of Mars. *Science, 319,* 1651–1654.

Oyama V. I., Berdahl B. J., and Carle G. C. (1977) Preliminary findings of the Viking gas exchange experiment and a model for martian surface chemistry. *Nature, 265,* 110–114.

Panikov N. S. and Sizova M. V. (2007) Growth kinetics of microorganisms isolated from Alaskan soil and permafrost in solid media frozen down to −35°C. *FEMS Microbiol. Ecol., 59,* 500–512.

Pavlov A. K., Shelegedin V. N., Vdovina M. A., and Pavlov A. A. (2010) Growth of microorganisms in martian-like shallow subsurface conditions: Laboratory modelling. *Intl. J. Astrobiol., 9,* 51–58.

Pavlov A. K., Blinov A. V., and Konstantinov A. N. (2002) Sterilization of martian surface by cosmic radiation. *Planet. Space Sci., 50,* 669–673.

Pla-Garcia J., Rafkin S. C. R., Kahre M., et al. (2016) The meteorology of Gale Crater as determined from Rover Environmental Monitoring Station observations and numerical modeling. Part I: Comparison of model simulations with observations. *Icarus, 280,* 103–113.

Pointing S. B. and Belnap J. (2012) Microbial colonization and controls in dryland systems. *Nature Rev. Microbiol., 10,* 654–654.

Pointing S. B., Chan Y., Lacap D. C., et al. (2009) Highly specialized microbial diversity in hyper-arid polar desert. *Proc. Natl. Acad. Sci. U.S.A., 106,* 19964–19969.

Potts M. (1994) Desiccation tolerance of prokaryotes. *Microbiol. Rev., 58,* 755–805.

Quinn R. C., Martucci H. F. H., Miller S. R., et al. (2013) Perchlorate radiolysis on Mars and the origin of martian soil reactivity. *Astrobiology, 13,* 515–520.

Rao M. V. S., Sreekantan B. V., and Venkata B. (1998) *Extensive Air Showers.* World Scientific, Singapore. 368 pp.

Renno N. O., Mehta M., Block B. P., and Braswell S. (2009) The discovery of liquid water on Mars and its implications for astrobiology. In *Instruments and Methods Astrobiology and Planetary Missions XII* (K. D. Retherford et al., eds.). SPIE Conf. Ser. 7441, Bellingham, Washington.

Richardson M. I. and Mischna M. A. (2005) Long-term evolution of transient liquid water on Mars. *J. Geophys. Res., 110,* E03003.

De los Ríos A., Valea S., Ascaso C., Davila A. F., Kastovsky J., McKay C. P., and Wierzchos J. (2010) Comparative analysis of the microbial communities inhabiting halite evaporites of the Atacama Desert. *Intl. Microbiol., 13,* 79–89.

Rivera-Valentín E. G., Gough R. V., Chevrier V. F., Primm K. M., et al. (2018) Constraining the potential liquid water environment at Gale Crater, Mars. *J. Geophys Res.–Planets, 123(5),* 1156–1167, DOI: 10.1002/2018JE005558.

Robinson C. K., Wierzchos J., Black C., Crits-Christoph A., Ma B., Ravel J., Ascaso C., et al. (2013) Microbial diversity and the presence of algae in halite endolithic communities are correlated to atmospheric moisture in the hyper-arid zone of the Atacama Desert. *Environ. Microbiol., 17(2),* 299–315.

Rummel J. D., Beaty D. W., Jones M. A., et al. (2014) A new analysis of Mars "special regions": Findings of the second MEPAG Special Regions Science Analysis Group (SR-SAG2). *Astrobiology, 14,* 887–968.

Sato N. and Murata N. (1980) Temperature shift-induced responses in lipids in the blue-green alga, *Anabaena Variabilis. Biochim. Biophys. Acta, 619,* 353–366.

Schon S. C. and Head J. W. (2011) Keys to gully formation processes on Mars: Relation to climate cycles and sources of meltwater. *Icarus, 213,* 428–432.

Schuerger A. and Nicholson W. (2006) Interactive effects of hypobaria, low temperature, and CO_2 atmospheres inhibit the growth of mesophilic *Bacillus* spp. under simulated martian conditions. *Icarus, 185,* 143–152.

Schuerger A. C., Moores J. E., Clausen C. A., et al. (2012) Methane from UV-irradiated carbonaceous chondrites under simulated martian conditions. *J. Geophys. Res., 117,* E08007.

Schuerger A. C., Ulrich R., Berry B. J., and Nicholson W. L. (2013) Growth of *Serratia Liquefaciens* under 7 mbar, 0°C, and CO_2-enriched anoxic atmospheres. *Astrobiology, 13,* 115–131.

Schulte M., Blake D., Hoehler T., and McCollom T. (2006) Serpentinization and its implications for life on the early Earth and Mars. *Astrobiology, 6,* 364–376.

Schulze-Makuch D., Fairén A. G., and Davila A. (2013) Locally targeted ecosynthesis: A proactive *in situ* search for extant life on other worlds. *Astrobiology, 13,* 674–678.

Schulze-Makuch D., Wagner D., Kounaves S. P., et al. (2018) Transitory microbial habitat in the hyperarid Atacama Desert. *Proc. Natl. Acad. Sci. U.S.A., 115,* 2670–2675.

Sears D. W. G. and Moore S. R. (2005) On laboratory simulation and the evaporation rate of water on Mars. *Geophys. Res. Lett., 32,* L16202.

Sherwood Lollar B., Onstott T. C., Lacrampe-Couloume G., and Ballentine C. J. (2014) The contribution of the Precambrian continental lithosphere to global H_2 production. *Nature, 516,* 379–382.

Smith M. D. (2004) Interannual variability in TES atmospheric observations of Mars during 1999–2003. *Icarus, 167,* 148–165.

Smith M. D. (2008) Spacecraft observations of the martian atmosphere. *Annu. Rev. Earth Planet. Sci., 36,* 191–219.

Smith P. H., Tamppari L. K., Arvidson R. E., et al. (2009) H_2O at the Phoenix landing site. *Science, 325,* 58–61.

Squyres S. W., Grotzinger J. P., Arvidson R. E., et al. (2004) *In situ* evidence for an ancient aqueous environment at Meridiani Planum, Mars. *Science, 306,* 1709–1714.

Stamenković V., Beegle L. W., Zacny K., Arumugam D. D., Baglioni P., Barba N., Baross J., et al. (2019) The next frontier for planetary and human exploration. *Nature Astron., 3,* 116–120, DOI: 10.1038/s41550-018-0676-9.

Stern J. C., Sutter B., Freissinet C., et al. (2015) Evidence for indigenous nitrogen in sedimentary and aeolian deposits from the Curiosity Rover investigations at Gale Crater, Mars. *Proc. Natl. Acad. Sci. U.S.A., 112,* 4245–4250.

Stevenson A., Burkhardt J., Cockell C. S., et al. (2015) Multiplication of microbes below 0.690 water activity: Implications for terrestrial and extraterrestrial life. *Environ. Microbiol., 17,* 257–277.

Stewart E. J. (2012) Growing unculturable bacteria. *J. Bacteriol., 194,* 4151–4160.

Stoker C. R., Zent A., Catling D. C., et al. (2010) Habitability of the Phoenix landing site. *J. Geophys. Res., 115,* 1–24.

Sutter B., McAdam A. C., Mahaffy P. R., Ming D. W., et al. (2017) Evolved gas analyses of sedimentary rocks and eolian sediment in Gale Crater, Mars: Results of the Curiosity Rover's sample analysis at Mars instrument from Yellowknife Bay to the Namib Dune. *J. Geophys. Res.–Planets, 122,* 2574–2609.

Tang I. N. (1980) Deliquescence properties and particle size change of hygroscopic aerosols. *Brookhaven National Laboratory,* Report No. BNL-27094.

Tarnas J. D., Mustard J. F., Lin H., Goudge T. A., Amador E. S., Bramble M. S., and Zhang X. (2019) Hydrated silica in the Jezero deltas. In *Lunar Planet. Sci. L,* Abstract #2551. Lunar and Planetary Institute, Houston.

Teodoro L., Davila A., Elphic R. C., et al. (2018) Habitability and biomarker preservation in the martian near-surface radiation environment. In *From Habitability to Life on Mars* (N. A. Cabrol and E. A. Grin, eds.), pp. 211–231. Elsevier, Netherlands.

Toner J. D., Catling D. C., and Light B. (2014) The formation of supercooled brines, viscous liquids, and low-temperature perchlorate glasses in aqueous solutions relevant to Mars. *Icarus, 233,* 36–47.

Viola D., McEwen A. S., Dundas C. M., and Byrne S. (2015) Expanded secondary craters in the Arcadia Planitia Region, Mars: Evidence for tens of Myr-old shallow subsurface ice. *Icarus, 248,* 190–204.

Wadsworth J. and Cockell C. S. (2017) Perchlorates on Mars enhance the bacteriocidal effects of UV light. *Sci. Rept., 7,* 4662.

Ward W. R. (1973) Large-scale variations in the obliquity of Mars. *Science, 181,* 260–262.

Warren-Rhodes K. A., Rhodes K. L., Pointing S. B., et al. (2006) Hypolithic cyanobacteria, dry limit of photosynthesis, and microbial ecology in the hyperarid Atacama Desert. *Microb. Ecol., 52,* 389–398.

Webster C. R., Mahaffy P. R., Atreya S. K., et al. (2015) Mars methane detection and variability at Gale Crater. *Science, 347,* 415–417.

Webster C. R., Mahaffy P. R., Atreya S. K., Moores J. E., Flesch G. J., Malespin C., McKay C. P., et al. (2018) Background levels of methane in Mars' atmosphere show strong seasonal variations. *Science, 360,* 1093–1096.

Wierzchos J. and Ascaso C. (2001) Life, decay, and fossilization of endolithic microorganisms from the Ross Desert, Antarctica. *Polar Biol., 24(11),* 863–868, DOI: 10.1007/s003000100296.

Wierzchos J. and Ascaso C. (2002) Microbial fossil record of rocks from the Ross Desert, Antarctica: Implications in the search for past life on Mars. *Intl. J. Astrobiol., 1(1),* 51–59, DOI: 10.1017/S1473550402001052.

Wierzchos J., Ascaso C., and McKay C. P. (2006) Endolithic cyanobacteria in halite rocks from the hyperarid core of the Atacama Desert. *Astrobiology, 6,* 415–422.

Wierzchos J., Cámara B., de los Ríos A., et al. (2011) Microbial colonization of Ca-Sulfate crusts in the hyperarid core of the Atacama Desert: Implications for the search for life on Mars. *Geobiology, 9,* 44–60.

Wierzchos J., Davila A. F., Sánchez-Almazo I. M. M., et al. (2012) Novel water source for endolithic life in the hyperarid core of the Atacama Desert. *Biogeosciences, 9,* 3071–3098.

Wierzchos J., de los Ríos A., and Ascaso C. (2012) Microorganisms in desert rocks: The edge of life on Earth. *Intl. Microbiol., 15,* 173–183.

Wilhelm M. B., Davila A. F., Parenteau M. N., et al. (2018) Constraints on the metabolic activity of microorganisms in Atacama surface soils inferred from refractory biomarkers: Implications for martian habitability and biomarker detection. *Astrobiology, 18,* 955–966.

Yung Y. L., Chen P., Nealson K., et al. (2018) Methane on Mars and habitability: Challenges and responses. *Astrobiology, 18,* 1221–1242.

Zent A. (2008) A historical search for habitable ice at the Phoenix landing site. *Icarus, 196,* 385–408.

Zent A. and McKay C. P. (1994) The chemical reactivity of the martian soil and implications for future missions. *Icarus, 108,* 146–157.

Zhu G., Wang S., Wang C., Zhou L., Zhao S., Li Y., Li F., Jetten M. S. M., Lu Y., and Schwark L. (2019) Resuscitation of Anammox bacteria after >10,000 years of dormancy. *ISME J., 13,* 1098–1109.

Zirkle R. E. (1954) Biological effects external X and gamma radiation, Part I. *Phys. Today, 7,* 20–22.

Zorzano M.-P., Mateo-Martí E., Prieto-Ballesteros O., et al. (2009) Stability of liquid saline water on present day Mars. *Geophys. Res. Lett., 36,* L20201.

Zurek R. W., Barnes J. R., Haberle R. M., et al. (1992) Dynamics of the atmosphere of Mars. In *Mars* (H. H. Kieffer, ed.), pp. 835–933. Univ. of Arizona, Tucson.

Schmidt B. E. (2020) The astrobiology of Europa and the jovian system. In *Planetary Astrobiology* (V. Meadows et al., eds.), pp. 185–215. Univ. of Arizona, Tucson, DOI: 10.2458/azu_uapress_9780816540068-ch008.

The Astrobiology of Europa and the Jovian System

B. E. Schmidt
Georgia Institute of Technology

Ocean worlds are a frontier in astrobiology — both as potential environments for past and present life in our solar system and as analogs for exotic types of planets that may be numerous around other stars. As our ability to explore these worlds and find potential ocean planets in other systems grows in parallel, understanding the processes that lead to habitable ocean worlds is critically important for astrobiologists of all disciplines, from astronomers to biologists. The jovian moons are a mini-planetary system orbiting Jupiter whose diverse characteristics and processes illustrate how critical the formation and evolution of planets is to their long-term habitability. Io and Callisto bracket the ends of the habitability spectrum: one much too hot, the other likely too cold and geologically inactive to support life. In between, Europa and Ganymede possess putative or known chemistry that is essential for life, as well as potential long-term geologic activity that could enable life to originate and continue. Europa in particular is among the most active worlds in the solar system and has likely been so since formation, which may also increase the chance for life to have emerged and been maintained. Both Europa and Ganymede are ocean worlds, but are fundamentally different from Earth. They represent an important concept: that habitability of any planet is a function of a complex system for which we have currently only one example of a known inhabited planet, and for which exotic configurations of different processes could potentially produce the same result. This chapter explores the historical and current inventory of potential long-term sources of energy for a biosphere in the sunless subsurface ice and oceans of the jovian moons, including observable signals of those processes. Thus, the jovian system provides many novel avenues for defining and adapting our concept of what planetary habitability means in any system, a sextant in our search for life on alien worlds, near and far.

"Behold, therefore, four stars . . . , and not of the common sort or multitude of the less notable fixed stars, but of the illustrious order of wandering stars, which, indeed make their journeys and orbits with a marvelous speed around the star of Jupiter, the most noble of them all, with mutually different motions, like children of the same family, while meanwhile all together, in mutual harmony, complete their great revolutions every twelve years about the center of the world, that is, about the Sun itself."
— *Galileo Galilei, 1610, Translated by A. Van Helden*

1. INTRODUCTION TO THE JOVIAN SYSTEM

Since Galileo's *Sidereus Nuncius* was published in 1610, the jovian system has been inextricably tied to humanity's perspective on the universe. When Galileo pointed the first telescope at the system, the orbital motions of the four innermost moons about Jupiter were discovered, demonstrating for the first time that a broader perspective was required for humans to understand the heavens, bolstering the Copernican revolution and driving the development of modern astronomy. Much later, among the primary discoveries made by the first observations with planetary spectroscopy would be ice on the surfaces of Europa, Ganymede, and Callisto and salts on Europa and Ganymede (*Johnson and McCord, 1971*). These observations opened the door to considering whether such worlds could host liquid reservoirs and even oceans (*Lewis,* 1971; *Cassen et al.,* 1979) that might make Europa habitable (*Reynolds et al.,* 1983; *Squyres et al.,* 1983), challenging the assumption that cold temperatures would result in solid ice-rock interiors, rendering these worlds lifeless buoys adrift around Jupiter. These observations helped launch the two Voyager spacecraft whose mission was to understand the vast reaches of the solar system. Voyager would reveal the jovian satellites as worlds with their own geologic processes, and set the stage for the Galileo mission to explore these worlds in depth for the first time. Galileo would achieve unprecedented views of Europa, showing its young surface, and detect the induced magnetic fields of Europa, Ganymede, and Callisto (*Khurana et al.,* 1998; *Kivelson et al.,* 2000), implying oceans beneath their icy surfaces. In a flash of poetic justice, observations made by the spacecraft bearing his name would show that the motions of the four moons that Galileo observed himself 400 years prior had potentially much more to say about life in the universe than could have been comprehended at the time: that oceans like our own persist across the solar system, maintained and invigorated by orbital dynamics. The conditions that arise due to, or are at the very least mediated by, the tidal evolution of the jovian system led to a diversity of moons. In particular Europa may represent the best potential to discover extant or recent life on another planet within the solar system.

Habitability, which is a term of now widespread use in the planetary, Earth, and astronomical scientific communities commonly describes the study of a planet's ability to promote the origin and sustenance of a biosphere (e.g., *Hays et al., 2015*). Habitability is a concept that has grown from its reductionist infancy — whereby liquid water and some elements of composition, in particular organics, were thought as the primary necessities for life, toward full system science that incorporates a planet's formation and geologic evolution as critically important for its suitability for life (e.g., *Des Marais et al., 2008; Hays et al., 2015; Sherwood-Lollar et al., 2018; National Academy of Sciences, 2019*). The jovian moons, with a spread in properties and processes, illustrate this concept well, both as potential locations for the search for life within this planetary system (on Europa and possibly Ganymede) and as reference points for what might define the suitability of planets and moons around other stars. Embedded in the following discussion is a focus on thinking of requirements for life not as a result of chemistry alone but one that depends on the orbital and internal dynamics that manifest during planetary evolution as potential long-term sources of energy for a biosphere.

Besides contributing to the maintenance of oceans beneath the ice, tides raised on the jovian moons by interactions with Jupiter and each other have provided a long-term source of energy with a wide range of effects: Io's immense and active volcanism (*Carr et al., 1979; Carr, 1986; Ojakangas and Stevenson, 1986; McEwen et al., 1998; Spencer et al., 2007*), which creates lava lakes that can be observed from the Earth; Europa's surprisingly young 30–90-m.y.-old (Ma) surface (*Zahnle et al., 2008; Bierhaus et al., 2009*) due to active resurfacing (e.g., *Pappalardo et al., 1998, 1999; Carr et al., 1998; Collins and Nimmo, 2009*) that may be ongoing even at present (e.g., *Schmidt et al., 2011; Roth et al., 2014a; Sparks et al., 2016; Jia et al., 2018*); Ganymede's surface structure and age dichotomy (e.g., *Showman and Malholtra, 1997; Patterson et al., 2010*); and potential resurfacing driven by internal heat ["endogenic" (e.g., *Hammond and Barr, 2014*)] seem to have derived from a period of high tidal forcing. Curiously, perhaps, Callisto remains the only satellite in the jovian system without its own known endogenic geologic processes, and it likely has not completely differentiated (melted and formed an interior stratified by density) (*Anderson et al., 1997; Schubert et al., 2004*), implying low internal energy. Some of these worlds share many similarities with our home planet in ways that

might be favorable to life (e.g., *Lewis, 1971; Reynolds et al., 1983; Chyba, 2000; Chyba and Phillips, 2001; Hand et al., 2009; Vance et al., 2016; Lunine, 2017*), and the jovian system provides an opportunity to investigate whether a "tidal Goldilocks" zone exists that could enable habitable ocean worlds far from a central star.

2. BUILDING A HABITABLE SYSTEM

The pioneering work of J. S. Lewis in 1971 discussed the concept that as a result of their formation conditions along with location in the solar nebula (and thus their composition), the moons of the giant planets could be ocean worlds. This work was contemporaneous with the first detections of ice on the surfaces of the jovian moons (*Johnson and McCord, 1971; Pilcher et al., 1972*). Calculations of protoplanetary disk and surface temperatures near Jupiter and simple thermal models assuming formation from chondritic materials led Lewis to conclude that the largest moons could preserve oceans of liquid water and some ammonia, covered with ice shells and with potentially differentiated silicate interiors. This early work provided a road map for considering how time of satellite formation, initial composition, and geophysical evolution depend on each other, which underpins how astrobiologists think about potential energy for a biosphere and our concept of habitability.

2.1. Formation

That the jovian system contains four large moons, a mini-planetary system in its own right (Fig. 1), is a consequence of how the planet itself formed. Giant planets, such as Jupiter and Saturn, form quickly after the formation of the star, and appear to form by rapid mass accretion to protoplanets larger than about 10× the mass of Earth (e.g., *Lunine et al., 2004; Lambrechts and Johansen, 2012*), forming gravitational perturbations that become subnebulas during the time that the protoplanetary disk is still gas rich. This in turn allows the planet to continue to grow rapidly through gas accretion funneled from the disk through the

Fig. 1. The formation and tidal history of the jovian system is written in the sizes, internal structures, and surfaces of the Galilean moons (left to right, and in increasing orbital distance): Io, Europa, Ganymede, and Callisto. The degree of tidal heating and tectonic resurfacing decreases and surface age increases from Io to Callisto. The bulk densities of the moons also decrease from Io to Callisto, and the innermost three are differentiated, while Callisto remains partially differentiated. Credit: NASA/JPL/DLR PIA1299.

system. Similar to instabilities in the protoplanetary disk, gravitational instabilities in the gas surrounding the growing planet drive a conveyor belt of material across the equatorial plane of the proto-giant-planet's disk, such that moons may be continually formed and devoured by the planet (e.g., *Canup and Ward,* 2002, 2006, 2009). This model makes testable predictions that are supported by the jovian system (Table 1): a hot and potentially dry inner disk represented by Io; a gradient in moon properties matched by Europa having less water than Ganymede; and cold outer disk temperatures, with the youngest moons toward the outer edge, which is consistent with Callisto having never differentiated despite its large size (*Canup and Ward,* 2006; *Barr and Canup,* 2010). This scenario predicts that the formation of the satellites likely occurs around the same time that the terrestrial planets formed, within 100 m.y. after the star forms, since moon accretion should cease shortly after the protoplanetary disk gas is exhausted. Further support for formation through gas drag in the jovian subnebula is that the same model also explains features of the saturnian system with a lower-mass host planet, where a diversity of moon sizes and properties also exists (*Canup and Ward,* 2006). Open questions remain as to the timing and duration of individual satellite formation, as well as to the exact details of whether the irregular satellites form from the planet's disk or are captured later.

An important implication for astrobiology is that if the formation of giant planet moon systems is fundamentally similar, then large moons around giant planets in any exoplanetary system should be relatively universal, broadening the number of potentially habitable worlds (e.g., *Reynolds et al.,* 1987; *Williams et al.,* 1997; *Heller and Pudritz,* 2015). Moreover, while the composition of the moons themselves would be difficult to measure directly, their general properties could be well constrained by the size and composition of the atmosphere of the host planet. The properties of the planet and its moons are linked to the time of formation and properties of the disk and would be a function of distance of the giant planet from the central star (e.g., *Chiang and Youdin,* 2010, *Heller and Pudritz,* 2015). Knowing when the moons formed and the properties of the host planet might reveal important constraints about the moons themselves that are otherwise difficult to know — for instance, changes in the relative abundance of volatiles or other components would mean drastically different ocean composition. Systems around Jupiter-mass planets might be dominated by spatially compressed larger moons due to deeper gravity wells and faster feeding rates, whereas Saturn-mass planets might possess more diverse distribution in moon size and location. As observing capabilities improve and exomoon systems become detectable and one day directly imaged, comparative studies will also provide additional data to understand satellite formation. Knowing the mass of the central planet alongside the mass and location of the moon would provide a useful tool for assessing how the moons formed and the likeliness that a barely detectable moon would be habitable, since this would constrain the density of the moon, its formation conditions, and potential for tidal evolution.

2.2. Tides and Tectonics

The jovian system is unique in its dynamics among the giant planet systems since the moon-moon interactions are particularly strong, which is critical to the astrobiological potential of the system. The jovian moons presently exist in a Laplace resonance, where the orbital periods of the bodies are integer numbers of each other. For each complete orbit of Ganymede, Europa orbits Jupiter twice, and Io orbits four times. Gravitational interactions between the moons prevent orbital circularization — they excite each other enough to maintain eccentric orbits despite the massive dissipation induced by Jupiter. The effect is that the three inner moons, but not Callisto, have a constant source of internal heat derived from massive tidal forces, which change the moon's shape over its orbit and generate energy via internal friction. It is also likely that the moons became resonant later in

TABLE 1. Properties of the Galilean satellites.

Body	Radius	Mean Density	Moment of Inertia	Hydrosphere Depth (km)
Io	1821.6 (0.0019)	3527.5 (2.9)	0.37824 (0.00022)	—
Europa	1565.0 (8.0)	2989 (46)	0.346 (0.005)	100–200
Ganymede	2631.2 (1.7)	1942.0 (4.8)	0.3115 (0.0028)	179–766*
Callisto	2410.3 (1.5)	1834.4 (3.4)	0.3549 (0.0042)	300

From *Schubert et al.* (2004). The axial moment of inertia is given by C/MR^2, where C is the moment of inertia, and describes the mass distribution within the planet. For reference, a sphere of uniform density would have a moment of inertia of 0.4. Here "hydrosphere" is the total thickness of the outermost ice layer and ocean.

* Taken from *Vance et al.* (2014).

the evolution of the system, which is estimated as ~2.6 Ga via the surface age of Ganymede. Ganymede's geology is dichotomous (Fig. 2) — consisting of regions of ~4 b.y.-old surface areas that are largely unevolved and saturated in craters, and areas of extreme tectonic deformation with ages as young as ~2.5 b.y., suggesting that a period of increased activity occurred at this time, consistent with entering into the orbital resonance (e.g., *Malhotra*, 1991; *Showman et al.*, 1997; *Bland et al.*, 2008). The exact orbital evolution depends on the initial states of the moons, but in any scenario the large excitation of orbital eccentricities imparts a strong pulse of energy into the bodies through dissipation, followed by slow damping of the orbits into their present state (e.g., *Tittemore*, 1990; *Hussman and Spohn*, 2004; *Moore and Hussman*, 2009; *Tobie et al.*, 2003, 2005). From this point on, the evolution of the interiors of the moons are coupled, where changes to orbital eccentricity can could cause cyclic thinning and thickening of Europa's ice shell from a few to many tens of kilometers. Ganymede may also have experienced shell thickness variations as a result of the excitation of orbital resonances (*Showman and Malhotra*, 1997; *Showman et al.*, 1997). From an astrobiological perspective, this is important for Europa in particular given that even if it had largely cooled after formation, from 2.6 Ga on, tidal energy dissipated in the ice shell and potentially in its silicate interior would increase dramatically and then stabilize or become cyclical. As will be discussed in greater detail below, Europa's habitability would be strongly affected by the changes such oscillations can induce, including changing ice shell and interior activity, and changing the nature of how material is exchanged between different energetic and compositional reservoirs within the moon.

2.3. Callisto

Within the context of the jovian system, Io and Callisto represent the two end members, one highly thermally evolved, and one relatively pristine, and for both the prospect of habitability is bleak. However, both of these moons belie the history of the system, one through volatile loss

(Io) and one through recording its impact history (Callisto), and provide a benchmark for potential states of worlds that rely on exogenic forces for energy. Callisto is a water world, the third largest and most massive moon in the solar system, and only slightly smaller than Mercury, and yet it has no evidence for internally driven surface processes (e.g., *Moore et al.*, 1999), which are found on moons that are much smaller and should have cooled faster once formed. Callisto likely has an ocean (*Khurana et al.*, 1998), but the ice shell is saturated with craters and weathering processes dominated by impact and CO_2 loss (*Moore et al.*, 1999), which demonstrates the importance of time of formation and tidal activity in satellite evolution. In the dynamical models described above, Callisto forms last (*Canup and Ward*, 2002), and was possibly denied short-lived radioisotopes that could contribute to melting the moon. Consequently the vast majority of its accretional heat likely went into melting ice that formed the ocean, thereby separating some of the ice from the inner core but leaving an undifferentiated interior (e.g., *Anderson et al.*, 1997; *Schubert et al.*, 2004, 2010). It is also possible that due to its position as the outermost large satellite, Callisto may have had a lower impact rate and lower impact velocities, also lowering the total amount of accretional energy for differentiation (*Barr and Canup*, 2010).

2.4. Io

Io, with a mean radius of 1866 km, is just larger than our Moon, and yet is the most volcanically active body in the solar system (Fig. 3) (e.g., *Peale et al.*, 1979; *Carr et al.*, 1979; *Carr*, 1986; *Ojakangas and Stevenson*, 1986; *McEwen et al.*, 1998; *Spencer et al.*, 2007). The Galileo spacecraft revealed a dynamic planet, dotted with hotspots (*Johnson et al.*, 1984; *Rathbun and Spencer*, 2010) with varying dynamic processes from lava fountains like Tvashtar that send magma tens of percent of the moon's radius into space (*Spencer et al.*, 2007) to lava lakes whose surfaces continually founder and expose new magma such as Loki Patera (*Rathbun et al.*, 2002; *Davies*, 2003). At extremely high temperatures (*McEwen et al.*, 1998), these regions possess enough thermal emission to be monitored from space and groundbased facilities on Earth (e.g., *Johnson et al.*, 1984; *Howell et al.*, 2001; *Marchis et al.*, 2002; *Rathbun and Spencer*, 2010).

Io's internal heat is so great that mantle convection may not be sufficient to cool the planet. Io's high temperatures might be expected to permit a deformable crust and low-viscosity mantle, leading to low topography and rapid plate tectonics, as has been suggested for super-Earths, for instance (e.g., *Valencia et al.*, 2007). Instead, Io may be the case example for heat pipe volcanism as a cooling mechanism (*Moore*, 2001). In this case, melt-through events pipe low-viscosity magma rapidly to the surface through a dense, dehydrated crust — where it spills out and cools, only to be replaced by additional magma (e.g., *Turcotte*, 1980, 1989). This rapid magmatism is fundamentally different than magma ocean dynamics and may also be relevant to planets

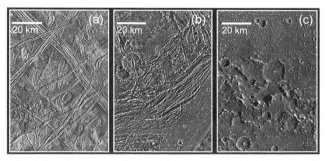

Fig. 2. Tectonized surfaces of (a) Europa and (b) Ganymede are significantly different from the ancient cratered surface of (c) Callisto. Ganymede's oldest surface terrain (~4 b.y. old) is similar to Callisto. Both Europa and Ganymede appear to have been differentiated during formation, while Callisto appears only partially differentiated. Credit: NASA/JPL/DLR PIA 01656.

Fig. 3. Volcanic activity has continued to be observed on Io by spacecraft with Voyager (left) continuing to Galileo (Fig. 1) and New Horizons (center, right) as well as from the ground. The left and center images show volcanic plumes on the limb and surface as viewed in visible wavelengths, and at right in the infrared showing thermal emission from multiple volcanic regions. Credit: NASA/JPL (left); NASA/JHUAPL/SWRI (center, right).

like Venus. Several lines of evidence support heat pipe volcanism on Io: Topography on Io is higher than expected if the crust was thin, and its surface is characterized by a crust with high mountains (*Schenk and Bullmer*, 1998; *Kirchoff and McKinnon*, 2009) and unmoving, constantly active regions (e.g., *McEwen et al.*, 1998; *Rathbun et al.*, 2002; *Rathbun and Spencer*, 2010). These observations indicate that at least since its entrance into the orbital resonance, Io has been melting and outgassing its mantle and crust constantly, producing a thick, rigid, and dry mantle.

For astrobiology, Io represents an important corollary for exotic states of terrestrial planet and moon evolution that demonstrates that internal heat is only one factor in the tectonic regime of a planet, and that other factors such as mantle hydration and strength are equally important (e.g., *O'Neil and Lenardic*, 2007). One such lesson is that super-Earths might be better approximated as super-Ios (e.g., *Barnes et al.*, 2009); this would imply tectonics governed by mantle rigidity and hydration state and that these planets may be subject to heat pipe volcanism to regulate their internal temperatures (e.g., *Moore et al.*, 2017). Second, Io might represent a transition state that planets pass through on their way to plate tectonics. Among the most difficult to answer questions regarding the initiation of plate tectonics is how and where the early crust might buckle to allow for plate motion to begin — given that either a gradient in strength, density, and compressive stresses across the surface is required. A provocative suggestion (e.g., *Moore and Webb*, 2013) is that in planets with stagnant lids, heat pipes could deliver deep mantle material to the surface, forming large deposits of magma that cause the crust to founder, in turn allowing for subduction to initiate and the "slab" of dense crust to pull the plate down with it. *Moore and Webb* (2013) argue that under early Earth conditions, the planet's water may have been largely absorbed within the mantle, and thus zircons recording contemporaneously subduction zone-like temperatures and high water content in ancient crustal materials might record that phenomena rather than water absorbed into minerals near established oceanic subduction zones (e.g., *Hopkins et al.*, 2010). Regardless of whether heat pipe volcanism is broadly relevant to tectonics, Io serves as an important reminder to astronomers and astro-

biologists that planetary evolution is a complex problem with multiple feedbacks. Io provides a highly-volcanically-active end member to consider along the spectrum of terrestrial planetary bodies from the inner solar system to early Earth to much larger terrestrial planets.

3. GANYMEDE: THE CLOSEST WATER WORLD

Ganymede is the largest moon in the solar system, a planet in its own right: larger than Mercury, possessing its own magnetic field (e.g., *Schubert et al.*, 1996; *Kivelson et al.*, 1997; *Jia et al.*, 2009; *Paty and Winglee*, 2004) likely powered by a molten core (e.g., *Anderson et al.*, 1996; *Hauck et al.*, 2006; *Kimura et al.*, 2009; *Bland et al.*, 2008; *Zhan and Schubert*, 2012), and a deep interior ocean sandwiched between low-pressure and high-pressure ices (e.g., *Kivelson et al.*, 2002), or perhaps multiple ocean layers interleaved between ice layers (*Vance et al.*, 2014). Outside of planetary bodies like Europa and Earth where liquid water is a relatively small layer on the outer edge of the planet, Ganymede represents the prototypical water world (*Grasset et al.*, 2013) with a layered interior of high-pressure ocean and ice layers that arise due to its size and gravity, which is relevant to water worlds in other planetary systems.

Like Europa and Io, the vast majority of data for Ganymede comes from the Galileo mission. Galileo revealed Ganymede to be two-faced: Its mostly water ice surface has two surface ages at 4 and ~2 b.y. (*Showman and Malhotra*, 1997; *Pappalardo et al.*, 2004; *Bland et al.*, 2009; *Patterson et al.*, 2010) and both dark [generally highly cratered and with CO_2 ices incorporated (*Hibbitts et al.*, 2000)] and bright surface units [generally the youngest, most tectonized surfaces (*Head et al.*, 2002; *Patterson et al.*, 2010)]. Craters and banded terrains are the most prevalent surface geology (Fig. 4). Craters on Ganymede, like on other icy satellites, are shallower than terrestrial craters, and complex craters have ring fractures rather than peak rings (e.g., *Schenk and McKinnon*, 1989, *Bray et al.*, 2012), which occurs since ice deforms more readily than rock. Many craters have fluidized ejecta (e.g., *Horner and Greeley*, 1982; *Boyce et al.*, 2010). The centers of Ganymede's large craters often host flat-floored structures with circular domes at their centers that may form as the result of refreezing of melted ice near the center of the crater (e.g., *Moore and Malin*, 1988; *Schenk*, 1993; *Bray et al.*, 2012), suggesting that at least temporary shallow reservoirs of water exist. A few irregular features have been suggested to derive from cryovolcanism — hypothesized cold water and ice volcanism — but few constraints exist on this process (e.g., *Showman et al.*, 2004; *Bland et al.*, 2009; *Patterson et al.*, 2010).

Ganymede's outer ice shell is very thick at ~100 km (e.g., *Kivelson et al.*, 2002; *Schubert et al.*, 2004; *Vance et al.*, 2014), thus it is possible that as the planet cools, the outer ice shell could thicken significantly and drive some fluids to the surface (*Showman and Malhotra*, 1997; *Pappalardo et al.*, 2004), although little direct evidence for such a pro-

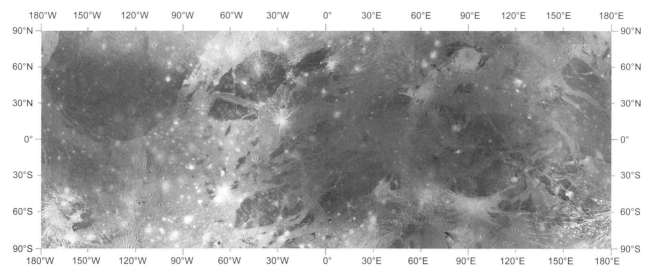

Fig. 4. Ganymede was studied extensively during the Voyager and Galileo missions, which revealed the dichotomous (two-component) surface shown here (global Mercator projection). The youngest terrain is bright and tectonized with an age ~2 b.y., having some similar properties to banded terrain on Europa, which is thought to indicate the point at which the Galilean moons entered their current orbital resonance. The oldest terrain is heavily cratered and with different or absent tectonism relative to the bright terrain, and an age of ~4 b.y. Credit: NASA/USGS/Patterson.

cess exists (*Patterson et al.,* 2010). However, the modern quiescence of Ganymede's icy shell as demonstrated by its surface age may limit any habitability in the ocean since crustal ice recycling would cease, e.g., limiting the delivery of oxidants formed on the surface to the subsurface. Given its immense internal pressure (up to ~1.2 GPa) (*Sotin and Tobie,* 2004; *Schubert et al.,* 2010; *Vance et al.,* 2014), ocean conditions on Ganymede are significantly different from Europa and Earth, and its internal pressures support the formation of high-pressure ice layers that could isolate the ocean from the core. While the outer layer of Ganymede's shell is either in the conductive or stagnant lid regime, its deeper ice layer may overturn under basal heating from the core (e.g., *Kalousová et al.,* 2018). Detailed calculation of the phase behavior of water-MgSO$_4$ salt mixtures under Ganymede conditions has now shown that depending on the composition and temperature of the ocean, the formation of multiple ocean layers separated by high-pressure ice III, V, and VI layers could occur, potentially with a deep reservoir of saline liquid above the silicate core (*Vance et al.,* 2014). Such prospects are exciting, since when coupled with evidence for a still active core, water may be co-located with warm or molten rock, which may increase the potential for water-rock reactions to be ongoing at Ganymede that could be relevant for life.

Ganymede is a planet-like satellite, and an example of what a water-rich planet might be like. Relevant to astrobiology, the most pertinent question is whether a sustained magnetic dynamo on Ganymede implies that significant heat preserves a liquid layer at the base of the lowermost ice layer above the core (*Vance et al.,* 2014), or if the hot core could deliver enough heat to the high-pressure ice layers to cause convective overturn of the ice or transport of liquids up through the ice shell (*Kalousová et al.,* 2018;

Choblet et al., 2017). If this activity is ongoing, then magmatic activity at the core-ice/ocean interface could drive high-pressure-reducing reactions, and some energy gradients for life's use might be sustained. It is thus unclear whether temperature-pressure-dependent isolation of the silicate interior from the majority of the ocean renders water worlds like Ganymede uninhabitable, and if it is important for habitability, where sources of oxidants could be derived in the absence of direct interactions with the surface. This may be a common limiting factor for habitability of water worlds, and pushes how far we can stretch our concept of Earth as an analog for other planetary environments. Given that ice-dominated Ganymede-like worlds could exist in the outer edges of solar systems, be common around giant planets and soon detectable (*Heller,* 2016), and be potentially numerous members of M-dwarf planetary systems (*Grimm et al.,* 2018), understanding the evolution of the silicate interface on large water worlds is important, and lessons from Ganymede therefore critical. Moreover, work to improve equation-of-state estimation for high-pressure ice-water systems and mixtures (e.g., *Vance et al.,* 2014), as well as coupled models of how these systems evolve for a range of conditions (e.g., *Noack et al.,* 2016), will be needed to better understand not only Ganymede but water worlds in other planetary systems. Ganymede thus offers a chance to benchmark these important studies for broader application, beyond its own intrinsic potential for habitability.

4. EUROPA: THE ARCHETYPE HABITABLE OCEAN WORLD?

Perhaps paradoxically, it is not how much water Europa has, but how little (relatively), that makes it special among the ocean worlds of the outer solar system. Oceans are not rare but rather may be everywhere in the outer solar system [see the chapters on Enceladus by Cable et al. and Titan by Lunine et al. in this volume, as well as *Hendrix et al.* (2019) for surveys on ocean worlds]. Among these worlds, Europa is more Earth-like in its bulk water-rock ratio and structure than it is like the other ocean worlds, which are characterized by relatively deeper oceans and ice shells. At 1565 km in radius and with a bulk density of 3000 kg m^{-3}, Europa is just smaller than Earth's Moon, but the outer ~100–150 km comprises an ice shell and ocean (*Anderson et al.,* 1997, 1998; *Kivelson et al.,* 2000) having 3× the volume of Earth's oceans. An important difference between Europa and potentially Ganymede, Callisto, and Titan is that Europa's ocean is in direct contact with its rocky interior (e.g., *Kivelson et al.,* 2002; *Schubert et al.,* 2004, 2010). There is not likely to be a significant difference between the initial bulk composition of the rocky materials that formed each of the jovian satellites except in their initial inventory of ^{26}Al and other short-lived radionuclides (*Canup and Ward,* 2002, 2006), depending on time of formation. It is possible that their impact history is different (*Barr and Canup,* 2010). But Europa's different interior is potentially due to its position in the jovian system contributing to its higher rock to ice ratio. As a result, Europa had more volumetric heating and per unit mass than Ganymede when it accreted, but with its smaller size its thermal insulation was lower. Having less water in total, or similarly less ice when it formed, means that less of Europa's accretion and radiogenic heat was spent on melting ice than the other two Galilean ocean moons.

Europa's relatively shallow ice shell and ocean, for an ocean world, has many important implications. There has been more volumetric heat available to melt the deep interior than on smaller or less rocky moons. Unlike the large ocean moons, Europa's ocean is in contact with the seafloor. Together these may drive water-rock reactions such as serpentinization that are known to occur on Earth and support chemosynthetic communities (e.g., *Reynolds et al.,* 1983; *Chyba,* 2000; *Chyba and Phillips,* 2001; *Zolotov and Shock,* 2004; *Kargel et al.,* 2000; *Vance and Goodman,* 2009; *Vance et al.,* 2007, 2016; *Preiner et al.,* 2018; *Egger et al.,* 2017). In addition, at a time when hydrothermal regions at the seafloor of Earth were pumping anoxic ocean waters through silicate chimneys that could have powered the origin of life (see chapter by Baross et al. in this volume), conditions in Europa's ocean may have been similarly conducive to life (e.g., *Barge and White,* 2017). Europa's present radiogenic heating and tidal heating provide less energy than Earth's thermal budget (e.g., *Davies and Davies,* 2010) by about two orders of magnitude (*Vance et al.,* 2016). However, between initial radiogenic heating, and tidal heating that stabilized by ~2 Ga (e.g., *Malhotra,* 1991; *Zahnle et al.,* 2003; *Moore and Hussman,* 2009), Europa likely had high internal heat during at least the first billion years of its history, and then at distinct intervals afterward, that could drive thermal gradients and energy production through seafloor hydrothermalism (e.g., *Travis et al.,* 2012). Thus, it is likely that both Europa and Earth have hosted hydrothermal reac-

tions between their oceans and seafloors over the history of the solar system. Yet there remain important questions as to whether or not Europa's long-term habitability gave rise to life or could sustain a biosphere. Can life originate in planetary oceans? Is life without surface environments or silicate weathering possible, given that these are important on modern and perhaps early Earth? Did Europa have enough geothermal activity to initiate life as we know it? What was the lowest level of internal heat generation for Europa and was it sufficient to support life? Would life on Europa be similar to our own?

4.1. Interior Structure

From an astrobiological perspective, the internal structure of Europa can help constrain how, if at all, vigorously convecting the early silicate mantle might have been, whether thermal evolution has ceased, and whether the pressures within the crust that can permit or limit long-term water circulation within Europa's silicate interior. The first models of Europa's potential interior emerged in the late 1970s and early 1980s, where the calculations of *Lewis* (1971) were expanded to demonstrate that rather than a thick ice shell over a rocky interior, the addition of tidal heating could maintain a long-lived ocean at depth (*Cassen et al.,* 1979; *Squyres et al.,* 1983), in advance of the Voyager 1979 flyby. When Voyager passed Europa, indications of a young, possibly active surface seemed evidence that a subsurface ocean was not only possible but likely. However, it would require extensive study by the Galileo mission to fully characterize Europa's interior.

4.1.1. Deep interior. *Anderson et al.* (1998) reported gravity data for Europa from the first two Galileo flybys, revealing data most consistent with a differentiated interior. They placed the first realistic constraints on its hydrosphere, suggesting a water and ice layer 150 km thick overlaying a silicate interior. By the end of the mission, measurements of Europa's moment of inertia were most consistent with internal differentiation including an iron or iron-sulfur core (e.g., *Schubert et al.,* 2004; *Moore and Hussman,* 2009). However, because gravity data are sensitive to material distribution within the planet (i.e., density of the internal layers), the small density difference between water and ice makes measuring the ice shell thickness via gravity alone difficult, and beyond the limits of the Galileo flyby data.

4.1.2. Ocean. The best constraints on the existence and properties of the ocean itself come from Europa's interactions with Jupiter's magnetic field. The jovian magnetosphere is the strongest planetary magnetic field in the solar system and rotates along with Jupiter every 10.1 hours. The magnetosphere is tilted with respect to the jovian pole, which means that the moons that are tightly co-planar along Jupiter's equatorial plane experience a changing magnetic field polarity that shifts with Jupiter's rotation period. Consequences of the jovian magnetic field are many. The field accelerates particles and plasma with a wide range of energies around the system, including material sputtered from the icy satellite surfaces as well as that ejected from Io's intense volcanism (e.g., *Carlson et al.,* 1999; *Paranicas et al.,* 2002; *Paranicas et al.,* 2009; *Nordheim et al.,* 2018). These impact the surface of the moons and interact with surface materials, producing new compounds as well as disassociating others. These materials also produce a strong asymmetry between the colors of the leading and trailing regions of Europa's surface and with the sub and anti-jovian surfaces [the so-called leading-trailing asymmetry (e.g., *Carlson et al.,* 2005, 2009; *Dalton et al.,* 2012, 2013; *Fischer et al.,* 2015; *Poston et al.,* 2017); see Fig. 5 and section 4.2]. From a diagnostic perspective, the changing jovian field strength and polarity that the moons — including Europa — experience in their orbits can induce a current within any electrically conductive layer of their interiors. During flybys of Europa, Ganymede, and Callisto, the magnetometer onboard Galileo measured the signature consistent with magnetic induction (*Kivelson et al.,* 1997, 2000; *Khurana et al.,* 1998). These measurements are made by assessing any changing magnetic field strength and direction as the spacecraft approaches the target. During flybys, the spacecraft begins embedded in magnetic field lines connected to Jupiter, but as it approaches moons with induced or internal fields, the spacecraft passes into regions where field lines connect to the moon. During close passes to Europa, the magnetic field polarity measured by Galileo's magnetometer would reverse while maintaining a nearly consistent strength, indicating that a shallow conductive layer indeed existed within the body. *Kivelson et al.* (2000) reported conclusive evidence for a subsurface ocean at Europa from six close flybys having multiple geometries and

different jovian field conditions in which the same observations were repeated, allowing them to conclude that Europa possessed an induced magnetic field consistent with a global ocean. Together with gravity, magnetic field data constrain the thickness of the ice shell, indicating that the ice-ocean interface is likely within a few tens of kilometers of the moon's surface (*Kivelson et al.,* 2000). The exact depth of Europa's ocean and its composition is as yet unknown, since while the magnetic and gravity field inversions can be coupled together to estimate ice shell and ocean properties, coupled solutions are non-unique. Nonetheless, the best solutions for all the present data include a subsurface ocean, and thus *Kivelson et al.* (2000) are generally credited with the discovery of Europa's ocean.

4.1.3. Ice shell. In general, two classes of ice shell thickness models are discussed with regard to Europa: a conductive or "thin" shell on the order of a few kilometers (e.g., *Greenberg et al.,* 1998), and a convective or a "thick" shell on the order of more than 10 kilometers (e.g., *Pappalardo et al.,* 1998; *Schenk,* 2002). *McKinnon* (1999) modeled in detail how Europa's thermal evolution impacts the ice shell state, and in particular tested the potential onset of convection within the ice shell as it grew thicker. This work suggests that once the shell reaches ~10 km in thickness, the ice shell must transition from a convective to conductive regime to remove heat from the interior. Geologic constraints on the ice shell thickness generally support one or the other of these models, in some cases depending on assumptions. A geologic map of Europa is found in Fig. 6. Impacts on Europa are few, especially large craters, but may hold clues

Fig. 5. See Plate 14 for color version. Europa's icy surface is characterized by a young geological age (having few impact craters) and both tectonic and endogenic geologic processes. Generally, young materials are dark in color, and age to brighter white with time. Both implanted materials from particles swept up by Europa (leading hemisphere) and from bombardment by ions trapped in Jupiter's rapidly rotating magnetosphere (trailing, mid-latitudes) can cause discoloration in addition to the emplacement of salt-rich materials from below. Credit: NASA/JPL-Caltech/SETI Institute PIA 19048.

toward the ice thickness. None of the impacts appear to have broken through the crust (e.g., *Turtle and Pierazzo*, 2001; *Schenk*, 2002; *Schenk and Turtle*, 2009; *Bray et al.*, 2014). *Bray et al.* (2014) concluded the shell could be as thin as 7–10 km based on impact morphology, while *Turtle and Pierazzo* (2001) and *Schenk* (2002) observed that the largest impact basins would be expected to create transient craters up to 20 km in depth, which they interpret as constraining the ice thickness at 25 km.

Fractures cross Europa's shell everywhere: They are both the youngest and oldest terrains on the surface and are globally distributed with a range of morphology (e.g., *Figueredo and Greeley*, 2004; *Doggett et al.*, 2009; *Leonard et al.*, 2018). Young fractures tend to have red coloration, while older fractures appear brighter. Formation models for the fractures and ridges differ between authors, from fully opening fractures (e.g., *Sullivan et al.*, 1998, *Greenberg et al.*, 2000) to tectonically reworked through strike-slip motion (e.g., *Nimmo and Gaidos*, 2002; *Nimmo and Schenk*, 2006; *Han and Showman*, 2008). It is possible that these features begin first as troughs, then proceed to form cracks that are then reworked to form ridges and ridge complexes (e.g., *Greeley et al.*, 1998; *Figueredo and Greeley*, 2004; *Nimmo and Gaidos*, 2002; *Nimmo and Schenk*, 2006; *Kattenhorn and Hurford*, 2009). Cycloids, series of large arcuit ridges (Fig. 7), were thought to represent rupture of the ice shell from diurnal tides, or "tidal walking," implying a very thin, few-kilometer shell (*Hoppa et al.*, 1999; *Greenberg et al.*, 1998). Later work has argued against diurnal tidal walking. For instance, within the same region cycloids may occur with similar age but with opposite facing arcs (*Kattenhorn*, 2004), possibly suggesting that cycloid geometry may be better fit by the propagation of fractures due to changing stress direction over longer than diurnal periods. And while the surface is covered in tectonic fractures, *Goldreich and Mitchell* (2010) have shown that tidal stresses are not able to break through Europa's ice shell if it is even just a few kilometers thick. Non-synchronous rotation may contribute a small amount to the stresses but appear still too small to fracture the ice shell (e.g., *Rhoden et al.*, 2011). Convective stresses between an overturning ductile shell and the brittle ice may provide additional mechanisms to break the ice (e.g., *Barr and Hammond*, 2015; *Weller et al.*, 2019), but more work is needed in this area. At present, fractures and ridges alone cannot discriminate ice shell thickness.

While there is some geology that has been argued to be evidence of a conductive or "thin" ice shell, the predominance of evidence from geology suggests a much thicker ice shell capable of solid-state heat transfer. Many authors have investigated the consequences of diapirism, convection, and tidal heating within the ice on geologic feature formation (e.g., *Ojakangas and Stevenson*, 1989; *McKinnon*, 1999; *Pappalardo and Barr*, 2004; *Showman and Han*, 2004; *Han and Showman*, 2005, 2010; *Peddinti and McNamara*, 2015, 2019). Chaos terrain, regions where pieces of the crust appear to have been broken, reoriented, and refrozen, were previously suggested to result from melt-through of the ice

shell by ocean plumes (e.g., *Carr et al.*, 1998; *O'Brien et al.*, 2002). However, the preponderance of evidence suggests that these regions form from melting within, or brine injection into, a thick ice shell forming 10–100-km-diameter brine-rich ice zones or even pockets of liquid water in the ice shell (e.g., *Pappalardo et al.*, 1998; *Collins et al.*, 2000; *Han and Showman*, 2005; *Sotin et al.*, 2002; *Schmidt et al.*, 2011). Until recently, there was only indirect evidence for plate tectonic resurfacing on Europa, indicated by the appearance of young "banded terrain" — regions of generally dark, ropy to flat lineated symmetrical sections of newer ice hundreds of kilometer long that bisect surface terrain that is contiguous on either side of the feature. These observations are consistent with spreading of the icy shell to produce upwelling new ice from a ductile layer below (*Nimmo et al.*, 2003; *Stempel et al.*, 2005; *Prockter and Patterson*, 2009; *Howell and Pappalardo*, 2018). With high-resolution data of Europa only available for about 10% of its surface (e.g., *Figueredo and Greeley*, 2004), resolving tectonic processes has proven difficult. However, *Kattenhorn and Prockter* (2014) reported observations of a region near Europa's pole where a geologic plate reconstruction demonstrated the removal of a ~99-km-wide swath of surface material, surrounded by other evidence of plate boundaries and subduction. There is no shallow or conductive shell hypothesis for such a process. This region, called a subsumption band, arguably represents the strongest evidence for plate tectonics on a planet other than Earth. This is an area for future work, since more information is needed to fully understand the processes that might underpin plate tectonics in ice (*Johnson et al.*, 2017; *Howell and Pappalardo*, 2019).

However thick the ice shell, the distance of the ocean from the surface is not necessarily key to Europa's habitability. However, a balance between composition and geophysical cycling that could maintain geochemical processes similar to those observed on Earth may be critical: Sources, sinks, and rates of nutrient and energy cycling all factor into habitability. Hence the existence of plate-tectonic-style recycling of the ice shell is potentially critical for maintaining Europa's habitability, and could be a key difference between Europa and other moons such as Enceladus with partially ancient surfaces (e.g., *Kirchoff and Schenk*, 2009).

4.2. Composition

Europa formed in the jovian subnebula, where presumably solar composition volatiles mixed with carbonaceous-chondrite-like material (*McKinnon and Zolensky*, 2003) were drawn into the gas-starved disk (*Canup and Ward*, 2002). Important for considering Europa's ability to host life, relative to Earth (which likely formed from drier ordinary chondrite rather than carbonaceous chondrite material, see the chapters in this volume by Zahnle and Carlson and Raymond et al.), is that Europa was likely enhanced in carbon, even though methane was not yet stable as a planet-forming solid at Europa's position in the solar system. Accretion of carbon from CO_2 ices was also likely, and Europa's initial

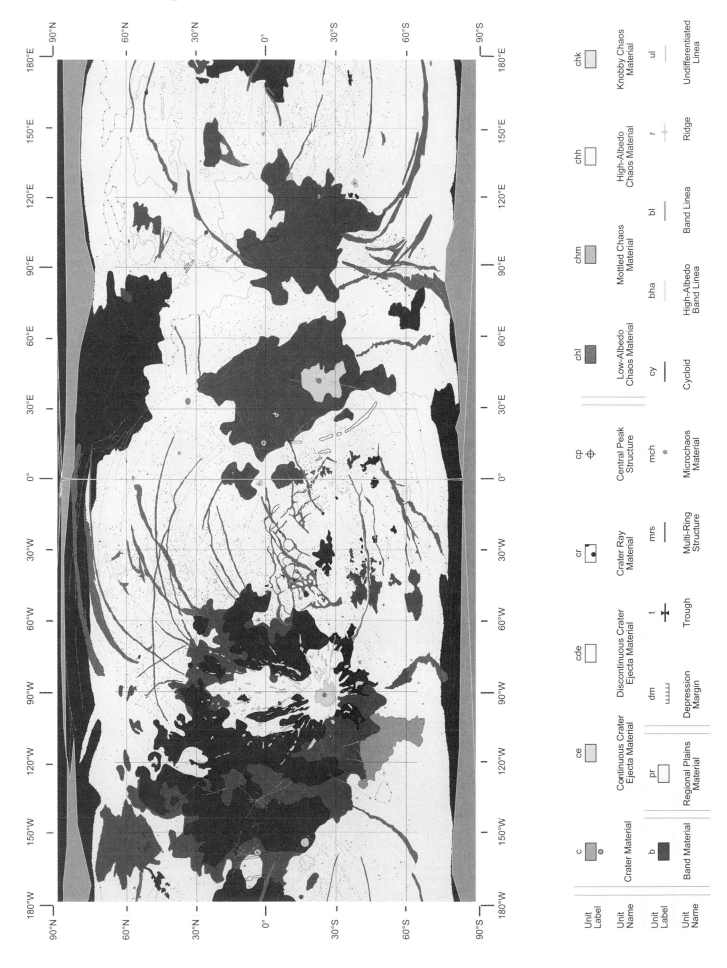

carbon inventory would be nearly solar, about 2.5× Earth (e.g., *Lewis,* 1971; *Zolotov and Shock,* 2001). CO_2 has been detected on the surface of Callisto (*Hibbitts et al.,* 2000) as well as Ganymede (*Hibbitts et al.,* 2003), where ozone has also been detected (*Noll et al.,* 1996). *Fanale et al.* (2001) suggest up to 0.28 wt.% of the ice could be CO_2. It is likely that the relatively low abundance of CO_2 on Europa's surface is due to its level of geologic activity, since reprocessing of the ice shell would mix the CO_2 into the ice, or remove any late veneer of CO_2 through warming or radiolytic processing. It is possible that CO_2 could be trapped in clathrates, but whether these could resist tectonic erosion is unclear. Absorption features consistent with hydrocarbons and S- and N-containing functional groups have been detected on Callisto (*McCord et al.,* 1997, 1998a), which argues that a veneer of organics has fallen on all the Galilean satellites.

Hand et al. (2009, and references therein) presented a detailed review of the elemental composition of Europa based on knowledge of the material from which it formed. The Jupiter system likely condensed with 0.25× the solar phosphorus, 3× the solar nitrogen, and 2.5× the solar carbon of Earth, and abundant sulfur, hydrogen, and oxygen (*Spencer and Calvin,* 2002). *Zolotov and Kargel* (2009) estimate the best constraints on phosphorus in the ocean as $0.34–4.2 × 10^{-6}$ kg, where Earth's ocean has $6 × 10^{-5}$ kg. In addition to the sources from formation, meteoritic infall of material could also be a source of bioavailable nutrients to the ocean. *Pierazzo and Chyba* (2002) calculated that $2 × 10^{11}$ to $3 × 10^{12}$ kg of N, $2 × 10^{10}$ to $3 × 10^{11}$ kg of P, $2 × 10^{11}$ to $2 × 10^{12}$ kg of S, and $9 × 10^{11}$ to $1 × 10^{13}$ kg of C could be delivered to the surface over the life of the solar system; these values could also be up to an order of magnitude larger (*Johnson et al.,* 2004).

4.2.1. Surface. Europa's surface is comprised of regions of relatively pure water ice, ice mixed with salts of many varieties, and deposits of highly concentrated material with little ice spectrally detected. Aside from water, there is evidence for both implanted materials from the Io torus concentrated on the leading and trailing sides, and endogenic materials. Exogenic materials from Io take two forms. High-energy particles, including sulfur, co-rotating within Jupiter's magnetic field sputter and implant in a large circular province on the trailing side, which can also form hydrates (e.g., *Carlson et al.,* 1999, 2009; *McCord et al.,*

Fig. 7. Cycloidal ridges may be evidence of changing tidal stress within the ice shell over time. Whether these features form due to a combination of diurnal and non-synchronous stresses (*Hoppa et al.,* 1999; *Hurford et al.,* 2007) or due to build up and release of tidal stress over longer than diurnal timescales (e.g., tail cracks) (*Kattenhorn,* 2004; *Kattenhorn and Hurford,* 2009) is still under debate. Credit: NASA/JPL/ University of Arizona.

1999; *Paranicas et al.,* 2009; *Dalton et al.,* 2012, 2013). The leading-hemisphere, non-magnetic particles ejected from Io are swept up by Europa's orbital motion. These mixed materials muddy our information on Europa's endogenic composition in those areas; however, other areas of the surface are more straightforward to interpret. Young geologic terrain, such as newer fractures, ridges, and chaos terrain (Fig. 5 and 8), all show reddish colors and lower albedos than the background surface (e.g., *Greeley et al.,* 1998; *Pappalardo et al.,* 1999; *Figueredo and Greeley,* 2004; *Doggett et al.,* 2009; *Leonard et al.,* 2018), indicating materials from the interior are exposed near the surface. Other compounds that are formed on Europa's surface and concentrated within geologically young terrain include hydrated sulfuric acid, sodium chloride, magnesium sulfate, and other salts at lower levels (e.g., *McCord et al.,* 1998b, 2002; *Zolotov and Shock,* 2001; *Carlson et al.,* 1999, 2002, 2009; *Shirley et al.,* 2016; *Dalton et al.,* 2012; *Brown and Hand,* 2013; *Fischer et al.,* 2015; *Hand and Carlson,* 2015; *Ligier et al.,* 2016). The composition provides windows into the body below, but the geochemical story is complex. For example, slight differences in the initial sulfur accreted by the moon can affect everything from core formation to the chemistry of the ocean, including the relative abundance of sodium and magnesium. Mixing models of these components allow us to better understand ice shell composition, processes, and sources. However, how these materials arrive at the surface requires models of not only geologic processes, but ice formation as well (*Buffo et al.,* 2018; *Buffo,* 2019).

Fig. 6. (facing page) See Plate 15 for color version. Global geologic map of Europa showing the distribution and juxtaposition of major terrains (*Leonard et al.,* 2018). The map was built by compiling images from Galileo and Voyager missions, and used the highest-resolution data to clarify local-scale crosscutting relationships. This is a draft of the global map; the final version will be Leonard et al. (in preparation) Geologic Map of Europa, U.S. Geological Survey Scientific Investigations Map, scale 1:15,000,000. Credit: NASA/Erin Leonard, Alex Patthoff, and Dave Senske, building on work by Ron Greeley, Thomas Doggett, and Melissa Bunte.

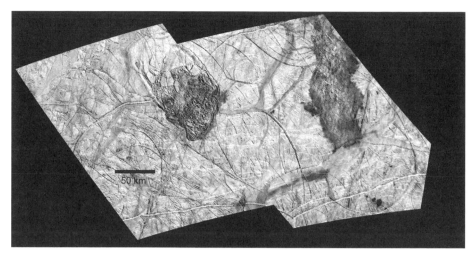

Fig. 8. See Plate 16 for color version. Thera Macula (left) and Thrace Macula (right) are two of Europa's largest chaos terrains. Multiple lines of evidence exist for liquid water and brines being involved in the formation of these features, including (1) their reddish color, which is associated with young age and high salt content; (2) the appearance of brines "soaking" from the features into surrounding terrain; and (3) their topography. Initially thought to be surface flows in Voyager spacecraft data, this image and locally-controlled topography reveal that Thera Macula is generally a large, circular depression with inward-facing scarps. The morphology suggests that the southern tip was the first area to rupture, and that the disruption of the ice may be proceeding northward and opening fractures creating new icebergs. Comparisons to Conamara Chaos fueled the formation of chaos evolution models (*Schmidt et al.,* 2011). Credit: NASA/JPL/University of Arizona PIA02099.

Relevant to astrobiology, Europa's surface represents the best source of oxidants available to mix into a presumably reductant-rich ocean produced through water rock reactions at the seafloor (*Chyba,* 2000; *Chyba and Phillips,* 2001), thereby providing chemical gradients for life. While the extreme radiation at the surface is likely destructive to life itself and can change or destroy biomarkers (e.g., *Paranicas et al.,* 2007, 2009; *Nordheim et al.,* 2018), the radiation also forms potentially useful byproducts. In particular, hydrogen peroxide (H_2O_2) is constantly produced on the surface and was detected in Galileo data (e.g., *McCord et al.,* 2002; *Paranicas et al.,* 2009), and it comprises 0.13% of the materials that are spectrally active on the surface. Oxidizing species are globally distributed, and therefore available to be recirculated into the interior by geophysical processes, including in clathrate hydrates (*Hand et al.,* 2006). In addition, surface carbon species that could be carried into the interior include formaldehyde (*Chyba,* 2000; *Chyba and Phillips,* 2001), methanol (*Bernstein et al.,* 1995), CO_2, and H_2CO_3 (*Nordheim et al.,* 2018). Nitrogen-bearing compounds with abundances between 5% and 10% that of Earth should have formed within Europa but would be easy to destroy radiolytically once at the surface (*Lewis,* 1971; *Loeffler et al.,* 2006). How the evidence of these cycles can be reconciled, including how and where biomarkers are produced and may survive (Fig. 9), will be key to interpreting Europa as a potentially habitable environment (*Figueredo et al.,* 2003; *Hand et al.,* 2009) that can be searched for life.

4.2.2. Ice shell and ocean. Because of Europa's complex interactions with the jovian environment and its own geology, teasing out the chemistry of the ice shell and ocean,

both possibly habitable niches, requires understanding of both chemical and physical processes that are ongoing. Most models of Europa's chemistry integrate assumptions about its initial rocky composition and water abundance (*Lewis,* 1971) with the detected compounds discovered thus far, both described above. *Zolotov and Shock* (2001, 2003, 2004), modeled the ice shell chemistry by assuming Europa formed from carbonaceous chondrites, and considering a wide range of factors affecting the alteration of this material through water-rock reactions during differentiation. They derive an ocean composition with bulk salinity of 12 ppt (best estimate) to 282 ppt (saturation). For comparison, 34 ppt is the salinity of Earth's ocean. A summary of their results relative to the Earth's composition is found in Table 2 (*Buffo,* 2019). *Hand and Chyba* (2007) show that depending on the ice shell thickness, magnetic field data can be consistent with anything from freshwater to nearly saturated salts.

Since ice forms from a liquid with efficient, but not complete, rejection of impurities, it reflects the conditions in the reservoir from which it forms. *Zolotov and Shock* (2001) estimated the composition of the ice shell by assuming the ocean composition they derived and calculating the solid-liquid phase boundary according to temperature. *Buffo* (2019) modeled the composition of the ice shell by incorporating multiphase transport into the solution, following polar ice analogs on Earth, and considered how porosity and drainage affect the end salinity and structure of the ice (*Buffo et al.,* 2018; *Buffo,* 2019). At the surface temperature of Europa, no liquids are likely. However, at the base of the ice shell, the temperature is expected to be at the freezing point of the ocean at a given pressure, somewhere around 270 K, similar to that on Earth. By a few kilometers depth in the ice shell, where the temperature reaches ~250–260 K (*McKinnon*, 1999, *Pappalardo and Barr*, 2004), the temperatures may be warm enough for melts of salt-rich materials under pressure ("eutectic" melt may become possible depending upon the salt compostion). *Buffo* (2019) derived a diffusive limit of impurity entrainment into Europa's ice shell as it forms by comparing it to sea ice that forms in high thermal gradient conditions — and thus could be analogous to the upper ice shell of Europa — and marine ice that forms in low thermal gradient conditions much more similar to expected conditions on Europa. Since the ice shell likely froze quickly down to a few kilometers and slowly after that, the upper ice shell is expected to have higher salinity than its base, by a few parts per trillion (*Buffo*, 2019; *Buffo et al.*, 2019). This work constrains the average salinity of Europa's ice shell as ranging from 1.0–14.8 ppt depending on whether the most likely ocean composition (12 ppt) or saturation (282 ppt) is assumed. For an ocean of identical composition to Earth's ocean, the bulk salinity for most of the shell would be 1.9 ppt (*Buffo*, 2019; *Buffo et al.*, 2019). The processes that govern how ice forms from water are also important for understanding geological processes, since water pockets or water-filled fractures within the ice shell would follow similar freezing behavior, affecting the composition and lifetime of shallow liquids within the ice shell, as well as the mechanical properties of the ice. High thermal gradients trap more salt in the ice, whereas low thermal gradients promote impurity rejection. Thus, fractures in the ice shell filled with salty water would entrain impurities from the liquid that could result in their red color and could become sources of local weakness from grain size effects and at contacts between pure and salty ice (*Buffo*, 2019).

Multiple lines of evidence exist for liquid water within the ice shell, which could be potential niches for life or regions of high biomarker preservation. The lifetime of water within the ice shell has been a topic of discussion, since due to the higher density of water, it has often been assumed that the water would drain through the ice (e.g., *Gaidos and Nimmo*, 2000; *Kalousova et al.*, 2014). The geology of chaos and lenticulae — domes, red material, and tilted icebergs without large amounts of missing material — suggests that brines from melt or injection may not drain, instead refreezing in place over time (*Schmidt et al.*, 2011). Schmidt et al. demonstrated that large reservoirs of water could form and remain within the ice, based on the geology and topography of chaos terrains and the observed behavior of sub- and englacial water on Earth. While water may be more dense than ice, observations of water and ice in the Earth's cryosphere suggests solid ice is impermeable to water if the porosity is less than about 5% (e.g., *Kovacs and Gow*, 1975; discussions in *Buffo et al.*, 2018, and *Buffo*, 2019). This is why, for instance, englacial brine (*Kovacs and Gow*, 1975) and supraglacial lakes and rivers (e.g., *Bell et al.*, 2017) exist on Earth even with the ice at relatively warm temperatures. If large-scale melting occurs within the ice shell as predicted (e.g., *Collins et al.*, 2000; *Sotin et al.*, 2002), with nowhere for the brines to escape, these pockets or zones of water can grow and refreeze in place over time (*Schmidt et al.*, 2011). Initially, the composition of the liquid would be the salinity of the ice that was melted; however, as the melt cools, inward solidification should occur, faster from the top. Convection would mix the liquid, and freezing would increase the salinity of the solution, such that fresher ice on top would give way to progressively more concentrated layers. Salts can preserve some liquid water for protracted lifetimes. Salt deposits have previously been hypothesized (e.g., *Han and Showman*, 2005; *Pappalardo and Barr*, 2004) but not constrained. At the end of the solidification, production of layers of salts may be possible, on the order of 1 m per kilometer of melted ice (*Buffo*, 2019), which both affect ice shell evolution and could preserve a record of the former liquid environment — potentially containing biomarkers.

From an astrobiological perspective, the chemistry of the ice has several implications. First, the ice shell is the most accessible window into the intrinsic composition of the ocean and deep interior. Thus, measuring the ice composition can constrain Europa's bulk composition if the processes that govern ice formation are known: If ice forms from ocean freezing, its composition is roughly 1% of the salinity of the ocean (*Buffo et al.*, 2018; *Buffo*, 2019). Ice on Earth

TABLE 2. Comparison of best estimate europan ocean and ice composition to that of Earth (after Buffo, 2019).

Species	Terrestrial Seawater (mol kg^{-1})	*Zolotov and Shock* (2001) Europan Ocean (mol kg^{-1})
Na^+	4.69×10^{-1}	4.91×10^{-2}
K^+	1.02×10^{-2}	1.96×10^{-3}
Ca^{2+}	1.03×10^{-2}	9.64×10^{-3}
Mg^{2+}	5.28×10^{-2}	6.27×10^{-2}
Cl^-	5.46×10^{-1}	2.09×10^{-2}
SO_4^{2-}	2.82×10^{-2}	8.74×10^{-2}
Bulk ocean salinity;	34	12.3
Bulk ice salinity (ppt)	1.9	1.0

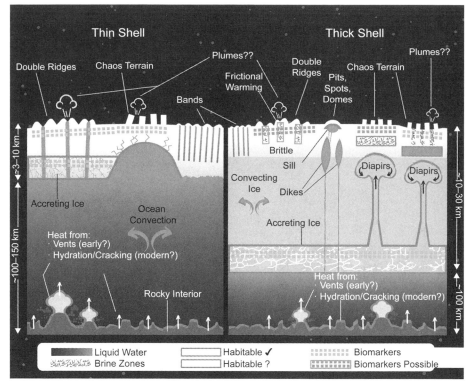

Fig. 9. See Plate 17 for color version. Schematic of habitability, biomarker production, and preservation on Europa as a function of geological processes for both a conductive (thin) ice shell and a convecting (thick) ice shell. All the colors and textures are chosen based on assessing hypothesized geological and physical processes for the corresponding feature. Locations within the system that have conditions hypothesized to be within the range of habitable conditions in ice on Earth are denoted as "Habitable" whereas regions that could once have had habitable conditions or for which the conditions are less favorable are marked "Habitable?" Regions where biomarkers could be produced and/or survive for long periods of time are marked "Biomarkers," and areas where the lifetime might be shorter or conditions relatively less favorable for producing biomarkers are marked "Biomarkers Possible." Inspired by *Figueredo et al.* (2003).

is a known host to life (e.g., *Priscu and Christner,* 2004; *Bowman et al.,* 2016; *Murray et al.,* 2012), thus Europa's ice shell may represent an important habitable niche, and materials within the ice comprise a critical source of energy with strong astrobiological implications. Areas where melting occurs within the shell should create reservoirs of brines with initially lower salinity than the ocean that may be suitable as habitats within the ice, particularly if the materials trapped within the ice from the ocean are reduced, given the likeliness that the ice shell is generally oxidizing. Second, on either end of melting within the ice shell, connected systems of brines within the ice may exist. Layers of ice with ~5% porosity could exist within the ice shell filled with brines. Life is known to exist even in low-porosity, relatively pure meteoric ice on Earth (e.g., *Priscu and Christner,* 2004), but these brine zones may be more akin to sea ice in terms of connectivity and habitability (e.g., *Bowman et al.,* 2016). Third, brines near the surface require high salinity to form or to survive for extended periods if injected from below and so will refreeze rapidly down to kilometers within the ice shell (*Buffo,* 2019). Thus, small and shallow liquid environments like fractures and sills may be short-lived. Finally, the formation and refreezing of melt lenses and brine zones should create cycles of salinity variation, from low to high,

as the area refreezes. The last stages of refreezing should produce extremely high salinity regions and perhaps precipitate concentrated salts and other materials, which could reactivate and encourage future melting.

A ramification of refreezing that creates high salinity, either near the surface or the middle ice shell, is the eventual formation of reservoirs with potentially very low water activity, depending on the exact composition of the ice and ocean to start. Water activity is an important consideration for the habitability of these areas and suggests that liquids within ice shells are not always habitats per se. Paradoxically, low-water activity solutions that are toxic to life can, however, preserve biological materials (*Touvila et al.,* 1987; *Borin et al.,* 2008). Given the potential for connectivity of some brine systems, and potentially long-lived reservoirs within the shell that end with a low-water activity state, these low-water activity zones and precipitates may well carry a concentrated, well-preserved record of the organisms within them. As such, water in the shell, and remnants in the form of salt deposits, are potentially important places to search for the record of past life on Europa, even if they are hostile to life at present.

4.2.3. Seafloor and deep interior. Not much is known directly about the composition of Europa's deep interior beyond assumptions of its starting material and later stage evolution. Gravity data are most consistent with the formation of a core, and the timing and location of the formation of Europa should have occurred within about 100 m.y. after the formation of the solar system from primarily P- and D-class outer-belt asteroids (*McKinnon and Zolensky*, 2003; *Canup and Ward*, 2009). Accordingly, its interior is likely best approximated by a chondritic composition. Whether or not the core is completely exchanged with the ocean, and whether ongoing geologic activity or other exchange is ongoing, are subjects of debate, since no data as yet can tell the difference between these models.

Reactions at the seafloor were likely ongoing early in Europa's history, and during tidally active periods, but whether these continue is as yet unknown. While *Kargel et al.* (2000), *Zolotov and Shock* (2001), and *Zolotov and Kargel* (2009) focused on bulk chemistry, other work has focused on ongoing water-rock reactions between the seafloor and the ocean. If magmatic activity is ongoing at Europa, Earth-like reactions similar to those near hydrothermal vents, cold seeps, and other similar environments are likely (see the chapter by Baross et al. in this volume). But even in the absence of magmatism, geothermal heat could contribute to hydration reactions like serpentinization that may be self-propagating, providing a source of reduced compounds to the ocean and pervasively fracturing and hydrating Europa's silicates (*Lowell and DuBose*, 2005; *Vance et al.*, 2007, 2016; *Vance and Goodman*, 2009). Serpentinization is a reaction between olivine species and water at low (>273 K) to moderate (<373 K) temperatures that produces H_2 as a byproduct that becomes bioavailable, so it is an important factor to consider for habitability. Only indirect measurements of hydrothermal byproducts have been made on Europa, including sulfur and magnesium, but these also have exogenic sources. The plume of Enceladus, however, does contain direct evidence of hydrothermal activity in the form of CH-compounds, silica nanoparticles,

(e.g., *Hsu et al.*, 2015; *Glein et al.*, 2015), and molecular hydrogen (*Waite et al.*, 2017), which is itself potentially useful for life (see the chapter by Cable et al. in this volume). Because serpentinizing reactions are both exothermic and expand the silicates as they hydrate, fractures can continue to propagate into the interior until pressure in the crust prevents further cracking (*Vance et al.*, 2007). *Vance et al.* (2016) place constraints on the depth of cracking and the potential rates of H_2 production within Europa's interior as a function of heating rate. Combined, fluxes of H_2 and O_2 for Europa span from two orders of magnitude lower up to nearly the lower limit for such fluxes on Earth (Fig. 10). Similarly low levels of activity are found in samples from accretion ice in Lake Vostok, Antarctica (e.g., *Christner et al.*, 2014) (Fig. 14).

5. PROSPECTS FOR HABITABILITY AND LIFE ON EUROPA

While no biosignatures have been detected on Europa as yet, there has been extensive work to try to bound the energy available for life, what life could exist there, what biomarkers would be expected, and whether these biomarkers would survive in the harsh conditions (Fig. 9). Geologic cycles could be important for habitability because they constantly reprocess planetary materials and set up chemical gradients of which life can take advantage (see the chapter by Hoehler et al. in this volume). On Earth, interactions between the atmosphere, continents, and ocean all provide sources of energy for life, and although the details of these processes have changed with time, their importance has remained. On Europa, it is unclear whether silicate weathering could occur, but cycles between the surface, ice shell, ocean, and seafloor are likely to be critical. Without an oxidizing atmosphere exposed to the ocean as on Earth, and with interactions between the ocean and seafloor potentially ongoing (e.g., *Zolotov and Shock*, 2001; *Vance et al.*, 2007, 2016), Europa's

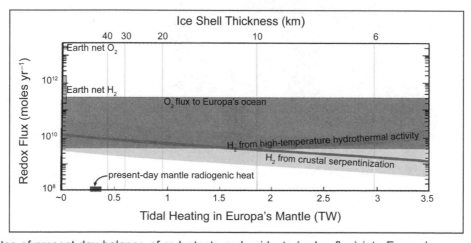

Fig. 10. Estimates of present-day balance of reductants and oxidants (redox flux) into Europa's ocean as a function of tidal heating input, after *Vance et al.* (2016). The oxidant flux is calculated from the presumed overturn rate of the surface based on its surface age, from 30 Ma to 90 Ma. The H_2 production is calculated based on the rate of serpentinization assuming that crack propagation can occur due to exothermic reactions down to 25 km depth.

ocean could become highly evolved through serpentinization and water-rock reactions, perhaps achieving high pH, as may have occurred on Enceladus (*Glein et al.,* 2015). However, the production of oxidants on its surface could provide a source to couple to reductants in the ocean (e.g., *Chyba,* 2000; *Chyba and Phillips,* 2001; *Hand et al.,* 2009, 2015). Although it is also possible that if the seafloor has no ongoing activity, the ocean could instead acidify (*Pasek and Greenberg,* 2012), active geologic cycles in the ice shell and subsurface could couple, and they could generate a long-term stable ocean environment with moderate to neutral pH conducive to life on Earth. Europa may be unique from the other solar system icy ocean worlds in that its combination of geologic activity and the duration over which it has continued may have given rise to habitable ocean conditions that existed over the lifetime of the solar system. Whether or not long periods of time are required for life to emerge, its support through geologic processes is likely on Earth, and may indeed be required elsewhere.

Important differences also occur between the oceans of Earth and Europa. While life is as yet found at all depths in Earth's ocean, life in Europa's deep ocean could be significantly different. For example, as pointed out in *Hand et al.* (2009), pressure differences between the oceans will affect carbonate production and dissolution, as CO_2 becomes more soluble with pressure below ~4 km ocean depth on Earth, and carbonates dissolve below that depth. The same would occur at about 40 km depth on Europa, well above the seafloor and a few to tens of kilometers from the base of the ice shell. If dissolution and precipitation form the carbon cycle on Europa, then carbon cycling may be linked directly to ocean mixing. Three-dimensional overturning circulation has been modeled by *Soderlund et al.* (2014) and *Soderlund* (2019). From a habitability perspective, these currents could dredge up carbon from deep in the ocean and cycle it toward the ice shell base, which would affect the stability of carbon

compounds and potentially enhance the habitability near this interface. Better understanding the differences and similarities between Europa's deep, global ocean and our own are key to understanding how the system works.

5.1. Surface-Subsurface Exchange

Galileo data revealed Europa as a recently active world, with a surface age of between 30 and 90 Ma (*Zahnle et al.,* 2003; *Bierhaus et al.,* 2009). This evidence suggests that the surface is constantly rebuilt, and indicates recent, if not ongoing, activity. This is supported by several pieces of evidence acquired during and since the Galileo mission that indicate recent surface/subsurface exchange. Cryovolcanism has been supposed (*Fagents et al.,* 2000; *Quick et al.,* 2013; *Sparks et al.,* 2017) but also challenged (*Fagents,* 2003; *Schmidt et al.,* 2011), in part due to terminology, since hydrology may better represent the physics of these systems where the melt or "magma" is water. A search for contemporary activity during Galileo was undertaken using the onboard cameras, but was inconclusive (*Phillips et al.,* 2000), although new work on other datasets from the mission have now revealed indications of potential activity.

Three methods of detection have shown potential plumes emitted from the surface of Europa. Hubble Space Telescope observations have detected potential plumes associated with southern high latitudes (Fig. 11) [*Roth et al.* (2014a) using UV emission data] and a region near the equator south of the Pwyll impact [*Sparks et al.* (2016, 2017) using UV transits of Jupiter]. In both cases, the detections are near the limit of the technique applied. While the southern plume has not been observed again (*Roth et al.,* 2014b), the potential equatorial source has been observed multiple times. Thus far, if they exist, the plumes do not appear to be tidally modulated. These telescopic detections sparked reanalysis of Galileo data. Galileo magnetometer data were thoroughly searched

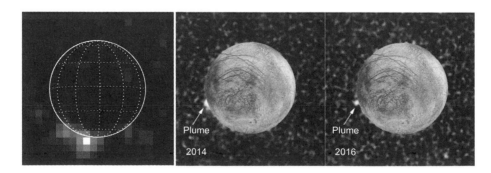

Fig. 11. Hubble Space Telescope (HST) images of potential plumes on Europa, after *Roth et al.* (2014a) (left) and *Sparks et al.* (2016) (right, with Galileo-derived globe superposed). *Roth et al.* (2014a) presented the first plume detection, observing 4σ enhancements of hydrogen and oxygen emission in HST Space Telescope Imaging Spectrometer (STIS) data, but detections at this location have not been repeated. *Sparks et al.* (2016, 2017) observed two instances of potential plumes near the equator of Europa during transits of the moon across Jupiter. Using Jupiter's UV illuminated disk as a background, extinction from putative plumes along the limb of Europa could be detected in STIS images. The location of this plume is just south of Pwyll crater, in a location where anomalous thermal inertia was observed in Galileo data, as well as where magnetic and plasma (*Jia et al.,* 2018) and ionosphere enhancements (*McGrath and Sparks,* 2017) were detected on the E6 flyby.

with comparisons with the plasma data, and are consistent with activity in the region near Pwyll during the mission (*Jia et al.,* 2018). A second potential plume visible in the magnetic field data has also been reported (*Arnold et al.,* 2019). While plumes themselves don't impact habitability directly (although they may be sources of surface-ocean or surface-liquid interaction), they are important indicators of contemporary activity, and provide a window into the subsurface composition. These plumes could be regions for future exploration by spacecraft, including sampling akin to that accomplished by Cassini at Enceladus (e.g., *Waite et al.,* 2017).

Potential surface exchange via fracture is possible although poorly constrained. The youngest cracks and ridges, characterized by a reddish-brown albedo, are likely regions of high salt content, and their color may require concentration mechanisms such as sourcing from highly concentrated brines or shallow refreezing. The ubiquitous surface fractures and tidal state of Europa's ice shell has led some authors to suggest that fractures could represent places where the ocean could come directly to the surface (e.g., *Greenberg et al.,* 1999) and thus these generally have been assumed to source the putative plumes. While perhaps intuitively satisfying, there are many arguments against this interpretation. The stresses expected from tides appear insufficient to open fractures beyond a few hundred meters (e.g., *Goldreich and Mitchell,* 2010), and models of basal fractures do not produce surface cracks (e.g., *Walker and Schmidt,* 2015; *Craft et al.,* 2016). Simple arguments using Archimedes' principle hold that, should a fracture open from the surface to the ocean, the ocean water can only reach within 10% of the surface, due to the density contrast between ice and the ocean, and thus for even a putative 5-km shell, the ocean reaches at most ~500 m from the surface. The bulk of its geology and most models suggest Europa's ice shell is tens of kilometers thick, thus a kilometer or more would separate the ocean from the surface. Clearly fractures exposed to the vacuum of space would produce plumes but would be unlikely to source surface flows, and holding these fractures open is also challenging. The extreme cold of the surface will freeze such fractures down to hundreds of meters within a few hours, much faster than the tidal cycle (*Buffo,* 2019). Some models have suggested that fractures could form via intrusions from the ocean or basal fractures (*Manga and Wang,* 2007; *Craft et al.,* 2016) or linear diapirism (buoyant ice plumes) (*Head et al.,* 1999). The fractures are surprisingly linear, and ridges have high angles to the surface, which has been interpreted as evidence for cryovolcanism (ice magmatism and associated plumes, e.g., *Fagents,* 2003; *Fagents et al.,* 2000; *Dameron and Burr,* 2018). However, no convincing cryomagmatic features have been confirmed, and ridge features can also be created through other processes: tail crack propagation (*Kattenhorn,* 2004), frictional processes (*Nimmo and Gaidos,* 2002; *Kalousova et al.,* 2016), and compression (*Culha et al.,* 2014). Nonetheless, while fractures may bring some material from the interior to the surface, it is unlikely that surface materials are brought downward. Banded terrain

is evidence of ice shell extension, akin to seafloor spreading (*Prockter and Patterson,* 2009), which would be evidence of part of a tectonic system analogous to the global crustal cycling on Earth where upward propagation of ice from the interior is likely (e.g., *Howell and Pappalardo,* 2018). This represents more upwelling from the interior, but no downward flux in these regions.

While upward transmission of material from the ocean has been thoroughly interrogated, evidence for whether and how material is returned to the ocean has been more elusive. Multiple models and hypotheses have considered convection within the ice shell, which would bring material from the ice-ocean interface to the base of the brittle portion of the ice shell and likely back down (e.g., *Pappalardo et al.,* 1998; *Pappalardo and Barr,* 2004; *Han and Showman,* 2005; *Mitri and Showman,* 2008; *Peddinti and McNamara,* 2015, 2019; *Barr and Hammond,* 2015). However, the brittle surface ice is likely porous and buoyant and as a result is difficult to draw down. Much work suggests water is formed within the ice shell above rising thermal plumes (*Head and Pappalardo,* 1999; *Sotin et al.,* 2002; *Han and Showman,* 2005; *Mitri and Showman,* 2008; *Schmidt et al.,* 2011; *Kalousova et al.,* 2014), either via tidal concentration of heating within the plume (*Han and Showman,* 2005) or eutectic melting (*Schmidt et al.,* 2011). These would potentially initiate full ice shell cycling including downward and lateral material propagation.

Chaos terrains may serve to cycle material through the ice shell. Comparing two of Europa's most prominent chaos features, Conamara Chaos and Thera Macula, *Schmidt et al.* (2011) showed that the geology of these features is most consistent with the formation of large shallow (a few kilometers deep) subsurface melt pockets within the ice shell. Thera Macula is a dark quasi-circular chaos terrain that constitutes an 80-km-wide surface depression of up to hundreds of meters deep, internal to which large ice rafts appear to float within a crushed matrix of ice. *Schmidt et al.* (2011) argued based on several geomorphic characteristics of older large chaos terrains, including Conamara, that the putative water body beneath Thera Macula was still liquid during the Galileo mission, implying large volumes of water were close to the surface, driving geologic activity that would likely be ongoing at present. Important for astrobiology, these would be areas where energetic mixing of the ice shell surface into the subsurface through fracture, disruption, and ice block capsize would deliver surface materials, probably including oxidants, to the deeper ice shell. These regions are characterized by kilometers-thick icebergs locally higher than crumbled ice matrix nearby (*Spaun et al.,* 1998; *Collins et al.,* 2000; *Schmidt et al.,* 2011) that are rotated and flipped, indicating that the surface disruption was initially violent. Similar events on Earth, the collapse of ice shelves, release kilotons of energy when they collapse and form icebergs surrounded by disintegrated ice (e.g., *Scambos et al.,* 2004; *Banwell et al.,* 2013), suggesting that the process on Europa should be similarly energetic and potentially mix surface material through the shell. Moreover, the geology surrounding these regions suggests horizontal transport of material during and after formation (Figs. 8 and 12) (*Schmidt et al., 2011*). Most work has assumed that once liquids form in the ice, the density contrast between the brines and the ice would lead to downward propagation of the liquids (*Gaidos and Nimmo, 2000; Kalousova et al., 2014*). However, since below critical porosity ice is impermeable, water formed within the ice shell could be trapped in place until frozen, or until fracturing or other process conducts it downward. In any case, as the water within chaos and other water pockets within the ice shell refreeze, these mixed materials would concentrate toward the bottom of the feature and could refreeze as denser layers atop or within convective regions of the ice shell. Thus, chaos terrains may be locations of upward and downward interaction between the surface and the ocean (Fig. 12), as well as an area of lateral transport, forming a conveyor belt even if brine drainage is not operating. Such environments may as a result cycle between periods of relatively mild and very extreme conditions that would impact their habitability, but once frozen out might preserve a record of the past environment.

Tectonic processes across the solar system are many; however, only Earth and possibly Europa have strong evidence for plate tectonics including processes that subsume material into the subsurface. Low spatial coverage of Eu-

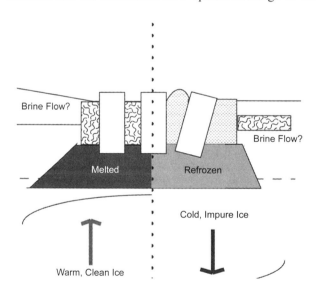

Fig. 12. Chaos terrain may be important for producing water within the ice shell as well as the horizontal and vertical transport of brines potentially important to habitability in the ice shell, and the mixing of surface materials into the deeper ice shell that can be transmitted to the ocean. Even the warm, clean ice shown here implies formation at the ice-ocean interface with some ocean-derived material entrained within that can be delivered to the surface and replaced by presumably oxidant rich material from the surface. Within the refrozen ice from the melt lens, increasing salinity with depth, as well as a possible sedimented salt layer, may occur that could preserve a detectable signal of the past water body, or pristine concentrated biomarkers that could be sampled relatively near the surface.

ropa's surface at high resolution makes evaluation of global crustal recycling complex, but several potential regions have been imaged (e.g., *Prockter and Patterson, 2009*). The most compelling and direct evidence for material exchange downward through Europa's ice shell is associated with so-called subsumption bands (*Kattenhorn and Prockter, 2014*). Near irregular-shaped bands in the northern hemisphere, regions of the surface can be reconstructed to reveal large portions of missing area, implying that this material has been consumed into the subsurface. Kattenhorn and Prockter build a provocative case for subduction of the ice, although how such a process would operate to entrain and draw down the ice is not detailed. Without topographic information, proving subduction is difficult. *Johnson et al.* (2017) showed that if there are locally dense regions of the ice shell, subduction might initiate, while *Howell and Pappalardo* (2018) have argued that buoyancy counteracts any subduction such that any process destroying surface ice at Europa is unlikely to be directly analogous to plate tectonics on Earth. However, no other models exist to explain the missing surface terrain, and the geologic evidence is convincing. Thus, while uncertainties remain, Europa is currently the only planet outside of Earth for which there is strong evidence of plate-tectonic-like activity. Future work should certainly investigate how ice tectonics may operate.

Rates of surface-subsurface exchange can be calculated based on Europa's surface age and the distribution of various kinds of terrain. The upper bound on modern surface recycling can be calculated from assuming the entire surface is eventually processed by subsumption of ice into the interior, which for a 30-Ma surface age implies 0.77 km^{-2} yr^{-1}, and 0.25 km^{-2} yr^{-1} for 90 Ma. For the best-constrained example of subsumption, an approximately 99-km-wide swath of ice was subsumed. Assuming that the recycling was ongoing at the time, a rate of ~1–3 km Ma^{-1} of subsumption would be the implied minimum rate of recycling. However, other features besides subsumption reprocesses the surface. Ridge and crack formation affects about 60% of the surface and likely does not transit material downward, while bands cover around 10% (*Figueredo and Greeley, 2004; Leonard et al., 2018*). Band formation, likely representing new surface created during tectonic resurfacing, could be the best constraint on the rate of subsumption, since bands are easier to observe than subsumption zones, implying a spreading and subsumption rate of 0.08 km^{-2} yr^{-1}. Chaos terrain represents between 14% and 25% of the surface (*Riley et al., 2000; Neish et al., 2012*), and could recycle 0.1–0.19 km^{-2} yr^{-1}.

Downward transmission of surface materials is potentially critically important for providing oxygen to the subsurface ocean in the form of hydrogen peroxide (H_2O_2), molecular oxygen (O_2), carbonic acid (H_2CO_3), and other surface materials. Downward transmission of phosphorus from micrometeorites could also be significant. Taking observed concentrations of hydrogen peroxide and oxygen discussed above, fluxes of these materials into the ocean could be high if the surface is rapidly drawn downward. However, the efficiency of the recycling of material is unknown. Given that not all of Europa's surface looks to have been recently subducted, somewhere between 10% (the approximate surface coverage of banded terrain) to 35% [the area of bands plus the areal coverage of chaos terrain at 14–25%, estimated by *Neish et al.* (2012)] would be reasonably expected to be mixed into the subsurface over the last <100 m.y. From here, downward convection could deliver materials to the subsurface trapped within the ice matrix. Models of ice shell convection suggest timescales of 10^4–10^5 yr for the full ice shell to overturn (e.g., *Pappalardo and Barr,* 2004; *Barr and Showman,* 2009), thus it is likely that the surface overturn is the limiting factor contributing to complete surface-subsurface exchange.

5.2. Analogs for Europan Life

Life on Europa cannot depend on photosynthesis, given Europa's distance from the Sun. On Earth, by about 7 m depth, no light penetrates through glacial ice (e.g., *Christner et al.,* 2018), thus at 5× further from the Sun, even a meter of ice would be effectively opaque at Europa. Therefore chemosynthesis, either powered by water-rock reactions or radiolysis, is likely to be the primary source of energy for putative europan life (Figs. 13 and 14). Ancient and modern seafloor hydrothermal vent communities are likely among the best analogs for potential europan communities. However, since the early oceans on Europa and Earth were both likely anoxic, and Europa's ocean today is also likely to be largely anoxic, ancient terrestrial hydro-

thermal systems are potentially better analogs for Europa (*Barge and White,* 2017). Importantly, if the origin of life on Earth took place at a hydrothermal vent, it would have been under anoxic conditions, similar to those expected on Europa. Assuming that seafloor pressures on Europa do not preclude such reactions, a seafloor origin of life might be supported. Several authors have also postulated on potential cold origins of life that could arise at strong chemical gradients within (e.g., *Price,* 2007) or below the ice (*Russell et al.,* 2017; *Vance et al.,* 2019); however, these are relatively new ideas and these scenarios have many challenges. Reaction rates at cold temperatures are likely much slower than within hydrothermal systems, and on Europa, such ice may be highly oxidized — rather than reduced — which may present challenges to the hypothesis of *Price* (2007). It is also likely that the ice contains an extremely dilute source of nutrients, since it forms via impurity rejection from the ocean, thus biogenic compounds may be extremely diffuse. Areas of cometary impacts could present locally higher concentrations of important elements for life such as carbon and phosphorus, and large impacts should form transient water pockets within the shell. However, the refreezing of these areas could be very fast and is unlikely to be cyclic. Melt-freeze cycles in the ice shell could be important for thermochemical cycling, but also are likely to be much slower than hydrothermal processes. Whether time is a key factor in the origin of life is still unknown. Structures formed as brine is rejected from sea ice — brinicles — have driven some interest as possible icy analogs of hydrothermal chimneys (*Vance et al.,* 2019). But brinicles are ephemeral, forming during rapid sea ice formation in areas of high

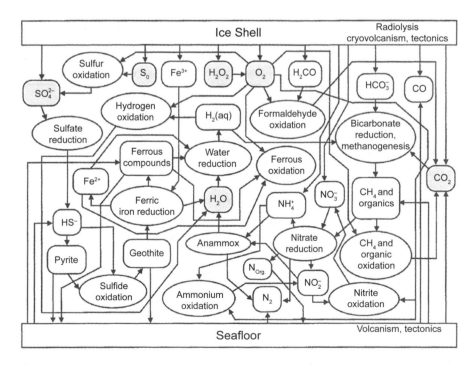

Fig. 13. Examples of geochemical and metabolic pathways that could support life on Europa. After *Zolotov and Shock* (2003, 2004); reproduced from *Hand et al.* (2009).

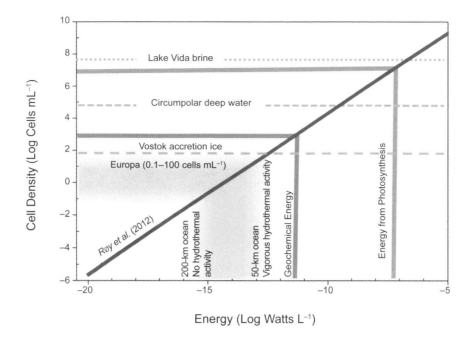

Fig. 14. Comparisons of available energy and number of cells found in a variety of Earth ecosystems compared to the energy expected for Europa. The bar signifying the Europa energy flux is bounded by that estimated in *Hand et al.* (2009) and *Vance et al.* (2016) (see Fig. 8). After *Hand et al.* (2017).

thermal gradients before shutting down (*Dayton and Martin,* 1971; *Perovich et al.,* 1995; *Kovac,* 1996). In Antarctica, brinicles have mostly been observed near coasts that enhance their formation by allowing for rapid drainage through tidal cracks open to the air, and that provide protection and low currents. The temperature gradients within Europa's ice shell after a few hundred meters of ice have become very shallow, which would likely preclude formation of such features; these are not observed under ice shelves where thermal gradients are more comparable to those at the ocean of Europa (e.g., *Buffo et al.,* 2018; *Buffo,* 2019). If periods of dramatic ice shell thickening do occur (e.g., *Figueredo and Greeley,* 2004; *Moore and Hussman,* 2009), enhanced salt flux may occur that could contribute to the formation of salt-rejection structures. Brinicle analogs could form if fractures rapidly drain saline fluids from the upper ice shell to the ocean, but there are as yet no strong indicators that such processes occur. Nonetheless, ice-formation driven chemical gradients could play a role in the europan system, while hydrothermalism still represents the most promising environment for an origin of life on Europa, as well as a sustained environment that might have been habitable since shortly after Europa's formation.

5.3. Life in Cryogenic Environments on Earth

If magmatic activity or deep cracking of the crust is ongoing on Europa at present, then hydrothermal vents or seeps might be possible at the seafloor to sustain life (e.g. ,*Lowell and DuBose,* 2005; see chapters in this volume by Barross et al. and Stüeken et al. for more details), depending on how tidal dissipation is operating in the interior. Regard-

less, if life once took hold on Europa, then the ice shell may represent a niche for life independent of tidal activity. On Earth, ice is a stronghold for life, both at interfaces with water and within the ice itself (e.g., *Priscu and Christner,* 2004; *Deming and Eicken,* 2007), and can be an excellent place to test strategies for life detection on ocean worlds like Europa (e.g., *Marion et al.,* 2003; *Gleeson et al.,* 2010, 2012; *Bulat et al.,* 2011; *Prieto-Ballesteros et al.,* 2011). The underside of sea ice, for example, represents the most concentrated zone of life along a vertical column thousands of meters in depth, from the atmosphere to the seafloor (*Thomas and Dieckmann,* 2008). The reason for this concentration of sea ice life is the convergence of energy and elements (nutrients) at the ice-water interface. If life exists on Europa, we may expect a similar scenario, with life concentrated near the first ice-water interface where energetic gradients established by ice-ocean interactions would be steepest. Pores, cracks, and grain boundaries in the ice near this interface may provide habitats for microbial life. Life in Earth's cryosphere (the ice component of the Earth system), while generally not independent from photosynthetic energy, provides constraints on potential activity and strategies for putative europan organisms (*Gaidos et al.,* 1999; *Chyba,* 2000; *Priscu and Christner,* 2004; *Deming and Eicken,* 2007, *Murray et al.,* 2012). Glacial melt water from the ice cap above subglacial volcano Grimsvotn was shown to be rich in cold-adapted organisms that were distinct from the water in the lake and the volcanic tephra (*Gaidos et al.,* 2004). The dry valleys of Antarctica host numerous shallow glacial water-fed or derived lakes with potentially important lessons for Europa. Data from Blood Falls (*Mikucki et al.,* 2009) has revealed communities of iron-reducing organisms within the subglacial water that spills out onto the ice, stressing the organisms as conditions go from reducing to highly oxidizing. *Murray et al.* (2012) discovered a bacterial community within the highly saline brines of permanently-ice-covered Lake Vida that were thriving in high concentrations of reduced metals, and other geochemical indicators that abiotic water-rock reactions were altering typical cryospheric metabolic niches. At Borup Fiord Pass, located on Ellesmere Island in the Canadian High Arctic, groundwater in permafrost is subject to glacial recharge, resulting in sulfide-rich artesian springs and an active sulfate-reducing microbial community (*Gleeson et al.,* 2011, 2012).

Work on deep subglacial lakes as well as under ice shelves has accelerated with the advent of improved ice-drilling capabilities. While not disconnected from the open ocean, water in some regions of ice shelves has long residence times, up to tens of years. This relative isolation from the open ocean inhibits local sources of energy below the ice and is likely to create low-energy conditions. Surprisingly, only three regions have been sampled under the Ross Ice Shelf, despite the fact that it is the largest ice shelf on Earth, spanning 1000 km from the open ocean to the grounding zone. Life is still seen in this relatively isolated subshelf water, and work at the grounding line (where the ice shelf meets land) of the Ross Ice Shelf at Whillans ice stream revealed small fish that must depend on the microbial environment for energy. About 300 km oceanward of the grounding zone, the first access through the shelf (400 m thick) at borehole J9 revealed strong evidence for nitrification as a community backbone (*Horrigan,* 1981). Toward the front of the shelf, the Coulman High region is another example of an ice-hosted community, where Edwardsiella Andrilliae anemones were discovered burrowing into the ice, and they are likely relying upon a microbial community for long-term survival (*Daly et al.,* 2013).

Just upstream of the Ross Ice Shelf/Whillans Ice Stream grounding line, drilling into subglacial Lake Whillans revealed a substrate-driven microbial community, with glacial melt and sedimentary-derived compounds contributing to the support of the ecosystem. Here, higher organic carbon was present than in ocean water, alongside high concentrations of ammonium and nitrate, with a community of organisms that oxidize nitrite or reduce nitrogen, iron, sulfur, and methane (*Christner et al.,* 2014; *Mikucki et al.,* 2016). Microbial communities within water and ice samples at subglacial Lake Vostok are also largely dependent on the rocks below (*Priscu et al.,* 1999; *Priscu and Christner,* 2004). However, subglacial lake communities do exist in regions with low energy availability. Within the accretion ice that has frozen from the top of Lake Vostok onto the base of the ice sheet, cell densities are as low as 200 cells mL^{-1}, making this among the lowest concentrations of life found on Earth. Subglacial lakes have mixed relevance for Europa. On one hand, they are deep, light-free environments experiencing long periods of isolation from the photosynthetic biosphere. In this way, they are adapted to low-energy environments

and chemosynthetic energy, and are therefore informative for Europa in this manner. However, the vast majority of the organisms in these environments are living off of sediments that are potentially rich in carbon. Thus, these may be more direct analogs for the seafloor of Europa (away from any hydrothermal vents), but not for communities within the ice and the ocean, which are potentially 100 km from significant volumes of silicates. The organisms under ice shelves are also under low-light conditions but have available energy sources that are also not expected on Europa, from melting of the meteoric ice, and are only partially disconnected from the open ocean.

6. FUTURE EXPLORATION OF EUROPA AND THE JOVIAN MOONS

Two orbiting spacecraft are already set to explore the moons of the jovian system, primarily focused on the icy moons Europa and Ganymede. NASA's Europa Clipper will launch no earlier than 2024, and the European Space Agency's (ESA's) JUpiter ICy moons Explorer (JUICE) will launch in 2022 but arrive after Europa Clipper, such that the two will explore Europa and the other moons during the late 2020–2030s. JUICE will make two close flybys of Europa before orbiting Ganymede. Europa Clipper will make a tour of the jovian system, interacting with Ganymede and Callisto and making distant flybys of Io, before entering an orbital phase of the mission that allows for globally distributed close flybys of Europa. While Europa Clipper has no official primary science goals at Ganymede, Callisto, Io, or Jupiter, its orbital tour and potentially long duration in the jovian environment should enable opportunistic science during the four-year prime mission and beyond, and early flybys of the satellites will allow Europa Clipper to calibrate its instruments. No missions are as yet planned for Io, although an Io orbiter mission was prioritized among the potential targets for NASA's New Frontiers 5 Program (*National Research Council,* 2012). An orbiting spacecraft at Io would revolutionize the study of tidal heating at its extremes for terrestrial planets, and reveal how the dynamics of planetary interiors interact with magnetic fields. Such a mission would also provide feed-forward data for exoplanets.

6.1. NASA's Europa Clipper

Europa Clipper is a Jupiter-orbiting spacecraft that leverages more than 40 close flybys of Europa to build up nearly global imaging, as well as regional coverage with other data. While orbiting Europa would also allow for global imaging coverage to be collected, due to the high radiation environment near Europa's orbit, spacecraft lifetimes would be limited. Orbiting at Europa also suffers from spending a significant part of the orbit in Europa eclipse, lowering the total amount of data that can be radioed back to Earth. Europa Clipper solves this issue by executing a unique orbit strategy, called "crank over the top" orbits, that allow the spacecraft to use Europa's gravity to redirect itself such that every orbit has a different periapsis and inclination. This will allow Europa Clipper to build up observations in a way that mimics the science of being in Europa orbit while allowing greater data to be returned and even a potentially longer mission duration. Europa Clipper will explore Europa through a diverse set of interdisciplinary investigations and is motivated by understanding its habitability (*Phillips and Pappalardo,* 2014). The mission will carry a suite of instruments, including a camera for geology and future landing site characterization, a visible and infrared imaging spectrometer to understand surface composition, an ice penetrating radar to image the subsurface of the ice shell and search for liquid water, an ion and neutral mass spectrometer and dust analyzers that will together constrain the composition and size of materials sputtered or emitted from the surface, a UV spectrometer, and a plasma and ion detector (see *https://europa.nasa.gov/*). Europa Clipper will also have the capability to measure Europa's internal structure using radio science and a magnetometer to improve measurements of Europa's interaction with the jovian magnetic field and better constrain the conductivity and thus salinity of the ocean. Europa Clipper represents the first major mission to target an ocean world and will obtain several orders of magnitude more data that will revolutionize our understanding of Europa and the jovian system.

6.2. ESA's JUICE

JUICE will focus on comparisons of Ganymede and Europa, with the primary goal of understanding how habitable worlds form around gas giants (*Grasset et al.,* 2013). This initiative is inspired by the synergy of studying the Galilean moons as habitable worlds, but also as they inform the study of exoplanets. Understanding the moons' natures and formation environment, as well as their relationship to their host giant planet, will enable JUICE to feed forward into studies of planetary systems, especially since gas giant planets are likely common elsewhere in the universe, and could host habitable moons at any orbital distance from the host star. JUICE will carry instruments that are complementary to those on Europa Clipper but also include systems designed to target Ganymede. These instruments include a visible camera system, an infrared mapping spectrometer, a laser altimeter, an ice-penetrating radar at 9 MHz that overlaps with the HF frequency on Europa Clipper, a magnetometer, a particle environment package, a radio interferometer and Doppler experiment, a submillimeter wave instrument, and a gravity investigation. JUICE will make two close flybys of Europa during its two-year tour into Ganymede orbit, and will spend a primary mission of one year in Ganymede orbit.

Europa Clipper and JUICE will transform the study of the Galilean moons. But the long-term prospects for landing and penetrating into and possible below the ice of Europa are tantalizing for astrobiology. Among the most critical aspects of the Europa Clipper mission will be to search for safe landing sites for future surface missions. Galileo imaging data only provided about 10% surface coverage at better

than 300 m per pixel (*Figueredo and Greeley,* 2004), and a handful of images at 10-m-scale resolution, making selection and certification of a landing site extremely challenging on the scale of a possible Europa lander (~1 m).

6.3. Landed Missions

Two Europa lander concept studies have been performed by NASA. The two landers have had similar goals: investigating Europa's habitability and seeking signs of life. The earlier concept studied a system that would ground truth measurements made by a Europa orbiter mission, while conducting a comprehensive study for biosignatures (*Pappalardo et al.,* 2013). The lander study planned imaging, spectral, seismic, magnetic, and sampling studies to understand the geophysical context of sampled materials. At that time, the envisioned mission would last a duration of six months on the surface. Comprehensive study of a smaller, shorter-duration lander was completed in 2016 that focused on the search for biosignatures. While this lander concept had a smaller mass/power/volume envelope and fewer instruments than the 2012 lander concept, the mission plan included a detailed assessment of how to search for signs of life based on organic, inorganic, and structural indicators, in concert with context from cameras and seismic instruments to understand the provenance of sampled materials that would be ingested into the lander (*Hand et al.,* 2017). Key to this mission was preparing science requirements based on the best-known analogs that might be expected for Europa. Measurement sensitivities were bounded by the lowest known concentrations of cells and biological material found in relevant environments on Earth, that of accretion ice formed at the top of Lake Vostok (*Priscu et al.,* 1999). This mission thus placed benchmarks for how to measure multiple lines of evidence that could pave the path toward detection of life at Europa.

No matter the lander, the fundamental limitation of landed investigations of Europa are that the most habitable areas and the highest concentration of biogenic materials are likely to be well below the surface. Aside from diffuse plumes that may be difficult to sample at Europa (*Lorenz,* 2016), the best constraints on shallow water on Europa place it at a few kilometers from the surface, with some localized areas where the water may reach up to ~100 m from the surface (*Schmidt et al.,* 2011). Thus, while rovers and surface missions are excellent ways to explore terrestrial planets with surface atmospheres and sedimentary geologic processes, melt probes and eventually underwater vehicles are the long-term goal for Europa and other ocean worlds. While swimming in the ocean may be decades off, a fundamental technology and exploration strategy is under development, and subsurface exploration could be on the near horizon.

6.4. Field Analog Exploration

Field studies of europan analog environments with technologies that act as testbeds for future mission technology have been underway since the late 1990s. Remotely operated vehicles (ROVs) have allowed *in situ* measurements to motivate decision-making strategies for autonomous underwater vehicles (AUVs) both on Earth and Europa. Hybrid vehicles that are capable of both directed and independent maneuvers allow exploration that has maximized science return while still making large steps forward in navigating below ice and without other navigation tools like GPS and ultra-stable beacon locators (USBL).

Melt probe missions to Europa have broad appeal (e.g., *Zimmerman et al.*, 2001) and significant terrestrial development. Valkyrie (*Stone et al.*, 2014) and IceDiver (*Winebrenner et al.*, 2016) have had successful deployments in glaciers in the U.S. and deployments for Valkyrie are planned for Alaska. IceMole (*Dachwald et al.*, 2014) has successfully penetrated into the brine channels that source Blood Falls, Antarctica, taking *in situ* measurements of the water chemistry and organisms. Recently, NASA has funded the first technology studies for melt probe missions, where teams are actively developing mission concepts and technologies for end-to-end components for such missions at the fidelity of benchtop and analog level function. Given sufficient funds, such missions could be slated for launch as soon as the 2030s, although the 2040s or 2050s are more likely given current mission flight timelines.

Ultimately, underwater vehicles are compelling platforms for navigating Europa's ocean or shallow water. The convergence of technological needs between polar ice and ocean science with that of the planetary oceanographic community is a newly expanding effort. Long-range under-ice vehicles such as Autosub 3 and Boaty McBoatface were not developed as Europa analog vehicles, but are still relevant accomplishments toward long-duration under-ice exploration. Autosub 3 established the record for under-ice-shelf operations, driving over 50 km beneath the Pine Island Glacier to map its oceanography and ice and seafloor bathymetry (*Jenkins et al.*, 2010; *Dutrieux et al.*, 2014). Ocean gliders have also

made significant progress in under-ice operations, operating continuously for a year in the open ocean and under sea ice (*Lee and Rudnick*, 2018), and recently under glacier ice during surveys under the Dotson Ice Shelf (*Boyle*, 2019). Polar under-ice vehicles specifically targeted for Europa-analog science include the Buoyant Rover for Under Ice Exploration (BRUIE) (*Berisford et al.*, 2013), hybrid remote-operated and autonomous under-ice vehicles (ROV/AUV) ENDURANCE (*Gulati et al.*, 2010), ARTEMIS (*Kimball et al.*, 2017), and Icefin (*Spears et al.*, 2016; *Meister et al.*, 2018), and AUV Nereid Under Ice (NUI) (*Jakuba et al.*, 2018; *German et al.*, 2019). These vehicles represent an investment in several different strategies for mobile under-ice vehicles. BRUIE is a small positively buoyant short-range vehicle that rolls along the ice surface and has operated under sea ice and lake ice in Alaska. ENDURANCE, ARTEMIS, and NUI are of a similar class of 500-kg or larger vehicles with round to oblate profiles and a few to tens of kilometers range. ENDURANCE operated under permanent lake ice (~4 m thick) in Lake Bonney, Antarctica, conducting sonar and conductivity, temperature, and depth (CTD) surveys along with *in situ* characterization of lake properties. ARTEMIS operated under the McMurdo Sea Ice (~5–8 m thick) and traversed 800 m under the McMurdo ice shelf (<15 m thick) in 2015 (*Kimball et al.*, 2017). NUI has operated under arctic sea ice exploring from the ice interface to hydrothermal vents at the seafloor. The hybrid AUV-ROV Icefin is a 100-kg, ~3-km-range vehicle that has operated under sea ice near McMurdo Station (2–13 m thick), over 1.2 km underneath the McMurdo Ice Shelf (where it was 10–40 m thick) and 1.5 km under the Erebus Glacier tongue (300 m thick; Fig. 15). Icefin explored the grounding zones of the Kamb Ice Stream and Thwaites Glacier, deploying through and below 600 m of ice in each case, in the austral summer 2019–2020. Under the ice at Thwaites Glacier, Icefin discovered a community of anemones that burrow into the ice and filter feed from the ocean (*Schmidt et al.*, 2020). At

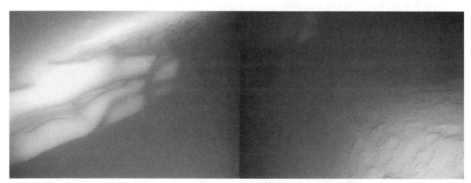

Fig. 15. Stills from a live video stream during the first exploration of the grounding zone of Erebus Glacier Tongue, Antarctica, by the Icefin hybrid remotely operated/autonomous vehicle in November 2018. The 24-cm-wide vehicle was navigating toward the grounding zone a few tens of meters upstream through a 2-m-wide channel in the ocean-carved ice (left) above a glacial flow-parallel few-meter high flute in the subglacial sediment (right). The left image shows 10–20-cm cups and ripples in the ice formed from melting by the ocean, and the smooth surfaces scraped by the contact of the ice and the seafloor. Icefin was built as a testbed for both science operations and instrumentation for future Europa and polar under ice exploration, but also with an eye toward better understanding how the ice and the ocean interact on Earth and other planets.

Kamb Ice Stream, Icefin observed *in situ* ice pumping forming marine ice accretion in a crevass, an analogous process thought to occur on Europa (*Lawrence et al., 2020*). These achievements push the bound of planetary analog research as well as our understanding of ice-ocean interactions on Earth, and the life it can support. Each of these vehicles has demonstrated an increase in science and technology fidelity that will need to be further developed to achieve missions one day on ocean worlds like Europa. Together, they represent progress toward better understanding oceans under the ice here on Earth, and pushing the boundaries of how data from other ocean worlds can be interpreted.

7. CONCLUSIONS AND FUTURE WORK

As the field of astrobiology has matured, progress toward systems science of habitable planets has emerged as a frontier in rethinking how planets support and maintain life (see also the chapter in this volume by Kopparapu et al. for terrestrial planets). The ocean worlds, and in particular the jovian moons, represent fertile new grounds for this line of inquiry. With similar and familiar compositions but different sizes and histories, these worlds offer a chance to interrogate the importance of planet formation and thermal evolution on the limits of habitability in any system. Having the four moons to compare provides both new lines of questions and a basis set to control future data from exoplanets. Just as the Cassini mission to Saturn and its moons was in many respects a Galileo-follow on with a similar but improved set of instruments, Europa Clipper represents a more sophisticated set of instruments leveraging what was learned from Cassini. These missions have allowed numerous cross-comparisons to be accomplished between studies of Europa, Enceladus, Titan, and Ganymede. With JUICE complementing Europa Clipper and expressly including studies relevant to the study of habitable giant planet systems, immense progress in this area will be made in the coming decades.

Key questions remain for understanding the jovian system itself, which is relevant to the individual worlds that make it up, but also to how we understand giant planet and waterworld formation. Our current paradigm of the jovian system interprets the formation of the moons as a continual process contemporary with the evolution of Jupiter, where many moons would have been formed and destroyed over time. As we gather better data about Jupiter's formation and evolution from the Juno mission, JUICE, and Europa Clipper, we will hope to constrain much more about the history of the system. But data from upcoming missions, combined with the growing sophistication of astronomical observations of distant planetary systems, will allow astrobiologists to simultaneously explore to what degree individual moons are the result of a general formation process. A few of the resulting questions are:

• How do moon systems form and evolve and how do the planet and its proto-moons interact?
  ◦ Will we observe more systems like Jupiter and Saturn?
  ◦ Where will the new discoveries happen?
  ◦ How important are impacts to the end state of moons (*Barr and Canup*, 2010; *Monteux et al.*, 2016)?
• Is there anything universal about the formation of ocean worlds?
  ◦ What was the early tidal history of the jovian system?
  ◦ Can icy ocean moons support geological processes without tidal influence?
  ◦ What did, or didn't, happen for Callisto to remain largely inactive, even when compared to smaller moons?

Io is an important reference for astrobiologists for many reasons — including as an exotic state of terrestrial planets as well as the only known terrestrial-planet-like body beyond the snow line. Much work remains to understand this violently active world. The challenge of operating spacecraft near Jupiter in intense radiation will make it difficult for the two currently planned Jupiter system missions to get up close to Io. Hopefully, an Io mission will be selected in a future competition. But in the meantime, continued observations from the ground and Earth orbit, including with the next generation of large observatories on the ground and in space, can allow the volcanic activity to be monitored. Together with improved geochemical and thermochronometry, Io may offer an important window into the diversity of terrestrial planet composition and processes. Work relevant to the future study of Io includes:

• Where's the water?
  ◦ Did Io possess an early hydrosphere?
  ◦ What tidal, geochemical, and thermal evolution conspired to result in Io's presently high level of activity?
• How does tectonics operate as a function of thermal, mechanical, and geochemical conditions?
  ◦ How do these factors couple together to produce a range of terrestrial bodies?
  ◦ What is the physical response of Io's interior to changing conditions, such as feedbacks between cyclic heating and cooling, melt production, and fractional crystallization and viscosity?
  ◦ How does plate tectonics begin and evolve, i.e., what makes the difference between Io, Earth, and Venus, for example?

With JUICE under development, science at Ganymede looks to expand dramatically in the coming decades. As the first of a different class of planetary body to be explored in detail, Ganymede represents the chance to achieve fundamental understanding of how thick ocean and ice worlds evolve and the implications for their habitability. With JUICE data, geologic processes will be better resolved, which promises to improve constraints on the entire jovian system. In particular, JUICE's data taken in orbit around Ganymede will provide the best constraints yet on the interior of an ocean world, making Ganymede a cornerstone for astrobiology both in the solar system and farther afield. Some areas of inquiry include:

• How do large water-rich planets function?

- Are multi-layered oceans possible?
- Do interior ocean layers separated by ice from each other or from the seafloor communicate with other layers within the planet?
- Are there geological sources of energy for life within large ocean worlds and how do they operate?
• What role do clathrates play in ocean world formation and evolution?
• Did Ganymede ever have an atmosphere?
• How has Ganymede maintained its internal dynamo, and what does that imply for water world evolution and habitability?

Europa, as an intermediate between a moon and a planet in its evolution, and as an intermediate between terrestrial and water worlds, will continue to be a critical place for stretching our concept of how habitability works. Europa's composition, to the degree we can measure presently, is not wildly different from Earth's; its most probable reservoirs for life are water-based, and it has experienced intense geological activity over the same duration of time as Earth. At the same time, it is geographically removed from Earth. As such, Europa represents arguably the best chance to understand a potential independent origin of life. Despite significant progress in understanding its formation and evolution, uncertainty remains given the relatively small amount of data available. But with the approach of Europa Clipper's launch there has already been a renaissance of Europa science, with ever-improving sophistication in the number and kind of processes and interactions that are being considered by planetary scientists and astrobiologists. Even on Earth, understanding how different reservoirs interact, and how processes at small scales affect those at large scales and vice versa, is a challenging task, but this is where Europa science is heading. Revolutionary data from Enceladus has been perfectly timed for thinking about how to formulate new questions about the europan system, and how to test them using existing and future data. The frontier for Europa is a systems-level approach to understanding the potential for life as a planetary process, and evaluating whether Europa can meet the requirements of a putative biosphere. This will be important for future Europa science as well as a model for how to conduct remote studies of habitability of worlds in other solar systems as a function of more than solar insolation, since the jovian moons represent more than just a representation of a giant planet moon system but also potential kinds of planets orbiting other stars. Some critical areas for future work include:

• What are the energy sources on Europa and on what processes do they depend?
  - How have geologic cycles changed over time?
  - Are nitrogen and phosphorus limiting, or are there available sources for life?
  - What are the impacts of ocean pressure on geochemical and geological processes and do they preclude or challenge life?
  - Is silicate weathering critical for life?
• How do geologic processes work in ice?

- How does plate tectonics work in ice, or does it?
- Is communication between the surface, ice shell, ocean, and interior of an ocean world important for life and how does it operate?
- Is there water in the ice shell and is it habitable?
• Would a Europa-like world around another star be habitable?
  - To what degree are the tidal evolution of the moon and surface irradiation critical as energy sources?
  - Does Europa have its own magnetic field suggesting an active core?
• Was there ever life on Europa, and does it still exist or leave a record of its existence? (*National Academies of Sciences, Engineering, and Medicine,* 2019)
  - How does biomarker production and preservation work in the ice?
  - Where and how can we best interrogate these possibilities?
  - Which if any conditions on Earth are required for life, or which can at least be replaced by something else playing the same role?
  - Can life originate within an ocean?

Much as it did in the time of Galileo, the jovian system, and Europa specifically, inspires us to rethink how our Earth-borne experience is both naïve and at the same time useful as we approach the field of astrobiology. How we ourselves pose questions for study, and how to best wrap our heads around what really bounds a habitable world, is important to consider in order to make progress understanding the probability for life beyond Earth.

REFERENCES

Anderson J. D., Lau E. L., Sjogren W. L., Schubert G., and Moore W. B. (1996) Gravitational constraints on the internal structure of Ganymede. *Nature, 384(6609),* 541.

Anderson J. D., Lau E. L., Sjogren W. L., Schubert G., and Moore W. B. (1997) Gravitational evidence for an undifferentiated Callisto. *Nature, 387(6630),* 264.

Anderson J. D., Schubert G., Jacobson R. A., Lau E. L., Moore W. B., and Sjogren W. L. (1998) Europa's differentiated internal structure: Inferences from four Galileo encounters. *Science, 281,* 2019–2022.

Arnold H., Liuzzo L., and Simon S. (2019) Magnetic signatures of a plume at Europa during the Galileo E26 flyby. *Geophys. Res. Lett., 46(3),* 1149–1157.

Banwell A. F., MacAyeal D. R., and Sergienko O. V. (2013) Breakup of the Larsen B Ice Shelf triggered by chain reaction drainage of supraglacial lakes. *Geophys. Res. Lett., 40(22),* 5872–5876.

Barge L. M. and White L. M. (2017) Experimentally testing hydrothermal vent origin of life on Enceladus and other icy/ocean worlds. *Astrobiology, 17(9),* 820–833.

Barnes R., Jackson B., Greenberg R., and Raymond S. N. (2009) Tidal limits to planetary habitability. *Astrophys. J. Lett., 700(1),* L30.

Barr A. C. and Canup R. M. (2010) Origin of the Ganymede-Callisto dichotomy by impacts during the late heavy bombardment. *Nature Geosci., 3(3),* 164.

Barr A. C. and Hammond N. P. (2015) A common origin for ridge-and-trough terrain on icy satellites by sluggish lid convection. *Phys. Earth Planet. Inter., 249,* 18–27.

Barr A. C. and Showman A. P. (2009) Heat transfer in Europa's icy shell. In *Europa* (R. T. Pappalardo et al., eds.), pp. 405–430. Univ. of Arizona, Tucson.

Bell R. E., Chu W., Kingslake J., Das I., Tedesco M., Tinto K. J., Zappa C. J., Frezzotti M., Boghosian A., and Lee W. S. (2017) Antarctic ice shelf potentially stabilized by export of meltwater in surface river.

Nature, 544(7650), 344.

Berisford D. F., Leichty J., Klesh A., and Hand K. P. (2013) Remote under-ice roving in Alaska with the buoyant rover for under-ice exploration. Abstract C13C-0684 presented at 2013 Fall Meeting, AGU, San Francisco, California, 9–13 December.

Bernstein M. P., Sandford S. A., Allamandola L. J., Chang S., and Scharberg M. A. (1995) Organic compounds produced by photolysis of realistic interstellar and cometary ice analogs containing methanol. *Astrophys. J., 454*, 327.

Bierhaus E. B., Zahnle K., Chapman C. R., Pappalardo R. T., McKinnon W. R., and Khurana K. K. (2009) Europa's crater distributions and surface ages. In *Europa* (R. T. Pappalardo et al., eds.), p. 161. Univ. of Arizona, Tucson.

Bland M. T., Showman A. P., and Tobie G. (2008) The production of Ganymede's magnetic field. *Icarus, 198(2)*, 384–399.

Bland M. T., Showman A. P., and Tobie G. (2009) The orbital-thermal evolution and global expansion of Ganymede. *Icarus, 200(1)*, 207–221.

Borin S., Crotti E., Mapelli F., Tamagnini I., Corselli C., and Daffonchio D. (2008) DNA is preserved and maintains transforming potential after contact with brines of the deep anoxic hypersaline lakes of the Eastern Mediterranean Sea. *Saline Syst., 4(1)*, 10.

Bowman J. S., Vick-Majors T. J., Morgan-Kiss R., Takacs-Vesbach C., Ducklow H. W., and Priscu J. C. (2016) Microbial community dynamics in two polar extremes: The lakes of the McMurdo Dry Valleys and the West Antarctic Peninsula marine ecosystem. *BioScience, 66(10)*, 829–847.

Boyce J., Barlow N., and Stewart S. (2010) Rampart craters on Ganymede: Their implications for fluidized ejecta emplacement. *Meteoritics & Planet. Sci., 45(4)*, 638–661.

Boyle A. (2019) Underwater robots survive a year probing climate change's effects on Antarctic ice. *GeekWire*, January 22, 2019, *https://www.geekwire.com/2019/seagoing-robots-survive-year-probing-climate-changes-effects-antarctic-ice/*.

Bray V. J., Schenk P. M., Melosh H. J., Morgan J. V., and Collins G. S. (2012) Ganymede crater dimensions — Implications for central peak and central pit formation and development. *Icarus, 217(1)*, 115–129.

Bray V. J., Collins G. S., Morgan J. V., Melosh H. J., and Schenk P. M. (2014) Hydrocode simulation of Ganymede and Europa cratering trends — How thick is Europa's crust? *Icarus, 231*, 394–406.

Brown M. E. and Hand K. P. (2013) Salts and radiation products on the surface of Europa. *Astron. J., 145(4)*, 110.

Bulat S. A., Alekhina I. A., Marie D., Martins J., and Petit J. R. (2011) Searching for life in extreme environments relevant to Europa: Lessons from subglacial ice studies at Lake Vostok (East Antarctica). *Adv. Space Res., 48(4)*, 697–701.

Buffo J. J. (2019) Multiphase reactive transport in planetary ices. Ph.D. thesis, Georgia Institute of Technology, *http://hdl.handle.net/1853/61767*.

Buffo J. J., Schmidt B. E., and Huber C. (2018) Multiphase reactive transport and platelet ice accretion in the sea ice of McMurdo Sound, Antarctica. *J. Geophys. Res.–Oceans, 123*, 324–345, DOI: 10.1002/2017JC013345.

Buffo J., Schmidt B. E., Huber C., and Walker C. C. (2019) Formation and dynamics of ocean-derived impurities within Europa's ice shell. *Sci. Adv.*, in review.

Canup R. M. and Ward W. R. (2002) Formation of the Galilean satellites: Conditions of accretion. *Astron. J., 124(6)*, 3404.

Canup R. M. and Ward W. R. (2006) A common mass scaling for satellite systems of gaseous planets. *Nature, 441(7095)*, 834.

Canup R. M. and Ward W. R. (2009) Origin of Europa and the Galilean satellites. In *Europa* (R. Pappalardo et al., eds.), pp. 59–83. Univ. of Arizona, Tucson.

Carlson R. W., Johnson R. E., and Anderson M. S. (1999) Sulfuric acid on Europa and the radiolytic sulfur cycle. *Science, 286*, 97–99.

Carlson R. W., Anderson M. S., Johnson R. E., Schulman M. B., and Yavrouian A. H. (2002) Sulfuric acid production on Europa: The radiolysis of sulfur in water ice. *Icarus, 157(2)*, 456–463.

Carlson R. W., Anderson M. S., Mehlman R., and Johnson R. E. (2005) Distribution of hydrate on Europa: Further evidence for sulfuric acid hydrate. *Icarus, 177(2)*, 461–471.

Carlson R. W., Calvin W. M., Dalton J. B., Hansen G. B., Hudson R. L., Johnson R. E., McCord T. B. and Moore M. H. (2009) Europa's surface composition. In *Europa* (R. T. Pappalardo et al., eds.), pp. 283–327. Univ. of Arizona, Tucson.

Carr M. H. (1986) Silicate volcanism on Io. *J. Geophys. Res.–Solid Earth, 91(B3)*, 3521–3532.

Carr M. H., Masursky H., Strom R. G., and Terrile R. J. (1979) Volcanic features of Io. *Nature, 280*, 729–733.

Carr M. H., Belton M. J. S., Chapman C. R., Davies M. E., Geissler P., Greenberg R., McEwen A. S., et al. (1998) Evidence for a subsurface ocean on Europa. *Nature, 391*, 363.

Cassen P., Reynolds R. T., and Peale S. J. (1979) Is there liquid water on Europa? *Geophys. Res. Lett., 6(9)*, 731–734.

Chiang E. and Youdin A. N. (2010) Forming planetesimals in solar and extrasolar nebulae. *Annu. Rev. Earth Planet. Sci., 38*, 493–522.

Choblet G., Tobie G., Sotin C., Kalousova K., and Grasset O. (2017) Heat transport in the high-pressure ice mantle of large icy moons. *Icarus, 285*, 252–262.

Christner B. C., Priscu J. C., Achberger A. M., Barbante C., Carter S. P., Christianson K., Michaud A. B., Mikucki J. A., Mitchell A. C., Skidmore M. L., Vick-Majors T. J., and the WISSARD Science Team (2014) A microbial ecosystem beneath the West Antarctic Ice Sheet. *Nature, 512*, 310–313.

Christner B. C., Lavender H. F., Davis C. L., Oliver E. E., Neuhaus S. U., Myers K. F., Hagedorn B., Tulaczyk S. M., Doran P. T., and Stone W. C. (2018) Microbial processes in the weathering crust aquifer of a temperate glacier. *Cryosphere, 12*, 3653–3669, DOI: 10.5194/tc-12-3653-2018.

Chyba C. F. (2000) Energy for microbial life on Europa. *Nature, 403(6768)*, 381.

Chyba C. F. and Phillips C. B. (2001) Possible ecosystems and the search for life on Europa. *Proc. Natl. Acad. Sci. U.S.A., 98(3)*, 801–804.

Collins G. and Nimmo F. (2009) Chaotic terrain on Europa. In *Europa* (R. Pappalardo et al., eds.), pp. 259–282. Univ. of Arizona, Tucson.

Collins G. C., Head J. W. III, Pappalardo R. T., and Spaun N. A. (2000) Evaluation of models for the formation of chaotic terrain on Europa. *J. Geophys. Res.–Planets, 105(E1)*, 1709–1716.

Craft K. L., Patterson G. W., Lowell R. P., and Germanovich L. (2016) Fracturing and flow: Investigations on the formation of shallow water sills on Europa. *Icarus, 274*, 297–313.

Culha C., Hayes A. G., Manga M., and Thomas A. M. (2014) Double ridges on Europa accommodate some of the missing surface contraction. *J. Geophys. Res.–Planets, 119(3)*, 395–403.

Dachwald B., Mikucki J., Tulaczyk S., Digel I., Espe C., Feldmann M., Francke G., Kowalski J., and Xu C. (2014) IceMole: A maneuverable probe for clean *in situ* analysis and sampling of subsurface ice and subglacial aquatic ecosystems. *Ann. Glaciol., 55(65)*, 14–22.

Dalton J. B., Shirley J. H., and Kamp L. W. (2012) Europa's icy bright plains and dark linea: Exogenic and endogenic contributions to composition and surface properties. *J. Geophys. Res.–Planets, 117(E3)*.

Dalton J. B. III, Cassidy T., Paranicas C., Shirley J. H., Prockter L. M., and Kamp L. W. (2013) Exogenic controls on sulfuric acid hydrate production at the surface of Europa. *Planet. Space Sci., 77*, 45–63.

Daly M., Rack F., and Zook R. (2013) *Edwardsiella andrillae*, a new species of sea anemone from Antarctic Ice. *PloS ONE, 8(12)*, e83476.

Dameron A. C. and Burr D. M. (2018) Europan double ridge morphometry as a test of formation models. *Icarus, 305*, 225–249.

Davies A. G. (2003) Temperature, age and crust thickness distributions of Loki Patera on Io from Galileo NIMS data: Implications for resurfacing mechanism. *Geophys. Res. Lett., 30(21)*.

Davies J. H. and Davies D. R. (2010) Earth's surface heat flux. *Solid Earth, 1(1)*, 5–24.

Dayton P. K. and Martin S. (1971) Observations of ice stalactites in McMurdo Sound, Antarctica. *J. Geophys. Res., 76(6)*, 1595–1599.

Deming J. W. and Eicken H. (2007) Life in ice. In *Planets and Life: The Emerging Science of Astrobiology* (W. T. Sullivan and J. A. Baross, eds.), pp. 292–312. Cambridge Univ., Cambridge.

Des Marais D. J., Nuth J. A. III, Allamandola L. J., Boss A. P., Farmer J. D., Hoehler T. M., Jakosky B. M., et al. (2008) The NASA astrobiology roadmap. *Astrobiology, 8(4)*, 715–730.

Doggett T., Greeley R., Figueredo P., and Tanaka K. (2009) Geologic stratigraphy and evolution of Europa's surface. In *Europa* (R. T. Pappalardo et al., eds.), pp. 137–160. Univ. of Arizona, Tucson.

Dutrieux P., De Rydt J., Jenkins A., Holland P. R., Ha H. K., Lee S. H., Steig E. J., Ding Q., Abrahamsen E. P., and Schröder M. (2014) Strong sensitivity of Pine Island ice-shelf melting to climatic variability. *Science, 343(6167)*, 174–178.

Egger M., Hagens M., Sapart C.J., Dijkstra N., van Helmond N. A., Mogollón J. M., Risgaard-Petersen N., van der Veen C., Kasten S., Riedinger N., and Böttcher M. E. (2017) Iron oxide reduction in methane-rich deep Baltic Sea sediments. *Geochim. Cosmochim. Acta, 207,* 256–276.

Fagents S. A. (2003) Considerations for effusive cryovolcanism on Europa: The post-Galileo perspective. *J. Geophys. Res.–Planets, 108(E12),* 5139.

Fagents S. A., Greeley R., Sullivan R. J., Pappalardo R. T., Prockter L. M., and the Galileo SSI Team (2000) Cryomagmatic mechanisms for the formation of Rhadamanthys Linea, triple band margins, and other low-albedo features on Europa. *Icarus, 144(1),* 54–88.

Fanale F. P., Li Y. H., De Carlo E., Farley C., Sharma S. K., Horton K., and Granahan J. C. (2001) An experimental estimate of Europa's "ocean" composition independent of Galileo orbital remote sensing. *J. Geophys. Res.–Planets, 106(E7),* 14595–14600.

Figueredo P. H. and Greeley R. (2004) Resurfacing history of Europa from pole-to-pole geological mapping. *Icarus, 167(2),* 287–312.

Figueredo P. H., Greeley R., Neuer S., Irwin L., and Schulze-Makuch D. (2003) Locating potential biosignatures on Europa from surface geology observations. *Astrobiology, 3(4),* 851–861.

Fischer P. D., Brown M. E., and Hand K. P. (2015) Spatially resolved spectroscopy of Europa: The distinct spectrum of large-scale chaos. *Astron. J., 150(5),* 164.

Gaidos E. J. and Nimmo F. (2000) Planetary science: Tectonics and water on Europa. *Nature, 405(6787),* 637.

Gaidos E. J., Nealson K. H., and Kirschvink J. L. (1999) Life in ice-covered oceans. *Science, 284(5420),* 1631–1633.

Gaidos E., Lanoil B., Thorsteinsson T., Graham A., Skidmore M., Han S. K., Rust T., and Popp B. (2004) A viable microbial community in a subglacial volcanic crater lake, Iceland. *Astrobiology, 4(3),* 327–344.

Galilei G. (1610) *Sidereus nuncius.* Thomas Baglioni, Venice.

German C. R. and Boetius A. (2019) Robotics-based scientific investigations at an ice-ocean interface: First results from Nereid Under Ice in the Arctic. In *Ocean Worlds* 4, Abstract #6039. LPI Contribution No. 2168, Lunar and Planetary Institute, Houston.

Gleeson D. F., Pappalardo R. T., Grasby S. E., Anderson M. S., Beauchamp B., Castaño R., Chien S. A., Doggett T., Mandrake L., and Wagstaff K. L. (2010) Characterization of a sulfur-rich Arctic spring site and field analog to Europa using hyperspectral data. *Remote Sens. Environ., 114(6),* 1297–1311.

Gleeson D. F., Williamson C., Grasby S. E., Pappalardo R. T., Spear J. R., and Templeton A. S. (2011) Low temperature S0 biomineralization at a supraglacial spring system in the Canadian High Arctic. *Geobiology, 9(4),* 360–375.

Gleeson D. F., Pappalardo R. T., Anderson M. S., Grasby S. E., Mielke R. E., Wright K. E., and Templeton A. S. (2012) Biosignature detection at an Arctic analog to Europa. *Astrobiology, 12(2),* 135–150.

Glein C. R., Baross J. A., and Waite J. H. Jr. (2015) The pH of Enceladus' ocean. *Geochim. Cosmochim. Acta, 162,* 202–219.

Goldreich P. M. and Mitchell J. L. (2010) Elastic ice shells of synchronous moons: Implications for cracks on Europa and non-synchronous rotation of Titan. *Icarus, 209(2),* 631–638.

Grasset O., Dougherty M. K., Coustenis A., Bunce E. J., Erd C., Titov D., Blanc M., Coates A., Drossart P., Fletcher L. N., and Hussmann H. (2013) JUpiter ICy moons Explorer (JUICE): An ESA mission to orbit Ganymede and to characterize the Jupiter system. *Planet. Space Sci., 78,* 1–21.

Greeley R., Sullivan R., Klemaszewski J., Homan K., Head J. W. III, Pappalardo R. T., Veverka J., Clark B. E., Johnson T. V., Klaasen K. P., and Belton M. (1998) Europa: Initial Galileo geological observations. *Icarus, 135(1),* 4–24.

Greenberg R., Geissler P. E., Hoppa G., Tufts B. R., Durda D. D., Pappalardo R., Head J. W., Greeley R., Sullivan R., and. Carr M. H. (1998) Tectonic processes on Europa: Tidal stresses, mechanical response, and visible features. *Icarus, 135,* 64–78.

Greenberg R., Hoppa G. V., Tufts B. R., Geissler P., Riley J., and Kadel S. (1999) Chaos on Europa. *Icarus, 141(2),* 263–286.

Greenberg R., Geissler P., Tufts B. R., and Hoppa G. V. (2000) Habitability of Europa's crust: The role of tidal-tectonic processes. *J. Geophys. Res.–Planets, 105(E7),* 17551–17562.

Grimm Simon L., Brice-Olivier D., Gillon M., Dorn C., Agol E., Burdanov A., Delrez L., et al. (2018) The nature of the TRAPPIST-1 exoplanets. *Astron. Astrophys., 613,* A68, DOI: 10.1051/0004-6361/201732233.

Gulati S., Richmond K., Flesher C., Hogan B. P., Murarka A., Kuhlmann G., Sridharan M., Stone W. C., and Doran P. T. (2010) Toward autonomous scientific exploration of ice-covered lakes — Field experiments with the ENDURANCE AUV in an Antarctic Dry Valley. In *2010 IEEE International Conference on Robotics and Automation,* pp. 308–315. IEEE, New York.

Hammond N. P. and Barr A. C. (2014) Formation of Ganymede's grooved terrain by convection-driven resurfacing. *Icarus, 227,* 206–209.

Han L. and Showman A. P. (2005) Thermo-compositional convection in Europa's icy shell with salinity. *Geophys. Res. Lett., 32(20),* L20201.

Han L. and Showman A. P. (2008) Implications of shear heating and fracture zones for ridge formation on Europa. *Geophys. Res. Lett., 35(3),* L03202.

Han L. and Showman A. P. (2010) Coupled convection and tidal dissipation in Europa's ice shell. *Icarus, 207,* 834–844.

Hand K. P. and Carlson R. W. (2015) Europa's surface color suggests an ocean rich with sodium chloride. *Geophys. Res. Lett., 42(9),* 3174–3178.

Hand K. P. and Chyba C. F. (2007) Empirical constraints on the salinity of the Europan ocean and implications for a thin ice shell. *Icarus, 189(2),* 424–438.

Hand K. P., Chyba C. F., Carlson R. W., and Cooper J. F. (2006) Clathrate hydrates of oxidants in the ice shell of Europa. *Astrobiology, 6(3),* 463–482.

Hand K. P., Chyba C. F., Priscu J. C., Carlson R. W., and Nealson K. H. (2009) Astrobiology and the potential for life on Europa. In *Europa* (R. T. Pappalardo et al., eds.), pp. 589–629. Univ. of Arizona, Tucson.

Hand K. P., Murray A. E., Garvin J. B., Brinkerhoff W. B., Edgett K. S., Ehlmann B. L., German C. R., Hayes A. G., Hoehler T. M., Horst S. M., Lunine J. I., Nealson K. H., Paranicas C., Schmidt B. E., Smith D. E., Rhoden A. R., Russell M. J., Templeton A. S., Willis P. A., Yingst R. A., Phillips C. B., Cable M. L., Craft K. L., Hofmann A. E., Nordheim T. A., Pappalardo R. T., and the Project Engineering Team (2017) *Report of the Europa Lander Science Definition Team.* JPL D-97667, National Aeronautics and Space Administration, Washington, DC, available online at *https://europa. nasa.gov/resources/58/europa-lander-study-2016-report/.*

Hauck S. A., Aurnou J. M., and Dombard A. J. (2006) Sulfur's impact on core evolution and magnetic field generation on Ganymede. *J. Geophys. Res.–Planets, 111(E9),* E09008.

Hays L. E., Archenbach L., Bailey J., Barnes R., Barros J., Bertka C., Boston P., et al. (2015) *Astrobiology Strategy.* National Aeronautics and Space Administration, Washington, DC.

Head J. W. III and Pappalardo R. T. (1999). Brine mobilization during lithospheric heating on Europa: Implications for formation of chaos terrain, lenticula texture, and color variations. *J. Geophys. Res.–Planets, 104(E11),* 27143–27155.

Head J. W., Pappalardo R. T., and Sullivan R. (1999) Europa: Morphological characteristics of ridges and triple bands from Galileo data (E4 and E6) and assessment of a linear diapirism model. *J. Geophys. Res.–Planets, 104(E10),* 24223–24236.

Head J., Pappalardo R., Collins G., Belton M. J., Giese B., Wagner R., Breneman H., Spaun N., Nixon B., Neukum G., and Moore J. (2002) Evidence for Europa-like tectonic resurfacing styles on Ganymede. *Geophys. Res. Lett., 29(24),* 4-1–4-4.

Heller R. (2016) Transits of extrasolar moons around luminous giant planets. *Astron. Astrophys., 588,* A34.

Heller R. and Pudritz R. (2015) Water ice lines and the formation of giant moons around super-jovian planets. *Astrophys. J., 806(2),* 181.

Hendrix A. R., Hurford T. A., Barge L. M., Bland M. T., Bowman J. S., Brinckerhoff W., Buratti B. J., et al. (2019) The NASA roadmap to ocean worlds. *Astrobiology, 19(1),* 1–27.

Hibbitts C. A., McCord T. B., and Hansen G. B. (2000) Distributions of CO_2 and SO_2 on the surface of Callisto. *J. Geophys. Res.–Planets, 105(E9),* 22541–22557.

Hibbitts C. A., Pappalardo R. T., Hansen G. B., and McCord T. B. (2003) Carbon dioxide on Ganymede. *J. Geophys. Res., 108,* DOI: 10.1029/2002JE001956.

Hopkins M. D., Harrison T. M., and Manning C. E. (2010) Constraints on Hadean geodynamics from mineral inclusions in>4 Ga zircons. *Earth Planet. Sci. Lett., 298(3–4),* 367–376.

Hoppa G., Tufts B. R., Greenberg R., and Geissler P. (1999) Strike-slip faults on Europa: Global shear patterns driven by tidal stress.

Icarus, 141, 287–298.

Horner V. M. and Greeley R. (1982) Pedestal craters on Ganymede. *Icarus, 51(3),* 549–562.

Horrigan S. G. (1981) Primary production under the Ross Ice Shelf, Antarctica. *Limnol Oceanogr., 26(2),* 378–382.

Howell S. M. and Pappalardo R. T. (2018) Band formation and ocean-surface interaction on Europa and Ganymede. *Geophys. Res. Lett., 45(10),* 4701–4709.

Howell S. M. and Pappalardo R. T. (2019) Can Earth-like plate tectonics occur in ocean world ice shells? *Icarus, 322,* 69–79.

Howell R. R., Spencer J. R., Goguen J. D., Marchis F., Prangé R., Fusco T., Blaney D. L., et al. (2001) Ground-based observations of volcanism on Io in 1999 and early 2000. *J. Geophys. Res.–Planets, 106(E12),* 33129–33139.

Hsu H. W., Postberg F., Sekine Y., Shibuya T., Kempf S., Horányi M., Juhász A., et al. (2015) Ongoing hydrothermal activities within Enceladus. *Nature, 519(7542),* 207.

Hurford T. A., Sarid A. R., and Greenberg R. (2007) Cycloidal cracks on Europa: Improved modeling and non-synchronous rotation implications. *Icarus, 186(1),* 218–233.

Hussmann H. and Spohn T. (2004) Thermal-orbital evolution of Io and Europa. *Icarus, 171(2),* 391–410.

Jakuba M. V., German C. R., Bowen A. D., Whitcomb L. L., Hand K., Branch A., Chien S., and McFarland C. (2018) Teleoperation and robotics under ice: Implications for planetary exploration. In *2018 IEEE Aerospace Conference,* pp. 1–14. DOI: 10.1109/AERO.2018.8396587. IEEE, New York.

Jenkins A., Dutrieux P., Jacobs S. S., McPhail S. D., Perrett J. R., Webb A. T., and White D. (2010) Observations beneath Pine Island Glacier in West Antarctica and implications for its retreat. *Nature Geosci., 3(7),* 468.

Jia X., Walker R. J., Kivelson M. G., Khurana K. K., and Linker J. A. (2009) Properties of Ganymede's magnetosphere inferred from improved three-dimensional MHD simulations. *J. Geophys. Res.–Space Phys., 114(A9).*

Jia X., Kivelson M. G., Khurana K. K., and Kurth W. S. (2018) Evidence of a plume on Europa from Galileo magnetic and plasma wave signatures. *Nature Astron., 2(6),* 459.

Johnson B. C., Sheppard R. Y., Pascuzzo A. C., Fisher E. A., and Wiggins S. E. (2017) Porosity and salt content determine if subduction can occur in Europa's ice shell. *J. Geophys. Res.–Planets, 122(12),* 2765–2778.

Johnson R. E., Carlson R. W., Cooper J. F., Paranicas C., Moore M. H., and Wong M. C. (2004) Radiation effects on the surfaces of the Galilean satellites. In *Jupiter: The Planet, Satellites and Magnetosphere* (F. Bagenal et al., eds.), pp. 485–512. Cambridge Univ., Cambridge.

Johnson T. V. and McCord T. B. (1971) Spectral geometric albedo of Galilean Satellites, 0.3 to 2.5 microns. *Astrophys. J., 169(3),* 589–594.

Johnson T. V., Morrison D., Matson D. L., Veeder G. J., Brown R. H., and Nelson R. M. (1984) Volcanic hotspots on Io: Stability and longitudinal distribution. *Science, 226(4671),* 134–137.

Kalousová K., Souček O., Tobie G., Choblet, G., and Čadek O. (2014) Ice melting and downward transport of meltwater by two-phase flow in Europa's ice shell. *J. Geophys. Res.–Planets, 119(3),* 532–549.

Kalousová K., Souček O., Tobie G., Choblet G., and Čadek O. (2016) Water generation and transport below Europa's strike-slip faults. *J. Geophys. Res.–Planets, 121(12),* 2444–2462.

Kalousová K., Sotin C., Choblet G., Tobie G., and Grasset O. (2018) Two-phase convection in Ganymede's high-pressure ice layer — Implications for its geological evolution. *Icarus, 299,* 133–147.

Kargel J. S., Kaye J. Z., Head J. W. III, Marion G. M., Sassen R., Crowley J. K., Ballesteros O. P., Grant S. A., and Hogenboom D. L. (2000) Europa's crust and ocean: Origin, composition, and the prospects for life. *Icarus, 148(1),* 226–265.

Kattenhorn S. A. (2004) Strike-slip fault evolution on Europa: Evidence from tailcrack geometries. *Icarus, 172(2),* 582–602.

Kattenhorn S. A. and Hurford T. (2009) Tectonics of Europa. In *Europa* (R. T. Pappalardo et al., eds.), pp. 199–236. Univ. of Arizona, Tucson.

Kattenhorn S. A. and Prockter L. M. (2014) Evidence for subduction in the ice shell of Europa. *Nature Geosci., 7(10),* 762.

Khurana K. K., Kivelson M. G., Stevenson D. J., Schubert G., Russell C. T., Walker R. J., and Polanskey C. (1998) Induced magnetic fields as evidence for subsurface oceans in Europa and Callisto.

Nature, 395(6704), 777.

Kimball P. W., Clark E. B., Scully M., Richmond K., Flesher C., Lindzey L. E., Harman J., et al. (2017) The ARTEMIS under-ice AUV docking system. *J. Field Robot., 35,* 299–308, DOI: 10.1002/rob.21740.

Kimura J., Nakagawa T., and Kurita K. (2009) Size and compositional constraints of Ganymede's metallic core for driving an active dynamo. *Icarus, 202(1),* 216–224.

Kirchoff M. R. and McKinnon W. B. (2009) Formation of mountains on Io: Variable volcanism and thermal stresses. *Icarus, 201(2),* 598–614.

Kirchoff M. R. and Schenk P. (2009) Crater modification and geologic activity in Enceladus' heavily cratered plains: Evidence from the impact crater distribution. *Icarus, 202,* 656–668.

Kivelson M. G., Khurana K. K., Coroniti F. V., Joy S., Russell C. T., Walker R. J., Warnecke J., Bennett L. and Polanskey C. (1997) The magnetic field and magnetosphere of Ganymede. *Geophys. Res. Lett., 24(17),* 2155–2158.

Kivelson M. G., Khurana K. K., Russell C. T., Volwerk M., Walker R. J., and Zimmer C. (2000) Galileo magnetometer measurements: A stronger case for a subsurface ocean at Europa. *Science, 289(5483),* 1340–1343.

Kivelson M. G., Khurana K. K., and Volwerk M. (2002) The permanent and inductive magnetic moments of Ganymede. *Icarus, 157,* 507–522.

Kovacs A. (1996) *Sea Ice. Part 1. Bulk Salinity Versus Ice Floe Thickness.* CRREL Report 96-7, Cold Regions Research and Engineering Lab, Hanover.

Kovacs A. and Gow A. J. (1975) Brine infiltration in the McMurdo Ice Shelf, McMurdo Sound, Antarctica. *J. Geophys. Res., 80(15),* 1957–1961.

Lambrechts M. and Johansen A. (2012) Rapid growth of gas-giant cores by pebble accretion. *Astron. Astrophys., 544,* A32.

Lawrence J., Schmidt B., Washam P., Meister M., Hurwitz B., Quartini E., Spears A., Mullen A., and Dichek D. (2020) Icefin observes super cooling and marine ice accretion in a basal crevasse beneath Ross Ice Shelf. To be presented at Scientific Committee on Antarctic Research Open Science Conference, Hobart, Australia, August 2020, in press.

Lee C. M. and Rudnick D. L. (2018) Underwater gliders. In *Observing the Oceans in Real Time* (R. Venkatesan et al., eds.), pp. 123–139. Springer, Cham.

Leonard E. J., Pappalardo R. T., and Yin A. (2018) Analysis of very-high-resolution Galileo images and implications for resurfacing mechanisms on Europa. *Icarus, 312,* 100–120.

Leonard E. J., Patthoff D. A., Senske D. A., and Collins G. C. (2018) The Europa Global Geologic Map. *Planetary Geologic Mappers Annual Meeting,* Abstract #7008. LPI Contribution No. 2066, Lunar and Planetary Institute, Houston.

Lewis J. S. (1971) Satellites of the outer planets: Their physical and chemical nature. *Icarus, 15(2),* 174–185.

Ligier N., Poulet F., Carter J., Brunetto R., and Gourgeot F. (2016) VLT/SINFONI observations of Europa: New insights into the surface composition. *Astron. J., 151(6),* 163.

Loeffler M. J., Raut U., Vidal R. A., Baragiola R. A., and Carlson R. W. (2006) Synthesis of hydrogen peroxide in water ice by ion irradiation. *Icarus, 180(1),* 265–273.

Lorenz R. D. (2016) Europa ocean sampling by plume flythrough: Astrobiological expectations. *Icarus, 267,* 217–219.

Lowell R. P. and DuBose M. (2005) Hydrothermal systems on Europa. *Geophys. Res. Lett., 32,* L05202.

Lunine J. I. (2017) Ocean worlds exploration. *Acta Astronaut., 131,* 123–130.

Lunine J. I., Coradini A., Gautier D., Owen T. C., and Wuchterl G. (2004) The origin of Jupiter. In *Jupiter: The Planet, Satellites and Magnetosphere* (F. Bagenal et al., eds.), pp. 19–34. Cambridge Univ., Cambridge.

Malhotra R. (1991) Tidal origin of the Laplace resonance and the resurfacing of Ganymede. *Icarus, 94(2),* 399–412.

Manga M. and Wang C. Y. (2007) Pressurized oceans and the eruption of liquid water on Europa and Enceladus. *Geophys. Res. Lett., 34,* L07202.

Marchis F., de Pater I., Davies A. G., Roe H. G., Fusco T., Le Mignant D., Descamps P., Macintosh B. A., and Prangé R. (2002) High-resolution Keck adaptive optics imaging of violent volcanic activity on Io. *Icarus, 160,* 124–131.

Marion G. M., Fritsen C. H., Eicken H. and Payne M. C. (2003) The search for life on Europa: Limiting environmental factors, potential habitats, and Earth analogues. *Astrobiology, 3(4),* 785–811.

McCord T. B., Carlson R. W., Smythe W. D., Hansen G. B., Clark R. N., Hibbitts C. A., Fanale F. P., et al. (1997) Organics and other molecules in the surfaces of Callisto and Ganymede. *Science, 278(5336),* 271–275.

McCord T. B., Hansen G. B., Clark R. N., Martin P. D., Hibbitts C. A., Fanale F. P., Granahan J. C., et al. (1998a) Non-water-ice constituents in the surface material of the icy Galilean satellites from the Galileo near-infrared mapping spectrometer investigation. *J. Geophys. Res.–Planets, 103(E4),* 8603–8626.

McCord T. B., Hansen G. B., Fanale F. P., Carlson R. W., Matson D. L., Johnson T. V., Smythe W. D., et al. (1998b) Salts on Europa's surface detected by Galileo's near infrared mapping spectrometer. *Science, 280(5367),* 1242–1245.

McCord T. B., Hansen G. B., Matson D. L., Johnson T. V., Crowley J. K., Fanale F. P., Carlson R. W., et al. (1999) Hydrated salt minerals on Europa's surface from the Galileo near-infrared mapping spectrometer (NIMS) investigation. *J. Geophys. Res.–Planets, 104(E5),* 11827–11851.

McCord T. B., Teeter G., Hansen G. B., Sieger M. T., and Orlando T. M. (2002) Brines exposed to Europa surface conditions. *J. Geophys. Res.–Planets, 107(E1),* DOI: 10.1029/2000JE001453.

McEwen A. S., Keszthelyi L., Spencer J. R., Schubert G., Matson D. L., Lopes-Gautier R., Klaasen K. P., et al. (1998) High-temperature silicate volcanism on Jupiter's moon Io. *Science, 281(5373),* 87–90.

McGrath M. A. and Sparks W. B. (2017) Galileo ionosphere profile coincident with repeat plume detection location at Europa. *Res. Notes AAS, 1,* 14.

McKinnon W. B. (1999) Convective instability in Europa's floating ice shell. *Geophys. Res. Lett., 26(7),* 951–954.

McKinnon W. B. and Zolensky M. E. (2003) Sulfate content of Europa's ocean and shell: Evolutionary considerations and some geological and astrobiological implications. *Astrobiology, 3(4),* 879–897.

Meister M., Dichek D., Spears A., Hurwitz B., Ramey C., Lawrence J. D., Philleo M., Lutz J., Lawrence J. P., and Schmidt B. E. (2018) Icefin: Redesign and 2017 Antarctic field deployment. In *OCEANS 2018 MTS/IEEE Charleston,* pp. 1–5. IEEE, New York.

Mikucki J. A., Pearson A., Johnston D. T., Turchyn A. V., Farquhar J., Schrag D. P., Anbar A. D., Priscu J. C., and Lee P. A. (2009) A contemporary microbially maintained subglacial ferrous "ocean." *Science, 324(5925),* 397–400.

Mikucki J. A., Lee P. A., Ghosh D., Purcell A. M., Mitchell A. C., Mankoff K. D., Fischer A. T., and the WISSARD Science Team (2016) Subglacial Lake Whillans biogeochemistry: A synthesis of current knowledge. *Philos. Trans. R. Soc., Ser. A, 374,* DOI: 10.1098/rsta.2014.0290.

Mitri G. and Showman A. P. (2008) A model for the temperature-dependence of tidal dissipation in convective plumes on icy satellites: Implications for Europa and Enceladus. *Icarus, 195(2),* 758–764.

Monteux J., Collins G. S., Tobie G., and Choblet G. (2016) Consequences of large impacts on Enceladus' core shape. *Icarus, 264,* 300–310.

Moore J. M. and Malin M. C. (1988) Dome craters on Ganymede. *Geophys. Res. Lett., 15(3),* 225–228.

Moore J. M., Asphaug E., Morrison D., Spencer J. R., Chapman C. R., Bierhaus B., Sullivan R. J., et al. (1999) Mass movement and landform degradation on the icy Galilean satellites: Results of the Galileo nominal mission. *Icarus, 140(2),* 294–312.

Moore W. B. (2001) The thermal state of Io. *Icarus, 154(2),* 548–550.

Moore W. B. and Hussmann H. (2009) Thermal evolution of Europa's silicate interior. In *Europa* (R. Pappalardo et al., eds.), p. 369. Univ. of Arizona, Tucson.

Moore W. B. and Webb A. A. G. (2013) Heat-pipe Earth. *Nature, 501(7468),* 501.

Moore W. B., Simon J. I., and Webb A. A. G. (2017) Heat-pipe planets. *Earth Planet. Sci. Lett., 474,* 13–19.

Murray A. E., Kenig F., Fritsen C. H., McKay C. P., Cawley K. M., Edwards R., Kuhn E., et al. (2012) Microbial life at −13°C in the brine of an ice-sealed Antarctic lake. *Proc. Natl. Acad. Sci. U.S.A., 109(50),* 20626–20631.

National Academies of Sciences, Engineering, and Medicine (2019) *An Astrobiology Strategy for the Search for Life in the Universe.* National Academies, Washington, DC. 188 pp.

National Research Council, Space Studies Board (2012) *Vision and Voyages for Planetary Science in the Decade 2013–2022.* National Academies, Washington, DC. 154 pp.

Neish C. D., Prockter L. M., and Patterson G. W. (2012) Observational constraints on the identification and distribution of chaotic terrain on icy satellites. *Icarus, 221(1),* 72–79.

Nimmo F. and Gaidos E. (2002) Strike-slip motion and double ridge formation on Europa. *J. Geophys. Res.–Planets, 107(E4),* 5–1.

Nimmo F. and Schenk P. (2006) Normal faulting on Europa: Implications for ice shell properties. *J. Structural Geol., 28(12),* 2194–2203.

Nimmo F., Pappalardo R. T., and Giese B. (2003) On the origins of band topography, Europa. *Icarus, 166(1),* 21–32.

Noack L., Höning D., Rivoldini A., Heistracher C., Zimov N., Journaux B., Lammer H., Van Hoolst T., and Bredehöft J. H. (2016) Water-rich planets: How habitable is a water layer deeper than on Earth? *Icarus, 277,* 215–236.

Noll K. S., Johnson R. E., Lane A. L., Domingue D. L., and Weaver H. A. (1996) Detection of ozone on Ganymede. *Science, 273,* 341–343.

Nordheim T. A., Hand K. P., and Paranicas C. (2018) Preservation of potential biosignatures in the shallow subsurface of Europa. *Nature Astron., 2(8),* 673.

O'Brien D. P., Geissler P., and Greenberg R. (2002) A melt-through model for chaos formation on Europa. *Icarus, 156(1),* 152–161.

Ojakangas G. W. and Stevenson D. J. (1986) Episodic volcanism of tidally heated satellites with application to Io. *Icarus, 66(2),* 341–358.

Ojakangas G. W. and Stevenson D. J. (1989) Thermal state of an ice shell on Europa. *Icarus, 81(2),* 220–241.

O'Neill C. and Lenardic A. (2007) Geological consequences of super-sized Earths. *Geophys. Res. Lett., 34,* L19204.

Pappalardo R. T. and Barr A. C. (2004) The origin of domes on Europa: The role of thermally induced compositional diapirism. *Geophys. Res. Lett., 31,* L01701.

Pappalardo R. T., Head J. W., Greeley R., Sullivan R. J., Pilcher C., Schubert G., Moore W. B., et al. (1998) Geological evidence for solid-state convection in Europa's ice shell. *Nature, 391(6665),* 365.

Pappalardo R. T., Belton M. J. S., Breneman H. H., Carr M. H., Chapman C. R., Collins G. C., et al. (1999) Does Europa have a subsurface ocean? Evaluation of the geological evidence. *J. Geophys. Res.–Planets, 104(E10),* 24015–24055.

Pappalardo R. T., Collins G. C., Head J. W., Helfenstein P., McCord T. B., Moore J. M., Prockter L. M., Schenk P. M., and Spencer J. R. (2004) Geology of Ganymede. In *Jupiter: The Planet, Satellites and Magnetosphere* (F. Bagenal et al., eds.), pp. 363–396. Cambridge Univ., Cambridge.

Pappalardo R. T., Vance S., Bagenal F., Bills B. G., Blaney D. L., Blankenship D. D., Brinckerhoff W. B., et al. (2013) Science potential from a Europa lander. *Astrobiology, 13(8),* 740–773.

Paranicas C., Ratliff J. M., Mauk B. H., Cohen C., and Johnson R. E. (2002) The ion environment near Europa and its role in surface energetics. *Geophys. Res. Lett., 29(5),* 18-1 to 18-4.

Paranicas C., Mauk B. H., Khurana K., Jun I., Garrett H., Krupp N., and Roussos E. (2007) Europa's near-surface radiation environment. *Geophys. Res. Lett., 34,* L15103.

Paranicas C., Cooper J. F., Garrett H. B., Johnson R. E., and Sturner S. J. (2009) Europa's radiation environment and its effects on the surface. In *Europa* (R. T. Pappalardo et al., eds.), pp. 529–544. Univ. of Arizona, Tucson.

Pasek M. A. and Greenberg R. (2012) Acidification of Europa's subsurface ocean as a consequence of oxidant delivery. *Astrobiology, 12(2),* 151–159.

Patterson G. W., Collins G. C., Head J. W., Pappalardo R. T., Prockter L. M., Lucchitta B. K., and Kay J. P. (2010) Global geological mapping of Ganymede. *Icarus, 207(2),* 845–867.

Paty C. and Winglee R. (2004) Multi-fluid simulations of Ganymede's magnetosphere. *Geophys. Res. Lett., 31,* L24806.

Peale S. J., Cassen P., and Reynolds R. T. (1979) Melting of Io by tidal dissipation. *Science, 203(4383),* 892–894.

Peddinti D. A. and McNamara A. K. (2015) Material transport across Europa's ice shell. *Geophys. Res. Lett., 42(11),* 4288–4293.

Peddinti D. A. and McNamara A. K. (2019) Dynamical investigation of a thickening ice-shell: Implications for the icy moon Europa. *Icarus, 329,* 251–269.

Perovich D. K., Richter-Menge J. A., and Morison J. H. (1995)

The formation and morphology of ice stalactites observed under deforming lead ice. *J. Glaciol., 41(138),* 305–312.

Phillips C. B. and Pappalardo R. T. (2014) Europa clipper mission concept: Exploring Jupiter's ocean moon. *Eos Trans. AGU, 95(20),* 165–167.

Phillips C. B., McEwen A. S., Hoppa G. V., Fagents S. A., Greeley R., Klemaszewski J. E., Pappalardo R. T., Klaasen K. P., and Breneman H. H. (2000) The search for current geologic activity on Europa. *J. Geophys. Res.–Planets, 105(E9),* 22579–22597.

Pierazzo E. and Chyba C. F. (2002) Cometary delivery of biogenic elements to Europa. *Icarus, 157(1),* 120–127.

Pilcher C. B., Ridgway S. T., and McCord T. B. (1972) Galilean satellites: Identification of water frost. *Science, 178(4065),* 1087–1089.

Poston M. J., Carlson R. W., and Hand K. P. (2017) spectral behavior of irradiated sodium chloride crystals under Europa-like conditions. *J. Geophys. Res.–Planets, 122(12),* 2644–2654.

Preiner M., Xavier J. C., Sousa F. L., Zimorski V., Neubeck A., Lang S. Q., Greenwell H. C., et al. (2018) Serpentinization: Connecting geochemistry, ancient metabolism and industrial hydrogenation. *Life, 8(4),* 41.

Price P. B. (2007) Microbial life in glacial ice and implications for a cold origin of life. *FEMS Microbiol. Ecol., 59(2),* 217–231.

Prieto-Ballesteros O., Vorobyova E., Parro V., Manfredi J. A. R., and Gómez F. (2011) Strategies for detection of putative life on Europa. *Adv. Space Res., 48(4),* 678–688.

Priscu J. C. and Christner B. C. (2004) Earth's icy biosphere. In *Microbial Diversity and Bioprospecting* (A. T. Bull, ed.), pp. 130–145. American Society for Microbiology, Washington, DC.

Priscu J. C., Adams E. E., Lyons W. B., Voytek M. A., Mogk D. W., Brown R. L., McKay C. P., et al. (1999) Geomicrobiology of subglacial ice above Lake Vostok, Antarctica. *Science, 286(5447),* 2141–2144.

Prockter L. M. and Patterson G. W. (2009) Morphology and evolution of Europa's ridges and bands. In *Europa* (R. T. Pappalardo et al., eds.), pp. 237–258. Univ. of Arizona, Tucson.

Quick L. C., Barnouin O. S., Prockter L. M., and Patterson G. W. (2013) Constraints on the detection of cryovolcanic plumes on Europa. *Planet. Space Sci., 86,* 1–9.

Rathbun J. A. and Spencer J. R. (2010) Ground-based observations of time variability in multiple active volcanoes on Io. *Icarus, 209(2),* 625–630.

Rathbun J. A., Spencer J. R., Davies A. G., Howell R. R., and Wilson L. (2002) Loki, Io: A periodic volcano. *Geophys. Res. Lett., 29(10),* DOI: 10.1029/2002GL014747.

Reynolds R. T., Squyres S. W., Colburn D. S., and McKay C. P. (1983) On the habitability of Europa. *Icarus, 56(2),* 246–254.

Reynolds R. T., McKay C. P., and Kasting J. F. (1987) Europa, tidally heated oceans, and habitable zones around giant planets. *Adv. Space Res., 7(5),* 125–132.

Rhoden A. R., Hurford T. A., and Manga M. (2011) Strike-slip fault patterns on Europa: Obliquity or polar wander? *Icarus, 211(1),* 636–647.

Riley J., Hoppa G. V., Greenberg R., Tufts B. R., and Geissler P. (2000) Distribution of chaotic terrain on Europa. *J. Geophys. Res.–Planets, 105(E9),* 22599–22615.

Roth L., Saur J., Retherford K. D., Strobel D. F., Feldman P. D., McGrath M. A., and Nimmo F. (2014a) Transient water vapor at Europa's south pole. *Science, 343(6167),* 171–174.

Roth L., Retherford K. D., Saur J., Strobel D. F., Feldman P. D., McGrath M. A., and Nimmo F. (2014b) Orbital apocenter is not a sufficient condition for HST/STIS detection of Europa's water vapor aurora. *Proc. Natl. Acad. Sci. U.S.A., 111(48),* E5123–E5132.

Russell M. J., Murray A. E., and Hand K. P. (2017) The possible emergence of life and differentiation of a shallow biosphere on irradiated icy worlds: The example of Europa. *Astrobiology, 17(12),* 1265–1273.

Scambos T. A., Bohlander J. A., Shuman C. U., and Skvarca P. (2004) Glacier acceleration and thinning after ice shelf collapse in the Larsen B embayment, Antarctica. *Geophys. Res. Lett., 31,* L18402.

Schenk P. M. (1993) Central pit and dome craters: Exposing the interiors of Ganymede and Callisto. *J. Geophys. Res.–Planets, 98(E4),* 7475–7498.

Schenk P. M. (2002) Thickness constraints on the icy shells of the Galilean satellites from a comparison of crater shapes. *Nature, 417,* 419–421.

Schenk P. M. and Bulmer M. H. (1998) Origin of mountains on Io by thrust faulting and large-scale mass movements. *Science, 279(5356),* 1514–1517.

Schenk P. M. and McKinnon W. B. (1989) Fault offsets and lateral crustal movement on Europa: Evidence for a mobile ice shell. *Icarus, 79(1),* 75–100.

Schenk P. M. and Turtle E. P. (2009) Europa's impact craters: Probes of the icy shell. In *Europa* (R. T. Pappalardo et al., eds.), pp. 181–198. Univ. of Arizona, Tucson.

Schmidt B. E., Blankenship D. D., Patterson G. W., and Schenk P. M. (2011) Active formation of 'chaos terrain' over shallow subsurface water on Europa. *Nature, 479(7374),* 502.

Schmidt B. E., Washam P., Davis P. E. D., Nicholls K. W., Lawrence J. D., Smith J., Dichek D. J. G., et al. (2020) The grounding zone of Thwaites Glacier explored by Icefin. To be presented at Scientific Committee on Antarctic Research Open Science Conference, Hobart, Australia, August 2020, in press.

Schubert G., Zhang K., Kivelson M. G., and Anderson J. D. (1996) The magnetic field and internal structure of Ganymede. *Nature, 384(6609),* 544.

Schubert G., Anderson J. D., Spohn T., and McKinnon W. B. (2004) Interior composition, structure and dynamics of the Galilean Satellites. In *Jupiter: The Planet, Satellites and Magnetosphere.* (F. Bagenal et al., eds.), pp. 281–306. Cambridge Univ., Cambridge.

Schubert G., Hussmann H., Lainey V., Matson D., McKinnon W., Sohl F., Sotin C., Tobie G., Turrini D., and VanHoolst T. (2010) Evolution of icy satellites. *Space Sci. Rev., 153,* 447–484.

Sherwood Lollar B., Atreya S. K., Boss A. P., Falkowski P. G., Farmer J. D., Guyon O., Joyce G. F., et al. (2018) Astrobiology science strategy for the search for life in the universe. Abstract #P23B-02 presented at 2018 Fall Meeting, AGU, Washington, DC, 10–14 December.

Shirley J. H., Jamieson C. S., and Dalton J. B III (2016) Europa's surface composition from near-infrared observations: A comparison of results from linear mixture modeling and radiative transfer modeling. *Earth Space Sci., 3(8),* 326–344.

Showman A. P. and Han L. (2004) Numerical simulations of convection in Europa's ice shell: Implications for surface features. *J. Geophys. Res.–Planets, 109(E1),* E01010, DOI: 10.1029/2003JE002103.

Showman A. P. and Malhotra R. (1997) Tidal evolution into the Laplace resonance and the resurfacing of Ganymede. *Icarus, 127,* 93–111.

Showman A. P., Stevenson D. J., and Malhotra R. (1997) Coupled orbital and thermal evolution of Ganymede. *Icarus, 129(2),* 367–383.

Showman A. P., Mosqueira I., and Head J. W. III (2004) On the resurfacing of Ganymede by liquid–water volcanism. *Icarus, 172(2),* 625–640.

Soderlund K. M. (2019) Ocean dynamics of outer solar system satellites. *Geophys. Res. Lett., 46,* 8700–8710, DOI: 10.1029/2018GL081880.

Soderlund K. M., Schmidt B. E., Wicht J., and Blankenship D. D. (2014) Ocean-driven heating of Europa's icy shell at low latitudes. *Nature Geosci., 7,* 16–19, DOI: 10.1038/ngeo2021.

Sotin C. and Tobie G. (2004) Internal structure and dynamics of the large icy satellites. *Compt. Rend. Phys., 5(7),* 769–780.

Sotin C., Head J. W. III, and Tobie G. (2002) Europa: Tidal heating of upwelling thermal plumes and the origin of lenticulae and chaos melting. *Geophys. Res. Lett., 29(8),* DOI: 10.1029/2001GL013844.

Sparks W. B., Hand K. P., McGrath M. A., Bergeron E., Cracraft M., and Deustua S. E. (2016) Probing for evidence of plumes on Europa with HST/STIS. *Astrophys. J., 829(2),* 121.

Sparks W. B., Schmidt B. E., McGrath M. A., Hand K. P., Spencer J. R., Cracraft M., and Deustua S. E. (2017) Active cryovolcanism on Europa? *Astrophys. J. Lett., 839(2),* L18.

Spaun N. A., Head J. W., Collins G. C., Prockter L. M., and Pappalardo R. T. (1998) Conamara Chaos region, Europa: Reconstruction of mobile polygonal ice blocks. *Geophys. Res. Lett., 25(23),* 4277–4280.

Spears A., West M., Meister M., Buffo J., Walker C., Collins T. R., Howard A., and Schmidt B. (2016) Under ice in Antarctica: The ICEFIN unmanned underwater vehicle development and deployment. *IEEE Robot. Autom. Mag., 23(4),* 30–41.

Spencer J. R. and Calvin W. M. (2002) Condensed O_2 on Europa and Callisto. *Astron. J., 124(6),* 3400.

Spencer J. R., Stern S. A., Cheng A. F., Weaver H. A., Reuter D. C., Retherford K., Lunsford A., et al. (2007) Io volcanism seen by New Horizons: A major eruption of the Tvashtar volcano. *Science, 318(5848),* 240–243.

Squyres S. W., Reynolds R. T., Cassen P. M., and Peale S. J. (1983) Liquid water and active resurfacing on Europa. *Nature, 301(5897),* 225.

Stempel M. M., Barr A. C., and Pappalardo R. T. (2005) Model constraints on the opening rates of bands on Europa. *Icarus, 177(2),* 297–304.

Stone W. C., Hogan B., Siegel V., Lelievre S., and Flesher C. (2014) Progress towards an optically powered cryobot. *Ann. Glaciol., 55(65),* 2–13.

Sullivan R., Greeley R., Homan K., Klemaszewski J., Belton M. J., Carr M. H., Chapman C. R., et al. (1998) Episodic plate separation and fracture infill on the surface of Europa. *Nature, 391(6665),* 371.

Thomas D. N. and Dieckmann G. S., eds. (2008) *Sea Ice: An Introduction to Its Physics, Chemistry, Biology and Geology.* Wiley, New York.

Tittemore W. C. (1990) Chaotic motion of Europa and Ganymede and the Ganymede-Callisto dichotomy. *Science, 250(4978),* 263–267.

Tobie G., Choblet C., and Sotin C. (2003) Tidally heated convection: Constraints on Europa's ice shell thickness. *J. Geophys. Res., 108(5124),* DOI: 10.1029/2003JE002099.

Tobie G., Mocquet A., and Sotin C. (2005) Tidal dissipation within large icy satellites: Applications to Europa and Titan. *Icarus, 177(2),* 534–549.

Travis B. J., Palguta J., and Schubert G. (2012) A whole-moon thermal history model of Europa: Impact of hydrothermal circulation and salt transport. *Icarus, 218,* 1006–1019.

Tuovila B. J., Dobbs F. C., LaRock P. A., and Siegel B. Z. (1987) Preservation of ATP in hypersaline environments. *Appl. Environ. Microbiol., 53(12),* 2749–2753.

Turcotte D. L. (1980) On the thermal evolution of the Earth. *Earth Planet. Sci. Lett., 48(1),* 53–58.

Turcotte D. L. (1989) A heat pipe mechanism for volcanism and tectonics on Venus. *J. Geophys. Res.–Solid Earth, 94(B3),* 2779–2785.

Turtle E. P. and Pierazzo E. (2001) Thickness of a Europan ice shell from impact crater simulations. *Science, 294(5545),* 1326–1328.

Valencia D., O'Connell R. J., and Sasselov D. D. (2007) Inevitability of plate tectonics on super-Earths. *Astrophys. J. Lett., 670(1),* L45.

Vance S. and Goodman J. (2009) Oceanography of an ice-covered moon. In *Europa* (R. T. Pappalardo et al., eds), pp. 459–484. Univ. of Arizona, Tucson.

Vance S., Harnmeijer J., Kimura J., Hussmann H., DeMartin B., and Brown J. M. (2007) Hydrothermal systems in small ocean planets. *Astrobiology, 7(6),* 987–1005.

Vance S., Bouffard M., Choukroun M., and Sotin C. (2014) Ganymede's internal structure including thermodynamics of magnesium sulfate oceans in contact with ice. *Planet. Space Sci., 96,* 62–70.

Vance S. D., Hand K. P., and Pappalardo R. T. (2016) Geophysical controls of chemical disequilibria in Europa. *Geophys. Res. Lett.,* 43(10), 4871–4879.

Vance S. D., Barge L. M., Cardoso S. S., and Cartwright J. H. (2019) Self-assembling ice membranes on Europa: Brinicle properties, field examples, and possible energetic systems in icy ocean worlds. *Astrobiology, 19(5),* 685–695.

Waite J. H., Glein C. R., Perryman R. S., Teolis B. D., Magee B. A., Miller G., Grimes J., et al. (2017) Cassini finds molecular hydrogen in the Enceladus plume: Evidence for hydrothermal processes. *Science, 356(6334),* 155–159.

Walker C. C. and Schmidt B. E. (2015) Ice collapse over trapped water bodies on Enceladus and Europa. *Geophys. Res. Lett., 42(3),* 712–719.

Weller M. B., Fuchs L., Becker T. W., and Soderlund K. M. (2019) Convection in thin shells of icy satellites: Effects of latitudinal surface temperature variations. *J. Geophys. Res.–Planets, 124,* 2029–2053. DOI: 10.1029/2018JE005799.

Williams D. M., Kasting J. F., and Wade R. A. (1997) Habitable moons around extrasolar giant planets. *Nature, 385(6613),* 234.

Winebrenner D. P., Elam W. T., Kintner P. M. S., Tyler S., and Selker J. S. (2016) Clean, logistically light access to explore the closest places on earth to Europa and Enceladus. Abstract # C51E-08 presented at 2016 Fall Meeting, AGU, San Francisco, California, 12–16 December.

Zahnle K., Schenk P., Levison H., and Dones L. (2003) Cratering rates in the outer solar system. *Icarus, 163(2),* 263–289. DOI: 10.1016/S0019-1035(03)00048-4.

Zahnle K., Alvarellos J. L., Dobrovolskis A., and Hamill P. (2008) Secondary and sesquinary craters on Europa. *Icarus, 194(2),* 660–674.

Zhan X. and Schubert G. (2012) Powering Ganymede's dynamo. *J. Geophys. Res.–Planets, 117,* E08011.

Zimmerman W., Bonitz R., and Feldman J. (2001) Cryobot: An ice penetrating robotic vehicle for Mars and Europa. In *2001 IEEE Aerospace Conference Proceedings Vol. 1,* Cat. No. 01TH8542. IEEE, New York.

Zolotov M. Y. and Kargel J. S. (2009) On the chemical composition of Europa's icy shell, ocean, and underlying rocks. In *Europa* (R. T. Pappalardo et al., eds.), pp. 431–457. Univ. of Arizona, Tucson.

Zolotov M. Y. and Shock E. L. (2001) Composition and stability of salts on the surface of Europa and their oceanic origin. *J. Geophys. Res.–Planets, 106(E12),* 32815–32827.

Zolotov M. Y. and Shock E. L. (2003) Energy for biologic sulfate reduction in a hydrothermally formed ocean on Europa. *J. Geophys. Res.–Planets, 108(5022).*

Zolotov M. Y. and Shock E. L. (2004) A model for low-temperature biogeochemistry of sulfur, carbon, and iron on Europa. *J. Geophys. Res.–Planets, 109,* E06003.

Cable M., Neveu M., Hsu H., and Hoehler T. (2020) Enceladus. In *Planetary Astrobiology* (V. Meadows et al., eds.), pp. 217–246. Univ. of Arizona, Tucson, DOI: 10.2458/azu_uapress_9780816540068-ch009.

Enceladus

M. L. Cable
NASA Jet Propulsion Laboratory

M. Neveu
NASA Goddard Space Flight Center/University of Maryland, College Park

H.-W. Hsu
University of Colorado

T. M. Hoehler
NASA Ames Research Center

Enceladus has recently come into the spotlight as one of the key targets for astrobiology exploration in the solar system. Discoveries that arose from data returned by the Voyager and Cassini spacecrafts from this tiny, intriguing moon provide tantalizing evidence of a global, subsurface ocean with water-rock interaction at its base, accessible via a sustained plume ejected from fissures at the south pole. Based on current understanding, the ocean of Enceladus is habitable. The next logical question to investigate is, does it have life? In this chapter we review the evidence for Enceladus' ocean and its habitability, identify constraints and outstanding questions on the extent of the supply of life's key ingredients, discuss how this supply might constrain the detectability of any life in Enceladus' interior, and present strategies to search for life in the specific context of Enceladus in the coming decades.

Life is flux.
— *Heraclitus (ca. 535–ca. 475 BCE)*

1. INTRODUCTION

Enceladus, a small moon in orbit around Saturn, has been full of surprises since it was first identified in 1789. The discovery of Enceladus is credited in part to William Herschel, who was fortuitously observing Saturn at equinox. During equinox, the Sun and Earth are aligned edge-on with the rings of Saturn, reducing the observed reflection of sunlight off the rings (ringshine), which is usually bright enough to mask Enceladus from most telescopes. Enceladus is named after one of the Titans, giants in Greek mythology, and contrary to its small size (504 km or 313 miles in diameter), it has lived up to this legacy.

Enceladus is a particularly exciting target for astrobiology in the solar system because it appears to be active and accessible: A recently discovered persistent water plume emanating from fissures near its south pole can provide a direct link to its global subsurface ocean. This ocean appears to meet the criteria for a habitable environment: extended regions of liquid water, conditions favorable for the assembly of complex organic molecules, and energy source(s) to sustain putative metabolism (*Des Marais et al.*, 2008). If these habitable conditions were present for an extended period over

Enceladus' history, the ocean could have supported life in the past, or might even host life today. Ocean material ejected from Enceladus through the plume eventually orbits Saturn as one of its outermost rings, the E ring.

The record of *in situ* observations of Enceladus spans decades. Given Enceladus' location within the E ring of Saturn and that the maximum intensity of this ring occurs near the orbit of Enceladus, it had been hypothesized from Voyager 1 observations that Enceladus was the primary source for E-ring particles (*Stone and Miner*, 1981). The orbital mean-motion resonance of Enceladus with another saturnian moon, Dione (a configuration that can promote noncircular orbits and therefore enhance tidal dissipation heating), led some to guess, even then, that these particles may be formed due to outgassing of liquid water from under a thin crust (*Terrile and Cook*, 1981). No backlit images of the Saturn system were obtained by Voyager 2 due to a malfunction of the scan platform (*Marchetto*, 1983); if it had not jammed, it is possible that Voyager 2 might have captured the first images of the Enceladus plume. Instead, confirmation of the plume, the connection between Enceladus and the E ring, and the presence of a global, subsurface ocean would require a dedicated mission to the Saturn system: Cassini-Huygens. [We note that recently a post-processed image taken by Voyager 1 appears to show a faint outline of the Enceladus plume (*Stryk, 2017*). However, as this image processing was not possible until recently, the Cassini-Huygens mission is still credited with discovery of the Enceladus plume.]

The Cassini-Huygens mission completely revolutionized our understanding of the saturnian system, with Enceladus as a prime example. The discovery of Enceladus' plume (*Porco et al., 2006; Dougherty et al., 2006; Hansen et al., 2006; Spahn et al., 2006; Tokar et al., 2006; Waite et al., 2006*) and its source — over 100 jets located along four surface fissures called "tiger stripes" in the south polar terrain (SPT) — provided strong evidence for a subsurface liquid water reservoir (*Schmidt et al., 2008; Postberg et al., 2009, 2011*). This reservoir is not a local sea but a global ocean, as demonstrated by two independent lines of evidence (*Hemingway and Mittal, 2019*). First, interpretation of gravity measurements (*McKinnon, 2015; Čadek et al., 2016; Beuthe et al., 2016*) suggests that the rapid (33-hr) spin of Enceladus affects the depth below which all pressures are hydrostatic (only due to gravity). The simplest explanation for this is a floating ice shell of variable thickness (Airy isostasy) sitting on top of a global ocean. Second, careful analysis of Enceladus images collected over seven years of Cassini observations revealed that Enceladus' rotation has a forced physical libration (wobble), too large to be consistent with a crust grounded to the core (*Thomas et al., 2016*). Again, the simplest explanation is a free-floating crust, which is only possible if there is a global ocean.

The implication of a global ocean as opposed to a regional sea is important for life because unlike a regional sea, a global ocean likely persists long enough for a biosphere to emerge [this took <1 b.y. on Earth (e.g., *Dodd et al., 2017*)]. A regional sea could have been formed by an impact and might only last millions of years (*Roberts and Stickle, 2017*). In contrast, a global ocean requires heating rates that on a small moon like Enceladus can only be generated by the dissipation of tides raised by a large body like Saturn. Tidal dissipation can be caused by orbital eccentricity (i.e., a noncircular orbit) driven by gravitational interactions with other moons in the saturnian system. Via this mechanism, tidal dissipation may take several hundred million years to subside, even once the orbital forcing ceases [the orbital circularization timescale (e.g., *Henning and Huford, 2014; Choblet et al., 2017*)]. Enceladus' orbit is currently forced to be eccentric due to interactions with Dione. The corresponding dissipation was thought to be too small to account for the high heat fluxes determined from Cassini measurements (*Meyer and Wisdom, 2007; Choblet et al., 2017*). However, Enceladus' orbit was recently found to expand at a surprisingly fast rate (*Lainey et al., 2017*), indicating that Saturn's tidal torque on Enceladus is higher than expected. This torque is likely able to transfer energy from Saturn's spin into heat inside Enceladus quickly enough to maintain the current ocean (*Nimmo et al., 2018*). Tidal dissipation could also have been higher in the geologically recent past (*Neveu and Rhoden, 2019*). In any case, with a global subsurface ocean, Enceladus beckons as one of the few places in the solar system where suitable conditions may have existed long enough for life to gain a foothold (Fig. 1). Because material exchange with the inner solar system is unlikely (*Worth et al., 2013*), any life on Enceladus would likely have emerged independently of life on Earth, and therefore would have much to teach us about which features of life are universal.

Enceladus might hold the key for discovering life in our own cosmic backyard, and it has provided a facile means to access it via the plume. With nearly 25 yr of observations of the E ring from the Voyager era (*Haff et al., 1983*), over 10 yr of direct observation of the plume by Cassini, and models that suggest that the plume is modulated by tidal dissipation (*Hedman et al., 2013; Běhounková et al., 2015; Patthoff and Kattenhorn, 2011*), we know that the plume is a long-lived phenomenon lasting at least tens to hundreds of years (section 2.4). This makes Enceladus the only confirmed place in the solar system where we can sample fresh material from the subsurface ocean without the need to dig or drill. [Europa may also have a plume (see *Sparks et al., 2017; Jia et al., 2018; Arnold et al., 2019*; and references therein; see also the chapter by Schmidt in this volume); the plumes of Neptune's moon Triton have not been conclusively linked to a subsurface ocean source (see *Kirk et al., 1995*).]

In this chapter, we address the astrobiological relevance of Enceladus from the outside-in. Section 2 describes the plume composition and structure. We address how this plume is connected to the subsurface ocean in section 3. In section 4, we delve deep into the ocean itself, and its interaction with the core. Section 5 takes these puzzle pieces and constructs a picture of Enceladus in the context of astrobiology. Finally, in section 6 we discuss how this picture informs the search for life at Enceladus: what biosignatures are most relevant, and where we might find them. However, our current picture of Enceladus is not complete; in section 7 we highlight some questions still unanswered for this fascinating moon.

2. PLUME COMPOSITION AND STRUCTURE

The plume of Enceladus is sourced from multiple (~100) jets and curtain eruptions located in surface hot spots along prominent parallel troughs called tiger stripes (*Porco et al., 2006, 2014; Spencer et al., 2006; Spitale et al., 2015*) located in Enceladus' SPT. These emissions are likely fed by coupled freezing and vaporization of liquid water beneath the icy crust (*Schmidt et al., 2008; Nakajima and Ingersoll, 2016; Kite and Rubin, 2016*), producing a plume that extends for thousands of kilometers and subsequently forms the E ring around Saturn (*Spahn et al., 2006; Kempf et al., 2010; Mitchell et al., 2015*).

2.1. Plume Composition

Enceladus' plume was sampled *in situ* by the Cassini spacecraft during multiple fly-throughs over the course of the mission (Fig. 2). Measurements have revealed that the plume is primarily composed of *gas* and *grains*.

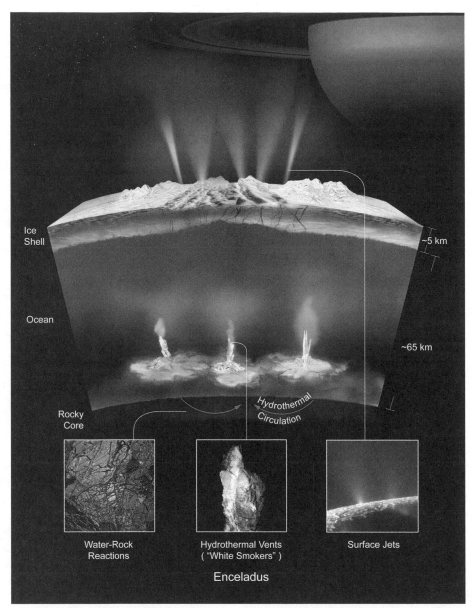

Fig. 1. See Plate 18 for color version. Enceladus' plume is sourced from a subsurface ocean in contact with the rocky core below at temperatures indicating hydrothermal activity. Credit: NASA/JPL-Caltech/SwRI.

The *gas* is mostly comprised of water vapor (Table 1), although importantly, significant amounts of methane and molecular hydrogen (H_2) have also been detected by the Cassini Ion and Neutral Mass Spectrometer (INMS). The methane may be too abundant to be formed solely from the decomposition of clathrate hydrates (methane trapped in water cages), indicating another process is at work (*Bouquet et al.*, 2015). This is supported by the detection of H_2 in the plume, strongly suggesting that the ocean of Enceladus is in contact with, and reacting with, the rocky core through hydrothermal processes (*Waite et al.*, 2017) (sections 3.2 and 4.3). These reactions drive the ocean out of chemical equilibrium, in a similar way to water around Earth's hydrothermal vents, potentially providing a source of chemical energy that life could exploit.

The plume *grains* have been detected and characterized by several instruments onboard Cassini, including the Cos-

mic Dust Analyzer (CDA) (*Spahn et al.*, 2006; *Postberg et al.*, 2009, 2011), the Cassini Plasma Spectrometer (CAPS) (*Jones et al.*, 2009), the Radio and Plasma Wave Science (RPWS) instrument (*Ye et al.*, 2014), and INMS (*Teolis et al.*, 2010, 2017). The closest Cassini fly-through was at an altitude of 25 km, so any grains too large or heavy to be lofted to this altitude were not sampled. The grains have been categorized based on composition (Table 2): water-rich (Type I), organic- or silicate-rich (Type II), and salt-rich (Type III).

Type I grains are almost pure water ice and tend to be small (typically <0.6 μm), implying that they are likely formed by homogeneous nucleation from the gas phase (*Postberg et al.*, 2011); this is supported by models of water vapor flow and condensation rates that are consistent with Cassini data of plume brightness and particle size distribution (*Schmidt et al.*, 2008). Salt-rich grains (Type III) tend to

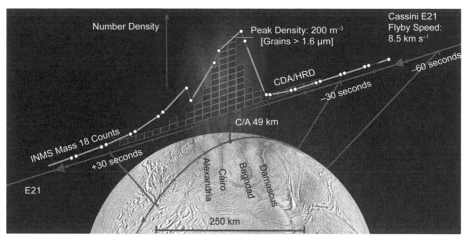

Fig. 2. See Plate 19 for color version. Plume ice grain and neutral gas profiles for the Cassini E21 fly-through of the Enceladus plume measured by the CDA High Rate Detector (HRD) (green boxes) and INMS (orange trace). The neutral gas profile shown is of water vapor (H_2O has a mass of 18 u). Closest approach (C/A) for this fly-through is indicated at 49 km altitude (this was the last and one of the closest plume crossings), with the time-delineated groundtrack outlined on Enceladus' surface in blue. In this image, the south pole of Enceladus is oriented up.

have 0.5–2% sodium and potassium salts by mass (*Postberg et al.*, 2011). These particles are typically larger (>0.6 μm), and are probably formed from flash freezing of salt water from the ocean (*Postberg et al.*, 2009).

Type II particles contain either organic or silicate impurities. Oxygen- and nitrogen-bearing volatile organic species have been identified in both the gas (*Magee and Waite*, 2017) and ice grain phases (*Khawaja et al.*, 2019), which could be synthesized to form more complex organic molecules through hydrothermal processes. Interestingly, a subset of these particles (~12% of Type II grains, ~3% of all E-ring particles) contain organic molecules with masses greater than 200 u (u = unified atomic mass units; as an example, the mass of glycine, the simplest amino acid, is 75 u; see Fig. 3). These organic molecules, called high mass organic cations (HMOC), are found in high abundance (~1% by mass) in these particles, and appear to be unsaturated (containing double and triple carbon-carbon and carbon-nitrogen bonds) and partially aromatic (containing cyclic structures). Significantly, analysis of CDA and INMS spectra collected at different velocities indicate that these species are probably fragments of even larger organic molecules, as the distribution of organics detected shifted to lower masses with increasing impact velocity (*Postberg et al.*, 2018; *Waite et al.*, 2009). However, Cassini's instruments lacked the mass range and resolution to characterize these fascinating organics any further. For more discussion of how such large, nonvolatile molecules might be expelled in the plume, see section 3.

2.2. Fate of the Plume Grains

The motion of grains in the plume is determined by gravity as well as other perturbing forces, whose ratio depends on the grain size. Smaller grains (<1 μm) are dominated by electromagnetic forces and can accumulate electric charges via collisions as the particles pass through the vents (*Jones et al.*, 2009). These grains could be picked up by Saturn's magnetosphere (*Meier et al.*, 2014; *Mitchell et al.*, 2015) and can escape with the plume gas to form the E ring (*Porco et al.*, 2006; *Kempf et al.*, 2010). A significant fraction of the larger grains (≥1 μm) fall back to the surface of Enceladus, covering craters and making the surface of Enceladus the whitest and most reflective in the solar system (*Southworth et al.*, 2019; *Porco et al.*, 2006, 2017; *Hedman et al.*, 2009; *Kempf et al.*, 2010; *Howett et al.*, 2010; *Verbiscer et al.*, 2007; *Schenk et al.*, 2011).

Once in the E ring, these ice grains are exposed to two important space weathering processes: sputtering erosion and plasma drag by the magnetospheric plasma ions (*Jurac et al.*, 2001; *Dikarev*, 1999). In addition to the effects on long-term E-ring dynamics (*Horányi et al.*, 2008), the silica (SiO_2) nanograins detected by the CDA instrument might be released from the E-ring grains by sputtering erosion. The high abundance and specific size distribution (particle radius <10 nm) of these nanograins indicate the pH (8.5–10.5, moderately alkaline) and temperature (≥90°C) at which they formed, most likely via homogeneous nucleation (*Hsu et al.*, 2015). These tiny sputtered nanograins provided the first evidence that hydrothermal reactions are occurring below the ocean floor of Enceladus.

TABLE 1. Major species present
in Enceladus' plume gas.

Constituent	Mixing Ratio (%)
H_2O	96 to 99
CO_2	0.3 to 0.8
CH_4	0.1 to 0.3
NH_3	0.4 to 1.3
H_2	0.4 to 1.4

Reproduced from *Waite et al.* (2017).

TABLE 2. Enceladus plume grain composition (number density) for particles >0.2 μm detected by CDA in the E ring or during plume fly-throughs at altitudes ≥25 km (*Postberg et al.,* 2008, 2009, 2011).

Grain Type	Plume Mixing Ratio (%)	E-Ring Mixing Ratio (%)	Composition
Type I	30	~70	Almost pure water ice (Na/H$_2$O mixing ratio ~10^{-7})
Type II	>30	25	Either organic-rich (hydrocarbons or polar organics) or silicate-rich (~1% by mass Mg-rich, Al-poor mineral inclusions in water ice)
Type III	<40	6*	0.5–2% by mass Na, K salts (mostly NaCl and NaHCO$_3$)

*The salt-rich Type III grains tend to be larger, and so are less likely to escape Enceladus' gravity and reach the E ring.

2.3. Plume Structure

The structure of the plume provides essential information on the interrelationship of the water vapor and ice grains in the vents, connecting to the subsurface ocean. After emerging from the surface, micrometer-sized ice grains soon collisionally decouple from the ascending gas flow. Their motion is mainly shaped under the force of Enceladus' gravity, following approximately ballistic trajectories. Thus, the Cassini measurements of the grain size distribution at various altitudes directly constrain the grains' ejection speed distribution and the source mechanism.

The depletion of larger and/or heavier grains at higher altitudes observed in near-infrared spectra (*Hedman et al.,* 2009) indicates that these grains are less likely to reach higher altitudes because of slower ejection speeds. This behavior is reproduced with a grain condensation model, in which ice grains form at nozzles in the subsurface vents when the vapor pressure reaches saturation, and are then transported upward by the gas flow as they grow through accretion (*Schmidt et al.,* 2008). However, the grains' ascent is hampered by wall collisions. After a collision, larger grains are accelerated more slowly back to the gas speed, because their surface-to-mass ratio is lower. The resulting effect is a size-dependent grain speed distribution when exiting the vents. Thus, the maximum height that a grain could reach depends on its size, which results in an altitude-dependent grain size distribution.

Plume stratification is also observed in the grain composition profile. The fraction of salt-rich ice grains (Type III) sharply increases from a few percent in the E ring to 40% near Enceladus' surface. This suggests that these grains, which are unable to escape into the E ring, are larger, likely flash-frozen water droplets (*Postberg et al.,* 2009, 2011).

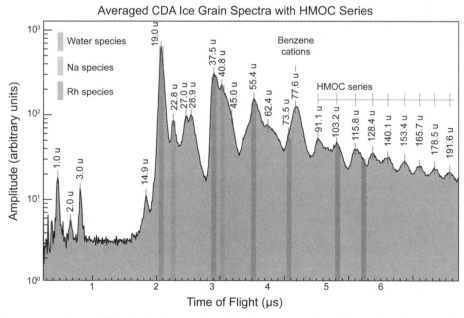

Fig. 3. Averaged mass spectrum of Enceladus ice grains with signatures of organic molecules seen by the Cassini CDA instrument (*Postberg et al.,* 2018). In addition to cations associated with the benzene ring at 77 and 79 u, a series of so-called high mass organic cations (HMOC) cover a mass range from 91 to 200 u, and likely continue further outside the instrument mass range, indicating the presence of large organic molecules in Enceladus' subsurface ocean. Na species are due to salt in the grains; Rh species are due to the rhodium metal target of the CDA instrument.

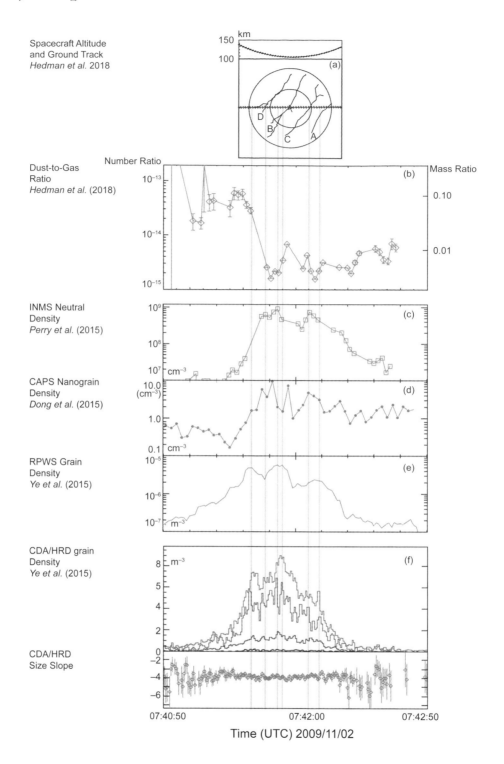

Fig. 4. Plume structure shown by the combined *in situ* gas and grain measurements during the Cassini E7 flyby on November 2, 2009. **(a)** Spacecraft altitude and ground track across the south polar terrain. Letters in gray are the first letter of each named tiger stripe, encountered in the flyby from left to right: Damascus (D), Baghdad (B), Cairo (C), and Alexandria (A). **(b)** The dust-to-gas number ratio and mass ratio as a function of time based on measurements shown in **(c)** and **(e)**. **(c)** Water vapor number density measured by the Cassini INMS instrument. **(d)** Negatively charged nanograin number density measured by the Cassini Plasma Spectrometer instrument. **(e)** Number density of micrometer-sized ice grains measured by the Cassini RPWS instrument. **(f)** Number density and size slope of micrometer-sized ice grains measured by the Cassini CDA instrument. The four different curves represent the number densities of grains with radii larger than 1.7, 2.2, 3.0, and 6.5 μm, respectively. The grain size distribution is consistent with a power-law distribution with a slope of -4. Vertical lines mark the time when the density peaks in either water vapor or micrometer-sized grain profiles. These figures are modified from *Ye et al.* (2014), *Dong et al.* (2015), *Perry et al.* (2015), and *Hedman et al.* (2018).

Stellar and solar occultation measurements (*Hansen et al.*, 2006, 2008, 2011, 2017) indicate that there are multiple narrow, supersonic gas jets embedded within a broad plume gas emission. The jet speed is roughly 1.5× the water vapor thermal speed (~500 m s^{-1} at 273 K) (*Tian et al.*, 2007; *Hansen et al.*, 2008; *Hurley et al.*, 2015), and the jet locations are spatially correlated with the grain density profiles as shown in Fig. 4 (*Hansen et al.*, 2011; *Ye et al.*, 2014; *Dong et al.*, 2015; *Perry et al.*, 2015; *Hedman et al.*, 2018). Two models, based on Cassini images, have been proposed to describe the emission of a mixture of water vapor and ice grains: the "jet model" of >100 discrete jets (*Porco et al.*, 2014), and a "curtain model" where broad, curtain-like eruptions may explain both the apparent jets and broad emission (*Spitale et al.*, 2015). The true plume environment at Enceladus may indeed be a combination of both. The dust emission through fast jets contributes significantly to replenishing the E ring (*Mitchell et al.*, 2015), while the low-speed gas emission, stemming from sublimation (*Goguen et al.*, 2013) or a layer at thermal equilibrium in contact with the fissure walls, also has a nonnegligible contribution (*Perry et al.*, 2015; *Teolis et al.*, 2017).

During the Cassini era, Enceladus' total mass output rate was estimated to be about 200 kg s^{-1}, ranging from 100 to 1000 kg s^{-1} (*Tian et al.*, 2007; *Burger et al.*, 2007; *Smith et al.*, 2010; *Hansen et al.*, 2011; *Dong et al.*, 2015; *Teolis et al.*, 2017). Most of the output mass is gas, with micrometer-sized grains comprising only tens of kilograms per second (*Gao et al.*, 2016; *Southworth et al.*, 2017, 2019; *Porco et al.*, 2017). Because of the previously described dynamical stratification, the local dust-to-gas mass density ratio could be closer to unity at altitudes below 25 km, such that grains may influence the dynamics of the ascending gas (*Hedman et al.*, 2018). The mass flux and dust-to-gas ratio also vary at different tiger stripes (Fig. 4) (*Kempf et al.*, 2008; *Jones et al.*, 2009; *Ye et al.*, 2014; *Dong et al.*, 2015; *Dhingra et al.*, 2017; *Hedman et al.*, 2018), implying different properties in their sources or in the connections to the subsurface reservoirs.

2.4. Variability

The plume varies with Enceladus' orbital phase. It is brightest near orbital apocenter (the furthest point from Saturn in Enceladus' elliptical orbit) by a factor of ~5 (*Hedman et al.*, 2013). This implies that tidal stresses regulate its emission (in addition to supplying energy required for Enceladus' activity; see section 4). Along Enceladus' slightly eccentric orbit, the level of crustal stress from deformation due to tides raised by Saturn varies. This is thought to modulate the width of fissures, and thus the ejection of grains (*Hedman et al.*, 2013; *Nimmo et al.*, 2014; *Perry et al.*, 2015; *Porco et al.*, 2018). Ultraviolet occultation measurements suggest a larger water emission in the jets when Enceladus is near its orbital apocenter. However, such enhancement does not result in a substantially higher plume water vapor density (*Hansen et al.*, 2017). The tidal control of plume activity

seems to be limited at the jets, modulating the emission of gas and dust across Enceladus' orbital phase (*Hedman et al.*, 2013; *Nimmo et al.*, 2014; *Hansen et al.*, 2017).

Temporal variabilities other than the diurnal modulation were also observed, from jets appearing in only a few minutes (*Spitale et al.*, 2017) to a factor of 4 decrease of the plume brightness (dust content) from 2005 to 2015 followed by a subsequent increase (*Ingersoll and Ewald*, 2017; *Porco et al.*, 2017). The mechanisms causing the plume's long-term variability remain unclear, but could be tied to orbital resonances with Dione (*Porco et al.*, 2017).

The plume must have been active since the detection of the OH torus around Saturn (*Shemansky et al.*, 1993; *Jurac et al.*, 2001) and the discovery of the E ring (*Feibelman*, 1967), and likely much earlier based on E-ring grain dynamics (at least a few hundred years) (*Horanyi et al.*, 2008) and the level of crater degradation (*Kirchoff and Schenk*, 2009; *Kempf et al.*, 2010). The central tiger stripe could be at least 0.1 to 1 m.y. old, based on a model in which the additional parallel stripes form because erupted ice grains that fall back onto the surface pile up along the sides of the central stripe, breaking long pieces of ice shell under their weight (*Hemingway et al.*, 2020). On geological timescales, Enceladus' activity may be episodic, as indicated by surface features and geophysical modeling (*O'Neill and Nimmo*, 2010; *Spencer and Nimmo*, 2013; *Crow-Willard and Pappalardo*, 2015). Considering heating from tidal dissipation in Enceladus' porous rocky core, a geophysical model suggests that the observed activity could be supported for tens of millions to billions of years (*Choblet et al.*, 2017). Over 1 b.y., a 200-kg s^{-1} plume would eject about 5% of the mass of Enceladus, with 10% of this mass falling back to the surface (*Schmidt et al.*, 2008).

3. TRACING THE PLUME TO THE OCEAN BELOW

3.1. Long, Narrow Channels with an Underground Source

As discussed in section 2, the plume composition and structure strongly indicate a subsurface ocean source. This is also supported by physical measurements and modeling. Cassini observations reveal that the tiger stripes are emitting gas at both high (up to Mach 10 or 6 km s^{-1}) and low (<1 km s^{-1}) speeds (*Teolis et al.*, 2017). The high-Mach gas is probably accelerated by the pressure gradient along the fissure length or through nozzle-like throats within the fissure. The low-Mach gas is either not accelerated in the fissures or has achieved thermal equilibrium by friction or thermal exchange with the fissure walls. Modeling indicates that high-Mach gas is best achieved by narrow jets with high aspect ratios (*Yeoh et al.*, 2017). Narrow spreading angles argue against sublimation from ice on the surface of Enceladus as the source of the plume, and instead suggest a deep underground source where the vapor has to move through long, narrow cracks before escaping. A turbulent dissipation model also suggests that the ocean-surface connection may be sustained on million-year timescales (*Kite and Rubin*, 2016).

Other aspects of the ocean-plume connection are much less constrained. Equilibrium thermodynamic and fluid mechanical modeling suggests that the vents are drawing from a deep subsurface liquid reservoir through a conduit, similar to explosive volcanic eruptions on Earth (*Mitchell et al.*, 2014). The throat of the vent is likely to be the point of peak pressure (*Mitchell et al.*, 2017). As ocean water and volatiles ascend through the conduit, the drop in temperature and pressure leads to boiling and freezing. These can occur together below the triple point pressure in the ocean water, according to the ratio of latent heat of evaporation to latent heat of fusion (*Mitchell et al.*, 2014). Modeling suggests that this can lead to a solid:vapor ratio similar to that observed in the plume jets [~6:1; note that only ~4% of the estimated 200 kg s^{-1} mass flux comes from the jets (*Porco et al.*, 2014)], but this model neglects exsolution (dissolved gases returning to the gas phase) by volatiles other than CO_2 and heat exchange during expansion. During ascent, gas and grains could also exchange momentum with each other and with the channel walls via sticking and nonsticking collisions. Salt grains could act as condensation nuclei and grow in size. Some particles may shrink in size as volatiles (CO_2, CH_4, etc.) sublimate. Triboelectric charging (accumulation of positive or negative charge due to contact, separation, or friction) could occur for small (nanograin) particles (section 2.2), possibly leading to chemical reactions (called tribochemistry) on their surfaces. Such processes should be taken into account when analyzing plume composition and inferring properties of the ocean. Further modeling efforts (including exsolution and sublimation effects of many volatile species, etc.) and experimental work would improve the understanding of the connection between the ocean and the vents.

Organic enrichment processes may be occurring at the interface between the ocean and the icy crust of Enceladus. The ~3% of E-ring grains bearing complex organic molecules (*Postberg et al.*, 2018) (see section 2) have a high abundance of these organics (about 1% by mass) but a very low abundance of salts, suggesting that the organic molecules are hydrophobic and/or sourced from a salt-poor matrix. In Earth's polar regions, clouds of ice particles enriched in organic material and depleted in salts are generated by sea-spray aerosols (also termed "atmospheric marine aerosol"). Wave breaking leads to the bursting of air bubbles at a sea-surface microlayer of organic material floating on the ocean surface (*Burrows et al.*, 2014). The submicrometer sea-spray aerosol thus generated is enriched 10 to 1000-fold in organic material. A similar process could occur on Enceladus, where aerosols formed at the water-ice interface due to rapid boiling beneath the jets could emerge from a boundary layer of organic material (*Postberg et al.*, 2018; *Porco et al.*, 2017). Such a process could also concentrate hydrophobic biosignatures such as lipids and fatty acids, if present (*Cable et al.*, 2017).

3.2. What the Plume Tells Us About the Ocean

Measurements of the plume can be combined with Cassini gravity measurements and models of Enceladus' internal structure to provide constraints on ocean composition, pH, and temperature. For ocean temperature, the presence of salts such as NaCl — at concentrations constrained by CDA measurements of the flash-frozen, salt-rich grains in the plume (*Postberg et al.*, 2009) — leads to freezing point depression, such that the minimum temperature is likely in the range of –1°C (*Glein et al.*, 2018). While other "antifreeze" agents such as NH_3 are present, they are not likely present in the ocean at sufficient concentrations to depress the temperature further (*Porco et al.*, 2006; *Glein et al.*, 2018).

The pH of the Enceladus ocean has been constrained to ~9–11 (Fig. 5), using a thermodynamic model of carbonate speciation based on three sets of measurements: (1) sodium chloride and carbonate abundances in plume particles measured by CDA, (2) the existence of silica nanograins (*Hsu et al.*, 2015; *Sekine et al.*, 2015), and (3) the amount of CO_2 in the plume gas measured by INMS (*Glein et al.*, 2015, 2018). The simplest model geochemical system consistent with these measurements is a Na-Cl-HCO_3/CO_3 solution with alkaline pH. The pH range of this observation-based study is consistent with the first-principles approach of *Zolotov* (2007), who assumed that Enceladus' ocean composition results from the aqueous alteration of carbonaceous chondrite material, a plausible analog of Enceladus' initial "rock" (nonvolatile) component (section 4.3). Notably, the pH of aqueous systems is temperature-dependent in a chemically closed system, so the hydrothermal fluid pH might be lower than the ocean, possibly in the range of 8–9 (*Sekine et al.*, 2015). However, this might not be the case if the ocean is chemically open to the icy crust through effective volatile exchange; in that instance, the hydrothermal fluids might have a pH close to or higher than the ocean. Similar geochemical modeling efforts with more quartz-rich compositions (quartz-magnesite-talc) constrain the pH to <10 and would be consistent with the observed silica nanograin abundance, while supporting a lower temperature (165°C) in the source region of the nanoparticles (*Glein et al.*, 2018).

Currently, efforts to fully describe the hydrothermal geochemistry of Enceladus are confounded by the discrepancy between the H_2 detection in the plume (meaning reaction with reduced minerals deep in the core, such as iron-bearing silicates) and the presence of silica nanograins (possibly sourced from quartz-rich rocks at shallower depths). No single water-rock interaction can produce both species: H_2 and SiO_2. *Glein et al.* (2018) propose a scenario where hydrothermal fluids react at depth with iron-bearing rocks, picking up dissolved H_2, and then flow up through quartz-rich carbonates to obtain dissolved silica

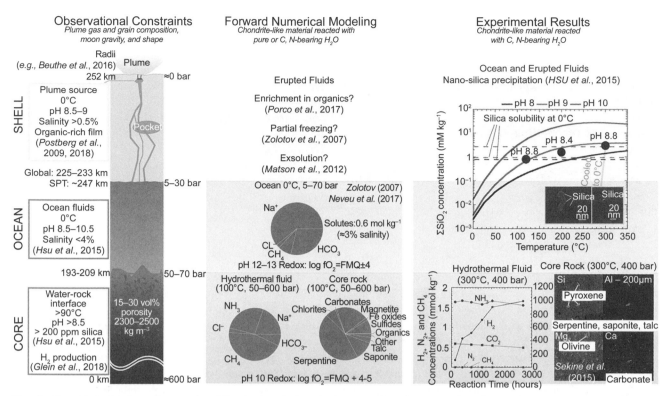

Fig. 5. See Plate 20 for color version. Physical and chemical connections between the plume, the ocean, and the core as evidenced by converging observational (section 3), modeling (section 4), and experimental results. SPT = south polar terrain.

before flowing into the ocean and eventually expressing both species in the plume. Further experimental work is needed, as well as *in situ* measurements at Enceladus, to fully understand the complex geochemistry of this moon.

The high pH of Enceladus' ocean has several implications. First, serpentinization (reaction of water with ultramafic rocks such as olivine and pyroxene, often resulting in the generation of H_2), which results in alkaline fluids, has most likely occurred or is still occurring on Enceladus (section 4). This is supported by the reported detection of H_2 in Enceladus' plume by INMS (*Waite et al.*, 2017). Second, alkaline conditions indicate that hydroxide (OH^-) should be fairly abundant, and (bi)carbonate (HCO_3^-/CO_3^{2-}) or sodium (bi)carbonate ($NaHCO_3/Na_2CO_3$) should be the dominant forms of dissolved inorganic carbon. Third, divalent metals (Ca^{2+}, Mg^{2+}) should be present only at low concentrations (insoluble). Finally, organic acids, if present, will be deprotonated (negatively charged) and should be frozen inside plume particles as sodium or potassium salts, and ocean-derived amines will be neutral and may be detectable in the plume gas if sufficiently volatile. These last two points might drive mission design or the selection of a future instrument suite to Enceladus (section 6).

4. CORE-OCEAN INTERACTION AND EVOLUTION

The potential for Enceladus to harbor life not just today, but also at any time since its accretion, is most likely controlled by the nature and extent of the interaction between liquid water and rock. By "rock," we refer to the nonice solids, which likely comprise mainly silicates, refractory organics, and metals including their sulfides and oxides. Water-rock interfaces provide, at least initially, locales of chemical disequilibrium between relatively oxidized (electron-poor) fluid and relatively reduced (electron-rich) rock. Reactions lead to reduction of fluid volatiles and oxidation of rock minerals and organics. In particular, H_2 results from the reduction of H_2O concomitant with the oxidation of Fe species (*Abrajano et al.*, 1990; *Sherwood Lollar et al.*, 1993; *Sleep et al.*, 2004). An example of a relevant reaction is (*McCollom and Bach*, 2009)

$$30\ Mg_{1.8}Fe_{0.2}SiO_4\ +\ 41\ H_2O\ \rightarrow$$
$$\text{olivine} \qquad\qquad \text{liquid water}$$

$$15\ Mg_3Si_2O_5(OH)_4 + 9\ Mg(OH)_2 + 2\ Fe_3O_4 + 2\ H_2$$
$$\text{serpentine} \qquad\qquad \text{brucite} \qquad \text{magnetite} \quad \text{hydrogen gas}$$

The synthesis of organic compounds of intermediate oxidation states from (bi)carbonate (*McCollom and Seewald*, 2007; *McDermott et al.*, 2015), can be achieved by reactions such as (here, for carbohydrates)

$$x\ HCO_3^- + 2x\ H_2 \rightarrow C_xH_{2x}O_x + x\ OH^- + x\ H_2O$$

The complete reduction to methane ($HCO_3^- + 4\ H_2 \rightarrow CH_4 + OH^- + 2\ H_2O$) proceeds slowly (*Sherwood Lollar et al.*, 2002) unless bio-mediated (*Taubner et al.*, 2018) or catalyzed, typically by metals (*McCollom*, 2016). Thus, core-ocean interaction likely governs the supply of chemical energy and key bioavailable elements, i.e., the habitability of Enceladus. To constrain Enceladus' astrobiological potential, it is therefore essential to understand to what extent the core and ocean interact, how long such interaction has taken place, and what chemical reactions can be predicted or inferred to be taking place.

4.1. Present-Day Setting and Extent of Water-Rock Interaction

Measurements of Enceladus' gravity (*Iess et al.*, 2014), shape, and libration (*Thomas et al.*, 2016) suggest that its interior comprises a rocky core surrounded by a global ocean thicker at the south pole, underlying an ice shell. The global ice shell thickness is estimated at 20 to 25 km (*Beuthe et al.*, 2016; *Čadek et al.*, 2016), but it is much thinner (2–15 km) at the south pole (*Čadek et al.*, 2019; *Patthoff et al.*, 2019; *Hemingway and Mittal*, 2019). The core density, estimated at 2300 to 2500 kg m^{-3} (*McKinnon*, 2015; *Čadek et al.*, 2016), is low compared to the densities of silicate grains: 2900–3000 kg m^{-3} for hydrated, serpentine-rich silicates, with some variation depending on the Fe/Mg ratio (*Tyburczy et al.*, 1991; *Auzende et al.*, 2006) and higher for dry silicates (*Chung*, 1971; *Yomogida and Matsui*, 1983; *Consolmagno and Britt*, 1998). Thus, even if the rock is hydrated silicate, the core also comprises material of lower density, such as 15–30 vol% of water, or >30 vol% of refractory organics with density 1500–1700 kg m^{-3} (*McKinnon et al.*, 2008; *Castillo-Rogez et al.*, 2012). The presence of water in the core is compatible with water-rock interaction as suggested by the chemistry of vented material (*Hsu et al.*, 2015; *Sekine et al.*, 2015; *Waite et al.*, 2017; *Glein et al.*, 2018). Similar core densities are expected for other icy worlds with a shape close to hydrostatic equilibrium (where gravity is balanced by internal pressure) and a radius below 800 km (Table 3), suggesting that waterlogged rocky cores may be the norm for such worlds.

The likely significant water content of Enceladus' core suggests that water-rock interaction is not restricted to the seafloor, but may instead be pervasive throughout the core. This is consistent with the expected low strength of the core because of low central pressures (<500 bar or 50 MPa) on this small, low-gravity world (*Neveu et al.*, 2014; *Klimczak et al.*, 2019). Two classes of models have been suggested for the water-rock interface: a core of cemented rock, but with porosity that either remains from accretion at low gravity or arises from fracturing processes (*Vance et al.*, 2007; *Neveu et al.*, 2014, 2015), or a core in which rock grains can move relative to one another with the water fluid (*Roberts*, 2015; *Choblet et al.*, 2017; *Bland and Travis*, 2017). The models are not mutually exclusive, as the rock fraction could increase with depth into the core to yield cemented, porous central regions surrounded by unconsolidated rock.

Regardless of the exact structure, a waterlogged core helps maintain the ocean because it is amenable to convective fluid circulation that both enhances tidal dissipation (*Choblet et al.*, 2017) and more efficiently transfers heat

out of the core compared to conduction alone (*Neveu and Rhoden*, 2019). As a result, core temperatures may be rather homogeneous with spatial variations not exceeding 100 K (*Chloblet et al.*, 2017; *Neveu and Rhoden*, 2019). Still, ≥90 K gradients must exist between the freezing interface at the bottom of the ice shell (–1°C) and the deep sites able to solubilize the aqueous silica detected as nanograins (>90°C) thought to have precipitated when mixing with colder water (*Hsu et al.*, 2015; *Sekine et al.*, 2015) (Fig. 5 and section 2.2). To sustain such gradients, Enceladus' core must not be too permeable, as higher permeabilities favor more vigorous fluid flows that tend to homogenize temperatures (*Phillips*, 1991; *Neveu et al.*, 2015).

4.2. Core-Ocean Interaction Through Time

The age of Enceladus' ocean is key to its ability to sustain life: too young, and there may not be enough time for life to emerge; too old, and chemical disequilibria (life's energy source) may have dissipated. The odds of us seeing Enceladus during a brief episode of unusually high activity are very low. Yet, how can this tiny world sustain such activity given its limited sources of endogenic (chiefly radioactivity) and exogenic energy (chiefly tidal dissipation) at the present day (*Meyer and Wisdom*, 2007), let alone through geologic time?

4.2.1. The origin of Enceladus. An upper bound on the age of Enceladus' ocean is the age of Enceladus itself. In the past few years, multiple new scenarios for the origin and evolution of Enceladus and Saturn's other mid-sized moons have been proposed. The result of these many plausible theories is that the spectrum of possible evolution scenarios for Enceladus and the other mid-sized moons of Saturn is vast (Fig. 6). Enceladus' ocean may have been interacting with the rocky core since the dawn of the solar system, or we may be witnessing a fluke lasting only decades to centuries.

In the canonical scenario, the moons formed in Saturn's subnebula at the dawn of the solar system 4.5 b.y. ago (*Mosqueira and Estrada*, 2003a,b). However, this might yield several larger (Titan-sized) moons, more akin to the Jupiter system than to what is seen at Saturn (*Canup and Ward*, 2006; *Crida and Charnoz*, 2012). Instead, Enceladus and its sibling moons could have formed from the debris of a Titan twin, disrupted after migrating too close to Saturn and dragged in by the remaining subnebular gas (*Canup*, 2010). Such a large moon would have already undergone ice-rock differentiation, and its core could have fallen into Saturn (*Canup*, 2010), leaving behind icier material to assemble the mid-sized moons we see today. If some metal-rock differentiation happened, metal-poor rock would be left over. Although this scenario reproduces the inner Saturn system, forming moons that are sufficiently rock-rich would require random delivery of rock by external impactors such as large comets (*Salmon and Canup*, 2017), which would provide further chemical mixing.

In yet a later formation scenario, Enceladus and its neighbors formed as ring material spread out (*Charnoz et al.*,

TABLE 3. Core densities of icy worlds estimated from spacecraft
measurements or telescope observations.

Object	Mean Radius (km)	Core Density (kg m^{-3})	Reference
Enceladus*	252	2300–2500	*Hemingway and Mittal* (2019)
Ceres	470	2426–2439	*Ermakov et al.* (2017)
Dione	561	2300–2575	*Beuthe et al.* (2016)
Rhea	763	1240–2350[†]	*Tortora et al.* (2016)
Haumea	798	2550–2750	*Dunham et al.* (2019)
Europa	1561	Unconstrained (2500–3900 near seafloor)	*Schubert et al.* (2009)
Titan	2576	Unconstrained (2500–4500)	*Coyette et al.* (2018)

* Because no gravity measurements have been made for Enceladus' neighboring moons Mimas and Tethys, the size of Mimas' core is unconstrained and Tethys' internal structure is unknown.
[†] Inferred from the reported range of possible core sizes (*Tortora et al., 2016*).

2011). This assumes that Saturn's rings predate the moons, which is debated (*Iess et al.*, 2019; *Crida et al.*, 2019), but in any case there may have been previous rings from disrupted moons. In this scenario, the moons form layered with a rock-rich core accreted first, being more resistant than ice to tidal disruption close to Saturn. Moreover, the closer-in moons (Mimas, Enceladus) are younger than their outer siblings (Rhea), possibly by billions of years.

A final scenario suggests Enceladus is staggeringly young, perhaps only a few tens to hundreds of millions of years old (*Cuk et al.*, 2016). This estimate is based on the absence of dynamical signatures of past moon-moon orbital

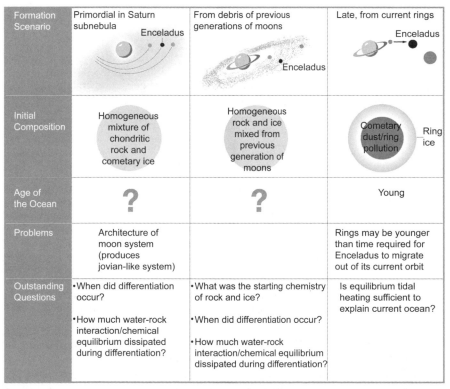

Fig. 6. Possible scenarios for Enceladus' origin and evolution. These control the timing, extent, and starting material composition of water-rock interaction, and therefore the extent of the habitability of Enceladus. These scenarios are not mutually exclusive and may have happened in sequence through time, from left to right, in which case Enceladus would be the product of the latest formation scenario.

resonances, indicating the moons formed later than when the resonances should have occurred, placing upper bounds on the ages of moons such as Enceladus.

4.2.2. Sustaining the ocean. Keeping water liquid inside Enceladus requires heat. Enceladus' main heat source likely arises from the dissipation of tides raised by Saturn (e.g., *Spencer and Nimmo,* 2013). Explanations for observed high levels of dissipation have been sought both in the way Enceladus' interior material responds to tides (*Tyler,* 2014; *Roberts,* 2015; *Choblet et al.,* 2017) and in variations of the orbital tidal forcing (*Travis and Schubert,* 2015; *Cuk et al.,* 2016). The main challenge in modeling Enceladus' internal evolution through time is that tidal dissipation couples geophysical and orbital processes, which act on vastly different timescales. Enceladus orbits Saturn with a period of just 32.9 hours (1.37 Earth days), and accurately simulating gravitational perturbations to its orbit from Saturn, the rings, and other moons requires computing interactions between each of these bodies and Enceladus hundreds of times per orbit (*Zhang and Nimmo,* 2009). This is not practical to simulate over the >4.5 b.y. of solar system history. Thus, in detailed dynamical simulations, the interior structure is usually simplified to a point mass with tidal dissipation and deformation averaged over the entire moon (e.g., *Meyer and Wisdom,* 2007; *Cuk et al.,* 2016). In detailed geophysical simulations, orbital properties are at least simplified (*Neveu and Rhoden,* 2019) and usually fixed (*Choblet et al.,* 2017), varied arbitrarily (*Travis and Schubert,* 2015), or ignored (*Schubert et al.,* 2007).

Another challenge is that essential quantities for Enceladus remain largely unknown. A dynamical example is the internal structure and propensity for tidal dissipation of Saturn, which determines how Enceladus' orbit and tidal forcing change over time (*Fuller et al.,* 2016; *Lainey et al.,* 2017; *Nimmo et al.,* 2018). A geophysical example is the estimation of mechanical properties of ice (*McCarthy and Cooper,* 2016; *Renaud and Henning,* 2018) and rock-ice mixtures (*Choblet et al.,* 2017) at relevant temperatures, stresses, and strain rates based on experiments (*Durham et al.,* 1992, 2010), necessarily limited in parameter space and timescale.

4.2.3. Observational clues. Perhaps a meaningful upper bound on the ocean age can be inferred from the fact that chemical reactions seem to still be happening (*Hsu et al.,* 2015; *Waite et al.,* 2017); chemical equilibrium has yet to be achieved. At the temperatures of 50° to 100°C required to keep silica soluble in order to then precipitate silica nanograins in colder fluids (*Hsu et al.,* 2015; *Sekine et al.,* 2015), chemical reactions should proceed very quickly relative to the age of the solar system. On Earth, water-rock reactions that complete the abiotic reduction of carbon all the way to methane seem to proceed on ~1-b.y. timescales below 100°C (*Etiope and Sherwood Lollar,* 2013). In this case, the path to equilibrium may be limited by how fast unreacted rock is exposed to ocean fluids. The distance traveled by a fluid front diffusing through rock increases with the square root of time (*McDonald and Fyfe,* 1985).

Equivalently, the timescale of alteration can be understood as the time taken by a fluid front to travel halfway between two adjacent water-rock interfaces (cracks or pores, or edges of rock chunks). Experiments suggest that the front travels about 1 m in about 10^6 yr, or 1 mm (the size of larger grains in chondrites) in a few years (*McDonald and Fyfe,* 1985). Thus, Enceladus' interior may not reach equilibrium readily if some of its core is cemented, shielding part of the rock from alteration. If adjacent fluid paths are no closer than ~20 m, some rock would remain unreacted over the age of the solar system. Comparatively, reactions do not seem limited by the availability of unreacted fluid, unless circulation through the core is much slower than fluid transport of silica and H_2 tracers of water-rock reactions, which are transported through the ocean and shell and vented to space within months (*Hsu et al.,* 2015; *Waite et al.,* 2017).

No matter when the current ocean was emplaced, Enceladus seems to have had episodic strong internal heating [>150 mW m^{-2} (*Bland et al.,* 2012)], akin to today's [~400 mW m^{-2} (*Howett et al.,* 2011)]. Evidence for episodic high heat flow is recorded in surface craters that are smoothed out (relaxed), yet relatively old because they are overprinted by younger craters and fractures (*Bland et al.,* 2012) (Fig. 7). The relative chronology is the following: (1) Enceladus' surface is impacted, leaving behind sharp craters; (2) Enceladus experiences high internal heating and the warm ice softens, smoothing the crater morphology; (3) the surface cools down enough to prevent the healing of subsequent craters and fractures; (4) steps 2 and 3 are repeated to the present day such that several sets of relaxed craters and fractures overlap. Unfortunately, it is not possible to tell how much time this sequence of events took. If Enceladus' craters were caused by debris in the Saturn system, they may have occurred only a few 10–100 m.y. ago, but if the impactors orbited the Sun (as those that generated craters on the terrestrial planets, the Moon, main-belt asteroids, and Pluto), the craters would be billions of years old, and so would Enceladus (*Kirchoff et al.,* 2018).

4.3. Chemistry of Water-Rock Interaction

Unlike the age of the ocean, the chemistry of water-rock interaction is relatively well constrained. Chemical reactions between liquid water and rock arise from chemical disequilibria at water-rock interfaces at the physical conditions (temperature, pressure, water-to-rock ratio) of the setting. Fundamental considerations (the rocky material, if analogous to meteoritic, cometary, or interplanetary dust matter, is much more reducing than water or fluids that would result from melting cometary ices), analogies with settings of water-rock interaction on Earth, and measurements of the Enceladus plume and stream particle compounds attributed to products of water-rock reactions (*Hsu et al.,* 2015; *Waite et al.,* 2017) all suggest that oxidizing (electron-poor) fluids are interacting with reduced (electron-rich) rock. The resulting oxidation-reduction (redox) disequilibrium could provide a long-term source of chemical energy that, on

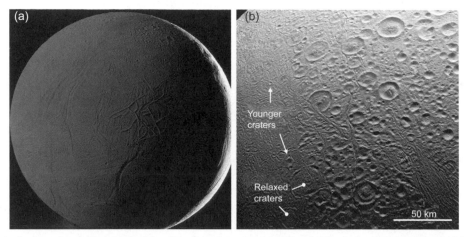

Fig. 7. (a) Enceladus' surface, illuminated by the Sun (crescent to the right) and sunlight reflected by Saturn, shows craters and fractures that overprint each other. Part of the south polar plume is visible at the bottom on the Saturn-lit side. **(b)** Zoom on a region near the north pole, showing how smooth craters overprinted by fractures and younger craters suggest past episodes of high heat flow (*Bland et al.,* 2012). Credit: NASA/JPL-Caltech/Space Science Institute/G. Ugarković.

Earth, can sustain ecosystems deprived of sunlight (e.g., *Schrenk et al.*, 2004).

The oxidized nature of Enceladus' fluids is suggested by the presence of of CO_2 in the plume (*Waite et al.*, 2006, 2009, 2017) and the compositions of cometary volatiles (*Bockelee-Morvan et al.*, 2004; *Mumma and Charnley,* 2011), which presumably represent the species accreted by icy moons (*Mousis and Alibert,* 2006). Additionally, the near surfaces of all airless bodies are subjected to photon and particle radiation, which can split H_2O to form reduced H_2 and oxidants such as H_2O_2 and O_2 (*Parkinson et al.*, 2008). H_2 escapes owing to its low molecular mass, leaving behind an oxidant-enriched shell. Although Enceladus' ice shell seems too thin to allow convective transport of near-surface oxidants to the ocean (*Barr and McKinnon,* 2007), and the shell would resist any convective motion driving subduction (*Howell and Pappalardo,* 2019), oxidants could be buried under accumulating plume fall-out, perhaps all the way down the few kilometers to the ocean (*Čadek et al.*, 2019; *Patthoff et al.*, 2019). Finally, oxidants can be produced by radiolysis of water via energetic particles resulting from the decay of potassium, thorium, and uranium radioisotopes in the core (*Bouquet et al.,* 2017). Potassium is also soluble in the ocean, as evidenced by the detection of K^+ in the plume (*Postberg et al.*, 2009), and could lead to oxidants via radiolysis in the ocean itself.

Enceladus' rocky core is likely to be much more reduced. Although the composition of the rock accreted by icy moons is unknown, a first guess is to extrapolate from recovered infall of primitive extraterrestrial material on Earth such as chondritic and/or carbonaceous (micro)meteorites and interplanetary dust particles. Chondritic material seems to have formed somewhat concurrently with planetesimals and giant planets in the first few million years of solar system history (e.g., *Desch et al.*, 2018, and references therein), and therefore might have been accreted by primordial icy moons, with several caveats. First, the composition of rock may

not have been homogeneous across the early solar system regions where chondritic parent bodies and giant planets formed, even though there could have been mixing due to planetary migration (e.g., *Walsh et al.*, 2011) and entrainment by nebular gas (*Johansen and Lambrechts,* 2017). Second, the giant planet subnebulae may have segregated material, potentially resulting in a different composition from material in the surrounding heliocentric (Sun-orbiting) protoplanetary disk (*Alibert et al.*, 2005). The proportions of subnebular and heliocentric material inside Enceladus are unknown because of the uncertainty on Enceladus' origin (section 4.2). Finally, the material recovered on Earth is likely biased not just by provenance region in the solar system, but also toward the materials able to survive ejection from the planetary body, Earth entry, and impact, and therefore against volatile-rich material (*Rivkin et al.*, 2014). Rock compositions in the Saturn system have also been informed by *in situ* analyses of ring particles made during the Grand Finale phase of the Cassini mission. These suggest that organic-rich ring particles are much more abundant than expected from earlier remote sensing (*Waite et al.*, 2018; *Hsu et al.*, 2018).

Another handle on possible fluid and rock compositions comes from numerical simulations of the interaction between fluid and rock initially at disequilibrium. Simulations of chemical equilibrium provide the end-member compositions toward which the system is driven, a necessary first step to understanding the fluid-rock system. These calculations, which seek to minimize the Gibbs free energy (amount of energy that can do work at given temperature and pressure), leverage measurements and extrapolations of thermodynamic properties for hundreds of species (e.g., *Johnson et al.,* 1992, and references therein), and their accuracy hinges on the comprehensiveness of these data (*Oelkers et al.*, 2009). Although most major species are covered, data for hydrated minerals are more uncertain (*Oelkers et al.*, 2009), especially for species that are rarer on Earth but potentially common on Enceladus such as ammonium-bearing clays that could sequester Enceladus' nitrogen (*Neveu et al.*, 2017). Whether equilibrium is achieved depends on kinetics along the path of reaction, which in turn depend on factors such as temperature, the extent of disequilibrium, concentrations of species such as H^+ (pH), mineral surface properties, or the action of catalysts (e.g., *Pokrovsky and Schott*, 1999; *Palandri and Kharaka*, 2004). Kinetic data are available for fewer species, and usually do not capture dependencies other than temperature. Consequently, forward modeling efforts of water-rock interactions inside Enceladus to date have involved equilibrium simulations and only qualitative discussion of how kinetic limitations might affect the results.

The equilibrium products of reactions between chondritic rock and water were determined numerically by *Zolotov* (2007) at Enceladus conditions inferred from plume analyses. In that study, the hydrogen fugacity (effective partial pressure), a measure of the redox state, was set to match the CO_2/CH_4 molar ratio of about 2 measured in the plume (Table 1). The system was assumed open with respect to H_2, to simulate H_2 removal by escape to space. This assumption, plausible given Enceladus' low gravity (0.11 m s^{-2} at the surface, about 1% of Earth's), leads to oxidized equilibrium fluid and rock. Predicted predominant minerals were, in order of volumetric abundance, saponite clay (e.g., $Mg_{0.17}$ $_5Fe_3Al_{0.35}Si_{3.65}O_{10}(OH)_2$), serpentine $(Fe,Mg)_3Si_2O_5(OH)_4$, iron oxides (goethite FeOOH or magnetite Fe_3O_4), iron sulfides (pyrrhotite FeS and/or pyrite FeS_2), chlorite clay $(Fe,Mg)_5Al_2Si_3O_{10}(OH)_8$, calcite $CaCO_3$, and nickel sulfide. Minor phosphates, chromite, and carbonates $(Ca,Mg,Fe,Mn)CO_3$ also result from simulations. Dominant solutes were Na^+, Cl^-, and (bi)carbonate, followed by K^+ and HS^-, in agreement with the findings of Na- and K-bearing salt grains in the plume (*Postberg et al.*, 2011), and indicating that the system was not so oxidized as to yield sulfate. This solution composition was not sensitive to water-to-rock mass ratios between 0.3 and 10 at 100°C. The fluid pH was alkaline, decreasing with temperature from 11 to 8, in agreement with the independent estimates of *Glein et al.* (2015, 2018) described in section 3.

These rock and fluid compositions are qualitatively similar to those determined by other studies (*Schulte and Shock*, 2004; *Neveu et al.*, 2017). Unlike the models of *Zolotov* (2007), these studies did not prescribe the redox evolution from plume data, but somewhat equivalently started from a carbonaceous chondrite mineralogy thought to have resulted from partial aqueous alteration and H_2 escape on the parent bodies of these meteorites (when starting from a less-altered mineralogy, much H_2 is produced). The agreement among models, albeit based in part on the same thermodynamic data, and with experimental results (*Sekine et al.*, 2015) and the limited plume measurements available, provides some confidence in our current understanding of Enceladus' interior chemistry (Fig. 6).

Nonetheless, even without new data, several improvements could be made to models or experiments seeking to reproduce the chemistry of Enceladus. First, in many models, the trapping of gases in clathrate hydrates (water cages) was ignored. However, conditions at Enceladus' seafloor are expected to be in the stability fields of several clathrates (e.g., *Mousis et al.*, 2015), which tend to be more stable at higher pressures and lower temperatures. Sequestration in clathrates would delay or prevent gas escape, affecting the types and amounts of species dissolved in the ocean waters and expressed in the plume.

Second, models and experiments could consider other possibilities for the starting ice (fluid) and rock compositions if Enceladus is not primordial. In the formation scenario of *Canup* (2010) described in section 4.2, Enceladus' rock would be less rich in iron (less mafic) and reducing, perhaps more akin to what is expected at the seafloor of Europa [which likely has experienced metal-rock differentiation (*Zolotov and Kargel*, 2009; *Schubert et al.*, 2009)]. If Saturn's inner moons formed from ring material (*Canup*, 2010; *Charnoz et al.*, 2011), *in situ* ring analyses indicate that more organic-rich compositions may be warranted (*Waite et al.*, 2018; *Hsu et al.*, 2018).

Finally, it is crucial to better model kinetic limitations in reaching equilibrium, because kinetics likely influence the fate of bioessential elements and their bioavailability. For example, reduction of fluid CO_2 to ultimately yield methane (CH_4) does not easily proceed to completion (*McCollom and Seewald*, 2001; *Sherwood Lollar et al.*, 2002; *Holland et al.*, 2013; *McCollom*, 2016), but rather stalls at metastable organic compounds of intermediate oxidation states (*Shock and McKinnon*, 1993, and references therein), promoting organic synthesis (*McCollom and Seewald*, 2007). If CH_4 formation is inhibited, forward models produce organic-rich fluids. Dominant C solute compounds are ethanol (CH_3CH_2OH) and acetate (CH_3COO^-), with less-abundant CH_3COOH, $NaCH_3COO$, NH_4CH_3COO, etc. (*Shock and McKinnon*, 1993; *Schulte and Shock*, 2004; *Neveu et al.*, 2017). Aqueous (bi)carbonate, aqueous CO_2, methanol CH_3OH, and formate $HCOO^-$ are slightly less abundant. There is much less H_2 production, as the reducing power is captured in the organics. Detection of diverse soluble organic compounds in the plume (*Khawaja et al.*, 2019) may support these processes.

In contrast and as evidenced by the detection of NH_3 and potentially H_2S in the plume (*Waite et al.*, 2009), N and S species may be less affected by kinetics (*Neveu et al.*, 2017). This may not seem intuitive. Indeed, reduction of N_2 to NH_3 involves breaking the strong N-N triple covalent bond, which is slow in the absence of catalysts (e.g., *Rodriguez et al.*, 2011; *Anderson et al.*, 2013). Sulfate can persist for months at high temperature in reducing conditions (*Mottl and Holland*, 1978; *Seyfried and Mottl*, 1982) and likely orders of magnitude longer in cold fluids (*Ohmoto and Lasaga*, 1982). However, much N and S may already be reduced initially, as evidenced by observations of NH_3 and H_2S in cometary gases [*Mumma and Charnley* (2011), but see *Rubin et al.* (2015) for a detection of oxidized N_2] and N-bearing organics and sulfide minerals in chondrites (*McKinnon and Zolensky*, 2003; *Alexander et al.*, 2017). Thus, the presence of aqueous NH_3 and NH_4^+ and sulfide minerals should not be impeded by kinetics. However, the release of N and S from organics (≤20% of the initial N and S pool in carbonaceous chondrites) could be inhibited (*Oh et al.*, 1988; *Maffei et al.*, 2012, 2013).

In summary, chemical disequilibria likely arose at the time of accretion. They could be replenished to a limited extent by burial of oxidants to the subsurface ocean and radiolysis (*Bouquet et al.*, 2017). The dissipation of disequilibria is likely limited by physical processes able to expose unreacted reduced rock to oxidizing fluids, and by chemical kinetics for some reactions such as those involving carbon oxidation/reduction. The dissipation of disequilibria causes alkaline conditions and promotes the synthesis of organics, which could provide available building blocks for prebiotic or biological synthesis.

5. ASTROBIOLOGICAL IMPLICATIONS

Observational and modeling constraints on conditions within Enceladus' ocean, and the physical and chemical processes that govern them, help to inform our understanding of the life-hosting potential of Enceladus — not only habitability but, in the case of an inhabited world, *detectability*.

A complete picture of biological potential should consider how processes and conditions would affect not only extant cellular life, but also its origin and evolution. The latter are discussed only briefly; our limited understanding of the origin of life and its environmental context, even on Earth, prohibit a more detailed discussion.

5.1. Origin of Life on Enceladus

Two prominent theories have different implications for the potential emergence of life on Enceladus. One theory posits that a three-phase (ocean-atmosphere-mineral) system favors, or may be essential for, the origin of life — in part because wetting-drying cycles can serve to concentrate organics, promote membrane self-assembly, and drive the formation of amino acid and RNA polymers (*Deamer and Georgiou*, 2015; *Deamer and Damer*, 2017). If this theory is correct, the origin of life on worlds that lack an atmosphere-water and atmosphere-mineral interface (e.g., Enceladus and other ocean-bearing icy moons) would be either disfavored or disallowed, depending on the stringency of the three-phase requirement.

Conversely, the "alkaline vent hypothesis" posits that life can and did originate in two-phase (ocean-mineral) systems (*Lane and Martin*, 2012; *Russell et al.*, 2014). In this scenario, alkaline, H_2-rich fluids produced by serpentinization yield conditions favorable for the origin of life as they emerge into an ocean that is more acidic; the resulting pH gradient provides an energy source (modern life utilizes intracellular pH gradients to store and release energy), along with reducing power and mineral catalysts capable of reducing CO_2 to organics that might be used by life. If this theory is correct, life could emerge on ice-covered ocean worlds that host the appropriate style of hydrothermal venting. As described in section 2, the presence of silica nanoparticles in Enceladus

plume materials is interpreted as evidence of hydrothermal venting (*Hsu et al.*, 2015), and alkaline and H_2-rich conditions in the ocean (*Glein et al.*, 2015; *Waite et al.*, 2017) would be consistent with alkaline venting specifically. An Earth example of an alkaline vent is the ultramafic "white smoker" hydrothermal system Lost City on the Mid-Atlantic Ridge (*Kelley et al.*, 2005), which is metabolically diverse; microbial cycling of sulfur and methane are the dominant active biogeochemical processes (*Brazelton et al.*, 2006).

Beyond data that are consistent with the presence of alkaline vents, it is challenging to bring observational data to bear on the potential for emergence of life on Enceladus. The alkaline conditions inferred for the present ocean (sections 3 and 4.3) could confound the ability to produce hydrolysable polymers such as those used by Earthly life for both catalysis (proteins) and information processing (RNA and DNA), particularly at the higher end of the suggested range of pH 9–11 (*Glein et al.*, 2018). Some organisms on Earth can tolerate pH as high as 13 [i.e., alkalithermophilic sulfate-reducing bacteria in Lake Calumet (*Slonczewski et al.*, 2009; *Roadcap et al.*, 2006)], but do so by virtue of cellular membranes and energy-expending mechanisms that make it possible to maintain intracellular pH at close to neutral levels and almost always <9 (*Slonczewski et al.*, 2009). In the absence of such mechanisms, exposure to alkaline pH would considerably diminish the stability of hydrolyzable polymers, or indeed the ability to form them in the first place. For example, free solution rates of RNA hydrolysis are about 5600× higher at pH 11 than at pH 7 (*Li and Breaker*, 1999). However, modern ocean conditions need not, and probably do not, represent the prevailing conditions over the lifetime of the ocean. A better understanding of the potential implications of alkaline pH for the origin of life will depend on the ability to constrain the possible history of ocean pH from its earliest stages.

5.2. Cellular Life on Enceladus

Relative to the requirements and limitations expressed by extant life on Earth (see the chapter by Hoehler et al. in this volume), our understanding of Enceladus from Cassini data portrays Enceladus as a habitable world. Earthly life requires liquid water, a source of light or chemical energy, a supply of the major biogenic elements (carbon, hydrogen, oxygen, nitrogen, phosphorus, and sulfur) and various metals, and physicochemical conditions that support a range of biological molecules, structures, and processes. Cassini data bear directly on the availability of liquid water, chemical energy, and C, H, N, and O, constrain salinity, and support inference of ocean pH. The moon hosts an extensive liquid water ocean. Sodium chloride and silica nanoparticles sourced in the plume materials are both indicative of significant interaction of the ocean with Enceladus' rocky mantle (section 4), which could serve to supply both energy and materials to the liquid water environment. The simultaneous presence of H_2 and CO_2 represents a chemical energy source that is utilized by microorganisms on Earth. Carbon is abundant in both oxidized and reduced forms, and nitrogen is abundant in a form — ammonia — that is readily oxidized by biology on Earth. While the physicochemical environment remains to be fully characterized, constraints on pH (*Glein et al.*, 2015, 2018; *Waite et al.*, 2017), temperature, and salinity (*Postberg et al.*, 2011) place those values within the biologically tolerated range (*Rothschild and Mancinelli*, 2001).

These data, of course, represent a qualitative "snapshot" of habitability on Enceladus that is applicable to extant, Earth-like life at the present moment: At least some organisms known on Earth would likely survive in the conditions that apparently now prevail on Enceladus. But habitability is an evolving, emergent property of the physical and chemical processes at work. These processes govern not only conditions, but *fluxes* of materials and energy that ultimately determine how large, diverse, and productive any biosphere on Enceladus could be. Although a truly quantitative picture of Enceladus' biological potential must await a more detailed understanding of the processes at work there, some quantitative constraints are suggested by observations and modeling work.

The size and productivity of any biosphere on Enceladus would likely not be limited by the availability of carbon, nitrogen, or sulfur. Average cell abundance in Earth's oceans is on the order of 10^6 cells cm^{-3} (*Whitman et al.*, 1998). Given the observed C, N, S, and P composition of typical aquatic cells (*Fagerbakke et al.*, 1996), the aqueous concentration of those elements that is equivalent to 10^6 cells cm^{-3} can be calculated and compared to their observed or inferred abundance in the plume or ocean of Enceladus (Table 4).

Reported abundances of CO_2 and NH_3 in the plume based on Cassini INMS data (*Waite et al.*, 2017) exceed biological requirements by 5–6 orders of magnitude. H_2S was reported in the plume at abundances of 12–34 ppm [vol:vol H_2O (*Waite et al.*, 2009)], about 5 orders of magnitude higher than the biological requirement, but the H_2S detection is considered ambiguous (*Magee and Waite*, 2017) because a range of other environmentally plausible compounds have nominally the same mass at the mass resolution of the INMS. H_2S abundance in the ocean has also been predicted via equilibrium geochemistry modeling (*Zolotov*, 2007), with a range, dependent on the temperature at which water-rock interaction occurs, of approximately 5×10^{-5} to 5×10^{-9} on a mol per mol H_2O basis. The low end of the range, corresponding to water-rock reaction temperature close to 273 K, still exceeds the biological requirement by more than an order of magnitude.

Possible phosphorus limitation of a biosphere on Enceladus has been suggested (*Lingham and Loeb*, 2018). The issue is not one of overall P abundance. Given the P content typical of aquatic cells on Earth [0.46–1.4 fg P cell^{-1} (*Fagerbakke et al.*, 1996)] and an Enceladus ocean volume of 2.7×10^{16} m^3 (*Steel et al.*, 2017), a global ocean biosphere averaging 10^6 cells cm^{-3} could be constructed from 3.5×10^{13} g P. Ordinary chondrites contain 0.8–2 g P kg^{-1} (*Mason*, 1979). Adopting the low (most conservative) end of the range, and assuming a silicate core density of 2400 kg m^{-3} (*Iess et al.*, 2014) and a core radius of 182 km (*Čadek et al.*, 2016), the needed P would be supplied by the upper 43 *mm* of the silicate core, whereas the observed

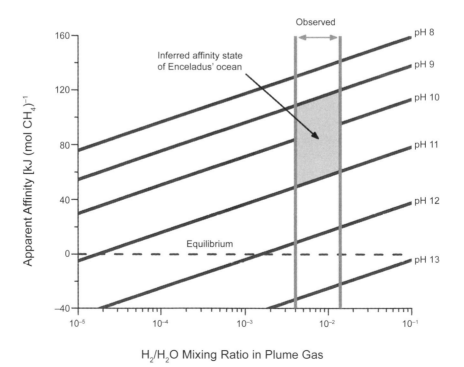

Fig. 8. Apparent affinity (Gibbs energy change) for methanogenesis in the ocean of Enceladus, constrained by ocean pH and the abundance of H_2. The shaded box lies above the dashed line for chemical equilibrium, meaning the ocean is in disequilibrium and therefore could support life. Reproduced from *Waite et al.* (2017).

TABLE 4. Comparison of elemental requirements for biology vs. abundances observed in the plume, or predicted via geochemical equilibrium modeling.

	C[¶]	N	P	S
Mass in a typical aquatic cell (fg)[*]	7–31	1.6–5	0.46–1.4	0.3–0.56
Required abundance for 10^6 cells cm^{-3} (mol per mol H$_2$O)[†]	1.1–4.7×10^{-8}	2.1–6.5×10^{-9}	2.7–8.2×10^{-10}	1.7–3.2×10^{-10}
Plume observation (vol per vol H$_2$O)[‡]	3–8.3×10^{-3}	4–13×10^{-3}	—	
Predicted ocean abundance (mol per mol H$_2$O)[§]			$<2 \times 10^{-12}$	5×10^{-5} to 5×10^{-9}

[*] *Fagerbakke et al.* (1996).
[†] Seawater substrate concentration, expressed as moles of solute per mole of H$_2$O, required to supply the elemental requirements of 10^6 cells cm^{-3}.
[‡] Elemental abundance expressed as volume of species per volume of H$_2$O, as calculated from mixing ratios of CO$_2$ and NH$_3$ (*Waite et al.,* 2017) in the volatile component of the Enceladus plume. Volume:volume units are directly comparable to mol:mol units if plume volatiles behave as ideal gases. The calculation neglects the contribution of plume ice particles to total water efflux (estimated at ~4%) (*Porco et al.,* 2017).
[§] Based on abundances plotted in Figs. 2 and 3 of *Zolotov* (2007).
[¶] As total inorganic carbon, the sum of CO$_2$, HCO$_3^-$, and CO$_3^{2-}$.

NaCl content implies much more extensive dissolution of rock. However, P as phosphate — the form of P in the biomolecules of life on Earth — partitions dominantly into the silicate mineral phase, with calculations suggesting that equilibrium phosphate concentrations should be in the range of 10^{-11}–10^{-10} mol kg^{-1} H$_2$O (10^{-13}–10^{-12} mol mol^{-1} H$_2$O) at water-rock reaction temperatures of 150°–300°C, and lower still at lower temperatures (*Zolotov,* 2007). These concentrations would be insufficient to supply the P required for an average cell density of 10^6 cells cm^{-3}, but could supply a 100-fold lower cell density. Importantly, low-equilibrium phosphate concentrations would not necessarily constitute a limit on biomass density, provided that ongoing water-rock processing can supply the ocean with a continuing, albeit small, flux of P that could be scavenged for the construction of biomass. However, it may stand as an important consideration for the origin of life, if the prebiotic synthesis of phosphate containing informational and energy-carrying molecules requires high concentrations of phosphate. This is an important area for progress in understanding the habitability of Enceladus.

In contrast to (non-P) bioessential element availability, energy availability may significantly constrain the possible size and productivity of a biosphere on Enceladus. Life on Earth is only known to utilize redox chemistry and near-infrared to visible light as sources of Gibbs (free) energy; other forms of energy are not known to be captured by life. Lacking direct access to solar radiation, any life within the Enceladus ocean would depend exclusively on whatever redox chemical energy the moon itself supplies.

Before considering how Enceladus may supply chemical energy for life, it is instructive to briefly review the two aspects of energy availability that impact life on Earth. First, chemical energy sources must represent a Gibbs energy change (ΔG) that makes their reaction not only chemically spontaneous ($\Delta G < 0$) but also suitable for driving the endergonic (energy-requiring) synthesis of the biological energy carrier, adenosine triphosphate (*Schink,* 1997; *Hoehler,* 2004). If the Gibbs energy requirement is met, metabolism using that energy source is possible. Theoretical and observational work place this requirement in the range of -10 to -20 kJ mol^{-1} (*Schink,* 1997; *Hoehler et al.,* 2001), and this provides a quantitative basis for evaluating potential redox energy sources for life. Biology is also heavily influenced by the *flux* of energy (how much energy through time), as life requires energy not only to construct new biomass, but also to maintain it through time (*Tijhuis et al.,* 1993; *Hoehler and Jørgensen,* 2013). The *biosynthetic* (biomass-building) energy requirement is quite well quantified in terran biology, varying within an order of magnitude as an understood function of metabolic and biochemical specifics (e.g., *McCollom and Amend,* 2005; *Thauer et al.,* 2008). This offers a means of equating energy fluxes with the amount of new biological material (e.g., cells, biosignatures, etc.) that can be created through time, even on global scales. The energy required to maintain existing biomass ("maintenance energy") is well understood on theoretical grounds, but attempts to quantify it yield values that range over several orders of magnitude [10^{-19}–10^{-13} kJ cell^{-1} d^{-1} (*Hoehler and Jørgensen,* 2013, and references therein)]. Comparing maintenance energy requirements to planetary energy flux (see below) constrains the size of a biosphere that could be supported on a given world.

Compositional data can bear directly on what redox couples may exist and whether they offer sufficient Gibbs energy change to satisfy biological requirements. In particular, INMS measurements of the plume composition have been used to constrain the abundance of all products and reactants in the known terran metabolism methanogenesis, i.e., based on the reaction CO$_2$ + 4H$_2$ → CH$_4$ + 2H$_2$O. Using these data, *Waite et al.* (2017) calculated the Gibbs energy change for methanogenesis, ΔG_{mp}, as a function of pH (Fig. 8). The calculations suggest that H$_2$-based methanogenesis would occur with ΔG_{mp} –120 to –45 kJ mol^{-1} in the pH range of 9–11, levels that are both thermodynamically favorable ($\Delta G_{mp} < 0$) and bioenergetically favorable ($\Delta G_{mp} < -10$ to –20 kJ mol^{-1}).

It is more difficult to constrain energy *flux* on the basis of compositional data. However, this important parameter determines how large and productive, and thus how potentially detectable, a biosphere could be. Estimates of energy flux have generally considered the flux of the reductant, hydrogen (denoted here as J_{H_2}), either from serpentinization (*Vance et al.*, 2016; *Steel et al.*, 2017; *Taubner et al.*, 2018) or radiolysis (*Bouquet et al.*, 2017) within the rocky core, into an ocean containing an excess of the oxidant, CO_2. The energy potentially available by virtue of this flux is then $(J_{H_2}/4) \times \Delta G_{mp}$. The factor 4 accounts for the 4:1 stoichiometry of H_2:CH_4 in the methanogenesis reaction, given that ΔG_{mp} is typically expressed per mole of methane.

The observed H_2:H_2O mixing ratio in the plume volatiles indicates that H_2 is being lost from the moon itself at a rate of 1–5 × 10^9 mol yr^{-1} (*Waite et al.*, 2017), but this need not represent the present flux of H_2 into the ocean from the core. Instead, it serves to indicate that the ocean contains abundant H_2 — suggesting a substantive time-integrated J_{H_2} — and, combined with constraints on pH, supports the hypothesis that serpentinization has served, or is presently serving, as a source of H_2. Efforts to constrain J_{H_2} thus depend on modeling the rate of serpentinization of the silicate core of Enceladus, with several approaches compared in Table 5:

1. *Vance et al.* (2016) calculated the J_{H_2} that would result from complete serpentinization of the silicate core at a continuous rate over the full age of the solar system. This estimate can be considered an upper-limit average if Enceladus' ocean has an age comparable to the age of the solar system, because it assumes full reaction of the silicate crust and also that H_2 production per unit of silicate rock is maximal in stoichiometric terms, whereas thermodynamic modeling indicates that H_2 yield per unit rock during ser-

pentinization can vary more than 10-fold as a function of reaction temperature (*McCollom and Bach*, 2009). However, full reaction over the course of a shorter lifetime for Enceladus' ocean would yield correspondingly higher J_{H_2}.

2. *Steel et al.* (2017) used (a) a heat flow model to estimate rates of hydrothermal fluid venting, with assumed vent fluid H_2 concentrations corresponding to those of serpentinizing systems on Earth, and (b) a model that assumes H_2 production is limited by the rate of exposure of fresh mineral surfaces to water, and determines that exposure rate as a function of the propagation of fractures through the crust as it cools (*Vance et al.*, 2016). J_{H_2} calculated via the fracture model was within the range of J_{H_2} determined via the heat flow model.

3. *Taubner et al.* (2018) modified the fracture-based modeling approach to account for potentially slow dissolution of minerals at newly produced fracture surfaces, which could serve to limit H_2 production to lower rates.

These approaches yield estimates of J_{H_2} that vary over 5 orders of magnitude (Table 5). For completeness, it should be noted that ongoing radioactive decay within the silicate crust can generate both H_2 and oxidants (e.g., H_2O_2). *Bouquet et al.* (2017) calculate that this process could augment J_{H_2} from serpentinization [relative to, e.g., the results of *Steel et al.* (2017)] by as much as 30%, with results highly dependent on assumed grain size and porosity within the core.

Taubner et al. (2018) used J_{H_2} to estimate global rates of biomass production and *Steel et al.* (2017) estimated both global biomass production and bulk ocean average cell abundance. Biomass production estimates differ to an even larger extent than do the corresponding estimates of J_{H_2}, owing to differing approaches in the energy-biomass conversion. Notably, despite this large range, even the highest cell densities predicted for the Enceladus ocean are about 1000-fold lower than those typical of Earth's ocean [10^6 cells cm^{-3} (*Watson et al.*, 1977)], as would be expected for a more energy-constrained system.

TABLE 5. Estimates of energy flux and biological potential in Enceladus' interior.

	J_{H2} (mol s^{-1})	J_{Energy} (W)	Biosynthesis (kg yr^{-1})	Cell abundance (cells cm^{-3})§
Vance et al. (2016)*	96†			
		1.1 × 10^6–2.9 × 10^6	9.9 × 10^5–5.4 × 10^6	0.6–890
Steel et al. (2017)*	0.6–34		4 × 10^4–2 × 10^6	80–4250
		6.1 × 10^3–1.9 × 10^6	6.1 × 10^3–1.9 × 10^6	0.004–340
Taubner et al. (2018)*	0.0009–0.013†		12.6–162‡	
		10.1–390	9.2–738	6 × 10^{-6}–0.12

* Values in upper panel (gray) were reported in the indicated paper. Values in the lower panel are recalculated based on reported J_{H_2}, and assumed values as follows: ΔG_{mp} = –120 to –45 kJ (mol CH_4)$^{-1}$ (*Waite et al.*, 2017); methanogen biomass yield = 1.3–7.2 (g dry biomass)·(mol CH_4)$^{-1}$ (*Thauer et al.*, 2008); maintenance energy = 1.2 × 10^{-19} to 6 × 10^{-17} W cell^{-1} (*Tijhuis et al.*, 1993; *Hoehler and Jørgensen*, 2013); average cell mass = 2 × 10^{-14} g (*Fagerbakke et al.*, 1996).
† Converted from values reported in units of mol yr^{-1}.
‡ Converted from values reported in units of C nmol g^{-1} L^{-1} yr^{-1}.
§ Calculated as a bulk ocean average, assuming an ocean volume of 2.7 × 10^{16} m^{-3} (*Steel et al.*, 2017).

To compare the implications of the three models for biospheric size and productivity on a common basis, we used the values of J_{H_2} reported in each study to (re)calculate global energy flux, global biomass production rate, and standing cell abundance (Table 5). For each, the values presented reflect the compound uncertainty in J_{H_2}, ΔG_{mp}, biosynthesis yield for methanogenesis, and cellular maintenance energy. The results of Table 5 demonstrate the presently large uncertainty in estimating biological potential for Enceladus. This is true even within a single study, and particularly regarding the potential of a system to support a standing biosphere. For example, *Steel et al.* (2017) estimated bulk ocean cell abundances of 80–4250 cells cm^{-3} by assuming an average cell lifetime of 1000 yr but no energetic cost for cell maintenance. The recalculated values, which include maintenance energy costs as observed in terran organisms, are 1–3 orders of magnitude lower and span an overall range of nearly 5 orders of magnitude. Better constrained estimates of both planetary energy flux and biomass-energy relations would help in narrowing the presently large range of possibility, as required to inform the strategies and measurement requirements of a search for evidence of life. This is an important area for advancement in further observations of Enceladus.

5.3. Unused Energy?

Exclusive of uncertainties in energy flux, constraints on plume composition indicate that energy is available in a known terran metabolism, methanogenesis, at levels that would satisfy terran Gibbs energy requirements (*Waite et al.*, 2018). Does the presence of such unused energy imply that no life is present to use it? This is a possibility, but the alternative cannot be ruled out.

"Unused energy" in plume chemistry could be consistent with an inhabited world if the plume sample fluids that have a residence time (the length of time the fluids are in the plume) that is short in comparison to the timescales associated with microbial growth and substrate utilization. The bulk ocean is estimated to have a residence time of about 200 m.y. with respect to recycling through warm crust and a mixing rate of kiloyears (*Steel et al.*, 2017), both much longer than the timescales associated with microbial growth. However, it has been suggested that buoyant vent fluid could be transported to the base of the ice shell in relatively coherent columns on a timescale of months (*Steel et al.*, 2017), which might be too short to allow for microbes to colonize and fully deplete available energy sources. If plume materials directly sampled such fluid columns, they could contain unused energy even on an inhabited world. This would also imply that plume measurements are reflective of vent fluid rather than bulk ocean chemistry.

Alternatively, physical or chemical factors might inhibit the origin or current presence of life, despite the presence of energy that could be used by biology. *Taubner et al.* (2018) addressed the sensitivity of methanogen growth and activity to a range of conditions and potentially inhibitory

substances that are associated with the Enceladus ocean. Those experiments demonstrated activity of at least some methanogens across much of the range of imposed conditions. Importantly, however, the pH of these experiments decreased due to the presence of CO_2 at high pressure to a range of pH 5–7, whereas Enceladus' ocean is thought to have an alkaline pH (*Glein et al.*, 2015; *Waite et al.*, 2017). Alkaline pH at the high end of the currently predicted range of 9–11 (*Glein et al.*, 2018) could prove inhibitory to biological growth or activity, as methanogen growth has thus far not been observed above pH 10.2 (*Taubner et al.*, 2015) and likewise could frustrate the formation of hydrolyzable polmers in a prebiotic stage of chemistry. High pH also decreases the solubilities of transition metals (Fe, Ni) that are typically utilized to catalyze key biochemical reactions in cells (i.e., in enzymes). However, on Earth this decreased solubility leads to precipitation of chimneys (i.e., Lost City) on the seafloor that concentrate minerals and focus redox/chemical gradients, potentially acting as incubators for biochemistry leading to life (*Barge and White*, 2017, and references therein). Further experimental and theoretical work is needed to fully understand the implications of pH and other factors on the energy budget of Enceladus.

5.4. Summary of Enceladus' Astrobiological Potential

Relative to the requirements and tolerances of modern terran life, Enceladus appears to be habitable. Qualitatively, it provides solvent, a supply of the biogenic elements, redox chemical energy, and physicochemical conditions in a range tolerated by at least some terran organisms. The impact of a potentially alkaline ocean pH on both the origin and modern-day activity of life remains as a key factor to understand. Quantitatively, any potential biosphere on Enceladus seems more likely to be limited by availability of energy than of the biogenic elements, with the caveat that the possibility for phosphorus limitation remains to be extensively considered. Present estimates of energy flux range over orders of magnitude, which affects the detectability of any biosphere, and how best to search for it. Whether or not Enceladus is, in fact, inhabited remains to be determined by future missions.

6. LIFE DETECTION ON ENCELADUS: BIOSIGNATURES AND MEASUREMENT REQUIREMENTS

Enceladus, which appears to meet current requirements for habitability, may hold the answer to one of the most profound questions: Does life exist elsewhere? Its ocean material can be accessed through the plume, circumventing the technological and planetary protection (see the chapter by Rummel in this volume) challenges of drilling through the icy crust.

Despite the ready sample access, a definitive answer to the life question will undoubtedly pose a significant challenge. Careful design of experiments, built upon the knowledge from the Cassini mission, may get us closer to eliminating abiotic hypotheses for the presence of potential biosignatures (*Neveu et al.*, 2018). A key step in this process is identifying biosignatures appropriate for this particular environment and where they might be found.

6.1. Biosignatures: Progress and Potential Pitfalls

Since the first "life" experiments on the Viking landers sent to Mars (see the chapter by Davila et al. in this volume), scientists and engineers have worked to design the most straightforward experiments to search for life while being as agnostic as possible to what that life might be. A few relevant biosignatures are discussed below, along with unique considerations for their detection in the Enceladus environment. While not a comprehensive list, the biosignatures described here illustrate the diverse approaches to life detection, as well as their potential weaknesses.

6.1.1. Amino acids. Amino acids, the building blocks of proteins, may serve as an excellent biosignature, given a few caveats. Amino acids can be generated both biotically and abiotically. They are commonly found in meteorites and are most likely present on Enceladus either due to primordial accretion (i.e., they were present in the carbonaceous chondrites from which Enceladus formed) and/or due to geochemical synthesis via serpentinization (*Truong et al.*, 2019). Biotic and abiotic amino acids can be distinguished in two ways: (1) enantiomeric excess (defined below) and (2) relative abundances compared to glycine.

Nearly all amino acids (except glycine) are chiral molecules: They and their mirror images are not superimposable (*Berg et al.*, 2012a). This means each amino acid has two forms called enantiomers, referred to as "L" or "D" based on the direction each rotates in plane-polarized light ("L" for levorotatory or counterclockwise, "D" for dextrorotatory or clockwise). In biochemistry, substituting a "D" for an "L" amino acid could cause a protein to fold differently, or prevent binding of an enzyme to a substrate. Thus, presumably to keep biological reactions simple, biology exhibits homochirality: All life as we know it uses only "L" amino acids (the question of why "L" and not "D" is debated). Importantly, most abiotic reactions do not exhibit chiral selectivity: Abiotic amino acids are an equal (50/50) mixture of "L" and "D" enantiomers. This is referred to as a racemic mixture, and any process that causes erasure of enantiomeric excess (a greater percentage of the "L" or "D" form relative to the other) is termed racemization.

Thus, an excess of either enantiomer would provide evidence against an abiotic source (*Summons et al.*, 2008, and references therein), and an excess of "D" amino acids would also be strong evidence that this life had evolved independently of life on Earth (*McKay*, 2004). However, accurate detection of enantiomeric excess can be confounded by an abundance of abiotic amino acids and racemization over time. Racemization in the Enceladus ocean may take place in just 10^4–10^7 yr (*Steel et al.*, 2017), so even biological activity that is recent in geological terms may not leave a detectable signal. Alternatively, if indeed an enantiomeric excess is detected at Enceladus, it would be strong evidence of a biosignature for recent life (within the last 10 m.y.).

Evidence from enantiomeric excess might be complemented by amino acid patterns. Abiotic generation of amino acids produces distributions consistent with thermodynamics: The simplest amino acids (glycine, alanine) are much more abundant than those that are larger or more complex. This is seen in meteorites, where glycine is typically more abundant than other amino acids by ~2 orders of magnitude (*Cronin and Pizzarello*, 1983; *Higgs and Pudriz*, 2009). In contrast, life requires more complex amino acids to induce protein folding motifs and create binding sites for metals and other substrates, essentially to form and operate the "molecular machinery" of life. Thus, environments influenced by biology tend to have amino acid patterns where serine, tyrosine, and other relatively complex amino acids are as abundant as glycine (*Davila and McKay*, 2014; *McKay*, 2004; *Glavin et al.*, 2010; *Creamer et al.*, 2017; *Cable et al.*, 2016).

However, with comparisons of amino acid abundances or mole fractions, abiotic production and destruction rates must be considered. Abiotic synthesis of the aromatic amino acid tryptophan was recently confirmed in clay-forming hydrothermal alteration of oceanic rocks (Fe-rich saponite) at depth beneath the Mid-Atlantic Ridge (*Menez et al.*, 2018), possibly ruling out this amino acid as a biosignature without a deeper understanding of the rates of abiotic production. *Truong et al.* (2019) calculated that the amino acids aspartic acid and threonine are particularly sensitive to decomposition in the Enceladus ocean, or the ocean of any hydrothermally active ocean world. While this implies that the ratio of these two amino acids to glycine might not have utility over longer timescales, it does provide another potential biosignature: Any aspartic acid or threonine detected in Enceladus' ocean must have been produced recently (within 1 m.y.) and cannot be primordial. Based on the model of *Truong et al.* (2019), any amino acids found at concentrations greater than 1 nM in the Enceladus ocean would indicate recent production via geochemical or biotic pathways; such concentrations could not be achieved with replenishment by cometary impact.

Simple amino acids probably exist as zwitterions (a molecule having separate positively and negatively charged groups, and thus a net neutral charge) or singly charged ions in the Enceladus ocean, depending on the equilibrium acid dissociation constant (pK_a) of their side-chain [e.g., the side chain of arginine is charged even at pH 10 (*Fitch et al.*, 2015)]. Those that have charge will probably bind to cations like Na^+; due to this and their relatively low volatilities, amino acids are most likely to be found in the grains of the Enceladus plume as opposed to the gas (*Glein et al.*, 2015) (section 3). This would also be true for proteins or protein fragments (polypeptides).

6.1.2. Lipids. A similar diagnostic test of life can be found in another set of molecules found in all known forms of life: lipids. These molecules typically have long, nonpolar, hydrophobic carbon chains that self-assemble into micelles and lipid bilayers, creating cell membranes. The carbon distribution in lipids used by life follows patterns. In bacteria and eukaryotes, lipid synthesis involves building chains two carbons at a time, yielding an excess of carbon chains of even length. In archaea, lipid chains are synthesized five carbons at a time (*Berg et al.*, 2012b). This is in stark contrast to abiotic synthesis of carbon chains (via Fischer-Tropsch-type reactions), which typically produces a standard distribution (Anderson-Schulz-Flory) of chain lengths with no addition pattern (*Dorn et al.*, 2011; *Fortsch et al.*, 2015). Thus, the detection of a pattern of repeating subunits in carbon chains of lipids may be a universal biosignature (*Summons et al.*, 2008; *Georgiou and Deamer*, 2014; *Parnell et al.*, 2007; *Cable et al.*, 2016). Furthermore, methanogens produce distinct ether lipids and isoprenoid hydrocarbons that could supplement the lipid pattern biosignature on Enceladus (*Taubner et al.*, 2018).

As with the amino acid biosignature, a thorough understanding of lipid degradation in the Enceladus environment is needed. Fortunately, degradation processes on Earth have been demonstrated to preserve the pattern (*Moldowan et al.*, 1985). As in the case of the amino acids, lipids and other long-chain carbon molecules are likely to be present in the plume grains, as they are not volatile enough to reach the gas phase. The detection of high-mass organic cations in the plume grains (*Postberg et al.*, 2018) may support this assessment.

6.1.3. Small molecules indicative of active biochemistry. Due to the alkalinity of the Enceladus ocean, hydroxide ions (OH^-) are likely to be abundant, opening up the potential for base-catalyzed reactions to influence the chemistry of organics in the ocean (*Glein et al.*, 2015). Molecules such as urea, nitriles, and acetyl thioesters, which hydrolyze rapidly under basic conditions, could serve as indicators of active biochemistry: If they are detected in such an environment, they must be produced at the present day. As with the biosignatures discussed above, evidence of these species is likely to be found frozen in the plume grains if they carry a charge, although neutral volatile reaction intermediates or products, such as amines, may be detectable in the plume gas.

6.1.4. Large molecules. Large, information-carrying macromolecules such as proteins, RNA, and DNA can serve as informative biosignatures. However, these are very Earth-life-centric. A more agnostic biosignature was proposed by *Benner* (2017): the presence of polymers with repeating charges in their backbones (the connecting chain in a polymer, also known as the main chain). These repeating charges force interstrand interactions in the polymer, and make it more likely to adopt an extended shape suitable for templating (neutral polymers tend to fold) to enable copying/replication of information. Repeating charges also ensure that the entire polymer will have similar physical properties, dominated by these charge interactions, even when the informational content of that polymer changes. Such large molecules, if they exist on Enceladus, could be present in the plume grains, either the salt-rich Type III or organic-rich

Type II grains (section 2), depending on whether they are free-floating molecules or entrained within hydrophobic cells or cell debris (next section).

6.1.5. Cells and cell assemblages. The chemical energy available in the ocean of Enceladus may be able to support cell densities up to ~10^3 cells cm^{-3}. However, this value is highly model-dependent (see Table 5) and represents a bulk ocean average; cell densities could be higher in certain regions (i.e., near vents). Cells could be ejected in the plume (*Porco et al.*, 2017), frozen within grains of sufficient size (most unicellular organisms on Earth are 0.2 μm or larger) (*Europa Lander Mission Concept Team,* 2016, and and references therein). The detection of cells within a plume grain depends on many factors, such as concentration mechanisms (sections 2 and 6.1.7), how the grains are sampled [capture above ~2 km s^{-1} can destroy cells, but encasing cells within ice provides protection (*Aksyonov and Williams,* 2001; *Burchell et al.*, 2014; *Porco et al.*, 2017)], and where the grains are sampled (large grains do not reach high altitudes; section 2).

More complex assemblages of microorganisms, such as biofilms or microbial mats, might also occur in areas where energy or nutrients are concentrated, such as at hydrothermal vents or the ice-ocean interface (*Russell et al.*, 2017). In the latter case, assuming they are preserved in the ice shell, evidence of life may be detectable *in situ* with microscopic or spectroscopic techniques (*Gleeson et al.*, 2012; *Storrie-Lombardi and Sattler,* 2009) or remotely via high-resolution imagery or key absorption features in the visible range if the deposit is substantial [water-ice tends to mask microbial absorptions in the near-infrared (*Poch et al.*, 2017)]. Sublimation may concentrate less-volatile biosignatures and improve detectability, in particular in Sun-exposed surface materials.

6.1.6. Biosignature concentration and preservation. Enceladus may host some unique conditions that could enhance the concentration and/or residence time of biosignatures. Organic material may become concentrated at the ocean-ice interface, similar to the formation of sea-spray aerosols on Earth, which can provide up to 1000-fold organic enrichment compared to the bulk ocean concentration (section 2). Moreover, any volatile exsolution leading to bubbles rising through the water column could scavenge organic material, including microorganisms, which would tend to attach to the bubble surface (*Porco et al.*, 2017). This process is most efficient with hydrophobic molecules and surfactants (which have a hydrophilic head and a hydrophobic tail), and has been shown to increase organic and microbial concentrations by orders of magnitude on Earth (*Carlucci and Williams,* 1965; *Blanchard and Syzdek,* 1972). However, this may not occur if turbulent mixing is occurring in the fractures. It is not clear whether these two enrichment processes are complementary or mutually exclusive, but if they occur on Enceladus, the potential for detecting organic biosignatures in the plume is much higher than may be expected based on estimates of bulk concentrations in the ocean alone.

The alkaline pH of the ocean, the nondetection of molecular oxygen by the Cassini Ultra Violet Imaging Spectrograph (UVIS) (*Hansen et al.*, 2005), and the lack of abundant sulfate in the plume (*Glein et al.*, 2015) all suggest that delivery of strong oxidants such as O_2 from the surface to the ocean of Enceladus has not been significant. In Earth's oceans, in the absence of O_2, organic matter is degraded much more slowly (e.g., *Cowie et al.*, 1995). Anoxia in the Enceladus ocean may therefore improve the preservation of organic material, including any biosignatures. However, *Glein et al.* (2015) note that the absence of sulfate could also be explained by robust oxidant delivery and rapid sulfate reduction by life, if enough organisms and reductants (e.g., H_2 from serpentinization) are present. A future mission to Enceladus may help resolve these questions.

6.2. Measurement Requirements

Cassini's instruments were designed to only interrogate simple molecules, because more complex compounds were unexpected in the Saturn system when Cassini was designed. The next missions to reach the Saturn system will undoubtedly have more advanced instrumentation. These instruments will likely be able to identify complex organic molecules, and hopefully distinguish biotic from abiotic compounds.

A path to life detection on an ocean world such as Enceladus is still in development (Fig. 9). *Sherwood* (2016) proposed three basic steps: (1) validation of habitability, (2) detection of biosignatures, and (3) confirmation of life. Arguably, the first step has been achieved at Enceladus with the Cassini mission. This mission has also helped constrain measurement requirements to achieve the second.

Many Enceladus mission concepts have been proposed at various levels of maturity (*Sherwood*, 2016). Architectures include a plume fly-through/orbiter, lander, sample return, fracture explorer/climber, and submersible. Measurement requirements for these mission concepts vary

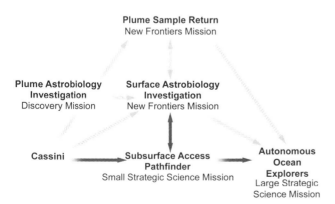

Fig. 9. Possible paths of investigation to search for life on Enceladus. Some paths may involve different mission classes (i.e., "New Frontiers" or "Strategic Science" missions in the NASA denomination). Dashed arrows indicate orbital/flyby/surface missions; solid arrows require subsurface access. Modified from *Sherwood* (2016).

based on which environment is targeted: the plume, the surface, or the ocean.

6.2.1. In situ plume sampling. Cassini mission operations demonstrated that plume gas and grains could be successfully sampled *in situ* at speeds of 8–18 km s^{-1} (these speeds are of the spacecraft relative to Enceladus; the plume velocity in the direction of the fly-through is negligible in comparison). A future plume fly-through mission concept could build on this work, using instruments sensitive to the energy of hypervelocity impacts [4–6 km s^{-1} has been reported as the optimal velocity range for detection of biomolecules such as amino acids, fatty acids, and peptides; at these speeds, biomolecules entrained in ice grains are ionized upon impact with minimal fragmentation, allowing analysis via mass spectrometry (*Jaramillo-Botero et al.*, 2012; *Klenner et al.*, 2020)]. Hypervelocity capture of cells is also possible (section 6.1.6), although more laboratory work and modeling would help characterize the relationship between capture speed and cell/ice ratio in preserving cell integrity.

An Enceladus orbiter could achieve fly-through speeds <1 km s^{-1}, enabling soft capture of grains for analysis. However, planetary protection requirements for the disposal of such a spacecraft are critical, as a significant amount of fuel would be needed to leave Enceladus orbit and impact a body of less astrobiological potential.

Given the half-angle (the angle from normal/perpendicular to the surface that defines the plume spread as it leaves the source) of 15° assumed for most plume models (*Hansen et al.*, 2008, *Kempf et al.*, 2010) and the distance between the tiger stripes, the material from the jets and curtains diffuses into a single plume at an altitude of ~45 km. Any future mission targeting the Enceladus plume would therefore have a high probability of collecting plume material in a flyby over the south polar terrain at this altitude, irrespective of whether the ground track is directly over an identified discrete jet.

A sample return architecture is also possible; analyses of returned samples are complementary to *in situ* measurements. Returned samples lack field environmental context and can undergo changes during sampling, caching, the return cruise, reentry, recovery, handling, and storage. However, sample return analyses are uniquely suited to addressing the need for complex, path-dependent analyses (*Tsou et al.*, 2012; *McKay et al.*, 2014). They therefore provide the ability to adapt to unexpected findings, which have been the hallmark of past searches for biosignatures (e.g., *Klein*, 1978; *McKay et al.*, 1996). Moreover, returned samples can be analyzed with the entire set of analytical techniques available on Earth, some of which cannot be flown, and none of which can be flown at a level of performance rivaling that achievable on Earth. Finally, samples can be archived for future analysis using techniques that we may not have conceived of yet.

Biosignature concentrations in the plume can be estimated based on (1) their predicted concentration in the ocean [i.e., 0.01–90 μmol (kg H_2O)$^{-1}$ for amino acids (*Steel et al.*, 2017)], (2) possible concentration before/during ejection (bubble scrubbing, enrichment via sea spray aerosol from a boundary layer, sublimation, etc.), and (3) models of plume gas and grain density over the fly-through sampling trajectory. Evidence suggests that the concentration of organic material in the plume is above detection limits of modern instruments ready or soon-to-be ready for flight: CDA onboard Cassini, an instrument built more than 20 yr ago, detected organic macromolecular compounds at concentrations well above its limit of detection, but did not have the mass resolution to identify their specific elemental compositions (*Postberg et al.*, 2018). For trace species, large collection apertures, cryotrapping, and/or multiple plume fly-throughs may be employed to obtain sufficient sample to meet instrument performance requirements. Other architectures such as a "plume hoverer" could also improve plume grain collection efficiency.

6.2.2. Sampling the surface. A lander to Enceladus, while much more complex than an orbital or fly-through architecture, offers unique advantages. A landed mission to the tiger stripes could access the largest grains emitted from the plume, since 68–93% of Enceladus plume particles fall back to the surface, including the largest grains more likely to contain whole cells (*Porco et al.*, 2017). A lander could also collect significantly greater sample volume than a plume fly-through mission could on a single pass.

However, significant obstacles exist for a lander. Planetary protection requirements must be considered, especially if the landed mission is powered by radioisotope thermoelectric generators. The probability of inadvertently melting through the ice, or any other mechanism by which microbial hitchhikers from Earth could reach the ocean, must be reduced to below 10^{-4} (*National Research Council*, 2012). In addition, a lander would be more costly, requiring a larger-class mission (*Sherwood*, 2016). Recent science and technology developments such as the Europa Lander mission concept (*Europa Lander Mission Concept Team*, 2016) and other programs, including the Concepts for Ocean worlds Life Detection Technology (COLDTech) and the Instrument Concepts for Exploration of Europa 2 (ICEE-2), could be leveraged to make such a mission possible in the coming years or decades. In addition, complementary efforts to these NASA programs are being undertaken worldwide.

6.2.3. Reaching the ocean. While not explicitly needed to answer the life question, absolute confirmation of microbial communities thriving near hydrothermal vents at the seafloor of Enceladus (or elsewhere in the ocean) would require direct access to the ocean. To achieve this, significant technological advancements are needed, ranging from melt probes or descent vehicles capable of navigating through the plume vents, to communication strategies to reach through ≥1 km of ice, to submersibles capable of autonomous navigation and with the appropriate scientific payload (*Sherwood*, 2016). Planetary protection requirements for such a mission would be appropriately more strict than for landers or plume flyby architectures. Much work is underway, including

funded programs in the U.S. and internationally, such as the Scientific Exploration of Subsurface Access Mechanism for Europa (SESAME) program, that would pave the way for exploration of many worlds in addition to Enceladus such as Europa, Titan, Ganymede, or perhaps even Triton.

7. OUTSTANDING QUESTIONS AND FURTHER READING

7.1. Outstanding Questions

Enceladus is one of the best places to look for life beyond Earth. Recognizing evidence of life in this moon, or interpreting the lack of life, will require answers to several outstanding questions. The list below covers three key areas of focus: the longevity of habitable conditions inside Enceladus, the modifications that material from possible subsurface habitats may undergo between its provenance and sampling locations, and the broader question of what evidence is sufficient to claim life detection.

How old is Enceladus? Much headway has been made in tackling this question in the past few years (*Canup,* 2010; *Charnoz et al.*, 2011; *Asphaug and Reufer,* 2013; *Cuk,* 2016; *Lainey et al.*, 2017). These studies have considerably broadened the fan of hypotheses for the origin of Enceladus and Saturn's other mid-sized moons. These new ideas on how moon systems might form and evolve, while exciting, show just how little is known about the history of these systems.

Can Enceladus' starting materials be better constrained? What ocean and core compositions result from starting compositions compatible with formation scenarios? Do they match measurements of surface and ejected material?

How does tidal heating work on Enceladus? How is it partitioned between the core, ocean, and ice shell? Can a better understanding of Enceladus inform these processes on other ocean worlds?

How long has the ocean existed? Accurate estimates by forward modeling require coupling detailed geophysical and orbital models that must be used to run simulations with about 10^{14} time steps (a hundred times a day over 4.5 b.y.). This is currently unachievable in any reasonable amount of time. Are there simplified modeling or observational means of constraining the age of the ocean?

To what extent have water and rock interacted? Is the core solid rock, loose sand and/or mud, or both? How extensive is the water-rock interface?

How close to chemical equilibrium is Enceladus? In principle, detailed forward modeling or experiments seeking to mimic mass spectrometric data collected by Cassini instruments could place tight constraints on the chemistry of Enceladus' water-rock interactions. One confounding factor is the potential for chemical fractionations as material is erupted and vented out, limiting for now robust compositional interpretations to refractory solids (*Postberg et al.*, 2011; *Hsu et al.*, 2015; *Sekine et al.*, 2015). In particular, detailed investigation of how exsolution and eruptive processes might affect H_2 abundances in the plume would be very informative, given its diagnostic power on the extent, progress, and speed of water-rock reactions and its use as a metabolic substrate by terran life. Studies of surface oxidant production and delivery to the ocean would constrain the oxidant supply.

How long has the plume been active? We know that the plume is at least decades old due to the presence of the E ring, but how much older is it? If it is episodic, what are its primary drivers?

How much is material modified or concentrated as it moves from the ocean to the plume? What are the characteristics of the conduits linking the ocean to the surface (e.g., aspect ratio, branching, chambers, wall roughness, lifetime), and how do these affect fluid ascent? Are the relative abundances of the chemical species seen in the plume equivalent to what is in the ocean, or have these ratios been modified en route? Is organic enrichment taking place? How much chemistry is happening in the vents?

How long has Enceladus been habitable? How much time has life had to arise on Enceladus? Is that time sufficient? If not, how far has prebiotic chemistry progressed? Will conditions persist long enough for life to emerge in the future? If life has just emerged on Enceladus, but has not yet become fully established (reached biological saturation based on nutrient availability), would any biosignatures still be above detection limits?

What would constitute convincing evidence of life on Enceladus? Would two independent biosignatures be enough? Would the scientific community be more easily convinced than the general public? Would seeing a motile cell or a squid be needed to settle any debate?

In the end, either discovering life on Enceladus or constraining its absence to below a biomass threshold would have profound implications for our understanding of life in a cosmic context. If life is present on Enceladus, it most likely emerged separately from life on Earth, owing to the low likelihood of material transfer between the two worlds (*Worth et al.*, 2013). This would imply that the emergence of life on habitable worlds is common, i.e., the f_l factor in the Drake equation (*Drake*, 2003) is high. Conversely, the absence of life on Enceladus seems impossible to ascertain unless each square micrometer of its material is analyzed. However, constraining the absence of biomass to below a threshold consistent with Enceladus' habitability would provide a solid example of an environment in which life is limited, bringing about a need to reconsider requirements for life in such environments. Either way, exploring Enceladus could revolutionize our understanding of how life works.

7.2. Further Reading

Spencer J. R. and Nimmo F. (2013) Enceladus: An active ice world in the Saturn system. *Annu. Rev. Earth Planet. Sci., 41*, 693–717.

Dougherty M., Esposito L., and Krimigis S., eds. (2009) *Saturn from Cassini-Huygens.* Springer, Berlin.

Schenk P. M., Clark R. N., Howett C. J. A., Verbiscer A. J., and Waite J. H., eds. (2018) *Enceladus and the Icy Moons of Saturn.* Univ. of Arizona, Tucson.

Hoehler T. M. and Jørgensen B. B. (2013) Microbial life under extreme energy limitation. *Nature Rev. Microbiol., 11*, 83–94.

The Saturn System: Through the Eyes of Cassini (2017) E-book, National Aeronautics and Space Administration (NASA)/Lunar and Planetary Institute (LPI), available online at *https://www.nasa.gov/connect/ebooks/the-saturn-system.html.*

Acknowledgments. We thank everyone on the Cassini mission, without whom Enceladus would have remained a mystery out of reach. We also thank C. Glein and C. J. A. Howett for their thorough and very helpful reviews of this chapter. Some of this research was carried out at the Jet Propulsion Laboratory, California Institute of Technology, under a contract with the National Aeronautics and Space Administration. Government sponsorship acknowledged.

REFERENCES

Abrajano T. A., Sturchio N. C., Kennedy B. M., Lyon G. L., Muehlenbachs K., and Bohlke J. K. (1990) Geochemistry of reduced gas related to serpentinization of the Zambales ophiolite, Philippines. *Appl. Geochem., 5(5)*, 625–630, DOI: 10.1016/0883-2927(90)90060-I.

Aksyonov S. A. and Williams P. (2001) Impact desolvation of electrosprayed microdroplets — a new ionization method for mass spectrometry of large biomolecules. *Rapid Commun. Mass Spectrom., 15*, 2001–2006, DOI:10.1002/rcm.470

Alexander C. M. O'D., Cody G. D., De Gregorio B. T., Nittler L. R., and Stroud R. M. (2017) The nature, origin and modification of insoluble organic matter in chondrites, the major source of Earth's C and N. *Chem. Erde–Geochem., 77(2)*, 227–256, DOI: 10.1016/j.chemer.2017.01.007.

Alibert Y., Mousis O., and Benz W. (2005) Modeling the jovian subnebula. I. Thermodynamic conditions and migration of proto-satellites. *Astron. Astrophys., 439*, 1205–1213.

Anderson J. S., Rittle J., and Peters J. C. (2013) Catalytic conversion of nitrogen to ammonia by an iron model complex. *Nature, 501(7465)*, 84–87, DOI: 10.1038/nature12435.

Asphaug E. and Reufer A. (2013) Late origin of the Saturn system. *Icarus, 223*, 544–565, DOI: 10.1016/j.icarus.2012.12.009.

Arnold H., Liuzzo L., and Simon S. (2019) Magnetic signatures of a plume at Europa during the Galileo E26 flyby. *Geophys. Res. Lett., 46*, 1149–1157.

Auzende A.-L., Pellenq R. J.-M., Devouard B., Baronnet A., and Grauby O. (2006) Atomistic calculations of structural and elastic properties of serpentine minerals: The case of lizardite. *Phys. Chem. Mineral., 33*, 266–275, DOI: 10.1007/s00269-006-0078-x.

Barge L. M. and White L. M. (2017) Experimentally testing hydrothermal vent origin of life on Enceladus and other icy/ocean worlds. *Astrobiology, 17(9)*, 820–833.

Barr A. C. and McKinnon W. B. (2007) Convection in ice I shells and mantles with self-consistent grain size. *J. Geophys. Res., 112*, E02012, DOI: 10.1029/2006JE002781.

Běhounková M., Tobie G., Čadek O., Choblet G., Porco C., and Nimmo F. (2015) Timing of water plume eruptions on Enceladus explained by interior viscosity structure. *Nature Geosci., 8*, 601–606, DOI: 10.1038/ngeo2475.

Benner S. A. (2017) Detecting Darwinism from molecules in the Enceladus plumes, Jupiter's moons, and other planetary water lagoons. *Astrobiology, 17(9)*, 840–851, DOI: 10.1089/ast.2016.1611.

Berg J. M., Tymoczko J. L., and Stryer L. (2012a) Protein composition and structure. In *Biochemistry*, 7th edition. Freeman, New York. 1050 pp.

Berg J. M., Tymoczko J. L., and Stryer L. (2012b) Fatty acid metabolism. In *Biochemistry*, 7th edition. Freeman, New York.

1050 pp.

Beuthe M., Rivoldini A., and Trinh A. (2016) Enceladus's and Dione's floating ice shells supported by minimum stress isostasy. *Geophys. Res. Lett.*, *43*, 10088–10096, DOI: 10.1002/2016GL070650.

Blanchard D. C. and Syzdek L. D. (1972) Concentration of bacteria in jet drops from bursting bubbles. *J. Geophys. Res.*, *77*, 5087–5099, DOI: 10.1029/JC077i027p05087.

Bland M. T., Singer K. N., McKinnon W. B., and Schenk P. M. (2012) Enceladus' extreme heat flux as revealed by its relaxed craters. *Geophys. Res. Lett.*, *39(17)*, L17204, DOI: 10.1029/2012GL052736.

Bland P. A. and Travis B. J. (2017) Giant convecting mud balls of the early solar system. *Sci. Adv.*, *3(7)*, e1602514, DOI: 10.1126/sciadv.1602514.

Bockelée-Morvan D., Crovisier J., Mumma M. J., and Weaver H. A. (2004) The composition of cometary volatiles. In *Comets II* (M. C. Festou et al., eds.), Univ. of Arizona, Tucson.

Bouquet A., Mousis O., Waite J. H., and Picaud S. (2015) Possible evidence for a methane source in En-celadus' ocean. *Geophys. Res. Lett.*, *42*, 1334–1339, DOI: 10.1002/2014GL063013.

Bouquet A., Glein C. R., Wyrick D., and Waite J. H. (2017) Alternative energy: Production of H_2 by radiolysis of water in the rocky cores of icy bodies. *Astrophys. J. Lett.*, *840*, L8, DOI: 10.3847/2041-8213/aa6d56.

Brazelton W. J., Schrenk M. O., Kelley D. S., and Baross J. A. (2006) Methane- and sulfur-metabolizing microbial communities dominate the Lost City hydrothermal field ecosystem. *Appl. Environ. Microbiol.*, *72*, 6257–6270.

Burchell M. J., Bowden S. A., Cole M., Price M. C., and Parnell J. (2014) Survival of organic materials in hypervelocity impacts of ice on sand, ice, and water in the laboratory. *Astrobiology*, *14(6)*, 473–485.

Burger M. H., Sittler E. C., Johnson R. E., Smith H. T., Tucker O. J., and Shematovich V. I. (2007) Understanding the escape of water from Enceladus. *J. Geophys. Res.*, *112*, A06219, DOI: 10.1029/2006JA012086.

Burrows S. M., Ogunro O., Frossard A. A., Russell L. M., Rasch P. J., and Elliott S. M. (2014) A physically based framework for modeling the organic fractionation of sea spray aerosol from bubble film Langmuir equilibria. *Atmos. Chem. Phys.*, *14*, 13601–13629, DOI: 10.5194/acp-14-13601-2014.

Cable M. L., Clark K., Lunine J. I., Postberg F., Reh K., Spilker L. J., and Waite J. H. (2016) Enceladus Life Finder: The search for life in a habitable moon. *IEEE Aerospace Conference*, Big Sky, Montana, 1–8, available online at *http://hdl.handle.net/2014/45905*.

Cable M. L., Postberg F., Lang S. Q., Aluqihare L., Huber J., Clark B., Spilker L. J., and Lunine J. I. (2017) Mechanisms for enrichment of organics in the Enceladus plume. *Astrobiology Science Conference*, abstract #3639.

Čadek O., Tobie G., Van Hoolst T., Massé M., Choblet G., Lefèvre A., Mitri G., Baland R.-M., Běhounková M., Bourgeois O., and Trinh A. (2016) Enceladus's internal ocean and ice shell constrained from Cassini gravity, shape, and libration data. *Geophys. Res. Lett.*, *43*, 5633–5660, DOI: 10.1002/2016GL068634.

Čadek O., Souček O., Běhounková M., Choblet G., Tobie G., and Hron J. (2019) Long-term stability of Enceladus' uneven ice shell. *Icarus*, *319*, 476–484.

Canup R. M. (2010) Origin of Saturn's rings and inner moons by mass removal from a lost Titan-sized satellite. *Nature*, *468*, 943–946, DOI: 10.1038/nature09661.

Canup R. M. and Ward W. R. (2006) A common mass scaling for satellite systems of gaseous planets. *Nature*, *44*, 834–839, DOI: 10.1038/nature04860.

Carlucci A. F. and Williams P. M. (1965) Concentration of bacteria from seawater by bubble scavenging. *ICES J. Marine Sci.*, *30*, 28–33.

Castillo-Rogez J. C., Johnson T. V., Thomas P. C., Choukroun M., Matson D. L., and Lunine J. I. (2012) Geophysical evolution of Saturn's satellite Phoebe, a large planetesimal in the outer solar system. *Icarus*, *219*, 86–109, DOI: 10.1016/j.icarus.2012.02.002.

Charnoz S., Crida A., Castillo-Rogez J. C., Lainey V., Dones L., Karatekin Ö., Tobie G., Mathis S., Le Poncin-Lafitte G., and Salmon J. J. (2011) Accretion of Saturn's mid-sized moons during the viscous spreading of young massive rings: Solving the paradox of silicate-poor rings versus silicate-rich moons. *Icarus*, *216*, 535–550, DOI: 10.1016/j.icarus.2011.09.017.

Choblet G., Tobie G., Sotin C., Běhounková M., Čadek O., Postberg F.,

and Souček O. (2017) Powering prolonged hydrothermal activity inside Enceladus. *Nature Astron.*, *1*, 841–847, DOI: 1038/s41550-017-0289-8.

Chung D. H. (1971) Pressure coefficients of elastic constants for porous materials: Correction for porosity and discussion on literature data. *Earth Planet. Sci. Lett.*, *10(3)*, 316–324.

Consolmagno G. J. and Britt D. T. (1998) The density and porosity of meteorites from the Vatican collection. *Meteoritics & Planet. Sci.*, *33(6)*, 1231–1241, DOI: 10.1111/j.1945-5100.1998.tb01308.x.

Cowie G. L., Hedges J. I., Prahl F. G., and de Lange G. J. (1995) Elemental and major biochemical changes across an oxidation front in a relict turbidite: An oxygen effect. *Geochim. Cosmochim. Acta*, *59*, 33–46, DOI: 10.1016/0016-7037(94)00329-K.

Coyette A., Baland R.-M., and Van Hoolst T. (2018) Variations in rotation rate and polar motion of a non-hydrostatic Titan. *Icarus*, *307*, 83–105, DOI: 10.1016/j.icarus.2018.02.003.

Creamer J. S., Mora M. F., and Willis P. A. (2017) Enhanced resolution of chiral amino acids with capillary electrophoresis for biosignature detection in extraterrestrial samples. *Anal. Chem.*, *89(2)*, 1329–1337.

Crida A. and Charnoz S. (2012) Formation of regular satellites from ancient massive rings in the solar system. *Science*, *338*, 1196–1199, DOI: 10.1126/science.1226477.

Crida A., Charnoz S., Hsu H.-W., and Dones L. (2019) Are Saturn's rings really young? *Nature Astron.*, *3*, 967-970.

Cronin J. and Pizzarello S. (1983) Amino acids in meteorites. *Adv. Space Res.*, *3(9)*, 5–18.

Crow-Willard E. N. and Pappalardo R. T. (2015) Structural mapping of Enceladus and implications for formation of tectonized regions. *J. Geophys. Res.–Planets*, *120*, 928–950, DOI: 10.1002/2015JE004818.

Cuk M., Dones L., and Nesvorny D. (2016) Dynamical evidence for a late formation of Saturn's moons. *Astrophys. J.*, *820*, 97, DOI: 10.3847/0004-637X/820/2/97.

Davila A. F. and McKay C. P. (2014) Chance and necessity in biochemistry: Implications for the search for extraterrestrial biomarkers in Earth-life environments. *Astrobiology*, *14*, 534–540.

Deamer D. W. and Damer B. (2017) Hydrothermal conditions and the origin of cellular life. *Astrobiology*, *17*, 834–839.

Deamer D. W. and Georgiou C. D. (2015) Hydrothermal conditions and the origin of cellular life. *Astrobiology*, *15*, 1091–1095.

Desch S. J., Kalyaan A., and Alexander C. M. O'D. (2018) The effect of Jupiter's formation on the distribution of refractory elements and inclusions in meteorites. *Astrophys. J. Suppl. Ser.*, *238*, 11.

Des Marais D. J., Nuth J. A., Allamandola L. J., Boss A. P., Farmer J. D., Hoehler T. M., Jakosky B. M., Meadows V. S., Pohorille A., Runnegar B., and Spormann A. M. (2008) The NASA Astrobiology Roadmap. *Astrobiology*, *8*, 715–730.

Dhingra D., Hedman M. M., Clark R. N., and Nicholson P. D. (2017) Spatially resolved near infrared observations of Enceladus' tiger stripe eruptions from Cassini VIMS. *Icarus*, *292*, 1–12.

Dikarev V. V. (1999) Dynamics of particles in Saturn's E ring: Effects of charge variations and the plasma drag force. *Astron. Astrophys.*, *346*, 1011–1019.

Dodd M. S., Papineau D., Grenne T., Slack J. F., Rittner M., Pirajno F., O'Neil J., and Little C. T. S. (2017) Evidence for early life in Earth's oldest hydrothermal vent precipitates. *Nature*, *543*, 60–64.

Dong Y., Hill T. W., and Ye S.-Y. (2015) Characteristics of ice grains in the Enceladus plume from Cassini observations. *J. Geophys. Res.–Space Physics*, *120*, 915–937, DOI: 10.1002/2014JA020288.

Dorn E. D., Nealson K. H., and Adami C. (2011) Monomer abundance distribution patterns as a universal biosignature: Examples from terrestrial and digital life. *J. Molec. Evol.*, *72*, 283–295.

Dougherty M. K., Khurana K. K., Neubauer F. M., Russell C. T., Saur J., Leisner J. S., and Burton N. E. (2006) Identification of a dynamic atmosphere at Enceladus with the Cassini magnetometer. *Science*, *311*, 1406–1409, DOI: 10.1126/science.1120985.

Drake F. (2003) The Drake equation revisited. *Astrobiology Magazine*, retrieved March 30, 2018, available online at *https://www.astrobio.net/alien-life/the-drake-equation-revisited-part-i*.

Dunham E., Desch S. J., and Probst L. (2019) Haumea's shape, composition, and internal structure. *Astrophys. J.*, *877*, 41.

Durham W. B., Kirby S. H., and Stern L. A. (1992) Effects of dispersed particulates on the rheology of water ice at planetary conditions. *J. Geophys. Res.*, *97(E12)*, 20883–20897, DOI: 10.1029/92JE02326.

Durham W. B., Prieto-Ballesteros O., Goldsby D. L., and Kargel J. S. (2010) Rheological and thermal properties of icy materials. *Space*

Sci. Rev., 153, 273–298, DOI: 10.1007/s11214-009-9619-1.

Ermakov A. I., Fu R. R., Castillo-Rogez J. C., Raymond C. A., Park R. S., Preusker F., Russell C. T., Smith D. E., and Zuber M. T. (2017) Constraints on Ceres' internal structure and evolution from its shape and gravity measured by the Dawn spacecraft. *J. Geophys. Res.–Planets, 122(11),* 2267–2293, DOI: 10.1002/2017JE005302.

Etiope G. and Sherwood Lollar B. (2013) Abiotic methane on Earth. *Rev. Geophys., 51,* 276–299.

Europa Lander Mission Concept Team (2016) *Europa Lander Study 2016 Report: Europa Lander Mission,* JPL D-97667, available online at *https://solarsystem.nasa.gov/docs/Europa_Lander_SDT_Report_2016.pdf.*

Fagerbakke K. M., Helgal M., and Norland S. (1996) Contents of carbon, oxygen, nitrogen, sulfur and phosphorus in native aquatic and cultured bacteria. *Aquat. Microb. Ecol., 10,* 15–27.

Feibelman W. (1967) Concerning the "D" ring of Saturn. *Nature, 214,* 793–794, DOI: 10.1038/214793a0.

Fitch C. A., Platzer G., Okon M., Garcia-Moreno B., and McIntosh L. P. (2015) Arginine: Its pKa value revisited. *Protein Sci., 24(5),* 752–761.

Fortsch D., Pabst K., and Gross-Hardt E. (2015) The product distribution in Fischer–Tropsch synthesis: An extension of the ASF model to describe common deviations. *Chem. Eng. Sci., 138,* 333–346.

Fuller J., Luan J., and Eliot Quataert E. (2016) Resonance locking as the source of rapid tidal migration in the Jupiter and Saturn moon systems. *Mon. Not. R. Astron. Soc., 458(4),* 3867–3879, DOI: 10.1093/mnras/stw609.

Gao P., Kopparla P., Zhang X., and Ingersoll A. P. (2016) Aggregate particles in the plumes of Enceladus. *Icarus, 264,* 227–238.

Georgiou C. D. and Deamer D. W. (2014) Lipids as universal biomarkers of extraterrestrial life. *Astrobiology,* 14, 541–549.

Glavin D. P., Callahan M. P., Dworkin J. P., and Elsila J. E. (2010) The effects of parent body processes on amino acids in carbonaceous chondrites. *Meteoritics & Planet. Sci., 45,* 1948–1972.

Gleeson D. F., Pappalardo R. T., Anderson M. S., Grasby S. E., Mielke R. E., Wright K. E., and Templeton A. S. (2012) Biosignature detection at an Arctic analog to Europa. *Astrobiology, 12(2),* 135–150.

Glein C. R., Baross J. A., and Waite J. H. (2015) The pH of Enceladus' ocean. *Geochim. Cosmochim. Acta, 162,* 202–219.

Glein C. R., Postberg F., and Vance S. D. (2018) The geochemistry of Enceladus: Compositon and controls. In *Enceladus and the Icy Moons of Saturn* (P. M. Schenk et al., eds.), pp. 39–56. Univ. of Arizona, Tucson.

Goguen J. D., Buratti B. J., Brown R. H., Clark R. N., Nicholson P. D., Hedman M. M., Howell R. R., Sotin C., Cruikshank D. P., Baines K. H., Lawrence K. J., Spencer J. R., and Blackburn D. G. (2013) The temperature and width of an active fissure on Enceladus measured with Cassini VIMS during the 14 April 2012 south pole flyover. *Icarus, 226(1),* 1128–1137, DOI: 10.1016/j.icarus.2013.07.012.

Haff P., Eviatar A., and Siscoe G. (1983) Ring and plasma: The enigmae of Enceladus. *Icarus, 56(3),* 426–438.

Hansen C. J., Hendrix A., Esposito L., Colwell J., Shemansky D., Pryor W., Stewart I., and West R. (2005) Cassini Ultra Violet Imaging Spectrograph (UVIS) observations of Enceladus' plume. *Eos Trans. AGU, 86(52),* Fall Meet. Suppl., Abstract P21F-04.

Hansen C. J., Esposito L., Stewart A. I. F, Colwell J., Hendrix A. P., Shemansky W. D., and West R. (2006) Enceladus' water vapor plume. *Science, 311,* 1422–1425, DOI: 10.1126/science.1121254.

Hansen C., Esposito L., Stewart A., et al. (2008) Water vapour jets inside the plume of gas leaving Enceladus. *Nature, 456,* 477–479, DOI: 10.1038/nature07542.

Hansen C. J. et al. (2011) The composition and structure of the Enceladus plume. *Geophys. Res. Lett.,* 38, L11202, DOI: 10.1029/2011GL047415.

Hansen C. J., Esposito L. W., Aye K.-M., Colwell J. E., Hendrix A. R., Portyankina G., and Shemansky D. (2017) Investigation of diurnal variability of water vapor in Enceladus' plume by the Cassini ultraviolet imaging spectrograph. *Geophys. Res. Lett., 44,* 672–677, DOI: 10.1002/2016GL071853.

Hedman M. M., Nicholson P. D., Showalter M. R., Brown R. H., Buratti B. J., and Clark R. N. (2009) Spectral observations of the Enceladus plume with Cassini-VIMS. *Astrophys. J., 693,* 1749–1762, DOI: 10.1088/0004-637X/693/2/1749.

Hedman M., Gosmeyer C., Nicholson P., et al. (2013) An observed correlation between plume activity and tidal stresses on Enceladus. *Nature, 500,* 182–184, DOI: 10.1038/nature12371.

Hedman M. M., Dhingra D., Nicholson P. D., Hansen C. J., Portyankina G., Ye S., and Dong Y. (2018) Spatial variations in the dust-to-gas ratio of Enceladus' plume. *Icarus, 305,* 123–138, DOI: 10.1016/j.icarus.2018.01.006.

Hemingway D. J. and Mittal T. (2019) Enceladus' ice shell structure as a window on internal heat production. *Icarus, 332,* 111–131.

Hemingway D. J., Rudolph M. L., and Manga M. (2020) Cascading parallel fractures on Enceladus. *ArXiV e-prints,* arXiv:1911.02730v1.

Henning W. G. and Hurford T. (2014) Tidal heating in multilayered terrestrial exoplanets. *Astrophys. J., 789,* 30.

Higgs P. G. and Pudriz R. E. (2009) A thermodynamic basis for prebiotic amino acid synthesis and the nature of the first genetic code. *Astrobiology,* 9, 483–490.

Hoehler T. M. (2004) Biological energy requirements as quantitative boundary conditions for life in the subsurface. *Geobiology, 2,* 205–215.

Hoehler T. M. and Jørgensen B. B. (2013) Microbial life under extreme energy limitation. *Nature Rev. Microbiol., 11,* 83–94.

Hoehler T. M., Alperin M. J., Albert D. B., and Martens C. S. (2001) Apparent minimum free energy requirements for methanogenic Archaea and sulfate-reducing bacteria in an anoxic marine sediment. *FEMS Microbiol. Ecol., 38,* 33–41.

Holland G., Sherwood Lollar B., Li L., Lacrampe-Couloume G., Slater G. F., and Ballentine C. (2013) Deep fracture fluids isolated in the crust since the Precambrian era. *Nature, 497(7449),* 357–360, DOI: 10.1038/nature12127.

Horányi M., Juhász A., and Morfill G. E. (2008) Large-scale structure of Saturn's E ring. *Geophys. Res. Lett., 35,* L04203.

Howell S. M. and Pappalardo R. T. (2019) Can Earth-like plate tectonics occur in ocean world ice shells? *Icarus, 322,* 69–79.

Howett C. J. A., Spencer J. R., Pearl J., and Segura M. (2010) Thermal inertia and bolometric Bond albedo values for Mimas, Enceladus, Tethys, Dione, Rhea and Iapetus as derived from Cassini/CIRS measurements. *Icarus, 206(2),* 573–593.

Howett C. J. A., Spencer J. R., Pearl J., and Segura M. (2011) High heat flow from Enceladus' south polar region measured using 10–600 cm^{-1} Cassini/CIRS data. *J. Geophys. Res.–Planets, 116(E3),* E03003, DOI: 10.1029/2010JE003718.

Hsu H. W., Postberg F., Sekine Y., Shibuya T., Kempf S., Horanyi M., Juhasz A., Altobelli N., Suzuki K., Masaki Y., Kuwatani T., Tachibana S., Sirono S. I., Moragas-Klostermeyer G., and Srama R. (2015) Ongoing hydrothermal activities within Enceladus. *Nature, 519,* 207–210.

Hsu H.-W., Schmidt J., Kempf S., Postberg F., Moragas-Klostermeyer G., Seiss M., Hoffmann H., Burton M., Ye S.-Y., Kurth W. S., Horanyi M., Khawaja N., Spahn F., Schirdewahn D., O'Donoghue J., Moore L., Cuzzi J., Jones G. H., and Srama R. (2018) *In situ* collection of dust grains falling from Saturn's rings into its atmosphere. *Science, 362(6410),* eaat3185.

Hurley D. M., Perry M. E., and Waite J. H. (2015) Modeling insights into the locations of density enhancements from the Enceladus water vapor jets. *J. Geophys. Res.–Planets, 120,* 1763–1773, DOI: 10.1002/2015JE004872.

Iess L., Stevenson D. J., Parisi M., Hemingway D., Jacobson R. A., Lunine J. I., Nimmo F., Armstrong J. W., Asmar S. W., Ducci M., and Tortora P. (2014) The gravity field and interior structure of Enceladus. *Science, 344(6179),* 78–80, DOI: 10.1126/science.1250551.

Iess L., Militzer B., Kaspi Y., Nicholson P., Durante D., Racioppa P., Anabtawi A., Galanti E., Hubbard W., Mariani M. J., Tortora P., Wahl S., and Zannoni M. (2019) Measurement and implications of Saturn's gravity field and ring mass. *Science, 364(6445),* eaat2965, DOI: 10.1126/science.aat2965.

Ingersoll A. P. and Ewald S. P. (2017) Decadal timescale variability of the Enceladus plumes inferred from Cassini images. *Icarus, 282,* 260–275, DOI: 10.1016/j.icarus.2016.09.018.

Jaramillo-Botero A., An Q., Cheng M.-J., Goddard W., Beegle L., and Hodyss R. (2012) Hypervelocity impact effect of molecules from Enceladus' plume and Titan's upper atmosphere on NASA's Cassini Spectrometer from reactive dynamics simulation. *Phys. Rev. Lett., 109,* 213201.

Jia X., Kivelson M. G., Khurana K. K., and Kurth W. S. (2018)

Evidence of a plume on Europa from Galileo magnetic and plasma wave signatures. *Nature Astron.*, 2, 459–464.

Johansen A. and Lambrechts M. (2017) Forming planets via pebble accretion. *Annu. Rev. Astron. Astrophys., 45,* 359–387.

Johnson J. W., Oelkers E. H., and Helgeson H. C. (1992) SUPCRT92: A software package for calculating the standard molal thermodynamic properties of minerals, gases, aqueous species, and reactions from 1 to 5000 bar and 0° to 1000°C. *Computers Geosci., 18(7),* 899–947, DOI: 10.1016/ 0098-3004(92)90029-Q.

Jones G. H., Arridge C. S., Coates A. J., Lewis G. R., Kanani S., Wellbrock A., Young D. T., Crary F. J., Tokar R. L., Wilson R. J., Hill T. W., Johnson R. E., Mitchell D. G., Schmidt J., Kempf S., Beckmann U., Russell C. T., Jia Y. D., Dougherty M. K., Waite J. H., and Magee B. A. (2009) Fine jet structure of electrically charged grains in Enceladus' plume. *Geophys. Res. Lett., 36,* L16204, DOI: 10.1029/2009GL038284.

Jurac S., Johnson R. E., and Richardson J. D. (2001) Saturn's E ring and production of the neutral torus. *Icarus, 149,* 384–396.

Kelley D. S., Karson J. A., Früh-Green G. L., Yoerger D. R., Shank T. M., Butterfield D. A., Hayes J. M., Schrenk M. O., Olson E. J., Proskurowski G., Jakuba M., Bradley A., Larson B., Ludwig K., Glickson D., Buckman K., Bradley A. S., Brazelton W. J., Roe K., Elend M. J., Delacour A., Bernasconi S. M., Lilley M. D., Baross J. A., Summons R. E., and Sylva S. P. (2005) A serpentinite-hosted ecosystem: The Lost City hydrothermal field. *Science, 307(5714),* 1428–1434.

Kempf S., Beckmann U., Moragas-Klostermeyer G., Postberg F., Srama R., Economou T., Schmidt J., Spahn F., and Grun E. (2008) The E ring in the vicinity of Enceladus I. Spatial distribution and properties of the ring particles. *Icarus, 193,* 420–437, DOI: 10.1016/j.icarus.2007.06.027.

Kempf S., Beckmann U., and Schmidt J. (2010) How the Enceladus dust plume feeds Saturn's E ring. *Icarus, 206,* 446–457, DOI: 10.1016/j.icarus.2009.09.016.

Khawaja N., Postberg F., Hillier J., Klenner F., Kempf S., Nolle L, Reviol R., Zou Z., and Srama R. (2019) Low mass nitrogen-, oxygen-bearing and aromatic compounds in Enceladean ice grains. *Mon. Not. R. Astron. Soc.,489(4),* 5231–5243.

Kirchoff M. R. and Schenk P. (2009) Crater modification and geologic activity in Enceladus' heavily cratered plains: Evidence from the impact crater distribution. *Icarus, 202(2),* 656–668.

Kirchoff M. R., Bierhaus E. B., Dones L., Robbins S. J., Singer K. N., Wagner R. J., and Zahnle K. J. (2018) Cratering histories in the saturnian system. In *Enceladus and the Icy Moons of Saturn* (P. M. Schenk et al., eds.), pp. 267–284. Univ. of Arizona, Tucson.

Kirk R. L., Soderblom L. A., Brown R. H., Kieffer S. W., and Kargel J. S. (1995) Triton's plumes: Discovery, characteristics, and models. In *Neptune and Triton* (D. P. Cruikshank, ed.), pp. 949–989. Univ. of Arizona, Tucson.

Kite E. S. and Rubin A. M. (2016) Sustained eruptions on Enceladus explained by turbulent dissipation in tiger stripes. *Proc. Natl. Acad. Sci. USA, 113,* 3972–3975.

Klein H. P. (1978) The Viking biological experiments on Mars. *Icarus, 34(3),* 666–674, DOI: 10.1016/0019-1035(78)90053-2.

Klenner F., Postberg F., Hillier J., Khawaja N., Reviol R., Stolz F., Cable M. L., Abel B., and Nölle L. (2020) Analog experiments for the identification of trace biosignatures in ice grains from extraterrestrial ocean worlds. *Astrobiology, 20,* 179–189.

Klimczak C., Byrne P. K., Regensburger P. V., Bohnenstiehl D. R., Hauck S. A., Dombard A. J., Hemingway D. J., Vance S. D., Melwani Daswani M., and Elder C. M. (2019) Strong ocean floors with Europa, Titan, and Ganymede limit geological activity there, Enceladus less so. *Lunar Planet Sci. L,* Abstract #2132. Lunar and Planetary Institute, Houston.

Lainey V., Jacobson R. A., Tajeddine R., Cooper N. J., Murray C., Robert V., Tobie G., Guillot T., Mathis S., Remus F., Desmars J., Arlot J.-E., De Cuyper J.-P., Dehant V., Pascu D., Thuillot W., LePoncin-Lafitte C., and Zahn J.-P. (2017) New constraints on Saturn's interior from Cassini astrometric data. *Icarus, 281,* 286–296, DOI: 10.1016/j.icarus.2016.07.014.

Lane N. and Martin W. F. (2012) The origin of membrane bioenergetics. *Cell, 151,* 1406–1416.

Li Y. and Breaker R. (1999) Kinetics of RNA degradation by specific base catalysis of transesterification involving the 2'-hydroxyl group. *J. Am. Chem. Soc., 121,* 5364–5372.

Lingham M. and Loeb A. (2018) Is extraterrestrial life suppressed on subsurface ocean worlds due to the paucity of bioessential elements? *Astron. J., 156,* 151, DOI: 10.3847/1538-3881/aada02.

Maffei T., Sommariva S., Ranzi E., and Faravelli T. (2012) A predictive kinetic model of sulfur release from coal. *Fuel, 91(1),* 213–223, DOI: 10.1016/j.fuel.2011.08.017.

Maffei T., Frassoldati A., Cuoci A., Ranzi E., and Faravelli T. (2013) Predictive one step kinetic model of coal pyrolysis for CFD applications. *Proc. Comb. Inst., 34(2),* 2401–2410, DOI: 10.1016/j.proci.2012.08.006.

Magee B. H. and Waite J. H. (2017) Neutral gas composition of Enceladus' plume — Model parameter insights from Cassini-INMS. *Lunar Planet. Sci. XLVIII,* Abstract #2974. Lunar and Planetary Institute, Houston.

Marchetto C. A. (1983) The Voyager 2 scan platform anomaly. *Guidance and Control, Proceedings of the Annual Rocky Mountain Conference,* 219–243. Keystone, Colorado, Feb. 5–9, AAS paper 83-046, available online at *https://ntrs.nasa.gov/search.jsp?R=19830062955.*

Mason B. (1979) Chapter B, Cosmochemistry, Part 1. Meteorites. In *Data of Geochemistry, 6th edition* (M. Fleischer, ed.), U.S. Geological Survey Professional Paper 440-B-1.

Mason D. L., Castillo-Rogez J. C., Davies A. G., and Johnson T. V. (2012) Enceladus: A hypothesis for bringing both heat and chemicals to the surface. *Icarus, 221(1),* 53–62.

McCarthy C. and Cooper R. F. (2016) Tidal dissipation in creeping ice and the thermal evolution of Europa. *Earth Planet. Sci. Lett., 443,* 185–194, DOI: 10.1016/j.epsl.2016.03.006.

McCollom T. M. (2016) Abiotic methane formation during experimental serpentinization of olivine. *Proc. Natl. Acad. Sci., 113(49),* 13965–13970.

McCollom T. M. and Amend J. P. (2005) A thermodynamic assessment of energy requirements for biomass synthesis by chemolithoautotrophic micro-organisms in oxic and anoxic environments. *Geobiology, 3,* 135–144.

McCollom T. M. and Bach W. (2009) Thermodynamic constraints on hydrogen generation during serpentinization of ultramafic rocks. *Geochim. Cosmochim. Acta, 73,* 856–875.

McCollom T. M. and Seewald J. S. (2001) A reassessment of the potential for reduction of dissolved CO_2 to hydrocarbons during serpentinization of olivine. *Geochim. Cosmochim. Acta, 65(21),* 3769–3778, DOI: 10.1016/S0016-7037(01)00655-X.

McCollom T. M. and Seewald J. S. (2007) Abiotic synthesis of organic compounds in deep-sea hydrothermal environments. *Chem. Rev., 107,* 382–401, DOI: 10.1021/cr0503660.

McDermott J. M., Seewald J. S., German C. R., and Sylva S. P. (2015) Pathways for abiotic synthesis at submarine hydrothermal fluids. *Proc. Natl. Acad. Sci., 112(25),* 7668–7672.

McDonald A. H. and Fyfe W. S. (1985) Rate of serpentinization in seafloor environments. *Tectonophysics, 116(1–2),* 123–135, DOI: 10.1016/0040-1951(85)90225-2.

McKay C. P. (2004) What is life — and how do we search for it in other worlds? *PLOS Biology, 2(9),* e302, DOI: 10.1371/journal.pbio.0020302.

McKay C. P., Anbar A. D., Porco C., and Tsou P. (2014) Follow the plume: The habitability of Enceladus. *Astrobiology, 14(4),* 352–355, DOI: 10.1089/ast.2014.1158.

McKay D. S., Gibson E. K., Thomas-Keprta K. L., and Vali H. (1996) Search for past life on Mars: Possible relic biogenic activity in martian meteorite ALH84001. *Science, 273(5277),* 924–930, DOI: 10.1126/science.273.5277.924.

McKinnon W. B. (2015) Effect of Enceladus's rapid synchronous spin on interpretation of Cassini gravity. *Geophys. Res. Lett., 42(7),* 2137–2143.

McKinnon W. B. and Zolensky M. E. (2003) Sulfate content of Europa's ocean and shell: Evolutionary considerations and some geological and astrobiological implications. *Astrobiology, 3(4),* 879–897, DOI: 10.1089/153110703322736150.

McKinnon W. B., Prialnik D., Stern S. A., and Coradini A. (2008) Structure and evolution of Kuiper belt objects and dwarf planets. In *The Solar System Beyond Neptune* (M. A. Barucci et al., eds.), pp. 213–241. Univ. of Arizona, Tucson.

Meier P., Kriegel H., Motschmann U., Schmidt J., Spahn F., Hill T. W., Dong Y., and Jones G. H. (2014) A model of the spatial and size distribution of Enceladus' dust plume. *Planet. Space Sci., 104,* 216–233, DOI: 10.1016/j.pss.2014.09.016.

Menez B., Pisapia C., Andreani M., Jamme F., Vanbellingen Q. P., Brunelle A., Richard L., Dumas P., and Refregiers M. (2018) Abiotic synthesis of amino acids in the recesses of the oceanic lithosphere. *Nature*, 564, 59–63.

Meyer J. and Wisdom J. (2007) Tidal heating in Enceladus. *Icarus*, 188, 535–539.

Mitchell C. J., Porco C. C., and Weiss J. W. (2015) Tracking the geysers of Enceladus into Saturn's E ring. *Astron. J.*, 149, 156, DOI: 10.1088/0004-6256/149/5/156.

Mitchell K. A., Ono M., Parcheta C., and Iacoponi S. (2017) Dynamic pressure at Enceladus' vents and implications for vent and conduit in-situ studies. *Lunar Planet Sci. XLVIII*, Abstract #2801. Lunar and Planetary Institute, Houston.

Mitchell K. L. (2014) Cryovolcanic conduit evolution and eruption on icy satellites. Abstract P52A-02 presented at 2014 Fall Meeting, AGU, San Francisco, California, 15–19 December.

Moldowan J. M., Seifert W. K., and Gallegos E. J. (1985) Relationship between petroleum composition and depositional environment of petroleum source rocks. *Am. Assoc. Petrol. Geol. Bull.*, 69, 1255–1268.

Mosqueira I. and Estrada P. R. (2003a) Formation of the regular satellites of giant planets in an extended gaseous nebula I: Subnebula model and accretion of satellites. *Icarus*, 163, 198–231.

Mosqueira I. and Estrada P. R. (2003b) Formation of the regular satellites of giant planets in an extended gaseous nebula II: Satellite migration and survival. *Icarus*, 163, 232–255.

Mottl M. J. and Holland H. D. (1978) Chemical exchange during hydrothermal alteration of basalt by seawater — I. Experimental results for major and minor components of seawater. *Geochim. Cosmochim. Acta*, 42(8), 1103–1115, DOI: 10.1016/ 0016-7037(78)90107-2.

Mousis O and Alibert Y. (2006) Modeling the jovian subnebula II. Composition of regular satellite ices. *Astron. Astrophys.*, 448, 771–778.

Mousis O., Chassefière E., Holm N. G., Bouquet A., Waite J. H., Geppert W. D., Picaud S., Aikawa Y., Ali-Dib M., Charlou J.-L., and Rousselot P. (2015) Methane clathrates in the solar system. *Astrobiology*, 15(4), 308–326, DOI: 10.1089/ ast.2014.1189.

Mumma M. J. and Charnley S. B. (2011) The chemical composition of comets — emerging taxonomies and natal heritage. *Annu. Rev. Astron. Astrophys.*, 49(1), 471–524, DOI: 10.1146/annurev-astro-081309-130811.

Nakajima M. and Ingersoll A. P. (2016) Controlled boiling on Enceladus. 1. Model of the vapor-driven jets. *Icarus*, 272, 309–318.

National Research Council (2012) *Assessment of Planetary Protection Requirements for Spacecraft Missions to Icy Solar System Bodies*. National Academies, Washington, DC, DOI: 10.17226/13401.

Neveu M. and Rhoden A. R. (2019) Evolution of Saturn's mid-sized moons. *Nature Astron.*, 3, 543–552.

Neveu M., Glein C. R., Anbar A. D., McKay C. P., Desch S. J., Castillo-Rogez J. C., and Tsou P. (2014) Enceladus' fully cracked core: Implications for habitability. *Workshop on the Habitability of Icy Worlds*, Abstract #4028. Lunar and Planetary Institute, Houston.

Neveu M., Desch S. J., and Castillo-Rogez J. C. (2015) Core cracking and hydrothermal circulation can profoundly affect Ceres' geophysical evolution. *J. Geophys. Res.–Planets*, 120, 123–154, DOI: 10.1002/2014JE004714.

Neveu M., Desch S. J., and Castillo-Rogez J. C. (2017) Aqueous geochemistry in icy world interiors: Equilibrium fluid, rock, and gas compositions, and fate of antifreezes and radionuclides. *Geochim. Cosmochim. Acta*, 212, 324–371, DOI: 10.1016/j.gca.2017.06.023.

Neveu M., Hays L. E., Voytek M. A., New M. H., and Schulte M. D. (2018) The Ladder of Life Detection. *Astrobiology*, 18, 1375–1402.

Nimmo F., Porco C. and Mitchell C. (2014) Tidally modulated eruptions on Enceladus: Cassini ISS observations and models. *Astron. J.*, 148, 46.

Nimmo F., Barr A. C., Běhounková M., and McKinnon W. B. (2018) The thermal and orbital evolution of Enceladus: Observational constraints and models. In *Enceladus and the Icy Moons of Saturn* (P. M. Schenk et al., eds.), pp. 79–94. Univ. of Arizona, Tucson.

Oelkers E. H., Bénézeth P., and Pokrovski G. S. (2009) Thermodynamic databases for water-rock interaction. *Rev. Min. Geochem.*, 70(1), 1–46, DOI: 10.2138/rmg.2009.70.1.

Oh M. S., Crawford R. W., Foster K. G., and Alcaraz A. (1988) Ammonia evolution from western and eastern oil shales. *Division of Petroleum Chemistry, Preprints (USA)*, 34(1), 94–102, American Chemical Society.

Ohmoto H. and Lasaga A. C. (1982) Kinetics of reactions between aqueous sulfates and sulfides in hydrothermal systems. *Geochim. Cosmochim. Acta*, 46(10), 1727–1745, DOI: 10.1016/0016-7037(82)90113-2.

O'Neill C. and Nimmo F. (2010) The role of episodic overturn in generating the surface geology and heat flow on Enceladus. *Nature Geosci.*, 3, 88–91, DOI: 10.1038/ngeo731.

Palandri J. L. and Kharaka Y. K. (2004) *A Compilation of Rate Parameters of Water-Mineral Interaction Kinetics for Application to Geochemical Modeling*. U.S. Geological Survey Open File Report 2004-1068, 71 pp., available online at *http://www.dtic.mil/dtic/tr/ fulltext/u2/a440035.pdf*.

Parkinson C. D., Liang M.-C., Yung Y. L., and Kirschivnk J. L. (2008) Habitability of Enceladus: Planetary conditions for life. *Orig. Life Evol. Biosph.*, 38(4), 355–369, DOI: 10.1007/s11084-008-9135-4.

Parnell J., Cullen D., Sims M. R., Bowden S., Cockell C. S., Court R., Ehrenfreund P., Gaubert F., Grant W., and Parro V. (2007) Searching for life on Mars: Selection of molecular targets for ESA's aurora ExoMars mission. *Astrobiology*, 7, 578–604.

Patthoff D. and Kattenhorn S. (2011) A fracture history on Enceladus provides evidence for a global ocean. *Geophys. Res. Lett.*, 38, L18201.

Patthoff D., Kattenhorn S. A., and Cooper C. M. (2019) Implications of nonsynchronous rotation on the deformational history and ice shell properties in the south polar terrain of Enceladus. *Icarus*, 321, 445–457.

Perry M. E., Teolis B. D., Hurley D. M., Magee B. A., Waite J. H., Brockwell T. G., Perryman R. S., and McNutt R. L. (2015) Cassini INMS measurements of Enceladus plume density. *Icarus*, 257, 139–162, DOI: 10.1016/j.icarus.2015.04.037.

Phillips O. M. (1991) *Flow and Reactions in Permeable Rocks*. Cambridge Univ., Cambridge.

Poch O., Joachim F., Isabel R., Antoine P., Bernhard J., and Nicolas T. (2017) Remote sensing of potential biosignatures from rocky, liquid or icy (exo)planetary surfaces. *Astrobiology*, 17(3), 231–252.

Pokrovsky O. S. and Schott J. (1999) Processes at the magnesium-bearing carbonates/solution interface. II. Kinetics and mechanism of magnesite dissolution. *Geochim. Cosmochim. Acta*, 63(6), 881–897, DOI: 10.1016/S0016-7037(99)00013-7.

Porco C. C., Helfenstein P., Thomas P. C., Ingersoll A. P., Wisdom J., West R., Neukum G., Denk T., Wagner R., Roatsch T., Kieffer S., Turtle E., McEwen A., Johnson T. V., Rathbun J., Veverka J., Wilson D., Perry J., Spitale J., Brahic A., Burns J. A., Del Genio A. D., Dones L., Murray C. D., and Squyres S. (2006) Cassini observes the active south pole of Enceladus. *Science*, 311(5766), 1393–1401, DOI: 10.1126/science.1123013.

Porco C. C., DiNino D., and Nimmo F. (2014) How the geysers, tidal stresses, and thermal emission across the south polar terrain of Enceladus are related. *Astron. J.*, 148(3), 24 pp.

Porco C. C., Dones L., and Mitchell C. (2017) Could it be snowing microbes on Enceladus? Assessing conditions in its plume and implications for future missions. *Astrobiology*, 17(9), 876–901.

Porco C., Mitchell C. J., Nimmo F., and Tiscareno M. S. (2018) Multiple long-period variations in the plume of Enceladus offer insights into eruption mechanism. Abstract P34F-3819 presented at 2018 Fall Meeting, AGU, Washington, DC, 10–14 December.

Postberg F., Kempf S., Hillier J. K., Srama R., Green S. F., McBride N., and Grün E. (2008) The E-ring in the vicinity of Enceladus II. Probing the moon's interior — the composition of E-ring particles. *Icarus*, 193, 438–454, DOI: 10.1016/j.icarus.2007.09.001.

Postberg F., Kempf S., Schmidt J., Brilliantov N., Beinsen A., Abel B., Buck Y., and Srama R. (2009) Sodium salts in E ring ice grains from an ocean below the surface of Enceladus. *Nature*, 459, 1098–1101, DOI: 10.1038/nature08046.

Postberg F., Schmidt J., Hillier J., Kempf S., and Srama R. (2011) A salt-water reservoir as the source of a compositionally stratified plume on Enceladus. *Nature*, 474(7353), 620–622, DOI: 10.1038/ nature10175.

Postberg F., Khawaja N., Abel B., Choblet G., Glein C. R., Gudipati M. S., Henderson B. L., Hsu H.-W., Kempf S., Klenner F., Moragas-Klostermeyer G., Magee B., Nolle L., Perry M., Reviol R., Schmidt J., Srama R., Stolz F., Tobie G., Trieloff M., and Waite J. H. (2018) Macromolecular organic compounds from the depths of Enceladus.

Nature, 558, 564–568.

Renaud J. P. and Henning W. G. (2018) Increased tidal dissipation using advanced rheological models: Implications for Io and tidally active exoplanets. *Astrophys. J., 857(2),* 98.

Rivkin A. S., Asphaug E., and Bottke W. S. (2014) The case of the missing Ceres family. *Icarus, 243,* 429–439, DOI: 10.1016/j.icarus.2014.08.007.

Roadcap G. S., Sanford R. A., Jin Q., Pardinas J. R., and Bethke C. M. (2006) Extremely alkaline (pH >12) ground water hosts diverse microbial community. *Groundwater, 44(4),* 511–517.

Roberts J. H. (2015) The fluffy core of Enceladus. *Icarus, 258,* 54–66, DOI: 10.1016/j.icarus.2015.05.033.

Roberts J. H. and Stickle A. M. (2017) Break the world's shell: An impact on Enceladus: Bringing the ocean to the surface. *Lunar Planet. Sci. XLVIII,* Abstract #1955. Lunar and Planetary Institute, Houston.

Rodriguez M. M., Bill E., Brennessel W. W., and Holland P. L. (2011) N_2 reduction and hydrogenation to ammonia by a molecular iron-potassium complex. *Science, 334(6057),* 780–783, DOI: 10.1126/science.1211906.

Rothschild L. and Mancinelli R. (2001) Life in extreme environments. *Nature, 409,* 1092–1101.

Rubin M., Altwegg K., Balsiger H., Bar-Nun A., Berthelier J.-J., Bieler A., Bochsler P., Briois C., Calmonte U., Combi M., De Keyser J., Dhooghe F., Eberhardt P., Fiethe B., Fuselier S. A., Gasc S., Gombosi T. I., Hansen K. C., Hässig M., Jäckel A., Kopp E., Korth A., Le Roy L., Mall U., Marty B., Mousis O., Owen T., Rème H., Sémon T., Tzou C.-Y., Waite J. H., and Wurz P. (2015) Molecular nitrogen in Comet 67P/Churyumov-Gerasimenko indicates a low formation temperature. *Science,* aaa6100, DOI: 10.1126/science.aaa6100.

Russell M. J., Barge L. M., Bhartia R., Bocanegra D., Bracher P. J., Branscomb E., Kidd R., McGlynn S., Meier D. H., Nitschke W., Shibuya T., Vance S., White L., and Kanik I. (2014) The drive to life on wet and icy worlds. *Astrobiology, 14,* 308–343.

Russell M. J., Murray A. E., and Hand K. P. (2017) The possible emergence of life and differentiation of a shallow biosphere on irradiated icy worlds: The example of Europa. *Astrobiology, 17(12),* 1265–1273.

Salmon J. J. and Canup R. M. (2017) Accretion of Saturn's inner mid-sized moons from a massive primordial ice ring. *Astrophys. J., 836,* 109, DOI: 10.3847/1538-4357/836/1/109.

Schenk P., Hamilton D. P, Johnson R. E., McKinnon W. B., Paranicas C., Schmidt J., and Showalter M. R. (2011) Plasma, plumes and rings: Saturn system dynamics as recorded in global color patterns on its midsize icy satellites. *Icarus, 211,* 740–757.

Schink B. (1997) Energetics of syntrophic cooperation in methanogenic degradation. *Microbiol. Molec. Biol. Rev., 61,* 262–280.

Schmidt J., Brilliantov N., Spahn F., and Kempf S. (2008) Slow dust in Enceladus' plume from condensation and wall collisions in tiger stripe fractures. *Nature, 451,* 685–688, DOI: 10.1038/nature06491.

Schrenk M. O., Kelley D. S., Bolton S. A., and Baross J. A. (2004) Low archaeal diversity linked to subseafloor geochemical processes at the Lost City Hydrothermal Field, Mid-Atlantic Ridge. *Environ. Microbiol., 6,* 1086–1095.

Schubert G., Anderson J. D., Travis B. J., and Palguta J. (2007) Enceladus: Present internal structure and differentiation by early and long-term radiogenic heating. *Icarus, 188,* 345–355, DOI: 10.1016/j.icarus.2006.12.012.

Schubert G., Sohl F., and Hussmann H. (2009) Interior of Europa. In *Europa* (R. T. Pappalardo et al., eds.), pp. 353–367. Univ. of Arizona, Tucson.

Schulte M. D. and Shock E. L. (2004) Coupled organic synthesis and mineral alteration on meteorite parent bodies. *Meteoritics & Planet. Sci., 39(9),* 1577–1590, DOI: 10.1111/j.1945-5100.2004.tb00128.x.

Seewald J. S. (2017) Detecting molecular hydrogen on Enceladus. *Science, 356(6334),* 132–133.

Sekine Y., Shibuya T., Postberg F., Hsu H.-W., Suzuki K., Masaki Y., Kuwatani T., Mori M., Hong P. K., Yoshizaki M., Tachibana S., and Sirono S. (2015) High-temperature water-rock interactions and hydrothermal environments in the chondrite-like core of Enceladus. *Nature Commun., 6(8604),* 1–8, DOI: 10.1038/ncomms9604.

Seyfried W. E. and Mottl M. J. (1982) Hydrothermal alteration of basalt by seawater under seawater-dominated conditions. *Geochim. Cosmochim. Acta, 46(6),* 985–1002, DOI: 10.1016/0016-7037(82)90054-0.

Shemansky D., Matheson P., Hall D., et al. (1993) Detection of the hydroxyl radical in the Saturn magnetosphere. *Nature, 363,* 329–331, DOI: 10.1038/363329a0.

Sherwood B. (2016) Strategic map for exploring the ocean-world Enceladus. *Acta Astron., 126,* 52–58.

Sherwood Lollar B., Frape S. K., Weise S. M., Fritz P., Macko S. A., and Welhan J. A. (1993) Abiogenic methanogenesis in crystalline rocks. *Geochim. Cosmochim. Acta, 57(23),* 5087–5097, DOI: 10.1016/0016-7037(93)90610-9.

Sherwood-Lollar B., Westgate T. D., Ward J. A., Slater G. F., and Lacrampe-Couloume G. (2002) Abiogenic formation of alkanes in the Earth's crust as a minor source for global hydrocarbon reservoirs. *Nature, 416(6880),* 522–524, DOI: 10.1038/416522a.

Shock E. L. and McKinnon W. B. (1993) Hydrothermal processing of cometary volatiles — applications to Triton. *Icarus, 106(2),* 464–477, DOI: 10.1006/icar.1993.1185.

Sleep N. H., Meibom A., Fridriksson T., Coleman R. G., and Bird D. K. (2004) H_2-rich fluids from serpentinization: Geochemical and biotic implications. *Proc. Natl. Acad. Sci. USA, 101(35),* 12818–12823, DOI: 10.1073/pnas.0405289101.

Slonczewski L., Fujisawa M., Dopson M., and Krulwich T. A. (2009) Cytoplasmic pH measurement and homeostasis in *Bacteria* and *Archaea. Adv. Microb. Physiol., 55,* 1–79.

Smith H. T., Johnson R. E., Perry M. E., Mitchell D. G., McNutt R. L., and Young D. T. (2010) Enceladus plume variability and the neutral gas densities in Saturn's magnetosphere. *J. Geophys. Res., 115,* A10252, DOI: 10.1029/2009JA015184.

Southworth B. S., Kempf S., Spitale J., Srama R., Schmidt J., and Postberg F. (2017) Resolving the mass production and surface structure of the Enceladus dust plume. *Lunar Planet. Sci. XLVIII,* Abstract #2904. Lunar and Planetary Institute, Houston.

Southworth B. S., Kempf S., and Spitale J. (2019) Surface deposition of the Enceladus plume and the zenith angle of emissions. *Icarus, 319,* 33–42.

Spahn F., Schmidt J., Albers N., Hörning M., Makuch M., Seiss M., Kempf S., Srama R., Dikarev V., Helfert S., Moragas-Klostermeyer G., Krivov A. V., Sremčević M., Tuzzolino A. J., Economou T., and Grün E. (2006) Cassini dust measurements at Enceladus and implications for the origin of the E ring. *Science, 311(5766),* 1416–1418, DOI: 10.1126/science.1121375.

Sparks W. B., Schmidt B. E., McGrath M. A., Hand K. P., Spencer J. R., Cracraft M., and Deustua S. E. (2017) Active cryovolcanism on Europa? *Astrophys. J. Lett., 839(2),* L18.

Spencer J. R. and Nimmo F. (2013) Enceladus: An active ice world in the Saturn system. *Annu. Rev. Earth Planet. Sci., 41,* 693–717, DOI: 10.1146/annurev-earth-050212-124025.

Spencer J. R., Pearl J. C., Segura M., Flasar F. M., Mamoutkine A., Romani P., Burratti B. J., Hendrix A. R., Spilker L. J., and Lopes R. M. C. (2006) Cassini encounters Enceladus: Background and the discovery of a south polar hot spot. *Science, 311(5766),* 1401–1405, DOI: 10.1126/science.1121661.

Spitale J. N., Hurford T. A., Rhoden A. R., Berkson E. E., and Platts S. S. (2015) Curtain eruptions form Enceladus' south-polar terrain. *Nature, 521,* 57–60.

Spitale J. N., Hurford T., and Rhoden A. R. (2017) Short-term variability in Enceladus' plume. *AAS/Division for Planetary Sciences Meeting Abstracts, 49,* 207.02.

Steel E., Davila A., and McKay C. P. (2017) Abiotic and biotic formation of amino acids in the Enceladus ocean. *Astrobiology, 17(9),* 862–875.

Stone E. C. and Miner E. D. (1981) Voyager 1 encounter with the saturnian system. *Science, 212(4491),* 159–163.

Storrie-Lombardi M. C. and Sattler B. (2009) Laser-induced fluorescence emission (L.I.F.E.): In situ nondestructive detection of microbial life in the ice covers of Antarctic lakes. *Astrobiology, 9(7),* 659–672.

Stryk T. (2017) Pre-discovery detection of the plumes of Enceladus. *Lunar Planet. Sci. XLVIII,* Abstract #1603. Lunar and Planetary Institute, Houston.

Summons R. E., Albrecht P., McDonald G., and Moldowan J. M. (2008) Molecular biosignatures. *Space Sci. Rev., 135,* 133–159.

Taubner R. S., Schleper C., Firneis M. G., and Rittmann S. K. M. R. (2015) Assessing the ecophysiology of methanogens in the context of recent astrobiological and planetological studies. *Life, 5,* 1652–

1686.

Taubner R. S., Pappenreiter P., Zwicker J., Smrzka D., Pruckner C., Kolar P., Bernacchi S., Seifert A. H., Krajete A., Bach W., Peckmann J., Paulik C., Firneis M. G., Schleper C., and Rittmann S. K. (2018) Biological methane productions under putative Enceladus-like conditions. *Nature Commun.*, *9*, 748.

Teolis B. D., Perry M. E., Magee B. A., Westlake J., and Waite J. H. (2010) Detection and measurement of ice grains and gas distribution in the Enceladus plume by Cassini's Ion Neutral Mass Spectrometer. *J. Geophys. Res., 115*, A09222.

Teolis B. D., Perry M. E., Hansen C. J., Waite J. H., Porco C. C., Spencer J. R., and Howett C. J. A. (2017) Enceladus plume structure and time variability: Comparison of Cassini observations. *Astrobiology, 17(9),* 926–940.

Terrile R. J. and Cook A. F. (1981) Enceladus: Evolution and possible relationship to Saturn's E-ring. *Lunar Planet. Sci. XII, Suppl. A, Satellites of Saturn*, p. 10. LPI Contribution No. 428, Lunar and Planetary Institute, Houston.

Thauer R. K., Kaster A. K., Seedorf H., Buckel W., and Hedderich R. (2008) Methanogenic archaea: Ecologically relevant differences in energy conservation. *Nature Rev. Microbiol., 6*, 579–591.

Thomas P. C., Tajeddine R., Tiscareno M. S., Burns J. A., Joseph J., Loredo T. J., Helfenstein P., and Porco C. (2016) Enceladus's measured physical libration requires a global subsurface ocean. *Icarus, 264*, 37–47, DOI: 10.1016/j.icarus.2015.08.037.

Tian F., Stewart A. I. F., Toon O. B., Larsen K. W., and Esposito L. W. (2007) Monte Carlo simulations of the water vapor plumes on Enceladus. *Icarus, 188(1),* 154–161, DOI: 10.1016/j.icarus.2006.11.010.

Tijhuis L., van Loosdrecht M. C. M., and Heijnen J. J. (1993) A thermodynamically based correlation for maintenance Gibbs energy requirements in aerobic and anaerobic chemotrophic growth. *Biotech. Bioeng., 42*, 509–519.

Tokar R. L., Johnson R. E., Hill T. W., Pontius D. H., Kurth W. S., Crary F. J., Young D. T., Thomsen M. F., Reisenfeld D. B., Coates A. J., Lewis G. R., Sittler E. C., and Gurnett D. A. (2006) The interaction of the atmosphere of Enceladus with Saturn's plasma. *Science, 311*, 1409–1412, DOI: 10.1126/science.1121061.

Tortora P., Zannoni M., Hemingway D., Nimmo F., Jacobson R. A., Iess L., and Parisi M. (2016) Rhea gravity field and interior modeling from Cassini data analysis. *Icarus, 264*, 264–273, DOI: 10.1016/j.icarus.2015.09.022.

Truong N., Monroe A. A., Glein C. R., Anbar A. D., and Lunine J. I. (2019) Decomposition of amino acids in water with application to in-situ measurements of Enceladus, Europa and other hydrothermally active icy ocean worlds. *Icarus, 329*, 140–147.

Travis B. J. and Schubert G. (2015) Keeping Enceladus warm. *Icarus, 250*, 32–42, DOI: 10.1016/j.icarus.2014.11.017.

Tsou P., Brownlee D. E., McKay C. P., Anbar A. D., Yano H., Altwegg K., Beegle L. W., Dissly R., Strange N. J., and Kanik I. (2012) LIFE: Life Investigation For Enceladus — A sample return mission concept in search for evidence of life. *Astrobiology, 12*, 730–742, DOI: 10.1089/ast.2011.0813.

Tyburczy J. A., Duffy T. S., Ahrens T. J., and Lange M. A. (1991) Shock wave equation of state of serpentine to 150 GPa: Implications for the occurrence of water in the Earth's lower mantle. *J. Geophys. Res.–Solid Earth, 96(B11),* 18011–18027, DOI: 10.1029/91JB01573.

Tyler R. (2014) Comparative estimates of the heat generated by ocean tides on icy satellites in the outer solar system. *Icarus, 243*, 358–385, DOI: 10.1016/j.icarus.2014.08.037.

Vance S., Harnmeijer J., Kimura J., Hussmann H., deMartin B., and Brown J. M. (2007) Hydrothermal systems in small ocean planets. *Astrobiology, 7(6),* 987–1005, DOI: 10.1089/ ast.2007.0075.

Vance S. D., Hand K. P., and Pappalardo R. T. (2016) Geophysical controls of chemical disequilibria in Europa. *Geophys. Res. Lett.*, 43, 4871–4879.

Verbiscer A., French R., Showalter M., and Helfenstein P. (2007) Enceladus: Cosmic graffiti artist caught in the act. *Science, 315(5813),* 815.

Waite J. H., Combi M. R., Ip W.-H., Cravens T. E., McNutt R. L., Kasprzak W., Yelle R., Luhmann J., Niemann H., Gell D., Magee B., Fletcher G., Lunine J. I., and Tseng W.-L. (2006) Cassini ion and neutral mass spectrometer: Enceladus plume composition and structure. *Science, 311(5766),* 1419–1422, DOI: 10.1126/science.1121290.

Waite J. H., Lewis W. S., Magee B. A., Lunine J. I., McKinnon W. B., Glein C. R., Mousis O., Young D. T., Brockwell T., Westlake J., Nguyen M.-J., Teolis B. D., Niemann H. B., McNutt R. L., Perry M., and Ip W.-H. (2009) Liquid water on Enceladus from observations of ammonia and ⁴⁰Ar in the plume. *Nature, 460(7254),* 487–490, DOI: 10.1038/nature08153.

Waite J. H., Glein C. R., Perryman R. S., Teolis B. D., Magee B. A., Miller G., Grimes J., Perry M. E., Miller K. E., Bouquet A., Lunine J. I., Brockwell T., and Bolton S. J. (2017) Cassini finds molecular hydrogen in the Enceladus plume: Evidence for hydrothermal processes. *Science, 356(6334),* 155–159, DOI: 10.1126/science.aai8703.

Waite J. H., Perryman R S., Perry M. E., Miller K. E., Bell J., Cravens T. E., Glein C. R., Grimes J., Hedman M., Cuzzi J., Brockwell T., Teolis B., Moore L., Mitchell D. G., Persoon A., Kurth W. S., Wahlund J.-E., Morooka M., Hadid L. Z., Chocron S., Walker J., Nagy A., Yelle R., Ledvina S., Johnson R., Tseng W., Tucker O. J., and Ip W.-H. (2018) Chemical interactions between Saturn's atmosphere and its rings. *Science, 362(6410),* eaat2382.

Walsh K. J., Morbidelli A., Raymond S. N., O'Brien D. P., and Mandell A. M. (2011) A low mass for Mars from Jupiter's early gas-driven migration. *Nature, 475*, 206–209, DOI: 10.1038/nature10201.

Watson S. W., Novitsky T. J., Quinby H. L., and Valois F. W. (1977) Determination of bacterial number and biomass in the marine environment. *Appl. Environ. Microbiol., 33(4),* 940–946.

Whitman W. B., Coleman D. C., and Wiebe W. J. (1998) Prokaryotes: The unseen majority. *Proc. Natl. Acad. Sci. USA, 95*, 6578–6583.

Worth R. J., Sigurdsson S., and House C. H. (2013) Seeding life on the moons of the outer planets via Lithopanspermia. *Astrobiology, 13*, 1155–1165, 10.1089/ast.2013.1028.

Ye S.-Y., Gurnett D. A., Kurth W. S., Averkamp T. F., Kempf S., Hsu H.-W., Srama R., and Grün E. (2014) Properties of dust particles near Saturn inferred from voltage pulses induced by dust impacts on Cassini spacecraft. *J. Geophys. Res.–Space Physics, 119(8),* 6294–6312.

Yeoh S. K., Li Z., Goldstein D. B., Varghese P. L., Levin D. A., and Trafton L. M. (2017) Constraining the Enceladus plume using numerical simulation and Cassini data. *Icarus, 281*, 357–378.

Yomogida K. and Matsui T. (1983) Physical properties of ordinary chondrites. *J. Geophys. Res.–Solid Earth, 88(B11),* 9513–9533, DOI: 10.1029/JB088iB11p09513.

Zhang K. and Nimmo F. (2009) Recent orbital evolution and the internal structures of Enceladus and Dione. *Icarus, 204(2),* 597–609, DOI: 10.1016/j.icarus.2009.07.007.

Zolotov M. Yu. (2007) An oceanic composition on early and today's Enceladus. *Geophys. Res. Lett., 34*, L23203, DOI: 10.1029/2007GL031234.

Zolotov M. Yu. and Kargel J. S. (2009) On the chemical composition of Europa's icy shell, ocean, and underlying rocks. In *Europa* (R. T. Pappalardo et al., eds.), pp. 431–458. Univ. of Arizona, Tucson.

Lunine J. I., Cable M. L., Hörst S. M., and Rahm M. (2020) The astrobiology of Titan. In *Planetary Astrobiology*
(V. Meadows et al., eds.), pp. 247–266. Univ. of Arizona, Tucson, DOI: 10.2458/azu_uapress_9780816540068-ch010.

The Astrobiology of Titan

Jonathan I. Lunine
Cornell University

Morgan L. Cable
NASA Jet Propulsion Laboratory

Sarah M. Hörst
Johns Hopkins University

Martin Rahm
Chalmers University of Technology

Titan's surface and interior provide a diverse range of locations to look for evidence of past and present life, while its atmosphere is the starting point for experiments in both biotic and prebiotic chemistry. An interior water ocean is the most Earth-like site for life. However, it is also the most remote and difficult to access. Transient sites of melting of the water ice crust, driven by cryovolcanism or hypervelocity impacts and suffused with organic molecules, may preserve the initial stages of chemical evolution toward life but are geologically short-lived (less than a million years). The organic-rich dunes, formed and driven by the wind, may host cycles of charging and discharging that could lead to some amount of chemical evolution. The environments around and within the methane-dominated seas are potential hosts of an exotic but as yet speculative biochemistry. These possibilities for life or its precursors provide ample rationale for the future exploration of Titan's surface.

1. WHAT WE KNOW OF TITAN RELEVANT TO ASTROBIOLOGY

Titan, Saturn's largest moon, is the only moon in the solar system that possesses a thick atmosphere. As with Earth, Titan's atmosphere is dominated by molecular nitrogen (N_2). In addition to N_2 (~98%), methane (CH_4) is also present (~2%), and photochemistry initiated by photons and energetic particles in Titan's upper atmosphere result in the formation of more complex organic compounds and eventually Titan's characteristic haze layer (see, e.g., *Hörst,* 2017). Titan's atmospheric inventory of relatively small molecules has been well documented through a combination of remote sensing and *in situ* observations from both Earth-based telescopes and the Voyager and Cassini-Huygens spacecraft. In addition to these molecules [such as acetyelene (C_2H_2), hydrogen cyanide (HCN), benzene (C_6H_6), vinyl cyanide (C_2H_3CN), etc.], the Cassini Plasma Spectrometer (CAPS) revealed the presence of very heavy ions in Titan's iono-sphere with mass to charge (m/z) up to 10,000 for negative ions (*Coates et al,*. 2007) and up to ~350 for positive ions (*Crary et al.,* 2009). Identification of the composition of these ions will require future measurements, but labora-tory experiments indicate that Titan's atmosphere is likely capable of producing molecular complexity that includes molecules of prebiotic interest such as nucleobases and amino acids (see, e.g., *Hörst et al.,* 2012). These very heavy ions are processed further in the atmosphere before reaching Titan's surface and potentially undergoing additional physi-cal and/or chemical modification.

Measurements from the Cassini-Huygens mission pro-vided substantial insights into the surface geology and atmosphere-surface interactions, revealing a world that is dominated by extensive sand seas at low latitudes, a mixture of enigmatic terrains at mid latitudes, and hydrocarbon lakes, seas, and rivers in the polar regions (filled in the case of the north pole, dry in the south). Although the surface appears to be carved by physical processes that are familiar from Earth and other terrestrial worlds, the composition of the surface is unique and poorly constrained. Titan's surface should be dominated by water ice (and potentially ammonia ice) and organic liquids and solids based on Titan's density, degree of differentiation, and atmospheric chemistry, but definitive surface composition measurements have proven elusive from orbit. The lack of composition information prevents surface geological processes from being well understood.

Due to Saturn's axial tilt (~26.7°), Titan experiences strong seasonal changes over its year (~29.5 Earth years), including changes in the large- and small-scale atmospheric structure, the altitude of the detached haze layer, the location of clouds, and presumably rain. The Cassini-Huygens mission lasted almost half a Titan year, allowing observation of these seasonal changes. Titan's methane cycle also experiences longer timescale cyclic variations as it appears that the distribution of currently filled lakes and seas (almost exclusively at the north pole) is the result of Croll-Milankovich cycles and that the location of these lakes and seas varies on 10–100-thousand-year cycles (*Aharonson et al.*, 2009; *Lora et al.*, 2014). Finally, the origin of Titan's methane remains unknown. The current atmospheric CH_4 is only stable against photochemistry for a few tens of millions of years (*Yung et al.*, 1984; *Wilson and Atreya*, 2004) and it is not clear if Titan's methane is resupplied constantly, episodically (over tens or hundreds of millions of years), or so sporadically that one must conclude the present state to be unusual or even unique in Titan's history.

Titan is often described as the "Earth-like moon," and indeed the balance of the various geologic and chemical processes that shaped Titan is more similar to Earth's than any other body in the solar system. Titan is noteworthy for its substantial organic inventory and complex chemistry. It possesses two different liquid environments: the subsurface liquid water ocean and the methane/ethane lakes and seas. There is ample solar energy to drive chemistry in the atmosphere. Titan therefore has the basic ingredients generally assumed to be required for life (*National Research Council*, 2007), and has been identified as an important astrobiological target since the 1970s (*Sagan*, 1974). However, Titan seems to lack other attributes that may be important for life, such as plate tectonics and a global-scale water-based ocean on the surface. Although there are liquids on the surface, they are methane with admixtures of ethane and nitrogen (*Mitchell et al.*, 2015; *Mastrogiuseppe et al.*, 2014, 2018). Biologically interesting molecules may have limited solubility in such liquids. The atmosphere absorbs much of the solar energy that reaches Titan and therefore most of the energy that reaches the surface is in the form of chemical bonds (in molecules like acetylene) rather than photons (see, e.g., *Lunine and Hörst*, 2011). Titan is therefore still an extremely interesting laboratory for the study of astrobiology, but there are numerous questions about whether, and what type of, life could originate and evolve on Titan.

Below we discuss four broad location types that are astrobiologically interesting: the subsurface liquid water ocean, frozen aqueous surface sites, the hydrocarbon seas,

and dunes (section 2). In section 3 we discuss the possible types of life forms and chemical processes that might support them, and in section 4 we discuss the measurements that are needed to more fully characterize Titan's ability to host life. Finally, in section 5 we present some lessons for methane-dominated exoplanets.

2. DESCRIPTION OF THE FOUR SITE TYPES FOR LIFE

2.1. Ocean

There are several pieces of evidence that suggest that Titan supports a subsurface water ocean, and hydrothermal activity at the ocean/interior interface may also occur. A subsurface ocean is inferred from measurements of the tidally-induced change in Titan's shape using Cassini's Radio Science Subsystem (RSS) (*Iess et al.*, 2012); the tidal response is so large that a liquid of density equal to or even slightly larger than that of water underlying a crust approximately 100 km thick is indicated (*Mitri et al.*, 2014). A second indicator of the liquid water ocean comes from interpreting the electric field measurements by the Huygens probe in terms of a "Schumann-like" resonance in a dielectric cavity bounded some 60 km below the surface by a high conductivity layer — consistent with a water ocean with salts, ammonia, or both (*Béghin et al.*, 2012). Finally, rotational data are also consistent with an ice shell decoupled from the interior (*Lorenz et al.*, 2008). The presence of ammonia in Titan's interior is supported by the Huygens-measured atmospheric Ar/N_2 ratio, which indicates ammonia as the primordial parent of the atmospheric molecular nitrogen (*Niemann et al.*, 2010), but the amount is poorly constrained by Cassini data (*Mitri et al.*, 2014). Interior models constrained by Titan's bulk properties and ratio of rock-to-ice perch the ocean between an ice-I (or methane clathrate) upper crust and a layer of high-pressure ice below (*Tobie et al.*, 2005, 2009). However, Titan's moment of inertia can be interpreted (non-uniquely) (*Gao and Stevenson*, 2013) as implying a partially hydrated rock core, which in turn leads to a thinner high-pressure ice layer (*Castillo-Rogez and Lunine*, 2010). That would allow for hydrothermal activity reaching the ocean.

Given how little we know of the ocean, speculation on the types of organisms one might find is exactly that: speculation. However, other than the basal pressure of the ocean, which exceeds 5 kbar in typical models (*Tobie et al.*, 2014), an ocean in contact with the rock beneath might offer a relatively Earth-like environment for microorganisms — saline waters at a temperature as high as 300 K (*Tobie et al.*, 2009). However, as described in section 3.1, it is possible in the laboratory to produce subclasses of microbes that exist at higher pressures (*Vanlint et al.*, 2011).

Surface expressions of the ocean may be difficult to find, given that the surface seems to be dominated by depositional processes. However, evidence in the form of warm diapirs rising through a thickening ice-I crust (*Tobie et al.*, 2006) might be sought by future missions. Whether organics might have interacted with the ocean depends on the thickness and stability of the crust, which in turn is a function of the thermal history of the interior.

2.2. Frozen-Aqueous Surface Sites

Although Titan's surface temperature is too cold for liquid water to be stable on geolologic timescales, impacts and cryovolcanism may result in the formation of transient bodies of liquid water. The organic molecules produced in Titan's atmosphere are not exposed to liquid water (and are exposed to very little gas phase water) at any point during their time in the atmosphere. Frozen aqueous surface sites are therefore particularly interesting because they represent locations where organics produced in the atmosphere have been further processed by aqueous reactions and are important for understanding the kinetics and products of abiotic chemistry.

As mentioned above, Titan's surface composition remains poorly constrained and therefore study of this aqueous processing has been restricted to the laboratory. Titan haze analogs (often called "tholins") (see, e.g., *Cable et al.,* 2012) have been subjected to hydrolysis in liquid water or water/ammonia mixtures at different temperatures by a number of different groups to look for potential hydrolysis products, reaction types, and reaction kinetics. Consistent among the different experiments is the transformation of laboratory haze analogs that are oxygen-poor by design into relative oxygen-rich products through various hydrolysis reactions. These experiments also consistently find the production of a number of proteinogenic amino acids and the nucleobases used by life on Earth (see, e.g., *Somogyi et al.,* 2005; *Neish et al.,* 2008, 2009, 2010; *Ramirez et al.,* 2010; *Poch et al.,* 2012; *Cleaves et al.,* 2014; *Brassé et al.,* 2017). These experiments all point to the fact that anywhere on Titan's surface where liquid water has come into contact with organics produced in Titan's atmosphere there are likely to be deposits of biologically interesting molecules.

Whether cryovolcanism has actually occurred on Titan anytime during its history remains an unanswered question (see, e.g., *Hörst,* 2017). The surface feature that most looks like a cryovolcanic remnant is the Sotra Patera region where what appear to be frozen flows are present that are tens of meters thick (*Lopes et al.,* 2013). Features of that thickness would freeze in ~1 year (*Neish et al.,* 2018). Even large cryovolcanic structures would likely freeze in a few hundred years (*Neish et al.,* 2006). If the source included ammonia, which would partly overcome the strong buoyancy penalty of liquid water, the flowing cryolava may have been close to its "peritectic" point where water ice and ammonia dihydrate coexist at ~176 K — and is therefore quite viscous (*Kargel,* 1992; *Neish et al.,* 2018). Laboratory hydrolysis experiments have not been conducted at such low temperatures. However, experiments conducted at 253 K with ammonia water mixtures found a half-life for reactions of a few days (*Neish et al.,* 2009). Using the

Arrhenius equation to extrapolate those rates to lower temperatures (section 3.3.1), it would take a few hundred years for them to complete at 176 K and therefore cryovolcanic flows may freeze completely before hydrolysis reactions occurring between Titan's organic inventory and water or ammonia water mixtures in the flow have run to completion (*Neish et al.,* 2018).

Titan's thick atmosphere shields the surface from small impactors, thus objects that survive to impact the surface are less frequent but larger. Indeed, Titan's surface exhibits a relatively small number of confirmed impact craters (*Wood et al.,* 2010; *Neish et al.,* 2016), which is likely due to a combination of this atmospheric shielding and a surface age of approximately 1 Ga (see, e.g., *Neish and Lorenz,* 2012; *Hörst,* 2017). Impacts into Titan's icy crust result in substantial melting and the amount of melt scales with impactor size. The freezing timescales range from hundreds to tens of thousands of years depending on the size of the impactor and the composition of the melt produced by the impact (*Thompson and Sagan,* 1992; *Artemieva and Lunine,* 2003; *O'Brien et al.,* 2005). Not only are these bodies of liquid likely to persist substantially longer than cryovolcanic constructs, but impact-generated melt pools can start out well above the water liquidus (273 K), rather than the ammonia water peritectic (176 K), allowing even more time for hydrolysis reactions to complete. Nonetheless, the meltwater survival timescales are short on geologic timescales, thus providing at best a strong test for the rate at which chemical complexification might occur in a prebiotic aqueous medium.

The amount of biologically interesting organic molecules trapped in frozen impact or cryolava sites is likely to be limited by production rates in the atmosphere to 1–10 M, less than the amounts derived from laboratory solubility studies (*Neish et al.,* 2018). These amounts will be difficult to detect by spectroscopy (even by a low-flying vehicle), and thus may require direct sampling (see section 6).

2.3. Methane Seas

2.3.1. Background on Titan's lakes and seas. The first high-resolution images of Titan's lakes were acquired by Cassini Synthetic Aperture Radar (SAR) in July 2006 — approximately 75 lacustrine ("lake-related") features were identified in the north polar region (*Stofan et al.,* 2007). To date, more than 650 seas (maria), lakes (lacūs), empty lake basins, and paleoseas have been identified, both in the north and south polar regions, with 577 of these presently filled with liquid (*Hayes,* 2016; *Hayes et al.,* 2008). Some candidate lakes have been identified in the equatorial (*Griffith et al.,* 2012) and mid-latitude (*Vixie et al.,* 2015) regions. If confirmed, the stability of these lakes might suggest the existence of a subsurface hydrocarbon reservoir, such as a "groundwater" table (*Hayes,* 2016; *Hayes et al.,* 2008; *Cornet et al.,* 2012), since methane rain appears to evaporate quickly at the equator after storms (*Turtle et al.,* 2011). As with Earth and early Mars, the size-frequency distribution of moderate to large lacustrine features follows a power law (*Hayes, 2016*).

Titan's three large seas — Kraken Mare, Ligeia Mare, and Punga Mare — comprise 80% of all liquid on Titan's surface by area (*Hayes et al., 2008*) and are all located in the north polar region (Table 1). Altimetry measurements by Cassini RADAR suggest all three are hydraulically connected (*Hayes et al., 2017*). Only one moderately-sized lake, Ontario Lacus, exists in the south polar region; this asymmetry is most likely due to orbital forcing from Saturn, as southern summer solstice (the longest day on Titan) matches almost perfectly with perihelion (the point in Saturn's orbit when it is closest to the Sun), which yields a peak solar flux that is 25% higher for southern summer compared to northern summer (*Aharonson et al., 2009*). If correct, this Croll-Milankovich type cycle will vary with Saturn's orbital parameters, leading to the lakes swapping poles in ~30,000 years with a full period of 125,000 years (*Lora et al., 2014*).

The lakes and seas of Titan are primarily comprised of liquid methane (*Mastrogiuseppe et al., 2014; Mitchell et al., 2015*), although liquid ethane has also been spectroscopically identified in the northern seas and in Ontario and Jingpo Lacūs (*Brown et al., 2008; Clark et al., 2010*). However, due to the absorption of near-infrared wavelengths within the first few millimeters of the lake surface, ethane concentration throughout the lakes and seas cannot be estimated accurately from this data [from the Visible and Infrared Mapping Spectrometer (VIMS) instrument]. Reported models vary widely in terms of methane- vs. ethane-rich compositions (*Cordier et al., 2009, 2012; Glein and Shock, 2013; Tan et al., 2013, 2015; Lorenz, 2014*). Recent laboratory work on the transparency of hydrocarbon mixtures at radio wavelengths (*Mitchell et al., 2015*), applied to Cassini RADAR altimetry data over the seas and one lake, suggests the northern seas are much richer in methane than ethane (*Mastrogiuseppe et al., 2016*) (Table 1). However, the presence of even minor concentrations of other species such as HCN and acetonitrile may affect the loss tangent, and therefore the predicted lake compositions (*Mastroguiseppe et al., 2018*). More laboratory work is needed on these species, dissolved individually and in various combinations in methane and ethane, to adequately address lake composition on Titan.

Some of the smaller lakes, termed sharp-edged depressions, are morphologically different from the larger lakes and seas (*Hayes et al., 2008, 2017*), and those that are filled appear to be compositionally different as well. SAR backscatter measurements suggest that either these depressions are extremely deep, or they are made up of more absorptive species (*Hayes et al., 2008*). The latter supports geophysical evidence that these depressions form via a dissolution-type mechanism, becoming saturated in organics such as butane and acetylene. However, raised rims on the order of 100 m for a subset of these depressions cannot be explained by dissolution erosion alone; other formation mechanisms such as sublimation could also be at work (*Birch et al., 2018*).

2.3.2. Lakes/seas as an environment for life. It is possible that life requires a liquid phase, to allow molecules to diffuse and interact (e.g., *Ball, 2001; Pross and Pascal, 2013*). The lakes and seas of Titan offer a unique test of whether the liquid phase supporting life must be water (or even, in a more general sense, polar), or if other solvents can also support their own biochemistries and life (*Ball, 2001; Benner et al., 2004*).

The time required for life to emerge on Earth is still largely unknown; most theories reported in the literature range from 0.1 to ~25 m.y. after conditions stabilized sufficiently to allow liquid water to persist (*Sleep et al., 2001; Lazcano and Miller, 1994; Anderson and Roth, 1977; Oberbeck and Fogleman, 1989*), and neglecting a protracted

TABLE 1. Size, depth, and composition of Titan's largest seas and lakes.

Sea/Lake	Surface Area (× 10^4 km^2)	Max. Depth (m)	Est. Volume (× 10^2 km^3)	Polar Region	Best-Fit Composition (vol.%)[‡]
Kraken Mare	50[*]	>190[§]	n.d.	North	n.d.
Ligeia Mare	13[*]	200[‡]	140[†]	North	71% CH_4, 12% C_2H_6, 17% N_2
Punga Mare	6.1[*]	103[§]	n.d.	North	Overlaps with Ligeia Mare
Ontario Lacus	n.d.	90[¶]	5.6[†]	South	59% CH_4, 28% C_2H_6, 13% N_2

[*] From *Hayes et al.* (2008).
[†] From *Hayes et al.* (2017).
[‡] From *Mastrogiuseppe et al.* (2016, 2018). Note that the uncertainties in these numbers are considerable and derive from multiple sources as discussed in the cited sources.
[§] From *Lorenz et al.* (2014); signal from the bottom not detected so the depth is a lower limit.
[¶] From *Hayes* (2016).

n.d. = no data

period in which sterilizing impacts frustrated the origin of life at any location (*Maher and Stevenson,* 1988). As mentioned above, evidence suggests that the lakes and seas of Titan are stable over timescales of tens of thousands of years, and move from south to north pole and back every 125,000 years (*Lora et al.,* 2014). Based on these very rough estimates, it may seem that a unique, nonaqueous origin of life on Titan might not have had enough time to form. However, the present-day lakes and seas may be a remnant of a once much larger and more stable global-scale ethane-methane sea (*Birch et al.,* 2017) and perhaps life within them evolved to withstand long dry periods as do spore-formers on Earth. Alternatively, the polar regions may host subsurface methane and ethane liquids (*Choukroun and Sotin,* 2012), enabling organisms to survive the "dry" part of Titan's Croll-Milankovich cycle.

Stagnant environments tend to be less conducive to supporting life than active ones; a general requirement for life is departure from thermodynamic equilibrium and hence access to a fresh supply of nutrients and a means to dispose of waste products. In this respect, the lakes and seas of Titan may serve putative life well. As noted above, the three largest seas in the north polar region appear to be hydraulically connected (*Lorenz,* 2014), enabling exchange of materials between them. Also, due to the fact that the lakes are comprised of multiple components, compositional stratification may occur (*Lu et al.,* 1969, 1970; *Hartwig et al.,* 2018) and has been observed in laboratory experiments (*Hanley et al.,* 2017). Sunlight absorbed near the surface can generate a warm liquid layer during summer that could cool in the winter and cause overturn (*Tokano,* 2005, 2009a). Such convective mixing could return the lake to a state of disequilibrium, and provide fresh access to material on a regular basis. In addition, methane fluvial (flowing) or pluvial (rain) events could lead to rapid exsolution of dissolved nitrogen (which is less soluble in methane than ethane; as the lake methane concentration increases, this would force the nitrogen out of solution and into the gas phase) and mixing within the lake (*Malaska et al.,* 2017).

Waves may enable periodic exchange or movement of material, but are unlikely to be the dominant process; Titan's low winds (≤1.5 m s⁻¹) mean waves rarely achieve heights over 40 cm (*Lorenz and Hayes,* 2012). The mysterious "Magic Islands," transient bright radar features observed on Ligeia and Kraken maria (*Hofgartner et al.,* 2014), indicate some type of activity is occurring at the sea surfaces; plausible explanations include bubbles, floating solids, or waves (*Hofgartner et al.,* 2014, 2016). We note that icebergs could not occur in Titan lakes, due to the higher densities of solid methane and ethane compared to their liquid phases (*Tokano,* 2009a). If bubbles are the primary mechanism for the "Magic Islands," physical processes such as bubble scrubbing may occur, transporting material from the lake floor up to the surface on the bubble walls. On Earth, this process can enrich surface materials in cells and hydrophobic molecules by several orders of magnitude (*Carlucci and Williams,* 1965; *Porco et al.,* 2017); on Titan,

the enrichment would likely be of molecules of the opposite sort — hydrophilic species.

Enrichment notwithstanding, putative biochemistry on Titan would have to get creative to work around a harsh reality based on thermodynamics: Most molecules do not dissolve in nonpolar solvents at cryogenic temperatures (see the chapter by Hoehler et al. in this volume). Life (as we know it) requires large molecules, such as DNA, RNA, and proteins, to store information and carry on the molecular machinery of life. However, solubility of any solute in any solvent is temperature-dependent, and favors higher temperature. Even a simple nonpolar molecule like benzene has a solubility of 18.5 mg L⁻¹ in liquid ethane at 94 K (*Malaska and Hodyss,* 2014; *Diez-y-Riega et al.,* 2014), which is three orders of magnitude lower than the ~66,000 Da globular protein bovine serum albumin (BSA) in water at room temperature (40 mg mL⁻¹ = 40,000 mg L⁻¹). Adding a second benzene ring to make naphthalene drops the solubility to 0.16 mg L⁻¹, equivalent to oil in water (palmitic acid solubility in water is 0.72 mg L⁻¹ at 20°C). So if life were to exist in a Titan lake, it might have to do so without large molecules to store genetic information or perform complex biochemical reactions. These examples refer to exposed, freely-floating molecules; for a discussion of possibilities that might improve solubility in liquid hydrocarbons through compartmentalization, see section 3.3.

Even if the bulk lake liquid is not conducive to putative life, a suitable environment might be found at the interface of liquid and solid — the "benthic layer" or "sludge" of the hydrocarbon seafloor (*Le Gall et al.,* 2016) — where large, insoluble molecules could be utilized while maintaining access to the liquid phase for nutrients and removal of waste. Or, perhaps life could subsist in the deposits the lakes leave behind. Lakes appear to evaporate, based on the appearance of "bathtub ring" margins that might be evaporites (*Barnes et al.,* 2009, 2011; *MacKenzie and Barnes,* 2016), although it is also possible that liquid is infiltrating underground. Small ponds at the margins of seas may also partially evaporate as tides periodically leave them isolated (*Tokano et al.,* 2014; *Vincent et al.,* 2018), akin to the evaporating ponds that Darwin imagined might be sites of life's origin on Earth.

Evaporites left behind by evaporating lakes or ponds could serve as areas where nutrients are concentrated, in minerals such as co-crystals or clathrates (*Maynard-Casely et al.,* 2018). These locations could provide chemical gradients that life could exploit. For example, as discussed in section 3.3, the metabolism of acetylene has been proposed as a mechanism for chemoautotrophy (deriving energy from a chemical source) on Titan (*Schulze-Makuch and Grinspoon,* 2005; *McKay and Smith,* 2005; *Tokano,* 2009b). Storage of acetylene in a Titan mineral such as a co-crystal may enable access to this resource, in a similar manner to the storage of carbon dioxide in carbonate deposits on Earth (*Cable et al.,* 2018). Further laboratory work would help elucidate possible mechanisms by which a Titan biome might subsist around pockets of chemical energy stored in the crust. Field studies of special sites on Earth where hydrocarbons mix with aqueous-based organisms (*Schulze-Makuch et al.,* 2011) might also provide some insight into analogous crustal interfaces on Titan.

2.4. Dunes

Titan has widespread regions of near-parallel, longitudinal dunes spanning the equatorial region (*Lorenz et al.,* 2006) and covering ~20% of the total surface (*Lorenz and Radebaugh,* 2009). Based on physical properties such as height, longitudinal symmetry, and superposition on other features, these dunes appear to be formed by depositional processes. Solid organic particles generated by complex photochemistry in the atmosphere are a likely source. The radar-dark appearance and high emissivity in the microwave range of the dunes (resulting in a low dielectric constant) are also consistent with this theory (*Lorenz et al.,* 2006).

The vast majority of the dunes appear based on morphology to be oriented nearly due eastward (*Lorenz and Radebaugh,* 2009). Varying orbital forcing may lead to changes in surface winds at low latitudes on Titan, causing dunes to reorient on a timescale of ~10^5 years (*Ewing et al.,* 2015). Methane rainstorms might also drive the net orientation (*Charnay et al.,* 2015). Due to Titan's low gravity and high atmospheric density, particles on the order of 0.18–0.25 mm (sand-sized) could be transported by winds (in a process called saltation) with a velocity as low as 0.1 m s^{-1} (*Lorenz et al.,* 2006). Transport in this manner could lead to triboelectric (frictional) charging of dune grains as they move against each other, which could lead to physical effects like aggregation (*Mendez-Harper et al.,* 2017), electric discharge, or even breaking/forming covalent bonds via tribochemistry (*Thomas and Beauchamp,* 2014). It is possible that life could take advantage of these energy gradients and mechanisms, although the idea that such dunes might host life may be more speculative than for the other sites considered in this section.

3. EXAMINATION OF POSSIBLE BIOLOGY ON TITAN

3.1. Deep Ocean

Modeling suggests that the subsurface liquid water ocean of Titan may have been stable over the entire 4.5-G.y. lifetime of Titan (*Grasset and Sotin,* 1996; *Grindrod et al.,* 2008), allowing ample time for life to have formed. Pressures in this ocean are likely to be in the range of 100–500 MPa (*Fortes,* 2000). Piezophilic (pressure-loving) organisms thrive at the ocean floors on Earth, where pressures reach 110 MPa (*Horikoshi,* 1998; *Kato and Bartlett,* 1998), and the most piezophilic organism isolated from the Mariana Trench so far can grow at pressures up to 130 MPa and possibly higher (*Yayanos et al.,* 1981; *Picard and Daniel,* 2013). A study of directed evolution of *E .coli*

produced a main population that survived pressures up to 700 MPa, and a subpopulation that survived up to 2 GPa (*Vanlint,* 2011). It is possible that organisms based on the biochemistry of life "as we know it" may thrive in the subsurface ocean of Titan.

If ammonia were present in the Titan ocean at a composition (~32%) corresponding to that at the minimum melting point, the pH could be as high as 11.4, but increasing pressure acts to shift the dissociation constant of water to reduce the pH to ~10.5 (*Fortes,* 2000). Alkaliphilic (high-pH-loving) organisms can survive up to pH 13.2 (*Rampelotto,* 2010, 2013; *Roadcap et al.,* 2006), well within this range.

3.2. Frozen-Remnants of Aqueous Prebiotic Experiments

The short timescales for liquid water in cryovolcanic or aqueous deposits, discussed in section 2.2, lead us to expect that preserved samples would be of prebiotic chemistry rather than of life itself. Of course, this should not be used to rule out testing such sites for life; it would indeed be a spectacular discovery if preserved cellular forms were found in such an environment. More likely is that one might recover the molecular remnants of aqueous organic chemistry that proceeded for some interval of time ranging from decades to millennia. Such natural chemical experiments would be of extremely high value given that it is difficult to perform such experiments in terrestrial laboratories over much shorter timescales than even a year, let alone centuries or millennia. Their utility would depend on some way to independently determine the duration of liquid water at the sampling site, as well as characterization of the aqueous environment (pH, ammonia content, etc.) that hosted the prolonged organic chemistry.

The organic products recovered could range from a diverse suite of prebiotic molecules such as amino acids and nucleic acid bases, along with other organic molecules, to more specific and organized product sets that might be suggestive of a system climbing the ladder toward biochemistry. Are the chiral molecules dominated by one or the other enantiomer? Is there evidence of production of preferred catalysts or information-carrying molecules in so-called autocatalytic cycles? Are the heaviest species so complex [defined by some measure as in, e.g., *Marshall et al.* (2017)] as to largely preclude abiotic production, thereby suggesting biochemistry? All these intriguing questions are the purview of such natural experiments, with great longevity relative to the laboratory but brief compared to geologic processes, and accessible only by surface sampling or (more likely) drilling.

3.3. Non-Aqueous Solid and Liquid Chemistry

3.3.1. Prerequisites for dynamic chemistry on the surface of Titan. The thermodynamic and kinetic requirements on chemical systems in order for them to constitute life is an active topic of debate (*Pascal,* 2012; *Pascal et al.,* 2013; *Benner et al.,* 2004). One basic requirement for

life is access to chemical processes that can proceed over reasonable timescales (kinetics). At the same time, life must exist in an out-of-thermodynamic equilibrium with respect to its waste products to allow for metabolism, replication, and information storage. On the thermodynamic side, partial reversibility in some underlying processes is likely important for pre-Darwinian exploration of chemical networks. Some degree of reversibility also exists in life as we know it, and is beneficial in, e.g., multi-step enzyme catalysis, where large exothermic steps tend to be avoided. Balanced kinetics and conditions favorable for thermodynamic reversibility are some basic requirements for thermally controlled dynamic chemistry, which arguably is one necessary stepping stone allowing for life's origin. At the same time, life may rely on thermodynamically irreversible chemical processes for its metabolism. Many chemical conditions for life have been discussed (*Benner et al.,* 2004; *Schulze-Makuch and Irwin,* 2006; see also the chapter by Hoehler et al. in this volume) and include, among others, the potential need for carbon-based scaffolding. Environmental aspects, such as available energy sources, and the often-assumed requirement for liquid water (*Ball,* 2013; *Lineweaver and Chopra,* 2012), are other potential showstoppers for life's development.

It is reasonable to assume that polymeric chemistry and some kind of macromolecules are a requirement for life (*Schulze-Makuch,* 2004; *Schulze-Makuch and Irwin,* 2006), including on Titan. Liquid water is central to current discussions on life's molecular origin on Earth (*Sutherland,* 2016; *Bolm et al.,* 2018), where biochemistry mostly occurs in the liquid state. Unfortunately, it is not likely that any conceivable polymer could exist in the liquid state under normal Titan surface conditions. With the exception of some very small non-polar molecules, such as acetylene and ethylene (*Singh et al.,* 2017), the solubility of most molecules is extremely low in 90–94 K methane/ethane (*Stevenson et al.,* 2015b; *Malaska and Hodyss,* 2014). What this de facto means is that under normal Titan conditions, i.e., barring special energy-providing events such as asteroid impacts

or cryovolcanism, chemistry is highly unlikely to proceed in a liquid state on the surface. With this in mind, just how far up the ladder of chemical complexity do we dare hope for on Titan? To attempt an answer to this question we will first discuss different types of interactions that might allow for the building of complexity through dynamic chemistry. After this, energy sources that could support metabolism on Titan will be introduced. Finally, we will discuss a few specific chemical structures and processes predicted using computations. The purpose of these discussions is not prediction per se, but rather to identify if any conceivable realistic chemistry can adhere to the stringent kinetic, thermodynamic, and environmental constraints believed to be required for life to occur on Titan.

The major impediment for chemistry occurring on the surface of Titan is not the lack of liquid water, it is the low surface temperature (90–94 K). In principle, there can be rich chemistry at any temperature as long as molecules are stable. However, because it affects reaction rates exponentially, temperature is the most important thermodynamic variable when it comes to controlling chemical reactivity; a reaction that proceeds on microsecond timescales under ambient Earth conditions would take years on Titan. For this reason, temperature imposes severe restrictions on the type of chemical interactions that may allow for dynamic chemistry on the surface of Titan.

Figure 1 compares ranges of activation energies required to support chemical reactions proceeding on timescales from milliseconds to years on Earth (298 K) and on Titan (94 K). These estimates have been made using transition-state theory and correspond to reaction half-lives assuming first-order reaction kinetics. For a reaction to proceed with a reasonable rate under the low temperatures on Titan, the free energy barrier to activation needs to be in the range 17–35 kJ mol⁻¹, significantly lower than the corresponding terrestrial value of 55–110 kJ mol⁻¹. Typical covalent carbon-carbon and carbon-oxygen single bonds have strengths of approximately 330 kJ mol⁻¹, which is why

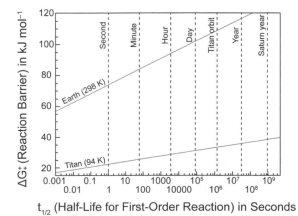 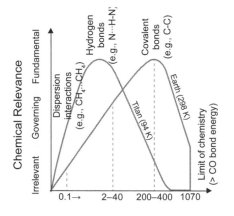

Fig. 1. (a) Reaction energy barriers required for chemical processes to occur over reasonable timescales (milliseconds to years) on Titan (17–35 kJ mol⁻¹) and on Earth (55-110 kJ mol⁻¹). **(b)** Approximate schematic showing differences in the types of chemical interactions that might allow for thermally driven dynamical processes.

life as we know it on Earth is reliant on efficient catalysts (enzymes) to lower barriers of activation. The much lower required free-energy barrier on Titan makes it difficult to imagine the breaking of covalent bonds as relevant to biochemistry there.

Nonetheless, the making and breaking of some covalent bonds can occur under Titan conditions. The presence of ammonia within Titan's interior and, early on, at its surface is inferred from noble gas and isotopic evidence (*Niemann et al.*, 2010; *Mandt et al.*, 2014); carbon dioxide is known to be present in the atmosphere and surface (see references in *Hörst*, 2017). Thus one example might be the reaction of CO_2 with amines (carbamation), known to proceed at very low temperatures (*Bossa et al.*, 2008, *Hodyss et al.*, 2017). Larger amines have not yet been directly observed, but they are one possible reaction product of the atmospheric chemistry. Still, dynamic chemistry on Titan's surface is likely to rely to a large extent on what we on Earth would consider weak (or secondary) interactions, in order to proceed at reasonable rates. As carbon-carbon bond breakage becomes increasingly unlikely, the importance of hydrogen bonds and dispersion interactions increase and might dominate any dynamic processes on Titan (Fig. 1). What can be achieved with these types of interactions?

Research fields such as supramolecular chemistry (chemistry beyond the molecule) and crystal engineering (e.g., *Stupp and Palmer*, 2014; *Wang and Yang*, 2013; *Braga et al.*, 2013), not to mention molecular biology, illustrate just how powerful understanding and exploitation of noncovalent interactions can be for the design and synthesis of molecular assemblies and extended structures with unique properties. Whereas such research has provided a good baseline of literature for the understanding of weak interactions in a large range of chemical environments, knowledge of such assemblies close to Titan temperatures is essentially non-existent.

3.3.1.1. Possible role of dispersion interactions on the surface of Titan: Dispersion (or Van der Waals) interactions arise due to fluctuations (temporary multipoles) in the electron density of molecules, and constitute the weakest meaningful type of interaction possible between molecules on the surface of Titan. Dispersion is responsible for the very weak (~0.1 kJ mol^{-1}) (*Li and Chao*, 2009) interactions between methane, ethane, and nitrogen that form Titan's seas and lakes. Dispersion interactions that are stronger by an order of magnitude can, for example, be found between benzene molecules. Benzene has been predicted and observed in the atmosphere of Titan, and we can expect it to have been deposited on the surface (*Coustenis et al.*, 2003; *Lavvas et al.*, 2008; *Krasnopolsky*, 2014). The dispersion interactions responsible for solidifying benzene are on the order of ~8 kJ mol^{-1} for an isolated benzene dimer (*Grimme*, 2004). Most dispersion interactions are non-directional, which makes them less useful for biological processes, compared to hydrogen bonding or covalent bonds. However, because dispersion interactions are additive in nature, larger conjugated molecules, such as naphthalene or coronene, can interact more strongly than benzene.

One class of dispersion-bound materials, which may be important on Titan, are co-crystals formed of organic molecules (Fig. 2). The self-assembly of formally nonbonded molecules into different ordered structures could greatly increase the diversity of materials available in the Titan environment. For example, if a sufficient variety of co-crystals exist on the surface, they might fill some roles analogous to those performed by minerals on Earth

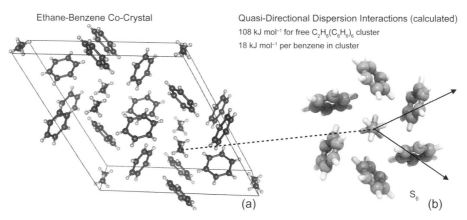

Ethane-Benzene Co-Crystal

Quasi-Directional Dispersion Interactions (calculated)
108 kJ mol^{-1} for free $C_2H_6(C_6H_6)_6$ cluster
18 kJ mol^{-1} per benzene in cluster

(a) S_6 (b)

Fig. 2. **(a)** Experimental crystal structure of a benzene-ethane co-crystal. This structure is thermodynamically favored relative to ethane and benzene, and is a potential evaporite material under Titan conditions (*Vu et al.*, 2014; *Cable et al.*, 2014; *Maynard-Casely et al.*, 2016). The possibility for such dispersion-driven cryogenic "mineralogy" demonstrates that even ethane, a small and non-polar molecule, and a major component of the Titan lakes and seas, can exist as a solid rather than a liquid at 94 K. This implies that chemical processes on the surface of Titan involving larger molecules are unlikely to proceed in a solvated state, and that they are more likely to occur inside and on the surfaces of solid-state agglomerates and crystals. **(b)** Quasi-directionality and cooperativity in dispersion-driven interactions might coordinate reactant molecules on crystal surfaces, and could be one driving force toward chirality in prebiotic chemistry. Calculations were performed at the B2PLYP/Def2-TZVPP level of theory (*Maynard-Casely et al.*, 2016). Binding energies have been corrected for zero-point energy vibrations.

(*Maynard-Casely et al.,* 2016, 2018). The crystal facets of different such "cryogenic minerals" could possibly act as structure-directing and templating agents in surface reactions. Analogies might then be drawn with the proposed catalytic role of crystalline surfaces for the origin of life on Earth (*Hazen and Sverjensky,* 2010). An example of directionality in dispersion-driven coordination is shown in Fig. 2b, where the topology of ethane facilitates the six-fold coordination of benzene and the formation of a rather strongly bound complex.

In addition to benzene with ethane (*Maynard-Casely et al.,* 2016; *Cable et al.,* 2014; *Vu et al.,* 2014), there are other examples of co-crystals that might be relevant to consider under Titan conditions. Acetylene can form a number of them: with ammonia (*Boese et al.,* 2009; *Cable et al.,* 2018), benzene, pyridine, acetonitrile, and water, and with several other aromatic systems (*Kirchner et al.,* 2010). Laboratory investigations into Titan petrology and mineralogy constitute a new field; many discoveries in this area are likely in the future (*Maynard-Casely et al.,* 2018).

3.3.1.2. The role of hydrogen bonding interactions on the surface of Titan:

Hydrogen bonding is a secondary interaction between a covalently bound hydrogen and second atom with a formal lone pair of electrons. The highly directional nature of hydrogen bonds makes them efficient in inducing structure, and on Earth they, for instance, govern protein folding and enable base pair recognition (cf. guanine-cytosine in Fig. 3). Relatively few molecular species with the potential to form hydrogen bonds have been detected on Titan. Hydrogen cyanide (HCN) is one of the major products of Titan's atmospheric photochemistry (*Teanby et al.,* 2007, 2006; *Griffith,* 2014). Adenine, an HCN-pentamer that might form under certain Titan conditions (*Hörst et al.,* 2012, *Raulin et al.,* 2012) and one of the building blocks of life as we know it, interacts too strongly to support dynamic chemistry on Titan. However, several other possibilities for C/N-H—N interactions lie in the right range (<35 kJ mol⁻¹) (cf. Fig. 1). More suitable interaction strengths can, for example, be found between imines and nitriles and between amines and HCN (Fig. 3).

3.3.2. Energy sources that might drive prebiotic chemistry, or life, on the surface of Titan.

In order for life to sustain an out-of-equilibrium state with respect to waste products and thermodynamic sinks, an external energy source is required. For most of life on Earth, and with the exception for volcanic vents and chemical energy stored in minerals, this energy source is, directly or indirectly, the Sun. Because Titan's great distance to the Sun, and its dense atmosphere, the solar flux impacting the surface of Titan is approximately 0.1% that of Earth (*McKay et al.,* 1991). This relatively low energy density is, however, still high enough to support photosynthesis in several life forms on Earth (*Raven et al.,* 2000). Higher in the atmosphere the solar radiation is about 1% as intense as that of Earth, and drives the photolysis of methane and nitrogen, and complex haze formation. Molecular hydrogen, acetylene (C_2H_2), and HCN are three of the most major products of the atmospheric photochemistry (e.g., *Yung et al.,* 1984). All these molecules carry a significant amount of chemical energy. At the surface, such chemical energy sources might drive exothermic reactions and, it has been speculated, possibly act as nutrition for some form of life (*McKay,* 2016; *McKay and Smith,* 2005;

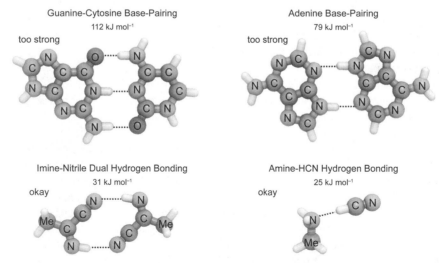

Fig. 3. See Plate 21 for color version. Examples of structure-directing hydrogen-bonds with varying interaction strength. Stacked guanine-cytosine base pairs are sufficiently strong to maintain the DNA double helix on Earth, but much too strong to allow for dynamic chemistry on Titan. On Titan, weaker C/N-H—N hydrogen bonds could be the primary linkages holding together biostructures. The interactions between adenine molecules, an HCN-pentamer that might form on Titan, exceeds the estimated 17–35 kJ mol⁻¹ range in which sufficiently rapid dynamics is possible. Interactions in a more favorable range is possible, for example, between amines, imines, nitriles, and HCN. Binding energies have been calculated at the DLPNO-CCSD(T)/aug-cc-pVTZ//wB97X-D3/Def2-SVPD level of theory, and are corrected for zero-point energy vibrations using the ORCA code (*Neese,* 2012).

Schulze-Makuch and Grinspoon, 2005). In particular, the reduction of acetylene and ethane to methane, and other similar hydrogenation reactions, have been discussed

$$C_2H_2 + 3H_2 \rightarrow 2CH_4 \qquad (1)$$
$$\Delta G_r^0 = -334 \text{ kJ mol}^{-1}$$

$$C_2H_6 + 2H_2 \rightarrow 2CH_4 \qquad (2)$$
$$\Delta G_r^0 = -57 \text{ kJ mol}^{-1}$$

Also, HCN is a conceivable energy source, if reacted/metabolized to molecular nitrogen

$$HCN + \frac{3}{2}H_2 \rightarrow 0.5N_2 + CH_4 \qquad (3)$$

Albeit singularly unlikely, such hydrogen-based metabolisms are tentatively in accord with a number of observations: The hydrogen mole fraction is significant, ~0.001 throughout the atmosphere (*Courtin et al.,* 2012). This, combined with the molecules' presumed high mobility in most liquid and solid media on the surface, would make the energy source accessible to any metabolic processes, despite the low temperature. The Huygens probe onboard Cassini measured H_2 throughout its descent and detected a drop in the H_2 mole fraction of $\sim 3 \times 10^{-5}$ over the last 3 km above the surface (*Niemann et al.,* 2010). Albeit small, this observation could indicate consumption by some chemical or biological process. More importantly, the atmosphere-to-surface H_2 flux is significant, on the order of $\sim 2 \times 10^{10}$ cm^{-2} s^{-1} (*Strobel,* 2010). This transport rate is not fully understood, but it significantly exceeds $\sim 10^8$ cm^{-2} s^{-1}, a suggested minimal estimate required to support methanogenic life (*McKay and Smith,* 2005). A further observation is a non-uniform latitudinal distribution of H_2 in the atmosphere, which has been interpreted as implying a significant downdraft of H_2 specifically in the north polar regions (*Courtin et al.,* 2012). A downdraft suggests the presence of some sort of chemical sink at the north pole. This is especially interesting in light of the fact that the north pole is the location of most of Titan's methane-rich lakes, which thus are composed mainly of what we regard for arguments sake as a metabolic waste product (equations (1)–(3)).

The two other speculated components of hydrogen-based metabolism, acetylene and hydrogen cyanide, are the two most abundant solid (on the surface) products of photochemistry in Titan's atmosphere (e.g., *Yung et al.,* 1984). As such, they can be expected to have formed significant deposits on the surface, over Titan's life span. While trace amounts of acetylene were detected by the Huygens Gas Chromatograph Mass Spectrometer (GCMS) after landing (*Niemann et al.,* 2010), no evidence for HCN crystallites have been found in surface spectra of Titan from the Cassini Visible and Infrared Mapping Spectrometer (VIMS) (*Clark*

et al., 2010). Chemical identification of the surface from space is hampered by atmospheric absorption, especially by methane, and scattering by haze particles. Nevertheless, the discrepancy between the solid products formed in the atmosphere and those found in surface deposits suggests that chemistry involving acetylene and HCN is ongoing on Titan's surface (*Rahm et al.,* 2016). This discrepancy and other possible biosignatures are discussed further in section 4.

Nitrogen and methane are the main components of Titan's atmosphere. If, somehow, biogenic nitrogen and methane were produced, then this would likely affect the atmospherically observable isotope ratios by enriching the lighter isotope. The magnitude of such an effect, and the ease by which it could be measured, would depend on the available biomass, the atmospheric photolysis rate, possible outgassing from the interior, and the influence of isotope fractionation and atmospheric escape.

Molecular nitrogen is the ultimate thermodynamic sink, and one of our hypothesized reaction products of HCN-based metabolism (equation (3)). The origin of molecular nitrogen in Titan's atmosphere is currently unknown. One possibility is endogenic production and outgassing of N_2 that might form from NH_3 reservoirs deep underground in reactions catalyzed by radioactive decay (*Glein,* 2015; *Lunine et al.,* 1999). The isotopic $^{14}N/^{15}N$ ratio (in N_2) has been reported as 167.7 (*Niemann et al.,* 2010). This low $^{14}N/^{15}N$ ratio is inconsistent with the terrestrial value of 272, and biological activity, but consistent with the initial source being ammonia in comets (*Mandt et al.,* 2014). At the same time, atmospheric HCN has a markedly lower $^{14}N/^{15}N$ ratio of 72 (*Molter et al.,* 2016). The reason for the large difference between N_2 and HCN is not known, but it is likely in part due to photolytic fractionation of N_2 in the upper atmosphere. Variations in the $^{14}N/^{15}N$ ratio with altitude is another possibility. We note that in a hypothetical scenario where HCN acts as a biological energy source on Titan (equation (3)), its lighter isotopomer is expected to be consumed faster and lower the HCN-based $^{14}N/^{15}N$ ratio in the atmosphere, in agreement with observations. At the same time, the metabolic conversion of HCN into N_2 would increase the N_2-based $^{14}N/^{15}N$ ratio relative to an abiotic reference situation. Current observations are inconclusive, but they do not exclude potential biological or prebiotic roles for HCN on the surface of Titan.

The supply of methane to Titan's surface is similarly unclear (*Cordier et al.,* 2016). The large amounts observed are thought to be unstable over longer timescales because of photodissociation and atmospheric escape (*Glein,* 2015), and a very small Kr/CH_4 molar ratio in Titan's atmosphere is evidence in favor of continuous methane replenishment by an unknown source. Whether any endogenic methane source could be geothermal outgassing, chemical, or even biological in nature, is an open question. Estimates to the $^{12}C/^{13}C$ ratios in Titan's atmosphere (91.4 ± 1.5) (*Niemann et al.,* 2010) are higher than the terrestrial standard value (89.4, NIST), and significantly higher than what has been found in the local interstellar medium (43 ± 4) (*Hawkins and Jura,* 1987). However, at the same time, Titan's $^{12}C/^{13}C$ ratio is also markedly similar to what has been measured in a range of different solar system objects (90 ± 5) (*Niemann et al.,* 2010).

3.3.3. Chemistry that might support prebiotic or biological processes on the surface of Titan.

We have discussed different types of chemical interactions (C/NH—N hydrogen bonding and dispersion interactions) that, based on their average interaction strength, might allow for dynamic chemistry. We have also mentioned some of the more widely available feedstock molecules that might drive metabolism (C_2H_2, HCN, H_2) on (or near) the surface of Titan. Arriving at more specific predictions of chemistry that might support dynamic processes under the kinetic, thermodynamic, and metabolic constraints discussed is not easy. Any one detail is likely to be in error, and we stress that what follows are speculations intended to drive the subject forward.

Suggesting chemistry unknown to experimental science should at minimum be validated by some form of calculations. The purpose of such theoretical pursuits should be to communicate as much realism as possible, ideally by predicting both the thermodynamics (can this happen?) and the kinetics (how fast can this happen?) of a given process and finally whether there might be an alternative route that is more favorable or that proceeds faster. The complexity involved in such efforts are significant, and as a consequence the field of computational molecular exobiology is very much in its infancy. However, we can touch on two concrete examples of computational predictions of possible relevance to prebiotic chemistry and life on Titan.

3.3.3.1. Compartmentalization: Cell membranes are essential prerequisite for life as we know it. Living systems on Earth rely on lipid bilayer membranes to protect the delicate molecular machinery needed for metabolism and replication. The cell membrane also has several other important functions ranging from selective uptake of nutrients and ions, to intercellular communication and waste disposal. By extrapolation, it is reasonable to assume that some sort of compartmentalization should be important also for life outside of Earth. This led researchers to suggests the fascinating possibility of so-called "azotosomes" on Titan (*Stevenson et al.,* 2015a)(Fig. 4). Azotosomes are defined as membranes made up of small organic molecules with a nitrile head group followed by a hydrocarbon end group. In contrast to normal lipid membranes in water, the azotosome membranes are conceived as displaying an inverted polarity, with hydrophobic groups on the outside. Molecular dynamics simulations in cryogenic methane have been used to predict that these structures would have

a similar elasticity as normal lipid bilayer membranes in aqueous solution if made from acrylonitrile (C_2H_3CN), a product of Titan's atmospheric chemistry definitively detected by ALMA at a level of a part per billion (*Vuitton et al.,* 2007, *Palmer et al.,* 2017). Simulations have also demonstrated that the proposed structure would be kinetically persistent near 90 K, with respect to the removal of one acrylonitrile monomer.

The possibility of azotosomes spurred great interest in discussions regarding the limits of life (*Schulze-Makuch et al.,* 2015), and even inspired the synthesis of different kinds of reversed surfactants operable under ambient Earth conditions (*Facchin et al.,* 2017). Molecular dynamics calculations suggest significant solubility of acrylonitrile in methane (*Stevenson et al.,* 2015b), with the potential for more than 10^7 cell membranes per cubic centimeter (*Palmer et al.,* 2017).

The basis for the "lipid world" hypothesis (*Segre et al.,* 2001), in which abiotic formation of membranes contributed to the emergence of life, is that lipids in water spontaneously self-assemble into supramolecular structures, such as membranes and micelles, above a critical concentration. As we have discussed, the expected thermodynamic ground state for any molecule larger than ethane near 94 K is a crystalline solid (molecular ice). Observing anything different from a small nitrile immersed in liquid methane would be remarkable. For the spontaneous self-assembly of azotosomes on Titan to occur, its structure would not only need to be kinetically persistent, as was shown computationally, it would also need to be thermodynamically lower in energy than the corresponding molecular ice. Recent calculations indicate that the azotosome structure is not thermodynamically favored, however (*Sandström and Rahm,* 2020).

Are cell membranes the only way to host the machinery of biochemistry on Titan? The main reason for membranes in life as we know it to ensure local entropy reduction, and to safeguard the precious contents of the cell from being diluted and rendered inactive in a much larger body of warm liquid water. On Titan, any life-bearing macromolecule or crucial machinery of a life form might exist in the solid state, and not risk destruction by dissolution. Already rendered immobile by the low temperature, it would rely on the diffusion of very small energetic molecules (e.g., H_2, C_2H_2, and/or HCN) in the atmosphere or the surrounding liquid to reach it in order for it to grow or replicate. Because of the high energetic cost associated with movement, any life form might rely on the natural cycling of liquids on Titan, such as rainfall, lake overturn, or seasonal fluctuation in lake and sea levels, to move solid materials along the surface when possible. Life existing under such conditions might

Fig. 4. See Plate 22 for color version. Proposed "azotosome" cell-membrane made from acrylonitrile (*Stevenson et al.,* 2015a).

lie largely dormant except near actively moving shorelines or during rainfall.

3.3.3.2. HCN polymerization as a route to biomolecules: The significant presence of HCN on Titan (*Teanby et al.,* 2007, 2006) is intriguing in particular because the molecule is considered a key precursor to the origin of life, capable of abiotic amino acid and nucleic acid synthesis (*Sutherland,* 2016; *Ruiz-Bermejo et al.,* 2013; *Seckbach,* 2005; *Patel et al.,* 2015; *Oro,* 1961). The polymerization of HCN (under Earth-laboratory timescales and conditions) is known to proceed under a variety of different conditions, and HCN can be the feedstock for reactions building significant chemical complexity, and a hydrogen-bonding rich environment consisting of carious heterocyclic aromatics, nitriles, imines, and amines (*He et al.,* 2012; *Bonnet et al.,* 2013; *Mamajanov and Herzfeld,* 2009; *Andersen et al.,* 2013; *de la Fuente et al.,* 2014). The structure of these complex materials is controversial, and the degree to which any specific structure might be attained under Titan-like conditions is still an open question. Experimental analysis of reaction products synthesized under anhydrous conditions have suggested that polyimine is one major component of base-catalyzed HCN polymerization (*He et al.,* 2012). Subsequent quantum chemical studies have highlighted this particular polymers' potential role in prebiotic chemistry on Titan (Fig. 5) (*Rahm et al.,* 2016).

Polyimine is fascinating for several reasons. First, and in contrast to the irreversibly exothermic polymerization of acetylene, the formation of polyimine from HCN is, in principle, thermodynamically reversible under Titan conditions ($\Delta G_{90K} \approx -20$ kJ mol^{-1}) (*Rahm et al.,* 2016). Due to a flexible backbone the polymer can exist in several different polymorphs, which are relatively close in energy. Under Earth ambient conditions the polymer is expected to behave as a random coil (and hydrolyze). However, on Titan more well-defined structures might compete in the low-temperature environment. The electronic and structural variability between them is extraordinary. The band gap, for example, changes over a 3eV range when moving from a planar sheet-like structure to increasingly coiled conformations (Fig. 5). Modulated by the surrounding physical and chemical environment, the material could, in principle, be capable of photon absorption in the infrared (IR), throughout the visible, and onward to the ultraviolet range. Calculations suggest that light absorption could induce local structural changes, as well as make protruding imine groups more reactive. Furthermore, several of the investigated low-energy polymorphs exhibit local domains capable of both accepting and donating hydrogen bonds, in a manner that resembles the catalytic sites of several enzymes. The potential for photochemical-driven catalysis and structural rearrangement makes polyimine one of the few concretely suggested macromolecules that might allow for dynamic chemistry under Titan's cryogenic conditions.

Understanding the pathways to forming various complex amorphous organics and ordered structures from HCN will likely be important for deciphering Titan's surface chemistry. Work is underway by a subset of the coauthors to evaluate explicit formation mechanisms and kinetics for the formation of polyimine.

Other possible HCN-based structures have been suggested as well (*Lv et al.,* 2017). The general challenge when considering Titan's potential surface and lake/sea-biochemistry is essentially one of shifting the window of relevant chemical interactions toward those of lower energy. While the most important chemical interactions for biology on Earth are strong covalent bonds and strong hydrogen bonds, the window of chemical interactions thermally accessible on Titan range from weak hydrogen bonds to dispersion interactions (Fig. 1). A broader understanding of these weaker interactions, and the variety

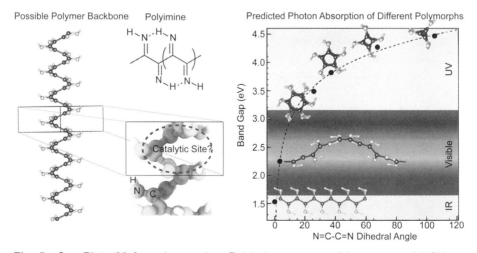

Fig. 5. See Plate 23 for color version. Polyimine, one possible product of HCN polymerization on Titan, might utilize inter- and intramolecular =N-H···N hydrogen bonds to govern the formation of different partially ordered structures, some of which may synergize with photon absorption and act catalytically. The band gaps of possible polyimine polymorphs span a range across the entire visible spectrum and into the ultraviolet. From *Rahm et al.* (2016); used with permission.

of structural motifs they might supply on Titan, would increase our ability to think of, rationalize, and detect any dynamic or prebiotic chemistry or non-aqueous forms of life. Photochemistry is an exception that may allow for the catalytic making and breaking of covalent linkages, and the construction of biomolecules even at 94 K. The identification of possible low-temperature reactions that can utilize Titan-abundant compounds is an important topic of future research. Hydrogen cyanide is one possible feedstock molecule whose polymerization may allow for dynamic chemistry, and formation of biomolecules capable of catalysis and photon absorption. Due to the low temperature, macromolecules on the surface of Titan will have a very low solubility in the hydrocarbon lakes, and predominantly exist in the solid state. Life under Titan's cryogenic surface conditions, encompassing macromolecules, might therefore rely on natural cycling of liquids for mobility and metabolism. As mentioned in section 2.3, the most environmentally dynamic locations on Titan are the shore lines, where liquid, solid, and gaseous phases meet. We believe these are the most likely regions for the unlikely chemical scenarios that may allow for life on the surface of Titan.

4. MEASUREMENTS AND INVESTIGATIONS NEEDED TO GO FROM "INTERESTING" TO "IS THERE LIFE?"

The current strategy for life detection on ocean worlds such as Titan follows the paradigm of (1) a habitability investigation, then (2) selection of habitable environments/locations where biosignatures might be concentrated, ultimately leading to (3) a mission with a payload specifically designed to search for life itself. Each stage of this model may utilize remote sensing techniques, *in situ* techniques, or both.

On Titan, the situation is unique, as life investigations may search for life as we know it (aqueous-based) or life as we don't know it (based on a non-polar solvent such as methane and/or novel organic chemistry).

4.1. Life As We Know It (The Aqueous Case)

A great deal of work has been done regarding the search for life that relies on water as its solvent. The terrestrial geologic and geochemical record, as well as the presence of prebiotic molecules in meteorites and throughout the solar system, provide some constraints on what to look for (*National Research Council*, 2002). Aqueous-based life in Titan' ocean or regions of transient liquid water — or natural prebiotic experiments in those environments — may be partially recorded in deposits of water ice extruded onto the surface and then exposed by erosion. The atmosphere of Titan limits subsequent processing of material due to photodegradation; very little high-energy radiation reaches the surface (*Lavvas et al.*, 2011). Similarly, smaller impactors are decelerated and broken up in the atmosphere

(*Artemieva and Lunine*, 2003; *Korycansky and Zahnle*, 2005), meaning biosignatures are also shielded from most impact gardening. So biosignatures generated in frozen-aqueous surface sites are most likely well-preserved for long periods of time, perhaps longer than on Earth and certainly Mars, and should be considered a top priority for any *in situ* Titan life investigation. Identification of regions with recent activity, possibly via detection of thermal anomalies or outgassing events, may help direct such *in situ* investigations.

Identification of areas where the subsurface ocean might be expressed on the surface would also be ideal targets to search for evidence of aqueous-based life. Regions consistent with cryovolcanic activity, such as Sotra Patera (formerly known as Sotra Facula) and Hotei Regio may be good targets in this respect (*Lopes et al.*, 2013; *Neish et al.*, 2006). However, understanding the provenance of organics in this case is important — the effects of temperature and pressure, for example, as biosignatures are moved from the subsurface through a cryovolcanic flow to the surface, should be considered. Chemical processing may alter the types or relative abundances of certain compound classes. Further laboratory work and modeling would help to address the chemical history of an *in situ* sample collected on Titan.

4.2. Life As We Don't Know It (The Non-Aqueous Case)

Some biosignatures for life in hydrocarbon solvents may be unique; however, some tried and true techniques can also be applied.

4.2.1. Isotopic distributions. As mentioned in section 3.3, methane is not stable in Titan's atmosphere; photoionization and photodissociation in the atmosphere would convert all methane to ethane and other HCN-containing species in 10^7 years (*Raulin and Owen*, 2002; *Lorenz et al.*, 2003). The fact that methane still exists in Titan's atmosphere today indicates it must be replenished in some way, either through periodic large outgassing events or some lower-yield, consistent process. Interestingly, Cassini INMS measured the carbon isotopic ratio in methane, and found it to be enriched in ^{12}C (*Waite et al.*, 2005). Highly specific reactions of enzymes tend to generate organic molecules enriched in lighter isotopes, as these can react more quickly than the heavier ^{13}C-containing isotopologs; this means organic molecules produced by life, such as methane, can be distinguished from abiotically-generated molecules based on isotopic composition (*Northrup*, 1981). The $^{12}C/^{13}C$ ratio is commonly used to identify biotic vs. abiotic sources of carbon on Earth (*Lollar et al.*, 2006; *Etiope and Lollar*, 2013), and has been proposed as a measurable indicator of life for the Europa Lander mission concept (*Hand et al.*, 2017). Such a measurement should also be applicable to the search for life on Titan, whether it is aqueous- or non-aqueous-based. Similarly, the $^{14}N/^{15}N$ ratio in various nitrogen-containing species (HCN, N_2, etc.) may be utilized to support or refute production/consumption of these molecules via biotic processes.

In situ measurements of larger isotope ratios in solid samples taken from surface deposits, compared to those of the atmosphere (i.e., $^{12}C/^{13}C > 91.4$) would indicate the presence of low-temperature reactions building the surface layer, but not necessarily be unequivocal evidence of life. However, when considered alongside other independent lines of evidence, isotopic measurements taken together could be considered a biosignature.

4.2.2. Chemical disequilibria. The presence of ^{12}C-enriched methane is not the only piece of evidence suggesting methanogenesis may be at work on Titan. The amount of acetylene on the surface is also much lower than expected based on models of photochemical production in the atmosphere (section 3.3); the ratio of acetylene to benzene, calculated to be on the order of 10^3 to 10^4, is <10 based on VIMS data (*Clark et al.,* 2010). One possible explanation is chemical: Acetylene will cyclize into benzene when exposed to heat, radiation such as UV or cosmic rays, or certain catalysts. Perhaps this chemical conversion, either in the atmosphere or on the surface, is more efficient than expected. Another possible explanation is physical: preferential transport of acetylene or benzene due to differences in material properties, either masking the surface acetylene signal from detection by remote sensing, or sequestering acetylene in the subsurface. The recently reported acetylene-ammonia co-crystal (*Cable et al.,* 2018), which is stable to pluvial or fluvial weathering, may support the hypothesis that acetylene could be trapped in underground reservoirs as a Titan mineral. More *in situ* measurements of Titan's surface are needed to address this discrepancy.

Concomitant with the missing acetylene is a high downward flux of H_2 to the surface, which is in excess of what would be needed for methanogenic organisms to subsist. If methanogenic life is consuming atmospheric hydrogen and acetylene to produce methane, it would decrease the H_2 mixing ratio (relative amount of H_2 in the air) in the troposphere, so the observed H_2 abundance with altitude is also consistent with an acetylene-powered biosphere (*Schulze-Makuch and Grinspoon,* 2005). However, as with acetylene, the H_2 mixing ratio could also be explained by abiotic processes (*Grinspoon and Schulze-Makuch,* 2010). It is possible that some of the less-saturated, complex hydrocarbon solids on the surface are somehow recombining with H_2 to reform CH_4, although such processes would likely be very slow or require a catalyst (*Strobel,* 2010).

Measurements to better constrain the abundances and distributions of key species, such as H_2, acetylene, and acetonitrile, in the atmosphere and on the surface would better establish the types of chemical disequilibria that may exist on Titan and perhaps be signs of life. A better understanding of surface composition (in particular the distribution of unsaturated complex hydrocarbons) would also determine if the source of methane is due to abiotic recombination of H_2 and surface organics, or if methanogenesis could be a viable

hypothesis. Any observed chemical disequilibrium must be considered in context, as Titan is a very complex system. Abiotic processes leading to a concentration gradient, such as the preferential sequestration of ethane over methane in clathrates in a subsurface "alkanofer" (*Mousis et al.,* 2016) or co-crystal layer (*Cable et al.,* 2014), must be eliminated before the claim of a biotic origin can be made.

4.2.3. Patterns in compound classes that are different from abiotic chemistry. Life as we know it requires certain building blocks, such as amino acids, to construct its molecular machinery. Work has demonstrated that patterns in these building blocks can serve as a biosignature (*Dorn et al.,* 2011). For example, abiotic synthesis of amino acids results in a significant amount of glycine (the simplest amino acid) compared to the more complicated amino acids that are most energetically expensive to form (*Higgs and Pudriz,* 2009; *Cronin and Pizzarello,* 1983). However, life as we know it requires a variety of amino acids to impart functionality to proteins, and so the pattern of amino acid abundances will emphasize the larger ones at the expense of glycine (*McKay,* 2004; *Davila and McKay,* 2014; *Glavin et al.,* 2010; *Creamer et al.,* 2017). Similar patterns are observed in lipids [i.e., the 2-carbon addition for bacteria and eukaryotes, or the 5-carbon addition for archaea (see *Georgiou and Deamer,* 2014)] and other organic molecules utilized by biology.

Non-aqueous life should also require key molecular building blocks to do its unique chemistry — the molecules may be different from aqueous-based life, but the patterns should still be detectable. A pattern of repeating subunits, such as $(CH_2)_m(HCN)_n$ (*Pernot et al.,* 2010), where the distribution of those molecules is inconsistent with purely abiotic mechanisms, could therefore be a biosignature in a non-aqueous system. Depending on the environment where this life subsides (lake liquids vs. the lake floors or solid phase — see section 2), these patterns may be detectable in the atmosphere, lake liquids, or solid materials comprising Titan's surface. As with the other potential biosignatures proposed here, full characterization of the chemical groups and structures of any (ideally all) of these environments, and how they change over time, would significantly improve the confidence that any pattern detected was due to biochemistry and not abiotic organic chemistry.

4.2.4. Morphological biosignatures. Life on Earth can form a variety of physical constructs that are often distinguishable from abiotic structures. These may include colonies of cells, biofabrics (concentrations of cells and spatial arrangement of minerals), or inclusions of entrapped biomass (*Hand et al.,* 2017). Importantly, this type of biosignature may remain long after the cells that produced it have disappeared — examples include stromatolites or other biofossils.

However, there are abiotic mechanisms that can create structures that might be confused with those made by biology. Therefore, detection of morphology consistent with life would probably require co-manifested chemical evidence to serve as a biosignature. On Titan, a more complete understanding of the organic chemistry occurring at the surface would help to eliminate false positives.

5. LESSONS FOR METHANE-DOMINATED EXOPLANETS

The exotic nature of Titan relative to Earth begs the question of possible "habitable" analogs around other stars, and indeed, the intensity of light Titan receives from the Sun is approximated by a planet at 1 AU from a spectroscopically "middle" M-dwarf star (*Lunine, 2009*). However, there are many differences. An M dwarf has a substantially different spectral energy distribution than a solar, G-type, star, and in particular the amount of UV radiation needed for photochemistry depends critically on whether the star is or is not a so-called "active" M dwarf (*Gilliam and McKay, 2011*). Furthermore, while it is conceivable that transits might detect such a body, a semimajor axis of 1 AU diminishes greatly the probability relative to the close-in planets that dominate Kepler data. Radial velocity detection, given projected capabilities and the limitations created by stellar surface noise, is likely to be limited to Earth-mass bodies. For these, the contribution of geothermal heating could be substantial (*Lunine, 2010*), again making for a divergence in the behavior of the surface-atmosphere system of such a body relative to Titan. The prolonged decline in brightness after reaching the main sequence also must be considered when drawing analogies with our solar system's Titan.

M dwarfs are the most abundant and long-lived main-sequence stars in the cosmos. However, they appear to be so active that close-in Earth analogs may not be habitable (*MacGregor et al., 2018*), leaving aside the complications that tidal locking and other issues that close proximity to the star present (*Lammer, 2007*). At distances on the order of 1 AU, where Titan analogs would be located, such effects are less important or absent. Hence, should Titan prove to be inhabited, in any of the astrobiological sites discussed in this chapter, it may serve as an avatar for the most common type of life-bearing body in the cosmos (*Lunine, 2010*).

6. FUTURE PROSPECTS AND CONCLUSIONS

Titan hosts a diversity of sites where biology might be present, or the remnants of prebiotic chemistry preserved for study. Additional analysis of Cassini and Huygens data will undoubtedly yield deeper insights and perhaps new discoveries relevant to Titan's astrobiological potential. Inevitably, with the end of the mission in September 2017, much of the community's focus has shifted to the future of Titan's exploration. Various concepts have been studied over more than a decade, ranging from balloons (*Lorenz, 2008*) to a Flagship-class armada of multiple *in situ* and orbital elements (*Coustenis et al., 2009*).

Titan Mare Explorer, a buoyant capsule to sample the Titan seas (*Stofan et al., 2013*), progressed to Phase A but was not selected for flight. In 2019 NASA selected Dragonfly as the New Frontiers 4 (NF4) mission. Dragonfly is an eight-bladed rotocraft lander designed to sample sites on the dunes and elsewhere on Titan, including where aqueous chemistry may have occurred (*Turtle et al., 2017; Lorenz et al., 2018*). This mission architecture is made possible due to maturation of automated sensing and control technologies related to multi-rotor drones, and is a more efficient means of travel on a world with a thick atmosphere than a standard driving rover platform. Planned for a 2026 launch, Dragonfly will arrive in 2034 to explore dozens of diverse locations, beginning with the dunes. It will cover tens to hundreds of kilometers over a planned 2.5-year prime mission. At each landing site, Dragonfly will conduct bulk elemental surface analyses, and will collect material for ingestion into a mass spectrometer to measure molecular composition. Meteorological, seismic, and other measurements round out the investigation of each landing site. At each landing site, Dragonfly recharges its batteries from a radioisotopic power source, to enable flight of up to a few hours to the next station. During flight, Dragonfly will make atmospheric profile measurements and image the surface to provide context for ground investigations and scout potential new landing sites.

Viewed from the perspective of the early years of planetary exploration, the importance that Titan represents as a target in the search for life and the exploration of prebiotic chemistry is one of the great surprises of the space age. With NASA's selection of Dragonfly, the exploration of this complex and in many ways still mysterious moon moves from remote sensing and a single-site lander to mobile surface operations. The Saturn system is a long way away and Titan is only one of many astrobiological targets. But it beckons, and perhaps within the next few decades we will know whether Titan is, or has been, an inhabited world.

Acknowledgments. All the authors are grateful to the book editor V. Meadows and series editor R. Binzel for the opportunity to address such a fascinating topic. We are grateful to H. Sandström of Chalmers University of Technology for pointing out an error in an earlier version of the manuscript. J.I.L. appreciates support from the Cassini Project. M.R. is thankful for support from the Swedish Research Council. Some of this research was carried out at the Jet Propulsion Laboratory, California Institute of Technology, under a contract with the National Aeronautics and Space Administration. Government sponsorship acknowledged.

REFERENCES

Aharonson O., Hayes A. G., Lunine J. I., Lorenz R. D., Allison M. D., and Elachi C. (2009) An asymmetric distribution of lakes on Titan as a possible consequence of orbital forcing. *Nature Geosci., 2,* 851–854.

Andersen J. L., Andersen T., Flamm C., et al. (2013) Navigating the chemical space of HCN polymerization and hydrolysis: Guiding graph grammars by mass spectrometry data. *Entropy, 15(10),* 4066–4083.

Anderson R. P. and Roth J. R. (1977) Tandem genetic duplications in phage and bacteria. *Annu. Rev. Microbiol., 31,* 473–505.

Artemieva N. and Lunine J. (2003) Cratering on Titan: Impact melt, ejecta, and the fate of surface organics. *Icarus, 164,* 471–480.

Ball P. (2001) *Life's Matrix: A Biography of Water.* Univ. of California, Berkeley. 417 pp.

Ball P. (2013) The importance of water. In *Astrochemistry and*

Astrobiology (I. W. M. Smith et al., eds.), pp. 169–210. Springer-Verlag, Berlin.

Barnes J. W., Brown R. H., Soderblom J. M., Soderblom L. A., Jaumann R., Jackson B., Le Mouelic S., Sotin C., Buratti B. J., Pitman K. M., Baines K. H., Clark R. N., Nicholson P. D., Turtle E. P., and Perry J. (2009) Shoreline features of Titan's Ontario Lacus from Cassini/VIMS observations. *Icarus, 201(1),* 217–225.

Barnes J. W., Bow J., Schwartz J., Brown R. H., Soderblom J. M., Hayes A. G., Vixie G., Le Mouelic S., Rodriguez S., Sotin C., Jaumann R., Stephan K., Soderblom L. A., Clark R. N., Buratti B. J., Baines K. H., and Nicholson P. D. (2011) Organic sedimentary deposits in Titan's dry lakebeds: Probable evaporite. *Icarus, 216(1),* 136–140.

Béghin C., Randriamboarison O., Hamelin M., Karkoschka E., Sotin C., Whitten R. C., Berthelier J-J., Grard R., and Simões F. (2012) Analytic theory of Titan's Schumann resonance: Constraints on ionospheric conductivity and buried water ocean. *Icarus, 218,* 1028–1042.

Benner S. A., Ricardo A., and Carrigan M. A. (2004) Is there a common chemical model for life in the universe? *Curr. Opin. Chem. Biol., 8(6),* 672–689.

Birch S. P. D., Hayes A. G., Dietrich W. E., Howard A. D., Bristow C. S., Malaska M. J., Moore J. M., Mastrogiuseppe M., Hofgartner J. D., Williams D. A., White O. L., Soderblom J. M., Barnes J. W., Turtle E. P., Lunine J. I., Wood C. A., Neish C. D., Kirk R. L., Stofan E. R., Lorenz R. D., and Lopes R. M. C. (2017) Geomorphologic mapping of Titan's polar terrains: Constraining surface processes and landscape evolution. *Icarus, 282,* 214–236.

Birch S. P. D., Hayes A. G., and Hofgartner J. D. (2018) The raised rims of Titan's small lakes. In *Lunar and Planetary Science XLIX,* Abstract #2083. Lunar and Planetary Institute, Houston.

Boese R., Blaser D., and Jansen G. (2009) Synthesis and theoretical characterization of an acetylene-ammonia cocrystal. *J. Am. Chem. Soc., 131(6),* 2104–2106.

Bolm C., Mocci R., Schumacher C., et al. (2018) Mechanochemical activation of iron cyano complexes: A prebiotic impact scenario for the synthesis of α-amino acid derivatives. *Angew. Chem., Int. Ed., 57(9),* 2423–2426.

Bonnet J.-Y., Thissen R., Frisari M., et al. (2013) Compositional and structural investigation of HCN polymer through high resolution mass spectrometry. *Intl. J. Mass Spectrom., 354–355,* 193–203.

Bossa J.-B., Borget F., Duvernay F., et al. (2008) Formation of neutral methylcarbamic acid ($CH_3NHCOOH$) and methylammonium methylcarbamate [$CH_3NH_3^+$][$CH_3NHCO_2^-$] at low temperature. *J. Phys. Chem. A, 112(23),* 5113–5120.

Braga D., Maini L., and Grepioni F. (2013) Mechanochemical preparation of co-crystals. *Chem. Soc. Rev., 42(18),* 7638–7648.

Brassé C., Buch A., Coll P., and Raulin F. (2017) Low-temperature alkaline pH hydrolysis of oxygen-free Titan tholins: Carbonates' impact. *Astrobiology, 17(1),* 8–26.

Brown R. H., Soderblom L. A., Soderblom J. M., Clark R. N., Jaumann R., et al. (2008) The identification of liquid ethane in Titan's Ontario Lacus. *Nature, 454,* 607–610.

Cable M. L., Hörst S. M., Hodyss R., Beauchamp P. M., Smith M. A., and Willis P. A. (2012) Titan tholins: Simulating Titan organic chemistry in the Cassini-Huygens era. *Chem. Rev., 112(3),* 1882–1909, DOI: 10.1021/cr200221x.

Cable M. L., Vu T. H., Hodyss R., Choukroun M., Malaska M., and Beauchamp P. (2014) Experimental determination of the kinetics of formation of the benzene-ethane co-crystal and implications for Titan. *Geophys. Res. Lett., 41(5),* 5396–5401.

Cable M. L., Vu T. H., Maynard-Casely H. E., Choukroun M., and Hodyss R. (2018) The acetylene-ammonia co-crystal on Titan. *ACS Earth Space Chem., 2 (4),* 366–375, DOI: 10.1021/acsearthspacechem.7b00135.

Carlucci A. F. and Williams P. M. (1965) Concentration of bacteria from seawater by bubble scavenging. *ICES J. Mar. Sci., 30,* 28–33.

Castillo-Rogez J. C. and Lunine J. I. (2010) Evolution of Titan's rocky core constrained by Cassini observations. *Geophys. Res. Lett., 37,* L20205, DOI: 10.1029/2010GL044398.

Charnay B., Barth E., Rafkin S., Narteau C., Lebonnois S., Rodriguez S., Courrech du Pont S., and Lucas A. (2015) Methane storms as a driver of Titan's dune orientation. *Nature Geosci., 8,* 362–366.

Choukroun M. and Sotin C. (2012) Is Titan's shape caused by its meteorology and carbon cycle? *Geophys. Res. Lett., 39,* L04201.

Clark R. N., Curchin J. M., Barnes J. W., et al. (2010) Detection and mapping of hydrocarbon deposits on Titan. *J. Geophys. Res.: Planets, 115(E10),* E10005.

Cleaves H. J., Neish C., Callahan M. P., Parker E., Fernández F. M., and Dworkin J. P. (2014) Amino acids generated from hydrated Titan tholins: Comparison with Miller-Urey electric discharge products. *Icarus, 237,* 182–189.

Coates A. J., Crary F. J., Lewis G. R., Young D. T., Waite J. H., and Sittler E. C. (2007), Discovery of heavy negative ions in Titan's ionosphere. *Geophys. Res. Lett., 34,* L22103, DOI: 10.1029/2007GL030978.

Cordier D., Mousis O., Lunine J. I., Lavvas P., and Vuitton V. (2009) An estimate of the chemical composition of Titan's lakes. *Astrophys. J. Lett., 707,* L128–L131.

Cordier D., Mousis O., Lunine J. I., Lebonnois S., Rannou P., et al. (2012) Titan's lakes chemical composition: Sources of uncertainties and variability. *Planet. Space Sci., 61,* 99–107.

Cordier D., Cornet T., Barnes J. W., et al. (2016) Structure of Titan's evaporites. *Icarus, 270,* 41–56.

Cornet T., Bourgeois O., Le Mouélic S., Rodriguez S., Lopez Gonzalez T., Sotin C., Tobie G., Fleurant C., Barnes J. W., Brown R. H., Baines K. H., Buratti B. J., Clark R. N., and Nicholson P. D. (2012) Geomorphological significance of Ontario Lacus on Titan: Integrated interpretation of Cassini VIMS, ISS, and RADAR data and comparison with the Etosha Pan (Namibia). *Icarus, 218(2),* 788–806.

Courtin R., Sim C. K., Kim S. J., et al. (2012) The abundance of H_2 in Titan's troposphere from the Cassini CIRS investigation. *Planet. Space Sci., 69(1),* 89–99.

Coustenis A., Salama A., Schulz B., Ott S., Lellouch E., Encrenaz T. H., Gautier D., and Feuchtgruber H. (2003) Titan's atmosphere from ISO mid-infrared spectroscopy. *Icarus, 161,* 383–403, DOI: 10.1016/S0019-1035(02)00028-3.

Coustenis A., Atreya S., Balint T., Brown R. H., Dougherty M., Ferri F., Fulchignoni M., Gautier D., Gowen R., Griffith C., Gurvits L., Jaumann R., Langevin Y., Leese M., Lunine J., McKay C. P., Moussas X., Müller-Wodarg I., Neubauer F., Owen T., Raulin F., Sittler E., Sohl F., Sotin C., Tobie G., Tokano T., Turtle E., Wahlund J.-E., Waite H., Baines K., Blamont J., Dandouras I., Krimigis T., Lellouch E., Lorenz R., Morse A., Porco C., Hirtizig M., Saur J., Coates A., Spilker T., Zarnecki J., and 113 co-authors (2009) TandEM: Titan and Enceladus mission. *Exp. Astron., 23,* 893–946, DOI: 10.1007/s10686-008-9103-z.

Crary F. J., Magee B. A., Mandt K., Waite J. H., Westlake J., and Young D. T. (2009) Heavy ions, temperatures, and winds in Titan's ionosphere: Combined Cassini CAPS and INMS observations. *Planet. Space Sci., 57,* 1847–1856, DOI: 10.1016/j.pss.2009.09.006.

Creamer J. S., Mora M. F., and Willis P. A. (2017) Enhanced resolution of chiral amino acids with capillary electrophoresis for biosignature detection in extraterrestrial samples. *Anal. Chem., 89(2),* 1329–1337.

Cronin J. and S. Pizzarello (1983) Amino acids in meteorites. *Adv. Space Res., 3(9),* 5–18.

Davila A. F. and McKay C. P. (2014) Chance and necessity in biochemistry: Implications for the search for extraterrestrial biomarkers in Earth-life environments. *Astrobiology, 14,* 534–540.

de la Fuente J. L., Ruiz-Bermejo M., Nna-Mvondo D., et al. (2014) Further progress into the thermal characterization of HCN polymers. *Polym. Degrad. Stab., 110,* 241–251.

Diez-y-Riega H., Camejo D., Rodriguez A. E., and Manzanares C. E. (2014) Unsaturated hydrocarbons in the lakes of Titan: Benzene solubility in liquid ethane and methane at cryogenic temperatures. *Planet. Space Sci., 99,* 28–35.

Dorn E. D., Nealson K. H., and Adami C. (2011) Monomer abundance distribution patterns as a universal biosignature: Examples from terrestrial and digital life. *J. Mol. Evol., 72,* 283–295.

Etiope G. and Lollar B. S. (2013) Abiotic methane on Earth. *Rev. Geophysics, 51,* 276–299.

Ewing R. C., Hayes A. G., and Lucas A. (2015) Sand dune patterns on Titan controlled by long-term climate cycles. *Nature Geosci., 8,* 15–19.

Facchin M., Scarso A., Selva M., et al. (2017) Towards life in hydrocarbons: Aggregation behaviour of "reverse" surfactants in cyclohexane. *RSC Adv., 7(25),* 15337–15341.

Fortes A. D. (2000) Exobiological implications of a possible ammonia-water ocean inside Titan. *Icarus, 146,* 144–452.

Gao P. and Stevenson D. J. (2013) Non-hydrostatic effects and the determination of icy satellites' moment of inertia. *Icarus, 226,*

1185–1191.

Georgiou C. D. and Deamer D. W. (2014) Lipids as universal biomarkers of extraterrestrial life. *Astrobiology, 14,* 541–549.

Gilliam A. E. and McKay C. P. (2011) Titan under a red dwarf star and as a rogue planet: requirements for liquid methane. *Planet. Space Sci., 59,* 835–839.

Glavin D. P., Callahan M. P., Dworkin J. P., and Elsila J. E. (2010) The effects of parent body processes on amino acids in carbonaceous chondrites. *Meteoritics & Planet. Sci., 45,* 1948–1972.

Glein C. R. (2015) Noble gases, nitrogen, and methane from the deep interior to the atmosphere of Titan. *Icarus, 250,* 570–586.

Glein C. R. and Shock E. L. (2013) A geochemical model of non-ideal solutions in the methane-ethane-propane-nitrogen-acetylene system on Titan. *Geochim. Cosmochim. Acta, 115,* 217–240.

Grasset O. and Sotin C. (1996) The cooling rate of a liquid shell in Titan's interior. *Icarus, 123,* 101–112.

Griffith C. A. (2014) Not just a storm in a teacup. *Nature, 514,* 40–41.

Griffith C. A., Lora J. M., Turner J., Penteado P. F., Brown R. H., et al. (2012) Possible tropical lakes on Titan from observations of dark terrain. *Nature, 486,* 237–239.

Grimme S. (2004) Accurate description of van der Waals complexes by density functional theory including empirical corrections. *J. Comput. Chem., 25(12),* 1463–1473.

Grindrod P. M., Fortes A. D., Nimmo F., Feltham D. L., Brodholt J. P., and Vocadlo L. (2008) The long-term stability of a possible aqueous ammonium sulfate ocean inside Titan. *Icarus, 197,* 137–151.

Grinspoon D. H. and Schulze-Makuch D. (2010) Possible niches for extant life on Titan in light of the first six years of Cassini/Huygens results. In *Bulletin of the American Astronomical Society, 42,* Abstract #61.09, American Astronomical Society, Washington, DC.

Hand K. P., Murray A. E., Garvin J. B., Brinckerhoff W. B., Christner B. C., Edgett K. S., Ehlmann B. L., German C. R., Hayes A. G., Hoehler T. M., Horst S. M., Lunine J. I., Nealson K. H., Paranicas C., Schmidt B. E., Smith D. E., Rhoden A. R., Russell M. J., Templeton A. S., Willis P. A., Yingst R. A., Phillips C. B., Cable M. L., Craft K. L., Hofmann A. E., Nordheim T. A., Pappalardo R. P., and the Project Engineering Team (2017) *Europa Lander Study 2016 Report.* Report of the Europa Lander Science Definition Team, JPL D-97667, Jet Propulsion Laboratory, Pasadena, California.

Hanley J. et al. (2017) Methane, ethane and nitrogen liquid stability on Titan. In *Lunar and Planetary Science XLVIII,* Abstract #1686. Lunar and Planetary Institute, Houston.

Hartwig J., Meyerhofer P., Lorenz R., and Lemmon E. (2018) An analytical solubility model for nitrogen-methane-ethane ternary mixtures. *Icarus, 299,* 175–186.

Hawkins I. and Jura M. (1987) The carbon ($^{12}C/^{13}C$) isotope ratio of the interstellar medium in the neighborhood of the Sun. *Astrophys. J., 317,* 926–950.

Hayes A. G. (2016) The lakes and seas of Titan. *Annu. Rev. Earth Planet. Sci., 44,* 57–83.

Hayes A., Aharonson O., Callahan P., Elachi C., Gim Y., et al. (2008) Hydrocarbon lakes on Titan: Distribution and interaction with a porous regolith. *Geophys. Res. Lett., 35,* L09204.

Hayes A. G., Birch S. P. D., Dietrich W. E., Howard A. D., Kirk R., et al. (2017) Topographic constraints on the evolution and connectivity of Titan's lacustrine basins. *Geophys. Res. Lett. 44(23),* 11745–11753.

Hazen R. M. and Sverjensky D. A. (2010) Mineral surfaces, geochemical complexities, and the origins of life. *Cold Spring Harbor Perspect. Biol., 2:a002162.*

He C., Lin G., Upton K. T., et al. (2012) Structural investigation of HCN polymer isotopomers by solution-state multidimensional NMR. *J. Phys. Chem. A, 116(19),* 4751–4759.

Higgs P. G. and Pudriz R. E. (2009) A thermodynamic basis for prebiotic amino acid synthesis and the nature of the first genetic code. *Astrobiology, 9,* 483–490.

Hodyss R., Piao S., Cable M. L., and Malaska M. (2017) Carbon dioxide chemistry at low temperatures. Presented at American Chemical Society National Meeting and Exposition, San Francisco, California, 2–6 April.

Hofgartner J. D., Hayes A. G., Lunine J. I., Zebker H., Stiles B. W., Sotin C., Barnes J. W., Turtle E. P., Baines K. H., Brown R. H., Buratti B. J., Clark R. N., Encrenaz P., Kirk R. D., Le Gall A., Lopes R. M., Lorenz R. D., Malaska M. J., Mitchell K. L., Nicholson P. D., Paillou P., Radebaugh J., Wall S. D., Wood C. (2014) Transient features in a Titan sea. *Nature Geosci, 7,* 493–496.

Hofgartner J. D., Hayes A. G., Lunine J. I., Zebker H., Lorenz R. D., Malaska M. J., Mastrogiuseppe M., Notarnicola C., and Soderblom J. M. (2016) Titan's "Magic Islands": Transient features in a hydrocarbon sea. *Icarus, 271,* 338–349.

Horikoshi H. (1998) Barophiles: Deep sea microorganisms adapted to an extreme environment. *Curr. Opin. Microbiol., 1(3),* 291–295.

Hörst S. M. (2017) Titan's atmosphere and climate. *J. Geophys. Res.-Planets, 122,* 432–482.

Hörst S. M. et al. (2012) Formation of amino acids and nucleotide bases in a Titan atmosphere simulation experiment. *Astrobiology, 12,* 809–817, DOI: 10.1089/ast.2011.0623.

Iess L., Jacobson R. A., Ducci M., Stevenson D. J., Lunine J. I., Armstrong J. W., Asmar S. W., Racioppa P., Rappaport N. J., and Tortora P. (2012) The tides of Titan. *Science, 337,* 457–459.

Kargel J. S. (1992) Ammonia-water volcanism on icy satellites: Phase relations at 1 atmosphere. *Icarus, 100,* 556–574.

Kato C. and Bartlett D. H. (1998) The molecular biology of barophilic bacteria. *Extremophiles, 1(3),* 111–116.

King M. D. and Marsh D. (1987) Head group and chain length dependence of phospholipid self-assembly studied by spin-label electron spin resonance. *Biochemistry, 26(5),* 1224–1231.

Kirchner M. T., Blaeser D., and Boese R. (2010) Co-crystals with acetylene: Small is not simple! *Chem. — Eur. J., 16(7),* 2131–2146.

Korycansky D. G. and Zahnle K. J. (2005) Modeling crater populations on Venus and Titan. *Planet. Space Sci., 53 (7),* 695–710.

Krasnopolsky V. A. (2014) Chemical composition of Titan's atmosphere and ionosphere: Observations and the photochemical model. *Icarus, 236,* 83–91.

Lammer H. (2007) M star planet habitability. *Astrobiology, 7,* 27–29.

Lavvas P. P., Coustenis A., and Vardavas I. M. (2008) Coupling photochemistry with haze formation in Titan's atmosphere, Part II: Results and validation with Cassini/Huygens data. *Planet. Space Sci., 56(1),* 67–99.

Lavvas P. et al. (2011) Energy deposition and primary chemical products in Titan's upper atmosphere. *Icarus, 213,* 233–251.

Lazcano A. and Miller S. L. (1994) How long did it take for life to begin and evolve to cyanobacteria? *J. Mol. Evol., 39,* 546–554.

Le Gall A., Malaska M., Lorenz R. D., Janssen M. A., Tokano T., Hayes A. G., Mastrogiuseppe M., Lunine J. I., Veyssière G., Encrenaz P., and Karatekin O. (2016) Composition, seasonal change, and bathymetry of Ligeia Mare, Titan, derived from its microwave thermal emission. *J. Geophys. Res.-Planets, 121,* 233–251, DOI: 10.1002/2015JE004920.

Li A. H.-T. and Chao S. D. (2009) Interaction energies of dispersion-bound methane dimer from coupled cluster method at complete basis set limit. *J. Mol. Struct.: THEOCHEM, 897(1),* 90–94.

Lineweaver C. H. and Chopra A. (2012) The habitability of our Earth and other Earths: astrophysical, geochemical, geophysical, and biological limits on planet habitability. *Annu. Rev. Earth Planet. Sci., 40,* 597–623.

Lollar B. S., Lacrampe-Couloume G., Slater G. F., Ward J. A., Moser D. P., Gihring T. M., Lin L.-H., and Onstott T. C. (2006) Unravelling abiogenic and biogenic sources of methane in the Earth's deep subsurface. *Chem. Geol., 226,* 328–339.

Lopes R. M. C., Kirk R. L., Mitchell K. L., Legall A., Barnes J. W., Hayes A., et al. (2013) Cryovolcanism on Titan: New results from Cassini RADAR and VIMS. *J. Geophys. Res.-Planets, 118,* 416–435.

Lora J. M., Lunine J. I., Russell J. L., and Hayes A. G. (2014) Simulations of Titan's paleoclimate. *Icarus, 243,* 264–273.

Lorenz R. D. (2008) A review of balloon concepts for Titan. *J. Br. Interplanet. Soc., 61,* 2–13.

Lorenz R. D. (2014) The flushing of Ligeia: Composition variations across Titan's seas in a simple hydrological model. *Geophys. Res. Lett., 41,* 5764–5770, DOI: 10.1002/ 2014GL061133.

Lorenz R. D. and Hayes A. G. (2012) The growth of wind-waves in Titan's hydrocarbon seas. *Icarus, 219,* 468–475.

Lorenz R. D. and Radebaugh J. (2009) Global pattern of Titan's dunes: Radar survey from the Cassini prime mission. *Geophys. Res. Lett., 36(3),* L03202.

Lorenz R. D., Kraal E., Asphaug E., and Thomson R. E. (2003) The seas of Titan. *Eos Trans. Am. Geophys. Union, 84,* 125, 131–132.

Lorenz R. D., Wall S., Radebaugh J., Boubin G., Reffet E., Janssen M., Stofan E., Lopes R., Kirk R., Elachi C., Lunine J., et al. (2006) The sand seas of Titan: Cassini RADAR observations of longitudinal dunes. *Science, 312 (5774),* 724–727.

Lorenz R. D., Stiles B. W., Kirk R. L., Allison M. D., Persi del Marmo P., Iess L., Lunine J. I., Ostro S. J., and Hensley S. (2008) Titan's rotation reveals an internal ocean and changing zonal winds. *Science, 319,* 1649–1651.

Lorenz R. D., Kirk R. I., Hayes A. G., Anderson Y. Z., Lunine J. I., Tokano T., Turtle E. P., Malaska M. J., Soderblom J. M., Lucas A., Karatekin Ö., and Wall S. D. (2014) A radar map of Titan seas: Tidal dissipation and ocean mixing through the throat of Kraken. *Icarus, 237,* 9–15.

Lorenz R. D., Turtle E. P., Barnes J. W., Trainer M. G., Adams D. S., Hibbard K. E., Sheldon C. Z., Zacny K., Peplowski P. N., Lawrence D. J., Ravine M. A., McGee T. G., Sotzen K. S., MacKenzie S. M., Langelaan J. W., Schmitz S., Wolfarth L. S., and Bedini P. (2018) Dragonfly: A Rotorcraft lander concept for scientific exploration at Titan. *Johns Hopkins APL Tech. Dig., 34(3).*

Lu B. C.-Y., Yu P., and Poon D. P. L. (1969) Formation of a third liquid layer in the nitrogen-methane-ethane system. *Nature, 222,* 768–769.

Lu B. C.-Y., Yu P., and Poon D. P. L. (1970) Liquid phase inversion. *Nature, 225,* 1128–1129.

Luspay-Kuti A., Chevrier V. F., Cordier D., Rivera-Valentin E. G., Singh S., Wagner A., and Wasiak F. C. (2015) Experimental constraints on the composition and dynamics of Titan's polar lakes. *Earth Planet. Sci. Lett., 410,* 75–83.

Lunine J. I. (2009) Saturn's Titan: A strict test for life's cosmic ubiquity. *Proc. Am. Philos. Soc., 153,* 403–418.

Lunine J. I. (2010) Titan and habitable planets around M dwarfs. *Faraday Discussions, 147,* 405–418.

Lunine J. I. and Hörst S. M. (2011) Organic chemistry on the surface of Titan. *Rend. Fis. Acc. Lincei, 22(3),* 183–189, DOI: 10.1007/s12210-011-0130-8.

Lunine J. I., Yung Y. L., and Lorenz R. D. (1999) On the volatile inventory of Titan from isotopic abundances in nitrogen and methane. *Planet. Space Sci., 47(10/11),* 1291–1303.

Lv K-P, Norman L., and Li Y-L. (2017) Oxygen-free biochemistry: The putative CHN foundation for exotic life in a hydrocarbon world? *Astrobiology, 17,* 1173–1181.

MacGregor M. A., Weinberger A. J., Wilner D. J., Kowalski A. F., and Cranmer S. R. (2018) Detection of a millimeter flare from Proxima Centauri. *Astrophys. J. Lett., 855(1),* L2.

MacKenzie S. M. and Barnes J. W. (2016) Compositional similarities and distinctions between Titan's evaporitic terrains. *Astrophys. J., 821(1),* 17.

Maher K. A. and Stevenson D. J. (1988) Impact frustration of the origin of life. *Nature, 331,* 612–614.

Malaska M. J. and Hodyss R. (2014) Dissolution of benzene, naphthalene, and biphenyl in a simulated Titan lake. *Icarus, 242,* 74–81.

Malaska M. J., Hodyss R., Lunine J. I., Hayes A. G., Hofgartner J. S., Hollyday G., and Lorenz R. D. (2017) Laboratory measurements of nitrogen dissolution in Titan lake fluids. *Icarus, 289,* 94–105.

Mamajanov I. and Herzfeld J. (2009) HCN polymers characterized by solid state NMR: Chains and sheets formed in the neat liquid. *J. Chem. Phys., 130(13),* 134503/1-134503/6.

Mandt K. E., Mousis O., Lunine J., and Gautier D. (2014) Protosolar ammonia as the unique source of Titan's nitrogen. *Astrophys. J. Lett., 788,* L24.

Marshall S. M., Murray A. R. G., and Cronin L. (2017) A probabilistic framework for identifying biosignatures using Pathway Complexity. *Philos. Trans. R. Soc. A, 375,* 20160342, DOI: 10.1098/rsta.2016.0342.

Mastrogiuseppe M., Poggiali V., Hayes A., Lorenz R., Lunine J. Picardi G., Seu R., Flamini E., Mitri G., Notarnicola C., Paillou P., and Zebker H. (2014) The bathymetry of a Titan sea. *Geophys. Res. Lett., 41,* DOI: 10.1002/2013GL058618.

Mastrogiuseppe M., Hayes A., Poggiali V., Seu R., Lunine J. I., and Hofgartner J. D. (2016) Radar sounding using the Cassini altimeter: Waveform modeling and Monte Carlo approach for data inversion of observations of Titan's seas. *IEEE Trans. Geosci. Remote Sens., 54,* 5646–5656, DOI: 10.1109/TGRS.2016.2563426.

Mastrogiuseppe M., Hayes A. G., Poggiali V., Seu R., Lunine J., and Hofgartner J. D. (2018) Bathymetry and composition of Titan's Ontario Lacus derived from Monte Carlo-based waveform inversion of Cassini RADAR altimetry data. *Icarus, 300,* 203–209.

Maynard-Casely H. E., Hodyss R., Cable M., L. et al. (2016) A co-crystal between benzene and ethane: A potential evaporate material for Saturn's moon Titan. *IUCrJ, 3(Pt 3),* 192–199, DOI: 10.1107/S2052252516002815.

Maynard-Casely H. E., Cable M. L., Malaska M. J., Vu T. H., Choukroun M., and Hodyss R. (2018) Prospects for mineralogy on Titan. *Am. Mineral., 103(3),* 343–349, DOI: 10.2138/am-2018-6259.

McKay C. P. (2004) What is life — and how do we search for it in other worlds? *PLoS Biol., 2(9),* e302, DOI: 10.1371/journal.pbio.0020302.

McKay C. P. (2016) Titan as the abode of life. *Life, 6(1),* 8.

McKay C. P. and Smith H. D. (2005) Possibilities for methanogenic life in liquid methane on the surface of Titan. *Icarus, 178(1),* 274–276.

McKay C. P., Pollack J. B., and Courtin R. (1991) The greenhouse and antigreenhouse effects on Titan. *Science, 253,* 1118–1121.

McKay C. P., Anbar A. D., Porco C., and Tsou P. (2014) Follow the plume: The habitability of Enceladus. *Astrobiology, 14(4),* 352–355, DOI: 10.1089/ast.2014.1158.

Méndez Harper J. S., McDonald G. D., Dufek J., Malaska M. J., Burr D. M., Hayes A. G., McAdams J., and Wray J. J. (2017) Electrification of sand on Titan and its influence on sediment transport. *Nature Geosci., 10,* 260–265.

Mitchell K. A., Barmatz M. B., Jamieson C. S., Lorenz R. D., and Lunine J. I. (2015) Laboratory measurements of cryogenic liquid alkane microwave absorptivity and implications for the composition of Ligeia Mare, Titan. *Geophys. Res. Lett., 42,* 1340–1345, DOI: 10.1002/2014GL059475.

Mitri G., Meriggiola R., Hayes A., Lefevre A., Tobie G., Genova A., Lunine J. I., and Zebker H. (2014) Shape, topography, gravity anomalies, and tidal deformation of Titan. *Icarus, 236,* 169–177.

Molter E. M., Nixon C. A., Cordiner M. A., Serigano J., Irwin P. G. J., Teanby N. A., Charnley S. B., and Lindberg J. E. (2016) Alma observations of HCN and its isotopologues on Titan. *Astron. J., 152(42),* 1–7.

Mousis O., Lunine J. I., Hayes A. G., and Hofgartner J. D. (2016) The fate of ethane in Titan's hydrocarbon lakes and seas. *Icarus, 270,* 37–40.

National Research Council (2002) *Signs of Life: A Report Based on the April 2000 Workshop on Life Detection Techniques.* National Academies, Washington, DC. 210 pp.

National Research Council (2007) *The Limits of Organic Life in Planetary Systems.* National Academies, Washington, DC. 100 pp.

Neese F. (2012) The ORCA program system. *Wiley Interdiscip. Rev.: Comput. Mol. Sci., 2(1),* 73–78.

Neish C. and Lorenz R. D. (2012) Titan's global crater population: A new assessment. *Planet. Space Sci., 60,* 26–33.

Neish C. D., Lorenz R. D., O'Brien D. P., and the Cassini RADAR Team (2006) The potential for prebiotic chemistry in the possible cryovolcanic dome Ganesa Macula on Titan. *Intl. J. Astrobiology, 5,* 57–65.

Neish C. D., Somogyi A., Imanaka H., Lunine J. I., and Smith M. A. (2008) Rate measurements of the hydrolysis of complex organic macromolecules in cold aqueous solutions: Implications for prebiotic chemistry on the early Earth and Titan. *Astrobiology, 8(2),* 273–287.

Neish C. D., Somogyi Á., Lunine J. I., and Smith M. A. (2009) Low temperature hydrolysis of laboratory tholins in ammonia-water solutions: Implications for prebiotic chemistry on Titan. *Icarus, 201,* 412–421.

Neish C. D., Somogyi Á., and Smith M. A. (2010) Titan's primordial soup: Formation of amino acids via low-temperature hydrolysis of tholins. *Astrobiology, 10,* 337–347, DOI: 10.1089/ast.2009.0402.

Neish C. D., Molaro J. L., Lora J. M., Howard A. D., Kirk R. L., Schenk P., et al. (2016) Fluvial erosion as a mechanism for crater modification on Titan. *Icarus, 270,* 114–129.

Neish C. D., Lorenz R. D., Turtle E. P., Barnes J. W., Trainer M. G., Stiles B., Kirk R., Hibbits C., and Malaska M. (2018) Strategies for detecting biological molecules on Titan. *Astrobiology, 18,* 571 585.

Niemann H. B., Atreya S. K., Demick J. E., et al. (2010) Composition of Titan's lower atmosphere and simple surface volatiles as measured by the Cassini-Huygens probe gas chromatograph mass spectrometer experiment. *J. Geophys. Res., 115,* E12006.

Oberbeck V. R. and Fogleman G. (1989) Estimates of the maximum time required to originate life. *Origins Life Evol. Biosph., 19,* 549–560.

O'Brien D. P., Lorenz R. D., and Lunine J. I. (2005) Numerical calculations of the longevity of impact oases on Titan. *Icarus, 173,* 243–253.

Oro J. (1961) Mechanism of synthesis of adenine from HCN under possible primitive Earth conditions. *Nature, 191,* 1193–1194.

Palmer M. Y, Cordiner M. A, Nixon C. A., et al. (2017) ALMA detection and astrobiological potential of vinyl cyanide on Titan. *Sci. Adv., 3(7),* e1700022.

Pascal R. (2012) Suitable energetic conditions for dynamic chemical complexity and the living state. *J. Syst. Chem., 3(3),* DOI: 10.1186/1759-2208-3-3.

Pascal R., Pross A., and Sutherland J. D. (2013) Towards an evolutionary theory of the origin of life based on kinetics and thermodynamics. *Open Biol., 3(11),* 130156.

Patel B. H., Percivalle C., Ritson D. J., et al. (2015) Common origins of RNA, protein, and lipid precursors in a cyanosulfidic protometabolism. *Nature Chem., 7(4),* 301–307.

Pernot P., Carrasco N., Thissen R., and Schmitz-Afonso I. (2010) Tholinomics — Chemical analysis of nitrogen-rich polymers. *Anal. Chem., 82(4),* 1371–1380.

Picard A. and Daniel I. (2013) Pressure as an environmental parameter for microbial life — A review. *Biophys. Chem., 183,* 30–41.

Poch O., Coll P., Buch A., Ramírez S. I., and Raulin F. (2012) Production yields of organics of astrobiological interest from H_2O-NH_3 hydrolysis of Titan's tholins. *Planet. Space Sci., 61,* 114–123, DOI: 10.1016/j.pss.2011.04.009.

Porco C. C., Dones L., and Mitchell C. (2017) Could it be snowing microbes on Enceladus? Assessing conditions in its plume and implications for future missions. *Astrobiology, 17,* 876–901.

Pross A. and Pascal R. (2013) The origin of life: What we know, what we can know, and what we will never know. *Open Biol., 3(3),* 120190, DOI: 10.1098/rsob.120190.

Rahm M, Lunine J. I., Usher D., et al. 2016. Polymorphism and electronic structure of polyimine and its potential significance for prebiotic chemistry on Titan. *Proc. Natl. Acad. Sci. U.S.A., 113,* 8121–8126.

Ramirez S. I., Coll P., Buch A., Brassé C., Poch O., and Raulin F. (2010) The fate of aerosols on the surface of Titan. *Faraday Discussions, 147,* 419–427.

Rampelotto P. H. (2010) Resistance of microorganisms to extreme environmental conditions and its contribution to astrobiology. *Sustainability, 2(6),* 1602–1623.

Rampelotto P. H. (2013) Extremophiles and extreme environments. *Life, 3,* 482–485.

Raulin F. and Owen T. (2002) Organic chemistry and exobiology on Titan. *Space Sci. Rev., 104,* 377–394.

Raulin F., Brasse C., Poch O., et al. (2012) Prebiotic-like chemistry on Titan. *Chem. Soc. Rev., 41(16),* 5380–5393.

Raven J. A., Kübler J. E., and Beardall J. (2000) Put out the light, and then put out the light. *J. Mar. Biol. Assoc. U. K., 80(1),* 1–25.

Reynolds J. A., Tanford C., and Stone W. L. (1977) Interaction of L-α-didecanoyl phosphatidylcholine with the AI polypeptide of high density lipoprotein. *Proc. Natl. Acad. Sci. U.S.A., 74(9),* 3796–3799.

Roadcap G. S., Sanford R. A., Jin Q., Pardinas J. R., and Bethke C. M. (2006) Extremely alkaline (pH > 12) ground water hosts diverse microbial community. *Groundwater, 44(4),* 511–517.

Ruiz-Bermejo M., Zorzano M-P., and Osuna-Esteban S. (2013) Simple organics and biomonomers identified in HCN polymers: An overview. *Life, 3(3),* 421–448.

Sagan C. (1974) The origin of life in a cosmic context. *Origins Life, 5(3–4),* 497–505.

Sandström H. and Rahm M. (2020) Can polarity-inverted membranes self-assemble on Titan? *Sci. Adv., 6(4),* eaax0272, DOI: 10.1126/sciadv.aax0272.

Schulze-Makuch D. and Grinspoon D. H. (2005) Biologically enhanced energy and carbon cycling on Titan? *Astrobiology, 5,* 560–567.

Schulze-Makuch D and Irwin L. N. (2004) *Life in the Universe: Expectations and Constraints.* Springer Nature Switzerland AG. 343 pp.

Schulze-Makuch D. and Irwin L. N. (2006) The prospect of alien life in exotic forms on other worlds. *Naturwissenschaften, 93(4),* 155–172.

Schulze-Makuch D., Haque S., Walther-Antonio M., Ali D., Hosein R., Song Y. C., and Hallam S. J. (2011) Microbial life in a liquid asphalt desert. *Astrobiology, 11(3),* 241–258, DOI: 10.1089/ast.2010.0488.

Schulze-Makuch D., Houtkooper J. M., et al. (2015) The physical, chemical and physiological limits of life. *Life, 5(3),* 1472–1486.

Seckbach J. (2005) *Origins: Genesis, Evolution and Diversity of Life.* Kluwer, Dordrecht. 709 pp.

Segre D., Ben-Eli D., Deamer D. W., et al. (2001) The lipid world. *Origins Life Evol. Biosph., 31(1–2),* 119–145.

Singh S., Combe J.-P., Cordier D., et al. (2017) Experimental determination of acetylene and ethylene solubility in liquid methane and ethane: Implications to Titan's surface. *Geochim. Cosmochim. Acta, 208,* 86–101.

Sleep N. H., Zahnle K., and Neuhoff P. S. (2001) Initiation of clement surface conditions on the early Earth. *Proc. Natl. Acad. Sci. U.S.A., 98,* 3666–3672.

Somogyi A., Oh C., Smith M. A., and Lunine J. I. (2005) Organic environments on Saturn's moon, Titan: Simulating chemical reactions and analyzing products by FT-ICR and ion-trap mass spectrometry. *J. Am. Soc. Mass Spectrom., 16(6),* 850–859.

Stevenson J., Clancy C., Lunine J. (2015a) Membrane alternatives in worlds without oxygen: Creation of an azotosome. *Sci. Adv., 1,* e1400067.

Stevenson J. M., Fouad W. A., David S., et al. (2015b) Solvation of nitrogen compounds in Titan's seas, precipitates, and atmosphere. *Icarus, 256,* 1–12.

Stofan E. R., Elachi C, Lunine J. I., Lorenz R. D., Stiles B., et al. (2007) The lakes of Titan. *Nature, 445,* 61–64.

Stofan E. R., Lorenz R. D., Lunine J. I., Bierhaus E., Clark B., Mahaffy P. R., and Ravine M. R. (2013) TiME: The Titan Mare Explorer. In *IEEE Aerospace Conference, Big Sky, MT, 2013,* pp. 1–10, DOI: 10.1109/AERO.2013.6497165.

Strobel D. F. (2010) Molecular hydrogen in Titan's atmosphere: Implications of the measured tropospheric and thermospheric mole fractions. *Icarus, 208(2),* 878–886.

Stupp S. I. and Palmer L. C. (2014) Supramolecular chemistry and self-assembly in organic materials design. *Chem. Mater., 26(1),* 507–518.

Sutherland J. D. (2016) The origin of life — Out of the blue. *Angew. Chem., Intl. Ed., 55,* 104–121.

Tan S. P., Kargel J. S., and Marion G. M. (2013) Titan's atmosphere and surface liquid: New calculation using statistical associating fluid theory. *Icarus, 222,* 53–72.

Tan S. P., Kargel J. S., Jennings D. E., Mastrogiuseppe M., Adidharma H., and Marion G. M. (2015) Titan's liquids: Exotic behavior and its implications on global fluid circulation. *Icarus, 250,* 64–75.

Teanby N. A., Irwin P. G. J., de Kok R., et al. (2006) Latitudinal variations of HCN, HC_3N, and C_2N_2 in Titan's stratosphere derived from Cassini CIRS data. *Icarus, 181(1),* 243–255.

Teanby N. A., Irwin P. G. J., de Kok R., et al. (2007) Vertical profiles of HCN, HC_3N, and C_2H_2 in Titan's atmosphere derived from Cassini/CIRS data. *Icarus, 186(2),* 364-384.

Thomas D. A. and Beauchamp J. L. (2014) Prebiotic chemistry on cryogenic worlds: Tribochemical reactions of organics and water in Titan dunes. In *Workshop on the Habitability of Icy Worlds,* Abstract #4089. LPI Contribution No. 1774, Lunar and Planetary Institute, Houston.

Thompson W. R. and Sagan C. (1992) Organic chemistry on Titan: Surface interactions. In *Proceedings of the Symposium on Titan,* pp. 167–176. ESA SP-338, Toulouse, France.

Tobie G., Grasset O., Lunine J. I., Mocquet J., and Sotin C. (2005) Titan's orbit provides evidence for a subsurface ammonia-water ocean. *Icarus, 175,* 496–502.

Tobie G., Lunine J. I., and Sotin C. (2006) Episodic outgassing as the origin of atmospheric methane on Saturn's moon Titan. *Nature, 440,* 61–64.

Tobie G., Choukroun M., Grasset O., Le Mouélic S., Lunine J. I., Sotin C., Bourgeois O., Gautier D., Hirtzig M., Lebonnois S., and Le Corre L. (2009) Evolution of Titan and implications for its hydrocarbon cycle. *Philos. Trans. R. Soc. A, 367,* 617–631.

Tobie G., Lunine J. I., Monteux J., Mousis O., and Nimmo F. (2014) The origin and evolution of Titan. In *Titan* (I. Muller-Wordag et al., eds.), pp. 29–62. Cambridge Univ., Cambridge.

Tokano T. (2005) Thermal structure of putative hydrocarbon lakes on Titan. *Adv. Space Res., 36,* 286–294.

Tokano T. (2009a) Impact of seas/lakes on polar meteorology of Titan: Simulation by a coupled GCM-Sea model. *Icarus, 204,* 619–636.

Tokano T. (2009b) Limnological structure of Titan's hydrocarbon lakes and its astrobiological implication. *Astrobiology, 9,* 147–164.

Tokano T., Lorenz R. D., and Van Hoolst T. (2014) Numerical simulation of tides and oceanic angular momentum of Titan's hydrocarbon seas. *Icarus, 242,* 188–201.

Turtle E., Perry J., Hayes A., Lorenz R., Barnes J., McEwen A., West R., Del Genio A., Barbara J., Lunine J., Schaller E., Ray T., Lopes R., and Stofan E. (2011) Rapid and extensive surface changes near Titan's equator: Evidence of April showers. *Science, 331,* 1414–1417.

Turtle E. P., Barnes J. W., Trainer M. G., Lorenz R. D., MacKenzie S. M., Hibbard K. E., Adams D., and Bedini P. (2017) Dragonfly: Exploring Titan's prebiotic organic chemistry. In *Lunar and Planetary Science XLVIII,* Abstract #1958. Lunar and Planetary Institute, Houston.

Vanlint D., Mitchell R., Bailey E., Meersman F., McMillan P. F., Michiels C. W., and Aertsen A. (2011) Rapid acquisition of gigapascal-high-pressure resistance by *Escherichia coli. mBio, 2(1),* e00130-10.

Vincent D., Karatekin O., Lambrechts J., Lorenz R. D., Dehant V., and Deleersnijder E. (2018) A numerical study of tides in Titan's northern seas, Kraken and Ligeia Mare. *Icarus, 310,* 105–126.

Vixie G, Barnes J. W., Jackson B, Rodriguez S, Le Mouelic S., et al. (2015) Possible temperate lakes on Titan. *Icarus, 257,* 313–323.

Vu T. H., Cable M. L., Choukroun M., Hodyss R., and Beauchamp P. (2014) Formation of a new benzene-ethane co-crystalline structure under cryogenic conditions. *J. Phys. Chem. A, 118(23),* 4087–4094.

Vuitton V., Yelle R. V., and McEwan M. J. (2007) Ion chemistry and N-containing molecules in Titan's upper atmosphere. *Icarus, 191,* 722–742.

Waite J. H. Jr., Niemann H., Yelle R. V., Kasprzak W. T., Cravens T. E., Luhmann J. G., McNutt R. L., Ip W. H., Gell D., De La Haye V., Müller-Wordag I., Magee B., Borggren N., Ledvina S., Fletcher G., Walter E., Miller R., Scherer S., Thorpe R., Xu J., Block B., and Arnett K. (2005) Ion neutral mass spectrometer results from the first flyby of Titan. *Science, 308,* 982–986.

Wang K. and Yang Y-W. (2013) Supramolecular chemistry. *Annu. Rept. Prog. Chem., Sect. B: Org. Chem., 109,* 67–87.

Wilson E. H. and Atreya S. K. (2004) Current state of modeling the photochemistry of Titan's mutually dependent atmosphere and ionosphere. *J. Geophys. Res., 109,* E06002, DOI: 10.1029/2003JE002181.

Wood C. A., Lorenz R., Kirk R., Lopes R., Mitchell K., and Stofan E. (2010) Impact craters on Titan. *Icarus, 206,* 334–344.

Yayanos A. A., Dietz A. S., and Van Boxtel R. (1981) Obligately barophilic bacterium from the Mariana trench. *Proc. Natl. Acad. Sci. U.S.A., 78,* 5212–5215.

Yoshihiro Y. and Ohashi Y. (1998) Crystal structure of acrylonitrile. *Bull. Chem. Soc. Japan, 71,* 345–348.

Yung Y. L., Allen M., and Pinto J. P. (1984) Photochemistry of the atmosphere of Titan — Comparison between model and observations. *Astrophys. J., 55,* 465–506, DOI: 10.1086/190963.

Rummel J. (2020) Planetary protection in planetary exploration missions. In *Planetary Astrobiology* (V. Meadows et al., eds.), pp. 267–284. Univ. of Arizona, Tucson, DOI: 10.2458/azu_uapress_9780816540068-ch011.

Planetary Protection in Planetary Exploration Missions

John D. Rummel
SETI Institute

Planetary protection (also known as planetary quarantine) is the name given to policies and practices that seek to avoid microbial and organic chemical contamination during space exploration and use. These policies, initiated at the very beginning of the space age (*Phillips, 1974; Meltzer, 2011*), intend to control both the contamination of other celestial bodies (forward contamination) and the possible contamination of Earth's biosphere by material returned from elsewhere (backward, or back contamination). The establishment and maintenance of these policies are an international endeavor, initially because of widespread agreement on the potential negative effects of biological and organic contamination on the science of what is now called astrobiology — for instance, misidentifying Earth organisms carried into space as extraterrestrial, or destroying existing planetary environments while they are being studied. These policies have also been viewed as part of a space agency's response to provisions against "harmful contamination" that are part of the 1967 United Nations Outer Space Treaty (*United Nations, 1967*), the activities of the Committee on Space Research (COSPAR) and the UN Committee on the Peaceful Uses of Outer Space (COPUOS). General agreement on how to address those concerns serves to enable international cooperative missions in astrobiology and solar system exploration. That general agreement, as of now, is described in this chapter, along with progress in ensuring robust science results. Future planetary protection challenges will include the rise of private commercial activities in space and on other planets, and the nature of planetary protection requirements associated with human missions to Mars.

> *"And scattered about . . . were the Martians — dead! — slain by the putrefactive and disease bacteria against which their systems were unprepared; slain as the red weed was being slain; slain, after all man's devices had failed, by the humblest things that God, in his wisdom, has put upon this earth."*
>
> — *H. G. Wells, 1898*

1. ORIGINS AND RATIONALE OF THE PLANETARY PROTECTION (QUARANTINE) CONCEPT

Considerations of the possible organic or biological contamination of other worlds were placed in the forefront of scientific discussion regarding space exploration as soon as Sputnik orbited Earth in 1957 (*Phillips, 1974*). Even earlier views of extraterrestrial life had laid the foundation for this immediate concern. For example, in the fourth century BCE, Metrodorus of Chios had stated that "to consider the Earth as the only populated world in infinite space is as absurd as to assert that in an entire field of millet, only one grain will grow" (*Guthrie, 1965*). In the late nineteenth century CE the views of Percival Lowell and Camille Flammarion pictured civilizations on Mars and likely other worlds — sometimes those were failing, but mostly they were superior to those on Earth (*Dick and Strick, 2005*). Such views also inspired the first specific, popular notice of the pitfalls of interplanetary biological contamination. In his novel *The War of the Worlds*, H. G. Wells (*Wells, 1898*) posits the invasion of Earth by a superior, malevolent martian civilization that (apparently) fails to take planetary protection into account, particularly the question of whether life on Earth might be dangerous to martians. Accordingly, the invaders are vanquished not by humanity's paltry military defenses, but by Earth's rich stock of pathogenic organisms. From this rich history of human experience, planetary protection has grown into a critical science for exploration. It is derived from both the ethical and scientific best practices needed to prevent ecosystems from both forward and backward contamination as we seek both robotic and human plans for the discovery and exploration of other solar system bodies, preventing such "invasions" as colloquially envisioned.

Microbial contamination *must* be taken into account when the tables are turned: The planet to be invaded is Mars, and we are the invaders. This important work prevents false positives and thus provides critical support for the scientific credibility of astrobiology as a field — we would not want to find that our contamination had prevented us from completing the scientific study of life on other worlds, one of the centerpieces of an astrobiology strategy (*SSB, 2019*) from before it was called "exobiology" (*Lederberg,* 1960). We certainly should have learned that from the long, sad history of human-enabled biological invasions on Earth. While popular, fictional treatments of the pitfalls of extraterrestrial contamination have more or less followed H. G. Wells [cf. Orson Welles' 1938 radio broadcast and Michael Crichton's *The Andromeda Strain* (*Crichton,* 1969)], the factual history of terrestrial biological invasions is there to give one pause. Whether the problem has been the release of European rabbits or American cactus to Australia, or the importation of a tree disease that wiped out the great chestnut forests of the eastern United States, or any of the hundreds of other examples of biological destruction (e.g., *Williamson,* 1996), the record on the cross-contamination of Earth organisms on Earth has been dismal — and continues to be, as invasive species displace native ones and upset the balance of Earth ecosystems on a large scale (e.g., *Vitousek et al.,* 1997).

2. THE EMERGENCE OF PLANETARY QUARANTINE POLICY AND AN INTERNATIONAL CONSENSUS THROUGH COSPAR

After the launch of Sputnik in 1957, the opportunity to use newfound capabilities in spaceflight to extend knowledge in the physical and chemical sciences was obvious to many, and as such, the scientific community started to organize its response. Thus, early in 1958 the International Council of Scientific Unions (ICSU; now the International Council for Science), at the urging of the U.S. National Academy of Sciences, had taken action to address the issues associated with contamination by forming its Committee on Contamination by Extraterrestrial Exploration (CETEX) (*News of Science,* 1958). After their first meeting in May 1958, CETEX made a report to ICSU that noted that the group believed that there was a real danger that space exploration could produce contamination of extraterrestrial bodies that might complicate or render impossible more detailed studies later, "when the technological problems of landing sensitive scientific instruments on the moon and planets have been solved." CETEX was concerned only with exploration intended to provide bona fide scientific data, and was charged with preventing cases where an experiment done for one purpose would make it impossible for other types of studies to be made later. CETEX considered various dangers of contamination, but did not propose a specific code of conduct — although they advocated for such a code to be defined and adopted "with a minimum of delay." By October 1958, ICSU formed an interdisciplinary Committee on Space Research (COSPAR) to engender progress in scientific research in space. The role previously expected of CETEX was assigned to the new COSPAR, to allow coordination among various purely scientific interests and to identify activities and studies that might interfere with themselves or the experiments of others. The governance of the COSPAR council was formed of both representatives of the various international scientific unions and representatives of national members, who were

generally appointed from a country's scientific academy (cf. *https://cosparhq.cnes.fr/about/origin-purpose-role*).

One of the hallmarks of the formation of COSPAR at that time was that it had the endorsement of both the U.S. and Soviet Academies of Science, and so, despite the Cold War, it immediately became a meeting ground where scientists from both the East and the West could plan for future research opportunities and compare results. At that point, a number of efforts were being made by national institutions to forward space science across multiple disciplines, which could expect an airing at COSPAR, and the life sciences and problems of biological contamination were among them (*Hall,* 1974). With the help of COSPAR's regular assembly as an international forum, the United States and the Soviet Union forged agreement on what the requirements for planetary protection should be, and how information on missions should be exchanged. From 1958 on, all missions going to *planetary* bodies had been launched and operated by either the United States or the Soviet Union, with international partners onboard from time to time.

Concomitant with the formation of COSPAR by ICSU — which focused on scientific studies as well as on the problems of conducting them — the United Nations formed its Committee on the Peaceful Uses of Outer Space (COPUOS), "to review the scope of international cooperation in peaceful uses of outer space, to devise programs in this field to be undertaken under United Nations auspices, to encourage continued research and the dissemination of information on outer space matters, and to study legal problems arising from the exploration of outer space" (*United Nations,* 1959). Early in its work, and to ensure it was kept informed of emerging knowledge about the space environment, COPUOS recognized COSPAR as permanent observer to its activities, along with the International Astronautical Federation, which focused on the technical challenges of space exploration.

From its inception, COSPAR's discussions on space biological research were not solely due to an interest in the effects of spaceflight on astronauts and cosmonauts (human or otherwise), but on the possibility that more could be learned about life in the universe in general. For example, in 1959 the Space Science Board (SSB, founded 1958; now Space Studies Board) of the U.S. National Academy of Sciences helped to support an activity known as the West Coast Committee on Extraterrestrial Life (WESTEX), which was focused on finding and characterizing extraterrestrial life through space exploration. The committee included such future luminaries as Joshua Lederberg (a Nobel Prize winner for his work in genetics), Melvin Calvin (a Nobel Prize winner for his work in photosynthesis), Harold Urey (a Nobel Prize winner in chemistry, and Stanley Miller's graduate advisor), and a young astronomer named Carl Sagan, among others. In their summary report (*SSB,* 1959) the committee stated that "the detection and analysis of planetary life is one of the major challenges of contemporary science and should be pre-eminent among the objectives of our space research programs," but they also noted that to do that research required spacecraft to avoid biological contamination altogether (at a probability of one in a million), and "reaffirm[ed] the necessity of maintaining strenuous precautions against such contamination." Looking forward to future sample return missions, they emphasized that avoiding contamination from Earth was essential to overcome the problems of understanding the actual dangers of biological interchange, noting "that the task of evaluating the potential hazard of a planetary biota will be multiplied if this has to be isolated from organisms inadvertently transferred from Earth" (*SSB*, 1959).

Despite the international understanding that organic and biological contamination in space exploration would be destructive, in the early years there was a naturally occurring disjunction between theory and practice. On the policy side, there was not enough of an understanding of extraterrestrial environments to properly gauge exactly how clean a spacecraft had to be to avoid their contamination. Planetary quarantine provisions called for "spacecraft sterilization" but it was not known if complete sterilization was needed, and it was not obvious that it was possible to achieve. A variety of sterilization methods were envisioned — from ethylene bromide gas treatment to "hard" radiation treatments — and eventually the use of dry heat treatment was found to be the most effective. Nonetheless, early spacecraft had parts and assemblies that were largely developed for other uses, and many were not able to withstand the rigors of the space environment — with or without an attempt to sterilize the spacecraft involved.

That being said, attempts at sterilization became for some an acceptable excuse for mission failures, whether or not those failures were in any way related to sterilization efforts. For example, between 1961 and 1964, the Jet Propulsion Laboratory (JPL) had six "Ranger" missions to the Moon fail, with the last four of those having failures that — despite evidence to the contrary — were attributed by laboratory personnel to sterilization efforts. Later in the program it was discovered that a failure of a particular component, not affected by heat, was a leading cause of the failures. Despite that disparity, the reaction of Ranger project managers was to attribute those failures to heat sterilization. As stated in a history of the Ranger program (*Hall*, 1977), "although lacking firm evidence that this requirement caused the equipment failures, JPL now requested and received more waivers from NASA Headquarters on heat sterilizing certain crucial components." Soon, however, NASA would build itself less flexibility and a more strict adherence to the concept of "sterilization when needed" with both the formalization of the SSB's advisory role in the mid-1960s, and a new international agreement.

Through the early 1960s, the SSB provided multidisciplinary scientific advice to NASA, including the reports *Conference on Potential Hazards of Back Contamination from the Planets* (*SSB*, 1964), *Conference on Hazard of Planetary Contamination Due to Microbiological Contamination in the Interior of Spacecraft Components*, (*SSB*, 1965), *Conference on Potential Hazards of Back Contamination from the Planets* (*SSB*, 1965*)*, and *Biology and the Exploration of Mars*

(*SSB*, 1966). Their recommendations to NASA on planetary quarantine were one of the critical subjects in their 1968 report *Planetary Exploration: 1968–1975* (*SSB*, 1968a) and their review of all NASA efforts in the biological sciences (NASA Biology Program) (*SSB*, 1968b).

With the adoption of the Outer Space Treaty in 1967 (OST) (*United Nations*, 1967), even though not specifically called out in the OST, it has been recognized that any space agency could conceivably contaminate another world or endanger Earth unilaterally. Thus, planetary protection has continued to be a required element in the international endeavor of space research and exploration. Specific to planetary protection, Article IX of the OST requires that parties pursuing studies of outer space and other celestial bodies should "conduct exploration of them so as to avoid their harmful contamination and also adverse changes in the environment of Earth resulting from the introduction of extraterrestrial matter." Furthermore, parties to the treaty "where necessary, shall adopt appropriate measures" for the purpose of avoiding harmful contamination and changes to Earth's environment. A particular facet of Article IX is the provision that states that if a party to the OST "has reason to believe that an activity or experiment planned by another State Party in outer space . . . would cause potentially harmful interference with activities in the peaceful exploration and use of outer space . . . may request consultation concerning the activity or experiment." Clearly, although the concern about forward contamination of other worlds pre-dates the OST, it does support consultation, at least, on those concerns. Note that the OST applies equally to both human and robotic missions, and under Article VI, to both governmental and non-governmental space activities.

By 1998, the launch of the Mars mission "Planet-B"/Nozomi by Japan's Institute of Space and Astronautical Science (ISAS), which did not plan to adhere to the then-established COSPAR policies for Mars planetary protection, indicated that the COSPAR policy guidelines were not well known, or being properly applied, by some agencies. In subsequent discussions, it was apparent that this was because the COSPAR policies themselves were not easily accessible by potential users. To enable COSPAR policy development, a more proactive approach was required. As such, a proposal was made to COSPAR by NASA at the COSPAR Assembly in 1998 in Nagoya to reestablish a Panel on Planetary Protection. The Panel then would be given the responsibility to coalesce the various COSPAR policy statements made since 1958, and combine them into a complete, stand-alone COSPAR Planetary Protection Policy. In March 1999, the COSPAR Bureau and Council approved the formation of that panel, which is still operating as of the writing of this chapter. Beginning in 2000 at the COSPAR Assembly in Warsaw (*Rummel et al.*, 2002a), and supplemented by a COSPAR/IAU workshop held in April 2002 in Williamsburg, Virginia (*Rummel*, 2002), the Panel on Planetary Protection assembled a single document that contained the entire COSPAR Planetary Protection Policy. That policy was approved by the COSPAR Bureau and Council at the

2002 World Space Congress, in Houston, Texas (*COSPAR*, 2003). COSPAR had not had a unified Planetary Protection Policy document since 1964.

Another outcome of the Williamsburg workshop and the unified COSPAR policy was less direct, but perhaps longer lasting. Within the first few years after that meeting, the European Space Agency (ESA) joined France's Centre National d'Études Spatiales (CNES) in establishing a formal planetary protection policy and apparatus for their agencies, and other agencies prepared draft policies for later implementation [e.g., Japan's Aerospace Exploration Agency's (JAXA) organization was solidified in 2017 after forming a Planetary Protection Safety Review Board in 2013]. Since ESA provided their planetary protection organization, located in the Independent Safety Office at the European Space Research and Technology Centre (ESTEC), ESA has been particularly active — both internationally and within ESA's Planetary Protection Working Group (PPWG) — in developing policy improvements and refinements in response to the challenges brought by new missions and new knowledge of extraterrestrial environments.

While a number of space agencies (e.g., NASA, ESA, CNES, JAXA) have established their own specific planetary protection policies to guide efforts in this area when conducting interplanetary missions, the origins of the field in the late 1950s resulted in a history of coordination and collaboration in this area prior to the development of the OST. For the NASA policy, the authorization under which planetary protection policies have been implemented continues to be the U.S. Government's Space Act of 1958 (US Public Law 85-568), as amended. The United States does not have a legal framework that directly implements the OST. Nonetheless, for the international community as a whole, the relationship to the OST can be considered an intrinsic feature of the COSPAR policy, and over time they have become closely aligned — for example, in 2017 COPUOS recognized "the long-standing role of COSPAR in maintaining the planetary protection policy as a reference standard for spacefaring nations and in guiding compliance with Article IX of the Outer Space Treaty" (*COPUOS*, 2017).

The future of COSPAR's role in providing an international consensus standard for applying planetary protection considerations to planetary missions is not firmly set, but work being done by both the International Institute of Space Law and The Hague International Space Resources Governance Working Group (*HISRGWG*, 2017) plan for future reference to "internationally agreed planetary protection policies" in their formulations. Such policies are exactly what COSPAR is intended to supply.

3. THE VIKING EXPERIENCE, EARLY 1970s

The Viking missions to Mars (see also the chapter by Davila et al. in this volume) were the first planetary exploration missions where planetary protection implementation was a major driver in both science and mission design because the Viking landers would be conducting life-detection experiments, trying to grow martian organisms in three different ways. The planetary quarantine (planetary protection) requirements that were applied to the Viking missions, particularly the Viking Landers, were driven by a mathematical calculation that linked the amount of biological material carried by the spacecraft, at launch, to the probability that the lander would contaminate the martian surface. The description of Viking planetary protection provisions given here depends chiefly on the work of *Daspit et al.* (1990) and *Stabekis and DeVincenzi* (1978), although helpful context is also given in the SSB's recent report on planetary protection policy development processes (*SSB*, 2018).

Under the regulations that applied to robotic missions in NASA at the time, each flight project was given a probability of contamination (P_C) allocation, consistent with overall policy objectives and an estimate of the probability of growth (P_g) in a particular planetary environment. For Viking, although the missions hoped to detect Mars life, the recommendation from the SSB — their estimate — was that the P_g should be 1×10^{-6} — one in a million. Also, NASA set the P_C for each mission in three categories: large impactables, ejecta-efflux, and lander sources. Using these basic categories, the Viking flight project performed a probabilistic analysis of all events and sources that was hoped to result in an optimized allocation of P_C among those categories to attain a total P_C of $<2 \times 10^{-4}$ for both missions. The specific calculation made is given in Fig. 1.

Because the Viking Landers also carried several life-detection instruments, additional requirements were applied to the spacecraft to provide additional protection against "false positive" indications of life on Mars that might result from direct contamination of those instruments. To meet this constraint, which was one chance of a false positive in one million, the Viking lander, within its enclosing "bioshield" capsule (see Fig. 2), had to be heated during terminal sterilization for durations exceeding the time required for planetary protection alone.

In preparing for the Viking missions, NASA made significant research investments into sterilization techniques and the appropriate values to be used in measuring potential planetary contamination (*Hall*, 1974). This research was overseen by a Planetary Quarantine Advisory Panel (PQAP),

---

The contamination equation for the Viking lander is

$$P_c = \Sigma\, N_0\, P_s\, P_r\, P_g$$

where

- P_c is the probability of contaminating Mars
- N_0 is the number of organisms present before sterilization
- P_s is the probability that a randomly selected organism will survive sterilization
- P_r is the probability of release
- P_g is the probability of growth

Fig. 1. The calculation of potential planetary contamination used in the Viking missions of the 1970s.

Fig. 2. The Viking Lander Capsule goes into the SAEF-2 oven at Cape Canaveral (*Daspit et al.,* 1990).

which established a base knowledge of planetary protection methods at the beginning of the NASA research program from which they could later measure progress over time. The PQAP, as it was known, had regular meetings that eventual Viking Project personnel could participate in (which they did, beginning two years before the project began), and from which they could learn the best practices and technologies to incorporate into the project.

Because of the intention to perform a terminal sterilization cycle (although not quite a full sterilization) using dry heat microbial reduction (all portions of the Viking lander to reach 111.7°C for at least 30 hours), careful attention was paid to the parts — especially the more than 60,000 electronic parts required for the missions — to ensure that they could survive those temperatures for that amount of

time, on one hand, and to lower the overall time required for the terminal sterilization process because of the credit the electronic parts received for the reduction of microbial contamination as a result of the temperatures obtained during the parts qualifications process.

It should be noted that because the entire Viking lander capsule was to be heated in an oven, with the heat from the oven soaking in from the outside, there was a concern that the outer portions of the lander were going to be heated hotter, and longer, than the rest. However, because each lander was to be powered by two radioisotope thermal electric generators (RTGs), the designers of the sterilization process used the waste heat from those RTGs, routed through "cooling" lines into the center of each lander, to shorten the overall time required in the oven (Fig. 3). When all was said and done, each lander was sealed in its bioshield, and subjected to a sterilization cycle that lasted between 45 and 54 hours (e.g., Fig. 4). The landers were not released from the bioshield until each mission reached Mars orbit and selected a landing site.

In order to monitor microbial contamination, NASA developed a culture-based process, now known as the NASA Standard Assay (*Puleo et al.,* 1977), to measure the number of aerobic spores in a microbial sample, as both a proxy for overall microbial contamination and because the microbes that were thought to be most resistant to the sterilization process at the time would be spore-forming microbes that exhibited great resistance to heat and other biocidal factors. The NASA Standard Assay thus became an important process in preparing clean spacecraft to the level of the Viking landers (cf. *SSB,* 1992), and as a standard it has lasted several decades longer than one might expect, given that any particular culture-based process can only detect approximately 1% or less of the total microbial contamination in a typical microbial environment [see, e.g., *La Duc et al.* (2004) for some conceptual updates].

When the Viking missions were flown, both lander spacecraft landed safely and performed their astrobiology-related

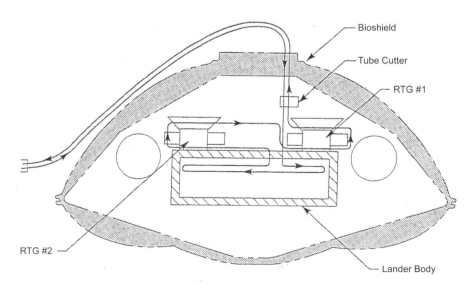

Fig. 3. Routing of the lander RTG cooling lines supported a shorter total lander heat treatment (*Daspit et al.,* 1990).

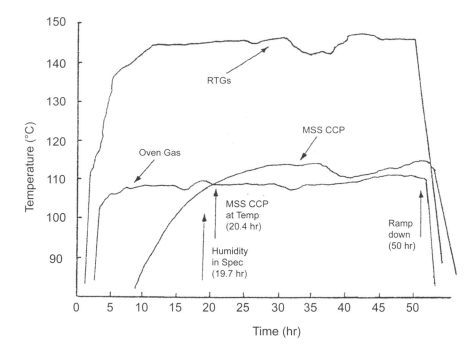

Fig 4. Details of the Viking Lander Capsule 1 sterilization cycle (*Daspit et al.,* 1990).

experiments without flaw (*Soffen,* 1976), and although their life-detection instruments did not detect accepted evidence of life on Mars, it was significant that there was never a case when the life-detection experiments conducted by the Vikings on Mars demonstrated any indications of Earth contamination carried on the spacecraft. Likewise, after a great deal of effort expended to achieve extremely low levels of organic contamination on the spaccecraft, there was no indication of the detection of any unknown organics inadvertently present when the spacecraft reached Mars.

4. EVOLUTION OF THE NASA AND COSPAR POLICIES AND THE CATEGORIZATION CONCEPT FOR ROBOTIC MISSIONS

The initial concerns of COSPAR were about how to address planetary contamination caused by spacecraft "forward contamination" (life or organic contamination carried to the planets from Earth), and that contamination was amenable to calculation with an estimate of the probability of growth for each given body. As such, the initial COSPAR planetary protection policy (*COSPAR,* 1964) focused on forward contamination alone, stating that COSPAR "accepts, as tentatively recommended interim objectives, a sterilization level such that the probability of a single viable organism aboard any spacecraft intended for planetary landing or atmospheric penetration would be less than 1×10^{-4}." That probability was later changed to 1×10^{-3}, over a specific timeframe (known as the period of biological exploration) of 20 years (*COSPAR,* 1969), and later 50 years (*COSPAR,* 1994). At that time, COSPAR did not establish any specific policy, let alone a probabilistic one, covering "backward contamination" (extraterrestrial

life carried back to Earth) and the potential for harmful contamination of Earth's biosphere.

While the Viking missions captured much of the attention of the exobiology community, they were by no means the only planetary spacecraft launched by NASA in the decade of the 1970s. The Pioneer missions to Jupiter and Pioneer Venus were successfully launched by NASA, as later on were the two Voyager missions that in combination would visit Jupiter, Saturn, Uranus, Neptune, and their satellites. For each of these missions, the allowable biological contamination that could be carried was the result of a specific formula that was governed, for the most part, by the number used for P_g, which itself was the subject of a recommendation to NASA by the SSB. The P_g for those missions (an estimate provided by the SSB) ranged from a P_g of 0 on the surface of Venus to 1×10^{-4} for the outer planets (*SSB,* 1972). A terminal heat sterilization cycle was not needed for any of those missions, although clean-room assembly and microbial monitoring (particularly regarding parts) was occasionally necessary. Generally speaking, those missions (except for Pioneer Venus) were not intended to land on, or even orbit, another planetary body, so a great deal of attention was paid chiefly to spacecraft reliability and navigation. During this period, where the SSB provided NASA with recommendations on the P_g for each solar system body, the resultant allowable biological contamination figure appeared to be derived quantitatively, but for many targets of exploration there was so little data about the body that it was impossible to make a reasonable estimate of that probability, making the result only appear to be quantitative. As additional data about the solar system became available, NASA continued to receive updated recommendations from the SSB. As the P_g terms provided by SSB study committees got smaller and smaller, the SSB and NASA became more and more uncomfortable with a qualitative way to distinguish between, for example, a P_g of 1×10^{-13} and one of 1×10^{-14}. Thus, the SSB was very supportive when NASA moved to a different system to evaluate planetary protection requirements for missions (*Cameron,* 1982). One of the last missions to be evaluated in the earlier way was the Galileo mission to Jupiter, then planned for launch on the space shuttle Atlantis in 1986.

In the early 1980s, then Planetary Quarantine Officer Donald DeVincenzi and his colleagues looked for a way to modify the NASA planetary protection regulations and later the COSPAR Planetary Protection Policy to remove this objection. The aim was to remove the numerically rigid, pseudo-quantitative policy of the time (see *SSB,* 1978) and instead tailor the various planetary protection requirements to specific mission/target-body combinations, with the potential to address specific concerns about various target bodies in a flexible manner, and to continue to avoid situations in which spacecraft contaminants might invalidate current or future scientific exploration of a body. Under the new scheme, each mission-target-body combination would be assigned a category associated with its perceived planetary protection risk (detailed in sections 5–6), and with different mission requirements.

The resultant COSPAR policy statement, as modified only slightly as of 2017, is:

> The conduct of scientific investigations of possible extraterrestrial life forms, precursors, and remnants must not be jeopardized. In addition, Earth must be protected from the potential hazard posed by extraterrestrial matter carried by a spacecraft returning from an interplanetary mission. Therefore, for certain space mission/target planet combinations, controls on contamination shall be imposed in accordance with issuances implementing this policy.

Table 1 provides a listing of mission categories and their associated requirements, while Table 2 provides the current listing of target bodies by their current category.

Note that while COSPAR adopted this policy (*DeVincenzi et al.,* 1983) in the early 1980s, and NASA applied the policy in draft form to both the Galileo and Mars Observer missions, NASA did not formally adopt this policy until the issuance of NASA Handbook (NHB) [later NASA Procedural Requirements (NPR)] 8020.12B in April 1999. That update dealt with concerns about both forward and backward contamination for robotic missions, but it would take until 2008 before the COSPAR policy would address both principles and guidelines for forward and backward contamination on human missions. NASA has yet to adopt those in any form.

About the same time that the COSPAR policy was being updated, DeVincenzi also determined that it was necessary to formally change the title of "Planetary Quarantine" and "Planetary Quarantine Officer" to the current "Planetary Protection" and "Planetary Protection Officer," as NASA's acceptance of a legal opinion that was first put forward by George Robinson of the Smithsonian Institution (*Robinson,* 1971) Robinson pointed out that particular NASA regulations [especially NASA Management Instruction (NMI) 8020.14] that had been adopted by NASA to allow the Planetary Quarantine Officer to arrest and quarantine people who were "extraterrestrially exposed" at the Manned Spacecraft Center, against their will, were problematic for NASA's purposes and unconstitutional overall, due to the unproven nature of the effects of such exposure. Since that arrest authority was pertinent only during the Apollo lunar quarantine activity (Apollos 11, 12, and 14), the change was readily, if slowly, accepted by NASA. The actual change was not formalized until the issuance of NMI 8020.7A in 1988 ("Biological Contamination Control for Outbound and Inbound Planetary Spacecraft") (*NASA,* 1988), which replaced the version of that policy issued (as a NASA Policy Document) in 1967 (see *Rummel,* 2019).

5. MISSION IMPLEMENTATION FOR CATEGORIES I–IV: OUTBOUND

Because the development of an international consensus regarding steps to be taken to prevent forward and backward contamination has practical benefits when flying joint missions with international partners, NASA and other international partners (e.g., ESA, CNES, etc.) all contribute to the development of the COSPAR planetary protection policies and ensure that, where appropriate, the COSPAR policies/guidelines are reflected in their own intrinsic planetary protection regulations. As such, it is possible to discuss mission implementation under these policies in a single narrative, currently focused on *Kminek et al.* (2017) for COSPAR, and on potential alterations that are proposed from time to time through COSPAR's Panel on Planetary Protection (see *Kminek and Fisk,* 2017; *SSB,* 2018). Overall, the COSPAR policy categorizes mission-target-body combinations for outbound missions according to first the scientific interest in the body relative to astrobiology, and second to whether the body is thought to be possible to contaminate by the introduction of Earth organisms or organic materials. There are four major categories for outbound missions, and one major category for missions returning to Earth.

Category I is reserved for bodies of the least concern to planetary protection. Such is the nature of human curiosity that NASA and others actually fly few missions to target bodies that result in a Category I classification of a mission, as Category I missions are to bodies that are "not of direct interest for understanding the process of chemical evolution or the origin of life." The NEAR Shoemaker mission to asteroid 433 Eros (launched 1996) was one such mission, given that a dry, small asteroid was not considered a likely locale for life or future life detection missions. Once a mission is assigned Category I, there are no further reporting requirements for planetary protection purposes.

Category II is the classification that is readily applied to missions to the majority of solar system bodies, including

TABLE 1. COSPAR categories for solar system bodies and types of missions (*Kminek et al.,* 2017).

	Category I	Category II	Category III	Category IV	Category V
Type of Mission	Any but Earth return	Any but Earth return	No direct contact (flyby, some orbiters)	Direct contact (lander, probe, some orbiters)	Earth return
Degree of Concern	None	Record of planned impact probability and contamination control measures	Limit on impact probability Passive bioburden control	Limit on probability of non-nominal impact Limit on bioburden (active control)	If *restricted* Earth return: • No impact on Earth or Moon • Returned hardware sterile • Containment of any sample
Representative Range of Requirements	None	Documentation only (all brief): • PP plan • Pre-launch report • Post-launch report • Post-encounter report • End-of-mission report	Documentation (Category II plus): • Contamination control • Organics inventory (as necessary) Implementing procedures such as: • Trajectory biasing • Cleanroom • Bioburden reduction (as necessary)	Documentation (Category II plus): • P_C analysis plan • Microbial reduction plan • Microbial assay plan • Organics inventory Implementing procedures such as: • Trajectory biasing • Cleanroom • Bioburden reduction • Partial sterilization of contacting hardware (as necessary) • Bioshield • Monitoring of bioburden via bioassay	*Outbound* Same category as target body/outbound mission *Inbound* If *restricted* Earth return: • Documentation (Category II plus) • P_C analysis plan • Microbial reduction plan • Microbial assay plan Implementing procedures such as: • Trajectory biasing • Sterile or contained returned hardware • Continual monitoring of project activities • Project advanced studies and research If unrestricted Earth return: • None

missions to asteroids, Venus, the Moon, comets, most planets, and most natural satellites, including Pluto, Ganymede, Callisto, etc. The common thread for these missions is that they are of "significant interest relative to the process of chemical evolution and the origin of life," but there is only a "remote chance that contamination carried by a spacecraft could compromise future investigations." The MESSENGER mission to Mercury (launched 2004) required a flyby of Venus en route, thus earning itself a Category II classification despite the fact that Mercury missions themselves might be considered Category I. Planetary protection requirements for Category II consist of short reports and the designation of the spacecraft location at the end of the mission, where

possible. Because it is such a broad classification, it is also very prone to minor modifications, such as the inclusion of an organic inventory or samples to be carried by the mission, etc. One such modification was attendant to the use of this classification when applied to the Galileo mission (launched 1989), when subsequent to the failure of the space shuttle Challenger, Galileo's propulsive characteristics were changed along with its trajectory to Jupiter to accommodate a less powerful upper stage and the Venus-Earth-Earth-Gravity-Assist (VEEGA) trajectory. The new mission design was assigned Category II as a mission to Jupiter, but the Planetary Protection Plan also provided for an assessment of the biological potential of the jovian satellites

TABLE 2. Category-specific listing of target body/mission types (*Kminek et al.,* 2017).

Category	Types of Mission	Target Bodies
Category I	Flyby, Orbiter, Lander	Undifferentiated, metamorphosed asteroids; Io; others to-be-defined (TBD)
Category II	Flyby, Orbiter, Lander	Venus; Moon (with organic inventory); comets; carbonaceous chondrite asteroids; Jupiter; Saturn; Uranus; Neptune; Ganymede*; Callisto; Titan*; Triton*; Pluto/Charon*; Ceres; Kuiper belt objects > 1/2 the size of Pluto*; Kuiper belt objects < 1/2 the size of Pluto; others TBD
Category III	Flyby, Orbiters	Mars; Europa; Enceladus; others TBD
Category IV	Landers	Mars; Europa; Enceladus; others TBD
Category V	Any Earth-return mission	
	"Restricted Earth return"	Mars; Europa; Enceladus; others TBD
	"Unrestricted Earth return"	Venus, Moon; others TBD

* The mission-specific assignment of these bodies to Category II must be supported by an analysis of the "remote" potential for contamination of the liquid-water environments that may exist beneath their surfaces (a probability of introducing a single viable terrestrial organism of $<1 \times 10^{-4}$), addressing both the existence of such environments and the prospects of accessing them.

(especially Europa) before the end of mission (EOM). Due to the exciting science results regarding potential oceans on Europa and the other jovian icy moons (e.g., *Pappalardo et al.,* 1998; *Kivelson et al.,* 2000), NASA and the Galileo project negotiated the EOM to preclude contamination of those satellites by the spacecraft. The Galileo spacecraft avoided an inadvertent collision with Europa and the other jovian satellites and ended its mission with a plunge into Jupiter on September 21, 2003.

In adopting the Category III and IV mission categorizations, the COSPAR policy originally assigned those categories to missions to any "target body of chemical evolution and/or origin of life interest and for which scientific opinion provides a significant chance of contamination which could compromise future investigations." Category III applies to flyby or orbital missions, while Category IV applies to landers or "probes." At the time of the adoption of these categories, only missions to Mars fit the definition of Category III or IV, and even Mars missions weren't assuredly thought to present a contamination hazard, given the assessment made by the SSB in the late 1970s (*SSB,* 1977; *SSB,* 1978).

Responding to a request from NASA, however, the SSB books were reopened on Mars in their 1992 report, *Biological Contamination of Mars: Issues and Recommendations* (*SSB,* 1992). From the results of that report, based on an emerging awareness of just how resilient Earth organisms can be, it was argued that missions to Mars should be retained as a Category III or IV, and that there is/was a real possibility that Earth organisms might be capable of contaminating Mars in ways not previously appreciated. In addition to their suggestions about how clean Mars-landing spacecraft should be (Viking pre-sterilization cleanliness as measured by the NASA Standard Assay), they also distinguished the level of cleanliness required if a mission was going to attempt life-detection tests of the sort seen with the Viking missions. Those SSB recommendations led to the establishment of two Category IV subcategories for Mars missions: Category IVa for lander missions not seeking to detect life (clean to the Viking pre-sterilization levels), and Category IVb for landers attempting to detect living organisms on Mars (clean to the Viking Lander levels, as flown). Meanwhile, flyby and orbital missions to Mars were to meet either Viking total contamination levels, or to adopt orbital lifetime guarantees ($\geq$20 years at 99% probability, >50 years at 95% probability).

Based on recommendations from the SSB report on Europa planetary protection (*SSB,* 2000), missions to Europa and to Saturn's moon, Enceladus, are also classified as Category III or Category IV (*Kminek et al.,* 2017). Because they are different in character from Mars, and in response to a specific recommendation from the SSB (*SSB,* 2000) their requirements are based on a probabilistic formulation that examines the probability of contamination of a mission. That formulation has been recommended for modification by the Planetary Protection of Outer Solar System (PPOSS) study, chiefly regarding the initial decision to adopt a probabilistic calculation in the first place (*Rettberg et al.,* 2019), although uncertainty about their precise environments makes even yes/no decisions difficult when dealing at the scale of a microorganism. Evolving thought and increased discussion between NASA and mission science and engineering teams has helped bring about a shift in how the probability of contamination can be calculated by including more realistic processes. For example, discussion and review during a workshop on planetary protection for Europa Clipper resulted in the period of biological exploration for Europa being reduced, for NASA purposes, from undefined — essentially requiring that the probability of contamination would have to be calculated without an end date — to instead to consider that probability for over only a period of 1000 years. Also, the same workshop accepted for the Europa Clipper missio, a 2.5-Mrad total ionizing dose (TID) as sterilization equivalence, a dose expected to kill all Earth organisms. These specific definitions and timelines make the calculation of risk more straightforward, and our understanding of potential contamination more explicit, enabling both cognizant planetary protection and exploration (*Pratt,* 2019).

6. MISSION IMPLEMENTATION FOR CATEGORY V: INBOUND ("RESTRICTED" VERSUS "UNRESTRICTED" EARTH RETURN)

Although the original paper making the COSPAR policy proposal (*DeVincenzi et al.,* 1983) had classified sample return missions to any solar system body as Category V, that paper also distinguished between sample return missions that were "safe for Earth return" and those that were "unsafe for Earth return." Those subcategory names were not sufficiently descriptive, so when the implementation of that proposal was made in the 2002, comprehensive, version of the COSPAR policy, Category V was instead divided (as it had been in the NASA procedural requirements document, NPR 8020.12) into "restricted Earth return" and "unrestricted Earth return," the latter of which was generally applicable to the return of samples from target bodies that would otherwise merit Category I or II on the outbound leg.

Two SSB studies that had been issued in the late 1990s were particularly influential in assisting NASA and COSPAR to provide metrics to distinguish among missions that were, or were not, allowed "unrestricted Earth return" status. The first focused specifically on sample return missions to Mars (*SSB,* 1997), where the strong recommendations of the SSB were that Mars sample return missions would be restricted:

- "Samples returned from Mars by spacecraft should be contained and treated as though potentially hazardous until proven otherwise. No uncontained martian materials, including spacecraft surfaces that have been exposed to the martian environment, should be returned to Earth unless sterilized.
- Integrity of containment should be maintained through reentry of the spacecraft and transfer of the sample to an appropriate receiving facility.
- Controlled distribution of unsterilized materials returned from Mars should occur only if rigorous analyses determine that the materials do not contain a biological hazard. If any portion of the sample is removed from containment prior to completion of these analyses, it should first be sterilized.
- The planetary protection measures adopted for the first

Mars sample return missions should not be relaxed for subsequent missions without thorough scientific review and concurrence by an appropriate independent body."

Second, when asked to recommend whether sample return missions from small bodies of the solar system should be either restricted or unrestricted, the *SSB* (1998) gave similar recommendations if a sampler were to be judged to have "biological potential," and provided some specific recommendations ("yes," "no," or "uncertain") regarding specific solar system bodies. Nonetheless, their longest lasting recommendation was that such a determination could be made by applying six questions that were developed to address the biological potential of most small bodies using the best scientific information at the time:

1. Does the preponderance of scientific evidence indicate that there was never liquid water in or on the target body?

2. Does the preponderance of scientific evidence indicate that metabolically useful energy sources were never present?

3. Does the preponderance of scientific evidence indicate that there was never sufficient organic matter (or CO_2 or carbonates and an appropriate source of reducing equivalents) in or on the target body to support life?

4. Does the preponderance of scientific evidence indicate that subsequent to the disappearance of liquid water, the target body has been subjected to extreme temperatures (i.e., $>160°C$)?

5. Does the preponderance of scientific evidence indicate that there is or was sufficient radiation for biological sterilization of terrestrial life forms?

6. Does the preponderance of scientific evidence indicate that there has been a natural influx to Earth, e.g., via meteorites, of material equivalent to a sample returned from the target body (*SSB,* 1998)?

If the answer to any one of these questions were "yes," the SSB recommended that strict containment and handling of the samples collected would not be necessary from a planetary protection perspective. In those questions, the SSB used the term "preponderance of scientific evidence" not in a legal sense, but rather "to connote a nonquantitative level of evidence compelling enough to research scientists in the field to support an informed judgment." In slightly modified form, those six questions are now part of the COSPAR (and NASA) Planetary Protection Policy (*Kminek et al.,* 2017).

Due to the nature of the timeline normally required to receive recommendations from the SSB (which may take as much as two years), the SSB advice has always been considered "strategic" rather than addressing most real-time implementation requirements. As such, it was notable that both the SSB report on Mars sample return (*SSB,* 1997) and the report on small-body sample return (*SSB,* 1998) made similar recommendations with respect to NASA's need for advice on how to plan and implement sample return missions. For example, from the Mars report:

A panel of experts, including representatives of relevant governmental and scientific bodies, should be established as soon as possible once serious planning for a Mars sample return mission has begun, to coordinate regulatory responsibilities and to advise NASA on the implementation of planetary protection measures for sample return missions (*SSB,* 1997).

and from the small-body sample return report:

> NASA should consult with or establish an advisory committee of experts from the scientific community when developing protocols and methods to examine returned samples for indicators of past or present extraterrestrial life forms (*SSB,* 1998).

Accordingly, NASA turned to its existing in-house advisory apparatus, the Space Science Advisory Committee (SScAC) of the NASA Advisory Council, which formed a Planetary Protection Task Force in 1999 as the first trial step toward the formation of a new advisory committee. After a year of operations by the Task Force the NASA Advisory Council formed the Planetary Protection Advisory Committee (PPAC), on par with the SScAC, but with a structure that explicitly included *ex officio* representatives from other U.S. Government organizations and from other space agencies around the world, balancing long-term familiarity against the need for short-timeframe decision making. One role of the PPAC was to recommend specific categorizations for missions that were proposed by NASA or its partners, in order to support the Planetary Protection Officer and demonstrate that NASA was meeting its commitment to its own Planetary Protection Policy, and by extension to its compatibility with the COSPAR policy guidelines. Especially important were both the joint compliance of NASA and its international partners, as well as the determination of whether a sample return mission should be judged a restricted or unrestricted Earth return mission.

Such determinations were made for NASA and NASA-partner missions, but they also serve double duty in areas where COSPAR's policy could be helpful. For example, in the spring of 2002, Japan's ISAS was planning a mission known as MUSES-C (later Hayabusa) which was intended to return a sample from the S-type asteroid 1998 SF36 (later 25143 Itokawa). However, the Australian government needed environmental approval prior to allowing MUSES-C to land at the Woomera test range, which was the preferred landing site for the sample return capsule for the mission. In that instance, it was decided by ISAS and the Australian government to seek COSPAR approval for the mission. Because the PPAC had recently (March 2002) recommended to NASA that MUSES-C was qualified as a Category V, "unrestricted Earth return" mission, the PPAC could make that same recommendation at the COSPAR/IAU workshop held in April 2002, which was attended by both ISAS and Australian government personnel. The COSPAR workshop could then adopt and forward that recommendation to the Australians in support of ISAS. This process resulted in a one-page decision directive from Environment Australia approving the return to Woomera, and after a public consultation, a 39-page directive from Biosecurity Australia providing its permission as well.

One of the innovations from the 2002 workshop was the tentative development of a definition for places on Mars "within which Earth-sourced organisms are likely to propagate," or are "interpreted to have high potential for extant martian life forms." These "Special Regions" on Mars were not explicitly defined at that time, but a new Category IV subcategory for Mars, Category IVc, was defined to be assigned to a mission that would seek to access a special region either by vertical or horizontal mobility (e.g., via a penetrator vs. a landed rover). The special region concept was endorsed by the SSB's PREVCOM report (*SSB,* 2006) on the forward contamination of Mars, and later work by *Beaty et al.* (2006) and *Kminek et al.* (2010) further defined Special Regions by the parameters of temperature and water activity. The most recent basic definition for Special Regions is provided by *Rummel et al.* (2014), with error margins designated in the current COSPAR policy reported by *Kminek et al.* (2017).

Unfortunately, as of this writing the NASA advisory committee that could provide advice on issues such as this has been disbanded, and no plan is presently in place for NASA to regain advice from a committee like the earlier PPAC. Such advice is particularly essential when envisioning the return of samples from Mars (or Europa, or Enceladus), and NASA is now dependent on forums like ESA's PPWG and COSPAR itself for informed discussion of the issues and policy implementation on future missions. The situation may be self-correcting, however, in that the most recent SSB *Review and Assessment of Planetary Protection Policy Development Processes* (*SSB,* 2018) made the specific recommendation "that NASA reestablish an independent advisory body and process to help guide formulation and implementation of planetary protection adequate to serve the best interests of the public, the NASA program, and the variety of new entrants that may become active in deep space operations in the years ahead."

For a variety of reasons and missions, that reestablishment will be critical for scoping mission implementation. One example of the need is tied directly to missions under study and proposed. Planning for a restricted Earth return mission is a complicated matter, and without an existing containment-capable receiving laboratory and protocol for assessing potential biohazards in the returned samples, a long lead time is required to include everything between planning, construction, and practice (Fig. 5). During the planning for a proposed NASA-CNES Mars sample return effort in the early 2000s, a draft version of what should be done and where was prepared and reviewed by a panel of experts (see *Rummel et al.,* 2002b). More recently, planning has begun for a future Mars sample return, including the planned launch of the Mars 2020 mission (Category V, restricted Earth return, to launch 2020), and need for a plan to implement a planetary protection strategy is called for in the iMars Phase 2 report (*Haltigin et al.,* 2018).

There is a clear need to have equal progress in the preparation for both the science to be done on a sample within a containment facility and the biohazard protocol to be developed to demonstrate that the sample is safe for sci-

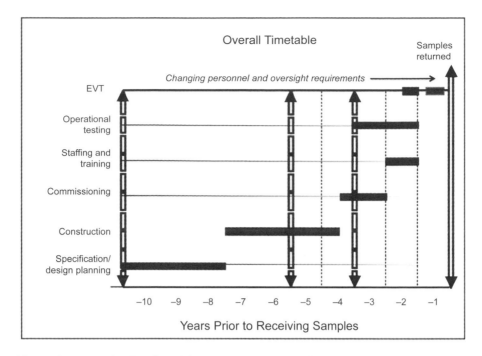

Fig 5. Planning timeline for a Mars sample receiving facility (*Rummel et al.,* 2002b).

entific study outside of containment (*Rummel and Kminek,* 2018). Indeed, a careful reading of the iMars Phase 2 report notes their recommendation that "detailed implementation plans should be put in place as soon as possible because of the projected timeline for designing, constructing, and operating a Sample Return Facility. As part of the plan, minimum requirements for safely receiving the samples on Earth should be clearly defined." Those activities will need to concern themselves with both the standard for the level of risk that is tolerable, as well as the dimensions (both conceptual and actual) of the Mars material to be contained (e.g., *Amman et al.,* 2012). Historically, such planning is generally difficult to put into place because of its long-lead-time nature, which generally precedes the commitment (and money) for a sample return mission, or suite of missions, by several years. But these procedures and facilities are needed for both robotic and human missions, and effective shortcuts in developing a sample receiving capability are not immediately evident.

7. MISSIONS WITH A HUMAN CREW

After extensive (although imperfect) planetary quarantine efforts for both lunar materials and astronauts during the Apollo program (Fig. 6) (*Allton et al.,* 1998), the focus of planetary exploration and planetary protection shifted to robotic missions. In the intervening years, robotic planetary protection policies and practices matured considerably, but while planetary protection is firmly established as a part of robotic mission planning, it needs to be incorporated into future human missions as well — especially for missions to Mars (e.g., *NASA,* 2007). Planetary protection provisions are needed to mitigate the potential for danger to astronauts and

to Earth, and the inclusion of planetary protection requirements need to be considered to be a critical aspect of all human mission system and subsystems development. For human missions, these concerns also include the possible immediate and long-term effects on the health of astronaut explorers from biologically-active materials encountered during exploration.

The primary goals of the COSPAR policy do not change when human explorers are involved. In developing preliminary guidelines for human missions to Mars in 2008, COSPAR noted that the greater capability of human explorers to contribute to the astrobiological exploration of Mars can be realized only if human-associated contamination is controlled and understood (*Kminek et al.,* 2017). To ensure human safety while conducting planetary exploration, consideration of planetary protection is essential. The unavoidable, and largely beneficial, association of humans with a huge diversity of commensal microbes means that tailored, appropriate implementation controls, different from those applied to robotic missions, will have to be developed (cf. *Stone,* 2009; *Voorhies and Lorenzi,* 2016).

Through organized workshops and interdisciplinary information exchanges, the planetary protection community, working with engineering and systems experts, has been studying the implementation of NASA and COSPAR planetary protection policies (e.g., *NASA,* 1999a; *Kminek et al.,* 2017) on numerous human associated activities and systems. A series of planetary protection studies and workshops have helped identify important data needs as well as priority research and development areas to support compatible astrobiology exploration with humans [cf. *Aeronautics and Space Engineering Board (ASEB) and Space Studies Board (SSB),* 2002; *Criswell et al.,* 2005;

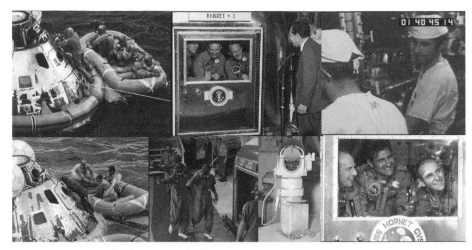

Fig. 6. Scenes from the Apollo back-contamination control effort. Apollo 11 is at the top, with Apollo 12 on the bottom. Note the Biological Isolation Garments donned by the Apollo 11 astronauts after landing, and the shirtsleeve motif for the Apollo 12 crew. Credit: NASA.

Hogan et al., 2006; Kminek et al., 2007; Race et al., 2008; NASA, 2007; NASA, 2015; etc.]. It is noteworthy that the science, technology, and legal considerations for planetary protection during long-duration human missions — especially for Mars — are significantly different than those applied during the Apollo program, or for human missions involving the International Space Station or other platforms in Earth orbit. Thus, it will be particularly important to continue discussions/interactions with space medicine, biomedical operations, and human factors communities to ensure that these areas are incorporated into the up-to-date implementation of planetary protection for future exploration beyond low-Earth orbit (LEO), and that the COSPAR principles and guidelines for the human exploration of Mars can be met.

A robust program of planetary protection, including forward contamination control, medical monitoring, spatial planning for human exploration, and precautions against back contamination, can be developed with the understanding that prior to human exploration there is a need for efforts to develop, rehearse, and refine planetary protection controls. Effectively, these principles involve "defense in depth" and the continuous evaluation throughout a mission of the contamination status of the crew and the planetary environments (surface and subsurface) they will explore and utilize.

8. FUTURE CONSIDERATIONS IN PLANETARY PROTECTION FOR COMMERCIAL AND OTHER MISSIONS

Over the modern lifetime of COSPAR's Panel on Planetary Protection, beginning in 1999, the Panel has effectively pursued an international consensus standard to avoid organic and biologic contamination during solar system exploration missions. Most, if not all, space agencies have followed that consensus standard, including NASA. Nonetheless, there is now significant interest by non-space-

agency actors (private and commercial) in the use of outer space, especially regarding the use of space resources such as metallic asteroids, lunar materials and oxygen, and water and/or water ice in many areas of the solar system. As such, there is an additional need to broadly communicate the rationale for the COSPAR policy to these other actors, and look for ways to preserve the scientific benefits of space exploration while acknowledging that there are other potential uses of the solar system that will continue to grow beyond science, and will also include other goals.

Whatever their interests, commercial users of outer space whose nations are signatories to the OST must comply with Article IX of the OST, avoiding "harmful contamination" and adhering to OST Article VI's authorization and continuing supervision provisions. Despite concerns about over-regulation of space entrepreneurs, it is anticipated that commercial entities can and will follow the international consensus standard for planetary protection maintained by COSPAR, and in the future have a voice in the crafting of that standard. Even so, for most of the solar system there are no obvious conflicts between adherence to the current COSPAR policy and the needs of commercial entities who are planning missions to the asteroids or the Moon. Those missions would not have operational requirements imposed on them — only recordkeeping would be required. In fact, for the foreseeable future, planetary protection conflicts in other than the science realm would be limited to the desire by some to begin human activities on and around Mars. There are many conflicting usage issues (science, human exploration and colonization, commercial use), and while at present a direct interest in Mars is limited to a few companies with the nascent capability of having their own missions, future exploration may be undertaken by a larger number of entities, similar to the current wave of international and commercial assets investigating LEO and the Moon.

While harvesting ice from lunar polar craters and shipping it to some useful location has significant technical challenges, those difficulties are nearly free of concern for biological contamination, and are likely to be compatible with the study of organic contributions to the lunar environment over the course of solar system history. On the other hand, the issues associated with future Mars missions are of a more complex nature. For example, if there remains the potential for a threat to Earth's biosphere from the importation of uncontained martian materials, then the back-contamination control provisions of Category V, restricted Earth return, should apply to all missions that return from Mars, with or without humans onboard. Likewise, there will be a need to understand the distribution and fate of human-associated microorganisms brought to Mars with either human explorers or tourists in order to understand Earth-sourced contamination when it is encountered. Of course, that is most important to the search for life on Mars on or near the surface. The *quid pro quo* is that an understanding of the potential for human explorers to encounter martian life is critical. In accordance with current COSPAR policies and the assessment of the scientific community, it would be both unethical and unwise to expose humans on Mars to alien life without planning for it, or testing for it as the human sphere of influence expands. And if, one day, it is decided that humans should permanently live on Mars, either in surface greenhouses or in a warmed-up martian environment, there will still be a need to control exposure to microbial populations — especially those brought from Earth.

Commercial spaceflight poses new challenges regarding the utilization of resources in space, space tourism, and others. These challenges include determining what committees or national agencies are responsible for enforcing policies regarding planetary protection. To this end, new workshops on planetary protection that include and engage the commercial spaceflight community are being undertaken, and NASA and others in the U.S. Government are seeking ways to include commercial space flight priorities and safeguards into the application of planetary protection concepts, given that States Parties to the OST are responsible for authorization and continued supervision of their nationals (Article VI), including avoiding "harmful contamination" of the Moon and other celestial bodies under Article IX of the treaty.

9. CONSIDERATIONS FOR FUTURE WORK IN PLANETARY PROTECTION RESEARCH

Many of the problems of implementing planetary protection on missions are due to either the lack of tools (technology) to do the job, or the lack of talent (experience) with the application of that technology. As such, planetary protection implementation on missions can take more time, and therefore money, to ensure mission success. All three aspects of doing the job — talent, tools, and time — can benefit from a solid science and technology research program. Current topics for research, technology development, and testing are discussed below.

9.1. Further Fundamental Knowledge on Microbial Limits of Life (and How to Kill Them)

Our understanding of environmental microbiology and extremophiles has expanded considerably, resulting in a greater awareness of the potential for the survival of terrestrial microbes in extreme environments, as well as the prospect for finding possible evidence of truly extraterrestrial life in other locations. It is essential to the proper implementation of planetary protection policy that habitability estimates for planetary environments be established conservatively, allowing for Earth-organism capabilities that are better than currently understood, and that appropriate measures are taken that allow for such capabilities and thus protect against contamination. Research on microbial diversity and adaptation to planetary environments are needed to inform planetary protection policies and their implementation.

Likewise, procedures that can be undertaken to end microbial viability without negatively affecting spacecraft systems are of great interest to space projects planned to explore solar system environments that could be contaminated by organisms from Earth. Both spacecraft sterilization and extremophile survival are closely related research topics.

For example, numerous studies have been conducted regarding the survival of microbes exposed to space conditions, where exposure to ultraviolet radiation in a vacuum has proven to be quickly fatal to both vegetative organisms and spores (e.g., *Rabbow et al.,* 2015). More recently, however, significant lethality has also been seen with the combined effects of heat and vacuum killing organisms under conditions when neither alone would have the same effect (*Schuerger et al.,* 2019).

In contrast, multiple factors may combine to provide a habitable environment for organisms in ways that are not apparent from Earth-focused experiments. Of particular note were the conditions at the Phoenix landing site in the north polar regions on Mars (*Smith et al.,* 2009; *Zent et al.,* 2010), where it was seen that water ice was present, and that overnight on some sols the relative humidity reached 100%. Combined with temperatures that on many days reached at least 253 K for over 4 hours, over the entire 24-hour sol the Phoenix site reached the conditions where it would be possible for an Earth organism capable of collecting water at night to use it for growth and reproduction during the day. To date, only growth has been recorded by the use of water vapor alone, and not reproduction (e.g., *Lange et al.,* 1986), but there does not appear to be any fundamental prohibition for both to occur.

9.2. Updating, Identifying, and Monitoring Potential Special Regions on Mars

Accordingly, a high priority for future research is updating the parameters and potential locations for naturally occurring Special Regions on Mars. More data about Mars environments should be available to help us understand and project where we might find Special Regions on

Mars — places where Earth organisms might be able to replicate, and where (eventually) we may gain insights on the potential for indigenous life on Mars. It is anticipated that new information and understanding about martian environments and terrestrial microbes will continue to be gained through a robotic program of Mars exploration, and an eventual sample return from Mars. This will require expanded knowledge of the limits of terrestrial microbial life, as well as the availability and action of water on Mars today, including specific features or depths in which places warm enough and wet enough for microbial activity might be found. We have only begun to understand the presence and availability of critical resources on Mars (e.g., *Rummel et al.,* 2014; *Orosei et al.,* 2018; *Goudge et al.,* 2017; see also the chapter by Davila et al. in this volume). If more fully addressed, such information would add powerful insights into Mars astrobiology and clarify planetary protection issues associated both with Special Regions themselves, and with the potential for Earth contamination on Mars to spread. Related research topics include:

- Further research into ice and the mixtures of ice and salt observed on Mars (e.g., at the Phoenix landing site)
- Extension of the coverage, resolution, and near-surface penetration of Mars radar surveys beyond those by Mars Advanced Radar for Subsurface and Ionosphere Sounding (MARSIS) and Mars SHAllow RADar sounder (SHA-RAD) to search for potential new habitats (see *Orosei et al.,* 2018, for an update)
- Further investigations into the conditions in caves on Mars (*Boston,* 2010; *Leveille and Datta,* 2010; *SSB,* 2019).

9.3. Knowledge Required to Better Understand Microbial Activity on Mars and Other Planets

We still are a long way from understanding the environmental limits to microbial reproduction on Earth. Because it is so difficult to discern what most microorganisms do, exactly, for a living, and on what aspects of the environment they may depend, it is even more difficult to understand what Earth organisms might do if transported to Mars (e.g., *Rummel et al.,* 2014; *SSB,* 2019). Mars appears to have water in places, and high-energy oxidants (perchlorates) on the surface in a number of locations. As such, there could very well be places on present-day Mars that would support Earth life. We do not currently know enough about the many places on Mars where life could occur to tell. Priority research needs in this area include:

- As noted above, improve our understanding of the synergy of multiple factors that enable enhanced microbial survival and growth, and mechanisms that may allow for temporal separation in microbial resource acquisition and use
- Conduct investigations into microbial activity at low water activity — additional physiological studies on the limits to microbial survival and replication
- Conduct investigations into microbial activity at the lower-temperature limits for life

9.4. Prepare for the Ocean Worlds

As noted in the recent SSB report, *Astrobiology Strategy for the Search for Life in the Universe* (*SSB,* 2019):

For the ocean worlds of the outer solar system, integrating the entire system into the discussion of habitability and the search for life allows the exploration of these worlds to be prioritized and planned. Over the past 20 years, the sophistication of understanding of these systems has grown dramatically. Knowledge of the system as a whole shifts focus from the search for water to the search for many parameters that are needed to support life, such as chemical energy and geologic activity.

Future exploration of the ocean worlds of the solar system, especially those showing positive surface-manifestations of a buried ocean, will be based on the currently evolving technology that is allowing oceanography on Earth to expand so rapidly. Due to investments by space agencies into ocean-world exploration technology and coupled analog studies, it is now possible to plan for missions that could penetrate the outer icy surface of those worlds and conduct oceanographic research at great distances from the oceans we are beginning to know on Earth (see *Rettberg et al.,* 2019).

From a planetary protection standpoint, such mission preparations benefit from the demands of hydrodynamic forces. It is far easier to surface-sterilize exploration tools that are smooth and circular in cross section than it is to deal with bare wire bundles on complex surface systems. And in some cases, notably Europa, the surface radiation environment can provide a means to reduce biological contamination on the surface or even interior of a landed or a long-orbiting spacecraft (e.g., *Hand,* 2017).

Unlike the situation on Mars, however, where it can be asserted that any biological contamination event will likely be transitory and isolated, the major concern regarding ocean worlds planetary protection is that any single contamination event could conceivably spread and contaminate a subsurface ocean. Extensive studies of ocean world dynamics and ocean access will be needed to positively deal with the challenges of providing proper planetary protection, especially in the forward direction.

9.5. Human-Associated Microbial Diversity and Distribution on Mars

Similarly, we have only recently recognized that humans themselves are a veritable scaffold upon which microbial ecosystems flourish. Powerful new analytical tools have become available to analyze and decipher such ecosystems and understand our human-associated microorganisms. Since these diverse microbial hitchhikers represent potential bio-contaminants during human exploration of the solar system, it is important to understand them to the fullest — their identities, abundance, and distribution, as well as their potential for dispersal, survival, and propagation as contaminants, and as markers in exploration environments, whether in habitat/work environments or exposed to the planet/moon itself.

Specific topics of relevance to the fundamental scientific understanding of human explorers in space include (but are not limited to):

- Development of a baseline inventory and understanding of human associated microbes, as relevant to the space environment
- Studies of human-associated microbes as potential contaminants, including their abundance, potential for release, and dispersal/survival/propagation during planetary exploration
- Understanding human-associated microbes as potential biomarkers of relevance, and their possible use as tracers of contamination
- Contamination transport models (near and far-field)
- Studies to better understand the contribution of ambient space environments toward passive mitigation of forward contaminant risks (radiation, temperature, etc.)

9.6. Lunar Testbeds for Technology Development and Operations

The Moon in particular is considered to be an excellent potential testbed to develop planetary protection procedures and practices in an environment sufficiently harsh to prove an adequate challenge, but isolated from the overwhelming background contamination of the terrestrial biosphere. Because the Moon is currently recognized as being of interest for understanding prebiotic chemistry and the origin of life, but is not hospitable to biological contamination by Earth life, the only planetary protection constraint for operations on the Moon is the requirement to document activity, per the requirements of COSPAR Category II. With no specific limits on contamination, the Moon can be an excellent place to test technologies developed for elsewhere in the solar system — in particular Mars. A coordinated lunar program addressing planetary protection issues could yield significant benefits (e.g., *NASA, 2007; LEAG, 2009*) such as providing valuable ground truth on *in situ* contamination of samples and the survival of microbes in the space environment; studying lunar habitat/spacesuit competency, containment, and leakage; and testing operational procedures associated with successful planetary protection implementation on a planetary surface. Depending on the specifics of its life-support system and in-space location, similar experiments could also be hosted by a human-occupied platform in deep space, near the Moon.

10. SUMMARY

1. Planetary protection is an important, cross-cutting concern of high relevance to astrobiology, and is required to ensure that space agencies are prepared to provide for safe solar system exploration and future utilization of space resources and locations.

2. For affected missions, whether robotic or human, planetary protection considerations need to be integrated throughout mission planning and systems development, requiring proactive coordination and collaboration between the planetary protection community and other experts from the earliest stages of mission development.

3. Numerous research and technology development areas relevant to planetary protection have been identified as important to implementing planetary protection controls on future robotic and human missions. These can provide the talent, tools, or time to ensure mission success.

4. Planetary protection is an active area of research on the science and engineering side as well, pitting a knowledge of the capabilities of Earth organisms against the availability of resources (at a microbial scale) on other worlds, and providing a cautionary tale to the search for life in the solar system.

Acknowledgments. I thank V. Meadows and B. Schmidt for their patience in receiving the manuscript for this chapter, V. Hipkin for her careful reading of it before submission, and the support of NASA through agreement NNX15AV46A to the SETI Institute (P.I. Margaret Race).

REFERENCES

Aeronautics and Space Engineering Board (ASEB) and Space Studies Board (SSB) (2002) *Safe On Mars: Precursor Measurements Necessary to Support Human Operations on the Martian Surface.* National Academies, Washington, DC.

Allton J. H., Bagby J. R. Jr., and Stabekis P. D. (1998) Lessons learned during Apollo lunar sample quarantine and sample curation. *Adv. Space Res., 22,* 373–382.

Ammann W., Baross J., Bennett A., Bridges J., Fragola J., Kerrest A., Marshall-Bowman K., Raoul H., Rettberg P., Rummel J., Salminen M., Stackebrandt E., and Walter N. (2012) *Mars Sample Return Backward Contamination — Report from the ESF-ESSC Study Group on MSR Planetary Protection Requirements.* European Science Foundation, Strasbourg. 59 pp.

Beaty D., Buxbaum K., Meyer M., Barlow N., Boynton W., Clark B., Deming J., et al. (2006) Findings of the Mars Special Regions Science Analysis Group. *Astrobiology, 6,* 677–732.

Boston P. J. (2010) Location, location, location! Lava caves on Mars for habitat, resources, and the search for life. *J. Cosmol., 12,* 3957–3979.

Cameron A. G. W. (1982) Letter as SSB Chair to Dr. Gerald A. Soffen, Director of NASA's Life Sciences Division, dated 18 May 1982.

COPUOS (2017) Final Report of the Committee on the Peaceful Uses of Outer Space, Sixtieth Session (7–16 June 2017), General Assembly, Official Records, Seventy-Second Session, Supplement No. 20, Paragraph 332 of UN document A/72/20, *http://www.unoosa.org/oosa/en/ourwork/copuos/2017/index.html.*

COSPAR (1964) Resolution 26.5. *COSPAR Inf. Bull., 20,* 25–26.

COSPAR (1969) COSPAR Decision No. 16. *COSPAR Inf. Bull., 50,* 15–16.

COSPAR (1994) COSPAR Decision No. 1/94. *COSPAR Inf. Bull., 131,* 30.

COSPAR (2003) COSPAR planetary protection policy 2002. *COSPAR Inf. Bull., 156,* 67–74.

Crichton M. (1969) *The Andromeda Strain.* Alfred A. Knopf, New York. 350 pp.

Criswell M., Race M., Rummel J., and Baker A., eds. (2005) *Planetary Protection Issues in the Human Exploration of Mars: Pingree Park Final Workshop Report.* NASA CP-2005-213461, A-0513375. NASA, Washington, DC. 88 pp.

Daspit L., Stern J., and Martin J. (1990) *Lessons Learned from the Viking Planetary Quarantine/Contamination Control Experience.* Bionetics Corporation Report for NASA Contract NASW-4355.

DeVincenzi D. L., Stabekis P. D., and Barengoltz J. B. (1983) A proposed new policy for planetary protection. *Adv. Space Res., 3,* 13–21.

Dick S. J. and Strick J. E. (2005) *The Living Universe: NASA and the Development of Astrobiology.* Rutgers Univ., New Brunswick. 308 pp.

Goudge T. A., Milliken R. E., Head J. W., Mustard J. F., and Fassett C. I. (2017) Sedimentological evidence for a deltaic origin of the western fan deposit in Jezero Crater, Mars and implications for future exploration. *Earth Planet. Sci. Lett., 458,* 357–365.

Guthrie W. K. C. (1965) *A History of Greek Philosophy, Volume II: The Presocratic Tradition from Parmenides to Democritus.* Cambridge Univ., Cambridge. 405 pp.

Hall L. (1974) Ten years of development of the planetary quarantine program of the United States. In *Life Sciences and Space Research: Proceedings of the Open Meeting of the Working Group on Space Biology of the Sixteenth Plenary Meeting of COSPAR* (P. H. A. Sneath, ed.), pp. 185–197, DOI: 10.1016/B978-0-08-021783-3.50030-1. Pergamon, Elmsford.

Hall R. C. (1977) *Lunar Impact: A History of Project Ranger.* NASA SP-4210. NASA, Washington, DC.

Haltigin T., Lange C., Mugnuolo R., Smith C., and the iMARS Working Group (2018) A draft mission architecture and science management plan for the return of samples from Mars: Phase 2 report of the International Mars Architecture for the Return of Samples (iMARS) Working Group. *Astrobiology, 18,* S1-1–S-131, DOI: 10.1089/ast.2018.29027.mars.

Hand K. P., Murray A. E., Garvin J. B., Brinkerhoff W. B., Edgett K. S., Ehlmann B. L., German C. R., Hayes A. G., Hoehler T. M., Horst S. M., Lunine J. I., Nealson K. H., Paranicas C., Schmidt B. E., Smith D. E., Rhoden A. R., Russell M. J., Templeton A. S., Willis P. A., Yingst R. A., Phillips C. B., Cable M. L., Craft K. L., Hofmann A. E., Nordheim T. A., Pappalardo R. T., and the Project Engineering Team (2017) *Report of the Europa Lander Science Definition Team,* JPL D-97667. NASA, Washington, DC.

HISRGWG (2017) *Building Blocks for an International Framework on Space Resource Activities,* September 2017. The Hague International Space Resources Governance Working Group, Univ. of Leiden, Netherlands.

Hogan J., Fisher J., Race M., Joshi J., and Rummel J., eds. (2006) *Life Support and Habitation and Planetary Protection Workshop Final Report.* NASA TM-2006-213485, NASA, Washington, DC.

Kivelson M. G., Khurana K. K., Russell C. T., Volwerk M., Walker R. J., and Zimmer C. (2000) Galileo magnetometer measurements: A stronger case for a subsurface ocean at Europa. *Science, 289(5483),* 1340–1343.

Kminek G. and Fisk L. A. (2017) Protecting our investment in the exploration and utilization of space. *Astrobiology, 17,* 955.

Kminek G., Rummel J., and Race M., eds. (2007) *Planetary Protection and Human System Research and Technology.* ESA-NASA Workshop Report, ESA WPP-276, ESTEC, Noordwijk.

Kminek G., Rummel J. D., Cockell C. S., Atlas R., Barlow N., Beaty D., Boynton W., Carr M., Clifford S., Conley C. A., Davila A. F., Debus A., Doran P., Hecht M., Heldmann J., Helbert J., Hipkin V., Horneck G., Kieft T. L., Klingelhoefer G., Meyer M., Newsom H., Ori G. G., Parnell J., Prieur D., Raulin F., Schulze-Makuch D., Spry J. A., Stabekis P. E., Stackebrand E., Vago J., Viso M., Voytek M., Wells L., and Westall F. (2010) Report of the COSPAR Mars Special Regions Colloquium. *Adv. Space Res., 46,* 811–829.

Kminek G., Conley C., Hipkin V., and Yano H. (2017) COSPAR's Planetary Protection Policy. *Space Res. Today, 200,* 12–25.

La Duc M. T., Kern K., and Venkateswaran K. (2004) Microbial monitoring of spacecraft and associated environments. *Microb. Ecol., 47,* 150–158.

Lange O. L., Kilian E., and Ziegler H. (1986) Water vapor uptake and photosynthesis of lichens: Performance differences in species with green and blue-green algae as phycobionts. *Oecologia, 71,* 104–110.

LEAG (2009) *The Lunar Exploration Roadmap: Exploring the Moon in the 21st Century: Themes, Goals, Objectives, Investigations, and Priorities, 2009.* Version 1.0. University of Notre Dame, Notre Dame. 8 pp.

Lederberg J. (1960) Exobiology: Approaches to life beyond the Earth. *Science, 132,* 393–400.

Leveille R. J. and Datta S. (2010) Lava tubes and basaltic caves as astrobiological targets on Earth and Mars: A review. *Planet. Space Sci., 58,* 592–598.

Meltzer M. (2011) *When Biospheres Collide: A History of NASA's Planetary Protection Programs.* NASA SP-2011-4234, Washington, DC.

NAS (U.S. National Academy of Science) (2019) Mission. National Academy of Science, *http://nasonline.org/about-nas/mission/,* accessed 12 April 2019.

NASA (1988) *Biological Contamination Control for Outbound and Inbound Planetary Spacecraft.* NASA Management Instruction (NMI) 8020.7A. NASA, Washington, DC.

NASA (1999a) *Biological Contamination Control for Outbound and Inbound Planetary Spacecraft.* NASA Policy Document (NPD) 8020.7G. NASA, Washington, DC. 10 pp.

NASA (1999b) *Planetary Protection Provisions for Robotic Extraterrestrial Missions.* NASA Handbook (NHB) 8020.12B. NASA, Washington, DC.

NASA (2007) *NASA Advisory Council Workshop on Science Associated with the Lunar Exploration Architecture,* available online at *http://www.lpi.usra.edu/meetings/LEAG/.*

NASA (2015) *Planetary Protection Knowledge Gaps for Human Extraterrestrial Missions: Workshop Report* (M. S. Race et al., eds.), pp. 1–73. NASA Ames, Moffett Field. 73 pp.

News of Science (1958) Development of international efforts to avoid contamination of extraterrestrial bodies. News of Science, 17 October 1958. *Science, 128,* 887–889, DOI: 10.1126/science.128.3329.887.

Orosei R., Lauro S. E., Pettinelli E., Cicchetti A., Coradini M., Cosciotti B., Di Paolo F., et al. (2018) Radar evidence of subglacial liquid water on Mars. *Science, 361(6401),* 490–493, DOI: 10.1126/science.aar7268.

Pappalardo R. T., Head J. W., Greeley R., Sullivan R. J., Pilcher C., Schubert G., Moore W. B., et al. (1998) Geological evidence for solid-state convection in Europa's ice shell. *Nature, 391(6665),* 365–368.

Phillips C. R. (1974) *The Planetary Quarantine Program, Origins and Achievements 1956–1973.* NASA SP-4902, NASA, Washington, DC.

Pratt L. (2019) *Clipper Planetary Protection Workshop.* Presentation at OPAG Meeting, April 2019, *https://www.lpi.usra.edu/opag/meetings/apr2019/presentations/Pratt.pdf.*

Puleo J. R., Fields N. D., Bergstrom S. L., Oxborrow G. S., Stabekis P. D., and Koukol R. (1977) Microbiological profiles of the Viking spacecraft. *Appl. Environ. Microbiol., 33,* 379–384.

Rabbow E., Rettberg P., Barczyk S., Bohmeier M., Parpart A., Panitz C., Horneck G., Burfeindt J., Molter F., Jaramillo E., and Pereira C. (2015) The astrobiological mission EXPOSE-R on board of the International Space Station. *Intl. J. Astrobiol., 14,* 3–16.

Race M., Kminek G., Rummel J., et al. (2008) Planetary protection and humans on Mars, NASA/ESA workshop results. *Adv. Space Res., 42,* 1128–1138.

Rettberg P., Antunes A., Brucato J., Cabezas P., Collins G., Haddaji A., Kminek G., Leuko S., McKenna-Lawlor S., Moissl-Eichinger C., Fellous J-L., Olsson-Francis K., Pearce D., Rabbow E., Royle S. l., Saunders M., Sephton M., Spry A., Walter N., Wimmer Schweingruber R., and Treuet J-C. (2019) Biological contamination prevention for outer solar system moons of astrobiological interest: What do we need to know? *Astrobiology, 19(8),* DOI: 10.1089/ast.2018.1996.

Robinson G. S. (1971) Earth exposure to extraterrestrial trial matter: NASA's quarantine regulations. *Intl. Lawyer, 5,* 219–248.

Rummel J. D. (2002) COSPAR/IAU Workshop on Planetary Protection. COSPAR, Paris.

Rummel J. D. (2019) From planetary quarantine to planetary protection: A NASA and international story. *Astrobiology, 19,* 624–627.

Rummel J. D. and Kminek G. (2018) It's time to develop a new "draft test protocol" for a Mars sample return mission (or two). *Astrobiology, 18,* 377–380.

Rummel J. D., Stabekis P. D., DeVincenzi D. L., and Barengoltz J. B. (2002a) COSPAR's planetary protection policy: A consolidated draft. *Adv. Space Res., 30,* 1567–1571.

Rummel J. D., Race M., DeVincenzi D. L., Schad P. J., Stabekis P. D., Viso M., and Acevedo S. (2002b) A draft test protocol for detecting possible biohazards in martian samples returned to Earth. NASA CP-2002-211842, Washington, DC.

Rummel J. D., Beaty D., et al. (2014) A new analysis of Mars "Special Regions": Findings of the second MEPAG Special Regions Science Analysis Group (SR-SAG2). *Astrobiology, 14,* 887–968.

Schuerger A. C., Moores J. E., Smith D. J., and Reitz G. (2019) A Lunar microbial survival model for predicting the forward contamination of the Moon. *Astrobiology, 19(6),* 730–756.

Smith P. H., Tamppari L. K., Arvidson R. E., Bass D., Blaney D., Boynton W. V., Carswell A., et al. (2009) H_2O at the Phoenix landing site. *Science, 325,* 58–61.

Soffen G. A. (1976) Scientific results of the Viking missions. *Science,*

194, 1274–1276.

SSB (1959) *Summary Report of WESTEX.* National Academies, Washington, DC. 174 pp.

SSB (1964) *Conference on Potential Hazards of Back Contamination from the Planets.* National Academies, Washington, DC. 86 pp.

SSB (1965) *Conference on Hazard of Planetary Contamination Due to Microbiological Contamination in the Interior of Spacecraft Components.* National Academies, Washington, DC. 18 pp.

SSB (1966) *Biology and the Exploration of Mars.* National Academies, Washington, DC. 532 pp.

SSB (1968a) *Planetary Exploration: 1968–1975.* National Academies, Washington, DC. 57 pp.

SSB (1968b) *Report on NASA Biology Program.* National Academies, Washington, DC. 34 pp.

SSB (1972) *Review of Planetary Quarantine Policy.* National Academies, Washington, DC. 5 pp.

SSB (1977) *Post-Viking Biological Investigations of Mars.* National Academies, Washington, DC. 35 pp.

SSB (1978) *Recommendations on Quarantine Policy for Mars, Jupiter, Saturn, Uranus, Neptune, and Titan.* National Academies, Washington, DC. 84 pp.

SSB (1992) *Biological Contamination of Mars: Issues and Recommendations.* National Academies, Washington, DC. 123 pp.

SSB (1997) *Mars Sample Return: Issues and Recommendations.* National Academies, Washington, DC. 58 pp.

SSB (1998) *Evaluating the Biological Potential in Samples Returned from Planetary Satellites and Small Solar System Bodies: Framework for Decision Making.* National Academies, Washington, DC. 116 pp.

SSB (2000) *Preventing the Forward Contamination of Europa.* National Academies, Washington, DC. 54 pp.

SSB (2006) *Preventing the Forward Contamination of Mars.* National Academies, Washington, DC. 166 pp.

SSB (2018) *Review and Assessment of Planetary Protection Policy Development Processes.* National Academies, Washington, DC. 138 pp.

SSB (2019) *An Astrobiology Strategy for the Search for Life in the Universe.* National Academies, Washington, DC. 188 pp.

Stabekis P. and DeVincenzi D. L. (1978) Planetary protection guidelines for outer planet missions. In *Life Sciences and Space Research, Volume XVI, Proceedings of the Open Meetings of the Working Group on Space Biology of the Twentieth Plenary Meeting of COSPAR, Tel Aviv, Israel, 7–18 June 1977* (R. Holmquist, ed.), pp. 39–44. Pergamon, Oxford.

Stone M. (2009) NIH builds substantial human microbiome project. *Microbe, 4(10),* DOI: 10.1128/microbe.4.451.1.

United Nations (1959) General Assembly Resolution 1472, International Co-operation in the Peaceful Uses of Outer Space, A/RES/14/72 (12 December 1959), available online at *unoosa.org/pdf/gares/ ARES_14_1472E.pdf.*

United Nations (1967) Treaty on Principles Governing the Activities of States in the Exploration and Use of Outer Space, Including the Moon and Other Celestial Bodies, A/RES/21/2222 (19 December 1966), available online at *unoosa.org/pdf/gares/ ARES_21_2222E.pdf.*

Vitousek P. M., D'Antonio C. M., Loope L. L., Rejmanek M., and Westbrooks R. (1997) Introduced species: A significant component of human-caused global change. *N. Z. J. Ecol., 21,* 1–16.

Voorhies A. A. and Lorenzi H. A. (2016) The challenge of maintaining a healthy microbiome during long-duration space missions. *Front. Astron. Space Sci., 3,* 23, DOI: 10.3389/fspas.2016.00023.

Wells H. G. (1898) *The War of the Worlds.* William Heinemann, London. 287 pp.

Williamson M. (1996) *Biological Invasions.* Chapman and Hall, London. 245 pp.

Zent A. P., Hecht M. H., Cobos D. R., Wood S. E., Hudson T. L., Milkovich S. M., DeFlores L. P., and Mellon M. T. (2010) Initial results from the thermal and electrical conductivity probe (TECP) on Phoenix. *J. Geophys. Res.–Planets, 115(E3),* DOI: 10.1029/2009JE003420.

Part 3:

The Solar System —
Exoplanet Synergy

Raymond S., Izidoro A., and Morbidelli A. (2020) Solar system formation in the context of extrasolar planets. In *Planetary Astrobiology* (V. Meadows et al., eds.), pp. 287–324. Univ. of Arizona, Tucson, DOI: 10.2458/azu_uapress_9780816540068-ch012.

Solar System Formation in the Context of Extrasolar Planets

Sean N. Raymond
Laboratoire d'Astrophysique de Bordeaux, Centre National de la Recherche Scientifique/Université de Bordeaux

Andre Izidoro
Universidade Estadual Paulista — Grupo de Dinmica Orbital Planetologia

Alessandro Morbidelli
Université de Nice Sophia-Antipolis, Centre National de la Recherche Scientifique/Observatoire de la Côte d'Azur

Exoplanet surveys have confirmed one of humanity's (and all teenagers') worst fears: We are weird. If our solar system were observed with present-day Earth technology — to put our system and exoplanets on the same footing — Jupiter is the only planet that would be detectable. The statistics of exo-Jupiters indicate that the solar system is unusual at the ~1% level among Sun-like stars (or ~0.1% among all main-sequence stars). But why are we different? This review focuses on global models of planetary system formation. Successful formation models for both the solar system and exoplanet systems rely on two key processes: orbital migration and dynamical instability. Systems of close-in "super-Earths" or "sub-Neptunes" cannot have formed *in situ*, but instead require substantial radial inward motion of solids either as drifting millimeter- to centimeter-sized pebbles or migrating Earth-mass or larger planetary embryos. We argue that, regardless of their formation mode, the late evolution of super-Earth systems involves migration into chains of mean-motion resonances anchored at the inner edge of the protoplanetary disk. The vast majority of resonant chains go unstable when the disk dissipates. The eccentricity distribution of giant exoplanets suggests that migration followed by instability is also ubiquitous in giant planet systems. We present three different models for inner solar system formation — the Low-Mass Asteroid Belt, Grand Tack, and Early Instability models — each of which invokes a combination of migration and instability. We discuss how each model may be falsified. We argue that most Earth-sized habitable zone exoplanets are likely to form much faster than Earth, with most of their growth complete within the disk lifetime. Their water contents should span a wide range, from dry rock-iron planets to water-rich worlds with tens of percent water. Jupiter-like planets on exterior orbits may play a central role in the formation of planets with small but non-zero, Earth-like water contents. Water loss during giant impacts and heating from short-lived radioisotopes like ^{26}Al may also play an important role in setting the final water budgets of habitable zone planets. Finally, we identify the key bifurcation points in planetary system formation. We present a series of events that can explain why our solar system is so weird. Jupiter's core must have formed fast enough to quench the growth of Earth's building blocks by blocking the flux of pebbles drifting inward through the gaseous disk. The large Jupiter/Saturn mass ratio is rare among giant exoplanets but may be required to maintain Jupiter's wide orbit. The giant planets' instability must have been gentle, with no close encounters between Jupiter and Saturn, also unusual in the larger (exoplanet) context. Our solar system is thus the outcome of multiple unusual, but not unheard of, events.

1. INTRODUCTION

The discovery of extrasolar planets demonstrated that the current solar system-inspired paradigm of planet formation was on the wrong track. Most extrasolar systems bear little resemblance to our well-ordered solar system. While the solar system is radially segregated by distance from the star, with small inner rocky worlds and more distant giant planets, few known exo-systems follow the same blueprint. Models designed with the goal of reproducing the solar system failed spectacularly to understand why other planetary systems looked different than our own.

Yet exoplanets represent a huge sample of outcomes of planet formation, and new ideas for solar system formation and evolution borrow liberally from models designed to explain the exoplanet population. While we are far from a complete picture, just a handful of processes may explain the broad characteristics of most exoplanet systems and the solar system.

We review the current thinking in how solar system formation fits in the larger context of extrasolar planetary systems. In section 1, we first review observational constraints on the frequency of solar system-like systems, pointing out specific characteristics of the solar system that don't fit within a simple formation picture. In section 2, we then briefly summarize the stages of planet formation — from dust to full-sized planets — as they are currently understood, with liberal references to more detailed recent reviews of different steps. Next, in section 3 we discuss current models for the different populations of extrasolar planets and how they match quantifiable constraints. We then turn our attention to the solar system (section 4). We present the empirical constraints and a rough timeline of events in solar system formation that includes a discussion of the Nice model for the solar system's (giant planet) dynamical instability. We discuss the classical model and its shortcomings, then present three newer competing models to match the important constraints of the inner solar system. A challenge for all current models is to explain the mass deficit interior to Venus' orbit. In section 5 we extrapolate to Earth-mass planets around other stars, discussing the various formation pathways for such planets and their expected water contents. We conclude that most exo-Earths are unlikely to be truly Earth-like. Finally, in section 6 we first synthesize these models into a large-scale picture of planetary system evolution, highlighting the key bifurcation points that may explain the observed diversity and the events that must have taken place to produce our own solar system. We lay out a path for future research by showing how to use theory and observations to test current models for both exoplanet and solar system formation.

1.1. How Common are Solar Systems?

To date, radial velocity (RV) and transit surveys have discovered thousands of extrasolar planets. Figure 1 shows a sample of the diversity of detected exo-systems. These surveys have determined occurrence rates of planets as a function of planet size/mass and orbital period around different types of stars (*Howard et al.,* 2010, 2012; *Mayor et al.,* 2011; *Fressin et al.,* 2013; *Dong and Zhu,* 2013; *Petigura et al.,* 2013; *Fulton et al.,* 2017). Meanwhile, gravitational microlensing and direct imaging surveys have placed constraints on the properties of outer planetary systems (*Cassan et al.,* 2012; *Biller et al.,* 2013; *Mróz et al.,* 2017; *Bowler and Nielsen,* 2018). [Note that microlensing may actually be the most sensitive method for detecting analogs to our solar system's giant planets. Indeed, microlensing observations have found a Jupiter-Saturn analog system (*Gaudi et al.,* 2008; *Bennett et al.,* 2010) as well as rough analogs to the ice giants (*Poleski et al.,* 2014; *Sumi et al.,* 2016). However, given that microlensing requires a precise alignment between a background source and the star whose planets can be found

(e.g., *Gould and Loeb,* 1992), it cannot be used to search for planets around a given star. Rather, its power is statistical in nature (see *Gould et al.,* 2010; *Clanton and Gaudi,* 2014, 2016; *Suzuki et al.,* 2016b). Nonetheless, upcoming microlensing surveys — especially spacebased surveys such as the Wide Field Infrared Survey Telescope (WFIRST) — are expected to find hundreds to thousands of planets in the Jupiter-Saturn regions of their stars (*Penny et al.,* 2019).]

To put the solar system on the same footing as the current sample of extrasolar planets we must determine what our system would look like when observed with present-day Earth technology. The outcome is somewhat bleak. The terrestrial planets are all too small and too low mass to be reliably detectable. Although sub-Earth-sized planets were discovered by Kepler (e.g., *Barclay et al.,* 2013), and approximately Earth-mass planets have been found by RV monitoring (e.g., *Anglada-Escudé et al.,* 2016), they were all on close-in orbits and many around stars much smaller than the Sun. Strong observational biases make it extremely challenging to detect true analogs to our terrestrial planets (e.g., *Charbonneau et al.,* 2007; *Fischer et al.,* 2014; *Winn,* 2018). However, a decade-long RV survey would detect Jupiter orbiting the Sun. Indeed, several Jupiter analogs have been discovered (e.g., *Wright et al.,* 2008). Saturn, Uranus, and Neptune are too distant to be within the reach of RV surveys. Figure 2 shows seven known Jupiter analog systems for scale.

The exo-solar system is therefore just the Sun-Jupiter system. Observations of the solar system as an exoplanet system would provide a decent measurement of Jupiter's mass (really, its m sin i, where i is the angle between our line of sight and its orbital plane) and semimajor axis, with modest constraints on its orbital eccentricity.

Based on current data, the Sun-Jupiter system is rare at the one-in-a-thousand level. The Sun — a member of the G-dwarf spectral class — is much more massive than most stars; among nearby stars only ~5% have similar masses (e.g., *Chabrier,* 2003). Roughly 10% of Sun-like stars have gas giant planets [defined as having masses $M \gtrsim 50\ M_\oplus$ (*Butler et al.,* 2006; *Udry and Santos,* 2007; *Cumming et al.,* 2008; *Mayor et al.,* 2011; *Clanton and Gaudi,* 2014)]. However, most have orbits that are either significantly closer-in or more eccentric, even after observational biases are accounted for. Using a relatively broad definition for Jupiter-like planets — as planets with orbital radii larger than 2 AU and orbital eccentricities below 0.1 — only 10% of giant exoplanets are Jupiter-like. This puts Jupiter as a 1% case among Sun-like stars, or ~0.1% overall.

The solar system's peculiarity can also be considered in terms of planets that are present in other systems but absent in ours. At least 30–50% of main-sequence stars have planets smaller than $4\ R_\oplus$ (or less massive than ~10–20 $M_\oplus$) on orbits closer-in than Mercury's (*Mayor et al.,* 2011; *Howard et al.,* 2012; *Fressin et al.,* 2013; *Petigura et al.,* 2013; *Dong and Zhu,* 2013; *Hsu et al.,* 2018; *Zhu et al.,* 2018). Recent modeling suggests that less than 8% of planetary systems have their innermost planet on an orbit as wide as Mercury's,

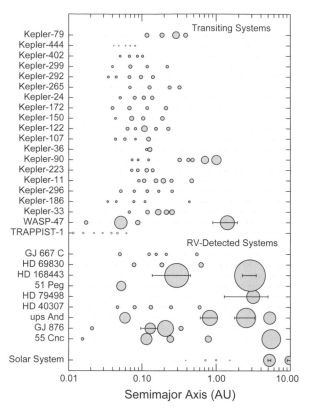

Fig. 1. A sample of extrasolar systems chosen to illustrate their diversity (but not the true distribution of systems; for instance, most super-Earth systems contain only a single detected planet). The top systems were discovered by transit surveys and the bottom systems by radial velocity, although some planets in the transit systems also have radial velocity constraints [e.g., WASP-47 (*Sinukoff et al., 2017*)] and some radial-velocity-detected systems host transiting planets [e.g., 55 Cnc e (*Demory et al., 2011*)]. The size of each planet is proportional to its true size (but not on the orbital scale); for planets with only mass (or m sin i) measurements, we used the $M \propto R^{2.06}$ scaling from *Lissauer et al.* (2011). For planets more massive than 50 $M_\oplus$ with eccentricities larger than 0.1 (as well as for Jupiter and Saturn), the horizontal error bar represents the planet's radial excursion over its orbit (from pericenter to apocenter). Given the logarithmic x axis, the separation between adjacent planets is a measure of their period ratio regardless of their orbital radii. This plot includes a number of systems of close-in super-Earths, including two — TRAPPIST-1 (*Gillon et al., 2017; Luger et al., 2017*) and Kepler-223 (*Mills et al., 2016*) — in which the planets have been shown to be in long chains of orbital resonances (note that the GJ 876 system also includes a three-planet Laplace resonance among more massive planets) (*Rivera et al., 2010*). There are systems with gas giants on eccentric orbits such as Ups And (*Ford et al., 2005*). The central stars vary dramatically in mass and luminosity for the different systems; for instance, the TRAPPIST-1 system orbits an ultracool dwarf star of just 0.08 $M_\odot$ (*Gillon et al., 2017*). Some systems include planets that are smaller than Earth [e.g., the Kepler-444 system (*Campante et al., 2015*)] and others include planets far more massive than Jupiter [e.g., HD 168443 (*Marcy et al., 2001*)]. There are systems with roughly Earth-sized planets in their star's habitable zones, notably Kepler-186 (*Quintana et al., 2014*), TRAPPIST-1 (*Gillon et al., 2017*), and GJ 667 C (*Anglada-Escudé et al., 2013*). Some of these planets are in multiple star systems [e.g., 55 Cnc (*Fischer et al., 2008*)].

and less than 3% have an innermost planet on an orbit as wide as Venus' (*Mulders et al.,* 2018). This reinforces the solar system's standing as an outsider.

The solar system's relative scarcity among exoplanet systems falls at an interesting level. We are not so rare that no solar system analogs have been found. Nor are we so common to be just a "face in the crowd" among exoplanets. Based on a single detectable planet, the solar system stands apart from the crowd but not alone. The question is, why?

1.2. Peculiarities of the Solar System

The orbital architecture of the solar system presents a number of oddities. But like a polka lover's musical pref-

erences, these oddities only become apparent when viewed within a larger context. The "classical model" (discussed in more detail in section 4.3) offers a convenient reference frame for the origin of the terrestrial planets. The classical model assumes that the planets formed mainly *in situ*, meaning from building blocks that originated at roughly their current orbital distances. It also assumes that terrestrial and giant planet formation can be considered separately.

The classical model invokes bottom-up planetary accretion. Starting from a distribution of solids, the planets that form retain a memory of their initial conditions (e.g., *Raymond et al.,* 2005). This motivated the "minimum mass solar nebula" model (*Weidenschilling,* 1977b; *Hayashi,* 1981), which uses the planets' present-day orbits to reconstruct

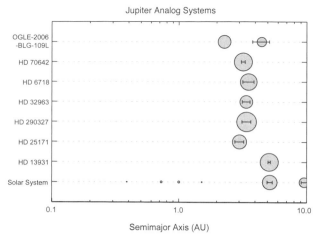

Jupiter Analog Systems

Fig. 2. A representative sample of seven extrasolar systems with Jupiter analogs. The top system (OGLE-2006-BLG-109L) was detected by gravitational microlensing and contains a pair of giant planets with broadly similar properties to Jupiter and Saturn orbiting a roughly half-$M_\odot$ star (*Gaudi et al.,* 2008; *Bennett et al.,* 2010). The next six Jupiter analog systems, all detected via the radial velocity method, each have a host star within 20% of the Sun's mass and contains a single detected planet: a gas giant with a mass between one-third and three times Jupiter's mass with a semimajor axis larger than 3 AU, and an orbital eccentricity of less than 0.1. The planet size scales with its mass$^{1/3}$ and the horizontal error bar denotes the planet's perihelion and aphelion distances. Data were downloaded from *exoplanets.org* (*Wright et al.,* 2011). The solar system is included for comparison. No directly-imaged planets are included in the figure, although there are two with orbits comparable in size to Saturn's: Beta Pictoris b (*Lagrange et al.,* 2009) and 51 Eridani b (*Macintosh et al.,* 2015).

a disk from which they may have formed (neglecting any large-scale motion toward or away from the star such as orbital migration, described in section 2). This style of growth leads to systems in which planets on adjacent orbits have similar sizes (e.g., *Kokubo and Ida,* 2002). The observed super-Earths do appear to have similar sizes within a given system (*Millholland et al.,* 2017; *Weiss et al.,* 2018), although it is hard to imagine a scenario for their formation that does not invoke large-scale radial drift of solids (see section 3.2).

When we compare the classical model blueprint with the actual solar system, several discrepancies emerge:

- Why is Mars so much smaller than Earth? Simulations of terrestrial planet formation from a smooth disk tend to produce Earth and Mars analogs with similar masses (e.g., *Wetherill,* 1978; *Chambers,* 2001; *Raymond et al.,* 2006b), in contrast with the actual 9:1 mass ratio between the two planets. This is called the "small Mars" problem, which was first pointed out by *Wetherill* (1991) and has motivated a number of models of terrestrial planet formation (see section 4).
- Why is Mercury so much smaller than Venus? Although it receives far less attention, the large Venus/Mercury mass

ratio is an even bigger problem than the "small Mars" problem. The Venus/Mercury mass ratio is 14:1 but simulations again tend to produce planets with similar masses and with more compact orbital configurations than the real one. It is worth noting that, in the context of extrasolar planetary systems (in which super-Earths are extremely common), the mass deficit interior in the very inner solar system is in itself quite puzzling.

- Why are the asteroid and Kuiper belts so low in mass yet dynamically excited? The asteroid and Kuiper belts contain very little mass: just $\sim5 \times 10^{-4}\ M_\oplus$ and $\sim0.01-0.1\ M_\oplus$, respectively (*DeMeo and Carry,* 2013; *Gladman et al.,* 2001). However, both belts are dynamically excited, with much higher eccentricities and inclinations than the planets' near circular, coplanar orbits. Yet the mass required to self-excite those belts exceeds the present-day mass by many orders of magnitude (e.g., *O'Brien et al.,* 2007). This apparent contradiction is a key constraint for solar system formation models.
- Why is Jupiter's orbit so wide, and why aren't all giant exoplanets in orbital resonance? Our wide orbit, non-resonant giant planets present an apparent contradiction when viewed through the lens of orbital migration. Migration is an inevitable consequence of planet formation. Given that planets form in massive gaseous disks, gravitational planet-disk interactions must take place. Migration is generally directed inward and the co-migration of multiple planets generically leads to capture in mean-motion resonances (e.g., *Kley and Nelson,* 2012; *Baruteau et al.,* 2014).

2. STAGES OF PLANET FORMATION

Global models of planet formation can be thought of as big puzzles. The puzzle pieces are the processes involved in planet formation, shaped by our current level of understanding. We now briefly review the stages and processes of planetary formation as envisioned by the current paradigm. We remain brief and refer the reader to recent reviews for more details.

2.1. Protoplanetary Disks

While high-resolution observations of disks around young stars show exquisite detail (e.g., *ALMA Partnership et al.,* 2015; *Andrews et al.,* 2016), the structure and evolution of the dominant, gaseous component of planet-forming disks remains uncertain (see discussion in *Morbidelli and Raymond,* 2016). Observations of disk spectra suggest that gas accretes from disks onto their stars (*Meyer et al.,* 1997; *Hartmann et al.,* 1998; *Muzerolle et al.,* 2003), and the occurrence rates of disks around stars in clusters of different ages suggest that disks dissipate on a timescale of a few million years (*Haisch et al.,* 2001; *Briceño et al.,* 2001; *Hillenbrand,* 2008; *Mamajek,* 2009). Disk models thus depend on mechanisms to transport angular momentum in order to generate large-scale radial gas motion (*Balbus and Hawley,* 1998; *Turner et al.,* 2014; *Fromang and Lesur,*

2017). Historically, models have assumed that disks are sufficiently ionized for the magneto-rotational instability to generate viscosity across the disk (*Lynden-Bell and Pringle,* 1974), often using the so-called alpha prescription (*Shakura and Sunyaev,* 1973). However, recent models accounting for electromagnetic effects (in particular the Hall effect and ambipolar diffusion) have found a fundamentally different structure and evolution than alpha disks (*Lesur et al.,* 2014; *Bai,* 2016; *Suzuki et al.,* 2016b), and this structure has implications for multiple stages of planet formation and migration (*Morbidelli and Raymond,* 2016). The final dissipation of the disk is thought to be driven by photo-evaporation from the central star (and in some cases by the external ultraviolet field) (*Hollenbach et al.,* 1994; *Adams et al.,* 2004). [For reviews of disk dynamics, structure, and dispersal, see *Armitage* (2011), *Turner et al.* (2014), *Alexander et al.* (2014), and *Ercolano and Pascucci* (2017).]

2.2. From Dust to Planetesimals

Based on observed infrared excesses in young protoplanetary disks, submicrometer-sized dust particles are inferred to be very abundant (e.g., *Briceño et al.,* 2001; *Haisch et al.,* 2001). Dust particles growing by coagulation in a gaseous disk encounter a number of barriers to growth such as fragmentation and bouncing (*Brauer et al.,* 2008; *Güttler et al.,* 2010). Once they reach roughly millimeter size (or somewhat larger), particles very rapidly drift inward, leading to what was historically called the "meter-sized barrier" to growth (*Weidenschilling,* 1977a; *Birnstiel et al.,* 2012). New models suggest that, if they are initially sufficiently concentrated relative to the gas, the streaming instability can produce clumps of drifting particles that are bound together by self-gravity and directly form 100-km-scale planetesimals (*Youdin and Goodman,* 2005; *Johansen et al.,* 2009, 2015; *Simon et al.,* 2016; *Yang et al.,* 2017). Planetesimals are the smallest macroscopic bodies that do not undergo rapid aerodynamic drift, and are often considered the building blocks of planets. Exactly where and when planetesimals form depends itself on the dynamics and structure of the disk (*Drążkowska et al.,* 2016; *Carrera et al.,* 2017), and some recent studies suggest that planetesimal growth may be favored near the inner edge of the disk (*Drążkowska et al.,* 2016) and just past the snow line (*Armitage et al.,* 2016; *Drążkowska and Alibert,* 2017; *Schoonenberg and Ormel,* 2017). [For reviews of dust growth/drift and planetesimal formation, see *Blum and Wurm* (2008), *Chiang and Youdin* (2010), *Johansen et al.* (2014), and *Birnstiel et al.* (2016).]

2.3. Pebble and Planetesimal Accretion

Planetesimals can grow by accreting other planetesimals (*Greenberg et al.,* 1978; *Wetherill and Stewart,* 1993; *Kokubo and Ida,* 2000) or pebbles that continually drift inward through the disk (*Ormel and Klahr,* 2010; *Johansen and Lacerda,* 2010; *Lambrechts and Johansen,* 2014). Pebbles are defined as particles for which the gas drag

timescale is similar to the orbital timescale, and are typically millimeter- to centimeter-sized in the terrestrial- and giant-planet-forming regions of disks (see *Johansen and Lambrechts,* 2017). Pebbles are thought to continually grow from dust and drift inward through the disk, such that growing planetesimals see a radial flux of pebbles across their orbits (*Lambrechts et al.,* 2014; *Chambers et al.,* 2016; *Ida et al.,* 2016). At low relative speeds, a large planetesimal efficiently accretes nearby small particles (either pebbles or small planetesimals) because the large planetesimal's gravity acts to increase its effective collisional cross section (a process known as gravitational focusing) (*Safronov,* 1969; *Rafikov,* 2004; *Chambers,* 2006). This triggers a phase of runaway growth (*Greenberg et al.,* 1978; *Wetherill and Stewart,* 1993; *Kokubo and Ida,* 1998). At later stages, growth by the accretion of other planetesimals is self-limited because the growing planetesimal excites the random velocities of nearby planetesimals, decreasing the efficiency of gravitational focusing (*Kokubo and Ida,* 2000; *Leinhardt and Richardson,* 2005). However, gas drag acts much more strongly on pebbles and maintains their low velocities relative to larger bodies. The efficiency of pebble accretion increases with the growing planetesimal's mass (*Lambrechts and Johansen,* 2012; *Morbidelli and Nesvorny,* 2012), and pebble accretion outpaces planetesimal accretion for bodies more massive than roughly 1 lunar mass (0.012 $M_\oplus$, although the exact value depends on the parameters of the disk) (*Johansen and Lambrechts,* 2017). Above roughly a lunar mass these objects are generally referred to as "planetary embryos." When an embryo reaches a critical mass, called the pebble isolation mass, it generates a pressure bump in the disk exterior to its orbit, which acts to block the inward flux of pebbles (*Morbidelli and Nesvorny,* 2012; *Lambrechts et al.,* 2014; *Bitsch et al.,* 2018). This acts to quench not only the embryo's growth but also the growth by pebble accretion of all objects interior to the embryo's orbit. Later growth must therefore rely on the accretion of planetesimals, other embryos or gas. We note that there is some debate about whether pebble accretion remains efficient for planets with significant gaseous envelopes below the pebble isolation mass [because pebbles may be ablated due to frictional heating upon entering the growing planets' primordial atmospheres (see *Alibert,* 2017; *Brouwers et al.,* 2018)]. [For reviews of pebble and planetesimal accretion, see *Johansen and Lambrechts* (2017) and *Kokubo and Ida* (2002) respectively.]

2.4. Gas Accretion and Giant Planet Growth

Once planetary embryos become sufficiently massive they accrete gas directly from the disk (*Pollack et al.,* 1996; *Ida and Lin,* 2004; *Alibert et al.,* 2005). Gas accretion operates to some degree for Mars-mass planetary embryos, and there is evidence from noble gases in Earth's atmosphere that a portion of the atmospheres of Earth's constituent embryos was retained during Earth's prolonged accretion (*Dauphas,* 2003). Gas accretion onto a growing planet depends on the gaseous envelope's opacity (*Ikoma et al.,* 2000; *Hubickyj et al.,* 2005; *Machida et al.,* 2010) and temperature, which is determined in part by the accretion rate of energy-depositing solid bodies (*Rice and Armitage,* 2003; *Broeg and Benz,* 2012). The dynamics of how gas is accreted onto a growing planet's surface is affected by small-scale gas flows in the vicinity of the planet's orbit (*Fung et al.,* 2015; *Lambrechts and Lega,* 2017) as well as the structure of the circumplanetary disk [if there is one (e.g., *Ayliffe and Bate,* 2009; *Szulágyi et al.,* 2016)]. When the mass in a planet's envelope is comparable to its solid core mass it undergoes a phase of runaway gas accretion during which the planet's expanding Hill sphere — the zone in which the planet's gravity dominates the star's — puts it in contact with ever more gas. This gas can be accreted, cool, and contract, allowing the planet to grow into a true gas giant (*Mizuno,* 1980; *Ida and Lin,* 2004; *Thommes et al.,* 2008c). This culminates with the carving of an annular gap in the disk, which slows the accretion rate (*Bryden et al.,* 1999; *Crida et al.,* 2006; *Lubow and D'Angelo,* 2006). [For reviews of giant planet growth, see *Lissauer and Stevenson* (2007) and *Helled et al.* (2014).]

2.5. Orbital Migration

Migration is an inevitable consequence of gravitational interactions between a growing planet and its natal gas disk. Planets launch density waves in the disk, whose flow is determined by the disk's dynamics. These density perturbations impart torques on the planets' orbits (*Goldreich and Tremaine,* 1980; *Ward,* 1986). These torques damp planets' eccentricities and inclinations (*Papaloizou and Larwood,* 2000; *Tanaka and Ward,* 2004) and also drive radial migration. For planets low enough in mass not to carve a gap in the disk (i.e., for planetary embryos and giant planet cores) the mode of migration is sometimes called *type I*. Multiple torques are at play. The *differential Lindblad torque* is almost universally negative, driving planets inward at a rate that is proportional to the planet mass (*Ward,* 1997; *Tanaka et al.,* 2002). In contrast, the *co-rotation torque* can in some situations be positive and overwhelm the differential Lindblad torque, leading to outward migration (*Paardekooper and Mellema,* 2006; *Kley and Crida,* 2008; *Masset and Casoli,* 2010; *Paardekooper et al.,* 2011; *Benítez-Llambay et al.,* 2015). The disk structure plays a central role in determining the regions in which outward migration can take place (*Bitsch et al.,* 2013, 2015a). Planets that carve a gap enter the *type II* migration regime, which is generally slower than type I. In this regime a planet's migration is determined in large part by the radial gas flow within the disk and thus its viscosity (*Ward,* 1997; *Dürmann and Kley,* 2015). Type II migration is directed inward in all but a few situations (*Masset and Snellgrove,* 2001; *Veras and Armitage,* 2004; *Crida et al.,* 2009; *Pierens et al.,* 2014). [For reviews of migration, see *Kley and Nelson* (2012) and *Baruteau et al.* (2014).]

2.6. Late-Stage Accretion

The gaseous component of planet-forming disks is observed to dissipate after a few million years (*Haisch et al.,* 2001; *Hillenbrand,* 2008). When the gas disk is gone, gas accretion, pebble accretion, gas-driven migration, and dust drift cease, and the final phase of growth begins. Late-stage accretion consists of a protracted phase of collisional sweep-up of remaining planetary embryos and planetesimals (*Wetherill,* 1978, 1996; *Chambers,* 2001; *Raymond et al.,* 2004). Whereas late accretion took place on all the rocky planets, the most energetic collisions are thought to have taken place during the growth of Earth and Venus. For these planets, this phase was characterized by giant impacts between Mars-mass or larger planetary embryos (e.g., *Agnor et al.,* 1999; *Stewart and Leinhardt,* 2012; *Quintana et al.,* 2016). The very last giant impact on Earth (and perhaps in the whole inner solar system) is thought to have been the Moon-forming impact (*Benz et al.,* 1986; *Canup and Asphaug,* 2001; *Ćuk and Stewart,* 2012). Dynamical friction from the gravitational interaction with remnant planetesimals slows down the larger planetary embryos but excites the smaller planetesimals (statistically speaking). This leads to an energy equipartition between planetesimals and embryos and keeps the orbits of embryos relatively circular during the early parts of late-stage accretion (*O'Brien et al.,* 2006; *Raymond et al.,* 2006b). While dynamical friction dwindles in importance as the planetesimal population is eroded by impacts and dynamical clearing, it is nonetheless important in setting the planets' final orbits. Growth generally proceeds inside to out. The timescale for solar-system-like systems to complete the giant impact phase is 100 m.y. (*Chambers,* 2001; *Raymond et al.,* 2009b; *Jacobson et al.,* 2014). Analogous terrestrial planet systems around different types of stars would have different durations of this phase (*Raymond et al.,* 2007b; *Lissauer,* 2007). However, some systems (e.g., those in resonant chains) may never undergo a late-stage phase of giant impacts. [For reviews of late-stage accretion, see *Morbidelli et al.* (2012), *Raymond et al.* (2014), and *Izidoro and Raymond* (2018).]

3. EXOPLANET FORMATION MODELS

We now review formation models for extrasolar planets. We focus on two specific categories of exoplanets: super-Earth systems and giant exoplanets. For each category we summarize the observational constraints, then present the relevant models. It is interesting to note that, despite the very different regimes involved, leading models for each category rely heavily on two processes: migration and dynamical instability.

3.1. Systems of Close-In Low-Mass Planets ("Super-Earths")

The abundance of close-in low-mass/small planets is one of the biggest surprises to date in exoplanet science. Both

RV and transit surveys find an occurrence rate of 30–50% for systems of planets with $R < 4$ $R_\oplus$ or $M < 10$–20 $M_\oplus$ and $P < 50$–100 days [see *Winn and Fabrycky* (2015) for a compilation of measured rates and a comparison between them]. These planets are commonly referred to as "super-Earths," even though a significant fraction appear to be gas-rich and thus closer to "sub-Neptunes" (*Rogers*, 2015; *Wolfgang et al.*, 2016; *Chen and Kipping*, 2017).

In this chapter we use the term super-Earth to encompass all planets smaller than 4 $R_\oplus$. This is partly for simplicity but also because, to date, formation models have focused primarily on determining the origin of the mass of these planets with little regard to reproducing the exact size distribution (but see, e.g., *Carrera et al.*, 2018). Whenever sub-Neptunes are discussed in the chapter it is to make a distinction between them and truly rocky super-Earths.

Super-Earth formation models are constrained by several lines of observations:

• Their high occurrence rate (*Howard et al.*, 2010, 2012; *Mayor et al.*, 2011; *Fressin et al.*, 2013; *Dong and Zhu*, 2013; *Petigura et al.*, 2013), which makes them the most common type of currently known planet in the galaxy.

• The distribution of orbital period ratios of adjacent planets (*Lissauer et al.*, 2011; *Fabrycky et al.*, 2014; *Steffen and Hwang*, 2015). Only a small fraction (perhaps 5–10%) of pairs or neighboring planets appear to be in resonance.

• The multiplicity distribution, or how many planets are detected around each star (*Lissauer et al.*, 2011; *Batalha et al.*, 2013; *Rowe et al.*, 2014; *Fabrycky et al.*, 2014). There is a large peak in the transit detection of singleton super-Earth systems compared with multiple planet systems [the so-called Kepler dichotomy (see, e.g., *Johansen et al.*, 2012; *Fang and Margot*, 2012; *Ballard and Johnson*, 2016)]. It remains debated whether this "dichotomy" is a signature of planet formation or simply an observational bias (e.g., *Izidoro et al.*, 2017, 2019; *Zhu et al.*, 2018).

• The distribution of planet sizes/mass and the size/mass ratios of adjacent planets. Recent analysis of the Kepler super-Earths and sub-Neptunes has found that adjacent planets tend to be similarly sized (*Millholland et al.*, 2017; *Weiss et al.*, 2018), consistent with the classical model but at odds with our own terrestrial planets.

• The distribution of physical densities of planets. Densities can be measured for transiting planets with good mass constraints from RV monitoring (e.g., *Fischer et al.*, 2014; *Marcy et al.*, 2014) or transit-timing variation analysis (for a review, see *Agol and Fabrycky*, 2017). Several analyses based on models of planetary interiors have concluded that planets smaller than roughly 1.2–2 $R_\oplus$ are predominantly rocky, whereas larger planets generally have thick gaseous envelopes (*Rogers*, 2015; *Wolfgang et al.*, 2016; *Chen and Kipping*, 2017). This marks the effective boundary between super-Earths and sub-Neptunes.

• The stellar mass-dependence of the occurrence of super-Earths and sub-Neptunes. Low-mass stars appear to have a similar overall abundance of close-in small planets but, compared with Sun-like stars, they have more super-Earths, fewer sub-Neptunes, and a higher average total mass in planets (*Dressing and Charbonneau*, 2015; *Mulders et al.*, 2015b,c).

All these constraints are naturally subject to observational bias, making a comparison with models challenging. For example, the period ratio distribution could be skewed if the middle planet in a three-planet system is not detected, leading to a detection of periods P_3 and P_1 but not P_2 and hence an inflated period ratio P_3/P_1. Alternately, the inferred distribution could be skewed by preferentially missing the transit of outer planets in pairs with large period ratios, i.e., by detecting P_2/P_1 but not P_3/P_2 if P_3/P_2 is much larger than P_2/P_1 (see discussion in *Izidoro et al.*, 2019).

A number of formation models for super-Earths existed prior to their discovery. *Raymond et al.* (2008b) compiled six different formation mechanisms for super-Earths (although in that paper they were referred to as "hot Earths"). Many of those mechanisms relied on the influence of gas giants in the system (e.g., *Fogg and Nelson*, 2005; *Zhou et al.*, 2005; *Raymond et al.*, 2006b; *Mandell et al.*, 2007) and are thus unable to match the general population of super-Earths (although those mechanisms could apply in select cases).

The simplest model of super-Earth formation — *in situ* accretion — was proposed by *Raymond et al.* (2008b), who subsequently discarded it as unrealistic (see also *Ogihara et al.*, 2015a). Given its recent revival (*Chiang and Laughlin*, 2013; *Hansen and Murray*, 2012, 2013) and surprising popularity in the exoplanet community, we think it worth explaining exactly why it cannot be considered a viable model. The simplest argument against *in situ* accretion is simply that it is not self-consistent. If super-Earths accreted *in situ* close to their stars, then their natal planet-forming disks must have been quite dense (*Raymond et al.*, 2008b; *Chiang and Laughlin*, 2013; *Schlichting*, 2014; *Schlaufman*, 2014). The timescales for accretion of these planets are very short because their disks are massive and the orbital time is short (*Safronov*, 1969). Simulations demonstrate that planets similar to the observed super-Earths indeed accrete on thousand- to hundred-thousand-year timescales (*Raymond et al.*, 2008b; *Hansen and Murray*, 2012; *Bolmont et al.*, 2014), long before the dispersal of gaseous disks. Thus super-Earths must have been massive enough to gravitationally interact with the gaseous disk, which itself must have been extremely dense to accommodate the planets' *in situ* growth. The planets must therefore have migrated (*Ogihara et al.*, 2015a). In fact, the disks required for *in situ* accretion are so dense that even aerodynamic drag alone would have caused their orbits to shrink (*Inamdar and Schlichting*, 2015).

The *in situ* model is thus caught in a logical impossibility. If the planets formed *in situ* then they must have migrated. But if they migrated, they did not form *in situ*. In other words, super-Earths simply cannot have formed *in situ*. Solids must have drifted relative to the gas, either at small scales (pebble drift) or large scales (migration). Meanwhile, there is abundant circumstantial evidence for inward planet migration. One extreme case is the existence of planets interior to the silicate sublimation radius (*Swift et al.*, 2013). Another piece of evidence is the existence of systems of super-Earths in resonant chains in systems such as Kepler-223 (*Mills et al.*, 2016) and TRAPPIST-1 (*Gillon et al.*, 2017; *Luger et al.*, 2017), given that it is extremely improbable for planets to end up in low-eccentricity resonant configurations without invoking migration.

Two models remain viable candidates to explain the origin of most super-Earths: the *drift* and *migration* models.

The drift model (*Boley and Ford*, 2013; *Chatterjee and Tan*, 2014, 2015; *Hu et al.*, 2016, 2017) proposes that inward-drifting pebbles are trapped in the inner parts of the disk, perhaps at a pressure bump associated with a region toward the inner edge of the disk where its properties (e.g., its viscosity) change abruptly. Pebbles accumulate at the pressure bump until a threshold is reached for them to form planetesimals (e.g., *Yang et al.*, 2017) and accrete into full-sized planets. *Chatterjee and Tan* (2014) proposed that the pressure bump itself would respond to the first planet's presence and retreat to an external orbit, providing a new nexus of super-Earth formation. While the model's predictions appear broadly consistent with observations, it has not been developed to the point of matching the observables laid out above. Below we argue that the late evolution of super-Earths in this model must also include migration.

A number of studies have placed themselves at the interface between the *in situ* accretion and drift models (e.g., *Hansen and Murray*, 2012, 2013; *Dawson et al.*, 2015, 2016; *Lee and Chiang*, 2016, 2017; *Moriarty and Ballard*, 2016). These studies assumed that the bulk of super-Earths' accretion happens close-in but within disks that have far less gas than would be present assuming that the local density of gas reflects the local density of solids (e.g., with a roughly 100:1 ratio, assuming solar metallicity). These studies thus inherently suppose that a previous process of solid enrichment took place within the inner disk, presumably by a mechanism such as dust/pebble drift. They further assume that planetesimals were uniformly distributed across the inner disk, usually as a simple, power-law profile. Such initial conditions — young disks with a broad close-in planetesimal distribution but little gas — are hard to reconcile with current thinking. If planetesimals formed quickly then the gas density should still be high and migration should be very fast. If planetesimals formed more slowly, presumably from pebbles that drifted inward to supplement the inner disk's solid reservoir, then they are unlikely to have a smooth radial distribution (e.g., *Chatterjee and Tan*, 2014; *Drążkowska et al.*, 2016). Thus, while these studies have provided interesting insights into various aspects of the accretion process, it seems unlikely that their starting conditions reflect reality. The mechanism by which inner disks are enriched in solids seems to be central to understanding the origins of super-Earths.

The migration model (*Terquem and Papaloizou*, 2007; *Ogihara and Ida*, 2009; *McNeil and Nelson*, 2010; *Ida and Lin*, 2010; *Cossou et al.*, 2014; *Izidoro et al.*, 2017, 2019) proposes that large planetary embryos form throughout the disk and migrate inward (see Fig. 3). Inward migration is counteracted by the positive surface density gradient at the disk's inner edge (*Masset et al.*, 2006). Embryos thus migrate inward and pile up into long resonant chains anchored at the disk's inner edge. Collisions are common during this phase, leading to a breaking of resonance followed by continued migration and rapid reformation of the resonant chain in a new configuration. After the gas disk dissipates many resonant chains become unstable, leading to a phase of giant impacts between growing planets that is not unlike the final phase of *in situ* accretion. If 90–95% of resonant chains become unstable, the resulting systems provide a quantitative match to the observed period ratio and multiplicity distributions (*Izidoro et al.*, 2017, 2019). In this model, the Kepler dichotomy is an observational artifact: The broad distribution of mutual inclinations in multiple super-Earth systems naturally produces a peak in systems with a single transiting planet. The resonant chains that remain stable are associated with observed multi-resonant systems such as TRAPPIST-1 (*Gillon et al.*, 2017; *Luger et al.*, 2017) and Kepler-223 (*Mills et al.*, 2016). [Note that the disk inner edge provides a built-in stopping mechanism for inward migration that plays a central role in the migration model. One may then wonder whether the *in situ* formation model could also have a mechanism for avoiding migration. Indeed, *in situ* growth within a region of slow or stopped migration — such as near the disk's inner edge — would naturally reduce the importance of migration. However, there are two important caveats. First, if planets grew *in situ* in slow-migration regions, this would not remove the importance of planet-disk interactions, which also affect the growing planets' eccentricities and inclinations (e.g., *Papaloizou and Larwood*, 2000; *Tanaka and Ward*, 2004). Second, regions of reduced migration are thought to be narrow (e.g., *Hasegawa and Pudritz*, 2011; *Bitsch et al.*, 2015a; *Baillié et al.*, 2015) such that even if some super-Earths did indeed grow more or less *in situ*, the bulk of growing super-Earths would still have experienced migration.]

Ormel et al. (2017) proposed a hybrid scenario in which planetesimals form first at the snow line, undergo pebble accretion, and then migrate inward. This idea is consistent with the migration model and also connects with dust growth and drift models, which find that planetesimals tend to form fastest just past the snow line (*Armitage et al.*, 2016; *Drążkowska and Alibert*, 2017; *Schoonenberg and Ormel*, 2017).

The late stages in super-Earth evolution should be the same for both the drift and migration models. Of course, the two models invoke different formation modes and feeding zones for the planets. However, once there is a population of planets massive enough to migrate, the subsequent evolution is independent of how the planets formed. Whatever the processes responsible for creating a population of such planets, they migrate. And the outcome of this migration is a well-studied problem. As long as there is a disk inner edge, a system of migrating planets invariably organizes itself into a chain of mean-motion resonances (e.g., *Snellgrove et al.*, 2001; *Lee and Peale*, 2002; *Papaloizou and*

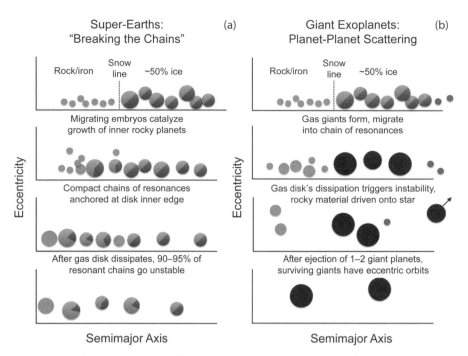

Fig. 3. How orbital migration and dynamical instabilities can explain the properties of exoplanet populations. **(a)** Evolution of the "breaking the chains" migration model for the origin of super-Earths (*Izidoro et al.,* 2017, 2019). Embryos within the snow line are entirely rocky and much smaller than those that form past the snow line, which also incorporate ice. Presumably ice-rich embryos migrate inward through the rocky material, catalyzing the growth of purely rocky planets interior to the ice-rich ones (*Raymond et al.,* 2018b). Planets migrate into long chains of mean-motion resonances, with the innermost planet at the inner edge of the disk. The vast majority (90–95%) of resonant chains become unstable when the gas disk dissipates. The resulting planets match the distributions of known super-Earths (*Izidoro et al.,* 2017, 2019). Given various loss processes (e.g., *Grimm and McSween,* 1993; *Genda and Abe,* 2005; *Marcus et al.,* 2010; *Monteux et al.,* 2018), the water/ice contents of these planets may be drastically overestimated. **(b)** Evolution of the planet-planet scattering model for the origin of giant exoplanets (e.g., *Adams and Laughlin,* 2003; *Chatterjee et al.,* 2008; *Raymond et al.,* 2010). Several embryos grow quickly enough to accrete gas and grow into gas giants. They subsequently migrate into a resonant chain without drastically affecting the orbits of nearby growing rocky planets (or outer planetesimal disks). After the disk dissipates, the vast majority (75–90%) of giant planet systems become unstable. The resulting systems match the correlated mass-eccentricity distribution of known giant exoplanets (e.g., *Ford and Rasio,* 2008; *Wright et al.,* 2009).

Terquem, 2006; *Cresswell et al.,* 2007). Magnetohydrodynamic simulations of disk accretion onto young stars show that for most plausible parameters, disks should indeed have inner edges (*Romanova et al.,* 2003; *Bouvier et al.,* 2007). Although the strength of the (positive) co-rotation torque depends on the local disk properties (*Masset and Casoli,* 2010; *Paardekooper et al.,* 2011), migration is likely to be directed inward during the late phases of disk evolution (*Bitsch et al.,* 2014, 2015a). Another factor is that a system of many super-Earths in resonance acts to excite their mutual eccentricities, decreasing the strength of the co-rotation torque (*Bitsch and Kley,* 2010; *Fendyke and Nelson,* 2014) and potentially leading to inward migration of the cohort (*Cossou et al.,* 2013).

We therefore argue that super-Earths must migrate regardless of how they formed (see Fig. 3). Late in the disk lifetime all formation scenarios converge on the evolution envisioned in the "breaking the chains" model of *Izidoro et al.* (2017, 2019). Migration should produce resonant chains with the innermost planets anchored at the inner edge of the disk. When the gas disk dissipates along with its stabiliz-

ing influence, the vast majority of resonant chains become unstable. While many details remain to be resolved, this evolution matches the key observed super-Earth constraints.

Raymond et al. (2008b) proposed that super-Earth formation models could be differentiated with two observables: the planets' compositions (via their densities) and the systems' orbital architectures. We have just argued that the late phases of the two viable models should converge to the same dynamical pathway, i.e., the breaking the chains model. We therefore do not expect the orbital architectures of super-Earth systems to provide a means of differentiation between models.

The compositions of super-Earths should in principle be different for the drift and migration models. In the drift model, all planet building takes place close in. Super-Earths should therefore be purely rocky, because temperatures so close in are so hot that the local building blocks should not contain any volatiles such as water. In contrast, a simple view of the migration model would predict ice-rich planets. Indeed, some models of dust growth and drift first produce planetesimals at or beyond the snow line, the radial distance beyond which water vapor can condense as ice (*Armitage et al.*, 2016; *Drążkowska and Alibert*, 2017). Immediately past the snow line, embryos are thought to grow faster and larger than closer in because pebbles may be somewhat larger and therefore easier to accrete, and may also be efficiently concentrated (*Ros and Johansen*, 2013; *Morbidelli et al.*, 2015a). This would suggest that ice-rich embryos from past the snow line are likely to be the first to migrate, and thus super-Earths should themselves be ice-rich in the migration model.

There are three problems with the idea of the migration model producing exclusively ice-rich super-Earths (see *Raymond et al.*, 2018b). First, in some cases rocky embryos may grow large enough to migrate (see *Lambrechts et al.*, 2014; *Izidoro et al.*, 2019). Indeed, in some models planetesimals form first in the terrestrial planet-forming region (*Drążkowska et al.*, 2016; *Surville et al.*, 2016), which would give rocky embryos a head-start in their growth. Second, migrating embryos must pass through the building blocks of terrestrial planets on their way to becoming super-Earths. Their migration can act to pile up rocky material in inner resonances with the migrating embryos (*Izidoro et al.*, 2014b) and catalyze the rapid formation of rocky super-Earths, which preferentially end up interior to the migrating ice-rich embryos. Indeed, simulations of the migration model show that the innermost super-Earths are often built entirely from inner planetary system material and should be purely rocky (*Raymond et al.*, 2018b). Third, it is not clear that embryos that migrate inward from beyond the snow line must be ice-rich. It is the fastest-forming embryos that are most likely to migrate, for simple timescale reasons. Yet rapid growth implies massive volatile loss. Thermal evolution models find that any planetesimals that form within 1 m.y. are completely dehydrated by strong ^{26}Al heating (*Grimm and McSween*, 1993; *Monteux et al.*, 2018). This short-lived radionuclide (half-life of ~700,000 yr) is thought to have been injected into the Sun's planet-forming disk from a nearby massive star (e.g., *Hester et al.*, 2004; *Gounelle and Meibom*, 2008; *Gaidos et al.*, 2009; *Ouellette et al.*, 2010) and to have played a central role in the thermal evolution of the fastest-forming planetesimals. In addition, giant collisions between ice-rich bodies preferentially strip outer icy mantles and leave behind rocky/iron cores (*Marcus et al.*, 2010). Likewise, later giant impacts can lead to substantial water loss for ocean planets (*Genda and Abe*, 2005).

It is the very closest-in planets that are easiest to characterize. Within the so-called "photo-evaporation valley," the atmospheres of any super-Earth-sized planets are thought to be rapidly evaporated by UV irradiation from the central star (*Lammer et al.*, 2003; *Baraffe et al.*, 2004; *Hubbard et al.*, 2007). Planets in this region should necessarily have lost their gaseous envelopes and thus be "naked" (*Lopez and Fortney*, 2013; *Owen and Wu*, 2013; *Sanchis-Ojeda et al.*, 2014), while planets just past the valley may have larger radii due to atmospheric heating (*Carrera et al.*, 2018). From their observed sizes and masses, these very close-in planets appear to be rocky in nature, not icy (*Owen and Wu*,

2017; *Lopez*, 2017; *Jin and Mordasini*, 2018). Some studies have suggested that the observed dip in planet occurrence between ~1.5 and 2 $R_{\oplus}$ (*Fulton et al.*, 2017; *Teske et al.*, 2018; *Fulton and Petigura*, 2018) may provide evidence that super-Earths are rocky (*Jin and Mordasini*, 2018; *Van Eylen et al.*, 2018). *Gupta and Schlichting* (2019) suggest that in this context, "rocky" planets should have less than 20% water by mass (which is comparable to Europa's water content by mass). Given that both the migration and drift models are consistent with rocky super-Earths (*Raymond et al.*, 2018b), observations cannot yet differentiate between the drift and migration models. Interpreting the mean densities of planets beyond the distance for atmospheric photoevaporation is challenging because of the degeneracies that arise once gas is included as a third potential constituent [along with rock/iron and water (see *Selsis et al.*, 2007; *Adams et al.*, 2008)]. The main difference between the migration and drift models is that the migration model predicts that some super-Earths should be ice-rich, in particular those planets that formed relatively late or in disks with little ^{26}Al.

Embryos embedded in the disk should also accrete gas (e.g., *Ikoma et al.*, 2000; *Rogers et al.*, 2011). Indeed, many sub-Neptunes are observed to have very low densities (e.g., *Marcy et al.*, 2014) and this has been interpreted as planets larger than ~1.5 $R_{\oplus}$ having gaseous envelopes that are typically 0.1–10% of their total mass (*Lopez and Fortney*, 2014; *Weiss and Marcy*, 2014; *Rogers*, 2015; *Wolfgang et al.*, 2016; *Chen and Kipping*, 2017).

Recent work has shown that gas accretion is far more complex than previously assumed. For few Earth-mass planets, currents of gas often pass within a small fraction of the planet's Hill sphere before exiting (*Fung et al.*, 2015; *Lambrechts and Lega*, 2017), casting doubts on the simple picture that gas within a planet's Hill sphere must simply cool sufficiently to approach the planet's surface. Nonetheless, growing super-Earths must accrete gas during the disk phase (*Lee et al.*, 2014; *Inamdar and Schlichting*, 2015) but should rarely reach the 50% gas-by-mass threshold for runaway gas accretion (*Pollack et al.*, 1996) because the occurrence rate of hot Jupiters is more than an order of magnitude smaller than that of super-Earths (*Howard et al.*, 2010; *Mayor et al.*, 2011). Super-Earths may be in a constant state of gas accretion (during the gas disk phase) (*Ikoma et al.*, 2000; *Lambrechts and Lega*, 2017) moderated by loss processes related to collisions (*Schlichting et al.*, 2015; *Inamdar and Schlichting*, 2016), as well as the dissipation of the disk itself (*Ikoma and Hori*, 2012; *Ginzburg et al.*, 2016). Models invoking atmospheric loss from a few large impacts — such as those characteristic of late instabilities in the breaking the chains model — can broadly match the observed distribution of gaseous envelope masses (*Inamdar and Schlichting*, 2016), although the initial, pre-impact atmospheric masses remain uncertain.

It is difficult to understand why close-in super-Earths exist in so many systems but not in all systems. Models of super-Earth formation struggle not to form them, whereas microlensing observations show that similar-mass planets are extremely abundant past the snow line (*Beaulieu et al.*, 2006; *Gould et al.*, 2010; *Suzuki et al.*, 2016a). A simple explanation is to invoke a timing constraint: If large embryos form too slowly then they would not have time to migrate all the way to the inner edge of the disk. However, most studies find that migration is fast [although new simulations by *McNally et al.* (2018) find that migration may be slower] so this requires a fine-tuned delay such that most super-Earths grow large just before the gas disk dissipates. And late accretion of smaller embryos beyond a few astronomical units after the dissipation of the disk would be quite inefficient (e.g., *Levison and Stewart*, 2001; *Thommes et al.*, 2003).

Lower disk masses also cannot explain why many stars do not host close-in super-Earths. There is an observed super-linear correlation between the disk mass and stellar mass (*Pascucci et al.*, 2016), albeit with large scatter at a given stellar mass (e.g., *Scholz et al.*, 2006; *Williams and Cieza*, 2011). This means that M-dwarf stars — with masses between roughly 8% and 60% of the Sun's — have on average significantly lower-mass disks than Sun-like stars. However, the occurrence rate of super-Earths around M dwarfs is at least as high as around Sun-like stars (*Howard et al.*, 2012; *Fressin et al.*, 2013; *Dressing and Charbonneau*, 2015), and systems of close-in low-mass planets contain on average a higher total mass around M dwarfs (*Mulders et al.*, 2015b,c). If lower-mass disks could not produce super-Earths, M dwarfs should naively have lower average occurrence rates.

Wide-orbit giant planets may potentially explain why some systems do not have close-in super-Earths. Once a planet accretes enough gas it carves an annular gap in the disk, slows its migration, and becomes a true gas giant (*Lin and Papaloizou*, 1986; *Bryden et al.*, 1999; *Crida et al.*, 2006). The inward migration of more distant large embryos are then blocked by the gas giant (*Izidoro et al.*, 2015b). In this context, Uranus and Neptune (and perhaps Saturn's core) may represent failed super-Earths, embryos whose migration was blocked by the young Jupiter (*Izidoro et al.*, 2015a). This predicts an anti-correlation between systems with many close-in super-Earths and those with wide-orbit gas giants (*Izidoro et al.*, 2015b). However, the occurrence rate of wide-orbit gas giants is 10% for Sun-like stars (*Cumming et al.*, 2008; *Mayor et al.*, 2011; *Wittenmyer et al.*, 2016; *Rowan et al.*, 2016) and is likely far lower for M dwarfs (*Johnson et al.*, 2007; *Lovis and Mayor*, 2007; *Dressing and Charbonneau*, 2015). For F-, G-, or K-type stars (FGK stars), this is a factor of 35 lower than the occurrence rate of super-Earths, and the problem is even worse for M dwarfs. Thus, the Jupiter migration barrier does not appear capable of explaining why most systems do not have close-in super-Earths.

Wide-orbit planets with masses comparable to the those of the ice giants (10–20 $M_\oplus$) may help solve this problem by stunting the growth of planetary embryos. The largest planetesimals are thought to represent the seeds of planetary embryos and to grow by accreting pebbles that drift inward through the disk (e.g., *Johansen and Lambrechts*, 2017) (see

section 2). Once an embryo reaches the pebble isolation mass it creates a pressure bump in the gas disk exterior to its orbit that acts to trap drifting pebbles and shut off the pebble flux (*Morbidelli and Nesvorny*, 2012; *Lambrechts et al.*, 2014; *Bitsch et al.*, 2018). This not only starves the embryo but all other embryos interior to its orbit, which may continue to accrete planetesimals (but not pebbles). For typical disk parameters the pebble-isolation mass is on the order of 20 $M_\oplus$ (*Lambrechts et al.*, 2014; *Bitsch et al.*, 2018). A fast growing wide-orbit planet with a mass similar to Neptune's (17 $M_\oplus$) may starve the inner disk and prevent closer-in embryos from reaching large enough masses to undergo rapid migration and follow the breaking the chains pattern discussed above. This mechanism is especially promising given that the abundance of wide-orbit ice-giant-mass planets inferred from microlensing is on the same order as the occurrence rate of close-in super-Earths (e.g., *Gould et al.*, 2010; *Petigura et al.*, 2013; *Clanton and Gaudi*, 2016; *Winn and Fabrycky*, 2015). However, given that pebble-blocking outer Neptunes must form quickly (to starve inner embryos), it is not clear how such planets could avoid migrating inward and becoming super-Earths themselves.

It is of course possible that migration may not always follow the pattern that we have laid out. In one type of disk model that invokes winds as an angular transport mechanism, the surface density in the inner 12 AU of the disk increases steeply with radius (*Suzuki et al.*, 2016b). In such a disk, type I migration is significantly slowed and may even be quenched (*Ogihara et al.*, 2015b), such that co-migrating super-Earths may not always form resonant chains (*Ogihara et al.*, 2018). Magnetic stresses in the mid-plane of certain disk models may generate positive torques that drive outward planet migration but only in specific situations, for example, if the star's spin vector is aligned with the magnetic field (*McNally et al.*, 2017, 2018). Such effects could in principle prevent inward migration in a subset of disks.

To conclude this subsection, we reiterate that the breaking the chains scenario, in which planetary embryos migrate into resonant chains that later become unstable (*Izidoro et al.*, 2017, 2019), provides a match to the period ratio and multiplicity distributions of observed super-Earth systems. The late evolution of that model should hold whether the bulk of the mass of super-Earths comes from pebbles that drifted inward (the drift model) (*Chatterjee and Tan*, 2014) or from cores that formed past the snow line (the migration model) (*Terquem and Papaloizou*, 2007). What remains unexplained is why so many — but not all — stars have close-in super-Earths.

3.2. Systems With Giant Exoplanets

Giant exoplanets are found around 10% of Sun-like stars after observational biases are accounted for [see discussion in section 1.2 (*Cumming et al.*, 2008; *Mayor et al.*, 2011; *Foreman-Mackey et al.*, 2016)]. Most giant exoplanets have orbital radii larger than 0.5–1 AU (*Butler et al.*, 2006; *Udry and Santos*, 2007) and only ~1% of stars have hot Jupiters

(*Howard et al.*, 2010; *Wright et al.*, 2012). Giant planet masses follow a roughly $dN/dM \propto M^{-1.1}$ distribution of minimum masses (*Butler et al.*, 2006). The median eccentricity of this population is ~0.25, roughly 5× larger than Saturn and Jupiter's long-term average eccentricities (*Quinn et al.*, 1991). More massive giant planets have statistically higher eccentricities than lower-mass giant planets (typically, the division between high-mass and low-mass giant planets is at roughly Jupiter's mass) (*Jones et al.*, 2006; *Ribas and Miralda-Escudé*, 2007; *Ford and Rasio*, 2008; *Wright et al.*, 2009).

As is the case for super-Earths, the population of giant exoplanets is thought to have been sculpted in large part by migration and instability. A subset of giant exoplanets has been found to be in mean-motion resonance. Strong resonances are thought to be a clear signature of migration (e.g., *Papaloizou and Terquem*, 2006; *Kley and Nelson*, 2012), whereas weaker resonances can be produced by instabilities (*Raymond et al.*, 2008a). In this context, the strength of a resonance can be measured by the amplitude of libration of specific resonant angles related to the relative orbital positions of the planets (for details, see *Murray and Dermott*, 1999). Notable examples of giant exoplanets in resonance include the GJ 876 system, in which three planets are locked in 4:2:1 Laplace resonance (*Rivera et al.*, 2010), and the HR 8799 system (*Marois et al.*, 2008, 2010), for which stability considerations indicate that the outer three or perhaps all four super Jupiter-mass planets may be in resonance (*Reidemeister et al.*, 2009; *Fabrycky and Murray-Clay*, 2010; *Goździewski and Migaszewski*, 2014; *Götberg et al.*, 2016). These systems are thought to represent signature cases of migration, and a recent analysis found that a significant fraction may be in resonance [25% of the 60 giant exoplanet systems studied in *Boisvert et al.* (2018)].

Giant planets are thought to form on circular orbits. However, a large fraction of giant exoplanets are found to have significant orbital eccentricities, which are thought to be an indicator of dynamical instability (see, e.g., *Ford and Rasio*, 2008). The eccentricity distribution of giant exoplanets can be reproduced by the planet-planet scattering model, which proposes that the observed planets are the survivors of system-wide instabilities (*Rasio and Ford*, 1996; *Weidenschilling and Marzari*, 1996; *Lin and Ida*, 1997). Giant planets are assumed to form in systems of two or more planets. After the gaseous disk dissipates the planets' orbits become dynamically unstable, leading to a phase of close gravitational scattering events. Scattering events involve orbital energy and angular momentum exchange between the planets and tend to increase their eccentricities and inclinations. During this phase of dramatic orbital excitation, nearby small bodies such as the building blocks of terrestrial planets are generally driven to such high eccentricities that they collide with the host star (*Veras and Armitage*, 2005, 2006; *Raymond et al.*, 2011, 2012; *Matsumura et al.*, 2013). More distant planetesimals are preferentially ejected (*Raymond et al.*, 2011, 2012, 2018a; *Marzari*, 2014). Giant planet

instabilities typically conclude with the ejection of one or more planets into interstellar space (*Veras and Raymond*, 2012). The surviving planets have eccentric orbits (*Ford et al.*, 2003; *Adams and Laughlin*, 2003; *Chatterjee et al.*, 2008; *Ford and Rasio*, 2008; *Raymond et al.*, 2010). The observed eccentricity distribution can be matched if at least 75% — and probably closer to 90% — of all giant exoplanet systems represent the survivors of instabilities (*Jurić and Tremaine*, 2008; *Raymond et al.*, 2010). Scattering can also match the observed orbital spacing of giant planet systems, with regard to their proximity to the analytically derived boundary for orbital stability (also called "Hill stability") (*Raymond et al.*, 2009a) and the coupled evolution of orbital eccentricities in multiple-planet systems (*Ford et al.*, 2005; *Timpe et al.*, 2013).

Gravitational scattering produces an energy equipartition among planets in the same system such that the lowest-mass planets have the highest eccentricities. This is in disagreement with observations, which show that higher-mass planets have higher eccentricities than lower-mass, with a statistically significant difference (*Ribas and Miralda-Escudé*, 2007; *Ford and Rasio*, 2008; *Wright et al.*, 2009). This discrepancy is resolved if systems with very massive giant planets ($M \gtrsim M_{Jup}$) systematically form multiple, very massive giant planets with near-equal masses (*Raymond et al.*, 2010; *Ida et al.*, 2013). The eccentricities of surviving planets are highest in systems with the most-massive, equal-mass planets (*Ford et al.*, 2001; *Raymond et al.*, 2010).

All that is needed to trigger instability is the formation and migration of 2–3 gas giants. [Wide binary stars can also trigger instabilities, as torques from passing stars and the galactic tidal field occasionally shrink their pericenters to approach the planetary region (*Kaib et al.*, 2013).] The timescale of instability is a function of the planets' initial separations (*Chambers et al.*, 1996; *Marzari and Weidenschilling*, 2002), so most studies simply started planets in unstable configurations to determine the outcome of the instability. A more self-consistent approach invokes a prior phase of orbital migration. While migration is often thought of as a dynamically calm process, several studies have shown that migration of multiple giant planets often generates instabilities after, or even during, the gaseous disk phase (*Moeckel et al.*, 2008; *Matsumura et al.*, 2010; *Marzari et al.*, 2010; *Moeckel and Armitage*, 2012; *Lega et al.*, 2013). The planets that emerge from migration-triggered instability match the eccentricity distribution of observed giant exoplanets (*Adams and Laughlin*, 2003; *Moorhead and Adams*, 2005).

Hot Jupiters present an interesting melding of migration and instability (see *Dawson and Johnson*, 2018, for a review). In recent years it has been debated whether hot Jupiters migrated in to their current locations (*Lin et al.*, 1996; *Armitage*, 2007) or were scattered to such high eccentricities (and such small pericenter distances) that tidal dissipation within the star (also called "tidal friction") shrank their orbits (*Nagasawa et al.*, 2008; *Beaugé and Nesvorný*,

2012). The Kozai effect — in which a perturbing planet or star on a highly-inclined orbit induces large-scale, anti-correlated oscillations in a planet's eccentricity and inclination — may play a role in generating the high eccentricities needed to produce hot Jupiters by tidal friction (*Fabrycky and Tremaine*, 2007; *Naoz et al.*, 2011), but only in situations in which such large mutual inclinations arise (e.g., after an instability). Of course, it is possible that hot Jupiters underwent rapid gas accretion close in and thus represent the rare hot super-Earths that grew fast enough to trigger runaway gas accretion (*Bodenheimer et al.*, 2000; *Boley et al.*, 2016; *Batygin et al.*, 2016). Yet even if hot Jupiters accreted their gas close in, the arguments presented in section 3.1 still indicate that their building blocks must have either migrated or drifted inward.

Another constraint comes from observations of the Rossiter-McLaughlin effect, which measures the projected stellar obliquity in systems with transiting planets (*Winn et al.*, 2005; *Gaudi and Winn*, 2007). This translates to a projection of the planet's orbital inclination with respect to the stellar equator. For a significant fraction of measured hot Jupiters, the orbital plane is measured to be strongly misaligned with the host stars' equators. More massive stars, for which the timescale for tidal dissipation are much longer (*Zahn*, 1977), are far more likely to host misaligned hot Jupiters (*Winn et al.*, 2010; *Triaud et al.*, 2010; *Albrecht et al.*, 2012). This may be explained by planets being scattered onto very eccentric and inclined orbits before their orbits are shrunk by tidal friction (*Fabrycky and Tremaine*, 2007; *Nagasawa et al.*, 2008; *Naoz et al.*, 2011; *Beaugé and Nesvorný*, 2012; *Lai*, 2012).

Migration predicts that hot Jupiters should remain aligned with their birth disks. However, disks themselves can be torqued into configurations that are misaligned with respect to the stellar equator (e.g., *Lai et al.*, 2011; *Batygin*, 2012). If planet-forming disks are themselves misaligned, then both the migration and close-in growth models can in principle explain hot Jupiters' misaligned orbits. In the Kepler-56 system two massive planets share an orbital plane that is misaligned with the stellar equator (*Huber et al.*, 2013). While this is suggestive of the planets having formed in a tilted disk, other dynamical mechanisms can plausibly explain such tilting (*Innanen et al.*, 1997; *Mardling*, 2010; *Kaib et al.*, 2011; *Boué and Fabrycky*, 2014; *Gratia and Fabrycky*, 2017). In addition, there is as yet no sign of debris disks — dust disks observed around older stars whose gas disks have already dissipated — that are misaligned with the equators of their host stars (*Greaves et al.*, 2014).

While the general picture of migration and instability appears to match the broad characteristics of giant exoplanets, questions remain. While slower than for low-mass planets, the timescale for type II migration — which is thought to be controlled in large part by the disk's viscosity (*Lin and Papaloizou*, 1986; *Ward*, 1997; *Dürmann and Kley*, 2015) — is still in many cases faster than the disk lifetime. Why, then, are there so few gas giants interior to 0.5–1 AU?

Photoevaporation of the disk produces inner cavities of roughly that size (*Alexander et al.*, 2014; *Ercolano and Pascucci*, 2017). The cavity is only generated late in the disk's lifetime, so if it is to explain the deficit of gas giants within 0.5–1 AU, this would require very slow migration and thus very low-viscosity disks [*Alexander and Pascucci* (2012); but see *Wise and Dodson-Robinson* (2018) and discussion in *Morbidelli and Raymond* (2016)].

To conclude, the general evolution of gas giant systems thus appears to follow a similar pattern as super-Earths' breaking the chains evolution. Gas giants form in cohorts and migrate into resonant configurations (e.g., *Kley and Nelson*, 2012). After the disk dissipates (or sometimes before), the vast majority of systems become unstable and undergo a violent phase of planet-planet scattering that generates the observed gas giant eccentricities (e.g., *Jurić and Tremaine*, 2008; *Raymond et al.*, 2010) and also disrupts the growth of any smaller planets in the systems (e.g., *Raymond et al.*, 2011). The origins of hot Jupiters remain debated, but viable formation models invoke a combination of migration and instability (see *Dawson and Johnson*, 2018).

4. MODELS FOR SOLAR SYSTEM FORMATION

We now turn our attention to the origin of the solar system. Given the astrobiological context of this chapter, we emphasize the origin of the inner solar system. However, it is important to keep in mind that dynamical perturbations from Jupiter during its growth, migration, and early evolution played an important part in shaping the terrestrial planets.

In this section we first lay out the observational constraints (section 4.1). Next, in section 4.2 we present a rough timeline of events, including a discussion of the late instability in the giant planets' orbits (the so-called "Nice model"). In section 4.3 we describe the so-called classical model of terrestrial planet formation and explain its shortcomings. In section 4.4 we present and contrast three models for the early evolution of the inner solar system. In section 4.5 we explore a feature of the solar system that remains hard to explain with all current models: the mass deficit in the very close-in solar system.

4.1. Solar System Constraints

A successful formation model must match the solar system's broad characteristics (see section 7.3 for a philosophical discussion). We now lay out the central constraints to be matched, in rough order of importance. We start with constraints related to the inner solar system (roughly from most to least stringent) and conclude with constraints related to the outer solar system (Jupiter and beyond).

4.1.1. Masses and orbits of the terrestrial planets. The terrestrial planets follow an odd pattern, with two large central planets (Venus and Earth) flanked by much smaller ones (Mercury and Mars). Roughly 90% of all the rocky material in the solar system is thus concentrated in a ring that is only 0.3 AU in width (encompassing the orbits of Venus and Earth). In addition, the terrestrial planets' orbits are remarkably close to circular. These constraints are often quantified using statistics on the radial mass concentration (*Chambers*, 2001) and degree of orbital excitation [often using the angular momentum deficit, defined as the fractional deficit in angular momentum of a system of planets relative to an identical system on perfectly circular, co-planar orbits (*Laskar*, 1997; *Chambers*, 2001)]. In addition, the Earth/Mars and Venus/Mercury mass ratios offer simple, surprisingly strong constraints.

4.1.2. Mass, orbital, and compositional structure of the asteroid belt. The main belt (between roughly 2 and 3.2 AU) covers a surface area more than 3× larger than that of the terrestrial planet region, yet the entire belt contains only ~4.5×10^{-4} $M_\oplus$ (*Krasinsky et al.*, 2002; *Kuchynka and Folkner*, 2013; *DeMeo and Carry*, 2013). Yet the asteroids' orbits are excited, with eccentricities that span from 0 to 0.3 and inclinations that extend above 20°. The belt contains a diversity of spectroscopically-distinct types of objects (*Bus and Binzel*, 2002). The belt is also radially segregated: the inner main belt (interior to ~2.7 AU) is dominated by S-type asteroids, whereas the main belt beyond 2.7 AU is dominated by C-types (*Gradie and Tedesco*, 1982; *DeMeo and Carry*, 2013, 2014). S-types are identified with ordinary chondrites, which are relatively dry (with water contents below 0.1% by mass), whereas C-types are associated with carbonaceous chondrites, which typically contain 10% water by mass (*Robert et al.*, 1977; *Kerridge*, 1985; *Alexander et al.*, 2018).

4.1.3. Cosmochemically-constrained growth histories of Earth and Mars. Isotopic analyses of different types of Earth rocks, and lunar and martian meteorites, constrain the growth histories of Earth and Mars respectively. Isotopic systems such as Hf-W with half-lives comparable to the planets' formation timescales are particularly useful (the half-life of radioactive ^{182}Hf is 9 m.y.) (see *Alexander et al.*, 2001). These studies indicate that Earth's core formation did not finish until at least ~40–100 m.y. after the start of planet formation (*Touboul et al.*, 2007; *Kleine et al.*, 2009). The final episode of core formation on Earth is generally assumed to have been the Moon-forming impact (*Benz et al.*, 1986; *Canup and Asphaug*, 2001). Mars' growth is directly constrained to have been far faster than Earth's (*Nimmo and Kleine*, 2007). Indeed, Mars' accretion was complete within 5–10 m.y. (*Dauphas and Pourmand*, 2011).

4.1.4. Abundance and isotopic signature of water on Earth. Despite being mostly dry and rocky, Earth nonetheless contains a small fraction of water by mass, which is thought to be essential for life. The exact amount of water on Earth remains only modestly well constrained. An "ocean" of water is defined as the total amount of water on Earth's surface, roughly 1.5×10^{24} g (or ~0.025 $M_\oplus$). The mantle is thought to contain between a few tenths of an ocean (*Hirschmann*, 2006; *Panero and Caracas*, 2017) and 5–10 oceans (*Lécuyer et al.*, 1998; *Marty*, 2012; *Halliday*, 2013). The core is generally thought to be very dry (*Badro*

et al., 2014), but *Nomura et al.* (2014) inferred a very large reservoir of water exceeding 50 oceans. Assuming a total water budget of four oceans, Earth's bulk water content is thus 0.1% by mass.

The isotopic signature of Earth's water — the D/H and ^{15}N/^{14}N ratios — is a key discriminant of different models of water delivery (see section 6). Earth's water is a good match to carbonaceous chondrite meteorites, specifically the CM subgroup (*Marty and Yokochi*, 2006; *Alexander et al.*, 2012). Earth's water is isotopically distinct from nebular and cometary sources [see data compiled in *Morbidelli et al.* (2000); note that there are two comets observed to have Earth-like D/H ratios but both have non-Earth-like ^{15}N/^{14}N ratios; see discussion in section 6].

4.1.5. Late veneer on Earth, Mars, and the Moon. Highly-siderophile elements are those that are thought to have a chemical affinity for iron rather than silicates. Most of these elements are thus thought to be sequestered in a planet's core during core-mantle segregation. All the highly-siderophile elements in Earth's crust must therefore have been delivered by impacts after the Moon-forming impact (*Kimura et al.*, 1974). From the abundance of highly-siderophile elements, and assuming the impactors to be chondritic in composition, it has been inferred that the last ~0.5% of Earth's accretion took place after the Moon-forming impact (*Day et al.*, 2007; *Walker*, 2009; *Morbidelli and Wood*, 2015). Meteorite constraints indicate that Mars accreted ~9× less material than Earth during this late veneer, and that the Moon accreted 200–1200× less than Earth (*Day et al.*, 2007; *Walker*, 2009).

4.1.6. Orbits and masses of the giant planets. The giant planets are radially segregated by mass, with the most massive planets closest-in. As discussed in section 1.1, Jupiter's orbit is wider than most known giant exoplanets' and it is only barely detectable by long-duration RV surveys. Jupiter and Saturn each have low-eccentricity but noncircular orbits, each with million-year-averaged eccentricities of ~0.05 (*Quinn et al.*, 1991). Uranus' average eccentricity is comparable to the gas giants' but Neptune's is only ~0.01. There are no mean-motion resonances among the giant planets. The Jupiter/Saturn, Saturn/Uranus, and Uranus/Neptune period ratios are 2.48, 2.85, and 1.95, respectively.

4.1.7. Total mass and orbital structure of the outer solar system's small-body populations. The Kuiper belt extends outward from just beyond Neptune's orbit. It has a complex orbital structure that includes a population of objects such as Pluto that are locked in mean-motion resonances with Neptune. The Kuiper belt has a broad eccentricity and inclination distribution and includes a population of very dynamically cold objects from 42 to 45 AU often called the cold classical belt. The total mass in the Kuiper belt has been estimated at a few percent up to 0.10 $M_\oplus$ (*Gladman et al.*, 2001). The scattered disk is a subset of Kuiper belt objects whose orbits cross those of the giant planets. The Oort cloud is the source of long-period (isotropic) comets and extends from roughly 1000 AU out to the Sun's ionization radius [currently at ~200,000 AU; see *Tremaine* (1993) for details].

4.2. Rough Timeline of Events

Theory and observations can combine to provide a rough timeline of the events that must have taken place in the solar system. Time zero is generally assumed to be the time of formation of calcium-aluminum-rich inclusions (CAIs), roughly millimeter-sized components of primitive (chondritic) meteorites that are well-dated to be 4.568 G.y. old (e.g., *Bouvier and Wadhwa*, 2010).

• Within 100,000 yr planetesimal formation was underway. CAIs and millimeter-scale chondrules had started to form (e.g., *Connelly et al.*, 2008; *Nyquist et al.*, 2009) and coalesce into larger objects (e.g., *Dauphas and Chaussidon*, 2011; *Johansen et al.*, 2015). [The origin of chondrules is hotly debated, and some models suggest that they are the outcomes of collisions between planetesimals and planetary embryos rather than their building blocks (e.g., *Asphaug et al.*, 2011; *Johnson et al.*, 2015; *Lichtenberg et al.*, 2018).]

• Within 1 m.y. large embryos had formed. Ages of iron meteorites indicate that embryos had formed in the inner solar system (e.g., *Halliday and Kleine*, 2006; *Kruijer et al.*, 2014; *Schiller et al.*, 2015). Meanwhile, the segregation of the parent bodies of carbonaceous and non-carbonaceous meteorites indicates that at least one ~10-$M_\oplus$ embryo — presumably Jupiter's core (*Kruijer et al.*, 2017) — had formed in the giant planet region. From this point onward, this core blocked the inward drift of pebbles and thus starved the inner solar system (e.g., *Bitsch et al.*, 2018).

• Within a few million years the gaseous planet forming disk had dissipated. Evidence for the timescale of disk dissipation comes from two sources. First, observations of hot dust — thought to trace the gas — around stars in young clusters with different ages indicate a typical dissipation timescale of 2–5 m.y. (*Haisch et al.*, 2001; *Hillenbrand*, 2008; *Pascucci et al.*, 2009; *Mamajek*, 2009). Second, given that all chondrule formation models require the presence of the gas disk, the latest-forming chondrules provide a lower limit on the gas disk's lifetime of 4–5 m.y. [the CB chondrites (see *Kita et al.*, 2005; *Krot et al.*, 2005; *Johnson et al.*, 2016)]. The existence of a hot Jupiter around a 2-m.y.-old T Tauri star (*Donati et al.*, 2016) demonstrates that giant planet formation and migration can happen on an even shorter timescale. The gas and ice giant planets were fully formed and likely in a compact resonant chain (*Morbidelli et al.*, 2007; *Izidoro et al.*, 2015a). By this time Mars was close to fully formed (*Nimmo and Kleine*, 2007; *Dauphas and Pourmand*, 2011), but Earth (and presumably Venus) was still actively accreting via planetesimal and embryo impacts.

• While the Sun was still in its birth cluster it is thought to have undergone a relatively close encounter with another star. Such an encounter has been invoked to explain the orbits of the Sednoids (*Morbidelli and Levison*, 2004; *Kenyon and Bromley*, 2004; *Jílková et al.*, 2015; *Pfalzner et al.*, 2018) — named after Sedna, the first one discovered (*Brown et al.*, 2004) — whose semimajor axes are greater than 250 AU and whose perihelia are detached from the planets'. The encounter may have either excited existing solar system planetesimals onto Sedna-like orbits or captured the objects from the passing star. The encounter distance was likely at a few hundred to 1000 AU. While the exact properties of the Sun's birth cluster remain a matter of debate (*Adams*, 2010; *Gounelle and Meynet*, 2012; *Portegies Zwart*, 2019), such encounters are expected to be a common occurrence (*Malmberg et al.*, 2011). It is interesting to note that the Sun must have left its parent cluster before the giant planet instability, since that is when the Oort cloud would have formed (*Brasser et al.*, 2013), and it would be much more compact had it formed in a cluster environment, given the stronger tidal field (*Tremaine*, 1993; *Kaib and Quinn*, 2008).

• Roughly 50–100 m.y. after CAIs, Earth suffered its final giant impact (*Touboul et al.*, 2007; *Kleine et al.*, 2009). This impact triggered Earth's final core formation event and the formation of the Moon (*Benz et al.*, 1986; *Canup and Asphaug*, 2001). Only ~0.5% of Earth's mass was accreted after this point (*Day et al.*, 2007; *Walker*, 2009; *Morbidelli and Wood*, 2015).

• Within 500 m.y. the outer solar system went unstable. The instability — thought to have been generated by interactions between the giant planets and an outer planetesimal disk, essentially the primordial Kuiper belt — is commonly referred to as the Nice model. Starting from a compact resonant chain originally formed as a consequence of a previous phase of migration in the gaseous disk (*Morbidelli et al.*, 2007), the giant planets underwent a series of close encounters. Interactions with the outer planetesimal disk caused the giant planets' orbits to radially spread out and destabilized the planetesimal disk (*Levison et al.*, 2011), which led to a phase of impacts throughout the solar system that was originally proposed to correspond to the so-called late heavy bombardment (*Tera et al.*, 1974; *Gomes et al.*, 2005), the event often associated with many of the oldest craters on Mercury, the Moon, and Mars. The instability can explain a number of features of the solar system, including the giant planets' present-day orbits (*Tsiganis et al.*, 2005; *Nesvorný and Morbidelli*, 2012), the orbital distribution of Jupiter's co-orbital asteroids (in particular their large inclinations) (*Morbidelli et al.*, 2005; *Nesvorný et al.*, 2013), and the characteristics of the giant planets' irregular satellites (*Nesvorný et al.*, 2007). Simulations that invoke that the young solar system had 1–2 additional ice giants that were ejected during the instability have a much higher success rate in matching the present-day solar system (*Nesvorný and Morbidelli*, 2012; *Batygin et al.*, 2012). The gas giants' relatively low eccentricities constitute a dynamical constraint: Jupiter and Saturn never underwent a close mutual encounter, although they must have had encounters with one or more ice giants (*Morbidelli et al.*, 2009).

While the existence of the instability remains in favor, the late timing has recently been challenged (*Boehnke and Harrison*, 2016; *Morbidelli et al.*, 2018; *Michael et al.*,

2018; *Nesvorný et al.*, 2018). An early giant planet instability is easier to understand from a dynamical perspective. The dispersal of the gaseous disk is the natural trigger for instabilities (e.g., *Matsumura et al.*, 2010) and simulations had substantial difficulty in delaying the onset of instability (*Gomes et al.*, 2005; *Levison et al.*, 2011). Simulations of giant planets interacting with outer planetesimal disks indeed show that most instabilities happen early, although there is a tail of instabilities that extends to much longer timescales (*Thommes et al.*, 2008b; *Raymond et al.*, 2010). A key input in such models — the inner edge location and orbital distribution of planetesimals in the outer primordial disk — remains poorly constrained.

The giant planet instability must have had a significant impact on the inner solar system. The changing dynamical environment caused by changes in Jupiter and Saturn's orbits caused secular resonances to sweep and/or jump across the inner solar system, exciting anything in their path (*Brasser et al.*, 2009; *Agnor and Lin*, 2012). A late instability tends to excite and often to destabilize the orbits of the already-formed terrestrial planets (*Brasser et al.*, 2013; *Roig et al.*, 2016; *Kaib and Chambers*, 2016). An early instability — triggered shortly after the gas disk's dispersal and before the final assembly of the terrestrial planets — has the potential to resolve this problem, and constitutes the basis for one of the terrestrial planet formation models we will discuss in section 4.4 [the Early Instability model of *Clement et al.* (2018)].

- For the past 4 b.y., the orbital architecture of the solar system has remained roughly constant. Most impacts on the terrestrial planets come from asteroids that are disrupted and whose fragments end in unstable resonances [often after drifting due to the Yarkovsky effect (see *Gladman et al.*, 1997; *Bottke et al.*, 2006b; *Granvik et al.*, 2017)]. The planets' orbits undergo secular oscillations due to long-range gravitational perturbations (e.g., *Quinn et al.*, 1991). The oscillations in Earth's orbit and spin are called Milankovitch cycles and play a key role in its climate evolution (e.g., *Berger et al.*, 2005). The inner solar system is chaotic with a Lyapunov timescale (for the divergence of nearby orbital trajectories) of a few million years (*Laskar*, 1990; *Batygin and Laughlin*, 2015). It remains unknown whether the outer solar system's evolution is chaotic or regular, as both types of solutions exist within the current error bars on the giant planets' positions (*Hayes*, 2007). Regardless of whether the outer solar system is chaotic or regular there is no chance of future instability. In contrast, the terrestrial planets have a 2% chance of becoming unstable before the Sun becomes a red giant in 4–5 G.y. (*Laskar and Gastineau*, 2009).

4.3. The Classical Model of Terrestrial Planet Formation

The so-called classical model of terrestrial planet formation was pioneered by a series of papers by George Wetherill spanning 2–3 decades (e.g., *Wetherill*, 1978, 1985, 1996). It

has succeeded in explaining a large number of features of the inner solar system, and its shortcomings have served to point newer models in the right direction. The classical model remains to this day the basis of comparison with more recent models (e.g., *Morbidelli et al.*, 2012; *Raymond et al.*, 2014; *Jacobson and Walsh*, 2015; *Izidoro and Raymond*, 2018).

The central assumption in the classical model is that giant planet formation can be considered separately from terrestrial accretion. At face value this appears to be a reasonable assumption. Gas-dominated protoplanetary disks are observed to dissipate in a few million years (*Haisch et al.*, 2001; *Hillenbrand*, 2008), setting an upper limit on the timescale of gas giant formation. In contrast, cosmochemical studies have demonstrated that Earth's accretion lasted 50–100 m.y. (*Kleine et al.*, 2009). Simulations of the classical model start from a population of rocky building blocks (planetary embryos and planetesimals) and fully-formed gas giants, inherently assuming that there was no prior interaction between these different populations and that the gas had already been removed.

Figure 4 shows the evolution of a characteristic simulation of the classical model (from *Raymond et al.*, 2006b). The simulation is gas-free and so its time zero effectively corresponds to the dissipation of the gaseous disk. The population of rocky embryos (initially approximately Ceres-to Moon-mass in this case) self-excites by mutual gravitational forcing from its inner regions outward, producing larger embryos with a characteristic spacing (*Kokubo and Ida*, 1998, 2000, 2002). The outer parts of the rocky disk are excited by secular and resonant forcing from Jupiter, and excited bodies transmit this disturbance through mutual gravitational scattering. There is a long chaotic phase characterized by excitation of planetesimals to high eccentricities; the embryos' eccentricities and inclinations are generally kept lower by dynamical friction (*O'Brien et al.*, 2006; *Raymond et al.*, 2006b). During this phase embryos grow by accreting planetesimals as well as other embryos, and given that other embryos are growing concurrently, the largest impacts tend to happen late (e.g., *Agnor et al.*, 1999). By roughly 100 m.y. after the start of the simulation, most remnant planetesimals have been cleared out and three planets have formed.

The simulation from Fig. 4 illustrates the successes of the classical model as well as its shortcomings. The two inner surviving planets bear a strong likeness to Venus and Earth. Their orbital separation and eccentricities are similar and their masses are reasonably close. In addition, their feeding zones are wide enough to extend into the outer asteroid belt and have accreted water-rich material (see *Morbidelli et al.*, 2000). Earth's accretion happened on a geochemically-appropriate timescale of ~100 m.y. and included late giant impacts suitable for Moon formation. However, the third planet bears little resemblance to Mars. Its orbit is somewhat wide of Mars' but the big problem is that the planet is as massive as Earth.

Mars is the classical model's Achilles heel. Simulations of the classical model systematically fail to match Mars'

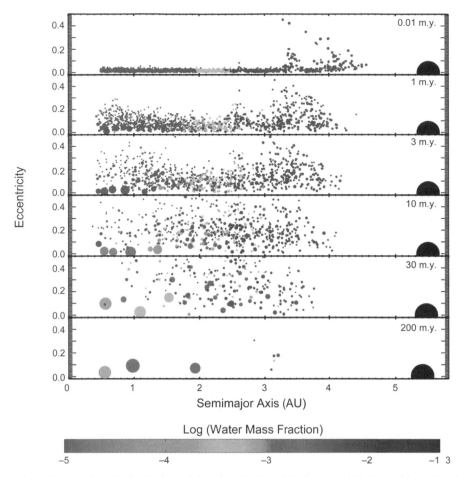

Fig. 4. See Plate 24 for color version. A simulation of the classical model of terrestrial planet formation (adapted from *Raymond et al., 2006b*). The simulation started from 1886 self-gravitating planetary embryos, represented as dots with size proportional to its mass$^{1/3}$. Jupiter is fully formed (large black dot) on a near-circular orbit at the start of the simulation. The color of each embryo represents its water content (see color bar at the bottom), with red objects being dry and the darkest blue containing 5% water by mass (see *Raymond et al., 2004*). This simulation produced quite good Earth and Venus analogs, a very poor Mars analog, and a plausible, albeit far too massive, asteroid belt. Note that time zero for this simulation corresponds to the dissipation of the gaseous disk, a few million years after CAIs. A movie of this simulation can be viewed at *https://youtu.be/m7hNlg9Gxvo*.

small mass and instead form Mars analogs that are a factor of 5–10 too massive (*Wetherill*, 1978; *Chambers*, 2001; *Raymond et al.*, 2006b, 2009b; *Morishima et al.*, 2010; *Fischer and Ciesla*, 2014; *Kaib and Cowan*, 2015). This was first pointed out by *Wetherill* (1991) and is commonly referred to as the "small Mars" problem.

The small Mars problem can be understood in a very simple way. If we assume that the disk of rocky building blocks extended smoothly from within 1 AU out to the giant planet region, then there was roughly the same amount of mass in Mars' feeding zone as in Earth's. In the absence of large perturbations, bottom-up accretion therefore produces Mars analogs that are as massive as Earth.

There are some circumstances under which the classical model can produce small Mars analogs. For example, if Jupiter and Saturn's orbits were more excited ($e_{Jup} \approx e_{Sat} \approx 0.07–0.1$) during terrestrial accretion than they are today, then secular resonances — orbits that precess at the same rate as a planet and therefore feel an enhanced gravita-

tional perturbation from that planet — would have been far stronger and could have acted to clear material from the Mars zone without depleting Earth's feeding zone [the "extra eccentric Jupiter and Saturn" (EEJS) configuration from *Raymond et al.* (2009b), *Morishima et al.* (2010), and *Kaib and Cowan* (2015)]. However, the EEJS setup has its own Achilles heel: Its initial conditions are not consistent with the evolution of Jupiter and Saturn in the gaseous disk. Simulations universally show that planet-disk interactions tend to drive the planets into resonance (in this case, specifically the 3:2 or 2:1 resonances) (*Pierens et al.*, 2014). However, if Jupiter and Saturn were in a resonant configuration, the location of their secular resonances within the terrestrial disk would not help to produce a small Mars (e.g., *Izidoro et al.*, 2016). A similar model invokes secular resonance sweeping during the dispersal of the gaseous disk to explain the depletion of the asteroid belt and Mars region (*Nagasawa et al.*, 2005; *Thommes et al.*, 2008a; *Bromley and Kenyon*, 2017). However, this model suffers from the same problem as the EEJS model: The gas giants' orbits are not consistent with the evolution of the disk, and using appropriate (generally lower-eccentricity, resonant) orbits removes the desired depletion. An early giant planet instability may, however, produce a giant planet configuration similar to the EEJS configuration, as we discuss in section 4.4.

The small Mars problem is inherently coupled with the asteroid belt's orbital excitation (*Izidoro et al., 2015c*). While very low in total mass, the asteroids' orbits are much more excited than the planets', with a broad range of eccentricities and inclinations. The current amount of mass in the belt cannot account for its excitation because there is not enough mass for gravitational self-stirring to be efficient (*Morbidelli et al., 2015b; Izidoro et al., 2016*). Yet a depleted region extending from Earth to the belt may explain why Mars is so small (*Izidoro et al., 2014a*). Indeed, the terrestrial planets' orbits are well-matched if they formed from a narrow ring of embryos that only extended from 0.7 to 1 AU (*Hansen, 2009; Walsh and Levison, 2016; Raymond and Izidoro, 2017b*). At face value, this means that a low-mass Mars implies an underexcited asteroid belt, and an appropriately excited asteroid belt implies a Mars that is far too massive (*Izidoro et al., 2015c*).

The small Mars and asteroid excitation problems are the primary shortcomings of the classical model. However, the model also cannot account for Mercury's small mass relative to Venus, although this remains a struggle for all models (see, e.g., *Lykawka and Ito, 2017*).

4.4. Viable Models for the Inner Solar System

We now discuss three successful models for the origin of the inner solar system: the Low-Mass Asteroid Belt, Grand Tack, and Early Instability models (summarized in Fig. 6). We explain the central assumptions of each model, what circumstances are required for the key mechanisms to operate, and how to test or falsify them. We order the models by when Mars' feeding zone was depleted, from earliest to latest.

4.4.1. Low-Mass Asteroid Belt model.
The Low-Mass Asteroid Belt model proposes that Mars is small simply because very few planetesimals formed between Earth's orbit and Jupiter's. Planetesimal formation has been shown to depend strongly on the gas disk's local properties (e.g., *Simon et al., 2016; Yang et al., 2017*). While gas disks are expected to have a relatively smooth radial distributions, Atacama Large Millimeter Array (ALMA) observations show that dust in young disks is concentrated into rings (*ALMA Partnership et al., 2015; Andrews et al., 2016*). It is not at all clear that planetesimals should form uniformly across the disk. Indeed, *Drążkowska et al.* (2016) modeled dust coagulation and drift in an evolving gas disk and found rings of planetesimals produced by the streaming instability centered at roughly 1 AU (see section 2). Additional mechanisms such as vortices can also act to strongly concentrate particles at ~1 AU to produce planetesimal rings (*Surville et al., 2016; Surville and Mayer, 2019*).

The Low-Mass Asteroid Belt model thus starts from a ring of planetesimals containing roughly 2 $M_\oplus$ centered between Venus' and Earth's present-day orbits. The terrestrial planets that accrete from such a planetesimal annulus provide a good match to the terrestrial planets' radial mass distribution (*Hansen, 2009; Kaib and Cowan, 2015; Walsh and Levison, 2016; Raymond and Izidoro, 2017b*). In this context, Mars' growth was stunted when it was scattered out of the dense ring of embryos. This naturally explains why Mars stopped accreting early (*Dauphas and Pourmand, 2011*). Earth's growth was more prolonged, lasting up to 100 m.y.

The compositional diversity of the asteroid belt in this scenario can be explained as a simple byproduct of the giant planets' growth [see Fig. 5 (from *Raymond and Izidoro, 2017a*)]. Jupiter's (and later, Saturn's) phase of rapid gas accretion invariably destabilized the orbits of nearby planetesimals and scattered them onto eccentric orbits. Gas drag acting on planetesimals with asteroid-belt-crossing or bits decreased their eccentricities, causing many to become trapped on stable, lower-eccentricity orbits in the belt. Scattered objects originated from across the outer solar system (out to 10–20 AU) and were preferentially trapped in the outer belt. The belt's radial structure can be matched by associating implanted planetesimals with C-type asteroids and assuming that a small amount of planetesimals native to the belt represent the S types. The giant planets' growth also scatters objects onto terrestrial planet-crossing orbits, providing a potential source of water for Earth (*Raymond and Izidoro, 2017a*). The efficiency with which planetesimals are scattered toward the terrestrial region is higher when gas drag is weaker and thus increases in efficiency as the disk dissipates (as well as for larger planetesimals). This process is universal and happens any time a giant planet forms (meaning that it happened several times in the solar system). The asteroid belt's excitation can be explained by processes such as chaotic excitation (*Izidoro et al., 2016*) or by secular excitation during the giant planet instability (*Deienno et al., 2018*).

An extreme version of the Low-Mass Asteroid Belt model invokes a completely empty belt in which absolutely no planetesimals formed between Earth's and Jupiter's orbits (*Raymond and Izidoro, 2017b*). Under that assumption, the terrestrial planets' orbits are naturally reproduced, and enough planetesimals are scattered out from the terrestrial planet region and implanted in the main belt to account for the total mass in S types. Given that the giant planets' growth invariably contributes C types (*Raymond and Izidoro, 2017a*), the "Empty Asteroid Belt" model thus proposes that all asteroids are refugees, implanted from across the solar system. However, there is a problem with the Empty Asteroid Belt model in that S-type asteroids [associated with ordinary chondrites (*Bus and Binzel, 2002*)] are compositionally distinct from Earth (e.g., *Warren, 2011*). In addition, the initial conditions for the simulations of *Raymond and Izidoro* (2017b) essentially invoke a single generation of planetesimals to explain the terrestrial planets and S types.

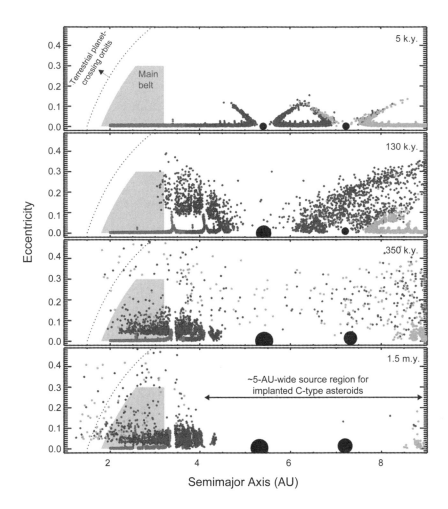

Fig. 5. See Plate 25 for color version. Injection of planetesimals into the asteroid belt and toward the terrestrial planet region as a consequence of giant planet formation (from *Raymond and Izidoro, 2017a*). This simulation starts with Jupiter and Saturn's 3 $M_\oplus$ cores embedded in a realistic gas disk [disk model from *Morbidelli and Crida* (2007), adapted to include gaps carved by the planets] including a population of 10^4 planetesimals assumed to be 100 km in diameter for the gas drag calculation. Jupiter grew (and carved a gap in the disk) from 100 to 200 k.y. and Saturn from 300 to 400 k.y., and the disk dissipated on a 200-k.y. exponential timescale (uniformly in radius). The colors of planetesimals represent their starting location. This case is the least dynamic possible scenario as it neglects migration of the gas giants and the formation and migration of the ice giants. Including those factors the source region of planetesimals implanted into the main belt extends out to 20 AU (*Raymond and Izidoro, 2017a*). An animation of this simulation can be viewed at *https://youtu.be/Ji5ZC7CP5to*.

However, measured ages imply that non-carbonaceous objects formed in several generations over the disk's lifetime (e.g., *Kruijer et al., 2017*). This problem may in principle be solved if another generation of planetesimals formed past Earth's orbit but still interior to the asteroid belt.

It is interesting to note that the implantation of planetesimals from the terrestrial planet-forming region into the asteroid belt happens regardless of the formation model. In the Empty Asteroid Belt model this represents the main source of volatile-depleted asteroids (*Raymond and Izidoro, 2017b*). However, classical model simulations have also found that terrestrial planetesimals are implanted into the main belt, and with a similar efficiency (*Bottke et al., 2006a; Mastrobuono-Battisti and Perets, 2017*). This suggests that the present-day belt must contain a population of leftovers

of terrestrial planet formation. It remains to be understood whether meteorites from such objects already exist in our collection or whether for unlucky reasons they are extremely rare (e.g., if there have not been any recent breakups of such asteroids).

The Low-Mass Asteroid Belt model's weakest point is its initial conditions. Planetesimal formation models in the context of dust concentration and streaming instability in disks with realistic structures struggle to produce planetesimal distributions consistent with the solar system. Some models do produce rings of planetesimals at ~1 AU suitable for terrestrial planet formation but no outer planetesimals that may have produced the giant planets' cores (*Drążkowska et al., 2016; Surville et al., 2016*). Other models produce planetesimals in outer planetary systems — in particular

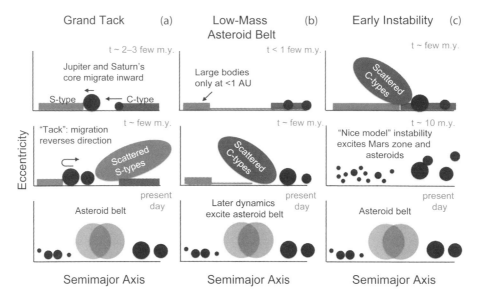

Fig. 6. Illustration of three models that can each match the large-scale properties of the inner solar system. **(a)** The Grand Tack model (*Walsh et al.*, 2011; *Raymond and Morbidelli*, 2014; *Brasser et al.*, 2016) invokes Jupiter's inward-then-outward migration to truncate the inner disk of rocky material, to explain the large Earth/Mars mass ratio. The asteroid belt is emptied then repopulated from two different source regions. S-types from interior to Jupiter's initial orbit were scattered outward during Jupiter's inward migration then back inward during its outward migration and implanted back onto similar orbits to their original ones, albeit with a low efficiency. C-types were implanted from exterior to Jupiter's initial orbit, also during the outward migration phase. **(b)** The Low-Mass Asteroid Belt model (*Hansen*, 2009; *Drążkowska et al.*, 2016; *Raymond and Izidoro*, 2017b), which proposes that the bulk of rocky planetesimals in the inner solar system formed within a narrow ring that only extended from roughly 0.7 to 1 AU. The ring produced terrestrial planets with the correct properties. Meanwhile, the asteroid belt was contaminated with C-type planetesimals implanted during the giant planets' growth (*Raymond and Izidoro*, 2017a). Later dynamical evolution — perhaps explained by chaotic evolution of Jupiter and Saturn (*Izidoro et al.*, 2016) or by secular forcing during the giant planet instability (*Deienno et al.*, 2018) — is required to explain the asteroids' level of excitation. **(c)** The Early Instability model (*Clement et al.*, 2018a), which is based on the giant planet instability happening within 10 m.y. after the dispersal of the gaseous disk. The instability acts to excite and deplete the asteroid belt and Mars' feeding zone, leading to realistic Earth/Mars mass ratios and terrestrial planets. Simulations find that the terrestrial system is best matched when the surviving giant planets match the real ones.

just past the snow line — but none in the terrestrial region (*Armitage et al.*, 2016; *Carrera et al.*, 2017; *Drążkowska and Alibert*, 2017). It remains unclear what conditions or processes are needed to produce planetesimal disks that are plausible precursors to the solar system. The abundance of new studies shows that this issue may be resolved in the near-term. From a geochemical standpoint, if the building blocks of the terrestrial planets were concentrated in a narrow ring then it is difficult to understand observed differences between Earth and Mars, as well as why Earth's chemistry is consistent with accretion from a heterogenous reservoir of material (*Rubie et al.*, 2011).

The Low-Mass Asteroid Belt model is robust to a modest degree of orbital migration of Jupiter and Saturn. Once the terrestrial ring of planetesimals has formed it is mostly separated from the giant planets' dynamical influence. In addition, there are disk-planet configurations for which Jupiter and Saturn's migration is slow or negligible (*Morbidelli and Crida*, 2007; *Pierens et al.*, 2014).

4.4.2. Grand Tack model. The Grand Tack model invokes large-scale migration of Jupiter to sculpt the inner solar system (*Walsh et al.*, 2011). The disk of rocky embryos and planetesimals is assumed to have extended smoothly from 0.5–0.7 AU out to the giant planet-forming region. Jupiter is assumed to have formed at 3–4 AU, opened a gap, and started to type II migrate inward, shepherding and scattering the rocky bodies in its path (as in previous studies focusing on exoplanets) (*Fogg and Nelson*, 2005; *Raymond et al.*, 2006a; *Mandell et al.*, 2007). Meanwhile, Saturn grew on an exterior orbit and migrated inward in the very rapid, gap-clearing type III regime (*Masset and Papaloizou*, 2003). Saturn became locked in Jupiter's exterior 2:3 resonance in a shared gap in the disk. This configuration changes the balance of disk torques felt by the coupled Jupiter-Saturn system, causing both planets to migrate outward while maintaining the resonance (*Masset and Snellgrove*, 2001; *Morbidelli and Crida*, 2007; *Pierens and Nelson*, 2008; *Zhang and Zhou*, 2010; *Pierens and Raymond*, 2011; *Pierens et al.*, 2014). If Jupiter's turnaround (or "tack") point was at 1.5–2 AU, then the disk of rocky material was truncated at ~1 AU, creating an edge reminiscent of the outer edge of embryos in the Low-Mass Asteroid Belt model. Jupiter and Saturn's migration continued until the disk itself started to dissipate, stranding the planets on resonant orbits consistent with the later giant planet instability (see *Nesvorný and Morbidelli*, 2012).

The disk of rocky material sculpted by Jupiter's migration can match the terrestrial planets' orbital and mass distributions and formation timescales (*Walsh et al., 2011; Jacobson et al., 2014; Jacobson and Morbidelli, 2014; Brasser et al., 2016*). Despite having traversed the asteroid belt twice, the belt is repopulated by both scattered inner disk material (linked here with S types) and implanted outer disk planetesimals (linked with C types) and its orbital distribution and total mass match the present-day belt (*Walsh et al., 2011, 2012; Deienno et al., 2017*). During the giant planets' outward migration, some C-type material is scattered inward past the asteroid belt onto terrestrial planet-crossing orbits, providing Earth with the appropriate water budget with the correct isotopic signature (*Walsh et al., 2011; O'Brien et al., 2014*) as well as atmophile elements (e.g., H, C, N, O, and the noble gases) (*Matsumura et al., 2016*). The Grand Tack can also match the terrestrial planets' compositions thanks to the early mixing of material originally formed over several astronomical units during Jupiter's inward migration phase (*Rubie et al., 2015*).

Hydrodynamical simulations of Jupiter-mass and Saturn-mass planets embedded in isothermal disks universally find that the two planets become locked in 3:2 resonance and migrate outward (*Masset and Snellgrove, 2001; Morbidelli and Crida, 2007; Pierens and Nelson, 2008; Zhang and Zhou, 2010; Pierens and Raymond, 2011*). Outward migration can extend across long stretches of planet-forming disks (*Crida et al., 2009*). When the disk mass and viscosity are varied and a more realistic thermal structure is accounted for, there is a spectrum of evolutionary pathways for the gas giants' orbits (*Pierens et al., 2014*). In disks with relatively low masses [smaller than the minimum-mass solar nebula model of *Weidenschilling* (1977b) and *Hayashi* (1981), introduced in section 1.2] and moderate viscosities [viscous stress parameters of $10^{-4} \lesssim \alpha \lesssim 10^{-2}$, where α is the poorly constrained parameter that controls the rate at which the disk evolves via angular momentum transport (*Shakura and Sunyaev*, 1973)] the gas giants are trapped in 2:1 resonance and maintain roughly stationary orbits. In somewhat more massive disks with moderate viscosities, Jupiter and Saturn are trapped in 3:2 resonance and migrate outward as in the isothermal case. In very low-viscosity disks ($\alpha \approx 10^{-5}$), Jupiter and Saturn are trapped in 2:1 resonance but migrate outward (*Pierens et al., 2014*).

It is unclear whether outward migration of Jupiter and Saturn can be maintained in the face of gas accretion (see discussion in *Raymond and Morbidelli, 2014*). This is the Grand Tack model's potential Achilles heel. In isothermal disks outward migration happens when two conditions are met (*Masset and Snellgrove, 2001*). First, Saturn must be at least half of its present-day mass in order to open a partial gap in the disk. Second, the Jupiter-to-Saturn mass ratio must be between 2 and 4. The outward migration of Jupiter and Saturn envisioned in the Grand Tack model spans a wide range in orbital distance (from ~1.5 to >5 AU for Jupiter) and takes a significant amount of time (likely on the order of 0.5–1 m.y. for the entire outward migra-tion phase). The question is, can the gas giants maintain the requisite conditions for outward migration during this entire phase when gas accretion onto the planets is taken into account? A definite answer to this question requires a good understanding of the disk's structure and evolution and of the planets' gas accretion. Both of these processes remain too poorly understood to allow for a clear evaluation at this point, although given the large amount of interest in the Grand Tack model, there is hope for progress in the near-term.

4.4.3. Early Instability model.
The Early Instability model proposes that terrestrial planet formation was strongly affected by the giant planet instability (*Clement et al., 2018*). This constrains the instability to have taken place within ~10 m.y. of the disk's dissipation (*Morbidelli et al., 2018*) in order to have an impact on stunting Mars' growth (*Nimmo and Kleine, 2007; Dauphas and Pourmand, 2011*). Within the inner solar system, an early instability effectively causes a rapid transition from a dynamically calm state to an excited one that bears a strong resemblance to the EEJS configuration discussed in section 4.2 (from *Raymond et al., 2009b*). An early instability thus explains how the giant planets could have reached EEJS-like orbits in a self-consistent way at an early enough time to make a difference for terrestrial accretion.

Simulations by *Clement et al.* (2018, 2019) show that an early instability acts to excite the asteroid belt and to clear out the Mars zone. This is in contrast to simulations of a late giant planet instability that often overexcite the terrestrial planets' orbits (or destabilize them entirely) (*Brasser et al., 2013; Roig et al., 2016; Kaib and Chambers, 2016*). With an early instability, the depletion is so strong that a significant fraction (~20%) of simulations leave Mars' present-day orbital region completely empty and the median Mars analog is close to Mars' true mass. In many simulations planets grow much larger than a Mars mass at Mars' orbital distance, are then excited to significant eccentricities by the instability, and then collide at pericenter with the growing Earth or Venus. Mars analogs in these simulations are often stranded embryos that avoid colliding with the larger embryos when they are scattered inward.

The Early Instability model can match the asteroid belt and Earth's water content. The belt is pre-excited before the instability by scattering of planetesimals by resident embryos (*Petit et al., 2001; Chambers and Wetherill, 2001; O'Brien et al., 2007*). During the instability the belt is depleted (*Morbidelli et al., 2010; Roig and Nesvorný, 2015; Clement et al., 2019*) and embryos are removed, and surviving planetesimals provide a decent match to the present-day belt's eccentricity and inclination distributions (*Clement et al., 2018, 2019*). As in the classical model, the belt's S-/C-type compositional dichotomy is assumed to be matched due to previous events [and recall that Jupiter's growth invariably implants C types into the outer main belt (*Raymond and Izidoro, 2017a*)]. Water-rich material is delivered to Earth by the same mechanism as in the classical model (*Morbidelli et al., 2000; Raymond et al.,*

2007a, 2009b) via impacts from water-rich planetesimals and embryos originating in the outer main belt.

An appealing aspect of the Early Instability model is that it simplifies the solar system's timeline. Rather than invoking separate mechanisms to sculpt the terrestrial and giant planet systems, a single key event can explain them both. In addition, the terrestrial planets are best reproduced when the giant planets reach their actual configuration [as measured by the giant planets' orbits and the strength of different secular modes (*Clement et al.,* 2018)].

The Early Instability model's weak points are related to the timing of the instability. Xenon isotopes are the first issue. Xenon in Earth's atmosphere is fractionated due to hydrodynamic escape and is enriched in heavier isotopes relative to xenon found in chondritic meteorites or in the solar wind (*Ozima and Podosek,* 2002). When this fractionation is taken into account, the isotopic signature of presumably primordial atmospheric xenon is still significantly different from that of xenon in the mantle. This suggests that another source of atmospheric xenon had to exist with a different isotopic signature than chondrites or the solar wind. Comet 67P/Churyumov-Gerasimenko matches the missing xenon signature and Earth's atmosphere can be matched with a 20:80 mixture of cometary and chondritic xenon (*Marty et al.,* 2017). In contrast, the mantle's xenon was purely chondritic (e.g., *Mukhopadhyay,* 2012; *Caracausi et al.,* 2016). The longstanding mystery of Earth's xenon thus appears to be solved if Earth's atmospheric xenon included an additional cometary component. In this scenario the mantle's xenon came from the inner solar system as Earth was accreting. The atmosphere's xenon came from a late bombardment of comets necessarily related to the giant planet instability, which represent a significant source of noble gases but not of water (*Marty et al.,* 2016). This elegant argument relates the relative timing of Earth's accretion and the cometary bombardment linked with the giant planet instability. In principle, a very early instability would cause a cometary bombardment at the same time as accretion such that one might expect Earth's mantle xenon to have the same isotopic signature as its fractionation-corrected atmospheric xenon. In contrast, a late cometary bombardment would only affect the atmospheric xenon signature.

At face value, the xenon constraint would seem to rule out a giant planet instability earlier than the Moon-forming impact. The relative timing of the instability and Moon-forming impact affects the ability of all terrestrial planet formation models to match the xenon constraint, not just the Early Instability model. However, there are two caveats. First, the particular xenon signature has only been measured in a single comet (although it is the only comet in which the signature could have been detected). Second, the repartition of the xenon signature between the mantle and atmosphere during Earth's impact-driven evolution is uncertain and certainly depends on factors related to Earth's growth history and chemical evolution.

Another potential conflict comes from the orbital structure of the Kuiper belt. *Nesvorný* (2015) found that the high-inclination classical Kuiper belt can be matched by a smooth phase of migration of Neptune that was interrupted by the instability. If true, this would restrict the earliest possible timing of the instability to be ~20 m.y. after gas disk dissipation, too late to stunt Mars' growth. This constraint is relatively indirect, as other models may potentially explain the high-inclination population.

4.5. An Outstanding Issue: The Mass Deficit in the Very Inner Solar System

The very low amount of mass in the very inner solar system (interior to Mercury's orbit) is hard to understand. This feature is important because it represents a divide between the solar system and exoplanet systems. Roughly half of all stars have Earth-sized or larger planets interior to Mercury's orbit (e.g., *Howard et al.,* 2010, 2012; *Mayor et al.,* 2011). However, our terrestrial planets are consistent with having formed from a narrow ring of planetesimals and embryos between Venus' and Earth's present-day orbits (*Hansen,* 2009; *Walsh and Levison,* 2016; *Raymond and Izidoro,* 2017b). Models such as those presented in section 4.4 can explain the outer edge of this ring of rocky material, for example, by invoking dynamical truncation by the migrating Jupiter (*Walsh et al.,* 2011). However, the inner edge of this ring — and the absence of other planets closer in than Mercury — remains challenging to explain. Of course, given its very large iron core and the reduced oxidation state of its crust and mantle, Mercury itself is a challenge to explain (e.g., *Ebel and Stewart,* 2018).

It was proposed by *Leake et al.* (1987) that a population of planetesimals formed on orbits interior to Mercury's and later bombarded Mercury. Indeed, there is a belt of dynamically stable orbits between 0.06 and 0.21 AU (sometimes called the "vulcanoid zone") (*Evans and Tabachnik,* 1999). However, a belt of planetesimals on such close-in orbits would undergo vigorous collisional grinding (*Stern and Durda,* 2000). Efficient removal of small bodies via radiative transport would remove the bulk of the population's mass, and the surviving planetesimals would themselves be further depleted by Yarkovsky effect-driven drift into unstable orbital configurations (*Vokrouhlický et al.,* 2000). Very few kilometer-scale planetesimals are expected to survive. Only planetesimals large enough to have significant self-gravity (D ≳ 100 km) would have survived, and to date none have been found. This suggests that a belt of 100-km-scale planetesimals typical of the streaming instability (*Simon et al.,* 2016; *Schäfer et al.,* 2017) did not form closer in than Mercury or, if it did, it was dynamically — not collisionally — removed.

Ida and Lin (2008) proposed that the rocky mass interior to roughly Venus' orbit was removed by inward migration. They assumed that accretion proceeds roughly as a wave sweeping outward in time and that the large embryos produced by accretion migrated inward and fell onto the young Sun. Embryos massive enough to undergo gas-driven orbital migration only had time to form interior to roughly Venus' orbit. While appealing, this model ignores the fact that disks have inner edges [as demonstrated by simulations of magnetic accretion onto young stars (see *Romanova et al.*, 2004)]. Embryos can migrate to the inner edge of the disk, where they are trapped by a strong positive torque (*Masset et al.*, 2006; *Romanova and Lovelace*, 2006), but they remain a great distance from the surface of the star (roughly one order of magnitude larger than the stellar radius).

Raymond et al. (2016) proposed that outward migration could explain the very inner solar system's mass deficit. Assuming planetesimals to form throughout the inner disk, they invoked the rapid formation of a large core of a few Earth masses close to the Sun. This object could plausibly have formed by trapping a fraction of inward-drifting pebbles, perhaps at a pressure bump in the very inner disk such as found in the disk models of *Flock et al.* (2017). For objects of a few Earth masses embedded in the inner parts of radiative viscous disks, migration is often directed outward (the details depend on parameters such as the disk metallicity and accretion rate; see *Bitsch et al.*, 2015a). The large core's outward migration through a population of planetesimals and embryos is analogous to the case of a large planet migrating inward through similar objects discussed in section 3.2 (see also *Fogg and Nelson*, 2005; *Raymond et al.*, 2006a; *Izidoro et al.*, 2014b). For outward migration timescales of ~10^5 yr, *Raymond et al.* (2016) found that the core shepherds embryos and planetesimals in exterior resonances and clears out the inner solar system. However, this mechanism typically broke down when the core reached 0.5–1 AU due to scattering between shepherded bodies. The core's migration would have continued to a zero-torque location past the snow line (*Bitsch et al.*, 2015a), often contaminating the primordial asteroid belt with material from the very inner solar system. In the context of this model, the migrating core represents Jupiter's core and therefore it is expected to contain a large fraction of rock. The weakness of this idea is the setup, as it requires (1) a large, close-in core that forms much faster than more distant, smaller embryos, and (2) a disk in which migration is directed outward. Yet this setup (at least point 1) appears to be plausible as many disk models have an inner pebble/dust trap (e.g., *Flock et al.*, 2017) and this idea is at the heart of the drift model for super-Earth formation (*Chatterjee and Tan*, 2014).

Morbidelli et al. (2016) proposed that the edge in the presumed disk of rocky building blocks at 0.7 AU corresponds to the location of the silicate condensation line at early times when the disk was hot. Rocky planetesimals that formed early could not have formed closer than the silicate condensation line. That early generation of planetesimals continued to grow by mutual collisions or by accreting pebbles, but the lack of closer-in material would be preserved if no other planetesimals formed closer in. There are two main uncertainties in this model. First, why did no planetesimals form closer in at later times? Given that several generations of planetesimals are thought to have formed in the inner solar system (e.g., *Kruijer et al.*, 2017),

it remains to be understood why none would form closer in. Second, what happens to the pebbles that drift inward past the growing planetesimals? Tens of Earth masses in pebbles are likely to have drifted inward past the rocky planetesimals (e.g., *Lambrechts and Johansen*, 2014). These are generally assumed to have reached the inner parts of the disk and simply sublimed. If even a small fraction is trapped, then it could lead to the formation of a large core as in the *Raymond et al.* (2016) model.

Two papers proposed that the early solar system contained a population of super-Earths that were destroyed. *Volk and Gladman* (2015) proposed that very energetic collisions ground the planets to dust. However, examination of the collisional parameters in the simulated collisions suggest that they are far below the catastrophic destruction threshold (*Leinhardt and Stewart*, 2012; *Wallace et al.*, 2017) and it is hard to understand how all of the planets' mass could have been removed. Unlike kilometer-scale planetesimals — which, as discussed above, would indeed be ground to dust and removed on orbits closer in than Mercury's (*Stern and Durda*, 2000; *Vokrouhlický et al.*, 2000) — the self-gravity of planets prevents their total destruction.

In contrast, *Batygin and Laughlin* (2015) invoked collisional debris generated by the Grand Tack model to push a population of primordial super-Earths onto the young Sun. While there are key issues related to the mechanism at play (see discussion in *Raymond et al.*, 2016), the main problem with this model is that, as described above, planets cannot simply migrate onto their stars. Rather, their migration is blocked by the inner edge of the disk (*Masset et al.*, 2006; *Romanova and Lovelace*, 2006) and their evolution should be similar to the breaking the chains model (*Izidoro et al.*, 2017, 2019) (see also Fig. 3 and discussion in section 3.1).

To conclude, in our minds the origin of the mass deficit closer in than Venus' orbit remains unexplained. While successful models to match the deficit do exist, there is no clear theory that does not have significant counterarguments or require specific assumptions. Based on our current understanding it is not plausible to invoke migration onto the Sun as a mechanism for losing close-in material. Rather, close-in material may have been swept outward by a migrating core (*Raymond et al.*, 2016), or the inner disk may simply never have produced planetesimals (*Morbidelli et al.*, 2016). It is also entirely possible that another mechanism may be responsible.

5. EXTRAPOLATION TO EXO-EARTHS: FORMATION TIMESCALES AND WATER CONTENTS

We now turn our attention to the more general question of Earth-like planets around other stars. Just how "Earth-like" should we expect Earth-sized planets in the habitable zones of their host stars to be? And how does this depend on other properties of these systems (in particular observable ones) such as the planetary system architecture, the planet size/mass, and the stellar type? In this section we address the formation and water contents of potentially habitable planets. We do not address the question of what conditions are needed for habitability. Rather we simply assume that approximately Earth-mass-sized planets in the habitable zones of their parent stars are viable candidates.

The solar system's terrestrial planets are thought to have accreted in steps (see section 2). First, planetesimals formed from drifting pebbles and dust. Then planetesimals grew into planetary embryos by accreting pebbles and other planetesimals. Embryos grew slowly enough and remained small enough not to have undergone any significant migration. Finally, there was an extended phase of giant collisions between embryos lasting ~100 m.y.

To generalize the formation of Earth-mass planets, we want to know how universal each of these steps is. Do all planet-forming disks follow the same general pattern as ours?

While great strides have been made in understanding how 100-km-scale planetesimals form, models disagree on where and when they form (e.g., *Drążkowska et al.,* 2016; *Carrera et al.,* 2017; *Drążkowska and Alibert,* 2017). We can imagine that planetesimal disks might be quite diverse in their structures; for instance, the Low-Mass Asteroid Belt model is based on a particular structure (see section 4.3). However, given our limited understanding of planetesimal formation in the solar system, we cannot reasonably extrapolate to other systems. For the purpose of this discussion we will simply assume that planetesimal formation is robust and has no strong radial dependence.

Embryo growth is a critical step. The solar system's terrestrial planets are consistent with having formed from a population of approximately Mars-mass embryos (e.g., *Morbidelli et al.,* 2012). The largest planetesimals undergo runaway accretion (of other planetesimals) and become Moon to Mars mass embryos (e.g., *Greenberg et al.,* 1978; *Wetherill and Stewart,* 1993; *Kokubo and Ida,* 2000). For a minimum-mass disk, embryos take 0.1–1 m.y. to grow at 1 AU, at which point they excite the orbits of nearby planetesimals, decrease the effects of gravitational focusing, and their growth from planetesimal accretion slows down drastically (e.g., *Kokubo and Ida,* 1998; *Leinhardt and Richardson,* 2005). Yet pebble accretion should continue and even accelerate (e.g., *Ormel and Klahr,* 2010; *Lambrechts and Johansen,* 2012).

Matching the terrestrial planets therefore requires a quenching of pebble accretion to prevent embryos at 1 AU from growing too massive [see *Lambrechts et al.* (2014) for simulations of the pebble-flux-governed bifurcation between terrestrial planets and rocky super-Earths]. This may have happened as a consequence of the growth of Jupiter's core to the pebble isolation mass, at which point it created a barrier to inward pebble drift (*Morbidelli and Nesvorný,* 2012; *Lambrechts et al.,* 2014; *Bitsch et al.,* 2018). *Kruijer et al.* (2017) used the temporal coexistence of meteorites with different nucleosynthetic signatures (carbonaceous vs. non-carbonaceous) to infer that Jupiter's

core did indeed provide a barrier within 1 m.y. after CAIs (see also *Desch et al.,* 2018). From that point on, pebble accretion was shut off in the inner solar system and the terrestrial planets grew by accreting planetesimals and embryos.

This line of thinking implies that the timing of the growth of Jupiter's core was critical (see, e.g., *Bitsch et al.,* 2015b, 2019). If it had grown much more slowly, pebble accretion would have generated more massive terrestrial embryos. These large embryos would then have migrated and likely followed the breaking the chains evolution discussed in section 3.1 (see also Fig. 3). If we assume that the growth of large outer cores varies significantly from disk to disk, these differences in timing can have big consequences. Systems with fast-growing cores may preserve their small inner rocky embryos, whereas in systems with slower-growing cores terrestrial embryos grow sufficiently fast that they cannot avoid migration.

Migration must play an important role in the formation of many habitable zone planets. This happens if (1) an outer core (analogous to Jupiter's core) formed slowly enough for large embryos to grow by pebble accretion, or (2) the central star is low enough in mass that the formation timescale in the habitable zone is very short. The accretion timescale depends on the local disk surface density (in planetesimals) and the orbital timescale (*Safronov,* 1969). Given the strong scaling of the habitable zone with stellar type (because of the strong stellar mass-luminosity scaling), the accretion timescales for planets in the habitable zones of low-mass stars are much shorter than for Sun-like stars (*Raymond et al.,* 2007b; *Lissauer and Stevenson,* 2007; *Dawson et al.,* 2015). Extrapolating from Earth's 50–100-m.y. formation timescale, planets in the habitable zones of stars less massive than one-half to one-third of a solar mass should form quickly, with most of the assembly taking place during the gas disk phase even in the absence of pebble accretion (*Raymond et al.,* 2007b).

In these systems embryos should follow the breaking the chains behavior described above: They should migrate into resonant chains anchored at the inner edge of the disk, most of which go unstable when the gas dissipates. Habitable zone planets would undergo a final phase of giant impacts shortly after disk dispersal (*Ogihara and Ida,* 2009; *Izidoro et al.,* 2017, 2019). The compositions of planets in these systems would vary (*Raymond et al.,* 2018b). Some could be made up entirely of large rocky embryos. However, given that embryos should still eventually grow large past the snow line, some planets would likely contain a significant fraction of their mass in volatiles.

What about systems with more massive central stars in which an outer core forms quickly? Fast-growing cores are themselves likely to migrate. If the disk properties are such that there is a zero-torque migration trap that lasts for close to the entire disk lifetime (*Lyra et al.,* 2010; *Bitsch et al.,* 2015a), the inner planetary system may be protected from the core's migration. In some cases cores may accrete gas to become gas giants and transition to slower, type II migration. In those cases terrestrial planet formation should in principle follow the same pattern as in the solar system provided the giant planets do not migrate all the way into the habitable zone (see *Fogg and Nelson,* 2005; *Raymond et al.,* 2006a). When migration is not stopped, the large core would plow into (or through, depending on the migration timescale) the growing terrestrial planets (*Izidoro et al.,* 2014b; *Raymond et al.,* 2018b). Planets can still form in the habitable zones in such systems but they are not rocky worlds like Earth. Rather, such planets should have compositions representative of the region past the snow line, presumably with large water contents.

Some simple analysis can hint at the distribution of outcomes. Low-mass (M) stars are the most common by number (e.g., *Chabrier,* 2003). The disk mass is observed to be a steeper than linear function of the stellar mass [scaling roughly as $M_{disk} \propto M_\star^{1.6}$ albeit with large scatter (e.g., *Scholz et al.,* 2006; *Pascucci et al.,* 2016)]. For low-mass stars, the snow line is farther removed from the habitable zone than for FGK stars, as measured simply by the snow line to habitable zone distance ratio (*Mulders et al.,* 2015a). Given that the growth timescale scales with the disk mass (e.g., *Safronov,* 1969), we expect outer cores to grow slowly around low-mass stars. Assuming that there is sufficient local material to build an Earth-mass planet *in situ* in the habitable zone, then rocky embryos should form quickly (*Raymond et al.,* 2007b; *Lissauer,* 2007; *Dawson et al.,* 2015), undergo pebble accretion, and migrate. Embryos from past the snow line are also likely to migrate later in the disk lifetime, as disks around low-mass stars are observed to have longer lifetimes than around Sun-like stars (*Pascucci et al.,* 2009). The location of the disk's inner edge depends on the rotation rate of young stars and is unlikely to be a strong function of the stellar mass. This means that the habitable zones of low-mass stars are closer to the inner edge of the disk than those of higher-mass stars. Rocky planets that form near the habitable zone do not necessarily migrate far away. Given the late instabilities characteristic of breaking the chains behavior, the final planets should be a mixture of embryos that started with terrestrial compositions and those with ice-rich compositions from past the snow line. This should result in a diversity of planetary compositions, from pure rock planets to planets with tens of percent ice by mass [neglecting various water/ice loss processes (e.g., *Grimm and McSween,* 1993; *Genda and Abe,* 2005; *Marcus et al.,* 2010; *Monteux et al.,* 2018)].

The main difference between this scenario for low-mass and higher-mass stars is the accretion timescale in the habitable zone (*Raymond et al.,* 2007b). For FGK stars, terrestrial embryos in the habitable zone are less likely to grow fast enough to migrate if they only accrete planetesimals. Compared with the same setting around low-mass stars, growing rocky planets in the habitable zones of FGK stars are more likely to be protected from pebble accretion by a

fast accreting core, and to have a large ice-rich core migrate into the terrestrial zone. Given the much faster core accretion compared with terrestrial accretion for FGK systems, close-in planets that result from the inward migration of large cores in these systems are likely to have higher average volatile contents than for low-mass stars. Of course, in some situations outer cores will accrete into gas giants and (in some cases) remain on wide orbits. In these cases, the terrestrial planets' accretion is protected, although the gas giants' growth may shower the terrestrial zone with volatile-rich planetesimals (*Raymond and Izidoro,* 2017a).

Compared with FGK stars, low-mass stars are found to have more super-Earths smaller than 2 $R_\oplus$ but fewer sub-Neptunes between 2 and 4 $R_\oplus$ and a higher total average planet mass on close-in orbits (*Mulders et al.,* 2015b,c). This can be explained by the reasoning presented above if the smaller super-Earths preferentially formed from migrating rocky embryos and larger sub-Neptunes formed mainly from migrating ice-rich embryos. Higher-mass stars have more gas giants (*Johnson et al.,* 2007; *Lovis and Mayor,* 2007). While growing gas giant cores block pebbles from drifting past (*Morbidelli and Nesvorný,* 2012; *Lambrechts et al.,* 2014; *Bitsch et al.,* 2018), gas giants themselves block embryos from migrating past (*Izidoro et al.,* 2015b). This might tilt the scales in favor of low-mass stars having a higher average total mass in close-in planets.

Let's put the pieces together. Given that M stars dominate by number, the formation pathway of their habitable zone planets likewise dominates. The habitable zones of low-mass stars are so close in that we expect planets to grow rapidly from both rocky and ice-rich material and to follow the breaking the chains evolution described in section 3.1. The bulk of these planets' growth took place during the gas disk phase, with last giant impacts happening during a late instability shortly after the dissipation of the disk. Their compositions are likely to span a wide range from pure rock to ice-rich depending on the objects' individual growth histories.

Earth-mass planets in the habitable zones of FGK stars are likely to have followed one of two pathways. In systems in which cores of 10–20 $M_\oplus$ or more grow quickly past the snow line, the flux of pebbles toward the inner system is shut off. In some systems (like our own) these large cores grow into gas giants, which migrate slowly and may remain isolated from the terrestrial zone. Further growth involves planetesimal and embryo accretion, and rocky embryos are unlikely to reach high enough masses to migrate within the gas disk's lifetime. In other systems outer, ice-rich cores do not become gas giants but instead migrate inward into the terrestrial zone. In that case habitable zone planets may typically be very volatile-rich (e.g., *Kuchner,* 2003).

How can we use observations to constrain these ideas? Systems that follow the breaking the chains evolution should commonly form habitable zone planets. However, the habitable zone planets themselves should have a diversity of compositions, encompassing systems in which rocky embryos grew large enough to migrate and those in which they did not. Low-mass stars should be more likely to have migrating rocky embryos and, since their habitable zones are likely to be closer to the inner edge of the disk, are also more likely to retain rocky embryos/planets in the habitable zone. Finally, systems with outer giant planets [beyond 2.5 and 3 AU for Sun-like stars (see *Raymond, 2006*)] and no inner giants are the best candidates for having habitable zone planets with small but non-zero, Earth-like water contents.

6. DISCUSSION

6.1. Central Processes that Sculpt Planetary Systems

Two processes appear to be ubiquitous in planet formation: migration and instability. These are essential ingredients in explaining the origins of exoplanet systems as well as the solar system. The breaking the chains model proposed by *Izidoro et al.* (2017, 2019) invokes inward migration of large embryos into long chains of mean-motion resonances anchored at the inner edge of the disk. The vast majority of resonant chains become unstable when the disk dissipates, leading to a late phase of giant collisions. Given that virtually all observed super-Earths are massive enough to undergo gas-driven migration, we argued in section 3.1 that, regardless of when and how they form, all super-Earth systems converge to the breaking the chains evolution.

The population of giant exoplanets may be explained by invoking the formation of multiple gas giants that also migrate into compact resonant configurations (see section 3.2). As for super-Earths, the vast majority of systems undergo instabilities when the disk dissipates or perhaps even during the late phases of the disk lifetime. The outcome of an instability correlates with the Safronov number, which is the ratio of the planets' escape speed to the local escape speed from the system (*Safronov, 1969*; *Ford and Rasio, 2008*). For high Safronov numbers the planets impart strong enough gravitational kicks that scattering is favored over collisions. Giant planet instabilities thus lead to planet-planet scattering and the surviving planets match the observed giant exoplanet eccentricity distribution [as well as the correlated mass-eccentricity distribution (*Raymond et al., 2010*)].

Solar system formation models likewise invoke different combinations of migration and instability. All current evolutionary pathways rely on the Nice model instability in the giant planets' orbits, although the timing of the instability is uncertain (*Morbidelli et al., 2018*). The Grand Tack model uses a Jupiter-specific migration path to deplete the asteroid belt and Mars' feeding zone (*Walsh et al., 2011*). In contrast, the Early Instability model proposes that an early giant planet instability was responsible for depleting the asteroid belt and Mars region (*Clement et al., 2018*).

Two additional processes are central in setting the stage: planetesimal formation and pebble accretion. They determine where and when embryos large enough to migrate can form. The Low-Mass Asteroid Belt model proposes that the mass depletion in the Mars and asteroid belt regions was inherited from the planetesimal formation stage (*Drążkowska et al., 2016*). Likewise, the timing of the formation of Jupiter's core — which acted to block the flux of pebbles to the inner solar system — was a key moment in keeping the terrestrial planets terrestrial (see discussion in section 5).

Different processes have different philosophical implications. Migration and instability both act to erase the initial conditions. Many formation pathways lead to a phase of migration, but once a system migrates it starts to forget its initial conditions, as all pathways converge to the same evolution. It is for that reason that density constraints on super-Earth compositions are so important: They are our only clue to the planets' origins, and even for accurate measurements density is a weaker diagnostic than one would like (see discussion in section 3.1). Likewise, the chaotic nature of instabilities makes it impossible to rewind the clock on a system of planets with eccentric orbits to uncover their pre-instability configuration [although statistical studies try to do just this (e.g., *Nesvorný and Morbidelli, 2012*)].

In contrast, bottom-up accretion retains a memory of its initial conditions. The planets that form from a disk with a given surface density still follow that same profile (*Raymond et al., 2005*). This is the central argument behind the "minimum-mass solar nebula" model (*Weidenschilling, 1977b*; *Hayashi, 1981*). This allows us to put strong — but not unique — constraints on the properties of the precursor disks that formed systems of small planets, assuming that accretion was the main process involved. For instance, a few different initial distributions of planetesimals and embryos can match the terrestrial planets but they all share common properties such as a strong mass deficit past Earth's orbit and interior to Venus' (*Hansen, 2009*; *Izidoro et al., 2015c*).

How well do planetesimal formation and pebble accretion remember their initial conditions? In the current paradigm, planetesimals form when drifting dust and pebbles are sufficiently concentrated to trigger the streaming instability (*Johansen et al., 2014*). Exactly where and when this happens depends on the underlying disk model (e.g., *Drążkowska et al., 2016*; *Carrera et al., 2017*). However, once planetesimal formation is triggered, objects form with a characteristic size distribution (*Simon et al., 2017*; *Schäfer et al., 2017*). Thus, while accretion may preserve a trace of where planetesimals formed and in what abundance, planetesimal formation itself does not retain a memory of dust coagulation and drift. Pebble accretion is too parameter-dependent to retain a memory of the historical pebble flux. If the pebble size (or Stokes number) and the initial planetesimal/embryo mass were known this might be possible.

6.2. Planet Formation Pathways and Bifurcations: How Did Our Solar System Get So Weird?

There are a few key bifurcation points in planetary system formation. At these points, small differences in outcome lead to very different evolution. We consider the key bifurcation points to be (1) disk properties, (2) planetesimal formation (where? when?), (3) giant planet formation, and (4) instability trigger (timing). We now go over each of these bifurcation points, then discuss which pathway the solar system must have followed.

A star's protoplanetary disk is its cradle, where its planetary system is born and raised. The characteristics of the disk and its evolution are perhaps the single most important factor in planet formation. While the detailed structure and evolution of disks are themselves poorly understood (see *Morbidelli and Raymond,* 2016), observations and theory demonstrate that there is a diversity in disk mass, structure, and lifetime (e.g., *Haisch et al.,* 2001; *Williams and Cieza,* 2011; *Bate,* 2018). The disk mass may itself be the key determinant of planetary system evolution (*Greaves et al.,* 2006; *Thommes et al.,* 2008c). More massive disks should more readily form planetesimals, embryos, and gas giants. Given their higher abundance of solids, higher-metallicity stars should also more readily form planetesimals and planets. This is borne out by the observed giant planet-metallicity correlation (*Fischer and Valenti,* 2005).

Where and when planetesimals form is of vital importance. If planetesimals form early then they are bathed in a flux of pebbles and can quickly grow into large embryos/cores and perhaps even gas giants (e.g., *Lambrechts and Johansen,* 2014; *Bitsch et al.,* 2015b). However, if planetesimals only form late — perhaps triggered by the dissipation of the gas disk and the accompanying increase in dust/gas ratio (*Throop and Bally,* 2005; *Carrera et al.,* 2017) — then the bulk of their growth must appeal to gas-free processes such as planetesimal accretion and no gas giants can form. The radial distribution of planetesimals is naturally of vital importance to planet formation. The Low-Mass Asteroid Belt model relies on planetesimals forming in a narrow ring in the inner solar system (*Drążkowska et al.,* 2016), whereas the Grand Tack (*Walsh et al.,* 2011) and Early Instability (*Clement et al.,* 2018) models were devised assuming that planetesimals did indeed form in the Mars region and asteroid belt and that a depletion mechanism was needed.

The formation of a giant planet is essentially two bifurcation points. Once a planet's core reaches the pebble isolation mass [of roughly $20\ M_\oplus$ at Jupiter's orbit for characteristic disk models (*Lambrechts et al.,* 2014; *Bitsch et al.,* 2018)], the pebble flux is blocked and the entire planetary system interior to the core is cut off from further pebble accretion. We argued in section 5 that the timing of the formation of a pebble-blocking core relative to the growth of inner embryos is the central parameter that determines whether most super-Earths are likely to be rocky or ice-dominated. After a core undergoes rapid gas accretion to become a gas giant and carves a gap in the disk, it also blocks the inward migration of any other large cores that form on exterior orbits (*Izidoro et al.,* 2015b). This is a second way in which a wide-orbit planet cuts off its inner planetary system from the inward-drifting/migrating mass. Of course, if the wide-orbit planet migrates inward then it itself becomes that inward-migrating mass.

Triggering instability is the final and perhaps most dramatic bifurcation point. While instability appears to be near-ubiquitous, the impact of instabilities can vary. For instance, the solar system's giant planets are thought to have undergone an instability but only a very weak one when compared with the instabilities in most exoplanet systems.

Indeed, the instabilities characteristic of giant exoplanet systems often drive the growing terrestrial planets into their host star (*Raymond et al.,* 2011, 2012). The late-stage accretion of the terrestrial planets also represents a form of instability that concluded with the Moon-forming impact. The breaking the chains (*Izidoro et al.,* 2017, 2019) evolution characteristic of super-Earth systems causes much more dramatic late instabilities that involve collisions between much larger (typically $\sim5\ M_\oplus$) objects and, given their close-in orbits, much higher collision speeds. Of course, a small minority of systems avoid instability. Stable systems such as Trappist-1 and Kepler-223 (see Fig. 1) are easily recognized by their chains of resonant orbits, which are systematically destroyed by instabilities [although scattering does generate resonances in a small fraction of cases (*Raymond et al.,* 2008a)].

What path must the solar system have taken with regard to these bifurcations? The Sun's planet-forming disk may have been somewhat more massive than average, with enough mass to form the cores of several gas giants (which total ~40–$50\ M_\oplus$), but not enough to form multiple Jupiter-mass planets. A few Earth-mass worth of rocky planetesimals must have formed early in the terrestrial planet region, either in a smooth disk or in one or more rings. Planetesimals also formed beyond the snow line and produced the giant planets' cores. Jupiter's core grew fast enough to starve the inner solar system of pebbles within ~1 m.y., keeping the precursors of carbonaceous and non-carbonaceous meteorites physically separated (*Kruijer et al.,* 2017) and preventing further growth of the terrestrial planets' constituent embryos (*Lambrechts et al.,* 2014). Jupiter's growth also stopped the ice giants and Saturn's core from migrating into the inner solar system (*Izidoro et al.,* 2015a,b). When the disk dissipated, the inner disk of terrestrial embryos entered its late instability, which lasted ~100 m.y. but during which Mars remained mostly isolated and protected. The giant planets' orbits became unstable sometime within the 500 m.y. following disk dissipation. While the disk's dispersal is the main natural trigger for instability, some geochemical arguments (e.g., the atmospheric xenon constraint; see section 4.3) point to a later trigger. While the instability did clear out the primordial Kuiper belt, it was far less dramatic than instabilities characteristic of extrasolar systems.

The solar system's presumed evolution contains multiple unusual occurrences. First, the gas giants' masses are quite different. The fact that the most massive giant exoplanets have the highest eccentricities (*Wright et al.,* 2009) indicates that massive gas giants (of roughly $1\ M_{Jup}$ or above) typically form in systems with other, roughly equal-mass gas giants that become unstable (*Raymond et al.,* 2010). Second,

Jupiter's orbit remained wide of the terrestrial region. This may be because of the dynamical influence of Saturn; the Jupiter-Saturn system can migrate outward or remain on near-stationary orbits depending on the disk properties (*Pierens et al.,* 2014). However, avoiding inward migration depends on a Jupiter/Saturn mass ratio close to its current value (*Masset and Snellgrove,* 2001), so this unusual occurrence may be intrinsically linked with the previous one. Third, the giant planet instability did not include any close encounters between Jupiter and Saturn. Simulations show that such an encounter would likely have ejected Saturn and stranded Jupiter with an eccentricity of ~0.2 (*Morbidelli et al.,* 2007). In other words, Jupiter's eccentricity would be typical of giant exoplanets if its instability had proceeded in typical fashion. However, in that case the terrestrial planets may well have been driven into the Sun (*Raymond et al.,* 2011).

We interpret these unusual occurrences as examples of why our solar system is weird. These evolutionary steps explain why Sun-Jupiter systems are rare within the known exoplanet sample (see section 1.1). This in turn implies that terrestrial planet systems like ours are also rare, although an understanding of the timescales of different processes is needed to assess that assertion in a more careful way (see section 5).

6.3. A Digression on the Significance of Models

A successful formation model is expected to match a planetary system (or a distribution of systems) in broad strokes and with a suitable success rate. But what exactly constitutes the "breadth" of the "strokes" needed for success? And at what rate is a model deemed successful? These inherently philosophical questions are central to models of planet formation. If a model matches the solar system in 1% of simulations, should we consider the problem solved? Or should we continue to test other models? And if model A provides a match in 30% of cases and model B in 10%, can we be confident that model A is truly preferred over model B?

We do not pretend to have a concrete solution, but we think it important to keep such considerations in mind. We expect that in the future, global planet formation modeling may make use of more rigorous statistical methods to address these issues.

6.4. Paths Forward

A plethora of outstanding problems in planet formation remain. As described in section 4.4, studies will strongly constrain (and may falsify some) models of solar system formation in the coming years. Nonetheless, we encourage the development of new models. NASA's Origins Spectral Interpretation Resource Identification and Security-Regolith Explorer (OSIRIS-REx) and JAXA's Hayabusa2 missions will return samples of carbonaceous asteroids (the B-type Bennu and the Cg-type Ryugu) that will certainly improve our understanding of the formation conditions of such objects and provide additional constraints on their origins.

NASA's Lucy mission will study Jupiter's co-orbital asteroids, thought to have been captured during the giant planet instability (*Morbidelli et al.,* 2005), and NASA's Psyche mission will study an apparently metallic asteroid that may have originated in the inner parts of a differentiated planetary embryo. Meanwhile, upcoming exoplanet-focused instruments such as NASA's Transiting Exoplanet Survey Satellite (TESS) and the European Space Agency's Planetary Transits and Oscillations of stars (PLATO) and Atmospheric Remote-sensing Infrared Exoplanet Large-survey (ARIEL) missions will deepen our understanding of the orbital architecture of planetary systems as well as their more detailed characteristics. This will provide additional constraints on exoplanet formation models.

Among our ever-increasing stockpile of extremely valuable data, we conclude this chapter by emphasizing the need for global models to connect the dots. Models should not be constrained by dogma or current paradigms. Of course, a model is only viable if it matches observations or measurements. A model is most useful if it is testable in the near term. And a model is most relevant when it lays the broadest possible foundation. This means putting solar system formation in the context of extrasolar planets.

Acknowledgments. We thank referees J. Chambers and K. Walsh and editor V. Meadows for their constructive reports, and are grateful to all of our colleagues who helped develop the ideas presented here. We each thank the Agence Nationale pour la Recherche for funding and support via grant ANR13BS050003002 (grant MOJO). A.I. acknowledges financial support from FAPESP (grants 16/126862 and 16/195567). S.N.R. also acknowledges NASA Astrobiology Institute's Virtual Planetary Laboratory Lead Team, funded under solicitation NNH12ZDA002C and cooperative agreement no. NNA13AA93A.

REFERENCES

Adams E. R., Seager S., and Elkins-Tanton L. (2008) Ocean planet or thick atmosphere: On the mass-radius relationship for solid exoplanets with massive atmospheres. *Astrophys. J., 673,* 1160.

Adams F. C. (2010) The birth environment of the solar system. *Annu. Rev. Astron. Astrophys., 48,* 47.

Adams F. C., Hollenbach D., Laughlin G., et al. (2004) Photoevaporation of circumstellar disks due to external far-ultraviolet radiation in stellar aggregates. *Astrophys. J., 611,* 360.

Adams F. C. and Laughlin G. (2003) Migration and dynamical relaxation in crowded systems of giant planets. *Icarus, 163,* 290.

Agnor C. B., Canup R. M., and Levison H. F. (1999) On the character and consequences of large impacts in the late stage of terrestrial planet formation. *Icarus, 142,* 219.

Agnor C. B. and Lin D. N. C. (2012) On the migration of Jupiter and Saturn: Constraints from linear models of secular resonant coupling with the terrestrial planets. *Astrophys. J., 745,* 143.

Agol E. and Fabrycky D. C. (2017) Transit-timing and duration variations for the discovery and characterization of exoplanets. In *Handbook of Exoplanets* (H. J. Deeg and J. A. Belmonte, eds.), p. 7. Springer Nature, Switzerland.

Albrecht S., Winn J. N., Johnson J. A., et al. (2012) Obliquities of hot Jupiter host stars: Evidence for tidal interactions and primordial misalignments. *Astrophys. J., 757,* 18.

Alexander C. M. O., Bowden R., Fogel M. L., et al. (2012) The provenances of asteroids, and their contributions to the volatile inventories of the terrestrial planets. *Science, 337,* 721.

Alexander C. M. O., Boss A. P., and Carlson R. W. (2001) The early evolution of the inner solar system: A meteoritic perspective.

Science, 293, 64.

Alexander C. M. O., McKeegan K. D., and Altwegg K. (2018) Water reservoirs in small planetary bodies: meteorites, asteroids, and comets. *Space Sci. Rev., 214,* 36.

Alexander R., Pascucci I., Andrews S., et al. (2014) The dispersal of protoplanetary disks. *Protostars and Planets VI* (H. Beuther et al., eds.), pp. 475–496. Univ of Arizona, Tucson.

Alexander R. D. and Pascucci I. (2012) Deserts and pile-ups in the distribution of exoplanets due to photoevaporative disc clearing. *Mon. Not. R. Astron. Soc., 422,* L82.

Alibert Y. (2017) Maximum mass of planetary embryos that formed in core-accretion models. *Astron. Astrophys., 606,* A69.

Alibert Y., Mordasini C., Benz W., et al. (2005) Models of giant planet formation with migration and disc evolution. *Astron. Astrophys., 434,* 343.

ALMA Partnership, Brogan C. L., Pérez L. M., et al. (2015) The 2014 ALMA long baseline campaign: First results from high angular resolution observations toward the HL Tau region. *Astrophys. J. Lett., 808,* L3.

Andrews S. M., Wilner D. J., Zhu Z., et al. (2016) Ringed substructure and a gap at 1 AU in the nearest protoplanetary disk. *Astrophys. J. Lett. 820,* L40.

Anglada-Escudé G., Amado P. J., Barnes J., et al. (2016) A terrestrial planet candidate in a temperate orbit around Proxima Centauri. *Nature, 536,* 437.

Anglada-Escudé G., Tuomi M., Gerlach E., et al. (2013) A dynamically-packed planetary system around GJ 667C with three super-Earths in its habitable zone. *Astron. Astrophys., 556,* A126.

Armitage P. J. (2007) Massive planet migration: Theoretical predictions and comparison with observations. *Astrophys. J., 665,* 1381.

Armitage P. J. (2011) Dynamics of protoplanetary disks. *Annu. Rev. Astron. Astrophys., 49,* 195.

Armitage P. J., Eisner J. A., and Simon J. B. (2016) Prompt planetesimal formation beyond the snow line. *Astrophys. J. Lett., 828,* L2.

Asphaug E., Jutzi M., and Movshovitz N. (2011) Chondrule formation during planetesimal accretion. *Earth Planet. Sci. Lett., 308,* 369.

Ayliffe B. A. and Bate M. R. (2009) Gas accretion on to planetary cores: Three-dimensional self-gravitating radiation hydrodynamical calculations. *Mon. Not. R. Astron. Soc., 393,* 49.

Badro J., Cote A., and Brodholt J. (2014) A seismologically consistent compositional model of Earth's core. *Proc. Natl. Acad. Sci. U S.A., 132,* 94.

Bai X.-N. (2016) Towards a global evolutionary model of protoplanetary disks. *Astrophys. J., 821,* 80.

Baillié K., Charnoz S., and Pantin E. (2015) Time evolution of snow regions and planet traps in an evolving protoplanetary disk. *Astron. Astrophys., 577,* A65.

Balbus S. A. and Hawley J. F. (1998) Instability, turbulence, and enhanced transport in accretion disks. *Rev. Mod. Phys., 70,* 1.

Ballard S. and Johnson J. A. (2016) The Kepler dichotomy among the M dwarfs: Half of systems contain five or more coplanar planets. *Astrophys. J., 816,* 66.

Baraffe I., Selsis F., Chabrier G., et al. (2004) *The effect of evaporation on the evolution of close-in giant planets. Astron. Astrophys., 419,* L13.

Barclay T., Rowe J. F., Lissauer J. J., et al. (2013) A sub-Mercury-sized exoplanet. *Nature, 494,* 452.

Baruteau C., Crida A., Paardekooper S.-J., et al. (2014) Planet-disk interactions and early evolution of planetary systems. In *Protostars and Planets VI* (H. Beuther et al., eds.), pp. 667–689. Univ. of Arizona, Tucson.

Batalha N. M., Rowe J. F., Bryson S. T., et al. (2013) Planetary candidates observed by Kepler. III. Analysis of the first 16 months of data. *Astrophys. J., Suppl. Ser., 204,* 24.

Bate M. R. (2018) On the diversity and statistical properties of protostellar discs. *Mon. Not. R. Astron. Soc., 475,* 5618.

Batygin K. (2012) A primordial origin for misalignments between stellar spin axes and planetary orbits. *Nature, 491,* 418.

Batygin K., Bodenheimer P. H., and Laughlin G. P. (2016) In situ formation and dynamical evolution of hot Jupiter systems. *Astrophys. J., 829,* 114.

Batygin K., Brown M. E., and Betts H. (2012) Instability-driven dynamical evolution model of a primordially five-planet outer solar system. *Astrophys. J. Lett., 744,* L3.

Batygin K. and Laughlin G. (2015) Jupiter's decisive role in the inner solar system's early evolution. *Proc. Natl. Acad. Sci. U.S.A., 112,*

4214.

Beaugé C. and Nesvorný D. (2012) Multiple-planet scattering and the origin of hot Jupiters. *Astrophys. J., 751,* 119.

Beaulieu J., Bennett D. P., Fouqué P. et al. (2006) Discovery of a cool planet of 5.5 Earth masses through gravitational microlensing. *Nature, 439,* 437.

Benítez-Llambay P., Masset F., Koenigsberger G., et al. (2015) Planet heating prevents inward migration of planetary cores. *Nature, 520,* 63.

Bennett D. P., Rhie S. H., Nikolaev S., et al. (2010) Masses and orbital constraints for the OGLE-2006-BLG-109Lb,c Jupiter/Saturn analog planetary system. *Astrophys. J., 713,* 837.

Benz W., Slattery W. L., and Cameron A. G. W. (1986) The origin of the Moon and the single-impact hypothesis. I. *Icarus, 66,* 515.

Berger A., Mélice J. L., and Loutre M. F. (2005) On the origin of the 100-kyr cycles in the astronomical forcing. *Paleoceanography, 20,* PA4019.

Biller B. A., Liu M. C., Wahhaj Z., et al. (2013) The Gemini/NICI planet-finding campaign: The frequency of planets around young moving group stars. *Astrophys. J., 777,* 160.

Birnstiel T., Fang M., and Johansen A. (2016) Dust evolution and the formation of planetesimals. *Space Sci. Rev., 205,* 41.

Birnstiel T., Klahr H., and Ercolano B. (2012) A simple model for the evolution of the dust population in protoplanetary disks. *Astron. Astrophys., 539,* A148.

Bitsch B. and Kley W. (2010) Orbital evolution of eccentric planets in radiative discs. *Astron. Astrophys., 523,* A30.

Bitsch B., Crida A., Morbidelli A., et al. (2013) Stellar irradiated discs and implications on migration of embedded planets. I. Equilibrium discs. *Astron. Astrophys., 549,* A124.

Bitsch B., Morbidelli A., Lega E. et al. (2014) Stellar irradiated discs and implications on migration of embedded planets. II. Accreting-discs. *Astron. Astrophys., 564,* A135.

Bitsch B., Johansen A., Lambrechts M., et al. (2015a) Formation of planetary systems by pebble accretion and migration: Growth of gas giants. *Astron. Astrophys., 575, A28.*

Bitsch B., Lambrechts M., and Johansen A. (2015b) The growth of planets by pebble accretion in evolving protoplanetary discs. *Astron. Astrophys., 582,* A112.

Bitsch B., Morbidelli A., Johansen A. et al. (2018) Pebble-isolation mass: Scaling law and implications for the formation of super-Earths and gas giants. *Astron. Astrophys., 612,* A30.

Bitsch B., Izidoro A., Johansen A., et al. (2019) Formation of planetary systems by pebble accretion and migration: growth of gas giants. *Astron. Astrophys., 623,* A88.

Blum J. and Wurm G. (2008) The growth mechanisms of macroscopic bodies in protoplanetary disks. *Annu. Rev. Astron. Astrophys., 46,* 21.

Bodenheimer P., Hubickyj O., and Lissauer J. J. (2000) Models of the in situ formation of detected extrasolar giant planets. *Icarus, 143,* 2.

Boehnke P. and Harrison T. M. (2016) Illusory Late Heavy Bombardments. *Proc. Natl. Acad. Sci. U.S.A., 113,* 10802.

Boisvert J. H., Nelson B. E., and Steffen J. H. (2018) Systematic mischaracterization of exoplanetary system dynamical histories from a model degeneracy near mean-motion resonance. *Mon. Not. R. Astron. Soc., 480,* 2846.

Boley A. C. and Ford E. B. (2013) The formation of systems with tightly-packed inner planets (STIPs) via aerodynamic drift. *ArXiV e-prints,* arXiv:1306.0566.

Boley A. C., Granados Contreras A. P., and Gladman B. (2016) The in situ formation of giant planets at short orbital periods. *Astrophys. J. Lett., 817,* L17.

Bolmont E., Raymond S. N., von Paris P., et al. (2014) Formation, tidal evolution, and habitability of the Kepler-186 system. *Astrophys. J., 793,* 3.

Bottke W. F., Nesvorný D., Grimm R. E., et al. (2006a) Iron meteorites as remnants of planetesimals formed in the terrestrial planet region. *Nature, 439,* 821.

Bottke Jr. W. F., Vokrouhlický D., Rubincam D. P., et al. (2006b) The Yarkovsky and YORP effects: Implications for asteroid dynamics. *Annu. Rev. Earth Planet. Sci., 34,* 157.

Boué G. and Fabrycky D. C. (2014) Compact planetary systems perturbed by an inclined companion. II. Stellar spin-orbit evolution. *Astrophys. J., 789,* 111.

Bouvier A. and Wadhwa M. (2010) The age of the solar system redefined by the oldest Pb-Pb age of a meteoritic inclusion. *Nature Geosci., 3,* 637.

Bouvier J., Alencar S. H. P., Harries T. J., et al. (2007) Magnetospheric accretion in classical T Tauri stars. In *Protostars and Planets V* (B. Reipurth et al., eds.), pp. 479–494. Univ. of Arizona, Tucson.

Bowler B. P. and Nielsen E. L. (2018) Occurrence rates from direct imaging surveys. In Handbook of Exoplanets (H. Deeg and J. Belmonte, eds.), pp. 1967–1983. Springer, Cham.

Brasser R., Matsumura S., Ida S., et al. (2016) Analysis of terrestrial planet formation by the Grand Tack Model: System architecture and tack location. *Astrophys. J., 821,* 75.

Brasser R., Morbidelli A., Gomes R., et al. (2009) Constructing the secular architecture of the solar system II: The terrestrial planets. *Astron. Astrophys., 507,* 1053.

Brasser R., Walsh K. J., and Nesvorný D. (2013) Constraining the primordial orbits of the terrestrial planets. *Mon. Not. R. Astron. Soc., 433,* 3417.

Brauer F., Dullemond C. P., and Henning T. (2008) Coagulation, fragmentation and radial motion of solid particles in protoplanetary disks. *Astron. Astrophys., 480,* 859.

Briceño C., Vivas A. K., Calvet N., et al. (2001) The CIDA-QUEST large-scale survey of Orion OB1: Evidence for rapid disk dissipation in a dispersed stellar population. *Science, 291,* 93.

Broeg C. H. and Benz W. (2012) Giant planet formation: Episodic impacts versus gradual core growth. *Astron. Astrophys., 538,* A90.

Bromley B. C. and Kenyon S. J. (2017) Terrestrial planet formation: Dynamical shake-up and the low mass of Mars. *Astron. J., 153,* 216.

Brouwers M. G., Vazan A., and Ormel C. W. (2018) How cores grow by pebble accretion. I. Direct core growth. *Astron. Astrophys., 611,* A65.

Brown M. E., Trujillo C., and Rabinowitz D. (2004) Discovery of a candidate inner Oort cloud planetoid. *Astrophys. J., 617,* 645.

Bryden G., Chen X., Lin D. N. C., et al. (1999) Tidally induced gap formation in protostellar disks: Gap clearing and suppression of protoplanetary growth. *Astrophys. J., 514,* 344.

Bus S. J. and Binzel R. P. (2002) Phase II of the Small Main-Belt Asteroid Spectroscopic Survey. A feature-based taxonomy. *Icarus, 158,* 146.

Butler R. P., Wright J. T., Marcy G. W., et al. (2006) Catalog of nearby exoplanets. *Astrophys. J., 646,* 505.

Campante T. L., Barclay T., Swift J. J., et al. (2015) An ancient extrasolar system with five sub-Earth-size planets. *Astrophys. J., 799,* 170.

Canup R. M. and Asphaug E. (2001) Origin of the Moon in a giant impact near the end of the Earth's formation. *Nature, 412,* 708.

Caracausi A., Avice G., Burnard P. G., et al. (2016) Chondritic xenon in the Earth's mantle. *Nature, 533,* 82.

Carrera D., Ford E. B., Izidoro A., et al. (2018) Identifying inflated super-Earths and photo-evaporated cores. *Astrophys. J., 866,* 104.

Carrera D., Gorti U., Johansen A., et al. (2017) Planetesimal formation by the streaming instability in a photoevaporating disk. *Astrophys. J., 839,* 16.

Cassan A., Kubas D., Beaulieu J.-P., et al. (2012) One or more bound planets per Milky Way star from microlensing observations. *Nature, 481,* 167.

Chabrier G. (2003) Galactic stellar and substellar initial mass function. *Publ. Astron. Soc. Pac., 115,* 763.

Chambers J. E. (2006) A semi-analytic model for oligarchic growth. *Icarus, 180,* 496.

Chambers J. E. (2001) Making more terrestrial planets. *Icarus, 152,* 205.

Chambers J. E. and Wetherill G. W. (2001) Planets in the asteroid belt. *Meteoritics & Planet. Sci., 36,* 381.

Chambers J. E., Wetherill G. W., and Boss A. P. (1996) The stability of multi-planet systems. *Icarus, 119,* 261.

Chambers K. C., Magnier E. A., Metcalfe N., et al. (2016) The Pan-STARRS1 surveys. *ArXiV e-prints,* arxiv:1612.05560.

Charbonneau D., Brown T. M., Burrows A., et al. (2007) When extrasolar planets transit their parent stars. In *Protostars and Planets V* (B. Reipurth et al., eds.), pp. 701–716. Univ of Arizona, Tucson.

Chatterjee S., Ford E. B., Matsumura S., et al. (2008) Dynamical outcomes of planet-planet scattering. *Astrophys. J., 686,* 580.

Chatterjee S. and Tan J. C. (2014) Inside-out planet formation. *Astrophys. J., 780,* 53.

Chatterjee S. and Tan J. C. (2015) Vulcan planets: Inside-out formation of the innermost super-Earths. *Astrophys. J. Lett., 798,* L32.

Chen J. and Kipping D. (2017) Probabilistic forecasting of the masses and radii of other worlds. *Astrophys. J., 834,* 17.

Chiang E. and Laughlin G. (2013) The minimum-mass extrasolar nebula: In situ formation of close-in super-Earths. *Mon. Not. R. Astron. Soc., 431,* 3444.

Chiang E. and Youdin A. N. (2010) Forming planetesimals in solar and extrasolar nebulae. *Annu. Rev. Earth Planet. Sci., 38,* 493.

Clanton C. and Gaudi B. S. (2014) Synthesizing exoplanet demographics from radial velocity and microlensing surveys. II. The frequency of planets orbiting M dwarfs. *Astrophys. J., 791,* 91.

Clanton C. and Gaudi B. S. (2016) Synthesizing exoplanet demographics: A single population of long-period planetary companions to M dwarfs consistent with microlensing, radial velocity, and direct imaging surveys. *Astrophys. J., 819,* 125.

Clement M. S., Kaib N. A., Raymond S. N., and Walsh K. J. (2018a) Mars' growth stunted by an early giant planet instability. *Icarus, 311,* 340–356.

Clement M. S., Raymond S. N., and Kaib N. A. (2018b) Excitation and depletion of the asteroid belt in the early instability scenario. *Astron. J., 157(1).*

Connelly J. N., Amelin Y., Krot A. N., et al. (2008) Chronology of the solar system's oldest solids. *Astrophys. J. Lett., 675,* L121.

Cossou C., Raymond S. N., Hersant F., et al. (2014) Hot super-Earths and giant planet cores from different migration histories. *Astron. Astrophys., 569,* A56.

Cossou C., Raymond S. N., and Pierens A. (2013) Convergence zones for type I migration: An inward shift for multiple planet systems. *Astron. Astrophys., 553,* L2.

Cresswell P., Dirksen G., Kley W., et al. (2007) On the evolution of eccentric and inclined protoplanets embedded in protoplanetary disks. *Astron. Astrophys., 473,* 329.

Crida A., Masset F., and Morbidelli A. (2009) Long range outward migration of giant planets, with application to Fomalhaut b. *Astrophys. J. Lett., 705,* L148.

Crida A., Morbidelli A., and Masset F. (2006) On the width and shape of gaps in protoplanetary disks. *Icarus, 181,* 587.

Ćuk M. and Stewart S. T. (2012) Making the Moon from a fast-spinning Earth: A giant impact followed by resonant despinning. *Science, 338,* 1047.

Cumming A., Butler R. P., Marcy G. W., et al. (2008) The Keck planet search: Detectability and the minimum mass and orbital period distribution of extrasolar planets. *Publ. Astron. Soc. Pac., 120,* 531.

Dauphas N. (2003) The dual origin of the terrestrial atmosphere. *Icarus, 165,* 326.

Dauphas N. and Chaussidon M. (2011) A perspective from extinct radionuclides on a young stellar object: The Sun and its accretion disk. *Annu. Rev. Earth Planet. Sci., 39,* 351.

Dauphas N. and Pourmand A. (2011) Hf-W-Th evidence for rapid growth of Mars and its status as a planetary embryo. *Nature, 473,* 489.

Dawson R. I., Chiang E., and Lee E. J. (2015) A metallicity recipe for rocky planets. *Mon. Not. R. Astron. Soc., 453,* 1471.

Dawson R. I. and Johnson J. A. (2018) Origins of hot Jupiters. *Annu. Rev. Astron. Astrophys., 56,* 175.

Dawson R. I., Lee E. J., and Chiang E. (2016) Correlations between compositions and orbits established by the giant impact era of planet formation. *Astrophys. J., 822,* 54.

Day J. M. D., Pearson D. G., and Taylor L. A. (2007) Highly siderophile element constraints on accretion and differentiation of the Earth-Moon system. *Science, 315,* 217.

Deienno R., Izidoro A., Morbidelli A., et al. (2018) Excitation of a primordial cold asteroid belt as an outcome of planetary instability. *Astrophys. J., 864,* 50.

Deienno R., Morbidelli A., Gomes R. S., et al. (2017) Constraining the giant planets' initial configuration from their evolution: Implications for the timing of the planetary instability, *Astron. J., 153,* 153.

DeMeo F. E. and Carry B. (2013) The taxonomic distribution of asteroids from multi-filter all-sky photometric surveys. *Icarus, 226,* 723.

DeMeo F. E. and Carry B. (2014) Solar system evolution from compositional mapping of the asteroid belt. *Nature, 505,* 629.

Demory B.-O., Gillon M., Deming D., et al. (2011) Detection of a transit of the super-Earth 55 Cancri e with warm Spitzer. *Astron. Astrophys., 533,* A114.

Desch S. J., Kalyaan A., and Alexander C. M. (2018) The effect of Jupiter's formation on the distribution of refractory elements and inclusions in meteorites. *Astrophys. J. Suppl.Ser., 238,* 11.

Donati J. F., Moutou C., Malo L., et al. (2016) A hot Jupiter orbiting a 2-million-year-old solar-mass T Tauri star. *Nature, 534,* 662.

Dong S. and Zhu Z. (2013) Fast rise of "Neptune-size" planets (4–8 R$_\oplus$) from P ~ 10 to ~250 days — Statistics of Kepler planet candidates up to ~0.75 AU. *Astrophys. J., 778,* 53.

Drążkowska J. and Alibert Y. (2017) Planetesimal formation starts at the snow line. *Astron. Astrophys., 608,* A92.

Drążkowska J., Alibert Y., and Moore B. (2016) Close-in planetesimal formation by pile-up of drifting pebbles. *Astron. Astrophys., 594,* A105.

Dressing C. D. and Charbonneau D. (2015) The occurrence of potentially habitable planets orbiting M dwarfs estimated from the full Kepler dataset and an empirical measurement of the detection sensitivity. *Astrophys. J., 807,* 45.

Dürmann C. and Kley W. (2015) Migration of massive planets in accreting disks. *Astron. Astrophys., 574,* A52.

Ebel D. S. and Stewart S. T. (2018) The elusive origin of Mercury. In *Mercury: The View after MESSENGER* (S. C. Solomon et al., eds.), pp. 497–515. Cambridge Univ., Cambridge.

Ercolano B. and Pascucci I. (2017) The dispersal of planet-forming discs: theory confronts observations. *R. Soc. Open Sci., 4,* 170114.

Evans N. W. and Tabachnik S. (1999) Possible long-lived asteroid belts in the inner solar system. *Nature, 399,* 41.

Fabrycky D. and Tremaine S. (2007) Shrinking binary and planetary orbits by Kozai cycles with tidal friction. *Astrophys. J., 669,* 1298.

Fabrycky D. C., Lissauer J. J., Ragozzine D., et al. (2014) Architecture of Kepler's multi-transiting systems. II. New investigations with twice as many candidates. *Astrophys. J., 790,* 146.

Fabrycky D. C. and Murray-Clay R. A. (2010) Stability of the directly imaged multiplanet system HR 8799: Resonance and masses. *Astrophys. J., 710,* 1408.

Fang J. and Margot J.-L. (2012) Architecture of planetary systems based on Kepler data: Number of planets and coplanarity. *Astrophys. J., 761,* 92.

Fendyke S. M. and Nelson R. P. (2014) On the corotation torque for low-mass eccentric planets. *Mon. Not. R. Astron. Soc., 437,* 96.

Fischer D. A., Howard A. W., Laughlin G. P., et al. (2014) Exoplanet detection techniques. In *Protostars and Planets VI* (H. Beuther et al., eds.), pp. 715–737. Univ. of Arizona, Tucson.

Fischer D. A., Marcy G. W., Butler R. P., et al. (2008) Five planets orbiting 55 Cancri. *Astrophys. J., 675,* 790.

Fischer D. A., and Valenti J. (2005) The planet-metallicity correlation. *Astrophys. J., 622,* 1102.

Fischer R. A. and Ciesla F. J. (2014) Dynamics of the terrestrial planets from a large number of N-body simulations. *Earth Planet. Sci. Lett., 392,* 28.

Flock M., Fromang S., Turner N. J., et al. (2017) 3D radiation nonideal magnetohydrodynamical simulations of the inner rim in protoplanetary disks. *Astrophys. J., 835,* 230.

Fogg M. J. and Nelson R. P. (2005) Oligarchic and giant impact growth of terrestrial planets in the presence of gas giant planet migration. *Astron. Astrophys., 441,* 791.

Ford E. B., Havlickova M., and Rasio F. A. (2001) Dynamical instabilities in extrasolar planetary systems containing two giant planets. *Icarus, 150,* 303.

Ford E. B., Lystad V., and Rasio F. A. (2005) Planet-planet scattering in the upsilon Andromedae system. *Nature, 434,* 873.

Ford E. B. and Rasio F. A. (2008) Origins of eccentric extrasolar planets: Testing the planet-planet scattering model. *Astrophys. J., 686,* 621.

Ford E. B., Rasio F. A., and Yu K. (2003) In *Scientific Frontiers in Research on Extrasolar Planets* (D. Deming and S. Seager, eds.), pp. 181–188. ASP Conf. Series 294, Astronomical Society of the Pacific, San Francisco.

Foreman-Mackey D., Morton T. D., Hogg D. W., et al. (2016) The population of long-period transiting exoplanets. *Astron. J., 152,* 206.

Fressin F., Torres G., Charbonneau D., et al. (2013) The false positive rate of Kepler and the occurrence of planets. *Astrophys. J., 766,* 81.

Fromang S. and Lesur G. (2017) Angular momentum transport in accretion disks: A hydrodynamical perspective. In *Astro Fluid 2016*, Abstracts Book, p. 65. Institut d'Astrophysique de Paris, arXiv:1705.03319.

Fulton B. J. and Petigura E. A. (2018) The California-Kepler survey. VII. Precise planet radii leveraging Gaia DR2 reveal the stellar mass dependence of the planet radius gap. *Astron. J., 156(6).*

Fulton B. J., Petigura E. A., Howard A. W., et al. (2017) The California-Kepler Survey. III. A gap in the radius distribution of small planets. *Astron. J., 154,* 109.

Fung J., Artymowicz P., and Wu Y. (2015) The 3D flow field around an embedded planet. *Astrophys. J., 811,* 101.

Gaidos E., Krot A. N., Williams J. P. et al. (2009) ^{26}Al and the formation of the solar system from a molecular cloud contaminated by Wolf-Rayet winds. *Astrophys. J., 696,* 1854.

Gaudi B. S., Bennett D. P., Udalski A., et al. (2008) Discovery of a Jupiter/Saturn analog with gravitational microlensing. *Science, 319,* 927.

Gaudi B. S. and Winn J. N. (2007) Prospects for the characterization and confirmation of transiting exoplanets via the Rossiter-McLaughlin effect. *Astrophys. J., 655,* 550.

Genda H. and Abe Y. (2005) Enhanced atmospheric loss on protoplanets at the giant impact phase in the presence of oceans. *Nature, 433,* 842.

Gillon M., Triaud A. H. M. J., Demory B.-O., et al. (2017) Seven temperate terrestrial planets around the nearby ultracool dwarf star TRAPPIST-1. *Nature, 542,* 456.

Ginzburg S., Schlichting H. E., and Sari R. (2016) Super-Earth atmospheres: Self-consistent gas accretion and retention. *Astrophys. J., 825,* 29.

Gladman B., Kavelaars J. J., Petit J., et al. (2001) The structure of the Kuiper belt: Size distribution and radial extent. *Astron. J., 122,* 1051.

Gladman B. J., Migliorini F., Morbidelli A., et al. (1997) Dynamical lifetimes of objects injected into asteroid belt resonances. *Science, 277,* 197.

Goldreich P. and Tremaine S. (1980) Disk-satellite interactions. *Astrophys. J., 241,* 425.

Gomes R., Levison H. F., Tsiganis K., et al. (2005) Origin of the cataclysmic Late Heavy Bombardment period of the terrestrial planets. *Nature, 435,* 466.

Götberg Y., Davies M. B., Mustill A. J., et al. (2016) Long-term stability of the HR 8799 planetary system without resonant lock. *Astron. Astrophys., 592,* A147.

Gould A., Dong S., Gaudi B. S., et al. (2010) Frequency of solar-like systems and of ice and gas giants beyond the snow line from high-magnification microlensing events in 2005–2008. *Astrophys. J., 720,* 1073.

Gould A. and Loeb A. (1992) Discovering planetary systems through gravitational microlenses. *Astrophys. J., 396,* 104.

Gounelle M. and Meibom A. (2008) The origin of short-lived radionuclides and the astrophysical environment of solar system formation. *Astrophys. J., 680,* 781.

Gounelle M. and Meynet G. (2012) Solar system genealogy revealed by extinct short-lived radionuclides in meteorites. *Astron. Astrophys., 545,* A4.

Goździewski K. and Migaszewski C. (2014) Multiple mean motion resonances in the HR 8799 planetary system. *Mon. Not. R. Astron. Soc., 440,* 3140.

Gradie J. and Tedesco E. (1982) Compositional structure of the asteroid belt. *Science, 216,* 1405.

Granvik M., Morbidelli A., Vokrouhlický D., et al. (2017) Escape of asteroids from the main belt. *Astron. Astrophys., 598,* A52.

Gratia P. and Fabrycky D. (2017) Outer-planet scattering can gently tilt an inner planetary system. *Mon. Not. R. Astron. Soc., 464,* 1709.

Greaves J. S., Fischer D. A., and Wyatt M. C. (2006) Metallicity, debris discs and planets. *Mon. Not. R. Astron. Soc., 366,* 283.

Greaves J. S., Kennedy G. M., Thureau N., et al. (2014) Alignment in star-debris disc systems seen by Herschel. *Mon. Not. R. Astron. Soc., 438,* L31.

Greenberg R., Hartmann W. K., Chapman C. R., et al. (1978) Planetesimals to planets — Numerical simulation of collisional evolution. *Icarus, 35,* 1.

Grimm R. E. and McSween H. Y. (1993) Heliocentric zoning of the asteroid belt by aluminum-26 heating. *Science, 259,* 653.

Gupta A. and Schlichting H. E. (2019) Sculpting the valley in the radius distribution of small exoplanets as a by-product of planet formation: The core-powered mass-loss mechanism. *Mon. Not. R. Astron. Soc., 487(1),* 24–33.

Güttler C., Blum J., Zsom A., et al. (2010) The outcome of protoplanetary dust growth: pebbles, boulders, or planetesimals?. I. Mapping the zoo of laboratory collision experiments. *Astron. Astrophys., 513,* A56.

Haisch K. E. Jr., Lada E. A., and Lada C. J. (2001) Disk frequencies and lifetimes in young clusters. *Astrophys. J. Lett., 553,* L153.

Halliday A. N. (2013) The origins of volatiles in the terrestrial planets.

Geochim. Cosmochim. Acta, 105, 146.

Halliday A. N. and Kleine T. (2006) Meteorites and the timing, mechanisms, and conditions of terrestrial planet accretion and early differentiation. In *Meteorites and the Early Solar System II* (D. S. Lauretta and H. Y. McSween Jr., eds.), pp. 775–801. Univ. of Arizona, Tucson.

Hansen B. M. S. (2009) Formation of the terrestrial planets from a narrow annulus. *Astrophys. J., 703,* 1131.

Hansen B. M. S. and Murray N. (2012) Migration then assembly: Formation of Neptune-mass planets inside 1 AU. *Astrophys. J., 751,* 158.

Hansen B. M. S. and Murray N. (2013) Testing in situ assembly with the Kepler planet candidate sample. *Astrophys. J., 775,* 53.

Hartmann L., Calvet N., Gullbring E., et al. (1998) Accretion and the evolution of T Tauri disks. *Astrophys. J., 495,* 385.

Hasegawa Y. and Pudritz R. E. (2011) The origin of planetary system architectures — I. Multiple planet traps in gaseous discs. *Mon. Not. R. Astron. Soc., 417,* 1236.

Hayashi C. (1981) Structure of the solar nebula, growth and decay of magnetic fields and effects of magnetic and turbulent viscosities on the nebula. *Progr. Theor. Phys. Suppl., 70,* 35.

Hayes W. B. (2007) Is the outer solar system chaotic? *Nature Phys., 3,* 689.

Helled R., Bodenheimer P., Podolak M., et al. (2014) Giant planet formation, evolution, and internal structure. In *Protostars and Planets VI* (H. Beuther et al., eds.), pp. 643–665. Univ. of Arizona, Tucson.

Hester J. J., Desch S. J., Healy K. R., et al. (2004) The cradle of the solar system. *Science, 304.*

Hillenbrand L. A. (2008) Disk-dispersal and planet-formation timescales. *Phys. Scr. Vol. T, 130,* 014024.

Hirschmann M. M. (2006) Water, melting, and the deep Earth H_2O cycle. *Annu. Rev. Earth Planet. Sci., 34,* 629.

Hollenbach D., Johnstone D., Lizano S., et al. (1994) Photoevaporation of disks around massive stars and application to ultracompact H II regions. *Astrophys. J., 428,* 654.

Howard A. W., Marcy G. W., Bryson S. T., et al. (2012) Planet occurrence within 0.25 AU of solar-type stars from Kepler. *Astrophys. J., Suppl. Ser., 201,* 15.

Howard A. W., Marcy G. W., Johnson J. A., et al. (2010) The occurrence and mass distribution of close-in super-Earths, Neptunes, and Jupiters. *Science, 330,* 653.

Hsu D. C., Ford E. B., Ragozzine D., and Morehead R. C. (2018) Improving the accuracy of planet occurrence rates from Kepler using approximate Bayesian computation. *Astron. J., 155(5).*

Hu X., Tan J. C., Zhu Z., et al. (2017) Inside-out planet formation. IV. Pebble evolution and planet formation timescales. *Astrophys. J., 857,* 1.

Hu X., Zhu Z., Tan J. C., et al. (2016) Inside-out planet formation. III. Planet-disk interaction at the dead zone inner boundary. *Astrophys. J., 816,* 19.

Hubbard W. B., Hattori M. F., Burrows A., et al. (2007) A mass function constraint on extrasolar giant planet evaporation rates. *Astrophys. J. Lett., 658,* L59.

Huber D., Carter J. A., Barbieri M., et al. (2013) Stellar spin-orbit misalignment in a multiplanet system. *Science, 342,* 331.

Hubickyj O., Bodenheimer P., and Lissauer J. J. (2005) Accretion of the gaseous envelope of Jupiter around a 5 10 Earth-mass core. *Icarus, 179,* 415.

Ida S., Guillot T., and Morbidelli A. (2016) The radial dependence of pebble accretion rates: A source of diversity in planetary systems. I. Analytical formulation. *Astron. Astrophys., 591,* A72.

Ida S. and Lin D. N. C. (2004) Toward a deterministic model of planetary formation. I. A desert in the mass and semimajor axis distributions of extrasolar planets. *Astrophys. J., 604,* 388.

Ida S. and Lin D. N. C. (2008) Toward a deterministic model of planetary formation. IV. Effects of type I migration. *Astrophys. J., 673,* 487.

Ida S. and Lin D. N. C. (2010) Toward a deterministic model of planetary formation. VI. Dynamical interaction and coagulation of multiple rocky embryos and super-Earth systems around solar-type stars. *Astrophys. J., 719,* 810.

Ida S., Lin D. N. C., and Nagasawa M. (2013) Toward a deterministic model of planetary formation. VII. Eccentricity distribution of gas giants. *Astrophys. J., 775,* 42.

Ikoma M. and Hori Y. (2012) In situ accretion of hydrogen-rich atmospheres on short-period super-Earths: Implications for the Kepler-11 planets. *Astrophys. J., 753,* 66.

Ikoma M., Nakazawa K., and Emori H. (2000) Formation of giant planets: Dependences on core accretion rate and grain opacity. *Astrophys. J., 537,* 1013.

Inamdar N. K. and Schlichting H. E. (2015) The formation of super-Earths and mini-Neptunes with giant impacts. *Mon. Not. R. Astron. Soc., 448,* 1751.

Inamdar N. K. and Schlichting H. E. (2016) Stealing the gas: Giant impacts and the large diversity in exoplanet densities. *Astrophys. J. Lett., 817,* L13.

Innanen K. A., Zheng J. Q., Mikkola S., et al. (1997) The Kozai mechanism and the stability of planetary orbits in binary star systems. *Astron. J., 113,* 1915.

Izidoro A. and Raymond S. N. (2018) Formation of terrestrial planets. In *Handbook of Exoplanets* (H. Deeg and J. Belmonte, eds.), Springer, Cham, DOI: 10.1007/978-3-319-55333-7_142.

Izidoro A., Haghighipour N., Winter O. C., et al. (2014a) Terrestrial planet formation in a protoplanetary disk with a local mass depletion: A successful scenario for the formation of Mars. *Astrophys. J., 782,* 31.

Izidoro A., Morbidelli A., and Raymond S. N. (2014b) Terrestrial planet formation in the presence of migrating super-Earths. *Astrophys. J., 794,* 11.

Izidoro A., Morbidelli A., Raymond S. N., et al. (2015a) Accretion of Uranus and Neptune from inward-migrating planetary embryos blocked by Jupiter and Saturn. *Astron. Astrophys., 582,* A99.

Izidoro A., Ogihara M., Raymond S. N., et al. (2017) Breaking the chains: Hot super-Earth systems from migration and disruption of compact resonant chains. *Mon. Not. R. Astron. Soc., 470,* 1750.

Izidoro A., Raymond S. N., Morbidelli A., et al. (2015b) Gas giant planets as dynamical barriers to inward-migrating super-Earths. *Astrophys. J. Lett., 800,* L22.

Izidoro A., Raymond S. N., Morbidelli A., et al. (2015c) Terrestrial planet formation constrained by Mars and the structure of the asteroid belt. *Mon. Not. R. Astron. Soc., 453,* 3619.

Izidoro A., Raymond S. N., Pierens A., et al. (2016) The asteroid belt as a relic from a chaotic early solar system. *Astrophys. J., 833,* 40.

Izidoro A., Bitsch B., Raymond S. N., et al. (2019) Formation of planetary systems by pebble accretion and migration: Hot super-Earth systems from breaking compact resonant chains. *ArXiV e-prints,* arXiv:1902.08772.

Jacobson S. A. and Morbidelli A. (2014) Lunar and terrestrial planet formation in the Grand Tack scenario. *Philos. Trans. R. Soc., A, 372,* 0174.

Jacobson S. A., Morbidelli A., Raymond S. N., et al. (2014) Highly siderophile elements in Earth's mantle as a clock for the Moon-forming impact. *Nature, 508,* 84.

Jacobson S. A. and Walsh K. J. (2015) Earth and terrestrial planet formation. In *The Early Earth: Accretion and Differentiation* (J. Badro and M. Walter, eds.), pp. 49–70. AGU Geophysical Monograph Series 212, American Geophysical Union, Washington DC.

Jílková L., Portegies Zwart S., Pijloo T., et al. (2015) How Sedna and family were captured in a close encounter with a solar sibling. *Mon. Not. R. Astron. Soc., 453,* 3157.

Jin S. and Mordasini C. (2018) Compositional imprints in density-distance-time: A rocky composition for close-in low-mass exoplanets from the location of the valley of evaporation. *Astrophys. J., 853,* 163.

Johansen A., Blum J., Tanaka H., et al. (2014) The multifaceted planetesimal formation process. In *Protostars and Planets VI* (H. Beuther et al., eds.), pp. 547–570. Univ. of Arizona, Tucson.

Johansen A., Davies M. B., Church R. P., et al. (2012) Can planetary instability explain the Kepler Dichotomy? *Astrophys. J., 758,* 39.

Johansen A. and Lacerda P. (2010) Prograde rotation of protoplanets by accretion of pebbles in a gaseous environment. *Mon. Not. R. Astron. Soc., 404,* 475.

Johansen A. and Lambrechts M. (2017) Forming planets via pebble accretion. *Annu. Rev. Earth Planet. Sci., 45,* 359.

Johansen A., Mac Low M.-M., Lacerda P., et al. (2015) Growth of asteroids, planetary embryos, and Kuiper belt objects by chondrule accretion. *Sci. Adv., 1,* 1500109.

Johansen A., Youdin A., and Klahr H. (2009) Zonal flows and long-lived axisymmetric pressure bumps in magnetorotational turbulence. *Astrophys. J., 697,* 1269.

Johnson B. C., Minton D. A., Melosh H. J., et al. (2015) Impact jetting as the origin of chondrules. *Nature, 517,* 339.

Johnson B. C., Walsh K. J., Minton D. A., et al. (2016) Timing of the formation and migration of giant planets as constrained by CB chondrites. *Sci. Adv., 2,* 12.

Johnson J. A., Butler R. P., Marcy G. W., et al. (2007) A new planet around an M dwarf: Revealing a correlation between exoplanets and stellar mass. *Astrophys. J., 670,* 833.

Jones H. R. A., Butler R. P., Tinney C. G., et al. (2006) High-eccentricity planets from the Anglo-Australian Planet Search. *Mon. Not. R. Astron. Soc., 369,* 249.

Jurić M. and Tremaine S. (2008) Dynamical origin of extrasolar planet eccentricity distribution. *Astrophys. J., 686,* 603.

Kaib N. A. and Chambers J. E. (2016) The fragility of the terrestrial planets during a giant-planet instability. *Mon. Not. R. Astron. Soc., 455,* 3561.

Kaib N. A. and Cowan N. B. (2015) The feeding zones of terrestrial planets and insights into Moon formation. *Icarus, 252,* 161.

Kaib N. A. and Quinn T. (2008) The formation of the Oort cloud in open cluster environments. *Icarus, 197,* 221.

Kaib N. A., Raymond S. N., and Duncan M. (2013) Planetary system disruption by galactic perturbations to wide binary stars. *Nature, 493,* 381.

Kaib N. A., Raymond S. N., and Duncan M. J. (2011) 55 Cancri: A coplanar planetary system that is likely misaligned with its star. *Astrophys. J. Lett., 742,* L24.

Kenyon S. J. and Bromley B. C. (2004) Detecting the dusty debris of terrestrial planet formation. *Astrophys. J. Lett., 602,* L133.

Kerridge J. F. (1985) Carbon, hydrogen and nitrogen in carbonaceous chondrites: Abundances and isotopic compositions in bulk samples. *Geochim. Cosmochim. Acta, 49,* 1707.

Kimura K., Lewis R. S., and Anders E. (1974) Distribution of gold and rhenium between nickel-iron and silicate melts: Implications for the abundance of siderophile elements on the Earth and Moon. *Geochim. Cosmochim. Acta, 38,* 683.

Kita N. T., Huss G. R., Tachibana S., et al. (2005) Constraints on the origin of chondrules and CAIs from short-lived and long-lived radionuclides. In *Chondrites and the Protoplanetary Disk* (A. N. Krot et al., eds.), p. 558. ASP Conf. Series 341, Astronomical Society of the Pacific, San Francisco.

Kleine T., Touboul M., Bourdon B., et al. (2009) Hf-W chronology of the accretion and early evolution of asteroids and terrestrial planets. *Geochim. Cosmochim. Acta, 73,* 5150.

Kley W. and Crida A. (2008) Migration of protoplanets in radiative discs. *Astron. Astrophys., 487,* L9.

Kley W. and Nelson R. P. (2012) Planet-disk interaction and orbital evolution. *Annu. Rev. Astron. Astrophys., 50,* 211.

Kokubo E. and Ida S. (1998) Oligarchic growth of protoplanets. *Icarus, 131,* 171.

Kokubo E. and Ida S. (2000) Formation of protoplanets from planetesimals in the solar nebula. *Icarus, 143,* 15.

Kokubo E. and Ida S. (2002) Formation of protoplanet systems and diversity of planetary systems. *Astrophys. J., 581,* 666.

Krasinsky G. A., Pitjeva E. V., Vasilyev M. V., et al. (2002) Hidden mass in the asteroid belt. *Icarus, 158,* 98.

Krot A. N., Amelin Y., Cassen P., et al. (2005) Young chondrules in CB chondrites from a giant impact in the early solar system. *Nature, 436,* 989.

Kruijer T. S., Burkhardt C., Budde C., et al. (2017) Age of Jupiter inferred from the distinct genetics and formation times of meteorites. *Proc. Natl. Acad. Sci. U.S.A., 114 (26),* 6712–6716.

Kruijer T. S., Touboul M., Fischer-Gödde M., et al. (2014) Protracted core formation and rapid accretion of protoplanets. *Science, 344,* 1150.

Kuchner M. J. (2003) Volatile-rich Earth-mass planets in the habitable zone. *Astrophys. J. Lett., 596,* L105.

Kuchynka P. and Folkner W. M. (2013) A new approach to determining asteroid masses from planetary range measurements. *Icarus, 222,* 243.

Lagrange A.-M., Gratadour D., Chauvin G., et al. (2009) A probable giant planet imaged in the β Pictoris disk. VLT/NaCo deep L'-band imaging. *Astron. Astrophys., 493,* L21.

Lai D. (2012) Tidal dissipation in planet-hosting stars: Damping of spin-orbit misalignment and survival of hot Jupiters. *Mon. Not. R. Astron. Soc., 423,* 486.

Lai D., Foucart F., and Lin D. N. C. (2011) Evolution of spin direction of accreting magnetic protostars and spin-orbit misalignment in exoplanetary systems. *Mon. Not. R. Astron. Soc., 412,* 2790.

Lambrechts M. and Johansen A. (2012) Rapid growth of gas-giant cores by pebble accretion. *Astron. Astrophys., 544,* A32.

Lambrechts M. and Johansen A. (2014) Forming the cores of giant planets from the radial pebble flux in protoplanetary discs. *Astron. Astrophys., 572,* A107.

Lambrechts M. and Lega E. (2017) Reduced gas accretion on super-Earths and ice giants. *Astron. Astrophys., 606,* A146.

Lambrechts M., Johansen A., and Morbidelli A. (2014) Separating gas-giant and ice-giant planets by halting pebble accretion. *Astron. Astrophys., 572,* A35.

Lambrechts M., Morbidelli A., Jacobson S., et al. (2019) Formation of planetary systems by pebble accretion and migration — How the radial pebble flux determines a terrestrial-planet or super-Earth growth mode. *Astron. Astrophys., 627,* A83.

Lammer H., Selsis F., Ribas I., et al. (2003) Atmospheric loss of exoplanets resulting from stellar X-ray and extreme-ultraviolet heating. *Astrophys. J. Lett., 598,* L121.

Laskar J. (1990) The chaotic motion of the solar system — A numerical estimate of the size of the chaotic zones. *Icarus, 88,* 266.

Laskar J. (1997) Large scale chaos and the spacing of the inner planets. *Astron. Astrophys., 317,* L75.

Laskar J. and Gastineau M. (2009) Existence of collisional trajectories of Mercury, Mars and Venus with the Earth. *Nature, 459,* 817.

Leake M. A., Chapman C. R., Weidenschilling S. J., et al. (1987) The chronology of Mercury's geological and geophysical evolution — The Vulcanoid hypothesis. *Icarus, 71,* 350.

Lécuyer C., Gillet P., and Robert F. (1998) The hydrogen isotope composition of sea water and the global water cycle. *Chem. Geol., 145,* 249.

Lee E. J. and Chiang E. (2016) Breeding super-Earths and birthing super-puffs in transitional disks. *Astrophys. J., 817,* 90.

Lee E. J. and Chiang E. (2017) Magnetospheric truncation, tidal inspiral, and the creation of short-period and ultra-short-period planets. *Astrophys. J., 842,* 40.

Lee E. J., Chiang E., and Ormel C. W. (2014) Make super-Earths, not Jupiters: Accreting nebular gas onto solid cores at 0.1 AU and beyond. *Astrophys. J., 797,* 95.

Lee M. H. and Peale S. J. (2002) Dynamics and origin of the 2:1 orbital resonances of the GJ 876 planets. *Astrophys. J., 567,* 596.

Lega E., Morbidelli A., and Nesvorný D. (2013) Early dynamical instabilities in the giant planet systems. *Mon. Not. R. Astron. Soc., 431,* 3494.

Leinhardt Z. M. and Richardson D. C. (2005) Planetesimals to protoplanets. I. Effect of fragmentation on terrestrial planet formation. *Astrophys. J., 625,* 427.

Leinhardt Z. M. and Stewart S. T. (2012) Collisions between gravity-dominated bodies. I. Outcome regimes and scaling laws. *Astrophys. J., 745,* 79.

Lesur G., Kunz M. W., and Fromang S. (2014) Thanatology in protoplanetary discs. The combined influence of Ohmic, Hall, and ambipolar diffusion on dead zones. *Astron. Astrophys., 566,* A56.

Levison H. F., Morbidelli A., Tsiganis K., et al. (2011) Late orbital instabilities in the outer planets induced by interaction with a self-gravitating planetesimal disk. *Astron. J., 142,* 152.

Levison H. F. and Stewart G. R. (2001) Remarks on modeling the formation of Uranus and Neptune. *Icarus, 153,* 224.

Lichtenberg T., Golabek G. J., Dullemond C. P., et al. (2018) Impact splash chondrule formation during planetesimal recycling. *Icarus, 302,* 27.

Lin D. N. C. and Papaloizou J. (1986) On the tidal interaction between protoplanets and the protoplanetary disk. III — Orbital migration of protoplanets. *Astrophys. J., 309,* 846.

Lin D. N. C. and Ida S. (1997) On the origin of massive eccentric planets. *Astrophys. J., 477,* 781.

Lin D. N. C., Bodenheimer P., and Richardson D. C. (1996) Orbital migration of the planetary companion of 51 Pegasi to its present location. *Nature, 380,* 606.

Lissauer J. J. (2007) Planets formed in habitable zones of M dwarf stars probably are deficient in volatiles. *Astrophys. J. Lett., 660,* L149.

Lissauer J. J., Ragozzine D., Fabrycky D. C., et al. (2011) Architecture and dynamics of Kepler's candidate multiple transiting planet systems. *Astrophys. J., Suppl. Ser., 197,* 8.

Lissauer J. J. and Stevenson D. J. (2007) Formation of giant planets. In *Protostars and Planets V* (B. Reipurth et al., eds.), pp. 591–606.

Univ. of Arizona, Tucson.

Lopez E. D. (2017) Born dry in the photoevaporation desert: Kepler's ultra-short-period planets formed water-poor. *Mon. Not. R. Astron. Soc., 472,* 245.

Lopez E. D. and Fortney J. J. (2013) The role of core mass in controlling evaporation: The Kepler radius distribution and the Kepler-36 density dichotomy. *Astrophys. J., 776,* 2.

Lopez E. D. and Fortney J. J. (2014) Understanding the mass-radius relation for sub-Neptunes: Radius as a proxy for composition. *Astrophys. J., 792,* 1.

Lovis C. and Mayor M. (2007) Planets around evolved intermediate-mass stars. I. Two substellar companions in the open clusters NGC 2423 and NGC 4349. *Astron. Astrophys., 472,* 657.

Lubow S. H. and D'Angelo G. (2006) Gas flow across gaps in protoplanetary disks. *Astrophys. J., 641,* 526.

Luger R., Sestovic M., Kruse E., et al. (2017) A seven-planet resonant chain in TRAPPIST-1. *Nature Astron., 1,* 0129.

Lykawka P. S. and Ito T. (2017) Terrestrial planet formation: Constraining the formation of Mercury. *Astrophys. J., 838,* 106.

Lynden-Bell D. and Pringle J. E. (1974) The evolution of viscous discs and the origin of the nebular variables. *Mon. Not. R. Astron. Soc., 168,* 603.

Lyra W., Paardekooper S.-J., and Mac Low M.-M. (2010) Orbital migration of low-mass planets in evolutionary radiative models: Avoiding catastrophic infall. *Astrophys. J. Lett., 715,* L68.

Machida M. N., Kokubo E., Inutsuka S.-I., et al. (2010) Gas accretion onto a protoplanet and formation of a gas giant planet. *Mon. Not. R. Astron. Soc., 405,* 1227.

Macintosh B., Graham J. R., Barman T., et al. (2015) Discovery and spectroscopy of the young jovian planet 51 Eri b with the Gemini Planet Imager. *Science, 350,* 64.

Malmberg D., Davies M. B., and Heggie D. C. (2011) The effects of fly-bys on planetary systems. *Mon. Not. R. Astron. Soc., 411,* 859.

Mamajek E. E. (2009) Initial conditions of planet formation: Lifetimes of primordial disk. In *American Institute of Physics Conference Series Vol. 1158* (T. Usuda et al., eds.), pp. 3–10. American Institute of Physics, College Park, Maryland.

Mandell A. M., Raymond S. N., and Sigurdsson S. (2007) Formation of Earth-like planets during and after giant planet migration. *Astrophys. J., 660,* 823.

Marcus R. A., Sasselov D., Stewart S. T., et al. (2010) Water/icy super-Earths: Giant impacts and maximum water content. *Astrophys. J. Lett., 719,* L45.

Marcy G. W., Butler R. P., Vogt S. S., et al. (2001) Two substellar companions orbiting HD 168443. *Astrophys. J., 555,* 418.

Marcy G. W., Isaacson H., Howard A. W., et al. (2014) Masses, radii, and orbits of small Kepler planets: The transition from gaseous to rocky planets. *Astrophys. J., Suppl. Ser., 210,* 20.

Mardling R. A. (2010) The determination of planetary structure in tidally relaxed inclined systems. *Mon. Not. R. Astron. Soc., 407,* 1048.

Marois C., Macintosh B., Barman T., et al. (2008) Direct imaging of multiple planets orbiting the star HR 8799. *Science, 322,* 1348.

Marois C., Zuckerman B., Konopacky Q. M., et al. (2010) Images of a fourth planet orbiting HR 8799. *Nature, 468,* 1080.

Marty B. (2012) The origins and concentrations of water, carbon, nitrogen and noble gases on Earth. *Earth Planet. Sci. Lett., 313,* 56.

Marty B., Altwegg K., Balsiger H., et al. (2017) Xenon isotopes in 67P/Churyumov-Gerasimenko show that comets contributed to Earth's atmosphere. *Science, 356,* 1069.

Marty B., Avice G., Sano Y., et al. (2016) Origins of volatile elements (H, C, N, noble gases) on Earth and Mars in light of recent results from the ROSETTA cometary mission. *Earth Planet. Sci. Lett., 441,* 91.

Marty B. and Yokochi R. (2006) Water in the early Earth. *Rev. Mineral. Geophys., 62,* 421.

Marzari F. (2014) Impact of planet-planet scattering on the formation and survival of debris discs. *Mon. Not. R. Astron. Soc., 444,* 1419.

Marzari F. and Weidenschilling S. J. (2002) Eccentric extrasolar planets: The jumping Jupiter model. *Icarus, 156,* 570.

Marzari F., Baruteau C., and Scholl H. (2010) Planet-planet scattering in circumstellar gas disks. *Astron. Astrophys., 514,* L4.

Masset F. and Snellgrove M. (2001) Reversing type II migration: Resonance trapping of a lighter giant protoplanet. *Mon. Not. R. Astron. Soc., 320,* L55.

Masset F. S. and Casoli J. (2010) Saturated torque formula for planetary migration in viscous disks with thermal diffusion: Recipe for

protoplanet population synthesis. *Astrophys. J., 723,* 1393.

Masset F. S., Morbidelli A., Crida A., et al. (2006) Disk surface density transitions as protoplanet traps. *Astrophys. J., 642,* 478.

Masset F. S. and Papaloizou J. C. B. (2003) Runaway migration and the formation of hot Jupiters. *Astrophys. J., 588,* 494.

Mastrobuono-Battisti A. and Perets H. B. (2017) The composition of solar system asteroids and Earth/Mars moons, and the Earth-Moon composition similarity. *Mon. Not. R. Astron. Soc., 469,* 3597.

Matsumura S., Brasser R., and Ida S. (2016) Effects of dynamical evolution of giant planets on the delivery of atmophile elements during terrestrial planet formation. *Astrophys. J., 818,* 15.

Matsumura S., Ida S., and Nagasawa M. (2013) Effects of dynamical evolution of giant planets on survival of terrestrial planets. *Astrophys. J., 767,* 129.

Matsumura S., Thommes E. W., Chatterjee S., et al. (2010) Unstable planetary systems emerging out of gas disks. *Astrophys. J., 714,* 194.

Mayor M., Marmier M., Lovis C., et al. (2011) The HARPS search for southern extra-solar planets XXXIV. Occurrence, mass distribution and orbital properties of super-Earths and Neptune-mass planets. *ArXiV e-prints,* arXiv:1109.2497.

McNally C. P., Nelson R. P., Paardekooper S.-J., et al. (2017) Low mass planet migration in magnetically torqued dead zones — I. Static migration torque. *Mon. Not. R. Astron. Soc., 472,* 1565.

McNally C. P., Nelson R. P., and Paardekooper S.-J. (2018) Low mass planet migration in magnetically torqued dead zones — II. Flow-locked and runaway migration, and a torque prescription. Mon. Not. R. Astron. Soc., 477(4), 4596–4614, DOI: 10.1093/mnras/sty905.

McNeil D. S. and Nelson R. P. (2010) On the formation of hot Neptunes and super-Earths. *Mon. Not. R. Astron. Soc., 401,* 1691.

Meyer M. R., Calvet N., and Hillenbrand L. A. (1997) Intrinsic near-infrared excesses of T Tauri stars: Understanding the classical T Tauri star locus. *Astron. J., 114,* 288.

Michael G., Basilevsky A., and Neukum G. (2018) On the history of the early meteoritic bombardment of the Moon: Was there a terminal lunar cataclysm? *Icarus, 302,* 80.

Millholland S., Wang S., and Laughlin G. (2017) Kepler multiplanet systems exhibit unexpected intra-system uniformity in mass and radius. *Astrophys. J. Lett., 849,* L33.

Mills S. M., Fabrycky D. C., Migaszewski C., et al. (2016) A resonant chain of four transiting, sub-Neptune planets. *Nature, 533,* 509.

Mizuno H. (1980) Formation of the giant planets. *Prog. Theor. Phys., 64,* 544.

Moeckel N. and Armitage P. J. (2012) Hydrodynamic outcomes of planet scattering in transitional discs. *Mon. Not. R. Astron. Soc., 419,* 366.

Moeckel N., Raymond S. N., and Armitage P. J. (2008) Extrasolar planet eccentricities from scattering in the presence of residual gas disks. *Astrophys. J., 688,* 1361.

Monteux J., Golabek G. J., Rubie D. C., et al. (2018) Water and the interior structure of terrestrial planets and icy bodies. *Space Sci. Rev., 214,* 39.

Moorhead A. V. and Adams F. C. (2005) Giant planet migration through the action of disk torques and planet planet scattering. *Icarus, 178,* 517.

Morbidelli A., Bitsch B., Crida A., et al. (2016) Fossilized condensation lines in the solar system protoplanetary disk. *Icarus, 267,* 368.

Morbidelli A., Brasser R., Gomes R., et al. (2010) Evidence from the asteroid belt for a violent past evolution of Jupiter's orbit. *Astron. J., 140,* 1391.

Morbidelli A., Brasser R., Tsiganis K., et al. (2009) Constructing the secular architecture of the solar system. I. The giant planets. *Astron. Astrophys., 507,* 1041.

Morbidelli A., Chambers J., Lunine J. I., et al. (2000) Source regions and time scales for the delivery of water to Earth. *Meteoritics & Planet. Sci., 35,* 1309.

Morbidelli A. and Crida A. (2007) The dynamics of Jupiter and Saturn in the gaseous protoplanetary disk. *Icarus, 191,* 158.

Morbidelli A., Lambrechts M., Jacobson S., et al. (2015a) The great dichotomy of the solar system: Small terrestrial embryos and massive giant planet cores. *Icarus, 258,* 418.

Morbidelli A. and Levison H. F. (2004) Scenarios for the Origin of the Orbits of the Trans-Neptunian Objects 2000 CR$_{105}$ and 2003 VB$_{12}$ (Sedna). *Astron. J., 128,* 2564.

Morbidelli A., Levison H. F., Tsiganis K., et al. (2005) Chaotic capture of Jupiter's Trojan asteroids in the early solar system. *Nature, 435,* 462.

Morbidelli A., Lunine J. I., O'Brien D. P. et al. (2012) Building terrestrial planets. *Annu. Rev. Earth Planet. Sci., 40,* 251.

Morbidelli A. and Nesvorný D. (2012) Dynamics of pebbles in the vicinity of a growing planetary embryo: Hydro-dynamical simulations. *Astron. Astrophys., 546,* A18.

Morbidelli A., Nesvorný D., Laurenz V., et al. (2018) The timeline of the lunar bombardment: Revisited. *Icarus, 305,* 262.

Morbidelli A. and Raymond S. N. (2016) Challenges in planet formation. *J. Geophys. Res.–Planets, 121,* 1962.

Morbidelli A., Tsiganis K., Crida A., et al. (2007) Dynamics of the giant planets of the solar system in the gaseous protoplanetary disk and their relationship to the current orbital architecture. *Astron. J., 134,* 1790.

Morbidelli A., Walsh K. J., O'Brien D. P., et al. (2015b) The dynamical evolution of the asteroid belt. In *Asteroids IV* (P. Michel et al., eds.), pp. 493–507. Univ. of Arizona, Tucson.

Morbidelli A. and Wood B. J. (2015) Late accretion and the late veneer. In *The Early Earth: Accretion and Differentiation* (J. Badro and M. Walter, eds.), p. 71. AGU Geophysical Monograph Series 212, American Geophysical Union, Washington DC.

Moriarty J. and Ballard S. (2016) The Kepler dichotomy in planetary disks: Linking Kepler observables to simulations of late-stage planet formation. *Astrophys. J., 832,* 34.

Morishima R., Stadel J., and Moore B. (2010) From planetesimals to terrestrial planets: N-body simulations including the effects of nebular gas and giant planets. *Icarus, 207,* 517.

Mróz P., Udalski A., Skowron J., et al. (2017) No large population of unbound or wide-orbit Jupiter-mass planets. *Nature, 548,* 183.

Mukhopadhyay S. (2012) Early differentiation and volatile accretion recorded in deep-mantle neon and xenon. *Nature, 486,* 101.

Mulders G. D., Ciesla F. J., Min M., et al. (2015a) The snow line in viscous disks around low-mass stars: Implications for water delivery to terrestrial planets in the habitable zone. *Astrophys. J., 807,* 9.

Mulders G. D., Pascucci I., and Apai D. (2015b) A stellar-mass-dependent drop in planet occurrence rates. *Astrophys. J., 798,* 112.

Mulders G. D., Pascucci I., and Apai D. (2015c) An increase in the mass of planetary systems around lower-mass stars. *Astrophys. J., 814,* 130.

Mulders G. D., Pascucci I., Apai D., et al. (2018) The exoplanet population observation simulator. I. The inner edges of planetary systems. *Astron. J., 156,* 24.

Murray C. D. and Dermott S. F. (1999) *Solar System Dynamics.* Cambridge Univ., Cambridge.

Muzerolle J., Hillenbrand L., Calvet N., et al. (2003) Accretion in young stellar/substellar objects. *Astrophys. J., 592,* 266.

Nagasawa M., Ida S., and Bessho T. (2008) Formation of hot planets by a combination of planet scattering, tidal circularization, and the Kozai mechanism. *Astrophys. J., 678,* 498.

Nagasawa M., Lin D. N. C., and Thommes E. (2005) Dynamical shake-up of planetary systems. I. Embryo trapping and induced collisions by the sweeping secular resonance and embryo-disk tidal interaction. *Astrophys. J., 635,* 578.

Naoz S., Farr W. M., Lithwick Y., et al. (2011) Hot Jupiters from secular planet-planet interactions. *Nature, 473,* 187.

Nesvorný D. (2015) Evidence for slow migration of Neptune from the inclination distribution of Kuiper belt objects. *Astron. J., 150,* 73.

Nesvorný D. and Morbidelli A. (2012) Statistical study of the early solar system's instability with four, five, and six giant planets. *Astron. J., 144,* 117.

Nesvorný D., Vokrouhlický D., Bottke W. F., et al. (2018) Evidence for very early migration of the solar system planets from the Patroclus-Menoetius binary Jupiter Trojan. *Nature Astron., 2,* 878.

Nesvorný D., Vokrouhlický D., and Morbidelli A. (2007) Capture of irregular satellites during planetary encounters. *Astron. J., 133,* 1962.

Nesvorný D., Vokrouhlický D., and Morbidelli A. (2013) Capture of Trojans by jumping Jupiter. *Astrophys. J., 768,* 45.

Nimmo F. and Kleine T. (2007) How rapidly did Mars accrete? Uncertainties in the Hf-W timing of core formation. *Icarus, 191,* 497.

Nomura R., Hirose K., Uesegi K., et al. (2014) Low core-mantle boundary temperature inferred from the solidus of pyrolite. *Science, 343,* 522.

Nyquist L. E., Kleine T., Shih C.-Y., et al. (2009) The distribution of short-lived radioisotopes in the early solar system and the chronology of asteroid accretion, differentiation, and secondary mineralization. *Geochim. Cosmochim. Acta, 73,* 5115.

O'Brien D. P., Morbidelli A., and Bottke W. F. (2007) The primordial excitation and clearing of the asteroid belt — Revisited. *Icarus, 191,* 434.

O'Brien D. P., Morbidelli A., and Levison H. F. (2006) Terrestrial planet formation with strong dynamical friction. *Icarus, 184,* 39.

O'Brien D. P., Walsh K. J., Morbidelli A., et al. (2014) Water delivery and giant impacts in the Grand Tack scenario. *Icarus, 239,* 74.

Ogihara M. and Ida S. (2009) N-body simulations of planetary accretion around M dwarf stars. *Astrophys. J., 699,* 824.

Ogihara M., Kokubo E., Suzuki T. K., et al. (2018) Formation of close-in super-Earths in evolving protoplanetary disks due to disk winds. *Astron. Astrophys., 615,* A63.

Ogihara M., Morbidelli A., and Guillot T. (2015a) A reassessment of the in situ formation of close-in super-Earths. *Astron. Astrophys., 578,* A36.

Ogihara M., Morbidelli A., and Guillot T. (2015b) Suppression of type I migration by disk winds. *Astron. Astrophys., 584,* L1.

Ormel C. W. and Klahr H. H. (2010) The effect of gas drag on the growth of protoplanets. Analytical expressions for the accretion of small bodies in laminar disks. *Astron. Astrophys., 520,* A43.

Ormel C. W., Liu B., and Schoonenberg D. (2017) Formation of TRAPPIST-1 and other compact systems. *Astron. Astrophys., 604,* A1.

Ouellette N., Desch S. J., and Hester J. J. (2010) Injection of supernova dust in nearby protoplanetary disks. *Astrophys. J., 711,* 597.

Owen J. E. and Wu Y. (2013) Kepler planets: A tale of evaporation. *Astrophys. J., 775,* 105.

Owen J. E. and Wu Y. (2017) The evaporation valley in the Kepler planets. *Astrophys. J., 847,* 29.

Ozima M. and Podosek F. (2002) *Noble Gas Geochemistry, 2nd edition.* Cambridge Univ., Cambridge. 286 pp.

Paardekooper S.-J. and Mellema G. (2006) Halting type I planet migration in non-isothermal disks. *Astron. Astrophys., 459,* L17.

Paardekooper S.-J., Baruteau C., and Kley W. (2011) A torque formula for non-isothermal type I planetary migration — II. Effects of diffusion. *Mon. Not. R. Astron. Soc., 410,* 293.

Panero W. R. and Caracas R. (2017) Stability of phase H in the $MgSiO_4H_2$-AlOOH-SiO_2 system. *Earth Planet. Sci. Lett., 463,* 171.

Papaloizou J. C. B. and Larwood J. D. (2000) On the orbital evolution and growth of protoplanets embedded in a gaseous disc. *Mon. Not. R. Astron. Soc., 315,* 823.

Papaloizou J. C. B. and Terquem C. (2006) Planet formation and migration. *Rept. Prog. Phys., 69,* 119.

Pascucci I., Apai D., Luhman K., et al. (2009) The different evolution of gas and dust in disks around Sun-like and cool stars. *Astrophys. J., 696,* 143.

Pascucci I., Testi L., Herczeg G. J., et al. (2016) A steeper than linear disk mass-stellar mass scaling relation. *Astrophys. J., 831,* 125.

Penny M. T., Gaudi B. S., Kerins E., et al. (2019) Predictions of the WFIRST microlensing survey I: Bound planet detection rates. *Astrophys. J. Suppl. Ser., 241(1),* DOI: 10.3847/1538-4365/aafb69.

Petigura E. A., Howard A. W., and Marcy G. W. (2013) Prevalence of Earth-size planets orbiting Sun-like stars. *Proc. Natl. Acad. Sci. U.S.A., 110,* 19273.

Petit J., Morbidelli A., and Chambers J. (2001) The primordial excitation and clearing of the asteroid belt. *Icarus, 153,* 338.

Pfalzner S., Bhandare A., Vincke K., et al. (2018) Outer solar system possibly shaped by a stellar fly-by. *Astrophys. J., 863,* 45.

Pierens A. and Nelson R. P. (2008) Constraints on resonant-trapping for two planets embedded in a protoplanetary disc. *Astron. Astrophys., 482,* 333.

Pierens A. and Raymond S. N. (2011) Two phase, inward-then outward migration of Jupiter and Saturn in the gaseous solar nebula. *Astron. Astrophys., 533,* A131.

Pierens A., Raymond S. N., Nesvorný D., et al. (2014) Outward migration of Jupiter and Saturn in 3:2 or 2:1 resonance in radiative disks: Implications for the Grand Tack and Nice models. *Astrophys. J. Lett., 795,* L11.

Poleski R., Skowron J., Udalski A., et al. (2014) Triple microlens OGLE-2008-BLG-092L: Binary stellar system with a circumprimary Uranus-type planet. *Astrophys. J., 795,* 42.

Pollack J. B., Hubickyj O., Bodenheimer P., et al. (1996) Formation of the Giant planets by concurrent accretion of solids and gas. *Icarus, 124,* 62.

Portegies Zwart S. (2019) The formation of solar system analogs in young star clusters. *Astron. Astrophys., 622,* A69, DOI:

10.1051/0004-6361/201833974.

Quinn T. R., Tremaine S., and Duncan M. (1991) A three million year integration of the Earth's orbit. *Astron. J., 101,* 2287.

Quintana E. V., Barclay T., Borucki W. J., et al. (2016) The frequency of giant impacts on Earth-like worlds. *Astrophys. J., 821,* 126.

Quintana E. V., Barclay T., Raymond S. N., et al. (2014) An Earth-sized planet in the habitable zone of a cool star. *Science, 344,* 277.

Rafikov R. R. (2004) Fast accretion of small planetesimals by protoplanetary cores. *Astron. J., 128,* 1348.

Rasio F. A. and Ford E. B. (1996) Dynamical instabilities and the formation of extrasolar planetary systems. *Science, 274,* 954.

Raymond S. N. (2006) The search for other Earths: Limits on the giant planet orbits that allow habitable terrestrial planets to form. *Astrophys. J. Lett., 643,* L131.

Raymond S. N., Armitage P. J., and Gorelick N. (2010) Planet-planet scattering in planetesimal disks. II. Predictions for outer extrasolar planetary systems. *Astrophys. J., 711,* 772.

Raymond S. N., Armitage P. J., Moro-Martín A., et al. (2011) Debris disks as signposts of terrestrial planet formation. *Astron. Astrophys., 530,* A62.

Raymond S. N., Armitage P. J., Moro-Martín A., et al. (2012) Debris disks as signposts of terrestrial planet formation. II. Dependence of exoplanet architectures on giant planet and disk properties. *Astron. Astrophys., 541,* A11.

Raymond S. N., Armitage P. J., Veras D., et al. (2018a) Implications of the interstellar object 1I/'Oumuamua for planetary dynamics and planetesimal formation. *Mon. Not. R. Astron. Soc., 476,* 3031.

Raymond S. N., Barnes R., Armitage P. J., et al. (2008a) Mean motion resonances from planet-planet scattering. *Astrophys. J. Lett., 687,* L107.

Raymond S. N., Barnes R., and Mandell A. M. (2008b) Observable consequences of planet formation models in systems with close-in terrestrial planets. *Mon. Not. R. Astron. Soc., 384,* 663.

Raymond S. N., Barnes R., Veras D. et al. (2009a) Planet-planet scattering leads to tightly packed planetary systems. *Astrophys. J. Lett., 696,* L98.

Raymond S. N., Boulet T., Izidoro A., et al. (2018b) Migration-driven diversity of super-Earth compositions. *Mon. Not. R. Astron. Soc., 479,* L81.

Raymond S. N. and Izidoro A. (2017a) Origin of water in the inner solar system: Planetesimals scattered inward during Jupiter and Saturn's rapid gas accretion. *Icarus, 297,* 134.

Raymond S. N. and Izidoro A. (2017b) The empty primordial asteroid belt. *Sci. Adv., 3,* e1701138.

Raymond S. N., Izidoro A., Bitsch B., et al. (2016) Did Jupiter's core form in the innermost parts of the Sun's protoplanetary disc? *Mon. Not. R. Astron. Soc., 458,* 2962.

Raymond S. N., Kokubo E., Morbidelli A., et al. (2014) Terrestrial planet formation at home and abroad. In *Protostars and Planets VI* (H. Beuther et al., eds.), pp. 595–618. Univ. of Arizona, Tucson.

Raymond S. N., Mandell A. M., and Sigurdsson S. (2006a) Exotic Earths: Forming habitable worlds with giant planet migration. *Science, 313,* 1413.

Raymond S. N. and Morbidelli A. (2014) The Grand Tack model: A critical review. *Proc. Intl. Astron. Union, 9*(S310), 194–203.

Raymond S. N., O'Brien D. P., Morbidelli A., et al. (2009b) Building the terrestrial planets: Constrained accretion in the inner solar system. *Icarus, 203,* 644.

Raymond S. N., Quinn T., and Lunine J. I. (2004) Making other Earths: Dynamical simulations of terrestrial planet formation and water delivery. *Icarus, 168,* 1.

Raymond S. N., Quinn T., and Lunine J. I. (2005) Terrestrial planet formation in disks with varying surface density profiles. *Astrophys. J., 632,* 670.

Raymond S. N., Quinn T., and Lunine J. I. (2006b) High-resolution simulations of the final assembly of Earth-like planets I. Terrestrial accretion and dynamics. *Icarus, 183,* 265.

Raymond S. N., Quinn T., and Lunine J. I. (2007a) High-resolution simulations of the final assembly of Earth-Like planets. 2. Water delivery and planetary habitability. *Astrobiology, 7,* 66.

Raymond S. N., Scalo J., and Meadows V. S. (2007b) A decreased probability of habitable planet formation around low-mass stars. *Astrophys. J., 669,* 606.

Reidemeister M., Krivov A. V., Schmidt T. O. B., et al. (2009) A possible architecture of the planetary system HR 8799. *Astron. Astrophys., 503,* 247.

Ribas I. and Miralda-Escudé J. (2007) The eccentricity-mass distribution of exoplanets: Signatures of different formation mechanisms? *Astron. Astrophys., 464,* 779.

Rice W. K. M. and Armitage P. J. (2003) On the formation timescale and core masses of gas giant planets. *Astrophys. J. Lett., 598,* L55.

Rivera E. J., Laughlin G., Butler R. P., et al. (2010) The Lick-Carnegie Exoplanet Survey: A Uranus-mass fourth planet for GJ 876 in an extrasolar Laplace configuration. *Astrophys. J., 719,* 890.

Robert F., Merlivat L., and Javoy M. (1977) Water and deuterium content in eight chondrites. *Meteoritics, 12,* 349.

Rogers L. A. (2015) Most 1.6 Earth-radius planets are not rocky. *Astrophys. J., 801,* 41.

Rogers L. A., Bodenheimer P., Lissauer J. J., et al. (2011) Formation and structure of low-density exo-Neptunes. *Astrophys. J., 738,* 59.

Roig F. and Nesvorný D. (2015) The evolution of asteroids in the jumping-Jupiter migration model. *Astron. J., 150,* 186.

Roig F., Nesvorný D., and DeSouza S. R. (2016) Jumping Jupiter can explain Mercury's orbit. *Astrophys. J. Lett., 820,* L30.

Romanova M. M. and Lovelace R. V. E. (2006) The magnetospheric gap and the accumulation of giant planets close to a star. *Astrophys. J. Lett., 645,* L73.

Romanova M. M., Ustyugova G. V., Koldoba A. V., et al. (2003) Three-dimensional simulations of disk accretion to an inclined dipole. I. Magnetospheric flows at different Θ. *Astrophys. J., 595,* 1009.

Romanova M. M., Ustyugova G. V., Koldoba A. V., et al. (2004) Three-dimensional simulations of disk accretion to an inclined dipole. II. Hot spots and variability. *Astrophys. J., 610,* 920.

Ros K. and Johansen A. (2013) Ice condensation as a planet formation mechanism. *Astron. Astrophys., 552,* A137.

Rowan D., Meschiari S., Laughlin G., et al. (2016) The Lick-Carnegie exoplanet survey: HD 32963 — A new Jupiter analog orbiting a Sun-like star. *Astrophys. J., 817,* 104.

Rowe J. F., Bryson S. T., Marcy G. W., et al. (2014) Validation of Kepler's multiple planet candidates. III. Light curve analysis and announcement of hundreds of new multi-planet systems. *Astrophys. J., 784,* 45.

Rubie D. C., Frost D. J., Mann U., et al. (2011) Heterogeneous accretion, composition and core-mantle differentiation of the Earth. *Earth Planet. Sci. Lett., 301,* 31.

Rubie D. C., Jacobson S. A., Morbidelli A., et al. (2015) Accretion and differentiation of the terrestrial planets with implications for the compositions of early-formed solar system bodies and accretion of water. *Icarus, 248,* 89.

Safronov V. S. (1969) *Evoliutsiia Doplanetnogo Oblaka.* Nauka, Moscow. 244 pp.

Sanchis-Ojeda R., Rappaport S., Winn J. N., et al. (2014) A study of the shortest-period planets found with Kepler. *Astrophys. J., 787,* 47.

Scalo J., Kaltenegger L., Segura A. G., et al. (2007) M stars as targets for terrestrial exoplanet searches and biosignature detection. *Astrobiology, 7,* 85.

Schäfer U., Yang C.-C., and Johansen A. (2017) Initial mass function of planetesimals formed by the streaming instability. *Astron. Astrophys., 597,* A69.

Schiller M., Connelly J. N., Glad A. C., et al. (2015) Early accretion of protoplanets inferred from a reduced inner solar system ^{26}Al inventory. *Earth Planet. Sci. Lett., 420,* 45.

Schlaufman K. C. (2014) Tests of in situ formation scenarios for compact multiplanet systems. *Astrophys. J., 790,* 91.

Schlichting H. E. (2014) Formation of close in super-Earths and mini-Neptunes: Required disk masses and their implications. *Astrophys. J. Lett., 795,* L15.

Schlichting H. E., Sari R., and Yalinewich A. (2015) Atmospheric mass loss during planet formation: The importance of planetesimal impacts. *Icarus, 247,* 81.

Scholz A., Jayawardhana R., and Wood K. (2006) Exploring brown dwarf disks: A 1.3 mm survey in Taurus. *Astrophys. J., 645,* 1498.

Schoonenberg D. and Ormel C. W. (2017) Planetesimal formation near the snowline: In or out? *Astron. Astrophys., 602,* A21.

Selsis F., Chazelas B., Bordé P., et al. (2007) Could we identify hot ocean-planets with CoRoT, Kepler and Doppler velocimetry? *Icarus, 191,* 453.

Shakura N. I. and Sunyaev R. A. (1973) Black holes in binary systems. Observational appearance. *Astron. Astrophys., 24,* 337.

Simon J. B., Armitage P. J., Li R., et al. (2016) The mass and size distribution of planetesimals formed by the streaming instability. I. The role of self-gravity. *Astrophys. J., 822,* 55.

Simon J. B., Armitage P. J., Youdin A. N., et al. (2017) Evidence for universality in the initial planetesimal mass function. *Astrophys. J. Lett., 847,* L12.

Sinukoff E., Howard A. W., Petigura E. A., et al. (2017) Mass constraints of the WASP-47 planetary system from radial velocities. *Astron. J., 153,* 70.

Snellgrove M. D., Papaloizou J. C. B., and Nelson R. P. (2001) On disc driven inward migration of resonantly coupled planets with application to the system around GJ 876. *Astron. Astrophys., 374,* 1092.

Steffen J. H. and Hwang J. A. (2015) The period ratio distribution of Kepler's candidate multiplanet systems. *Mon. Not. R. Astron. Soc., 448,* 1956.

Stern S. A. and Durda D. D. (2000) Collisional evolution in the Vulcanoid region: Implications for present-day population constraints. *Icarus, 143,* 360.

Stewart S. T. and Leinhardt Z. M. (2012) Collisions between gravity-dominated bodies. II. The diversity of impact outcomes during the end stage of planet formation. *Astrophys. J., 751,* 32.

Sumi T., Udalski A., Bennett D. P., et al. (2016) The first Neptune analog or super-Earth with a Neptune-like orbit: MOA-2013-BLG-605Lb. *Astrophys. J., 825,* 112.

Surville C. and Mayer L. (2019) Dust-vortex instability in the regime of well-coupled grains. *Astrophys. J., 883(2),* 176, DOI: 10.3847/1538-4357/ab3e47.

Surville C., Mayer L., and Lin D. N. C. (2016) Dust capture and long-lived density enhancements triggered by vortices in 2D protoplanetary disks. *Astrophys. J., 831,* 82.

Suzuki D., Bennett D. P., Sumi T., et al. (2016a) The exoplanet mass-ratio function from the MOA-II Survey: Discovery of a break and likely peak at a Neptune mass. *Astrophys. J., 833,* 145.

Suzuki T. K., Ogihara M., Morbidelli A., et al. (2016b) Evolution of protoplanetary discs with magnetically driven disc winds. *Astron. Astrophys., 596,* A74.

Swift J. J., Johnson J. A., Morton T. D., et al. (2013) Characterizing the cool KOIs. IV. Kepler-32 as a prototype for the formation of compact planetary systems throughout the galaxy. *Astrophys. J., 764,* 105.

Szulágyi J., Masset F., Lega E., et al. (2016) Circumplanetary disc or circumplanetary envelope? *Mon. Not. R. Astron. Soc., 460,* 2853.

Tanaka H., Takeuchi T., and Ward W. R. (2002) Three-dimensional interaction between a planet and an isothermal gaseous disk. I. Corotation and Lindblad torques and planet migration. *Astrophys. J., 565,* 1257.

Tanaka H. and Ward W. R. (2004) Three-dimensional interaction between a planet and an isothermal gaseous disk. II. Eccentricity waves and bending waves. *Astrophys. J., 602,* 388.

Tera F., Papanastassiou D. A., and Wasserburg G. J. (1974) Isotopic evidence for a terminal lunar cataclysm. *Earth Planet. Sci. Lett., 22,* 1.

Terquem C. and Papaloizou J. C. B. (2007) Migration and the formation of systems of hot super-Earths and Neptunes. *Astrophys. J., 654,* 1110.

Teske J. K., Ciardi D. R., Howell S. B., et al. (2018) The effects of stellar companions on exoplanet radius distributions. *Astron. J., 156(6),* 292, DOI: 10.3847/1538-3881/aaed2d.

Thommes E., Nagasawa M., and Lin D. N. C. (2008a) Dynamical shake-up of planetary systems. II. N-body simulations of solar system terrestrial planet formation induced by secular resonance sweeping. *Astrophys. J., 676,* 728-739.

Thommes E. W., Bryden G., Wu Y., et al. (2008b) From mean motion resonances to scattered planets: Producing the solar system, eccentric exoplanets, and late heavy bombardments. *Astrophys. J., 675,* 1538.

Thommes E. W., Duncan M. J., and Levison H. F. (2003) Oligarchic growth of giant planets. *Icarus, 161,* 431.

Thommes E. W., Matsumura S., and Rasio F. A. (2008c) Gas disks to gas giants: Simulating the birth of planetary systems. *Science, 321,* 814.

Throop H. B. and Bally J. (2005) Can photoevaporation trigger planetesimal formation? *Astrophys. J. Lett., 623,* L149.

Timpe M., Barnes R., Kopparapu R., et al. (2013) Secular behavior of exoplanets: Self-consistency and comparisons with the planet-planet scattering hypothesis. *Astron. J., 146,* 63.

Touboul M., Kleine T., Bourdon B., et al. (2007) Late formation and prolonged differentiation of the Moon inferred from W isotopes in lunar metals. *Nature, 450,* 1206.

Tremaine S. (1993) The distribution of comets around stars. In *Planets Around Pulsars* (J. A. Phillips et al., eds.), pp. 335–344. ASP Conference Series 36, Astronomical Society of the Pacific, San Francisco.

Triaud A. H. M. J., Collier Cameron A., Queloz D., et al. (2010) Spin-orbit angle measurements for six southern transiting planets. New insights into the dynamical origins of hot Jupiters. *Astron. Astrophys., 524,* A25.

Tsiganis K., Gomes R., Morbidelli A., et al. (2005) Origin of the orbital architecture of the giant planets of the solar system. *Nature, 435,* 459.

Turner N. J., Fromang S., Gammie C., et al. (2014) Transport and accretion in planet-forming disks. In *Protostars and Planets VI* (H. Beuther et al., eds.), pp. 411–432. Univ. of Arizona, Tucson.

Udry S. and Santos N. C. (2007) Statistical properties of exoplanets. *Annu. Rev. Astron. Astrophys., 45,* 397.

Van Eylen V., Agentoft C., Lundkvist M. S., et al. (2018) An asteroseismic view of the radius valley: Stripped cores, not born rocky. *Mon. Not. R. Astron. Soc., 479,* 4786.

Veras D. and Armitage P. J. (2004) Outward migration of extrasolar planets to large orbital radii. *Mon. Not. R. Astron. Soc., 347,* 613.

Veras D. and Armitage P. J. (2005) The influence of massive planet scattering on nascent terrestrial planets. *Astrophys. J. Lett., 620,* L111.

Veras D. and Armitage P. J. (2006) Predictions for the correlation between giant and terrestrial extrasolar planets in dynamically evolved systems. *Astrophys. J., 645,* 1509.

Veras D. and Raymond S. N. (2012) Planet-planet scattering alone cannot explain the free-floating planet population. *Mon. Not. R. Astron. Soc., 421,* L117.

Vokrouhlický D., Farinella P., and Bottke W. F. (2000) The depletion of the putative vulcanoid population via the Yarkovsky effect. *Icarus, 148,* 147.

Volk K. and Gladman B. (2015) Consolidating and crushing exoplanets: Did It happen here? *Astrophys. J. Lett., 806,* L26.

Walker R. J. (2009) Highly siderophile elements in the Earth, Moon and Mars: Update and implications for planetary accretion and differentiation. *Chem. Erde–Geochem., 69,* 101.

Wallace J., Tremaine S., and Chambers J. (2017) Collisional fragmentation is not a barrier to close-in planet formation. *Astron. J., 154,* 175.

Walsh K. J. and Levison H. F. (2016) Terrestrial planet formation from an annulus. *Astron. J., 152,* 68.

Walsh K. J., Morbidelli A., Raymond S. N., et al. (2011) A low mass forMars from Jupiter's early gas-driven migration. *Nature, 475,* 206.

Walsh K. J., Morbidelli A., Raymond S. N., et al. (2012) Populating the asteroid belt from two parent source regions due to the migration of giant planets — "The Grand Tack." *Meteoritics & Planet. Sci., 47,* 1941.

Ward W. R. (1986) Density waves in the solar nebula — Differential Lindblad torque. *Icarus, 67,* 164.

Ward W. R. (1997) Protoplanet migration by nebula tides. *Icarus, 126,* 261.

Warren P. H. (2011) Stable-isotopic anomalies and the accretionary assemblage of the Earth and Mars: A subordinate role for carbonaceous chondrites. *Earth Planet. Sci. Lett., 311,* 93.

Weidenschilling S. J. (1977a) Aerodynamics of solid bodies in the solar nebula. *Mon. Not. R. Astron. Soc., 180,* 57.

Weidenschilling S. J. (1977b) The distribution of mass in the planetary system and solar nebula. *Astrophys. Space Sci., 51,* 153.

Weidenschilling S. J. and Marzari F. (1996) Gravitational scattering as a possible origin for giant planets at small stellar distances. *Nature, 384,* 619.

Weiss L. M. and Marcy G. W. (2014) The mass-radius relation for 65 exoplanets smaller than 4 Earth radii. *Astrophys. J. Lett., 783,* L6.

Weiss L. M., Marcy G. W., Petigura E. A., et al. (2018) The California-Kepler survey. V. Peas in a pod: Planets in a Kepler multi-planet system are similar in size and regularly spaced. *Astron. J., 155,* 48.

Wetherill G. W. (1978) Accumulation of the terrestrial planets. In *Protostars and Planets* (T. Gehrels, ed.), pp. 565–598. Univ. of Arizona, Tucson.

Wetherill G. W. (1985) Occurrence of giant impacts during the growth of the terrestrial planets. *Science, 228,* 877.

Wetherill G. W. (1991) Why isn't Mars as big as Earth? In *Lunar and Planetary Science XXII,* Abstract #1743. Lunar and Planetary

Institute, Houston.

Wetherill G. W. (1996) The formation and habitability of extra-solar planets. *Icarus, 119,* 219.

Wetherill G. W. and Stewart G. R. (1993) Formation of planetary embryos — Effects of fragmentation, low relative velocity, and independent variation of eccentricity and inclination. *Icarus, 106,* 190.

Williams J. P. and Cieza L. A. (2011) Protoplanetary disks and their evolution. *Annu. Rev. Astron. Astrophys., 49,* 67.

Winn J. N. (2018) Planet occurrence: Doppler and transit surveys. In *Handbook of Exoplanets* (H. J. Deeg and J. A. Belmonte, eds.), pp. 1949–1966. Springer Nature, Switzerland.

Winn J. N., Fabrycky D., Albrecht S. et al. (2010) Hot stars with hot Jupiters have high obliquities. *Astrophys. J. Lett., 718,* L145.

Winn J. N. and Fabrycky D. C. (2015) The occurrence and architecture of exoplanetary systems. *Annu. Rev. Astron. Astrophys., 53,* 409.

Winn J. N., Noyes R. W., Holman M. J., et al. (2005) Measurement of spin-orbit alignment in an extrasolar planetary system. *Astrophys. J., 631,* 1215.

Wise A. W. and Dodson-Robinson S. E. (2018) Photoevaporation does not create a pileup of giant planets at 1 au. *Astrophys. J., 855,* 145.

Wittenmyer R. A., Butler R. P., Tinney C. G., et al. (2016) The Anglo-Australian Planet Search XXIV: The frequency of Jupiter analogs. *Astrophys. J., 819,* 28.

Wolfgang A., Rogers L. A., and Ford E. B. (2016) Probabilistic mass-radius relationship for sub-Neptune-sized planets. *Astrophys. J., 825,* 19.

Wright J. T., Marcy G. W., Butler R. P., et al. (2008) The Jupiter twin HD 154345b. *Astrophys. J. Lett., 683,* L63.

Wright J. T., Upadhyay S., Marcy G. W., et al. (2009) Ten new and updated multiplanet systems and a survey of exoplanetary systems. *Astrophys. J., 693,* 1084.

Wright J. T., Fakhouri O., Marcy G. W., et al. (2011) The exoplanet orbit database. *Publ. Astron. Soc. Pac., 123(902),* 412, DOI: 10.1086/659427.

Wright J. T., Marcy G. W., Howard A. W., et al. (2012) The frequency of hot Jupiters orbiting nearby solar-type stars. *Astrophys. J., 753,* 160.

Yang C.-C., Johansen A., and Carrera D. (2017) Concentrating small particles in protoplanetary disks through the streaming instability. *Astron. Astrophys., 606,* A80.

Youdin A. N. and Goodman J. (2005) Streaming instabilities in protoplanetary disks. *Astrophys. J., 620,* 459.

Zahn J.-P. (1977) Tidal friction in close binary stars. *Astron. Astrophys., 57,* 383.

Zhang H. and Zhou J.-L. (2010) On the orbital evolution of a giant planet pair embedded in a gaseous disk. I. Jupiter-Saturn configuration. *Astrophys. J., 714,* 532.

Zhou J.-L., Aarseth S. J., Lin D. N. C., et al. (2005) Origin and ubiquity of short-period Earth-like planets: Evidence for the sequential accretion theory of planet formation. *Astrophys. J. Lett., 631,* L85.

Zhu W., Petrovich C., Wu Y., et al. (2018) About 30% of Sun-like stars have Kepler-like planetary systems: A study of their intrinsic architecture. *Astrophys. J., 860,* 101.

Meech K. and Raymond S. (2020) Origin of Earth's water: Sources and constraints. In *Planetary Astrobiology* (V. Meadows et al., eds.), pp. 325–353. Univ. of Arizona, Tucson, DOI: 10.2458/azu_uapress_9780816540068-ch013.

Origin of Earth's Water: Sources and Constraints

Karen Meech
Institute for Astronomy

Sean N. Raymond
Laboratoire d'Astrophysique de Bordeaux, CNRS, and Université de Bordeaux

The origin of Earth's water is a longstanding question in the fields of planetary science, planet formation, and astrobiology. In this chapter we review our current state of knowledge. Empirical constraints on the origin of Earth's water come from chemical and isotopic measurements of solar system bodies and of Earth itself. Dynamical models have revealed potential pathways for the delivery of water to Earth during its formation, most of which are anchored to specific models for terrestrial planet formation. Meanwhile, disk chemical models are focused on determining how the isotopic ratios of the building blocks of planets varied as a function of radial distance and time, defining markers of material transported along those pathways. Carbonaceous chondrite meteorites — representative of the outer asteroid belt — provide a good match to Earth's bulk water content (although mantle plumes have been measured at a lower D/H). What remains to be understood is how this relationship was established. Did Earth's water originate among the asteroids (as in the classical model of terrestrial planet formation)? Or, more likely, was Earth's water delivered from the same parent population as the hydrated asteroids (e.g., external pollution, as in the Grand Tack model)? We argue that the outer asteroid belt — at the boundary between the inner and outer solar system — is the next frontier for new discoveries. The outer asteroid belt is icy, as shown by its population of icy bodies and volatile-driven activity seen on 12 main-belt comets (MBCs), seven of which exhibit sublimation-driven activity on repeated perihelion passages. Measurements of the isotopic characteristics of MBCs would provide essential missing links in the chain between disk models and dynamical models. Finally, we extrapolate to water delivery to rocky exoplanets. Migration is the only mechanism likely to produce very-water-rich planets with more than a few percent water by mass (and even with migration, some planets are purely rocky). While water-loss mechanisms remain to be studied in more detail, we expect that water should be delivered to the vast majority of rocky exoplanets.

1. INTRODUCTION

We have only one example of an inhabited world, namely Earth, with its thin veneer of water, the solvent essential to known life. Is a terrestrial planet like Earth that lies in the habitable zone and has the ingredients of habitability a common outcome of planet formation or an oddity that relied on a unique set of stochastic processes during the growth and subsequent evolution of our solar system?

Ultimately, water originates in space, likely formed on dust grain surfaces via reactions with atoms inside cold molecular clouds (*Tielens and Hagen,* 1982; *Jing et al.,* 2011). Grain surface water ice then provides a medium for the chemistry that forms molecules of biogenic relevance (*Herbst and van Dishoeck,* 2009). Both water and organ-

ics are abundant ingredients in protoplanetary disks out of which solar systems form. In the disk, water is present both as a gas and a solid and is likely processed during planet formation. Earth formed in our solar system's inner protoplanetary disk. The sequence of events that led to the growth of a rocky Earth with a thin veil of water have yet to be fully understood. Both thermal processing — in particular, desiccation of the earliest generations of planetesimals due to ^{26}Al heating (*Grimm and McSween,* 1993; *Monteux et al.,* 2018; *Alexander,* 2019a) — and Earth's orbital location within the planet-forming disk likely played important roles. Water only exists as ice past the disk's snow line (the radial distance from the star beyond which the temperatures are low enough for gases to condense as ice), which itself shifts inward in time as the disk evolves, and which may have been located interior to Earth's orbit during part of Earth's formation. Reconciling Earth's water content with meteoritic constraints and models for solar system formation remains a challenge.

While there has been significant work to investigate the origin of Earth's water — the wellspring of life — we still do not know if it came mostly from the inner disk or if it was delivered from the outer disk. No one knows if our solar system, with a planet possessing the necessary ingredients for life within the habitable zone, e.g., sufficient water and organics, is a cosmic rarity. Nor do we know whether the gas giants in our solar system aided or impeded the delivery of essential materials to the habitable zone. The answers to these questions are contained in volatiles (compounds that are typically found as gases or ices in the outer solar system) unaltered since the formation of the giant planets. To access this record, we need (1) a population of bodies that faithfully records the history of volatile migration in the early solar system, (2) a source of volatiles that we can access, (3) knowledge that the volatiles were not altered by aqueous interaction with their parent body, and (4) measurements from multiple chemical markers with sufficient precision to distinguish between original volatile reservoirs.

Comets were long thought to be the most likely "delivery service" of Earth's water based on a comparison of their D/H ratios with that of Earth's ocean (*Delsemme*, 1992; *Owen and Bar-Nun*, 1995). But new models and data, including the Rosetta mission's survey of Comet 67P/Churyumov-Gerasimenko, have cast doubt on this source. The relative abundances of 67P's volatile isotopes (D/H, N, C, and noble gases) do not match those of Earth (*Altwegg et al.*, 2015; *Marty et al.*, 2016). Furthermore, comets will not provide an answer to the original source of Earth's water because they are dynamically unstable and cannot be traced back to where they formed (*Levison and Duncan*, 1997; *Brasser et al.*, 2013). The only way to learn where Earth's water came from is to match the chemical fingerprints of inner solar system volatiles to a location in the protoplanetary disk. Such data can distinguish between competing models of solar system formation to specify where the water came from and how it was delivered.

Our solar system is quantifiably unusual compared to the thousands of known exoplanet systems (*Martin and Livio*, 2015; *Mulders et al.*, 2018; see also the chapter by Raymond et al. in this volume). After taking observational biases into account, only ~10% of Sun-like stars have gas giants (*Cumming et al.*, 2008; *Mayor et al.*, 2011; *Fernandes et al.*, 2019), and only ~10% of outer gas giants have low-eccentricity orbits like Jupiter (*Butler et al.*, 2006; *Udry et al.*, 2007), putting the solar system's orbital architecture in an ~1% minority (for a discussion, see the chapter by Raymond et al.). Many exoplanetary systems have Earth-sized planets in or near their host star's habitable zones [e.g., in the TRAPPIST-1 system (*Gillon et al.*, 2017)], and a handful of systems have outer gas giants and inner small planets [e.g., the Kepler-90 8-planet system (*Cabrera et al.*, 2014)]. It remains unclear whether these planets are actually

habitable. Liquid water is certainly a key ingredient, and it remains to be seen whether the solar system's formation and the planetary arrangement itself were critical in setting the conditions needed for life, in terms of the quantity and timing of volatile delivery. However, it is important to ascertain the mechanisms for water delivery to the inner planets, as this is directly relevant to the habitability of those planets, and thus bears on the age-old question, "Are we alone?"

1.1. What We Know About Earth's Water

The inner solar system is relatively dry (*Abe et al.*, 2000; *van Dishoeck et al.*, 2014). Mercury is dry except for surface ice seen at the poles, likely deposited from recent exogenous impacts (*Deutsch et al.*, 2019). Venus was once wet (*Donahue et al.*, 1982), but experienced major water loss through hydrodynamic escape or impact-driven desiccation (*Kasting and Pollack*, 1983; *Kurosawa*, 2015). Likewise, Mars may have once had significant water, but has also lost a large fraction of its atmosphere to space via sputtering (*Jakosky et al.*, 2017). While images of Earth from space suggest that it is a blue oasis in space, rich in water, the bulk Earth is in fact relatively dry. However, its precise water content is uncertain. Assessments of the amount of water stored as hydrated silicates within the mantle vary between roughly one and ten "oceans" (*Hirschmann*, 2006; *Mottl et al.*, 2007; *Marty*, 2012; *Halliday*, 2013), where one ocean is defined as the amount of water on Earth's surface (by mass, roughly 2.5×10^{-4} $M_\oplus$). The amount of water in Earth's core is also uncertain (*Badro et al.*, 2014; *Nomura et al.*, 2014), although recent laboratory experiments suggest that it is quite dry and that the bulk of Earth's water resides in the mantle and on the surface (*Clesi et al.*, 2018).

Earth's water was incorporated during its formation. Planets form around young stars in disks of gas and dust whose characteristics may vary (*Bate*, 2018). Disks are ~99% gas by mass. The solid component of the disk starts as submicrometer-sized dust grains inherited from the interstellar medium. Water is provided to the disk as ice coatings on grains. What is not clear is how much recycling of water there is in the disk. The disk is flared, i.e., the thickness of the disk increases with radius, so the disk top and bottom surfaces are exposed to radiation from the star. This heats the surfaces, producing a vertical variation in temperature (in addition to a general decrease in temperature with distance from the star), as shown schematically in Fig. 1. Matter is accreted from the disk onto the star, adding to the star's UV and X-ray flux. This can alter water chemistry and isotopic composition as a function of position and time. Isotopes can preserve the signature of these kinds of disk processes and will be the key to tracing the volatiles in the early solar system.

Cosmochemical and geochemical evidence provides clues that Earth's water arrived early (Fig. 2) (see also the discussion in the chapter by Zahnle and Carlson). The condensation of the first solar system solids (the calcium- and aluminum-rich inclusions, or CAIs) sets the

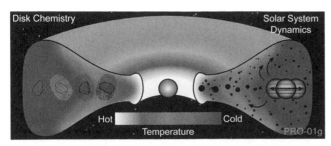

Fig. 1. See Plate 26 for color version. Protoplanetary disk chemical signatures are implanted on planetesimals as ices freeze near the midplane. This fingerprint is a sensitive measure of the planetesimal formation location. Subsequent planetesimal scattering as the giant planets grew will scramble this signature. The key to understanding where inner solar system volatiles originated is to measure the isotopes in a primitive reservoir of material that records the dynamical history during formation.

time "zero" for the birth of the solar system at 4.567 Ga (*Amelin et al.,* 2002; *Krot and Bizzarro,* 2009; *Bouvier and Wadhwa,* 2010). As discussed in section 3, it is believed that water (1) could have arrived at Earth locally as gas that was adsorbed onto the surface of dust grains (*King et al.,* 2010; *Asaduzzaman et al.,* 2014), or (2) could have formed on Earth's surface if a magma ocean was in contact with a primordial hydrogen atmosphere (*Ikoma and Genda,* 2006), or (3) could have been dynamically delivered from the ice-rich outer solar system (*Walsh et al.,* 2011; *Raymond and Izidoro,* 2017a). Both observations and models of protoplanetary disks show that the lifetimes of gaseous protoplanetary disks are very short, a few million years (*Haisch et al.,* 2001; *Ercolano and Pascucci,* 2017), so local processes would have had to have occurred early. Likewise, Earth's magma ocean probably could have lasted anywhere from 100 k.y. to more than ~100 m.y. (*Elkins-Tanton* 2012; *Hamano et al.,* 2013; *Monteux et al.,* 2016). Estimates of the final formation stages of Earth (the core formation and the Moon-forming impact event) show that this did not extend beyond 140 m.y. after the formation of the first solids (*Rubie et al.,* 2007; *Fischer and Nimmo,* 2018). Evidence from Hadean zircons show that oceans likely existed on Earth within ~150–250 m.y. of the start of solar system formation (*Mojzsis et al.,*

2001; *Wilde et al.,* 2001). This indicates that water (and accompanying organics and other volatiles) must have arrived as Earth was forming, consistent with models for Earth's chemical evolution during its accretion (*Wood et al.,* 2010; *Dauphas,* 2017).

This chapter reviews current paradigms for the origins of Earth's water. As water delivery is inherently linked with planet formation, we summarize the geochemical, cosmochemical, astrochemical, astronomical, and dynamical evidence for how habitable worlds form, addressing in particular the origin of water on rocky planets. Isotopes and noble gases are powerful tracers of the early solar system chemical processes in the protoplanetary disk. Interpreting these fingerprints requires an understanding of how the planetesimals grew, moved around, and were eventually incorporated into habitable planets (see Fig. 1). The remainder of section 1 introduces the history of exploring the origin of water through the isotopic record. Section 2 summarizes the dynamical and chemical constraints and processes for solar system formation, section 3 summarizes models for the delivery of water to the inner solar system, and section 4 discusses the implications.

1.2. Deterium/Hydrogen as a Tracer of Earth's Water

Deuterated molecules have a highly-temperature-sensitive chemistry that can provide information about the physical conditions at the time of their formation. The original D/H ratio set in the Big Bang (*Spergel et al.,* 2003) and altered by stellar nucleosynthesis is measured from interstellar absorption lines as starlight passes through diffuse gases. Deuterium is enriched in interstellar ices in cold dense regions in molecular clouds via complex gas phase and gas-grain chemistry reaction networks (*Millar et al.,* 1989).

It had long been known that the D/H ratio of Earth's oceans (e.g., standard mean ocean water, D/H_{SMOW} = 15.576 × 10⁻⁵) (*Lécuyer et al.,* 1998) was significantly elevated (by a factor of 6.4) above that of the expected protosolar value (*Geiss and Gloeckler,* 1998) (see Table 1). The measurement of a similarly elevated D/H value in water (2 × SMOW) for Comet Halley by the Giotto mission

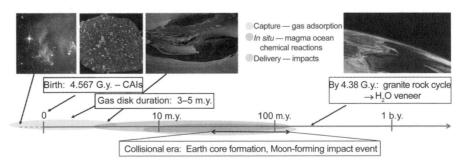

Fig. 2. See Plate 27 for color version. Earth received its water early; by 4.38 Ga Earth had oceans. Credit: ESA and the Hubble Heritage team (STScI/auRA)-ESA/Hubble Collaboration; AMNH-Creative commons license, 2.0 (*https://creative-commons.org/licenses/by/2.0/deed.en*); NASA.

TABLE 1. Astronomical and solar system D/H values.

Source	D/H Value		Note
	($\times 10^{-5}$)	δD (‰)[*]	
Big Bang	2.62 ± 0.19	-832 ± 12	1
Local ISM gas	2.3 ± 0.24	-852 ± 15	2
ISM/YSO ice	10–100	$-350 \rightarrow +5419$	3
Protosolar	2.0 ± 0.35	-872 ± 22	4
SMOW	15.57	0	5
Earth mantle	<12.2	<–217	6
Giant planets	1.7–4.4	$-890 \rightarrow -717$	7
Mars mantle	<19.9	<278	8
C chondrites	8.3–18	$-467 \rightarrow +156$	9
Rings, moons	14.3–130	$-82 \rightarrow +7346$	10
Comets	16–53	$+27 \rightarrow +2402$	11

Notes: [1] *Spergel et al.* (2003); [2] *Linsky* (2007); [3] *Cazaux et al.* (2011), *Coutens et al.* (2014); [4] *Geiss and Gloeckler* (1998); [5] *Lécuyer et al.* (1998); [6] *Hallis et al.* (2015); [7] As measured in H_2, *Lellouch et al.* (2001), *Feuchtgruber et al.* (2013), *Pierel et al.* (2017); [8] *Hallis* (2017); [9] *Alexander et al.* (2012b); [10] *Waite et al.* (2009), *Clark et al.* (2019); [11] *Bockelée-Morvan et al.* (2015), *Altwegg et al.* (2015).

[*]δD in per mil (‰) = $[(D/H)_{sample}/(D/H)_{SMOW}-1] \times 1000$.

(*Eberhardt et al.,* 1987, 1995) led to the idea that comets could have been the source of Earth's water (*Owen and Bar-Nun,* 1995) and that D/H could serve as the "fingerprint" to identify the origin of inner solar system water.

However, as more measurements of D/H in comets were obtained (mostly of long-period comets, or LPCs; see Fig. 3) showing that the values were all elevated above that of SMOW (*Mumma and Charnley,* 2011), it was clear that comets could not be the only source of Earth's water.

Initially, D/H measurements from astronomical sources were compared to Earth's oceans. However, since the oceans likely do not represent the bulk of Earth's water inventory, the comparison should be made to Earth's bulk D/H. Earth's oceans are not a closed system; they interact with the atmosphere, and water is mixed into the mantle via subduction. One scenario invoking large-scale loss of an early H-rich atmosphere proposes that the atmospheric D/H may have increased over the age of the solar system by a factor of 2–9 (*Genda and Ikoma,* 2008). Determining

the Earth's primordial D/H ratio thus requires sampling a primitive undegassed mantle source. Magma ocean crystallization models (*Elkins-Tanton,* 2008) suggest that there may have been small volumes of late-solidifying material in the first 30–75 m.y. of Earth's history that still exist near the core mantle boundary. Mantle plumes in Hawai'i, Iceland, and Baffin Island appear to have tapped into undegassed, deep mantle sources, based on their He isotope ratios (*Starkey et al.,* 2009; *Stuart et al.,* 2003; *Jackson et al.,* 2010). Measurements from the Icelandic plume show a measurement for Earth's mantle that has a D/H lower than that of the oceans (*Hallis et al.,* 2015) (see section 3.1.2), although this may not be representative of the whole mantle. Among solar system water reservoirs, only carbonaceous chondrites have D/H ratios that span the estimates for Earth's ocean and interior (*Marty and Yokochi,* 2006; *Marty,* 2012; *Hallis et al.,* 2015; *Alexander et al.,* 2012b). While chondritic water is isotopically lighter than Earth's water, chondritic organics are heavier

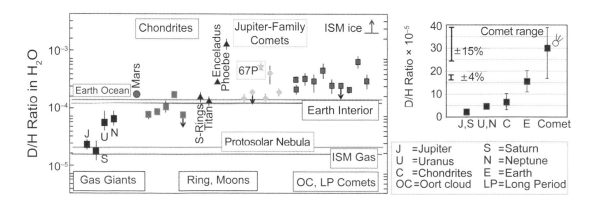

Fig. 3. See Plate 28 for color version. D/H measurements from a variety of solar system reservoirs. The values and references are shown in Table 1. While high-precision measurements of D/H can discriminate between source reservoirs, some reservoirs have similar D/H values. D/H alone cannot identify the source of inner solar system water.

(*Alexander et al.,* 2012a). These two reservoirs (water and organics) are expected to have equilibrated during accretion, and it is the bulk D/H isotopic composition that should be considered.

1.3. Oxygen Isotopic Variations

Isotopic measurements of D/H alone are insufficient to uniquely ascertain a formation location in the disk because of the complexity of disk chemical models (see section 1.6). Oxygen has three isotopes whose variation in the disk depends on different physical processes, thus the variation can be used as an independent tracer.

The oxygen isotopic composition of the Sun inferred from samples of the solar wind returned by the Genesis spacecraft is ^{16}O-rich (*McKeegan et al.,* 2011), whereas nearly all solar system solids are ^{16}O-depleted relative to the Sun's value (see Fig. 4). Exceptions are refractory inclusions that formed from a gas of approximately solar composition and are ^{16}O-rich (*Scott and Krot,* 2014). Most physical and chemical fractionation processes depend on mass, and on an oxygen three-isotope plot of $\delta^{17}O$ vs. $\delta^{18}O$ samples should fall on a line with a slope of ~0.52. Most samples from Earth (excluding the atmosphere) fall along this line, called the terrestrial fractionation line (TFL). However, the compositions of chondrules and refractory inclusions in the primitive (unmetamorphosed and unaltered) carbonaceous chondrites fall along a mass-independent fractionation line with a slope of ~1 (see section 2.3.2). The physical processes controlling the distribution of oxygen isotopes in primitive solar system bodies are different from those for D/H, thus the oxygen isotopic composition of water can also provide clues about its origins (see section 2.3.2).

1.4. Nitrogen Isotopic Variations

Nitrogen isotopes can provide additional constraints for planetesimal formation distances (*Alexander et al.,* 2018a). There is a general trend of increasing $\delta^{15}N$ with increasing distance of formation from the Sun, although again, we only have these measurements for a small number of comets. While many carbonaceous chondrites have nitrogen isotopic ratios that match that of Earth, cometary values are very different from the telluric value. The large cometary ^{15}N excess relative to Earth's atmosphere indicates that their volatiles underwent isotopic fractionation at some point in the early solar system. Within the uncertainties on estimates of volatile budgets, *Marty* (2012) and *Alexander et al.* (2012b) have proposed that Earth's inventory of H, C, and noble gases may be matched with a few percent CM/CI chondritic material (see Fig. 5). However, Earth's bulk nitrogen content appears to be depleted by an order of magnitude relative to those elements. It remains uncertain whether Earth preferentially lost much of its nitrogen during giant impacts or whether there remains an as-yet-unidentified reservoir of nitrogen in Earth's interior.

1.5. Noble Gases

While Earth's D/H may be a match for carbonaceous chondrites, the noble gases are not a match (*Owen and Bar-Nun,* 1995, 2000) and an additional solar component is required. Ideas for how this solar component was acquired include (1) that it was accreted directly from the nebula, (2) that it was solar-wind-implanted into the surfaces of small objects accreted by Earth, or (3) that it was delivered by cometary ices (see also the discussion in the chapter by Zahnle and Carlson). The noble gas fractionation patterns for the martian and terrestrial atmospheres are similar, both depleted in Xe relative to the other noble gases compared to carbonaceous chondrites. The first detection of noble gases in comets was from the Rosetta mission and showed that Ar and Kr were solar (*Rubin et al.,* 2018), but there were deficits in the heavy Xe isotopes (*Marty et al.,* 2017). This suggests that there was at least a small contribution of

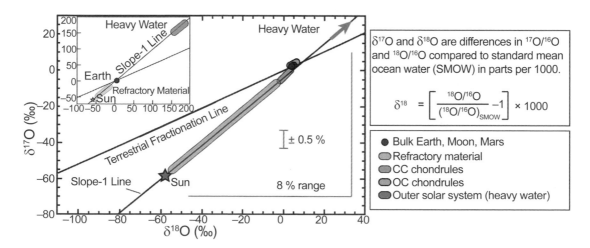

Fig. 4. See Plate 29 for color version. The range of oxygen isotopic variation in the solar system is small, so distinguishing reservoirs requires very-high-precision measurements.

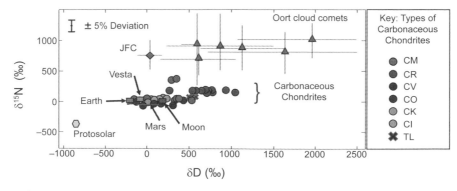

Fig. 5. See Plate 30 for color version. Combining nitrogen and hydrogen isotope ratios helps discriminate between reservoirs. The N-isotope difference between the group defined by Earth, Vesta, the Moon, Mars, chondrites, the Oort cloud comets, and the protosolar value is large. The difference between the CR chondrites and the other chondrites is 15%. The cometary values are measured in a variety of different molecules including water, HCN, and NH_3. It is challenging to infer the bulk isotopic compositions.

cometary volatiles that delivered little of Earth's water but a significant fraction of its (atmospheric) noble gases (*Marty et al.*, 2016, 2017).

1.6. Multiple Fingerprints Are Needed

Generally, volatiles that have been heated and reequilibrated with inner solar system gas will have a low (protosolar) D/H value, and bodies formed in the distant solar system will have high D/H. Most of the early comet D/H measurements were from LPCs, which were thought to form closer to the Sun than the Jupiter-family comets (JFCs). Based on the expectation that the JFCs likely formed in the colder outer disk and the LPCs formed in the giant planet region closer to the Sun (*Meech and Svoren*, 2004), it was expected that the JFCs would have an even higher D/H ratio than the LPCs. However, the D/H measurement in the EPOXI mission target, JFC 103P/Hartley 2, matched that of Earth's oceans, leading to claims that the JFC comet reservoir could have delivered Earth's water (*Hartogh et al.*, 2011).

More recently, the high D/H measurements from the Rosetta mission target (also a JFC) (see Fig. 3) have been interpreted to imply that comets did not bring Earth's water (*Altwegg et al.*, 2015). However, there are two key uncertainties. First, cometary isotopic ratios were only measured in certain molecules and may not represent the bulk isotopic compositions [recall that while carbonaceous chondrites match Earth's bulk D/H, chondritic water is isotopically lighter, whereas chondritic organics are isotopically heavier (*Alexander et al.*, 2012a)]. Second, there may be a correlation between measured D/H values and the level of cometary activity, with more active comets having lower D/H (*Lis et al.*, 2019). This effect is likely to be small, however, and models show that this should not be a factor if comets are observed at perihelion (*Podolak et al.*, 2002).

Many of these D/H measurements have stimulated the development of new disk chemical models when data did not match predicted trends (*Aikawa et al.*, 2002; *Willacy and Woods* 2009; *Jacquet and Robert*, 2013; *Yang et al.*, 2013).

The real issue is that the predicted D/H variation along the mid-plane from chemical models of protosolar disks is complex, and D/H alone is not sufficient to determine a formation distance to compare to dynamical models. This was seen with the Rosetta D/H measurement (*Altwegg et al.*, 2015), which was consistent with several disk chemical and dynamical models (see Fig. 6).

There is an additional complication. The D/H of water in ordinary and R chondrites is higher than in carbonaceous chondrites (*Alexander et al.*, 2012b), possibly due to either higher presolar water abundances in the inner solar system, or oxidation of metal by water and subsequent isotopic fractionation as the generated H_2 is lost (*Alexander*, 2019a,b).

Finally, because of the complex chemistry and dynamical mixing as the planets grew, the chemical fingerprints are "smeared." Multiple reservoirs can also have the same fingerprint. Thus, to understand the origins of inner solar system volatiles, we need to measure multiple isotopes, comparing these to the dynamical models.

2. SOLAR SYSTEM FORMATION CONSTRAINTS AND PROCESSES

Given that the origin of Earth's water cannot be decoupled from its formation, we must consider the large-scale constraints on solar system formation. In this section, we briefly summarize the empirical constraints and the key processes of planet formation. For more detail we refer the reader to recent in-depth reviews focused on dynamical modeling and solar system formation (*Morbidelli and Raymond*, 2016; *O'Brien et al.*, 2018; see also chapters in this volume by Zahnle and Carlson and by Raymond et al.).

2.1. Planet Formation Model Empirical Constraints

The following observations represent the fossil evidence of our solar system's formation, which successful planet formation scenarios must reproduce (see *Chambers*, 2001; *Raymond et al.*, 2009).

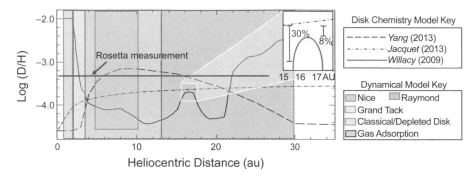

Fig. 6. See Plate 31 for color version. Measuring only D/H cannot uniquely determine an origin location. The Rosetta measurement from Comet 67P/Churyumov-Gerasimenko is consistent with at least two disk chemical models, predicting formation at several heliocentric distances compatible with several dynamical models. Furthermore, disk chemical models can have similar D/H at different distances, and different dynamical models can scatter from overlapping regions.

2.1.1. The planets' orbits. Most of the mass in terrestrial planets is concentrated in a narrow ring between the orbits of Venus and Earth. The terrestrial planets themselves have near-circular, coplanar orbits, which for years posed a problem for models of terrestrial accretion. Meanwhile, the giant planets' low-eccentricity but spread-out orbits may indicate an orbital instability in our solar system's distant past (*Tsiganis et al.*, 2005), albeit one that was far weaker than those inferred for exoplanet systems (*Raymond et al.*, 2010).

2.1.2. Small-body populations. Solar system small bodies are the leftovers of planet formation, although there is no reason to think that they are a representative sample. The asteroid belt is very low in mass but has an excited orbital distribution. In broad strokes, the inner belt is dominated by dry objects (such as the S types) and the outer belt by hydrated asteroids (e.g., the C types) (*Gradie and Tedesco*, 1982; *DeMeo and Carry*, 2013). In addition to the asteroids, volatile-rich bodies include comets, small satellites, and Kuiper belt objects. The Kuiper belt contains a total of ~0.1 Earth masses ($M_\oplus$), (*Gladman et al.*, 2001), and the Oort cloud up to a few Earth masses (*Boe et al.*, 2019, and references therein).

For a long time it was believed that comets formed in distinct regions, e.g., that the LPCs formed in the giant planet region and were scattered to the Oort cloud during formation, and that the JFCs were formed further out in the region of the Kuiper belt — eventually getting perturbed inward, during which time they became Centaurs, until their orbits were influenced by Jupiter (*Meech and Svoren*, 2004). However, while there are clear trends in comet chemistry (*A'Hearn et al.*, 1995; *Mumma and Charnley*, 2011), they have not been clearly tied to dynamical class. Comets likely formed over a range of distances outside the solar system's snow line and have experienced significant dynamical scattering. In considering possible sources for Earth's water we must therefore consider these objects as a population because individual comets cannot be traced back in time; rather, they sample the entire disk outside the snow line.

2.1.3. Meteorite constraints on growth timescales. Isotopic analyses of different types of meteorites provide vital constraints on formation timescales of different types of objects. CAIs, the oldest dated solids to have formed in the solar system, are generally used as "time zero" for planet formation. Age estimates of iron meteorites provide upper limits to the formation timescales of differentiated bodies (*Kruijer et al.*, 2014). The existence of two types of chondritic meteorites with different isotopic anomalies (non-carbonaceous and carbonaceous) but similar ages has been interpreted as evidence for the rapid growth of Jupiter's core, which would have provided a barrier between these populations (*Kruijer et al.*, 2017; *Desch et al.*, 2018a; see also the chapter in this volume by Zahnle and Carlson). Finally, the Hf/W system provides estimates of the timing of core formation, and suggest that Mars' growth was rapid (*Nimmo and Kleine*, 2007; *Dauphas and Pourmand*, 2011), whereas Earth's was prolonged (*Kleine et al.*, 2009; *Jacobson et al.*, 2014).

These constraints are inherently tied to the conditions of planet formation. For example, numerical experiments have shown that a smooth disk of solids extending from Mercury's orbit out to Jupiter's generally fails to reproduce the solar system because (1) the terrestrial planets' orbits are overly excited (*Chambers*, 2001; *O'Brien et al.*, 2006; *Raymond et al.*, 2006b), (2) Mars is too massive and grows too slowly (*Raymond et al.*, 2009; *Fischer et al.*, 2014; *Izidoro et al.*, 2015b), and (3) the asteroids' orbits are underexcited (*Izidoro et al.*, 2015b). However, we note that these problems may be solved if the giant planet instability happens *during* terrestrial planet formation (*Clement et al.*, 2018) (see section 3 and Fig. 11 for a comparison between models).

2.2. Key Planet Formation Processes

Planet formation models are built of processes. Each process can be thought of as a puzzle piece that must be assembled into a global picture of planetary growth (see the chapter in this volume by Raymond et al. for a review dedicated entirely to this endeavor for the solar system and exoplanet systems). We now very briefly summarize the key planet formation processes from the ground up.

2.2.1. Disk structure and evolution. The underlying structure and dynamics of protoplanetary disks remain poorly understood (*Morbidelli and Raymond,* 2016). An essential piece of the story is how angular momentum is transported within disks (*Turner et al.,* 2014). The radial surface density of gas and dust sets the stage for planet formation. Within a disk, the gas is subject to hydrodynamic pressure forces as well as gravity, and its motion deviates from pure Keplerian motion, with radial velocities that are generally slightly slower than the Keplerian velocity. Dust grows and drifts within the disk (*Birnstiel et al.,* 2012). Dust accumulates at pressure bumps, narrow rings at which the gas velocity matches the Keplerian velocity such that the drag force disappears (*Haghighipour and Boss,* 2003). Exterior to a pressure bump the gas velocity is sub-Keplerian such that dust particles feel a headwind, lose orbital energy, and drift radially inward. Just interior to pressure bumps the gas velocity is super-Keplerian so particles feel a tailwind, driving them back outward. Very small dust particles remain strongly coupled to the gas, and large bodies have enough inertia to drift slowly; the fastest-drifting particles are "pebbles" (e.g., *Ormel and Klahr,* 2010; *Lambrechts and Johansen,* 2012). As disks evolve they cool down, and locations associated with specific temperatures — e.g., the *snow line,* the radial distance beyond which a volatile (such as water) may condense as ice — move inward. Gaseous disks are observed to dissipate on a characteristic timescale of a few million years (*Haisch et al.,* 2001; *Hillenbrand et al.,* 2008).

2.2.2. Planetesimal formation. Planetesimals are the smallest macroscopic bodies for which gravity dominates over hydrodynamical forces. Their origin has long been difficult to reproduce with formation models because any growth model must traverse the size scale at which particles start to decouple from the gas motion. These intermediate-sized particles then experience the headwind that causes them to rapidly spiral inward, preventing aggregation into larger bodies. This occurs at centimeter to meter sizes, so this is sometimes called the "meter barrier" (*Weidenschilling,* 1977). New models have demonstrated that millimeter-sized particles can be concentrated and clump directly into planetesimals via processes such as the streaming instability (*Youdin and Goodman,* 2005; *Johansen et al.,* 2009; *Simon et al.,* 2016; *Yang et al.,* 2017), thus jumping over the meter barrier. The conditions for triggering the streaming instability vary in time and position within a given disk (*Drążkowska and Alibert,* 2017; *Carrera et al.,* 2017).

2.2.3. Pebble accretion. Once planetesimals form, they may grow by accreting other planetesimals as well as pebbles drifting inward through the disk (*Johansen and Lambrechts,* 2017). Here, "pebbles" are taken to be particles that drift rapidly through the gas and are typically millimeter- to centimeter-sized for typical disk parameters (*Ormel and Klahr,* 2010; *Lambrechts and Johansen,* 2012). Pebble accretion can be extremely fast under some conditions and objects can quickly grow to many Earth masses in the giant planet region if there is a sufficient reservoir of pebbles (*Lambrechts and Johansen,* 2014; *Morbidelli et al.,* 2015). Pebble accretion is self-limiting, as above a given mass

(typically 10–20 $M_\oplus$ at Jupiter's orbit) (*Bitsch et al.,* 2018) a core generates a pressure bump exterior to its orbit that holds back the inward-drifting pebbles.

2.2.4. Gas accretion. Cores that grow large enough and fast enough accrete gas from the disk. The *core-accretion* scenario for giant planet formation (*Pollack et al.,* 1996) envisions the growth of ~10 $M_\oplus$ cores followed by a slow phase of gas accretion. When the mass in the gaseous envelope is comparable to the core mass, gas accretion can accelerate and quickly form Saturn- to Jupiter-mass gas giants. During this rapid accretion phase, the orbits of nearby planetesimals are destabilized and many are scattered inward, contaminating the inner planetary system (*Raymond and Izidoro* 2017b). Given that most cores are not expected to grow into gas giants, this model predicts a much higher abundance of ice giant-mass planets relative to gas giants, which has been confirmed by exoplanet statistics (*Gould et al.,* 2010; *Mayor et al.,* 2011; *Petigura et al.,* 2013). On the other hand, some giant exoplanets — in particular those at large orbital radii — may form rapidly by direct gravitational collapse (*Boss,* 1997; *Mayer et al.,* 2002; *Boley,* 2009).

2.2.5. Orbital migration. Gravitational interactions between a growing planet and its nascent gaseous disk generate density perturbations that torque the planet's orbit and cause it to shrink or grow [i.e., to *migrate* inward or outward (*Kley and Nelson* 2012; *Baruteau et al.,* 2014)]. Migration matters for planets more massive than ~0.1–1 $M_\oplus$. In most cases migration is directed inward but the corotation torque, which depends on the local disk conditions, can in some instances be positive and strong enough to drive outward migration (*Kley and Crida* 2008; *Paardekooper et al.,* 2011). In the context of the entire disk, in some regions planets migrate toward a common location, although these convergence zones themselves shift as disks evolve (*Lyra et al.,* 2010; *Bitsch et al.,* 2015). Above a critical mass, a planet clears an annular gap in the gaseous disk and migration transitions to "type 2" [as opposed to "type 1" for planets that do not open gaps (*Lin and Papaloizou,* 1986; *Ward,* 1997; *Crida et al.,* 2006)]. Type 2 migration is generally slower than type 1 migration and is again directed inward in most instances.

2.2.6. Giant impacts. Impacts between similar-sized massive objects are thought to be common in planet formation. The late phases of terrestrial planet growth are attributed to a small number of ever-larger giant impacts between growing rocky bodies (*Wetherill,* 1991; *Agnor et al.,* 1999). Giant impacts among large ice-rich cores have been invoked to explain the large obliquities of Uranus and Neptune (*Benz et al.,* 1989; *Izidoro et al.,* 2015a). The final giant impact on Earth is believed to be the one that led to the formation of the Moon (*Benz et al.,* 1986; *Canup and Asphaug,* 2001).

2.2.7. Late accretion. Giant impacts are thought to be energetic enough to trigger core formation events, which sequester siderophile ("iron-loving") elements in the planet's core (*Harper and Jacobsen,* 1996). Highly-siderophile elements (HSEs) in a planet's mantle and crust are, therefore, considered to have been delivered by impacts with planetesimals *after* the giant impact phase (*Kimura et al.,* 1974; *Day et al.,* 2007; *Walker,* 2009). This is called late accretion or the late veneer. Earth's HSE budget implies that roughly 0.5% of an Earth mass was delivered in late accretion (*Walker et al.,* 2015; *Morbidelli and Wood,* 2015).

2.2.8. Dynamical instability. After the disappearance of the gaseous disk, systems of planets may become unstable. This applies both to systems of low-mass planets (i.e., super-Earths) and gas giants (see the chapter by Raymond et al. in this volume). Our solar system's giant planets are thought to have undergone such an instability (*Tsiganis et al.,* 2005; *Morbidelli et al.,* 2007; *Levison et al.,* 2011). This instability can explain the giant planets' orbits and a multitude of characteristics of small-body populations (for a review, see *Nesvorný et al.,* 2018a). This "Nice model" was originally conceived to explain the Late Heavy Bombardment (*Gomes et al.,* 2005), a perceived spike in the impact rate on the Moon starting roughly 500 m.y. after planet formation (*Tera et al.,* 1974). However, a new interpretation of the evidence has led to the conclusion that there was probably no delayed bombardment but rather a smooth decline in the impact rate in the inner solar system (*Boehnke and Harrison,* 2016; *Zellner,* 2017; *Morbidelli et al.,* 2018; *Hartmann,* 2019). The instability is still thought to have occurred but may have taken place anytime in the first ~100 m.y. of solar system history (*Nesvorný et al.,* 2018b). The broad eccentricity distribution of exoplanets implies that instabilities are common (*Chatterjee et al.,* 2008; *Ford and Rasio,* 2008; *Jurić and Tremaine,* 2008) (the median eccentricity of giant exoplanets is ~0.25). Instabilities in most giant planet systems are likely to have been far more violent than in our own solar system (*Raymond et al.,* 2010; *Ida et al.,* 2013), and to have often disrupted growing terrestrial exoplanets and outer planetesimal disks (*Veras and Armitage,* 2006; *Raymond et al.,* 2011, 2012).

2.3. Protoplanetary Disk Chemistry

Disk chemistry is responsible for the key chemical markers that can be used to trace the transport of water in the disk, and this is transferred to the planetesimals as ice freezes on dust grains. The isotopic composition is imprinted on planetesimals when they form. However, given the widespread planetesimal scattering and dynamical rearrangement during and after planetary formation (e.g., *Levison et al.,* 2008; *Walsh et al.,* 2011; *Raymond and Izidoro,* 2017a), it is important to recognize that the present-day orbits of solar system bodies may not reflect their formation locations. For example, while Jupiter is often used as the boundary between the inner and outer solar system, recent models suggest that the parent bodies of the carbonaceous chondrites likely originated beyond Jupiter (*Walsh et al.,* 2011; *Kruijer et al.,* 2017; *Raymond and Izidoro,* 2017a).

Unique chemical signatures are thus imprinted on the icy material that is incorporated in the planetesimals that grow to form planets. Planet formation models incorporate volatile condensation onto grains (*Grossman,* 1972) and water

transport and condensation beyond the snow line (*Stevenson and Lunine,* 1988; *Garaud and Lin,* 2007), inside which the temperatures are too warm for water ice to remain stable.

Modern protoplanetary-disk-evolution models are based on chemical networks developed for the interstellar medium (ISM), and include disk structure, isotopic fractionation, and gas transport and incorporate the physics of grain growth, settling, and radial migration. Although models are always incomplete and remain crude representations of reality, recent observations have provided key constraints to help refine the models. Models and observations (*Pontoppidan et al.,* 2014; *van Dishoeck et al.,* 2014) reveal radial and vertical variations in thermal and chemical disk structure (Fig. 1). Infrared observations from Spitzer and Herschel (*Zhang et al.,* 2013; *Du and Bergin,* 2014) and new spatially resolved Atacama Large Millimeter Array (ALMA) telescope observations constrain the models (*Qi et al.,* 2013). ALMA observations also set limits on ionization in protoplanetary disks (*Cleeves et al.,* 2014a, 2015), providing constraints on models of deuterium-enrichment in water (*Cleeves et al.,* 2014b) due to ion-molecule reactions. These state-of-the-art disk-chemistry models (*Willacy and Woods,* 2009; *Jacquet and Robert,* 2013; *Yang et al.,* 2013) make very different testable predictions of radial isotope distributions in protoplanetary disks.

2.3.1. Deuterium chemistry. Observations and detailed models provide a good understanding of the complex chemistry of cold interstellar clouds in which new stars form (*Bergin and Tafalla,* 2007). For example, low-temperature ion-molecule reactions drive deuterium fractionation (*Millar et al.,* 1989). Water ice becomes enriched in deuterium, with a D/H ratio of 0.001 to 0.02 compared to a cosmic abundance of 2.6×10^{-5}. Physical processes in the disk control the temperature and radiation-field dependent chemistry. In the hot inner region of the disk, isotopic exchange reactions between water vapor and hydrogen gas reduce the D/H ratio to $\sim 2 \times 10^{-5}$, the "protosolar" value (see Table 1). In the nebula's outer disk, D/H evolved from an initial supply of water from the molecular cloud ("ISM ice" in Fig. 3) that was highly enriched in deuterium via ion-molecule and gas-grain reactions (*Herbst,* 2003; *van Dishoeck et al.,* 2013) at temperatures <30 K. Indeed, high D/H water probably cannot be produced within disks themselves and must be inherited (*Cleeves et al.,* 2014a, 2016). Sharp gradients in the disk's D/H isotopic composition arise from mixing between inherited water and water that had reequilibrated by isotopic exchange with hydrogen in the hot inner disk [sometimes called the *protosolar nebula* (*Geiss and Gloeckler,* 1998; *Lellouch et al.,* 2001; *Yang et al.,* 2013; *Jacquet and Robert,* 2013)]. The radial extent of equilibration is uncertain, as stellar outbursts (also called FU Orionis outbursts) can strongly heat the disk for short periods (years to decades) and drive the snow line out to tens of astronomical units (au) (*Cieza et al.,* 2016).

2.3.2. Oxygen isotope fractionation. The most widely accepted mechanism for explaining the oxygen isotopic diversity among solar system materials is CO self-shielding

(*Lyons and Young*, 2005). The three oxygen isotopes (^{16}O, ^{17}O, ^{18}O) have dramatically different abundances (~2500, 1, 5, respectively). The wavelengths necessary to dissociate $^{12}C^{16}O$, $^{12}C^{17}O$, and $^{12}C^{18}O$ are distinct, and the number of photons at each wavelength is similar in the UV continuum. At the edge of a dense molecular cloud or accretion disk, UV light dissociates the same fraction of all the three isotopologues. But as the light penetrates into the cloud or disk, the photons that dissociate the $^{12}C^{16}O$ are depleted by absorption, so deeper in the cloud only $^{12}C^{17}O$ and $^{12}C^{18}O$ are dissociated. The resulting oxygen ions can either recombine into CO or combine with H_2 to form H_2O. Deeper in the cloud, the H_2O will be enriched in ^{17}O and ^{18}O. If the solar system started out with the composition of the Sun, self-shielding would have produced isotopically heavy water in the outer parts of the disk. *Yurimoto and Kuramoto* (2004) also suggest that this self-shielding could occur in the pre-solar molecular cloud, and this material was transported into the solar nebula by icy dust grains during the cloud collapse. As they drifted in toward the Sun and sublimated, this enriched the inner disk gas.

An alternative mechanism to explain the oxygen isotopic diversity is the Galactic Chemical Evolution (GCE) model (*Krot et al.*, 2010), although there are reasons why this model may not work (*Alexander et al.*, 2017). According to the GCE model, the solids and gas in the protosolar molecular cloud had different ages and average compositions; the solids were younger and ^{16}O-depleted relative to the gas. According to the CO self-shielding model, oxygen isotope compositions of the primordial and thermally processed solids must follow a slope 1.0 line, whereas there is no *a priori* reason to believe that the GCE model results in the formation of solid and gaseous reservoirs falling on a slope 1.0 line (*Lugaro et al.*, 2012). Patterns of oxygen isotope fractionation can be compared against these two models.

Models combined with Genesis observations (*McKeegan et al.*, 2011) indicate that primordial dust and gas had the same ^{16}O-rich composition as the Sun. The result is an array of points with a slope ~1 on an oxygen three-isotope plot (Fig. 4), with the initial CO plotted at the lower left (marked "Sun" on the diagram) and the ^{17}O-, ^{18}O-rich water plotted at the upper right (marked "heavy water") (*Clayton et al.*, 1973; *Yurimoto et al.*, 2008). Outside the snow line, the heavy water froze on the surface of dust and settled to the disk mid-plane. Other compositions in the diagram can be produced by combining isotopically "heavy" water with isotopically "light" condensates with compositions similar to the Sun (*Lyons and Young*, 2005; *Yurimoto and Kuramoto*, 2004). The total range in oxygen isotope variation seen in the solar system is small (see range marked in Fig. 4), so distinguishing different reservoirs and formation distances requires high-precision oxygen isotope information. Self-shielding depends on UV intensity and gas densities, and therefore relates to solar distance and time.

2.3.3. Nitrogen chemistry. Nitrogen fractionation in the disk is dominated by different physical mechanisms than for D/H and oxygen, thus providing a third independent tracer. The primordial nebula and the Sun are significantly depleted in ^{15}N relative to Earth's atmosphere (*Owen et al.*, 2001; *Meibom et al.*, 2007; *Marty et al.*, 2010; *Anders and Grevesse*, 1989). Other reservoirs (CN and HCN in comets, some carbonaceous chondrite organics, and Titan) are enriched in ^{15}N (Fig. 5). This large fractionation may be inherited from the protosolar molecular cloud, where it is attributed to low-temperature (<10 K) ion-molecule reactions (*Anders and Grevesse*, 1989) or it could have resulted from photochemical self-shielding effects in the protoplanetary disk (*Heays et al.*, 2014; *Lyons*, 2012); the radial dependence on isotopic composition will differ significantly between these mechanisms. We expect the radial dependence of nitrogen isotopes will be similar to those of oxygen. However, the radial dependence is additionally subject to low-temperature effects, active in the outer disk that is exposed to X-rays, making it distinguishable from oxygen.

2.3.4. Noble gases as a thermometer. When water ice condenses from a gas at temperatures less than 100 K, it condenses in the amorphous form and can trap other volatiles. The amount of the trapped volatiles and their isotopic fractionation is a sensitive function of trapping temperature and pressure (*Bar-Nun et al.*, 1985; *Yokochi et al.*, 2012; *Rubin et al.*, 2018). Over the age of the solar system, solar insolation and impacts can heat many small bodies above 137 K, the amorphous ice crystallization temperature. When the ice crystallizes, a fraction of the trapped gases is retained in the water ice [possibly in the form of clathrates (*Rubin et al.*, 2018; *Laufer et al.*, 2017)] and is released only when the ice sublimates. This is seen consistently in laboratory experiments (*Bar-Nun et al.*, 1988). Measurements of the $^{84}Kr/^{36}Ar$ ratio can provide a sensitive indication of temperature at which the gases were trapped (*Mousis et al.*, 2018), and this can be linked through disk chemistry models to location in the protoplanetary disk.

3. MODELS FOR THE ORIGIN OF EARTH'S WATER AND THEIR CONTEXT IN PLANET FORMATION

We now delve into the depths of models that describe exactly how Earth may have acquired its water. There are two categories of models: those that propose that Earth's water could have been sourced locally, and those that require the *delivery* of water from more distant regions.

Below we explain how each model works and its inherent assumptions. For some models that requires going into the dynamics they invoke to sculpt the solar system. We then confront each model with the empirical constraints laid out in section 2.1.

3.1. Local Accretion

To date, two models have been proposed that advocate for a local source of water on the terrestrial planets. Here we describe those models and their challenges.

3.1.1. Adsorption onto grains.
It has been suggested that water could be "in-gassed" into growing planets (e.g., *Sharp*, 2017). *Drake* (2004) and *Muralidharan et al.* (2008a) proposed that water vapor could be adsorbed onto silicate grains. This would allow for a *local* source of water at 1 au within the Sun's planet-forming disk. Density functional theory calculations have shown that water vapor can indeed be adsorbed onto forsterite (*Muralidharan et al.,* 2008b) or olivine (*Stimpfl et al.,* 2006; *Asaduzzaman et al.,* 2014) grains. In principle, this mechanism can explain the accretion of multiple oceans of water onto Earth.

There are four apparent issues with this model. First, it cannot account for the delivery of other volatiles to Earth, including carbon, nitrogen, and the noble gases. In fact, this model would require an additional, hydrogen-depleted source for these species. Second, while the mechanism may accrete a water budget of perhaps a few oceans, it does not account for collisional loss of water as planetesimals grow, which can be substantial (*Genda and Abe,* 2005). Second, it begs the question of why the enstatite chondrite meteorites, which offer a close match to Earth's composition [albeit with some important differences, such that Earth cannot be made entirely of enstatite condrites (e.g., *Dauphas,* 2017)], failed to incorporate any water and are extremely dry. Third, this mechanism implies that Earth's water should have the same isotopic signature as the nebular gas. However, the D/H ratio of the disk gas is likely to have been the same as the Sun's, which is a factor of 6.4 lower than Earth's oceans [see *Geiss and Gloeckler* (1998) and section 1]. Finally, disk-evolution models find that an individual parcel of gas in the vicinity of the snow line moves radially much faster than the snow line does (*Morbidelli et al.,* 2016). This means that the water vapor in the inner disk likely does not remain but rather moves inward and is accreted onto the star much more quickly than it can be replenished. The inner disk's gas is therefore dry for a large fraction of the disk's lifetime.

3.1.2. Oxidation of a primordial hydrogen-rich atmosphere.
Ikoma and Genda (2006) also proposed that Earth's water could have been acquired locally, but as a result of reactions between Earth's primordial atmosphere and surface. During the magma ocean phase, atmospheric hydrogen can be oxidized by gas-rock interactions and produce water. A magma ocean phase that meets the criteria for this mechanism is expected for planets more massive than $\sim$0.3 $M_\oplus$ (*Ikoma and Genda,* 2006), although the duration of a magma ocean is also a strong function of orbital distance (*Hamano et al.,* 2013).

This mechanism produces terrestrial water directly from nebular gas. Yet the D/H ratio of nebular gas is a factor of $\sim$6.4 lower than Earth's (*Geiss and Gloeckler,* 1998). However, most of this early hydrogen-rich atmosphere had to escape (Jean's escape), and the escape process leads to fractionation and an increase in the D/H ratio. *Genda and Ikoma* (2008) showed that the D/H of Earth's water can increase to its present-day value for certain values of the efficiency and timescale of hydrogen escape. However, the collateral effects of this loss on the isotopic fractionation

of other species such as the noble gases remains unaddressed. Given that nitrogen should not be fractionated by this process, it appears to imply that Earth should have a solar nitrogen isotopic composition, which is not the case (*Marty,* 2012).

Measurements of water thought to be sourced from an isolated deep mantle reservoir yield D/H values closer to the nebular one (*Hallis et al.,* 2015). This indicates that at least a fraction of Earth's water may have a nebular origin. There is also circumstantial evidence from noble gases to support the idea that Earth had a primordial nebular atmosphere (*Dauphas,* 2003; *Williams and Mukhopadhyay,* 2019).

It remains to be seen whether nebular gas can explain the origin of the bulk of Earth's water. This mechanism clearly requires Earth to have accreted and later lost a thick hydrogen envelope. While such a process is a likely outcome of growth within the gaseous disk and there is evidence that many $\sim$Earth-sized exoplanets have thick hydrogen atmospheres (e.g. *Wolfgang et al.,* 2016), it seems a cosmic coincidence for the parameters to have been right for Earth to end up with the same D/H ratio (and $^{15}N/^{14}N$ ratio) as carbonaceous chondrites (*Marty,* 2012). Carbonaceous chondrite-like objects thus seem a more likely source for Earth's water.

3.2. Water Delivery

An alternate explanation for the origin of Earth's water is delivery from the outer solar system. The following models invoke sources of water that are separated from Earth's orbit. Hence water is "delivered" from these regions to Earth. These models are driven by a number of processes, including a combination of the moving snow line and drifting pebbles (section 3.2.1), widening of terrestrial planets' feeding zones (section 3.2.2), gravitational scattering of planetesimals during the growth and migration of the giant planets (section 3.2.3), and inward migration of large planetary embryos (section 3.2.4).

3.2.1. Pebble "snow."
As disks evolve and thin out, they cool down. Condensation fronts move inward (*Dodson-Robinson and Bodenheimer,* 2010), including the water-ice snow line, located at T $\sim$ 150 K. The exact evolution of the snow line depends on the thermal evolution of the disk and therefore on the assumed heating mechanisms, the most important of which are stellar irradiation and viscous heating. For alpha-viscosity disk models, the snow line starts in the Jupiter-Saturn region and moves inside 1 au within 1 m.y. or so (*Sasselov and Lecar,* 2000; *Lecar et al.,* 2006; *Kennedy and Kenyon,* 2008). The snow line also moves inward and generally ends up inside 1 au in more complex disk models (*Oka et al.,* 2011; *Martin and Livio,* 2012; *Bitsch et al.,* 2015, 2019). However, it is worth noting that radial variations in viscosity may under certain assumptions keep the snow line outside 1 au for the disk lifetime (*Kalyaan and Desch,* 2019). Figure 7 shows how a planet forming in a close-in, rocky part of the disk can become hydrated as the snow line sweeps inward (*Sato et al.,* 2016).

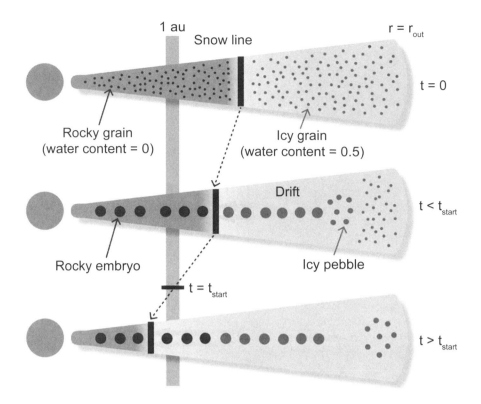

Fig. 7. Snapshots in time of the evolution of a disk, showing how the snow line sweeps inward as the disk cools. Water may thus be delivered to rocky planets at 1 AU by pebbles as they drift inward. We refer to this mechanism in the text as *pebble snow*. The region bounding embryos and pebbles outside the snowline represents the "pebble production front," the outward-moving radial location in the disk where pebbles grow from dust and start to drift inward. From *Sato et al.* (2016).

Planetesimals are thought to form from dust and pebbles that are locally concentrated by drifting within the gas disk, followed by a phase of further concentration (e.g., by the streaming instability) to produce gravitationally bound objects (see, e.g., the review by *Johansen et al.,* 2014). When planetesimals form, their compositions "lock in" the local conditions at that time. Yet most of the mass in solid bodies remains in dust and pebbles, which are small enough that their compositions likely change as they drift inward, in particular by losing their volatiles. Planetesimals continue to grow, in part by accreting pebbles (see Fig. 8 for a cartoon representation of the different phases of growth). The planetesimal and pebble phases overlap for the entire gas disk lifetime.

The evolution of the pebble flux controls the water distribution within the solids in the disk. When the snow line sweeps inward, the source of water is not condensing gas but inward-drifting particles [i.e., pebbles (*Morbidelli et al.,* 2016)]. This is because the gas' radial motion is faster than the speed at which the snow line moves. As the snow line moves inward, it does not sweep over water vapor that can condense. Rather, the gas interior to the snow line is dry (or mostly dry) simply because it moves more quickly than the snow line itself. The source of water at the snow line is instead in the form of ice-rich pebbles that drift inward from farther out in the disk. There are two ways in which

the flux of pebbles drifting inward through the disk can drop significantly: The pebble supply can be exhausted or the pebble flow can be blocked.

Dust within the disk coagulates to sizes large enough to drift inward rapidly (e.g., *Birnstiel et al.,* 2012, 2016). However, dust grows into pebbles faster closer-in to the star, where accretion timescales are short and densities high. So the dust is consumed faster in the disk interior, resulting in an outward-moving front at which pebbles are produced, after which the pebbles drift inward (*Lambrechts and Johansen,* 2014; *Ida et al.,* 2016). When this *pebble production front* reaches the outer edge of the disk, the pebble flux drops drastically, as the source of pebbles is exhausted. Taking this into account limits the degree to which an inward-sweeping snow line can hydrate rocky planets because the mass in water-bearing pebbles drops off drastically in time. Nonetheless, in many cases this mechanism can deliver Earth-like water budgets (*Ida et al.,* 2019).

The pebble flux can also be blocked, either by growing planets or by structures within the disk itself. Drifting particles follow the local pressure gradient (*Haghighipour and Boss,* 2003), which drives pebbles monotonically inward in a perfectly smooth disk. However, pressure "bumps" act as traps for inward-drifting particles. These are radially-confined regions in which the gas pressure gradient becomes high enough that the gas orbits at the Keplerian speed, thus

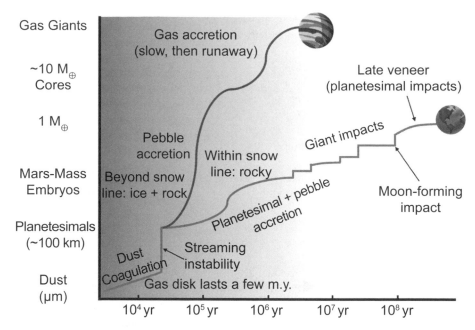

Fig. 8. A rough summary of the current understanding of the growth history of rocky and gas giant planets, illustrating the various planet formation processes. Images of Earth and Jupiter from *www.kissclipart.com*.

eliminating the headwind and associated drag forces on pebbles. Such traps may exist naturally within the disk, or they can be produced by growing planets. Once a planet reaches a critical mass [of roughly 20 M$_\oplus$ at Jupiter's orbit for typical disk parameters (*Lambrechts and Johansen*, 2014)], it generates a pressure bump exterior to its orbit, which acts as a trap for inward-drifting pebbles (*Morbidelli and Nesvorny*, 2012; *Lambrechts and Johansen*, 2014; *Bitsch et al.*, 2018). This not only starves the planet itself but also all other planets interior to its orbit.

When the pebble flux is blocked by a growing planet, it renders the concept of the snow line ambiguous. Given that inward-drifting pebbles are the source of water, the location at which the temperature drops below 150 K continues to move inward in the disk but does not bring any water along with it (recall that the gas is dry because it moves much faster than the snow line). In this way, the water distribution within a disk is "fossilized" at the time when an outer planet first grew large enough to block the pebble flux (*Morbidelli et al.*, 2016). This fossilization is analogous to the snow line on a mountain, which marks the location at which the temperature reached zero degrees Celsius while it was snowing (i.e., while pebbles were drifting). Once it stops snowing, the snow line on a mountain is no longer linked with the local temperature (as long as it does not warm up past the freezing point). After the inward drift of pebbles is cut off, redistribution of water within the disk requires dynamical processes that transport objects at larger size scales (e.g., via migration or gravitational scattering as invoked by the other water delivery mechanisms).

Can the pebble snow mechanism explain the origin of Earth's water? The *pebble snow* mechanism can produce planets with water contents similar to Earth's (*Ida et al.*, 2019). However, understanding whether Earth could have

accreted a large enough contribution from inward-drifting pebbles requires an understanding of the chemical properties of Earth's building blocks.

Nucleosynthetic isotope differences are seen in a number of elements between the two main classes of meteorites: carbonaceous and non-carbonaceous (*Warren*, 2011; *Kruijer et al.*, 2017). These two populations appear to have been sourced from different reservoirs within the planet-forming disk whose origins remain debated (*Nanne et al.*, 2019). The rapid growth of Jupiter's core has been invoked as a mechanism to keep the two populations separate, by preventing the drift of outer, carbonaceous pebbles into the inner solar system, which is thought to have been dominated by non-carbonaceous material (*Budde et al.*, 2016; *Kruijer et al.*, 2017).

Earth's water's D/H ratio is well-matched by carbonaceous chondrite meteorites (see section 1.2 and Fig. 3). The pebble snow model would thus invoke carbonaceous pebbles as the source of Earth's water, delivered late enough that the disk had cooled to the point that the snow line was located inside Earth's orbit.

However, the pebble snow model does not appear to match empirical constraints. The latest-forming chondrites have ages that extend to roughly 4 m.y. after CAIs (*Sugiura and Fujiya*, 2014; *Alexander et al.*, 2018a; *Desch et al.*, 2018b), likely the full length of the disk lifetime. This seems to indicate that the two reservoirs remained spatially separated throughout. In other words, there are no known classes of chondritic meteorites with nucleosynthetic anomalies that lie in between, which would be a signature of the mixing.

This early separation between carbonaceous and non-carbonaceous pebble reservoirs seems to preclude drifting pebbles as the source of Earth's water. If water-rich carbonaceous pebbles drifted inward to deliver water to Earth, then it should follow that the two reservoirs should have mixed and some meteorites should exist with intermediate compositions, which is not the case.

Dynamical models naturally implant carbonaceous planetesimals from Jupiter's orbit and beyond into the asteroid belt and terrestrial planet region [discussed in detail in section 3.2.3 (*Walsh et al.,* 2011; *Raymond and Izidoro,* 2017a)]. This may indicate that the non-carbonaceous S-type asteroids formed in the inner solar system, whereas the carbonaceous C-types were implanted as planetesimals.

3.2.2. Wide feeding zones. In the well-studied framework of the *classical* model of terrestrial planet formation, the late phases of terrestrial accretion occur after the formation of the giant planets (*Wetherill,* 1991, 1996; *Chambers,* 2001; *Raymond et al.,* 2006b, 2009, 2014; *O'Brien et al.,* 2006; *Morishima et al.,* 2010; *Fischer et al.,* 2014; *Izidoro et al.,* 2015b; *Kaib and Cowan,* 2015). In Fig. 8, late-stage accretion starts after dispersal of the gaseous disk, during the giant impact phase. The late stage of terrestrial accretion (essentially starting from the "giant impacts" phase shown in

Fig. 8) thus starts from a population of roughly Mars-mass planetary embryos embedded in a sea of planetesimals, under the dynamical influence of the already-formed giant planets.

Figure 9 shows an example simulation of the classical model. The compositional gradient in the present-day solar system is assumed to represent the initial conditions for planet formation: Inside roughly 2.5 au objects are dry, and between 2.5 au and Jupiter's orbit they have water contents of 5–10%, similar to carbonaceous chondrites (*Morbidelli et al.,* 2000; *Raymond et al.,* 2004).

In the inner disk, there is an effective wave of growth that sweeps outward in time, driven by self-gravity among the growing embryos (*Kokubo and Ida,* 2000). In the asteroid region, eccentricities are excited by Jupiter via secular and resonant forcing. Dynamical friction keeps the most massive planetary embryos on near-circular orbits while planetesimals and smaller embryos often have eccentric and inclined orbits (*O'Brien et al.,* 2006; *Raymond et al.,* 2006b).

Collisional growth continues for 10–100 m.y. as the planets grow by giant impacts. At the end of the simulation, three

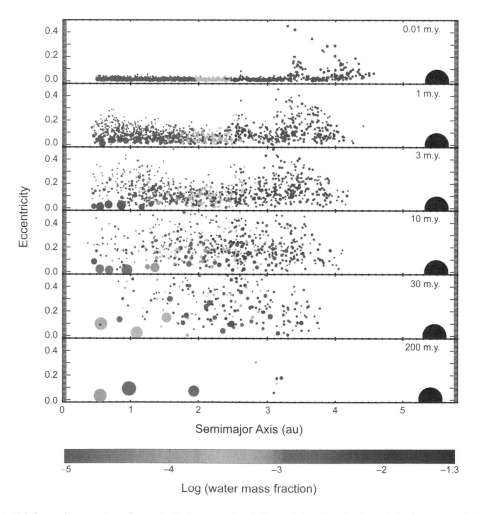

Fig. 9. See Plate 32 for color version. Snapshots from a simulation of the classical model of terrestrial planet formation. Jupiter was included from the start, on a low-eccentricity orbit at 5.5 au (large black circle). Almost 2000 planetary embryos are represented by their relative sizes (proportional to their masses$^{1/3}$) and their water contents, which were imposed at the start of the simulation (red = dry, black = 5% water by mass; see color bar). Three terrestrial planets formed, and each acquired material from beyond 2.5 au that delivered its water. Adapted from *Raymond et al.* (2006b).

terrestrial planets have formed (see *Raymond et al.*, 2006b, for details). These include reasonable analogs to Venus and Earth and a planet close to Mars' orbit that is roughly 10× more massive than the actual planet.

The feeding zones of all three planets included a tail that extended into the outer asteroid belt. The water content of each planet in this simulation was thus sourced from the outer asteroid belt.

The wide feeding zones of the planets in Fig. 9 are a generic feature of late-stage accretion. Eccentricity excitation implies that any planet's building blocks sample a wide region. Water delivery is therefore a robust outcome of classical model-type accretion.

But the classical model has an Achilles heel: Mars. Classical model simulations systematically form Mars "analogs" that are 5–10× more massive than the real planet (quantified by *Raymond et al.*, 2009; *Morishima et al.*, 2010; *Fischer et al.*, 2014; *Izidoro et al.*, 2015b; *Kaib and Cowan*, 2015). The problem is not Mars' absolute mass but the fact that it is so much less massive than Earth. Accretion tends to form systems in which neighboring planets have comparable masses (e.g., *Lissauer*, 1987). Models that succeed in reproducing the inner solar system invoke mechanisms to deplete Mars' feeding zone relative to Earth's (summarized in Fig. 11).

The *Early Instability* model (*Clement et al.*, 2018, 2019a,b) matches the terrestrial planets (including the Earth/Mars mass ratio) while preserving many of the assumptions of the classical model. It assumes that the giant planet instability [sometimes referred to as the Nice model instability because it was developed in the French town of Nice (*Tsiganis et al.*, 2005; *Morbidelli et al.*, 2007)] took place shortly after the dissipation of the gaseous disk.

Perturbations during the giant planets' instability act to strongly excite and deplete the asteroid belt and Mars region without strongly affecting the region within 1 au. Simulations match the terrestrial planets' mass distribution and the rate of success in matching the inner solar system correlates with that in matching the outer solar system (*Clement et al.*, 2018).

In the Early Instability model, water is delivered to the growing terrestrial planets from the outer asteroid region in the same way as in Fig. 9. That water would presumably have the same chemical fingerprint as today's C-type asteroids, represented by carbonaceous chondrites, and thus match Earth.

The possibility that Earth's water could be a result of its wide feeding zone therefore rests on the viability of the Early Instability model itself. To date there are three successful models that can explain the early evolution of the inner solar system (see the chapter by Raymond et al. in this volume, as well as Fig. 11). Future studies will use empirical and theoretical arguments to evaluate these models.

3.2.3. External pollution. Water may be delivered by a relatively low-mass population of volatile-rich planetesimals that "rain down" onto the terrestrial planet-forming region. In this scenario, the terrestrial planets would have formed predominantly from local rocky material but with a small amount of *external water-bearing pollution*. The difference between this model and the classical model is that the polluting planetesimals are not simply an extension of the planets' feeding zones but rather were dynamically injected from more distant regions of the planet-forming disk. These water-bearing planetesimals would have been scattered on high-eccentricity orbits by the growth and/or migration of the giant planets during the late parts of the gaseous disk phase.

To date, two mechanisms have been proposed to produce a population of high-eccentricity planetesimals. The first is a general mechanism that applies to every instance of giant planet growth (*Raymond and Izidoro*, 2017a). The second is inherently tied to the Grand Tack model (*Walsh et al.*, 2011).

In the core-accretion model (*Pollack et al.*, 1996), gas giant planets grow in two steps (see Fig. 8). First they accrete large solid cores of 5–20 $M_\oplus$ [likely by pebble accretion (e.g., *Lambrechts and Johansen*, 2012)]. Then they accrete gas from the disk. Gas accretion proceeds slowly until a critical threshold is reached (likely the point at which the mass in the gaseous envelope is comparable to the solid core mass), after which accretion accelerates and the planet rapidly grows into a Saturn- to Jupiter-mass planet (e.g., *Lissauer et al.*, 2009) and carves an annular gap in the disk (*Crida et al.*, 2006).

Figure 10 shows how the growth of gas giant planets affects nearby planetesimals (from *Raymond and Izidoro*, 2017a). When a planet such as Jupiter undergoes a phase of rapid gas accretion it destabilizes the orbits of nearby objects. Planetesimals undergo close encounters with the growing Jupiter and are scattered onto eccentric orbits across the solar system. Meanwhile, gas drag acts to damp the planetesimals' eccentricities. Planetesimals scattered inward can thus become decoupled from Jupiter as their aphelia decrease, and can be trapped onto stable orbits in the inner solar system (*Raymond and Izidoro* 2017a; *Ronnet et al.*, 2018). Saturn's growth has a similar effect, although planetesimals must be scattered first by Saturn and then by Jupiter to reach the inner solar system. Because Saturn is thought to have grown later than Jupiter when the planet-forming disk had evolved and was lower in mass than it was during Jupiter's formation, many planetesimals are scattered inward *past* the asteroid belt to the terrestrial planet region. Many planetesimals destabilized by Saturn may also be captured in Jupiter's circumplanetary disk (*Ronnet et al.*, 2018).

The same population of scattered planetesimals end up crossing the terrestrial planets' orbits and populating the outer asteroid belt (*Raymond and Izidoro*, 2017a). This mechanism naturally explains why carbonaceous chondrites (from C-type asteroids) are a chemical match to Earth's water.

The balance between planetesimals implanted into the belt and scattering toward the terrestrial zone depends on unknown parameters. The most important parameter is simply the strength of gas drag, which depends on a combination of the planetesimal size and the gas surface density in the inner disk at the time of planetesimal scattering. It is likely that there were many generations of planetesimal scattering

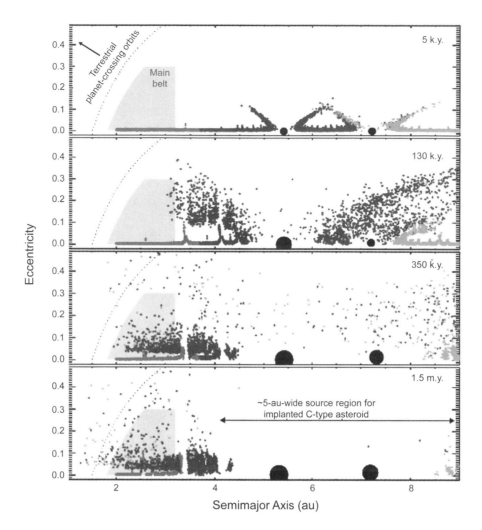

Fig. 10. See Plate 33 for color version. Dynamical injection of planetesimals into the inner solar system during Jupiter and Saturn's growth. This figure shows snapshots of a simulation in which 100-km planetesimals interact gravitationally with the growing gas giants and by gas drag with the disk. Jupiter's mass was increased from a core to its final mass from 100 to 200 k.y. and Saturn from 300 to 400 k.y. The colors of planetesimals serve to indicate their starting location. A large number of planetesimals were captured on stable orbits in the outer asteroid belt, providing a good match to the C-type asteroids. Many planetesimals were scattered onto high-eccentricity orbits that cross the growing terrestrial planets' and may have delivered water to Earth. The source region for implanted planetesimals was from 4 to 9 au in this example, but can extend out to 20 au when migration and the ice giants are considered. From *Raymond and Izidoro* (2017a).

into the inner solar system: during Jupiter and Saturn's growth and possible migration and the ice giants' growth and migration. In the example simulation from Fig. 10, planetesimals are implanted from a 5-au-wide swath of the disk. However, this is a *minimum* width, as taking migration and the ice giants into account can extend the source region to past 20 au (*Raymond and Izidoro,* 2017a).

The mechanism illustrated in Fig. 10 is generic and applies to any instance of giant planet formation (*Raymond and Izidoro,* 2017a). It may thus explain the initial conditions of the classical model. This mechanism has also been shown to be robust to a number of migration histories for the giant planets (*Raymond and Izidoro,* 2017a; *Ronnet et al.,* 2018; *Pirani et al.,* 2019).

This mechanism also naturally provides a source of C types and terrestrial water for the *Low-Mass Asteroid Belt*

model, another viable model for terrestrial planet formation. The Low-Mass Asteroid Belt model proposes that the terrestrial planets did not form from a broad disk of rocky material but rather from a narrow annulus (for details see the chapter by Raymond et al. in this volume and Fig. 11). In this model, the large Earth/Mars mass ratio is a simple consequence of a primordial mass deficit in the Mars region (*Hansen* 2009; *Kaib and Cowan,* 2015; *Raymond and Izidoro,* 2017b) (this would also explain the large Venus/Mercury mass ratio). Perhaps planetesimal formation was simply more efficient at 1 au and 5 au than in between. ALMA has indeed found a number of young circumstellar disks containing rings of dust (*ALMA Partnership et al.,* 2015; *Andrews et al.,* 2018). Given that planetesimal formation is only triggered in regions of high dust density (*Carrera et al.,* 2015; *Yang et al.,* 2017), it is plausible to imagine that planetesimals

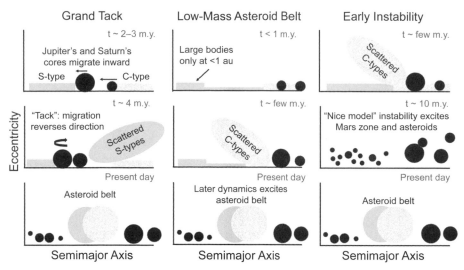

Fig. 11. Illustration of the evolution of three models that can match the inner solar system. From *Raymond et al.* (2018b).

also form in rings. The Low-Mass Asteroid Belt model can plausibly match the terrestrial planets and asteroid belt and is on the same footing as the Early Instability and Grand Tack models.

The second mechanism of water delivery by external pollution depends on *outward* migration in the framework of the Grand Tack model (*Walsh et al.*, 2011; *Jacobson and Morbidelli*, 2014; *Raymond and Morbidelli*, 2014; *Brasser et al.*, 2016) (see also Fig. 8). The growing Jupiter would have carved a gap in the gaseous disk and migrated inward in the type 2 regime. Saturn would have grown later farther out, migrated inward, and become trapped in mean-motion resonance with Jupiter (*Morbidelli et al.*, 2007; *Pierens and Nelson*, 2008). Two planets are in mean-motion resonance when their orbital periods form the ratio of small integers; e.g., in the 3:2 resonance the inner planet orbits the star three times for every two orbits of the outer planet, at which time the planets realign. After this point the two planets would have shared a common gap in the disk and tilted the torque balance so as to migrate outward together (*Masset and Snellgrove*, 2001; *Morbidelli et al.*, 2007; *Crida et al.*, 2009). This outward migration mechanism operates when two planets share a common gap with the innermost being more massive. Hydrodynamical simulations find that Jupiter and Saturn can migrate outward in the 3:2 or 2:1 resonances (*Masset and Snellgrove*, 2001; *Morbidelli et al.*, 2007; *Zhang and Zhou*, 2010; *Pierens and Raymond*, 2011; *Pierens et al.*, 2014). Jupiter and Saturn would thus have migrated outward together until the disk dissipated.

Jupiter's migration would have dramatically sculpted small-body populations (*Walsh et al.*, 2011, 2012). During its inward migration, Jupiter pushed most of the inner rocky material further inward by resonant shepherding (see *Fogg and Nelson*, 2005; *Raymond et al.*, 2006a), which acted to compress a broad disk of rocky material into a narrow annulus (similar to the one invoked for the Low-Mass Asteroid Belt model). A fraction of rocky planetesimals were scattered onto wider orbits, and some were scattered out to the Oort

cloud (*Meech et al.*, 2016). At the turnaround point of its migration Jupiter was 1.5–2 au from the Sun (*Walsh et al.*, 2011; *Brasser et al.*, 2016). Then, during Jupiter and Saturn's outward migration, the giant planets first encountered the scattered rocky planetesimals and then pristine outer-disk planetesimals. As the giant planets migrated through these small bodies most were ejected, but a small fraction were scattered inward and left behind on stable orbits once the planets migrated past. This mechanism is less dependent on the planetesimal size than the one illustrated in Fig. 10 because the orbits of scattered planetesimals are stabilized by the giant planets migrating away rather than by gas drag. In the Grand Tack model, the surviving planetesimals are trapped in the asteroid belt with a similar orbital distribution to the observed one (*Deienno et al.*, 2016).

In the Grand Tack model, water-rich planetesimals are scattered into the terrestrial planet-forming region from beyond Jupiter's orbit (*Walsh et al.*, 2011; *O'Brien et al.*, 2014). The mass in water-delivering planetesimals can be calibrated to the mass in planetesimals trapped in the asteroid belt. Taking into account later depletion of the belt (*Minton and Malhotra*, 2010; *Nesvorný*, 2015), the mass in polluting planetesimals is a few to 10% of a $M_\oplus$. The growing Earth accretes enough water to account for its current water budget (*Walsh et al.*, 2011; *O'Brien et al.*, 2014).

External pollution represents a viable scenario for water delivery. Inward scattering of planetesimals is an inevitable byproduct of giant planet formation (see Fig. 10) (*Raymond and Izidoro*, 2017a). The Grand Tack model matches a number of characteristics of the inner solar system. The main uncertainty lies in the outward migration mechanism (see *Raymond and Cossou*, 2014, for a discussion), which requires that Jupiter remain substantially more massive than Saturn during the entire accretion phase (*Masset and Snellgrove*, 2001).

The astute reader may ask themself how the external pollution mechanism differs from the old comet-delivery model. That model proposed that Earth grew locally and later received its water via a bombardment of comets (e.g., *Delsemme*, 1992; *Owen and Bar-Nun,* 1995). In contrast with the cometary model, the external pollution model invokes (1) self-consistent, global dynamical scenarios that explain the origin of Earth's water in the context of models that match the architecture of the inner solar system; and (2) does not invoke comets but rather polluting objects that originate from the same reservoir as C-type asteroids, which match Earth's D/H and $^{15}N/^{14}N$ ratios (comets do not match the Earth's $^{15}N/^{14}N$ ratio; see Fig. 5).

3.2.4. Inward migration. Orbital migration is a ubiquitous process in planet formation. Given that planets form in gaseous disks, gas-driven migration is simply inevitable once planets reach a critical mass (*Kley and Nelson,* 2012; *Baruteau et al.,* 2014). Migration is almost certainly a key process in the formation of so-called *super-Earths* and *sub-Neptunes,* which exist around 30–50% of main-sequence stars (*Howard et al.,* 2012; *Mayor et al.,* 2011; *Petigura et al.,* 2013; *Mulders et al.,* 2018). For virtually any disk profile, these planets should have migration timescales that are far shorter than the disk lifetime (e.g., *Ogihara et al.,* 2015). Models that invoke the migration of growing planetary embryos can quantitatively match the observed super-Earth distributions (e.g., *Ida and Lin,* 2010; *Cossou et al.,* 2014; *Izidoro et al.,* 2017, 2019; *Ogihara et al.,* 2018).

There is good reason to think that planetary embryos massive enough to undergo long-range migration form past the snow line. Planetesimals may form by the streaming instability most readily just past the snow line (*Armitage et al.,* 2016; *Drążkowska and Alibert,* 2017). Pebble accretion is also much more efficient past the snow line; by the time 10 $M_\oplus$ ice-rich cores have formed at 5 au, rocky planetary embryos in the inner disk may only reach ~0.1 $M_\oplus$ (*Morbidelli et al.,* 2015). This fits nicely with our picture of solar system formation, which requires a population of ~Mars-mass rocky embryos in the inner disk and a handful of giant planet cores in the outer disk.

If embryos large enough to migrate do indeed form past the snow line, then many are likely to be ice-rich. However, if embryos form at the snow line and migrate inward while continuing to accrete, they may only be ~5–10% water by mass, a far cry from the 50% that is often assumed. That assumption is questionable given that the most water-rich meteorites (including both components, the chondrites and matrix) have water-to-rock ratios of ~0.4 despite appearing to be outer solar system objects (*Alexander,* 2019a,b). In addition, while the inward migration of icy super-Earths perturbs the growth of terrestrial planets (*Izidoro et al.,* 2014), it may lead to the formation of very close-in planets that are entirely rocky (*Raymond et al.,* 2018). By all these avenues, migration tends to produce planets whose feeding zones are disconnected from their final orbital radii (e.g., *Kuchner,* 2003).

Could inward migration explain the origin of Earth water? Probably not. Multiple lines of (admittedly circumstantial) evidence suggest that the building blocks of the terrestrial planets were roughly Mars-sized (see *Morbidelli and Nes-*

vorny, 2012). Mars is below the mass required for long-range migration. In addition, if Earth or its building blocks did migrate inward then it is hard to understand why they would have stopped where they did rather than migrating closer to the Sun.

However, migration may indeed play a role in delivering water to Earth-like planets in other systems. This may be particularly important for planets orbiting low-mass stars (see section 4.1).

4. DISCUSSION

As discussed in section 3, a number of different scenarios can in principle match the amount and isotopic composition of Earth's water. However, some of them do not fit in a clear way within a self-consistent picture of the dynamical and chemical evolution of the solar system. On the other hand, other mechanisms are essentially inevitable, as they are simple byproducts of planet formation [for instance, the growth of a giant planet invariably pollutes its inner regions with water-rich planetesimals (*Raymond and Izidoro,* 2017a)].

4.1. Evaluation of Water Models

We now evaluate critically the six scenarios from section 3. We find that the external pollution mechanism is currently the most likely candidate.

While physically motivated, it is hard to see how the two scenarios for local water accretion could fit in a bigger picture of solar system formation. The adsorption of water vapor onto grains struggles because (1) the gas in the inner solar system was likely mainly dry; (2) the D/H ratio of adsorbed water should in principle be nebular, not Earth-like; and (3) the existence of dry chondritic meteorites (e.g., enstatite chondrites) restricts the plausible parameter space for the mechanism to operate. However, it should be noted that the D/H ratio for the primordial mantle material is only an upper limit and it could be lower, or the measurements did not measure the D/H of the original/indigenous water (*Hallis et al.,* 2015), so more measurements are needed.

Earth and its constituent planetary embryos may have accreted primordial hydrogen-rich atmospheres. Oxidation during the magma ocean phase may have produced water (*Ikoma and Genda,* 2006), after which extensive atmospheric loss could have increased the D/H ratio to Earth ocean-like values (*Genda and Ikoma,* 2008). However, it seems a great coincidence for the surviving water on Earth to match carbonaceous chondrites in their D/H ratios. It is also unclear whether this model could explain Earth's $^{15}N/^{14}N$ ratio given that the nitrogen would have come from a different source.

The Sun's snow line may have been interior to Earth's orbit during a significant fraction of the disk lifetime (e.g., *Oka et al.,* 2011; *Martin and Livio* 2012). However, it seems unlikely that inward-drifting pebbles provided Earth's water. The age distributions of the two isotopically-distinct classes of meteorites — carbonaceous and non-carbonaceous — suggest that those reservoirs were kept separate as of ~1 m.y.

after CAIs, and this segregation has been interpreted as being caused by the growing giant planets (perhaps Jupiter's core) blocking the inward drift of carbonaceous pebbles (*Budde et al.*, 2016; *Kruijer et al.*, 2017; *Desch et al.*, 2018a). Since the carbonaceous pebbles represent the source of water, it is hard to imagine how they could have delivered water to Earth without producing a population of meteorites intermediate in composition between the two known classes. Yet the ordinary and R chondrites formed at ~2 m.y. after CAIs but show signs of having accreted water ice (*Alexander et al.*, 2018b). The origin of that ice is hard to understand, and may be linked with pebble recycling in the inner disk or Jupiter's core acting as an imperfect pebble barrier (e.g., *Morbidelli et al.*, 2016).

In the classical model of solar system formation (*Wetherill*, 1992), Earth's water is a byproduct of its broad feeding zone (*Morbidelli et al.*, 2000; *Raymond et al.*, 2004, 2007a). However, the classical model cannot easily match the large Earth/Mars mass ratio (e.g., *Raymond et al.*, 2009; *Morishima et al.*, 2010) and is therefore suspect. The Early Instability model can reproduce Mars' mass and also delivers water to Earth due to its broad feeding zone (*Clement et al.*, 2018, 2019a). Yet the processes that shaped the initial water distribution remain unexplained by such models.

At present, the external pollution model provides the most complete explanation for the origin of Earth's water. A population of planetesimals on high-eccentricity orbits crossing the terrestrial zone is naturally produced by the giant planets' growth (*Raymond and Izidoro*, 2017a) and migration (*Walsh et al.*, 2011; *O'Brien et al.*, 2014). This fits within the Grand Tack and Low-Mass Asteroid Belt models for terrestrial planet formation (see Fig. 11). The same dynamical processes also implant objects into the outer asteroid belt. The objects that delivered water to Earth should then have had the same chemical signature as carbonaceous chondrites, which do indeed provide a good match to Earth's isotopic composition in terms of water and nitrogen (*Lécuyer et al.*, 1998; *Marty*, 2012). The amount of water delivered depends on unconstrained parameters (e.g., the disk properties and planetary migration rates) but plausible values can match Earth. Of course, these dynamical models remain a matter of debate (see the chapter by Raymond et al. in this volume). Nonetheless, there are no obvious problems with this delivery mechanism.

Finally, inward-migrating planetary embryos can indeed deliver water to inner rocky planets [or themselves become inner ice-rich planets (e.g., *Terquem and Papaloizou*, 2007; *Izidoro et al.*, 2019; *Bitsch et al.*, 2019)]. However, the building blocks of our solar system's terrestrial planets are likely to have been ~Mars-mass (see *Morbidelli and Nesvorny*, 2012), below the mass for substantial orbital migration.

4.2. Water Loss Processes

Several mechanisms exist that may significantly dry out planetesimals and planets that were not accounted for in the models presented in section 3.

The short-lived radionuclide ^{26}Al (half-life of ~700,000 yr) provided a huge amount of heat to the early solar system. As a result, any planetesimals that formed within roughly 2 m.y. of CAIs would have been completely dehydrated (*Grimm and McSween*, 1993; *Monteux et al.*, 2018). In addition, the presence of ^{26}Al can lead to a bifurcation between very water-rich planets with tens of percent of water by mass and relatively dry, rocky planets like Earth (*Lichtenberg et al.*, 2019). Because ^{26}Al is produced in massive stars, its abundance in planet-forming disks may vary considerably (*Hester et al.*, 2004; *Gounelle and Meibom*, 2008; *Gaidos et al.*, 2009; *Lichtenberg et al.*, 2016), leading to a diversity in the water contents of terrestrial exoplanets.

Impacts may also strip planets of water. Ice-rich mantles can be stripped in giant impacts (*Marcus et al.*, 2010). Given that the impact impedance of water is lower than that of rock, giant impacts preferentially remove water from the surfaces of water-rich planets (*Genda and Abe*, 2005). Planetesimal impacts may erode planetary atmospheres (*Svetsov*, 2007; *Schlichting et al.*, 2015), causing water loss if a large fraction of planets' water is in the atmosphere.

Improving our understanding of water loss may be as important as improving our understanding of water delivery.

4.3. Extrapolation to Exoplanets

The bulk of extrasolar planetary systems look very different than our own (see section 1 and discussion in the chapter by Raymond et al. in this volume). Yet we think that the same fundamental processes govern the formation of all systems. In this section we evaluate which of the mechanisms from section 3 are likely to be dominant in exoplanet systems.

Our analysis leads to two conclusions. First, we expect water to be delivered to virtually all rocky planets. Any completely dry planets are likely to have *lost* their water. Second, while pebble snow, broad feeding zones, and external pollution are likely to play a role, we expect migration to be the dominant mechanism of water delivery. Migration is also the mechanism that should produce planets with the highest water contents. This conclusion is based on overwhelming empirical and theoretical evidence that the population of super-Earths form predominantly during the gaseous disk lifetime, leading to the inescapable conclusion that planet-disk interactions — and therefore orbital migration — must have played a role in shaping this population (e.g., as in the model of *Izidoro et al.*, 2017, 2019). Indeed, current thinking invokes migration as one of a handful of essential ingredients in planet formation models (see the chapter by Raymond et al.).

Imagine a planet growing in the hotter regions of a planet-forming disk, in its star's habitable zone. Ironically, while a planet in this region can maintain liquid water on its surface (with the right atmosphere), the local building blocks are dry. Given the diversity of water-delivery mechanisms, it is hard to imagine such a planet remaining dry. Of course, if water is sourced locally — from adsorption onto grains or from the oxidation of a primordial H-rich envelope — then it will be hydrated anyway. If the planet accretes from a smooth disk of solid material then its feeding zone will widen in time to encompass more distant, volatile-rich bodies. Even if the planet is growing from a ring of material, the snow line moves inward as the gaseous disk cools, such that pebbles can "snow" down and deliver water. Pebble snow can be shut down by the growth of a large outer planet, but that planet would produce a rain of water-rich planetesimals that pollute the inner rocky zone. A large outer planet might even migrate inward and deliver water in bulk (and perhaps become the ocean-covered seed of a habitable zone planet). Most stars are lower in mass than the Sun (*Chabrier*, 2003), with lower luminosities and correspondingly closer-in habitable zones (*Kasting et al.,* 1993). Certain factors argue that habitable zone planets around low-mass stars should be drier than around Sun-like stars. In dynamical terms, the snow line is farther away from the habitable zone around low-mass stars (*Kennedy and Kenyon,* 2008; *Mulders et al.,* 2015a). This would presumably reduce the efficiency of pebble snow and also of external pollution [which is also reduced by the lower frequency of gas giants around low-mass stars (*Johnson et al.,* 2007)]. Since their habitable zones are closer in, accretion timescales are shorter around low-mass stars (compared with Sun-like stars) and impact speeds higher (*Lissauer and Stevenson,* 2007; *Raymond et al.,* 2007b). Other factors are ambiguous with regard to water delivery. While fast accretion timescales imply an increase in the efficiency of migration, the masses of planet-forming disks appear to be lower around low-mass stars (*Pascucci et al.,* 2016). Low-mass stars are observed to have more "super-Earths" and fewer "mini-Neptunes" than Sun-like stars (*Mulders et al.,* 2015b) but it remains unclear how this connects with planet formation and water delivery.

Migration is likely a dominant process in determining planetary water contents. Many models for the origin of super-Earths invoke large-scale migration of large planetary embryos (*Terquem and Papaloizou,* 2007; *McNeil and Nelson,* 2010; *Ida and Lin,* 2010; *Cossou et al.,* 2014; *Ogihara et al.,* 2015; *Izidoro et al.,* 2017, 2019). Given that embryos are thought to form faster past the snow line [by pebble accretion (*Lambrechts et al.,* 2014; *Morbidelli et al.,* 2015)], most of the super-Earths formed by migration should be ice-rich (but see *Raymond et al.,* 2018). Some large embryos may grow close to the inner edge of the disk by accumulating drifting pebbles (*Chatterjee and Tan,* 2014, 2015). Embryos that grow large past the snow line before migrating inward have water contents of ten or more percent (*Bitsch et al.,* 2019; *Izidoro et al.,* 2019), whereas those that grow close-in should be purely rocky. Despite the preferential stripping of icy mantles during giant impacts (*Marcus et al.,* 2010), collisions between these two populations will likely maintain the bimodal distribution of very water-rich planets and very dry ones. Planets with intermediate water contents would only form after one ice-rich embryo underwent a series of giant impacts with pure rock embryos.

None of the other water delivery mechanisms is likely to produce planets with more than ~1% water by mass. Pebble snow is susceptible to being deactivated by the growth of any large core on an exterior orbit (*Lambrechts and Johansen,* 2014; *Bitsch et al.,* 2018). In addition, the pebble flux may drop substantially when the pebble production front reaches the outer edge of the disk, limiting the overall amount of water that can be delivered (*Ida et al.,* 2019). The sweet spot for pebble snow may thus be early, before the growth of large cores and while the pebble flux is high. Very water-rich planets may grow from ice-rich pebbles (*Lambrechts and Johansen,* 2012; *Morbidelli et al.,* 2015), and these are the planets likely to migrate inward (e.g., *Ormel et al.,* 2017; *Bitsch et al.,* 2019).

Water contents similar to Earth's (~0.1% water by mass) are not detectable with present-day techniques. Our knowledge of the water contents of exoplanets is very limited. Mass and radius measurements exist for dozens of known close-in low-mass planets (e.g., *Batalha et al.,* 2013; *Marcy et al.,* 2014). These data have revealed a dichotomy between small, solid *super-Earths* ($R \lesssim 1.5$–2 $R_\oplus$) and larger, gas-rich *sub-Neptunes* (*Rogers,* 2015; *Wolfgang et al.,* 2016; *Chen and Kipping,* 2017). Some well-measured super-Earths have densities high enough to be scaled-up versions of Earth or even Mercury (*Howard et al.,* 2013; *Santerne et al.,* 2018; *Bonomo et al.,* 2019). However, constraining water contents from density measurements is fraught with uncertainty (e.g., *Adams et al.,* 2008; *Selsis et al.,* 2007; *Dorn et al.,* 2015). The gap in the size distribution of super-Earths (*Fulton et al.,* 2017) has been interpreted as an indication that most super-Earths are rocky (*Lopez,* 2017; *Owen and Wu,* 2017; *Jin and Mordasini,* 2018) but there is considerable debate (e.g., *Zeng et al.,* 2019). Given measurement uncertainties, the meaning of "rocky" vs. "ice-rich" is ambiguous; some models can match the radius gap with planets with anything up to 20% water by mass (*Gupta and Schlichting,* 2019).

4.4. What New Measurements Are Needed?

What measurements do we need to move forward? As discussed previously, the external pollution mechanism is currently the most likely candidate. However, this still leaves a wide range of possible distances from where the volatiles may have originated. The source location in the disk will depend upon where and when Jupiter's core formed. The key will be to match the signatures from volatiles in primitive objects that preserve the early solar system record and whose dynamical provenance is reasonably well understood. This region is the "wet" outer asteroid belt. Material scattered into the inner solar system will be both implanted into the asteroid belt, and impact the growing terrestrial planets. While the material that built the terrestrial planets no longer exists, the material scattered inward still exists today in the asteroid belt, and volatiles have been preserved in the outer belt.

Direct evidence for water in the outer asteroid belt stems from the Herschel detection of water vapor from Ceres

(*Küppers et al.,* 2014) coupled with water ice detections on Ceres' surface by the Dawn mission (*Combe et al.,* 2016), and a possible H$_2$O-frost signature on the asteroids 90 Antiope (*Hargrove et al.,* 2015) and 24 Themis (*Campins et al.,* 2010; *Rivkin and Emery,* 2010). Main-belt comets (MBCs) are the primitive representatives of accessible ice from the outer asteroid belt. Orbiting in the outer asteroid belt since the early stages of solar system formation (*Hsieh and Jewitt,* 2006; *Jewitt,* 2012), they exhibit comet-like tails, attributed to outgassing of volatiles from ices in their interiors. The ices, preserved by a layer of dust, carry a signature of the early solar system.

All objects whose semimajor axes are less than Jupiter's, with an orbit dynamically decoupled from Jupiter (*Vaghi,* 1973; *Kresak,* 1980), and exhibiting mass loss with a cometary appearance, are termed "active asteroids." They are dynamically unrelated to comets (*Levison et al.,* 2006; *Haghighipour,* 2009). There are 25 currently known. Several causes for their comet-like activity have been postulated, including electrostatic ejection (*Criswell and De,* 1977), mass shedding from collisional and radiation torques (YORP) (*Drahus et al.,* 2011; *Jacobson and Scheeres,* 2011), rotational instability, and volatile sublimation (*Jewitt,* 2012). Only a thermal process (e.g., sublimation) can explain the observed repeated perihelion activity. The 12 believed to be water-ice sublimation-driven [7 with repeat activity (*Hsieh et al.,* 2018a)] are termed MBCs (*Snodgrass et al.,* 2017), of value for their primordial ices. At the distance of the asteroid belt, the interior temperatures are warm enough that only water and its trapped volatiles remain over the age of the solar system (*Prialnik and Rosenberg,* 2009). MBC sublimation is believed to be triggered by impacts of meter-sized objects that remove part of the surface layer, allowing heat to reach the ice (*Hsieh et al.,* 2009; *Haghighipour et al.,* 2016). Models predict activity will continue for hundreds of orbits, and will show significant fading before ceasing (*Prialnik and Rosenberg,* 2009). Conversely, non-MBC active asteroids exhibit very different, short-lived dust structures (*Jewitt,* 2012; *Hainaut et al.,* 2012; *Kleyna et al.,* 2013). All attempts to directly detect gas (H$_2$O and its more read-

ily observable proxy, CN, a minor species that fluoresces strongly and is dragged out of the nucleus with water as it sublimates) have been unsuccessful — unsurprising, given that the detection limit for a 10-m telescope is 1 to 2 orders of magnitude above the amount of gas required to lift the observed dust. Indeed, Kuiper belt and Oort cloud comets at the same distance routinely show no spectral evidence for gas, although their dust comae are strong.

Main-belt comets are small (radii < 2 km), with low albedos and flat featureless spectra in the visible. Most MBCs are related to collisional asteroid families (*Hsieh et al.,* 2018b) since this is an excellent way to bring interior ices closer to the surface. For example, the Themis family represents the collisional remnants of an icy protoplanet a few hundred kilometers in size (Fig. 12). A weak or absent hydration signature suggests limited aqueous alteration in the Themis parent interior (*Marsset et al.,* 2016), consistent with thermal models predicting a thick, unaltered outer layer of primitive ice and rock (*Castillo-Rogez and Schmidt,* 2010). Diversity in the Themis family spectral properties is interpreted as a gradient in the parent-body composition, supporting that scenario (*Fornasier et al.,* 2016). Stripping the Themis outer layer formed a large fraction of the family members, including the MBCs, without reprocessing their volatile content (*Durda et al.,* 2007; *Rivkin et al.,* 2014) (Fig. 12).

Main-belt comets are samples from the unexplored icy asteroids that are small enough and/or formed late enough to have escaped complete hydrothermal processing and still have primitive outer layers (Fig. 12), unlike Ceres, an icy asteroid that has suffered intensive aqueous alteration (*Ammannito et al.,* 2016). An alternate explanation for the origin of some MBCs is that they are primordial planetesimals that never accreted into larger bodies. In both cases, MBCs offer accessible pristine volatiles. Thermal models show that these ices can survive over the age of the solar system due to an insulating dust layer (*Prialnik and Rosenberg,* 2009; *Schorghofer,* 2008). Thus, these represent the best source of material from which measurements can help distinguish between solar system formation models and the origin of

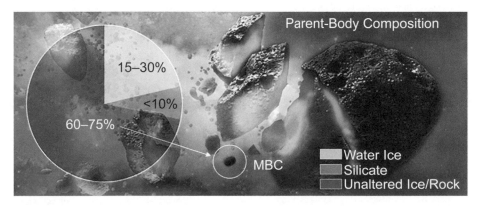

Fig. 12. See Plate 34 for color version. Thermal models show that MBC parent bodies preserve unaltered materials throughout 60–75% of the body for their history (*Castillo-Rogez and Schmidt,* 2010). MBCs are from this unaltered fraction (*Marsset et al.,* 2016).

inner solar system water. The key measurements to make will be to obtain the multiple volatile isotopic fingerprints: D/H in water, $^{17}O/^{16}O$, $^{18}O/^{16}O$, $^{15}N/^{14}N$, and $^{84}Kr/^{36}Ar$.

The next generation of submillimeter telescopes envisioned for the next decade will have 10× the sensitivity and resolution of the exquisite performance of ALMA (*Murphy et al.*, 2018). These facilities will be able probe planet formation inside 10 au (*Andrews et al.*, 2018), unveiling the chemistry and structures in the habitable zones in nearby protoplanetary disks (*Ricci et al.*, 2018; *McGuire et al.*, 2018). Careful isotopic measurements of primitive volatiles in the outer asteroid belt gives us the ability to explore these past processes in our own solar system.

Acknowledgments. We thank referees H. Genda and C. Alexander for helpful reports that greatly improved this review. We acknowledge extensive discussions that shaped our thinking over the past many years with researchers in the field, including E. Bergin, G. Huss, A. Krot, A. Morbidelli, J. Lunine, B. Marty, A. Izidoro, J. Chambers, K. Walsh, S. Jacobson, and C. Alexander. K.J.M. acknowledges support from the NASA Astrobiology Institute under Cooperative Agreement No. NNA09DA77A issued through the Office of Space Science, NASA Grant 80-NSSC18K0853, and a grant from the National Science Foundation, AST1617015. S.N.R. thanks NASA Astrobiology Institute's Virtual Planetary Laboratory Lead Team, funded under solicitation NNH12ZDA002C and Cooperative Agreement No. NNA13AA93A. We would particularly like to thank J. E. Polk (Jet Propulsion Laboratory) for helping to edit this chapter.

REFERENCES

Abe Y., Ohtani E., Okuchi T., Righter K., and Drake M. (2000) Water in the early Earth. In *Origin of the Earth and Moon* (R. M. Canup and K. Righter, eds.), pp. 413–433. Univ. of Arizona, Tucson.

Adams E. R., Seager S., and Elkins-Tanton L. (2008) Ocean planet or thick atmosphere: On the mass-radius relationship for solid exoplanets with massive atmospheres. *Astrophys. J.*, 673, 1160–1164.

Agnor C. B., Canup R. M., and Levison H. F. (1999) On the character and consequences of large impacts in the late stage of terrestrial planet formation. *Icarus*, 142, 219–237.

A'Hearn M. F., Millis R. C., Schleicher D. O., Osip D. J., and Birch P. V. (1995) The ensemble properties of comets: Results from narrowband photometry of 85 comets, 1976–1992. *Icarus*, 118, 223–270.

Aikawa Y., van Zadelhoff G. J., van Dishoeck E. F., and Herbst E. (2002) Warm molecular layers in protoplanetary disks. *Astron. Astrophys.*, 386, 622–632.

Alexander C. M. O. (2019a) Quantitative models for the elemental and isotopic fractionations in chondrites: The carbonaceous chondrites. *Geochim. Cosmochim. Acta*, 254, 277–309.

Alexander C. M. O. (2019b) Quantitative models for the elemental and isotopic fractionations in the chondrites: The non-carbonaceous chondrites. *Geochim. Cosmochim. Acta*, 254, 246–276.

Alexander C. M. O., Bowden R., Fogel M. L., Howard K. T., Herd C. D. K., and Nittler L. R. (2012a) The provenances of asteroids, and their contributions to the volatile inventories of the terrestrial planets. *Science*, 337, 721.

Alexander C. M. O., Bowden R., Fogel M. L., Howard K. T., Herd C. D. K., and Nittler L. R. (2012b) The provenances of asteroids, and their contributions to the volatile inventories of the terrestrial planets. *Science*, 337, 721.

Alexander C. M. O., Nittler L. R., Davidson J., and Ciesla F. J. (2017) Measuring the level of interstellar inheritance in the solar protoplanetary disk. *Meteoritics & Planet. Sci.*, 52, 1797–1821.

Alexander C. M. O., McKeegan K. D., and Altwegg K. (2018a) Water reservoirs in small planetary bodies: Meteorites, asteroids, and

comets. *Space Sci. Rev.*, 214, 36.

Alexander C. M. O., McKeegan K. D., and Altwegg K. (2018b) Water reservoirs in small planetary bodies: Meteorites, asteroids, and comets. *Space Sci. Rev.*, 214, 36.

ALMA Partnership, Brogan C. L., Pérez L. M., Hunter T. R., Dent W. R. F., Hales A. S., Hills R. E., et al. (2015) The 2014 ALMA Long Baseline Campaign: First results from high angular resolution observations toward the HL Tau region. *Astrophys. J. Lett.*, 808, L3.

Altwegg K., Balsiger H., Bar-Nun A., Berthelier J. J., Bieler A., Bochsler P., Briois C., Calmonte U., Combi M., and De Keyser J. (2015) 67P/Churyumov-Gerasimenko, a Jupiter family comet with a high D/H ratio. *Science*, 347, 1261952.

Amelin Y., Krot A. N., Hutcheon I. D., and Ulyanov A. A. (2002) Lead isotopic ages of chondrules and calcium-aluminum-rich inclusions. *Science*, 297, 1678–1683.

Ammannito E., DeSanctis M. C., Ciarniello M., Frigeri A., Carrozzo F. G., Combe J. P., Ehlmann B. L., Marchi S., McSween H. Y., and Raponi A. (2016) Distribution of phyllosilicates on the surface of Ceres. *Science*, 353, aaf4279.

Anders E. and Grevesse N. (1989) Abundances of the elements: Meteoritic and solar. *Geochim. Cosmochim. Acta*, 53, 197–214.

Andrews S. M., Wilner D. J., Macías E., Carrasco-González C., and Isella A. (2018) Resolved substructures in protoplanetary disks with the ngVLA. In *Science with a Next Generation Very Large Array* (E. Murphy, ed.), p. 137. ASP Conf. Ser. 517, Astronomical Society of the Pacific, San Francisco.

Armitage P. J., Eisner J. A., and Simon J. B. (2016) Prompt planetesimal formation beyond the snow line. *Astrophys. J. Lett.*, 828, L2.

Asaduzzaman A. M., Zega T. J., Laref S., Runge K., Deymier P. A., and Muralidharan K. (2014) A computational investigation of adsorption of organics on mineral surfaces: Implications for organics delivery in the early solar system. *Earth Planet. Sci. Lett.*, 408, 355–361.

Badro J., Côté A. S., and Brodholt J. P. (2014) A seismologically consistent compositional model of Earth's core. *Proc. Natl. Acad. Sci. U.S.A.*, 111, 7542–7545.

Bar-Nun A., Herman G., Laufer D., and Rappaport M. L. (1985) Trapping and release of gases by water ice and implications for icy bodies. *Icarus*, 63, 317–332.

Bar-Nun A., Kleinfeld I., and Kochavi E. (1988) Trapping of gas mixtures by amorphous water ice. *Phys. Rev. Ser. B*, 38, 7749–7754.

Baruteau C., Crida A., Paardekooper S.-J., Masset F., Guilet J., Bitsch B., Nelson R., Kley W., and Papaloizou J. (2014) Planet-disk interactions and early evolution of planetary systems. In *Protostars and Planets VI* (H. Beuther et al., eds.), pp. 667–689. Univ. of Arizona, Tucson.

Batalha N. M., Rowe J. F., Bryson S. T., Barclay T., Burke C. J., Caldwell D. A., Christiansen J. L., et al. (2013) Planetary candidates observed by Kepler. III. Analysis of the first 16 months of data. *Astrophys. J. Suppl.*, 204, 24.

Bate M. R. (2018) On the diversity and statistical properties of protostellar discs. *Mon. Not. R. Astron. Soc.*, 475, 5618–5658.

Benz W., Slattery W. L., and Cameron A. G. W. (1986) The origin of the moon and the single-impact hypothesis I. *Icarus*, 66, 515–535.

Benz W., Cameron A. G. W., and Melosh H. J. (1989) The origin of the moon and the single impact hypothesis III. *Icarus*, 81, 113–131.

Bergin E. A. and Tafalla M. (2007) Cold dark clouds: The initial conditions for star formation. *Annu. Rev. Astron. Astrophys.*, 45, 339–396.

Birnstiel T., Klahr H., and Ercolano B. (2012) A simple model for the evolution of the dust population in protoplanetary disks. *Astron. Astrophys.*, 539, A148.

Birnstiel T., Fang M., and Johansen A. (2016) Dust evolution and the formation of planetesimals. *Space Sci. Rev.*, 205, 41–75.

Bitsch B., Johansen A., Lambrechts M., and Morbidelli A. (2015) The structure of protoplanetary discs around evolving young stars. *Astron. Astrophys.*, 575, A28.

Bitsch B., Morbidelli A., Johansen A., Lega E., Lambrechts M., and Crida A. (2018) Pebble-isolation mass: Scaling law and implications for the formation of super-Earths and gas giants. *Astron. Astrophys.*, 612, A30.

Bitsch B., Raymond S. N., and Izidoro A. (2019) Rocky super-Earths or waterworlds: The interplay of planet migration, pebble accretion, and disc evolution. *Astron. Astrophys.*, 624, A109.

Bockelée-Morvan D., Calmonte U., Charnley S., Duprat J., Engrand C., Gicquel A., Hässig M., et al. (2015) Cometary isotopic measurements. *Space Sci. Rev.*, 197, 47–83.

Boe B., Jedicke R., Meech K. J., Wiegert P., Weryk R. J., Chambers
K. C., Denneau L., et al. (2019) The orbit and size-frequency
distribution of long period comets observed by Pan-STARRS1.
Icarus, 333, 252–272.

Boehnke P. and Harrison T. M. (2016) Illusory late heavy
bombardments. *Proc. Natl. Acad. Sci. U.S.A., 113*, 10802–10806.

Boley A. C. (2009) The two modes of gas giant planet formation.
Astrophys. J. Lett., 695, L53–L57.

Bonomo A. S., Zeng L., Damasso M., Leinhardt Z. M., Justesen A. B.,
Lopez E., Lund M. N., Malavolta L., Silva Aguirre V., and Buchhave
L. A. (2019) A giant impact as the likely origin of different twins in
the Kepler-107 exoplanet system. *Nature Astron., 3*, 416–423.

Boss A. P. (1997) Giant planet formation by gravitational instability.
Science, 276, 1836–1839.

Bouvier A. and Wadhwa M. (2010) The age of the solar system
redefined by the oldest Pb-Pb age of a meteoritic inclusion. *Nature
Geosci., 3*, 637–641.

Brasser R., Walsh K. J., and Nesvorný D. (2013) Constraining the
primordial orbits of the terrestrial planets. *Mon. Not. R. Astron. Soc.,
433(4)*, 3417–3427.

Brasser R., Matsumura S., Ida S., Mojzsis S. J., and Werner S. C. (2016)
Analysis of terrestrial planet formation by the Grand Tack model:
System architecture and tack location. *Astrophys. J., 821*, 75.

Budde G., Burkhardt C., Brennecka G. A., Fischer-Gödde M., Kruijer
T. S., and Kleine T. (2016) Molybdenum isotopic evidence for the
origin of chondrules and a distinct genetic heritage of carbonaceous
and non-carbonaceous meteorites. *Earth Planet. Sci. Lett., 454*,
293–303.

Butler R. P., Wright J. T., Marcy G. W., Fischer D. A., Vogt S. S.,
Tinney C. G., Jones H. R. A., et al. (2006) Catalog of nearby
exoplanets. *Astrophys. J., 646*, 505–522.

Cabrera J., Csizmadia S., Lehmann H., Dvorak R., Gandolfi D., Rauer
H., Erikson A., Dreyer C., Eigmüller P., and Hatzes A. (2014) The
planetary system to KIC 11442793: A compact analogue to the solar
system. *Astrophys. J., 781*, 18.

Campins H., Hargrove K., Pinilla-Alonso N., Howell E. S., Kelley
M. S., Licandro J., Mothé-Diniz T., Fernández Y., and Ziffer J.
(2010) Water ice and organics on the surface of the asteroid 24
Themis. *Nature, 464*, 1320–1321.

Canup R. M. and Asphaug E. (2001) Origin of the Moon in a giant
impact near the end of the Earth's formation. *Nature, 412*, 708–712.

Carrera D., Johansen A., and Davies M. B. (2015) How to form
planetesimals from mm-sized chondrules and chondrule aggregates.
Astron. Astrophys., 579, A43.

Carrera D., Gorti U., Johansen A., and Davies M. B. (2017) Planetesimal
formation by the streaming instability in a photoevaporating disk.
Astrophys. J., 839, 16.

Castillo-Rogez J. C. and Schmidt B. E. (2010) Geophysical evolution of
the Themis family parent body. *Geophys. Res. Lett., 37*, L10202.

Cazaux S., Caselli P., and Spaans M. (2011) Interstellar ices as witnesses
of star formation: Selective deuteration of water and organic
molecules unveiled. *Astrophys. J. Lett., 741*, L34.

Chabrier G. (2003) Galactic stellar and substellar initial mass function.
Publ. Astron. Soc. Pac., 115, 763–795.

Chambers J. E. (2001) Making more terrestrial planets. *Icarus, 152*,
205–224.

Chatterjee S. and Tan J. C. (2014) Inside-out planet formation.
Astrophys. J., 780, 53.

Chatterjee S. and Tan J. C. (2015) Vulcan planets: Inside-out formation
of the innermost super-Earths. *Astrophys. J. Lett., 798*, L32.

Chatterjee S., Ford E. B., Matsumura S., and Rasio F. A. (2008)
Dynamical outcomes of planet-planet scattering. *Astrophys. J., 686*,
580–602.

Chen J. and Kipping D. (2017) Probabilistic forecasting of the masses
and radii of other worlds. *Astrophys. J., 834*, 17.

Cieza L. A., Casassus S., Tobin J., Bos S. P., Williams J. P., Perez
S., Zhu Z., et al. (2016) Imaging the water snow-line during a
protostellar outburst. *Nature, 535*, 258–261.

Clark R. N., Brown R. H., Cruikshank D. P., and Swayze G. A. (2019)
Isotopic ratios of Saturn's rings and satellites: Implications for the
origin of water and Phoebe. *Icarus, 321*, 791–802.

Clayton R. N., Grossman L., and Mayeda T. K. (1973) A component of
primitive nuclear composition in carbonaceous meteorites. *Science,
182*, 485–488.

Cleeves L. I., Bergin E. A., and Adams F. C. (2014a) Exclusion of
cosmic rays in protoplanetary disks. II. Chemical gradients and

observational signatures. *Astrophys. J., 794*, 123.

Cleeves L. I., Bergin E. A., Alexander C. M. O. D., Du F., Graninger D.,
Öberg K. I., and Harries T. J. (2014b) The ancient heritage of water
ice in the solar system. *Science, 345*, 1590–1593.

Cleeves L. I., Bergin E. A., Qi C., Adams F. C., and Öberg K. I. (2015)
Constraining the X-ray and cosmic-ray ionization chemistry of the
TW Hya protoplanetary disk: Evidence for a sub-interstellar cosmic-
ray rate. *Astrophys. J., 799*, 204.

Cleeves L. I., Bergin E. A., Alexander C. M. O. D., Du F., Graninger
D., Öberg K. I., and Harries T. J. (2016) Exploring the origins of
deuterium enrichments in solar nebular organics. *Astrophys. J., 819*,
13.

Clement M. S., Kaib N. A., Raymond S. N., and Walsh K. J. (2018)
Mars' growth stunted by an early giant planet instability. *Icarus,311*,
340-356.

Clement M. S., Kaib N. A., Raymond S. N., Chambers J. E., and
Walsh K. J. (2019a) The early instability scenario: Terrestrial
planet formation during the giant planet instability, and the effect of
collisional fragmentation. *Icarus, 321*, 778–790.

Clement M. S., Raymond S. N., and Kaib N. A. (2019b) Excitation and
depletion of the asteroid belt in the early instability scenario. *Astron.
J., 157*, 38.

Clesi V., Bouhifd M. A., Bolfan-Casanova N., Manthilake G., Schiavi
F., Raepsaet C., Bureau H., Khodja H., and Andrault D. (2018) Low
hydrogen contents in the cores of terrestrial planets. *Sci. Adv., 4*,
e1701876.

Combe J.-P., McCord T. B., Tosi F., Ammannito E., Carrozzo F. G., De
Sanctis M. C., Raponi A., Byrne S., Landis M. E., and Hughson
K. H. G. (2016) Detection of local H_2O exposed at the surface of
Ceres. *Science, 353*, aaf3010.

Cossou C., Raymond S. N., Hersant F., and Pierens A. (2014) Hot super-
Earths and giant planet cores from different migration histories.
Astron. Astrophys., 569, A56.

Coutens A., Vastel C., Hincelin U., Herbst E., Lis D. C., Chavarría L.,
Gérin M., van der Tak F. F. S., Persson C. M., Goldsmith P. F., and
Caux E. (2014) Water deuterium fractionation in the high-mass star-
forming region G34.26+0.15 based on Herschel/HIFI data. *Mon. Not.
R. Astron. Soc., 445*, 1299–1313.

Crida A., Morbidelli A., and Masset F. (2006) On the width and shape
of gaps in protoplanetary disks. *Icarus, 181*, 587–604.

Crida A., Masset F., and Morbidelli A. (2009) Long range outward
migration of giant planets, with application to Fomalhaut b.
Astrophys. J. Lett., 705, L148–L152.

Criswell D. R. and De B. R. (1977) Intense localized photoelectric
charging in the lunar sunset terminator region, 2. Supercharging at
the progression of sunset. *J. Geophys. Res., 82*, 1005.

Cumming A., Butler R. P., Marcy G. W., Vogt S. S., Wright J. T., and
Fischer D. A. (2008) The Keck planet search: Detectability and the
minimum mass and orbital period distribution of extrasolar planets.
Publ. Astron. Soc. Pac., 120, 531–554.

Dauphas N. (2003) The dual origin of the terrestrial atmosphere. *Icarus,
165*, 326–339.

Dauphas N. (2017) The isotopic nature of the Earth's accreting material
through time. *Nature, 541*, 521–524.

Dauphas N. and Pourmand A. (2011) Hf-W-Th evidence for rapid
growth of Mars and its status as a planetary embryo. *Nature, 473*,
489–492.

Day J. M. D., Pearson D. G., and Taylor L. A. (2007) Highly siderophile
element constraints on accretion and differentiation of the Earth-
Moon system. *Science, 315*, 217–219.

Deienno R., Gomes R. S., Walsh K. J., Morbidelli A., and Nesvorný
D. (2016) Is the Grand Tack model compatible with the orbital
distribution of main belt asteroids? *Icarus, 272*, 114–124.

Delsemme A. H. (1992) Cometary origin of carbon, nitrogen, and water
on the earth. *Origins Life Evol. Biosph., 21*, 279–298.

DeMeo F. E. and Carry B. (2013) The taxonomic distribution of
asteroids from multi-filter all-sky photometric surveys. *Icarus, 226*,
723–741.

Desch S. J., Kalyaan A., and Alexander C. M. O. D. (2018a) The effect
of Jupiter's formation on the distribution of refractory elements and
inclusions in meteorites. *Astrophys. J. Suppl., 238*, 11.

Desch S. J., Kalyaan A., and Alexander C. M. O. D. (2018b) The effect
of Jupiter's formation on the distribution of refractory elements and
inclusions in meteorites. *Astrophys. J. Suppl., 238*, 11.

Deutsch A. N., Head J. W., and Neumann G. A. (2019) Age constraints
of Mercury's polar deposits suggest recent delivery of ice. *Earth*

Planet. Sci. Lett., 520, 26–33.

Dodson-Robinson S. E. and Bodenheimer P. (2010) The formation of Uranus and Neptune in solid-rich feeding zones: Connecting chemistry and dynamics. *Icarus, 207,* 491–498.

Donahue T. M., Hoffman J. H., Hodges R. R., and Watson A. J. (1982) Venus was wet: A measurement of the ratio of deuterium to hydrogen. *Science, 216,* 630–633.

Dorn C., Khan A., Heng K., Connolly J. A. D., Alibert Y., Benz W., and Tackley P. (2015) Can we constrain the interior structure of rocky exoplanets from mass and radius measurements? *Astron. Astrophys., 577,* A83.

Drahus M., Jewitt D., Guilbert-Lepoutre A., Waniak W., Hoge J., Lis D. C., Yoshida H., Peng R., and Sievers A. (2011) Rotation state of comet 103P/Hartley 2 from radio spectroscopy at 1 mm. *Astrophys. J. Lett., 734,* L4.

Drake M. J. (2004) Origin of water in the terrestrial planets. *Meteoritics & Planet. Sci. Suppl., 39,* 5031.

Drążkowska J. and Alibert Y. (2017) Planetesimal formation starts at the snow line. *Astron. Astrophys., 608,* A92.

Du F. and Bergin E. A. (2014) Water vapor distribution in protoplanetary disks. *Astrophys. J., 792,* 2.

Durda D. D., Bottke W. F., Nesvorný D., Enke B. L., Merline W. J., Asphaug E., and Richardson D. C. (2007) Size-frequency distributions of fragments from SPH/ N-body simulations of asteroid impacts: Comparison with observed asteroid families. *Icarus, 186,* 498–516.

Eberhardt P., Dolder U., Schulte W., Krankowsky D., Lammerzahl P., Hoffman J. H., Hodges R. R., Berthelier J. J., and Illiano J. M. (1987) The D/H ratio in water from comet P/Halley. *Astron. Astrophys., 187,* 435.

Eberhardt P., Reber M., Krankowsky D., and Hodges R. R. (1995) The D/H and $^{18}O/^{16}O$ ratios in water from comet P/Halley. *Astron. Astrophys., 302,* 301.

Elkins-Tanton L. T. (2008) Linked magma ocean solidification and atmospheric growth for Earth and Mars. *Earth Planet. Sci. Lett., 271,* 181–191.

Elkins-Tanton L. T. (2012) Magma oceans in the inner solar system. *Annu. Rev. Earth Planet. Sci., 40,* 113–139.

Ercolano B. and Pascucci I. (2017) The dispersal of planet-forming discs: Theory confronts observations. *R. Soc. Open Sci., 4,* 170114.

Fernandes R. B., Mulders G. D., Pascucci I., Mordasini C., and Emsenhuber A. (2019) Hints for a turnover at the snow line in the giant planet occurrence rate. *Astrophys. J., 874,* 81.

Feuchtgruber H., Lellouch E., Orton G., de Graauw T., Vandenbussche B., Swinyard B., Moreno R., Jarchow C., Billebaud F., Cavalié T., Sidher S., and Hartogh P. (2013) The D/H ratio in the atmospheres of Uranus and Neptune from Herschel-PACS observations. *Astron. Astrophys., 551,* A126.

Fischer R. A. and Nimmo F. (2018) Effects of core formation on the Hf-W isotopic composition of the Earth and dating of the Moon-forming impact. *Earth Planet. Sci. Lett., 499,* 257–265.

Fischer D. A., Howard A. W., Laughlin G. P., Macintosh B., Mahadevan S., Sahlmann J., and Yee J. C. (2014) Exoplanet detection techniques. In *Protostars and Planets VI* (H. Beuther et al., eds.), pp. 715–737. Univ. of Arizona, Tucson.

Fogg M. J. and Nelson R. P. (2005) Oligarchic and giant impact growth of terrestrial planets in the presence of gas giant planet migration. *Astron. Astrophys., 441,* 791–806.

Ford E. B. and Rasio F. A. (2008) Origins of eccentric extrasolar planets: Testing the planet-planet scattering model. *Astrophys. J., 686,* 621–636.

Fornasier S., Lantz C., Perna D., Campins H., Barucci M. A., and Nesvorný D. (2016) Spectral variability on primitive asteroids of the Themis and Beagle families: Space weathering effects or parent body heterogeneity? *Icarus, 269,* 1–14.

Fulton B. J., Petigura E. A., Howard A. W., Isaacson H., Marcy G. W., Cargile P. A., Hebb L., et al. (2017) The California-Kepler Survey. III. A gap in the radius distribution of small planets. *Astron. J., 154,* 109.

Gaidos E., Krot A. N., Williams J. P., and Raymond S. N. (2009) ^{26}Al and the formation of the solar system from a molecular cloud contaminated by Wolf-Rayet winds. *Astrophys. J., 696,* 1854–1863.

Garaud P. and Lin D. N. C. (2007) The effect of internal dissipation and surface irradiation on the structure of disks and the location of the snow line around Sun-like stars. *Astrophys. J., 654,* 606–624.

Geiss J. and Gloeckler G. (1998) Abundances of deuterium and helium-3 in the protosolar cloud. *Space Sci. Rev., 84,* 239–250.

Genda H. and Abe Y. (2005) Enhanced atmospheric loss on protoplanets at the giant impact phase in the presence of oceans. *Nature, 433,* 842–844.

Genda H. and Ikoma M. (2008) Origin of the ocean on the Earth: Early evolution of water D/H in a hydrogen-rich atmosphere. *Icarus, 194,* 42–52.

Gillon M., Triaud A. H. M. J., Demory B.-O., Jehin E., Agol E., Deck K. M., Lederer S. M., et al. (2017) Seven temperate terrestrial planets around the nearby ultracool dwarf star TRAPPIST-1. *Nature, 542,* 456–460.

Gladman B., Kavelaars J. J., Petit J., Morbidelli A., Holman M. J., and Loredo T. (2001) The structure of the Kuiper belt: Size distribution and radial extent. *Astron. J., 122,* 1051–1066.

Gomes R., Levison H. F., Tsiganis K., and Morbidelli A. (2005) Origin of the cataclysmic Late Heavy Bombardment period of the terrestrial planets. *Nature, 435,* 466–469.

Gould A., Dong S., Gaudi B. S., Udalski A., Bond I. A., Greenhill J., Street R. A., et al. (2010) Frequency of solar-like systems and of ice and gas giants beyond the snow line from high-magnification microlensing events in 2005–2008. *Astrophys. J., 720,* 1073–1089.

Gounelle M. and Meibom A. (2008) The origin of short-lived radionuclides and the astrophysical environment of solar system formation. *Astrophys. J., 680,* 781–792.

Gradie J. and Tedesco E. (1982) Compositional structure of the asteroid belt. *Science, 216,* 1405–1407.

Grimm R. E. and McSween H. Y. (1993) Heliocentric zoning of the asteroid belt by aluminum-26 heating. *Science, 259,* 653–655.

Grossman L. (1972) Condensation in the primitive solar nebula. *Geochim. Cosmochim. Acta, 36,* 597–619.

Gupta A. and Schlichting H. E. (2019) Sculpting the valley in the radius distribution of small exoplanets as a by-product of planet formation: The core-powered mass-loss mechanism. *Mon. Not. R. Astron. Soc., 487,* 24–33.

Haghighipour N. (2009) Dynamical constraints on the origin of main belt comets. *Meteoritics & Planet. Sci., 44,* 1863–1869.

Haghighipour N. and Boss A. P. (2003) On pressure gradients and rapid migration of solids in a nonuniform solar nebula. *Astrophys. J., 583,* 996–1003.

Haghighipour N., Maindl T. I., Schäfer C., Speith R., and Dvorak R. (2016) Triggering sublimation-driven activity of main belt comets. *Astrophys. J., 830,* 22.

Hainaut O. R., Kleyna J., Sarid G., Hermalyn B., Zenn A., Meech K. J., Schultz P. H., Hsieh H., Trancho G., and Pittichová J. (2012) P/2010 A2 LINEAR. I. An impact in the asteroid main belt. *Astron. Astrophys., 537,* A69.

Haisch K. E. Jr., Lada E. A., and Lada C. J. (2001) Disk frequencies and lifetimes in young clusters. *Astrophys. J. Lett., 553,* L153–L156.

Halliday A. N. (2013) The origins of volatiles in the terrestrial planets. *Geochim. Cosmochim. Acta, 105,* 146–171.

Hallis L. J. (2017) D/H ratios of the inner solar system. *Philos. Trans. R. Soc., A, 375,* 20150390.

Hallis L. J., Huss G. R., Nagashima K., Taylor G. J., Halldórsson S. A., Hilton D. R., Mottl M. J., and Meech K. J. (2015) Evidence for primordial water in Earth's deep mantle. *Science, 350,* 795–797.

Hamano K., Abe Y., and Genda H. (2013) Emergence of two types of terrestrial planet on solidification of magma ocean. *Nature, 497,* 607–610.

Hansen B. M. S. (2009) Formation of the terrestrial planets from a narrow annulus. *Astrophys. J., 703,* 1131–1140.

Hargrove K. D., Emery J. P., Campins H., and Kelley M. S. P. (2015) Asteroid (90) Antiope: Another icy member of the Themis family? *Icarus, 254,* 150–156.

Harper C. L. and Jacobsen S. B. (1996) Evidence for ^{182}Hf in the early solar system and constraints on the timescale of terrestrial accretion and core formation. *Geochim. Cosmochim. Acta, 60,* 1131–1153.

Hartmann W. K. (2019) The collapse of the terminal cataclysm paradigm . . . and where we go from here? In *Lunar Planet. Sci. L,* Abstract #1064. Lunar and Planetary Institute, Houston.

Hartogh P., Lis D. C., Bockelée-Morvan D., de Val-Borro M., Biver N., Küppers M., Emprechtinger M., Bergin E. A., Crovisier J., and Rengel M. (2011) Ocean-like water in the Jupiter-family comet 103P/Hartley 2. *Nature, 478,* 218–220.

Heays A. N., Visser R., Gredel R., Ubachs W., Lewis B. R., Gibson S. T., and van Dishoeck E. F. (2014) Isotope selective photodissociation of N_2 by the interstellar radiation field and cosmic rays. *Astron. Astrophys., 562,* A61.

Herbst E. (2003) Isotopic fractionation by ion-molecule reactions. *Space Sci. Rev., 106,* 293–304.

Herbst E. and van Dishoeck E. F. (2009) Complex organic interstellar molecules. *Annu. Rev. Astron. Astrophys., 47,* 427–480.

Hester J. J., Desch S. J., Healy K. R., and Leshin L. A. (2004) The cradle of the solar system. *Science, 304(5674),* 1116-1117.

Hillenbrand L. A., Carpenter J. M., Kim J. S., Meyer M. R., Backman D. E., Moro-Martín A., Hollenbach D. J., Hines D. C., Pascucci I., and Bouwman J. (2008) The complete census of 70 μm-bright debris disks within "The Formation and Evolution of Planetary Systems" Spitzer Legacy Survey of Sun-like stars. *Astrophys. J., 677,* 630–656.

Hirschmann M. M. (2006) Water, melting, and the deep Earth H$_2$O cycle. *Annu. Rev. Earth Planet. Sci., 34,* 629–653.

Howard A. W., Marcy G. W., Bryson S. T., Jenkins J. M., Rowe J. F., Batalha N. M., Borucki W. J., et al. (2012) Planet occurrence within 0.25 AU of solar-type stars from Kepler. *Astrophys. J. Suppl., 201,* 15.

Howard A. W., Sanchis-Ojeda R., Marcy G. W., Johnson J. A., Winn J. N., Isaacson H., Fischer D. A., Fulton B. J., Sinukoff E., and Fortney J. J. (2013) A rocky composition for an Earth-sized exoplanet. *Nature, 503,* 381–384.

Hsieh H. H. and Jewitt D. (2006) A population of comets in the main asteroid belt. *Science, 312,* 561–563.

Hsieh H. H., Jewitt D., and Ishiguro M. (2009) Physical properties of main-belt comet P/2005 U1 (Read). *Astron. J., 137,* 157–168.

Hsieh H. H., Ishiguro M., Kim Y., Knight M. M., Lin Z.-Y., Micheli M., Moskovitz N. A., Sheppard S. S., Thirouin A., and Trujillo C. A. (2018a) The 2016 reactivations of the main-belt comets 238P/Read and 288P/(300163) 2006 VW139. *Astron. J., 156,* 223.

Hsieh H. H., Novaković B., Kim Y., and Brasser R. (2018b) Asteroid family associations of active asteroids. *Astron. J., 155,* 96.

Ida S. and Lin D. N. C. (2010) Toward a deterministic model of planetary formation. VI. Dynamical interaction and coagulation of multiple rocky embryos and super-Earth systems around solar-type stars. *Astrophys. J., 719,* 810–830.

Ida S., Lin D. N. C., and Nagasawa M. (2013) Toward a deterministic model of planetary formation. VII. Eccentricity distribution of gas giants. *Astrophys. J., 775,* 42.

Ida S., Guillot T., and Morbidelli A. (2016) The radial dependence of pebble accretion rates: A source of diversity in planetary systems. I. Analytical formulation. *Astron. Astrophys., 591,* A72.

Ida S., Yamamura T., and Okuzumi S. (2019) Water delivery by pebble accretion to rocky planets in habitable zones in evolving disks. *Astron. Astrophys., 624,* A28.

Ikoma M. and Genda H. (2006) Constraints on the mass of a habitable planet with water of nebular origin. *Astrophys. J., 648,* 696–706.

Izidoro A., Morbidelli A., and Raymond S. N. (2014) Terrestrial planet formation in the presence of migrating super-Earths. *Astrophys. J., 794,* 11.

Izidoro A., Morbidelli A., Raymond S. N., Hersant F., and Pierens A. (2015a) Accretion of Uranus and Neptune from inward-migrating planetary embryos blocked by Jupiter and Saturn. *Astron. Astrophys., 582,* A99.

Izidoro A., Raymond S. N., Morbidelli A., and Winter O. C. (2015b) Terrestrial planet formation constrained by Mars and the structure of the asteroid belt. *Mon. Not. R. Astron. Soc., 453,* 3619–3634.

Izidoro A., Ogihara M., Raymond S. N., Morbidelli A., Pierens A., Bitsch B., Cossou C., and Hersant F. (2017) Breaking the chains: Hot super-Earth systems from migration and disruption of compact resonant chains. *Mon. Not. R. Astron. Soc., 470,* 1750–1770.

Izidoro A., Bitsch B., Raymond S. N., Johansen A., Morbidelli A., Lambrechts M., and Jacobson S. A. (2019) Formation of planetary systems by pebble accretion and migration: Hot super-Earth systems from breaking compact resonant chains. *Astron. Astrophys., 623,* A88.

Jackson M. G., Carlson R. W., Kurz M. D., Kempton P. D., Francis D., and Blusztajn J. (2010) Evidence for the survival of the oldest terrestrial mantle reservoir. *Nature, 466,* 853–856.

Jacobson S. A. and Morbidelli A. (2014) Lunar and terrestrial planet formation in the Grand Tack scenario. *Philos. Trans. R. Soc., A, 372,* 0174.

Jacobson S. A. and Scheeres D. J. (2011) Dynamics of rotationally fissioned asteroids: Source of observed small asteroid systems. *Icarus, 214,* 161–178.

Jacobson S. A., Morbidelli A., Raymond S. N., O'Brien D. P., Walsh K. J., and Rubie D. C. (2014) Highly siderophile elements in Earth's mantle as a clock for the Moon-forming impact. *Nature, 508,* 84–87.

Jacquet E. and Robert F. (2013) Water transport in protoplanetary disks and the hydrogen isotopic composition of chondrites. *Icarus, 223,* 722–732.

Jakosky B. M., Slipski M., Benna M., Mahaffy P., Elrod M., Yelle R., Stone S., and Alsaeed N. (2017) Mars' atmospheric history derived from upper-atmosphere measurements of ^{38}Ar/^{36}Ar. *Science, 355,* 1408–1410.

Jewitt D. (2012) The active asteroids. *Astron. J., 143,* 66.

Jin S. and Mordasini C. (2018) Compositional imprints in density-distance-time: A rocky composition for close-in low-mass exoplanets from the location of the valley of evaporation. *Astrophys. J., 853,* 163.

Jing D., He J., Brucato J., De Sio A., Tozzetti L., and Vidali G. (2011) On water formation in the interstellar medium: Laboratory study of the O+D reaction on surfaces. *Astrophys. J. Lett., 741,* L9.

Johansen A. and Lambrechts M. (2017) Forming planets via pebble accretion. *Annu. Rev. Earth Planet. Sci., 45,* 359–387.

Johansen A., Youdin A., and Klahr H. (2009) Zonal flows and long-lived axisymmetric pressure bumps in magnetorotational turbulence. *Astrophys. J., 697,* 1269–1289.

Johansen A., Blum J., Tanaka H., Ormel C., Bizzarro M., and Rickman H. (2014) The multifaceted planetesimal formation process. In *Protostars and Planets VI* (H. Beuther et al., eds.), pp. 547–570. Univ. of Arizona, Tucson.

Johnson J. A., Butler R. P., Marcy G. W., Fischer D. A., Vogt S. S., Wright J. T., and Peek K. M. G. (2007) A new planet around an M dwarf: Revealing a correlation between exoplanets and stellar mass. *Astrophys. J., 670,* 833–840.

Jurić M. and Tremaine S. (2008) Dynamical origin of extrasolar planet eccentricity distribution. *Astrophys. J., 686,* 603–620.

Kaib N. A. and Cowan N. B. (2015) The feeding zones of terrestrial planets and insights into Moon formation. *Icarus, 252,* 161–174.

Kalyaan A. and Desch S. J. (2019) Effect of different angular momentum transport mechanisms on the distribution of water in protoplanetary disks. *Astrophys. J., 875,* 43.

Kasting J. F. and Pollack J. B. (1983) Loss of water from Venus. I. Hydrodynamic escape of hydrogen. *Icarus, 53,* 479–508.

Kasting J. F., Whitmire D. P., and Reynolds R. T. (1993) Habitable zones around main sequence stars. *Icarus, 101,* 108–128.

Kennedy G. M. and Kenyon S. J. (2008) Planet formation around stars of various masses: The snow line and the frequency of giant planets. *Astrophys. J., 673,* 502–512.

Kimura K., Lewis R. S., and Anders E. (1974) Distribution of gold and rhenium between nickel-iron and silicate melts: Implications for the abundance of siderophile elements on the Earth and Moon. *Geochim. Cosmochim. Acta, 38,* 683–701.

King H. E., Stimpfl M., Deymier P., Drake M. J., Catlow C. R. A., Putnis A., and de Leeuw N. H. (2010) Computer simulations of water interactions with low-coordinated forsterite surface sites: Implications for the origin of water in the inner solar system. *Earth Planet. Sci. Lett., 300,* 11–18.

Kleine T., Touboul M., Bourdon B., Nimmo F., Mezger K., Palme H., Jacobsen S. B., Yin Q.-Z., and Halliday A. N. (2009) Hf-W chronology of the accretion and early evolution of asteroids and terrestrial planets. *Geochim. Cosmochim. Acta, 73,* 5150–5188.

Kley W. and Crida A. (2008) Migration of protoplanets in radiative discs. *Astron. Astrophys., 487,* L9–L12.

Kley W. and Nelson R. P. (2012) Planet-disk interaction and orbital evolution. *Annu. Rev. Astron. Astrophys., 50,* 211–249.

Kleyna J., Hainaut O. R., and Meech K. J. (2013) P/2010 A2 LINEAR. II. Dynamical dust modelling. *Astron. Astrophys., 549,* A13.

Kokubo E. and Ida S. (2000) Formation of protoplanets from planetesimals in the solar nebula. *Icarus, 143,* 15–27.

Kresak L. (1980) Dynamics, interrelations, and evolution of the systems of asteroids and comets. *Earth Moon Planets, 22,* 83–98.

Krot A. N. and Bizzarro M. (2009) Chronology of meteorites and the early solar system. *Geochim. Cosmochim. Acta, 73,* 4919–4921.

Krot A. N., Nagashima K., Ciesla F. J., Meyer B. S., Hutcheon I. D., Davis A. M., Huss G. R., and Scott E. R. D. (2010) Oxygen isotopic composition of the Sun and mean oxygen isotopic composition of the protosolar silicate dust: Evidence from refractory inclusions. *Astrophys. J., 713,* 1159–1166.

Kruijer T. S., Touboul M., Fischer-Gödde M., Bermingham K. R.,

Walker R. J., and Kleine T. (2014) Protracted core formation and rapid accretion of protoplanets. *Science, 344,* 1150–1154.

Kruijer T. S., Burkhardt C., Budde C., and Kleine T. (2017) Age of Jupiter inferred from the distinct genetics and formation times of meteorites. *Proc. Natl. Acad. Sci. U.S.A., 114(26),* 6712-6716.

Kuchner M. J. (2003) Volatile-rich Earth-mass planets in the Habitable Zone. *Astrophys. J. Lett., 596,* L105–L108.

Küppers M., O'Rourke L., Bockelée-Morvan D., Zakharov V., Lee S., von Allmen P., Carry B., Teyssier D., Marston A., and Müller T. (2014) Localized sources of water vapour on the dwarf planet Ceres. In *Asteroids, Comets, Meteors 2014, Book of Abstracts* (K. Muinonen et al., eds.), p. 298. Univ. of Helsinki, Finland.

Kurosawa K. (2015) Impact-driven planetary desiccation: The origin of the dry Venus. *Earth Planet. Sci. Lett., 429,* 181–190.

Lambrechts M. and Johansen A. (2012) Rapid growth of gas-giant cores by pebble accretion. *Astron. Astrophys., 544,* A32.

Lambrechts M. and Johansen A. (2014) Forming the cores of giant planets from the radial pebble flux in protoplanetary discs. *Astron. Astrophys., 572,* A107.

Lambrechts M., Johansen A., and Morbidelli A. (2014) Separating gas-giant and ice-giant planets by halting pebble accretion. *Astron. Astrophys., 572,* A35.

Laufer D., Bar-Nun A., and Ninio Greenberg A. (2017) Trapping mechanism of O_2 in water ice as first measured by Rosetta spacecraft. *Mon. Not. R. Astron. Soc., 469,* S818–S823.

Lecar M., Podolak M., Sasselov D., and Chiang E. (2006) On the location of the snow line in a protoplanetary disk. *Astrophys. J., 640,* 1115–1118.

Lécuyer C., Gillet P., and Robert F. (1998) The hydrogen isotope composition of seawater and the global water cycle. *Chem. Geol., 145,* 249–261.

Lellouch E., Bézard B., Fouchet T., Feuchtgruber H., Encrenaz T., and de Graauw T. (2001) The deuterium abundance in Jupiter and Saturn from ISO-SWS observations. *Astron. Astrophys., 370,* 610–622.

Levison H. F. and Duncan M. J. (1997) From the Kuiper belt to Jupiter-family comets: The spatial distribution of ecliptic comets. *Icarus, 127,* 13–32.

Levison H. F., Terrell D., Wiegert P. A., Dones L., and Duncan M. J. (2006) On the origin of the unusual orbit of comet 2P/Encke. *Icarus, 182,* 161–168.

Levison H. F., Morbidelli A., Vanlaerhoven C., Gomes R., and Tsiganis K. (2008) Origin of the structure of the Kuiper belt during a dynamical instability in the orbits of Uranus and Neptune. *Icarus, 196,* 258–273.

Levison H. F., Morbidelli A., Tsiganis K., Nesvorný D., and Gomes R. (2011) Late orbital instabilities in the outer planets induced by interaction with a self-gravitating planetesimal disk. *Astron. J., 142,* 152.

Lichtenberg T., Parker R. J., and Meyer M. R. (2016) Isotopic enrichment of forming planetary systems from supernova pollution. *Mon. Not. R. Astron. Soc., 462,* 3979–3992.

Lichtenberg T., Golabek G. J., Burn R., Meyer M. R., Alibert Y., Gerya T. V., and Mordasini C. (2019) A water budget dichotomy of rocky protoplanets from ^{26}Al-heating. *Nature Astron., 3,* 307–313.

Lin D. N. C. and Papaloizou J. (1986) On the tidal interaction between protoplanets and the protoplanetary disk. III — Orbital migration of protoplanets. *Astrophys. J., 309,* 846–857.

Linsky J. L. (2007) D/H and nearby interstellar cloud structures. *Space Sci. Rev., 130,* 367–375.

Lis D. C., Bockelée-Morvan D., Güsten R., Biver N., Stutzki J., Delorme Y., Durán C., Wiesemeyer H., and Okada Y. (2019) Terrestrial deuterium-to-hydrogen ratio in water in hyperactive comets. *Astron. Astrophys., 625,* L5.

Lissauer J. J. (1987) Timescales for planetary accretion and the structure of the protoplanetary disk. *Icarus, 69,* 249–265.

Lissauer J. J. and Stevenson D. J. (2007) Formation of giant planets. In *Protostars and Planets V* (B. Reipurth et al., eds.), pp. 591–606. Univ. of Arizona, Tucson.

Lissauer J. J., Hubickyj O., D'Angelo G., and Bodenheimer P. (2009) Models of Jupiter's growth incorporating thermal and hydrodynamic constraints. *Icarus, 199,* 338–350.

Lopez E. D. (2017) Born dry in the photoevaporation desert: Kepler's ultra-short-period planets formed water-poor. *Mon. Not. R. Astron. Soc., 472,* 245–253.

Lugaro M., Liffman K., Ireland T. R., and Maddison S. T. (2012) Can galactic chemical evolution explain the oxygen isotopic variations in the solar system? *Astrophys. J., 759,* 51.

Lyons J. R. (2012) Isotope signatures in organics due to CO and N_2 self-shielding. In *Lunar Planet. Sci. XLIII,* Abstract #2858. Lunar and Planetary Institute, Houston.

Lyons J. R. and Young E. D. (2005) CO self-shielding as the origin of oxygen isotope anomalies in the early solar nebula. *Nature, 435,* 317–320.

Lyra W., Paardekooper S.-J., and Mac Low M.-M. (2010) Orbital migration of low-mass planets in evolutionary radiative models: Avoiding catastrophic infall. *Astrophys. J. Lett., 715,* L68–L73.

Marcus R. A., Sasselov D., Stewart S. T., and Hernquist L. (2010) Water/icy super-Earths: Giant impacts and maximum water content. *Astrophys. J. Lett., 719,* L45–L49.

Marcy G. W., Isaacson H., Howard A. W., Rowe J. F., Jenkins J. M., Bryson S. T., Latham D. W., et al. (2014) Masses, radii, and orbits of small Kepler planets: The transition from gaseous to rocky planets. *Astrophys. J. Suppl., 210,* 20.

Marsset M., Vernazza P., Birlan M., DeMeo F., Binzel R. P., Dumas C., Milli J., and Popescu M. (2016) Compositional characterisation of the Themis family. *Astron. Astrophys., 586,* A15.

Martin R. G. and Livio M. (2012) On the evolution of the snow line in protoplanetary discs. *Mon. Not. R. Astron. Soc., 425,* L6–L9.

Martin R. G. and Livio M. (2015) The solar system as an exoplanetary system. *Astrophys. J., 810,* 105.

Marty B. (2012) The origins and concentrations of water, carbon, nitrogen and noble gases on Earth. *Earth Planet. Sci. Lett., 313,* 56–66.

Marty B. and Yokochi R. (2006) Water in the early Earth. In *Water in Nominally Anhydrous Minerals* (H. Keppler and J. R. Smyth, eds.), pp. 421–450. Reviews in Mineralogy and Geochemistry, Vol. 62, Mineralogical Society of America, Washington.

Marty B., Zimmermann L., Burnard P. G., Wieler R., Heber V. S., Burnett D. L., Wiens R. C., and Bochsler P. (2010) Nitrogen isotopes in the recent solar wind from the analysis of Genesis targets: Evidence for large scale isotope heterogeneity in the early solar system. *Geochim. Cosmochim. Acta, 74,* 340–355.

Marty B., Avice G., Sano Y., Altwegg K., Balsiger H., Hässig M., Morbidelli A., Mousis O., and Rubin M. (2016) Origins of volatile elements (H, C, N, noble gases) on Earth and Mars in light of recent results from the ROSETTA cometary mission. *Earth Planet. Sci. Lett., 441,* 91–102.

Marty B., Altwegg K., Balsiger H., Bar-Nun A., Bekaert D. V., Berthelier J. J., Bieler A., Briois C., Calmonte U., and Combi M. (2017) Xenon isotopes in 67P/Churyumov-Gerasimenko show that comets contributed to Earth's atmosphere. *Science, 356,* 1069–1072.

Masset F. and Snellgrove M. (2001) Reversing type II migration: Resonance trapping of a lighter giant protoplanet. *Mon. Not. R. Astron. Soc., 320,* L55–L59.

Mayer L., Quinn T., Wadsley J., and Stadel J. (2002) Formation of giant planets by fragmentation of protoplanetary disks. *Science, 298,* 1756–1759.

Mayor M., Marmier M., Lovis C., Udry S., Ségransan D., Pepe F., Benz W., et al. (2011) The HARPS search for southern extra-solar planets XXXIV. Occurrence, mass distribution and orbital properties of super-Earths and Neptune-mass planets. *ArXiV e-prints,* arXiv:1109.2497.

McGuire B. A., Bergin E., Blake G. A., Burkhardt A. M., Cleeves L. I., Loomis R. A., Remijan A. J., Shingledecker C. N., and Willis E. R. (2018) Observing the effects of chemistry on exoplanets and planet formation. In *Science with a Next Generation Very Large Array* (E. Murphy, ed.), p. 217. ASP Conf. Ser. 517, Astronomical Society of the Pacific, San Francisco.

McKeegan K. D., Kallio A. P. A., Heber V. S., Jarzebinski G., Mao P. H., Coath C. D., Kunihiro T., Wiens R. C., Nordholt J. E., and Moses R. W. (2011) The oxygen isotopic composition of the Sun inferred from captured solar wind. *Science, 332,* 1528.

McNeil D. S. and Nelson R. P. (2010) On the formation of hot Neptunes and super-Earths. *Mon. Not. R. Astron. Soc., 401,* 1691–1708.

Meech K. J. and Svoren J. (2004) Using cometary activity to trace the physical and chemical evolution of cometary nuclei. In *Comets II* (M. C. Festou et al., eds.), p. 317–335. Univ. of Arizona, Tucson.

Meech K. J., Yang B., Kleyna J., Hainaut O. R., Berdyugina S., Keane J. V., Micheli M., Morbidelli A. R., and Wainscoat R. J. (2016) Inner solar system material discovered in the Oort cloud. *Sci. Adv., 2,* e1600038.

Meibom A., Krot A. N., Robert F., Mostefaoui S., Russell S. S., Petaev

M. I., and Gounelle M. (2007) Nitrogen and carbon isotopic composition of the sun inferred from a high-temperature solar nebular condensate. *Astrophys. J. Lett.*, *656*, L33–L36.

Millar T. J., Bennett A., and Herbst E. (1989) Deuterium fractionation in dense interstellar clouds. *Astrophys. J.*, *340*, 906.

Minton D. A. and Malhotra R. (2010) Dynamical erosion of the asteroid belt and implications for large impacts in the inner solar system. *Icarus*, *207*, 744–757.

Mojzsis S. J., Harrison T. M., and Pidgeon R. T. (2001) Oxygen-isotope evidence from ancient zircons for liquid water at the Earth's surface 4,300 Myr ago. *Nature*, *409*, 178–181.

Monteux J., Andrault D., and Samuel H. (2016) On the cooling of a deep terrestrial magma ocean. *Earth Planet. Sci. Lett.*, *448*, 140–149.

Monteux J., Golabek G. J., Rubie D. C., Tobie G., and Young E. D. (2018) Water and the interior structure of terrestrial planets and icy bodies. *Space Sci. Rev.*, *214*, 39.

Morbidelli A. and Nesvorný D. (2012) Dynamics of pebbles in the vicinity of a growing planetary embryo: Hydro-dynamical simulations. *Astron. Astrophys.*, *546*, A18.

Morbidelli A. and Raymond S. N. (2016) Challenges in planet formation. *J. Geophys. Res.–Planets*, *121*, 1962–1980.

Morbidelli A. and Wood B. J. (2015) Late accretion and the late veneer. In *The Early Earth: Accretion and Differentiation* (J. Badro and M. Walter, eds.), pp. 71–82. Geophys. Monogr. Ser., Vol. 212, American Geophysical Union, Washington, DC.

Morbidelli A., Chambers J., Lunine J. I., Petit J. M., Robert F., Valsecchi G. B., and Cyr K. E. (2000) Source regions and time scales for the delivery of water to Earth. *Meteoritics & Planet. Sci.*, *35*, 1309–1320.

Morbidelli A., Tsiganis K., Crida A., Levison H. F., and Gomes R. (2007) Dynamics of the giant planets of the solar system in the gaseous protoplanetary disk and their relationship to the current orbital architecture. *Astron. J.*, *134*, 1790–1798.

Morbidelli A., Lambrechts M., Jacobson S., and Bitsch B. (2015) The great dichotomy of the solar system: Small terrestrial embryos and massive giant planet cores. *Icarus*, *258*, 418–429.

Morbidelli A., Bitsch B., Crida A., Gounelle M., Guillot T., Jacobson S., Johansen A., Lambrechts M., and Lega E. (2016) Fossilized condensation lines in the solar system protoplanetary disk. *Icarus*, *267*, 368–376.

Morbidelli A., Nesvorný D., Laurenz V., Marchi S., Rubie D. C., Elkins-Tanton L., Wieczorek M., and Jacobson S. (2018) The timeline of the lunar bombardment: Revisited. *Icarus*, *305*, 262–276.

Morishima R., Stadel J., and Moore B. (2010) From planetesimals to terrestrial planets: N-body simulations including the effects of nebular gas and giant planets. *Icarus*, *207*, 517–535.

Mottl M., Glazer B., Kaiser R., and Meech K. (2007) Water and astrobiology. *Chem. Erde–Geochem.*, *67*, 253–282.

Mousis O., Ronnet T., Lunine J. I., Luspay-Kuti A., Mandt K. E., Danger G., Pauzat F., Ellinger Y., Wurz P., and Vernazza P. (2018) Noble gas abundance ratios indicate the agglomeration of 67P/Churyumov-Gerasimenko from warmed-up ice. *Astrophys. J. Lett.*, *865*, L11.

Mulders G. D., Ciesla F. J., Min M., and Pascucci I. (2015a) The snow line in viscous disks around low-mass stars: Implications for water delivery to terrestrial planets in the habitable zone. *Astrophys. J.*, *807*, 9.

Mulders G. D., Pascucci I., and Apai D. (2015b) A stellar-mass-dependent drop in planet occurrence rates. *Astrophys. J.*, *798*, 112.

Mulders G. D., Pascucci I., Apai D., and Ciesla F. J. (2018) The exoplanet population observation simulator. I. The inner edges of planetary systems. *Astron. J.*, *156*, 24.

Mumma M. J. and Charnley S. B. (2011) The chemical composition of comets — Emerging taxonomies and natal heritage. *Annu. Rev. Astron. Astrophys.*, *49*, 471–524.

Muralidharan K., Deymier P., Stimpfl M., de Leeuw N. H., and Drake M. J. (2008a) Origin of water in the inner solar system: A kinetic Monte Carlo study of water adsorption on forsterite. *Icarus*, *198*, 400–407.

Muralidharan K., Deymier P., Stimpfl M., de Leeuw N. H., and Drake M. J. (2008b) Origin of water in the inner solar system: A kinetic Monte Carlo study of water adsorption on forsterite. *Icarus*, *198*, 400–407.

Murphy E. J., Bolatto A., Chatterjee S., Casey C. M., Chomiuk L., Dale D., de Pater I., Dickinson M., Francesco J. D., and Hallinan G. (2018) The ngVLA science case and associated science requirements.

In *Science with a Next Generation Very Large Array* (E. Murphy, ed.), p. 3. ASP Conf. Ser. 517, Astronomical Society of the Pacific, San Francisco.

Nanne J. A. M., Nimmo F., Cuzzi J. N., and Kleine T. (2019) Origin of the non-carbonaceous-carbonaceous meteorite dichotomy. *Earth Planet. Sci. Lett.*, *511*, 44–54.

Nesvorný D. (2015) Evidence for slow migration of Neptune from the inclination distribution of Kuiper belt objects. *Astron. J.*, *150*, 73.

Nesvorný D., Vokrouhlický D., Bottke W. F., and Levison H. F. (2018a) Evidence for very early migration of the solar system planets from the Patroclus-Menoetius binary Jupiter Trojan. *Nature Astron.*, *2*, 878–882.

Nesvorný D., Vokrouhlický D., Bottke W. F., and Levison H. F. (2018b) Evidence for very early migration of the solar system planets from the Patroclus-Menoetius binary Jupiter Trojan. *Nature Astron.*, *2*, 878–882.

Nimmo F. and Kleine T. (2007) How rapidly did Mars accrete? Uncertainties in the Hf-W timing of core formation. *Icarus*, *191*, 497–504.

Nomura R., Hirose K., Uesegi K., Ohishi Y., Tsuchiyama A., Miyake A., and Ueno Y. (2014) Low core-mantle boundary temperature inferred from the solidus of pyrolite. *Science*, *343*, 522–525.

O'Brien D. P., Morbidelli A., and Levison H. F. (2006) Terrestrial planet formation with strong dynamical friction. *Icarus*, *184*, 39–58.

O'Brien D. P., Walsh K. J., Morbidelli A., Raymond S. N., and Mandell A. M. (2014) Water delivery and giant impacts in the Grand Tack scenario. *Icarus*, *239*, 74–84.

O'Brien D. P., Izidoro A., Jacobson S. A., Raymond S. N., and Rubie D. C. (2018) The delivery of water during terrestrial planet formation. *Space Sci. Rev.*, *214*, 47.

Ogihara M., Morbidelli A., and Guillot T. (2015) A reassessment of the *in situ* formation of close-in super-Earths. *Astron. Astrophys.*, *578*, A36.

Ogihara M., Kokubo E., Suzuki T. K., and Morbidelli A. (2018) Formation of close-in super-Earths in evolving protoplanetary disks due to disk winds. *Astron. Astrophys.*, *615*, A63.

Oka A., Nakamoto T., and Ida S. (2011) Evolution of snow line in optically thick protoplanetary disks: Effects of water ice opacity and dust grain size. *Astrophys. J.*, *738*, 141.

Ormel C. W. and Klahr H. H. (2010) The effect of gas drag on the growth of protoplanets. Analytical expressions for the accretion of small bodies in laminar disks. *Astron. Astrophys.*, *520*, A43.

Ormel C. W., Liu B., and Schoonenberg D. (2017) Formation of TRAPPIST-1 and other compact systems. *Astron. Astrophys.*, *604*, A1.

Owen J. E. and Wu Y. (2017) The evaporation valley in the Kepler planets. *Astrophys. J.*, *847*, 29.

Owen T. and Bar-Nun A. (1995) Comets, impacts, and atmospheres. *Icarus*, *116*, 215–226.

Owen T. C. and Bar-Nun A. (2000) Volatile contributions from icy planetesimals. In *Origin of the Earth and Moon* (R. M. Canup and K. Righter, eds.), pp. 459–471. Univ. of Arizona, Tucson.

Owen T., Mahaffy P. R., Niemann H. B., Atreya S., and Wong M. (2001) Protosolar nitrogen. *Astrophys. J. Lett.*, *553*, L77–L79.

Paardekooper S.-J., Baruteau C., and Kley W. (2011) A torque formula for non-isothermal Type I planetary migration — II. Effects of diffusion. *Mon. Not. R. Astron. Soc.*, *410*, 293–303.

Pascucci I., Testi L., Herczeg G. J., Long F., Manara C. F., Hendler N., Mulders G. D., et al. (2016) A steeper than linear disk mass-stellar mass scaling relation. *Astrophys. J.*, *831*, 125.

Petigura E. A., Howard A. W., and Marcy G. W. (2013) Prevalence of Earth-size planets orbiting Sun-like stars. *Proc. Natl. Acad. Sci. U.S.A.*, *110*, 19273–19278.

Pierel J. D. R., Nixon C. A., Lellouch E., Fletcher L. N., Bjoraker G. L., Achterberg R. K., Bézard B., Hesman B. E., Irwin P. G. J., and Flasar F. M. (2017) D/H ratios on Saturn and Jupiter from Cassini CIRS. *Astron. J.*, *154*, 178.

Pierens A. and Nelson R. P. (2008) Constraints on resonant-trapping for two planets embedded in a protoplanetary disc. *Astron. Astrophys.*, *482*, 333–340.

Pierens A. and Raymond S. N. (2011) Two phase, inward-then-outward migration of Jupiter and Saturn in the gaseous solar nebula. *Astron. Astrophys.*, *533*, A131.

Pierens A., Raymond S. N., Nesvorný D., and Morbidelli A. (2014) Outward migration of Jupiter and Saturn in 3:2 or 2:1 resonance in radiative disks: Implications for the Grand Tack and Nice models.

Astrophys. J. Lett., *795*, L11.

Pirani S., Johansen A., Bitsch B., Mustill A. J., and Turrini D. (2019) Consequences of planetary migration on the minor bodies of the early solar system. *Astron. Astrophys.*, *623*, A169.

Podolak M., Mekler Y., and Prialnik D. (2002) Is the D/H ratio in the comet coma equal to the D/H ratio in the comet nucleus? *Icarus*, *160*, 208–211.

Pollack J. B., Hubickyj O., Bodenheimer P., Lissauer J. J., Podolak M., and Greenzweig Y. (1996) Formation of the giant planets by concurrent accretion of solids and gas. *Icarus*, *124*, 62–85.

Pontoppidan K. M., Salyk C., Bergin E. A., Brittain S., Marty B., Mousis O., and Öberg K. I. (2014) Volatiles in protoplanetary disks. In *Protostars and Planets VI* (H. Beuther et al., eds.), p. 363. Univ. of Arizona, Tucson.

Prialnik D. and Rosenberg E. D. (2009) Can ice survive in main-belt comets? Long-term evolution models of comet 133P/Elst-Pizarro. *Mon. Not. R. Astron. Soc.*, *399*, L79–L83.

Qi C., Öberg K. I., Wilner D. J., D'Alessio P., Bergin E., Andrews S. M., Blake G. A., Hogerheijde M. R., and van Dishoeck E. F. (2013) Imaging of the CO snow line in a solar nebula analog. *Science*, *341*, 630–632.

Raymond S. N. and Cossou C. (2014) No universal minimum-mass extrasolar nebula: Evidence against *in situ* accretion of systems of hot super-Earths. *Mon. Not. R. Astron. Soc.*, *440*, L11–L15.

Raymond S. N. and Izidoro A. (2017a) Origin of water in the inner solar system: Planetesimals scattered inward during Jupiter and Saturn's rapid gas accretion. *Icarus*, *297*, 134–148.

Raymond S. N. and Izidoro A. (2017b) The empty primordial asteroid belt. *Sci. Adv.*, *3*, e1701138.

Raymond S. N. and Morbidelli A. (2014) The Grand Tack model: A critical review. In *Complex Planetary Systems* (Z. Knežević and A. Lemaître, eds.), pp. 194–203. Proceedings of the International Astronomical Union, Vol. 310, Cambridge Univ., Cambridge.

Raymond S. N., Quinn T., and Lunine J. I. (2004) Making other Earths: Dynamical simulations of terrestrial planet formation and water delivery. *Icarus*, *168*, 1–17.

Raymond S. N., Mandell A. M., and Sigurdsson S. (2006a) Exotic Earths: Forming habitable worlds with giant planet migration. *Science*, *313*, 1413–1416.

Raymond S. N., Quinn T., and Lunine J. I. (2006b) High-resolution simulations of the final assembly of Earth-like planets. I. Terrestrial accretion and dynamics. *Icarus*, *183*, 265–282.

Raymond S. N., Quinn T., and Luninc J. I. (2007a) High resolution simulations of the final assembly of Earth-like planets. 2. Water delivery and planetary habitability. *Astrobiology*, *7*, 66–84.

Raymond S. N., Scalo J., and Meadows V. S. (2007b) A decreased probability of habitable planet formation around low-mass stars. *Astrophys. J.*, *669*, 606–614.

Raymond S. N., O'Brien D. P., Morbidelli A., and Kaib N. A. (2009) Building the terrestrial planets: Constrained accretion in the inner solar system. *Icarus*, *203*, 644–662.

Raymond S. N., Armitage P. J., and Gorelick N. (2010) Planet-planet scattering in planetesimal disks. II. Predictions for outer extrasolar planetary systems. *Astrophys. J.*, *711*, 772–795.

Raymond S. N., Armitage P. J., Moro-Martín A., Booth M., Wyatt M. C., Armstrong J. C., Mandell A. M., Selsis F., and West A. A. (2011) Debris disks as signposts of terrestrial planet formation. *Astron. Astrophys.*, *530*, A62.

Raymond S. N., Armitage P. J., Moro-Martín A., Booth M., Wyatt M. C., Armstrong J. C., Mandell A. M., Selsis F., and West A. (2012) Debris disks as signposts of terrestrial planet formation. II. Dependence of exoplanet architectures on giant planet and disk properties. *Astron. Astrophys.*, *541*, A11.

Raymond S. N., Kokubo E., Morbidelli A., Morishima R., and Walsh K. J. (2014) Terrestrial planet formation at home and abroad. In *Protostars and Planets VI* (H. Beuther et al., eds.), pp. 595–618. Univ. of Arizona, Tucson.

Raymond S. N., Boulet T., Izidoro A., Esteves L., and Bitsch B. (2018) Migration-driven diversity of super-Earth compositions. *Mon. Not. R. Astron. Soc.*, *479*, L81–L85.

Ricci L., Isella A., Liu S., and Li H. (2018) Imaging planetary systems in the act of forming with the ngVLA. In *Science with a Next Generation Very Large Array* (E. Murphy, ed.), p. 147. ASP Conf. Ser. 517, Astronomical Society of the Pacific, San Francisco.

Rivkin A. S. and Emery J. P. (2010) Detection of ice and organics on an asteroidal surface. *Nature*, *464*, 1322–1323.

Rivkin A. S., Asphaug E., and Bottke W. F. (2014) The case of the missing Ceres family. *Icarus*, *243*, 429–439.

Rogers L. A. (2015) Most 1.6 Earth-radius planets are not rocky. *Astrophys. J.*, *801*, 41.

Ronnet T., Mousis O., Vernazza P., Lunine J. I., and Crida A. (2018) Saturn's formation and early evolution at the origin of Jupiter's massive moons. *Astron. J.*, *155*, 224.

Rubie D. C., Nimmo F., and Melosh H. J. (2007) Formation of Earth's core. In *Treatise on Geophysics, 1st edition, Vol. 9: Evolution of the Earth* (G. Schubert, ed.), pp. 51–90. Elsevier, Amsterdam.

Rubin M., Altwegg K., Balsiger H., Bar-Nun A., Berthelier J.-J., Briois C., Calmonte U., Combi M., De Keyser J., and Fiethe B. (2018) Krypton isotopes and noble gas abundances in the coma of comet 67P/Churyumov-Gerasimenko. *Sci. Adv.*, *4*, eaar6297.

Santerne A., Brugger B., Armstrong D. J., Adibekyan V., Lillo-Box J., Gosselin H., Aguichine A., Almenara J. M., Barrado D., and Barros S. C. C. (2018) An Earth-sized exoplanet with a Mercury-like composition. *Nature Astron.*, *2*, 393–400.

Sasselov D. D. and Lecar M. (2000) On the snow line in dusty protoplanetary disks. *Astrophys. J.*, *528*, 995–998.

Sato T., Okuzumi S., and Ida S. (2016) On the water delivery to terrestrial embryos by ice pebble accretion. *Astron. Astrophys.*, *589*, A15.

Schlichting H. E., Sari R., and Yalinewich A. (2015) Atmospheric mass loss during planet formation: The importance of planetesimal impacts. *Icarus*, *247*, 81–94.

Schorghofer N. (2008) The lifetime of ice on main belt asteroids. *Astrophys. J.*, *682*, 697–705.

Scott E. R. D. and Krot A. N. (2014) Chondrites and their components. In *Treatise on Geochemistry, 2nd edition, Vol. 1: Meteorites and Cosmochemical Processes* (A. M. Davis, ed.), pp. 65–137. Elsevier, Amsterdam.

Selsis F., Chazelas B., Bordé P., Ollivier M., Brachet F., Decaudin M., Bouchy F., et al. (2007) Could we identify hot ocean-planets with CoRoT, Kepler and Doppler velocimetry? *Icarus*, *191*, 453–468.

Sharp Z. D. (2017) Nebular ingassing as a source of volatiles to the terrestrial planets. *Chem. Geol.*, *448*, 137–150.

Simon J. B., Armitage P. J., Li R., and Youdin A. N. (2016) The mass and size distribution of planetesimals formed by the streaming instability. I. The role of self-gravity. *Astrophys. J.*, *822*, 55.

Snodgrass C., Agarwal J., Combi M., Fitzsimmons A., Guilbert-Lepoutre A., Hsieh H. H., Hui M.-T., Jehin E., Kelley M. S. P., and Knight M. M. (2017) The main belt comets and ice in the solar system. *Astron. Astrophys. Rev.*, *25*, 5.

Spergel D. N., Verde L., Peiris H. V., Komatsu E., Nolta M. R., Bennett C. L., Halpern M., et al. (2003) First-year Wilkinson Microwave Anisotropy Probe (WMAP) observations: Determination of cosmological parameters. *Astrophys. J. Suppl.*, *148*, 175–194.

Starkey N. A., Stuart F. M., Ellam R. M., Fitton J. G., Basu S., and Larsen L. M. (2009) Helium isotopes in early Iceland plume picrites: Constraints on the composition of high ^{3}He/^{4}He mantle. *Earth Planet. Sci. Lett.*, *277*, 91–100.

Stevenson D. J. and Lunine J. I. (1988) Rapid formation of Jupiter by diffusive redistribution of water vapor in the solar nebula. *Icarus*, *75*, 146–155.

Stimpfl M., Walker A. M., Drake M. J., de Leeuw N. H., and Deymier P. (2006) An ångström-sized window on the origin of water in the inner solar system: Atomistic simulation of adsorption of water on olivine. *J. Cryst. Growth*, *294*, 83–95.

Stuart F. M., Lass-Evans S., Godfrey Fitton J., and Ellam R. M. (2003) High ^{3}He/^{4}He ratios in picritic basalts from Baffin Island and the role of a mixed reservoir in mantle plumes. *Nature*, *424*, 57–59.

Sugiura N. and Fujiya W. (2014) Correlated accretion ages and ε^{54}Cr of meteorite parent bodies and the evolution of the solar nebula. *Meteoritics & Planet. Sci.*, *49*, 772–787.

Svetsov V. V. (2007) Atmospheric erosion and replenishment induced by impacts of cosmic bodies upon the Earth and Mars. *Solar System Res.*, *41*, 28–41.

Tera F., Papanastassiou D. A., and Wasserburg G. J. (1974) Isotopic evidence for a terminal lunar cataclysm. *Earth Planet. Sci. Lett.*, *22*, 1.

Terquem C. and Papaloizou J. C. B. (2007) Migration and the formation of systems of hot super-Earths and Neptunes. *Astrophys. J.*, *654*, 1110–1120.

Tielens A. G. G. M. and Hagen W. (1982) Model calculations of the molecular composition of interstellar grain mantles. *Astron.*

Astrophys., 114, 245–260.

Tsiganis K., Gomes R., Morbidelli A., and Levison H. F. (2005) Origin of the orbital architecture of the giant planets of the solar system. *Nature, 435,* 459–461.

Turner N. J., Fromang S., Gammie C., Klahr H., Lesur G., Wardle M., and Bai X.-N. (2014) Transport and accretion in planet-forming disks. In *Protostars and Planets VI* (H. Beuther et al., eds.), pp. 411–432. Univ. of Arizona, Tucson.

Udry S., Bonfils X., Delfosse X., Forveille T., Mayor M., Perrier C., Bouchy F., et al. (2007) The HARPS search for southern extra-solar planets. XI. Super-Earths (5 and 8 $M_\oplus$) in a 3-planet system. *Astron. Astrophys., 469(3),* L43–L47.

Vaghi S. (1973) Orbital evolution of comets and dynamical characteristics of Jupiter's family. *Astron. Astrophys., 29,* 85.

Van Dishoeck E. F., Herbst E., and Neufeld D. A. (2013) Interstellar water chemistry: From laboratory to observations. *Chem. Rev., 113,* 9043–9085.

Van Dishoeck E. F., Bergin E. A., Lis D. C., and Lunine J. I. (2014) Water: From clouds to planets. In *Protostars and Planets VI* (H. Beuther et al., eds.), p. 835. Univ. of Arizona, Tucson.

Veras D. and Armitage P. J. (2006) Predictions for the correlation between giant and terrestrial extrasolar planets in dynamically evolved systems. *Astrophys. J., 645,* 1509–1515.

Waite J. H. Jr., Lewis W. S., Magee B. A., Lunine J. I., McKinnon W. B., Glein C. R., Mousis O., Young D. T., Brockwell T., and Westlake J. (2009) Liquid water on Enceladus from observations of ammonia and ^{40}Ar in the plume. *Nature, 460,* 1164.

Walker R. J. (2009) Highly siderophile elements in the Earth, Moon, and Mars: Update and implications for planetary accretion and differentiation. *Chem. Erde–Geochem., 69,* 101–125.

Walker R. J., Bermingham K., Liu J., Puchtel I. S., Touboul M., and Worsham E. A. (2015) In search of late-stage planetary building blocks. *Chem. Geol., 411,* 125–142.

Walsh K. J., Morbidelli A., Raymond S. N., O'Brien D. P., and Mandell A. M. (2011) A low mass for Mars from Jupiter's early gas-driven migration. *Nature, 475,* 206–209.

Walsh K. J., Morbidelli A., Raymond S. N., O'Brien D. P., and Mandell A. M. (2012) Populating the asteroid belt from two parent source regions due to the migration of giant planets — "The Grand Tack." *Meteoritics & Planet. Sci., 47,* 1941–1947.

Ward W. R. (1997) Protoplanet migration by nebula tides. *Icarus, 126,* 261–281.

Warren P. H. (2011) Stable-isotopic anomalies and the accretionary assemblage of the Earth and Mars: A subordinate role for carbonaceous chondrites. *Earth Planet. Sci. Lett., 311,* 93–100.

Weidenschilling S. J. (1977) Aerodynamics of solid bodies in the solar nebula. *Mon. Not. R. Astron. Soc., 180,* 57–70.

Wetherill G. W. (1991) Why isn't Mars as big as Earth? In *Lunar Planet. Sci. XXII,* Abstract #1743. Lunar and Planetary Institute, Houston.

Wetherill G. W. (1992) An alternative model for the formation of the asteroids. *Icarus, 100,* 307–325.

Wetherill G. W. (1996) The formation and habitability of extra-solar planets. *Icarus, 119,* 219–238.

Wilde S. A., Valley J. W., Peck W. H., and Graham C. M. (2001) Evidence from detrital zircons for the existence of continental crust and oceans on the Earth 4.4 Gyr ago. *Nature, 409,* 175–178.

Willacy K. and Woods P. M. (2009) Deuterium chemistry in proto-planetary disks. II. The inner 30 AU. *Astrophys. J., 703,* 479–499.

Williams C. D. and Mukhopadhyay S. (2019) Capture of nebular gases during Earth's accretion is preserved in deep-mantle neon. *Nature, 565,* 78–81.

Wolfgang A., Rogers L. A., and Ford E. B. (2016) Probabilistic mass-radius relationship for sub-Neptune-sized planets. *Astrophys. J., 825,* 19.

Wood B. J., Halliday A. N., and Rehkämper M. (2010) Volatile accretion history of the Earth. *Nature, 467,* E6.

Yang C.-C., Johansen A., and Carrera D. (2017) Concentrating small particles in protoplanetary disks through the streaming instability. *Astron. Astrophys., 606,* A80.

Yang L., Ciesla F. J., and Alexander C. M. O. D. (2013) The D/H ratio of water in the solar nebula during its formation and evolution. *Icarus, 226,* 256–267.

Yokochi R., Marboeuf U., Quirico E., and Schmitt B. (2012) Pressure dependent trace gas trapping in amorphous water ice at 77 K: Implications for determining conditions of comet formation. *Icarus, 218,* 760–770.

Youdin A. N. and Goodman J. (2005) Streaming instabilities in protoplanetary disks. *Astrophys. J., 620,* 459–469.

Yurimoto H. and Kuramoto K. (2004) Molecular cloud origin for the oxygen isotope heterogeneity in the solar system. *Science, 305,* 1763–1766.

Yurimoto H., Krot A. N., Choi B. G., Aleon J., Kunihiro T., and Brearley A. J. (2008) Oxygen isotopes of chondritic components. *Rev. Mineral. Geochem., 68,* 141–186.

Zellner N. E. B. (2017) Cataclysm no more: New views on the timing and delivery of lunar impactors. *Origins Life Evol. Biosph., 47,* 261–280.

Zeng L., Jacobsen S. B., Sasselov D. D., Petaev M. I., Vanderburg A., Lopez-Morales M., Perez-Mercader J., et al. (2019) Growth model interpretation of planet size distribution. *Proc. Natl. Acad. Sci. U.S.A., 116(20),* 9723-9728.

Zhang H. and Zhou J.-L. (2010) On the orbital evolution of a giant planet pair embedded in a gaseous disk. I. Jupiter-Saturn configuration. *Astrophys. J., 714,* 532–548.

Zhang K., Pontoppidan K. M., Salyk C., and Blake G. A. (2013) Evidence for a snow line beyond the transitional radius in the TW Hya protoplanetary disk. *Astrophys. J., 766,* 82.

Arney G. N. and Kane S. (2020) Venus as an analog for hot Earths. In *Planetary Astrobiology* (V. Meadows et al., eds.), pp. 355–378. Univ. of Arizona, Tucson, DOI: 10.2458/azu_uapress_9780816540068-ch014.

Venus as an Analog for Hot Earths

Giada N. Arney
NASA Goddard Space Flight Center

Stephen Kane
University of California at Riverside

This chapter evaluates our nearest planetary neighbor, Venus, as an example of the runaway greenhouse state that bounds the inner edge of the habitable zone. Despite its current hellish surface environment, Venus may once have been habitable with oceans of surface liquid water. Over time, it lost its potentially clement environment as the Sun brightened. Today, it represents an end-state of habitable planet evolution, and it therefore provides valuable lessons on habitability as a planetary process. Beyond the solar system, exo-Venus analogs may be common types of planets, and we likely have already discovered many of Venus' sisters orbiting other stars. Furthermore, our near-future exoplanet detection and characterization methods are biased toward observing these types of hot worlds. Therefore, it is instructive to consider what Venus can teach us about exo-Venus analogs. By observing exo-Venus planets of differing ages and in different astrophysical contexts, these distant hot terrestrial worlds may likewise allow us to witness processes that occurred on Venus in the past.

Here is no water but only rock
Rock and no water and the sandy road
The road winding above among the mountains
Which are mountains of rock without water
If there were water we should stop and drink
Amongst the rock one cannot stop or think
Sweat is dry and feet are in the sand
If there were only water amongst the rock
— Excerpt from "The Waste Land" by T. S. Elliot

1. INTRODUCTION

A major focus of astrobiology is the search for habitable conditions and life beyond Earth. We have scrutinized Earth intensely to understand how we might search for habitable conditions elsewhere. In many ways, Venus is the most Earth-like planet in the solar system with its similar mass, radius, and bulk density. Indeed, it is likely that Venus and Earth had very similar starting conditions in terms of their relative compositions of both volatiles and refractory compounds. Yet Earth has been habitable since at least the start of the Archean geological eon (about 3.8 b.y. ago) and possibly during the Hadean eon (before 3.8 b.y. ago) (*Bell et al.,* 2015), while Venus' timescale for habitability is much more uncertain. At some point, the evolution of these two planets diverged dramatically, and Venus is now one of the most uninhabitable planets we might imagine. In the same way that we can study Earth's history to understand how biospheres may evolve over time and co-evolve with their environments, Venus can teach us about an equally

fundamental process: how habitability evolves through nonbiological forcing and how planets reach the end-state of habitability. While Earth shows how self-regulating negative feedbacks (e.g., the carbonate silicate cycle) work to maintain and preserve habitability on geological timescales, Venus offers us an example of a world whose self-regulation mechanisms failed catastrophically.

To paraphrase an observation by climate and planetary scientist François Forget from the 2017 Habitable Worlds meeting in Laramie, Wyoming: Would we ever imagine a Venus-like exoplanet if we did not have an example of one nearby? Without Venus in our solar system, we might not have developed the theory of the runaway greenhouse, and might be more inclined to consider the size and mass of Earth as fundamental factors affecting habitable conditions in terms of both geological and atmospheric evolution. We might anticipate that terrestrial exoplanets closer to their stars are hotter versions of Earth rather than fundamentally different kinds of worlds. Yet Venus shows us that a similar planetary structure to Earth can produce an environment that may be considered at the opposite end of the habitability scale. Therefore, in our searches for worlds circling distant stars, an incomplete understanding of the evolution of Venus and the processes that govern its atmosphere and interior today will hinder our ability to interpret observations of hot terrestrial exoplanets, and will also hinder our ability to understand processes that govern planetary habitability more generally.

Many basic parameters of Venus are still unknown. These include the relative size of the core to its mantle, the rate of current volcanic activity, its moment of inertia, and the identity of the unknown ultraviolet (UV) absorber in the atmosphere. Yet Venus accounts for 40% of the mass of the rocky solar system planets and is the closest planet in the universe to Earth. When we work to understand the basic properties of exoplanets, we must have a robust understanding of our local planetary neighborhood to guide us. For example, comparisons between Earth and solar compositions are being used to predict the interiors of exoplanets based on the stellar abundances of the host star (e.g., *Hinkel and Unterborn,* 2018). Such inferences of planetary abundances will benefit enormously from a more detailed understanding of those abundances present within Venus and other terrestrial planets of our solar system.

Important for future observations of exoplanets, Venus analogs represent one of the most readily observable types of terrestrial planets for the transit transmission observations that will become possible in the near future with the James Webb Space Telescope and large groundbased telescopes (*Kane and von Braun,* 2008). This is because Venus-analog worlds orbit their stars closely and therefore enjoy a higher transit probability and frequency of transit events compared to planets in the habitable zone (HZ). Indeed, many terrestrial exoplanets already discovered, and undoubtedly many that will be uncovered by the recently-launched Transiting Exoplanet Survey Telescope (TESS), are likely to be more Venus-like than Earth-like.

In this chapter, we will tell the story of Venus, starting with a description of its present diabolical state, continuing to its possible history and evolution that may include past habitable conditions, and ending with considerations of "pale yellow dots" orbiting around distant stars.

2. THE CURRENT STATE OF VENUS

Understanding the current state of Venus is a necessary prerequisite to understanding Venus-like exoplanets. Venus is a complex and intriguing planet, and its surface conditions were veiled in mystery for many decades, prompting wild speculations of swampy oases beneath its dense butter-colored clouds (e.g., *Arrhenius,* 1918). A hint of the hot Venus surface temperature came from millimeter-wave observations in the late 1950s (*Mayer et al.,* 1958), but confirmation of the true nature of the blistering venusian surface environment had to wait for spacecraft visits in the 1960s (e.g., *Dickel,* 1966).

We now know that the surface temperature of Venus is hot enough to melt lead, so it is of little surprise that the few landers that have been to the venusian surface have survived for no more than about an hour at most. Yet the challenges of studying the Venus environment are also opportunities that invite us to consider how and why this planet, which is in many ways similar to Earth in terms of size (0.94× Earth's radius), mass (0.81× Earth's mass), and bulk composition, had such a different evolutionary outcome. These questions are even more important in light of evidence suggesting that Venus was once more Earth-like. Was it once habitable, and if so, how did Venus lose its habitability?

Venus is a planet of many extremes. At its surface, a crushing 92 bar of pressure bear down, equivalent to the pressure at 1.6 km underwater on Earth. The clouds are composed not of water but of a concentrated solution (75–85%) of sulfuric acid (H_2SO_4) and water (*Sill,* 1972). Even Venus' rotation is unusual in the solar system. Venus' sidereal rotation period is 243 Earth days — longer than its year of 225 Earth days — while its length of day is 116 Earth days (*Dyce et al.,* 1967). A combination of Earth-Sun body tides and solar atmospheric tides may have led to its present rotation rate (*Dobrovolskis and Ingersoll,* 1980), which is retrograde relative to the other planets. *Correia and Laskar* (2003) suggest that its rotational axis could have flipped 180° due to internal core-mantle friction and tides from the atmosphere. They suggest that for planets with thick atmospheres like Venus, most initial conditions can lead to Venus-like rotation regimes via two mechanisms: The spin axis can flip directions, or a planet with prograde rotation can become retrograde as the obliquity decreases toward zero. This slow rotation may be at least partially responsible for Venus' lack of an internally-generated magnetic field (*Stevenson,* 2003), although other factors inhibiting thermal convection — such as its lack of plate tectonics — have also been implicated (*Nimmo,* 2002).

Perhaps the most notable characteristic of Venus is its high surface temperature. Its globally averaged surface temperature is 735 K (*Marov et al.,* 1973; *Seiff et al.,* 1985), although higher and lower elevations are colder and hotter, respectively. The surface is so hot that a faint red incandescence emanates from the ground (*Bullock and Grinspoon,* 2013). The solar flux (i.e., the amount of solar radiation a planet receives per unit area) at Venus is almost 2× as much as Earth's solar constant (2610 W m^{-2} compared to Earth's 1360 W m^{-2}), but this is not the only cause of its hot surface temperatures. Indeed, with its high albedo due to its brightly reflective H_2SO_4 clouds, Venus actually reflects almost 80% of the incident solar radiation back to space. Due to the dense atmosphere and highly reflected cloud cover, only about 2.5% of the total incoming solar radiation actually reaches the surface, as the rest is absorbed by the atmosphere. The high surface temperatures Venus experiences are due to its highly efficient greenhouse effect.

In this section, we will describe characteristics of modern Venus that may be accessible on exoplanets or relevant to understanding their observable features: its atmosphere, its clouds, geological activity on the surface, and the speculative possibility of extant life in the Venus clouds.

2.1. The Atmosphere

The atmosphere of an exoplanet is one of its most readily accessible characteristics. Here, we provide an overview of the current state of the gases in the venusian atmosphere: Exo-Venus planets may host similar suites of gases. The atmospheric composition of Venus is summarized in Fig. 1, which shows the vertical profiles of key trace gases from the Venus International Reference Atmosphere (VIRA) (*Moroz and Zasova, 1997*).

The Venus atmosphere is extremely efficient at trapping heat: The dense 92-bar atmosphere is composed of 96.5% CO_2 (*Oyama et al., 1979*), producing an almost impenetrable greenhouse enabled by strong, pressure-broadened, infrared absorption bands (*Sagan, 1961; Pollack, 1971; Bullock and Grinspoon, 2013*). Carbon dioxide produces its strongest absorption bands near 4.3 and 15 µm. On Earth, the 15-µm CO_2 band is important for our planet's greenhouse effect. With its elevated temperature, Venus still emits significant amounts of radiation at wavelengths as short as 4.3 µm, and so this band is more important for the venusian greenhouse than it is for Earth's (*Bullock and Grinspoon, 2013*). Collisional broadening of these CO_2 bands increases their extent in wavelength. Gaps between these strong bands where radiation could otherwise escape are effectively closed by pressure-broadened H_2O and SO_2 bands from trace amounts of water vapor and sulfur dioxide in the atmosphere, as well as the sulfuric acid clouds. A relatively narrow spectral region in the near-infrared between roughly 1 and 2.7 µm, where the clouds are mostly non-absorptive (see section 2.2), allows thermal radiation from the surface and lower atmosphere to leak out to space between CO_2 and H_2O bands.

Venus appears to have degassed its bulk inventory of CO_2 and N_2 to the atmosphere. If Earth's crustal CO_2 — most of which is locked up in carbonates on continental margins — was liberated into the atmosphere, our planet would have about 100 bar of CO_2 (*Sleep and Zahnle, 2001; Zahnle et al., 2007*), similar to Venus. Molecular nitrogen is the second most abundant gas, comprising 3.5% of the total atmosphere on Venus. Although this is a small fraction of the whole venusian atmosphere, it is about 4× as much total N_2 as Earth's atmosphere contains (*Hoffman et al., 1980; Von Zahn et al., 1983*). On Earth, nitrogen cycles between the atmosphere and mantle are driven by various biotic and abiotic processes (e.g., *Canfield et al., 2010; Wordsworth, 2016*). On Venus, it has been proposed that oxygen liberated by H_2O photolysis and water loss (described in section 3) could have been absorbed by the mantle, increasing its oxygen fugacity and releasing N_2 to the atmosphere (*Wordsworth and Pierrehumbert, 2014*).

At altitudes starting at about 10 km and higher, the Venus atmosphere superrotates, meaning that it rotates faster than the surface. Extreme winds in the cloud layers travel from east to west at speeds 60× faster than the surface rotation rate (*Schubert, 1983*). These speeds reach 120 m s⁻¹ near the equatorial cloud tops. Meanwhile, wind speeds at the surface are only about 1 m s⁻¹ (*Marov et al., 1973*). The complete mechanism that causes the Venus superrotation is not well understood, but evidence suggests angular momentum is transferred between the planetary surface and upper atmosphere (*Mendonça and Read, 2016*).

Studies of atmospheric waves propagating in the venusian atmosphere may be important to understanding the nature of its superrotation (e.g., *Peralta et al., 2009, 2014; Piccialli et al., 2014*). Studies of these waves have been aided by the Japanese Aerospace Exploration Agency (JAXA) Akatsuki mission, which is currently in orbit around Venus. Akatsuki was intended to go into orbit around Venus in 2010, but the insertion attempt failed. Instead, it remained in a heliocentric orbit until 2015, when a second close to pass to Venus resulted in a successful orbital insertion (*Nakamura et al., 2016*).

Akatsuki data has revealed an unusual stationary wave in the upper cloud deck (*Fukuhara et al., 2017*), suggesting complex atmospheric dynamics at play. Normal wind speeds at these altitudes whip across the planet at roughly 100 m s⁻¹, but this UV-bright, bow-shaped feature remains stationary relative to the surface far below. The center of the feature is located above the western slope of highland region Aphrodite Terra and may be a stationary atmospheric gravity wave associated with lower atmosphere wind flows over this feature.

This same type of feature was observed in 1985 Vega Venus Balloon observations: Stationary waves generated by Aphrodite Terra topography were observed to propagate upward to the middle cloud layers where they perturbed atmospheric motions at balloon altitudes of 50–55 km (*Young et al., 1987*). Additionally, such waves have also been seen in analyses of Venus Express (VEx) data, which showed them concentrated over highland surface features (*Peralta et al., 2017*). While these waves are clearly correlated to topography, latitude and time of day also appear to play a role, suggesting the influence of solar heating on atmospheric

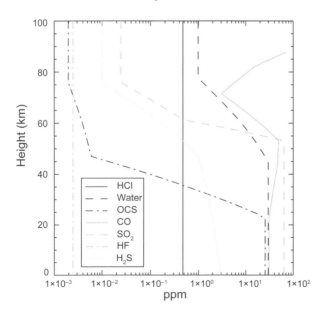

Fig. 1. Profiles of trace gases in the Venus atmosphere. The bulk gases are not shown here and include CO_2 (96.5% of the Venus atmosphere) and N_2 (3.5%).

wave phenomena (*Kouyama et al.,* 2017). Recent observations using a cloud-tracking pipeline to study atmospheric motions have revealed that the middle and lower atmosphere is unexpectedly variable, and low latitude equatorial winds can exceed 80 m s^{-1} (*Horinouchi et al.,* 2017). This equatorial jet phenomenon may be caused by vertical or horizontal momentum transfer driven by atmospheric waves, although the precise mechanism is not yet clear.

Possibly, the propagation of these waves is aided by amplification caused by the lower density of the atmosphere at higher altitudes and resonance driven by variations in static stability and zonal winds as suggested by *Young et al.* (1987). On Earth, gravity waves in the troposphere can be generated when wind flows over mountainous regions; waves propagating through the atmosphere at higher altitudes where the air is thinner (i.e., over mountains) can experience nonlinear wave-breaking effects, affecting the dynamics of Earth's middle atmosphere. On Venus, the propagation of such mountain waves to the cloud tops may be difficult as convection between the cloud top and ground can disturb wave features (e.g., *Seiff et al.,* 1980).

Atmospheric mountain waves may even alter the rotation rate of Venus. An analysis of VEx data suggests Venus' rotation rate may have varied slightly between Magellan measurements taken from 1990 to 1992 (rotation period = 243.0185 ± 0.0001 days) and VEx measurements between 2006 and 2008 (rotation period = 243.023 ± 0.023 days). A recent study suggests that mountain waves, together with thermal tides and baroclinic waves, can affect the solid-body rotation rate by up to two minutes per day (*Navarro et al.,* 2018), which may contribute to the discrepancy in rotation rates observed by Magellan and more recent observations.

Seasonal variability in the venusian atmosphere should be minor because its axial tilt is only 1.7°, compared to Earth's 23.5° obliquity. Indeed, the temperature gradient at the surface inferred from latitudes <60° is ~4 K, with little annual variation (*Seiff et al.,* 1980), although there are no deep atmosphere or surface measurements poleward of 60°. Despite this apparent lack of seasonality, surprising variability has been observed in the distributions of gases in the Venus atmosphere, which vary on unknown timescales due to unknown mechanisms. Hemispherical asymmetries in the abundances of trace gases like H_2O, CO, SO_2, and OCS have been observed to vary on poorly constrained timescales (*Arney et al.,* 2014; *Collard et al.,* 1993; *Krasnopolsky,* 2010a; *Marcq et al.,* 2006). For example, *Arney et al.* (2014) found time-varying hemispherical asymmetries in the distributions of CO, H_2O, OCS, SO_2, and cloud droplet H_2SO_4 concentration using groundbased observations of the Venus subcloud atmosphere in observations from 2009 and 2010. Seasonal hemispherical dichotomies have been observed on other solar system bodies such as Saturn (*Sinclair et al.,* 2013) and Titan (e.g., *Coustenis and Bezard,* 1995; *Teanby et al.,* 2008). However, the small venusian obliquity makes seasonally-driven variations difficult to invoke as an explanation for the observed variable asymmetries. In principle, orbital eccentricity could also drive seasonal variability,

but Venus' orbital eccentricity is even smaller than Earth's (0.0067 vs. Earth's 0.017).

Some gases in the Venus atmosphere exhibit striking spatial variability. For instance, CO and OCS display marked anticorrelation in their distributions in the lower atmosphere, and the apparent opposing hemispherical asymmetries of these gases may relate to chemical transformation between these species (*Krasnopolsky,* 2007; *Yung et al.,* 2009). The distribution of CO in the lower atmosphere varies with latitude due to atmospheric circulation processes. CO is thought to be produced from CO_2 photolysis above the cloud deck, and it freshly sinks to lower altitudes near 60° north and south. Atmospheric circulation carries it to the equator at lower elevations, and reactions with surface minerals or other thermochemical reactions with SO_2 in the atmosphere can transform it into OCS. Indeed, observations suggest that CO and OCS are broadly anticorrelated to each other, with CO abundances peaking at higher latitudes and reaching minima at lower latitudes, while OCS displays the opposite behavior (*Marcq et al.,* 2005, 2006, 2008; *Arney et al.,* 2014).

Other variations in the Venus atmosphere have been observed. Decade-long variations in upper-atmosphere SO_2 abundance were seen in data spanning multiple spacecraft visits to Venus [described in section 2.2 (*Marcq et al.,* 2013)], leading to speculation of transport of SO_2 from the lower atmosphere through atmospheric dynamical processes, or by volcanism. In addition, diurnal variations of the atmosphere may be present (*Sandor et al.,* 2010), cloud top wind speeds were observed to increase between 2006 and 2012 in VEx observations (*Khatuntsev et al.,* 2013), and periodicity in wind speed oscillations tracked by cloud motion has been reported (*Kouyama et al.,* 2013). None of these processes are well understood.

2.2. The Clouds

Clouds and other types of aerosols can obscure information about the deep planetary environment on exoplanets, as they do on Venus. The Venus cloud deck hides the surface at most wavelengths, making historical observations of the planet's subcloud environment challenging. Cloud droplets on Venus are composed of a concentrated H_2SO_4/H_2O solution and are typically 75–85% H_2SO_4 (*Sill,* 1972; *Young,* 1975; *Hansen and Hovenier,* 1974; *Barstow et al.,* 2012; *Cottini et al.,* 2012). The H_2SO_4 forms photochemically at altitudes around 60 km (160 mbar) through reactions involving SO_2 and trace amounts of water vapor (*Yung and DeMore,* 1982).

$$SO_2 + h\nu \rightarrow SO + O$$

$$SO_2 + O \rightarrow SO_3$$

$$SO_3 + H_2O + M \rightarrow H_2SO_4 + M$$

The H_2SO_4 condenses into droplets with H_2O. At high altitudes, the particles exist as a fine haze. As condensation continues, the particles coalesce and settle into the main cloud deck below (*Gao et al.*, 2014).

Models of the cloud structure on Venus have been proposed by several authors (e.g., *Crisp*, 1986; *Grinspoon et al.*, 1993; *Pollack et al.*, 1993; *Tomasko et al.*, 1985), and more recent studies have worked to better constrain the cloud structure and properties (e.g., *Barstow et al.*, 2012; *Satoh et al.*, 2009).

The main cloud deck can be divided into three major layers characterized by different sized particles, plus a fine haze above and below the main clouds (*Knollenberg and Hunten*, 1980; *Ragent and Blamont*, 1980; *Crisp*, 1986). These layers are summarized in Table 1. The upper layers (>70 km) are composed of a fine haze of "mode 1" particles approximately 0.4 μm in radius. Below this, the upper cloud (57–70 km) contains "mode 2" particles, approximately 1.4 μm in radius; mode 1 particles are also in the upper cloud. In the middle (50–57 km) and lower clouds (48–50 km), the cloud particles consist of mode 2 and mode 3 particles, which have an average radius of 3.85 μm (*Knollenberg and Hunten*, 1980). These modes are summarized in Table 2. The large mode 3 particles are responsible for the bulk of the opacity and mass of the cloud layers (*Crisp*, 1986). Mode 3 particles may have a crystalline component (*Esposito et al.*, 1983; *Knollenberg and Hunten*, 1980), requiring the presence of some species other than H_2SO_4. It has also been proposed that mode 3 particles represent the large end of the particle size distribution for the mode 2 particles (*Toon et al.*, 1984). In this case, the particles would be noncrystalline, but their optical properties would disagree with Pioneer Venus nephelometer (*Ragent and Blamont*, 1980) and solar flux radiometer data (*Tomasko et al.*, 1980).

The altitude at which the H_2SO_4 vapor pressure exceeds its saturation vapor pressure sets the height of the bottom of the cloud deck at about 47 km (approximately 2 bar). A fine haze of H_2SO_4 extends below the cloud deck. Below the clouds, H_2SO_4 can exist in vapor phase down to about 38 km (about 4 bar), below which it is thermochemically decomposed at temperatures around 430 K (*Bullock and Grinspoon*, 2001) through the reaction $H_2SO_4 \rightarrow SO_3 + H_2O$.

The clouds themselves are important to the Venus greenhouse effect. Climate modeling shows that if the cloud deck were removed on Venus, the surface temperature would decrease by 142.8 K because the clouds are strongly absorptive at thermal wavelengths (Fig. 2) [compare to a decrease of 422.7 K if CO_2 is removed, and 68.8 K if H_2O is removed (*Bullock and Grinspoon*, 2001)].

For wavelengths shorter than about 2.5 μm, the clouds are very weakly absorbing (Fig. 2), although they are still highly scattering. The low absorption in this wavelength range allows thermal radiation to leak out to space between strong, pressure-broadened CO_2 and H_2O bands (section 2.1), although this radiation is still highly scattered. When observed on the planet's nightside, because there is no confounding scattered sunlight, nightside thermal emission from the subcloud atmosphere and surface can be observed directly, even from Earth-based observatories. These "spectral windows" were discovered when *Allen and Crawford* (1984) found an excess of radiation near 1.74 and 2.3 μm on the Venus nightside, and subsequent studies confirmed that these spectral ranges sense thermal emission emanating from below the cloud layer (*Allen*, 1987; *Bézard et al.*, 1990). Subsequently, additional windows at shorter wavelengths were discovered at 1.0, 1.1, 1.18, 1.27, and 1.31 μm (*Allen*, 1990; *Bézard et al.*, 1990; *Carlson et al.*, 1991; *Crisp et al.*, 1991) that sense the surface and lowest atmospheric scale height (~16 km) (*Pollack et al.*, 1993; *Lecacheux et al.*, 1993; *Meadows and Crisp*, 1996). Additional weaker windows were discovered at 0.85 and 0.9 μm (*Carlson et al.*, 1991; *Baines et al.*, 2000) and 1.51, 1.55, 1.78, and 1.81 μm (*Erard et al.*, 2009).

Also in the clouds is an "unknown UV absorber" (e.g., *Ross*, 1928; *Barker*, 1979; *Esposito*, 1980; *Molaverdikhani et al.*, 2012; *Markiewicz et al.*, 2014), which absorbs wavelengths from 320 to 500 nm and is responsible for the characteristic yellowish color of Venus. The absorber appears to be distributed in the upper cloud layer at 58 to 62 km (*Ekonomov et al.*, 1984). It also appears to be distributed heterogeneously (*Molaverdikhani et al.*, 2012) and has been observed to be correlated to SO_2 (*Titov et al.*, 2008; *Esposito and Travis*, 1982).

Understanding at what altitudes the UV absorber is located may aid in understanding its composition. Recent Akatsuki observations at 365 and 283 nm (the unknown UV absorber and an SO_2 band, respectively) show faster zonal winds in the 283-nm (SO_2) channel (*Horinouchi et al.*, 2018), suggesting its features are produced at higher altitudes than those in the 365-nm channel (UV absorber). *Lee et al.* (2017) similarly found that phase angle dependency in 365-nm observations can most easily be explained by the UV absorber distributed below the cloud top, not above.

A number of candidates have been suggested for the identity of this absorber, including dimers of disulfur dioxide

TABLE 1. Venus cloud layer properties.

Layer name	Altitude (km)	Temperature (K)	Pressure (bar)	Particle mode types (see Table 2)	Optical depth at 0.63 μm
Upper haze	70–90	255–190	0.0267–0.0028	1	0.2–1
Upper cloud	56.5–70	286–255	0.406–0.0267	1, 2	6–8
Middle cloud	50.5–56.5	245–286	0.981–0.406	1, 2', 3	8–10
Lower cloud	47.5–50.5	367–345	1.391–0.981	1, 2, 3	6–12
Lower haze	38–47.5	430–367	4–1.39	1	0.1–0.2

Derived from *Esposito et al.* (1983).

TABLE 2. Venus cloud particle sizes from *Crisp* (1986).

Mode	Effective Radius (µm)	Variance
1	0.49	0.22
2	1.04	0.19
2′	1.4	0.207
3	3.85	0.262

(OSSO); polysulfur; and solutions of $FeCl_3$ in H_2SO_4, Cl_2, SCl_2, and S_2O (e.g., *Esposito et al.,* 1983, 1997; *Zasova et al.,* 1981; *Frandsen et al.,* 2016; *Zhang et al.,* 2012). Current leading candidates include polysulfur compounds and OSSO.

Polysulfur, S_x, can form through reactions involving OCS photolysis around the cloud tops (*Zhang et al.,* 2012). *Young* (1977) suggested S_8 absorption could account for the UV absorber if present in sufficient quantities, while *Toon et al.* (1982) discussed how S_3 and S_4 allotropes have strong absorptions centered near 400 nm and 530 nm and convert to S_8 in hours to days, which could explain the UV absorber's short lifetime. A more recent study by *Zhang et al.* (2012) used a photochemical model to study sulfur chemistry in the Venus atmosphere and found that S_2, S_3, S_4, and S_5 are supersaturated at various altitudes in and above the cloud deck.

The OSSO dimer provides a good spectral match to the absorber at UV wavelengths. The high concentration of SO implies that OSSO dimers should be present in the atmosphere, and concentrations of OSSO are expected to be similar to SO concentrations at altitudes of 58–70 km (*Frandsen et al.,* 2016). Recently, *Pérez-Hoyos et al.* (2018) analyzed MErcury Surface, Space ENvironment, GEochemistry, and Ranging (MESSENGER) spectral data taken during a Venus flyby between 0.3 and 1.5 µm and determined that the absorption spectrum of the UV absorber is best matched by OSSO, and is also reasonably matched by S_2O. Conversely, Pérez-Hoyos et al. found other candidates such as SCl_2, Cl_2, and $FeCl_3$ have absorption features that are too narrow to match the observations.

It is possible that the UV absorber may in fact be a combination of species. Indeed, Pérez-Hoyos et al. pointed out that while the absorption they observe for the UV absorber is most closely matched by OSSO, this compound does not produce sufficient absorption near 0.4 µm to match observations. In this case, the authors suggest that S_4 is an excellent candidate to add opacity in this spectral region.

The lifetime of the H_2SO_4 cloud deck is likely to be short, given loss of SO_2 from the atmosphere. This implies replenishment of SO_2 from ongoing volcanic processes. Gaseous SO_2 required to form the Venus clouds can be lost from the atmosphere through reactions with surface minerals: *Fegley and Prinn* (1989) showed that at Venus-like conditions, SO_2 should react with calcite on the surface to form anhydrite.

$$CaCO_3 + SO_2 \leftrightarrow CaSO_4 + CO$$

Based on this reaction, all the SO_2 in the Venus atmosphere could be lost in only 2 m.y. (*Fegley and Prinn,* 1989). However, this calculation did not account for diffusion of SO_2 through surface rock layers to reach fresh calcite, which would increase the atmospheric lifetime of SO_2. *Bullock and Grinspoon* (2001) applied a model that included diffusion to the problem and estimated the atmospheric lifetime of SO_2 to be 10× longer than the *Fegley and Prinn* (1989) calculation. They estimate the lifetime for SO_2 to be 20 m.y. *Bullock and Grinspoon* (2001) estimated that the amount of SO_2 currently in the atmosphere is in excess of equilibrated conditions by a factor of 100. If SO_2 concentrations drop below 10 ppmv, the clouds should dissipate (*Bullock and Grinspoon,* 2001). Thus, the existence of a long-lived cloud deck implies ongoing volcanic processes generating SO_2, and the existence of the clouds on modern Venus require SO_2 injections to the atmosphere within the past 20 m.y. The most abundant compounds outgassed by typical basaltic volcanism on Earth are CO_2, H_2O, and SO_2 (*Kaula and Phillips,* 1981). *Fegley and Prinn* (1989) estimate that between 0.4 and 11 km³ of fresh magma per year are needed to maintain the cloud deck.

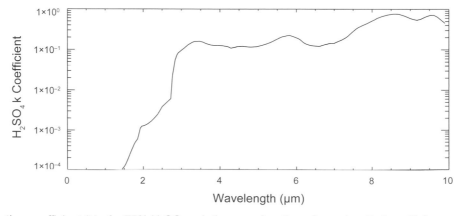

Fig. 2. The extinction coefficient (k) of a 75% H_2SO_4 solution as a function of wavelength from *Palmer and Williams* (1975). The extinction coefficient, which measures how absorptive the cloud particles are, decreases by orders of magnitude for wavelengths shorter than about 2.5 µm, allowing thermal radiation from the lower atmosphere to escape through the cloud at these wavelengths. At longer wavelengths, the clouds are highly absorptive.

Strikingly, variations in SO_2 at the cloud top level have been observed across multiple decades of observations by NASA's Pioneer Venus (in orbit from 1978 to 1992) and the European Space Agency's (ESA's) VEx (its tenure at Venus started in 2006 and ended in 2015 when it was deorbited in the upper atmosphere), suggesting that there may be long-term cycling mechanisms transporting SO_2 from the lower atmosphere to the upper layers (*Marcq et al.*, 2013). This mechanism could be related to atmospheric dynamical and transport processes. The subcloud atmosphere contains roughly 130–180 ppm SO_2 (*Bezard et al.*, 1993; *Pollack et al.*, 1993; *Marcq et al.*, 2008; *Arney et al.*, 2014), while the atmosphere above the clouds contains only a few hundred parts per billion (ppb) (*Marcq et al.*, 2013; *Krasnopolsky*, 2010b), so transport of the relatively SO_2-rich lower layers to higher levels could cause measurable changes. The variations in SO_2 seen above the clouds could also be related to volcanic injections of SO_2 (*Esposito*, 1984), rather than purely atmospheric dynamical processes, which is intriguing in light of the necessity of volcanism to maintain the cloud deck, and other evidence for volcanism (section 2.3).

2.3. The Surface and Possible Volcanism

Much of what is known about the venusian atmosphere can be linked to surface geological activity. While volcanism on a Venus-like exoplanet may be difficult to detect, inferences of exoplanet geological activity might be gleaned from the existence of outgassed atmospheric species (e.g., SO_2 and H_2SO_4 as discussed in section 2.2). Understanding the geological history of the venusian surface may also have important implications for its history of transition between a possibly habitable world to the hot planet it is today.

The modern bulk Venus atmosphere points to a long history of outgassing. On Earth, CO_2 outgassed into the atmosphere can be removed through reactions with silicates in Earth's crust. These materials are eventually cycled back into the mantle via plate tectonics and subsequently outgassed back into the atmosphere, completing the cycle. Indeed, this recycling mechanism, known as the carbonate silicate cycle, has been used to explain Earth's relatively stable climate over geological timescales (e.g., *Walker et al.*, 1981) because the rate of CO_2-silicate rock reactions is temperature-dependent, resulting in a stabilizing negative feedback process. On Venus, the lack of plate tectonics means that CO_2 outgassed into the atmosphere cannot return to the interior of the planet and simply accumulates over time. The fact that Venus' atmosphere appears to contain a comparable amount of CO_2 to Earth's total CO_2 budget locked in carbonate rocks (*Sleep and Zahnle*, 2001) suggests that Venus may have started with a comparable volatile inventory and outgassed all or most of its mantle CO_2 budget.

Little was known about the surface and possible volcanic processes of Venus prior to the Magellan mission, which launched in 1989. Earlier observations using, e.g., the 85-ft Goldstone radio dish in the early 1960s (*Carpenter*, 1964, 1966; *Muhleman*, 1961) could discern the venusian solid-body rotation rate and its radius. Magellan's radar images mapped nearly the whole surface at a resolution of 200 m, unveiling a complex basaltic landscape. Most of the Venus surface is rolling plains splattered with a variety of volcanic features and lava flows but without plate tectonic activity. For instance, rift valleys exist on Venus (e.g., in the Alta Regio, Beta Regio, and western Estalia Regio areas), and they may have formed from localized volcanic activity that stretched and deformed parts of the crust (*Campbell et al.*, 1984; *Kiefer and Swafford*, 2006).

Remote searches for active volcanism on the surface of Venus have been enabled by Venus' near-infrared spectral windows (*Allen and Crawford*, 1984). However, the scattering footprint of surface radiation escaping through the cloud deck is about 100 km² (*Drossart et al.*, 2007), so smaller-scale features are smeared out.

Initial searches for volcanism on Venus did not prove fruitful: An analysis of VEx data obtained with the Venus Monitoring Camera (VMC) between October 31, 2007, and June 15, 2009, did not reveal any hot spots that might be indicative of volcanism (*Shalygin et al.*, 2012). However, more recently, *Smrekar et al.* (2010) discussed the discovery of nine emissivity anomalies identified in VEx Visible and Infrared Thermal Imaging Spectrometer (VIRTIS) data as sites of possible volcanic activity that may be 2.5 m.y. old, and more likely as young as 250,000 years old. These age estimates are derived from estimated weathering rates of fresh basalt. Strengthening the case for a volcanic interpretation, Magellan gravity data suggests the emissivity anomalies are associated with surface regions likely to have a thin, elastic lithosphere. More evidence for modern volcanism came in 2015 via new analyses of VMC observations. *Shalygin et al.* (2015) observed four temporally variable hotspots at volcanos Ozza Mons and Maat Mons in the Ganiki Chasma rift zone, suggesting fresh volcanic activity.

The cratering record on Venus implies a relatively young surface, which may suggest significant geologic activity in the recent past. The entire surface of Venus has only about 1000 randomly distributed impact craters largely unaffected by tectonism or volcanism (*Schaber et al.*, 1998). This cratering record suggests that almost the whole surface of Venus is roughly 300–800 m.y. old (*Strom et al.*, 1994; *Phillips et al.*, 1992; *Namiki and Solomon*, 1994; *Herrick and Rumpf*, 2011). This unusual finding has been interpreted by some authors to suggest a global resurfacing event (*Strom et al.*, 1994) that obliterated older crust from the first 80% of Venus' history. This hypothesis is generally referred to as the catastrophic or episodic resurfacing scheme. Such an event could be driven by rising mantle temperatures caused by heat trapped by a thick lithosphere that cannot be released because Venus has no plate tectonics. The rising temperatures weaken the crust until widespread subduction occurs and the crust is completely recycled over a relatively short time period (<100 m.y.) (*Schaber et al.*, 1992; *Strom et al.*, 1994; *Turcotte et al.*, 1999). This resurfacing event could have caused a 100 K change in temperature at the surface (*Bullock and Grinspoon,* 2001), which would have put stresses on the lithosphere over the next 100 m.y., possibly creating observed wrinkle ridge features (*Solomon et al.,* 1999).

There may be regions of the Venus surface that did not participate in this global resurfacing event, and these regions might someday yield clues about the ancient venusian surface. Studies have suggested that certain regions on the surface called tesserae, which are highly deformed elevated surface regions (e.g., Aphrodite Terra), pre-date the resurfacing event (*Romeo and Turcotte,* 2008; *Ivanov and Head,* 1996, 2013). The word "tessera" means "floor tiles" in Greek (such as used to construct a mosaic), and the textures of these regions are unique to Venus in the solar system. These tesserae could record a distinct tectonic regime from the venusian past (*Gilmore et al.,* 1997; *Brown and Grimm,* 1997). Models suggest that they may have formed by mantle up/downwelling processes (*Ghent et al.,* 2005), or instead they may represent deformed lava planes (*Hansen et al.,* 2000). Venus Express data hints that tesserae may have a lower emissivity than the low-lying basaltic planes (*Basilevsky et al.,* 2012). This could be indicative of composition differences indicative of silica-rich minerals (*Hashimoto and Sugita,* 2003) possibly formed through ancient continent building processes, or alternatively the emissivity variations could suggest, e.g., grain-size differences in tesserae (*Basilevsky et al.,* 2004, 2007).

An alternative to the global resurfacing hypothesis is continuous local volcanic activity, which can also match the crater data. This is sometimes referred to as the equilibrium or evolutionary resurfacing hypothesis (*Phillips et al.,* 1992; *Hansen and Young,* 2007). If the equilibrium resurfacing hypothesis is correct, significant parts of Venus' surface may actually be more than 1 b.y. old, implying a rich geological history may be preserved on Venus.

In this scheme, the uniform crater distribution on the venusian surface is suggested to have resulted from random tectonic or volcanic events that destroy localized impact craters (e.g., *Phillips et al.,* 1992; *Guest and Stofan,* 1999). A number of studies have suggested that Venus' population of impact craters exhibit geological characteristics that are challenging to explain through the catastrophic resurfacing hypothesis (e.g., *Herrick et al.,* 1995; *Hansen and Young,* 2007; *Hansen and López,* 2010; *Herrick and Rumpf,* 2011). *Hansen and Young* (2007), for example, discuss that the parts of Venus' surface hypothesized to be geologically ancient and pre-dating the resurfacing event on elevated plateaus do not necessarily correlate with the oldest surfaces indicated by crater counting and crater morphology records. While some Monte Carlo modeling efforts have suggested that equilibrium resurfacing is unable to reproduce Venus' impact crater characteristics (*Bullock et al.,* 1993; *Strom et al.,* 1994), *Bjonnes et al.* (2012) more recently performed suites of Monte Carlo modeling experiments and suggested that some configurations of equilibrium resurfacing are consistent with the crater data.

Beneath the surface, there is evidence that Venus may have a liquid core, even though it lacks a planetary magnetic field. Evidence for the liquid core comes from the tidal Love number measured from Magellan and Pioneer Venus data (*Konopliv and Yoder,* 1996; *Yoder,* 1997). Despite this, the planet lacks a magnetic field (*Russell et al.,* 1980). If Venus had a magnetic field in the past, it may have been lost if the planet transitioned from a tectonic regime to a sluggish mantle or stagnant lid regime that is inefficient at allowing internal heat to escape. Alternatively, it is thought that the core dynamo on Earth is partially driven by compositional convection caused by the inner core solidifying. If the venusian core has not cooled enough to allow this, that may stymie the generation of a dynamo (*Stevenson et al.,* 1983; *Nimmo,* 2002).

2.4. Suggestions of Extant Life

Although numerous questions remain about the nature of current Venus, it is abundantly clear that Earth's closest planetary neighbor (in terms of distance and bulk properties) appears to be one of the least likely terrestrial worlds of the solar system to support life. However, it has been suggested that there may be possible habitats for microbes in the cloud layers where temperatures and pressures are more amenable to life (*Schulze-Makuch et al.,* 2002, 2004). For instance, at altitudes of around 50 km, pressures are 1 bar and temperatures are 300–350 K. It is hypothesized venusian microbes could have taken refuge in this relatively clement atmospheric layer after the surface of Venus was rendered uninhabitable. On Earth, microbes can exist in the upper atmosphere (*Smith,* 2013). For instance, stratospheric samples from 20 km over the Pacific Ocean have revealed the presence of bacterial species and the *Penicillium* fungus genus (*Smith et al.,* 2010), but the atmospheric residence time for these samples is only 7–10 days, so this discovery does not indicate an independent aerial ecosystem on Earth. It has been hypothesized that the large "mode 3" particles in the Venus clouds (*Grinspoon,* 1997) or the unknown UV absorber (*Limaye et al.,* 2018) might even be biological, but this possibility remains highly speculative.

3. THE HISTORY AND EVOLUTION OF VENUS

The hellish current conditions of present day Venus stand in stark contrast to possible warm, watery conditions that have been suggested for its deep past. If it was indeed habitable, how was this paradise lost? Moreover, if Venus had surface liquid water, how much was there, and how long did it last? These questions do not have definitive answers, yet there have been numerous studies investigating aspects of these problems, and they have sketched the outlines of a doomed once-habitable world just next door. Understanding Venus is vital to understanding processes that shape habitability in our solar system and elsewhere in the universe.

Despite their vast differences today, Venus may have formed with similar bulk composition and initial volatile inventory to Earth. For instance, *Morbidelli et al.* (2012)

suggests that Venus and Earth are unlikely to have formed with greater than 5 orders of magnitude difference in their water inventories, and some accretional modeling scenarios suggest Venus received 5–30 Earth oceans worth of water from accretion of volatile-rich bodies (*Chassefière et al.,* 2012). In some formation scenarios, most of the water may have been lost in the first 100 m.y. of the planet's history (*Hamano et al.,* 2013). However, even early Earth may also have lost much of its volatiles in the first 100 m.y. (*Finlay et al.,* 2016; *Pujol et al.,* 2013), and volatile loss would also have occurred in the Moon-forming impact (*Cameron et al.,* 1976). Later water delivery may have occurred during the late veneer (*Frank et al.,* 2012).

3.1. Evidence for Lost Water

A tantalizing clue of the past nature of Venus lies in the deuterium-to-hydrogen (D/H) ratio in its atmosphere, which may indicate early water loss. If water vapor is photolyzed, and hydrogen escapes, the lighter-weight H is lost to space more easily than the heavier D, leading to an excess of deuterium and thus increasing the D/H ratio of the atmosphere. Measurements of D/H ratios are often quoted in comparison to Earth's standard mean ocean water (SMOW); i.e., a planetary D/H of 100× SMOW would indicate a planet's D/H ratio is 100× higher than Earth's, possibly implying past atmospheric loss to cause this enrichment. Interestingly, the D/H ratio of SMOW is close to that of chondritic water, which may suggest the origin of some of Earth's water may be chondritic asteroids (*Alexander et al.,* 2012).

The Pioneer Venus Mass Spectrometer directly sampled the venusian atmosphere and its results yielded venusian D/H = 0.016 ± 0.002 (*Donahue et al.,* 1982), or 157 ± 30× terrestrial SMOW (*Donahue et al.,* 1982, 1997), although interpretation of these measurements is somewhat complicated by the clogging of the mass spectrometer inlet. Thus, better measurements of the bulk D/H ratio of the Venus atmosphere will aid future studies of its potential early water inventory and loss processes. Complementary remote-sensing measurements from the nightside infrared spectral windows suggest D/H is ~127× Earth's SMOW in the Venus near-surface environment (*Bézard et al.,* 2011).

Despite the desiccated conditions of Venus today, the D/H ratio suggests that early Venus may have been more water rich. Currently, the amount of H_2O on Venus is 6×10^{15} kg, which is considerably smaller than Earth's 1.4×10^{21} kg (*Lecuyer et al.,* 2000). Below the clouds, the atmosphere contains only about 30 ppmv of water vapor (*Meadows and Crisp,* 1996; *Bézard et al.,* 2011; *Chamberlain et al.,* 2013; *Arney et al.,* 2014), and higher layers are even drier (Fig. 1). Estimates from D/H suggest early Venus may have had 4 m to 525 m of liquid water if spread evenly over the surface (e.g., *Donahue and Russell,* 1997), although other estimates suggest Venus formed with less water (*Raymond et al.,* 2006). An alternative explanation for the elevated D/H ratio in the Venus atmosphere is that it may reflect steady-state evolution where fractionating water loss is balanced by cometary or volcanic input fluxes (*Grinspoon,* 1993; *Donahue and Hodges,* 1992).

Even at the start of its history, the current orbit of Venus would have been closer to the Sun than the inner edge of the "conservative" HZ in the solar system (*Kopparapu et al.,* 2013), so the potential existence of an early liquid water ocean is challenging to reconcile in light of this. However, highly reflective water clouds may have helped maintain habitable conditions on Venus for hundreds of millions to billions of years after the planet's formation. The importance of high-albedo water clouds for maintaining habitable conditions on Venus was shown by *Grinspoon and Bullock* (2007), who found that if the atmosphere of young Venus was cloud-free, it would have only been able to stave off catastrophic loss of habitability for roughly 500 m.y. after its formation. With 50% cloud coverage, climate stability can be maintained for up to 2 b.y. after formation.

Recent three-dimensional atmospheric modeling efforts have helped illuminate mechanisms that could have supported a habitable early Venus enabled by these cloud effects. On slowly rotating planets, atmospheric circulation patterns generate strong rising motion on the dayside, which can lead to thick substellar cloud decks. *Way et al.* (2016) show that for Venus' slow rotation period of 116 days, atmospheric convection patterns would have generated thick water vapor clouds near the substellar point that may have produced substantial surface cooling by increasing the planet's dayside albedo (*Yang et al.,* 2013; *Kopparapu et al.,* 2017). Whether Venus' early rotation rate was as slow as its current rate is debatable, as it may have required atmospheric tides in a current Venus mass atmosphere to slow it to its present rotation rate (*Dobrovolskis and Ingersoll,* 1980). However, more recent work suggests that even a 1-bar atmosphere can create this tidally-induced rotation slowing (*Leconte et al.,* 2015). *Way et al.* (2016) show that for a modern Earth-like atmosphere, rotation rates slower than 16× modern Earth's rotation rate can produce habitable conditions for early Venus, and at its present-day rotation rate, habitable surface conditions are possible for up to at least 0.715 b.y. ago through the substellar cloud cooling effect. This means that Venus could have been habitable up to the time of its purported global resurfacing event. These simulations even show that snowfall could have occurred on the Venus nightside billions of years ago.

3.2. Loss of Habitability

Even if early Venus was habitable through mechanisms aided by its slow rotation rate, thick clouds, or other processes, it could not escape its inevitable loss of habitability. All main-sequence stars steadily brighten as they age, pushing the boundaries of their HZs ever outward. As the Sun inexorably increased in luminosity, habitable conditions on Venus eventually could not be supported.

How did Venus go from a potentially clement world to our scalding planetary neighbor? As temperatures increased on Venus, driven by the brightening solar luminosity, increasing amounts of water would have evaporated into the atmosphere. Water vapor is a greenhouse gas, so this would have created a positive feedback loop: Higher temperatures caused the evaporation of more water, which in turn drove temperatures even higher, which in turn caused more water to evaporate, and so on (*Ingersoll,* 1969).

A runaway greenhouse occurs when a planet's atmosphere becomes saturated with water vapor, which limits outgoing radiation and heats the planetary surface. As increasing quantities of greenhouse gases like water vapor are added to a planetary atmosphere, the outgoing thermal radiation emitted to space (also known as "outgoing longwave radiation," or OLR) for a given surface temperature will decrease. Thus, to balance the radiation budget of the atmosphere, the surface temperature must increase. There is a limit to the maximum amount of OLR a planet can radiate to space in a moist atmosphere. When the atmosphere is saturated with water, so that atmospheric spectral windows through which energy can be radiated to space become opaque, a maximally efficient greenhouse known as the "runaway greenhouse" is achieved. This radiation limit for the troposphere was found as early as 1927 (*Simpson,* 1928) with subsequent work by *Nakajima et al.* (1992). A stratospheric radiation limit, known as the Komyabashi-Ingersoll limit (*Kombayashi,* 1967; *Ingersoll,* 1969) has been studied as the maximum OLR a planet can radiate to space. These radiation limits have been reviewed in *Goldblatt et al.* (2012). The limit on outgoing thermal radiation has been estimated at 310 W m^{-2} (*Kasting,* 1988), although a study using more modern spectral line lists suggest a lower radiation limit of 282 W m^{-2} (*Goldblatt et al.,* 2013). In principle, a planet that has undergone a runaway greenhouse can overcome this OLR limit by heating the surface up to such extreme temperatures that it is able to radiate in the near-infrared and even visible part of the spectrum, where water vapor is not a good greenhouse gas and radiation can again escape to space.

In practice, before a planet enters the runaway greenhouse regime, it can enter a state at lower temperatures called the "moist greenhouse" (*Kasting,* 1988). On Earth, most of the water vapor in our atmosphere is "cold trapped" in the troposphere. The top of the troposphere, the tropopause, marks the region where water vapor condenses into clouds and rains back out to the surface. On Earth, this occurs at a pressure level of roughly 0.1 bar (e.g., *Robinson and Catling,* 2014), or 9–20 km in altitude, depending on latitude and season. Layers of the atmosphere above the tropopause are comparatively dry. However, as planetary surface temperatures increase and the atmosphere grows more moist with evaporated water vapor, the cold trap will lift to higher altitudes. As this occurs, the stratosphere of the planet can become "moist" with orders of magnitude more water vapor than these altitudes would normally contain when the cold trap was lower.

The moist stratosphere present in the moist and runaway greenhouses can result in water loss and D/H fractionation. The timescale for water loss approaches the age of Earth when the stratospheric water vapor mixing ratio is 3 × 10^{-3}, which occurs at a surface temperature of 340 K (*Koppar-*

apu et al., 2013). D/H fractionation only occurs efficiently for atoms transported to the exobase of the atmosphere; otherwise, relatively nonfractionating hydrodynamic escape processes will operate. Over time, as CO_2 liberated from rocks and volcanic outgassing entered the atmosphere, this more massive atmosphere could transport deuterium and hydrogen to higher altitudes where fractionating thermal and nonthermal escape processes could occur. In this scenario, this fractionating escape occurred only during the loss of the final 4 bar of H_2O (*Kasting,* 1988), so the D/H ratio remnant that is detectable today may not reflect the total initial water inventory. Kasting concluded that Venus could have maintained its oceans for 600 m.y. after its formation before losing them to the moist greenhouse, but did not include the effect of clouds that lengthen the interval of possible habitability (section 3.1).

While free hydrogen can be easily lost to space, the fate of the liberated oxygen atoms from water photolysis is more ambiguous. The current Venus atmosphere does not host a significant quantity of free oxygen (section 4.3). The oxygen produced by water photolysis may have been lost to reactions with crustal and mantle materials. *Chassefière* (1996) calculated that Venus' surface and mantle could have absorbed O_2 for initial water inventories up to 0.45 of an Earth ocean. Loss of a larger ocean may have left O_2 behind in the atmosphere (section 4.3), although surface and mantle sequestration of O_2 is still not well understood (*Rosenqvist and Chassefière,* 1995). Unfortunately, because much of the surface of Venus may be less than a billion years old, geological evidence of oxidation processes has likely been lost. At present, the modern Venus crust does not appear to be as oxidized as Mars (*Pieters et al.,* 1986).

In addition to surface and mantle sequestration of O_2, some fraction of the oxygen produced by water photolysis may have been lost to space. A strong magnetic field on Venus could have dramatically slowed loss of water to space in the planet's past (*Driscoll and Berkovici,* 2013). However, it does not currently have a global magnetic field, and gaseous species are still escaping despite having an induced magnetic field generated by solar wind and ionosphere interactions (*Luhmann et al.,* 2004). The Analyzer for Space Plasmas and Energetic Atoms (ASPERA-4) instrument onboard VEx has showed that dominant ions escaping from the modern atmosphere include H$^+$, He$^+$, and O$^+$. Stoichiometrically, water is still being lost from Venus: The H$^+$/O$^+$ escape rate ratio measured by ASPERA-4 is 2:1 (*Barabash et al.,* 2007).

There is evidence that atmospheric loss processes appear to be more efficient on Venus when compared to Earth. Analysis of VEx measurements have shown that Venus has an "electric wind" that can efficiently strip even heavy ions like oxygen to space (*Collinson et al.,* 2016). In a planetary ionosphere containing electrons and ions, the lighter electrons can be more easily stripped to space as they move along magnetic field lines. When electrons move outward from a planet, the heavier ions pull on them through Coulomb forces, the result being an "ambipolar" electric field or "electric wind" that slows the movement of the electrons and accelerates the ions outward toward the electrons. Surprisingly, the "electric wind" surrounding Venus is 5× stronger than the field in Earth's ionosphere. It is unclear why the Venus field is unexpectedly strong, but it may be because Venus is closer to the Sun than Earth and thus is buffeted by larger amounts of solar ionizing radiation. This "electric wind" appears to be persistent, stable, and global in extent, and it can carry even heavy O^+ ions away to space (*Collinson et al.,* 2016). This may have implications for atmospheric loss processes on exoplanets orbiting closely to energetic stars as discussed in section 4.3.

3.3. Noble Gases as Records of the Past

Aspects of Venus' evolutionary history can be revealed by noble gas measurements. Noble gases are useful for constraining early states of planetary environments since they do not react with other gases or surface materials, so they can record the early history of terrestrial planet evolution and formation (*Zahnle,* 1993). Such gases include helium (He), xenon (Xe), krypton (Kr), and argon (Ar).

Noble gases may help to distinguish between the catastrophic and equilibrium resurfacing models invoked to explain the crater distribution on Venus by placing constraints on Venus' history of volcanic outgassing (section 2.3). Isotopes of noble gases including ^{40}Ar, ^4He, ^3He, ^{129}Xe, and ^{136}Xe are incompatible with crustal minerals and so will escape into to the atmosphere during volcanic events. Radiogenic ^{40}Ar and ^4He are created by the decay of K, U, and Th. Helium, being lightweight, escapes to space easily with a lifetime of hundreds of millions of years (*Prather and McElroy,* 1983), and therefore can be used to constrain the rate of recent volcanic outgassing events. Meanwhile, argon's high atomic mass means ^{40}Ar will tend to remain in an atmosphere without escaping. Measurements of ^{40}Ar may therefore constrain the long outgassing history of a planet. Venus has 25% less ^{40}Ar than Earth (*Turcotte and Schubert,* 1988). Because ^{40}Ar is created by the decay of ^{40}K, this could imply Venus has less K than Earth, or that it has outgassed less, possibly because of a significant decrease in the rate of degassing 1 b.y. after formation (*Baines et al.,* 2013). The decay of ^{129}I and ^{244}Pu generate ^{129}Xe and ^{136}Xe, and these isotopes could also be used to constrain mantle-degassing rates.

Noble gases might reveal massive impact events in the history of Venus and energetic atmospheric processing. For instance, the initial Ar inventory of all of the terrestrial planets is thought to be the same as the Sun, but modern Earth and Mars have smaller $^{36}Ar/^{38}Ar$ ratios than the Sun, indicating atmospheric loss (*Atreya et al.,* 2013). Existing constraints on venusian $^{36}Ar/^{38}Ar$ are insufficient to distinguish between primordial and processed states. Meanwhile, Xe isotopes observed on Mars and Earth exhibit strong mass fractionation, which may be evidence of significant atmospheric erosion processes (e.g., impacts) that occurred during these planets' formation and early histories (*Pepin,* 1991; *Zahnle,* 1993). An upper limit for Xe on Venus was measured by Pioneer

Venus (120 ppb) (*Donahue et al.,* 1981), but this is insufficient to shed light on early energetic processes. If Venus has a similar Xe fractionation to Earth and Mars, this could suggest that all three bodies were delivered volatiles by the same type of source. Radiogenic isotopes of ^{129}Xe on Earth can be used to date the age of its atmosphere to 95 m.y. after the formation of the solar system, which is interpreted to be when the Moon-forming impact occurred (*Tolstikhin et al.,* 2014). Measurements of ^{129}Xe on Venus might similarly indicate whether such a large impact occurred in its early history. Additionally, Pioneer Venus measurements suggest that Venus has an unusually high amount of Ne, ^{36}Ar, and ^{38}Ar (*Pepin,* 1991), which may suggest a large impact event. For example, a large planetesimal endowed with noble gases from the solar wind could have impacted Venus and affected its abundances of these noble gases (*McElroy and Prather,* 1981). The Ar and Ne observed on Venus might also have been delivered by a large comet (*Owen et al.,* 1992).

3.4. The Future of Earth

Venus may serve as a magic mirror that reflects the future of Earth's climate evolution back to forward-thinking observers. As discussed, all main-sequence stars brighten slowly over their lifetimes, and the inner edge of the solar system's HZ will someday sweep past Earth. The solar luminosity as a function of time can be estimated with the following relation (*Gough,* 1981).

$$L_t = \left[1 + \frac{2}{5} \left(1 - \frac{t}{t_0} \right) \right]^{-1} L_0$$

Here, L_0 is the solar constant in units of the modern Sun luminosity ($L_0 = 1$), and t_0 is the age of the solar system today, i.e., 4.6 b.y. Extrapolating into the future, *Wolf and Toon* (2015) find that stable climates on Earth can be maintained for solar insolations <21% greater than today and surface temperatures <368.2 K. Climate modeling by Wolf and Toon suggests these conditions will be met about 2 b.y. into the future, although the planet would become uninhabitable to human life before ocean loss is triggered due to extreme hot temperatures. Along with water loss and surface heating, a shutdown of plate tectonics should occur in Earth's future. Subsequent volcanic outgassing of CO_2 and SO_2, coupled to photochemical processes that generate H_2SO_4 clouds, could lead to a state closely resembling Venus on Earth.

4. VENUS: THE EXOPLANET LABORATORY NEXT DOOR

Venus-like exoplanets may be one of the most common types of terrestrial worlds, and our current and near-future detection and characterization techniques are biased toward detecting exo-Venus analogs over exo-Earth analogs. Because of this detection bias, much of the modeling work that has yet been undertaken on exoplanet environments has focused on understanding hot worlds (e.g., *Schaefer and Fegley,* 2009, 2011; *Leger et al.,* 2009; *Miguel et al.,* 2011; *Treiman and Bullock,* 2012). Very hot terrestrial planets (i.e., $T_{surf} > 1500$ K) are likely to be in a regime even beyond Venus, with atmospheres composed of exotic compounds including SiO, and clouds made of K and Na compounds (*Schaefer and Fegley,* 2009). Hot planets in a less-extreme temperature regime may be more like Venus, with CO_2 atmospheres and other compounds that may include N_2, H_2O, SO_2, HCl, HF, OCS, and CO (*Schaefer and Fegley,* 2011; *Treiman and Bullock,* 2012), all of which occur in the atmosphere of Venus. A review of Venus in the context of exoplanets can be found in *Kane et al.* (2019).

4.1. Exo-Venus Analogs

Recent results from exoplanet demographics studies suggest that Venus-like planets might be relatively common, and possibly more common than Earth-like worlds. Using estimates of the inner and outer edges of the habitable zone (HZ) (*Kasting et al.,* 1993; *Kopparapu et al.,* 2013, 2014) and the detection of Earth-sized, close-in planets, we have been able to estimate the fraction of stars with rocky planets within the HZ (referred to as η_{Earth}). Recent calculations of η_{Earth} allow us to predict how many potentially Earth-like planets we may detect with future observatories. Many of these recent calculations of η_{Earth} use Kepler statistics to provide a sufficiently large sample to perform a meaningful statistical analysis (*Dressing and Charbonneau,* 2013, 2015; *Kopparapu,* 2013; *Petigura et al.,* 2013; *Burke et al.,* 2015). (Kepler was a spacebased mission searching for exoplanets transiting their host stars.) Since Kepler has preferentially detected planets interior to the HZ, these worlds are therefore

more likely to be potential Venus analogs than Earth analogs.

Kane et al. (2014) defined the "Venus zone" (VZ) as a target selection tool to identify terrestrial planets whose atmospheres could be pushed into a runaway or moist greenhouse, producing environmental conditions similar to those on Venus. Figure 3 shows the VZ and HZ for stars of different temperatures. The outer boundary of the VZ is demarcated by the "runaway greenhouse" line, which is calculated using climate models of Earth's atmosphere. The inner boundary (dashed line) is estimated based on where the stellar radiation from the star would cause complete atmospheric erosion. This inner boundary of the VZ is empirically derived from solar system observations of atmospheric density as a function of incident flux and surface gravity (*Catling and Zahnle,* 2009, 2013), referred to as the "cosmic shoreline" (*Zahnle and Catling,* 2013). The images of Venus shown in this region represent planet candidates detected by Kepler that may be Venus-analogs. *Kane et al.* (2014) calculated an occurrence rate of VZ terrestrial planets as $32^{+0.05}_{-0.07}$% for low-mass stars and $45^{+0.06}_{-0.09}$% for Sun-like stars. Note however that, like the HZ, the boundaries of the VZ should be considered a testable hypothesis of star-planet separation and other characteristics that lead to a Venus-like runaway state. Detection and analysis of exoplanet atmospheric properties will aid in determining if runaway greenhouse conditions could occur beyond the calculated VZ boundary (*Foley,* 2015; *Luger and Barnes,* 2015) (section 4.2).

The prevalence of Venus analogs will become increasingly relevant in the era of forthcoming exoplanet missions. The TESS mission is expected to detect hundreds of terrestrial planets orbiting bright host stars (*Sullivan et al.,* 2015; *Barclay et al.,* 2018), many of which will lie within their stars' VZ. Hot terrestrial planets are already being discovered by TESS (*Dragomir et al.,* 2019; *Kostov et al.,* 2019; *Vanderspek et al.,* 2019). The PLAnetary Transits and Oscillations

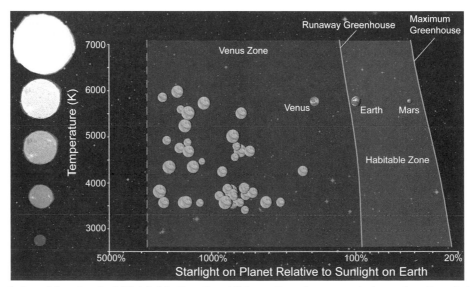

Fig. 3. The boundaries of the classical "Habitable Zone" and "Venus Zone," represented as a function of stellar temperature and insolation flux relative to Earth. Shown on the plot are the location of the solar system planets (Venus, Earth, Mars) and the potential Venus analogs identified by *Kane et al.* (2014). Credit: Chester Harman; planet images from NASA/JPL.

of stars (PLATO) mission will add further to the inventory of candidate Venus analogs orbiting bright stars and potentially extend to longer orbital period sensitivity than TESS (*Rauer et al.,* 2016). These new discoveries will provide key opportunities for transmission spectroscopy follow-up observations using the James Webb Space Telescope (JWST) (*Morley et al.,* 2017; *Lincowski et al.,* 2018), among other facilities, such as the Atmospheric Remote-sensing Exoplanet Large-survey (ARIEL) mission (*Zingales et al.,* 2018). Observationally confirming the division between Earths and Venuses will allow us to place empirical constraints on the interior edge of the HZ. Quantifying the actual occurrence rate of true Venus analogs (η_{Venus}) will help us to decode why the atmosphere of Venus so radically diverged from its sister planet, Earth, and will help constrain the frequency of these evolutionary outcomes.

We may already have detected several of Venus' sisters in the exoplanet population. These worlds include terrestrial-sized planets such as TRAPPIST-1b and c (*Gillon et al.,* 2016, 2017), which receive 4.3× and 2.3× Earth's insolation; Kepler-1649b (*Angelo et al.,* 2017), which receives 2.3× Earth's insolation; GJ 1132b (*Berta-Thompson et al.,* 2015), which receives 19× Earth's insolation; Ross 128b (*Bonfils et al.,* 2017), which receives 1.38× Earth's insolation; and L 98-59 b/c/d, which receive 23.9×/12.4×/4.85× Earth's insolation (*Kostov et al.,* 2019).

True Venus analogs (i.e., with a thick Venus-like cloud layer) would display relatively featureless spectra at many wavelengths, both in transit transmission and direct imaging (*Morley et al.,* 2017; *Meadows et al.,* 2018a; *Lincowski et al.,* 2018). So, while these worlds may be cosmically ubiquitous, they may be challenging to observe and interpret. This underscores the importance of allowing Venus to guide our understanding of exo-Venus worlds. On the other hand, since the presence of Venus' clouds requires ongoing volcanic outgassing processes, trace amounts of water vapor (for H_2SO_4 photochemistry), and temperatures conducive to H_2SO_4 condensation, Venus-like worlds with less volcanic activity, that are more desiccated, or that are simply too hot, may host no or thinner cloud layers (*Lincowski et al.,* 2018). Figure 4 shows the spectrum of a planet exactly like Venus for reflected, transmitted, and thermally emitted light orbiting an M dwarf the size of TRAPPIST-1 (stellar radius of 0.12 R_{Sun}).

In reflected light (Fig. 4a), absorption features can be discerned from CO_2, and for wavelengths <0.5 μm, the unknown UV absorber is prominent, darkening the spectrum. At wavelengths >2.5 μm, H_2SO_4 becomes significantly absorptive (Fig. 2), causing the spectral falloff in the reflected light spectrum. Although water vapor is a trace species in the Venus atmosphere, it is present throughout the atmospheric column, and water vapor features can be seen in the reflected light spectrum.

In transit transmission, the cloud deck truncates the strength of gaseous absorption features, but it may still be possible to observe CO_2 features. H_2SO_4 itself also produces absorption features that might be observed in transmission observations, notably near 10 μm. These features might be detectable with JWST, although prelaunch estimates suggest JWST may encounter a noise floor of 20 ppm in the near-infrared and 50 ppm for mid-infrared instruments (*Greene et al.,* 2016). The actual detectability of exoplanet spectral features with JWST will not be known until after telescope commissioning and early observations are completed. Fortunately, since planets orbiting smaller stars like TRAPPIST-1 will produce larger transit depths, some features may exceed 100 ppm (*Morley et al.,* 2017; *Lincowski et al.,* 2018) for CO_2-dominated, Venus-like but clear atmospheres. Clouds will decrease the strength of absorption features, but strong features (e.g., the CO_2 bands near 4 and 15 μm) might still be detectable on exo-Venuses. Because observations with JWST will be costly in terms of observing time, careful target selection is absolutely critical (*Morgan et al.,* 2018).

Because clouds dramatically impact the spectrum of Venus — including producing spectral features of their own (Fig. 4) — it will be important to better understand and predict the potential for clouds to form in exo-Venus atmospheres to anticipate their impact on observations. Because the Venus clouds consist of H_2SO_4, which is generated in Venus' atmosphere by photochemistry, the UV spectrum of the star will impact how much H_2SO_4, and therefore clouds, are generated on Venus-like exoplanets. Possibly, planets around stars with higher UV outputs may develop thicker cloud decks (*Schafer and Fegley,* 2011), which will have important implications for the detectability of spectral features on exo-Venus analogs since thicker clouds, and higher cloud decks, lead to diminished gaseous spectral features. Furthermore, some exoplanets may be too hot to support Venus-like cloud decks: *Lincowski et al.* (2018) suggest that a Venus-like TRAPPIST-1b may be too hot to condense a H_2SO_4 cloud layer.

Venus-like planets may also be observable through thermal phase curves (*Morley et al.,* 2017; *Meadows et al.,* 2018a). Such observations may be possible through photometry using the JWST Mid Infrared Instrument (MIRI). The planet-star contrast ratio increases at longer mid-infrared wavelengths, making the planet more observable. Observations of the 15-μm CO_2 band at thermal wavelengths (Fig. 4c) may provide information that could help to discriminate between different types of planetary environments, and also act as a discriminant between planets with atmospheres and those without (*Meadows et al.,* 2018a). If the planet's emitted flux at 15 μm is lower than the expected thermal emission continuum level based on other wavelengths, this is evidence for CO_2.

4.2. Exo-Venus Analogs in the Habitable Zone

In addition to exoplanets in the VZ, some exoplanets in their stars' HZs may in fact be more Venus-like than Earth-like for worlds orbiting M dwarfs. Planets in the HZs of M dwarfs may experience significantly different evolutionary histories compared to planets orbiting solar-type stars. M dwarfs, especially younger stars, are highly active, producing frequent flares (*West et al.,* 2015; *MacGregor et al.,* 2018) and high amounts of X-ray radiation that can erode atmospheres significantly (e.g., *Owen and Mohanty,* 2016; *Airapetian et al.,* 2017; *Garcia-Sage et al.,* 2017; *Dong et al.,* 2017). M dwarfs are also inherently dim compared to solar-type stars, and planets must huddle close to them for warmth in the HZ, making them vulnerable to this extreme activity and strong tidal forces (*Barnes et al.,* 2013).

Even before planets orbiting M dwarfs can settle into their long lifetimes in the HZ, they must contend with their stars' extended superluminous pre-main-sequence phase caused by longer Kelvin-Helmholtz contraction timescales for lower-mass stars (e.g., *Baraffe et al.,* 1998; *Reid and Hawley,* 2013; *Dotter et al.,* 2008). A 1-$M_\odot$ star contracts to the main sequence in <50 m.y. (*Baraffe et al.,* 1998), while M dwarfs can spend hundreds of millions of years to up to a

billion years contracting. During this phase, stars can be up to 2 orders of magnitude more luminous than their eventual main-sequence brightness, so planets that orbit within the ultimate HZ of M dwarfs may be well interior to it for an extended period of time. Studies suggest that rocky planets form in 10—100 m.y. (*Chambers,* 2004; *Raymond et al.,* 2007, 2013; *Kleine et al.,* 2009), so planets in M-dwarf HZs are subjected to extreme luminosities early on for an extended period of time.

This lengthy extreme luminosity could cause these planets to enter runaway greenhouse states before their stars stop contracting and the HZ boundaries settle to their main-sequence-phase positions. Therefore, planets orbiting in the HZs of M dwarfs may actually be desiccated exo-Venus worlds in disguise, despite orbital parameters suggesting potential habitability (*Luger and Barnes,* 2015; *Ramirez and Kaltenegger,*

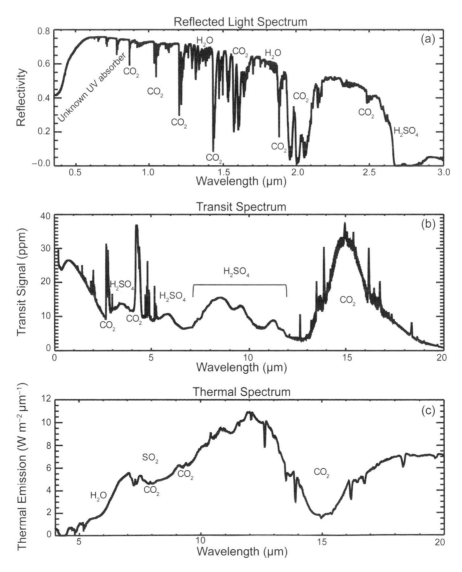

Fig. 4. (a) The reflected light spectrum, **(b)** a transit transmission spectrum of Venus orbiting the M dwarf TRAPPIST-1 at the orbital separation of TRAPPIST-1c, and **(c)** thermal emission from the same planet. This figure does not represent a true Venus-like exoplanet, which may have a different composition due to evolution and/or photochemistry; rather, it is provided as an example of typical Venus spectral features accessible to different observing techniques. For transmission and emission spectra of more self-consistent TRAPPIST-1 exo-Venuses, see *Lincowski et al.* (2018).

2014; *Meadows et al.,* 2018a; *Lincowski et al.,* 2018). Models show that planets enduring the extended superluminous pre-main-sequence phase of an M dwarf could lose several Earth ocean equivalents of water through evaporation and hydrodynamic escape (*Bolmont et al.,* 2016). Therefore, the "Venus zone" may extend into, and possibly beyond, the HZ for some M dwarfs (*Lincowski et al.,* 2018). Because the length of the pre-main-sequence phase scales inversely with stellar mass, planets in the HZs of higher-mass M dwarfs are less likely to become exo-Venus planets than worlds orbiting lower-mass M dwarfs. Some of these potential exo-Venus HZ planets include TRAPPIST-1d–g (M8V), Proxima Centauri b (M5.5V), and LHS 1140b (M4.5V). Others will surely be discovered by TESS and other facilities.

Planets subjected to the superluminous pre-main-sequence phase could still be habitable. It is possible that habitability could be restored to planets desiccated by the pre-main-sequence phase if late volatile delivery occurs (*Morbidelli et al.,* 2000). Alternatively, planets could migrate into the HZ from farther out in the system where solid ices could endow them with abundant volatiles, and complete desiccation might be averted. Indeed, the resonant orbits of the planets in the TRAPPIST-1 system suggests that radial migration within a gaseous disk occurred (*Luger et al.,* 2017). The migration timescale for this system was likely much shorter than the pre-main-sequence phase of TRAPPIST-1 (up to 1 b.y.) (*Bolmont et al.,* 2016), so a "late" migration that might avoid being close to the star during the damaging pre-main-sequence is unlikely. But if the TRAPPIST-1 planets are endowed with a significant volatile fraction acquired at their more distant formation orbits (*Unterborn et al.,* 2018), this may have allowed them to avoid total desiccation.

In any case, the extended pre-main-sequence phase for M dwarfs suggests that Venus-like states may be possible for HZ planets, depending on initial planetary volatile inventory (provided these planets can retain atmospheres despite the high activity levels typical of M dwarf stars). The types of scenarios that can "save" these planets from desiccation and how frequently they occur are just beginning to be explored, and observations of HZ exoplanets orbiting M dwarfs should provide valuable insights.

Exo-Venus planets may also shed light on processes that occurred on Venus in its past. As discussed in section 3.1, Venus' slow rotation rate may have enabled habitability for an extended period of time by driving the formation of a thick layer of brightly reflective water clouds on the dayside. Beyond the solar system, slow rotation may be likely for synchronously-rotating exoplanets within or closer to the stars than the HZs of M-dwarf stars. Similar thick substellar cloud decks may affect the potential habitability of these worlds, cooling their surface temperatures and possibly allowing for habitable conditions interior to the inner edge of the traditional HZ (*Kopparapu et al.,* 2017; *Fujii et al.,* 2017). Such planets may even be in "habitable moist greenhouse" states with moist stratospheres gradually losing water to space, but with habitable surface temperatures. If water loss is slow, habitable conditions may exist on these planets

for a significant period of time (e.g., hundreds of millions to several billion years) (*Kopparapu et al.,* 2017). By studying the history of Venus, we can gain a better understanding of how these processes may operate elsewhere. Likewise, by observing slowly-rotating exoplanets in the HZ, this can help us to understand the potential for this "habitable moist greenhouse" process to have occurred on Venus in the past. In light of the difficulty of obtaining data from the early history of Venus, these observations of analog exoplanets, especially at different ages, are particularly valuable for illuminating the possible history of Venus (section 2.3).

4.3. Exo-Venus Analogs and False Positive Oxygen

One consequence of the extreme water loss that planets endure during a runaway or moist greenhouse could be the accumulation of significant quantities of O_2 in planets' atmospheres (*Luger and Barnes,* 2015; *Tian,* 2015). Depending on the initial water inventory, hundreds or thousands of bars of O_2 could be generated. Photolysis and hydrogen loss from one Earth ocean would produce about 240 bar of O_2 (*Kasting,* 1997). The amount of O_2 that actually remains in the atmosphere would be a function of a number of planetary and stellar characteristics and processes including stellar activity, the original quantity of water, the mass of the planet, oxygen sinks and loss processes, and the planet-star distance.

Oxygen is one of the longest-standing exoplanet biosignatures (e.g., *Hitchcock and Lovelock,* 1967; *Meadows,* 2017; *Meadows et al.,* 2018b), so understanding the nature and likelihood of planetary mechanisms for abiotic generation of O_2 (false positives) is critical to interpreting whether observed O_2 is biological. Any "false positive" processes that generate O_2 abiotically are important to consider when planning for future observations that are designed to search for signs of life on potentially habitable exoplanets. In light of this, several previous studies have discussed O_2 biosignature false positive mechanisms (*Hu et al.,* 2012; *Domagal-Goldman et al.,* 2014; *Tian et al.,* 2014; *Gao et al.,* 2015; *Harman et al.,* 2015; *Schwieterman et al.,* 2016; *Wordsworth and Pierrehumbert,* 2014; *Schaefer et al.,* 2016; *Meadows,* 2017; *Harman et al.,* 2018).

One of the appealing features of O_2 as a biosignature is that atmospheres with abundant oxygen produce spectral features from the UV to the mid-infrared from O_2 itself, from its photochemical→ byproduct ozone (O_3), and from collisional pairs of O_2 molecules (O_2-O_2 or O_4). Collisionally-induced O_4 features are density-dependent, and appear most strongly in atmospheres with significant quantities of O_2. O_2 itself produces strong spectral features in the visible and near-infrared (NIR) at 0.628 μm (the "B band"), 0.762 μm (the "A band"), and the $\alpha^1\Delta_g$ band near 1.27 μm (the designations "B band" and "A band" were first given by Joseph von Fraunhofer in his pioneering observations of the spectrum of sunlight passing through Earth's atmosphere) (e.g., *Thomas,* 1991). Of these features, the "A band" is the most prominent (*Rothman et al.,* 2013). Ozone produces strong absorption in the UV (0.2–0.350 μm, the "Hartley-

Huggins band"), the visible (0.5–0.7 μm, the "Chappuis band"), and the mid-infrared (9.6 μm). Of these, the UV "Hartley-Huggins band" is the strongest and detectable for the lowest quantities of O_2 and O_3. Prominent O_4 features are produced between 0.3 and 0.7 μm (*Hermans et al.,* 1999; *Thalman and Volkamer,* 2013), at 1.06 μm (*Greenblatt et al.,* 1990), and at 1.269 μm (*Maté et al.,* 1999) (coinciding with the $\alpha^1 \Delta_g$ feature).

The type of massive O_2 accumulation that could occur during the loss of oceans' worth of water should reveal itself as a biosignature false positive through the appearance of O_4 features because these collisionally-induced absorption features only become prominent when a substantial amount of O_2 is present (i.e., several bars). In transit transmission observations, for planets with up to 10 bar of pressure but with small O_2 atmospheric fractions (<20%), O_4 features are challenging to discern (*Misra et al.,* 2014a). However, for an ~1-bar atmosphere dominated by O_2, the NIR O_4 features are stronger (*Schwieterman et al.,* 2016). For the extreme case of a 100-bar O_2 atmosphere (which could result from the loss of 0.4× Earth's ocean volume), the NIR O_4 features become stronger than the O_2 features themselves, a clear spectral indicator that implies the abiotic nature of the oxygen (*Schwieterman et al.,* 2016; *Meadows et al.,* 2018b). In direct imaging, the O_4 features at both visible and NIR wavelengths can be detected. In transit transmission, Rayleigh scattering tends to mask visible wavelength O_4 features due to the longer path lengths through the atmosphere.

As one case study of a potential extreme kind of exo-Venus, *Schaefer et al.* (2016) modeled ocean loss for GJ 1132b in the presence of a magma ocean. This planet is much closer to its star than the HZ and receives considerable more stellar flux than Venus (roughly 19× Earth's insolation, or 9.5× Venus' insolation). Although this planet's temperature regime is even higher than Venus, it is still instructive to consider the possibility of abiotic O_2 on it as an extreme scenario. Schaefer et al. suggest that the most likely outcome of water loss on GJ 1132b is a "thin" (i.e., pressure less than a few bar) O_2 atmosphere, although large initial water abundances could produce atmospheres with thousands of bar of O_2.

Besides the superluminous pre-main-sequence phase driving massive water loss and the potential subsequent generation of O_2, other oxygen false positive scenarios have been proposed. These include the following scenarios, all of which can be better understood when considered in the context of Venus atmospheric processes.

Low non-condensable gas inventories (e.g., low p_{N_2}) can generate false positive O_2. This mechanism permits enhanced water vapor abundance at high altitudes, because in atmospheres with low non-condensable gas inventories, the atmospheric "cold trap" that would otherwise prevent water from reaching upper layers is less effective (*Wordsworth and Pierrehumbert,* 2014). As we have discussed in the context of Venus' water loss, water vapor in the upper atmosphere is vulnerable to photolysis followed by loss. If the O_2 generated by H_2O photolysis is not itself lost, it could produce detectable false positive spectral signatures. For a planet with modern Earth's surface temperature, reducing the atmospheric partial pressure of N_2 by an order of magnitude could allow for a moist upper atmosphere, and photochemical production of abiotic O_2 could follow (*Wordsworth and Pierrehumbert,* 2014). There is evidence that the pressure of Earth's atmosphere may have been as low as ±0.23 bar at 2.7 b.y. ago (*Som et al.,* 2016), and analysis of 2.7-b.y.-old meteorites may suggest abundant stratospheric oxygen at this time (*Tomkins et al.,* 2016).

Abiotic oxygen can also be produced by other photochemical processes. Carbon dioxide photolysis can produce oxygen.

$$CO_2 + h\nu \rightarrow CO + O$$

This also occurs on Venus. Direct recombination of CO and O back into CO_2 is spin forbidden and proceeds very slowly. So, O produced by this mechanism could react to form O_2 and O_3 (*Harman et al.,* 2015; *Domagal-Goldman et al.,* 2014; *Tian et al.,* 2014; *Hu et al.,* 2012; *Gao et al.,* 2015). Therefore, atmospheres with abundant CO_2 and efficient CO_2 photolysis (which occurs via photons with wavelength <175 nm) may show signs of false positive oxygen. However, catalysts present in an atmosphere may allow the rapid recombination of CO and O back to CO_2, reducing the formation of abiotic O_2. Lightning-generated NO catalysts can drive the recombination of CO and O created by CO_2 photolysis (*Harman et al.,* 2018). Water vapor photolysis that generates OH and HOx chemistry can also enable this through several reactions (*Meadows,* 2017).

$$O + O_2 + M \rightarrow HO_2 + M$$
$$O + HO_2 \rightarrow OH + O_2$$
$$OH + CO \rightarrow CO_2 + H$$

The net result of this is $CO + O \rightarrow CO_2$. *Gao et al.* (2015) pointed out that extremely desiccated atmospheres (<1 ppm H_2O, less than Venus) can enhance abiotic oxygen accumulation by inhibiting this catalytic recombination to CO_2. A lack of H_2O features in combination with detectable oxygen and abundant CO_2 might suggest the production of O_2 from a highly desiccated planet (and also would indicate a planet that is uninhabited as water is considered fundamentally important for life). However, water vapor may be difficult to detect in transmission in general since water is normally cold-trapped in the troposphere on habitable planets, which may be inaccessible in transit observations due to the effects of refraction (*Bétrémieux and Kaltenegger,* 2013; *Misra et al.,* 2014b) and clouds. Detection of abundant CO might therefore be a better discriminant of this type of false positive scenario (*Gao et al.,* 2015; *Schwieterman et al.,* 2016).

Besides water vapor, oceans can also provide sinks to remove oxygen from the atmosphere: O_2 and CO in the ocean can reform CO_2 through aqueous chemistry (*Harman et al.,* 2015). However, stars that can efficiently photolyze CO_2 may still generate false positive O_2 even for planets with oceans if other O_2 sinks such as reduced volcanic gases (e.g., CH_4, H_2S) are minor components of the atmosphere. These types of false positive planets could be discriminated from planets with biological O_2 through the presence of strong CO_2 and CO features, and weak or nonexistent spectral features from reduced gases.

Venus's atmospheric loss processes and photochemistry have a vital role to play in guiding our understanding of the potential for the false positive O_2 scenarios discussed here to operate on exoplanets. Venus may have undergone massive water loss in its past, and currently undergoes rapid CO_2 photolysis, yet it does not exhibit an O_2-rich atmosphere today. Venus' "electric wind" has illuminated what may be a key process for highly-irradiated planets and that may diminish the prevalence of high-O_2 atmospheres on exoplanets (*Collinson et al.,* 2016) (section 3.2). If exoplanets have comparable or stronger "winds," they too could lose heavy ions like O^+ to space regardless of whether they possess a magnetic field. The unexpected strength of the venusian electric wind has been suggested to be due to Venus' closer proximity to the Sun than Earth, causing it to receive more ionizing radiation (*Collinson et al.,* 2016). Planets orbiting closely to M dwarfs may have even stronger "electric winds" when considering their proximity to their stars and the high levels of stellar activity typical for M dwarfs, diminishing their potential to retain oxygen ions. The potential for electric winds and other atmospheric loss processes to operate on exoplanets must be investigated in detail to assess their impact on atmospheric retention, habitability, and biosignature false positive scenarios.

On Venus today, atmospheric CO_2 is photolytically dissociated at very high rates (*McElroy et al.,* 1973), yet Venus does not have a significant amount of O_2 and O_3 in its atmosphere, which has significant implications for our understanding of O_2 false positive mechanisms. *Montmessin et al.* (2011) have reported small amounts of ozone (1.5×10^{-4} times Earth's ozone abundance) on the Venus nightside. Also on the nightside, recombination of excited O_2 produces spatially- and temporally-variable airglow (Fig. 5) that can be detected near 1.27 μm (*Connes et al.,* 1979; *Crisp et al.,* 1996). Free oxygen produced by CO_2 photolysis on the planet's dayside is transported rapidly to the planet's nightside and recombines as excited O_2 ($\alpha^1\Delta_g$) in the mesosphere. The excited O_2 quickly relaxes to the ground state and emits the 1.27-μm photons seen in observations. The airglow produced by O_2 emission on Venus is highly variable, both spatially and temporally. Despite the presence of O_2 airglow from excited oxygen, ground-state O_2 has never been detected on Venus, with upper limits suggesting <2 ppm above 60 km (*Trauger and Lunine,* 1983; *Mills,* 1999). The implication is that the oxygen formed by CO_2 photolysis is rapidly removed from the atmosphere. Chlorine catalysts have been suggested as a mechanism to scrub O_2 from Venus' atmosphere (*Yung and DeMore,* 1982); laboratory experiments (*Pernice et al.,* 2004) and measurements of venusian O_3 abundance (*Montmessin et al.,* 2011) suggest this mechanism is viable, but intermediate species involved have not been detected in Venus' atmosphere. Alternatively, or in addition to catalytic chlorine chemistry, heterogeneous aerosol chemistry may help to scrub O_2 from the venusian atmosphere (*Mills and Allen,* 2007).

The catalytic mechanisms that remove O_2 from the Venus atmosphere have not been included in studies examining the possibility of abiotic O_2 production on exoplanets as a biosignature false positive. This is in large part because these processes are still not well understood. Therefore, better understanding the processes involving O_2 in the venusian atmosphere are vitally important for understanding the viability of many proposed oxygen false positive mechanisms for exoplanets.

March 2, 2009

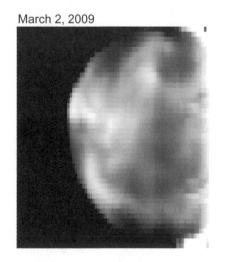

March 3, 2009

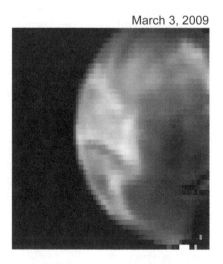

Fig. 5. Variable O_2 airglow at 1.27 μm can be seen illuminating the Venus nightside in these observations from March 2, 2009, and March 3, 2009, from the Apache Point Observatory 3.5-m telescope TripleSpec instrument. The dayside crescent (not shown) is to the right in these images.

5. CONCLUDING REMARKS

Venus stands as a stark reminder that habitability is not a static state that planets remain in throughout their entire lives. Habitability can be lost, and the runaway greenhouse is the catastrophic final resting place of once-watery worlds. Understanding habitability as a planetary process crucially depends on understanding what happened to the putative lost water of Venus. Although the fanciful humid swamps imagined by early researchers (*Arrhenius*, 1918) could not be farther from the desiccated reality of the current venusian environment, perhaps they contain a grain of truth about past Venus.

Venus is extremely valuable in the context of exoplanets. We can make models that seek to understand exoplanet atmospheres and environments more robust by validating them across the diverse and complex atmospheres that exist in the solar system (e.g., Venus, Earth, Mars, etc.). Given that the planetary atmosphere, surface, and interior models that we extrapolate and apply to exoplanets are based on *in situ* data acquired within our solar system, it is imperative that the planetary objects within reach are explored to the fullest extent. As a world of extremes in temperature and pressure, Venus is particularly useful for model validation across a range of conditions. And because exo-Venus planets may be cosmically ubiquitous, it is particularly important to better understand the world next door so that we may be able to improve our interpretation of future observations of analog worlds. Venus may even teach us something about detecting life on exoplanets by guiding our understanding of oxygen biosignature false positives. Likewise, just as Venus can teach us about exoplanets, studying hot exoplanets of varying ages and in varied astrophysical contexts may shed light on processes that have operated on Venus in the past.

The Nobel-prize-winning physicist and chemist Svante Arrhenius concluded his vision of Venus' environment in the book *The Destinies of the Stars* by advocating for comparative planetology, with the concept of the other planets providing "a prediction of the fate that once, after milliards of years perhaps, will befall the later descents of the present generation" (*Arrhenius*, 1918). These musings seem startlingly prescient when considering that exoplanets may someday allow us to witness planetary states lost to our solar system's past, and Venus may show us the possible future of Earth.

Acknowledgments. We thank reviewers David Crisp and Zachary Berta-Thompson for their careful reviews and constructive comments and suggestions which greatly improved this chapter. This work was performed as part of the NASA Virtual Planetary Laboratory, a member of the Nexus for Exoplanet System Science Research Coordination Network, and funded by the NASA Astrobiology Program under grant 80NSSC18K0829.

REFERENCES

Airapetian V. S., Glocer A., Khazanov G. V., Loyd R. O. P., France K., Sojka J., Danchi W. C., and Liemohn M. W. (2017) How hospitable are space weather affected habitable zones? The role of ion escape. *Astrophys. J. Lett.*, *836(1)*, L3.

Allen D. A. (1987) The dark side of Venus. *Icarus*, *69(2)*, 221–229.

Allen D. A. (1990) Venus. *International Astronomical Union Circular 4962: 1989c1; EF Eri; Venus.*

Allen D. A. and Crawford J. W. (1984) Cloud structure on the dark side of Venus. *Nature*, *307(5948)*, 222.

Alexander C. O. D., Bowden R., Fogel M. L., Howard K. T., Herd C. D. K., and Nittler L. R. (2012) The provenances of asteroids, and their contributions to the volatile inventories of the terrestrial planets. *Science*, *337(6095)*, 721–723.

Angelo I. et al. (2017) Kepler-1649b: An exo-Venus in the solar neighborhood. *Astron. J.*, *153*, 162.

Arney G. et al. (2014) Spatially-resolved measurements of H_2O, HCl, CO, OCS, SO_2, cloud opacity, and acid concentration in the Venus near-infrared spectral windows. *J. Geophys. Res.–Planets*, *119(8)*, 1860–1891, DOI: 10.1002/2014JE004662.

Arrhenius S. (1918) *The Destinies of the Stars.* Knickerbocker, New York. 330 pp.

Atreya S. K., Trainer M. G., Franz H. B., Wong M. H., Manning H. L., Malespin C. A., Mahaffy P. R., Conrad P. G., Brunner A. E., Leshin L. A., and Jones J. H. (2013) Primordial argon isotope fractionation in the atmosphere of Mars measured by the SAM instrument on Curiosity and implications for atmospheric loss. *Geophys. Res. Lett.*, *40(21)*, 5605–5609.

Baines K. H., Bellucci G., Bibring J. P., Brown R. H., Buratti B. J., Bussoletti E., Capaccioni F., Cerroni P., Clark R. N., Coradini A., and Cruikshank D. P. (2000) Detection of sub-micron radiation from the surface of Venus by Cassini/VIMS. *Icarus*, *148(1)*, 307–311.

Baines K. H., Atreya S. K., Bullock M. A., Grinspoon D. H., Mahaffy P., Russell C. T., and Zahnle K. (2014) The atmospheres of the terrestrial planets: Clues to the origins and early evolution of Venus, Earth, and Mars. In *Comparative Climatology of Terrestrial Planets* (S. J. Mackwell et al., eds.), pp. 137–162. Univ. of Arizona, Tucson.

Barabash S., Sauvaud J. A., Gunell H., Andersson H., Grigoriev A., Brinkfeldt K., Holmström M., Lundin R., Yamauchi M., Asamura K., and Baumjohann W. (2007) The Analyser of Space Plasmas and Energetic Atoms (ASPERA-4) for the Venus Express mission. *Planet. Space Sci.*, *55(12)*, 1772–1792.

Baraffe I., Chabrier G., Allard F., and Hauschildt P. (1998) Evolutionary models for solar metallicity low-mass stars: Mass-magnitude relationships and color-magnitude diagrams. *Astron. Astrophys.*, *337(2)*, 403–412.

Barclay T., Pepper J., and Quintana E. V. (2018) A revised exoplanet yield from the transiting exoplanet survey satellite (TESS). *Astrophys. J. Suppl. Ser.*, *239(1)*, 2.

Barker E. S. (1979) Detection of SO_2 in the UV spectrum of Venus. *Geophys. Res. Lett.*, *6(2)*, 117–120.

Barnes R., Mullins K., Goldblatt C., Meadows V. S., Kasting J. F. and Heller R. (2013) Tidal Venuses: Triggering a climate catastrophe via tidal heating. *Astrobiology*, *13(3)*, 225–250.

Barstow J. K. et al. (2012) Models of the global cloud structure on Venus derived from Venus Express observations. *Icarus*, *217(2)*, 542–560, DOI: 10.1016/j.icarus.2011.05.018.

Basilevsky A. T., Head J. W., and Abdrakhimov A. M. (2004) Impact crater air fall deposits on the surface of Venus: Areal distribution, estimated thickness, recognition in surface panoramas, and implications for provenance of sampled surface materials. *J. Geophys. Res.–Planets*, *109(E12)*.

Basilevsky A. T., Ivanov M. A., Head J. W., Aittola M., and Raitala J. (2007) Landing on Venus: Past and future. *Planet. Space Sci.*, *55(14)*, 2097–2112.

Basilevsky A. T., Shalygin E. V., Titov D. V., Markiewicz W. J., Scholten F., Roatsch T., Kreslavsky M. A., Moroz L. V., Ignatiev N. I., Fiethe B., and Osterloh B. (2012) Geologic interpretation of the near-infrared images of the surface taken by the Venus Monitoring Camera, Venus Express. *Icarus*, *217(2)*, 434–450.

Bell E. A., Boehnke P., Harrison T. M., and Mao W. L. (2015) Potentially biogenic carbon preserved in a 4.1 billion-year-old zircon. *Proc. Natl. Acad. Sci.*, *112(47)*, 14518–14521.

Berta-Thompson Z. K., Irwin J., Charbonneau D., Newton E. R., Dittmann J. A., Astudillo-Defru N., Bonfils X., Gillon M., Jehin E., Stark A. A., and Stalder B. (2015) A rocky planet transiting a nearby low-mass star. *Nature*, *527(7577)*, 204.

Bétrémieux Y. and Kaltenegger L. (2013) Transmission spectrum of Earth as a transiting exoplanet from the ultraviolet to the near-infrared. *Astrophys. J. Lett.*, *772(2)*, L31.

Bézard B., De Bergh C., Crisp D., and Maillard J. P. (1990) The deep atmosphere of Venus revealed by high-resolution nightside spectra. *Nature*, *345(6275)*, 508.

Bézard B., De Bergh C., Fegley B., Maillard J. P., Crisp D., Owen T., Pollack J. B. and Grinspoon D. (1993) The abundance of sulfur dioxide below the clouds of Venus. *Geophys. Res. Lett.*, *20(15)*, 1587–1590.

Bézard B., Fedorova A., Bertaux J. L., Rodin A., and Korablev O. (2011) The 1.10- and 1.18-μm nightside windows of Venus observed by SPICAV-IR aboard Venus Express. *Icarus*, *216(1)*, 173–183.

Bjonnes E. E., Hansen V. L., James B., and Swenson J. B. (2012) Equilibrium resurfacing of Venus: Results from new Monte Carlo modeling and implications for Venus surface histories. *Icarus*, *217(2)*, 451–461.

Bolmont E., Selsis F., Owen J. E., Ribas I., Raymond S. N., Leconte J., and Gillon M. (2016) Water loss from terrestrial planets orbiting ultracool dwarfs: Implications for the planets of TRAPPIST-1. *Mon. Not. R. Astron. Soc.*, *464(3)*, 3728–3741.

Bonfils X., Astudillo-Defru N., Díaz R., Almenara J. M., Forveille T., Bouchy F., Delfosse X., Lovis C., Mayor M., Murgas F., Santos N. C., Ségransan D., Udry S., Wünsche A., and Pepe F. (2017) A temperate exo-Earth around a quiet M dwarf at 3.4 parsecs. *Astron. Astrophys*, *613*, A25.

Brown C. D. and Grimm R. E. (1997) Tessera deformation and the contemporaneous thermal state of the plateau highlands, Venus. *Earth Planet. Sci. Lett.*, *147(1–4)*, 1–10.

Bullock M. and Grinspoon D. (2001) The recent evolution of climate on Venus. *Icarus*, *150(1)*, 19–37, DOI: 10.1006/icar.2000.6570.

Bullock M. A. and Grinspoon D. H. (2013) The atmosphere and climate of Venus. In *Comparative Climatology of Terrestrial Planets* (S. J. Mackwell et al., eds.), pp. 19–54. Univ. of Arizona, Tucson.

Bullock M. A., Grinspoon D. H., and Head J. W. (1993) Venus resurfacing rates: Constraints provided by 3-D Monte Carlo simulations. *Geophys. Res. Lett.*, *20(19)*, 2147–2150.

Burke C. J., Christiansen J. L., Mullally F., Seader S., Huber D., Rowe J. F., Coughlin J. L., Thompson S. E., Catanzarite J., Clarke B. D., and Morton T. D. (2015) Terrestrial planet occurrence rates for the Kepler GK dwarf sample. *Astrophys. J.*, *809(1)*, 8.

Catling D. C. and Zahnle K. J. (2009) The planetary air leak. *Sci. Am.*, *300(5)*, 36–43.

Catling D. C. and Zahnle K. J. (2013) An impact stability limit controlling the existence of atmospheres on exoplanets and solar system bodies. In *Lunar. Planet. Sci. XLIV*, Abstract #2665. Lunar and Planetary Institute, Houston.

Cameron A. G. and Ward W. R. (1976) The origin of the Moon. In *Proc. Lunar Planet. Sci. Conf. 7th*, pp. 120–122.

Campbell D. B., Head J. W., Harmon J. K., and Hine A. A. (1984) Venus: Volcanism and rift formation in Beta Regio. *Science*, *226(4671)*, 167–170.

Canfield D. E., Glazer A. N., and Falkowski P. G. (2010) The evolution and future of Earth's nitrogen cycle. *Science*, *330(6001)*, 192–196, DOI: 10.1126/science.1186120.

Carlson R. O., Baines K. H., Encrenaz T., Taylor F. W., Drossart P., Kamp L. W., Pollack J. B., Lellouch E., Collard A. D., Calcutt S. B., and Grinspoon D. (1991) Galileo infrared imaging spectroscopy measurements at Venus. *Science*, *253(5027)*, 1541–1548.

Carpenter R. L. (1964) Symposium on radar and radiometric observations of Venus during the 1962 conjunction: Study of Venus by CW radar. *Astron. J.*, *69*, 2.

Carpenter R. L. (1966) Study of Venus by CW radar — 1964 results. *Astron. J.*, *71*, 142.

Chamberlain S., Bailey J., Crisp D., and Meadows V. (2013) Ground-based near-infrared observations of water vapour in the Venus troposphere. *Icarus*, *222(1)*, 364–378.

Chambers J. E. (2004) Planetary accretion in the inner solar system. *Earth Planet. Sci. Lett.*, *223(3–4)*, 241–252.

Chassefière E. (1996) Hydrodynamic escape of hydrogen from a hot water-rich atmosphere: The case of Venus. *J. Geophys. Res.-Planets*, *101(E11)*, 26039–26056.

Chassefière E., Wieler R., Marty B., and Leblanc F. (2012) The evolution of Venus: Present state of knowledge and future exploration. *Planet. Space Sci.*, *63*, 15–23.

Collard A. D. et al. (1993) Latitudinal distribution of carbon monoxide in the deep atmosphere of Venus. *Planet. Space Sci.*, *41(7)*, 487–494.

Collinson G. A., Frahm R. A., Glocer A., Coates A. J., Grebowsky J. M.,

Barabash S., Domagal-Goldman S. D., Fedorov A., Futaana Y., Gilbert L. K., and Khazanov G. (2016) The electric wind of Venus: A global and persistent "polar wind"-like ambipolar electric field sufficient for the direct escape of heavy ionospheric ions. *Geophys. Res. Lett.*, *43(12)*, 5926–5934.

Connes P., Noxon J. F., Traub W. A., and Carleton N. P. (1979) O_2 (1Δ) emission in the day and night airglow of Venus. *Astrophys. J. Lett.*, *233*, L29–L32.

Correia A. C. M. and Laskar J. (2003) Long-term evolution of the spin of Venus. II. Numerical simulations. *Icarus*, *163(1)*, 24–25, DOI: 10.1016/S0019-1035(03)00043-5.

Cottini V. et al. (2012) Water vapor near the cloud tops of Venus from Venus Express/VIRTIS dayside data. *Icarus*, *217(2)*, 561–569, DOI: 10.1016/j.icarus.2011.06.018.

Coustenis A. and Bezard B. (1995) Titan's atmosphere from Voyager infrared observations: IV. Latitudinal variations of temperature and composition. *Icarus*, *115(1)*, 126–140.

Crisp D. (1986) Radiative forcing of the Venus mesosphere. *Icarus*, *514(67)*, 484–514.

Crisp D., Allen D. A., Grinspoon D. H., and Pollack J. B. (1991) The dark side of Venus: Near-infrared images and spectra from the Anglo-Australian Observatory. *Science*, *253(5025)*, 1263–1266.

Crisp D., Meadows V. S., Bézard B., Bergh C. D., Maillard J. P., and Mills F. P. (1996) Ground-based near-infrared observations of the Venus nightside: 1.27-μm O_2 (a 1Δ g) airglow from the upper atmosphere. *J. Geophys. Res.–Planets*, *101(E2)*, 4577–4593.

Dickel J. (1966) Measurement of the temperature of Venus at a wavelength of 3.75 cm for a full cycle of planetary phase. *Icarus*, *5*, 305–208.

Dobrovolskis A. R. and Ingersoll A. P. (1980) Atmospheric tides and the rotation of Venus: 1. Tidal theory and the balance of torques. *Icarus*, *41*, 1–17.

Domagal-Goldman S. D., Segura A., Claire M. W., Robinson T. D., and Meadows V. S. (2014) Abiotic ozone and oxygen in atmospheres similar to prebiotic Earth. *Astrophys. J.*, *792(2)*, 90.

Donahue T. M. and Hodges R. R. (1992) Past and present water budget of Venus. *J. Geophys. Res.–Planets*, *97(E4)*, 6083–6091.

Donahue T. M. and Russell C. T. (1997) The Venus atmosphere and ionosphere and their interaction with the solar wind: An overview. In *Venus II: Geology, Geophysics, Atmosphere, and Solar Wind Environment* (S. W. Bougher et al., eds.), pp. 3–32. Univ. of Arizona, Tucson.

Donahue T. M., Hoffman J. H., and Hodges R. R. (1981) Krypton and xenon in the atmosphere of Venus. *Geophys. Res. Lett.*, *8(5)*, 513–516.

Donahue T. M., Hoffman J. H., Hodges R. R., and Watson A. J. (1982) Venus was wet: A measurement of the ratio of deuterium to hydrogen. *Science*, *216(4546)*, 630–633.

Donahue T. M., Grinspoon D. H., Hartle R. E., and Hodges R. R. Jr. (1997) Ion/neutral escape of hydrogen and deuterium: Evolution of water. In *Venus II — Geology, Geophysics, Atmosphere, and Solar Wind Environment* (S. W. Bougher et al., eds.), pp. 385–414. Univ. of Arizona, Tucson.

Dong C., Lingam M., Ma Y., and Cohen O. (2017) Is Proxima Centauri b habitable? A study of atmospheric loss. *Astrophys. J. Lett.*, *837(2)*, L26.

Dotter A., Chaboyer B., Jevremović D., Kostov V., Baron E., and Ferguson J. W. (2008) The Dartmouth stellar evolution database. *Astrophys. J. Suppl.*, *178(1)*, 89.

Dragomir D., Teske J., Günther M. N., Ségransan D., Burt J. A., Huang C. X., Vanderburg A., Matthews E., Dumusque X., Stassun K. G., and Pepper J. (2019) TESS delivers its first Earth-sized planet and a warm sub-Neptune. *Astrophys. J. Lett.*, *875(2)*, L7.

Dressing C. D. and Charbonneau D. (2013) The occurrence rate of small planets around small stars. *Astrophys. J.*, *767(1)*, 95.

Dressing C. D. and Charbonneau D. (2015) The occurrence of potentially habitable planets orbiting M dwarfs estimated from the full Kepler dataset and an empirical measurement of the detection sensitivity. *Astrophys. J.*, *807*, 45.

Driscoll P. and Bercovici D. (2013) Divergent evolution of Earth and Venus: Influence of degassing, tectonics, and magnetic fields. *Icarus*, *226(2)*, 1447–1464.

Drossart P., Piccioni G., Adriani A., Angrilli F., Arnold G., Baines K. H., Bellucci G., Benkhoff J., Bézard B., Bibring J. P., and Blanco A. (2007) Scientific goals for the observation of Venus by VIRTIS on ESA/Venus Express mission. *Planet. Space Sci.*, *55(12)*, 1653–1672.

Dyce B. R., Pettengill G. H., and Shapiro I. I. (1967) Radar determination of the rotations of Venus and Mercury. *Astron. J., 72,* 351.

Ekonomov A. P., Moroz V. I., Moshkin B. E., Gnedykh V. I., Golovin Y. M., and Crigoryev A. V. (1984) Scattered UV solar radiation within the clouds of Venus. *Nature, 307(5949),* 345.

Erard S., Drossart P., and Piccioni G. (2009) Multivariate analysis of visible and infrared thermal imaging spectrometer (VIRTIS) Venus express nightside and limb observations. *J. Geophys. Res.–Planets, 114(E9).*

Esposito L. W. (1980) Ultraviolet contrasts and the absorbers near the Venus cloud tops. *J. Geophys. Res.–Space Phys., 85(A13),* 8151–8157.

Esposito L. W. (1984) Sulfur dioxide: Episodic injection shows evidence for active Venus volcanism. *Science, 223(4640),* 1072–1074.

Esposito L. W. and Travis L. D. (1982) Polarization studies of the Venus UV contrasts: Cloud height and haze variability. *Icarus, 51(2),* 374–390.

Esposito L. W. et al. (1983) The clouds and hazes of Venus. In *Venus* (D. M. Hunten et al., eds.), pp. 484–564. Univ. of Arizona, Tucson.

Esposito L. W., Bertaux J. L., Krasnopolsky V., Moroz V. I., and Zasova L. V. (1997) Chemistry of lower atmosphere and clouds. In *Venus II — Geology, Geophysics, Atmosphere, and Solar Wind Environment* (S. W. Bougher et al., eds.), pp. 415–458. Univ. of Arizona, Tucson.

Fegley B. Jr. and Prinn R. G. (1989) Estimation of the rate of volcanism on Venus from reaction rate measurements. *Nature, 337(6202),* 55.

Finlay M. S., Mark D. F., Gandanger P., and McConville P. (2016) Earth-atmosphere evolution based on new determination of Devonian atmosphere Ar isotopic composition. *Earth Planet. Sci. Lett., 446,* 21–26.

Foley B. J. (2015) The role of plate tectonic-climate coupling and exposed land area in the development of habitable climates on rocky planets. *Astrophys. J., 812(1),* 36.

Frandsen B. N., Wennberg P. O., and Kjaergaard H. G. (2016) Identification of OSSO as a near-UV absorber in the venusian atmosphere. *Geophys. Res. Lett., 43*(21), 11146–11155.

Frank E. A., Maier W. D., and Mojzsis S. J. (2012) The 'late veneer' on Earth: Evidence from Eoarchean ultramafic schists (metakomatiites). In *Lunar Planet. Sci. XLIII,* Abstract #2890, Lunar and Planetary Institute, Houston.

Fujii Y., Del Genio A. D., and Amundsen D. S. (2017) NIR-driven moist upper atmospheres of synchronously rotating temperate terrestrial exoplanets. *Astrophys. J., 848(2),* 100.

Fukuhara T. et al. (2017) Large stationary gravity wave in the atmosphere of Venus. *Nature Geosci., 10(2),* 85–88, DOI: 10.1038/ngeo2873.

Gao P., Zhang X., Crisp D., Bardeen C. G., and Yung Y. L. (2014) Bimodal distribution of sulfuric acid aerosols in the upper haze of Venus. *Icarus, 231,* 83–98.

Gao P., Hu R., Robinson T. D., Li C., and Yung Y. L. (2015) Stability of CO_2 atmospheres on desiccated M dwarf exoplanets. *Astrophys. J., 806(2),* 249.

Garcia-Sage K., Glocer A., Drake J. J., Gronoff G., and Cohen O. (2017) On the magnetic protection of the atmosphere of Proxima Centauri b. *Astrophys. J. Lett., 844(1),* L13.

Gough D. O. (1981) Solar interior structure and luminosity variations. In *Physics of Solar Variations* (V. Domingo, ed.), pp. 21–34. Springer, Dordrecht.

Gillon M., Jehin E., Lederer S. M., Delrez L., de Wit J., Burdanov A., van Grootel V., Burgasser A. J., Triaud A. H., Opitom C., and Demory B. O. (2016) Temperate Earth-sized planets transiting a nearby ultracool dwarf star. *Nature, 533(7602),* 221–224.

Gillon M., Triaud A. H., Demory B. O., Jehin E., Agol E., Deck K. M., Lederer S. M., De Wit J., Burdanov A., Ingalls J. G. and Bolmont E. (2017) Seven temperate terrestrial planets around the nearby ultracool dwarf star TRAPPIST-1. *Nature, 542(7642),* 456.

Goldblatt C. and Watson A. J. (2012) The runaway greenhouse: Implications for future climate change, geoengineering and planetary atmospheres. *Philos. Trans. R. Soc. London A, 370(1974),* 4197–4216.

Goldblatt C., Robinson T. D., Zahnle K. J., and Crisp D. (2013) Low simulated radiation limit for runaway greenhouse climates. *Nature Geosci., 6(8),* 661.

Greenblatt G. D., Orlando J. J., Burkholder J. B., and Ravishankara A. R. (1990) Absorption measurements of oxygen between 330 and 1140 nm. *J. Geophys. Res.–Atmos., 95(D11),* 18577–18582.

Greene T. P., Line M. R., Montero C., Fortney J. J., Lustig-Yaeger J., and Luther K. (2016) Characterizing transiting exoplanet atmospheres with JWST. *Astrophys. J., 817(1),* 17.

Ghent R. R., Phillips R. J., Hansen V. L., and Nunes D. C. (2005) Finite element modeling of short-wavelength folding on Venus: Implications for the plume hypothesis for crustal plateau formation. *J. Geophys. Res.–Planets, 110(E11).*

Gilmore M. S., Ivanov M. A., Head J. W., and Basilevsky A. T. (1997) Duration of tessera deformation on Venus. *J. Geophys. Res.–Planets, 102(E6),* 13357–13368.

Grinspoon D. H. (1993) Implications of the high D/H ratio for the sources of water in Venus' atmosphere. *Nature, 363(6428),* 428.

Grinspoon D. H. (1997) *Venus Revealed.* Addison-Wesley, Boston. 355 pp.

Grinspoon D. H. and Bullock M. A. (2007) Astrobiology and Venus exploration. In *Exploring Venus as a Terrestrial Planet* (L. W. Esposito et al., eds.), pp. 191–206. American Geophysical Union, Washington, DC.

Grinspoon D. H. et al. (1993) Probing Venus cloud structure with Galileo NIMS. *Planet. Space Sci., 41(7),* 515–542, DOI: 10.1016/0032-0633(93) 90034-Y.

Guest J. E. and Stofan E. R. (1999) A new view of the stratigraphic history of Venus. *Icarus, 139(1),* 55–66.

Hamano K., Abe Y., and Genda H. (2013) Emergence of two types of terrestrial planet on solidification of magma ocean. *Nature, 497(74),* 607–610

Hamano K., Abe Y., and Genda H. (2013) Emergence of two types of terrestrial planet on solidification of magma ocean. *Nature, 497(74),* 607–610.

Hansen J. E. and Hovenier J. W. (1974) Interpretation of the polarization of Venus. *J. Atmos. Sci., 31(4),* 1137–1160.

Hansen V. L. and Young D. A. (2007) Venus's evolution: A synthesis. In *Convergent Margin Terranes and Associated Regions: A Tribute to W. G. Ernst* (M. Cloos et al., eds.), p. 255. GSA Special Paper 419, Geological Society of America.

Hansen V. L. and López I. (2010) Venus records a rich early history. *Geology, 38(4),* 311–314.

Hansen V. L., Phillips R. J., Willis J. J., and Ghent R. R. (2000) Structures in tessera terrain, Venus: Issues and answers. *J. Geophys. Res.–Planets, 105(E2),* 4135–4152.

Harman C.E., Schwieterman E. W., Schottelkotte J. C., and Kasting J. F. (2015) Abiotic O_2 levels on planets around F, G, K, and M stars: Possible false positives for life? *Astrophys. J., 812(2),* 137.

Harman C. E., Felton R., Hu R., Domagal-Goldman S. D., Segura A., Tian F., and Kasting J. F. (2018) Abiotic O_2 levels on planets around F, G, K, and M stars: Effects of lightning-produced catalysts in eliminating oxygen false positives. *Astrophys. J., 866(1),* 56.

Hashimoto G. L. and Sugita S. (2003) On observing the compositional variability of the surface of Venus using nightside near-infrared thermal radiation. *J. Geophys. Res.–Planets, 108(E9),* DOI: 10.1029/2003JE002082.

Hermans C., Vandaele A. C., Carleer M., Fally S., Colin R., Jenouvrier A., Coquart B., and Mérienne M. F. (1999) Absorption cross-sections of atmospheric constituents: NO_2, O_2, and H_2O. *Environ. Sci. Pollut. Res., 6(3),* 151–158.

Herrick R. R. and Rumpf M. E. (2011) Postimpact modification by volcanic or tectonic processes as the rule, not the exception, for venusian craters. *J. Geophys. Res.–Planets, 116(E2),* DOI: 10.1029/2010JE003722.

Herrick R. R., Izenberg N., and Phillips R. J. (1995) Comment on "The global resurfacing of Venus" by R. G. Strom, G. G. Schaber, and D. D. Dawson. *J. Geophys. Res.–Planets, 100(E11),* 23355–23359, DOI: 10.1029/95JE02293.

Hinkel N. R. and Unterborn C. T. (2018) The star-planet connection. I. Using stellar composition to observationally constrain planetary mineralogy for the 10 closest stars. *Astrophys. J., 853,* 83, DOI: 10.3847/1538-4357/aaa5b4.

Hitchcock D. R. and Lovelock J. E. (1967) Life detection by atmospheric analysis. *Icarus, 7(1–3),* 149–159.

Hoffman J. et al. (1980) Composition of the Venus lower atmosphere from the Pioneer Venus mass spectrometer. *J. Geophys. Res., 85(80),* 7882–7890, DOI: 10.1029/JA085iA13p07882.

Horinouchi T., Murakami S. Y., Satoh T., Peralta J., Ogohara K., Kouyama T., Imamura T., Kashimura H., Limaye S. S., McGouldrick K., and Nakamura M. (2017) Equatorial jet in the lower to middle cloud layer of Venus revealed by Akatsuki. *Nature Geosci.*, *10(9)*, 646.

Horinouchi T., Kouyam T., Lee Y. J., Murakami S. Y., Ogohara K., Takagi M., Imamura T., Nakajima K., Peralta J., Yamazaki A., and Yamada M. (2018) Mean winds at the cloud top of Venus obtained from two-wavelength UV imaging by Akatsuki. *Earth Planets Space*, *70(1)*, 10.

Hu R., Seager S., and Bains W. (2012) Photochemistry in terrestrial exoplanet atmospheres. I. Photochemistry model and benchmark cases. *Astrophys. J.*, *761(2)*, 166.

Ingersoll A. P. (1969) The runaway greenhouse: A history of water on Venus. *J. Atmos. Sci.*, *26(6)*, 1191–1198.

Ivanov M. A. and Head J. W. (1996) Tessera terrain on Venus: A survey of the global distribution, characteristics, and relation to surrounding units from Magellan data. *J. Geophys. Res.–Planets*, *101(E6)*, 14861–14908.

Ivanov M. A. and Head J. W. (2013) The history of volcanism on Venus. *Planet. Space Sci.*, *84*, 66–92.

Kane S. R. and von Braun K. (2008) Constraining orbital parameters through planetary transit monitoring. *Astrophys. J.*, *689*, 492.

Kane S. R., Kopparapu R. K., and Domagal-Goldman S. D. (2014) On the frequency of potential Venus analogs from Kepler data. *Astrophys. J. Lett.*, *794*, L5.

Kane S. R., Arney G., Crisp D., Domagal-Goldman S., Glaze L. S., Goldblatt C., Grinspoon D., Head J. W., Lenardic A., Unterborn C., and Way M. J. (2019) Venus as a laboratory for exoplanetary science. *J. Geophys. Res.–Planets*, *124(8)*, 2015–2028, DOI: 10.1029/2019JE005939.

Kaula W. M. and Phillips R. J. (1981) Quantitative tests for plate tectonics on Venus. *Geophys. Res. Lett.*, *8(12)*, 1187–1190.

Kasting J. F. (1988) Runaway and moist greenhouse atmospheres and the evolution of Earth and Venus. *Icarus*, *74(3)*, 472–494.

Kasting J. F. (1997) Habitable zones around low mass stars and the search for extraterrestrial life. In *Planetary and Interstellar Processes Relevant to the Origins of Life* (D. C. B. Whittet, ed.), pp. 291–307. Springer, Dordrecht.

Kasting J. F., Whitmire D. P., and Reynolds R. T. (1993) Habitable zones around main sequence stars. *Icarus*, *101(1)*, 108–128.

Khatuntsev I. V. et al. (2013) Cloud level winds from the Venus Express Monitoring Camera imaging. *Icarus*, *226(1)*, 140–158, DOI: 10.1016/j.icarus.2013.05.018.

Kiefer W. S. and Swafford L. C. (2006) Topographic analysis of Devana Chasma, Venus: Implications for rift system segmentation and propagation. *J. Structural Geol.*, *28(12)*, 2144–2155.

Kleine T., Touboul M., Bourdon B., Nimmo F., Mezger K., Palme H., Jacobsen S. B., Yin Q. Z., and Halliday A. N. (2009) Hf-W chronology of the accretion and early evolution of asteroids and terrestrial planets. *Geochim. Cosmochim. Acta*, *73(17)*, 5150–5188.

Knollenberg R. and Hunten D. (1980) The microphysics of the clouds of Venus: Results of the Pioneer Venus particle size spectrometer experiment. *J. Geophys Res.*, *85(A13)*, 8039–8058, DOI: 10.1029/JA085iA13p08039.

Komabayasi M. (1967) Discrete equilibrium temperatures of a hypothetical planet with the atmosphere and the hydrosphere of one component-two phase system under constant solar radiation. *J. Meteorological Soc. Japan, Ser. II*, *45(1)*, 137–139.

Konopliv A. S. and Yoder C. F. (1996) Venusian k_2 tidal Love number from Magellan and PVO tracking data. *Geophys. Res. Lett.*, *23(14)*, 1857–1860.

Kopparapu R. K. (2013) A revised estimate of the occurrence rate of terrestrial planets in the habitable zones around Kepler M-dwarfs. *Astrophys. J. Lett.*, *767(1)*, L8.

Kopparapu R. K. et al. (2013) Habitable zones around main-sequence stars: New estimates. *Astrophys. J.*, *765(2)*, 131.

Kopparapu R. K. et al. (2014) Habitable zones around main-sequence stars: Dependence on planetary mass. *Astrophys. J. Lett.*, *787*, L29.

Kopparapu R. K., Wolf E. T., Arney G., Batalha N. E., Haqq-Misra J., Grimm S. L., and Heng K. (2017) Habitable moist atmospheres on terrestrial planets near the inner edge of the habitable zone around M-dwarfs. *Astrophys. J.*, *845(1)*, 5.

Kostov V. B., Schlieder J. E., Barclay T., Quintana E. V., Colón K. D., Brande J., and Kreidberg L. (2019) The L 98-59 system: Three transiting, terrestrial-size planets orbiting a nearby M dwarf. *Astron. J.*, *158(1)*, 32.

Kouyama T. et al. (2013) Long-term variation in the cloud-tracked zonal velocities at the cloud top of Venus deduced from Venus Express VMC images. *J. Geophys. Res.–Planets*, *118(1)*, 37–46, DOI: 10.1029/2011JE004013.

Kouyama T., Imamura T., Taguchi M., Fukuhara T., Sato T. M., Yamazaki A., Futaguchi M., Murakami S., Hashimoto G. L., Ueno M., and Iwagami N. (2017) Topographical and local time dependence of large stationary gravity waves observed at the cloud top of Venus. *Geophys. Res. Lett.*, *44(24)*, 12098–12105.

Krasnopolsky V. A. (2007) Chemical kinetic model for the lower atmosphere of Venus. *Icarus*, *191(1)*, 25–37.

Krasnopolsky V. A. (2010a) Spatially-resolved high-resolution spectroscopy of Venus 1. Variations of CO_2, CO, HF, and HCl at the cloud tops. *Icarus*, *208(2)*, 539–547, DOI: 10.1016/j.icarus.2010.02.012.

Krasnopolsky V. A. (2010b) Spatially-resolved high-resolution spectroscopy of Venus 2. Variations of HDO, OCS, and SO_2 at the cloud tops. *Icarus*, *209(2)*, 314–322.

Lecacheux J., Drossart P., Laques P., Deladerriere F., and Colas F. (1993) Detection of the surface of Venus at 1.0 μm from ground-based observations. *Planet. Space Sci.*, *41(7)*, 543–549.

Leconte J., Wu H., Menou K., and Murray N. (2015) Asynchronous rotation of Earth-mass planets in the habitable zone of lower-mass stars. *Science*, *347(6222)*, 632–635.

Lecuyer C., Simon L., and Guyot F. (2000) Comparison of carbon, nitrogen and water budgets on Venus and the Earth. *Earth Planet. Sci. Lett.*, *181(1–2)*, 33–40.

Léger A., Rouan D., Schneider J., Barge P., Fridlund M., Samuel B., Ollivier M., Guenther E., Deleuil M., Deeg H. J., and Auvergne M. (2009) Transiting exoplanets from the CoRoT space mission VIII. CoRoT-7b: The first super-Earth with measured radius. *Astron. Astrophys.*, *506(1)*, 287–302.

Lee Y. J., Yamazaki A., Imamura T., Yamada M., Watanabe S., Sato T. M., Ogohara K., Hashimoto G. L., and Murakami S. (2017) Scattering properties of the venusian clouds observed by the UV imager on board Akatsuki. *Astron. J.*, *154(2)*, 44.

Limaye S. S., Mogul R., Smith D. J., Ansari A. H., Słowik G. P., and Vaishampayan P. (2018) Venus' spectral signatures and the potential for life in the clouds. *Astrobiology*, *18 (9)*, DOI: 10.1089/ast.2017.1783.

Lincowski A. P., Meadows V. S., Crisp D., Robinson T. D., Luger R., Lustig-Yaeger J., and Arney G. N. (2018) Evolved climates and observational discriminants for the TRAPPIST-1 planetary system. *Astrophys. J.*, *867(1)*, 76.

Luger R. and Barnes R. (2015) Extreme water loss and abiotic O_2 buildup on planets throughout the habitable zones of M dwarfs. *Astrobiology*, *15(2)*, 119–143.

Luger R., Sestovic M., Kruse E., Grimm S. L., Demory B. O., Agol E., Bolmont E., Fabrycky D., Fernandes C. S., van Grootel V., and Burgasser A. (2017) A seven-planet resonant chain in TRAPPIST-1. *Nature Astron.*, *1(6)*, 0129.

Luhmann J. G., Ledvina S. A., and Russell C. T. (2004) Induced magnetospheres. *Adv. Space Res.*, *33(11)*, 1905–1912.

MacGregor M. A., Weinberger A. J., Wilner D. J., Kowalski A. F., and Cranmer S. R. (2018) Detection of a millimeter flare from Proxima Centauri. *Astrophys. J. Lett.*, *855(1)*, L2.

Markiewicz W. J., Petrova E., Shalygina O., Almeida M., Titov D. V., Limaye S. S., Ignatiev N., Roatsch T., and Matz K. D. (2014) Glory on Venus cloud tops and the unknown UV absorber. *Icarus*, *234*, 200–203.

Marcq E. et al. (2005) Latitudinal variations of CO and OCS in the lower atmosphere of Venus from near-infrared nightside spectro-imaging. *Icarus*, *179(2)*, 375–386, DOI: 10.1016/j.icarus.2005.06.018.

Marcq E. et al. (2006) Remote sensing of Venus lower atmosphere from ground-based IR spectroscopy: Latitudinal and vertical distribution of minor species. *Planet. Space Sci.*, *54(13–14)*, 1360–1370, DOI: 10.1016/j.pss.2006.04.024.

Marcq E. et al. (2008) A latitudinal survey of CO, OCS, H_2O, and SO_2 in the lower atmosphere of Venus: Spectroscopic studies using VIRTIS-H. *J. Geophys Res.*, *113(E5)*, DOI: 10.1029/2008JE003074.

Marcq E., Bertaux J. L., Montmessin F., and Belyaev D. (2013) Variations of sulphur dioxide at the cloud top of Venus's dynamic atmosphere. *Nature Geosci.*, *6(1)*, 25.

Marov M. Y. et al. (1973) Measurements of temperature, pressure, and wind velocity on the illuminated side of Venus. *J. Atmos. Sci.*, *30(6)*, 1210–1214.

Maté B., Lugez C., Fraser G. T., and Lafferty W. J. (1999) Absolute intensities for the O_2 1.27 μm continuum absorption. *J. Geophys Res.–Atmos.*, *104(D23)*, 30585–30590.

Mayer C. H., McCullough T. P., and Sloanaker R. M. (1958) Observations of Venus at 3.15-cm wave length. *Astrophys. J.*, *127(1)*, 1–10, DOI: 10.1086/146433.

McElroy M. B. and Prather M. J. (1981) Noble gases in the terrestrial planets. *Nature*, *293(5833)*, 535.

McElroy M. B., Dak Sze N., and Ling Yung Y. (1973) Photochemistry of the Venus atmosphere. *J. Atmos. Sci.*, *30(7)*, 1437–1447.

Meadows V. S. (2017) Reflections on O_2 as a biosignature in exoplanetary atmospheres. *Astrobiology*, *17(10)*, 1022–1052.

Meadows V. S. and Crisp D. (1996) Ground-based near-infrared observations of the Venus nightside: The thermal structure and water abundance near the surface. *J. Geophys. Res.–Planets*, *101(E2)*, 4595–4622.

Meadows V. S., Arney G. N., Schwieterman E. W., Lustig-Yaeger J., Lincowski A. P., Robinson T., Domagal-Goldman S. D., Deitrick R., Barnes R. K., Fleming D. P., and Luger R. (2018a) The habitability of Proxima Centauri b: Environmental states and observational discriminants. *Astrobiology*, *18(2)*, 133–189.

Meadows V. S., Reinhard C. T., Arney G. N., Parenteau M. N., Schwieterman E. W., Domagal-Goldman S. D., Lincowski A. P., Stapelfeldt K. R., Rauer H., DasSarma S., and Hegde S.(2018b) Exoplanet biosignatures: Understanding oxygen as a biosignature in the context of its environment. *Astrobiology*, *18(6)*, 630–662.

Mendonça J. M. and Read P. L. (2016) Exploring the Venus global super-rotation using a comprehensive general circulation model. *Planet. Space Sci.*, *134*, 1–18, DOI: 10.1016/j.pss.2016.09.001.

Miguel Y., Kaltenegger L., Fegley B., and Schaefer L. (2011) Compositions of hot super-Earth atmospheres: Exploring Kepler candidates. *Astrophys. J. Lett.*, *742(2)*, L19.

Mills F. P. (1999) A spectroscopic search for molecular oxygen in the Venus middle atmosphere. *J. Geophys. Res.–Planets*, *104(E12)*, 30757–30763.

Mills F. P. and Allen M. (2007) A review of selected issues concerning the chemistry in Venus' middle atmosphere. *Planet. Space Sci.*, *55(12)*, 1729–1740.

Misra A., Meadows V., Claire M., and Crisp D. (2014a) Using dimers to measure biosignatures and atmospheric pressure for terrestrial exoplanets. *Astrobiology*, *14(2)*, 67–86.

Misra A., Meadows V., and Crisp D. (2014b) The effects of refraction on transit transmission spectroscopy: Application to Earth-like exoplanets. *Astrophys. J.*, *792(1)*, 61.

Molaverdikhani K., McGouldrick K., and Esposito L. W. (2012) The abundance and vertical distribution of the unknown ultraviolet absorber in the venusian atmosphere from analysis of Venus Monitoring Camera images. *Icarus*, *217(2)*, 648–660.

Montmessin F., Bertaux J. L., Lefèvre F., Marcq E., Belyaev D., Gérard J. C., Korablev O., Fedorova A., Sarago V., and Vandaele A. C. (2011) A layer of ozone detected in the nightside upper atmosphere of Venus. *Icarus*, *216(1)*, 82–85.

Morbidelli A., Chambers J., Lunine J. I., Petit J. M., Robert F., Valsecchi G. B., and Cyr K. E. (2000) Source regions and timescales for the delivery of water to the Earth. *Meteoritics & Planet. Sci.*, *35(6)*, 1309–1320.

Morbidelli A., Lunine J. I., O'Brien D. P., Raymond S. N., and Walsh K. J. (2012) Building terrestrial planets. *Annu. Rev. Earth Planet. Sci.*, *40*, 251–275.

Morgan E., Kerins E., Awiphan S., McDonald I., Hayes J., Komonjinda S., Mkritchian D., and Sanguansak N. (2018) Exoplanetary atmosphere target selection in the era of comparative planetology. *ArXiV e-prints*, arXiv: 1802.05645 [astro-ph.EP].

Morley C. V., Kreidberg L., Rustamkulov Z., Robinson T., and Fortney J. J. (2017) Observing the atmospheres of known temperate Earth-sized planets with JWST. *Astrophys. J.*, *850(2)*, 121.

Moroz V. I. and Zasova L. V (1997) VIRA-2: A review of inputs for updating the Venus International Reference Atmosphere. *Adv. Space Res.*, *19(8)*, 1191–1201, DOI: 10.1016/S0273-1177(97)00270-6.

Muhleman D. O. (1961) Early results of the 1961 JPL Venus radar experiment. *Astron. J.*, *66*, 292.

Nakajima S., Hayashi Y. Y., and Abe Y. (1992) A study on the "runaway greenhouse effect" with a one-dimensional radiative-convective equilibrium model. *J. Atmos. Sci.*, *49(23)*, 2256–2266.

Nakamura M., Imamura T., Ishii N., Abe T., Kawakatsu Y., Hirose C., Satoh T., Suzuki M., Ueno M., Yamazaki A., and Iwagami N. (2016) AKATSUKI returns to Venus. *Earth Planets Space*, *68(1)*, 75.

Namiki N. and Solomon S. C. (1994) Impact crater densities on volcanoes and coronae on Venus: Implications for volcanic resurfacing. *Science*, *265(5174)*, 929–933.

Navarro T., Schubert G., and Lebonnois S. (2018) Atmospheric mountain wave generation on Venus and its influence on the solid planet's rotation rate. *Nature Geosci.*, *11*, 487–491.

Nimmo F. (2002) Why does Venus lack a magnetic field? *Geology*, *30(11)*, 987–990.

Owen J. E. and Mohanty S. (2016) Habitability of terrestrial-mass planets in the HZ of M dwarfs — I. H/He-dominated atmospheres. *Mon. Not. R. Astron. Soc.*, *459(4)*, 4088–4108, DOI: 10.1093/mnras/stw959.

Owen T., Bar-Nun A., and Kleinfeld I. (1992) Possible cometary origin of heavy noble gases in the atmospheres of Venus, Earth and Mars. *Nature*, *358(6381)*, 43.

Oyama V. I., Carle G. C., Woeller F., and Pollack J. B. (1979) Venus lower atmospheric composition: Analysis by gas chromatography. *Science*, *203(4382)*, 802–805.

Palmer K. F. and Williams D. (1975) Optical constants of sulfuric acid; application to the clouds of Venus? *Appl. Opt.*, *14(1)*, 208–219.

Pepin R. O. (1991) On the origin and early evolution of terrestrial planet atmospheres and meteoritic volatiles. *Icarus*, *92(1)*, 2–79.

Peralta J. et al. (2009) Characterization of mesoscale gravity waves in the upper and lower clouds of Venus from VEX-VIRTIS images. *J. Geophys Res.–Planets*, *114(5)*, 1–12, DOI: 10.1029/2008JE003185.

Peralta J. et al. (2014) Analytical solution for waves in planets with atmospheric superrotation. I. Acoustic and inertia-gravity waves. *Astrophys. J. Suppl.*, *213(1)*, DOI: 10.1088/0067-0049/213/1/17.

Peralta J., Hueso R., Sánchez-Lavega A., Lee Y. J., Muñoz A. G., Kouyama T., Sagawa H., Sato T. M., Piccioni G., Tellmann S., and Imamura T. (2017) Stationary waves and slowly moving features in the night upper clouds of Venus. *Nature Astron.*, *1*, article #0187.

Pérez-Hoyos S., Sánchez-Lavega A., García-Muñoz A., Irwin P. G. J., Peralta J., Holsclaw G., McClintock W. M., and Sanz-Requena J. F. (2018) Venus upper clouds and the UV absorber from MESSENGER/MASCS observations. *J. Geophys. Res.–Planets*, *123(1)*, 145–162.

Pernice H., Garcia P., Willner H., Francisco J. S., Mills F. P., Allen M., and Yung Y. L. (2004) Laboratory evidence for a key intermediate in the Venus atmosphere: Peroxychloroformyl radical. *Proc. Natl. Acad. Sci.*, *101(39)*, 14007–14010.

Petigura E. A., Howard A. W., and Marcy G. W. (2013) Prevalence of Earth-size planets orbiting Sun-like stars. *Proc. Natl. Acad. Sci.*, *110*, 19273.

Phillips R. J., Raubertas R. F., Arvidson R. E., Sarkar I. C., Herrick R. R., Izenberg N., and Grimm R. E. (1992) Impact craters and Venus resurfacing history. *J. Geophys. Res.–Planets*, *97(E10)*, 15923–15948.

Piccialli A., Titov D. V., and Sanchez-Lavega A. (2014) High latitude gravity waves at the Venus cloud tops as observed by the Venus Monitoring Camera on board Venus Express. *Icarus*, *227*, 94–111, DOI: 10.1016/j.icarus.2013.09.012.

Pieters C. M., Head J. W., Pratt S., Patterson W., Garvin J., Barsukov V. L., Basilevsky A. T., Khodakovsky I. L., Selivanov A. S., Panfilov A .S., and Gektin Y. M. (1986) The color of the surface of Venus. *Science*, *234(4782)*, 1379–1383.

Pollack J. B. (1971) A nongrey calculation of the runaway greenhouse: Implications for Venus' past and present. *Icarus*, *14(3)*, 295–306.

Pollack J. et al. (1993) Near-infrared light from Venus' nightside: A spectroscopic analysis. *Icarus*, *103*, 1–42.

Prather M. J. and McElroy M. B. (1983) Helium on Venus: Implications for uranium and thorium. *Science*, *220(4595)*, 410–411.

Pujol M., Marty B., Burgess R., Turner G., and Philippot P. (2013) Argon isotopic composition of Archaean atmosphere probes early Earth geodynamics. *Nature*, *498(7452)*, 87–90.

Ragent B. and Blamont J. (1980) The structure of the clouds of Venus: Results of the Pioneer Venus nephelometer experiment. *J. Geophy. Res.–Space Phys.*, *85(A13)*, 8089–8105, DOI: 10.1029/JA085iA13p08089.

Ramirez R. M. and Kaltenegger L. (2014) The habitable zones of pre-main-sequence stars. *Astrophys. J. Lett., 797(2)*, L25.

Rauer H. et al. (2016) The PLATO mission. *Astron. Nachr., 337*, 961.

Raymond S. N., Quinn T., and Lunine J. I. (2006) High-resolution simulations of the final assembly of Earth-like planets I. Terrestrial accretion and dynamics. *Icarus, 183(2)*, 265–282.

Raymond S. N., Scalo J., and Meadows V. S. (2007) A decreased probability of habitable planet formation around low-mass stars. *Astrophys. J., 669(1)*, 606.

Raymond S. N., Kokubo E., Morbidelli A., Morishima R., and Walsh K. J. (2014) Terrestrial planet formation at home and abroad. In *Protostars and Planets VI* (H. Beuther et al., eds.), pp. 595–618. Univ. of Arizona, Tucson.

Reid N. I. and Hawley S. L. (2013) *New Light on Dark Stars: Red Dwarfs, Low-Mass Stars, Brown Dwarfs.* Springer, Berlin. 470 pp.

Robinson T. D. and Catling D. C. (2014) Common 0.1 bar tropopause in thick atmospheres set by pressure-dependent infrared transparency. *Nature Geosci., 7(1)*, 12.

Rosenqvist J. and Chassefière E. (1995) Inorganic chemistry of O_2 in a dense primitive atmosphere. *Planet. Space Sci., 43(1–2)*, 3–10.

Romeo I. and Turcotte D. L. (2008) Pulsating continents on Venus: An explanation for crustal plateaus and tessera terrains. *Earth Planet. Sci. Lett., 276(1–2)*, 85–97.

Ross F. E. (1928) Photographs of Venus. *Astrophys. J., 68*, 57.

Rothman L. S., Gordon I. E., Babikov Y., Barbe A., Benner D. C., Bernath P. F., Birk M., Bizzocchi L., Boudon V., Brown L. R., and Campargue A. (2013) The HITRAN2012 molecular spectroscopic database. *J. Quant. Spectrosc. Radiat. Transfer, 130*, 4–50.

Russell C. T., Elphic R. C., and Slavin J. A. (1980) Limits on the possible intrinsic magnetic field of Venus. *J. Geophys Res.–Space Phys., 85(A13)*, 8319–8332.

Sagan C. (1961) The planet Venus. *Science, 133(3456)*, 849–858.

Sandor B. J. et al. (2010) Sulfur chemistry in the Venus mesosphere from SO_2 and SO microwave spectra. *Icarus, 208(1)*, 49–60, DOI: 10.1016/j.icarus.2010.02.013.

Satoh T. et al. (2009) Cloud structure in Venus middle-to-lower atmosphere as inferred from VEX/VIRTIS 1.74 μm data. *J. Geophys. Res.–Planets, 114(E9)*, DOI: 10.1029/2008JE003184.

Schaber G. G., Strom R. G., Moore H. J., Soderblom L. A., Kirk R. L., Chadwick D. J., Dawson D. D., Gaddis L. R., Boyce J. M., and Russell J. (1992) Geology and distribution of impact craters on Venus: What are they telling us? *J. Geophys. Res.–Planets, 97(E8)*, 13257–13301.

Schaber G. G., Kirk R. L., and Strom R. G. (1998) *Data Base of Impact Craters on Venus Based on Analysis of Magellan Radar Images and Altimetry Data.* USGS Open File Report 98-104, U.S. Geological Survey, DOI: 10.3133/ofr98104.

Schaefer L. and Fegley B. Jr. (2009) Chemistry of silicate atmospheres of evaporating super-Earths. *Astrophys. J. Lett., 703(2)*, L113.

Schaefer L. and Fegley B. Jr. (2011) Atmospheric chemistry of Venus-like exoplanets. *Astrophys. J., 729(1)*, 6.

Schaefer L., Wordsworth R. D., Berta-Thompson Z., and Sasselov D. (2016) Predictions of the atmospheric composition of GJ 1132b. *Astrophys. J., 829(2)*, 63.

Shalygin E. V., Basilevsky A. T., Markiewicz W. J., Titov D. V., Kreslavsky M. A., and Roatsch T. (2012) Search for ongoing volcanic activity on Venus: Case study of Maat Mons, Sapas Mons and Ozza Mons volcanoes. *Planet. Space Sci., 73(1)*, 294–301.

Shalygin E. V., Markiewicz W. J., Basilevsky A. T., Titov D. V., Ignatiev N. I., and Head J. W. (2015) Active volcanism on Venus in the Ganiki Chasma rift zone. *Geophys. Res. Lett., 42(12)*, 4762–4769.

Schubert G. (1983) General circulation and the dynamical state of the Venus atmosphere. In *Venus* (D. M. Hunten et al., eds.), pp. 681–765. Univ. of Arizona, Tucson.

Schulze-Makuch D. and Irwin L. N. (2002) Reassessing the possibility of life on Venus: Proposal for an astrobiology mission. *Astrobiology, 2(2)*, 197–202.

Schulze-Makuch D., Grinspoon D. H., Abbas O., Irwin L. N., and Bullock M. A. (2004) A sulfur-based survival strategy for putative phototrophic life in the venusian atmosphere. *Astrobiology, 4(1)*, 11–18.

Schwieterman E. W., Meadows V. S., Domagal-Goldman S. D., Deming D., Arney G. N., Luger R., Harman C. E., Misra A., and Barnes R. (2016) Identifying planetary biosignature impostors: Spectral features of CO and O_4 resulting from abiotic O_2/O_3 production.

Astrophys. J. Lett., 819(1), L13.

Seiff A. et al. (1980) Measurements of thermal structure and thermal contrasts in the atmosphere of Venus and related dynamical observations: Results from the four pioneer Venus probes. *J. Geophys Res., 85(A13)*, 7903–7933, DOI: 10.1029/JA085iA13p07903.

Seiff A. et al. (1985) Models of the structure of the atmosphere of Venus from the surface to 100 kilometers altitude. *Adv. Space Res., 5(11)*, 3–58.

Simpson G. C. (1928) Some studies in terrestrial radiation. *Mem. R. Meteoritical Soc., 2(16)*, DOI: 10.1002/qj.49705522908.

Sill G. T. (1972) Sulfuric acid in the Venus clouds. *Commun. Lunar Planet. Lab., 166–171(9)*, 191–198.

Sinclair J. et al. (2013) Seasonal variations of temperature, acetylene and ethane in Saturn's atmosphere from 2005 to 2010, as observed by Cassini-CIRS. *Icarus, 225(1)*, 257–271, DOI: 10.1016/j.icarus.2013.03.011.

Sleep N. H. and Zahnle K. (2001) Carbon dioxide cycling and implications for climate on ancient Earth. *J. Geophys. Res.–Planets, 106(E1)*, 1373–1399.

Smith D. J., Griffin D. W., and Schuerger A. C. (2010) Stratospheric microbiology at 20 km over the Pacific Ocean. *Aerobiologia, 26(1)*, 35–46.

Smith D. J. (2013) Microbes in the upper atmosphere and unique opportunities for astrobiology research. *Astrobiology, 13(10)*, 981–990.

Smrekar S. E., Stofan E. R., Mueller N., Treiman A., Elkins-Tanton L., Helbert J., Piccioni G., and Drossart P. (2010) Recent hotspot volcanism on Venus from VIRTIS emissivity data. *Science, 328(5978)*, 605–608.

Solomon S. C., Bullock M. A., and Grinspoon D. H. (1999) Climate change as a regulator of tectonics on Venus. *Science, 286(5437)*, 87–90.

Som S. M., Buick R., Hagadorn J. W., Blake T. S., Perreault J. M., Harnmeijer J. P., and Catling D. C. (2016) Earth's air pressure 2.7 billion years ago constrained to less than half of modern levels. *Nature Geosci., 9*, 448–451.

Stevenson D. J. (2003) Planetary magnetic fields. *Earth Planet. Sci. Lett., 208(1–2)*, 1–11.

Stevenson D. J., Spohn T., and Schubert G. (1983) Magnetism and thermal evolution of the terrestrial planets. *Icarus, 54(3)*, 466–489.

Strom R. G., Schaber G. G., and Dawson D. D. (1994) The global resurfacing of Venus. *J. Geophys. Res.–Planets, 99(E5)*, 10899–10926.

Sullivan P.W., Winn J. N., Berta-Thompson Z. K., Charbonneau D., Deming D., Dressing C. D., Latham D. W., Levine A. M., McCullough P. R., Morton T., and Ricker G. R. (2015) The Transiting Exoplanet Survey Satellite: Simulations of planet detections and astrophysical false positives. *Astrophys. J., 809(1)*, 77, DOI: 10.1088/0004-637X/809/1/77.

Teanby N. A. et al. (2008) Global and temporal variations in hydrocarbons and nitriles in Titan's stratosphere for northern winter observed by Cassini/CIRS. *Icarus, 193(2)*, 595–611, DOI: 10.1016/j.icarus.2007.08.017.

Thalman R. and Volkamer R. (2013) Temperature dependent absorption cross-sections of O_2-O_2 collision pairs between 340 and 630 nm and at atmospherically relevant pressure. *Phys. Chem. Chem. Phys., 15(37)*, 15371–15381.

Thomas N. C. (1991) The early history of spectroscopy. *J. Chem. Educ., 68(8)*, 631, DOI: 10.1021/ed068p631.

Tian F. (2015) History of water loss and atmospheric O_2 buildup on rocky exoplanets near M dwarfs. *Earth Planet. Sci. Lett., 432*, 126–132.

Tian F., France K., Linsky J. L., Mauas P. J., and Vieytes M. C. (2014) High stellar FUV/NUV ratio and oxygen contents in the atmospheres of potentially habitable planets. *Earth Planet. Sci. Lett., 385*, 22–27.

Titov D. V., Taylor F. W., Svedhem H., Ignatiev N. I., Markiewicz W. J., Piccioni G., and Drossart P. (2008) Atmospheric structure and dynamics as the cause of ultraviolet markings in the clouds of Venus. *Nature, 456(7222)*, 620.

Tolstikhin I., Marty B., Porcelli D., and Hofmann A. (2014) Evolution of volatile species in the Earth's mantle: A view from xenology. *Geochim. Cosmochim. Acta, 136*, 229–246.

Tomasko M. G. et al. (1980) Measurements of the flux of sunlight in the atmosphere of Venus. *J. Geophys. Res.–Atmos., 85(80)*, 8167–8186,

DOI: 10.1029/JA085iA13p08167.

Tomasko M. G., Doose L. R., and Smith P. H. (1985) The absorption of solar energy and the heating rate in the atmosphere of Venus. *Adv. Space Res., 5(9),* 71–79, DOI: 10.1016/0273-1177(85)90272-8.

Tomkins A. G., Bowlt L., Genge M., Wilson S. A., Brand H. E., and Wykes J. L. (2016) Ancient micrometeorites suggestive of an oxygen-rich Archaean upper atmosphere. *Nature, 533(7602),* 235.

Toon O. B., Turco R. P., and Pollack J. B. (1982) The ultraviolet absorber on Venus: Amorphous sulfur. *Icarus, 51(2),* 358–373.

Toon O. et al. (1984) Large, solid particles in the clouds of Venus: Do they exist? *Icarus, 57,* 143–160.

Trauger J. T. and Lunine J. I. (1983) Spectroscopy of molecular oxygen in the atmospheres of Venus and Mars. *Icarus, 55(2),* 272–281.

Treiman A. H. and Bullock M. A. (2012) Mineral reaction buffering of Venus' atmosphere: A thermochemical constraint and implications for Venus-like planets. *Icarus, 217(2),* 534–541.

Turcotte D. L. and Schubert G. (1988) Tectonic implications of radiogenic noble gases in planetary atmospheres. *Icarus, 74(1),* 36–46.

Turcotte D. L., Morein G., Roberts D., and Malamud B. D. (1999) Catastrophic resurfacing and episodic subduction on Venus. *Icarus, 139(1),* 49–54.

Unterborn C. T., Desch S. J., Hinkel N. R., and Lorenzo A. (2018) Inward migration of the TRAPPIST-1 planets as inferred from their water-rich compositions. *Nature Astron., 2,* 297–302, DOI: 10.1038/s41550-018-0411-6.

Vanderspek R., Huang C. X., Vanderburg A., Ricker G. R., Latham D. W., Seager S., Winn J. N., Jenkins J. M., Burt J., Dittmann J., and Newton E. (2019) TESS discovery of an ultra-short-period planet around the nearby M dwarf LHS 3844. *Astrophys. J. Lett., 871(2),* L24.

Von Zahn U. et al. (1983) Composition of the Venus atmosphere. In *Venus* (D. M. Hunten et al., eds.), pp. 299–430. Univ. of Arizona, Tucson.

Walker J. C., Hays P. B., and Kasting J. F. (1981) A negative feedback mechanism for the long-term stabilization of Earth's surface temperature. *J. Geophys Res.–Oceans, 86(C10),* 9776–9782.

Way M. J., Del Genio A. D., Kiang N. Y., Sohl L. E., Grinspoon D. H., Aleinov I., Kelley M., and Clune T. (2016) Was Venus the first habitable world of our solar system? *Geophys. Res. Lett., 43(16),* 8376–8383.

West A. A., Weisenburger K. L., Irwin J., Berta-Thompson Z. K., Charbonneau D., Dittmann J., and Pineda J. S. (2015) An activity-rotation relationship and kinematic analysis of nearby mid-to-late-type M dwarfs. *Astrophys. J., 812(1),* 3.

Wolf E. T. and Toon O. B. (2015) The evolution of habitable climates under the brightening Sun. *J. Geophys Res.–Atmos., 120(12),* 5775–5794.

Wordsworth R. D. (2016) Atmospheric nitrogen evolution on Earth and Venus. *Earth Planet. Sci. Lett., 447,* 103–111, DOI: 10.1016/j.epsl.2016.04.002.

Wordsworth R. and Pierrehumbert R. (2014) Abiotic oxygen-dominated atmospheres on terrestrial habitable zone planets. *Astrophys. J. Lett., 785(2),* 2–5, DOI: 10.1088/2041-8205/785/2/L20.

Yang J., Cowan N. B., and Abbot D. S. (2013) Stabilizing cloud feedback dramatically expands the habitable zone of tidally locked planets. *Astrophys. J. Lett., 771(2),* L45.

Yoder C. F. (1997) Venusian spin dynamics. In *Venus II — Geology, Geophysics, Atmosphere, and Solar Wind Environment* (S. W. Bougher et al., eds.), pp. 1087–1124. Univ. of Arizona, Tucson.

Young A. T. (1975) The clouds of Venus. *J. Atmos. Sci., 32(6),* 1125–1132.

Young A. T. (1977) An improved Venus cloud model. *Icarus, 32(1),* 1–26.

Young R. E., Walterscheid R. L., Schubert G., Seiff A., Linkin V. M., and Lipatov A. N. (1987) Characteristics of gravity waves generated by surface topography on Venus: Comparison with the VEGA balloon results. *J. Atmos. Sci., 44(18),* 2628–2639.

Yung Y. and DeMore W. (1982) Photochemistry of the stratosphere of Venus : Implications for atmospheric evolution. *Icarus, 51,* 199–247.

Yung Y. L., Liang M. C., Jiang X., Shia R. L., Lee C., Bézard B., and Marcq E. (2009) Evidence for carbonyl sulfide (OCS) conversion to CO in the lower atmosphere of Venus. *J. Geophys. Res.–Planets, 114(E5).*

Zahnle K. (1993) Planetary noble gases. In *Protostars and Planets III* (E. H. Levy and J. I. Lunine, eds.), pp. 1305–1338. Univ. of Arizona, Tucson.

Zahnle K. J. and Catling D. C. (2013) The cosmic shoreline. In *Lunar Planet. Sci. XLIV,* Abstract #2787. Lunar and Planetary Institute, Houston.

Zahnle K., Arndt N., Cockell C., Halliday A., Nisbet E., Selsis F., and Sleep N. H. (2007) Emergence of a habitable planet. *Space Sci. Rev., 129(1–3),* 35–78.

Zasova L. V., Krasnopolsky V. A., and Moroz V. I. (1981) Vertical distribution of SO_2 in upper cloud layer of Venus and origin of UV-absorption. *Adv. Space Res., 1(9),* 13–16.

Zhang X., Liang M. C., Mills F. P., Belyaev D. A., and Yung Y. L. (2012) Sulfur chemistry in the middle atmosphere of Venus. *Icarus, 217(2),* 714–739.

Zingales T. et al. (2018) The ARIEL mission reference sample. *Exp. Astron., 46(1),* 67–100.

Robinson T. and Reinhard T. (2020) Earth as an exoplanet. In *Planetary Astrobiology* (V. Meadows et al., eds.), pp. 379–416. Univ. of Arizona, Tucson, DOI: 10.2458/azu_uapress_9780816540068-ch015.

Earth as an Exoplanet

Tyler D. Robinson
Northern Arizona University

Christopher T. Reinhard
Georgia Institute of Technology

Earth is the only planet known to harbor life and, as a result, the search for habitable and inhabited planets beyond the solar system commonly focuses on analogs to our planet. However, Earth's atmosphere and surface environment have evolved substantially in the last 4.5 b.y. A combination of *in situ* geological and biogeochemical modeling studies of our planet have provided glimpses of environments that, while technically belonging to our Earth, are seemingly alien worlds. For modern Earth, observations from groundbased facilities, satellites, and spacecraft have yielded a rich collection of data that can be used to effectively view our planet within the context of exoplanet characterization. Application of planetary and exoplanetary remote sensing techniques to these datasets then enables the development of approaches for detecting signatures of habitability and life on other worlds. In addition, an array of models have been used to simulate exoplanet-like datasets for the distant Earth, thereby providing insights that are often complementary to those from existing observations. Understanding the myriad ways Earth has been habitable and inhabited, coupled with remote sensing approaches honed on the distant Earth, provides a key guide to recognizing potentially life-bearing environments in other planetary systems.

*Look again at that dot. That's here. That's home. That's us. [. . .] [E]very saint
and sinner in the history of our species lived there — on a mote of dust suspended
in a sunbeam. — Carl Sagan*

1. INTRODUCTION

The quest for both habitable and inhabited worlds beyond Earth is key to understanding the potential distribution of life in the universe. This ongoing search seeks to answer profound questions: Are we alone? How unique is Earth? Should the hunt for life beyond Earth uncover a multitude of habitable worlds and few (if any) inhabited ones, humanity would begin to understand just how lonely and fragile our situation is. On the other hand, if our hunt yields a true diversity of inhabited worlds, then we would learn something fundamental about the tenacity of life in the cosmos.

But how will we recognize a distant habitable world, and how would we know if this environment hosts some form of life? A key opportunity for understanding the remote characterization of habitability and life comes from studying our own planet — Earth will always be our best example of a habitable and inhabited world. Thus, by studying our planet within the context of exoplanet exploration and characterization, we develop ideas, approaches, and tools suitable for remotely detecting the signs of (near) global surface habitability and a vigorous planet-wide biosphere.

While habitable exoplanets are unlikely to look exactly like Earth, these worlds will probably share some important characteristics with our own, including the presence of oceans, clouds, surface inhomogeneities, and, potentially, life. Studying globally-averaged observations of Earth within the context of remote sensing therefore provides insights into the ideal measurements to identify planetary habitability from data-limited exoplanet observations.

Of course, Earth is not a static environment. Life emerged on our planet into an environment completely unlike the Earth we understand today. The subsequent evolution of our planet — an intimate coupling between life and geochemical processes — produced worlds seemingly alien to modern Earth. Ranging from ice-covered "Snowball Earth" scenarios to the likely oxygen-free and, potentially, intermittently hazy atmosphere of the Archean [3.8–2.5 giga-annum (Ga)], each evolutionary stage of our planet offers a unique opportunity to understand habitable, life-bearing worlds distinct from the present Earth.

The chapter presented here summarizes studies of Earth within the context of exoplanet characterization. Following a brief synopsis of the current state of exoplanet science, we review our understanding of the evolution of Earth, and its associated appearance, over the last 4 b.y. Then, using this understanding of Earth through time, we review how key remotely-detectable biosignatures for our planet may have changed over geological timescales. We then shift our emphasis to modern Earth, where existing observational datasets and modeling tools can be used to explore ideas related to characterizing Earth-like planets from a distance. Finally, we present an overview of what has been learned by studying Earth as an exoplanet, summarizing approaches to remote characterization of potentially habitable or inhabited worlds. For further reading, we note that an entire book on studies of the distant Earth has been published by *Vázquez et al.* (2010).

1.1. Current State of Exoplanet Science

Following the first detection of an exoplanet around a Sun-like star (*Mayor and Queloz,* 1995) and of an exoplanet atmosphere (*Charbonneau et al.,* 2002) over a decade ago, the field of exoplanetary science has been marked by two clear trends: the steady discovery of increasingly smaller worlds on longer-period orbits, and the ever-increasing quality of observational data suitable for characterizing worlds around other stars. Due to advances in exoplanet detection using a variety of techniques, we now know that, on average, every star in the Milky Way galaxy hosts at least one exoplanet (*Cassan et al.,* 2012). Furthermore, due in large part to the success of the Kepler mission (*Borucki et al.,* 2010), we understand that occurrence rates of potentially Earth-like worlds orbiting within the habitable zone of main-sequence stars are relatively large, with estimates spanning roughly 10–50% (*Dressing and Charbonneau,* 2013, 2015; *Petigura et al.,* 2013; *Batalha,* 2014; *Foreman-Mackey et al.,* 2014; *Burke et al.,* 2015; *Kopparapu et al.,* 2018). Excitingly, and especially for low-mass stellar hosts, surveys have revealed a number of nearby potentially Earth-like exoplanets, such as Proxima Centauri b (*Anglada-Escudé et al.,* 2016) or the worlds in the TRAPPIST-1 system (*Gillon et al.,* 2016, 2017).

The subsequent characterization of exoplanet atmospheres has largely been accomplished using transit and/or secondary eclipse spectroscopy (for a review, see *Kreidberg,* 2018). The former relies on the wavelength-dependent transmittance of an exoplanet atmosphere (*Seager and Sasselov,* 2000; *Brown,* 2001; *Hubbard et al.,* 2001), which causes a transiting world to block more (for lower transmittance) or less (for higher transmittance) light when crossing the disk of its host. By comparison, secondary eclipse spectroscopy measures the planet-to-star flux ratio by observing the combined star and exoplanet spectrum prior to the planet disappearing behind its host star (i.e., secondary eclipse). As with any burgeoning field, some findings related to exoplanet atmospheres remain controversial or have undergone substantial revision (*Line et al.,* 2014; *Hansen et al.,* 2014; *Diamond-Lowe et al.,* 2014). Nevertheless, using these techniques astronomers have

probed the atmospheres of a striking variety of exoplanets, spanning so-called hot Jupiters (*Grillmair et al.,* 2008; *Swain et al.,* 2008, 2009; *Pont et al.,* 2008; *Sing et al.,* 2009; *Madhusudhan and Seager,* 2009), as well as mini-Neptunes and Earth-like planets orbiting within the habitable zone of their Sun-like hosts, due to the long orbital periods, small transit probabilities, and low signal sizes for such worlds. ("Sun-like" typically refers to main-sequence stars with spectral type F, G, or K. Such stars range from 30% larger and 6% more luminous than our Sun down to 30% smaller and 13% less luminous than the Sun. Sun-like stars are often contrasted to "late-type" stars, which are red, cool main-sequence M stars that can be smaller than 10% the size of our Sun and can have luminosities as small as 1–2% that of the Sun.) Here, direct (or high-contrast) imaging will likely be the leading observational approach, and, as a result, the material below focuses primarily on directly observing Earth (in both reflected light and thermal emission). For exoplanets, direct imaging involves blocking the light of a bright central host star in order to resolve and observe faint companions to that star (*Traub and Oppenheimer,* 2010). Both internal and external occulting technologies are under active study (*Guyon et al.,* 2006; *Cash et al.,* 2007; *Shaklan et al.,* 2010; *Mawet et al.,* 2012), and groundbased telescopes equipped with coronagraphs already enable the characterization of hot gas giant exoplanets orbiting young, nearby stars (*Marois et al.,* 2008; *Skemer et al.,* 2012; *Macintosh et al.,* 2015).

1.2. The Future of Rocky Exoplanets

A number of planned or under-study missions will improve and expand our ability to characterize exoplanet atmospheres and surfaces. Foremost among these is NASA's James Webb Space Telescope (JWST) (*Gardner et al.,* 2006), which is expected to provide high-quality transit and secondary eclipse spectra of many tens of targets over the duration of its designed five-year mission (*Beichman et al.,* 2014). Some of these observations will probe lower-mass, potentially rocky exoplanets (*Deming et al.,* 2009; *Batalha et al.,* 2015). Critically, JWST may even be capable of characterizing temperate Earth-sized planets orbiting low-mass stars (*Kaltenegger and Traub,* 2009; *Cowan et al.,* 2015; *Barstow et al.,* 2016), although the ability to conduct such studies depends largely on the behavior and size of systematic noise sources (*Greene et al.,* 2016).

Following JWST, NASA will launch the Wide-Field InfraRed Survey Telescope (WFIRST) (*Spergel et al.,* 2013). It is anticipated that WFIRST will be equipped with a Coronagraphic Instrument (CGI) capable of visible-light imaging and, potentially, spectroscopy or spectrophotometry of exoplanets (*Noecker et al.,* 2016). Key outcomes of this mission will include a demonstration of high-precision coronagraphy in space, as well as the study of a small handful of cool, gas giant exoplanets (*Marley et al.,* 2014; *Hu,* 2014; *Burrows,* 2014; *Lupu et al.,* 2016; *Traub et al.,* 2016;

Nayak et al., 2017). However, the planned capabilities of WFIRST/CGI will make observations of Earth-like planets extremely unlikely (*Robinson et al.,* 2016).

Exoplanet direct imaging missions that could build on the technological successes of WFIRST are already under investigation. Included here are the WFIRST starshade rendezvous concept (*Seager et al.,* 2015), the Habitable Exoplanet (HabEx) imaging mission (*Mennesson et al.,* 2016; *Gaudi et al.,* 2018), and the Large Ultraviolet-Visible-InfraRed (LUVOIR) explorer (*Peterson et al.,* 2017; *Roberge and Moustakas,* 2018). While the scope and capabilities of these mission concepts are varied (*Stark et al.,* 2016), a central goal unites these designs: to detect and characterize "pale blue dots" around our nearest stellar neighbors.

2. EARTH AS AN EVOLVING EXOPLANET

Earth, however, has not always been the pale blue dot we see today. Indeed, the Earth system has evolved considerably over time (Fig. 1). These changes have in turn impacted both the habitability of Earth surface environments (e.g., *Kasting and Catling,* 2003) and the remote detectability of Earth's biosphere (*Kaltenegger et al.,* 2007; *Meadows,* 2008; *Reinhard et al.,* 2017a; *Rugheimer and Kaltenegger,* 2018). In particular, the atmospheric abundances of almost all potential biosignature gases (e.g., CH_4, O_2, O_3, N_2O, CO_2) have changed by many orders of magnitude throughout Earth's history. The timing and magnitude of these changes have been controlled by often complex interactions between biological, geologic, and

stellar factors. At the same time, Earth's climate system and surface habitability have changed significantly, as influenced by both long-term trends in solar energy flux, catastrophic climate destabilization during low-latitude "Snowball Earth" glaciations, and major impact events.

Despite these dramatic changes, all life on Earth appears to share a single origin that is perhaps nearly as ancient as Earth itself (*Fox et al.,* 1980; *Hedges,* 2016). Earth's history thus allows us to explore the long-term evolutionary factors controlling the production and maintenance of remotely detectable signatures of habitability and life against the backdrop of a continuously inhabited planet. Fully illuminating this history requires integration of geologic observations, geochemical data, and constraints from theoretical models — and provides a unique opportunity to develop predictive frameworks that can be leveraged in the search for living planets beyond Earth.

In this section, our focus is on observations and models aimed at constraining surface habitability and atmospheric composition through time on Earth, with a particular eye toward key habitability indicators and atmospheric biosignatures. Critically, data and models can be used to constrain the surface habitability and atmospheric composition of Earth through time, enabling an assessment of the strength of different biosignature features and habitability markers and how these may change through planetary evolution. In addition, Earth's geologic history provides a series of empirical tests of our understanding of Earth-like planets as integrated systems — with different periods of Earth's evolutionary history serving as analogs for alien, yet habitable, worlds for which we have biological and geological constraints.

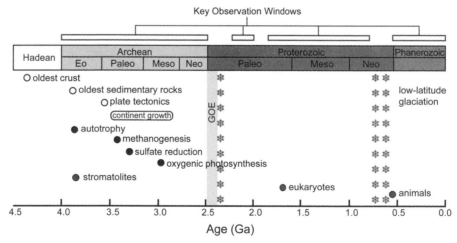

Fig. 1. Summary timeline of geologic history for Earth, with major divisions of the geologic timescale (top) related to quantitative ages (bottom) according to billions of years before the present (Ga). Major early geologic events are shown by open white circles, including the development of the oldest crust and sedimentary rocks, the emergence of plate tectonics, and an interval of significant growth/exposure of continental crust above sea level. Filled circles denote approximate geological and geochemical constraints on major biological innovations and events. These include carbon fixation (autotrophy), methane production, sulfate reduction, oxygenic photosynthesis, the first putative stromatolites (layered sedimentary structures induced by microorganisms), the emergence of eukaryotes (large, complex cells with a membrane-bound nucleus), and the first multicellular animal fossils. Also shown are the initial oxygenation of Earth's atmosphere — the Great Oxidation Event (GOE) — and three occurrences of unusually intense low-latitude glaciation ("Snowball Earth" events). Above the geologic timeline we summarize our key "observation windows" examined in Fig. 2 and Table 1.

2.1. Geological Constraints on Evolving Climate

Understanding the evolution of Earth's climate system is critical for diagnosing how potential observational discriminants of Earth's habitability may have changed with time. Interestingly, with some notable exceptions (see below), Earth's climate appears to have been clement for the vast majority of its history. For instance, isotopic evidence from the oldest minerals on Earth suggest the presence of liquid water at Earth's surface by 4.3 b.y. ago (Ga) (*Mojzsis et al.,* 2001; *Ushikubo et al.,* 2008), and a consistent if fragmentary marine sedimentary rock record attests to a large-scale fluid-mediated rock cycle for the last 3.8 b.y. (e.g., *Rosing et al.,* 1996). Other lines of evidence (*Knauth and Epstein,* 1976; *Robert and Chaussidon,* 2006; *Gaucher et al.,* 2008) have been used to suggest that surface temperatures during much of the Archean Eon (3.8–2.5 Ga) were hot, perhaps as high as 70°C. More recent estimates suggest much cooler (but still quite warm) temperatures between 25° and 40°C (*Hren et al.,* 2009; *Blake et al.,* 2010).

Geologic evidence for glacial deposits can provide first-order information about evolving climate, particularly if the approximate location and altitude of glacial activity can be constrained. The most well-established sedimentary evidence for early glacial activity near sea level is found in the Mozaan Group, South Africa at ~2.9 Ga (*Young et al.,* 1998). Given existing paleolatitude constraints of around 45°–50° (*Kopp et al.,* 2005), these deposits suggest a climate similar to or colder than that of the Pleistocene Earth [a relatively cold epoch of repeated glaciations spanning 2.59–0.012 mega-annum (Ma)]. More recently, sedimentological evidence of glacial activity has been reported from the ~3.5-Ga Overwacht Group, Barberton Greenstone Belt, South Africa (*de Wit and Furnes,* 2016), although their glacial origin is less definitive than that of the Mozaan Group sediments (e.g., *Viljoen and Viljoen,* 1969). The reconstructed paleolatitudes of these deposits are between ~20° and 40°, so if they are indeed glaciogenic in origin they would imply a relatively cold early Archean climate.

Firm evidence for glacial activity near sea level does not reappear until after the Archean-Proterozoic boundary, with a series of glacial deposits observed in North America (*Young,* 2001) and South Africa (*Rasmussen et al.,* 2013) between ~2.4 and 2.3 Ga. Glaciogenic deposits found on the Kaapval craton, South Africa, recently dated to 2.426 ± 0.003 Ga (*Gumsley et al.,* 2017), show evidence for being deposited at low latitudes (*Evans et al.,* 1997), leading to the suggestion that these deposits record a Paleoproterozoic "Snowball Earth" — classically envisaged as a catastrophic destabilization of the climate system during which runaway ice-albedo feedback causes the advance of ice sheets to the tropics and a virtual shutdown of the hydrologic cycle (*Budyko,* 1969; *Sellers,* 1969). The temporal correspondence between this apparently intense ice age and the initial accumulation of O_2 in Earth's atmosphere (Fig. 1) has led to the intriguing suggestion that the climate system was transiently destabilized by a sharp drop in atmospheric CH_4 attendant to rising atmospheric O_2 (*Kasting,* 2005).

Recent evidence suggests that these intense ice ages were followed by a transient period of atmospheric oxygenation, after which the climate system appears to have been relatively stable with little firm evidence for glaciation between ~1.8 and 0.8 b.y. ago (here referred to as the "mid-Proterozoic"). A notable exception to this comes in the form of putative glacial deposits from the Vazante Group, east-central Brazil (*Azmy et al.,* 2008; *Geboy et al.,* 2013), although their age is somewhat enigmatic (*Geboy et al.,* 2013; *Rodrigues et al.,* 2012). These deposits indicate that at least portions of the mid-Proterozoic were not entirely ice-free, but their deposition at relatively high paleolatitude (*Tohver et al.,* 2006) renders their broader climatic implications somewhat difficult to interpret.

The close of the Proterozoic Eon (2.5–0.541 Ga) bore witness to perhaps the most severe climate perturbations in Earth's history, the Neoproterozoic "Snowball Earth" events (*Hoffman et al.,* 1998). Recent high-resolution geochronology delineates two major glacial episodes, the protracted Sturtian glaciation (lasting between 717 and 660 Ma) and the shorter Marinoan glaciation (terminating at 635 Ma), with a relatively brief interglacial period lasting less than 25 m.y. (*Rooney et al.,* 2015). While understanding the intensity, dynamics, and biogeochemical impacts of these glaciations remain areas of active research [extensively reviewed in *Hoffman and Schrag* (2002), *Pierrehumbert et al.* (2011), and *Hoffman et al.* (2017)], it is clear that this period marks a dramatic perturbation to planetary climate and would have represented a significant and protracted shift in the remotely detectable indicators of Earth's habitability.

The Phanerozoic Eon (e.g., the last 541 m.y.) has been marked by at least three large-scale ice ages (*Delabroye and Vecoli,* 2010; *Veevers and Powell,* 1987; *Zachos et al.,* 2001). These events have been linked with faunal turnover and mass extinction (*Raymond and Metz,* 2004), and in some cases reflect major milestones in the evolution of Earth's biosphere such as the earliest colonization of the land surface by simple plant life at ~470 Ma (*Lenton et al.,* 2012) and the extensive production and burial of organic matter by burgeoning terrestrial ecosystems around 300 Ma (*Feulner,* 2017). At the same time, the most recent half-billion years of Earth's history show evidence for significant transient perturbations to Earth's carbon cycle and climate system on a wide range of timescales (*Zachos et al.,* 2008; *Hönisch et al.,* 2012), often associated with dramatic changes to the diversity and abundance of macroscopic life (*Erwin,* 1994; *Payne et al.,* 2004). Nevertheless, despite large changes to carbon fluxes into and out of the ocean-atmosphere system, Phanerozoic Earth's climate has consistently avoided the sort of catastrophic climate destabilization witnessed during the late Proterozoic.

In sum, Earth's geologic record suggests that the establishment of a robust hydrosphere, with liquid water oceans and low-temperature aqueous weathering of exposed crust, occurred very shortly after Earth's formation. In addition, surface temperatures have generally been stable and relatively warm for the vast majority of Earth's history, despite long-term changes in solar insolation and dramatic changes to atmospheric composition (see below). However, this history also highlights the importance of internal feedbacks within the climate system in structuring planetary habitability on Earth (and thus the likelihood of remote detection) over time. For example, the "Snowball Earth" glaciations suggest that an Earth-like planet that spends its lifetime safely within the habitable zone of its host star can still undergo catastrophic climate destabilization, and that both the timing and duration of these events can be unpredictable (e.g., *Rooney et al.,* 2015). This is in marked contrast to the regular, periodic mode of climate instability predicted for Earth-like planets near the outer edge of the habitable zone (*Haqq-Misra et al.,* 2016). In addition, the contrasting timescales of the two Neoproterozoic glaciations imply a wide range of potential effects on the long-term maintenance of remotely detectable biosignatures, placing significant impetus on better understanding the large-scale biogeochemistry of "Snowball Earth" conditions (see below).

2.2. Geological Constraints on Evolving Atmospheric Chemistry

Geologic and geochemical data also provide a window into the dramatic evolution of Earth's atmospheric chemistry, with implications for both the habitability of surface environments and the remote detectability of atmospheric biosignatures. For the last ~2 m.y., atmospheric composition on Earth can be tracked directly by analyzing the composition of volatiles trapped in ice (*Wolff and Spahni,* 2007). Prior to this, biogeochemists and planetary scientists seeking to reconstruct the composition of Earth's atmosphere must rely on some form of "proxy" — an indirect indicator of atmospheric composition. For example, in many species of plant the spatial density of cells on the leaf that are used to exchange gases with the environment — "stomata" — scales coherently with the abundance of CO_2 in the growth environment (e.g., *Royer,* 2001). Paleobotanists can thus use the stomatal density of fossil plant leaves as a proxy for atmospheric CO_2 abundance in Earth's past. Our focus in this chapter is on major changes to atmospheric gas species that are important for regulating global climate (e.g., CO_2, CH_4, N_2, and possibly H_2) and species that are potentially promising biosignature gases (e.g., O_2, O_3, CH_4, N_2O). Some species, most notably methane and organic hazes, serve dual roles as both arbiters of climate and potential biosignatures.

There are four broad temporal intervals of Earth's evolutionary history that are relevant to our purposes: the Archean (4.0–2.5 Ga), the early Paleoproterozoic (more specifically the interval between 2.2 and 2.0 Ga), the mid-Proterozoic (specifically the interval between 1.8 and 0.8 Ga), and the Phanerozoic (roughly 0.5 Ga to the present) (see Fig. 1). It is important to bear in mind that these all represent extremely long periods of time, and that there is likely to be higher-order variability within each interval., Nevertheless,

model-derived spectra of Earth during these key evolutionary stages demonstrate how varying atmospheric compositions (Fig. 2) have led to dramatically different spectral appearances for our planet over time (Fig. 3).

2.2.1. Atmospheric oxygen (O_2) and methane (CH_4) on Earth through time. The atmospheric abundances of O_2 CH_4 on Earth have changed considerably over time (Fig. 2). In particular, the abundant O_2 in Earth's modern atmosphere is a relatively recent phenomenon: Atmospheric O_2 abundance has increased over Earth's history by many orders of magnitude, from the very low levels of the Hadean/Archean, through a period of intermediate values during the Proterozoic, eventually rising to the high levels we observe on Earth today. Because both O_2 and CH_4 are largely generated and recycled through biological processes, and are spectrally active, they both represent potentially useful atmospheric biosignatures. Levels of O_2 and CH_4 are also linked mechanistically through the redox state of Earth's atmosphere, such that periods of Earth's history characterized by low atmospheric O_2 tend to feature elevated atmospheric CH_4, and vice versa (Fig. 2). In this section, we discuss the evolution of atmospheric O_2 and CH_4 on Earth as constrained by a range of geologic and geochemical indicators, with an eye toward better understanding how the detectability of Earth's biosphere has changed over time.

Sulfur (S) isotope distributions in marine sedimentary rocks of Archean age suggest extremely low atmospheric O_2 levels, constrained to an upper limit of ~10^{-6} bar (*Farquhar et al.,* 2000; *Pavlov and Kasting,* 2002), but most likely below ~10^{-8} bar (*Claire et al.,* 2006; *Zahnle et al.,* 2006). These anomalies also suggest high abundance of some reducing gas for effective production of S_8 in the atmosphere, with CH_4 as the most likely candidate (*Zahnle et al.,* 2006). More recently, coherent time-dependent changes in S isotope systematics have been tied to the transient production and breakdown of atmospheric organic hazes during the late Archean (*Zerkle et al.,* 2012; *Izon et al.,* 2017). Given the surface CH_4 fluxes required to maintain persistent organic hazes, their presence may serve as an effective biosignature in reducing planetary atmospheres such as that of the Archean Earth (*Arney et al.,* 2018). This isotopic evidence for a broadly reducing, low-O_2 ocean-atmosphere system is consistent with a wide range of other geochemical and sedimentological observations (*Rye and Holland,* 1998; *Rasmussen and Buick,* 1999; *Planavsky et al.,* 2010; *Crowe et al.,* 2014).

Sedimentological and isotopic evidence records the initial accumulation of O_2 in Earth's atmosphere during the "Great Oxidation Event" (GOE) at ~2.3 b.y. ago (e.g., *Holland,* 1984, 2002; *Luo et al.,* 2016) (Fig. 1). More recently, geologic and geochemical evidence has led to the hypothesis of a protracted, but ultimately transient, period of ocean-atmosphere oxygenation following the GOE (reviewed in *Lyons et al.,* 2014). In particular, marine sedimentary carbonate rocks record an extended interval of ^{13}C enrichment — the so-called "Lomagundi Event" (*Karhu and Holland,* 1996; *Bekker,* 2001; *Melezhik et al.,* 2007) — which implies a

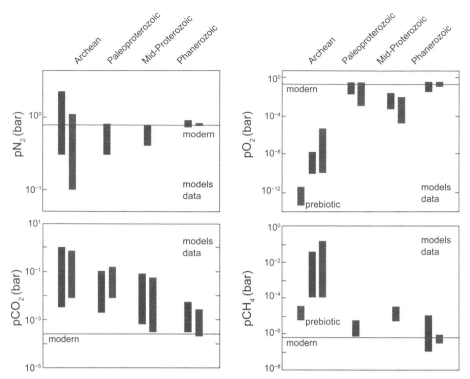

Fig. 2. See Plate 35 for color version. Summary of theoretical and empirical constraints on the abundances of N_2, O_2, CO_2, and CH_4 in Earth's atmosphere for the four major time periods discussed in the text (Archean, Paleoproterozoic, mid-Proterozoic, and Phanerozoic). Blue bars show reconstructions from models, while red bars show inferences based on empirical data. Also shown for the Archean are model-based estimates of prebiotic O_2 and CH_4 levels (gray bars). The ranges are meant to be inclusive, and some of the variability in a given time period should be considered to arise from time-dependent variability rather than uncertainty (e.g., *Olson et al., 2018b*). Constraints are as described in Table 1.

massive release of O_2 to the ocean-atmosphere system according to conventional models of Earth's carbon cycle (*Kump and Arthur,* 1999). This period also records the earliest extensive marine deposits of sulfate-bearing evaporite minerals (*Schröder et al.,* 2008), increasingly large S isotope fractionations (*Canfield,* 2005), a dramatic increase in the ratio of oxidized iron in marine shales (*Bekker and Holland,* 2012), significant enrichments of redox-sensitive metals in anoxic marine sediments (*Canfield et al.,* 2013; *Partin et al.,* 2013; *Reinhard et al.,* 2013), and the first economic phosphorite deposits (*Lepland et al.,* 2014).

A wide range of geochemical proxies point to a subsequent return to relatively low ocean-atmosphere oxygen levels during the mid-Proterozoic, between ~1.8 and 0.8 b.y. ago, perhaps to levels that would have rendered O_2 difficult to detect remotely (see section 3). In particular, the disappearance of sulfate evaporites and large phosphorite deposits from the rock record (*Schröder et al.,* 2008; *Reinhard et al.,* 2017b), a drop in ferric to total iron ratios in marine shales (*Bekker and Holland,* 2012), the S isotope systematics of marine sedimentary rocks (*Planavsky et al.,* 2012; *Scott et al.,* 2014), and the incomplete retention of Fe and Mn in ancient soil horizons (*Zbinden et al.,* 1988) all point to a decrease in ocean-atmosphere oxygen levels following the Lomagundi Event. More recent geochemical observations have buttressed this view (*Planavsky et al.,* 2014b; *Cole et al.,* 2016; *Hardisty et al.,* 2017; *Tang et al.,* 2016a; *Partin*

et al., 2013; *Reinhard et al.,* 2013; *Sheen et al.,* 2018; *Bellefroid et al.,* 2018). The majority of geochemical observations are consistent with a background pO_2 value at or well below ~10^{-2} bar (*Lyons et al,* 2014). However, S isotopic anomalies in sedimentary rocks characteristic of the Archean do not return during this interval, suggesting that atmospheric pO_2 remained above ~10^{-6} bar, atmospheric pCH_4 remained well below ~10^{-2}–10^{-3} bar, or both (e.g., *Zahnle et al.,* 2006). Precisely quantifying atmospheric pO_2 during this period remains a significant outstanding challenge, and Archean oxygen levels, perhaps paradoxically, are perhaps better constrained than those of the Proterozoic.

The late Proterozoic bore witness to significant changes to ocean-atmosphere redox, before, during, and after the "Snowball Earth" glaciation events (reviewed in *Lyons et al.,* 2014). Indeed, there is some evidence for a shift in ocean-atmosphere redox immediately preceding the first low-latitude glaciation (*Planavsky et al.,* 2014b; *Thomson et al,* 2015), implicating time-dependent changes to Earth's oxygen cycle as a potentially important component of climate destabilization in both the Paleoproterozoic and Neoproterozoic (Fig. 1). The ultimate result of these upheavals appears to have been an oxygenation of the ocean-atmosphere system to a degree approaching that of the modern Earth. For most of Phanerozoic time (541 m.y. ago to the present), atmospheric pO_2 appears to have remained within the fire window of between ~0.15 and 0.35 bar (*Belcher and McElwain,*

2008; *Glasspool and Scott,* 2010), although atmospheric pO_2 during the Paleozoic prior to the rise of land plants is somewhat poorly constrained and may have been below 0.1 bar (e.g., *Bergman,* 2004; *Lenton et al.,* 2018). [The "fire window" refers to a range of atmospheric pO_2 levels that are constrained by widespread charcoal in the Phanerozoic geologic record as well as the persistence of plant life in this same record. At the lower bound of the window, burning of plant material would be rare and charcoal would not be widely produced. At the upper bound of the window, plant burning would be global and could not be extinguished (*Scott and Glasspool,* 2006).]However, essentially all available geologic, geochemical, and biological observations are consistent with a well-oxygenated ocean-atmosphere system for the last 500–600 m.y. (*Lyons et al.,* 2014).

2.2.2. Atmospheric carbon dioxide (CO_2) on Earth through time.
The abundance of CO_2 in Earth's atmosphere has also changed by orders of magnitude throughout Earth's history (Fig. 2). These changes are deeply intertwined with Earth's overall habitability through the carbonate-silicate cycle, or "Walker feedback" (*Walker et al.,* 1981), which is hypothesized to regulate atmospheric CO_2 levels through temperature-dependent weathering of the silicate crust. The operation of such a feedback is central to our understanding of planetary habitability, and indeed forms the basis for the "habitable zone" concept (*Kasting et al.,* 1993). Earth's history provides some evidence in favor of this framework, with atmospheric CO_2 abundance generally decreasing over time in the face of a long-term increase in solar luminosity

(Fig. 2). However, more precisely quantifying the strength and transient dynamics of this feedback, and the boundary conditions under which it may break down, are critical ongoing tasks for researchers with an interest in planetary habitability. In this section, we discuss geologic and geochemical constraints on atmospheric CO_2 abundance on Earth through time, while section 2.4 below discusses models of the long-term carbon cycle and the impact of evolving atmospheric CO_2 on global climate.

Archean atmospheric pCO_2 is not very well-constrained, although all existing data are consistent with values that were elevated above those of the modern Earth, perhaps by 2–3 orders of magnitude. Mineral assemblages observed within riverine sediments provide a lower limit on atmospheric pCO_2 of roughly 10^{-3} bar at 3.2 Ga (*Hessler et al.,* 2004), but these observations are also consistent with CO_2 levels an order of magnitude or more higher than this. Similarly, secondary mineral assemblages in ancient soil horizons (paleosols) that formed between ~2.7 and 2.5 Ga have been interpreted to indicate pCO_2 values between ~10^{-3} and 10^{-2} bar during the late Archean (*Rye et al.,* 1995; *Sheldon,* 2006). A more recent model suggests much higher Archean pCO_2, up to or exceeding ~10^{-1} bar (*Kanzaki and Murakami,* 2015), although estimates according to this method can vary over many orders of magnitude at any given time due to uncertainties in assumed soil formation timescales.

Broadly, geologic observations tend to suggest a drop in atmospheric CO_2 levels between the Paleoproterozoic and the mid-Proterozoic. Paleosols formed between 2.5

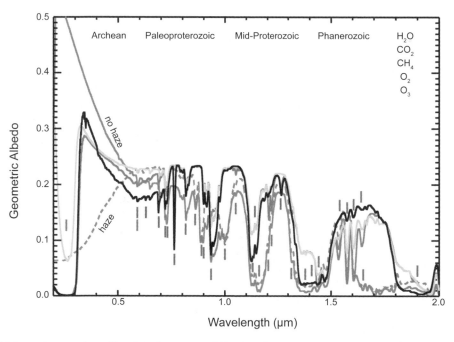

Fig. 3. See Plate 36 for color version. Simulated spectra of Earth at key evolutionary stages. Colors indicate time period: Archean (orange), Paleoproterozoic (dark gray), mid-Proterozoic (light gray), and Phanerozoic (black). Both hazy and haze-free Archean models are shown, and all models include fractional water cloud coverage. Key absorption features are indicated. Original sources for spectra are *Arney et al.* (2016) (Archean), *Robinson et al.* (2011) (Phanerozoic), and E. Schwieterman (personal communication) (Paleoproterozoic and mid-Proterozoic).

and 1.8 Ga yields a somewhat more consistent picture, with estimates from both of the most recent pCO_2 reconstruction techniques yielding values on the order of $\sim 10^{-3}$–10^{-2} bar (*Sheldon*, 2006; *Kanzaki and Murakami,* 2015), although the estimates of *Kanzaki and Murakami* (2015) for similarly aged paleosols are in general a factor of 2–5 higher than those of *Sheldon* (2006). There is only one well-studied paleosol near the Archean-Proterozoic boundary, making it difficult to establish with confidence whether signifcant changes in atmospheric pCO_2 occurred transiting the Archean-Proterozoic boundary. Moving into the mid-Proterozoic, reconstructions based on paleosols (*Sheldon*, 2013), C isotopes (*Kaufman and Xiao*, 2003), and fossils of cyanobacteria (*Kah and Riding*, 2007) are all broadly consistent with atmospheric pCO_2 values on the order of $\sim 3 \times 10^{-3}$ bar, although individual estimates vary between values roughly equivalent to the modern Earth to high values of $\sim 10^{-2}$ bar.

A much wider range of potential pCO_2 proxies exists for the Phanerozoic, including a higher-resolution paleosol record, the carbon (C) isotope compositions of plant and phyto-plankton fossils, the density of stomata on fossilized leaves, and geochemical proxies of ocean pH (reviewed in *Royer,* 2014). Although these approaches are all undergoing continual refinement, they generally point to a range for atmospheric pCO_2 during the Phanerozoic between roughly $\sim 10^{-4}$ and 10^{-3} bar, with significant time-dependent shifts associated with major biospheric innovations and changes in climate (see above). An important caveat to this record is that constraints become very patchy during early Paleozoic time (e.g., prior to around 400 m.y. ago), a period during which Earth system models indicate atmospheric CO_2 levels were higher than at any other time during the Phanerozoic (*Berner and Kothavala,* 2001). However, recent model inversions suggest that atmospheric pCO_2 has never been significantly above $\sim 5 \times 10^{-3}$ bar during the last 500 m.y. (*Royer et al.,* 2014; *Lenton et al.,* 2018) (see section 2.3).

2.2.3. Atmospheric nitrogen (N_2) on Earth through time. Earth's modern atmosphere is $\sim 80\%$ molecular nitrogen (N_2) by volume. A range of geologic processes could potentially lead to large changes in atmospheric N_2 on Earth, but the empirical trajectory of N_2 levels is poorly known at present. This is important, because on Earth the abundance of N_2 in the atmosphere largely controls overall atmospheric pressure, which can significantly impact overall atmospheric opacity to infrared radiation through pressure-induced broadening of absorption lines for greenhouse gases, linking atmospheric pressure and global surface temperature (e.g., *Goldblatt et al.,* 2009). In addition, one of the few known mechanisms for producing an abiotic O_2-rich planetary atmosphere, and thus a potential biosignature "false positive" for O_2 (*Meadows,* 2017), relies on low atmospheric pressure (*Wordsworth and Pierrehumbert,* 2014). As a result, there is strong impetus to better understand the evolutionary history of N_2 on Earth as a means to better understand the factors controlling planetary N_2 cycling on other Earth-like planets.

A current frontier in reconstructing the evolution of Earth's atmosphere is developing constraints on atmospheric pressure, as linked most directly with changes in atmospheric N_2 abundance. As discussed in section 2.3, mass balance calculations suggest that N_2 levels may have varied significantly from the present level of ~ 0.8 bar, with potentially non-trivial impacts on climate (*Goldblatt et al.,* 2009). Recent approaches toward reconstructing overall atmospheric pressure have included estimating air density based on the diameter of fossilized raindrop imprints in a 2.7-b.y.-old tuff from the Ventersdorp Supergroup, South Africa (*Som et al.,* 2012), and estimating total barometric pressure via the size distribution of vesicles in a lava flow from roughly the same age preserved in the Pilbara Craton, Australia (*Som et al.,* 2016). The technique based on raindrop imprints implies that atmospheric pressure at 2.7 Ga was between ~ 0.5 and 2.0 bar, although placing an upper limit with this method is difficult (*Kavanagh and Goldblatt,* 2015). The vesicle approach provides a much more stringent upper limit of around 0.5 bar. *Marty et al.* (2013) attempted to estimate pN_2 directly by analyzing the isotopic composition of nitrogen (N) and argon (Ar) in 3.5 and 3.0-b.y.-old fluids trapped in hydrothermal quartz from the Pilbara Craton, Australia, deriving mixing arrays between end-member hydrothermal fluids of variable composition with a single end member for air-saturated Archean seawater. Their analysis indicates that pN_2 was not significantly above ~ 0.5–1.0 bar during the early Archean. *Nishizawa et al.* (2007) provide a pN_2 estimate of ~ 3 bar from fluid inclusions in the same unit, but the $N_2/^{36}Ar$ values from their samples indicate that their data do not capture the low-N_2 end member analyzed by *Marty et al.* (2013) and thus likely overestimate ambient pN_2. In any case, uncertainties in all current approaches and the fragmentary nature of the archives required for their application are such that current data are consistent with atmospheric N_2 abundance a factor of 2 or more above/below modern, rendering the potential climate impacts of N_2 somewhat enigmatic at present but important to consider (see Table 1 below).

2.3. Model Constraints on Evolving Climate

Standard stellar evolution models (*Gough,* 1981) predict that the Sun was 20–30% less luminous than it is today during the Hadean and Archean. With a planetary greenhouse effect equivalent to that of the modern Earth, this would lead to below-freezing global average surface temperatures prior to ~ 2 Ga, in stark contrast to observations from Earth's rock record — an inconsistency often referred to as the "faint young Sun paradox" (*Sagan and Mullen,* 1972; *Feulner,* 2012). The observation of a generally clement or even warm climate during the Hadean and Archean (see above) thus implies that the composition of Earth's atmosphere was very different from that of the modern (barring major changes in orbital parameters or non-chemical albedo effects). Indeed, as discussed above,

TABLE 1. Empirical and theoretical constraints on Earth's evolving atmospheric composition.

Period	Case	pN_2 (bar)	pO_2 (bar)	pCO_2 (bar)	pCH_4 (bar)	Reference(s)
Prebiotic	model high	—	$3.0 \times 10^{-12*}$	—	3.0×10^{-5}	[1,2]
	model low	—	$3.0 \times 10^{-14\dagger}$	—	5.0×10^{-6}	[1,3]
	data high	—	—	—	—	—
	data high	—	—	—	—	—
Archean	model high	2.3	2.0×10^{-8}	1.0×10^{0}	3.5×10^{-2}	[4–12]
	model low	0.3	1.0×10^{-10}	3.0×10^{-3}	1.0×10^{-4}	[5–13]
	data high	1.1	2.0×10^{-6}	7.0×10^{-1}	$1.4 \times 10^{-1\ddagger}$	[14–16]
	data low	$0.1^{\S}$	—	7.0×10^{-3}	1.0×10^{-4}	[7,17–19]
Paleoproterozoic	model high	0.8	3.0×10^{-1}	1.0×10^{-1}	5.0×10^{-6}	[8–10,13,20,21]
	model low	0.3	2.0×10^{-2}	2.0×10^{-3}	7.0×10^{-7}	[8–10,13,21]
	data high	—	3.0×10^{-1}	1.5×10^{-1}	—	[16,22]
	data low	—	1.0×10^{-3}	7.0×10^{-3}	—	[18,22,23]
Mid-Proterozoic	model high	0.8	2.0×10^{-2}	8.0×10^{-2}	3.0×10^{-5}	[8–10,13,24–26]
	model low	0.4	6.0×10^{-4}	6.0×10^{-4}	$5.0 \times 10^{-6\P}$	[8–10,13,25,27]
	data high	—	8.0×10^{-3}	5.5×10^{-2}	—	[23,28,29]
	data low	—	2.0×10^{-5}	3.0×10^{-4}	—	[30,31]
Phanerozoic	model high	0.9	3.0×10^{-1}	5.5×10^{-3}	1.0×10^{-5}	[32–36]
	model low	0.7	4.0×10^{-2}	2.8×10^{-4}	1.0×10^{-7}	[32–36]
	data hig	0.8	1.5×10^{-1}	2.8×10^{-3}	8.0×10^{-7}	[37–39]
	data low	0.8	3.0×10^{-1}	1.9×10^{-4}	3.5×10^{-7}	[39–41]

All values are approximate. See primary references for assumptions and caveats not noted here. References: [1] *Haqq-Misra et al.* (2011); [2] *Emmanuel and Ague* (2007); [3] *Tian et al.* (2011); [4] *Goldblatt et al.* (2009); [5] *Claire et al.* (2006); [6] *Zahnle et al.* (2006); [7] *Kurzweil et al.* (2013); [8] *Kasting* (1987); [9] *Halevy and Bachan* (2017); [10] *Krissansen-Totton et al.* (2018); [11] *Kharecha et al.* (2005); [12] *Ozaki et al.* (2018); [13] *Stüeken et al.* (2016); [14] *Marty et al.* (2013); [15] *Pavlov and Kasting* (2002); [16] *Kanzaki and Murakami* (2015); [17] *Som et al.* (2016); [18] *Sheldon* (2006); [19] *Driese et al.* (2011); [20] *Bachan and Kump* (2015); [21] *Harada et al.* (2015); [22] *Bekker and Holland* (2012); [23] *Rye and Holland* (1998); [24] *Laakso and Schrag* (2017); [25] *Catling et al.* (2007); [26] *Zhao et al.* (2018); [27] *Olson et al.* (2016); [28] *Kaufman and Xiao* (2003); [29] *Kah and Riding* (2007); [30] *Planavsky et al.* (2014b); [31] *Sheldon* (2013); [32] *Berner* (2006); [33] *Royer* (2014); [34] *Lenton et al.* (2018); [35] *Bartdorff et al.* (2008); [36] *Beerling et al.* (2009); [37] *Belcher and McElwain* (2008); [38] *Royer* (2014); [39] *Wolff and Spahni* (2007); [40] *Glasspool and Scott* (2010); [41] *Galbraith and Eggleston* (2017).

*Assumes pCO_2 = 2 bar.
†Assumes pCO_2 = 0.02 bar.
‡Assuming data high pCO_2 and CH_4/CO_2 = 0.2.
§Assuming predominantly N_2 atmosphere.
¶Assuming $pO_2 = 10^{-3}$ bar, $[SO_4^{2-}]$ = 500 mol kg^{-1}.

there is persuasive geological and geochemical evidence that the composition of Earth's atmosphere was very different during the Hadean, Archean, and Proterozoic.

The most prominent solutions to this problem invoke a larger inventory of greenhouse gases in Earth's early atmosphere. *Sagan and Mullen* (1972) explored a reducing NH_3–CH_4–H_2S–H_2O–CO_2 greenhouse, with a dominant role for NH_3. However, the rapid photolysis of NH_3 in the upper atmosphere would have required a very large source at Earth's surface, and would have in turn resulted in the production of rather extreme amounts of N_2 on geologically rapid timescales (*Kuhn and Atreya,* 1979). More recently, it has been suggested that the photolysis of NH_3 in the upper atmosphere may have been mitigated somewhat by absorption of UV photons by a fractal organic haze (*Wolf and Toon,* 2010), an idea that warrants additional scrutiny of the relative altitudes of peak NH_3 photolysis and haze absorption in future work (*Wolf and Toon,* 2010). In any

case, subsequent work has tended to focus on CH_4–CO_2–H_2O greenhouses and, more recently, the possible radiative effects of high H_2 (*Kasting,* 2005; *Haqq-Misra et al.,* 2008).

The factors regulating planetary climate on the prebiotic Earth are not particularly well understood. Direct constraints are few, but models predict that the abundance of CH_4 in Earth's prebiotic atmosphere would have been low (*Kasting,* 2005; *Emmanuel and Ague,* 2007), while a recent inversion using a geologic carbon cycle model yields a median pCO_2 estimate of 0.3 bar with a 95% confidence interval of 0.03–1.0 bar (*Krissansen-Totton et al.,* 2018). Greenhouses dominated by H_2O and CO_2 with pCO_2 values on the lower end of this range would be unlikely to exhibit clement surface temperatures under Hadean or early Archean solar luminosity, but both one-dimensional radiative-convective and three-dimensional global climate models predict that values at the upper end of this range would result in surface temperatures well above freezing under early and late Archean solar luminosity (*Kasting and Ackerman*, 1986; *Kasting*, 1987; *Charnay et al.*, 2013; *Wolf and Toon*, 2013, 2014). Significant additional warming may have been provided by collision-induced absorption by H_2–N_2 under plausible prebiotic conditions (*Wordsworth and Pierrehumbert*, 2013), although the strength of this would have depended strongly on atmospheric H_2 and N_2 abundance, both of which are poorly constrained for the prebiotic atmosphere.

The emergence of a biosphere on Earth would have had a significant impact on atmospheric chemistry and climate. In particular, primitive microbial metabolisms such as methanogenesis, acetogenesis, and anoxygenic photosynthesis would have dramatically increased fluxes of CH_4 to Earth's atmosphere. The implications of this for climate are two-fold. First, CH_4 is an important greenhouse gas in its own right, providing another means toward offsetting decreased solar luminosity that would be particularly effective in a reducing atmosphere (*Pavlov et al.*, 2000; *Haqq-Misra et al.*, 2008). Second, the potential for large biogenic CH_4 fluxes introduces an additional feedback on climate via formation of an organic haze in the atmosphere (*Zahnle*, 1986; *Pavlov et al.*, 2001; *Haqq-Misra et al.*, 2008). Photochemical models (*Haqq-Misra et al.*, 2008; *Zerkle et al.*, 2012; *Arney et al.*, 2016) and laboratory experiments (*DeWitt et al.*, 2009; *Trainer et al.*, 2004, 2006) predict that once the CH_4/CO_2 ratio of the atmosphere increases beyond ~0.1 hydrocarbon aerosols will begin to form, with optical thicknesses at visible and UV wavelengths that increase rapidly at CH_4/CO_2 values between ~0.1 and 1 (*Domagal-Goldman et al.*, 2008; *Haqq-Misra et al.*, 2008). The formation of these hazes can lead to significant cooling, even at relatively low CH_4/CO_2 values of ~0.1–0.2 (*Arney et al.*, 2016). These combined effects would have been important in regulating climate and surface temperature during both the Archean and Proterozoic.

Taken together, models suggest that clement or even warm surface temperatures could have been maintained on the prebiotic Earth by a CO_2–H_2O greenhouse, potentially supplemented by collision- and pressure-induced warming at greater H_2 and N_2 abundance (*Kasting*, 1987; *Goldblatt et al.*, 2009; *Wordsworth and Pierrehumbert*, 2013; *Krissansen-Totton et al.*, 2018). Following the emergence of a surface biosphere, one-dimensional radiative-convective and three-dimensional global climate models, coupled ecosystem-biogeochemistry models, and the geologic record are all consistent in suggesting that surface temperatures well above freezing could have been maintained throughout the Archean with a CH_4–H_2O–CO_2 greenhouse and optically thin haze (*Haqq-Misra et al.*, 2008; *Zerkle et al.*, 2012; *Charnay et al.*, 2013; *Wolf and Toon*, 2013; *Izon et al.*, 2017; *Ozaki et al.*, 2018; *Krissansen-Totton et al.*, 2018). However, additional factors beyond changes to the atmospheric greenhouse may also have been important, and together could lead to significant warming. For example, changes to the average size of cloud droplets attendant to fewer cloud condensation nuclei (CCN) in the early atmosphere would have resulted in more effective rainout and fewer low clouds at low-mid latitudes (*Charnay et al.*, 2013; *Wolf and Toon*, 2014) and possibly thinner clouds overall (*Goldblatt and Zahnle*, 2011; *Charnay et al.*, 2013), both of which would have resulted in significant warming. Maintaining a habitable climate during Earth's earliest history is thus not particularly challenging, as all of these factors were potentially in play. However, achieving the highest estimates of Archean temperature with plausible atmospheric CO_2 and CH_4 levels remains a challenge.

Existing models of the long-term carbon cycle and surface temperature are broadly consistent with geochemical evidence for elevated pCO_2 during the Paleoproterozoic (e.g., *Halevy and Bachan*, 2017; *Krissansen-Totton et al.*, 2018). However, the initial rise of atmospheric O_2 at ~2.3 Ga would have destabilized the Archean atmospheric greenhouse (*Claire et al.*, 2006; *Zahnle et al.*, 2006; *Haqq-Misra et al.*, 2008), potentially leading to dramatic effects on climate during the earliest Paleoproterozoic (see above). Understanding climate dynamics and coherently modeling climate and biogeochemistry during and after the GOE and through the Lomagundi Event remain outstanding challenges, but existing models suggest large changes to Earth surface temperature, the global carbon cycle, and atmospheric composition (*Claire et al.*, 2006; *Harada et al.*, 2015).

There is limited geologic evidence for glacial conditions during the mid-Proterozoic (see above), despite a solar luminosity ~10% lower than that of the modern Earth (*Gough*, 1981). Recent coupled three-dimensional climate modeling indicates that global glaciation should occur under these conditions if atmospheric pCO_2 drops to around 10^{-3} bar (*Fiorella and Sheldon*, 2017), and extensive glaciation at high and middle latitudes should occur even above pCO_2 values an order of magnitude higher than the modern Earth unless surface temperatures are buffered by some other greenhouse gas. Nitrous oxide (N_2O) would be unlikely to provide the requisite radiative forcing on its own, given the relatively low atmospheric pO_2 during the mid-Proterozoic and the large biological N fluxes required (*Roberson et al.*, 2011), but may have a marginal impact on surface temperatures (*Buick*, 2007; *Stanton et al.*, 2018). Methane (CH_4) is another candidate, but three-dimensional models of ocean biogeochemistry (*Olson et al.*, 2016) interpreted in light of three-dimensional climate modeling (*Fiorella and Sheldon*, 2017) indicate that an ocean-only CH_4 cycle would have been unable to maintain an ice-free climate during the mid-Proterozoic. One possible solution to this would be a significant microbial CH_4 flux from terrestrial ecosystems (*Zhao et al.*, 2018). Alternatively, atmospheric pCO_2 may have been somewhat higher than geochemical proxies suggest (*Krissansen-Totton et al.*, 2018), or other changes to factors like cloud droplet radius or surface albedo may have contributed to stabilizing relatively warm temperatures (*Fiorella and Sheldon*, 2017). Lastly, at least periods of the mid-Proterozoic may not have been entirely ice-free (see above). In any case, climate models, geochemical proxies, and models of marine/terrestrial biogeochemistry yield a picture of a relatively weak H_2O–CO_2 greenhouse buffered by CH_4 levels on the order of $\sim 10^{-5}$ bar or slightly less, depending on the importance of terrestrial CH_4 cycling.

The dynamics of Earth's climate during the intense ice ages of the late Proterozoic are much more well-studied than their Paleoproterozoic counterparts, and the reader is here referred to two recent comprehensive reviews on the subject (*Pierrehumbert et al.,* 2011; *Hoffman et al.,* 2017). However, the biogeochemical dynamics associated with these perturbations are more poorly understood, particularly with regard to Earth's O_2 and CH_4 cycles, and this represents an important topic of future work. For example, although impacts to the Earth's "oxidized" carbon cycle have been explored in a range of models (*Le Hir et al.,* 2008a,b; *Mills et al.,* 2011), it remains unclear what role Earth's CH_4 cycle may have played in the inception or recovery from low-latitude glaciation, if any (*Schrag et al.,* 2002; *Pierrehumbert et al.,* 2011; *Olson et al.,* 2016). In any case, the Sturtian glacial episode in particular is estimated to have lasted for roughly 50 m.y. (*Rooney et al.,* 2014), with the attendant impacts on atmospheric biosignatures and the remote detectability of any surviving biosphere almost completely unknown.

The Phanerozoic climate system, although perhaps relatively stable in the scope of Earth's entire history, has been extremely dynamic, with at least three major ice ages and intervening periods of relatively warm, largely ice-free conditions (see above). Although they differ in their tectonic boundary conditions, scope/intensity, and overall biological impact, these climate shifts are generally considered to have been driven primarily by variations in atmospheric CO_2, together with internal climate system feedbacks and modulated by changes in Earth's orbital parameters (*Zachos et al.,* 2001; *Royer et al.,* 2004; *Herrmann et al.,* 2004; *Montañez and Poulsen,* 2013). The Phanerozoic climate system is thus thought to have been controlled largely by an H_2O–CO_2 greenhouse, buffered by volcanic outgassing, organic carbon weathering under a high-O_2 atmosphere, and solar luminosity roughly equivalent to that of the modern Earth. A notable exception to this may have occurred during the Carboniferous "coal swamp" era roughly 300 m.y. ago, during which biogeochemical and climate models predict significant radiative forcing from atmospheric CH_4 (*Bartdorff et al.,* 2008; *Beerling et al.,* 2009). Taken together, low-order and three-dimensional climate models are consistent with geologic and geochemical records in indicating that Earth's surface has been consistently habitable for the last $\sim$500 m.y., with long-term average surface temperatures largely within the range of $\sim$15°–25°C (*Royer et al.,* 2004; *Lenton et al.,* 2018).

2.4. Model Constraints on Evolving Atmospheric Chemistry

Theoretical models have also provided a great deal of insight into the evolving redox state and major background gas composition of Earth's atmosphere. Earth's prebiotic atmosphere is generally thought to have been mildly reducing, composed predominantly of N_2–CO_2–H_2O, with variable H_2 and CH_4 and only trace amounts of species like O_2, O_3, and N_2O. Photochemical models assuming that hydrogen escaped the Archean atmosphere at the diffusion limit (e.g., *Hunten,* 1973) predict that ground-level atmospheric O_2 would have been on the order of $\sim 10^{-14}$–10^{-12} bar, with H_2 on the order of $\sim 10^{-4}$–10^{-3} bar depending on the assumed volcanic outgassing rate (*Kasting and Walker,* 1981; *Haqq-Misra et al.,* 2011). However, it has been suggested that lower exobase temperatures in an O_2-poor, CO_2-rich atmosphere would have significantly decreased the efficiency of thermal (Jeans) escape at the top of the atmosphere, with the result that rates of hydrogen escape would have been controlled instead by extreme ultraviolet (EUV) energy fluxes from the young Sun (*Tian et al.,* 2005). Balancing these "energy-limited" escape rates with plausible volcanic H_2 outgassing rates results in H_2 mixing ratios as high as $\sim$0.3 bar, with important ramifications for prebiotic chemistry (*Tian et al.,* 2005) and early climate (*Wordsworth and Pierrehumbert,* 2013). There remains some debate as to whether heating of the exobase by gases other than O_2 and/or other nonthermal escape processes could promote more efficient escape (*Catling,* 2006; *Tian et al.,* 2006). Nevertheless, it remains plausible that Earth's prebiotic atmosphere was relatively H_2-rich.

The evolution of the earliest biosphere would have dramatically transformed atmospheric chemistry. In particular, the emergence of microbial methanogenesis and the evolution of the most primitive forms of anoxygenic photosynthesis, both of which are thought to be very ancient based on geochemical (*Tice and Lowe,* 2004; *Ueno et al.,* 2006) and phylogenetic (*Xiong et al.,* 2000; *Wolfe and Fournier,* 2018) evidence, would have consumed H_2 through reactions such as

$$CO_2 + 4H_2 \rightarrow CH_4 + 2H_2O$$
$$2H_2 + CO_2 \rightarrow CH_2O + H_2O$$

Both processes would have the net effect of decreasing the atmospheric H_2/CH_4 ratio, to an extent that would be limited by the availability of energy and nutrients (*Kharecha et al.,* 2005; *Ozaki et al.,* 2018). In particular, globally integrated rates of anoxygenic photosynthesis would most likely have been limited by the availability of electron donors (H_2, Fe^{2+}, H_2S), in contrast to oxygenic photosynthesis, which can use water as an electron donor (see below). Models of the Earth's primitive biosphere can produce extremely high atmospheric CH_4 levels, on the order of $\sim 10^{-2}$ bar (Fig. 2), depending on electron donor flux, levels of available nutrients, and photosynthetic community assemblage (*Kharecha et al.,* 2005; *Ozaki et al.,* 2018). However, these models neglect microbial anaerobic oxidation of methane (AOM), under the presumption that Archean seawater sulfate levels were extremely low (e.g., *Crowe et al.,* 2014), and to some extent plausible upper limits on atmospheric CH_4 are constrained by the climate effects of hydrocarbon haze formation at elevated CH_4/CO_2 ratios such that atmospheric CH_4 should not be treated in isolation (*Ozaki et al.,* 2018). Nevertheless, existing models consistently predict that the emergence of Earth's earliest biosphere would have dramatically shifted the atmospheric CH_4/H_2 ratio, readily supporting atmospheric CH_4 levels that would have the potential for remote observation (*Reinhard et al.,* 2017a; *Arney et al.,* 2018).

Mass balance calculations (*Goldblatt et al.,* 2009) and time-dependent biogeochemical models (*Stüeken et al.,* 2016) are consistent with elevated atmospheric N_2 throughout the Archean, but can also accommodate long-term decrease in pN_2 through the Hadean and Archean depending on the assumed history of CO_2 outgassing and the mechanics coupling organic C and N burial (*Stüeken et al.,* 2016). Similarly, existing models of the carbon cycle that are coupled to long-term stellar evolution are broadly consistent with geochemical data in indicating high atmospheric pCO_2 and a secular decline through the Hadean and Archean (*Kasting,* 1987; *Sleep and Zahnle,* 2001; *Halevy and Bachan,* 2017; *Charnay et al.,* 2017; *Krissansen-Totton et al.,* 2018). The particular trajectories depend on assumptions regarding secular changes in heat flow, the mechanisms regulating seafloor weathering, and the timing and magnitude of major changes to the Earth surface CH_4 cycle. Fully understanding the impact of atmospheric N_2 and CO_2 on the detectability of habitability markers and atmospheric biosignatures on the Hadean/early Archean Earth will require both more precise geochemical/paleobarometric constraints and further development of approaches that effectively couple models of long-term biogeochemistry and climate to dynamic models of ocean-atmosphere redox.

The evolution of oxygenic photosynthesis resulted in an autotrophic biosphere that could use water as an electron donor, freeing global autotrophy from electron donor limitation and dramatically increasing the potential energy flux through the biosphere

$$CO_2 + H_2O \rightarrow CH_2O + O_2$$

The timing of this event is still debated, but is likely to have occurred at some point prior to the late Archean (*Kurzweil et al.,* 2013; *Planavsky et al.,* 2014a; *Magnabosco et al.,* 2018). Following the emergence of biological oxygen production, atmospheric chemistry would have been controlled largely by the balance of fluxes between O_2 and CH_4 produced by the surface biosphere and the consumption of O_2 by reducing volcanic/metamorphic gases and weathering reactions with reduced phases in Earth's upper crust (*Catling and Claire,* 2005). Low-order biogeochemical models and one-dimensional models of atmospheric chemistry are consistent with constraints from the geologic record (discussed above) in suggesting low atmospheric pO_2, on the order of $\sim 10^{-10}$–10^{-8} bar, and relatively high atmospheric pCH_4, around 10^{-4}–10^{-3} bar, on the Archean Earth after the emergence of an oxygenic biosphere (*Gold-*

blatt et al., 2006; *Claire et al.,* 2006; *Zahnle et al.,* 2006; *Daines and Lenton,* 2016).

A number of potential drivers have been suggested for the GOE, including secular tectonic processes (*Kump and Barley,* 2007; *Holland,* 2009; *Gaillard et al.,* 2011), changes in global biological fluxes (*Kopp et al.,* 2005; *Konhauser et al.,* 2009), and time-integrated hydrogen escape from the atmosphere (*Catling et al.,* 2001; *Claire et al.,* 2006). It is likely that to some extent all of these factors were important. Atmospheric pO_2 may have changed by many orders of magnitude moving across the GOE, with most models predicting a geologically instantaneous rise from $\sim 10^{-8}$ bar to as high as $\sim 10^{-2}$ bar regardless of the underlying mechanism (*Claire et al.,* 2006; *Goldblatt et al.,* 2006). Both the timing and magnitude of this event are consistent with existing isotopic records (e.g., *Luo et al.,* 2016). However, biogeochemical models suggest that the imbalance in the global redox budget required to transit the GOE need not necessarily have been large (*Claire et al.,* 2006; *Goldblatt et al.,* 2006).

The GOE effectively represented a shift in the trace redox gas in Earth's atmosphere from O_2 to CH_4. Although models do indeed predict an initial drop in atmospheric pCH_4 during the GOE, the ultimate establishment of a substantial stratospheric O_3 layer attendant to rising ground-level pO_2 is predicted to shield CH_4 from destruction in the troposphere. This allows atmospheric CH_4 levels to rebound in photochemical models to $\sim 10^{-4}$ bar in the Proterozoic (*Claire et al.,* 2006; *Goldblatt et al.,* 2006). However, models that include ocean biogeochemistry and microbial consumption of CH_4 with O_2 and SO_4^{2-} generally result in lower steady-state atmospheric pCH_4 following the GOE, typically on the order of $\sim 10^{-5}$ bar or lower (*Catling et al.,* 2007; *Daines and Lenton,* 2016; *Olson et al.,* 2016). These results depend strongly on assumed atmospheric pO_2 and the ocean reservoir of SO_4^{2-} (*Olson et al.,* 2016), and are not currently equipped to deal with the potentially important impacts of a terrestrial biosphere (*Zhao et al.,* 2018). A full exploration of this problem will require coupled, open-system models of photochemistry and ocean/terrestrial microbial metabolism. Nevertheless, the abundance of CH_4 in Earth's atmosphere following the GOE was likely significantly lower than that of the Archean Earth, and in particular would have been orders of magnitude below that of the earliest Archean and Hadean Earth prior to the evolution of oxygenic photosynthesis (*Kharecha et al.,* 2005; *Ozaki et al.,* 2018). Unfortunately, a quantitative geologic or geochemical indicator of atmospheric pCH_4 at levels below those of the Archean has not yet been developed, and atmospheric pCH_4 is very difficult to track empirically throughout the remainder of Earth's history.

Although the differences in mean state before and after the GOE are relatively well understood, the dynamics of climate and atmospheric chemistry in the immediate aftermath of the GOE are not. Some models predict that this change to Earth's surface redox balance would have had significant climate impacts (*Claire et al.,* 2006; *Haqq-Misra et al.,* 2008), one result of which may have been an ultimately transient but quantitatively dramatic elevation in atmospheric pO_2 (*Harada et al.,* 2015). This scenario would be consistent with emerging geochemical evidence for elevated atmospheric O_2 (and thus O_3), possibly for 100-m.y. timescales, during the Paleoproterozoic (see above). The protracted, but ultimately transient, rise in atmospheric pO_2 implies a significant drop in atmospheric CH_4 levels (*Harada et al.,* 2015), and potentially a substantial drop in atmospheric pCO_2 unless buffered by a sedimentary rock cycle very different from that of the modern Earth (e.g., *Bachan and Kump,* 2015). In any case, long-term (e.g, mean state) carbon cycle and climate models that are entirely uncoupled or only implicitly coupled to the O_2 cycle are broadly consistent with the current geochemical constraints for atmospheric pCO_2 during the Paleoproterozoic discussed above (*Sleep and Zahnle,* 2001; *Halevy and Bachan,* 2017; *Krissansen-Totton et al.,* 2018).

The initial accumulation of O_2 in Earth's atmosphere appears to have been followed by a subsequent return to relatively low ocean-atmosphere oxygen levels during the mid-Proterozoic (between ~1.8 and 0.8 b.y. ago). However, the absence of non-mass-dependent S isotope fractionations in marine sediments and the apparent absence of reduced detrital minerals in fluvial settings indicate atmospheric pO_2 remained above ~10^{-6} bar. On long timescales, the modern atmospheric O_2 level is maintained dynamically by the balance between net O_2 sources (principally the burial of organic C and reduced S into the Earth's upper crust) and net O_2 sinks (largely the subsequent exhumation and oxidative weathering of organic C and reduced S, along with reactions between O_2 and reduced metamorphic and volcanic gases). However, there are strong nonlinearities in the scaling relationships between these fluxes and the amount of O_2 in the atmosphere. In addition, the major sink fluxes on the modern Earth decrease in magnitude as atmospheric pO_2 drops, while the major source fluxes increase. As a result, not all atmospheric pO_2 values are equally stable, and understanding the internal processes and feedbacks capable of maintaining atmospheric O_2 levels above those characteristic of the Archean but well below those of the modern Earth remains an outstanding question (*Lyons et al.,* 2014; *Daines et al.,* 2017).

As discussed above, models of mid-Proterozoic climate and biogeochemistry suggest a relatively weak H_2O–CO_2 greenhouse buffered by modest CH_4 levels (Fig. 2, Table 1). In particular, long-term carbon cycle models suggest atmospheric pCO_2 of around ~10^{-3}–10^{-2} bar during the mid-Proterozoic, while models of marine/terrestrial biogeochemistry suggest atmospheric pCH_4 values on the order of ~10^{-6}–10^{-5} bar, together consistent with most geochemical constraints and a largely ice-free climate state (*Fiorella and Sheldon,* 2017). That said, some paleosol reconstructions approach roughly modern pCO_2 values (e.g., *Sheldon,* 2013), which is difficult to reconcile with evidence for a largely ice-free Earth surface for most of the mid-Proterozoic unless Earth's greenhouse was impacted strongly by fluxes of CH_4 from a terrestrial microbial biosphere (*Zhao et al.,* 2018). A full picture of Earth's mid-Proterozoic atmosphere awaits a comprehensive model that couples open-system carbon cycling with a balanced redox budget and dynamic O_2-CH_4 cycle, but existing data and models are consistent with this period of Earth's history representing a potential "false negative" for conventional biosignature techniques (e.g., *Reinhard et al.,* 2017a) — a period through which the spectral features of most canonical biosignature gases would have been relatively weak, perhaps for geologic timescales.

Although there is accumulating geologic and geochemical evidence that the extreme low-latitude glaciations of the late Proterozoic were associated with significant changes to ocean-atmosphere redox and atmospheric chemistry (*Hoffman et al.,* 1998, 2017; *Canfield et al.,* 2007; *Sahoo et al.,* 2012; *Cox et al.,* 2013; *Planavsky et al.,* 2014b; *Thomson et al.,* 2015), the relative timing and mechanistic links remain somewhat obscure. Simple biogeochemical models indicate that low-latitude glacial episodes can readily drive a secular transition from low- to high-oxygen steady states at sufficiently high pCO_2 thresholds for deglaciation (*Laakso and Schrag,* 2017). For example, a deglaciation threshold of pCO_2 ~0.1 bar is sufficient to drive a permanent transition in atmospheric pO_2 from 10^{-3} to 10^{-1} bar (*Laakso and Schrag,* 2017) during deglaciation. Glacial CO_2 levels of this order are readily achievable even in models that allow for efficient ocean-atmosphere gas exchange and seafloor weathering (e.g., *Le Hir et al.,* 2008b). Efforts to better understand the temporal polarity and mechanistic details linking climate destabilization and nonlinear changes to atmospheric chemistry during both the Paleoproterozoic and Neoproterozoic represent an important avenue of future work. Nevertheless, significant changes in the redox state of Earth's ocean-atmosphere system are strongly implicated as having been both cause and consequence of sporadic perturbations to Earth's habitability.

Biogeochemical models are generally consistent in suggesting a high-O_2, low-CH_4, and moderate CO_2 atmosphere throughout the Phanerozoic (541 Ma to the present). Both atmospheric O_2 and atmospheric CO_2 have been controlled by the combined effects of roughly modern solar luminosity, time-dependent variability in volcanic degassing, rock uplift, changes to the major ion chemistry of seawater, and the emergence and expansion of terrestrial ecosystems (*Berner,* 1991, 2006; *Royer et al.,* 2014; *Lenton et al.,* 2018). Long-term atmospheric CH_4 levels have been controlled largely by the evolutionary and climate dynamics controlling biogenic CH_4 fluxes from terrestrial ecosystems (*Bartdorff et al.,* 2008). Despite some discrepancies between different models in estimates of atmospheric pO_2 during the earliest part of the Phanerozoic, most models indicate ranges for atmospheric pO_2, pCO_2, and pCH_4 between ~0.1 and 0.3 bar, ~10^{-4} and 10^{-3} bar, and ~10^{-7} and 10^{-5} bar, respectively, all of which are consistent with existing geologic and geochemical constraints (Fig. 2, Table 1). Similar models for atmospheric pN_2 through time suggest values close to that of the modern Earth for most of the Phanerozoic, although direct geologic constraints on this are lacking for all but the most recent periods of Earth's history.

Observations from the geologic record and results from quantitative models are united in suggesting extensive changes to the Earth system over time, including the chemistry of the ocean-atmosphere system, the dynamics of long-term climate, and the size and scope of Earth's biosphere. The contours of this evolution provide important information for exoplanet characterization efforts. In particular, simulating and predicting observations across the spectrum of habitable worlds represented in Earth's evolutionary history provides a series of test cases for evaluating putative discriminants of habitability and life on Earth-like planets beyond our solar system, and can potentially provide important insight into the challenges associated with deciphering exo-Earth observations.

3. REMOTE DETECTABILITY OF EARTH'S BIOSPHERE THROUGH TIME

How would Earth's evolving climate and atmospheric chemistry have appeared to a remote observer? We focus here on a subset of the most prominent biosignatures that may be remotely detectable — namely, atmospheric oxygen (O_2), its photochemical byproduct ozone (O_3), methane (CH_4), hydrocarbon haze, and nitrous oxide (N_2O) (e.g., *Schwieterman et al.,* 2018). We also include two major habitability indicators — water vapor (H_2O) and carbon dioxide (CO_2) —the latter of which is required for climate system stabilization via the carbonate-silicate geochemical cycle (*Walker et al.,* 1981). The true detectability of any particular biosignature or habitability indicator will depend on the magnitude of the signal produced (related to, e.g., a species' atmospheric abundance and the atmospheric opacity produced by a specific feature), the parameters controlling observational precision (e.g., stellar host, distance to the target, instrument and astrophysical noise sources), and the wavelength range accessible to the instrument being used. Our discussion is thus not meant to be exhaustive or definitive, but is instead meant to provide some context for motivating remote observations of the modern Earth and simulated observations of different periods of Earth's evolutionary history. As a guide to this discussion, Table 2 contains a detailed listing of the wavelengths, widths, and strengths of spectral features for key biosignature and habitability indicator gases.

The most prominent features of molecular oxygen (O_2) are at 0.76 and 0.69 µm, the Fraunhofer A and B bands respectively. The Fraunhofer A band is the stronger of the two bands, but is likely to be relatively weak unless atmospheric pO_2 is above the few percent level (*Des Marais et al.,* 2002; *Reinhard et al.,* 2017a; *Schwieterman et al.,* 2018). When considered in the context of Earth's evolution, it is clear that O_2 spectral features were likely non-existent during the Archean, and may have been weak during much of the Proterozoic. However, absorption by O_2 at both the Fraunhofer A and B bands would have been relatively strong through essentially all of Phanerozoic time and possibly during a protracted interval in the Paleoproterozoic.

Ozone (O_3), which is produced photochemically by O_2 and thus scales with atmospheric O_2 abundance, is an extremely useful biosignature in the context of Earth evolution. Ozone offers strong spectral features across a range of wavelengths, including diagnostic features in the mid-infrared (at 9.6 µm), the visible/near-infrared (the Chappuis band between 0.55 and 0.65 µm), and the ultraviolet (the Hartley-Huggins bands centered near 0.26 µm). The latter of these is of particular note, as it is sensitive to extremely low peak O_3 abundances of ~1 ppm or less. Although all these features would be extremely weak at the vanishingly low atmospheric O_2/O_3 levels characteristic of the Archean, the Hartley-Huggins band would have produced relatively strong, although not saturated, absorption at atmospheric pO_2 values approaching the lowest inferred for the mid-Proterozoic, while absorption at both the Hartley-Huggins and mid-infrared bands would have been relatively strong at the upper end of mid-Proterozoic pO_2 estimates (*Segura et al.,* 2003; *Reinhard et al.,* 2017a; *Rugheimer and Kaltenegger,* 2018; *Olson et al.,* 2018a). All these features would have been relatively strong for the last ~500 m.y. of Earth's evolutionary history, and possibly during the early Paleoproterozoic.

Methane (CH_4) has a number of spectral features, including many relatively weak features spanning the visible wavelength range; stronger near-infrared features at 1.65, 2.3, and 2.4 µm; and a strong mid-infrared band at 7.7 µm. The visible wavelength features between 0.6 and 1.0 µm only have appreciable depth for atmospheric CH_4 abundances above ~10^{-3} bar, suggesting that they may have been relatively strong during the Archean and may have been particularly promising biosignatures prior to the evolution of oxygenic photosynthesis (*Kharecha et al.,* 2005; *Ozaki et al.,* 2018). The stronger near-infrared methane features would likely have been prominent for the vast majority of the Archean (*Reinhard et al.,* 2017a). For most of Earth's history subsequent to the Archean, absorption by CH_4 in the mid-infrared at 7.7 µm would have been relatively strong. Indeed, this feature is apparent even at the very low atmospheric CH_4 abundance of the modern Earth (*Des Marais et al.,* 2002). However, overlap with a significant H_2O band may render this feature challenging to detect in some cases.

Hydrocarbon hazes — which can be produced in reducing atmospheres with CH_4/CO_2 ratios above ~0.1 — also produce strong features and could represent a biosignature "proxy" for biotic CH_4 production (*Arney et al.,* 2016, 2018). Indeed, there is isotopic evidence for at least sporadic haze production on the Archean Earth (see section 2.2.1), and photochemical models suggest that the surface CH_4 fluxes required to maintain both haze production and the clement climate state implied by Earth's rock record are most consistent with biospheric CH_4 production. The most prominent features for haze include a broad ultraviolet/visible absorption feature and a band near 6.5 µm in the mid-infrared. Both may have been relatively strong during the Archean, and the shortwave feature could have

TABLE 2. Spectral feature details for key biosignature and habitability indicator gases.

Species	λ^* (μm)	$\Delta\lambda^\dagger$ (μm)	$\lambda/\Delta\lambda$	Opacity‡ (cm^2 molecule^{-1})	Optical Depth§
O_2	0.629	3.0×10^{-3}	210	1.5×10^{-25}	6.7×10^{-1}
O_2	0.689	4.1×10^{-3}	170	3.9×10^{-24}	$1.7 \times 10^{+1}$
O_2	0.762	5.1×10^{-3}	150	5.8×10^{-23}	$2.6 \times 10^{+2}$
O_2	0.865	6.6×10^{-3}	130	1.6×10^{-27}	7.3×10^{-3}
O_2	1.07	4.3×10^{-3}	250	2.7×10^{-27}	1.2×10^{-2}
O_2	1.27	5.8×10^{-3}	220	7.8×10^{-25}	$3.5 \times 10^{+0}$
O_2	6.30	3.2×10^{-1}	19	3.2×10^{-27}	1.4×10^{-2}
			—		
O_3	0.256	3.9×10^{-2}	6.5	1.2×10^{-17}	$1.1 \times 10^{+2}$
O_3	0.600	1.2×10^{-1}	5.0	4.8×10^{-21}	4.4×10^{-2}
O_3	2.48	2.1×10^{-3}	1100	1.5×10^{-21}	1.3×10^{-2}
O_3	3.27	6.9×10^{-3}	470	1.2×10^{-20}	1.1×10^{-1}
O_3	3.59	4.7×10^{-2}	77	1.8×10^{-21}	1.6×10^{-2}
O_3	4.74	8.7×10^{-2}	54	7.6×10^{-20}	6.9×10^{-1}
O_3	5.80	1.6×10^{-1}	37	3.8×10^{-21}	3.4×10^{-2}
O_3	9.58	3.5×10^{-1}	27	7.3×10^{-19}	$6.7 \times 10^{+0}$
O_3	14.3	$1.0 \times 10^{+0}$	14	2.6×10^{-20}	2.3×10^{-1}
			—		
CH_4	0.510	5.0×10^{-3}	100	4.3×10^{-27}	1.5×10^{-7}
CH_4	0.542	5.0×10^{-3}	110	4.6×10^{-26}	1.5×10^{-7}
CH_4	0.576	8.0×10^{-3}	72	1.3×10^{-26}	1.5×10^{-7}
CH_4	0.598	8.0×10^{-3}	75	9.0×10^{-27}	1.5×10^{-7}
CH_4	0.619	8.0×10^{-3}	77	2.2×10^{-25}	1.5×10^{-7}
CH_4	0.667	1.5×10^{-2}	44	5.6×10^{-26}	1.5×10^{-7}
CH_4	0.703	1.1×10^{-2}	64	1.1×10^{-25}	1.5×10^{-7}
CH_4	0.726	1.0×10^{-2}	73	1.4×10^{-24}	1.5×10^{-7}
CH_4	0.798	2.5×10^{-2}	32	4.4×10^{-25}	1.5×10^{-7}
CH_4	0.840	1.0×10^{-2}	84	3.3×10^{-25}	1.5×10^{-7}
CH_4	0.861	1.1×10^{-2}	78	1.9×10^{-24}	1.5×10^{-7}
CH_4	0.887	1.8×10^{-2}	49	1.1×10^{-23}	1.5×10^{-7}
CH_4	1.00	3.7×10^{-2}	27	6.4×10^{-24}	1.5×10^{-7}
CH_4	1.13	1.3×10^{-2}	87	3.2×10^{-22}	1.1×10^{-2}
CH_4	1.16	7.4×10^{-3}	160	7.7×10^{-22}	2.6×10^{-2}
CH_4	1.33	1.4×10^{-4}	9400	1.9×10^{-21}	2.6×10^{-2}
CH_4	1.65	1.3×10^{-2}	120	1.8×10^{-20}	6.2×10^{-1}
CH_4	1.67	1.4×10^{-3}	1200	1.4×10^{-20}	4.9×10^{-1}
CH_4	1.68	1.3×10^{-2}	130	8.1×10^{-21}	2.8×10^{-1}
CH_4	2.20	1.2×10^{-3}	1800	8.1×10^{-21}	2.8×10^{-1}
CH4	2.31	6.6×10^{-2}	35	2.9×10^{-20}	$1.0 \times 10^{+0}$
CH_4	2.37	2.3×10^{-2}	100	3.7×10^{-20}	$1.3 \times 10^{+0}$
CH_4	2.59	1.9×10^{-2}	130	2.5×10^{-21}	8.7×10^{-2}
CH_4	3.32	9.4×10^{-2}	35	2.1×10^{-18}	$7.1 \times 10^{+1}$
CH_4	7.66	2.3×10^{-1}	34	7.7×10^{-19}	$2.6 \times 10^{+1}$
			—		

TABLE 2. (*continued*)

Species	λ^* (μm)	$\Delta\lambda^\dagger$ (μm)	$\lambda/\Delta\lambda$	Opacity‡ (cm^2 molecule^{-1})	Optical Depth§
N_2O	1.52	1.1×10^{-2}	140	1.2×10^{-22}	7.9×10^{-4}
N_2O	1.67	1.3×10^{-2}	130	8.0×10^{-23}	5.1×10^{-4}
N_2O	1.70	1.4×10^{-2}	120	2.9×10^{-23}	1.8×10^{-4}
N_2O	1.77	1.5×10^{-2}	120	7.5×10^{-23}	4.8×10^{-4}
N_2O	1.96	1.8×10^{-2}	110	2.2×10^{-22}	1.4×10^{-3}
N_2O	1.99	2.0×10^{-2}	100	2.3×10^{-22}	1.5×10^{-3}
N_2O	2.04	2.1×10^{-2}	100	4.0×10^{-23}	2.6×10^{-4}
N_2O	2.11	2.1×10^{-2}	100	3.3×10^{-21}	2.1×10^{-2}
N_2O	2.16	2.4×10^{-2}	91	4.6×10^{-22}	2.9×10^{-3}
N_2O	2.26	2.5×10^{-2}	90	5.2×10^{-21}	3.3×10^{-2}
N_2O	2.46	3.1×10^{-2}	78	2.2×10^{-22}	1.4×10^{-3}
N_2O	2.61	3.3×10^{-2}	80	6.6×10^{-21}	4.2×10^{-2}
N_2O	2.67	3.5×10^{-2}	75	3.1×10^{-21}	2.0×10^{-2}
N_2O	2.87	4.0×10^{-2}	71	1.5×10^{-19}	9.5×10^{-1}
N_2O	2.97	4.6×10^{-2}	65	7.3×10^{-21}	4.7×10^{-2}
N_2O	3.58	5.6×10^{-2}	63	2.5×10^{-20}	1.6×10^{-1}
N_2O	3.90	7.6×10^{-2}	52	1.1×10^{-19}	6.8×10^{-1}
N_2O	4.06	8.3×10^{-2}	49	2.5×10^{-20}	1.6×10^{-1}
N_2O	4.31	9.9×10^{-2}	43	2.2×10^{-21}	1.4×10^{-2}
N_2O	4.50	9.9×10^{-2}	45	4.4×10^{-18}	$2.8 \times 10^{+1}$
N_2O	5.32	8.7×10^{-2}	61	1.6×10^{-20}	1.0×10^{-1}
N_2O	5.72	1.4×10^{-1}	42	6.4×10^{-22}	4.1×10^{-3}
N_2O	6.12	2.0×10^{-1}	31	4.1×10^{-22}	2.6×10^{-3}
N_2O	7.78	3.1×10^{-1}	25	7.4×10^{-19}	$4.7 \times 10^{+0}$
N_2O	8.56	3.7×10^{-1}	25	2.6×10^{-20}	1.6×10^{-1}
N_2O	9.47	4.7×10^{-1}	20	2.7×10^{-23}	1.7×10^{-4}
N_2O	10.7	5.5×10^{-1}	19	1.4×10^{-22}	8.8×10^{-4}
N_2O	14.4	9.6×10^{-1}	15	7.4×10^{-22}	4.7×10^{-3}
N_2O	17.0	$1.1 \times 10^{+0}$	15	6.4×10^{-19}	$4.1 \times 10^{+0}$
			—		
H_2O	0.653	1.0×10^{-2}	65	1.2×10^{-23}	$1.1 \times 10^{+0}$
H_2O	0.722	1.1×10^{-2}	64	1.3×10^{-22}	$1.2 \times 10^{+1}$
H_2O	0.823	1.5×10^{-2}	55	1.7×10^{-22}	$1.7 \times 10^{+1}$
H_2O	0.940	2.1×10^{-2}	45	2.2×10^{-21}	$2.1 \times 10^{+2}$
H_2O	1.14	2.9×10^{-2}	39	4.9×10^{-21}	$4.8 \times 10^{+2}$
H_2O	1.38	4.3×10^{-2}	32	6.6×10^{-20}	$6.5 \times 10^{+3}$
H_2O	1.89	8.9×10^{-2}	21	1.0×10^{-19}	$9.8 \times 10^{+3}$
H_2O	2.65	1.8×10^{-1}	15	8.5×10^{-19}	$8.3 \times 10^{+4}$
H_2O	3.17	3.0×10^{-1}	11	6.4×10^{-21}	$6.3 \times 10^{+2}$
H_2O	3.68	2.2×10^{-1}	17	5.2×10^{-23}	$5.0 \times 10^{+0}$
H_2O	6.27	$1.1 \times 10^{+0}$	5.7	1.2×10^{-18}	$1.2 \times 10^{+5}$
			—		
CO_2	1.43	9.8×10^{-3}	150	2.6×10^{-22}	$1.8 \times 10^{+0}$
CO_2	1.58	1.1×10^{-2}	140	7.6×10^{-23}	5.3×10^{-1}

TABLE 2. (*continued*)

Species	λ^* (µm)	$\Delta\lambda^\dagger$ (µm)	$\lambda/\Delta\lambda$	Opacity‡ (cm^2 molecule^{-1})	Optical Depth§
CO_2	1.61	1.2×10^{-2}	130	7.6×10^{-23}	5.3×10^{-1}
CO_2	1.96	1.7×10^{-2}	110	1.9×10^{-21}	$1.3 \times 10^{+1}$
CO_2	2.01	1.9×10^{-2}	110	5.6×10^{-21}	$3.9 \times 10^{+1}$
CO_2	2.06	1.7×10^{-2}	120	1.2×10^{-21}	$8.3 \times 10^{+0}$
CO_2	2.69	3.4×10^{-2}	79	2.6×10^{-19}	$1.8 \times 10^{+3}$
CO_2	2.77	3.6×10^{-2}	76	1.7×10^{-19}	$1.2 \times 10^{+3}$
CO_2	4.26	8.6×10^{-2}	50	1.6×10^{-17}	$1.1 \times 10^{+5}$
CO_2	7.31	2.5×10^{-1}	29	2.9×10^{-24}	2.1×10^{-2}
CO_2	7.94	3.0×10^{-1}	27	2.6×10^{-24}	1.8×10^{-2}
CO_2	9.40	4.2×10^{-1}	22	1.3×10^{-22}	9.4×10^{-1}
CO_2	10.4	5.1×10^{-1}	20	8.5×10^{-23}	0.6×10^{-1}
CO_2	15.0	9.6×10^{-1}	16	4.4×10^{-18}	$3.1 \times 10^{+4}$

*Feature central wavelength.
†Feature full-width at half-max (assuming p = 1 bar and T = 288 K).
‡Peak opacity at assumed pressure and temperature.
§Assuming modern Earth (vertical) column densities and adopting peak opacity values.

caused Earth to present as a "pale orange dot" early in its history (*Arney et al.,* 2016). However, it is unlikely that Earth's CH_4/CO_2 ratio has been high enough to produce haze after ~2.5 Ga.

Nitrous oxide (N_2O) is a potential atmospheric biosignature, in addition to being a powerful greenhouse gas and an important component of stratospheric ozone chemistry (*Prather and Hsu,* 2010). On the modern Earth, natural (non-anthropogenic) sources of N_2O are dominated by microbial activity in terrestrial soils and productive regions of the surface ocean (*Matson and Vitousek,* 1990; *Hirsch et al.,* 2006). Under low oxygen concentrations, N_2O can be produced biologically during the metabolic oxidation of ammonium (NH_4^+) and nitrite (NO_2^-) and during the reduction of nitrate (NO_3^-) during incomplete denitrification (*Bianchi et al.,* 2012; *Freing et al.,* 2012). The inorganic reaction of nitric oxide (NO) with dissolved ferrous iron (Fe^{2+}) can also produce N_2O in a process referred to as "chemodenitrification" (*Wullstein and Gilmour,* 1966). The source NO for this reaction can be derived from either biological N fixation or abiotically through the breakdown of atmospheric N_2 by lightning. Still, N_2O is considered a putative biosignature because production of N_2O from the latter is likely to be relatively small (*Schumann and Huntrieser,* 2007). Importantly, the photochemical stability of N_2O is a relatively strong function of atmospheric O_2 (*Levine et al.,* 1979; *Kasting and Donahue,* 1980) — for example, sustaining a modern atmospheric N_2O abundance at a plausible Proterozoic atmospheric pO_2 of ~1% of the present atmospheric level requires a surface N_2O flux of roughly 30× that on the modern Earth for a Sun-like star (*Roberson et al.,* 2011; *Stanton et al.,* 2018). This suggests that through Earth's history atmospheric N_2O abundance would broadly have tracked the abundance of

atmospheric O_2 and O_3. Because even modern-Earth-like N_2O abundances produce relatively weak spectral features (*Schwieterman et al.,* 2018, and Table 2), N_2O is often not a primary focus of biosignature detectability studies.

Finally, H_2O and CO_2, while essential to many biological functions, are also critical signposts of a habitable world. The former will be present in the lower atmosphere of any world with stable surface liquid water, and the latter is key to maintaining surface habitability through its properties as a greenhouse gas and its role in the carbonate-silicate cycle. The spectrum of modern Earth is strongly sculpted by H_2O absorption features throughout the red-visible, near-infrared, and mid-infrared, and these same features would have produced strong spectral features on Earth during all non-snowball periods of its evolution. However, even in extremely cold "Snowball Earth" scenarios, the H_2O bands spanning the near-infrared at 6.3 µm would remain apparent. Carbon dioxide has several strong features at longer near-infrared wavelengths (most notably at 4.3 µm) and in the mid-infrared (at 15 µm). Earlier in Earth's history, where higher levels of CO_2 would have been required to maintain habitable surface conditions, CO_2 features at shorter near-infrared wavelengths would have appeared much stronger (e.g., bands near 1.6 and 2 µm) (*Meadows,* 2008; *Arney et al.,* 2016; *Rugheimer and Kaltenegger,* 2018).

4. OBSERVING EARTH FROM AFAR

Observational data for the distant Earth provide a critical opportunity to study the spectral appearance of an ocean-bearing, inhabited world. As is discussed below, such data enable investigations into the remote characterization of the physical and chemical state of our planet

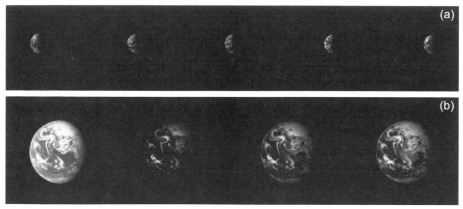

Fig. 4. **(a)** Time sequence of Earth and the Moon acquired with the SSI "infrared" filter (961–1011 nm) after the second Galileo flyby (December 1992). **(b)** Images of Earth acquired with the SSI from the first Galileo flyby (December 1990). Filters from left to right are violet (382–427 nm), green (527–592 nm), red (641–701 nm), and "infrared." The subobserver point in each image is approximately identical, and Australia is the landmass near the center of the images. For all images south is oriented "up," as it was in the true flyby geometry.

through applications of solar system planetary exploration techniques. Zooming even further out, though, observations where Earth is treated as a single pixel — a pale blue dot — provide insights into how we will, one day, study photometric and spectroscopic observations of potentially Earth-like exoplanets for signs of habitability and life.

4.1. Earth as a Planet: The Galileo Experiment

The Pioneer 10/11 (launched 1972) (*Baker et al.,* 1975; *Gehrels,* 1976; *Ingersoll et al.,* 1976; *Kliore and Woiceshyn,* 1976) and Voyager 1/2 missions (launched 1977) (*Kohlhase and Penzo,* 1977; *Hanel et al.,* 1977) enabled the initial exploration of key approaches to analyzing spacecraft flyby data for many solar system worlds. Instruments and techniques for either acquiring or interpreting spatially-resolved observations of planets and moons using photometry or spectroscopy at wavelengths spanning the ultraviolet through the infrared were among the important developments. From these observations planetary scientists were able to infer details about atmospheric chemistry and composition, cloud and aerosol formation and distribution, atmospheric thermal structure and circulation, and surface chemical and thermal properties (for worlds with thin atmospheres), as well as planetary energy balance.

Flybys of Earth by the Galileo spacecraft (launched 1989) (*Johnson et al.,* 1992) in December 1990 and 1992 afforded planetary scientists the first opportunity to analyze our planet using the same tools and techniques that had been (and would be) applied throughout the solar system. During the flybys, data were acquired using the Near-Infrared Mapping Spectrometer (NIMS) (*Carlson et al.,* 1992), the Solid-State Imaging system (SSI) (*Belton et al.,* 1992), the Ultraviolet Spectrometer (UVS) (*Hord et al.,* 1992), and the Plasma Wave Subsystem (PWS) (*Gurnett et al.,* 1992). Critically, spatially-resolved imagery from the SSI was provided across eight filters (*Belton et al.,* 1992, their Table II), and spatially-resolved spectra were acquired by NIMS over 0.7–5.2 μm

(at a wavelength resolution, $\Delta\lambda$, of 0.025 μm longward of 1 μm, and 0.0125 μm shortward of 1 μm).

Figure 4 shows a sampling of SSI images of Earth from the first Earth flyby and a time sequence of SSI images of Earth and the Moon that were acquired after the second Earth flyby. Many images from the second Earth flyby suffered from saturation defects. Reconstructed NIMS images from both Earth flybys are shown in Fig. 5, where certain instrument and mapping defects can be seen. Here, the 4.0-μm images highlight a wavelength range with relatively little atmospheric opacity and where thermal emission dominates. By contrast, the 2.75-μm image (located within a H_2O absorption band) contains both reflected and thermal contributions, which, for example, results in reflective clouds only being seen in the sunlit portions of the images.

In a landmark study, *Sagan et al.* (1993) used the Galileo flyby data to ascertain key details about the surface and atmospheric state of our planet. Spectra from the NIMS instrument contain information about surface and atmospheric chemistry, and can also indicate surface thermal conditions at longer wavelengths. Thus, Sagan et al. argued that reflective polar caps seen in SSI images were water ice, and that the surface of the planet spanned the freezing point of water (covering at least 240–290 K). Additionally, the darkest regions in the SSI images showed signs of specular refection, indicating that these regions were liquid oceans. Finally, clear sky soundings of H_2O absorption features indicated a surface with large relative humidity (i.e., near the condensation point). Taken altogether, these lines of evidence clearly indicate that the world under investigation is habitable, or capable of maintaining liquid water on its surface.

Drossart et al. (1993) retrieved abundances of key atmospheric constituents (CO_2, H_2O, CO, O_3, CH_4, and N_2O) via simple parameterized fits to resolved NIMS observations (see Fig. 6). Here, model spectra were generated by adopting scaled Earth-like profiles for trace atmospheric species. Thermal structure profiles were derived using the

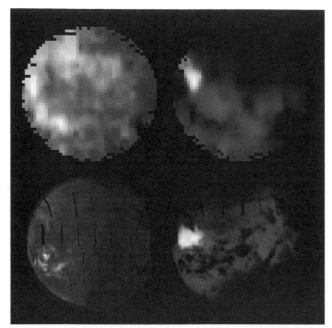

Fig. 5. Reconstructed images of Earth using NIMS observations from the first (top) and second (bottom) Galileo Earth flybys. Images in the left column are at 2.75 μm while images in the right column are at 4.0 μm. The bright (warm) source seen in both 4.0-μm images is likely Australia. The spacecraft was nearer to Earth in the second flyby dataset, thereby providing better resolution across the disk.

4.3-μm CO_2 band, which rely on this gas having a well-mixed vertical profile (i.e., a near-constant mixing ratio with altitude). For a well-mixed gas, variation in an infrared

absorption band can be attributed to thermal structure rather than variation in abundance with altitude.

Observations from the NIMS instrument also indicated large column densities of O_2 in Earth's atmosphere (*Sagan et al.,* 1993). It was estimated that diffusion-limited escape of hydrogen (produced from H_2O photolysis) would require many billions of years to build up atmospheric oxygen to the observed levels, implying an alternative source. Using the abundances derived in *Drossart et al.* (1993), it was shown that atmospheric CH_4 was in a state of extreme disequilibrium with no known geological source that could supply CH_4 at the rate required to maintain the observed concentrations. Certain surface regions of the planet imaged in multiple SSI filters demonstrated a sharp increase in reflectivity at wavelengths beyond 700 nm, known from ground-truth investigations to be the vegetation "red edge" (a rapid increase in reflectivity near 0.7 μm that is related to pigments). Thus, multiple lines of evidence point toward the potential for biological activity to be shaping the surface and atmospheric properties of Earth. [The Galileo Earth dataset has also been investigated for signs of shadowing caused by trees (*Doughty and Wolf,* 2016), whose vertical structures create distinct bi-directional reflectance distribution functions that have been proposed as a biosignature for Earth-like exoplanets (*Doughty and Wolf,* 2010).]

Of course, the strongest evidence for Earth being inhabited came from the PWS dataset (*Sagan et al.,* 1993). Here, radio emissions from our planet were monitored as a function of time and frequency throughout the Galileo encounter. Narrow-band emissions between 4 and 5 MHz (i.e., in the high-frequency portion of the radio spectrum

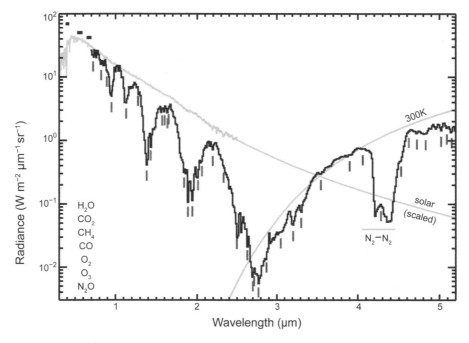

Fig. 6. See Plate 37 for color version. Spatially-resolved spectrum of Earth from the second Galileo flyby, acquired for a small region of the planet in the Indian Ocean and near the terminator at the time of observation. Visible SSI photometry (in the violet, green, and red filters) and NIMS spectroscopy are both shown, and key absorption bands are indicated. Reflected light contributes significantly below about 3 μm, while thermal emission dominates at longer wavelengths. Light gray curves indicate a scaled solar spectrum and a 300 K blackbody.

where a variety of radio communications occur), isolated in both frequency and time, were interpreted as radio transmissions from an intelligent species on our planet. In other words, an observational approach championed in the search for extraterrestrial intelligence (SETI) — listening at radio frequencies — yielded the least ambiguous evidence for the inhabitance of Earth from the Galileo flyby dataset.

In the words of *Sagan et al.* (1993), the Galileo Earth datasets offered a "unique control experiment on the ability of flyby spacecraft to detect life at various stages of evolutionary development." Combining lines of evidence that spanned the ultraviolet, visible, infrared, and radio spectral regimes, the Galileo observations indicated a habitable planet with a diversity of surface environments and whose atmosphere (and thus spectrum) is strongly influenced by life. In the context of exoplanets, however, the key question becomes which habitability and life signatures are lost when Earth is studied not as a resolved source but as a distant, unresolved target.

4.2. Observing the Pale Blue Dot

A more accurate model for future observations of Earth-like exoplanets is not the Galileo flyby observations, but instead the famous "pale blue dot" photograph of Earth taken by the Voyager 1 spacecraft (see Fig. 7). Here — even for future large telescopes — observations will not be able to spatially resolve features on the disk of an exoplanet [although time-domain data can be used to obtain some spatial resolution (e.g., *Cowan et al.,* 2009;

Majeau et al., 2012)]. The resolution of a telescope is limited by the physics of light diffraction to an angular size of roughly λ/D, where λ is wavelength and D is telescope diameter. For a 10-m-class telescope observing at visible wavelengths (i.e., near 500 nm), the angular resolution is at best 5×10^{-8} rad (or about 10 milli-arcsec). Even for our nearest stellar neighbors (e.g., α Centauri at 1.3 parsec) this corresponds to a spatial resolution of 2×10^6 km, or about 3× the radius of our Sun.

Unresolved observations of planets are sometimes referred to as being "disk-integrated." Here, the entire three-dimensional complexity of a planet is effectively collapsed into a single pixel. For worlds like Earth, this means that cloud-free regions of the planet are blended with cloudy regions, warm equatorial zones are observationally mixed with cold polar caps, and continents become unresolved from oceans. Additionally, viewing geometry plays an important role, as the portions of the planet near the limb will contribute less overall flux to an observation (since these areas are of smaller solid angular size), and, in reflected light, regions near the day/night terminator will also contribute relatively little flux owing to their lower insolation.

If we wish to extend the analysis techniques applied to the Galileo Earth flyby observations to an unresolved pale blue dot, we must look toward either observational datasets for an unresolved Earth, or toward datasets or products that can mimic an unresolved Earth. There are, in general, three approaches to obtaining (or constructing) such disk-integrated observations of our planet. First, one can observe light reflected from Earth from the portion of the Moon that is not illuminated by the Sun (i.e., so-called "Earthshine"

Fig. 7. The famous "Pale Blue Dot" image of Earth, acquired by the Voyager 1 spacecraft from a distance of 40 AU. Credit: NASA/JPL-Caltech.

observations) (*Danjon,* 1928; *Dubois,* 1947; *Woolf et al.,* 2002; *Pallé et al.,* 2003; *Turnbull et al.,* 2006). Second, one can use higher spatial resolution observations from satellites in Earth orbit to piece together a more integrated view of Earth (*Hearty et al.,* 2009; *Macdonald and Cowan,* 2019). Finally, spacecraft observations of the distant Earth — like those acquired by Galileo — can be integrated over the planetary disk to yield exoplanet-like datasets. We discuss each of these approaches below, highlighting the advantages and disadvantages of each technique.

4.2.1. Earthshine. Using Earthshine from the dark portion of the Moon — which is illuminated by Earth but not the Sun — has a long history of revealing key details about our planet. In the *Dialogue Concerning the Two Chief World Systems,* Galileo used Earthshine to deduce that "seas would appear darker, and [. . .] land brighter" when observed from a distance (*Galilei,* 1632). In the first multi-year Earthshine monitoring experiment, described in *Danjon* (1928) and continued by *Dubois* (1947), the broadband visual reflectivity of Earth was shown to vary by several tens of percent at a given phase angle (i.e., the planet-star-observer angle), and cloud variability was identified as the likely driver of these variations.

Modern Earthshine observations (*Goode et al.,* 2001; *Woolf et al.,* 2002; *Pallé et al.,* 2004b) have reached an impressive level of precision. Achieving this precision requires corrections for airmass effects, the lunar phase function (i.e., how the Moon scatters light into varied directions), and variations in reflectivity across the lunar surface. Nevertheless, it is now common for Earthshine measurements to achieve 1% precision on a given night (*Qiu et al.,* 2003). Such high-quality photometric observations have

revealed variability in the visible reflectivity of Earth at daily, monthly, seasonal, and decadal timescales (*Goode et al.,* 2001; *Pallé et al.,* 2003, 2004a, 2009a, 2016). Figure 8 shows a collection of phase-dependent visual (400–700 nm) apparent albedo measurements from Earthshine measurements. Apparent albedo (A_{app}) is defined by normalizing an observed planetary flux to that from a perfectly reflecting Lambert sphere (i.e., a sphere whose surface reflects light equally well into all directions) observed at the same phase angle, or

$$A_{app} = \frac{3}{2} \frac{F_p}{F_s} \frac{\pi}{\sin\alpha + (\pi - \alpha)\cos\alpha}$$

where α is the star-planet-observer (i.e., phase) angle, F_p is the emergent planetary flux at the top of the atmosphere, F_s is the solar/stellar flux at normal incidence on the top of the planetary atmosphere, and the flux quantities can either be wavelength-dependent (resulting in a wavelength-dependent apparent albedo) or integrated. (Note that the factor of 3/2 comes from the conversion between geometric and spherical albedo.) Thus, a Lambert sphere would have a constant apparent albedo as a function of phase. Critically, then, the non-constant apparent albedo of Earth demonstrated in Fig. 8 reveals a weak back-scattering peak at small phase angles, a region of Lambert-like scattering at intermediate phase angles, and a strong forward-scattering peak at large phase angles.

Spectroscopic studies of Earthshine (*Woolf et al.,* 2002) offer additional insights into Earth as an exoplanet (Fig. 9), beyond those obtained through photometric Earthshine

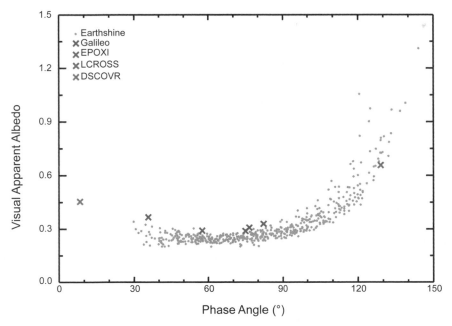

Fig. 8. See Plate 38 for color version. Measurements of the phase-dependent visual (400–700 nm) apparent albedo of Earth from Earthshine data spanning several years (from *Pallé et al.,* 2003), and from several spacecraft missions. Apparent albedo values larger than unity indicate stronger directional scattering than can be produced by a sphere whose surface reflects light isotropically (i.e., a Lambert sphere). The DSCOVR datapoint is derived from the available four narrowband channels that span the visible range since an integrated 400–700-nm observation cannot be produced from DSCOVR data.

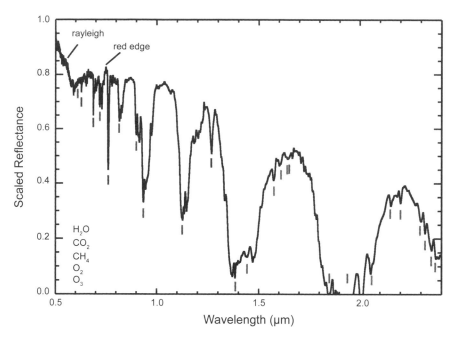

Fig. 9. See Plate 39 for color version. Scaled reflectance spectrum of Earth at visible and near-infrared wavelengths measured from Earthshine. Key absorption and refection features are indicated. Data courtesy of M. Turnbull from *Turnbull et al.* (2006).

investigations. Using spectroscopic Earthshine data collected over several weeks or months, *Arnold et al.* (2002) and *Seager et al.* (2005) showed that the aforementioned vegetation red edge signature is variable in the reflectance spectrum of Earth, and can lead to sharp reflectivity increases at the 10% level in the 600–800-nm range. A red-edge-focused Earthshine study by *Montañés-Rodríguez et al.* (2006) found no strong signature in spectroscopic data from a single night, which highlighted the importance of cloud cover both in setting the overall brightness of Earth and in masking surface reflectance features. In Earthshine observations that spanned 0.7–2.4 µm, *Turnbull et al.* (2006) noted a plethora of absorption features that were indicative of life, habitability, and geological activity. Also, after accounting for how the lunar surface depolarizes radiation, polarization-sensitive spectroscopic Earthshine observations have explored the degree to which a spectrum of Earth can be polarized (0–20%, depending on wavelength) as well as the impact of cloud cover on this signature (*Sterzik et al.,* 2012; *Miles-Páez et al.,* 2014). Finally, by investigating the Earthshine spectrum at extremely high spectral resolution ($\lambda/\Delta\lambda$; also referred to as the spectral resolving power), *González-Merino et al.* (2013) uncovered narrow spectral features due to atomic sodium in Earth's atmosphere that are either of terrestrial or meteoritic origin.

Of course, Earth-like planets around other stars may not be solely investigated using reflected-light techniques (e.g., *Snellen et al.,* 2013), especially in the case of potentially habitable worlds orbiting M-dwarf hosts where transit or secondary eclipse observations would be the preferred approach. Impressively, observational techniques developed

for Earthshine data collection have been repurposed to enable observations of the transmission spectrum of Earth's atmosphere. By observing the Moon during a lunar eclipse, *Pallé et al.* (2009b) were able to measure light that had been transmitted through our atmosphere and reflected by the lunar surface. These observations revealed signatures of key atmospheric and biosignature gases, and even included narrow features due to ionized calcium as well as broad pressure-induced features from O_2 and N_2 (the latter of which is typically difficult to detect due to its general lack of rotational-vibrational features). A follow-up analysis of these data by *García Muñoz et al.* (2012) showed that refractive effects in transit spectra of Earth twins would limit the atmospheric depths probed (during mid-transit) to be above about 10 km, thus providing limited information from the surface and tropospheric environments. Also, additional high-resolution transmission spectra acquired using Earthshine-related techniques have revealed variability in the depths of H_2O absorption features (*Yan et al.,* 2015), likely tied to the condensible nature of this gas in our atmosphere.

Finally, while Earthshine techniques have been proven to be both powerful and versatile, this approach does have its shortcomings. First, due to the groundbased nature of the observations, full diurnal cycles in the reflectivity of Earth cannot be observed except during polar night (*Briot et al.,* 2013). Second, it is often difficult to calibrate Earthshine observations in a fashion that reveals the absolute brightness of our planet. Thus, some Earthshine datasets are only reported as a scaled reflectance value, and these products are of lower utility when it comes to exoplanet detectability and characterization studies. Finally, Earthshine cannot be used to observe thermal emission from Earth since the Moon is also self-luminous at infrared wavelengths.

4.2.2. Orbit. A large suite of satellites are continuously monitoring the Earth system from space. While most of these Earth-observing satellites only resolve a small patch of our planet in any individual observation, the collective dataset from these satellites benefits from extensive temporal, spatial, and spectral coverage. Thus "stitching" together spatially-resolved radiance measurements from one (or several) observing platform(s) can enable a view of the entire disk of Earth. This approach was pioneered by *Hearty et al.* (2009), who used spatially-resolved thermal radiance observations from the Atmospheric Infrared Sounder (AIRS) instrument (onboard NASA's Aqua satellite) (*Aumann et al.,* 2003) to create disk-integrated infrared spectra of Earth (Fig. 10).

The practice of stitching together resolved radiance measurements from an Earth-observing satellite is, unfortunately, not straightforward. Temporal gaps sometimes exist in these datasets where a given latitude/longitude patch of Earth has not been observed in a given 24-hr period. Thus, if the goal is to produce a snapshot of Earth at a given time, an interpolation of existing radiance observations across time must be performed. As the Earth climate system (as well as top-of-atmosphere radiances) is non-linear, this interpolation introduces some uncertainties.

The greater challenge to deriving whole-Earth views from resolved satellite observations, though, stems from viewing geometry constraints. Most Earth-observing satellites are designed to acquire observations in the nadir (i.e., direct downward) direction. For satellites observing in reflected light, the range of solar incidence angles can also be limited, especially over any given several-day period. Thus, when stitching together a whole-disk observation, data for certain viewing geometries (e.g., patches located near the limb, where the observing geometry is quite distinct from nadir-looking) may not exist. In this case, assumptions must be adopted for how radiance will vary with the emission angle and/or the solar incidence angle. For example, *Hearty et al.* (2009) adopted a limb darkening law to transform radiances acquired at nadir to radiances appropriate for other emission angles.

An alternative approach to using directly-observed radiances from satellites is, instead, to adopt a collection of satellite-derived "scene" models. These scene models describe the viewing geometry-dependent brightness of different surface categories for Earth (e.g., ocean or desert) under different cloud coverage scenarios. In other words, these models specify the bi-directional reflectance distribution functions for a large variety of surface type and cloud cover- age combinations. Scene models can be derived from satellite observations (*Suttles et al.,* 1988) or can be designed to fit satellite observations (*Manalo-Smith et al.,* 1998). Combining data that describe the time-dependent distribution of clouds, snow, and ice on Earth with a set of scene models then enables the recreation of whole-disk views of our planet (*Ford et al.,* 2001; *Pallé et al.,* 2003; *Oakley and Cash,* 2009). Integrating these three-dimensional models over the planetary disk then yields the brightness (or reflectivity) of the pale blue dot. One

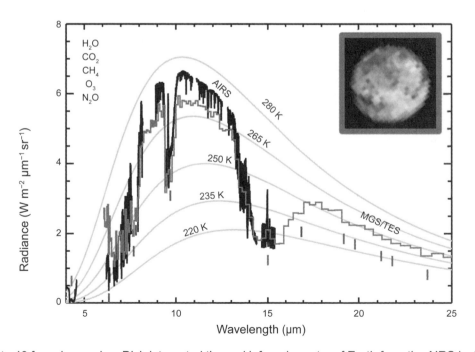

Fig. 10. See Plate 40 for color version. Disk-integrated thermal infrared spectra of Earth from the AIRS instrument (*Hearty et al.,* 2009) and from the Mars Global Surveyor Thermal Emission Spectrometer (MGS/TES) (*Christensen and Pearl,* 1997), where differences are due to the combined effects of seasons, climate, clouds, and viewing geometry. Key features are labeled and blackbody spectra at different emitting temperatures are shown. Inset is a broadband thermal infrared (6–10 μm) image of Earth from the LCROSS mission (*Robinson et al.,* 2014).

key shortcoming of the scene model approach, however, is that such models are rarely spectrally resolved, and instead specify a broadband reflectivity or brightness. Thus, such models cannot produce spectra of the disk-integrated Earth, and, instead, focus on computing broadband lightcurves for the pale blue dot.

Solar occultation observations acquired from Earth orbit provide a direct measurement of the transmittance along a slant path through the atmosphere. Initially such datasets (*Abrams et al.,* 1999; *Bovensmann et al.,* 1999; *Bernath et al.,* 2005) provided an excellent model validation for tools designed to simulate transit spectra of Earth-like exoplanets (*Kaltenegger and Traub,* 2009; *Misra et al.,* 2014b). However, as was recognized by *Robinson et al.* (2014) and *Dalba et al.* (2015), occultation observations from orbit can be directly translated into transit spectra. Using data from the Canadian Atmospheric Chemistry Experiment–Fourier Transform Spectrometer (ACE-FTS) (*Bernath et al.,* 2005), *Schreier et al.* (2018) created transit spectra of Earth spanning 2.2–13.3 μm and demonstrated that signatures of chlorofluorocarbons appeared in the occultation-derived transit observations, in addition to more-standard features of H_2O, CO_2, CH_4, N_2O, N_2, NO_2, and O_2 (see also *Macdonald and Cowan,* 2019).

4.2.3. Spacecraft. The ideal approach for mimicking direct observations of Earth-like exoplanets is, of course, to acquire photometry and/or spectroscopy for a truly distant Earth. Such observations must be taken from distances beyond low-Earth or geostationary orbit, as the entire disk of the planet is not entirely visible from these vantages (e.g., only about 85% of the disk is observable from geostationary orbit). Thus, views of Earth from spacecraft at lunar distances or from Earth-Sun Lagrange points, or observations from interplanetary spacecraft, are all excellent sources. Until the recent launch of the Deep Space Climate Observatory (DSCOVR) (*Biesecker et al.,* 2015) mission to the Earth-Sun L1 point, no dedicated mission existed for observing Earth from a great distance. Thus, the majority of the spacecraft observations relevant to Earth as an exoplanet came from missions sent to other solar system worlds.

While spacecraft observations of the distant Earth are ideal for exoplanet-themed investigations, this approach is not without its shortcomings. First, it is difficult to find time during the main phase of a mission to dedicate toward observations of non-primary targets such as Earth. This means that the temporal coverage of spacecraft datasets for the distant Earth is poor, with many of these datasets acquired during the cruise phase of a mission. Second, and most unfortunately, spacecraft datasets for the distant Earth often remain unpublished. In these circumstances, the data may have been acquired only for press or outreach purposes, or it might be that analysis and publication of these data are seen as a distraction from the main goals of a mission. Unpublished datasets are known to exist (both from private communications and press releases) for a number of other missions, including Cassini, Clementine, Lunar Reconnaissance Orbiter, Mars Express, Mars Recon-

TABLE 3. Published spacecraft datasets for Earth as an exoplanet.

Spacecraft	Date[*]	Phase Angle(s)	Wavelength (μm)	Resolution[†]	Source(s)
Galileo	1990-12-10	35°	0.38–5.2	$\Delta\lambda = 0.01$ 0.44 μm (vis)	*Sagan et al.* (1993); *Drossart et al.* (1993)
	1992-12-09	82°			
	1992-12-16	89°		$\Delta\lambda = 0.025$ μm (NIR)	
MGS/TES	1996-11-23	n/a	6–50	$\lambda/\Delta\lambda = 15$ 170	*Christensen and Pearl* (1997)
EPOXI	2008-03-18	58°	0.37–4.54	$\Delta\lambda$ H 0.1 μm (vis)	*Livengood et al.* (2011); *Cowan et al.* (2011); *Fujii et al.* (2011); *Robinson et al.* (2011)
	2008-05-28	75°			
	2008-06-04	77°		$\lambda/\Delta\lambda = 215$ 730 (NIR)	
	2009-03-27	87°			
	2009-10-04	86°			
LCROSS	2009-08-01	23°	0.26–13.5	$\lambda/\Delta\lambda = 300$ 800 (vis)	*Robinson et al.* (2014)
	2009-08-17	129°		$\Delta\lambda = 0.3, 0.8$ μm (NIR)	
	2009-09-18	75°		$\Delta\lambda - 4, 7.5$ μm (thermal)	
DSCOVR	ongoing	4–12°	0.318–0.780	$\Delta\lambda = 1$ 3 nm	*Biesecker et al.* (2015); *Yang et al.* (2018)

[*]Based on UT at start of observations.

[†]Abbreviating visible range (~0.4–1 μm) as "vis" and near-infrared range (~1–5 μm) as "NIR."

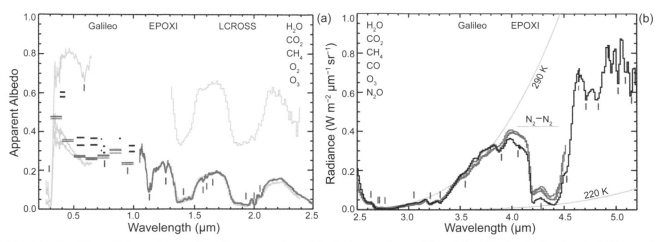

Fig. 11. See Plate 41 for color version. Summary of published observations of the distant Earth at ultraviolet, visible, and near-infrared wavelengths. The first figure presents spectra of Earth's apparent albedo from Galileo (*Sagan et al.,* 1993), EPOXI (*Livengood et al.,* 2011), and LCROSS (*Robinson et al.,* 2014). A crescent-phase observation from LCROSS is marked by large apparent albedo, which was driven primarily by forward scattering from a glint spot. The second figure presents near-infrared emission observations from Galileo and EPOXI, and blackbody spectra are provided. Key absorption features are indicated in both figures.

naissance Orbiter, MESSENGER, OSIRIS-REx, SELENE/ Kaguya, and Venus Express.

A detailing of published spacecraft-acquired datasets that are relevant to Earth as an exoplanet is shown in Table 3, emphasizing photometric and spectroscopic observations that span the ultraviolet, visible, and infrared wavelengths. Beyond the previously-discussed Galileo Earth flyby observations, key datasets also come from a snapshot thermal infrared spectrum acquired by the Mars Global Surveyor Thermal Emission Spectrometer (MGS/ TES), visible photometry and near-infrared spectroscopy spanning 24 hr on five separate dates from the EPOXI mission (which repurposed the Deep Impact flyby spacecraft), visible spectroscopy and infrared photometry and spectroscopy taken over brief intervals on three separate dates by the Lunar CRater Observation and Sensing Satellite (LCROSS), and the aforementioned DSCOVR data (which include images taken in 10 narrowband channels spanning ultraviolet and visible wavelengths, and bolometric measurements in several channels spanning 0.2–100 µm). The EPOXI dataset has been used to analyze key spectral features for Earth in the near-infrared range and to quantify the vegetation red-edge signature in disk-integrated observations (*Livengood et al.,* 2011; *Robinson et al.,* 2011), and to investigate mapping techniques for unresolved objects (*Cowan et al.,* 2009; *Fujii et al.,* 2011; *Cowan et al.,* 2011). In *Robinson et al.* (2014), the LCROSS Earth observations were used to quantify the impact of ocean glint and ozone absorption on phase-dependent disk-integrated visible spectroscopic data for the pale blue dot. A digest of ultraviolet, visible, and near-infrared observations is shown in Fig. 11.

Finally, while existing datasets have made many valuable contributions to our understanding of the appearance of the pale blue dot, major gaps still exist in our observational coverage. Regarding Table 3, it is obvious

that spacecraft observations of Earth in reflected light at crescent phases (i.e., phase angles from roughly 145°–180°) are lacking — only a single dataset, from the LCROSS mission (*Robinson et al.,* 2014), exists for all phase angles beyond quadrature (which occurs at a phase angle of 90°, where the planet is half illuminated). Thermal infrared observations at moderate to high spectral resolution are also not represented. Additionally, no datasets span a continuous timeframe of longer than roughly 24 hr, which hinders studies of rotational variability. Lastly, excepting the few LCROSS pointings (*Robinson et al.,* 2014), existing visible-wavelength datasets only contain photometry, so spectroscopy below about 1 µm is not well represented.

5. MODELING THE PALE BLUE DOT

Techniques for simulating observations of the distant Earth provide a complementary approach to spacecraft, orbital, and Earthshine observations. Especially once validated against observational datasets, models of the disk-integrated Earth enable the exploration of the pale blue dot across a wide range of wavelengths and spectral resolutions, and can also fill in the various gaps that exist between different observational approaches. Currently, a hierarchy of Earth models exists, spanning simple reflectance tools to complex three-dimensional models whose outputs cover the ultraviolet through the far-infrared.

5.1. One-Dimensional Approaches

One-dimensional models of the pale blue dot capture the vertical structure of Earth's atmosphere, but omit any latitudinal or longitudinal structure in the atmosphere and surface. Such simplifications enable these one-dimensional approaches to be computationally efficient, and often allow for higher spectral resolution in model outputs. Nevertheless, key details about the fractional distribution of clouds and various surface types on Earth must be accounted for, either through data-informed weighting factors or through tuning parameters.

Traub and Jucks (2002) presented one of the earliest models of the pale blue dot. This one-dimensional tool spanned the ultraviolet through thermal infrared, and included absorption and emission from key atmospheric species. Radiation multiple scattering was neglected, and modeled observations in reflected light were generated by linearly combining spectral components (including Rayleigh, clear sky, high cloud, and others). At visible wavelengths, disk-integrated observations were simulated using a single solar zenith angle (i.e., the Sun was placed at a zenith angle of 60° over a plane-parallel atmosphere). Both *Woolf et al.* (2002) and *Turnbull et al.* (2006) used the *Traub and Jucks* (2002) model to analyze Earthshine spectra. By fitting the reflected-light spectral components in the *Traub and Jucks* (2002) model to the Earthshine data, these authors determined that the most important aspects of their reflected-light observations were a clear sky component, a gray high cloud continuum, and Rayleigh scattering. More recently, the Traub and Jucks model has been used to study the spectral evolution of Earth through time (*Kaltenegger et al.*, 2007; *Rugheimer and Kaltenegger*, 2018), including a comparison to the previously mentioned EPOXI dataset (*Rugheimer et al.*, 2013).

A multiple-scattering one-dimensional model, developed by *Martín-Torres et al.* (2003), was adopted by *Montañés-Rodríguez et al.* (2006) to help understand the signature of the vegetation red edge in Earthshine spectra. In this work, the good match between the Earthshine data and the simulations was attributed to the scattering treatment within the model. Also, the *Montañés-Rodríguez et al.* (2006) study developed a sophisticated approach to capturing the latitudinal and longitudinal distribution of clouds and surface types on Earth. Specifically, disk-averaged cloud and surface coverage maps were derived from Earth science data products, including appropriate weighting factors for the solar and lunar geometry.

5.2. Three-Dimensional Models

In general, three-dimensional models of the pale blue dot compute the spatially-resolved radiance over the planetary disk, and then integrate this radiance over solid angle to produce a disk-integrated quantity. More formally, three-dimensional models of Earth aim to compute the integral of the projected area weighted intensity in the direction of an observer, which is written as

$$F_\lambda(\hat{o},\hat{s}) = \frac{R_E^2}{d^2} \int_{2\pi} I_\lambda(\hat{n},\hat{o},\hat{s}) \, (\hat{n}\cdot\hat{o}) \, d\omega$$

where F_λ is the disk-integrated specific flux density received from a world of radius R_E at a distance d from the observer, $I_\lambda(\hat{n}, \hat{o}, \hat{s})$ is the location-dependent specific intensity in the direction of the observer, $d\omega$ is an infinitesimally small unit of solid angle on the globe, $\hat{n}$ is a surface normal unit vector for the portion of the surface corresponding to $d\omega$, and $\hat{o}$ and $\hat{s}$ are unit vectors in the direction of the observer and the Sun, respectively (see Fig. 12). The integral in equation (2) is over the entire observable hemisphere (2π steradians) and the dot product at the end of the expression ensures that an element of area $R_E^2 \, d\omega$ near the limb is weighted less than an element of equal size near the subobserver point. Note that, for reflected light, I_λ will be zero at locations on the night side of the world (i.e., where $\hat{n} \cdot \hat{s} < 0$), but is non-zero at all locations when considering thermal emission.

The most straightforward (quasi) three-dimensional models use empirical bi-directional reflectance distribution functions [e.g., the previously-mentioned scene models from *Manalo-Smith et al.* (1998)] to specify the reflectivity of a patch on the disk as a function of viewing geometry. These three-dimensional tools can either be spectrally-resolved (*Ford et al.*, 2001) or broadband (*McCullough*, 2006; *Pallé et al.*, 2003, 2008; *Williams and Gaidos*, 2008; *Oakley and Cash*, 2009). Atmospheric effects (e.g., gas absorption and scattering) are typically omitted, although *Fujii et al.* (2010) produced a three-dimensional reflectance model that blended wavelength-dependent bi-directional reflectance distribution functions from a variety of sources and also included an additive atmospheric Rayleigh scattering term. Time-dependent distributions of clouds and surface types are derived from Earth science datasets, such as the International Satellite Cloud Climatology Project (ISCCP) (*Schiffer and Rossow*, 1983).

The most complex three-dimensional tools for simulating observations of the distant Earth solve the full plane-

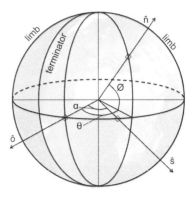

Fig. 12. Geometry for modeling disk-integrated Earth observations. The surface normal vector, and the vectors in the direction of the observer and Sun are $\hat{n}$, $\hat{o}$, and $\hat{s}$ respectively. The angle α is the phase angle, while ϕ and θ are the coordinates of latitude and longitude, respectively. Earth view generated by the Earth and Moon Viewer, first implemented by J. Walker (*http://www.fourmilab.ch/cgi-bin/Earth*).

parallel, multiple-scattering radiative transfer equation to determine the emergent radiance over the planetary disk. By including realistic atmospheric radiative effects, these fully multiple-scattering tools can more self-consistently capture gas and cloud absorption and scattering (*Tinetti et al.*, 2006; *Fujii et al.*, 2011; *Robinson et al.*, 2011; *Feng et al.*, 2018) as well as polarization effects (*Stam*, 2008). Like the previously-discussed reflectance models, cloud and surface type coverages are typically derived from Earth science datasets, while cloud optical thicknesses must also be adopted from similar datasets to include in the multiple-scattering calculation. Such sophisticated three-dimensional models can serve as virtual "laboratories" for studying the pale blue dot across a wide range of timescales, wavelengths, and viewing geometries (see Fig. 13), insofar as they are validated against observations (that, admittedly, do not span all possible combinations of planetary phase, wavelength coverage, and spectral resolution; see section 4). In any case, our observational analysis and models of the pale blue dot must also be extended to confront the realization that Earth's major characteristics and remotely observable properties have changed considerably throughout the course of planetary evolution.

6. DECIPHERING EXO-EARTH OBSERVATIONS

The observations and models discussed in previous sections provide insights into remote sensing approaches to understanding distant habitable worlds. At wavelengths

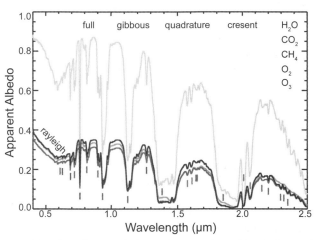

Fig. 13. See Plate 42 for color version. Simulations of Earth's phase- and wavelength-dependent apparent albedo (from *Robinson et al.*, 2010). Models are averaged over a full rotation at each phase, and the angles of the given phases are 0° (full), 45° (gibbous), 90° (quadrature), and 135° (crescent). Large apparent albedos at crescent phase are primarily due to ocean glint and cloud forward scattering. The similarity in apparent albedo scales for the quadrature, gibbous, and full spectra indicate that Earth largely scatters like a Lambert sphere across these phase angles. A slight enhancement in apparent albedo at full phase is due to cloud back scattering.

spanning the ultraviolet through the infrared, and for both broadband photometry and spectroscopy across a range of resolutions, data (or simulated data) for the distant Earth — at any of its evolutionary stages — contain a great deal of information about the planetary environment. The sections below discuss the information content of observations of reflected light and of thermal emission. Additional information can be found in reviews by *Meadows* (2008), *Kaltenegger et al.* (2010), *Kaltenegger et al.* (2012), and *Robinson* (2018).

6.1. Visible Photometry

Single-instance broadband photometry of a distant Earth-like world provides limited information about the planetary environment. As Earthshine observations have shown (*Qiu et al.*, 2003; *Pallé et al.*, 2003), photometric data could constrain planetary albedo — which is central to an understanding of planetary energy balance — as long as planetary size and phase are known. (If the planetary radius is unknown, the planetary reflectivity and size are degenerate, although see section 6.3 for a discussion of using thermal infrared observations to constrain the planetary radius.) Additionally, broad absorption features can be detected using photometric observations [e.g., as was the case for the 950-nm H_2O band in EPOXI observations (*Livengood et al.*, 2011)], although constraining atmospheric abundances from low-resolution observations is extremely challenging (*Lupu et al.*, 2016). Planetary color derived from broadband observations has been suggested as a means of identifying exo-Earth candidates (*Traub*, 2003), and, regarding Fig. 3, visible photometry could differentiate a hazy Archean Earth from a non-hazy Earth at all other evolutionary stages, even at low signal-to-noise. Distinguishing our planet from certain non-Earth-like planets might be more problematic for Earth at any geological phase other than the Phanerozoic (*Krissansen-Totton et al.*, 2016b) and could be confused by planetary phase effects as well as our lack of knowledge on realistic colors of temperate or cool exoplanets.

Disk-integrated photometric observations that resolve the rotation of an exo-Earth yield much more powerful diagnostics than single-instance photometry (Fig. 14). When acquired over multiple days (rotations), the rotation rate of the planet can be determined from diurnal variability [which is typically 10–20% (*Ford et al.*, 2001; *Livengood et al.*, 2011)], even in the presence of evolving weather patterns (*Pallé et al.*, 2008; *Oakley and Cash*, 2009). Once the rotation rate of an exo-Earth is known, the correspondence between time and subobserver longitude enables longitudinally-resolved mapping (*Cowan et al.*, 2009; *Fujii et al.*, 2010, 2011; *Kawahara and Fujii*, 2010; *Fujii and Kawahara*, 2012; *Cowan and Strait*, 2013; *Lustig-Yaeger et al.*, 2018), although degeneracies can occur in mapping approaches, especially with regard to spectral unmixing (*Fujii et al.*, 2017). Additionally, studying photometric variability inside absorption bands of well-mixed gases (e.g., O_2) as

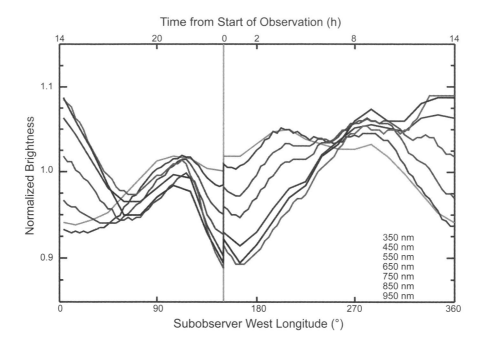

Fig. 14. See Plate 43 for color version. Full-rotation lightcurves for Earth from the March 2008 EPOXI dataset in all shortwave filters. Filter bandpasses are 100 nm wide, and filter center wavelengths are indicated. Overall brightness is driven largely by clouds at shorter wavelengths and by both clouds and continents at longer wavelengths (*Robinson et al.*, 2011). Variability is primarily due to Earth's rotation; however, differences in brightness after a full rotation are due to longer-term evolution of cloud patterns.

compared to variability inside bands of other species (e.g., H_2O) can reveal condensation processes in planetary atmospheres (*Fujii et al.*, 2013). Depending on the optical thickness of a potential haze in the atmosphere of the Archean Earth (*Arney et al.*, 2016), it might be necessary to push photometric observations to red or near-infrared wavelengths to have surface and near-surface sensitivity in lightcurves.

Photometric exo-Earth observations resolved at both rotational and orbital timescales could reveal additional information about the planetary surface. Due to the obliquity of the planetary rotational axis [that could be constrained from lightcurves (*Schwartz et al.*, 2016; *Kawahara*, 2016)], and depending on orbital inclination, maps resolved in latitude and longitude could be produced from high-quality data (*Fujii and Kawahara*, 2012; *Cowan et al.*, 2013). Even in the absence of rotationally-resolved photometry, surface oceans — whose presence directly confirms the habitability of an exoplanet — could be detected via the effect of specular reflectance on a planetary phase curve (*McCullough*, 2006; *Williams and Gaidos*, 2008), especially at red and near-infrared wavelengths where observations at large phase angles are less strongly impacted by Rayleigh scattering and, thus, have better surface sensitivity (*Robinson et al.*, 2010; *Zugger et al.*, 2011) [observations at these wavelengths would also be less influenced by any hazes on the Archean Earth (*Arney et al.*, 2016)]. Additionally, scattering at the Brewster angle will maximize the polarization signature from an exo-ocean and could be detected in the polarization phase curve of an exo-Earth (*McCullough*,

2006; *Stam*, 2008; *Williams and Gaidos*, 2008; *Zugger et al.*, 2010). Finally, polarization and reflectance phase curves for Earth-like exoplanets can also reveal cloud properties through scattering effects (*Bailey*, 2007; *Karalidi et al.*, 2011, 2012; *Karalidi and Stam*, 2012), although little work has been done to understand how photochemical hazes (e.g., like those that may have been present on the Archean Earth) would impact polarimetric observations.

6.2. Visible Spectroscopy

Spectroscopic observations in reflected light provide powerful information about the atmospheric and surface environment of the pale blue dot at any stage in its evolution. In addition to the insights offered from photometry (as spectra can always be degraded to lower resolution), observations at even moderate spectral resolution enable the detection of trace atmospheric gases. For example, *Drossart et al.* (1993) used near-infrared Galileo data to constrain the abundances of CO_2, H_2O, CO, O_3, CH_4, and N_2O in the atmosphere of Earth. While the *Drossart et al.* (1993) study used spatially-resolved observations, the same gaseous absorption features appear in the disk-integrated EPOXI dataset (*Livengood et al.*, 2011).

Beyond trace gas detection, spectroscopic reflected-light observations can also constrain atmospheric pressure — a key determinant of habitability — through Rayleigh scattering effects, broadening of gas absorption lines and bands, and through collision-induced absorption and dimer features.

Pressure-induced absorption features due to O_2 and N_2 occur throughout the near-infrared [and into the mid-infrared (*Misra et al.*, 2014a; *Schwieterman et al.*, 2015)]. Of course, interpretation of Rayleigh scattering features and pressure-broadened absorption bands is not straightforward. The former depends on surface gravity and atmospheric mean molecular weight and can be masked by surface or haze absorption at blue wavelengths (see Fig. 3), while the latter is impacted by the composition of the background atmosphere (e.g., *Hedges and Madhusudhan*, 2016). However, despite difficulties associated with detection, constraining O_2 and N_2 levels in the atmosphere of an Earth-like exoplanet would be key to deciphering the disequilibrium signature of N_2-O_2-H_2O (*Krissansen-Totton et al.*, 2016a).

Feng et al. (2018) investigated retrievals of planetary and atmospheric properties for the modern pale blue dot from visible-wavelength (0.4–1.0 μm) spectroscopy at a variety of spectral resolutions and signal-to-noise ratios. Here, firm constraints on key gas mixing ratios (for H_2O, O_3, and O_2), total atmospheric pressure, and planetary radius could be achieved with simulated Earth observations at a V-band signal-to-noise ratio of 20 and spectral resolution of 140. Thus, observations of this quality for modern exo-Earth twin could be sufficient to indicate that the planet is either super-Earth or Earth-sized and that O_2 is a major atmospheric constituent, which is strong evidence that the planet may be inhabited.

Low-resolution, low signal-to-noise ultraviolet observations of an Earth-like planet could rapidly distinguish the ozone-free Archean Earth from Earth at different evolutionary stages, and the ultraviolet Hartley-Huggins band may have exhibited dramatic seasonal variations during the Proterozoic (*Olson et al.*, 2018a). For the Archean, spectral models have demonstrated strong features due to methane and haze (*Arney et al.*, 2016), but retrieval investigations have yet to show how observations of different quality and wavelength coverage could be used to infer methane and haze concentrations for an Archean Earth-like exoplanet. Here, and as opposed to a haze-free Earth, planetary radius may be difficult to constrain from reflected light observations as the size determination relies strongly on measuring a Rayleigh scattering feature. (Since the Rayleigh scattering properties of a gas are well-defined, the planet-to-star flux ratio in a Rayleigh scattering feature is dependent primarily on the planetary size and orbital distance.)

6.3. Thermal Infrared Observations

Owing to the great technical challenges posed by techniques for observing Earth-like planets around Sun-like stars at long wavelengths, relatively little attention has been focused on understanding disk-integrated observations of our planet at thermal infrared wavelengths. Nevertheless, infrared spectra of the distant Earth — even at relatively low spectral resolution — provide a great deal of information about the atmospheric and surface environment. Most fundamentally, and unlike reflected-light data, infrared

observations can directly constrain the radius of an exo-Earth. The flux received from a true blackbody depends on its temperature, size, and distance. If the distance to a target star is known, and with the temperature constrained via Wien's displacement law, the size of a planet can then be determined from low-resolution thermal infrared observations. Of course, Earth does not emit like a true blackbody, which would introduce some uncertainty into a fitted planetary radius.

Beyond planetary size, infrared gas absorption features, by definition, reveal the key greenhouse gases of a planetary atmosphere. For Earth, observations (Fig. 10) plainly reveal signatures of CO_2, H_2O, O_3, CH_4, and N_2O (*Christensen and Pearl*, 1997; *Hearty et al.*, 2009). Regarding our evolving planet, the 9.7-μm ozone band becomes apparent in Earth's emission spectrum after the rise of oxygen and the CO_2 dioxide bands at 9.4, 10.4, and 15 μm strongly track decreasing atmospheric CO_2 levels with time (*Meadows*, 2008; *Rugheimer and Kaltenegger*, 2018). Critically, these would be observable at modest resolving powers and characteristic spectral signal-to-noise ratios of 5 (*Rugheimer and Kaltenegger*, 2018). Additionally, pressure-induced absorption features can be used to indicate bulk atmospheric composition and pressure, and one such feature from N_2 has been detected in observations of the distant Earth near 4 μm (*Schwieterman et al.*, 2015).

Critically, as molecular absorption bands are pressure broadened, and because high-opacity regions of a molecular band probe lower atmospheric pressures than do low-opacity regions, infrared observations can be used to probe the thermal structure of the atmosphere and surface of an exo-Earth. This idea would apply for Earth at any stage in its evolution, even for a hazy Archean Earth as such hydrocarbon aerosols are typically transparent at infrared wavelengths (*Arney et al.*, 2016). Finally, thermal infrared lightcurves could also reveal variability due to weather (and associated condensational processes), rotation, and seasons (*Hearty et al.*, 2009; *Selsis et al.*, 2011; *Robinson*, 2011; *Gómez-Leal et al.*, 2012; *Cowan et al.*, 2012).

Using spatially-resolved Galileo/NIMS Earth observations, along with adopted *a priori* knowledge of total atmospheric pressure and the CO_2 mixing ratio, *Drossart et al.* (1993) derived the thermal structures of cloud-free regions on Earth from the 4.3-μm CO_2 band. More recently, *von Paris et al.* (2013) used retrieval techniques on simulated infrared observations of a distant, modern Earth to show that low-resolution ($\lambda/\Delta\lambda = 20$) observations at signal-to-noise ratios of 10–20 could constrain thermal structure and atmospheric composition reasonably well. However, like the *Drossart et al.* (1993) retrievals, the results from *von Paris et al.* (2013) emphasize a cloud-free atmosphere. Thus, it remains unclear how realistic patchy clouds would influence our ability to understand the atmosphere and surface of an exo-Earth from thermal infrared observations, and how this might impact attempts to characterize exo-Earths in different thermal states (e.g., a snowball state vs. a clement or hothouse state).

7. SUMMARY

Exoplanetary science is rapidly progressing toward its long-term goal of discovering and characterizing Earth-like planets around our nearest stellar neighbors. We now know that exoplanets, including potentially habitable Earth-sized worlds, are quite common, and small worlds orbiting within the habitable zone of nearby cool stars have already been discovered. Advances in observational technologies, especially with regards to exoplanet direct imaging techniques, will enable the detection of pale blue dots around other stars, potentially in the not-too-distant future.

Flybys of Earth by the Galileo spacecraft in the early 1990s enabled the remote detection of habitability and life on our planet using planetary science remote sensing techniques. A combination of spatially-resolved visible and near-infrared spectral observations argued conclusively for the presence of liquid water on Earth's surface. These same datasets indicated an atmosphere that was in a state of strong chemical disequilibrium — a sign of life — and observations at radio wavelengths contained features that indicated the presence of intelligent organisms.

More recently, a variety of observational approaches have yielded datasets that effectively allow us to view Earth as a distant exoplanet. While observations from spacecraft at or beyond the Moon's orbit are ideal for understanding habitability and life signatures from the pale blue dot, such data are rarely acquired. Critically, satellite and Earthshine observations complement, and fill in certain gaps between, spacecraft data for the distant Earth.

Beyond observational datasets, models have proved effective tools for simulating and characterizing Earth as an exoplanet. These tools span a wide range of complexities, including one-dimensional (vertical) spectral simulators, simple reflectance tools that capture the broadband reflectivity of Earth at visible wavelengths, and complex three-dimensional models that can simulate observations of the distant Earth at arbitrary viewing geometry across wavelengths that span the ultraviolet through the thermal infrared. Especially in the absence of frequent spacecraft observations, models of the pale blue dot can serve as testing grounds for proposed approaches to detecting and characterizing Earth-like exoplanets.

Geological and biogeochemical studies of the long-term evolution of Earth reveal a world that, while being continuously habitable and inhabited, has progressed through a variety of surface and atmospheric states. Abundances of key atmospheric constituents, including biosignature gases, have varied by many orders of magnitude. As these gases imprint information about their concentrations on spectra of Earth, applying the aforementioned spectral simulation tools to the pale blue dot at different geologic epochs reveals planetary spectra (and associated biosignatures) quite distinct from modern Earth. Especially for the Archean Earth, the term "pale blue dot" may not even apply.

Combining an understanding of remote sensing techniques relevant to exoplanets with knowledge of the conditions on the current and ancient Earth yields insights into approaches for detecting and studying Earth-like worlds around other stars. Broadband observations have the potential to reveal habitable environments on ocean-bearing exoplanets, and time-resolved photometry can be used to extract spatial information from spatially-unresolved data. More powerfully, spectroscopic observations at moderate resolutions can uncover key details about the surface and atmospheric state on a potentially Earth-like planet, including fundamental details relevant to life detection. Only by uncovering the key signatures that indicate the habitability and inhabitance of Earth — at any point in its evolution — can we properly design the observational tools needed to discover and fully characterize other Earths.

Acknowledgments. This research has made use of the Planetary Data System (PDS) and USGS Integrated Software for Imagers and Spectrometers (ISIS). TR gratefully acknowledges support from NASA through the Sagan Fellowship Program executed by the NASA Exoplanet Science Institute and through the Exoplanets Research Program (award #80NSSC18K0349). Both TR and CR would like to acknowledge support from the NASA Astrobiology Institute, both through a grant to the Virtual Planetary Laboratory (under Cooperative Agreement No. NNA13AA93A) and to the University of California, Riverside "Alternative Earths" team. The results reported herein benefitted from collaborations and/or information exchange within NASA's Nexus for Exoplanet System Science (NExSS) research coordination network sponsored by NASA's Science Mission Directorate. We thank M. Turnbull, P. Christensen, T. Hearty, E. Pallé, E. Schwieterman, and G. Arney for openly sharing data used in this chapter. Both authors also thank V. Meadows, G. Arney, N. Cowan, and an anonymous reviewer for detailed and constructive comments on versions of this manuscript.

REFERENCES

Abrams M. C., Goldman A., Gunson M. R., Rinsland C. P., and Zander R. (1999) Observations of the infrared solar spectrum from space by the ATMOS experiment. *Appl. Opt., 35,* 2747.

Anglada-Escudé G., Amado P. J., Barnes J., Berdiñas Z. M., Butler R. P., Coleman G. A. L., de la Cueva I., et al. (2016) A terrestrial planet candidate in a temperate orbit around Proxima Centauri. *Nature, 536,* 437–440.

Arney G. N., Domagal-Goldman S. D., Meadows V. S., Wolf E. T., Schwieterman E., Charnay B., Claire M., Hébrard E., and Trainer M. G. (2016) The pale orange dot: The spectrum and habitability of hazy Archean Earth. *Astrobiology, 16,* 873–899.

Arney G. N., Domagal-Goldman S. D., and Meadows V. S. (2018) Organic haze as a biosignature in anoxic Earth-like atmospheres. *Astrobiology, 18,* 311–329.

Arnold L., Gillet S., Lardière O., Riaud P., and Schneider J. (2002) A test for the search for life on extrasolar planets. Looking for the terrestrial vegetation signature in the Earthshine spectrum. *Astron. Astrophys., 392,* 231–237.

Aumann H. H., Chahine M. T., Gautier C., Goldberg M. D., Kalnay E., McMillin L. M., Revercomb H., et al. (2003) AIRS/AMSU/HSB on the Aqua mission: Design, science objectives, data products, and processing systems. *IEEE Trans. Geosci. Electron., 41,* 253–264.

Azmy K., Kendall B., Creaser R. A., Heaman L., and de Oliveira T. F. (2008) Global correlation of the Vazante Group, São Francisco Basin, Brazil: Re-Os and U-Pb radiometric age constraints.

Precambrian Res., 164, 160–172.

Bachan A. and Kump L. R. (2015) The rise of oxygen and siderite oxidation during the Lomagundi Event. *Proc. Natl. Acad. Sci. U.S.A., 112,* 6562–6567.

Bailey J. (2007) Rainbows, polarization, and the search f or habitable planets. *Astrobiology, 7,* 320–332.

Baker A. L., Baker L. R., Beshore E., Blenman C., Castillo N. D., Chen Y.-P., Doose L. R., Elston J. P., Fountain J. W., and Goffeen D. L. (1975) The imaging photopolarimeter experiment on Pioneer 11. *Science, 188,* 468–472.

Barstow J. K., Aigrain S., Irwin P. G. J., Kendrew S., and Fletcher L. N. (2016) Telling twins apart: Exo-Earths and Venuses with transit spectroscopy. *Mon. Not. R. Astron. Soc., 458,* 2657–2666.

Bartdorff O., Wallmann K., Latif M., and Semenov V. (2008) Phanerozoic evolution of atmospheric methane. *Global Biogeochem. Cycles, 22,* GB1008.

Batalha N. M. (2014) Exploring exoplanet populations with NASA's Kepler Mission. *Proc. Natl. Acad. Sci. U.S.A., 111,* 12647–12654.

Batalha N., Kalirai J., Lunine J., Clampin M., and Lindler D. (2015) Transiting exoplanet simulations with the James Webb Space Telescope. *ArXiv e-prints,* arXiv:1507.02655.

Bean J. L., Kempton E. M.-R., and Homeier D. (2010) A ground-based transmission spectrum of the super-Earth exoplanet GJ 1214b. *Nature, 468(7324),* 669–672.

Beerling D., Berner R. A., Mackenzie F. T., Harfoot M. B., and Pyle J. A. (2009) Methane and the CH_4 related greenhouse effect over the past 400 million years. *Am. J. Sci., 309,* 97–113.

Beichman C., Benneke B., Knutson H., Smith R., Lagage, P.-O., Dressing, C., Latham D., et al. (2014) Observations of transiting exoplanets with the James Webb Space Telescope (JWST) *Publ. Astron. Soc. Pac., 126(946),* 1134–1173.

Bekker A. (2001) Chemostratigraphy of the Paleoproterozoic Duitschland Formation, South Africa: Implications for coupled climate change and carbon cycling. *Am. J. Sci., 301,* 261–285.

Bekker A. and Holland H. D. (2012) Oxygen overshoot and recovery during the early Paleoproterozoic. *Earth Planet. Sci. Lett., 317,* 295–304.

Belcher C. M. and McElwain J. C. (2008) Limits for combustion in low O_2 redefine paleoatmospheric predictions for the Mesozoic. *Science, 321,* 1197.

Bellefroid E. J., Hood A. V. S., Hoffman P. F., Thomas M. D., Reinhard C. T., and Planavsky N. J. (2018) Constraints on Paleoproterozoic atmospheric oxygen levels. *Proc. Natl. Acad. Sci. U.S.A., 115,* 8104–8109.

Belton M. J. S., Klaasen K. P., Clary M. C., Anderson J. L., Anger C. D., Carr M. H., Chapman C. R., Davies M. E., Greeley R., and Anderson D. (1992) The Galileo Solid-State Imaging experiment. *Space Sci. Rev., 60,* 413–455.

Bergman N. M. (2004) COPSE: A new model of bio-geochemical cycling over Phanerozoic time. *Am. J. Sci., 304,* 397–437.

Bernath P. F., McElroy C. T., Abrams M. C., Boone C. D., Butler M., Camy-Peyret C., Carleer M., et al. (2005) Atmospheric Chemistry Experiment (ACE): Mission overview. *Geophys. Res. Lett., 32,* L15S01.

Berner R. A. (1991) A model for atmospheric CO_2 over Phanerozoic time. *Am. J. Sci., 291,* 339–376.

Berner R. A. (2006) Geological nitrogen cycle and atmospheric N_2 over Phanerozoic time. *Geology, 34,* 413.

Berner R. A. and Kothavala Z. (2001) GEOCARB III: A revised model of atmospheric CO_2 over Phanerozoic time. *Am. J. Sci., 301(2),* 182–204.

Bianchi D., Dunne J. P., Sarmiento J. L., Galbraith E. D. (2012) Data-based estimates of suboxia, denitrification, and N_2O production in the ocean and their sensitivities to dissolved O_2. *Global Biogeochem. Cycles, 26(2).*

Biesecker D. A., Reinard A., Cash M. D., Johnson J., Burek M., de Koning C. A., Szabo A., et al. (2015) Presenting DSCOVR: The first NOAA mission to leave Earth orbit. Abstract SH12A–06 presented at 2015 Fall Meeting, AGU, San Francisco, California, 14–18 December.

Blake R. E., Chang S. J., and Lepland A. (2010) Phosphate oxygen isotopic evidence for a temperate and biologically active Archaean ocean. *Nature, 464,* 1029–1032.

Borucki W. J., Koch D., Basri G., Batalha N., Brown T., Caldwell D., Caldwell J., et al. (2010) Kepler planet-detection mission:

Introduction and first results. *Science, 327,* 977–980.

Bovensmann H., Burrows J. P., Buchwitz M., Frerick J., Noël S., Rozanov V. V., Chance K. V., and Goede A. P. H. (1999) SCIAMACHY: Mission objectives and measurement modes. *J. Atmos. Sci., 56,* 127–150.

Briot D., Arnold L., and Jacquemoud S. (2013) Present and future observations of the Earthshine from Antarctica. In *Astrophysics from Antarctica* (M. G. Burton et al., eds.), pp. 214–217. IAU Symp. 288, Cambridge Univ., Cambridge.

Brown T. M. (2001) Transmission spectra as diagnostics of extrasolar giant planet atmospheres. *Astrophys. J., 553(2),* 1006.

Budyko M. I. (1969) The effect of solar radiation variations on the climate of the Earth. *Tellus, 21,* 611–619.

Buick R. (2007) Did the Proterozoic Canfield ocean cause a laughing gas greenhouse? *Geobiology, 5(2),* 97–100.

Burke C. J., Christiansen J. L., Mullally F., Seader S., Huber D., Rowe J. F., Coughlin J. L., et al. (2015) Terrestrial planet occurrence rates for the Kepler GK dwarf sample. *Astrophys. J., 809,* 8.

Burrows A. (2014) Scientific return of coronagraphic exoplanet imaging and spectroscopy using WFIRST. *ArXiv e-prints,* arXiv:1412.6097.

Canfield D. E. (2005) The early history of atmospheric oxygen: Homage to Robert M. Garrels. *Annu. Rev. Earth Planet. Sci., 33,* 1–36.

Canfield D. E., Poulton S. W., and Narbonne G. M. (2007) Late-Neoproterozoic deep-ocean oxygenation and the rise of animal life. *Science, 315,* 92.

Canfield D. E., Ngombi-Pemba L., Hammarlund E. U., Bengtson S., Chaussidon M., Gauthier-Lafaye F., et al. (2013) Oxygen dynamics in the aftermath of the Great Oxidation of Earth's atmosphere. *Proc. Natl. Acad. Sci. U.S.A., 110,* 16736–16741.

Carlson R. W., Weissman P. R., Smythe W. D., and Mahoney J. C. (1992) Near-Infrared Mapping Spectrometer experiment on Galileo. *Space Sci. Rev., 60,* 457–502.

Cash W., Schindhelm E., Arenberg J., Lo A., Polidan R., Kasdin J., Vanderbei R., Kilston S., and Noecker C. (2007) External occulters for direct observation of exoplanets: An overview. In *UV/Optical/IR Space Telescopes: Innovative Technologies and Concepts III* (H. A. MacEwen and J. B. Breckinridge, eds.), pp. 668–712. SPIE Conf. Ser. 6687, Bellingham, Washington.

Cassan A., Kubas D., Beaulieu J. P., Dominik M., Horne K., Greenhill J., Wambsganss J., et al. (2012) One or more bound planets per Milky Way star from microlensing observations. *Nature, 481(7380),* 167–169.

Catling D. C. (2006) Comment on "A hydrogen-rich early Earth atmosphere." *Science, 311,* 38a.

Catling D. C. and Claire M. W. (2005) How Earth's atmosphere evolved to an oxic state: A status report. *Earth Planet. Sci. Lett., 237,* 1–20.

Catling D. C., Zahnle K. J., and McKay C. P. (2001) Biogenic methane, hydrogen escape, and the irreversible oxidation of early Earth. *Science, 293,* 839–843.

Catling D. C., Claire M. W., and Zahnle K. J. (2007) Anaerobic methanotrophy and the rise of atmospheric oxygen. *Philos. Trans. R. Soc., A, 365,* 1867–1888.

Charbonneau D., Brown T. M., Noyes R. W., and Gilliland R. L. (2002) Detection of an extrasolar planet atmosphere. *Astrophys. J., 568,* 377–384.

Charnay B., Forget F., Wordsworth R., Leconte J., Millour E., Codron F., and Spiga A. (2013) Exploring the faint young Sun problem and the possible climates of the Archean Earth with a 3-D GCM. *J. Geophys. Res.–Atmos., 118,* 10.

Charnay B., Le Hir G., Fluteau F., Forget F., and Catling D. C. (2017) A warm or a cold early Earth? New insights from a 3-D climate-carbon model. *Earth Planet. Sci. Lett., 474,* 97–109.

Christensen P. R. and Pearl J. C. (1997) Initial data from the Mars Global Surveyor thermal emission spectrometer experiment: Observations of the Earth. *J. Geophys. Res., 102,* 10875–10880.

Claire M. W., Catling D. C., and Zahnle K. J. (2006) Biogeochemical modeling of the rise in atmospheric oxygen. *Geobiology, 4,* 239–269.

Cole D. B., Reinhard C. T., Wang X., Gueguen B., Halverson G. P., Gibson T., Hodgskiss M. S. W., McKenzie N. R., Lyons T. W., and Planavsky N. J. (2016) A shale-hosted Cr isotope record of low atmospheric oxygen during the Proterozoic. *Geology, 44,* 555–558.

Cowan N. B. and Strait T. E. (2013) Determining reflectance spectra of surfaces and clouds on exoplanets. *Astrophys. J. Lett., 765,* L17.

Cowan N. B., Agol E., Meadows V. S., Robinson T., Livengood T. A.,

Deming D., Lisse C. M., A'Hearn M. F., Wellnitz D. D., Seager S., and Charbonneau D. (the EPOXI Team) (2009) Alien maps of an ocean-bearing world. *Astrophys. J., 700,* 915–923.

Cowan N. B., Robinson T., Livengood T. A., Deming D., Agol E., A'Hearn M. F., Charbonneau D., et al. (2011) Rotational variability of Earth's polar regions: Implications for detecting snowball planets. *Astrophys. J., 731(76).*

Cowan N. B., Voigt A., and Abbot D. S. (2012) Thermal phases of Earth-like planets: Estimating thermal inertia from eccentricity, obliquity, and diurnal forcing. *Astrophys. J., 757,* 80.

Cowan N. B., Fuentes P. A., and Haggard H. M. (2013) Light curves of stars and exoplanets: Estimating inclination, obliquity, and albedo. *Mon. Not. R. Astron. Soc., 434,* 2465–2479.

Cowan N. B., Greene T., Angerhausen D., Batalha N. E., Clampin M., Colón K., Crossfeld I. J. M., et al. (2015) Characterizing transiting planet atmospheres through 2025. *Publ. Astron. Soc. Pac., 127,* 311–327.

Cox G. M., Halverson G. P., Minarik W. G., Le Heron D. P., Macdonald F. A., Bellefroid E. J., and Strauss J. V. (2013) Neoproterozoic iron formation: An evaluation of its temporal, environmental and tectonic significance. *Chem. Geol., 362,* 232–249.

Crowe S. A., Paris G., Katsev S., Jones C., Kim S.-T., Zerkle A. L., Nomosatryo S., et al. (2014) Sulfate was a trace constituent of Archean seawater. *Science, 346,* 735–739.

Daines S. J. and Lenton T. M. (2016) The effect of widespread early aerobic marine ecosystems on methane cycling and the Great Oxidation. *Earth Planet. Sci. Lett., 434,* 42–51.

Daines S. J., Mills B. J. W., and Lenton T. M. (2017) Atmospheric oxygen regulation at low Proterozoic levels by incomplete oxidative weathering of sedimentary organic carbon. *Nature Commun., 8,* 14379.

Dalba P. A., Muirhead P. S., Fortney J. J., Hedman M. M., Nicholson P. D., and Veyette M. J. (2015) The transit transmission spectrum of a cold gas giant planet. *Astrophys. J., 814,* 154.

Danjon A. (1928) Recherches sur la photométrie de la lumiere cendrée et l'albedo de la terre. *Ann. Obs. Strasbourg, 2,* 165–180.

Delabroye A. and Vecoli M. (2010) The end-Ordovician glaciation and the Hirnantian stage: A global review and questions about Late Ordovician event stratigraphy. *Earth-Sci. Rev., 98,* 269–282.

Deming D., Seager S., Winn J., Miller-Ricci E., Clampin M., Lindler D., Greene T., et al. (2009) Discovery and characterization of transiting super Earths using an all-sky transit survey and follow-up by the James Webb Space Telescope. *Publ. Astron. Soc. Pac., 121,* 952–967.

Des Marais D. J., Harwit M. O., Jucks K. W., Kasting J. F., Lin D. N. C., Lunine J. I., Schnieder J., Seager S., Traub W. A., and Woolf N. J. (2002) Remote sensing of planetary properties and biosignatures on extrasolar terrestrial planets. *Astrobiology, 2,* 153–181.

de Wit M. J. and Furnes H. (2016) 3.5-Ga hydrothermal fields and diamictites in the Barberton Greenstone Belt — Paleoarchean crust in cold environments. *Sci. Adv., 2,* e1500368.

de Wit J., Wakeford H. R., Lewis N. K., Delrez L., Gillon M., Selsis F., Leconte J., et al. (2018) Atmospheric reconnaissance of the habitable-zone Earth-sized planets orbiting TRAPPIST-1. *Nature Astron., 2,* 214–219.

DeWitt H. L., Trainer M. G., Pavlov A. A., Hasenkopf C. A., Aiken A. C., Jimenez J. L., McKay C. P., Toon O. B., and Tolbert M. A. (2009) Reduction in haze formation rate on prebiotic Earth in the presence of hydrogen. *Astrobiology, 9,* 447–453.

Diamond-Lowe H., Stevenson K. B., Bean J. L., Line M. R., and Fortney J. J. (2014) New analysis indicates no thermal inversion in the atmosphere of HD 209458b. *Astrophys. J., 796,* 66.

Domagal-Goldman S. D., Kasting J. F., Johnston D. T., and Farquhar J. (2008) Organic haze, glaciations, and multiple sulfur isotopes in the Mid-Archean era. *Earth Planet. Sci. Lett., 269,* 29–40.

Doughty C. E. and Wolf A. (2010) Detecting tree-like multicellular life on extrasolar planets. *Astrobiology, 10,* 869–879.

Doughty C. E. and Wolf A. (2016) Detecting 3D vegetation structure with the Galileo Space Probe: Can a distant probe detect vegetation structure on Earth? *PLoS ONE, 11,* e0167188.

Dressing C. D. and Charbonneau D. (2013) The occurrence rate of small planets around small stars. *Astrophys. J., 767,* 95.

Dressing C. D. and Charbonneau D. (2015) The occurrence of potentially habitable planets orbiting M dwarfs estimated from the full Kepler dataset and an empirical measurement of the detection sensitivity. *Astrophys. J., 807,* 45.

Driese S. G., Jirsa M. A., Ren M., Brantley S. L., Sheldon N. D., Parker D., and Schmitz M. (2011) Neoarchean paleoweathering of tonalite and metabasalt: Implications for reconstructions of 2.69 Ga early terrestrial ecosystems and paleoatmospheric chemistry. *Precambrian Res., 189,* 1–17.

Drossart P., Rosenqvist J., Encrenaz T., Lellouch E., Carlson R. W., Baines K. H., Weissman P. R., Smythe W. D., Kamp L. W., and Taylor F. W. (1993) Earth global mosaic observations with NIMS-Galileo. *Planet. Space Sci., 41,* 551–561.

Dubois J. (1947) Sur l'albedo de la Terre. *Bull. Astron., 13,* 193–196.

Ehrenreich D., Bonfils X., Lovis C., Delfosse X., Forveille T., Mayor M., Neves V., Santos N. C., Udry S., and Ségransan D. (2014) Near-infrared transmission spectrum of the warm-Uranus GJ 3470b with the Wide Field Camera-3 on the Hubble Space Telescope. *Astron. Astrophys., 570,* A89.

Emmanuel S. and Ague J. J. (2007) Implications of present-day abiogenic methane fluxes for the early Archean atmosphere. *Geophys. Res. Lett., 34,* L15810.

Erwin D. H. (1994) The Permo-Triassic extinction. *Nature, 367,* 231–236.

Evans D. A., Beukes N. J., and Kirschvink J. L. (1997) Low-latitude glaciation in the Palaeoproterozoic era. *Nature, 386,* 262–266.

Farquhar J., Bao H., and Thiemens M. (2000) Atmospheric influence of Earth's earliest sulfur cycle. *Science, 289,* 756–758.

Feng Y. K., Robinson T. D., Fortney J. J., Lupu R. E., Marley M. S., Lewis N. K., Macintosh B., and Line M. R. (2018) Characterizing Earth analogs in reflected light: Atmospheric retrieval studies for future space telescopes. *Astron. J., 155,* 200.

Feulner G. (2012) The faint young Sun problem. *Rev. Geophys., 50(2),* RG2006.

Feulner G. (2017) Formation of most of our coal brought Earth close to global glaciation. *Proc. Natl. Acad. Sci. U.S.A., 114,* 11333–11337.

Fiorella R. P. and Sheldon N. D. (2017) Equable end Mesoproterozoic climate in the absence of high CO_2. *Geology, 45,* 231–234.

Ford E. B., Seager S., and Turner E. L. (2001) Characterization of extrasolar terrestrial planets from diurnal photometric variability. *Nature, 412,* 885–887.

Foreman-Mackey D., Hogg D. W., and Morton T. D. (2014) Exoplanet population inference and the abundance of Earth analogs from noisy, incomplete catalogs. *Astrophys. J., 795,* 64.

Fox G. E., Stackebrandt E., Hespell R. B., Gibson J., Maniloff J., Dyer T. A., Wolfe R. S., et al. (1980) The phylogeny of prokaryotes. *Science, 209,* 457–463.

Fraine J., Deming D., Benneke B., Knutson H., Jordán A., Espinoza N., Madhusudhan N., Wilkins A., and Todorov K. (2014) Water vapour absorption in the clear atmosphere of a Neptune-sized exoplanet. *Nature, 513,* 526–529.

Freing A., Wallace D. W. R., and Bange H. W. (2012) Global oceanic production of nitrous oxide. *Philos. Trans. R. Soc., B, 367,* 1245–1255.

Fujii Y. and Kawahara H. (2012) Mapping Earth analogs from photometric variability: Spin-orbit tomography for planets in inclined orbits. *Astrophys. J., 755,* 101.

Fujii Y., Kawahara H., Suto Y., Taruya A., Fukuda S., Nakajima T., and Turner E. L. (2010) Colors of a second Earth: Estimating the fractional areas of ocean, land, and vegetation of Earth-like exoplanets. *Astrophys. J., 715,* 866–880.

Fujii Y., Kawahara H., Suto Y., Fukuda S., Nakajima T., Livengood T. A., and Turner E.L. (2011) Colors of a second Earth. II. Effects of clouds on photometric characterization of Earth-like exoplanets. *Astrophys. J., 738,* 184.

Fujii Y., Turner E. L., and Suto Y. (2013) Variability of water and oxygen absorption bands in the disk-integrated spectra of Earth. *Astrophys. J., 765,* 76.

Fujii Y., Lustig-Yaeger J., and Cowan N. B. (2017) Rotational spectral unmixing of exoplanets: Degeneracies between surface colors and geography. *Astron. J., 154,* 189.

Gaillard F., Scaillet B., and Arndt N. T. (2011) Atmospheric oxygenation caused by a change in volcanic degassing pressure. *Nature, 478,* 229–232.

Galbraith E. D. and Eggleston S. (2017) A lower limit to atmospheric CO_2 concentrations over the past 800,000 years. *Nature Geosci., 10,* 295–298.

Galilei G. (1632) *Dialogue Concerning the Two Chief Worlds Systems.* Giovanni Battista Landini, Florence.

García Muñoz A., Zapatero Osorio M. R., Barrena R., Montañés-Rodríguez P., Martín E. L., and Pallé E. (2012) Glancing views of the Earth: From a lunar eclipse to an exoplanetary transit. *Astrophys. J., 755,* 103.

Gardner J. P., Mather J. C., Clampin M., Doyon R., Greenhouse M. A., Hammel H. B., Hutchings J. B., et al. (2006) The James Webb Space Telescope. *Space Sci. Rev., 123,* 485–606.

Gaucher E. A., Govindarajan S., and Ganesh O. K. (2008) Palaeotemperature trend for Precambrian life inferred from resurrected proteins. *Nature, 451,* 704–707.

Gaudi B. S., Seager S., Mennesson B., Kiessling A., Warfeld K. R., and the Habitable Exoplanet Observatory Science and Technology Definition Team (2018) The Habitable Exoplanet Observatory. *Nature Astron., 2,* 600–604.

Geboy N. J., Kaufman A. J., Walker R. J., Misi A., de Oliviera T. F., Miller K. E., Azmy K., Kendall B., and Poulton S. W. (2013) Re-Os age constraints and new observations of Proterozoic glacial deposits in the Vazante Group, Brazil. *Precambrian Res., 238,* 199–213.

Gehrels T. (1976) The results of the imaging photopolarimeter on Pioneers 10 and 11. In *Jupiter* (T. Gehrels, ed.), pp. 531–563. Univ. of Arizonz, Tucson.

Gillon M., Jehin E., Lederer S. M., Delrez L., de Wit J., Burdanov A., Van Grootel V., et al. (2016) Temperate Earth-sized planets transiting a nearby ultracool dwarf star. *Nature, 533,* 221–224.

Gillon M., Triaud A. H. M. J., Demory B.-O., Jehin E., Agol E., Deck K. M., Lederer S. M., et al. (2017) Seven temperate terrestrial planets around the nearby ultracool dwarf star TRAPPIST-1. *Nature, 542,* 456–460.

Glasspool I. J. and Scott A. C. (2010) Phanerozoic concentrations of atmospheric oxygen reconstructed from sedimentary charcoal. *Nature Geosci., 3,* 627–630.

Goldblatt C. and Zahnle K. J. (2011) Clouds and the faint young Sun paradox. *Climate Past, 7,* 203–220.

Goldblatt C., Lenton T. M., and Watson A. J. (2006) Bistability of atmospheric oxygen and the Great Oxidation. *Nature, 443,* 683–686.

Goldblatt C., Claire M. W., Lenton T. M., Matthews A. J., Watson A. J., and Zahnle K. J. (2009) Nitrogen-enhanced greenhouse warming on early Earth. *Nature Geosci., 2,* 891–896.

Gómez-Leal I., Pallé E., and Selsis F. (2012) Photometric variability of the disk-integrated thermal emission of the Earth. *Astrophys. J., 752,* 28.

González-Merino B., Pallé E., Motalebi F., Montañés-Rodríguez P., and Kissler-Patig M. (2013) Earthshine observations at high spectral resolution: Exploring and detecting metal lines in the Earth's upper atmosphere. *Mon. Not. R. Astron. Soc., 435,* 2574–2580.

Goode P. R., Qiu J., Yurchyshyn V., Hickey J., Chu M., Kolbe E., Brown C. T., and Koonin S. E. (2001) Earth-shine observations of the Earth's reflectance. *Geophys. Res. Lett., 28,* 1671–1674.

Gough D. O. (1981) Solar interior structure and luminosity variations. *Solar Phys., 74,* 21–34.

Greene T. P., Line M. R., Montero C., Fortney J. J., Lustig-Yaeger J., and Luther K. (2016) Characterizing transiting exoplanet atmospheres with JWST. *Astrophys. J., 817(1),* 17.

Grillmair C. J., Burrows A., Charbonneau D., Armus L., Stauffer J., Meadows V., van Cleve J., von Braun K., and Levine D. (2008) Strong water absorption in the dayside emission spectrum of the planet HD 189733b. *Nature, 456,* 767–769.

Gumsley A. P., Chamberlain K. R., Bleeker W., Söderlund U., de Kock M. O., Larsson E. R., and Bekker A. (2017) Timing and tempo of the Great Oxidation Event. *Proc. Natl. Acad. Sci. U.S.A., 114,* 1811–1816.

Gurnett D. A., Kurth W. S., Shaw R. R., Roux A., Gendrin R., Kennel C. F., Scarf F. L., and Shawhan S. D. (1992) The Galileo plasma wave investigation. *Space Sci. Rev., 60,* 341–355.

Guyon O., Pluzhnik E. A., Kuchner M. J., Collins B., and Ridgway S. T. (2006) Theoretical limits on extrasolar terrestrial planet detection with coronagraphs. *Astrophys. J. Suppl., 167,* 81–99.

Halevy I. and Bachan A. (2017) The geologic history of seawater pH. *Science, 355,* 1069–1071.

Hanel R., Conrath B., Kunde V., Lowman P., Maguire W., Pearl J., Pirraglia J., Gautier D., Gierasch P., and Kumar S. (1977) The Voyager infrared spectroscopy and radiometry investigation. *Space Sci. Rev., 21,* 129–157.

Hansen C. J., Schwartz J. C., and Cowan N. B. (2014) Features in the broad-band eclipse spectra of exoplanets: Signal or noise? *Mon. Not. R. Astron. Soc., 444,* 3632–3640.

Haqq-Misra J. D., Domagal-Goldman S. D., Kasting P. J., and Kasting J. F. (2008) A revised, hazy methane greenhouse for the Archean Earth. *Astrobiology, 8,* 1127–1137.

Haqq-Misra J., Kasting J. F., and Lee S. (2011) Availability of O_2 and H_2O_2 on pre-photosynthetic Earth. *Astrobiology, 11,* 293–302.

Haqq-Misra J., Kopparapu R. K., Batalha N. E., Harman C. E., and Kasting J. F. (2016) Limit cycles can reduce the width of the habitable zone. *Astrophys. J., 827,* 120.

Harada M., Tajika E., and Sekine Y. (2015) Transition to an oxygen-rich atmosphere with an extensive overshoot triggered by the Paleoproterozoic snowball Earth. *Earth Planet. Sci. Lett., 419,* 178–186.

Hardisty D. S., Lu Z., Bekker A., Diamond C. W., Gill B. C., Jiang G., Kah L. C., et al. (2017) Perspectives on Proterozoic surface ocean redox from iodine contents in ancient and recent carbonate. *Earth Planet. Sci. Lett., 463,* 159–170.

Hearty T., Song I., Kim S., and Tinetti G. (2009) Mid-infrared properties of disk averaged observations of Earth with AIRS. *Astrophys. J., 693,* 1763–1774.

Hedges S. B. (2016) The origin and evolution of model organisms. *Nature Rev. Genet., 3,* 838–849.

Hedges C. and Madhusudhan N. (2016) Effect of pressure broadening on molecular absorption cross sections in exoplanetary atmospheres. *Mon. Not. R. Astron. Soc., 458,* 1427–1449.

Herrmann A. D., Patzkowsky M. E., and Pollard D. (2004) The impact of paleogeography, pCO_2, poleward ocean heat transport and sea level change on global cooling during the Late Ordovician. *Palaeogeogr., Palaeoclimatol., Palaeoecol., 206,* 59–74.

Hessler A. M., Lowe D. R., Jones R. L., and Bird D. K. (2004) A lower limit for atmospheric carbon dioxide levels 3.2 billion years ago. *Nature, 428,* 736–738.

Hirsch A. I., Michalak A. M., Bruhwiler L. M., Peters W., Dlugokencky E. J., and Tans P. P. (2006) Inverse modeling estimates of the global nitrous oxide surface flux from 1998–2001. *Global Biogeochem. Cycles, 20.*

Hoffman P. F., Abbot D. S., Ashkenazy Y., Benn D. I., Brocks J. J., Cohen P. A., Cox G. M., et al. (2017) Snowball Earth climate dynamics and Cryogenian geology-geobiology. *Sci. Adv., 3,* e1600983.

Hoffman P. F., Kaufman A. J., Halverson G. P., and Schrag D. P. (1998) A Neoproterozoic Snowball Earth. *Science, 281,* 1342.

Hoffman P. F. and Schrag D. P. (2002) The snowball Earth hypothesis: Testing the limits of global change. *Terra Nova, 14,* 129–155.

Holland H. (1984) *The Chemical Evolution of the Atmosphere and Oceans.* Princeton Univ., Princeton.

Holland H. D. (2002) Volcanic gases, black smokers, and the great oxidation event. *Geochim. Cosmochim. Acta, 66,* 3811–3826.

Holland H. D. (2009) Why the atmosphere became oxygenated: A proposal. *Geochim. Cosmochim. Acta, 73,* 5241–5255.

Hönisch B., Ridgwell A., Schmidt D. N., Thomas E., Gibbs S. J., Sluijs A., Zeebe R., et al. (2012) The geological record of ocean acidification. *Science, 335,* 1058.

Hord C. W., McClintock W. E., Stewart A. I. F., Barth C. A., Esposito L. W., Thomas G. E., Sandel B. R., Hunten D. M., Broadfoot A. L., and Shemansky D. E. (1992) Galileo Ultraviolet Spectrometer experiment. *Space Sci. Rev., 60,* 503–530.

Hren M. T., Tice M. M., and Chamberlain C. P. (2009) Oxygen and hydrogen isotope evidence for a temperate climate 3.42 billion years ago. *Nature, 462,* 205–208.

Hu R. (2014) Ammonia, water clouds and methane abundances of giant exoplanets and opportunities for super-Earth exoplanets. *ArXiv e-prints,* arXiv:1412.7582.

Hubbard W., Fortney J., Lunine J., Burrows A., Sudarsky D., and Pinto P. (2001) Theory of extrasolar giant planet transits. *Astrophys. J., 560(1),* 413.

Hunten D. M. (1973) The escape of light gases from planetary atmospheres. *J. Atmos. Sci., 30,* 1481–1494.

Ingersoll A. P., Muench G., Neugebauer G., and Orton G. S. (1976) Results of the infrared radiometer experiment on Pioneers 10 and 11. In *Jupiter* (T. Gehrels, ed.), pp. 197–205. Univ. of Arizona, Tucson.

Izon G., Zerkle A. L., Williford K. H., Farquhar J., Poulton S. W., and Claire M. W. (2017) Biological regulation of atmospheric chemistry en route to planetary oxygenation. *Proc. Natl. Acad. Sci. U.S.A., 114,* E2571–E2579.

Johnson T. V., Yeates C. M., and Young R. (1992) *Space Science Reviews* volume on Galileo mission overview. *Space Sci. Rev., 60,* 3–21.

Kah L. C. and Riding R. (2007) Mesoproterozoic carbon dioxide levels inferred from calcified cyanobacteria. *Geology, 35,* 799.

Kaltenegger L. and Traub W. (2009) Transits of Earth-like planets. *Astrophys. J., 698,* 519.

Kaltenegger L., Traub W. A., and Jucks K. W. (2007) Spectral evolution of an Earth-like planet. *Astrophys. J., 658,* 598–616.

Kaltenegger L., Selsis F., Fridlund M., Lammer H., Beichman C., Danchi W., Eiroa C., et al. (2010) Deciphering spectral fingerprints of habitable exoplanets. *Astrobiology, 10,* 89–102.

Kaltenegger L., Miguel Y., and Rugheimer S. (2012) Rocky exoplanet characterization and atmospheres. *Intl. J. Astrobiol., 11,* 297–307.

Kanzaki Y. and Murakami T. (2015) Estimates of atmospheric CO_2 in the Neoarchean-Paleoproterozoic from paleosols. *Geochim. Cosmochim. Acta, 159,* 190–219.

Karalidi T. and Stam D. M. (2012) Modeled flux and polarization signals of horizontally inhomogeneous exoplanets applied to Earth-like planets. *Astron. Astrophys., 546,* A56.

Karalidi T., Stam D. M., and Hovenier J. W. (2011) Flux and polarisation spectra of water clouds on exoplanets. *Astron. Astrophys., 530,* A69.

Karalidi T., Stam D. M., and Hovenier J.W. (2012) Looking for the rainbow on exoplanets covered by liquid and icy water clouds. *Astron. Astrophys., 548,* A90.

Karhu J. A. and Holland H. D. (1996) Carbon isotopes and the rise of atmospheric oxygen. *Geology, 24,* 867.

Kasting J. F. (1987) Theoretical constraints on oxygen and carbon dioxide concentrations in the Precambrian atmosphere. *Precambrian Res., 34,* 205–229.

Kasting J. (2005) Methane and climate during the Precambrian era. *Precambrian Res., 137,* 119–129.

Kasting J. F. and Ackerman T. P. (1986) Climatic consequences of very high carbon dioxide levels in the Earth's early atmosphere. *Science, 234,* 1383–1385.

Kasting J. F. and Catling D. (2003) Evolution of a habitable planet. *Annu. Rev. Astron. Astrophys., 41,* 429–463.

Kasting J. F. and Donahue T. M. (1980) The evolution of atmospheric ozone. *J. Geophys. Res., 85,* 3255–3263.

Kasting J. F. and Walker J. C. G. (1981) Limits on oxygen concentration in the prebiological atmosphere and the rate of abiotic fixation of nitrogen. *J. Geophys. Res., 86,* 1147–1158.

Kasting J. F., Whitmire D. P., and Reynolds R. T. (1993) Habitable zones around main sequence stars. *Icarus, 101,* 108–128.

Kaufman A. J. and Xiao S. (2003) High CO_2 levels in the Proterozoic atmosphere estimated from analyses of individual microfossils. *Nature, 425,* 279–282.

Kavanagh L. and Goldblatt C. (2015) Using raindrops to constrain past atmospheric density. *Earth Planet. Sci. Lett., 413,* 51–58.

Kawahara H. (2016) Frequency modulation of directly imaged exoplanets: Geometric effect as a probe of planetary obliquity. *Astrophys. J., 822,* 112.

Kawahara H. and Fujii Y. (2010) Global mapping of Earthlike exoplanets from scattered light curves. *Astrophys. J., 720,* 1333–1350.

Kharecha P., Kasting J. K., and Siefert J. (2005) A coupled atmosphere-ecosystem model of the early Archean Earth. *Geobiology, 3,* 53–76.

Kliore A. J. and Woiceshyn P. M. (1976) Structure of the atmosphere of Jupiter from Pioneer 10 and 11 radio occultation measurements. In *Jupiter* (T. Gehrels, ed.), pp. 216–237. Univ. of Arizona, Tucson.

Knauth L. P. and Epstein S. (1976) Hydrogen and oxygen isotope ratios in nodular and bedded cherts. *Geochim. Cosmochim. Acta, 40,* 1095–1108.

Knutson H. A., Benneke B., Deming D., and Homeier D. (2014a) A featureless transmission spectrum for the Neptune-mass exoplanet GJ 436b. *Nature, 505,* 66–68.

Knutson H. A., Dragomir D., Kreidberg L., Kempton E. M.-R., McCullough P. R., Fortney J. J., Bean J. L., Gillon M., Homeier D., and Howard A. W. (2014b) Hubble Space Telescope near-IR transmission spectroscopy of the super-Earth HD 97658b. *Astrophys. J., 794,* 155.

Kohlhase C. E. and Penzo P. A. (1977) Voyager mission description.

Space Sci. Rev., 21, 77–101.

Konhauser K. O., Pecoits E., Lalonde S. V., Papineau D., Nisbet E. G., Barley M. E., Arndt N. T., Zahnle K., and Kamber B. S. (2009) Oceanic nickel depletion and a methanogen famine before the Great Oxidation Event. *Nature, 458,* 750–753.

Kopp R.E., Kirschvink J. L., Hilburn I. A., and Nash C. Z. (2005) The Paleoproterozoic snowball Earth: A climate disaster triggered by the evolution of oxygenic photosynthesis. *Proc. Natl. Acad. Sci. U.S.A., 102,* 11131–11136.

Kopparapu R. K., Hebrard E., Belikov R., Batalha N. M., Mulders G. D., Stark C., Teal D., Domagal-Goldman S., and Mandell A. (2018) Exoplanet classification and yield estimates for direct imaging missions. *Astrophys. J., 856,* 122.

Kreidberg L. (2018) Exoplanet atmosphere measurements from transmission spectroscopy and other planet star combined light observations. In *Handbook of Exoplanets* (H. J. Deeg and J. A. Belmonte, eds.), pp. 1–23. Springer, Cham.

Kreidberg L., Bean J. L., Désert J.-M., Benneke B., Deming D., Stevenson K. B., Seager, S., Berta-Thompson Z., Seifahrt A., and Homeier D. (2014) Clouds in the atmosphere of the super-Earth exoplanet GJ 1214b. *Nature, 505(7481),* 69–72.

Krissansen-Totton J., Bergsman D. S., and Catling D. C. (2016a) On detecting biospheres from chemical thermodynamic disequilibrium in planetary atmospheres. *Astrobiology, 16,* 39–67.

Krissansen-Totton J., Schwieterman E. W., Charnay B., Arney G., Robinson T. D., Meadows V., and Catling D. C. (2016b) Is the Pale Blue Dot unique? Optimized photometric bands for identifying Earth-like exoplanets. *Astrophys. J., 817,* 31.

Krissansen-Totton J., Arney G. N., and Catling D. C. (2018) Constraining the climate and ocean pH of the early Earth with a geological carbon cycle model. *Proc. Natl. Acad. Sci. U.S.A., 115,* 4105–4110.

Kuhn W. R. and Atreya S. K. (1979) Ammonia photolysis and the greenhouse effect in the primordial atmosphere of the Earth. *Icarus, 37,* 207–213.

Kump L. R. and Arthur M. A. (1999) Interpreting carbon-isotope excursions: Carbonates and organic matter. *Chem. Geol., 161,* 181–198.

Kump L. R. and Barley M. E. (2007) Increased subaerial volcanism and the rise of atmospheric oxygen 2.5 billion years ago. *Nature, 448,* 1033–1036.

Kurzweil F., Claire M., Thomazo C., Peters M., Hannington M., and Strauss H. (2013) Atmospheric sulfur rearrangement 2.7 billion years ago: Evidence for oxygenic photosynthesis. *Earth Planet. Sci. Lett., 366,* 17–26.

Laakso T. A. and Schrag D. P. (2017) A theory of atmospheric oxygen. *Geobiology, 415,* 366–384.

Le Hir G., Goddéris Y., Donnadieu Y., and Ramstein G. (2008a) A geochemical modelling study of the evolution of the chemical composition of seawater linked to a "snowball" glaciation. *Biogeosciences, 5,* 253–267.

Le Hir G., Ramstein G., Donnadieu Y., and Goddéris Y. (2008b) Scenario for the evolution of atmospheric pCO2 during a snowball Earth. *Geology, 36,* 47–50.

Lenton T. M., Crouch M., Johnson M., Pires N., and Dolan L. (2012) First plants cooled the Ordovician. *Nature Geosci., 5,* 86–89.

Lenton T. M., Daines S. J., and Mills B. J. W. (2018) COPSE reloaded: An improved model of biogeochemical cycling over Phanerozoic time. *Earth-Sci. Rev., 33,* 1–28.

Lepland A., Joosu L., Kirsimäe, K., Prave A. R., Romashkin A. E., Črne A. E., Martin A. P., et al. (2014) Potential infuence of sulphur bacteria on Palaeoproterozoic phosphogenesis. *Nature Geosci., 7,* 20–24.

Levine J. S., Hays P. B., and Walker J. C. G. (1979) The evolution and variability of atmospheric ozone over geological time. *Icarus, 39,* 295–309.

Line M. R., Knutson H., Deming D., Wilkins A., and Désert J.-M. (2013) A near-infrared transmission spectrum for the warm Saturn HAT-P-12b. *Astrophys. J., 778,* 183.

Line M. R., Knutson H., Wolf A. S., and Yung Y. L. (2014) A systematic retrieval analysis of secondary eclipse spectra. II. A uniform analysis of nine planets and their C to O ratios. *Astrophys. J., 783,* 70.

Livengood T. A., Deming L. D., A'Hearn M. F., Charbonneau, D., Hewagama T., Lisse C. M., McFadden L. A., et al. (2011) Properties of an Earth-like planet orbiting a Sun-like star: Earth observed by the EPOXI mission. *Astrobiology, 11,* 907–930.

Luo G., Ono S., Beukes N. J., Wang D. T., Xie S., and Summons R. (2016) The loss of mass-independent fractionation in sulfur due to a Paleoproterozoic collapse of atmospheric methane. *Sci. Adv., 2.*

Lupu R. E., Marley M. S., Lewis N., Line M., Traub W. A., and Zahnle K. (2016) Developing atmospheric retrieval methods for direct imaging spectroscopy of gas giants in reflected light. I. Methane abundances and basic cloud properties. *Astron. J., 152,* 217.

Lustig-Yaeger J., Meadows V. S., Tovar Mendoza G., Schwieterman E. W., Fujii Y., Luger R., and Robinson T. D. (2018) Detecting ocean glint on exoplanets using multiphase mapping. *Astron. J., 156,* 301.

Lyons T. W., Reinhard C. T., and Planavsky N. J. (2014) The rise of oxygen in Earth's early ocean and atmosphere. *Nature, 506,* 307–315.

Macdonald E. J. R. and Cowan N. B. (2019) An empirical infrared transit spectrum of Earth: Opacity windows and biosignatures. *Mon. Not. R. Astron. Soc., 489(1),* 196–204.

Macintosh B., Graham J. R., Barman T., De Rosa R. J., Konopacky Q., Marley M. S., Marois C., et al. (2015) Discovery and spectroscopy of the young jovian planet 51 Eri b with the Gemini Planet Imager. *Science, 350,* 64–67.

Madhusudhan N. and Seager S. (2009) A temperature and abundance retrieval method for exoplanet atmospheres. *Astrophys. J., 707,* 24–39.

Magnabosco C., Moore K. R., Wolfe J. M., and Fournier G. P. (2018) Dating phototrophic microbial lineages with reticulate gene histories. *Geobiology, 16,* 179–189.

Majeau C., Agol E., and Cowan N. B. (2012) A two-dimensional infrared map of the extrasolar planet HD 189733b. *Astrophys. J. Lett., 747,* L20.

Manalo-Smith N., Smith G. L., Tiwari S. N., and Staylor W. F. (1998) Analytic forms of bidirectional reflectance functions for application to Earth radiation budget studies. *J. Geophys. Res., 103,* 19733–19752.

Marley M., Lupu R., Lewis N., Line M., Morley C., and Fortney J. (2014) A quick study of the characterization of radial velocity giant planets in reflected light by forward and inverse modeling. *ArXiv e-prints,* arXiv:1412.8440.

Marois C., Macintosh B., Barman T., Zuckerman B., Song I., Patience J., Lafrenière D., and Doyon R. (2008) Direct imaging of multiple planets orbiting the star HR 8799. *Science, 322,* 1348.

Martín-Torres F. J., Kutepov A., Dudhia A., Gusev O., and Feoflov A. G. (2003) Accurate and fast computation of the radiative transfer absorption rates for the infrared bands in the atmosphere of Titan. *EGS-AGU-EUG Joint Assembly Abstracts,* 7735.

Marty B., Zimmermann L., Pujol M., Burgess R., and Philippot P. (2013) Nitrogen isotopic composition and density of the Archean atmosphere. *Science, 342,* 101–104.

Matson P. A. and Vitousek P. M. (1990) Ecosystem approach to a global nitrous oxide budget. *BioScience, 40,* 667–672.

Mawet D., Pueyo L., Lawson P., Mugnier L., Traub W., Boccaletti A., Trauger J. T., et al. (2012) Review of small-angle coronagraphic techniques in the wake of ground-based second-generation adaptive optics systems. In *Space Telescopes and Instrumentation 2012: Optical, Infrared, and Millimeter Wave* (M. C. Clampin et al., eds.), p. 844204. SPIE Conf. Ser. 8442, Bellingham, Washington.

Mayor M. and Queloz D. (1995) A Jupiter-mass companion to a solar-type star. *Nature, 378,* 355–359.

McCullough P. R. (2006) Models of polarized light from oceans and atmospheres of Earth-like extrasolar planets. *ArXiv e-prints,* arXiv:astro-ph/0610518.

Meadows V. S. (2008) Planetary environmental signatures for habitability and life. In *Exoplanets* (J. W. Mason, ed.), p. 259. Springer, Berlin.

Meadows V. S. (2017) Reflections on O_2 as a biosignature in exoplanetary atmospheres. *Astrobiology, 17,* 1022–1052.

Melezhik V. A., Huhma H., Condon D. J., Fallick A. E., and Whitehouse M. J. (2007) Temporal constraints on the Paleoproterozoic Lomagundi-Jatuli carbon isotopic event. *Geology, 35,* 655.

Mennesson B., Gaudi S., Seager S., Cahoy K., Domagal-Goldman S., Feinberg L., Guyon O., et al. (2016) The Habitable Exoplanet (HabEx) imaging mission: Preliminary science drivers and technical requirements. In *Space Telescopes and Instrumentation 2016: Optical, Infrared, and Millimeter Wave* (H. A. MacEwen et al., eds.),

p. 99040L. SPIE Conf. Ser. 9904, Bellingham, Washington.

Miles-Páez P. A., Pallé E., and Zapatero Osorio M. R. (2014) simultaneous optical and near-infrared linear spectropolarimetry of the Earthshine. *Astron. Astrophys., 562,* L5.

Mills B., Watson A. J., Goldblatt C., Boyle R., and Lenton T. M. (2011) Timing of Neoproterozoic glaciations linked to transport-limited global weathering. *Nature Geosci., 4,* 861–864.

Misra A., Meadows V., Claire M., and Crisp D. (2014a) Using dimers to measure biosignatures and atmospheric pressure for terrestrial exoplanets. *Astrobiology, 14,* 67–86.

Misra A., Meadows V., and Crisp D. (2014b) The effects of refraction on transit transmission spectroscopy: Application to Earth-like exoplanets. *Astrophys. J., 792,* 61.

Mojzsis S. J., Harrison T. M., and Pidgeon R. T. (2001) Oxygen-isotope evidence from ancient zircons for liquid water at the Earth's surface 4,300 Myr ago. *Nature, 409,* 178–181.

Montañez I.P. and Poulsen C. J. (2013) The late Paleozoic ice age: An evolving paradigm. *Annu. Rev. Earth Planet. Sci., 41,* 629–656.

Montañés-Rodríguez P., Pallé E., Goode P. R., and Martín-Torres F. J. (2006) Vegetation signature in the observed globally integrated spectrum of Earth considering simultaneous cloud data: Applications for extrasolar planets. *Astrophys. J., 651,* 544–552.

Nayak M., Lupu R., Marley M. S., Fortney J. J., Robinson T., and Lewis N. (2017) Atmospheric retrieval for direct imaging spectroscopy of gas giants in reflected light. II. Orbital phase and planetary radius. *Publ. Astron. Soc. Pac., 129(3),* 034401.

Nishizawa M., Sano Y., Ueno Y., and Maruyama S. (2007) Speciation and isotope ratios of nitrogen in fluid inclusions from seafloor hydrothermal deposits at 3.5 Ga. *Earth Planet. Sci. Lett., 254,* 332–344.

Noecker M. C., Zhao F., Demers R., Trauger J., Guyon O., and Kasdin N. J. (2016) Coronagraph instrument for WFIRST-AFTA. *J Astron. Telesc. Instrum. Syst., 2(1),* 011001.

Oakley P. H. H. and Cash W. (2009) Construction of an Earth model: Analysis of exoplanet light curves and mapping the next Earth with the New Worlds Observer. *Astrophys. J., 700,* 1428–1439.

Olson S. L., Reinhard C. T., and Lyons T. W. (2016) Limited role for methane in the mid-Proterozoic greenhouse. *Proc. Natl. Acad. Sci. U.S.A., 113,* 11447–11452.

Olson S. L., Schwieterman E. W., Reinhard C. T., Ridgwell A., Kane S. R., Meadows V. S., and Lyons T. W. (2018a) Atmospheric seasonality as an exoplanet biosignature. *Astrophys. J. Lett., 858,* L14.

Olson S. L., Schwieterman E. W., Reinhard C. T., and Lyons T. W. (2018b) Earth: Atmospheric evolution of a habitable planet. In *Handbook of Exoplanets* (H. J. Deeg and J. A. Belmonte, eds.), pp. 1–37. Springer, Cham.

Ozaki K., Tajika E., Hong P. K., Nakagawa Y., and Reinhard C. T. (2018) Effects of primitive photosynthesis on Earth's early climate system. *Nature Geosci., 11,* 55–59.

Pallé E., Goode P. R., Yurchyshyn V., Qiu J., Hickey J., Montañés Rodriguez P., Chu M., Kolbe E., Brown C. T., and Koonin S. E. (2003) Earthshine and the Earth's albedo: 2. Observations and simulations over 3 years. *J. Geophys. Res.–Atmos., 108,* 4710.

Pallé E., Goode P. R., Montañés-Rodríguez P., and Koonin S. E. (2004a) Changes in Earth's reflectance over the past two decades. *Science, 304,* 1299–1301.

Pallé E., Montañés Rodriguez P., Goode P. R., Qiu J., Yurchyshyn V., Hickey J., Chu M.-C., Kolbe E., Brown C. T., and Koonin S. E. (2004b) The Earthshine Project: Update on photometric and spectroscopic measurements. *Adv. Space Res., 34,* 288–292.

Pallé E., Ford E. B., Seager S., Montañés-Rodríguez P., and Vazquez M. (2008) Identifying the rotation rate and the presence of dynamic weather on extrasolar Earthlike planets from photometric observations. *Astrophys. J., 676,* 1319–1329.

Pallé E., Goode P. R., and Montañés-Rodríguez P. (2009a) Interannual variations in Earth's reflectance 1999–2007. *J. Geophys. Res.–Atmos., 114,* D00D03.

Pallé E., Zapatero Osorio M. R., Barrena R., Montañés-Rodríguez P., and Martín E. L. (2009b) Earth's transmission spectrum from lunar eclipse observations. *Nature, 459,* 814–816.

Pallé E., Goode P. R., Montañés-Rodríguez P., Shumko A., Gonzalez-Merino B., Lombilla C. M., Jimenez-Ibarra F., et al. (2016) Earth's albedo variations 1998–2014 as measured from ground-based Earthshine observations. *Geophys. Res. Lett., 43,* 4531–4538.

Partin C. A., Bekker A., Planavsky N. J., Scott C. T., Gill B. C., Li C., Podkovyrov V., et al. (2013) Large-scale fluctuations in Precambrian atmospheric and oceanic oxygen levels from the record of U in shales. *Earth Planet. Sci. Lett., 369,* 284–293.

Pavlov A. A. and Kasting J. F. (2002) Mass-independent fractionation of sulfur isotopes in Archean sediments: Strong evidence for an anoxic Archean atmosphere. *Astrobiology, 2,* 27–41.

Pavlov A. A., Kasting J. F., Brown L. L., Rages K. A., and Freedman R. (2000) Greenhouse warming by CH_4 in the atmosphere of early Earth. *J. Geophys. Res., 105,* 11981–11990.

Pavlov A. A., Brown L. L., and Kasting J. F. (2001) UV shielding of NH_3 and O_2 by organic hazes in the Archean atmosphere. *J. Geophys. Res., 106,* 23267–23288.

Payne J. L., Lehrmann D. J., Wei J., Orchard M. J., Schrag D. P., and Knoll A. H. (2004) Large perturbations of the carbon cycle during recovery from the end-Permian extinction. *Science, 305,* 506–509.

Peterson B. M., Fischer D., and LUVOIR Science and Technology Definition Team (2017) The Large Ultraviolet/Optical/Infrared Surveyor (LUVOIR). *AAS Meeting Abstracts, 229,* Abstract #405.04. American Astronomical Society, Washington, DC.

Petigura E. A., Marcy G. W., and Howard A. W. (2013) A plateau in the planet population below twice the size of Earth. *Astrophys. J., 770,* 69.

Pierrehumbert R. T., Abbot D. S., Voigt A., and Koll D. (2011) Climate of the Neoproterozoic. *Annu. Rev. Earth Planet. Sci., 39,* 417–460.

Planavsky N., Bekker A., Rouxel O. J., Kamber B., Hofmann A., Knudsen A., and Lyons T. W. (2010) Rare Earth element and yttrium compositions of Archean and Paleoproterozoic Fe formations revisited: New perspectives on the significance and mechanisms of deposition. *Geochim. Cosmochim. Acta, 74,* 6387–6405.

Planavsky N. J., Bekker A., Hofmann A., Owens J. D., and Lyons T. W. (2012) Sulfur record of rising and falling marine oxygen and sulfate levels during the Lomagundi event. *Proc. Natl. Acad. Sci. U.S.A., 109,* 18300–18305.

Planavsky N. J., Asael D., Hofmann A., Reinhard C. T., Lalonde S. V., Knudsen A., Wang X., et al. (2014a) Evidence for oxygenic photosynthesis half a billion years before the Great Oxidation Event. *Nature Geosci., 7,* 283–286.

Planavsky N. J., Reinhard C. T., Wang X., Thomson D., McGoldrick P., Rainbird R. H., Johnson T., Fischer W. W., and Lyons T. W. (2014b) Low mid-Proterozoic atmospheric oxygen levels and the delayed rise of animals. *Science, 346,* 635–638.

Pont F., Knutson H., Gilliland R., Moutou C., and Charbonneau D. (2008) Detection of atmospheric haze on an extrasolar planet: The 0.55–1.05 µm transmission spectrum of HD 189733b with the Hubble Space Telescope. *Mon. Not. R. Astron. Soc., 385(1),* 109–118.

Prather M. J., and Hsu J. (2010) Coupling of nitrous oxide and methane by global atmospheric chemistry. *Science, 330,* 952–954.

Qiu J., Goode P., Pallé E., Yurchyshyn V., Hickey J., Montanés-Rodriguez P., Chu M., Kolbe E., Brown C., and Koonin S. (2003) Earthshine and the Earth's albedo: 1. Earthshine observations and measurements of the lunar phase function for accurate measurements of the Earth's bond albedo. *J. Geophys. Res, 108(4709),* 1999–2007.

Rasmussen B. and Buick R. (1999) Redox state of the Archean atmosphere: Evidence from detrital heavy minerals in ca. 3250–2750 Ma sandstones from the Pilbara Craton, Australia. *Geology, 27,* 115.

Rasmussen B., Bekker A., and Fletcher I. R. (2013) Correlation of Paleoproterozoic glaciations based on U-Pb zircon ages for tuff beds in the Transvaal and Huronian Supergroups. *Earth Planet. Sci. Lett., 382,* 173–180.

Raymond A. and Metz C. (2004) Ice and its consequences: Glaciation in the Late Ordovician, Late Devonian, Pennsylvanian-Permian, and Cenozoic compared. *J. Geol., 112,* 655–670.

Reinhard C. T., Planavsky N. J., Robbins L. J., Partin C. A., Gill B. C., Lalonde S. V., Bekker A., Konhauser K. O., and Lyons T. W. (2013) Proterozoic ocean redox and biogeochemical stasis. *Proc. Natl. Acad. Sci. U.S.A., 110,* 5357–5362.

Reinhard C. T., Olson S. L., Schwieterman E. W., and Lyons T. W. (2017a) False negatives for remote life detection on ocean-bearing planets: Lessons from the early Earth. *Astrobiology, 17,* 287–297.

Reinhard C. T., Planavsky N. J., Gill B. C., Ozaki K., Robbins L. J., Lyons T. W., Fischer W. W., Wang C., Cole D. B., and Konhauser K. O. (2017b) Evolution of the global phosphorus cycle. *Nature,*

541, 386–389.

Roberge A. and Moustakas L. A. (2018) The Large Ultraviolet/Optical/Infrared Surveyor. *Nature Astron., 2,* 605–607.

Roberson A. L., Roadt J., Halevy I., and Kasting J. F. (2011) Greenhouse warming by nitrous oxide and methane in the Proterozoic eon. *Geobiology, 9,* 313– 320.

Robert F. and Chaussidon M. (2006) A palaeotemperature curve for the Precambrian oceans based on silicon isotopes in cherts. *Nature, 443,* 969–972.

Robinson T. D. (2011) Modeling the infrared spectrum of the Earth-Moon system: Implications for the detection and characterization of Earthlike extrasolar planets and their Moonlike companions. *Astrophys. J., 741,* 51.

Robinson T. D. (2018) Characterizing exoplanets for habitability. In *Handbook of Exoplanets* (H. J. Deeg and J. A. Belmonte, eds.), pp. 3137-3157 Springer, Cham.

Robinson T. D., Meadows V. S., and Crisp D. (2010) Detecting oceans on extrasolar planets using the glint effect. *Astrophys. J. Lett., 721,* L67–L71.

Robinson T. D., Meadows V. S., Crisp D., Deming D., A'Hearn M. F., Charbonneau D., Livengood T. A., et al. (2011) Earth as an extrasolar planet: Earth model validation using EPOXI Earth observations. *Astrobiology, 11,* 393–408.

Robinson T. D., Maltagliati L., Marley M. S., and Fortney J. J. (2014a) Titan solar occultation observations reveal transit spectra of a hazy world. *Proc. Natl. Acad. Sci. U.S.A., 111(25),* 9042–9047.

Robinson T. D., Ennico K., Meadows V. S., Sparks W., Bussey D. B. J., Schwieterman E. W., and Breiner J. (2014b) Detection of ocean glint and ozone absorption using LCROSS Earth observations. *Astrophys. J., 787,* 171.

Robinson T. D., Stapelfeldt K. R., and Marley M. S. (2016) Characterizing rocky and gaseous exoplanets with 2 m class space-based coronagraphs. *Publ. Astron. Soc. Pac., 128(2),* 025003.

Rodrigues J. B., Pimentel M. M., Buhn B., Matteini M., Dardenne M. A., Alvarenga C. J. S., and Armstrong R. A. (2012) Provenance of the Vazante Group: New U-Pb, Sm-Nd, Lu-Hf isotopic data and implications for the tectonic evolution of the Neoproterozoic Brasília Belt. *Gondwana Res., 21,* 439–450.

Rooney A. D., Macdonald F. A., Strauss J. V., Dudás F. Ö., Hallmann C., and Selby D. (2014) Re-Os geochronology and coupled Os-Sr isotope constraints on the Sturtian snowball Earth. *Proc. Natl. Acad. Sci. U.S.A., 111,* 51–56.

Rooney A. D., Strauss J. V., Brandon A. D., and Macdonald F. A. (2015) A Cryogenian chronology: Two long-lasting synchronous Neoproterozoic glaciations. *Geology, 43,* 459–462.

Rosing M. T., Rose N. M., Bridgwater D., and Thomsen H. S. (1996) Earliest part of Earth's stratigraphic record: A reappraisal of the >3.7 Ga Isua (Greenland) supracrustal sequence. *Geology, 24,* 43.

Royer D. L. (2001) Stomatal density and stomatal index as indicators of paleoatmospheric CO_2 concentration. *Rev. Palaeobot. Palynol., 114,* 1–28.

Royer D. L. (2014) Atmospheric CO_2 and O_2 during the Phanerozoic: Tools, patterns, and impacts. In *Treatise on Geochemistry* (H. D. Holland and K. K. Turekian, eds.), pp. 251–267. Elsevier, Netherlands.

Royer D. L., Berner R. A., Montanez I. P., Tabor N. J., and Beerling D. J. (2004) CO_2 as a primary driver of Phanerozoic climate. *GSA Today, 14,* 4–10.

Royer D. L., Donnadieu Y., Park J., Kowalczyk J., and Godderis Y. (2014) Error analysis of CO_2 and O_2 estimates from the long-term geochemical model GEOCARBSULF. *Am. J. Sci., 314,* 1259–1283.

Rugheimer S. and Kaltenegger L. (2018) Spectra of Earth-like planets through geological evolution around FGKM stars. *Astrophys. J., 854,* 19.

Rugheimer S., Kaltenegger L., Zsom A., Segura A., and Sasselov D. (2013) Spectral fingerprints of Earth-like planets around FGK stars. *Astrobiology, 13,* 251–269.

Rye R. and Holland H. D. (1998) Paleosols and the evolution of atmospheric oxygen; a critical review. *Am. J. Sci., 298,* 621–672.

Rye R., Kuo P. H., and Holland H. D. (1995) Atmospheric carbon dioxide concentrations before 2.2 billion years ago. *Nature, 378,* 603–605.

Sagan C. and Mullen G. (1972) Earth and Mars: Evolution of atmospheres and surface temperatures. *Science, 177,* 52–56.

Sagan C., Thompson W. R., Carlson R., Gurnett D., and Hord C. (1993)

A search for life on Earth from the Galileo spacecraft. *Nature, 365,* 715–721.

Sahoo S. K., Planavsky N. J., Kendall B., Wang X., Shi X., Scott C., Anbar A. D., Lyons T. W., and Jiang G. (2012) Ocean oxygenation in the wake of the Marinoan glaciation. *Nature, 489,* 546–549.

Schiffer R. A. and Rossow W. B. (1983) The International Satellite Cloud Climatology Project (ISCCP): The first project of the World Climate Research Programme. *Bull. Am. Meteor. Soc., 64,* 779.

Schrag D. P., Berner R. A., Hoffman P. F., and Halverson G. P. (2002) On the initiation of a snowball Earth. *Geochem., Geophys., Geosyst., 3.*

Schreier F., Städt S., Hedelt P., and Godolt M. (2018) Transmission spectroscopy with the ACE-FTS infrared spectral atlas of Earth: A model validation and feasibility study. *Mol. Astrophys., 11,* 1–22.

Schröder S., Bekker A., Beukes N. J., Strauss H., and van Niekerk H. S. (2008) Rise in seawater sulphate concentration associated with the Paleoproterozoic positive carbon isotope excursion: Evidence from sulphate evaporites in the 2.2–2.1 Gyr shallow-marine Lucknow Formation, South Africa. *Terra Nova, 20,* 108–117.

Schumann U. and Huntrieser H. (2007) The global lightning-induced nitrogen oxides source. *Atmos. Chem. Phys., 7,* 3823–3907.

Schwartz J. C., Sekowski C., Haggard H. M., Pallé E., and Cowan N. B. (2016) Inferring planetary obliquity using rotational and orbital photometry. *Mon. Not. R. Astron. Soc., 457,* 926–938.

Schwieterman E. W., Robinson T. D., Meadows V. S., Misra A., and Domagal-Goldman S. (2015) Detecting and constraining N_2 abundances in planetary atmospheres using collisional pairs. *Astrophys. J., 810,* 57.

Schwieterman E. W., Kiang N. Y., Parenteau M. N., Harman C. E., DasSarma S., Fisher T. M., Arney G. N., et al. (2018) Exoplanet biosignatures: A review of remotely detectable signs of life. *Astrobiology, 18,* 663–708.

Scott A. C. and Glasspool I. J. (2006) The diversification of Paleozoic fire systems and fluctuations in atmospheric oxygen concentration. *Proc. Natl. Acad. Sci. U.S.A., 103,* 1086–10865.

Scott C., Wing B. A., Bekker A., Planavsky N. J., Medvedev P., Bates S. M., Yun M., and Lyons T. W. (2014) Pyrite multiple-sulfur isotope evidence for rapid expansion and contraction of the early Paleoproterozoic seawater sulfate reservoir. *Earth Planet. Sci. Lett., 389,* 95–104.

Seager S. and Sasselov D. (2000) Theoretical transmission spectra during extrasolar giant planet transits. *Astrophys. J., 537(2),* 916.

Seager S., Turner E. L., Schafer J., and Ford E. B. (2005) Vegetation's red edge: A possible spectroscopic biosignature of extraterrestrial plants. *Astrobiology, 5,* 372–390.

Seager S., Turnbull M., Sparks W., Thomson M., Shaklan S. B., Roberge A., Kuchner M., et al. (2015) The Exo-S Probe Class Starshade Mission. In *Techniques and Instrumentation for Detection of Exoplanets VII* (S. Shaklan, ed.), p. 96050W. SPIE Conf. Ser. 9605, Bellingham, Washington.

Segura A., Krelove K., Kasting J. F., Sommerlatt D., Meadows V., Crisp D., Cohen M., and Mlawer E. (2003) Ozone concentrations and ultraviolet fluxes on Earth-like planets around other stars. *Astrobiology, 3,* 689–708.

Sellers W. D. (1969) A global climatic model based on the energy balance of the Earth-atmosphere system. *J. Appl. Meteorol., 8,* 392–400.

Selsis F., Wordsworth R. D., and Forget F. (2011) Thermal phase curves of nontransiting terrestrial exoplanets. I. Characterizing atmospheres. *Astron. Astrophys., 532,* A1.

Shaklan S. B., Noecker M. C., Glassman T., Lo A. S., Dumont P. J., Kasdin N. J., Cady E. J., Vanderbei R., and Lawson P. R. (2010) Error budgeting and tolerancing of starshades for exoplanet detection. In *Space Telescopes and Instrumentation 2010: Optical, Infrared, and Millimeter Wave* (J. M. Oschmann Jr. et al., eds.), p. 77312G. SPIE Conf. Ser. 7731, Bellingham, Washington.

Sheen A. I., Kendall B., Reinhard C. T., Creaser R. A., Lyons T. W., Bekker A., Poulton S. W., and Anbar A. D. (2018) A model for the oceanic mass balance of rhenium and implications for the extent of Proterozoic ocean anoxia. *Geochim. Cosmochim. Acta, 227,* 75–95.

Sheldon N. D. (2006) Precambrian paleosols and atmospheric CO_2 levels. *Precambrian Res., 147,* 148–155.

Sheldon N. D. (2013) Causes and consequences of low atmospheric pCO_2 in the Late Mesoproterozoic. *Chem. Geol., 362,* 224–231.

Sing D., Désert J.-M., Lecavelier Des Etangs A., Ballester G., Vidal-

Madjar A., Parmentier V., Hebrard G., and Henry G. (2009) Transit spectrophotometry of the exoplanet HD 189733b. I. Searching for water but finding haze with HST NICMOS. *Astron. Astrophys., 505,* 891–899.

Sing D. K., Fortney J. J., Nikolov N., Wakeford H. R., Kataria T., Evans T. M., Aigrain S., et al. (2016) A continuum from clear to cloudy hot-Jupiter exoplanets without primordial water depletion. *Nature, 529,* 59–62.

Skemer A. J., Hinz P. M., Esposito S., Burrows A., Leisenring J., Skrutskie M., Desidera S., et al. (2012) First Light LBT AO images of HR 8799 bcde at 1.6 and 3.3 μm: New discrepancies between young planets and old brown dwarfs. *Astrophys. J., 753,* 14.

Sleep N. H. and Zahnle K. (2001) Carbon dioxide cycling and implications for climate on ancient Earth. *J. Geophys. Res., 106,* 1373–1400.

Snellen I. A. G., de Kok R. J., Le Poole R., Brogi M., and Birkby J. (2013) Finding extraterrestrial life using ground-based high-dispersion spectroscopy. *Astrophys. J., 764,* 182.

Som S. M., Catling D. C., Harnmeijer J. P., Polivka P. M., and Buick R. (2012) Air density 2.7 billion years ago limited to less than twice modern levels by fossil raindrop imprints. *Nature, 484,* 359–362.

Som S. M., Buick R., Hagadorn J. W., Blake T. S., Perreault J. M., Harnmeijer J. P., and Catling D. C. (2016) Earth's air pressure 2.7 billion years ago constrained to less than half of modern levels. *Nature Geosci., 9,* 448–451.

Spergel D., Gehrels N., Breckinridge J., Donahue M., Dressler A., Gaudi B., Greene T., et al. (2013) Wide-field infrared survey telescope-astrophysics focused telescope assets WFRST-AFTA final report. *ArXiv e-prints,* arXiv:1305.5422.

Stam D. M. (2008) Spectropolarimetric signatures of Earth-like extrasolar planets. *Astron. Astrophys., 482,* 989–1007.

Stanton C. L., Reinhard C. T., Kasting J. F., Ostrom N. E., Haslun J. A., Lyons T. W., and Glass J. B. (2018) Nitrous oxide from chemodenitrifcation: A possible missing link in the Proterozoic greenhouse and the evolution of aerobic respiration. *Geobiology, 16(6),* 597–609.

Stark C. C., Cady E. J., Clampin M., Domagal-Goldman S., Lisman D., Mandell A. M., McElwain M. W., et al. (2016) A direct comparison of exoEarth yields for starshades and coronagraphs. In *Space Telescopes and Instrumentation 2016: Optical, Infrared, and Millimeter Wave* (H. A. McEwen et al., eds.), p. 99041U. SPIE Conf. Ser. 9904, Bellingham, Washington.

Sterzik M. F., Bagnulo S., and Pallé E. (2012) Biosignatures as revealed by spectropolarimetry of Earthshine. *Nature, 483,* 64–66.

Stevenson K. B. (2016) Quantifying and predicting the presence of clouds in exoplanet atmospheres. *Astrophys. J. Lett., 817,* L16.

Stevenson K. B., Harrington J., Nymeyer S., Madhusudhan N., Seager S., Bowman W.C., Hardy R. A., Deming D., Rauscher E., and Lust N. B. (2010) Possible thermochemical disequilibrium in the atmosphere of the exoplanet GJ 436b. *Nature, 464,* 1161–1164.

Stüeken E. E., Kipp M. A., Koehler M. C., Schwieterman E. W., Johnson B., and Buick R. (2016) Modeling pN_2 through geological time: Implications for planetary climates and atmospheric biosignatures. *Astrobiology, 16,* 949–963.

Suttles J. T., Green R. N., Minnis P., Smith G. L., Staylor W. F., Wielicki B. A., Walker I., Young D., Taylor V., and Stowe L., eds. (1988) *Angular Radiation Models for Earth-Atmosphere System: Volume 1 — Shortwave Radiation.* NASA Scientific and Technical Information Division Reference Publication 1184.

Swain M. R., Vasisht G., and Tinetti G. (2008) The presence of methane in the atmosphere of an extrasolar planet. *Nature, 452,* 329–331.

Swain M. R., Vasisht G., Tinetti G., Bouwman J., Chen P., Yung Y., Deming D., and Deroo P. (2009) Molecular signatures in the near-infrared dayside spectrum of HD 189733b. *Astrophys. J. Lett., 690,* L114–L117.

Tang D., Shi X., Wang X., and Jiang G. (2016a) Extremely low oxygen concentration in mid-Proterozoic shallow seawaters. *Precambrian Res., 276,* 145–157.

Tang M., Chen K., and Rudnick R. L. (2016b) Archean upper crust transition from mafic to felsic marks the onset of plate tectonics. *Science, 351,* 372–375.

Thomson D., Rainbird R. H., Planavsky N., Lyons T. W., and Bekker A. (2015) Chemostratigraphy of the Shaler Supergroup, Victoria Island, NW Canada: A record of ocean composition prior to the Cryogenian glaciations. *Precambrian Res., 263,* 232–245.

Tian F., Toon O. B., Pavlov A. A., and De Sterck H. (2005) A hydrogen-rich early Earth atmosphere. *Science, 308,* 1014–1017.

Tian F., Toon O. B., and Pavlov A. A. (2006) Response to comment on "A hydrogen-rich early Earth atmosphere." *Science, 311,* 38b.

Tian F., Kasting J. F., and Zahnle K. (2011) Revisiting HCN formation in Earth's early atmosphere. *Earth Planet. Sci. Lett., 308,* 417–423.

Tice M. M. and Lowe D. R. (2004) Photosynthetic microbial mats in the 3,416-Myr-old ocean. *Nature, 431,* 549–552.

Tinetti G., Meadows V. S., Crisp D., Fong W., Fishbein E., Turnbull M., and Bibring J.-P. (2006) Detectability of planetary characteristics in disk-averaged spectra. I: The Earth model. *Astrobiology, 6,* 34–47.

Tohver E., D'Agrella-Filho M. S., and Trindade R. I. F. (2006) Paleomagnetic record of Africa and South America for the 1200–500 Ma interval, and evaluation of Rodinia and Gondwana assemblies. *Precambrian Res., 147,* 193–222.

Trainer M. G., Pavlov A. A., Curtis D. B., McKay C. P., Worsnop D. R., Delia A. E., Toohey D. W., Toon O. B., and Tolbert M. A. (2004) Haze aerosols in the atmosphere of early Earth: Manna from heaven. *Astrobiology, 4,* 409–419.

Trainer M. G., Pavlov A. A., Dewitt H. L., Jimenez J. L., McKay C. P., Toon O. B., and Tolbert M. A. (2006) Inaugural article: Organic haze on Titan and the early Earth. *Proc. Natl. Acad. Sci. U.S.A., 103,* 18035–18042.

Traub W. A. (2003) The colors of extrasolar planets. In *Scientific Frontiers in Research on Extrasolar Planets* (D. Deming and S. Seager, eds.), pp. 595–602. ASP Conf. Ser. 294, Astronomical Society of the Pacific, San Francisco.

Traub W. A. and Jucks K. W. (2002) A possible aeronomy of extrasolar terrestrial planets. In *Atmospheres in the Solar System: Comparative Aeronomy* (M. Mendillo et al., eds.), p. 369. American Geophysical Union, Washington, DC.

Traub W. A., Breckinridge J., Greene T. P., Guyon O., Kasdin N. J., and Macintosh B. (2016) Science yield estimate with the Wide-Field Infrared Survey Telescope coronagraph. *J. Astron. Telesc. Instrum. Syst., 2(1),* 011020.

Traub W. A. and Oppenheimer B. R. (2010) Direct imaging of exoplanets. In *Exoplanets* (S. Seager, ed.), pp. 111–156. Univ. of Arizona, Tucson.

Turnbull M. C., Traub W. A., Jucks K. W., Woolf N. J., Meyer M. R., Gorlova N., Skrutskie M. F., and Wilson J. C. (2006) Spectrum of a habitable world: Earthshine in the near-infrared. *Astrophys. J., 644,* 551–559.

Ueno Y., Yamada K., Yoshida N., Maruyama S., and Isozaki Y. (2006) Evidence from fluid inclusions for microbial methanogenesis in the early Archaean era. *Nature, 440,* 516–519.

Ushikubo T., Kita N. T., Cavosie A. J., Wilde S. A., Rudnick R. L., and Valley J. W. (2008) Lithium in Jack Hills zircons: Evidence for extensive weathering of Earth's earliest crust. *Earth Planet. Sci. Lett., 272,* 666–676.

Vázquez M., Pallé E., and Rodríguez P. (2010) *The Earth as a Distant Planet: A Rosetta Stone for the Search of Earth-Like Worlds.* Springer-Verlag, New York. 422 pp.

Veevers J. J. and Powell C. M. (1987) Late Paleozoic glacial episodes in Gondwanaland reflected in transgressive-regressive depositional sequences in Euramerica. *Bull. Geol. Soc. Am., 98,* 475.

Viljoen M. J. and Viljoen R. P. (1969) Evidence for the existence of a mobile extrusive peridotitic magma from the Komati Formation of the Overwacht Group. *Geol. Soc. S. Afr., Spec. Publ., 2,* 87–112.

Von Paris P., Hedelt P., Selsis F., Schreier F., and Trautmann T. (2013) Characterization of potentially habitable planets: Retrieval of atmospheric and planetary properties from emission spectra. *Astron. Astrophys., 551,* A120.

Walker J. C. G., Hays P. B., and Kasting J. F. (1981) A negative feedback mechanism for the long-term stabilization of Earth's surface temperature. *J. Geophys. Res., 86,* 9776–9782.

Williams D. M. and Gaidos E. (2008) Detecting the glint of starlight on the oceans of distant planets. *Icarus, 195,* 927–937.

Wolf E. T. and Toon O. B. (2010) Fractal organic hazes provided an ultraviolet shield for early Earth. *Science, 328,* 1266.

Wolf E. T. and Toon O. B. (2013) Hospitable Archean climates simulated by a general circulation model. *Astrobiology, 13,* 656–673.

Wolf E. T. and Toon O. B. (2014) Controls on the Archean climate system investigated with a global climate model. *Astrobiology, 14,* 241–253.

Wolfe J. M. and Fournier G. P. (2018) Horizontal gene transfer constrains the timing of methanogen evolution. *Nature Ecol. Evol., 2,* 897–903.

Wolff E. and Spahni R. (2007) Methane and nitrous oxide in the ice core record. *Philos. Trans. R. Soc., A, 365,* 1775–1792.

Woolf N. J., Smith P. S., Traub W. A., and Jucks K. W. (2002) The spectrum of Earthshine: A pale blue dot observed from the ground. *Astrophys. J., 574,* 430–433.

Wordsworth R. and Pierrehumbert R. (2013) Hydrogen-nitrogen greenhouse warming in Earth's early atmosphere. *Science, 339,* 64.

Wordsworth R. and Pierrehumbert R. (2014) Abiotic oxygen-dominated atmospheres on terrestrial habitable zone planets. *Astrophys. J. Lett., 785(2).*

Wullstein L. H. and Gilmour C. M. (1966) Non-enzymatic formation of nitrogen gas. *Nature, 210,* 1150–1151.

Xiong J., Fischer W. M., Inoue K., Nakahara M., and Bauer C. E. (2000) Molecular evidence for the early evolution of photosynthesis. *Science, 289,* 1724–1730.

Yan F., Fosbury R. A. E., Petr-Gotzens M. G., Zhao G., Wang W., Wang L., Liu Y., and Pallé E. (2015) High-resolution transmission spectrum of the Earth's atmosphere: Seeing Earth as an exoplanet using a lunar eclipse. *Intl. J. Astrobiol., 14,* 255–266.

Yang W., Marshak A., Várnai T., and Knyazikhin Y. (2018) EPIC spectral observations of variability in Earth's global reflectance. *Remote Sens., 10,* 254.

Young G. (2001) Paleoproterozoic Huronian basin: Product of a Wilson cycle punctuated by glaciations and a meteorite impact. *Sediment. Geol., 141,* 233–254.

Young G. M., Brunn V. V., Gold D. J. C., and Minter W. E. L. (1998) Earth's oldest reported glaciation: Physical and chemical evidence from the Archean Mozaan Group (~2.9 Ga) of South Africa. *J. Geol., 106,* 523–538.

Zachos J., Pagani M., Sloan L., Thomas E., and Billups K. (2001) Trends, rhythms, and aberrations in global climate 65 Ma to present. *Science, 292,* 686–693.

Zachos J. C., Dickens G. R., and Zeebe R. E. (2008) An early Cenozoic perspective on greenhouse warming and carbon-cycle dynamics. *Nature, 451,* 279–283.

Zahnle K. J. (1986) Photochemistry of methane and the formation of hydrocyanic acid (HCN) in the Earth's early atmosphere. *J. Geophys. Res., 91,* 2819–2834.

Zahnle K. J., Claire M. W., and Catling D. C. (2006) The loss of mass-independent fractionation in sulfur due to a Paleoproterozoic collapse of atmospheric methane. *Geobiology, 4,* 271–283.

Zbinden E. A., Holland H. D., Feakes C. R., and Dobos S. K. (1988) The sturgeon falls paleosol and the composition of the atmosphere 1.1 Ga BP. *Precambrian Res., 42,* 141–163.

Zerkle A. L., Claire M. W., Domagal-Goldman S. D., Farquhar J., and Poulton S. W. (2012) A bistable organic-rich atmosphere on the Neoarchaean Earth. *Nature Geosci., 5,* 359–363.

Zhao M., Reinhard C. T., and Planavsky N. (2018) Terrestrial methane fluxes and Proterozoic climate. *Geology, 46,* 139–142.

Zugger M. E., Kasting J. F., Williams D. M., Kane T. J., and Philbrick C. R. (2010) Light scattering from exoplanet oceans and atmospheres. *Astrophys. J., 723,* 1168–1179.

Zugger M. E., Kasting J. F., Williams D. M., Kane T. J., and Philbrick C R. (2011) Searching for water Earths in the near-infrared. *Astrophys. J., 739,* 12.

Part 4:

Synthesis

Del Genio A. D., Brain D., Noack L., and Schaefer L. (2020) The inner solar system's habitability through time. In *Planetary Astrobiology* (V. Meadows et al., eds.), pp. 419–447. Univ. of Arizona, Tucson, DOI: 10.2458/azu_uapress_9780816540068-ch016.

The Inner Solar System's Habitability Through Time

Anthony D. Del Genio
NASA Goddard Institute for Space Studies

David Brain
University of Colorado

Lena Noack
Freie Universität Berlin

Laura Schaefer
Stanford University

Earth, Mars, and Venus, irradiated by an evolving Sun, have had fascinating but diverging histories of habitability. Although only Earth's surface is considered to be habitable today, all three planets might have simultaneously been habitable early in their histories to microbial life. We consider how physical processes that have operated similarly or differently on these planets determined the creation and evolution of their atmospheres and surfaces over time. These include the geophysical and geochemical processes that determined the style of their interior dynamics and the presence or absence of a magnetic field, the surface-atmosphere exchange processes that acted as a source or sink for atmospheric mass and composition, the Sun-planet interactions that controlled escape of gases to space, and the atmospheric processes that interacted with these to determine climate and habitability. The divergent evolutions of the three planets provide an invaluable context for thinking about the search for life outside the solar system.

1. INTRODUCTION

One of the greatest challenges in thinking about life elsewhere is that we have only one known example of a planet with life. We therefore use the criteria for life as we know it on Earth as a guide to thinking about other planets: liquid water as a solvent for biochemical reactions, an energy source such as sunlight, chemical elements present in all life forms (C, H, N, O, P, S), and an environment suitable for life to exist and reproduce (*Hoehler,* 2007; *Cockell et al.,* 2016; see the chapter by Hoehler et al. in this volume).

Earth may or may not be the only planet in the solar system that supports life now. We do know, however, that the conditions for life have continually evolved since the planets were formed. Life has existed for much of Earth's history (*Mojzsis et al.,* 1996; *Bell et al.,* 2015; *Dodd et al.,* 2017; *Tashiro et al.,* 2017; see the chapter by Baross et al. in this volume), yet the physical and chemical conditions and types of life have varied over time. Mars and Venus, currently inhospitable for surface life, have their own fascinating histories that offer clues to possible habitability in earlier times.

A broader perspective on planetary habitability is obtained by considering these three neighboring planets together, at different points in their evolutions, giving us in effect five to ten different planets to study rather than just one. This chapter focuses on external influences on all three planets over time, and how processes evolved similarly or differently on each planet due to their size and/or interactions with the Sun to cause habitability to vary from ancient to modern times. As such, this chapter lies at the intersection of Earth science, planetary science, heliophysics, and astrophysics. It builds on detailed descriptions of each planet's habitability elsewhere in this book (see the chapters in this volume by Zahnle et al., Amador and Ehlmann, Davila et al., and Arney et al.). It also links to discussions of exoplanet properties (see the chapter in this volume by Kopparapu et al.) in both directions: solar system rocky planets through time as relatively well-explored prototypes of habitable/uninhabitable planets that inform thinking about poorly constrained exoplanets, and the larger population and diversity of exoplanets as a context for common vs. atypical aspects of our solar system. *Baines et al.* (2013) and *Lammer et al.* (2018) provide complementary reviews of the formation and early evolution of Earth, Mars, and Venus.

The apparent link between solar composition and the building blocks of the solar system as observed in carbonaceous chondrites [apart from volatile elements, specifically N, C, and O (*Ringwood, 1979*)] can also help to predict possible compositions (and hence interior structures) of rocky exoplanets, given observations of the stellar spectrum. The elemental abundances (at least of major, non-volatile elements such as Fe, Mg, and Si) should resemble those of the accretion disk and therefore of planets around other stars (*Bond et al., 2010*). This theory is being tested with solar system rocky planets (*Dorn et al., 2015; Unterborn and Panero, 2017*). The mass-radius ratio of Mars, Earth, and Venus can be explained based on solar abundances — derived from chondritic meteorites and the solar spectrum — within an error interval, whereas Mercury is an exception, having had a more complicated formation history than assumed for the other inner planets (see the chapter by Raymond et al. in this volume). Some of the most abundant elements in the solar spectrum (H, O, C) condense further out beyond the snowlines of the various icy compounds they form (pure ices, hydrates, clathrates). Planetesimals from these regions of the planetary disk contribute to the formation of icy moons and the enrichment of volatiles and heavy elements in gas giant atmospheres (*Johnson et al., 2012*). The fraction of these abundant elements on the terrestrial planets depends on the extent to which they occur in minerals with high melting points as opposed to volatile ices. For this reason, O is the most abundant element in Earth's crust and the second most abundant element (after Fe) on the planet (*McDonough and Sun, 1995*).

Exoplanets on close-in orbits, for which the mass and radius are known with a sufficiently high accuracy, seem to confirm that a relation exists between element abundances in stellar spectra and planet building blocks, as observed via the densities of the planets CoRoT-7b, Kepler-10b, and Kepler-93b (*Santos et al., 2015*). Some degeneracy in the prediction of exoplanet composition cannot be avoided, though, since the migration of planets during their formation as well as the accretion history matter. This can be seen for Mercury, for which iron content is in excess of what would be predicted from the solar composition, plausibly caused by preferential removal of silicate material during a giant impact (*Benz et al., 2008*). However, additional processes near the inner edge of the protoplanetary disk may further act to fractionate silicates and metals in close-in planets. This may occur by preferential loss of lower-density planetesimals (*Weidenschilling, 1978*) or photophoresis (*Wurm et al., 2013*), or by later fractionation through atmospheric escape from lithophile-rich magma ocean atmospheres (*Fegley et al., 2016*), i.e., atmospheres formed at high temperature during the planet accretion stage when giant impacts melt a significant portion of the silicates and metals form most of the planet (*Elkins-Tanton, 2012*).

The planets of the inner solar system probably formed from a relatively, although not completely, homogeneous protoplanetary disk of gas and dust, as indicated by isotopes of various refractory elements (see, e.g., *Boss, 2004; Larson*

et al., 2011; Pringle et al., 2013). Thus, many of the differences among them today are likely to be explained by location-dependent differences in their subsequent evolution. *Baines et al.* (2013) (their Table 1) extensively document abundance and isotopic ratio differences among the terrestrial planets and the clues these provide about similarities and differences in their evolutions. We discuss several of these later in this chapter. Two major differences among the planets specifically relevant to habitability are the small size of Mars relative to Earth and Venus, which has implications for its present climate, and the timing and magnitude of water delivery to each planet (summarized by *O'Brien et al., 2018*; see also the chapters by Raymond et al. and Meech et al. in this volume).

The other major influence on the habitability of Earth, Venus, and Mars has been the evolution of the Sun. Solar luminosity has increased by ~30% since it first reached the main sequence at ~4.6 Ga, due to hydrogen fusion, core contraction, and the resulting increase in core temperature. Accompanying this is a slight shift of the Sun's spectrum toward shorter wavelengths (Fig. 4 of *Claire et al., 2012*). Claire et al. present parameterizations of solar flux evolution for an average G-type star, summarized in terms of the flux incident at Earth, Mars, and Venus in Fig. 1. The short-wavelength solar flux (X-ray and UV) for such a star decreases over time from a very active "young Sun" to its more quiescent behavior now. There is, however, considerable uncertainty in the Sun's early history. *Tu et al.* (2015)

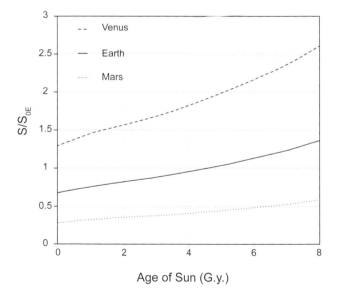

Fig. 1. Temporal evolution of the broadband solar flux incident on Earth, Venus, and Mars (S) normalized by that incident on Earth today (S_{0E}), based on results of Claire et al. (2012). Values of S/S_{0E} early in solar system history, when Mars was probably habitable and Venus might have been, are the basis for traditional estimates of the width of the habitable zone (see the chapter by Kopparapu et al. in this volume).

show that depending on the Sun's initial rotation rate, a large range of X-ray luminosities in its first billion years is possible. The uncertain evolution of the Sun has implications for photochemistry and escape processes, as we discuss later. It is also a reminder that exoplanets we observe now are snapshots in their own evolving but uncertain climate history and that the lifetime of any planet within the habitable zone is finite (*Rushby et al.,* 2013).

The net result of the initial state of each planet, the brightening of the Sun in the visible and infrared, and the decrease in X-ray and UV activity has been that the composition, climate, and habitability of Earth, Mars, and Venus have diverged over geologic time from early in solar system history when all three may have been habitable to the present epoch, in which Venus is clearly uninhabitable and Mars is either marginally habitable or uninhabitable (Fig. 2). It is sometimes assumed in exoplanet research that atmospheric thickness and composition are a function primarily of planet mass (e.g., *Kopparapu et al.,* 2014). While this may be true in a statistical sense over the broad ranges of size and mass that differentiate the largest planets that retained their primordial thick H_2-He envelope and the smallest planets that lost any primordial atmosphere, it is clearly not the rule for small rocky planets with modest size differences. Consider, e.g., Venus, which is almost as large as Earth but with a surface pressure 92× that of Earth, or Titan, a body smaller than Mars but with a 1.5-bar atmosphere compared to Mars' 6 mbar.

These seeming inconsistencies point to more complex histories influenced by a variety of processes: (1) exogenous sources that deliver different amounts of volatiles to planets as a function of location within the solar system or stochastically, determined by dynamical interactions among planets and planetesimals as the solar system evolved; (2) outgassing of volatiles from planet interiors and sequestration of other volatiles in the interiors as geodynamic and geochemical processes evolve toward surface-atmosphere

chemical equilibrium; and (3) atmospheric loss processes, depending on the type and age of the planet's host star, its distance from that star, the presence or absence of a planetary magnetic field, and dynamic/thermodynamic/chemical processes within the atmosphere. Thus the surface habitability of any planet depends on the combined evolution of its atmosphere, surface, and interior, as well as that of its star (Fig. 2). It is a great challenge to anticipate the histories of exoplanets given the meager observational constraints. Thus the more well-observed terrestrial planets of the solar system are an invaluable source of information about how these processes have played out over time for a fairly narrow range of distances in a single planetary system. We assume that the divergent evolutions of the three planets are mostly deterministic functions of their size and distance from the Sun, but stochastic sources and sinks early in their histories are likely, and bistability and bifurcations in evolution from similar initial states are possible as well (*Lenardic et al.,* 2016a).

This chapter compares processes operating in the interiors of the three terrestrial planets (section 2), at their surface-atmosphere interfaces (section 3), and at their atmosphere-space interfaces (section 4) to produce the conditions that have determined their climates and habitability through geologic time (section 5). The chapter closes with a discussion of the perspectives these planets provide for the search for life outside our solar system (section 6).

2. PLANETARY INTERIOR PROCESSES

Planetary interior processes drive the evolution of the atmospheric composition, surface tectonics, and magnetic dynamo generation, and so play multiple key roles in planetary habitability. The interior structure and composition of the planet are set during the earliest evolution of a rocky planet, including the formation of the core that is needed for

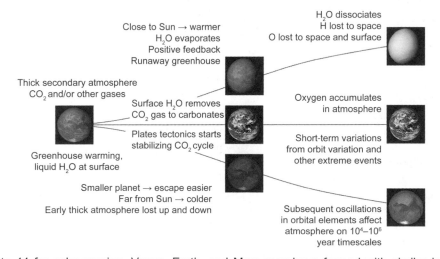

Fig. 2. See Plate 44 for color version. Venus, Earth, and Mars may have formed with similar inventories of chemical elements and could conceivably have had similar surface conditions early in solar system history. Differences in their sizes and/or distances from the Sun, as well as the emergence of life on Earth (see the chapter by Baross et al. in this volume), are the most likely causes of their subsequent divergent evolutions and explain their disparate present climates.

a long-term dynamo generation and the mantle chemistry. The redox state of the mantle influences the behavior of volatiles when rocks start to melt, and so the composition of gases released from the interior via volcanism that will ultimately contribute to a planet's atmosphere and climate. The interior structure and composition also affect the planet's thermal structure and rate of heat loss, which in turn factor into the tectonic mode of the planet and whether or not plate tectonics, which may help buffer climate, can occur, and how long a planet can stay geologically active. Below we describe these key characteristics of the planetary interior that are relevant to habitability.

2.1. Core Formation and Fractionation of Elements

Information about exoplanet interiors beyond rough density estimates for planets for which both minimum mass and radius are known is difficult to obtain, except in rare instances in which dust tails from disintegrating planets can be observed (e.g., *van Lieshout et al.,* 2014, 2016; *Bodman et al.,* 2018). Thus, knowledge about the interiors of solar system rocky planets is crucial for constraining ideas about exoplanet formation and climate evolution. Each of the terrestrial planets of the solar system has a central, metallic core, surrounded by a dominantly silicate mantle and crust. Earth's metallic core is the source of its magnetic field (section 2.2), which shields the planet from harmful high-energy particles and may protect the atmosphere from certain escape processes (section 4). Differentiation during planet formation, in which denser elements such as Fe sink to the core and lighter elements form the mantle and crust, determines the temperature, size, and composition of the core. This process also influences the oxidation state of the mantle (section 2.3) as well as the abundances of both minor (e.g., Ni, Co, C, S, etc.) and major (e.g., O, Si, Fe) elements in the primitive silicate mantle. Later silicate differentiation further fractionates elements between the mantle and crust. The size and density of Earth's inner and outer core are well known from seismic measurements and can be used to infer information about the core composition in conjunction with knowledge of siderophile-element abundances in the silicate mantle. In contrast, the size, composition, and phase of the cores of Venus and Mars must be inferred through indirect observations, meteoritic abundances (for Mars), and models.

Metal and sulfide-loving (siderophile and chalcophile) trace elements will preferentially partition into the core-forming phase and be removed from the mantle (e.g., *Ringwood,* 1959; *Li and Agee,* 1996). This leads to an overall depletion of, in particular, highly siderophile elements in the observable silicate planet. Other elements may partition into the core to a greater or lesser degree depending on the conditions during separation, e.g., temperature, pressure, oxygen fugacity (a measure of the amount of free or uncombined oxygen available for chemical reactions), etc. Late-stage accretion of terrestrial planets likely occurred through impacts of objects that had already undergone differentiation, suggesting that material was added to the core of the

growing planet in discrete intervals. Partial reequilibration of the cores of the accreting objects with the mantle occurs at progressively higher pressures and temperatures as the planet grows (e.g., *Rubie et al.,* 2015). Measurements of siderophile-element abundances in the silicate mantle can be used to constrain the conditions of core formation for Earth (e.g., *Rubie et al.,* 2015; *Fischer et al.,* 2015) and for Mars based on meteoritic data (e.g., *Righter et al.,* 2015).

The cores of Earth and Mars are thought to contain significant amounts of lighter elements. For Earth, a density deficit in the outer core compared to pure Fe liquid indicates that this light element(s) must make up ~10 wt% of the outer core (*Birch,* 1952), therefore indicating an abundant (in the cosmochemical sense) element must be present. Many models suggest combinations of Si, O, S, C, and H in Earth's core (*Poirier,* 1994; *Hillgren et al.,* 2000). Addition of Si and O, in particular, to Earth's core may influence the oxidation state of the mantle. Recent models suggest up to 8.5 wt% Si and 1.6 wt% O in Earth's core (*Fischer et al.,* 2015). The inner core, however, is thought to contain only Fe, Ni alloy, causing an enrichment of the light element in the liquid outer core that may help power the geodynamo (section 2.5). Mars' core is thought to have very large amounts of S (10–16 wt%) (*Gaillard et al.,* 2013; *Lodders and Fegley,* 1997; *Sanloup et al.,* 1999; *Wänke and Dreibus,* 1988), although a recent study of martian meteorites suggests lower values <5–10 wt% (*Wang and Becker,* 2017). However, the core mass fraction (0.21–0.24) and core radius (1673–1900 km) of Mars are relatively poorly known (*Rivoldini et al.,* 2011). A high proportion of S likely leads to the crystallization of a sulfide such as Fe_3S rather than pure Fe when the inner core nucleates. However, it remains unclear at this time if Mars has a solid inner core at the present day (*Helffrich,* 2017). NASA's Interior Exploration using Seismic Investigations, Geodesy and Heat Transport (InSight) mission (*Banerdt and Russell,* 2017) is designed to determine these properties and more.

The core of Venus is even more poorly known. Internal structure models for Venus rely on measurements of the moment of inertia and the Love number (a set of dimensionless numbers that characterize the rigidity of a body and thus how easily its shape can change), which have proven more difficult to make than for Mars. The moment of inertia of Venus is poorly known due to the slow spin rate of the planet, the effect of drag of the dense atmosphere on orbiting spacecraft, and the short lifetime of surface landers. The potential Love number k_2 was derived from Magellan and Pioneer Venus spacecraft data (*Konopoliv and Yoder,* 1996). It indicates a partially or fully liquid core, but reanalysis of the data using a viscoelastic rheology (rather than fully elastic) suggests that a fully solid core cannot be ruled out (*Dumoulin et al.,* 2017). The mass and radius of Venus' core cannot be fully determined from the Love number alone, and therefore many models of Venus' internal structure use a scaled version of Earth (e.g., *Mocquet et al.,* 2011; *Aitta,* 2012). Using a range of compositional models derived from planet formation models and different internal temperature profiles, *Dumoulin et al.* (2017) find a range of core radii of 2941–3425 km (0.48–0.57 R_{Venus}), with core-mantle boundary pressures of 103–127 GPa, which are all consistent with the measured total mass and radius of Venus. The assumption of a scaled version of the density of Earth's core, implying a similar abundance of light elements, must be tested by future models and measurements. Inferences about Venus' internal thermal history and core cooling and the possible existence of a magnetodynamo before the present day therefore remain speculative until additional constraints on the internal structure of Venus can be obtained.

2.2. Magnetic Field Development

The magnetic field may be a contributing factor to the habitability of Earth, as it shields life at the surface from harmful radiation, and may have helped to limit atmosphere erosion in the early solar system, when the Sun emitted strong EUV radiation leading to thermal and non-thermal escape processes (*Tian et al.,* 2008) (but see section 4.4 for a more thorough discussion and some caveats). Earth is not the only body in the solar system with an active magnetic dynamo leading to a magnetosphere. Mercury, the smallest planet in the solar system, has a weak magnetic field of about 200 nT (*Anderson et al.,* 2011) (compared to more than 50,000 nT field strength for present-day Earth). In contrast, neither Venus nor Mars have an active dynamo creating a measurable magnetic field. Mars' magnetic dynamo stopped about 4.1 G.y. ago (*Morschhauser et al.,* 2018), whereas Venus shows no evidence of past dynamo activity.

A planetary magnetic dynamo needs a fluid with high electrical conductivity, convective motion, and planetary rotation [although even slow rotation like that of Venus can sustain a magnetic field if the other conditions are satisfied (*López-Morales et al.,* 2012)]. The conductive fluid could be, e.g., liquid iron (as for rocky planets in the solar system), metallic hydrogen (as is the case for Jupiter), or melt in a magma ocean. Rotation leads to the Coriolis effect, which organizes flow into rolls aligned in the direction of the rotation axis. For terrestrial planets with a similar interior structure as Earth — i.e., divided into metal core, solid rock mantle, and surface crust — dynamo action is driven by convection in the liquid part of the core. Different scenarios are possible, which can explain the existence or absence of a dynamo in the rocky planets of the solar system.

A thermal dynamo refers to convection in the core driven by a strong, super-adiabatic heat flux from core into mantle. Such a dynamo would be expected in the early evolution of planetary bodies (e.g., Mars and the Moon), where after core formation and magma ocean solidification the temperatures are much higher in the metal core than in the silicate mantle, driving a strong heat flux at the core-mantle boundary. In contrast, a compositional (sometimes called chemical) dynamo refers to convection driven by density differences from heterogenous core composition — either due to freezing of the inner core (e.g., Earth), which leads to an enrichment of lighter elements above the inner core boundary, or by snow-

like iron precipitate starting at the core-mantle boundary, as suggested for Ganymede and Mercury (*Hauck et al.,* 2006; *Chen et al.,* 2008; *Rückriemen et al.,* 2018).

The absence of a magnetic dynamo on present-day Venus may best be explained by a core without an inner solid part and/or by insufficient heat flux at the core-mantle boundary, since Venus' mantle may be much hotter than Earth's mantle due to the absence of a cooling mechanism such as plate tectonics in the more recent past. A fully solid core would not allow for a magnetic field today. In that case, early Venus would likely have had higher interior temperatures, such that a dynamo could have been possible but would also have ended relatively early (e.g., *Stevenson et al.,* 1983). A present-day solid core on Venus would imply significantly lower light-element concentration in the metal than for Earth, since lighter elements (e.g., sulfur) strongly decrease the metal melting temperature. This is difficult to reconcile with core-formation models. It would be difficult to explain the lack of a present-day dynamo on Venus for a partially molten core, except by a recent change in tectonic style (e.g., cessation of plate tectonics) or episodic lid tectonics to suppress core heat flux (e.g., *Nimmo,* 2002; *Armann and Tackley,* 2012) and halt inner core solidification. A fully molten core is consistent with the lack of a present-day dynamo, but does not prohibit an early and potentially long-lived dynamo driven by thermal convection (*Stevenson et al.,* 1983; *Nimmo,* 2002; *Driscoll and Bercovici,* 2014). The nucleation of a solid inner core in the future might allow for another period of dynamo activity on Venus. The presence of magnetite below its Curie temperature in the highlands of Venus (*Starukhina and Kreslavsky,* 2002) could provide a record of a previous geodynamo, but detecting such remnant magnetism would require a surface lander.

2.3. Redox State of Mantle

The redox state of a terrestrial planet's mantle, which is linked to the volatile content of mantle rocks, can be inferred from trace-element compositions of melt products that reached the surface (e.g., *Smythe and Brenan,* 2016). The redox state at least of Earth's early Archean uppermost mantle was similar to the present-day state (*Canil,* 1997; *Delano,* 2001; *Li and Lee,* 2004). However, the mid-ocean ridge basalt (MORB) samples dating back to 3 Ga show a different picture of reduced mantle conditions with decreasing oxygen fugacity with depth, which would have led to reduced outgassing from melt originating from these depths (*Aulbach and Stagno,* 2016). A later oxidation of the mantle may be linked to the initiation of plate tectonics (*Mikhail and Sverjensky,* 2014). The change of reduced Hadean crust to oxidized conditions may be linked to the decline in chondritic addition of reduced species by comets (*Yang et al.,* 2014a). On the other hand, the early Archean (and possibly Hadean) already having an oxidized upper mantle (*Delano,* 2001) could reflect volatile recycling processes, even though these would be expected to have been less efficient in oxidizing Earth's mantle compared to subduction of hydrated crust. The redox state of the upper mantle may therefore not reflect the deep mantle's oxygen fugacity.

The initial redox state depends on the building blocks of Earth and on the volatiles that were able to be contained in the mantle when the last magma ocean, produced by the large impact that formed the Moon, solidified (*Hier-Majumder and Hirschmann,* 2017; *Mukhopadhyay,* 2012). During the evolution of Earth, the redox state can change, for example, due to melting processes (*Parkison and Arculus,* 1999) or chemical mineral changes (*Frost et al.,* 2004; *Galimov,* 2005), as well as subduction of volatiles into the mantle (*Mikhail and Sverjensky,* 2014).

The redox state of the mantle influences the formation of carbonates in the melt and therefore has a crucial influence on the outgassing products at the surface (*Mikhail and Sverjensky,* 2014). Figure 3 shows the amount of CO_2 that can be dissolved in basaltic melt assuming equilibrium with graphite and an oxygen fugacity at the iron-wüstite (IW) buffer (*Holloway et al.,* 1992; *Grott et al.,* 2011). The melt fraction directly reflects the temperature. Melting temperatures increase strongly with increasing pressure. A melt fraction of 0 represents here the solidus melting temperature, where the first minerals of the rock assemblage begin to melt; a melt fraction of 100% represents the liquidus melting temperature, where all minerals would be molten. Increasing pressure leads to a strong reduction in CO_2 concentration in the melt, whereas an increase in temperature leads to higher concentrations. The oxygen fugacity is directly reflected in the amount of outgassing, as a redox state increased by one (e.g., IW + 1) leads to outgassing that is stronger by an order of magnitude. Changing oxygen fugacity in Earth's mantle at the depths where uprising melt is produced therefore strongly impacts the outgassing products. The redox state of the melt also influences the composition of the gas that is released into

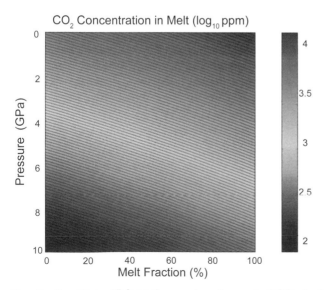

Fig. 3. See Plate 45 for color version. Amount of CO_2 that can be dissolved in graphite-saturated, tholeiitic melt for an oxygen fugacity at the iron-wustite buffer, based on *Holloway et al.* (1992) and *Grott et al.* (2011).

the atmosphere — an oxidized mantle would outgas species such as N_2, CO_2, H_2O, and SO_2, whereas from a reduced mantle NH_3, CH_4, CO, H_2, and H_2S would be degassed.

The redox state of Mars' mantle has been inferred from geochemical analysis of SNC meteorites and has been derived to be moderately reduced (e.g., *Wadhwa,* 2001; *Herd et al,* 2002; *Grott et al.,* 2011). The oxygen fugacity of the Moon was measured in rock samples of the Apollo 12, 14, and 15 missions (*Sato et al.,* 1973). The lunar mantle seems to be more reduced in comparison to Earth, with an oxygen fugacity below the iron-wüstite buffer at around IW–1. Even more reduced conditions are expected for Mercury's mantle based on high S and low FeO abundances, with an oxygen fugacity several orders of magnitude below the iron-wüstite buffer (*McCubbin et al.,* 2012).

2.4. Heat Transport from Interior to Surface

After the accretion and magma ocean phase (section 3.1), enormous amounts of heat are stored in the interior of planets that are slowly, over geological timescales, transported to the surface, where the overlying atmosphere can radiate it to space. On modern Earth, the surface heat flow is estimated to be ~45–49 TW (e.g., *Davies and Davies,* 2010). This heat flux is attributed to plate tectonics, one of the processes that regulates atmospheric CO_2 and thus habitability (section 3.3), helping to efficiently cool the mantle. Heat can be transported through a medium via thermal conduction, convection of material, or radiation. Inside rocky planets, only the first two mechanisms transport heat from the core to the surface. Convective currents can occur both in the liquid core and in the solid mantle, but on very different time scales. Liquid iron has a small viscosity comparable to water on the order of few millipascals per second, and convection velocities are on the order of millimeters per second. In the solid mantle, where the viscosity of rocks is ~20 or more orders of magnitude higher, convective velocities are on the order of tens of centimeters per year. Convection in the mantle only occurs if sufficiently large variations in the density occur (with hot, less-dense material accumulating at the core-mantle boundary and cold, denser material at the top of the mantle) to trigger an instability of the system. The efficiency of convection is described by the non-dimensional Rayleigh number Ra, which expresses the ratio between driving and resisting factors for convection. Estimating the Rayleigh number for different planets can give a first idea of the efficiency of convective heat transport (*Breuer,* 2009). In the inner solar system, the highest mantle Ra (~10^8) is for Earth, closely followed by Venus (~10^7–10^8). Mars's Ra is 1–2 orders of magnitude smaller [mostly due to its smaller mantle thickness d ~ 1600 km (*Rivoldini et al.,* 2011)], since Ra α d³, and hence we expect less vigorous convection. Mercury, with a thin ~400-km mantle (*Dumberry and Rivoldini,* 2015), might not experience active mantle convection anymore. The Galilean moon Io is comparable to Earth's Moon in size but shows strong volcanic resurfacing, implying strong convection in the interior. It should be noted, though, that Io is heated constantly by tidal dissipation due to its close-in orbit around Jupiter together with tidal forces acting on Io from its neighboring moons. The surface heat flux is about 3× higher than on present-day Earth (*Veeder et al.,* 1994), and material is efficiently extracted to the surface — which may have been similar on early Earth (*Moore and Webb,* 2013).

Ra only indicates how efficient material (and hence heat) transport is over a fully convecting layer. For Archean Earth, however, mantle temperatures were higher than at present by ~150–200°C (*van Hunen and Moyen,* 2012). Higher temperatures decrease the stiffness of the mantle, and might therefore trigger stronger convection.

The other bodies of the inner solar system do not have a crust separated into smaller plates; instead, the mantle is covered by an immobile lithosphere, which is referred to as a stagnant lid. For Venus, this may have been different during its earlier evolution — a time from which no evidence persists at the surface today as far as we know (section 3). If plate tectonics occurs, then the surface participates in the convection cycle and leads to efficient cooling of the mantle, which may be important to maintain Earth's magnetic dynamo (section 2.2).

While plate tectonics may have been important for the development of complex life on Earth, stagnant lid planets seem to be more abundant (given the few examples we have in the solar system). It is therefore worthwhile to investigate the potential of stagnant lid planets with respect to surface habitability (*Noack et al.,* 2017; *Foley and Smye,* 2018; *Tosi et al.,* 2017), as they may also be abundant elsewhere in the galaxy. While planets lacking subduction and efficient resurfacing may be limited in their potential to regulate climate over long, geological timescales, planets lacking plate tectonics may still be habitable for timespans long enough for life to begin. This may have implications for the frequency with which exoplanets near the outer edge of the habitable zone can maintain habitable conditions after their early magma ocean outgassing phase.

2.5. Plate Tectonics and Emergence of Continents

As a result of plate tectonics, most of Earth's ancient surface rocks have been eroded and subducted over time. The rock record thus starts at only ~4 Ga, leaving no direct evidence of conditions during most of the Hadean Eon, and surviving samples are found only in a few places and make up only a tiny portion of Earth's present-day surface. Oceanic crust is continuously recycled back into the mantle on timescales of tens to hundreds of millions of years and is on average very young. Continental, granitic crust has a lighter composition than oceanic crust, and is more resistant against erosion and subduction. It contains the oldest outcropped rocks (for example in Greenland, Canada, western Australia, and southern Africa) (*Papineau,* 2010), and floats on Earth's lithospheric mantle as part of the Wilson cycle, where continental crust regularly accumulates to form supercontinents, only to break up into smaller continents spread over the entire planet's surface on geological timescales. Continental crust is formed by remelting of previously extracted crust, for example, in subduction zone settings, where water released from a subducting slab hydrates the mantle wedge and — by reducing the melting temperature of the hydrated rocks compared to dry rocks — triggers rising melt and remelting of the base of the crust. Different ideas have been advanced for the first felsic crust (i.e., crust that has been remelted, is less dense than primitive basaltic crust, and therefore makes up the continental crust). This crust formed during the Hadean Eon under hydrated conditions, as shown by zircon minerals dating back 4.4 G.y., which were then exposed later in Archean rocks (*Harrison,* 2009; *Arndt and Nisbet,* 2012; *Trail et al.,* 2013), as well as Archean felsic crust containing tonalite-trondhjemite-granodiorite (TTG) inclusions. These ideas include early plate tectonics, intraplate melting, or melting at plate-like boundaries formed by uprising plumes without accompanying subduction processes (e.g., *Marschall et al.,* 2010; *Moyen and Martin,* 2012; *Rozel et al.,* 2017; *Harrison,* 2009).

Different estimates exist for the evolution of continental crustal surface coverage over time (*Hawkesworth et al.,* 2010) — ranging from an almost complete outcropping of present-day volumes of continental crust during the Archaean (*Fyfe,* 1978; *Reymer and Schubert,* 1984), to slow, increasing continental growth starting mostly at the end of the Archaean at 2.5 Ga (*Taylor and McLennan,* 1995; *Breuer and Spohn,* 1995), or to intermediate models with more or less steady continental growth with time (e.g., *Belousova et al.,* 2010). The rise and growth of continents, if assumed to have taken place at the end of the Archaean, has been linked to plate tectonics (*Höning et al.,* 2014) and the Great Oxygenation Event (GOE) (e.g., *Gaillard et al.,* 2011), which in turn has been linked to the increase of biovolume of single organisms (*Payne et al.,* 2009). *Rosing et al.* (2006) on the other hand suggested that energy harvested by life through photosynthesis may have played a crucial role in Earth's energy cycle and could have influenced the rise of continents. It is not clear, however, how much continental crust already existed before the GOE, and when plate tectonics initiated. Estimates go from the end of the Proterozoic Eon (~1 Ga) to as far back as the Hadean Eon (>4 Ga); the majority of studies suggest the geological transition of Earth's rock mineralogy observed throughout the mid-Archean to be evidence for plate tectonics initiation (see *Korenaga,* 2013, for an in-depth review on this topic).

Earth is the only rocky body in the solar system for which evidence for currently active plate tectonics has been identified. Icy bodies such as Europa may experience a similar tectonic feature, allowing for resurfacing of the icy crust (*Kattenhorn and Prockter,* 2014), similarly to how buoyant crust is subducted on a lava lake on Earth. Venus, which today probably has surface temperatures too high to allow for active plate tectonics (*Landuyt and Bercovici,* 2009), may have experienced plate tectonics in the past, or may even be trapped in an episodic change between resurfacing and stagnant phases, determined by the interior and surface temperature evolution (*Noack et al., 2012; Gillmann and Tackley,* 2014). Any signs of past plate tectonics — if it ever existed — have been erased by recent recycling of the crust in the last 300–1100 m.y. (*Hansen and Lopez,* 2010). It is also not clear if Venus possesses any continental crust with a composition notably different from the lowland crust, which would be comparable to the continental-oceanic crust dichotomy that we see on Earth. The oldest crust of Venus' past, the tessera terrains, could well be felsic in composition (*Müller et al.,* 2008), but continental crust may be formed without plate tectonics, and maybe even without liquid water at the surface, and therefore even detection of felsic, continental crust on Venus does not tell us anything about its past potential habitability. Models for the early history of Venus range from Earth-like scenarios to a hellish planet from day one, with a dense runaway greenhouse atmosphere from the formation stage of the planet (section 5).

The surface of Mars is much older in contrast to Venus and Earth, and unveils more of its early history (*Frey,* 2006). If plate tectonics or a similar crustal recycling mechanism was ever active on Mars, it must have been in the first few hundreds of millions of years [within 100 m.y. following *Debaille et al.* (2009)] after the magma ocean solidification stage (~20 m.y.) (*Bovier et al.,* 2018), since the apparently younger lowlands in the northern hemisphere are still at least 4 G.y. old (*Frey,* 2006). Also, it has been suggested that part of the crust on Mars, for example in the Gale crater (*Sautter et al.,* 2015), is of continental-like composition, even though it is only 3.61 G.y. old, when plate tectonics on Mars definitely was not active. This suggests that also on Mars, continental-like crust may not have been produced by plate tectonics, but by other processes such as intrusive melt pockets leading to melting of existing crust as has been suggested for early Earth (see discussion above).

The examples of Venus and Mars suggest that the emergence of felsic, continent-like regions is not necessarily linked to plate tectonics. Continuous subduction of hydrated plates as on Earth may yield an increased production of continental crust, but this effect may be balanced by the expected increased erosion of continental crust, leading possibly to a steady-state continental crust amount exposed at Earth's surface.

3. SURFACE-ATMOSPHERE INTERACTIONS

3.1. Magma Ocean-Atmosphere Exchange: Implications for Early Water

N-body simulations suggest that giant impacts were very common during the formation of predominantly rocky planets (*Agnor et al.,* 1999; *Quintana et al.,* 2016; see also the chapter by Zahnle et al. in this volume). Giant impacts are energetic enough to melt a significant fraction of the parent body, creating a magma ocean (*Tonks and Melosh,* 1993; *Canup,* 2008; *Nakajima and Stevenson,* 2015; *deVries et al.,* 2016). Common volatiles such as water, CO_2, and N_2, as well as more minor species such as CH_4, H_2, and NH_3, are soluble in silicate melts (e.g., *Holloway and Blank,* 1994; *Papale,* 1997; *Mysen et al.,* 2008; *Hirschmann et al.,* 2012; *Ardia et al.,* 2013). Therefore exchange with the atmosphere throughout the solidification process determines both the amount of volatiles that become trapped in the solidified interior, and the composition and abundance of volatiles in the atmosphere after the magma ocean solidifies (e.g., *Abe and Matsui,* 1985, 1988; *Elkins-Tanton,* 2008; *Salvador et al.,* 2017). Early oceans may result from collapse of a steam-dominated magma ocean atmosphere.

Early atmospheres for the terrestrial planets have long been thought to be dominated by water vapor and CO_2. H_2O is more soluble in silicate melts than CO_2, so the abundance of water vapor in the atmosphere varies more as the magma ocean cools. In contrast, carbon partitions more readily into the metallic phase and the atmosphere. Some models suggest that for a magma ocean in contact with a metallic liquid, carbon could be pumped out of the atmosphere into the core (*Hirschmann,* 2012). The mantle oxidation state of Mars is more reduced than that of Earth (*Wadhwa,* 2008), which may favor a more reduced early atmosphere with substantial H_2 (*Ramirez et al.,* 2014; *Batalha et al.,* 2015; *Wordsworth et al.,* 2017; *Sholes et al.,* 2017), although the implications of such an atmosphere for the magma ocean and subsequent climatic evolution are still being explored.

The timing of the last giant impact, and presumably last magma ocean stage, on Earth has been constrained by Hf-W dating of early Earth materials as well as lunar materials. These results suggest that the Moon-forming impact must have occurred after the lifetime of Hf, so no earlier than ~60 m.y. after the formation of Ca-Al-rich inclusions (CAIs), which are the earliest datable materials formed in the solar system (*Kleine et al.,* 2009). Similar studies of Hf-W and Sm-Nd systematics for whole rock analyses of martian meteorites suggests that the martian magma ocean cooled off within ~10–15 m.y. after CAIs, with crustal formation happening no more than 20 m.y. later (*Kruijer et al.,* 2017). New measurements of U-Pb ages and Hf systematics of individual zircons extracted from martian meteorites supports an early magma ocean crystallization age and first crust formation at ~20 m.y. after CAIs (*Bouvier et al.,* 2018). No direct timing constraints exist for a venusian magma ocean since no physical samples of Venus are available for laboratory study.

Indirect evidence for a venusian magma ocean comes from measurements of the D/H ratio in the atmosphere, which is highly enriched in D relative to Earth and chondrites (see the chapter by Arney et al. in this volume). This suggests that significant water escape could have occurred from the atmosphere of Venus during the magma ocean stage. Magma ocean models by *Gillmann et al.* (2009) and *Hamano et al.* (2013) suggest that a magma ocean stage on Venus would be significantly prolonged compared to magma oceans on Earth or Mars, due to its closer orbital proximity to the Sun. Photolysis and escape of H from water vapor during this prolonged runaway greenhouse magma ocean stage results in significant depletion of the water reservoir and eventual solidification of the magma ocean. Thus magma ocean models suggest that it is plausible that Venus has been in its present hot, dessicated state for almost its entire lifetime. *Hamano et al.* (2013) find that planets beyond a critical distance — close to Venus' present orbital location — cool rapidly enough for the water envelope to collapse into an early ocean. Such would be the case for early Earth and Mars. If Venus was instead somewhat outside the critical distance early in its history, then enough water might have remained after the magma ocean crystallized to form an initial liquid water ocean, which then would have been lost via a later leaky moist greenhouse stage as the Sun brightened. Evidence for surface carbonates in the older regions of Venus' surface would suggest early liquid water, but the likelihood of carbonates surviving on Venus' surface is small (section 3.2).

Crust formation on Earth and Mars was delayed toward the end of the magma ocean stage. At sufficiently high water abundance and temperature, significant rocky elements may be soluble in the water vapor atmosphere (*Fegley et al.,* 2016). Models suggest that early crust may be generated by condensation of this rocky vapor atmosphere (*Baker and Sofonio,* 2017). Other models suggest very early hydration and weathering of the earliest basaltic crust on Earth by the hot early ocean or steam atmosphere (*Abe and Matsui,* 1985; *Matsui and Abe,* 1986, 1987; *Zahnle et al.,* 1988; *Sleep et al.,* 2001). However, no record of this earliest crust on Earth exists, except perhaps in the form of ancient Jack Hills zircons (*Trail et al.,* 2011), which support the early presence of liquid water at or near the surface. The earliest crust of Mars, however, is relatively well preserved. Recent experimental work by *Cannon et al.* (2017) shows that reaction between a basaltic crust and a supercritical steam atmosphere or hot early ocean is rapid and may form thick phyllosilicate layers. Those authors simulate later crustal reworking by impacts and volcanism to show that this early clay layer matches observations of deep clay exposures and could be preserved to the present as a deep, disrupted layer in the martian crust. The crust of Venus, in contrast, appears on the whole to be much younger (*Strom et al.,* 1994; *McKinnon et al.,* 1997) and therefore is unlikely to preserve direct signatures of the magma ocean stage.

Therefore, the case for early condensed (solid or liquid) water on Earth and Mars at the end of a magma ocean stage is relatively robust in models and seems to be well supported by the available evidence. The case for early water on Venus is harder to make if a magma ocean stage occurred, which appears difficult to avoid in planet formation models. However, the desiccation of a magma ocean atmosphere depends on the early evolution of the solar flux, which is somewhat uncertain, and radiative transfer models of thick H_2O-CO_2 atmospheres do not universally find this early dessication (see, e.g., *Kasting,* 1988; *Lammer et al.,* 2018; see also the discussion in the chapter by Arney et al. in this volume). In fact, water loss can be quite sensitive to CO_2 abundance (*Wordsworth and Pierrehumbert,* 2013b). Therefore it re-

mains uncertain whether Venus could have had early water oceans. Future characterization of the atmospheres of some of the known exoplanets slightly inside the inner edge of the traditional habitable zone (*Kane et al.,* 2014) may provide some perspective on the potential of such planets to lose vs. retain any early water. We consider both scenarios for Venus in the following sections.

3.2. Surface Weathering Reactions

Reactions of the atmosphere with the surface of a planet are a major sink of atmospheric gases and can therefore strongly impact its climate. The present-day crusts of Mars and Venus are dominantly basaltic with small regions (<10%) of possible felsic material on both planets (*Ehlmann and Edwards,* 2014; *Gilmore et al.,* 2017). Present-day Earth has a bimodal crustal distribution with felsic continents of variable age and very young basaltic oceanic crust. The growth rate of the felsic continental crust remains a highly debated topic (*Armstrong,* 1981; *Belousova et al.,* 2010; *Dhuime et al.,* 2012; *Korenaga,* 2018). Some models use ages derived from different elements to estimate the crust age distribution, while others estimate the evolution of depletion of incompatible elements from the mantle. The range of results from these two types of models prevents differentiating between the conclusion that recycling of the crust was insignificant or that it has been an important part of crustal evolution over the age of Earth. Surface weathering rates depend on surface temperature (*Walker et al.,* 1981), as well as the composition of the surface; typically mafic minerals weather more quickly than felsic minerals (*Kump et al.,* 2000). Therefore surface-atmosphere reactions will vary with planet age based on the evolution of the crustal composition as well as climate.

The most important weathering reactions involve CO_2 and water (both gaseous and liquid), which can produce carbonates, as well as hydrous and oxidized minerals. The reaction of CO_2 with silicates to produce carbonate minerals creates a negative feedback due to the temperature dependence of the reaction and the warming greenhouse behavior of CO_2 in the atmosphere (*Walker et al.,* 1981; *Kump et al.,* 2000; *Sleep and Zahnle,* 2001). The formation of hydrated silicates will sequester water out of the atmosphere and ocean, which may be eventually transported to the upper mantle (see section 3.3). Hydration reactions may also produce H_2 gas as a byproduct that can potentially escape and allow progressive oxidation of the planet, but may also produce greenhouse warming (e.g., *Wordsworth and Pierrehumbert,* 2013a). Additional weathering reactions involve trace gases such as SO_2, HCl, and HF, which can produce sulfates as well as chloride and fluoride salts, and are important on both Mars and Venus, but these are expected to have less of a role in climate, so we refer the reader to the reviews by *Zolotov* (2015, 2018) for further details.

Following the magma ocean period on Earth, CO_2 is expected to have been very abundant in the atmosphere due to its low solubility in silicate melts, followed by a period of very rapid weathering (*Zahnle et al.,* 2010). However, rapid early drawdown of CO_2 on Earth is expected to pose a problem for clement surface conditions due to the faint young Sun paradox unless other greenhouse gases were also present (*Sleep and Zahnle,* 2001) (section 5). Unfortunately, no geologic evidence other than remnant zircon mineral grains remain from the Hadean, making it difficult to constrain the atmospheric composition and surface conditions on Earth for the first ~500 m.y.

On Venus, in contrast, the abundance of CO_2 in the atmosphere is consistent with the amount of CO_2 locked in sedimentary rocks on present-day Earth due to this cycle, suggesting that the majority of Venus' CO_2 inventory has been outgassed (*Lecuyer et al.,* 2000). At present, the carbonate-silicate cycle does not operate on Venus due to the high surface temperature and low water abundance. Rather, Venus may at present have a high enough surface temperature for mineral reactions to buffer the abundances of some atmospheric gases, although this model has been disputed in recent years (see reviews in *Gilmore et al.,* 2017; *Zolotov,* 2015, 2018) and there is at present no evidence for major gas sinks in Venus' lower atmosphere. However, the past history of surface-atmosphere interactions on Venus is not well studied, in the wet early Venus scenario.

Venus' present-day surface appears to be dominated by basaltic plains, but observed tessera, which are structurally deformed materials often found in plateaus, could be composed of either mafic or felsic material (e.g., *Romeo and Turcotte,* 2008). Felsic silicates are produced by melting of other silicate materials in the presence of water (*Campbell and Taylor,* 1983), therefore the possible existence of massive felsic terrains on Venus would imply that they formed while water was still relatively abundant. If early Venus retained water, even if it resided mostly in the atmosphere, massive carbonate minerals would likely have formed. Venus' surface is sufficiently young that massive carbonates are not expected to have survived resurfacing, although they could potentially be preserved in the tessera terrain if it is truly ancient. However, carbonates are expected to be unstable on the surface of Venus today due to reaction with atmospheric sulfur compounds, and therefore any carbonates would likely have decomposed and contributed CO_2 to the greenhouse warming of the planet (*Gilmore et al.,* 2017; *Zolotov,* 2015, 2018).

There is ample evidence for the presence of water on early Mars, based on the abundance of clays (*Ehlmann et al.,* 2011; *Carter et al.,* 2015), the presence of valley networks (*Carr and Clow,* 1981, *Hynek et al.,* 2010), deltaic fan deposits (*Malin and Edgett,* 2003), and conglomerates (*Williams et al.,* 2013). However, there is evidence for only limited carbonate formation on early Mars (*Edwards and Ehlmann,* 2015; *Niles et al.,* 2013), whereas massive carbonates should be expected if Mars had an active hydrologic cycle and massive CO_2 atmosphere. There is a possibility that carbonate layers may be deeply layered in the crust (*Michalski and Niles,* 2010). *Batalha et al.* (2016) suggest that the outgassing of CO_2 on Mars was unable to keep up with the rate of carbonate formation, causing major climatic swings from glaciated to clement in Mars' early history. In order to produce clement conditions, this model relies on H_2 produced by either volcanic outgassing or serpentinization of the crust to provide additional warming (*Batalha et al.,* 2015).

3.3. Deep Volatile Recycling

Fluxes of volatiles into and out of the mantle on geologic timescales play a key role in the stability of habitable conditions at the surface by regulating greenhouse gases and replenishing volatiles from the mantle, as described in section 3.2. Volatile fluxes out of the mantle are dictated by the rate of volcanism, whereas fluxes into the mantle depend strongly on the tectonic style of the planet as well as the weathering reactions discussed in the previous section that are responsible for removal of atmospheric gases. Mobile lid planets such as Earth may actively transport these trapped volatiles into the mantle through subduction (e.g., *Sleep and Zahnle,* 2001; *McGovern and Schubert,* 1989), whereas stagnant lid or episodic lid planets may have more sporadic or limited transport of volatiles into the mantle (e.g., *Morschhauser et al.,* 2011).

Earth's mantle may hold ~1–10× as much water as the surface oceans (see, e.g., Table 1 of *Bounama et al.,* 2001), although more recent work discounts larger values in favor of values from 0.5 to 2.5 ocean masses of water (*Hirschmann and Kohlstedt,* 2012; *Korenaga et al.,* 2017), equivalent to 170–870 ppm by mass. *Hirschmann and Dasgupta* (2009) constrain the H/C ratio in the mantle to be 0.99 ± 0.42, and the water abundance in the mantle to be 70–570 ppm. This yields a C abundance in Earth's mantle of 90–740 ppm. The abundance of water in Venus' mantle has been estimated to be ~50 ppm at present (*Smrekar and Sotin,* 2012). Models by *Elkins-Tanton et al.* (2007) suggest that even extensive melting of the mantle should not be able to completely deplete Venus' mantle of water or CO_2. However, there is no estimate of the amount of CO_2 in Venus' mantle, and most of it is assumed to have outgassed based on analogy with Earth's total CO_2 inventory (*Lecuyer et al.,* 2000). Martian meteorites have been used to constrain the abundance of water in the mantle to 10–100 ppm over geologic time (*McCubbin et al.,* 2016; *Weis et al.,* 2017). The carbon content is less certain and is assumed to have been altered more significantly than water by outgassing.

Figure 4 illustrates how material exchange including volatile recycling into the mantle could occur under present conditions on Earth compared to Archaean Earth. For current Earth, Fig. 4a shows material moving into a subduction zone, where water is released after serpentinites become unstable in the upper mantle (*Irifune et al.,* 1998; *Höning et al.,* 2014). The water flows upward, leading to locally lower viscosities (*Hirth and Kohlstedt,* 2003) and reduced melting temperatures (*Katz et al.,* 2003) as well as more buoyant melt (*Jing and Karato,* 2009).

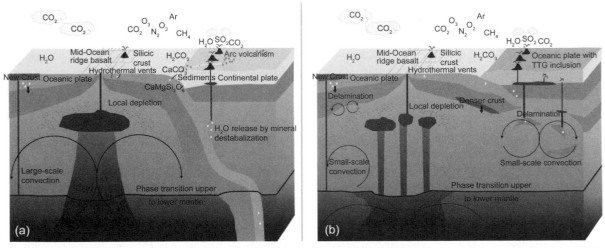

Fig. 4. See Plate 46 for color version. Sketch of **(a)** Earth's present-day volatile cycles, including outgassing of greenhouse and trace gases, recycling of carbonates and water, and deep water release due to destabilization of serpentinites, in comparison to **(b)** less-efficient material recycling mechanisms that have been suggested for early Earth in the absence of (present-day-like) plate tectonics and with a possible layering of convection between upper and lower mantle (as discussed by *Faccenda and DalZilio*, 2017). The depth profile of the sketch is logarithmic.

On early Earth (Fig. 4b), before plate tectonics initiated, the hydrated crustal material was probably recycled less efficiently via several different possible processes. Flat subduction under a lighter (possibly proto-continental) crust could have played a role (*van Hunen and Moyen*, 2012). Furthermore, resurfacing would have been possible by adding new crust on top of the hydrated crust, thus pushing the hydrated minerals deeper into the mantle (*Kamber et al.*, 2005; *Gorczyk and Vogt*, 2018). Sagduction, which refers to crustal material sinking into the mantle as drop-shaped material, or subduction-like behavior on short timescales due to formation of denser minerals such as eclogite, may also have occurred (e.g., *Aoki and Takahashi*, 2004; *Rozel et al.*, 2017).

Whereas Earth has very active recycling of oceanic crust, there is no evidence for crustal recycling on Mars, which suggests that it has always been in the stagnant lid regime. Models for Mars therefore assume no volatile recycling into the mantle (e.g., *Grott et al.*, 2011; *Morschhauser et al.*, 2011; *Sandu and Kiefer*, 2012), so gases emitted into the atmosphere by volcanic outgassing remain there or on the surface until/unless they escape. In comparison, the young age of most of Venus' crust indicates at least one episode of crustal recycling. *Elkins-Tanton et al.* (2007) propose that Venus may have active recycling in the form of plume volcanism and lithospheric delamination providing a range of melting environments and source regions and a mechanism to recycle volatiles within the mantle. Using a coupled mantle-atmosphere model, *Noack et al.* (2012) find intermittent recycling of the lithosphere of Venus following periods of increased surface temperature. This would lead to return of any volatiles sequestered in the crust (perhaps massive carbonates?) back into the mantle. In a similar model, *Gillmann and Tackley* (2014) find that the CO_2 abundance in Venus' atmosphere did not vary significantly over time,

but atmospheric water vapor is very sensitive to outgassing events, which may cause significant surface temperature fluctuations. However, these models do not allow for water oceans on Venus so the implications for volatile recycling on Venus must still be explored.

4. ATMOSPHERIC LOSS PROCESSES

The ability of a planet to retain an atmosphere influences whether water can be stable as a liquid at the planet's surface, and therefore strongly affects habitability. In broad terms, an atmosphere is retained when some fraction of the constituent particles is neither removed to the surface/subsurface nor removed to space. In the former case, particles can be removed through weathering (section 3.2), adsorption, or simple surface deposition/condensation. In many cases these processes are reversible; i.e., atmospheric particles removed at a planet's surface may be restored to the atmosphere if conditions near the surface change. Particles removed to space from the top of the atmosphere, however, are irreversibly lost from the system.

An individual atmospheric particle can be removed to space if it (1) is traveling upward, (2) is unlikely to collide with other atmospheric particles, and (3) has sufficient energy to escape. These three conditions must be met regardless of the mechanism that leads to their escape. The second condition requires that particles escape from the exobase region (or higher), defined as the location where the mean free path between collisions is approximately equal to the local atmospheric scale height. Regardless, a rather obvious point is that escape proceeds from the tops of planetary atmospheres. The third condition requires that a particle's velocity exceed the escape velocity $v_{esc} = (2GM/R)^{1/2}$ for the planet, where G is the universal gravitational constant, M the planet mass, and R the planet radius. Earth (~ 11 km s^{-1})

and Venus (~10 km s⁻¹) have similar escape velocities due to their similar size, while smaller Mars (~5 km s⁻¹) has a smaller escape velocity. Note that more massive species must have greater kinetic energy to achieve escape velocity. An oxygen atom requires ~10 eV to escape from Venus or Earth, but a hydrogen ion requires only ~0.5 eV (these atoms require ~2 eV and ~0.1 eV, respectively, to escape from Mars). Viewed in this light, escape should be more effective at removing the atmospheres of smaller planets and lighter species. However, there are a number of potentially complicating factors, including the mechanisms for escape, whether sufficient energy or upper atmospheric particles are available to enable escape (e.g., *Zahnle and Catling,* 2017), and whether a planet's magnetic field limits escape (e.g., *Moore and Horwitz,* 2007; *Brain et al.,* 2013). How these effects combine for particular stars and planets has implications for whether a given rocky exoplanet can retain an atmosphere for a sufficiently long time to be habitable (e.g., *Airapetian et al.,* 2017; *Dong et al.,* 2017a,b, 2018).

Atmospheric loss to space has occurred and continues to occur on all solar system planets via a variety of different mechanisms — all of which supply energy to upper atmospheric particles. In this section we discuss these mechanisms as they have operated at Venus, Earth, and Mars over time, and in the context of planetary surface habitability. In the three following subsections we discuss atmospheric blowoff, which operated early in the histories of the terrestrial planets; contemporary escape processes (including non-thermal escape); energy- and diffusion-limited escape; and the role of planetary magnetic fields in regulating escape.

4.1. Atmospheric Blowoff

Shortly after their formation the terrestrial planet atmospheres likely contained substantial amounts of hydrogen and helium captured from the surrounding disk (*Baines et al.,* 2013; *Lammer et al.,* 2018). At the same time, the EUV/XUV output of the Sun, responsible for heating upper atmospheric particles, may have been much higher than it is today due to faster rotation [which is not certain (see *Tu et al.,* 2015)] and resultant stronger magnetic activity of the young Sun (*Ribas et al.,* 2005). The combination of low atomic mass of atmospheric species and high energy input makes it likely that early in solar system history a significant fraction of particles in the upper atmospheres of the terrestrial worlds had enough energy to escape. Light species escaped the planets in sufficient quantities that they behaved as a fluid, or an escaping "wind." This process is sometimes referred to as hydrodynamic escape, and sometimes as blowoff (*Hunten,* 1973; *Watson et al.,* 1981).

Whether hydrodynamic escape occurs on a given planet depends on the temperature of atmospheric particles and the energy required to remove them. An "escape parameter" λ can be considered for a planet that is the ratio of gravitational potential energy to the thermal energy of the gas. This can be written as $\lambda = GMm/[kT(R + z)]$, where m is the particle mass, T is temperature, and z is altitude. A value of $\lambda \sim 3$ is typically taken as indicating that blowoff will occur (i.e., thermal energy is at least one-third of gravitational potential energy). Importantly, the escaping light atmospheric particles behave as a fluid and therefore can also entrain and carry away more massive particles. Thus, species with insufficient thermal energy to escape may be removed via collisions.

The primordial H_2 and He atmospheres of Venus, Earth, and Mars were lost to space early in their history via hydrodynamic escape (see discussion in *Lammer et al.,* 2008). While the timing and duration of the hydrodynamic escape period is not well-constrained, there is general consensus that the CO_2 and N_2 atmospheres of these planets today are secondary, produced primarily by outgassing from the interiors. The more massive particles and declining EUV/XUV flux early in solar system history left the atmospheres with too little thermal energy per particle to escape as a fluid. Increased attention to the topic of atmospheric loss from exoplanets has spurred a substantial amount of work on hydrodynamic escape from planets of all kinds and the role it plays in atmospheric evolution (e.g., *Yelle,* 2004; *Tian et al.,* 2005; *Owen and Jackson,* 2012; *Lammer et al.,* 2014; *Johnstone et al.,* 2015; *Fossati et al.,* 2017).

The above discussion highlights a simple but important idea when considering atmospheric evolution and habitability: A planet that has lost its atmosphere via blowoff may still develop a substantial (and habitable) atmosphere at a future time if it remains geologically active (section 3).

4.2. Contemporary Escape Processes

Several different mechanisms have been responsible for removing atmospheric particles of the terrestrial planets to space over time (Table 1). The most straightforward of these is thermal escape, whereby the portion of the velocity distribution of upper atmospheric particles in excess of the escape velocity is able to leave the planet. Hydrodynamic escape is an extreme form of thermal escape, where the fraction of the distribution above the escape velocity is large (achieved as the escape velocity approaches the average thermal velocity

TABLE 1. Contemporary loss mechanisms and their relevance for the escape of different upper atmospheric species at each of the three main terrestrial planets today.

	Thermal	Photochemical	Ion	Sputtering
Venus	—	—	O, C, H?	Ar?
Earth	H	—	O, H	—
Mars	H	O, N, C	O, C, H?	Ar?

of the distribution). Thermal escape is negligible for species heavier than hydrogen on all the terrestrial planets today. However, it is the dominant mechanism by which hydrogen is removed from the martian atmosphere (*Chassefière and Leblanc*, 2004), and is significant for Earth as well (approximately one-third of hydrogen loss is via thermal escape). Temperatures near the Venus exobase are too low, because of cooling via CO_2 thermal emission, for thermal escape to be significant (*Lammer et al.*, 2008).

Energy to drive escape also comes from photochemical reactions in the upper atmosphere. Molecular species (e.g., O_2, N_2, CO) can be "photoionized" by interaction with UV light, producing an ion and free electron (e.g., hv + O_2 → O_2^+ + e). They then dissociate when they recombine, giving the resulting atoms sufficient velocity to escape. Photochemical reactions are the dominant mechanism by which oxygen is removed from the atmosphere of Mars (e.g., *Cravens et al.*, 2016; *Lillis et al.*, 2017). These same reactions at Venus and Earth, however, do not produce atoms with velocities above the escape velocity.

Atmospheric particles are removed as ions as well, and electric fields provide the energy to enable them to escape. Ions escape from Earth's atmosphere near the poles, where the planet's global magnetic field lines are oriented vertically and are open to the solar wind. In these locations vertical electron pressure gradients are the source of electric fields that accelerate ions (O^+, H^+, etc.) upward.

Two additional sources of electric field are relevant at the unmagnetized planets Venus and Mars. First, the flowing solar wind and its embedded interplanetary magnetic field (IMF) combine to create a motional (v × B) electric field that accelerates planetary ions above the exobase region. The ions are picked up by the solar wind flow and carried downstream from the planet. This electric field is important at Venus and Mars because the flowing solar wind penetrates much closer to the planet than at magnetized Earth, so that exospheric ions actually experience this field (e.g., *Luhmann and Schwingenschuh*, 1990; *Brain et al.*, 2016).

Second, interplanetary magnetic field lines drape around the conducting ionospheres of Venus and Mars, and are carried "up and over" the planetary obstacle by the flowing solar wind. However, the solar wind far from the planet is able to flow much more quickly than the solar wind that encounters the ionosphere, so that the draped magnetic field lines being carried by the solar wind can be highly curved or bent close to the planet. The bent magnetic field lines will make current flow near the location of the bend, and a corresponding Hall (J × B) electric field will accelerate planetary ions downstream. In some sense, the curved magnetic field lines "scoop and slingshot" planetary ions away. Again, this process is relevant for Venus and Mars because the interplanetary magnetic field lines penetrate closer to the planet than at Earth, and therefore encounter regions where significant planetary ions are available to be accelerated (e.g., *Halekas et al.*, 2017).

We note that all three electric field terms can contribute to atmospheric escape at a planet, and even for a single planetary ion. For example, pressure gradients and/or J × B electric fields can accelerate a particle at low altitudes where the contribution from pickup is negligible, and v × B can provide additional acceleration at high altitudes where the other terms are small.

A final process of sputtering is considered for the unmagnetized planets Venus and Mars, although it has yet to be unambiguously observed (*Leblanc et al.*, 2018). Sputtering occurs when particles near the exobase region are removed via collisions with incident energetic particles — usually recently accelerated planetary ions. The contribution of sputtering to atmospheric escape at Venus and Mars today is likely small, but may have been larger in the past. Sputtering is also likely to be the dominant removal mechanism for species such as Ar that are massive (so thermal escape is unlikely) and do not react chemically (so photochemical escape cannot occur).

When considering the role that atmospheric escape plays in habitability it is important to consider the influence that the contemporary loss processes discussed above have on both overall thickness of an atmosphere (i.e., surface pressure or total atmospheric abundance) and its composition (e.g., reducing vs. oxidizing atmosphere, partial pressure of water, etc.). It is thus important to keep in mind that many distinct mechanisms remove atmospheric particles to space today, and that the different mechanisms are important for different species.

Argon isotopic ratios ($^{38}Ar/^{36}Ar$) measured in Mars' upper atmosphere indicate that Mars has lost most of its early secondary atmosphere (*Jakosky et al.*, 2017), consistent with geologic evidence for ancient clement surface temperatures and liquid water (section 5). Extrapolation back in time of the present-day escape rate estimates for each process, accounting for a more active early Sun, suggests that escape to space has been a dominant process in the evolution of the martian atmosphere, and that 0.5 bar (or more) of CO_2 was lost to space (*Jakosky et al.*, 2018). The presence of a large CO_2 atmosphere poses problems for some photochemical models, however, which suggest that additional greenhouse gases are required to keep such an atmosphere stable (*Zahnle et al.*, 2008), and even then large loss rates would result (*Tian et al.*, 2009). *Tian et al.* (2009) suggest that the high early escape rates created a thin, cold early Noachian Mars atmosphere and that the subsequent decrease in escape allowed outgassing to thicken the CO_2 atmosphere, explaining the timing of the valley networks. *Kurokawa et al.* (2018), who assume a cooler exobase temperature, estimate at least a 0.5-bar atmosphere at this time. Other escape processes such as impacts (e.g., *Brain and Jakosky*, 1998; *de Niem et al.*, 2012) or sequestration of CO_2 in the subsurface (e.g., *Dobrovolskis and Ingersoll*, 1975; *Stewart and Nimmo*, 2002; *Zent et al.*, 1987) may also have been important.

There is still considerable disagreement over the present-day mass loss rate at Earth (*Yau et al.*, 1988; *Seki et al.*, 2001; *Slapak et al.*, 2017), as well as the timing and total amount of mass loss. Atmospheric escape has occurred in large enough quantities, however, to leave Xe mass fractionated. This has presented a puzzle for Earth, since it is assumed that the Xe was removed via hydrodynamic escape, yet other noble gases such as Kr and Ar are not as strongly fractionated as Xe despite being less massive (and therefore more easily removed, one might guess) (*Ozima and Podosek,* 1983). One recent solution that has been proposed involves the escape of Xe in ionized form, since Xe is more easily ionized than most atmospheric species (*Zahnle et al.,* 2019), with the consequence that substantial atmospheric escape persisted for long periods of time as opposed to subsiding early in Earth's history. Still, models suggest that escape would have been robust early on; *Airapetian and Usmanov* (2016) estimate from a three-dimensional model that the solar wind at 1 AU 3.95 Ga was 2× as fast, 50× denser, and 2× as hot as today, leading to early mass loss rates 1–2 orders of magnitude greater than today. Compared to Mars, however, Earth's atmospheric evolution has been much more complex due to tectonic activity (section 2) and the resulting surface-interior exchange of gases over time (section 3). In fact, whether Earth's atmosphere is thicker or thinner today than in its past is not agreed upon (section 6).

Even less is known about Venus' atmospheric evolution: It is marginally close enough to the Sun that it may have experienced an extended hot magma ocean phase and catastrophic and hydrodynamic loss of its early atmosphere (sections 3, 5). A subsequent volcanic period (sections 2, 3) is likely the proximate source of its current thick atmosphere. Venus' current atmospheric mass loss appears to be due primarily to nonthermal solar wind-induced escape (*Barabash et al.,* 2007a, 2007b; *Brain et al.,* 2016).

4.3. Upper Limits on Escape: Energy-Limited vs. Diffusion-Limited

The processes described in section 4.2 all result in the loss of atmospheric particles to space, and the difference between them on some level comes down to the differing physics of the mechanisms. When considering the flux of particles of a given species from an atmosphere, however, it is useful to ask whether the planet is in an energy-limited or diffusion-limited regime.

In an energy-limited situation, there is insufficient energy arriving at the atmosphere (in the form of solar EUV/X-ray and solar wind) to remove all particles of a given species near the exobase. Thus, an increase in energy input to the atmosphere (from a solar storm, for example, or over the course of a solar cycle) increases atmospheric escape rates. To first order, one can expect that the major atmospheric species of Earth, Mars, and Venus and their dissociation products are energy-limited because their abundance is so high at the top of the atmosphere that it is difficult to supply enough energy to remove them all.

In a diffusion-limited situation, there is ample energy to drive escape of a given species, so that a particle transported from below to the exobase region will be quickly removed from the atmosphere. In this limit the escape rates for specific escape mechanisms cease to be important; instead the rate of supply of species to the upper atmosphere controls the escape rate. One expects trace atmospheric gases to be diffusion-limited, applying similar logic to the paragraph above. These gases, such as hydrogen in a terrestrial planet atmosphere, can be critically important for surface habitability. For example, as insolation increases on a planet with surface liquid water, moist convective storms in the troposphere may deepen and increasingly inject water into the stratosphere, where photodissociation can occur and the resulting H atoms can escape to space. The extent to which this occurs depends on the strength of the tropopause cold trap, which depends on the partial pressure of non-condensing gases in the atmosphere (*Wordsworth and Pierrehumbert,* 2014; *Kleinböhl et al.,* 2018). Thus diffusion-limited escape of H can be tied to tropospheric moist convective storms [or, in the case of Mars, dust storms (*Chaffin et al.,* 2017)] on the terrestrial planets. The same may not be true, however, of planets orbiting cool stars, where radiatively driven large-scale ascent supplies stratospheric water instead (*Fujii et al.,* 2017).

Kasting et al. (1993) estimate that diffusion-limited escape would remove an Earth ocean's worth of water over the age of the solar system once the molar mixing ratio reached $\sim 3 \times 10^{-3}$. This is the basis for the "moist greenhouse" definition of the inner edge of the habitable zone (see the chapter by Kopparapu et al. in this volume). Modern Earth is far from this state, but it may be Earth's fate in ~ 1–2 G.y., as solar luminosity continues to increase (Fig. 1) (see also *Wolf and Toon,* 2015). Whether the same process played an important role at some point in Venus' past, given its present absence of water, depends on whether it ever had a surface ocean (see section 5). *Zahnle et al.* (2008) argue that hydrogen escape on modern Mars is at least close to the diffusion-limited rate.

4.4. Role of a Planetary Magnetic Field

Earth today differs dramatically from Venus and Mars in that it has a strong magnetic field. This implies that its upper atmosphere interacts less directly with its local plasma environment than the currently unmagnetized Mars and Venus (*Brain et al.,* 2016). A debate has arisen in the past decade about the importance of planetary magnetic fields in determining atmospheric escape rates (*Moore and Horwitz,* 2007; *Strangeway et al.,* 2010; *Brain et al.,* 2013). The resolution of the debate will have large consequences for our understanding of habitability of both the terrestrial planets in our own solar system and likely candidates for habitable exoplanets.

The debate arose when it was noticed that the escape rates of ions from Venus, Earth, and Mars were the same, to within 1–2 orders of magnitude, despite the fact that Earth possesses a global dynamo magnetic field and Venus and Mars do not (*Strangeway et al.,* 2010). It was proposed that, contrary to conventional wisdom, magnetic fields do not substantially reduce atmospheric escape — that Earth's magnetic field does not actually shield the atmosphere from being stripped by the solar wind. Instead, energy from the solar wind (i.e., Poynting flux) is transferred to the upper atmosphere along magnetic field lines, and is concentrated at the poles in magnetic cusp regions. The energy then drives ion outflow from the cusps, resulting in escape rates comparable to the situation if Earth had no global magnetic field.

At present the debate is unresolved, although at least four lines of argument suggest that a planetary magnetic field has at least some effect on ion escape rates. First, measurements of ion escape from Earth and Mars as a solar storm passed both objects suggest that Mars responded much more strongly to the event than Earth (*Wei et al.,* 2012). However, solar storms evolve as they propagate away from the Sun and vary laterally, so that Earth and Mars likely did not experience identical driving conditions. Second, both models and *in situ* spacecraft observations show that the ion escape rate from Mars varies as it rotates. Mars has strong crustal magnetic fields (*Acuña et al.,* 1999) that form "minimagnetospheres" in some geographic locations. Variation in ion escape as Mars rotates suggests that the crustal magnetic fields reduce the escape of atmospheric particles overall, but that the effect is different when the strong field is on the dayside vs. the nightside (e.g., *Ma et al.,* 2014; *Fang et al.,* 2015; *Ramstad et al.,* 2017). However, the magnitude of this effect ranges from ~10% to a factor of 2.5, depending on the analysis. Third, recent work employing theory and simple models suggests that a global magnetic field does influence escape rates (*Blackman and Tarduno,* 2018; *Gunell et al.,* 2018), although in one case a magnetic field actually increases escape (*Gunell et al.,* 2018). However, the arguments are still mostly conceptual at this stage, and wait to be followed up with rigorous models. Fourth, recent global plasma models show that ion escape rates vary as the strength of a global magnetic field is increased (*Sakai et al.,* 2018; *Egan et al.,* 2019). All the above assume that the magnetic field extends above the atmosphere; if the magnetic field is weak, or the atmosphere is substantially inflated (*Lichtenegger et al.,* 2010; *Lammer et al.,* 2018), then the magnetic field may play a minimal role.

We note two additional caveats that make obtaining an answer challenging. First, Venus, Earth, and Mars are different sizes and orbit at different distances from the Sun. Thus, one would not necessarily expect the three planets to have comparable escape rates even if they all had magnetic fields (or not) and identical atmospheres. Mars might offer a preferable alternative control experiment since it has both magnetized and unmagnetized regions on a planet of a single size, atmospheric composition, and distance from the Sun. Second, the discussion here has focused solely on ion escape, whereas atmospheric escape results from processes that affect neutral particles as well. These neutral processes are, to first order, insensitive to magnetic fields, suggesting that if the escape of a given species from a planet is dominated by one of these mechanisms then the presence or absence of a global magnetic field should not much matter.

From the perspective of planetary habitability, it would be exciting to have an answer to this question. Then we would better understand whether Earth's magnetic field played an important role in making it habitable, in contrast to Venus and Mars. And, as it becomes possible to detect exoplanetary magnetic fields, we would have an additional lever arm to use in identifying which planets in their habitable zone are most likely to be able to support life.

5. CLIMATE EVOLUTION

5.1. Planetary Energy Balance and Climate

The terrestrial planets probably formed from similar parts of the solar nebula (section 1). Thus the diverging habitability of Earth, Mars, and Venus over time are due to differences in (1) incident solar flux; (2) partitioning in the abundances of molecules between the atmosphere, ocean, and interior due to chemical and internal processes (sections 2, 3); (3) escape of gases to space (section 4); (4) how these processes combined to determine how much of the incident sunlight is absorbed rather than reflected back to space by each planet; and (5) how the temperature at the surface relates to that at higher altitudes where a planet with an atmosphere emits heat to space.

Mars and Venus have never been illuminated as strongly and weakly, respectively, as Earth has at any time, and the solar heating of Earth has changed by ~30% over its history (Fig. 1). Yet Earth has been habitable for most of this time (see the chapters by Zahnle et al. and Hoehler et al. in this volume), Mars had an early habitable phase (chapter by Amador and Ehlmann), and even ancient Venus might have once been habitable (chapter by Arney et al.). Thus incident stellar flux by itself (one of the only pieces of information currently available to astronomers for their initial assessments of exoplanet climates) is very limited as a guide to whether a given planet is habitable, and if so, what its climate might be like. To understand habitability, we need to consider the series of processes by which incident stellar flux ultimately determines the surface temperature of a planet.

In equilibrium, a planet's absorbed solar flux equals its emitted thermal flux. This balance can be expressed as $[S_O(1-A)/4d^2] = \sigma T_e^4$, where S_O is the solar constant, A the planetary (Bond) albedo, d the planet-Sun distance in AU, T_e the equilibrium temperature (i.e., the blackbody temperature that is consistent with the planet's thermal emission to space), and σ the Stefan-Boltzmann constant (see the chapter by Kopparapu et al. in this volume). All these quantities are measured for modern Earth, Mars, and Venus, but A is not known for these planets at different times in their past and therefore must be modeled. (The same is true for exoplanets, for which astronomers just assume a value of A to estimate T_e in the absence of reflected light phase curves.) T_e is not the quantity of interest for habitability, however, because a planet with an atmosphere containing absorbing gases emits most of its heat to space from higher altitudes within the atmosphere rather than from the surface. The surface temperature $T_s > T_e$ by an amount that depends on the extent to which surface emitted heat is absorbed by the atmosphere and re-radiated downward. This can be expressed as $\sigma T_s^4 = \sigma T_e^4 + G(\tau)$, where G, the greenhouse effect of the atmosphere, increases with τ, its thermal infrared optical thickness (*Raval and Ramanathan,* 1989). An optically thin atmosphere ($\tau \lesssim 1$) may be in radiative equilibrium, with T_s determined only by radiative processes. In an optically thick atmosphere ($\tau \gg 1$), convection sets in to moderate T_s, producing radiative-convective equilibrium (*Manabe and Strickler,* 1964). Thus to know the surface temperature, the incident stellar flux, the fraction of that flux reflected to space by the planet, and the composition and pressure of the atmosphere (which determine how opaque it is to thermal radiation) must all be constrained.

The roles of individual greenhouse gases depend on their chemical stability, temperature dependence, and wavelengths of their absorption bands. Climate is controlled by greenhouse gases that do not condense at the temperatures typically found in a given atmosphere, regardless of the relative warming by each gas. Thus, Earth's temperature is regulated by CO_2 (which does not condense at temperatures encountered on Earth) even though H_2O contributes more warming. This occurs because atmospheric H_2O vapor concentration increases strongly with temperature via the Clausius-Clapeyron equation (modulated by weather-related changes in relative humidity). If CO_2 and its greenhouse effect were removed, the climate would cool, causing water to condense and the H_2O vapor concentration to decrease, producing a weaker greenhouse effect and thus further cooling, leading to further reduction in H_2O vapor, further cooling, and so on. In other words, H_2O vapor cannot act on its own as a greenhouse gas on Earth — it only affects climate as a feedback that amplifies CO_2 warming, which can be considered an external forcing at temperatures at which it does not condense (*Lacis et al.,* 2010). For planets warmer than Earth, with atmospheres where H_2O condensation is unlikely, then H_2O dominates the greenhouse effect. For planets colder than Earth, gases that can avoid condensation at temperatures that would freeze H_2O (e.g., CO_2, CH_4, H_2) become more important than H_2O.

Bond albedo depends on the clear atmosphere, condensate clouds, other atmospheric particulates ("aerosols" or "hazes"), and the surface (see discussion in *Del Genio,* 2013). Rayleigh scattering by gases is a minor contributor except for cold planets with very thick atmospheres. Aerosols can be a minor (Earth) or major (Venus) contributor to albedo. Condensates account for approximately two-thirds of Earth's Bond albedo (*Stephens et al.,* 2012), but their role varies from one planet to another depending on the water (and other condensable) abundance and circulation regime. Surface albedo can be a major or minor contributor to Bond albedo depending on masking by the atmosphere. Earth's surface albedo contributes only ~12% of its total ~0.3 Bond albedo (*Donohoe and Battisti,* 2011). On Venus, it contributes virtually nothing because of a planet-wide thick sulfuric acid haze. On cold planets or planets with thin atmospheres (e.g., Mars), the surface is important to the radiative balance. General circulation model (GCM) simulations indicate that habitable exoplanets orbiting G stars that are more strongly or more weakly irradiated than modern Earth will more likely than not have higher Bond albedos because of the enhanced contributions of clouds or sea ice/snow, respectively, while most habitable planets orbiting M stars will have lower Bond albedos than modern Earth because of the enhanced absorption of the mostly near-IR incident radiation by water vapor and sea ice (*Shields et al.,* 2013; *Del Genio et al.,* 2019).

5.2. Implications for the Evolution of Climate on Earth, Mars, and Venus

5.2.1. The faint young Sun problem (or opportunity).
Earth's surface sustained liquid water and life even in the early Archean Eon (3.8–2.5 Ga), when the Sun was ~25% dimmer than today (see the chapter by Zahnle et al. in this volume). *Sagan and Mullen* (1972) noted the paradox between the faint Sun and the origin of life, estimating that a modern Earth-like Archean atmosphere would have a surface temperature below freezing, and thus a larger greenhouse effect was needed to maintain liquid water on the surface. *Walker et al.* (1981) proposed the carbonate-silicate cycle feedback as a reason for elevated Archean CO_2 to explain this (section 3). This is also assumed to be relevant to the habitability of ancient Mars and to define the outer edge of the habitable zone (see the chapter by Kopparapu et al. in this volume). Models often just assume that more weakly irradiated planets have higher CO_2, but some represent the carbonate-silicate cycle control on CO_2 interactively (*Menou,* 2015; *Batalha et al.,* 2016; *Haqq-Misra et al.,* 2016; *Charnay et al.,* 2017; *Krissansen-Totton et al.,* 2018). The ancient Earth and Mars climate problems raise the question of what CO_2 levels might plausibly be expected on weakly irradiated exoplanets. This has led several groups to explore other controls on the weathering sink, such as planet rotation rate (*Jansen et al.,* 2019), and on the outgassing source, such as planet size and tectonic regime (*Noack et al.,* 2017; *Rushby et al.,* 2018) (see also section 2.4). It has also led to proposals to derive statistical constraints on the feedback by observationally constraining CO_2 abundances on a large number of exoplanets with a range of incident stellar fluxes (*Bean et al.,* 2017).

As a result, it is fair to say that the faint young Sun paradox no longer exists for Earth, although it cannot yet be considered solved because the observational evidence on our own planet is inadequate to constrain whether Archean climate was cool or very warm (*Feulner,* 2012). Coupled climate-carbon cycle models predict a high (relative to modern Earth) ~100 mbar of CO_2 (*Charnay et al.,* 2017; *Krissansen-Totton et al.,* 2018). Observational inferences of Archean CO_2 vary by orders of magnitude (*Olson et al.,* 2018), from ~0.9 mbar (*Rosing et al.,* 2010) to hundreds of millibars (*Kanzaki and Murakami,* 2015). The absence of an $^{14}N/^{15}N$ isotopic anomaly in Earth's atmosphere (e.g., *Mandt et al.,* 2015) suggests that its exobase was never high enough for significant escape of N to occur. This in turn implies that sufficient CO_2, much higher than the *Rosing et al.* (2010) estimate, was present on early Earth to radiatively cool the upper atmosphere and keep the exobase altitude sufficiently low to protect it from an enhanced early EUV flux (*Lammer et al.,* 2018).

CH_4 is also poorly constrained; the Archean pre-dates the rise of O_2 and thus CH_4 was more chemically stable and probably much more abundant than in today's oxidizing atmosphere. Plausible CH_4 concentrations (*Olson et al.,* 2018) add ~10°–15°C to estimated global temperatures (*Charnay et al.,* 2013; *Wolf and Toon,* 2013), but the actual role of CH_4 at this time depends on the strength of biological fluxes. Depending on how much more vigorous tectonics

was on early Earth and thus how efficient the carbonate-silicate cycle was (section 3), CH_4 (or another greenhouse gas) might have been needed not only in the early Archean but even in the prior Hadean Eon to maintain warm temperatures after Earth's magma ocean phase (*Sleep and Zahnle,* 2001). One possibility is outgassed H_2 (*Wordsworth and Pierrehumbert,* 2013a), depending on its early rate of escape (section 4). Another is photochemical N_2O produced from energetic particles associated with strong coronal mass ejections (*Airapetian et al.,* 2016), but whether this process could affect the troposphere enough to matter for the climate is not yet known.

Figure 5 shows the effect of uncertainty in Archean CO_2 on its predicted climate. With only 0.9 mbar of CO_2

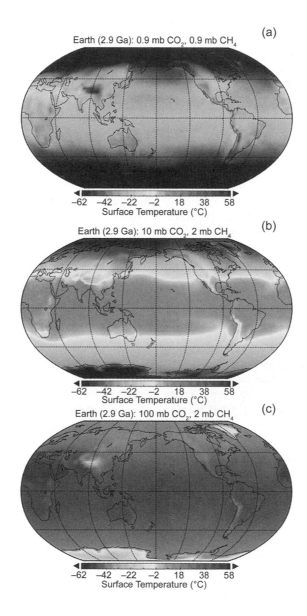

Fig. 5. See Plate 47 for color version. 2.9-Ga Archean Earth surface temperature simulated by the ROCKE-3D GCM (*Way et al.,* 2017) assuming atmospheric compositions for cases **(a)** A, **(b)** B, and **(c)** C of *Charnay et al.* (2013) and the solar spectrum from *Claire et al.* (2012). Uncertainties due to the unknown land-ocean distribution are a few degrees. Simulations courtesy of Michael Way.

(Fig. 5a), the entire surface is glaciated, even with CH_4, because of the positive sea-ice–albedo feedback. In fact, if $[CH_4]/[CO_2] > 0.2$, an organic haze would have formed and further cooled the surface (*Sagan and Chyba*, 1997; *Wolf and Toon*, 2010; *Arney et al.*, 2016). With 10 mbar of CO_2 (Fig. 5b), an above-freezing tropical "waterbelt" is sustained, consistent with geologic and oxygen isotope data implying glacial conditions at subtropical to middle latitudes (*de Wit and Furnes*, 2016). CO_2 levels as high as those inferred by *Kanzaki and Murakami* (2015) (Fig. 5c) produce a hot Archean climate, warmer than modern Earth and with ice only on polar land masses (if there were any at that time; section 2.5). Bond albedo changes magnify the effects of the increased CO_2 and CH_4 since the planet darkens as it warms. This is the net result of decreases in sea ice and thicker, brighter clouds.

If explaining a habitable Archean Earth under the faint young Sun is a challenge, what then to make of ancient Mars, which was illuminated <40% as strongly as modern Earth (Fig. 1) yet almost surely had surface liquid water (see the chapter by Amador and Ehlmann in this volume; *Ehlmann et al.*, 2016; *Wordsworth*, 2016a)? Clearly this requires larger greenhouse gas concentrations than are needed to explain Archean Earth's climate. Today's thin CO_2 Mars atmosphere is thought to be largely due to escape of an early thick CO_2-H_2O atmosphere supplied by volcanism (*Jakosky and Phillips*, 2001) (section 4). How much of the early CO_2 was instead sequestered as carbonate beneath the surface is highly uncertain, though. The carbonate-silicate cycle feedback might support a thicker CO_2 atmosphere on a cool ancient Mars with moderate liquid water given a less efficient chemical weathering sink than that on modern Earth (section 3), perhaps with cyclic behavior (*Batalha et al.*, 2016). Evidence for surface carbonates from weathering is less than expected (*Edwards and Ehlmann*, 2015), although carbonates may be more widespread than originally thought (*Wray et al.*, 2016).

The thickness of Mars' early atmosphere is thus poorly constrained. *Kurokawa et al.* (2018) estimate a lower limit of 0.5 bar at 4 Ga, although *Tian et al.* (2009) reach a different conclusion with different model assumptions. Higher pressures (up to ~1–2 bar) during the valley network-forming period of the late Noachian (~3.8–3.6 Ga) are possible (*Kite et al.*, 2014; *Hu et al.*, 2015). However, a thicker early Mars CO_2 atmosphere is not enough. The maximum greenhouse effect of CO_2 occurs at ~8 bar (*Kasting et al.*, 1993); at higher pressure Rayleigh scattering increases the Bond albedo sufficiently to prevent further warming. Even at pressures this high, CO_2 and H_2O alone do not produce above-freezing temperatures on early Mars (*Kasting*, 1991).

Consequently, understanding the ancient Mars climate requires other warming agents. For example, cold CO_2 atmospheres form CO_2 ice clouds. These were initially thought to oppose warming because condensation reduces the temperature lapse rate and thus weakens the CO_2 greenhouse effect (*Kasting*, 1991). On the other hand, downward scattering of thermal radiation by CO_2 clouds enhances

warming (*Forget and Pierrehumbert*, 1997; *Mischna et al.*, 2000; *Colaprete and Toon*, 2003). *Kitzmann* (2016), though, finds that more accurate radiative calculations produce a much smaller warming than estimated by earlier studies. Furthermore, in three-dimensional model simulations, CO_2 clouds do not completely cover the planet, further reducing any warming impact they might have (*Forget et al.*, 2013; *Wordsworth et al.*, 2013).

Other greenhouse gases may provide the additional warming, e.g., CH_4 and H_2 (*Ramirez et al.*, 2014; *Lasue et al.*, 2015; *Chassefière et al.*, 2016; *Kite et al.*, 2017a; *Wordsworth et al.*, 2017). New estimates of CO_2-H_2 and CO_2-CH_4 collisionally induced absorption (*Wordsworth et al.*, 2017) suggest that ~1% levels of each gas could lift an ancient Mars with >0.5 bar of CO_2 above freezing for geologically brief periods that might explain the valley networks and lakes. If so, then the question is whether ancient Mars had a warm climate with a long-lived northern hemisphere ocean (*Ramirez and Craddock*, 2018), or a generally cold and icy climate without an ocean but with enough episodic warming due to obliquity variations, impacts, or volcanic events to occasionally melt ice at high elevations (*Wordsworth*, 2016a). Clearly, a faint young Sun paradox for Mars still exists.

Furthermore, it is not even certain that the thick early Mars atmosphere was purely CO_2. The model of *Zahnle et al.* (2008) predicts that CO_2 should be photochemically unstable on early Mars under the cold conditions produced by the faint young Sun. Including the effect of O escape, their model predicts gradual reduction of the CO_2 atmosphere to CO. Depending on the timescale for O escape, as well as other factors such as volcanic activity and the composition of impact material, the timescale for conversion might be as short as tens to hundreds of millions of years. Thus it is possible that inferences of surface pressure on ancient Mars from surface features could be biased by the assumption of a CO_2 atmosphere, rather than a mixture of CO_2 and CO of unknown proportions.

To some extent we know more about ancient Mars' climate than we do about that of ancient Earth. Earth's surface has been completely reworked by plate tectonics, volcanism, and weathering of rocks (sections 2, 3). Thus, we do not know exactly when and where continents emerged above the ocean surface; samples of rocks from the Archean or the earlier Hadean Eons are relatively rare, and what evidence exists may or may not reflect the climate at the location at which the evidence was found. By comparison, Mars did not have plate tectonics (section 2) and lost much of its atmosphere in its first billion years (section 4), preserving its history of habitable conditions via its surface geomorphology (section 2) and geochemistry (section 3). These provide not only evidence that surface liquid water was present but regional features that constrain the atmospheric general circulation and regional water cycle (*Wordsworth*, 2016a).

Venus represents the opposite extreme, since its resurfacing period (sections 2, 3) left little or no geomorphological evidence of its ancient climate. Venus' current 92-bar CO_2 atmosphere and strong insolation produce a runaway greenhouse climate that makes directly exploring its surface a challenge (see the chapter by Arney et al. in this volume). Thus no direct evidence of past surface liquid water on Venus exists; we only know that today it is dry. If water was delivered to the terrestrial planets stochastically by embryos from the outer solar system, Venus' initial inventory may have been less than 1 Earth ocean (*Raymond et al.,* 2006). After the hot magma ocean and thick H_2O steam + CO_2 phase on Earth and Venus (section 3), did their evolutions diverge immediately or much later? *Hamano et al.* (2013) and *Lupu et al.* (2014) find that the incident solar flux at Earth allows a magma ocean to cool and solidify fast enough for a water ocean to form. At Venus, though, the stronger solar flux produces slower cooling and more water loss to space. Thus, Venus may or may not have ever formed a liquid water ocean.

The deuterium/hydrogen (D/H) ratio provides indirect evidence that Venus may have once had liquid water, although it might also be explained by an extended magma ocean phase with no surface water (section 3.1). D/H is several orders of magnitude higher on Venus than on Earth (*Donahue et al.,* 1982). If this is due to water isotope fractionation after photodissociation due to preferential escape of lighter H atoms, D/H implies a putative global water ocean depth of 4–525 m (*Donahue and Russell,* 1997; *Chassefière et al.,* 2012). For even a weak or moderately active young Sun, *Lichtenegger et al.* (2016) estimate that enhanced XUV fluxes were strong enough for hydrogen to escape efficiently; D and H escape rates would thus have been similar during the early blow-off phase, implying that today's observed D/H provides only a lower limit to Venus' possible past water loss. If water was replenished from later impactors or volcanism, though, the observed D/H does not require an early surface ocean (*Grinspoon,* 1993).

Lichtenegger et al. (2016) also show that oxygen does not escape as efficiently as hydrogen, implying that considerable O_2 should have been left behind on Venus after its first 100 m.y. Similar inferences have been made for habitable zone exoplanets orbiting M stars (*Luger and Barnes,* 2015). However, Venus' atmosphere today has little O_2 (see the chapter by Arney et al. in this volume). Thus, the missing oxygen must have been removed by interactions between the atmosphere and solid planet. Different possible scenarios are discussed by *Lammer et al.* (2018), who suggest that the lack of oxygen today is easiest to explain if a liquid water ocean never formed and oxygen was simply incorporated into Venus' magmatic crust and oxidized the mantle during its first 100 m.y.

If primitive Venus did cool enough for liquid water to form, the faint young Sun might have been an opportunity for it to be an early habitable planet depending on the evolution of its spin. The "conservative" cloud-free habitable zone (see the chapter by Kopparapu et al.) predicts that a solar flux only slightly higher than modern Earth's would turn it into a "moist greenhouse" or "runaway greenhouse." *Yang et al.* (2014b) show that a planet that rotates slowly enough creates rising motion and thick dayside clouds that raise the Bond albedo and shield the planet. *Way et al.* (2016) esti-

mate that a planet rotating at current Venus' 243-d sidereal period 2.9 Ga could have maintained surface temperatures similar to modern Earth despite an incident solar flux 1.4× that received by modern Earth.

We do not know whether Venus always rotated slowly or evolved to its slow current value via the solar gravitational torque on Venus' atmospheric thermal tide (*Dobrovolskis and Ingersoll,* 1980; *Correia and Laskar,* 2003; *Leconte et al.,* 2015). *Yang et al.* (2014b) and *Way et al.* (2016) find that a rotation period of ~1–2 months is slow enough to produce the stabilizing dayside cloud deck required for a habitable climate. This is determined by the radiative relaxation time of the atmosphere, since radiative heating/cooling of the dayside/nightside must be strong enough to drive upward/downward atmospheric motion. Thus, several possible scenarios exist: (1) Venus never had a water ocean because it lost its hydrogen in its cool-down phase or initially rotated too rapidly to prevent an early runaway greenhouse; (2) Venus cooled rapidly enough to form an initial ocean that transformed within the first 100 m.y. into a runaway greenhouse, leading to escape of its hydrogen and part of its oxygen, with the remaining oxygen being oxidized into the hot surface; (3) Venus cooled rapidly enough and rotated slowly early enough in its history, or had enough water delivered later, to form a transient ocean, thick clouds, and a temperate climate for a finite time period, eventually losing its oxygen to the surface after it evolved into something closer to its current state. [Oxygen loss to space from Venus due to the solar wind is thought to have been modest (*Kulikov et al.,* 2006), although it may be supplemented by processes such as ion acceleration by electric fields due to pressure gradients and gravitational ion-electron separation (see *Brain et al.,* 2016, for a review of relevant processes)].

5.2.2. From ancient to modern times. Earth, Mars, and Venus have not warmed monotonically since their original cool-down after formation. Because of its smaller size, much of Mars' secondary atmosphere has escaped over time (section 4), so its climate is colder today than in its past despite the brightening Sun. At least Earth and Mars appear to have also fluctuated between warm and cold periods throughout their histories. These climate fluctuations are of two fundamentally different types.

On Earth, the first type consists of occasional catastrophic cooling events that caused glacial conditions to extend equatorward, sometimes fully glaciating the planet (see the chapters in this volume by Zahnle et al. and Robinson et al.). The oldest such "snowball Earth" period occurred ~2.2 Ga; there may have been regional events as early as ~2.9 Ga (*Kopp et al.,* 2005). The most well-studied snowball periods are the Sturtian (~715 Ma) and Marinoan (~650 Ma), although it is not clear whether Earth was globally glaciated at these times (*Kirschvink,* 1992; *Hoffman et al.,* 1998; *Hoffman and Schrag,* 2002) or had a tropical liquid "waterbelt" (e.g., *Sohl and Chandler,* 2007).

External climate forcing cannot explain the snowball events. The 2.2-Ga glaciation occurred shortly after the "Great Oxidation Event" (GOE), a sharp rise in O_2 now thought to be merely the most dramatic change in a long history of the rise of O_2 (*Lyons et al.*, 2014; *Olson et al.*, 2018; see also the chapters in this volume by Zahnle et al. and Robinson et al.). Perhaps the release of large amounts of O_2 into the atmosphere chemically destroyed CH_4 and reduced its greenhouse effect (see the chapter by Robinson et al.). If so, this sets limits on late Archean CO_2. Figure 5 shows that a 2.9-Ga Earth with 100 mbar of CO_2 is far too warm to later glaciate Earth by removing CH_4. A midrange scenario with ~10 mbar of CO_2 gives a cold mean climate but with open tropical oceans, a situation more susceptible to full glaciation if CH_4 is destroyed. If the putative ~2.9-Ga event is related to one of the "whiffs" of O_2 prior to the GOE (*Planavsky et al.*, 2014), this would also argue for modest CO_2 levels.

An equally interesting question is how Earth recovered from these snowball periods. *Walker et al.* (1981) had previously suggested in general that if a decrease in insolation had ever glaciated Earth, the precipitation weathering CO_2 sink component of the carbonate-silicate cycle feedback mechanism would largely cease, allowing CO_2 to gradually build up via volcanic activity. *Kirschvink* (1992) first suggested that the observed Marinoan glaciation, although not driven by insolation change, might have ended via a similar volcanic CO_2 buildup. It is not clear whether inferred CO_2 levels were high enough to deglaciate a full snowball; even small areas of open ocean might provide enough of a sink to limit CO_2 buildup (*LeHir et al.*, 2008). Dust has been proposed as one possible mechanism assisting snowball deglaciation (*Abbot and Halevy*, 2010; *Abbot and Pierrehumbert*, 2010; *LeHir et al.*, 2010). *Abbot et al.* (2012) and *Abbot* (2014) have suggested that naturally occurring low-altitude stratocumulus clouds would provide a net warming effect over bright sea ice by enhancing downward longwave radiative fluxes to the surface. This process regulates Arctic sea ice temperatures on modern Earth (*Stramler et al.*, 2011) and may reduce the CO_2 needed to emerge from a snowball state.

Earth has also been anomalously warm in its past, e.g., in the Cretaceous (145–66 Ma), Paleocene-Eocene Thermal Maximum (56 Ma), and early Eocene (50 Ma) (*Lunt et al.*, 2017). These may have been caused by high fluxes of CO_2 and/or CH_4 from the ocean or interior, e.g., due to enhanced volcanism or destabilization of seafloor methane clathrates. These were equable climates, when the equator-pole temperature gradient almost vanished and tropical conditions existed at the poles. *Heller and Armstrong* (2014) argue that Earth is not an optimal habitable planet, and that "superhabitable" planets with more extensive, easily detectable biospheres may exist. Equable climate periods in Earth's past may be a prototype for superhabitable exoplanets.

The second, more modest type of climate fluctuation in Earth's geologic record is shorter-term (~10^4–10^5 yr) advances and retreats of the ice sheets, such as the Last Glacial Maximum or "Ice Age" (~20 Ka). *Milankovitch* (1941) calculated that Earth's orbital eccentricity, obliquity, and precession would vary with periods of ~100,000, 41,000, and 23,000 yr, respectively, due to gravitational interactions

with other solar system objects. *Hays et al.* (1976) showed that temperatures inferred from deep-sea sediment cores were consistent with the Milankovitch cycles. Orbital variations do not primarily change total insolation, but rather its seasonal and latitudinal distribution. This causes sea-ice albedo and water vapor feedbacks and ocean-atmosphere exchanges of CO_2 that amplify the forced climate change (*Hansen et al.*, 2008). Milankovitch cycles still affect Earth's climate but are now dwarfed by anthropogenic greenhouse warming. For some rocky exoplanets, obliquity excursions over time might be much larger, with significant ramifications for their climates and thus for the location of the outer edge of the habitable zone (*Deitrick et al.*, 2018a,b; *Kilic et al.*, 2017, 2018; *Colose et al.*, 2019).

The valley networks and lakes that document liquid water and clement temperatures on Mars were created during the mid-late Noachian (~3.9–3.7 Ga), after a strong early escape period at ~4 Ga (*Tian et al.*, 2009) had ended and allowed Mars' CO_2 atmosphere to remain fairly thick (*Irwin et al.*, 2005; *Fassett and Head*, 2008, 2011) until the Noachian-Hesperian boundary (~3.6 Ga). During this time, CO_2 may have been photochemically unstable and the atmosphere may have become enriched in CO (*Zahnle et al.*, 2008). The relative timing of important events is uncertain, but they include the cessation of Mars' magnetic field in the early Noachian (~4.1 Ga); subsequent atmospheric loss due to impact erosion, sputtering, and sequestration (*Brain and Jakosky*, 1998) (section 4); and the eventual collapse of the cooling martian atmosphere as the CO_2 ice caps began to grow (*Jakosky and Phillips*, 2001; *Soto et al.*, 2015; *Kite et al.*, 2017a). After this the Mars climate was cold and arid for much of the Hesperian period (~3.4–3.1 Ga) and unfavorable for life, similar to present-day Mars.

Despite this, the evolution of Mars' post-Noachian climate was not monotonic. Alluvial fan deposits indicate that from the late Hesperian to early Amazonian (~3.4–2.8 Ga), wetter, habitable conditions occurred intermittently for ~10^8 yr (*Kite et al.*, 2017b). Even in the mostly cold, hyperarid, uninhabitable late Amazonian period evidence exists for occasional liquid water on the surface that implies climate changes (*Morgan et al.*, 2011).

Like Earth, Mars exhibits Milankovitch cycles. Earth's obliquity variation is limited by its large Moon (*Lissauer et al.*, 2011). Mars' obliquity varies chaotically and over a larger range (*Armstrong et al.*, 2004; *Laskar et al.*, 2004; *Brasser and Walsh*, 2011). At low obliquity, Mars' ice caps expand and its climate cools; at high obliquity, melting takes place, the ice caps recede, and the climate warms, due to the lower albedo of Mars with less ice and because sublimation of CO_2 increases the atmospheric pressure and greenhouse effect. CO_2 snow migrates to low latitudes when obliquity is large (*Haberle et al.*, 2003; *Forget et al.*, 2006; *Mischna et al.*, 2013), where it can melt episodically. Obliquity fluctuations may explain several features of Mars' past climate and habitability, e.g., the "icy highlands" scenario for Noachian valley network formation (*Wordsworth et al.*, 2015) and, in combination with proposed "methane bursts" due to CH_4-

clathrate destabilization, sporadic early Amazonian lake formation (*Lasue et al.,* 2015; *Chassefière et al.,* 2016; *Kite et al.,* 2017a; *Wordsworth et al.,* 2017).

Since we have little information about Venus' past, little can be said about its evolution. If Venus cooled too slowly for a water ocean to develop, it was always uninhabitable. Major issues include understanding the surface-atmosphere and escape processes (sections 3, 4) that kept Venus from accumulating O_2 (*Kasting and Pollack,* 1983; *Gillmann et al.,* 2009; *Hamano et al.,* 2013; *Lichtenegger et al.,* 2016), which also has implications for exoplanets near and inside the inner edge of the habitable zone (*Luger and Barnes,* 2015; *Harman et al.,* 2015); a possible transition from an early tectonic to a stagnant lid geophysical regime (*Lenardic et al.,* 2008, 2016b); and the timing of major volcanism and plains emplacement (section 2). There is furthermore no guarantee that Venus' evolution was destined to diverge from Earth's. The coupled climate-tectonic system may exhibit bistability (*Lenardic et al.,* 2016a): Internal dynamical perturbations due to unknowable inhomogeneities, or transient external surface temperature anomalies like those Earth is known to have experienced through its history, might conceivably make the difference between an evolutionary path that leads to plate tectonics vs. one that maintains stagnant lid conditions, with different consequences for the subsequent buildup of atmospheric CO_2 and thus the climate. If so, then Venus' and Earth's eventual fate might not always be replicated by rocky exoplanets in other stellar systems irradiated at similar levels.

If instead Venus formed a primitive ocean, are there plausible scenarios for an extended habitable period before the transition to today's thick CO_2 atmosphere and hot climate? A wet, slowly rotating early Venus could have remained habitable even to the present because of cloud shielding (*Way et al.,* 2018). This did not occur, so an ocean at the low end of D/H inferences, coupled with elevated CO_2 as for Archean Earth, might explain why. In this case the ocean evaporates and H_2O is transported to the stratosphere (where photolysis and escape set in) more quickly. This leaves a dry atmosphere and subsequent warming and CO_2 buildup, as weathering ceases, surface carbonates decompose, the mantle partially melts and outgasses, and perhaps the interior changes its tectonic regime, producing its modern state (*Noack et al.,* 2012; *Lenardic et al.,* 2016b) (sections 2, 3.2).

6. CONCLUSION

The inner solar system appears to have had a rich history of habitability. At least one, most probably two, and possibly three planets were habitable soon after they formed and cooled down. Habitability did not evolve in a straight line but with fascinating ups and downs that are possible in a climate system that couples an atmosphere to an ocean, a land surface, a dynamic interior, a dynamic heliosphere, and a star that increases in brightness. The future looks to be less interesting, though. Venus' habitable period, if it existed at all, is long gone and will not return. Mars' habitable era has also either come to a

close or is nearing that stage. Even Earth only has ~1–2 G.y. left before the Sun's luminosity increases by ~20% and it enters a runaway greenhouse. By then it may already have long lost its ocean, mostly through transport to the stratosphere, photolysis, and hydrogen escape (*Wolf and Toon,* 2015) and secondarily via subduction into the mantle (*Bounama et al.,* 2001). Of course the immediate problem for the inhabitants of Earth is the lifetime of the intelligent species that reads book chapters such as this one (*Frank and Sullivan,* 2014).

Taken together, Earth, Mars, and Venus provide an invaluable, but incomplete, template for thinking about the range of conditions under which planets can be habitable. Mars is a case study of the narrow window of habitability of small-mass planets. It forces us to confront our fairly primitive understanding of the factors controlling habitability, as we struggle to explain what we can see through the eyes of spacecraft instruments. Mars provides a good lesson about our limited imaginations as we begin to anticipate the possibilities on hundreds of planets in other stellar systems that will only yield their secrets slowly.

On the other hand, Mars is a missed opportunity as well. What if it had formed with a mass comparable to Earth's? Would a weakly illuminated planet with surface liquid water, a magnetic field, and a thick CO_2 atmosphere, more of which is retained rather than lost, have been continuously habitable over 4.5 b.y.? As we attempt decades from now to interpret potential exoplanet biosignatures and ask what the probability of life is given a habitable planet (e.g., *Catling et al.,* 2018), how might our thinking have changed if two rather than one of two habitable planets in our solar system had developed and maintained life? How much farther along would our understanding be if 1-bar N_2 atmospheres with trace amounts of CO_2 and CH_4 were not our only prototype for life-bearing planets?

In fact, a 1-bar atmosphere is not even indicative of our own planet's evolution: Earth may have seen large changes in N_2 and total pressure in past eons, although whether pressure was greater or less in the distant past is debated (*Marty et al.,* 2013; *Johnson and Goldblatt,* 2015, 2018; *Som et al.,* 2016; *Stüeken et al.,* 2016; *Zerkle and Mikhail,* 2017). These changes are due to both biotic and abiotic mechanisms (*Wordsworth,* 2016b), emphasizing the fact that the habitability and inhabitance of a planet are intertwined (*Goldblatt,* 2016). The clearest example of this is the unintended experiment that humanity is currently conducting on the habitability of our own planet — so much so that we are now considered to be in a new geologic epoch, the Anthropocene (*Steffen et al.,* 2007).

Venus is an equally valuable example for our exoplanet thinking, although in different ways. Given our Earth-centric view, could we have imagined the runaway greenhouse end state of evolution for our so-called "sister" planet if we had not seen an example? Before we knew about Venus' crushing pressure and harsh surface conditions, astronomers sometimes imagined it to be an Earth-like planet with possibly abundant water and ice clouds producing its high albedo (e.g., *Bottema et al.,* 1965). Now that we have discovered many exoplanets in the "Venus zone" located inside the inner edge of the traditional habitable zone (*Kane et al.,* 2014), might we find a few to be habitable if we do not rule them out without making the effort to characterize them?

A key feature of Earth's evolution is that it has remained inhabited since life first appeared. During periods when its global mean temperature was below freezing, it may have retained equatorial regions of liquid surface water. During more extreme snowball periods when it was fully glaciated, life still persisted in oceanic niches. Thus the simplistic idea of a habitable exoplanet sustaining surface liquid water is invalidated by our own planet's history (see the chapters in this volume by Zahnle et al., Baross et al., Hoehler et al., Stüeken et al., and Johnson et al.). Such cold, but habitable and inhabited, periods in our history may be of less practical interest in the search for life on exoplanets, because of the problems they pose for biosignature detection (*Reinhard et al.,* 2017; see also the chapter by Robinson et al. in this volume). Nonetheless, the inner solar system's lessons about the temporal evolution of habitability provide needed perspective for our thinking about other stellar systems: The parched desert planet we find with a future telescope may have teemed with life eons ago, while a frozen, apparently lifeless planet we detect may blossom a billion years into the future, just as our time as a habitable planet is coming to an end.

Acknowledgments. We thank Helmut Lammer, Robin Wordsworth, and Vikki Meadows for insightful comments that helped us improve the manuscript. This chapter was made possible by collaborations supported by the NASA Astrobiology Program through the Nexus for Exoplanet System Science, by the Sellers Exoplanet Environments Collaboration, by the NASA Planetary Atmospheres Program, and by the NASA Psyche mission. This work was also funded by the Deutsche Forschungsgemeinschaft (SFB-TRR 170, subproject C06). This is TRR 170 Publication No. 67.

REFERENCES

Abbot D. S. (2014) Resolved Snowball Earth clouds. *J. Climate, 27,* 4391–4402.

Abbot D. S. and Halevy I. (2010) Dust aerosol important for Snowball Earth deglaciation. *J. Climate, 23,* 4121–4132.

Abbot D. S. and Pierrehumbert R. T. (2010) Mudball: Surface dust and Snowball Earth deglaciation. *J. Geophys. Res.–Atmospheres, 115,* D03104, doi:10.1029/2009JD012007.

Abbot D. S., Voigt A., Branson M., et al. (2012) Clouds and Snowball Earth deglaciation. *Geophys. Res. Lett., 39,* L20711, DOI: 10.1029/2012GL052861.

Abe Y. and Matsui T. (1985) The formation of an impact-generated H_2O atmosphere and its implications for the early thermal history of the Earth. *Proc. Lunar Planet. Sci. Conf. 15th,* in *J. Geophys. Res., 90,* C545–C559.

Abe Y. and Matsui T. (1988) Evolution of an impact-generated H_2O-CO_2 atmosphere and formation of a hot proto-ocean on Earth. *J. Atmos. Sci., 45,* 3081–3101.

Acuña M. H., Connerney J. E. H., Ness N. F., et al. (1999) Global distribution of crustal magnetization discovered by the Mars Global Surveyor MAG/ER experiment. *Science, 284,* 790–793.

Agnor C. B., Canup R. M., and Levison H. F. (1999) On the character and consequences of large impacts in the late stage of terrestrial planet formation. *Icarus, 142,* 219–237.

Airapetian V. S. and Usmanov A. V. (2016) Reconstructing the solar wind from its early history to current epoch. *Astrophys. J. Lett.,*

817, L24.

Airapetian V. S., Glocer A., Gronoff G., Hébrard E., and Danchi W. (2016) Prebiotic chemistry and atmospheric warming of early Earth by an active young Sun. *Nature Geosci., 9,* 452–455.

Airapetian V. S., Glocer A., Khazanov G. V., et al. (2017) How hospitable are space weather affected habitable zones? The role of ion escape. *Astrophys. J. Lett., 836,* L3.

Aitta A. (2012) Venus' internal structure, temperature and core composition. *Icarus, 218,* 967–974.

Anderson B. J., Johnson C. L., Korth H., Purucker M. E., Winslow R. M., Slavin J. A., Solomon S. C., McNutt Jr. R. L., Raines J. M., and Zurbuchen T. H. (2011) The global magnetic field of Mercury from MESSENGER orbital observations. *Science, 333,* 1859–1862.

Aoki I. and Takahashi E. (2004) Density of MORB eclogite in the upper mantle. *Phys. Earth Planet. Inter., 143–144,* 129–143.

Ardia P., Hirschmann M. M., Withers A. C., and Stanley B. D. (2013) Solubility of CH_4 in a synthetic basaltic melt, with applications to atmosphere-magma ocean-core partitioning of volatiles and to the evolution of the martian atmosphere. *Geochim. Cosmochim. Acta, 114,* 52–71.

Armann M. and Tackley P. J. (2012) Simulating the thermochemical magmatic and tectonic evolution of Venus's mantle and lithosphere: Two-dimensional models. *J. Geophys. Res., 117,* E12003, DOI: 10.1029/2012JE004231.

Armstrong J. C., Leovy C. B., and Quinn T. (2004) A 1 Gyr climate model for Mars: New orbital statistics and the importance of seasonally resolved polar processes. *Icarus, 171,* 255-271.

Armstrong R. L. (1981) Radiogenic isotopes: The case for crustal recycling on a near-steady-state no-continental-growth Earth. *Philos. Trans. R. Soc. A, 301,* 443–472.

Arndt N. T. and Nisbet E. G. (2012) Processes on the young Earth and the habitats of early life. *Annu. Rev. Earth Planet. Sci., 40,* 521–549.

Arney G., Domagal-Goldman S. D., Meadows V. S., Wolf E. T., Schwieterman E., Charnay B., Claire M., Hébrard E., and Trainer M. G. (2016) The spectrum and habitability of hazy Archean Earth. *Astrobiology, 16,* 873–899.

Aulbach S. and Stagno V. (2016) Evidence for a reducing Archean ambient mantle and its effects on the carbon cycle. *Geology, 44,* 751–754.

Baines K. H., Atreya S. K., Bullock M. A., et al. (2013) The atmospheres of the terrestrial planets: Clues to the origins and early evolution of Venus, Earth, and Mars. In *Comparative Climatology of Terrestrial Planets* (S. J. Mackwell et al., eds.), pp. 137 160. Univ. of Arizona, Tucson.

Baker D. R. and Sofonio K. (2017) A metasomatic mechanism for the formation of Earth's earliest evolved crust. *Earth Planet. Sci. Lett., 463,* 48–55.

Banerdt W. B. and Russell C. T. (2017) Editorial on: Topical collection on InSight mission to Mars. *Space Sci. Rev., 211,* 1–3.

Barabash S., Fedorov A., Sauvaud J. J., et al. (2007a) The loss of ions from Venus through the plasma wake. *Nature, 450,* 650–653.

Barabash S., Sauvaud J.-A., Gunell H., et al. (2007b) The Analyser of Space Plasmas and Energetic Atoms (ASPERA-4) for the Venus Express mission. *Planet. Space Sci., 55,* 1772–1792.

Batalha N., Domagal-Goldman S. D., Ramirez R., and Kasting J. F. (2015) Testing the early Mars H_2-CO_2 greenhouse hypothesis with a 1-D photochemical model. *Icarus, 258,* 337–349.

Batalha N. E., Kopparapu R. K., Haqq-Misra J., and Kasting J. F. (2016) Climate cycling on early Mars caused by the carbonate-silicate cycle. *Earth Planet. Sci. Lett., 455,* 7–13.

Bean J. L., Abbot D. S., and Kempton E. M.-R. (2017) A statistical comparative planetology approach to the hunt for habitable exoplanets and life beyond the solar system. *Astrophys. J. Lett., 841,* L24.

Bell E. A., Boehnke P., Harrison T. M., and Mao W. L. (2015) Potentially biogenic carbon preserved in a 4.1 billion-year-old zircon. *Proc. Natl. Acad. Sci. U.S.A., 112,* 14518–14521.

Belousova E. A., Kostitsyn Y. A., Griffin W. L., Begg G. C., O'Reilly S. Y., and Pearson N. J. (2010) The growth of the continental crust: Constraints from zircon Hf-isotope data. *Lithos, 119,* 457–466.

Benz W., Anic A., Horner J., and Whitby J. A. (2008) The origin of Mercury. In *Mercury* (A. Balogh et al., eds.), pp. 7–20. Space Sciences Series of ISSI, Vol. 26, Springer-Verlag, New York.

Birch F. (1952) Elasticity and constitution of the Earth's interior. *J. Geophys. Res., 57,* 227–286.

Blackman E. G. and Tarduno A. J. (2018) Mass, energy, and momentum capture from stellar winds by magnetized and unmagnetized planets: Implications for atmospheric erosion and habitability. *Mon. Not. R. Astron. Soc., 481,* 5146–5155.

Bodman E. H. L., Wright J. T., Desch S. J., and Lisse C. M. (2018) Inferring the composition of disintegrating planet interiors from dust tails with future James Webb Space Telescope observations. *Astron. J., 156,* 173.

Bond J. C., O'Brien D. P., and Lauretta D. S. (2010) The compositional diversity of extrasolar terrestrial planets. I. *In situ* simulations. *Astrophys. J., 715,* 1050–1070.

Boss A. P. (2004) Evolution of the solar nebula. VI. Mixing and transport of isotopic heterogeneity. *Astrophys. J., 616,* 1265–1277.

Bottema M., Plummer W., Strong J., and Zander R. (1965) The composition of the Venus clouds and implications for model atmospheres. *J. Geophys. Res., 70,* 4401–4402.

Bounama C., Franck S., and von Bloh W. (2001) The fate of Earth's ocean. *Hydrol. Earth Syst. Sci., 5,* 569–576.

Bouvier L. C., Costa M. M., Connelly J. N., Jensen N. K., Wielandt D., Storey M., Nemchin A. A., Whitehouse M. J., Snape J. F., Bellucci J. J., Moynier F., Agranier A., Gueguen B., Schonbachler M., and Bizzarro M. (2018) Evidence for extremely rapid magma ocean crystallization and crust formation on Mars. *Nature, 558,* 586–589.

Brain D. A. and Jakosky B. M. (1998) Atmospheric loss since the onset of the martian geologic record: Combined role of impact erosion and sputtering. *J. Geophys. Res., E10,* 22689–22694.

Brain D. A., Leblanc F., Luhmann J. G., Moore T. E., and Tian F. (2013) Planetary magnetic fields and climate evolution. In *Comparative Climatology of Terrestrial Planets* (S. J. Mackwell et al., eds.), pp. 487–501. Univ. of Arizona, Tucson.

Brain D. A., Bagenal F., Ma Y.-J., Nilsson H., and Wieser G. S. (2016) Atmospheric escape from unmagnetized bodies. *J. Geophys. Res.–Planets, 121,* 2364–2385.

Brasser R. and Walsh K. J. (2011) Stability analysis of the martian obliquity during the Noachian era. *Icarus, 213,* 423–427.

Breuer D. (2009) Dynamics and thermal evolution. In *Solar System: Landolt-Börnstein — Group VI Astronomy and Astrophysics (Numerical Data and Functional Relationships in Science and Technology), Vol. 4B.* Springer, Berlin, Heidelberg, DOI: 10.1007/978-3-540-88055-4_19.

Breuer D. and Spohn T. (1995) Possible flush instability in mantle convection at the Archaean-Proterozoic transition. *Nature, 378,* 608–610.

Campbell I. H. and Taylor S. R. (1983) No water, no granites — No oceans, no continents. *Geophys. Res. Lett., 10,* 1061–1064.

Canil D. (1997) Vanadium partitioning and the oxidation state of Archaean komatiite magmas. *Nature, 389,* 842–845.

Cannon K. M., Parman S. W., and Mustard J. F. (2017) Primordial clays on Mars formed beneath a steam or supercritical atmosphere. *Nature, 552,* 88–91.

Canup R. M. (2008) Accretion of the Earth. *Philos. Trans. R. Soc. A, 366,* 4061–4075.

Carr M. H. and Glow G. D. (1981) Martian channels and valleys — Their characteristics, distribution, and age. *Icarus, 48,* 91–117.

Carter J., Loizeau D., Mangold N., Poulet F., and Bibring J.-P. (2015) Widespread surface weathering on early Mars: A case for a warmer and wetter climate. *Icarus, 248,* 373–382.

Catling D. C., Krissansen-Totton J., Kiang N. Y., et al. (2018) Exoplanet biosignatures: A framework for their assessment. *Astrobiology, 19,* 709–738.

Chaffin M. S., Deighan J., Schneider N. M., and Stewart A. I. F. (2017) Elevated atmospheric escape of atomic hydrogen from Mars induced by high-altitude water. *Nature Geosci., 10.,* 174–178.

Charnay B., Forget F., Wordsworth R., Leconte J., Millour E., Codron F., and Spiga A. (2013) Exploring the faint young Sun problem and the possible climates of the Archean Earth with a 3-D GCM. *J. Geophys. Res.–Atmospheres, 118,* DOI: 10.1002/jgrd.50808.

Charnay B., Le Hir G., Fluteau F., Forget F., and Catling D. C. (2017) A warm or a cold early Earth? New insights from a 3-D climate-carbon model. *Earth Planet Sci. Lett., 474,* 97–109.

Chassefière E. and Leblanc F. (2004) Mars atmospheric escape and evolution; interaction with the solar wind. *Planet. Space Sci., 52,* 1039–1058.

Chassefière E., Wieler R., Marty B., and Leblanc F. (2012) The evolution of Venus: Present state of knowledge and future exploration. *Planet. Space Sci., 63–64,* 15–23.

Chassefière E., Lasue J., Langlais B., and Quesnel Y. (2016) Early Mars serpentinization-derived CH_4 reservoirs, H_2-induced warming and paleopressure evolution. *Meteoritics & Planet. Sci., 51,* 2234–2245.

Chen B., Li J., and Hauck S. A. II (2008) Non-ideal liquidus curve in the Fe-S system and Mercury's snowing core. *Geophys. Res. Lett., 35,* L07201, DOI: 10.1029/2008GL033311.

Claire M. W., Sheets J., Cohen M., Ribas I., Meadows V. S., and Catling D. C. (2012) The evolution of solar flux from 0.1 nm to 160 μm. *Astrophys. J., 757,* 95.

Cockell C. S., Bush. T., Bryce C., et al. (2016) Habitability: A review. *Astrobiology, 16,* 89–117.

Colaprete A. and Toon O. B. (2003) Carbon dioxide clouds in an early dense martian atmosphere. *J. Geophys. Res., 108, E4,* 5025, DOI: 10.1029/2002JE001967.

Colose C. M., Del Genio A. D., and Way M. J. (2019) Enhanced habitability on high obliquity bodies near the outer edge of the habitable zone of Sun-like stars. *Astrophys. J., 884,* 138.

Correia A. C. M. and Laskar J. (2003) Long-term evolution of the spin of Venus. II. Numerical simulations. *Icarus, 163,* 24–45.

Cravens T. E, Rahmati A., Fox J. L., et al. (2016) Hot oxygen escape from Mars: Simple scaling with solar EUV irradiance. *J. Geophys. Res.–Space Physics, 122,* 1102–1116.

Davies J. H. and Davies D. R. (2010) Earth's surface heat flux. *Solid Earth, 1,* 5–24.

Debaille V., Brandon A. D., O'Neill C., Yin Q. Z., and Jacobsen B. (2009) Early martian mantle overturn inferred from isotopic composition of nakhlite meteorites. *Nature Geosci., 2,* 548–552.

Deitrick R., Barnes R., Quinn T. R., Armstrong J., Charnay B., and Wilhelm C. (2018a) Exo-Milankovitch cycles. I. Orbits and rotation states. *Astron. J., 155,* 60.

Deitrick R., Barnes R., Bitz C., Fleming D., Charnay B., Meadows V., Wilhelm C., Armstrong J., and Quinn T. R. (2018b) Exo-Milankovitch cycles. II. Climates of G-dwarf planets in dynamically hot systems. *Astron. J., 155,* 266.

Delano J. W. (2001) Redox history of the Earth's interior since ~3900 Ma: Implications for prebiotic molecules. *Origins Life Evol. Biospheres, 31,* 311–341.

Del Genio A. D. (2013) Physical processes controlling Earth's climate. In *Comparative Climatology of Terrestrial Planets* (S. J. Mackwell et al., eds.), pp. 3–18. Univ. of Arizona, Tucson.

Del Genio A. D., Kiang N. Y., Way M. J., Amundsen D. A., Sohl L. E., Fujii Y., Chandler M., Aleinov A., Colose C. M., Guzewich S. D., and Kelley M. (2019) Albedos, equilibrium temperatures, and surface temperatures of habitable planets. *Astrophys. J., 884,* 75.

de Vries J., Nimmo F., Melosh H. J., Jacobson S. A., Morbidelli A., and Rubie D. C. (2016) Impact induced melting during accretion of the Earth. *Prog. Earth Planet. Sci., 3(7),* DOI: 10.1186/s40645-016-0083-8.

de Niem D., Kührt E., Morbidelli A., and Motschmann U. (2012) Atmospheric erosion and replenishment indued by impacts upon the Earth and Mars during a heavy bombardment. *Icarus, 221,* 495–507.

de Wit M. J. and Furnes H. (2016) 3.5-Ga hydrothermal fields and diamictites in the Barberton Greenstone Belt — Paleoarchean crust in cold environments. *Sci. Adv., 2(2),* e1500368.

Dhuime B., Hawkesworth C. J., Cawood P. A., and Storey C. D. (2012) A change in the geodynamics of continental growth 3 billion years ago. *Science, 335,* 1334–1336.

Dobrovolskis A. and Ingersoll A. P. (1975) Carbon dioxide — Water clathrate as a reservoir of CO_2 on Mars. *Icarus, 26,* 353–357.

Dobrovolskis A. R. and Ingersoll A. P. (1980) Atmospheric tides and the rotation of Venus. 1. Tidal theory and the balance of toques. *Icarus, 40,* 1–17.

Dodd M. S., Papineau D., Grenne T., Slack J. F., Rittner M., Pirajno F., O'Neil J., and Little C. T. S. (2017) Evidence for early life in Earth's oldest hydrothermal vent precipitates. *Nature, 543,* 60–64.

Donahue T. M. and Russell C. T. (1997) The Venus atmosphere and ionosphere and their interaction with the solar wind: An overview. In *Venus II* (S. W. Bougher et al., eds.), pp. 3–31. Univ. of Arizona, Tucson.

Donahue T. M., Hoffman J. H., Hodges R. R. Jr., and Watson A. J. (1982) Venus was wet: A measurement of the ratio of deuterium to hydrogen. *Science, 216,* 630–633.

Dong C., Huang Z., Lingam M., et al. (2017a) The dehydration of water worlds via atmospheric losses. *Astrophys. J., Lett., 847,* L4.

Dong C., Lingam M., Ma Y., and Chen O. (2017b) Is Proxima Centauri b habitable? A study of atmospheric loss. *Astrophys. J., Lett., 837,* L26.

Dong C., Jin M., Lingam M., et al. (2018) Atmospheric escape from the TRAPPIST-1 planets and implications for habitability. *Proc. Natl. Acad. Sci. U.S.A., 115,* 260–265.

Donohoe A. and Battisti D. S. (2011) Atmospheric and surface contributions to planetary albedo. *J. Climate, 24,* 4402–4418.

Dorn C., Khan A., Heng K., Connolly J. A. D., Alibert Y., Benz W., and Tackley P. (2015) Can we constrain the interior structure of rocky exoplanets from mass and radius measurements? *Astron. Astrophys., 577,* A83, DOI: 10.1051/0004-6361/201424915.

Driscoll P. and Bercovici D. (2014) On the thermal and magnetic histories of Earth and Venus: Influences of melting, radioactivity, and conductivity. *Phys. Earth Planet. Inter., 236,* 36–51.

Dumberry M. and Rivoldini A. (2015) Mercury's inner core size and core-crystallization regime. *Icarus, 248,* 254–268.

Dumoulin C., Tobie G., Verhoeven O., Rosenblatt P., and Rambaux N. (2017) Tidal constraints on the interior of Venus. *J. Geophys. Res.–Planets, 122,* 1338–1352.

Edwards C. S. and Ehlmann B. L. (2015) Carbon sequestration on Mars. *Geology, 43,* 863–866.

Egan H., Ma Y., Dong C., et al. (2019) Comparison of global martian plasma models in the context of MAVEN observations. *J. Geophys. Res.–Space Phys., 123,* 3714–3726.

Ehlmann B. L., Mustard J. F., and Murchie S. L. (2011) Subsurface water and clay mineral formation during the early history of Mars. *Nature, 479,* 53–60.

Ehlmann B. L. and Edwards C. S. (2014) Mineralogy of the martian surface. *Annu. Rev. Earth Planet. Sci., 42,* 291–315.

Ehlmann B. L., Anderson F. S., Andrews-Hanna J., et al. (2016) The sustainability of habitability on terrestrial planets: Insights, questions, and needed measurements from Mars for understanding the evolution of Earth-like worlds. *J. Geophys. Res.–Planets, 121,* 1927–1961.

Elkins-Tanton L. T. (2008) Linked magma ocean solidification and atmospheric growth for Earth and Mars. *Earth Planet. Sci. Lett., 271,* 181–191.

Elkins-Tanton L. T. (2012) Magma oceans in the inner solar system. *Annu. Rev. Earth Planet. Sci., 40,* 113–139.

Elkins-Tanton L. T., Smrekar S. E., Hess P. C., and Parmentier E. M. (2007) Volcanism and volatile recycling on a one-plate planet: Applications to Venus. *J. Geophys. Res., 112,* E04S06.

Faccenda M. and Dal Zilio L. (2017) The role of solid-solid phase transitions in mantle convection. *Lithos, 268–271,* 198–224.

Fang X., Ma Y., Brain D., Dong Y., and Lillis R. (2015) Control of Mars global atmospheric loss by the continuous rotation of the crustal magnetic field: A time-dependent MHD study. *J. Geophys. Res.–Space Phys., 120(12),* 10926–10944.

Fassett C. I. and Head J. W. III (2008) The timing of martian valley network activity: Constraints from buffered crater counting. *Icarus, 195,* 61–89.

Fassett C. I. and Head J. W. III (2011) Sequence and timing of conditions on early Mars. *Icarus, 211,* 1204–1214.

Fegley B. Jr., Jacobson N. S., Williams K. B., Plane J. M. C., Schaefer L., and Lodders K. (2016) Solubility of rock in steam atmospheres of planets. *Astrophys. J., 824,* 2.

Feulner G. (2012) The faint young Sun problem. *Rev. Geophys., 50,* RG2006.

Fischer R. A., Nakajima Y., Campbell A. J., et al. (2015) High pressure metal-silicate partitioning of Ni, Co, V, Cr, Si, and O. *Geochim. Cosmochim. Acta, 167,* 177–194.

Foley B. J. and Smye A. J. (2018) Carbon cycling and habitability of Earth-size stagnant lid planets. *Astrobiology, 18,* 873–896.

Forget F. and Pierrehumbert R. T. (1997) Warming early Mars with carbon dioxide clouds that scatter infrared radiation. *Science, 278,* 1273–1276.

Forget F., Haberle R. M., Montmessin F., Levrard B., and Head J. W. (2006) Formation of glaciers on Mars by atmospheric precipitation at high obliquity. *Science, 311,* 368–371.

Forget F., Wordsworth R., Millour E., Madeleine J.-B., Kerber L., Leconte J., Marcq E., and Haberle R. M. (2013) 3D modeling of the early martian climate under a denser CO_2 atmosphere: Temperatures and CO_2 ice clouds. *Icarus, 222,* 81–99.

Fossati L., Erkaev N. V., Lammer H., et al. (2017) Aeronomical constraints to the minimum mass and maximum radius of hot low-mass planets. *Astron. Astrophys., 598,* A90.

Frank A. and Sullivan W. (2014) Sustainability and the astrobiological perspective: Framing human futures in a planetary context. *Anthropocene, 5,* 32–41.

Frey H. V. (2006) Impact constraints on the age and origin of the lowlands of Mars. *Geophys. Res. Lett., 33,* L08S02.

Frost D. J., Liebske C., Langenhorst F., McCammon C. A., Trønnes R. G., et al. (2004) Experimental evidence for the existence of iron-rich metal in the Earth's lower mantle. *Nature, 248,* 409–412.

Fujii Y., Del Genio A. D., and Amundsen D. S. (2017) NIR-driven moist upper atmospheres of synchronously rotating temperate terrestrial planets. *Astrophys. J., 848,* 100.

Fyfe W. S. (1978) The evolution of the Earth's crust: Modern plate tectonics to ancient hot spot tectonics? *Chem. Geol., 23,* 89–114.

Gaillard F., Scaillet B., and Arndt N. T. (2011) Atmospheric oxygenation caused by a change in volcanic degassing pressure. *Nature, 478,* 229–232.

Gaillard F., Michalski J., Berger G., McLennan S. M., and Scaillet B. (2013) Geochemical reservoirs and timing of sulfur cycling on Mars. *Space Sci. Rev., 174,* 251–300.

Galimov E. M. (2005) Redox evolution of the Earth caused by a multi-stage formation of its core. *Earth Planet. Sci. Lett., 233,* 263–276.

Gillmann C., Chassefière E., and Lognonné P. (2009) A consistent picture of early hydrodynamic escape of Venus atmosphere explaining present Ne and Ar isotopic ratios and low oxygen atmospheric content. *Earth Planet. Sci. Lett., 286,* 503–513.

Gillmann C. and Tackley P. (2014) Atmosphere/mantle coupling and feedbacks on Venus. *J. Geophys. Res.–Planets, 119,* 1189–1217.

Gilmore M., Treiman A., Helbert J., and Smrekar S. (2017) Venus surface composition constrained by observation and experiment. *Space Sci. Rev., 212,* 1511–1540.

Goldblatt C. (2016) The inhabitance paradox: How habitability and inhabitancy are inseparable. In *NASA Conference Proceedings of Comparative Climates of Terrestrial Planets II,* Moffett Field, 2015, arXiv:1603.00950.

Gorczyk W. and Vogt K. (2018) Intrusion of magmatic bodies into the continental crust: 3D numerical models. *Tectonics, 37,* 705–723.

Grinspoon D. H. (1993) Implications of the high D/H ratio for the sources of water in Venus' atmosphere. *Nature, 363,* 428–431.

Grott M., Morschhauser A., Breuer D., and Hauber E. (2011) Volcanic outgassing of CO_2 and H_2O on Mars. *Earth Planet. Sci. Lett., 308,* 391–400.

Gunell H., Maggiolo R., Nilsson H., et al. (2018) Why an intrinsic magnetic field does not protect a planet against atmospheric escape. *Astron. Astrophys., 614,* L3.

Haberle R. M., Murphy J. R., and Schaeffer J. (2003) Orbital change experiments with a Mars general circulation model. *Icarus, 161,* 66–89.

Halekas J. S., Brain D. A., Luhmann J. G., et al. (2017) Flows, fields, and forces in the Mars-solar wind interaction. *J. Geophys. Res.–Space Phys., 122,* 11320–11341.

Hamano K., Abe Y., and Genda H. (2013) Emergence of two types of terrestrial planet on solidification of magma ocean. *Nature, 497,* 607–610.

Hansen J., Sato M., Kharecha P., et al. (2008) Target atmospheric CO_2: Where should humanity aim? *Open Atmos. Sci. J., 2,* 217–231.

Hansen V. L. and López I. (2010) Venus records a rich early history. *Geology, 38,* 311–314.

Haqq-Misra J., Kopparapu R. K., Batalha N. E., Harman C. E., and Kasting J. F. (2016) Limit cycles can reduce the width of the habitable zone. *Astrophys. J., 827,* 120.

Harman C. E., Schwieterman E. W., Schottelkotte J. C., and Kasting J. F. (2015) Abiotic O_2 levels on planets around F, G, K, and M stars: Possible false positives for life? *Astrophys. J., 812,* 137.

Harrison M. (2009) The Hadean crust: Evidence from >4 Ga zircons. *Annu. Rev. Earth Planet. Sci., 37,* 479–505.

Hauck S. A. II, Aurnou J. M., and Dombard A. J. (2006) Sulfur's impact on core evolution and magnetic field generation on Ganymede. *J. Geophys. Res.–Planets, 111,* E9008, DOI: 10.1029/2005JE002557.

Hawkesworth C. J., Dhuime B., Pietranik A. B., Cawood P. A., Kemp A. I. S., and Storey C. D. (2010) The generation and evolution of the continental crust. *J. Geol. Soc., 167,* 229–248.

Hays J. D., Imbrie J., and Shackleton N. J. (1976) Variations in the Earth's orbit: Pacemaker of the ice ages. *Science, 194,* 1121–1132.

Helffrich G. (2017) Mars core structure — Concise review and anticipated insights from InSight. *Prog. Earth Planet. Sci., 4,* 24.

Heller R. and Armstrong J. (2014) Superhabitable worlds. *Astrobiology, 14,* 50–66.

Herd C. D. K., Borg L. E., Jones J. J., and Papike J. J. (2002) Oxygen fugacity and geochemical variations in the martian basalts: Implications for martian basalt petrogenesis and the oxidation state of the upper mantle of Mars. *Geochim. Cosmochim. Acta, 66*, 2025–2036.

Hier-Majumder S. and Hirschmann M. M. (2017) The origin of volatiles in the Earth's mantle. *Geochem. Geophys. Geosyst., 18*, 3078–3092.

Hillgren V. J., Gessmann C. K., and Li J. (2000) An experimental perspective on the light element in Earth's core. In *Origin of the Earth and Moon* (R. M. Canup et al., eds.), pp. 245–263. Univ. of Arizona, Tucson.

Hirschmann M. M. (2012) Magma ocean influence on early atmosphere mass and composition. *Earth Planet. Sci. Lett., 341–344*, 48–57.

Hirschmann M. M. and Dasgupta R. (2009) The H/C ratios of Earth's near-surface and deep reservoirs, and consequences for deep Earth volatile cycles. *Chem. Geol., 262*, 4–16.

Hirschmann M. and Kohlstedt D. (2012) Water in Earth's mantle. *Phys. Today, 65*, 40–45.

Hirschmann M. M., Withers A. C., Ardia P., and Foley N. T. (2012) Solubility of molecular hydrogen in silicate melts and consequences for volatile evolution of terrestrial planets. *Earth Planet. Sci. Lett., 345–348*, 38–48.

Hirth G. and Kohlstedt D. L. (2003) Rheology of the upper mantle and mantle wedge: A view from experimentalists. In *Inside the Subduction Factory* (J. Eiler, ed.), pp. 83–105. AGU Geophys. Monogr. 138, American Geophysical Union, Washington DC.

Hoehler T. M. (2007) An energy balance concept for habitability. *Astrobiology, 7*, 824–838.

Hoffman P. F. and Schrag D. P. (2002) The snowball Earth hypothesis: Testing the limits of global change. *Terra Nova, 14*, 129–155.

Hoffman P. F., Kaufman A. J., Halverson G. P., and Schrag D. P. (1998) A Neoproterozoic snowball Earth. *Science, 281*, 1342–1346.

Holloway J. R. and Blank J. G. (1994) Application of experimental results to C-O-H species in natural melts. *Rev. Mineral., 30*, 185–230.

Holloway J. R., Pan V., and Gudmundsson G. (1992) High-pressure fluid-absent melting experiments in the presence of graphite: Oxygen fugacity, ferric/ferrous ratio, and dissolved CO_2. *Eur. J. Mineral., 4*, 105–114.

Höning D., Hansen-Goos H., Airo A., and Spohn T. (2014) Biotic vs. abiotic Earth: A model for mantle hydration and continental coverage. *Planet. Space Sci., 98*, 5–13.

Hu R., Kass D. M., Ehlmann B. L., and Yung Y. L. (2015) Tracing the fate of carbon and the atmospheric evolution of Mars. *Nature Commun., 6*, 10003.

Hunten D. M. (1973) The escape of light gases from planetary atmospheres. *J. Atmos. Sci., 30*, 1481–1494.

Hynek B. M., Beach M., and Hoke M. R. T. (2010) Updated global map of Martian valley networks and implications for climate and hydrologic processes. *J. Geophys. Res., 115*, E09008.

Irifune T., Kobu N., Isshiki M., and Yamasaki Y. (1998) Phase transformations in serpentine and transportation of water into the lower mantle. *Geophys. Res. Lett., 25*, 203–206.

Irwin R. P. III, Howard A. D., Craddock R. A., and Moore J. M. (2005) An intense terminal epoch of widespread fluvial activity on early Mars: 2. Increased runoff and paleolake development. *J. Geophys. Res., 110*, E12S15, DOI: 10.1029/2005JE002460.

Jakosky B. M. and Phillips R. J. (2001) Mars' volatile and climate history. *Nature, 412*, 237–244.

Jakosky B. M., Slipski M., Benna M., et al. (2017) Mars' atmospheric history derived from upper-atmosphere measurements of $^{38}Ar/^{36}Ar$. *Science, 355*, 1408–1410.

Jakosky B. M., Brain D., Chaffin M., et al. (2018) Loss of the martian atmosphere to space: Present-day loss rates determined from MAVEN observations and integrated loss through time. *Icarus, 315*, 146–157.

Jansen T., Scharf C., Way M., and Del Genio A. (2019) Climates of warm Earth-like planets II: Rotational 'Goldilocks' zones for fractional habitability and silicate weathering. *Astrophys. J., 875*, 79.

Jing Z. and Karato S. (2009) The density of volatile bearing melts in the Earth's deep mantle: The role of chemical composition. *Chem. Geol., 262*, 100–107.

Johnson B. and Goldblatt C. (2015) The nitrogen budget of Earth. *Earth-Sci. Rev., 148*, 150–173.

Johnson B. W. and Goldblatt C. (2018) EarthN: A new Earth system nitrogen model. *Geochem. Geophys. Geosys., 19*, 2516–2542.

Johnson T. V., Mousis O., Lunine J. I., and Madhusudhan N. (2012) Planetesimal compositions in exoplanet systems. *Astrophys. J, 757*, 192.

Johnstone C. P., Güdel M., Stökl A., et al. (2015) The evolution of stellar rotation and the hydrogen atmospheres of habitable-zone terrestrial planets. *Astrophys. J. Lett., 815*, L12.

Kamber B. S., Whitehouse M. J., Bolhar R., and Moorbath S. (2005) Volcanic resurfacing and the early terrestrial crust: Zircon U-Pb and REE constraints from the Isua Greenstone Belt, southern West Greenland. *Earth Planet. Sci. Lett., 240(2)*, 276–290.

Kane S. R., Kopparapu R. K., and Domagal-Goldman S. D. (2014) On the frequency of potential Venus analogs from Kepler data. *Astrophys. J. Lett., 794*, L5.

Kanzaki Y. and Murakami T. (2015) Estimates of atmospheric CO_2 in the Neoarchean-Paleoproterozoic from paleosols. *Geochim. Cosmochim. Acta, 159*, 190–219.

Kasting J. F. (1988) Runaway and moist greenhouse atmospheres and the evolution of Earth and Venus. *Icarus, 74*, 472–494.

Kasting J. F. (1991) CO_2 condensation and the climate of early Mars. *Icarus, 94*, 1–13.

Kasting J. F. and Pollack J. B. (1983) Loss of water from Venus. I. Hydrodynamic escape of hydrogen. *Icarus, 53*, 479–508.

Kasting J. F., Whitmire D. P., and Reynolds R. T. (1993) Habitable zones around main sequence stars. *Icarus, 101*, 108–128.

Kattenhorn S. A. and Prockter L. M. (2014) Evidence for subduction in the ice shell of Europa. *Nature Geosci., 7*, 762–767.

Katz R. F., Spiegelman M., and Langmuir C. H. (2003) A new parameterization of hydrous mantle melting. *Geochem. Geophys. Geosyst., 4*, 1073.

Kilic C., Raible C. C., and Stocker T. F. (2017) Multiple climate states of habitable exoplanets: The role of obliquity and irradiance. *Astrophys. J., 844*, 147.

Kilic C., Lunkeit F., Raible C. C., and Stocker T. F. (2018) Stable equatorial ice belts at high obliquity in a coupled atmosphere-ocean model. *Astrophys. J., 864*, 106.

Kirschvink J. L. (1992) Late Proterozoic low-latitude global glaciation: The snowball Earth. In *The Proterozoic Biosphere, A Multidisciplinary Study* (J. W. Schopf and C. Klein, eds.), pp. 51–52. Cambridge Univ., New York.

Kite E. S., Williams J.-P., Lucas A., and Aharonson O. (2014) Low palaeopressure of the martian atmosphere estimated from the size distribution of ancient craters. *Nature Geosci., 7*, 335–339.

Kite E. S., Gao P., Goldblatt C., Mischna M. A., Mayer D. P., and Yung Y. (2017a) Methane bursts as a trigger for intermittent lake-forming climates on post-Noachian Mars. *Nature Geosci., 10*, 737–740.

Kite E. S., Sneed J., Mayer D. P., and Wilson S. A. (2017b) Persistent or repeated surface habitability on Mars during the late Hesperian-Amazonian. *Geophys. Res. Lett., 44*, 3991–3999.

Kitzmann D. (2016) Revisiting the scattering greenhouse effect of CO_2 ice clouds. *Astrophys. J. Lett., 817*, L18.

Kleinböhl A., Willacy K., Friedson A. J., Chen P., and Swain M. R. (2018) Buildup of abiotic oxygen and ozone in moist atmospheres of temperate terrestrial exoplanets and its impact on the spectral fingerprint in transit observations. *Astrophys. J., 862*, 92.

Kleine T., Touboul M., Bourdon B., Nimmo F., Mezger K., Palme H., Jacobsen S. B., Yin Q.-Z., and Halliday A. (2009) Hf-W chronology of the accretion and early evolution of asteroids and terrestrial planets. *Geochim. Cosmochim. Acta, 73*, 5150–5188.

Konopliv A. S. and Yoder C. F. (1996) Venusian k_2 tidal Love number from Magellan and PVO tracking data. *Geophys. Res. Lett., 23*, 1857–1860.

Kopp R. E., Kirschvink J. L., Hilburn I. A., and Nash C. Z. (2005) The Paleoproterozoic snowball Earth: A climate disaster triggered by the evolution of oxygenic photosynthesis. *Proc. Natl. Acad. Sci. U.S.A., 102*, 11131–11136.

Kopparapu R. K., Ramirez R. M., and Kotte J. S. (2014) Habitable zones around main-sequence stars: Dependence on planetary mass. *Astrophys. J. Lett., 787*, L29.

Korenaga J. (2013) Initiation and evolution of plate tectonics on Earth: Theories and observations. *Annu. Rev. Earth Planet. Sci., 41*, 117–151.

Korenaga J. (2018) Estimating the formation age distribution of continental crust by unmixing zircon ages. *Earth Planet. Sci. Lett., 482*, 388–395.

Korenaga J., Planavsky N. J., and Evans D. A. D. (2017) Global water cycle and the coevolution of the Earth's interior and surface environment. *Philos. Trans. R. Soc. A, 375*, 20150393.

Krissansen-Totton J., Arney G., and Catling D. C. (2018) Constraining the climate and ocean pH of the early Earth with a geological carbon cycle model. *Proc. Natl. Acad. Sci. U.S.A., 115*, 4105–4110.

Kruijer T. S., Kleine T., Borg L. E., Brennecka G. A., Irving A. J., Bischoff A., and Agee C. B. (2017) The early differentiation of Mars inferred from Hf-W chronometery. *Earth Planet. Sci. Lett., 474*, 345–354.

Kulikov Y. N., Lammer H., Lichtenegger H. I. M., et al. (2006) Atmospheric and water loss from early Venus. *Planet. Space Sci., 54*, 1425–1444.

Kump L. R., Brantley S. L., and Arthur M. A. (2000) Chemical weathering, atmospheric CO_2, and climate. *Annu. Rev. Earth Planet. Sci., 28*, 611–667.

Kurokawa H., Kurosawa K., and Usui T. (2018) A lower limit of atmospheric pressure on early Mars inferred from nitrogen and argon isotopic compositions. *Icarus, 299*, 443–459.

Lacis A. A., Schmidt G. A., Rind D. A., and Ruedy R. A. (2010) Atmospheric CO_2: Principal control knob governing Earth's temperature. *Science, 330*, 356–359.

Lammer H., Kasting J. F., Chassefière E., Johnson R. E., Kulikov Y. N., and Tian F. (2008) Atmospheric escape and evolution of terrestrial planets and satellites. *Space Sci. Rev., 139*, 399–436.

Lammer H., Stokl A., Erkaev N. V., et al. (2014) Origin and loss of nebula-captured hydrogen envelopes from 'sub-' to 'super-Earths' in the habitable zone of Sun-like stars. *Mon. Not. R. Astron. Soc., 439*, 3225–3238.

Lammer H., Zerkle A. H., Gebauer S., et al. (2018) Origin and evolution of the atmospheres of early Venus, Earth, and Mars. *Astron. Astrophys. Rev., 26*, 2.

Landuyt W. and Bercovici D. (2009) Variations in planetary convection via the effect of climate on damage. *Earth Planet. Sci. Lett., 277*, 29–37.

Larsen K. K., Trinquier A., Paton C., et al. (2011) Evidence for magnesium isotope heterogeneity in the solar protoplanetary disk. *Astrophys. J. Lett., 735*, L37.

Laskar J., Correia A. C. M., Gastineau M., Joutel F., Levrard B., and Robutel P. (2004) Long term evolution and chaotic diffusion of the insolation quantities of Mars. *Icarus, 170*, 343–364.

Lasue J., Quesnel Y., Langlais B., and Chassefière E. (2015) Methane storage capacity of the early martian cryosphere. *Icarus, 260*, 205–214.

Leblanc F., Martinez A., Chaufray J. Y., et al. (2018) On Mars' atmospheric sputtering after MAVEN's first martian year of measurements. *Geophys. Res. Lett., 45*, 4685–4691.

Leconte J., Wu H., Menou K., and Murray N. (2015) Asynchronous rotation of Earth-mass planets in the habitable zone of lower-mass stars. *Science, 347*, 632–635.

Lecuyer C., Simon L., and Guyot F. (2000) Comparison of carbon, nitrogen, and water budgets on Venus and the Earth. *Earth Planet. Sci. Lett., 181*, 33–40.

Le Hir G., Ramstein G., Donnadieu Y., and Goddé (2008) Scenario for the evolution of atmospheric pCO_2 during a snowball Earth. *Geology, 36*, 47–50.

Le Hir G., Donnadieu Y., Krinner G., and Ramstein G. (2010) Toward the snowball Earth deglaciation. *Climate Dynam., 35*, 285–297.

Lenardic A., Jellinek A. M., and Moresi L.-N. (2008) A climate induced transition in the tectonic style of a terrestrial planet. *Earth Planet. Sci. Lett., 271*, 34–42.

Lenardic A., Crowley J. W., Jellinek A. M., and Weller M. (2016a) The solar system of forking paths: Bifurcations in planetary evolution and the search for life-bearing planets in our galaxy. *Astrobiology, 16*, 551–559.

Lenardic A., Jellinek A. M., Foley B., O'Neill C., and Moore W. B. (2016b) Climate-tectonic coupling: Variations in the mean, variations about the mean, and variations in mode. *J. Geophys. Res., 121*, 1831–1864.

Li J. and Agee C. B. (1996) Geochemistry of mantle-core differentiation at high pressure. *Nature, 381*, 686–689.

Li Z. X. A and Lee C. T. A. (2004) The constancy of upper mantle fO_2 through time inferred from V/Sc ratios in basalts. *Earth Planet. Sci. Lett., 228*, 483–493.

Lichtenegger H. I. M., Lammer H., Griessmeier J.-M., et al. (2010) Aeronomical evidence for higher CO_2 levels during Earth's Hadean epoch. *Icarus, 210*, 1–7.

Lichtenegger H. I. M., Kislyakova K. G., Odert P., et al. (2016) Solar XUV and ENA-driven water loss from early Venus' steam atmosphere. *J. Geophys. Res.–Space Phys., 121*, 4718–4732.

Lillis R. J., Deighan J., Fox J. L., et al. (2017) Photochemical escape of oxygen from Mars: First results from MAVEN *in situ* data. *J. Geophys. Res.–Space Phys., 122*, 3815–3836.

Lissauer J. J., Barnes J. W., and Chambers J. E. (2011) Obliquity variations of a moonless Earth. *Icarus, 217*, 77–87.

Lodders K. and Fegley B. (1997) An oxygen isotope model for the composition of Mars. *Icarus, 126*, 373–394.

López-Morales M., Gómez-Pérez N., and Ruedas T. (2012) Magnetic fields in Earth-like exoplanets and implications for habitability around M-dwarfs. *Origins Life Evol. Biospheres, 41*, 533–537.

Luger R. and Barnes R. (2015) Extreme water loss and abiotic O_2 buildup on planets throughout the habitable zones of M dwarfs. *Astrobiology, 15*, 119–143.

Luhmann J. G. and Schwingenschuh K. (1990) A model of the energetic ion environment of Mars. *J. Geophys. Res.–Space Phys, 95*, 939–945.

Lunt D. J., Huber M., Anagnostou E., et al. (2017) The DeepMIP contribution to PMIP4: Experimental designs for model simulations of the EECO, PETM, and pre-PETM (version 1.0). *Geosci. Model Dev., 10*, 889–901.

Lupu R. E., Zahnle K., Marley M. S., Schaefer L., Fegley B., Morley C., Cahoy K., Freedman R., and Fortney J. J. (2014) The atmospheres of Earthlike planets after giant impact events. *Astrophys. J., 784*, 27.

Lyons T. W., Reinhard C. T., and Planavsky N. J. (2014) The rise of oxygen in Earth's early ocean and atmosphere. *Nature, 506*, 307–315.

Ma Y., Fang X., Russell C. T., et al. (2014) Effects of crustal field rotation on the solar wind plasma interaction with Mars. *Geophys. Res. Lett., 41*, 6563–6569.

Malin M. C. and Edgett K. S. (2003) Evidence for persistent flow and aqueous sedimentation on early Mars. *Science, 302*, 1931–1934.

Manabe S. and Strickler R. F. (1964) Thermal equilibrium of the atmosphere with a convective adjustment. *J. Atmos. Sci., 21*, 361–385.

Mandt K., Mousis O., and Chassefière E. (2015) Comparative planetology of the history of nitrogen isotopes in the atmospheres of Titan and Mars. *Icarus, 254*, 259–261.

Marschall H. R., Hawkesworth C. J., Storey C. D., Dhuime B., Leat P. T., Meyer H.-P., and Tamm-Buckle S. (2010) The Annandagstoppane granite, East Antarctica: Evidence for Archaean intracrustal recycling in the Kaapvaal–Grunehogna Craton from zircon O and Hf isotopes. *J. Petrol., 51*, 2277–2301.

Marty B., Zimmermann L., Pujol M., Burgess R., and Philippot P. (2013) Nitrogen isotopic composition and density of the Archean atmosphere. *Science, 342*, 101–104.

Matsui T. and Abe Y. (1986) Formation of a 'magma ocean' on the terrestrial planets due to the blanketing effect of an impact-induced atmosphere. *Earth Moon Planets, 34*, 223–230.

Matsui T. and Abe Y. (1987) Evolutionary tracks of the terrestrial planets. *Earth Moon Planets, 39*, 207–214.

McCubbin F. M., Riner M. A., Vander Kaaden K. E., and Burkemper L. K. (2012) Is Mercury a volatile-rich planet? *Geophys. Res. Lett., 39*, L09202, DOI: 10.1029/2012GL051711.

McCubbin F. M., Boyce J. W., Srinivasan P., Santos A. R., Elardo S. M., Filiberto J., Steele A., and Shearer C. K. (2016) Heterogeneous distribution of H_2O in the martian interior; implications for the abundance of H_2O in depleted and enriched mantle sources. *Meteoritics & Planet. Sci., 51*, 2036–2060.

McDonough W. F. and Sun S.-S. (1995) The composition of the Earth. *Chem. Geol., 120*, 223–253.

McGovern P. J. and Schubert G. (1989) Thermal evolution of the Earth: Effects of volatile exchange between atmosphere and interior. *Earth Planet. Sci. Lett., 96*, 27–37.

McKinnon W. B., Zahnle K. J., Ivanov B. I., and Melosh H. J. (1997) Cratering on Venus: Models and observations. In *Venus II* (S. W. Bougher et al., eds.), pp. 969–1014. Univ. of Arizona, Tucson.

Menou K. (2015) Climate stability of habitable Earth-like planets. *Earth Planet. Sci. Lett., 429*, 20–24.

Michalski J. R. and Niles P. B. (2010) Deep crustal carbonate rocks exposed by meteor impact on Mars. *Nature Geosci., 3,* 751–755.

Mikhail S. and Sverjensky D. A. (2014) Nitrogen speciation in upper mantle fluids and the origin of Earth's nitrogen-rich atmosphere. *Nature Geosci., 7,* 816–819.

Milankovitch M. (1941) *Canon of Insolation and the Ice-Age Problem: Special Publications of the Royal Serbian Academy, Vol. 132.* Israel Program for Scientific Translations, Jerusalem. 484 pp.

Mischna M. A., Kasting J. F., Pavlov A., and Freedman R. (2000) Influence of carbon dioxide clouds on early martian climate. *Icarus, 145,* 546–554.

Mischna M. A., Baker V., Milliken R., Richardson M., and Lee C. (2013) Effects of obliquity and water vapor/trace gas greenhouses in the early martian climate. *J. Geophys. Res.–Planets, 118,* 560–576.

Mojzsis S. J., Arrhenius G., McKeegan K. D., Harrison T. M., Nutman A. P., and Friend C. R. L. (1996) Evidence for life on Earth before 3,800 million years ago. *Nature, 384,* 55–59.

Moore T. E. and Horwitz J. L. (2007) Stellar ablation of planetary atmospheres. *Rev. Geophys., 45,* RG3002, DOI: 10.1029/2005RG000194.

Moore W. B. and Webb A. A. G. (2013) Heat-pipe Earth. *Nature, 501,* 501–505.

Mocquet A., Rosenblatt P., Dehant V., and Verhoeven O. (2011) The deep interior of Venus, Mars, and the Earth: A brief review and the need for planetary surface-based measurements. *Planet. Space Sci., 59,* 1048–1061.

Morgan G. A., Head J. W., and Marchant D. R. (2011) Preservation of Late Amazonian Mars ice and water-related deposits in a unique crater environment in Noachis Terra: Age relationships between lobate debris tongues and gullies. *Icarus, 211,* 347–365.

Morschhauser A., Grott M., and Breuer D. (2011) Crustal recycling, mantle dehydration, and the thermal evolution of Mars. *Icarus, 212,* 541–558.

Morschhauser A., Vervelidou F., Thomas P., Grott M., Lesur V., and Gilder S. A. (2018) Mars' crustal magnetic field. In *Magnetic Fields in the Solar System* (H. Lühr et al., eds.), pp. 331–356. Springer Nature, Switzerland.

Moyen J.-F. and Martin H. (2012) Forty years of TTG research. *Lithos, 148,* 312–336.

Müller N., Helbert J., Hashimoto G. L., Tsang C. C. C., Erard S., Piccioni G., and Drossart P. (2008) Venus surface thermal emission at 1 mm in VIRTIS imaging observations: Evidence for variation of crust and mantle differentiation conditions. *J. Geophys. Res., 113,* E00B17, DOI: 10.1029/2008JE003118.

Mukhopadhyay S. (2012) Early differentiation and volatile accretion recorded in deep-mantle neon and xenon. *Nature, 486,* 101–104.

Mysen B. O., Yamashita S., and Chertkova N. (2008) Solubility and solution mechanisms of NOH volatiles in silicate melts at high pressure and temperature-amine groups and hydrogen fugacity. *Am. Mineral., 93,* 1760–1770.

Nakajima M. and Stevenson D. J. (2015) Melting and mixing states of the Earth's mantle after the Moon-forming impact. *Earth Planet. Sci. Lett., 427,* 286–295.

Niles P. B., Catling D. C., Berger G., Chassefiere E., Ehlmann B. L., Michalski J. R., Morris R., Ruff S. W., and Sutter B. (2013) Geochemistry of carbonates on Mars: Implications for climate history and nature of aqueous environment. *Space Sci. Rev., 174,* 301–328.

Nimmo F. (2002) Why does Venus lack a magnetic field? *Geology, 30,* 987–990.

Noack L., Breuer D., and Spohn T. (2012) Coupling the atmosphere with interior dynamics: Implications for the resurfacing of Venus. *Icarus, 217,* 484–498.

Noack L., Rivoldini A., and Van Hoolst T. (2017) Volcanism and outgassing of stagnant-lid planets: Implications for the habitable zone. *Phys. Earth Planet. Inter., 269,* 40–57.

O'Brien D. P., Izidoro A., Jacobson S. A., Raymond S. N., and Rubie D. C. (2018) The delivery of water during terrestrial planet formation. *Space Sci. Rev., 214,* 47.

Olson S. L., Schwieterman E. W., Reinhard C. T., and Lyons T. W. (2018) Earth: Atmospheric evolution of a habitable planet. In *Handbook of Exoplanets* (H. Deeg and J. Belmonte, eds.), pp. 1–37. Springer Nature, Switzerland.

Owen J. E. and Jackson A. P. (2012) Planetary evaporation by UV and X-ray radiation: Basic hydrodynamics. *Mon. Not. R. Astron. Soc., 425,* 2931–2947.

Ozima M. and Podosek F. A. (1983) *Noble Gas Geochemistry.* Cambridge Univ., New York. 367 pp.

Papale P. (1997) Modeling of the solubility of a one-component H_2O or CO_2 fluid in silicate liquids. *Contrib. Mineral. Petrol., 126 (3),* 237–251.

Papineau D. (2010) Mineral environments on the earliest Earth. *Elements, 6,* 25–30.

Parkinson I. J. and Arculus R. J. (1999) The redox state of subduction zones: Insights from arc-peridotites. *Chem. Geol., 160,* 409–423.

Payne J. L., Boyer A. G., Brown J. H., et al. (2009) Two-phase increase in the maximum size of life over 3.5 billion years reflects biological innovation and environmental opportunity. *Proc. Natl. Acad. Sci. U.S.A, 106,* 24–27.

Planavsky N. J., Asael D., Hofmann A., et al. (2014) Evidence for oxygenic photosynthesis half a billion years before the Great Oxidation Event. *Nature Geosci., 7,* 283–286.

Poirier J.-P. (1994) Light elements in the Earth's outer core: A critical review. *Phys. Earth Planet. Inter., 85,* 319–337.

Pringle E. A., Savage P. S., Jackson M. G., Barrat J.-A., and Moynier F. (2013) Si isotope homogeneity of the solar nebula. *Astrophys. J., 779,* 123.

Quintana E. V., Barclay T., Borucki W. J., Rowe J. F., and Chambers J. E. (2016) The frequency of giant impacts on Earth-like worlds. *Astrophys. J., 821,* 126.

Ramirez R. M. and Craddock R. A. (2018) The geological and climatological case for a warmer and wetter early Mars. *Nature Geosci., 11,* 230–237.

Ramirez R. M., Kopparapu R., Zugger M. E., Robinson T. D., Freedman R., and Kasting J. F. (2014) Warming early Mars with CO_2 and H_2. *Nature Geosci., 7,* 59–63.

Ramstad R., Barabash S., Futaana Y., Nilsson H., and Holmström M. (2017) Effects of the crustal magnetic fields on the martian atmospheric ion escape rate. *Geophys. Res. Lett., 43,* 10574–10579.

Raval A. and Ramanathan V. (1989) Observational determination of the greenhouse effect. *Nature, 342,* 758–761.

Raymond S. N., Quinn T., and Lunine J. I. (2006) High-resolution simulations of the final assembly of Earth-like planets I. Terrestrial accretion and dynamics. *Icarus, 183,* 265–282.

Reinhard C. T., Olson S. L., Schwieterman E. W., and Lyons T. W. (2017) False negatives for remote life detection on ocean-bearing planets: Lessons from the early Earth. *Astrobiology, 17,* 287–297.

Reymer A. and Schubert G. (1984) Phanerozoic addition rates to the continental crust and crustal growth. *Tectonics, 3,* 63–77.

Ribas I., Guinan E. F., Güdel M., and Audard M. (2005) Evolution of the solar activity over time and effects on planetary atmospheres. I. High-energy irradiances (1–1700 Å). *Astrophys. J., 622,* 680–694.

Righter K., Danielson L. R., Pando K. M., et al. (2015) Highly siderophile element (HSE) abundances in the mantle of Mars are due to core formation at high pressure and temperature. *Meteoritics & Planet. Sci., 50,* 604–631.

Ringwood A. E. (1959) On the chemical evolution and densities of the planets. *Geochim. Cosmochim. Acta, 15,* 257–283.

Ringwood A. E. (1979) Composition and origin of the Earth. In *The Earth: Its Origin, Structure and Evolution* (M. W. McElhinny, ed.), pp. 1–58. Academic, San Diego.

Rivoldini A., Van Hoolst T., Verhoeven O., Mocquet A., and Dehant V. (2011) Geodesy constraints on the interior structure and composition of Mars. *Icarus, 213,* 451–472.

Romeo I. and Turcotte D. L. (2008) Pulsating continents on Venus: An explanation for crustal plateaus and tessera terrains. *Earth Planet. Sci. Lett., 276,* 85–97.

Rosing M. T., Bird D. K., Sleep N. H., Glassley W., and Albarede F. (2006) The rise of continents — An essay on the geologic consequences of photosynthesis. *Palaeogeogr., Palaeoclimatol., Palaeoecol., 232,* 99–113.

Rosing M. T., Bird D. K., Sleep N. H., and Bjerrum C. J. (2010) No climate paradox under the faint early Sun. *Nature, 464,* 744–747.

Rozel A. B., Golabek G. J., Jain C., Tackley P. J., and Gerya T. (2017) Continental crust formation on early Earth controlled by intrusive magmatism. *Nature, 545,* 332–335.

Rubie D. C., Jacobson S. A., Morbidelli A., O'Brien D. P., Young E. D., de Vries J., Nimmo F., Palme H., and Frost D. J. (2015) Accretion and differentiation of the terrestrial planets with implications for the compositions of early-formed solar system bodies and accretion of water. *Icarus, 248,* 89–108.

Rückriemen T., Breuer D., and Spohn T. (2018) Top-down freezing in a Fe-FeS core and Ganymede's present-day magnetic field. *Icarus, 307,* 172–196.

Rushby A. J., Claire M. W., Osborn H., and Watson A. J. (2013) Habitable zone lifetimes of exoplanets around main sequence stars. *Astrobiology, 13,* 833–849.

Rushby A. J., Johnson M., Mills B. J. W., Watson A. J., and Claire M. W. (2018) Long-term planetary habitability and the carbonate-silicate cycle. *Astrobiology, 18,* 469–480.

Sagan C. and Chyba C. (1997) The early faint Sun paradox: Organic shielding of ultraviolet-labile greenhouse gases. *Science, 276,* 1217–1221.

Sagan C. and Mullen G. (1972) Earth and Mars: Evolution of atmospheres and surface temperatures. *Science, 177,* 52–56.

Sakai S., Seki K., Terada N., et al. (2018) Effects of a weak intrinsic magnetic field on atmospheric escape from Mars. *Geophys. Res. Lett., 45,* 9336–9343.

Salvador A., Massol H., Davaille A., Marcq E., Sarda P., and Chassefiere E. (2017) The relative influence of H_2O and CO_2 on the primitive surface conditions and evolution of rocky planets. *J. Geophys. Res., 122,* 1458–1486.

Sandu C. and Kiefer W. S. (2012) Degassing history of Mars and the lifespan of its magnetic dynamo. *Geophys. Res. Lett., 39,* L02301.

Sanloup C., Jambon A., and Gillet P. (1999) A simple chondritic model of Mars. *Phys. Earth Planet. Inter.,112,* 43–54.

Santos N. C., Adibekyan V., Mordasini C., et al. (2015) Constraining planet structure from stellar chemistry: The cases of CoRoT-7, Kepler-10, and Kepler-93. *Astron. Astrophys., 580,* L13.

Sato M., Hickling N. L., and McLane J. E. (1973) Oxygen fugacity values of Apollo 12, 14, and 15 lunar samples and reduced state of lunar magmas. *Proc. Lunar Planet. Sci. Conf. 4th,* pp. 1061–1079.

Sautter V., Toplis M. J., Wiens R. C., et al. (2015) *In situ* evidence for continental crust on early Mars. *Nature Geosci., 8,* 605–609.

Seki K., Elphic R. C., Hirahara M., Terasawa T., and Mukai T. (2001) On atmospheric loss of oxygen ions from Earth through magnetospheric processes. *Science, 291,* 1939–1941.

Shields A. L., Meadows V. S., Bitz C. M., Pierrehumbert R. T., Joshi M. M., and Robinson T. D. (2013) The effect of host star spectral energy distribution and ice-albedo feedback on the climate of extrasolar planets. *Astrobiology, 13,* 715–739.

Sholes S. F., Smith M. L., Claire M. W., Zahnle K. J., and Catling D. C. (2017) Anoxic atmospheres on Mars driven by volcanism: Implications for past environments and life. *Icarus, 290,* 46–62.

Slapak R., Schillings A., Nilsson H., et al. (2017) Atmospheric loss from the dayside open polar region and its dependence on geomagnetic activity: Implications for atmospheric escape on evolutionary timescales. *Ann. Geophys., 35,* 721–731.

Sleep N. H. and Zahnle K. (2001) Carbon dioxide cycling and implications for climate on ancient Earth. *J. Geophys. Res., 106,* 1373–1399.

Sleep N. H., Zahnle K., and Neuhoff P. S. (2001) Initiation of clement surface conditions on the earliest Earth. *Proc. Natl. Acad. Sci. U.S.A., 98,* 3666–3672.

Smrekar S. E. and Sotin C. (2012) Constraints on mantle plumes on Venus: Implications for volatile history. *Icarus, 217,* 510–523.

Smythe D. J. and Brenan J. M. (2016) Magmatic oxygen fugacity estimated using zircon-melt partitioning of cerium. *Earth Planet. Sci. Lett., 453,* 260–266.

Sohl L. E. and Chandler M. A. (2007) Reconstructing Neoproterozoic palaeoclimates using a combined data/modeling approach. In *Deep-Time Perspectives on Climate Change: Marrying the Signal from Computer Models and Biological Proxies* (M. Williams et al., eds.), pp. 61–80. Micropalaeontological Society Special Publication #2, Geological Society, London.

Som S. M., Buick R., Hagadorn J. W., Blake T. S., Perreault J. M., Harnmeijer J. P., and Catling D. C. (2016) Earth's air pressure 2.7 billion years ago constrained to less than half of modern levels. *Nature Geosci., 9,* 448–451.

Soto A., Mischna M., Schneider T., Lee C., and Richardson M. (2015) Martian atmospheric collapse: Idealized GCM studies. *Icarus, 250,* 553–569.

Starukhina L. V. and Kreslavsky M. A. (2002) Radiophysical properties of Venusian highlands: Possible role of magnetic effects. In *Lunar Planet. Sci. XXXIII,* Abstract #1559. Lunar and Planetary Institute, Houston.

Steffen W., Crutzen P. J., and McNeill J. R. (2007) The Anthropocene: Are humans now overwhelming the great forces of nature. *Ambio, 36,* 614–621.

Stephens G. L., Li J., Wild M., Clayson C. A., Loeb N., Kato S., L'Ecuyer T., Stackhouse Jr. P. W., Lebsock M., and Andrews T. (2012) An update on Earth's energy balance in light of the latest global observations. *Nature Geosci., 5,* 691–696.

Stevenson D. J., Spohn T., and Schubert G. (1983) Magnetism and thermal evolution of the terrestrial planets. *Icarus, 54,* 466–489.

Stewart S. T. and Nimmo F. (2002) Surface runoff features on Mars: Testing the carbon dioxide formation hypothesis. *J. Geophys. Res.–Planets, 107(E9),* 5069.

Stramler K., Del Genio A. D., and Rossow W. B. (2011) Synoptically driven Arctic winter states. *J. Climate, 24,* 1747–1762.

Strangeway R. J., Russell C. T., Luhmann J. G., et al. (2010) Does a planetary-scale magnetic field enhance or inhibit ionospheric plasma outflows? Abstract SM33B-1893 presented at 2010 Fall Meeting, AGU, San Francisco, Calif., 13–17 Dec.

Strom R. G., Schaber G. G., and Dawson D. D. (1994) The global resurfacing of Venus. *J. Geophys. Res., 99,* 10899–10926.

Stüeken E. E., Kipp M. A., Koehler M. C., Schwieterman E. W., Johnson B., and Buick R. (2016) Modeling pN_2 through geologic time: Implications for planetary climates and atmospheric biosignatures. *Astrobiology, 16,* 949–963.

Tashiro T., Ishida A., Hori M., Igisu M., Koike M., Méjean P., Takahata N., Sano Y., and Komiya T. (2017) Early trace of life from 3.95 Gya sedimentary rocks in Labrador, Canada. *Nature, 549,* 516–518.

Taylor S. R. and McLennan S. M. (1995) The geochemical evolution of the continental crust. *Rev. Geophys., 33,* 241–265.

Tian F., Toon O. B., Pavlov A. A., and DeSterck H. (2005) Transonic hydrodynamic escape of hydrogen from extrasolar planetary atmospheres. *Astrophys. J., 621,* 1049–1060.

Tian F., Kasting J. F., Lin H., and Roble R. G. (2008) Hydrodynamic planetary thermosphere model: 1. The response of the Earth's thermosphere to extreme solar EUV conditions and the significance of adiabatic cooling. *J. Geophys. Res., 113,* E05008, DOI: 10.1029/2007JE002946.

Tian F., Kasting J. F., and Solomon S. C. (2009) Thermal escape of carbon from the early martian atmosphere. *Geophys. Res. Lett., 36,* L02205, DOI: 10.1029/2008GL036513.

Tonks W. B. and Melosh H. J. (1993) Magma ocean formation due to giant impacts. *J. Geophys. Res.–Planets, 98,* 5319–5333.

Tosi N., Godolt M., Stracke B., et al. (2017) The habitability of a stagnant-lid Earth. *Astron. Astrophys., 605,* A71.

Trail D., Watson E. B., and Tailby N. D. (2011) The oxidation state of Hadean magmas and implications for early Earth's atmosphere. *Nature, 480,* 79–82.

Trail D., Watson E. B., and Tailby N. D. (2013) Insights into the Hadean Earth from experimental studies of zircon. *J. Geol. Soc. India, 81,* 605–636.

Tu L., Johnstone C. P., Güdel M., and Lammer H. (2015) The extreme ultraviolet and X-ray Sun in time: High-energy evolutionary tracks of a solar-like star. *Astron. Astrophys., 577,* L3.

Unterborn C. T. and Panero W. R. (2017) The effects of Mg/Si on the exoplanetary refractory oxygen budget. *Astrophys. J., 845,* 61.

van Hunen J. and Moyen J-F. (2012) Archean subduction: Fact or fiction? *Annu. Rev. Earth Planet. Sci., 40,* 195–219.

van Lieshout R., Min M., and Dominik C. (2014) Dusty tails of evaporating exoplanets. I. Constraints on the dust composition. *Astron. Astrophys., 572,* A76.

van Lieshout R., Min M., Dominik C., et al. (2016) Dusty tails of evaporating exoplanets. II. Physical modeling of the KIC 12557548b light curve. *Astron. Astrophys., 596,* A32.

Veeder G. J., Matson D. L., Johnson T. V., Blaney D. L., and Gougen J. D. (1994) Io's heat flow from infrared radiometry: 1983–1993. *J. Geophys. Res., 99,* 17095–17162.

Wadhwa M. (2001) Redox state of Mars' upper mantle and crust from Eu anomalies in shergottite pyroxenes. *Science, 291,* 1527–1530.

Wadhwa M. (2008) Redox conditions on small bodies, the Moon and Mars. *Rev. Mineral. Geochem., 68,* 493–510.

Walker J. C. G., Hays P. B., and Kasting J. F. (1981) A negative feedback mechanism for the long-term stabilization of Earth's surface temperature. *J. Geophys. Res., 86,* 9776–9782.

Wang Z. and Becker H. (2017) Chalcophile elements in martian meteorites indicate low sulfur content in the martian interior and a volatile element-depleted late veneer. *Earth Planet. Sci. Lett., 463,* 56–68.

Wänke H. and Dreibus G. (1988) Chemical composition and accretion history of terrestrial planets. *Philos. Trans. R. Soc. A, 325,* 545–557.

Watson A. J., Donahue T. M., and Walker J. C. G. (1981) The dynamics of a rapidly escaping atmosphere: Applications to the evolution of Earth and Venus. *Icarus, 48,* 150–166.

Way M. J., Del Genio A. D., Kiang N. Y., Sohl L. E., Grinspoon D. H., Aleinov I., Kelley M., and Clune T. (2016) Was Venus the first habitable world of our solar system? *Geophys. Res. Lett., 43,* 8376–8383.

Way M. J., Aleinov I., Amundsen D. S., Chandler M. A., Clune T. L., Del Genio A. D., Fujii Y., Kelley M., Kiang N. Y., Sohl L., and Tsigaridis T. (2017) Resolving Orbital and Climate Keys of Earth and Extraterrestrial Environments with Dynamics (ROCKE-3D) 1.0: A general circulation model for simulating the climates of rocky planets. *Astrophys. J., Suppl. Ser., 231,* 12.

Way M. J., Del Genio A., and Amundsen D. S. (2018) Modeling Venus-like worlds through time. *Proceedings of the Venera-D Modeling Workshop,* Moscow, Russia, arXiv:1802.05434.

Wei Y., Fraenz M., Dubinin E., et al. (2012) Enhanced atmospheric oxygen outflow on Earth and Mars driven by a corotating interaction region. *J. Geophys. Res.–Space Phys., 117,* A03208, DOI: 10.1029/2011JA017340.

Weidenschilling S. J. (1978) Iron/silicate fractionation and the origin of Mercury. *Icarus, 35,* 99–111.

Weis F. A., Bellucci J. J., Skogby H., Stadler R., Nemchin A. A., Whitehouse M. J. (2017) Water content in the martian mantle: A Nakhla perspective. *Geochim. Cosmochim. Acta, 212,* 84–98.

Williams R. M. E. et. al. (2013) Martian fluvial conglomerates at Gale Crater. *Science, 340,* 1068–1072.

Wolf E. T. and Toon O. B. (2010) Fractal organic hazes provided an ultraviolet shield for early Earth. *Science, 328,* 1266–1268.

Wolf E. T. and Toon O. B. (2013) Hospitable Archean climates simulated by a general circulation model. *Astrobiology, 13,* 1–18.

Wolf E. T. and Toon O. B. (2015) The evolution of habitable climates under the brightening Sun. *J. Geophys. Res.–Atmospheres, 120,* 5775–5794.

Wordsworth R. D. (2016a) The climate of early Mars. *Annu. Rev. Earth Planet. Sci., 44,* 1–31.

Wordsworth R. D. (2016b) Atmospheric nitrogen evolution on Earth and Venus. *Earth Planet. Sci. Lett., 447,* 103–111.

Wordsworth R. and Pierrehumbert R. (2013a) Hydrogen-nitrogen greenhouse warming in Earth's early atmosphere. *Science, 339,* 64–67.

Wordsworth R. and Pierrehumbert R. (2013b) Water loss from terrestrial planets with CO_2-rich atmospheres. *Astrophys. J., 778,* 154.

Wordsworth R. and Pierrehumbert R. (2014) Abiotic oxygen-dominated atmospheres on terrestrial habitable zone planets. *Astrophys. J. Lett., 785,* L20.

Wordsworth R, Forget F., Millour E., Head J. W., Madeleine J.-B., and Charnay B. (2013) Global modeling of the early martian climate under a denser CO_2 atmosphere: Water cycle and ice evolution. *Icarus, 222,* 1–19.

Wordsworth R. D., Kerber L., Pierrehumbert R. T., Forget F., and Head J. W. (2015) Comparison of "warm and wet" and "cold and icy" scenarios for early Mars in a 3-D climate model. *J. Geophys. Res.–Planets, 120,* 1201–1219.

Wordsworth R., Kalugina Y., Lokshtanov S., Vigasin A., Ehlmann B., Head J., Sanders C., and Wang H. (2017) Transient reducing greenhouse warming on early Mars. *Geophys. Res. Lett., 44,* 665–671.

Wray J. J., Muchie S. L., Bishop J. L., et al. (2016) Orbital evidence for more widespread carbonate-bearing rocks on Mars. *J. Geophys. Res.–Planets, 121,* 652–677.

Wurm G., Trieloff M., and Rauer H. (2013) Photophoretic separation of metals and silicates: The formation of Mercury-like planets and metal depletion in chondrites. *Astrophys. J., 769,* 78.

Yang X., Gaillard F., and Scaillet B. (2014a) A relatively reduced Hadean continental crust and implications for the early atmosphere and crustal rheology. *Earth Planet. Sci. Lett., 393,* 210–219.

Yang J., Boué, Fabrycky D. C., and Abbot D. S. (2014b) Strong dependence of the inner edge of the habitable zone on planetary rotation rate. *Astrophys. J. Lett., 787,* L2.

Yau A. W., Peterson W. K., and Shelley E. G. (1988) Quantitative parametrization of energetic ionospheric ion outflow. In *Modeling Magnetospheric Plasma* (T. E. Moore et al., eds.), pp. 211–217. American Geophysical Union, Washington, DC.

Yelle R. V. (2004) Aeronomy of extra-solar giant planets at small orbital distances. *Icarus, 170,* 167–179.

Zahnle K. J. and Catling D. C. (2017) The cosmic shoreline: The evidence that escape determines which planets have atmospheres, and what this may mean for Proxima Centauri b. *Astrophys. J., 843,* 122.

Zahnle K. J., Kasting J. F., and Pollack J. B. (1988) Evolution of a steam atmosphere during Earth's accretion. *Icarus, 74,* 62–97.

Zahnle K., Haberle R. M., Catling D. C., and Kasting J. F. (2008) Photochemical instability of the ancient martian atmosphere. *J. Geophys. Res., 113,* E11004, DOI: 10.1029/2008JE003160.

Zahnle K. J., Schaefer L., and Fegley B. Jr. (2010) Earth's earliest atmospheres. *Cold Spring Harbor Perspect. Biol., 2:a004895,* DOI: 10.1101/cshperspect.a004895.

Zahnle K. J., Gacesa M., and Catling D. C. (2019) Strange messenger: A new history of hydrogen on Earth, as told by xenon. *Geochim. Cosmochim. Acta, 244,* 56–85.

Zent A. P., Fanale F. P., and Postawko S. E. (1987) Carbon dioxide: Adsorption on palagonite and partitioning in the martian regolith. *Icarus, 71,* 241–249.

Zerkle A. L. and Mikhail S. (2017) The geobiological nitrogen cycle: From microbes to the mantle. *Geobiology, 15,* 343–352.

Zolotov M. Y. (2015) Solid planet-atmosphere interactions. In *Treatise on Geophysics,* 2nd edition (G. Schubert, ed.), pp. 411–427. Elsevier, Amsterdam.

Zolotov M. Y. (2018) Gas-solid interactions on Venus and other solar system bodies. *Rev. Mineral. Geochem., 84,* 351–392.

Kopparapu R. K., Wolf E. T., and Meadows V. S. (2020) Characterizing exoplanet habitability. In *Planetary Astrobiology* (V. Meadows et al., eds.), pp. 449–476. Univ. of Arizona, Tucson, DOI: 10.2458/azu_uapress_9780816540068-ch017.

Characterizing Exoplanet Habitability

Ravi Kumar Kopparapu
NASA Goddard Space Flight Center

Eric T. Wolf
University of Colorado, Boulder

Victoria S. Meadows
University of Washington

Habitability is a measure of an environment's potential to support life, and a habitable exoplanet supports liquid water on its surface. However, a planet's success in maintaining liquid water on its surface is the end result of a complex set of interactions between planetary, stellar, planetary system, and even galactic characteristics and processes, operating over the planet's lifetime. In this chapter, we describe how we can now determine which exoplanets are most likely to be terrestrial, and the research needed to help define the habitable zone under different assumptions and planetary conditions. We then move beyond the habitable zone concept to explore a new framework that looks at far more characteristics and processes, and provide a comprehensive survey of their impacts on a planet's ability to acquire and maintain habitability over time. We are now entering an exciting era of terrestrial exoplanet atmospheric characterization, where initial observations to characterize planetary composition and constrain atmospheres is already underway, with more powerful observing capabilities planned for the near and far future. Understanding the processes that affect the habitability of a planet will guide us in discovering habitable, and potentially inhabited, planets.

There are countless suns and countless earths all rotating around their suns in exactly the same way as the seven planets of our system. We see only the suns because they are the largest bodies and are luminous, but their planets remain invisible to us because they are smaller and non-luminous. The countless worlds in the universe are no worse and no less inhabited than our earth. — Giordano Bruno, 1584 C.E.

1. INTRODUCTION

Statistical studies of the thousands of known exoplanets suggest that the majority of stars host planetary systems (*Cassan et al.,* 2012; *Dressing and Charbonneau,* 2015; *Gaidos et al.,* 2016; *Winn,* 2018), and so it seems inconceivable that Earth is the only habitable world in the universe, even though that may indeed be true. One of the primary goals of both exoplanet science and astrobiology is to search for and identify a potentially habitable, and possibly inhabited, planet orbiting another star. For an exoplanet, habitability is defined as the ability to support and maintain liquid water on the planetary surface. There are several extrasolar planets that are currently considered to be prime candidates for follow-up observations to determine their habitability potential, and future discoveries may yield even more candidate habitable worlds, increasing the odds of finding life outside our solar system. New NASA mission concepts currently under consideration are designed to have the capability to characterize the most promising planets for signs of habitability and life. We are at an exhilarating point in human history where the answer to the question "Are we alone?" lies within our scientific and technological grasp.

To understand habitability more broadly for exoplanets, however, we need to better understand how stars both like and unlike the Sun impact planetary environments. The habitability potential of a planet critically depends upon the host stars characteristics, which can include stellar spectral energy distribution, activity, stellar winds, age, X-ray/ultraviolet (UV) emission, magnetic field, and stellar multiplicity. Several of these factors may also change with the age of the star, consequently affecting habitability of a planet over time. Many of these factors become particularly critical for M-dwarf habitable zone planets, which orbit much closer to their parent stars than the Earth does to the Sun.

In addition to host star properties, habitability of a planet is also influenced by the properties and processes of the planet itself, which include but are not limited to atmospheric composition, atmospheric escape/retention, volatile inventory and delivery, cycling of elements between surface and interior, planetary magnetic field, planet mass and size, orbital architecture of planets in the system, and the presence of giant planets. Life itself may also have an influence on the habitability of a planet (*Nisbet et al.,* 2007; see also the chapter by Stüeken et al. in this volume). Within our solar system there is a diversity of planetary environmental conditions, with Earth as the only known planet with surface liquid water. Our closest neighbors, Mars and Venus, seem to have taken different evolutionary paths than Earth, primarily in response to the influence of our changing Sun over the last 4.5 b.y., but also due to geological factors. There is evidence that Mars had flowing water on its surface 3.5 G.y. ago (*Fassett and Head,* 2008), and it is hypothesized that Venus may have had liquid water; however, the evidence remains unclear (*Donahue et al.,* 1982; *Grinspoon,* 1993; *Kulikov et al.,* 2006; *Hamano et al.,* 2013; *Way et al.,* 2016). Nevertheless, the implication that the solar system may have had at least two planets with liquid water on their surface (and so perhaps potentially habitable) at some point in the ancient past raises an interesting possibility of similar history on planets around other stars.

In this chapter, we will address some of the requirements for understanding and assessing planetary habitability, emphasizing that this assessment is specifically focused on exoplanets. Consequently, the surveys and measurements needed to explore the habitability potential of a planet are quite different than solar system planets, which are discussed in earlier chapters. Without a great technological leap forward, we cannot send satellites and landers to study exoplanets at close range, as has been done for virtually all the major objects in our solar system. All knowledge of habitable exoplanets must be obtained via astronomical observations, and our understanding of the planetary and stellar factors that control habitability must be used to interpret these data.

2. IDENTIFYING POTENTIALLY HABITABLE EXOPLANETS

For exoplanets, a habitable planet is defined as one that can support liquid water on its surface. This "surface liquid water" criterion has been used to define the habitable zone (HZ) (*Hart,* 1978, 1979) as that range of distances from a parent star in which an Earth-like planet could maintain liquid water on its surface (*Kasting et al.,* 1993; *Kopparapu et al.,* 2013) and so potentially host a surface biosphere. Although subsurface liquid water is entirely possible and may even be common (as suggested by the interior oceans of the icy moons in our solar system), detecting that water, and any subsurface biosphere supported by it, is far less likely with remote-sensing telescopic observations (for more detailed discussions of the habitability potential of the surface and subsurface environments of Mars and the solar system's icy moons, see the chapters in this volume by Amador et al., Davila et al., Schmidt, and Cable et al.). Consequently, the search for habitability and life on exoplanets will focus

on telescopic observations of planetary atmospheres and surfaces, where a surface biosphere will be more apparent.

The HZ is therefore designed as a useful concept to identify for remote-sensing studies that region around a star where an orbiting planet has the highest probability of being *detectably* habitable. Although we do not currently have a means of observing markers of surface habitability on exoplanets, these capabilities are expected in the near future (see section 7). Arguments that the HZ is somehow too limited, because it does not encompass the subsurface habitability exemplified by the solar system's icy moons (e.g., *Stevenson,* 2018; *Tasker et al.,* 2017), do not take into account the definition and purpose of the HZ.

In the search for habitable exoplanets, it is an important first step to be able to identify those planets that are most likely to be habitable. These planets will become the highest-priority targets for future telescopes that will be able to observationally confirm whether or not a planet supports liquid surface water. A first-order assessment of potential habitability would be to find a planet that (1) has the solid surface needed to support an ocean and (2) resides within the HZ, so that liquid surface water is more likely to be possible. This initial assessment can be made with three readily observable characteristics: the planet's mass or size, the type of star it orbits, and its distance from that star.

However, as the field of astrobiology develops, it is becoming clearer that multiple factors, characteristics, and processes can impact whether a planet is able to acquire and maintain liquid water on its surface. These include the properties of the planet, star, and planetary system and how these interact over time (*Meadows and Barnes,* 2018). Finding a terrestrial-type rocky planet in the HZ can then be thought of as a two-dimensional slice through a far more complex, interdisciplinary, and multi-dimensional parameter space. Moreover, a planet's position in the HZ does not guarantee habitability, because aspects of its formation or evolution may preclude habitability. For example, the planet could have formed with little or no water (*Raymond et al.,* 2004, 2007), or lost that water in the first billion years of the star's evolution (*Ramirez and Kaltenegger,* 2014; *Luger et al.,* 2015; *Tian and Ida,* 2015). In the rest of this section we discuss the larger context of types of exoplanets and how we now feel confident we can identify those planets most likely to be terrestrial, and will also expand our discussion of how the HZ is defined. In subsequent sections we review the many factors that can impact exoplanet habitability more broadly, and conclude with a discussion of future work in this area.

2.1. The Search for Terrestrial Exoplanets

Exoplanet discoveries have revealed a diversity of exoplanets that span a broad range of mass/radius and orbital distance, exceeding the types of planets seen in our solar system, and arrayed in planetary system architectures that are often completely unlike our own (*Fulton et al.,* 2017; *Winn,* 2018). While direct analogs of gas giants, ice giants, and terrestrial planets likely exist in other systems, the exoplanet population has also revealed hot Jupiters, jovian planets in extremely short orbital periods (approximately a few hours to days); hot Earths, planets that are likely rocky and receive many more times the insolation received by Mercury (e.g., *Berta-Thompson et al.,* 2015); and warm Neptunes, which have ice giant sizes and densities but reside within the inner planetary system. Perhaps the most unexpected discovery has been that of the sub-Neptune population of exoplanets. These planets are smaller than Neptune and often larger than Earth, and are currently seen in relatively close orbits (~200 days or less). The sub-Neptunes are a type of planet that has no analog in our solar system, and they are extremely common, currently comprising the largest fraction of the known population of exoplanets. These sub-Neptunes appear to consist of two subgroups divided by composition and potentially formation mechanisms: mini-Neptunes that are ice-dominated, and super-Earths that have densities more consistent with rock and therefore may well be terrestrial exoplanets (*Rogers,* 2015; *Fulton et al.,* 2017).

To date, the vast majority of exoplanets have been discovered by indirect detection, i.e., the presence of the planet is inferred from the behavior of the star — which may dim, brighten, or move under the influence of its orbiting planet. For a review of the four principal indirect detection techniques, including radial velocity, transit, astrometry, and microlensing, see *Fischer et al.* (2014) and *Wright* (2018). Here we will only further discuss the two current principal indirect detection techniques — radial velocity and transit — as well as direct detection techniques that isolate photons from the planet itself, such as direct imaging and secondary eclipse.

Radial velocity (RV) was the first exoplanet detection technique to discover multiple planets orbiting main-sequence stars. The RV technique detects the presence of the planet when the planet and star orbit a mutual center of mass, which causes the star to appear to move toward and away from the observer. To detect the star's radial motion (toward and away), high-resolution spectroscopy is used with ultrastable spectral reference frames, e.g., iodine gas cells, to detect the star's tiny shifts in RV. Because the RV amplitude is proportional to the mass of the planet and inversely proportional to the planet's orbital distance and the mass of the star, the RV technique is particularly sensitive to large planets close to small parent stars. Radial velocity measurements can reveal planetary orbital period and eccentricity, and can put constraints on planetary mass, deriving a minimum mass — which is uncertain due to the often-unknown orbital inclination of the planet with respect to the observer. Radial velocity was initially the most successful planet detection technique, and in 1995 it was used to detect the first exoplanet around a main-sequence star, 51 Pegasus b (*Mayor and Queloz,* 1995). Given the detection sensitivity biases of this technique, it should be no surprise that 51 Pegasus b was a hot Jupiter, a large planet close to its parent star. Many other hot Jupiters were initially discovered by the RV technique, again due to the sensitiv-

ity bias, even though we now know that systems housing hot Jupiters are rare, comprising only approximately 1% of planetary systems (*Wright et al.,* 2012). The RV technique has continued to push down to smaller masses, and Earth-sized planets were eventually discovered orbiting M dwarfs using this technique, perhaps the most notable being Proxima Centauri b, a 1.3-$M_\oplus$ minimum mass planet orbiting the nearest star to our Sun (*Anglada-Escudé et al.,* 2016). Radial velocity currently lacks the sensitivity to be able to detect an Earth-like planet orbiting a Sun-like star, but new community initiatives in extreme precision radial velocity (EPRV) will tackle this challenge in the coming decade.

The launch of the Kepler spacecraft in 2009 ushered in the heyday of planet detection using the transit technique (*Borucki et al.,* 2010). Unlike the RV technique, which looks for tiny motions of the star, the transit technique detects planets via observation of periodic dimming of the parent star as a planet passes in front of it along our line of sight. Like the RV technique, transit is most sensitive to larger planets (which can block more light) orbiting smaller stars (so that a larger percentage of light is blocked), and it also favors planets that are closer to the parent star. The latter attribute is valuable both because closer planets have a higher probability of appearing to transit their star relative to the observer and because closer planets have shorter orbital periods and therefore multiple transits, which make for a more robust detection, occur in shorter intervals of time. The transit technique can measure the orbital period and size of the planet. If the transiting planet can also be detected with RV, then the orbital inclination of the planet with respect to the observer, as determined by the transit geometry, removes the ambiguity on the RV mass, allowing a true mass to be inferred. With mass from RV and size from transit, density can be calculated for the planet, which can be used to help constrain planetary bulk composition. In multi-planet transiting systems, gravitational interactions between planets can delay or accelerate the time of subsequent transits, resulting in transit timing variations (TTVs) that can also be used to infer planetary mass (*Agol et al.,* 2005; *Holman and Murray,* 2005; *Winn,* 2010). The first detection of a transiting planet was made in 1999, the hot Jupiter HD209458 b (*Charbonneau et al.,* 2000). In the subsequent decade, groundbased telescopes continued to make tens of transiting planet discoveries, detecting planets ranging in size from hot Jupiters to the mini-Neptune GJ1214b (*Charbonneau et al.,* 2009). The launch of the dedicated transit-detection Kepler space telescope in 2009 pushed exoplanet detection into the thousands, rapidly eclipsing the planets detected by the RV technique. The transit technique can also find terrestrial-sized planets orbiting in the HZ of their parent M dwarfs, with the TRAPPIST-1 system of seven Earth-like planets orbiting a late-type (the smaller and cooler end of the spectral class) M dwarf as a key example (*Gillon et al.,* 2016, 2017; *Luger et al.,* 2017a). However, transit also finds detecting Earth-like planets orbiting more Sun-like stars more challenging, although the main objective of the upcoming PLATO mission is to determine the bulk properties and ages of small planets, including those in the HZ of Sun-like stars (*Rauer et al.,* 2014).

Direct detection provides another suite of techniques that can be used to both detect and characterize exoplanets, and can study planets beyond the inner planetary regions favored by RV and transit measurements. In direct detection, photons from the planet are separated from the star either spatially, with direct imaging, or temporally, using secondary eclipse, where a planet passes behind its parent star. In direct imaging, telescopes of sufficient size have the ability to angularly separate the planet from the star on the sky, and some form of starlight suppression technique is used to reduce the glare from the parent star so that the planet and star can be seen as two separate points of light. This allows studies of the planet using both direct reflected light photometry and spectroscopy of the planet's atmosphere and surface, if it has one. To date, direct imaging has been successful only for tens of young (and therefore still hot and self-luminous) jovian planets in the outer regions of planetary systems (*Marois et al.,* 2008, 2010; *Rajan et al.,* 2017). Future observations of Neptune-sized objects closer to the star may be possible with coronagraphs onboard the James Webb Space Telescope (JWST) (*Beichman et al.,* 2019). Direct imaging of terrestrial planets in the HZ of M dwarfs may be possible for a handful of the nearest M-dwarf planets in the near term (*Quanz et al.,* 2015; *Crossfield,* 2016; *López-Morales et al.,* 2019). On longer timescales, the imaging of terrestrial planets in the HZs of more Sun-like stars, and of the cool jovians that characterize our own planetary system, will require large-aperture spacebased telescopes like the Habitable Exoplanet Observatory (HabEx) and Large UV/Optical/IR Surveyor (LUVOIR) mission concepts (see *https://www.greatobservatories.org*).

Secondary eclipse is a means of separating a transiting planet's emitted photons from the parent star's without the planet and star needing to be spatially resolved. The secondary eclipse technique uses observations of the unresolved star and planet and then subtracts an observation of the star alone, taken when the planet is behind the star. This isolates radiation from the planet, and this technique is most effective at mid-infrared wavelengths where the emitted contrast between planet and star is relatively larger, and thus easier to differentiate. Because emitted radiation is being measured, secondary eclipse is sensitive to planetary temperature, and emission spectroscopy can also be used to measure atmospheric molecules.

The statistics provided by the many exoplanet detection techniques, and by Kepler in particular, have enabled many seminal exoplanet discoveries. The majority of planets detected by Kepler are larger than our Earth but smaller than Neptune, and reside close to their host stars (*Winn,* 2018). Note that this does not necessarily mean that sub-Neptunes larger than Earth are the most common type of planet in the galaxy, as the Kepler survey is not sensitive enough to detect smaller terrestrial planets that are potentially more numerous. We have also learned that many planetary systems are not like our solar system, either because they contain hot Jupiters or sub-Neptunes close to the star, or because they are systems where multiple planets are packed much closer to the star than Mercury is to the Sun (*Lissauer et al.,* 2011).

In the past few years, astronomers have made significant progress in understanding the nature of the sub-Neptunes, and most importantly for astrobiology, in identifying likely terrestrial planets in this population. Using the sample of small Kepler planets that also have RV measurements, such that the mass, radius, and density were known, researchers have applied Bayesian statistics to help identify the dividing line in radius that corresponds to a higher likelihood that a planet smaller than that radius has a rocky composition, whereas one larger is likely to be dominated by ice or gas (*Weiss and Marcy,* 2014; *Rogers,* 2015; *Wolfgang et al.,* 2016; *Chen and Kipping,* 2017). The radius below which an exoplanet is more likely to be composed of rock and metal, and therefore could support a surface ocean, is between 1.5 and 1.7 $R_\oplus$ (*Rogers,* 2015). This upper limit for terrestrial planet size was supported by more precise groundbased measurements of stellar — and therefore planetary — radii for over 1000 Kepler planet-hosting stars. This more precise dataset found a gap in the radius distribution that had previously been washed out by the larger errors on planetary radii that now divided sub-Neptune planets into two populations with R < 1.5 $R_\oplus$ and R = 2.0–3.0 $R_\oplus$ (*Fulton et al.,* 2017).

These two populations are inferred to be terrestrial, rocky planets, and a cohort of larger planets that have rocky cores augmented by significant gas envelopes. This inference is due in part to the previous research showing that planets smaller than 1.5 $R_\oplus$ had densities consistent with terrestrial planets, but also because a gap in the radius distribution of sub-Neptune planets was predicted as a result of photoevaporation of planetary envelopes by X-ray and extreme ultraviolet (XUV) radiation (*Fortney et al.,* 2013; *Owen and Wu,* 2013; *Jin et al.,* 2014; *Chen and Rogers,* 2016). The observed gap therefore also lends credibility to the idea that photoevaporation is a key process that sculpts the population of sub-Jovian class planets, although core-powered mass loss — where the luminosity of a cooling rocky core erodes thin H_2 envelopes, but preserves thicker ones — is a feasible alternative explanation (e.g., *Ginzburg et al.,* 2018). Subsequent research has also shown that for lower-mass stars, this bimodal distribution for sub-Neptune-sized planets has a gap that shifts to smaller sizes, consistent with smaller stars producing smaller planet cores (*Fulton and Petigura,* 2018). These studies also indicate that there are a comparable number of worlds in the terrestrial and mini-Neptune classes, with the proviso that we do not have the necessary sensitivity to detect the smallest component of the terrestrial class, or completeness for planets on orbits longer than 200 days. Nonetheless, Kepler has shown that the universe is indeed teeming with terrestrial-sized worlds, most of which are likely to have the higher densities associated with terrestrial worlds in our solar system.

The suite of different types of exoplanets have expanded our knowledge beyond the subset seen in our solar system. Figure 1 shows a schematic for a potential classification

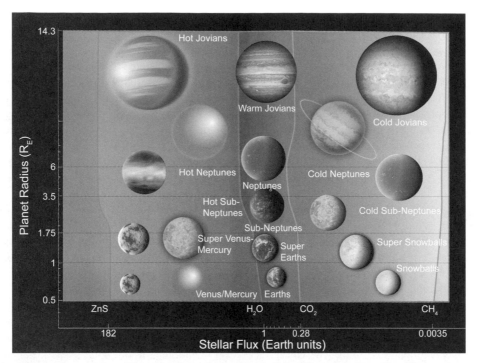

Fig. 1. See Plate 48 for color version. Classification of exoplanets into different categories (*Kopparapu et al.,* 2018). The boundaries of the boxes represent the regions where different chemical species are condensing in the atmosphere of that particular-sized planet at that stellar flux, according to equilibrium chemistry calculations. The radius division is from *Fulton et al.* (2017) for super-Earths and sub-Neptunes, and from *Chen and Kipping* (2017) for the upper limit on jovians.

scheme for these worlds based on planetary radius, stellar flux received, and the corresponding boundaries for which condensates will form clouds in these atmospheres (*Kopparapu et al.,* 2018). The planetary radius bins start with a lower limit for terrestrial atmospheric retention (*Zahnle and Catling,* 2017), and end with the radius past which planets transition to brown dwarf stars (*Chen and Kipping,* 2017), with subcategories for terrestrials including super-Earths, mini-Neptunes [following the *Fulton et al.* (2017) distributions], Neptunes, and jovians. The incident stellar flux then divides them into hot, warm, or cold examples of their size class, with corresponding condensates. In hot exoplanet atmospheres, ZnS mineral clouds have been considered as possible condensates (*Morley et al.,* 2012; *Charnay et al.,* 2015). Moving further away from the star, H_2O starts condensing in the atmosphere of more temperate worlds. At lower incident stellar fluxes, CO_2 and CH_4 condensates bracket the final boundaries. In this broader scheme the HZ can be thought of as that region between instellations (stellar incident flux) at which liquid water clouds (~1 stellar flux) and carbon dioxide clouds (~0.3 stellar flux) form (*Abe et al.,* 2011). Terrestrial planets found within this instellation range are more likely to be habitable than planets in other regions of this diagram. In the next section, we discuss the calculation of the circumstellar HZ in more detail.

2.2. Predicting the Habitable Zone

As discussed in the introduction, the HZ of a star is the circumstellar region where a terrestrial radius and mass

planet can maintain liquid water on its surface. In other words, the HZ identifies a range of orbital distances where a planet is more likely to support habitable conditions on its surface, and thus can be detectable and characterizable by astronomical observations. While numerous studies have sought to understand the impacts of planetary properties and other factors on the limits of the HZ, the most useful region is still likely to be where all HZ estimates overlap for a given stellar type, as that will indicate our best understanding of the region of highest probability for surface liquid water. However, all these models, despite their commonality in many cases, are still predictions, and are based on our understanding of processes working on Earth. These theoretical predictions of the HZ will be subject to revision once observational data on the atmospheric compositions and habitability of terrestrial exoplanets are obtained over the next five years to decades.

Traditionally, one-dimensional climate models were used to estimate the position of the HZs around different stars (*Huang,* 1959; *Hart,* 1978; *Kasting et al.,* 1993; *Selsis et al.,* 2007; *Pierrehumbert et al.,* 2011; *Kopparapu et al.,* 2013; *Zsom et al.,* 2013; *Ramirez and Kaltenegger,* 2017a). These models assume Earth-like planets with CO_2, H_2O, and N_2 atmospheres and an active carbonate-silicate cycle, which provides a negative feedback to buffer atmospheric CO_2 as a function of surface temperature (*Walker et al.,* 1981). This latter assumption will result in "Earth-like" planets that have less CO_2 than Earth near the inner boundary, and significantly more at the outer boundary of the HZ. The one-dimensional nature of the model comes from the atmosphere being approximated as a single column extending from the surface to about ~100 km in altitude (~0.1 mbar), divided into numerous levels where radiative transfer calculations are performed. Such one-dimensional column models are meant to capture planet-wide average conditions, in a simple and efficient package. They are often run cloud-free, or with simplistic approximations for the radiative effect of clouds (*Kasting et al.,* 1993), and with atmospheres where H_2O and CO_2 are the only greenhouse gases.

The width of the HZ is defined by inner and outer edges, which are bounded by climate catastrophes. The models simulate where surface liquid water is no longer stable if one pushes the planet closer to the star, increasing the incident stellar flux (inner HZ, or IHZ), or away from the star decreasing the stellar flux (outer HZ, or OHZ; see Fig. 2). The IHZ proposes two habitability limits: (1) A moist greenhouse limit, where the stellar radiation warms the atmosphere sufficiently so that the stratospheric water vapor volume mixing ratio becomes $>10^{-3}$ (Earth's H_2O mixing ratio is ~10^{-6} at 1 mbar), causing the planet to lose water by photolysis and then subsequent escape of free hydrogen to space; and (2) a runaway greenhouse limit, whereby surface water is vaporized and the atmosphere becomes opaque to outgoing thermal radiation due to excess amounts of H_2O in the atmosphere, heating uncontrollably perhaps to beyond 1500 K (*Ingersoll,* 1969; *Goldblatt et al.,* 2013). While the runaway greenhouse is the more violent and catastrophic end, the moist greenhouse is the more proximal. Habitability could potentially be terminated via the moist greenhouse process long before a thermal runaway occurs. However, some studies found that moist greenhouse may be inhibited under certain conditions, because of the subsaturation resulting in cooler stratospheres (*Leconte et al.,* 2013b). The moist

greenhouse limit depends entirely on a planet's inventory of non-condensing gases. A planet with negligible N_2 and Ar would enter the moist greenhouse limit even if it were in a snowball state (*Wordsworth and Pierrehumbert,* 2014).

The OHZ is defined by the maximum greenhouse limit, where the warming provided by the build up of atmospheric CO_2 (due to the active carbonate-silicate cycle) is maximum. Models indicate that this occurs with ~6 to 10 bars of CO_2 in the atmosphere. For thick CO_2 atmospheres, the enhancement of the greenhouse effect from adding more CO_2 begins to saturate, while the reflectivity of the atmosphere due to Rayleigh scattering from a thick atmosphere continues to increase. Beyond a certain amount of atmospheric CO_2, increases in scattering win out over the increases to the greenhouse effect, causing the planet to experience cooling instead of warming. This turning point marks the maximum CO_2 greenhouse outer edge limit to the HZ.

To guide our search for liquid water on a planetary surface, the conservative HZ uses the runaway greenhouse inner limit and the maximum greenhouse outer limit, but a more optimistic HZ can be defined empirically based on phenomena in our solar system. For the optimistic IHZ one can define a recent Venus limit, based on geological evidence that Venus has not had liquid water on its surface for at least the past 1 b.y. (*Solomon and Head,* 1991). If we assume Venus was habitable right up until 1 b.y. ago, then the recent Venus limit is the equivalent distance from our modern Sun that would have matched the insolation at Venus 1 b.y. ago under a fainter Sun. For the outer edge, there is a corresponding early Mars empirical estimate, based on geological evidence that suggests that Mars had liquid water on its surface 3.8 b.y. ago. These optimistic empirical limits, and the conservative limits calculated using climate

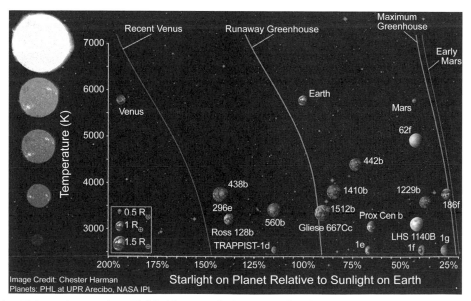

Fig. 2. See Plate 49 for color version. Habitable zone limits for an Earth-like planet around stars with different stellar temperatures (vertical axis) in terms of incident stellar flux (horizontal axis) on the planet, from a one-dimensional climate model. The "conservative HZ" is the region between the runaway greenhouse and maximum greenhouse limits. The "optimistic HZ" is the region between recent Venus and early Mars limits. See text for HZ definitions. Currently confirmed terrestrial exoplanets, along with the solar system ones, are also shown.

models, can be used as a first-order means of identifying habitable planet candidates for follow-up observations. All the currently known terrestrial exoplanets that are in their host stars' HZs are shown in Fig. 2.

Although one-dimensional climate models are relatively fast to run and can include reasonable physics and chemistry (*Lincowski et al.*, 2018), there are instances where complex three-dimensional climate models are particularly needed to understand the impacts of planetary circulation, rotation rate, and cloud formation on climate and habitability. In particular, three-dimensional global circulation models (GCMs) can address the impacts of ice-albedo feedback (*Joshi et al.*, 1997; *Joshi*, 2003; *Shields et al.*, 2013) on planetary volatile abundance, circulation, and climate (*Abe et al.*, 2011; *Wordsworth et al.*, 2010; *Edson et al.*, 2011; *Pierrehumbert et al.*, 2011; *Pierrehumbert and Ding*, 2016; *Way et al.*, 2016); the impacts of cloud formation on the inner edges of the HZ (*Leconte et al.*, 2013b; *Yang et al.*, 2013, 2014; *Wolf and Toon*, 2015; *Way et al.*, 2015; *Godolt et al.*, 2015; *Kopparapu et al.*, 2016, 2017; *Haqq-Misra et al.*, 2018); and the effect of feedbacks between ice formation, atmospheric composition, and surface temperature for the outer edge (*Turbet et al.*, 2017a,b).

Estimates of the HZ for Sun-like stars (F, G, K dwarfs) from three-dimensional climate models are within ~5–7% of the predictions of one-dimensional models. However, three-dimensional models for ocean-covered planets orbiting late K- and M-dwarf stars predict significantly expanded HZs, which is due in part to planetary dynamical spin states. Planets near and within the IHZ of these later-type stars are close enough to the star that tidal locking is a likely dynamical outcome (*Ribas et al.*, 2016). If the planet's orbital eccentricity is small, this can result in synchronous rotation, where the rotation period of the planet equals its orbital period (*Dole*, 1964; *Peale*, 1977; *Dobrovolskis*, 2009; *Leconte et al.*, 2015; *Barnes*, 2017), producing permanent day- and nightsides. Tidally-locked planets are more likely to have slower rotational periods than Earth. This slower rotation diminishes the atmospheric Coriolis force and changes atmospheric circulation, affecting relative humidity, clouds, heat transport, and ultimately the climate. On more rapidly rotating planets, like Earth, the Coriolis force defects air parcels to the right in the northern hemisphere and to the left in the southern hemisphere, producing latitudinally-banded cloud patterns with a cloudy (reflective) equator and clearer subtropics. However, for slowly rotating planets the Coriolis force is too weak, and instead strong and persistent convection occurs at the substellar region, creating a stationary and optically thick cloud deck. This causes a strong increase in the planetary albedo, cooling the planet and stabilizing the climate against a thermal runaway for large incident stellar fluxes. Thus, an ocean-covered planet may be able to maintain clement global-mean surface temperatures (~280 K) around M-dwarf stars at much higher stellar fluxes than predicted by one-dimensional models. This in turn extends the inner edge of the HZ closer to the star,

increasing the width of the HZ (*Yang et al.*, 2013, 2014; *Kopparapu et al.*, 2017).

Modeling to better understand limits at the outer edge of the HZ have also been undertaken. Global circulation model simulations indicate that planets at the outer edge of the HZ around M dwarfs are less susceptible to snowball climates due to the lower snow/ice albedo at near-infrared (NIR) wavelengths, which, interacting with the M-dwarf's red/NIR incident spectrum, causes surface ice to melt more easily compared to under a Sun-like incident spectra (*Shields et al.*, 2013). This may extend the outer edge of the HZ around M dwarfs to lower stellar fluxes compared to models that do not include the ice-albedo feedback. However, oscillations between ice-free and globally glaciated states, called limit cycles, could occur on planets with volcanic outgassing rates that are too low to sustain a CO_2-warmed climate (*Kadoya and Tajika*, 2014, 2015; *Menou*, 2015; *Haqq-Misra et al.*, 2016). Planets orbiting Sun-like stars, and F-type stars in particular, may be more susceptible to limit cycles due to the stronger ice-albedo feedback with the strongly blue spectrum of the F-type star, reducing the extent of the HZ for Sun-like stars. However, for cases where volcanic outgassing is more pronounced, and/or the atmosphere includes various cocktails of other greenhouse gases such as H_2 and CH_4 in addition to high amounts of CO_2, then the OHZ may be extended (*Pierrehumbert and Gaidos*, 2011; *Wordsworth and Pierrehumbert*, 2013a; *Seager*, 2014; *Ramirez and Kaltenegger*, 2017b). Determining which of the above processes, if any, actually govern climates near the edges of the HZ awaits near-term observations of terrestrial atmospheres with JWST and groundbased telescopes that may be able to identify greenhouse gas compositions and search for signs of runaway greenhouse processes or ocean loss (*Morley et al.*, 2017; *Lincowski et al.*, 2018; *Lustig-Yaeger et al.*, 2019; *Turbet et al.*, 2019).

Single stars are the primary focus in the search for habitable planets, but exploration of the HZ for binaries is being undertaken in anticipation of the eventual discovery of terrestrial planet candidates orbiting binary stars. Binary stars are common, with nearly half of all Sun-like stars residing in binary (and higher multiple star) systems. At the time of this writing there are 6 confirmed planets orbiting one member of a sub-20-AU binary stellar system [i.e., circumprimary planets or S-type systems (*Kley and Haghighipour*, 2014)] and 12 confirmed planets orbiting within 3 AU of both members of sub-AU binary star systems [circumbinary planets or P-type systems (e.g., *Welsh et al.*, 2015; *Kostov et al.*, 2016)]. Based on known circumbinary systems, estimates suggest a 1–10% occurrence rate of Neptune- to Jupiter-sized planets (e.g., *Armstrong et al.*, 2014; *Kostov et al.*, 2016; *Welsh et al.*, 2015). Almost half of known circumbinary planets (planets orbiting both the stars of a binary stellar system) reside in the HZ (*Doyle et al.*, 2011; *Orosz et al.*, 2012a,b; *Welsh et al.*, 2015; *Kostov et al.*, 2013, 2016), but these planets are not terrestrials and so are likely not habitable. Discovering transiting planets orbiting binaries is challenging, and is usually done by eye, because their transits are strongly aperiodic. Promising new techniques are being developed to automate the search, and increase our chances of eventually finding smaller terrestrial planets (*Windemuth et al., 2019*).

Meanwhile, there has been some progress in predicting the HZs of Earth-like planets around binary stars (*Kaltenegger and Haghighipour, 2013; Haghighipour and Kaltenegger, 2013; Eggl et al., 2012, 2013; Forgan, 2014; Kane and Hinkel, 2013; Forgan, 2016; Popp and Eggl, 2017; Wang and Cuntz, 2019*), although this is also challenging. Stellar insolation received by HZ planets in circumbinary systems can change by up 50% due to the oscillations of the host stars. These changes in stellar insolation occur on timescales of approximately tens to approximately hundreds of Earth days, and can drive extreme weather and seasonality on circumbinary planets. This could affect prospects for habitability on such worlds. Currently, there are efforts by various groups to simulate such systems using hierarchical models of one-dimensional, energy-balance models (EBM) and GCMs.

2.3. Occurrence Rates for Potentially Habitable Worlds

Now that we understand the size range most likely associated with a terrestrial-type world, and have a working estimate for the limits of the HZ, we can identify those known planets that are more likely to be habitable (see Fig. 2) and can calculate initial estimates of the occurrence rate of potentially habitable planets (likely terrestrials in the HZ). This quantity, $\eta_\oplus$, is defined as the fraction of stars that have at least one planet in the HZ. Current estimates of $\eta_\oplus$ from the Kepler data for Sun-like stars range from 0.22 ± 0.08 (*Petigura et al., 2013*) to 0.36 ± 0.14 (*Mulders et al., 2018*), with some estimates as low as 0.02 (*Foreman-Mackey et al., 2014*). However, realizing this large variation, and to help achieve a useful community-wide consensus of occurrence rates for FGK stars, the NASA-funded Exoplanet Exploration Program Analysis Group (ExoPAG) led Study Analysis Group 13 (SAG13). Study Analysis Group 13 standardized a grid of period and planet radius, interpolated most published occurrence rates to this grid, collected additional contributions from the community, and compiled the results to form a community-wide average with uncertainties. Integrating the SAG13 occurrence rates over the boundaries of stellar and planetary parameters to which the community agreed gives $\eta_\oplus = 0.24^{+0.46}_{-0.16}$ (*Stark et al., 2019*). This value of $\eta_\oplus$ was used to calculate the FGK exo-Earth yields for mission concept studies LUVOIR and HabEX. For M-dwarf stars $\eta_\oplus$ is estimated to be $0.16^{+0.1}_{-0.07}$ for conservative HZ and $0.24^{+0.18}_{-0.08}$ for the optimistic HZ (*Dressing and Charbonneau, 2015*).

Another potential source of habitable worlds are exomoons of HZ jovian planets. Since the launch of the Kepler telescope, exomoon candidates have received increased attention (*Kipping et al., 2012; Szabó et al., 2013; Simon et al., 2015; Agol et al., 2015*). Particularly, the habitability of exomoon candidates has been explored both with theoretical models (*Heller, 2012; Hinkel and Kane, 2013; Forgan and Kipping, 2013; Heller and Barnes, 2013, 2015; Lammer et al., 2014; Forgan and Dobos, 2016; Dobos et al., 2017; Haqq-Misra and Heller, 2018*) and observational efforts to discover exomoon candidates in the HZ of their host stars (*Kipping et al., 2013, 2014, 2015; Forgan, 2017*). Recent occurrence rate estimates of giant planets ($3-25$ R_E) within the optimistic HZ of Kepler stars find a frequency of $(6.5 \pm 1.9)\%$ for G-type stars, $(11.5 \pm 3.1)\%$ for K-type stars, and $(6 \pm 6)\%$ for M-type stars (*Hill et al., 2018*). If one assumes that each giant planet has one large terrestrial moon, then these moons are less likely to exist in the HZ than terrestrial planets. However, if each giant planet holds more than one moon, then the occurrence rates of moons in the HZ would be comparable to that of terrestrial planets and could potentially exceed them. Although there is presently no robust detection of exomoons, there are tentative detections (*Teachey and Kipping, 2018; Rodenbeck et al., 2018*) indicating that a confirmed exomoon discovery is imminent.

2.4. Moving Beyond the Habitable Zone: Factors Affecting Habitability

Although the HZ provides an excellent first-order means to quickly assess the potential habitability of a newly discovered planet, and will be even more powerful if observationally confirmed, there is a growing realization that the HZ is in many ways too simplistic. Although it provides a zeroth-order assessment of whether the planet may be able to support liquid water on its surface now, it does not take into account the formation and subsequent evolution of the planet or the diversity of characteristics or interactions between planet, star, and planetary system that can shape whether or not the planet was able to acquire or maintain liquid on its surface. Planetary habitability is now recognized as the interdisciplinary, multi-factorial outcome of a planet's evolution and planetary system environment.

Factors — characteristics and processes — that impact habitability can be identified in three major areas: planetary characteristics, stellar characteristics, and planetary system characteristics. Habitability is influenced by these properties but also by the interactions that occur between these components as a function of time that allow a planet to acquire and maintain liquid water on its surface (Fig. 3). While each of these factors is important, only a subsample are potentially observable (denoted by the blue text in Fig. 3), and so could be used in the near-term to help characterize habitability. Nonetheless, continued theoretical work into understanding how each of these factors impacts habitability will better prepare us to search for habitable planets and life, and to interpret upcoming data on terrestrial exoplanet environments.

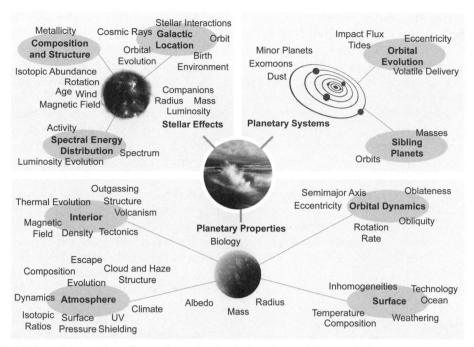

Fig. 3. See Plate 50 for color version. Currently understood planetary, stellar, and planetary system properties that may impact planetary habitability. The larger the number of these factors that can be determined for a given HZ candidate, the more robust our assessment of habitability will be. Colors of type denotes characteristics that could be observed directly with sufficiently powerful telescopes (blue); those that require modeling interpretation, possibly constrained by observations (green); and the properties or processes that are accessible primarily through theoretical modeling (orange). From *Meadows and Barnes* (2018).

3. PLANETARY CHARACTERISTICS FOR HABITABILITY

The planet's environment, mass, radius, orbit, interior, surface, and atmosphere set the stage for habitability. Once life has evolved on a habitable world, it becomes a planetary process that can also impact its environment (*Lovelock and Margulis*, 1974; *Goldblatt et al.*, 2009; *Tziperman et al.*, 2011; *Lenton et al.*, 2012). A detailed discussion of life as a planetary process on Earth and its co-evolution with our environment over Earth's history can be found in the chapter by Stüeken et al. in this volume. Below we describe several of the key planetary characteristics and processes that support habitability and briefly describe our likely ability to observe these characteristics on exoplanets.

3.1. Effect of Mass and Radius on Habitability

While true limits on planetary radii or masses for habitable planets are currently unknown, the radius/mass range within which a planet is more likely to be habitable can be constrained. As discussed above, observations of small Kepler planets, for which the mass, radius, and density are known, have suggested that 1.5-$R_\oplus$ radii is the upper limit for an exoplanet to be more likely to have a predominantly rocky composition (*Weiss et al.*, 2016; *Rogers*, 2015; *Fulton et al.*, 2017). Above this limit, planetary densities drop significantly, suggesting rocky cores with thick hydrogen envelopes: mini-Neptunes, which would be much less

likely to be habitable (*Owen and Mohanty*, 2016). The lower mass limit for which a planet is likely to have sufficient radiogenic heating to drive plate tectonics and atmospheric replenishment via outgassing has been theoretically calculated as 0.3 $M_\oplus$ for an Earth-like composition (0.7 $R_\oplus$ for an object of Earth's density) (*Raymond et al.*, 2007; *Williams et al.*, 1997).

A planet's mass impacts planetary habitability in multiple ways. It provides radiogenic heating from long-lived radionuclides to drive internal heating and tectonics (*Lenardic and Crowley*, 2012) as well as generation of a magnetic field (*Driscoll and Barnes*, 2015), which is a key parameter that determines atmospheric retention (*Chassefiere et al.*, 2007; *Lammer*, 2012; *Egan et al.*, 2019). Planetary mass, via planetary gravity, also controls atmospheric scale height, which can change the rate the planet radiates to space and modify its climate and the limits of the HZ (*Kopparapu et al.*, 2014). Mass is also a key parameter in atmospheric retention, which is dependent on the interplay of planetary mass, radius, and insolation (*Zahnle and Catling*, 2017). Even though Mars lies within the HZ, at 0.1 $M_\oplus$ it has not been able to retain or replenish a sufficiently large atmosphere to maintain liquid water on its surface, and therefore is below the habitable mass limit. If the locations of Venus and Mars were swapped, it is possible that the solar system might have supported two habitable planets.

Planetary radii and masses can be relatively straightforward to measure, depending on the technique used to detect the planet. Exoplanet radius is straightforward to measure if the planet transits and the stellar radius is well known, since planet size is derived from the drop in measured flux as the planet passes in front and the size of the star (*Borucki et al.*, 2010; *Batalha et al.*, 2011). Size is extremely challenging to observe if the planet does not transit due to an inherent size-albedo degeneracy at visible-NIR wavelengths, which can be broken with observations in the thermal infrared (*Des Marais et al.*, 2002). However, exoplanet masses can be measured using transit timing variations (*Deck and Agol*, 2015; *Agol and Fabrycky*, 2018), astrometry (*Benedict et al.*, 2006), or RV measurements (*Mayor and Queloz*, 2012) combined with planetary system inclinations derived from transit-duration observations (*Borucki et al.*, 2010) or high-resolution-spectroscopy measurements of exoplanet orbital velocity (*Snellen et al.*, 2010; *Luger et al.*, 2017b).

3.2. Planetary Orbit, Obliquity, and Rotation Rate

The planetary orbital parameters, such as semimajor axis, eccentricity, obliquity, and rotation rate, have a significant influence on planetary habitability through their control on the stellar radiation received by a planet over its orbit as well as associated feedbacks on the climate system. Fundamentally, the time-averaged amount of stellar radiation received by a planet is determined by its distance from the host star and thus its semimajor axis. Stellar radiation is the primary source of energy for planetary atmospheres. First-order assessments of planetary habitability and the HZ typically rely on the received stellar flux as a primary metric (Figs. 1 and 2). However, the combination of eccentricity, obliquity, and planet rotation rate contribute to complicated temporal and spatially dependent variations of the stellar radiation received by a planet (*Berger et al.*, 1993; *Shields et al.*, 2016). Orbital system parameters may also evolve over time (*Armstrong et al.*, 2014). These characteristics of orbital systems can affect the prospects for habitability, sometimes strongly.

Planets on eccentric orbits receive significant variations in stellar radiation over the course of their orbits as the star-planet distance changes between aphelion and perihelion. If the eccentricity is large, this can result in seasonal changes to a planet's surface temperature. However, Earth-like planets (those with significant oceans and atmospheres) have a large thermal inertia and thus can buffer time-varying changes in the stellar radiation. The long-term climate stability of eccentric planets is determined primarily by the average stellar flux received over the course of their orbit, and not by the extremes received at aphelion and perihelion respectively (*Williams and Pollard*, 2002; *Bolmont et al.*, 2016; *Dressing et al.*, 2010; *Way and Georgakarakos*, 2017; *Adams et al.*, 2019). Still, seasonal temperature extremes and fluctuating environmental conditions could pose significant challenges for the evolution and adaptation of complex life (*Sherwood and Huber*, 2010). Generally, planets with thicker and wetter atmospheres are better able to buffer time-dependent changes in stellar radiation, while planets with thinner and drier atmospheres will be more susceptible climate oscillations driven by time-varying stellar radiation.

Earth has a small but non-zero eccentricity. However, Earth's seasons are not driven by its eccentricity, but rather by its obliquity. Obliquity is the axial tilt of the planet's rotational axis relative to the star-planet plane. A planet's obliquity determines the latitudinal variation in the received stellar radiation. As the planet orbits the star, a non-zero obliquity results in a meridional migration of the substellar point north and south of the equator. Still, for low-obliquity planets like Earth, the annually averaged stellar radiation remains centered at the equator, with meridional excursions of the substellar point creating the seasons. However, for high-obliquity planets (>54°), the time-averaged pattern of the stellar radiation reverses, with a maximum flux received at polar regions and a minimum at the equator (*Jenkins*, 2000). This peculiar pattern of stellar radiation creates unique climate states where ice belts may accumulate around the equator while the poles remain temperate and habitable (*Spiegel et al.*, 2009, 2010; *Kilic et al.*, 2017). For high-obliquity planets, bistability thresholds between habitable temperate climates and uninhabitable snowball climates are notably altered compared to thresholds identified for low-obliquity worlds (*Linsenmeier et al.*, 2015; *Rose et al.*, 2017; *Colose et al.*, 2019).

Planetary rotation rate controls the diurnal period (the length of day). Unique among orbital parameters, the planetary rotation rate imparts a significant impact on the circulation state of the atmosphere through the action of the Coriolis effect. The Coriolis effect is a fictitious force that arises due to Earth's rotation, and defects large-scale atmospheric motions relative to Earth's surface. Changes to the planet's rotation rate, and thus the Coriolis effect, can trigger the emergence of different atmospheric circulation regimes (*Carone et al.*, 2015; *Noda et al.*, 2017; *Haqq-Misra et al.*, 2018). The atmospheric circulation state affects horizontal heat transport and the spatial distribution of clouds, each of which can significantly affect the climate and habitability of a planet (*Yang et al.*, 2013; *Kopparapu et al.*, 2017; *Wolf et al.*, 2019; *Komacek and Abbot*, 2019).

For slowly rotating planets like Venus or planets found around M-dwarf stars where tidal locking is expected, the Coriolis effect is weak and the atmospheric circulation regime supports the creation of thick stationary clouds at the substellar point that effectively reflect sunlight and permit a planet to be habitable at higher insolation levels (*Yang et al.*, 2013; *Kopparapu et al.*, 2016; *Way et al.*, 2016). For more rapidly rotating planets, like Earth, a stronger Coriolis effect leads to predominately zonal circulation and the creation of zonally banded cloud decks, which are less efficient at reflecting sunlight (*Yang and Abbot*, 2014; *Kopparapu et al.*, 2017).

For planets that are tidally-locked and synchronously rotating (in a 1:1 spin-orbit resonance), one side of the planet always faces the star and the other side of the planet is in permanent darkness. On such worlds, the day-night temperature differences can be enhanced, leading to the possibility of the atmosphere freezing out or "collapsing" onto the nightside (*Joshi et al.*, 1997; *Joshi*, 2003; *Turbet et al.*, 2016; *Leconte et al.*, 2013b). This is true especially for colder and thinner Mars-like atmospheres; however, modeling has shown that thicker Earth-like atmospheres can maintain suffcient day-to-night heat transport to prevent collapse (*Joshi et al.*, 1997; *Wordsworth*, 2015; *Kopparapu et al.*, 2016; *Wolf et al.*, 2019). Note that synchronous rotation is not guaranteed for tidally-locked planets around M-dwarf stars, however, as trapping into spin-orbit resonances (like Mercury's) are also possible (*Hut*, 1981; *Rodríguez et al.*, 2012; *Ribas et al.*, 2016). A large atmosphere may also prevent synchronization (*Gold and Soter*, 1969; *Leconte et al.*, 2015). The tidal damping of the rotation rate into a synchronous state is model dependent (*Ferraz-Mello et al.*, 2008; *Barnes*, 2017) and depends on the planet's structure (*Henning and Hurford*, 2014) and, if present, the tidal dissipation in a planet's ocean (*Egbert and Ray*, 2000; *Green et al.*, 2017).

Planetary orbital properties are generally amenable to observations. Semimajor axes can be observationally determined using transit, RV, and astrometry. Eccentricity is also straightforward to measure with RV and astrometry but more challenging with transit, and in the latter case more accurate if both the primary and secondary eclipse can be observed (*Demory et al.*, 2007). However, determining obliquity and rotation rate for a terrestrial planet will require time-dependent mapping using direct imaging observations (*Fujii et al.*, 2018; *Lustig-Yaeger et al.*, 2017; *Kawahara and Fujii*, 2010; *Cowan et al.*, 2009). By observing the thermal emission as a function of orbital phase, different climate states may be discernable (*Leconte et al.*, 2013b; *Yang et al.*, 2013; *Haqq-Misra et al.*, 2018; *Wolf et al.*, 2019; *Adams et al.*, 2019).

3.3. Planetary Interior and Geological Activity

The interior of a planet plays a critical role in determining the habitability of a planet. An active and dynamic interior can drive the generation of a magnetic field (*Olson and Christensen*, 2006; *Driscoll and Bercovici*, 2013) and outgassing (*Driscoll and Bercovici*, 2014), which are key for producing and maintaining a secondary atmosphere.

The atmospheres of terrestrial planets in our solar system are a product of outgassing. Primordial atmospheres, if they ever existed for terrestrial planets, are dominated by H_2 and He with traces of Ar and Ne accumulated during the formation of the solar system from the gaseous nebular disk (*Sekiya et al.*, 1980; *Lammer et al.*, 2018). However, such an atmosphere will have only a fleeting existence on a newly formed terrestrial planet, because the temperatures are hot enough for the lighter gases (H_2 and He) to escape the low-gravity well of the planet (a simple relation between mean kinetic energy and the internal energy due to the temperature of the gas). The heavier icy materials (H_2O, CH_4, NH_3), on the other hand, combine with the rocky materials (like iron and olivine) and get integrated into the crust and the mantle. If the terrestrial planet is big enough to maintain the formation heat, it can sustain an active tectonic activity, which results in volcanism, which in turn releases these trapped icy materials producing secondary atmospheres.

Tectonic activity on a planet can influence the habitability of a terrestrial planet through cycles of volcanic outgassing and consequent weathering of the released gases. Tectonic activity also creates weatherable topography, which has a long-term impact on the evolution of the climate (*Lenardic et al.*, 2016) in terms of negative feedback between silicate weathering (the loss process for atmospheric CO_2) and surface temperature (*Walker et al.*, 1981). Furthermore, tectonic activity similar to that of Earth facilitates an efficient water cycling between the surface and the interior, sustaining oceans on the planet (*Sandu et al.*, 2011; *Cowan and Abbot*, 2014; *Schaefer and Sasselov*, 2015; *Komacek and Abbot*, 2016). Tectonic and volcanic activity cycles that operate beyond the characteristics of Earth may occur on exoplanets of varying mass and composition, such as stagnant lid (*Solomatov and Moresi*, 2000), episodic tectonics (*Lenardic et al.*, 2016), and heat pipe (*Moore and Webb*, 2013; *Moore et al.*, 2017). On the other hand, coupling melt models (*Katz et al.*, 2003) with mantle convection models can simulate volcanic outgassing, which depends upon the composition and the internal temperature of the planet.

Augmenting a planet's primordial internal heat for extended periods of time can be possible by radionuclide decay (*Dye*, 2012) or tidal stress (*Jackson et al.*, 2008; *Barnes et al.*, 2009). Radionuclide decay releases high-energy particles, which can be absorbed in the planetary interior. The nature of a planet's radiogenic sources is predetermined during the formation of the system. Different radio isotopes decay at different rates. For example, ^{26}Al is a short-lived isotope whose half-life is just ~700,000 years, which is suspected to have been present during the solar system's formation. Aluminum-26 is produced in supernova explosions of massive stars, which provide the ingredients and the initial "fuse" (through shock waves) for forming a planetary system. Short-lived isotopes drive the differentiation of elements inside a planet through their radiogenic heating during the primordial stages of their formation. Once the differentiation begins, the frictional heat of partitioning elements sustains the internal heat deep inside the planet.

On the other hand, ^{40}K is a long-lived radio isotope with a half-life of more than a billion years. Long-lived isotopes generally concentrate near the crust and the mantle, providing heat at these layers, due to their large size preventing dense packing deep within Earth.

The internal heat energy needed for tectonic activity can also be generated by tidal heating, where the differential gravity of the planet due to a companion (usually the host star or nearby planet) causes internal stress, and energy is deposited by friction (*Jackson et al.*, 2008; *Driscoll and Barnes*, 2015). A key area of future research for the impact of planetary interiors on terrestrial planet evolution and habitability will be understanding degassing from terrestrial planets of different composition, including the potentially volatile-rich migrated terrestrial planets found orbiting M dwarfs (*Gillon et al.,* 2017; *Luger et al.,* 2017a; *Grimm et al.,* 2018).

Planetary interior properties will be challenging to determine observationally. However, precise characterization of the planetary system's orbital state could theoretically be used to yield constraints on planetary interior structure — including determination of the rigidity of the planetary body and its susceptibility to tidal deformation (*Buhler et al.,* 2016; *Becker and Batygin,* 2013). Constraints on the planet's interior structure and composition could also be gleaned from a combination of knowledge of the star's composition, the planet's mass and radius, and planet-formation models (*Dorn et al.,* 2015; *Unterborn et al.,* 2016). For multi-planet transiting systems, density measurements to constrain interior composition could also be obtained from observations of planetary radius from transit and of planetary mass from RV, astrometry, or transit timing variations. Hints of the planet's interior composition may also be obtained from measurements of atmospheric and cloud composition, which may point to a steady-state volcanic outgassing source, as it does for the clouds of Venus (*Bullock,* 1997). Transmission or direct imaging observations may also reveal transient compositional changes in atmospheric gases (*Kaltenegger and Sasselov,* 2009; *Kaltenegger et al.,* 2010) or aerosols (*Misra et al.,* 2015) that are indicative of ongoing volcanic activity.

The redox state of a planet also influences its habitability. The escape of hydrogen is important as it can drive water loss, and can also alter the planet's surface redox state to more oxidizing (*Catling et al.,* 2001; *Catling and Claire,* 2005; *Kump,* 2008; *Armstrong et al.,* 2019). Similarly, the degree of iron segregation to the core and/or iron redox disproportionation sets the redox state of the mantle (*Frost and McCammon,* 2008), which determines whether reducing or oxidizing gases get outgassed by volcanos. These gases have a large impact on the composition of the secondary terrestrial planetary atmosphere (*Wordsworth and Pierrehumbert,* 2013b). Highly reducing conditions have climate consequences but they are also fundamental to driving prebiotic chemistry, particularly when abundant HCN is present (*Ferris and Hagan,* 1984; *Orgel,* 2004). Conversely, highly oxidizing surface conditions are not only a roadblock to the origin of life (see the chapters by Hoehler et al. and Baross et al. in this volume), they can also make conditions toxic for complex life altogether if O_2 levels are high enough (*Baker et al.,* 2017).

3.4. Magnetic Fields

Magnetic fields are an important factor when considering the habitability of a planet, as they may protect planets from losing volatiles (such as water) through stellar wind interactions (*Chassefiere et al.,* 2007; *Lundin et al.,* 2007; *Lammer,* 2012; *Driscoll and Bercovici,* 2013; *do Nascimento et al.,* 2016; *Driscoll,* 2018). However, this "magnetic umbrella" hypothesis is still debated, as the magnetic field may also increase the interaction area with the solar wind, which could drive increased escape (*Brain et al.,* 2013; *Egan et al.,* 2019). Although atmospheric escape was often assumed to be due to thermal processes, and independent of the magnetic field (e.g., *Hunten and Donahue,* 1976; *Watson et al.,* 1981; *Lammer et al.,* 2008), modeling had suggested that the magnetic-limited escape rate does indeed decrease with increasing planetary magnetic moment (*Driscoll and Bercovici,* 2013). However, recent modeling suggests that the situation is more complex, and that whether a magnetic field decreases or increases atmospheric escape is dependent on multiple factors including the strength of the planet's intrinsic magnetic field and the incoming solar wind pressure (*Egan et al.,* 2019).

Convection in the iron-rich core maintains the planetary magnetic field, and is tied to the interior thermal evolution, which can reveal the energetic state and history of the interior (*Stanley and Glatzmaier,* 2010; *Stevenson,* 2010; *Schubert and Soderlund,* 2011). A magnetic dynamo is generated via convection in the outer core (*Olson and Christensen,* 2006). This process is influenced by the planetary rotation rate and core material properties, and enhanced by buoyancy driven by the core cooling rate, which is in turn controlled by the overlying mantle. Plate tectonics cool the planet's interior, and Venus' lack of plate tectonics may explain its lack of a magnetic field (*Nimmo,* 2002), although as a counterpoint Mercury and Ganymede maintain dynamos below their stagnant lids (*Ness,* 1978; *Kivelson et al.,* 1997). Observations and models of volatile-loss rates for planets and satellites with or without magnetic fields will provide additional insight into the generation of magnetic dynamos and the extent of magnetoprotection of atmospheres. Detecting the presence of a magnetic field on an exoplanet will be challenging; however, recent observations have inferred magnetic fields around hot Jupiters (*Cauley et al.,* 2019). Magnetic star-planet interactions involve the release of energy stored in the stellar and planetary magnetic fields. These signals thus offer indirect detections of exoplanetary magnetic fields. Large planetary magnetic field strengths may produce observable electron cyclotron maser radio emission by preventing the maser from being quenched by the planet's ionosphere (*Ergun et al.,* 2000). Intensive radio monitoring of exoplanets will help to confirm these fields and inform the generation mechanism of magnetic fields. Constraints on magnetospheric strength might be gained in the near future with the detection of auroral lines in high-resolution spectra of exoplanet such as Proxima Centauri b (*Luger et al.,* 2017c), although caution will be needed in discriminating these from more diffuse, globally-prevalent airglow lines. Radio emission may also indicate a planetary magnetic field, with coherent emission frequency providing a constraint on the strength of the field itself (*Zarka,* 2007; *Hess and Zarka,* 2011) and characteristic radio emission from the star due to its interaction with a magnetized planet (*Driscoll and Olson,* 2011; *Turnpenney et al.,* 2018).

3.5. Atmospheric Properties

While the atmosphere of a terrestrial planet generally constitutes only a minuscule fraction of the planet's overall mass and radius, atmospheric properties play an outsized role in determining habitability. For example, while Earth and the Moon each receive the same amount of stellar radiation, their surface environments are decidedly different because Earth has an atmosphere and the Moon does not. An atmosphere is an envelope of gas that surrounds a planet, and is retained due to the force of gravity. An atmosphere is a necessary condition for a planet to have liquid water at its surface because water can only remain stable as a liquid under a relatively narrow range of temperatures at a given pressure. Without adequate pressure, surface liquid water would irreversibly evaporate or sublimate away, as it would on Mars and the Moon despite each residing within the HZ (*Kopparapu et al.,* 2013). Adequate atmospheric pressure is particularly important for synchronously rotating planets where day-night temperature differences can grow large, potentially resulting in atmospheric collapse onto the nightside (*Wordsworth,* 2015; *Turbet et al.,* 2016, 2017b). However, atmospheric collapse can be countered by denser atmospheres, which promote efficient heat transport. Of course, atmospheres that are too dense may negatively impact habitability, by becoming opaque to stellar radiation. This may be particularly problematic for planets that do not lose their primordial H_2 atmospheres (*Owen and Mohanty,* 2016) and for habitable planets near the outer edge of the HZ, which require thick atmospheres to stay warm.

Appropriate surface temperatures for liquid water are maintained through a delicate balance between absorbed incoming radiation from the star, emitted thermal radiation from the planet, and horizontal heat transports (*Trenberth et al.,* 2009). The constituents of planetary atmospheres, including gases, clouds, and aerosols, critically modulate a planet's energy balance and thus its climate and ultimately its local surface temperatures (*Read et al.,* 2015). Maintaining the right surface temperatures for liquid water to exist requires having just the right amount and combination of atmospheric gases. CO_2 is perhaps the most familiar greenhouse gas, and helps maintain clement temperatures on Earth. The long-term regulation of CO_2 in the atmosphere via the silicate weathering cycle is thought to keep planets habitable despite large differences in their received stellar flux (*Walker et al.,* 1981). Planets at large distances from their star may be kept habitable by the strong greenhouse effect provided by several bars of CO_2 (*Kasting et al.,* 1993; *Kopparapu et al.,* 2013; *Selsis et al.,* 2007). However, as the atmosphere becomes thick, Rayleigh (molecular) scattering increases, reflecting stellar energy away from the planet. Other greenhouse gases may also help keep planets sufficiently warm in the HZ, including CH_4, N_2O, and NH_3, as well as collision-induced absorption of N_2 and H_2 (*Wordsworth and Pierrehumbert,* 2013a; *Pierrehumbert and Gaidos,* 2011; *Ramirez and Kaltenegger,* 2018; *Koll and Cronin,* 2019).

With respect to the climate of habitable planets, H_2O is perhaps the most interesting atmospheric constituent, and not because it is a prerequisite for life. On a robustly habitable planet like Earth, the expected surface and atmospheric temperature variations allow water to exist in all three thermodynamic phases simultaneously in the atmosphere, in the oceans, and on the surface. Each phase of water contributes strong competing feedbacks on a planet's climate. Water vapor is a strong greenhouse gas and NIR absorber and acts to warm a planet. High-altitude ice water clouds (i.e., cirrus clouds) act also as strong greenhouse agents and warm a planet. Liquid water clouds (e.g., stratus clouds) are highly reflective, raising the albedo and cooling a planet. Finally, water that condenses on the surface as snow and ice also raises the albedo and cools a planet. The water vapor greenhouse feedback and ice-albedo feedbacks are both positive climate feedbacks, meaning that they will amplify climate perturbations, potentially leading to climate catastrophes of a runaway greenhouse and runaway glaciation and the end of habitability. While water is of course critical for the existence of life, water has an inherently destabilizing force on the climate system.

Beyond the regulation of climate, atmospheres play other important roles that factor into habitability. For instance, for oxygen-rich planets, stratospheric ozone plays a major role in maintaining surface habitability by shielding harmful UV fluxes from reaching the surface (*Segura et al.,* 2003, 2005; *Rugheimer et al.,* 2015). Alternatively, anoxic planets may form Titan-like photochemical hazes that are created in the upper atmosphere and also constitute a significant UV shield that could protect life on the surface (*Sagan and Chyba,* 1997; *Wolf and Toon,* 2010; *Arney et al.,* 2016, 2017). Although, while O_3 has little effect on a planet's surface temperature, if sufficiently thick a photochemical haze layer could significantly cool a planet's surface and threaten habitability (*McKay et al.,* 1999; *Haqq-Misra et al.,* 2008). Absorbing species in the atmosphere can also affect habitability by modifying the atmospheric thermal structure, which in turn can either help or hinder water loss via photolysis in the stratosphere (*Wordsworth and Pierrehumbert,* 2014; *Fujii et al.,* 2017). Planetary surface characteristics such as the presence of global oceans and the locations of continents also can affect habitability through the modulation of ocean heat transports and their effect on the overall climate (*Hu and Yang,* 2014; *Del Genio et al.,* 2019; *Yang et al.,* 2019).

Atmospheric properties will be probed via transit transmission spectroscopy, thermal phase curves, secondary eclipse, and direct imaging spectroscopy (*Meadows et al.,* 2018). Transit transmission spectroscopy can help us identify gas species in the upper atmosphere of planets (*Morley et al.,* 2017; *Lincowski et al.,* 2018; *Lustig-Yaeger et al.,* 2019). The detectability of a molecule depends on its atmospheric abundance and the strength of spectral features, as well as the wavelength range observed, with some molecules more likely to be observed than others (*Schwieterman et al.,* 2015). However, the presence of condensates, which generally present featureless spectra, may sharply obscure our ability to observe underlying gases with transmission spectroscopy (*Kreidberg et al.,* 2015; *Lincowski et al.,* 2018; *Morley et al.,* 2017; *Lustig-Yaeger et al.,* 2019). Thermal emission and reflected light phase curves may yield clues to atmospheric composition and the presence of clouds and aerosols, and may allow temperature mapping (*Yang et al.,* 2014; *Koll and Abbot,* 2016; *Wolf et al.,* 2019; *Kreidberg et al.,* 2019). The temperature structure of atmospheres is more challenging to observe, but could be derived from thermal infrared spectroscopy that encompasses the 15-µm CO_2 band.

4. STELLAR CHARACTERISTICS FOR HABITABILITY

The host star's characteristics have a huge influence on a planet's environment and habitability. Stellar mass and radius determine many of the star's fundamental characteristics, such as temperature and lifetime. Stellar luminosity evolution drives strong climate change and may result in atmospheric or ocean loss, which is a compositional change and often a threat to habitability. The stellar spectrum and activity levels influence atmospheric escape and climate, provide the most abundant surface energy source for the majority of HZ planets, and photochemically modify the planet's atmospheric composition.

4.1. Luminosity, Age, and Spectral Energy Distribution

The energy emitted by the host star and received by the planet plays a primary role in determining whether a planet can be habitable. The luminosity of the star is a measure of the total energy it emits per unit time, and it depends on the star's size and emitting temperature. The stellar luminosity controls the energy received by a planet and in large part determines the semimajor axis of its HZ. Stars have a finite lifetime, determined by their rate of fuel consumption. Smaller, cooler stars like M dwarfs have much lower luminosities than larger, hotter F dwarfs, and have nearly fully convective interiors that can deliver more fuel to the reacting core, so they burn at a low rate, but for longer. More massive stars support their higher luminosities by burning their atomic fuel at a much higher rate, but can't convect additional fuel to the core as efficiently as smaller stars, and so have significantly shorter life spans. While our Sun, a G dwarf, may live for 10 b.y., an A dwarf that is twice as massive as the Sun would remain on the main sequence for only 2 b.y. This is significantly less time than it took for oxygen to rise to even 10% of the current atmospheric level on our planet (*Lyons et al.,* 2014) and so produce a detectable biosphere. M-dwarf stars are small and dim, and can spend hundreds of billions of years on the main sequence (*Rushby et al.,* 2013), far longer than the 13.8-G.y. age of the universe.

Stars brighten over their main-sequence lifetimes, thus the radiation received by an orbiting planet increases over time. For instance, in Earth's early history it received 25% less stellar energy than it does today. Still, other factors controlling the atmospheric composition and the greenhouse effect allowed Earth to maintain continuously habitable surface temperatures despite this change in stellar irradiance over time. Our Sun will continue brightening at a rate of 1% every 100 m.y. (*Gough,* 1981). The rate of luminosity evolution depends on stellar mass, with larger hotter stars naturally brightening more rapidly than smaller, cooler stars.

Finally, the spectral energy distribution (SED) is the amount of radiation emitted by the star as a function of wavelength. The SED is strongly dependent on the emitting temperature of the stellar photosphere. The wavelength of peak stellar emission is inversely proportional to the emitting temperature. G dwarfs like our Sun have peak emission at visible light wavelengths, while cooler stars like M dwarfs emit primarily NIR radiation. The relative SED of the star in turn can strongly influence the climate system as surface reflectivity, NIR gas absorption, and Rayleigh scattering are all sensitive to changes in the SED. The relative SED may also strongly influence biology, as photosynthesis plays a dominate role in Earth's biosphere.

Stars evolve in luminosity, from a superluminous phase as they collapse down to their main-sequence sizes, through gradual brightening as they fuse hydrogen into helium in their cores, increasing the core temperature and accelerating fusion. For lower-mass stars the pre-main-sequence phase can be as long as 2.5 G.y., but is only 10 m.y. for a G dwarf like our Sun (*Baraffe et al.,* 2015). This superluminous pre-main-sequence phase would subject planets that form in what will become the M-dwarf's main-sequence HZ to very large amounts of radiation early on, increasing the chance that these planets will experience ocean and atmospheric loss (*Luger and Barnes,* 2015; *Ramirez et al.,* 2014; *Tian and Ida,* 2015). For more Sun-like stars, the pre-main-sequence phase is relatively short, and the star's luminosity then increases strongly during its main-sequence phase (the Sun will undergo an 80% increase in luminosity in its lifetime). For the smaller M-dwarf stars, the long pre-main-sequence phase fades to an almost constant main-sequence luminosity over trillions of years. While a star's current luminosity can be straightforwardly measured, its luminosity evolution is determined primarily using models that are validated against observed luminosities as a function of spectral type, mass, and age (*Baraffe et al.,* 2015).

4.2. Activity Levels

The stellar mass and age of the star will also affect the level of stellar activity, which can produce UV and shorter-wavelength radiation that is potentially damaging to planetary atmospheres, ozone layers, and surface life (*Wheatley et al.,* 2019; *Tilley et al.,* 2017; *Segura et al.,* 2010). Stellar activity, including sunspots and fares, is produced by stellar magnetic field interactions, which are a function of the internal convection of the star and its rotation rate. For solar-type stars a stellar magnetic field is generated via shearing due to differential rotation at the boundary between the radiative inner zone and the convective outer zone of the star's interior. The field generated at this boundary rises buoyantly through the star's convective zone to emerge as magnetic loops on the stellar surface, which eventually release their magnetic energy in stellar fares. For high-mass F dwarfs, the outer convective layer is too shallow for much field to be generated, but by M3V (stars ≤ 0.3 M$_\odot$) stars become fully convective, and the magnetic field is generated instead by a turbulent dynamo, which can produce large-scale fields and strong stellar activity. The generation of the magnetic field is intimately linked to the stellar rotation, and this evolves as the star ages. Stars form with relatively high angular momentum and spin down over the course of their lifetimes so that young stars are more magnetically active than older stars, up until a characteristic saturation spin velocity at which the observed activity appears to level out (*White et al.*, 2007; *Güdel*, 2004; *West and Basri*, 2009). While early M dwarfs have solar-like spindown times and are inactive in just under a billion years (*West et al.*, 2008), fully-convective later-type M dwarfs have much longer spindown times, extending up to 8 G.y. for M7V stars (*Hawley et al.*, 2000; *Gizis et al.*, 2002). Stellar activity levels and frequency are relatively easily measured using broadband photometry, but spectral information on the fares requires UV spectroscopy from spacebased platforms such as the Galaxy Evolution Explorer (GALEX) or the Hubble Space Telescope (HST). There is a need for more extreme ultraviolet (EUV or XUV) observations by future space missions to fully understand the impact of stellar activity on planetary habitability.

5. PLANETARY SYSTEM CHARACTERISTICS FOR HABITABILITY

In addition to the star-planet interaction, the planet may also interact with other components of the planetary system during formation and subsequent evolution. When assessing habitability, these planetary system components will need to be inventoried to better understand planet formation, volatile delivery, and orbital modification.

Other components of a planetary system, such as jovian planets, asteroid and Kuiper belts, and nearby sibling planets can all impact the potential habitability of terrestrial planets and provide clues to their formation and evolutionary history (*Raymond et al.*, 2008). The masses and orbits of jovian planets in particular should be characterized, as they can affect volatile delivery to forming terrestrial planets. Eccentric jovians could result in the formation of water-poor terrestrials (*Raymond et al.*, 2004), whereas jovians that remain on wide orbits protect terrestrial planet formation in the inner planetary system while potentially enriching it with volatiles (*Raymond and Izidoro*, 2017) (for an in-depth discussion of the impacts of planet formation on terrestrial planet characteristics and habitability, see the chapters in this volume by Raymond et al. and by Meech and Raymond). The presence of debris disks may indicate that an eccentric jovian is not present (*Raymond et al.*, 2012). Nearby sib-

ling planets can also modify orbital parameters including eccentricity and obliquity, and help maintain tidally-locked planets in 3:2 resonances, rather than the 1:1 resonance of synchronous rotation, although this may also occur if the planet has a very low "triaxiality" and is more truly spherical (*Ribas et al.*, 2016). Sibling perturbation from circular orbits and synchronous rotation may occur even for the closely packed systems seen orbiting M dwarfs, of which TRAPPIST-1 is a well-known example (*Gillon et al.*, 2016, 2017; *Luger et al.*, 2017a). The seven TRAPPIST-1 planets are found closely packed together in a resonant chain that implies migration from more distant birth orbits (*Luger et al.*, 2017a), and their mutual gravitational interactions produce transit timing variations (TTVs) that have been used to determine their masses and densities and constrain their orbital eccentricities to be less than 0.08 in most cases (*Gillon et al.*, 2017).

Other planets in the planetary system can be observed in transit, via TTVs, with RV, astrometry, or direct imaging. Belts of minor planets analogous to our asteroid and Kuiper belts serve as a reservoir for water-rich bodies, and the disk's dust distribution can reveal collisions among these smaller bodies or the gravitational signature of unseen planets. These features of the planetary system may be detectable as infrared excesses or directly imaged (*Kraus et al.*, 2017). Exomoons can also influence planetary habitability by damping large obliquity oscillations for habitable worlds, but remain challenging to detect. Future observations may look for changes in the center of the planet-exomoon composite image (*Agol et al.*, 2015), or additional transit timing signals (Kipping 2011).

6. STAR-PLANET-PLANETARY SYSTEM INTERACTIONS AND HABITABILITY

The intrinsic properties of planets, stars, and planetary systems described above exert significant controls on planetary habitability. However, the interactions among a planet, its host star, and its planetary system constitute another category of factors that in part determine whether a planet is and can remain habitable. Radiative interactions with the host star can modify planetary atmospheric compositions by driving the photochemical production of aerosols or gas species. These modifications of the atmosphere subsequently affect planetary climate and the UV flux incident at the planet's surface, both of which directly affect habitability. In more extreme cases, oceans of water vapor and/or the atmosphere itself can be stripped away to space by EUV/XUV radiation and the stellar wind. Gravitational interactions between the host star, planet, and system can modify orbital properties that in turn modulate insolation levels and therefore climate. Gravitational interactions may also be responsible for late volatile deliveries from comets that defected into the inner part of stellar systems. Tidal interactions between bodies in the system can influence planetary interiors, controlling the magnetic dynamo and plate tectonics, both of which play significant roles in the maintenance and retention of secondary atmospheres on terrestrial planets. These processes will be better understood with an interdisciplinary systems approach to modeling terrestrial exoplanet environments. Below we describe the significant planetary processes that are impacted by these interactions in more detail.

6.1. Atmospheric and Ocean Loss and Replenishment

Unlike the gas giants, none of the four terrestrial planets in our solar system have retained their primordial H_2-dominated atmospheres. Instead they exhibit secondary atmospheres, composed of fractionated remnants of their primordial atmospheres augmented by outgassed volatiles from their interiors and volatiles delivered from comets and asteroids (*Pepin*, 2006). The loss of a primordial H_2-dominated atmosphere is probably beneficial for planetary habitability (*Luger et al.*, 2015; *Owen and Mohanty*, 2016). For instance, the gas giants in our solar system have retained their primordial atmospheres, and as a consequence have dense and opaque atmospheres, immense pressures, and no true surfaces. Any extant life on a gas giant would probably need to live among the clouds (e.g., *Morowitz and Sagan*, 1967). Still, the retention of a planet's secondary atmosphere remains of significant concern for the long-term habitability of terrestrial planets. In our own solar system, we observe that Mars' secondary atmosphere has been stripped away over time (*Jakosky et al.*, 2011), leaving insufficient atmospheric pressure remaining to keep Mars warm today despite it being located within the HZ. On the other hand, Venus has retained a thick secondary atmosphere of outgassed CO_2, but has lost its water to space over time (*Kasting and Pollack*, 1983; *Donahue et al.*, 1982), rendering it water-poor and thus uninhabitable.

Atmospheric escape can be driven by several processes, including EUV/XUV radiation from the star, stellar wind interactions, and erosion by impact events (*Ahrens*, 1993; *Quintana et al.*, 2016; *Genda and Abe*, 2005). Smaller planets, hotter planets, and planets without a protective magnetic field will be more prone to atmospheric loss processes. An analysis of the planets and moons in our solar system, along with known exoplanets, suggest that bodies with and without atmospheres may be divided as a function of stellar insolation and escape velocity (*Zahnle and Catling*, 2017). Large-mass planets have stronger gravity and higher escape velocities, and thus are better able to resist atmospheric escape to space due to stellar EUV/XUV radiation and impacts. Planets with magnetic fields are able to shield their atmospheres from stripping by the solar wind by deflecting charge particles around the planet.

Planets around M-dwarf stars may be particularly vulnerable to atmospheric loss via hydrodynamic escape processes due to their long (>1 G.y.) superluminous pre-main-sequence phase and high activity levels (*Lammer et al.*, 2008; *Luger and Barnes*, 2015; *Meadows et al.*, 2018; *Barnes et al.*, 2018) as well as ion pickup (*Ribas et al.*, 2016). The strong stellar magnetic fields of M dwarfs also can reduce the size of

planetary magnetospheres, exposing more of the planet's atmosphere to erosion by the stellar wind (*Vidotto et al.*, 2013). For magnetized planets orbiting M dwarfs, the calculated polar wind losses for a 1-bar Earth-like atmosphere suggested a lifetime for that atmosphere of less than 400 m.y., although the calculated loss rate of H_2^+ and O_2^+ did not exceed Earth's current replenishment rate via outgassing and volatile delivery, and therefore maintenance of the atmosphere might be possible (*Garcia-Sage et al.*, 2017). However, planets without a magnetic field will be more vulnerable to atmospheric loss, as may be the case for older synchronously-rotating M-dwarf planets in the HZ (*Driscoll and Barnes*, 2015). If enough atmosphere is lost via interaction with the star, then the entire atmosphere could potentially condense onto the cold nightside (*Joshi et al.*, 1997; *Leconte et al.*, 2013a; *Turbet et al.*, 2017b) or at the poles of the planet (*Turbet et al.*, 2017b). Planets that have completely lost their atmospheres could be identifiable by the presence of extreme day-night temperature differences (*Kreidberg et al.*, 2019).

Venus, on the other hand, poses the scenario where many oceans of water can be lost to space over time, even when other atmospheric constituents remain. Deuterium/hydrogen ratios suggest that Venus has lost significant amounts of water to space over its lifetime (*Donahue et al.*, 1982; *Donahue*, 1999), while climate modeling studies suggest that Venus could have had clement surface conditions during the early history of our solar system when the Sun was dimmer (*Yang et al.*, 2014; *Way et al.*, 2016; see also the chapter by Arney and Kane in this volume). Independent of the bulk stripping of the atmosphere, water loss to space for Venus and other planets occurs when the climate enters a moist or runaway greenhouse state (*Kasting*, 1988), both of which allow large quantities of water vapor to permeate a planet's stratosphere where it can be photolyzed and the freed H atoms then irreversibly escape to space. Hydrogen atoms, being the lightest of all atoms, can escape to space most easily while heavier molecular gases like N_2 and CO_2 remain bound to the planet. Habitable planets around M-dwarf stars may be at greater risk of water loss during moist greenhouse states due to their particular patterns of atmospheric circulation (*Kopparapu et al.*, 2017; *Fujii et al.*, 2017); however, water loss rates may ultimately depend on the level of stellar activity (*Chen et al.*, 2019).

The loss of oceans to space could significantly modify the composition of a planet's atmosphere (*Luger and Barnes*, 2015). Photolysis of H_2O and the subsequent removal of H results in the buildup of approximately 250 bars of O_2 for each Earth-ocean equivalent of water lost. The amount of O_2 that remains in an atmosphere over long timescales depends on further loss processes, including atmospheric loss, and losses to sequestration in a magma ocean or via other surface processes (*Schaefer et al.*, 2016; *Wordsworth et al.*, 2018). This potential buildup of abiotic O_2 is a potential false positive biosignature (*Harman et al.*, 2015).

While ocean loss processes are themselves irreversible, the effects of ocean loss during moist and runaway greenhouse climate states may be mitigated by planets forming with higher initial water abundances (*Tian and Ida*, 2015), planetary migration from beyond the snowline (*Luger et al.*, 2017a), the presence of a dense protective H_2 atmospheres early on (*Luger et al.*, 2015; *Barnes et al.*, 2018), and subsequent cometary delivery or outgassing (*Albarede*, 2009; *Meadows et al.*, 2018). For example, despite Earth having lost 80–95% of its volatiles within the first 50–500 m.y. (*Turner*, 1989), large quantities of water may have remained sequestered in the mantle and were then slowly outgassed over time (*Sleep et al.*, 2012). Volatile cycling between the mantle and atmosphere may take billions of years to reach a steady-state, allowing exoplanets to regain surface volatiles that were lost during the early phases in their history (*Komacek and Abbot*, 2016). If M-dwarf terrestrial exoplanets can similarly acquire volatiles after an initial loss of water and atmosphere, they may, over billions of years, accumulate a surface ocean and atmosphere from volcanic outgassing or volatile delivery after the M dwarf has settled into its more benign main-sequence phase.

6.2. Photochemistry

The composition of terrestrial exoplanet atmospheres can be strongly affected by photochemistry (*Segura et al.*, 2005; *Rugheimer et al.*, 2015; *Meadows et al.*, 2018; *Lincowski et al.*, 2018). Photochemistry refers to gas-phase chemistry that occurs in a planet's atmosphere and is driven by light of the host star. Photochemistry is driven by the stellar UV spectrum, and the resultant gases and aerosols depend on the initial atmospheric composition and temperature structure, the total amount of UV emitted by the star, and the wavelength dependence of the star's UV output (*Rugheimer et al.*, 2015). Upper-atmospheric chemistry can also be driven by stellar energetic particle precipitation from coronal mass ejections (*Airapetian et al.*, 2016). Some molecules can be directly photolyzed, as is the case for CH_4, N_2O, and H_2O, while the abundance of other molecules (including CH_4) can be highly sensitive to the presence of catalysts. The concentration of catalysts may be independently controlled by both photochemical (*Segura et al.*, 2005) and dynamical processes (*Schoeberl and Hartmann*, 1991).

Cooler main-sequence stars often emit less UV and thus are often not as efficient as the Sun at photolyzing water vapor. This has a variety of consequences, including the reduced production of the $O(^1D)$ catalyst that is efficient at chemically destroying CH_4 (*Segura et al.*, 2005) and the reduced production of OH radicals required for CO_2 recombination (*Harman et al.*, 2015). Planets around quiet M-dwarf stars may experience only limited water loss from moist greenhouse climate states due to inefficient photolysis; however, water loss rates may be critically dependent on the level of stellar activity (*Chen et al.*, 2019). Photochemically produced O_3 in oxygen-rich atmospheres and Titan-like hydrocarbon hazes in methane-rich atmospheres can both provide UV shielding of the planetary surface by absorbing UV radiation high in the atmosphere (*Arney et al.*, 2016). However, note that

some photochemical modifications to the atmosphere can be harmful. For instance, planets around M-dwarf stars might build up significant abundances of CO in their atmospheres, which is a poisonous gas for complex life as we know it (*Schwieterman et al.*, 2019).

6.3. Climate

As discussed above, if water is to remain liquid at the surface, a planet must maintain a climate state that supports surface temperatures that lie in a fairly narrow range. A planet's climate is determined through a delicate balance between absorbed incoming stellar radiation and emitted thermal radiation, modulated by the presence of greenhouse gases, clouds, and aerosols (*Trenberth et al.*, 2009). However, star, planet, and planetary system interactions can have important influences on both patterns of stellar insolation and on the composition of planetary atmospheres.

Gravitational interactions can also drive changes to planetary climate via orbital modifications including migration and synchronous rotation. Long-term gravitational interactions can drive planet migration, which changes the mean insolation received by a planet. Giant planet migration may also be responsible for disturbing the orbits of comets and asteroids, triggering a bombardment and delivery of fresh volatiles to the inner terrestrial planets, shaping the composition of their secondary atmospheres (*Mojzsis et al.*, 2019). Gravitational interactions among neighboring bodies can drive oscillations in planetary eccentricity and obliquity (*Spiegel et al.*, 2010; *Armstrong et al.*, 2014; *Brasser et al.*, 2014; *Deitrick et al.*, 2018), which can affect the seasonal variability of planetary surface temperatures (*Bolmont et al.*, 2016; *Rose et al.*, 2017; *Kilic et al.*, 2017; *Way and Georgakarakos*, 2017; *Adams et al.*, 2019; *Colose et al.*, 2019). While the time-mean climates of highly eccentric planets may be stable (*Bolmont et al.*, 2016), dramatic seasonal swings in surface temperature or ice cover could make sustained surface habitability challenging (*Sherwood and Huber*, 2010). Furthermore, planets in the HZs of M-dwarf stars likely experience tidal locking where the planet maintains a synchronous or resonant orbit, thus governing the planet's rotation rate and significantly changing atmospheric circulation patterns and climate (*Yang et al.*, 2014).

Photochemistry, driven by the stellar energy distribution and stellar activity level of the host star, can cause the chemical modification of terrestrial planet atmospheres (*Segura et al.*, 2010; *Arney et al.*, 2017), which will impact temperature structure, climate, and water loss. Photochemistry impacts planetary climate and thus habitability by destroying or creating greenhouse gases, by creating absorbing aerosol species, and by driving planetary water loss process by removing H from high-altitude water vapor (thus permitting H to escape to space). The photolysis and ultimate destruction of water vapor in terrestrial planet atmospheres can yield a desiccated planet that has very different climate dynamics compared to water-rich planets (*Abe et al.*, 2011; *Leconte et al.*, 2013a; *Kodama et al.*, 2015). Photochemistry can also drive the creation or destruction of different novel greenhouse gas species in a planetary atmosphere such as CH_4, C_2H_6, NH_3, or N_2O, which all can act as greenhouse gases and cause warming of the planet (*Segura et al.,* 2005; *Haqq-Misra et al.,* 2008; *Airapetian et al.,* 2016; *Meadows et al.,* 2018). Conversely, photochemically produced upper atmospheric hazes can have a potentially significant cooling effect on climate (*McKay et al.,* 1999; *Haqq-Misra et al.,* 2008; *Arney et al.,* 2016). The addition of greenhouse gases or atmospheric hazes can either help or hinder planetary habitability depending upon the initial climate state.

6.4. Tidal Effects

Planets close to their host stars are expected to be affected by the differential gravitational force and experience tides, which can affect a planet's habitability (*Rasio et al.,* 1996; *Jackson et al.,* 2008). In particular, HZ planets around low-mass stars, such as M dwarfs, may experience tidal forces that can deform their solid bodies and also cause changes in angular momenta (rotation) and energy that affect their atmospheric dynamics (*Yang et al.,* 2013a; *Kopparapu et al.,* 2016; *Wolf,* 2017; *Haqq-Misra et al.,* 2018).

In addition to tidal locking (*Dole,* 1964; *Barnes,* 2017), which can result in synchronous rotation, tides may also impact habitability via orbital circularization, orbital migration, obliquity erosion, and tidal heating. Tides will also drive planetary obliquities toward 0 or 180 (*Goldreich,* 1966; *Heller et al.,* 2011), which changes the insolation pattern and may lead to atmospheric collapse at the poles (*Joshi et al.,* 1997). Near the end state of tidal evolution, planets may become tidally locked, and once their eccentricity and obliquity are both near zero they may synchronously rotate with one side of the planet always facing the star. However, close-in planets can also be found in non-circular orbits, because either they have not yet been tidally circularized, or they have had their eccentricity or obliquity maintained over long time periods via perturbation by another planet — which counteracts the tendency of tides to damp eccentricity and obliquity to zero (*Barnes et al.,* 2010). Planets in non-circular orbits can be heated by friction induced by the changing deformation of the planet (*Jackson et al.,* 2008; *Barnes et al.,* 2009). Planets in the HZs of M dwarfs may experience Io-like levels of surface heat flux, which could significantly change their internal properties and outgassing rates and potentially induce a runaway greenhouse (*Barnes et al.,* 2013). Tidal heating can also induce a prolonged magma ocean stage for young, close-in planets before they circularize (*Driscoll and Barnes,* 2015), and dense atmospheres may prolong this magma ocean phase (*Hamano et al.,* 2013). This tidal heating of the mantle can promote core cooling and generation of an early magnetic dynamo, which is maintained as the planet circularizes and loses its tidal heating component. However, if eccentricity is maintained for prolonged periods the mantle cannot cool, which results in core solidification

and loss of the magnetic dynamo, exposing the planet to stellar wind erosion. For these hotter eccentric planets, massive melt eruptions may also render them uninhabitable (*Driscoll and Barnes,* 2015).

6.5. Galactic Effects

The planetary system that a habitable planet forms in also interacts with the galaxy, and these interactions may affect the composition of the host star and protoplanetary disk, the dynamical stability of the planetary system, and the radiation received by the planet [for a more detailed review of galactic impacts on habitability, see *Kaib* (2018)]. Initial observations of a correlation between stars with enhanced metallicity (elements heavier than hydrogen) and the formation of giant planets (*Fischer and Valenti,* 2005) implied that perhaps planets would be more frequent in the inner, higher-metallicity regions of the galaxy. However, subsequent studies showed no such correlation for planets with $R < 4\ R_\oplus$ (*Buchhave et al.,* 2012; *Buchhave and Latham,* 2015), which are found around stars with a range of metallicities that are much wider than expected based on the hot Jupiter results. Consequently, there is no clear constraint on the location of terrestrial planet formation in our galaxy as a function of stellar metallicity and, by extension, the metallicity of the interstellar medium.

Galactic metallicity may also be a less important factor in supporting a planet with active plate tectonics, which is part of the carbonate-silicate cycle that buffers planetary climate (*Walker et al.,* 1981). Planet formation in a region of the galaxy that does not have recent star formation and regular type-II supernovae, which generate the $^{235,\ 238}U$, ^{232}Th, and ^{40}K unstable isotopes thought to drive radiogenic internal energy for tectonics, was also thought to limit the formation of terrestrial planets with sufficient long-lived energy for plate tectonics. However, neutron star mergers, which are less tied to very recent star formation, and for which gravitational wave signatures were recently detected (*Abbott et al.,* 2017), are now recognized to be a dominant contributor to the production of a more constant supply of unstable isotopes of U and Th. In addition, there is still uncertainty as to the relative contributions to Earth's current internal heat, and radiogenic decay may produce a heat flux that is roughly equal to the Earth's primordial heat of formation (*Korenaga,* 2008; *Dye,* 2012), such that even a complete lack of heating from radionuclides may not significantly drop the surface heat flow.

Gravitational interactions with the galactic environment can perturb planetary systems and possibly contribute to significant migration of the host star from its birth environment, ultimately affecting habitability. While still in their formation clusters, stars are closer to each other and have lower relative velocities than they will out in the field, and therefore gravitational encounters between stars are more likely to perturb planetary orbits, including ejecting planets (*Laughlin and Adams,* 1998; *Spurzem et al.,* 2009). However, this is unlikely to be an issue for most HZ planets, as an extremely close pass by another star would be needed to eject an inner planet, and most star-forming clusters disperse on 10-m.y. timescales, which lowers the probability of such an encounter. Although it would be challenging to eject a HZ planet, outer planets are more likely to be perturbed and the modification to their orbits could be transferred to inner planets via planet-planet interactions (*Malmberg et al.,* 2011). After the open cluster phase, stellar interactions would normally be less frequent, but if the host star has a distant stellar companion, that companion can act as an antenna for gravitational encounters with other stars. This can ultimately destabilize the planetary system, and make habitability for planets in widely space binaries less likely over long time periods (*Kaib et al.,* 2013). On an even larger scale, stars on nearly circular orbits co-rotating with the galaxy's spiral structure may be susceptible to large radial migrations (*Sellwood and Binney,* 2002; *Roškar et al.,* 2012). Sun-like stars could have migrated up to 6 kpc within the galactic disk, from the inner galaxy outward, significantly changing the radiation environment over time as well as the stellar encounter frequency (*Wielen et al.,* 1996; *Roškar et al.,* 2011). Increased stellar encounters when the Sun was younger and closer to the center of the galaxy could have influenced the volatile delivery and impact history of Earth by modifying the Oort cloud and the flux of comets through the inner solar system (*Kaib et al.,* 2011). These interactions could have reduced or destroyed reservoirs of distant icy bodies or injected outer planetary system icy bodies into habitable orbits.

Finally, the galactic environment can affect the planet directly, via interaction with radiation and particles from highly-energetic events. Proposed mechanisms include the action of gamma-ray bursts (*Melott et al.,* 2004; *Atri et al.,* 2014), supernovae (*Gehrels et al.,* 2003), or kilonovae (*Abbott et al.,* 2017), which could erode the ozone layer on a habitable planet. However, models of the effect of a nearby (8 pc distant) supernova suggest that surface UV flux would only increase by a factor of 2, which may not be catastrophic. Moreover, such a nearby supernova would likely only happen once every 8 G.y. (*Gehrels et al.,* 2003). Gamma-ray bursts (GRBs) are far more energetic, and could indeed produce mass extinction events if a habitable planet were exposed to one within 1–2 kpc (*Thomas et al.,* 2005). However, GRBs are most often observed in metal-poor (less than 10% of the Sun's metallicity) galaxies (*Piran and Jimenez,* 2014), and if their progenitors are similarly metal poor, very few stars would have been subjected to these GRBs, and those would be in the more metal-poor outskirts of the galaxy (*Gowanlock,* 2016). Consequently, the impact of gamma-ray bursts and supernovae on planetary habitability may be relatively modest, especially when compared to processes that are likely to have a higher impact, e.g., stellar activity, within a planetary system.

7. CURRENT AND NEAR-TERM OBSERVATIONS OF HABITABLE ZONE PLANETS

We are now entering a new era of terrestrial exoplanet characterization that is providing a glimpse into the environments of possibly habitable worlds. Although models may provide valuable information about the probability of habitability, it is observations, albeit often interpreted by models, that will ultimately be used to assess whether or not a planet is habitable. In the past few years, the very first attempts at determining planetary bulk composition and observing the atmospheres of likely terrestrial exoplanets in the HZs of M dwarfs have been undertaken. A handful of good targets are known, including the HZ transiting planets TRAPPIST-1 e, f, and g in the seven-planet TRAPPIST-1 system (*Gillon et al.,* 2016, 2017), LHS 1140 b (*Dittmann et al.,* 2017), the RV-detected Proxima Centauri b (*Anglada-Escudé et al.,* 2016), and Ross 128 b (*Bonfils et al.,* 2018). To complement these HZ planets, other likely terrestrials include the exo-Venuses TRAPPIST-1 b, c, and d (*Gillon et al.,* 2016, 2017) and GJ1132 b (*Berta-Thompson et al.,* 2015). For the transiting planets, observations of both the size and inclination from transit, combined with masses from RV or TTVs, have produced densities that provide our first steps toward characterizing habitability. The densities of nearby M-dwarf terrestrial planets [e.g., GJ1132 b, 6.0 ± 2.5 g cm^{-3} (*Berta-Thompson et al.,* 2015), and LHS1140 b, 12.5 ± 3.4 g cm^{-3} (*Dittmann et al.,* 2017)] are comparable to the densities of Earth (5.5 g cm^{-3}) or Venus (5.3 g cm^{-3}), consistent with mixtures of silicate rock and iron. Initial estimates suggested that TRAPPIST-1 planets have densities that span 0.6 to 1.0× Earth's density (i.e., 3.3–5.5 g cm^{-3}) (*Grimm et al.,* 2018), and innovative techniques that used data from the seven planets together to probe their interiors suggested that the measurement errors were constant with a consistent or increasing water mass fraction with semimajor axis (*Dorn et al.,* 2018). However, new measurements with better precision suggest that the densities of most of the TRAPPIST-1 planets are in fact similar to each other and are closer in value, albeit with higher water fractions, to our solar system terrestrials. The generally lower densities of the TRAPPIST-1 planets (*Gillon et al.,* 2017; *Grimm et al.,* 2018), along with their resonant orbits, suggest that the planets have formed at larger distances from the star in a more volatile-rich birth environment and migrated inward (*Luger et al.,* 2017b; *Unterborn et al.,* 2018). So despite predictions of extreme atmospheric and ocean loss for these planets, it is likely that their interiors remain volatile-rich.

Spectra and photometry have also been obtained for several of the TRAPPIST-1 planets, and although these spectra appear to be featureless, they do help in attempts to constrain the planetary atmospheric properties. The HST Wide Field Camera 3 transmission spectroscopy has ruled out H_2-dominated atmospheres for the innermost six TRAPPIST-1 planets (*de Wit et al.,* 2016, 2018). In addition, laboratory data and models suggest that it is unlikely that the flat spectra observed are due to suspended aerosols, and instead may be high mean-molecular-weight secondary outgassed atmospheres (*Moran et al., 2018*), of as-yet-unconstrained composition (*Delrez et al., 2018; Lincowski et al., 2018*). However, whether the planets have high molecular weight atmospheres (e.g., CO_2- or O_2-dominated) or no atmospheres requires observations with future facilities.

In the next 5–10 years the JWST — scheduled for launch early in 2021 — and next-generation groundbased extremely large telescopes will provide additional capabilities to study the atmospheres of HZ terrestrial exoplanets. The best targets for JWST will likely come from ground-based exoplanet detection surveys that focus on late-type M dwarfs, as these small stars produce excellent atmospheric signals for transiting planets (*Morley et al., 2017; Lustig-Yaeger et al., 2019*). Planets discovered by the Transiting Exoplanet Survey Satellite (TESS) mission are more likely to be orbiting brighter, earlier type M dwarfs that produce relatively weak differential signals from planetary atmospheres and so will not be the best targets for JWST. However, for systems like TRAPPIST-1, which is orbiting a late-type M8 dwarf, JWST can likely detect the presence of a terrestrial atmosphere (containing CO_2 but no aerosols) for each of the planets using transmission spectroscopy, by co-adding data from fewer than 10 transits for each planet (*Morley et al., 2017; Lustig-Yaeger et al., 2019*). For cloudy atmospheres the integration time will be up to 20–30 transits (*Lustig-Yaeger et al., 2019*). For characterizing the nature of the atmosphere and the planet's habitability, JWST will be much more adept at proving ocean loss, or a loss of habitability, than confirming that an ocean is present. This is because signs of ocean loss, such as strong O_2-O_2 collisional absorption from a massive O_2 atmosphere (*Meadows, 2017; Lincowski et al., 2018*), as well as potential signatures from enhanced D/H from atmospheric loss, are potentially detectable in 2–11 transits for the inner planets. The presence of gases that are normally soluble in water, such as SO_2, or the presence of an H_2SO_4-H_2O haze layer with SO_2 gas (*Lincowski et al., 2018; Loftus et al., 2019*) are more observationally challenging (*Lustig-Yaeger et al., 2019; Loftus et al., 2019*) but would point to a surface environment almost entirely devoid of water. On the other hand, water vapor, at the relatively desiccated altitudes that transit probes, will take upward of 60 transits for the HZ TRAPPIST-1 planets (*Lustig-Yaeger et al., 2019; Lincowski et al., 2019*), and even if detected it is not definitive proof that surface liquid water exists. Biosignatures like O_2 will be extremely challenging for JWST and unlikely to be detected due to the poor sensitivity of the instruments at wavelengths shortward of 1.3 μm.

Groundbased extremely large telescopes (ELTs) may also be capable of probing M-dwarf planetary atmospheres starting in 2025. The ELTs can probe a handful of HZ planets using high-resolution spectroscopy for transiting (*Rodler and López-Morales, 2014*) and reflected light observations (*Snellen et al., 2015; Lovis et al., 2017*).

These telescopes may also be able to take direct imaging mid-infrared (MIR) observations of planets orbiting G-dwarf stars (*Quanz et al., 2015*). For the best targets, these facilities may have the precision required to undertake the first spectroscopic search for atmospheric water vapor and biosignature gases, such as O_2 and CH_4 (*Lovis et al., 2017; Rodler and López-Morales, 2014; López-Morales et al., 2019*), but these groundbased measurements will also be unable to directly detect surface water on an exoplanet.

In the more distant future, large space-based coronagraphic direct imaging mission concepts currently under consideration by NASA may obtain spectra of hundreds of HZ planets. These planets will be orbiting stars of spectral type from F down through M and will provide a large statistical sample for observational determination of the HZ and detection of biosignature gases. Constraints on CO_2 levels on the larger samples of planets obtained by these telescopes could be used to test for evidence of the carbonate-silicate cycle directly (*Bean et al., 2017*). These telescopes may also have the capability to map the planetary surface as it rotates under the observer and directly detect glint from exoplanet oceans (*Robinson et al., 2010; Cowan et al., 2012; Lustig-Yaeger et al., 2018*), providing a definitive detection of habitability. For a more detailed discussion of how to observationally determine exoplanet habitability, see the chapter by Robinson and Reinhard in this volume.

8. SUMMARY AND CONCLUSIONS

The characteristics and processes relevant to the maintenance of surface liquid water on a terrestrial planet are broad, interdisciplinary, and interconnected, and both modeling and observations will be needed to understand them. To date, our best first-order assessment method for whether or not a planet is likely to be habitable has been to check whether a newly discovered exoplanet is in the size range that is likely to be terrestrial, and is in the HZ of its parent star. However, we now know that habitability is maintained via the interplay of planetary, stellar, and planetary system characteristics over the planet's lifetime. Within this new framework, the HZ can be seen as a two-dimensional slice in stellar type and semimajor axis through a multi-dimensional parameter space. Planets that form in the HZ may also not be habitable due to either a low initial volatile inventory or loss of volatiles over time. Atmosphere and ocean loss is of particular concern for M-dwarf terrestrial planets, which orbit close to a star that spent its early life in a superluminous state and maintains high stellar activity for an extended period of time. On the other hand, terrestrial planetary atmospheres are most likely secondary, outgassed from the interior. Understanding how the balance between outgassing and atmospheric escape sculpts the resulting terrestrial planet atmosphere, and potentially replenishes an ocean, will be an important new frontier in terrestrial exoplanet evolution and habitability. While some of the characteristics and processes that inform planetary habitability may be observable in the coming decades, many will instead be explored via modeling, or a combination of modeling and observations. An interdisciplinary system science approach will be needed to fully explore the depth and complexity of planetary habitability. An improved understanding of the factors that affect habitability will enable identification of those exoplanets that are most likely to be habitable, and inform our interpretation of upcoming exoplanet data to be used to search for life beyond the Earth.

Acknowledgments. We sincerely thank reviewers D. Abbot and R. Wordsworth for their constructive comments and suggestions, which greatly improved the review chapter. This chapter benefitted from discussions with A. Lincowski and the assistance of E. Davis. This work was performed by NASA's Virtual Planetary Laboratory, a member of the Nexus for Exoplanet System Science Research Coordination Network, and funded by the NASA Astrobiology Program under grant 80NSSC18K0829.

REFERENCES

Abbott B. P., Abbott R., Abbott T. D., Acernese F., Ackley K., Adams C., Adams T., et al. (2017) GW170817: Observation of gravitational waves from a binary neutron star inspiral. *Phys. Rev. Lett.*, *119(16)*, 161101.

Abe Y., Abe-Ouchi A., Sleep N. H., and Zahnle K. J. (2011) Habitable zone limits for dry planets. *Astrobiology*, *11*, 443–460.

Adams A., Boos W., and Wolf E. (2019) Aquaplanet models on eccentric orbits: Effects of the rotation rate on observables. *Astrophys. J.*, *157(189)*, 19.

Agol E. and Fabrycky D. C. (2018) Transit-timing and duration variations for the discovery and characterization of exoplanets. In *Handbook of Exoplanets* (H. Deeg and J. Belmonte, eds.), pp. 797–816. Springer, Cham.

Agol E., Steffen J., Sari R., and Clarkson W. (2005) On detecting terrestrial planets with timing of giant planet transits. *Mon. Not. R. Astron. Soc.*, *359(2)*, 567–579.

Agol E., Jansen T., Lacy B., Robinson T. D., and Meadows V. (2015) The center of light: Spectroastrometric detection of exomoons. *Astrophys. J.*, *812(1)*, 5.

Ahrens T. J. (1993) Impact erosion of terrestrial planetary atmospheres. *Annu. Rev. Earth Planet. Sci.*, *21(1)*, 525–555.

Airapetian V. S., Glocer A., Gronoff G., Hebrard E., and Danchi W. (2016) Prebiotic chemistry and atmospheric warming of early Earth by an active young Sun. *Nature Geosci.*, *9*, 452–455.

Albarede F. (2009) Volatile accretion history of the terrestrial planets and dynamic implications. *Nature*, *461(7268)*, 1227.

Anglada-Escudé G., Amado P. J., Barnes J., Berdiñas Z. M., Butler R. P., Coleman G. A., de la Cueva I., et al. (2016) A terrestrial planet candidate in a temperate orbit around Proxima Centauri. *Nature*, *536(7617)*, 437–440.

Armstrong J., Barnes R., Domagal-Goldman S., Breiner J., Quinn T., and Meadows V. (2014) Effects of extreme obliquity variations on the habitability of exoplanets. *Astrobiology*, *14(4)*, 277–291.

Armstrong K., Frost D. J., McCammon C. A., Rubie D. C., and Boffa Ballaran T. (2019) Deep magma ocean formation set the oxidation state of Earth's mantle. *Science*, *365(6456)*, 903–906.

Arney G., Domagal-Goldman S. D., Meadows V. S., Wolf E. T., Schwieterman E., Charnay B., Claire M., Hébrard E., and Trainer M. G. (2016) The pale orange dot: The spectrum and habitability of hazy Archean Earth. *Astrobiology*, *16(11)*, 873–899.

Arney G. N., Meadows V. S., Domagal-Goldman S. D., Deming D., Robinson T. D., Tovar G., Wolf E. T., and Schwieterman E. (2017) Pale orange dots: The impact of organic haze on the habitability and detectability of Earthlike exoplanets. *Astrophys. J.*, *836(1)*, 49.

Atri D., Melott A. L., and Karam A. (2014) Biological radiation dose from secondary particles in a Milky Way gamma-ray burst. *Intl. J. Astrobiol.*, *13*, 224–228.

Baker S. J., Hesselbo S. P., Lenton T. M., Duarte L. V., and Belcher C. M. (2017) Charcoal evidence that rising atmospheric oxygen terminated Early Jurassic ocean anoxia. *Nature Commun.*, *8*, 15018.

Baraffe I., Homeier D., Allard F., and Chabrier G. (2015) New evolutionary models for pre-main sequence and main sequence low-mass stars down to the hydrogen-burning limit. *Astron. Astrophys.*, *577*, A42.

Barnes R. (2017) Tidal locking of habitable exoplanets. *Cel. Mech. Dyn. Astron.*, *129(4)*, 509–536.

Barnes R., Jackson B., Greenberg R., and Raymond S. N. (2009) Tidal limits to planetary habitability. *Astrophys. J. Lett.*, *700(1)*, L30.

Barnes R., Raymond S. N., Greenberg R., Jackson B., and Kaib N. A. (2010) CoRoT-7b: Super-Earth or super-Io? *Astrophys. J. Lett.*, *709(2)*, L95.

Barnes R., Mullins K., Goldblatt C., Meadows V. S., Kasting J. F., and Heller R. (2013) Tidal Venuses: Triggering a climate catastrophe via tidal heating. *Astrobiology*, *13(3)*, 225– 250.

Barnes R., Deitrick R., Luger R., Driscoll P. E., Quinn T. R., Fleming D. P., Guyer B., et al. (2018) The habitability of Proxima Centauri b I: Evolutionary scenarios. *ArXiv e-prints*, arXiv:1608.06919v2.

Batalha N. M., Borucki W. J., Bryson S. T., Buchhave L. A., Caldwell D. A., Christensen-Dalsgaard J., Ciardi D., et al. (2011) Kepler's first rocky planet: Kepler-10b. *Astrophys. J.*, *729(1)*, 27.

Bean J., Abbot D., and Kempton E. M.-R. (2017) A statistical comparative planetology approach to the hunt for habitable exoplanets and life beyond the solar system. *Astrophys. J. Lett.*, *841*, L24.

Becker J. C. and Batygin K. (2013) Dynamical measurements of the interior structure of exoplanets. *Astrophys. J.*, *778*, 100.

Beichman C., Barrado D., Belikov R., Biller B., Boccaletti A., Burrows A., Danielski C., et al. (2019) Direct imaging and spectroscopy of exoplanets with the James Webb Space Telescope. *Bull. Am. Astron. Soc.*, *51(3)*, 58.

Benedict G. F., McArthur B. E., Gatewood G., Nelan E., Cochran W. D., Hatzes A., Endl M., Wittenmyer R., Baliunas S. L., and Walker G. A. H. (2006) The extrasolar planet ε Eridani b: Orbit and mass. *Astrophys. J.*, *132(5)*, 2206–2218.

Berger A., Loutre M.-F., and Tricot C. (1993) Insolation and Earth's orbital periods. *J. Geophys. Res.–Atmos.*, *98(D6)*, 10341–10362.

Berta-Thompson Z. K., Irwin J., Charbonneau D., Newton E. R., Dittmann J. A., Astudillo-Defru N., Bonfils X., et al. (2015) A rocky planet transiting a nearby low-mass star. *Nature*, *527(7577)*, 204–207.

Bolmont E., Libert A.-S., Leconte J., and Selsis F. (2016) Habitability of planets on eccentric orbits: The limits of the mean flux approximation. *Astron. Astrophys.*, *591*, A106.

Bonfils X., Astudillo-Defru N., Díaz R., Almenara J. M., Forveille T., Bouchy F., Delfosse X., et al. (2018) A temperate exo-Earth around a quiet M dwarf at 3.4 parsec. *Astron. Astrophys.*, *613*, A25.

Borucki W. J., Koch D., Basri G., Batalha N., Brown T., Caldwell D., Caldwell, J., et al. (2010) Kepler planet-detection mission: Introduction and first results. *Science*, *327(5968)*, 977–980.

Brain D. A., Leblanc F., Luhmann J. G., Moore T. E., and Tian F. (2013) *Planetary Magnetic Fields and Climate Evolution*. Univ. of Arizona, Tucson. 708 pp.

Brasser R., Ida S., and Kokubo E. (2014) A dynamical study on the habitability of terrestrial exoplanets — II. The super-Earth HD 40307 g. *Mon. Not. R. Astron. Soc.*, *440*, 3685–3700.

Buchhave L. A. and Latham D. W. (2015) The metallicities of stars with and without transiting planets. *Astrophys. J.*, *808(2)*, 187.

Buchhave L. A., Latham D. W., Johansen A., Bizzarro M., Torres G., Rowe J. F., Batalha N. M., et al. (2012) An abundance of small exoplanets around stars with a wide range of metallicities. *Nature*, *486(7403)*, 375–377.

Buhler P. B., Knutson H. A., Batygin K., Fulton B. J., Fortney J. J., Burrows A., and Wong I. (2016) Dynamical constraints on the core mass of hot Jupiter HAT-P-13b. *Astrophys. J.*, *821*, 26.

Bullock M. A. (1997) The stability of climate on Venus. Ph.D. thesis, Univ. of Colorado, Boulder, available online at *http://libraries. colorado.edu/record=b2784318~S3*.

Carone L., Keppens R., and Decin L. (2015) Connecting the dots — II. Phase changes in the climate dynamics of tidally locked terrestrial exoplanets. *Mon. Not. R. Astron. Soc.*, *453(3)*, 2412–2437.

Cassan A., Kubas D., Beaulieu J. P., Dominik M., Horne K., Greenhill J., Wambsganss J., et al. (2012) One or more bound planets per Milky Way star from microlensing observations. *Nature*, *481(7380)*, 167–169.

Catling D. C. and Claire M. W. (2005) How Earth's atmosphere evolved to an oxic state: A status report. *Earth Planet. Sci. Lett., 237(1-2),* 1–20.

Catling D. C., Zahnle K. J., and McKay C. P. (2001) Biogenic methane, hydrogen escape, and the irreversible oxidation of early Earth. *Science, 293(5531),* 839–843.

Cauley P. W., Shkolnik E. L., Llama J., and Lanza A. F. (2019) Magnetic field strengths of hot Jupiters from signals of star-planet interactions. *Nature Astron., 3,* 1128–1134.

Charbonneau D., Brown T. M., Latham D.W., and Mayor M. (2000) Detection of planetary transits across a Sun-like star. *Astrophys. J. Lett., 529(1),* L45–L48.

Charbonneau D., Berta Z. K., Irwin J., Burke C. J., Nutzman P., Buchhave L. A., Lovis C., et al. (2009) A super-Earth transiting a nearby low-mass star. *Nature, 462(7275),* 891–894.

Charnay B., Meadows V., Misra A., Leconte J., and Arney G. (2015) 3D modeling of GJ1214b's atmosphere: Formation of inhomogeneous high clouds and observational implications. *Astrophys. J. Lett., 813(1),* L1.

Chassefiere E., Leblanc F., and Langlais B. (2007) The combined effects of escape and magnetic field histories at Mars. *Planet. Space Sci., 55(3),* 343–357.

Chen J. and Kipping D. (2017) Probabilistic forecasting of the masses and radii of other worlds. *Astrophys. J., 834(1),* 17.

Chen H. and Rogers L. A. (2016) Evolutionary analysis of gaseous sub-Neptune-mass planets with MESA. *Astrophys. J., 831(2),* 180.

Chen H., Wolf E. T., Zhan Z., and Horton D. E. (2019) Habitability and spectroscopic observability of warm M dwarf exoplanets with a 3D chemistry-climate model. *Astrophys. J., 886(1).*

Colose C., Del Genio A., and Way M. (2019) Enhanced habitability of high obliquity bodies near the outer edge of the habitable zone of Sun-like stars. *Astrophys. J., 884(2),* DOI: 10.3847/1538-4357/ab4131.

Cowan N. B. and Abbot D. S. (2014) Water cycling between ocean and mantle: Super-Earths need not be waterworlds. *Astrophys. J., 781(1),* 27.

Cowan N. B., Agol E., Meadows V. S., Robinson T., Livengood T. A., Deming D., Lisse C. M., et al. (2009) Alien maps of an ocean-bearing world. *Astrophys. J., 700(2),* 915.

Cowan N. B., Abbot D. S., and Voigt A. (2012) A false positive for ocean glint on exoplanets: The latitude-albedo effect. *Astrophys. J. Lett., 752(1),* L3.

Crossfeld I. J. (2016) Exoplanet atmospheres and giant ground-based telescopes. *ArXiv e-prints,* arXiv:1604.06458.

de Wit J., Wakeford H. R., Gillon M., Lewis N. K., Valenti J. A., Demory B.-O., Burgasser A. J., et al. (2016) A combined transmission spectrum of the Earth-sized exoplanets TRAPPIST-1 b and c. *Nature, 537,* 69–72.

de Wit J., Wakeford H. R., Lewis N. K., Delrez L., Gillon M., Selsis F., Leconte J., et al. (2018) Atmospheric reconnaissance of the habitable-zone Earth-sized planets orbiting TRAPPIST-1. *Nature Astron., 2,* 214–219.

Deck K. M. and Agol E. (2015) Measurement of planet masses with transit timing variations due to synodic "chopping" effects. *Astrophys. J., 802,* 116.

Deitrick R., Barnes R., Quinn T. R., Armstrong J., Charnay B., and Wilhelm C. (2018) Exo-Milankovitch cycles. I. Orbits and rotation states. *Astron. J., 155,* 60.

Del Genio A. D., Way M. J., Amundsen D. S., Aleinov I., Kelley M., Kiang N. Y., and Clune T. L. (2019) Habitable climate scenarios for Proxima Centauri b with a dynamic ocean. *Astrobiology, 19(2),* 1–27.

Delrez L., Gillon M., Triaud A. H. M. J., Demory B. O., de Wit J., Ingalls J. G., Agol E., et al. (2018) Early 2017 observations of TRAPPIST-1 with Spitzer. *Mon. Not. R. Astron. Soc., 475(3),* 3577–3597.

Demory B.-O., Gillon M., Barman T., Bonfils X., Mayor M., Mazeh T., Queloz D., et al. (2007) Characterization of the hot Neptune GJ 436 b with Spitzer and ground-based observations. *Astron. Astrophys., 475(3),* 1125–1129.

Dittmann J. A., Irwin J. M., Charbonneau D., Bonfils X., Astudillo-Defru N., Haywood R. D., Berta-Thompson Z. K., et al. (2017) A temperate rocky super-Earth transiting a nearby cool star. *Nature, 544(7650),* 333–336.

Do Nascimento J.-D. Jr., Vidotto A. A., Petit P., Folsom C., Castro M., Marsden S. C., Morin J., et al. (2016) Magnetic field and wind of Kappa Ceti: Toward the planetary habitability of the young Sun when life arose on Earth. *Astrophys. J. Lett., 820(1),* L15.

Dobos V., Heller R., and Turner E. L. (2017) The effect of multiple heat sources on exomoon habitable zones. *Astron. Astrophys., 601,* A91.

Dobrovolskis A. R. (2009) Insolation patterns on synchronous exoplanets with obliquity. *Icarus, 204(1),* 1–10.

Dole S. H. (1964) *Habitable Planets for Man.* Blaisdell, New York. 158 pp.

Donahue T. M. (1999) New analysis of hydrogen and deuterium escape from Venus. *Icarus, 141(2),* 226–235.

Donahue T. M., Hoffman J. H., Hodges R. R., and Watson A. J. (1982) Venus was wet: A measurement of the ratio of deuterium to hydrogen. *Science, 216,* 630–633.

Dorn C., Khan A., Heng K., Connolly J. A., Alibert Y., Benz W., and Tackley P. (2015) Can we constrain the interior structure of rocky exoplanets from mass and radius measurements? *Astron. Astrophys., 577,* A83.

Dorn C., Mosegaard K., Grimm S. L., and Alibert Y. (2018) Interior characterization in multiplanetary systems: TRAPPIST-1. *Astrophys. J., 865(1),* 20.

Doyle L. R., Carter J. A., Fabrycky D. C., Slawson R. W., Howell S. B., Winn J. N., Orosz J. A., et al. (2011) Kepler-16: A transiting circumbinary planet. *Science, 333(6049),* 1602.

Dressing C. D. and Charbonneau D. (2015) The occurrence of potentially habitable planets orbiting M dwarfs estimated from the full Kepler dataset and an empirical measurement of the detection sensitivity. *Astrophys. J., 807(1),* 45.

Dressing C. D., Spiegel D. S., Scharf C. A., Menou K., and Raymond S. N. (2010) Habitable climates: The influence of eccentricity. *Astrophys. J., 721(2),* 1295.

Driscoll P. E. (2018) Planetary interiors, magnetic fields, and habitability. In *Handbook of Exoplanets* (H. Deeg and J. Belmonte, eds.), pp. 2917-2935. Springer, Cham.

Driscoll P. E. and Barnes R. (2015) Tidal heating of Earth-like exoplanets around M stars: Thermal, magnetic, and orbital evolutions. *Astrobiology, 15,* 739–760.

Driscoll P. and Bercovici D. (2013) Divergent evolution of Earth and Venus: Influence of degassing, tectonics, and magnetic fields. *Icarus, 226,* 1447–1464.

Driscoll P. and Bercovici D. (2014) On the thermal and magnetic histories of Earth and Venus: Influences of melting, radioactivity, and conductivity. *Phys. Earth Planet. Inter., 236,* 36–51.

Driscoll P. and Olson P. (2011) Optimal dynamos in the cores of terrestrial exoplanets: Magnetic field generation and detectability. *Icarus, 213,* 12–23.

Dye S. T. (2012) Geoneutrinos and the radioactive power of the Earth. *Rev. Geophys., 50(3),* RG3007.

Edson A., Lee S., Bannon P., Kasting J. F., and Pollard D. (2011) Atmospheric circulations of terrestrial planets orbiting low-mass stars. *Icarus, 212(1),* 1–13.

Egan H., Jarvinen R., Ma Y., and Brain D. (2019) Planetary magnetic field control of ion escape from weakly magnetized planets. *Mon. Not. R. Astron. Soc., 488(2),* 2108–2120.

Egbert G. D. and Ray R. D. (2000) Significant dissipation of tidal energy in the deep ocean inferred from satellite altimeter data. *Nature, 405,* 775–778.

Eggl S., Pilat-Lohinger E., Georgakarakos N., Gyergyovits M., and Funk B. (2012) An analytic method to determine habitable zones for S-type planetary orbits in binary star systems. *Astrophys. J., 752(1),* 74.

Eggl S., Pilat-Lohinger E., Funk B., Georgakarakos N., and Haghighipour N. (2013) Circumstellar habitable zones of binary-star systems in the solar neighbourhood. *Mon. Not. R. Astron. Soc., 428(4),* 3104–3113.

Ergun R. E., Carlson C. W., McFadden J. P., Delory G. T., Strangeway R. J., and Pritchett P. L. (2000) Electron-cyclotron maser driven by charged-particle acceleration from magnetic field-aligned electric fields. *Astrophys. J., 538(1),* 456–466.

Fassett C. I. and Head J. W. (2008) Valley network-fed, open-basin lakes on Mars: Distribution and implications for Noachian surface and subsurface hydrology. *Icarus, 198,* 37–56.

Ferraz-Mello S., Rodríguez A., and Hussmann H. (2008) Tidal friction in close-in satellites and exoplanets: The Darwin theory re-visited. *Cel. Mech. Dyn. Astron., 101,* 171–201.

Ferris J. P. and Hagan W. J. Jr. (1984) HCN and chemical evolution: The possible role of cyano compounds in prebiotic synthesis.

Tetrahedron, 40(7), 1093–1120.

Fischer D. A. and Valenti J. (2005) The planet-metallicity correlation. *Astrophys. J., 622(2)*, 1102–1117.

Fischer D. A., Howard A. W., Laughlin G. P., Macintosh B., Mahadevan S., Sahlmann J., and Yee J. C. (2014) Exoplanet detection techniques. In *Protostars and Planets VI* (H. Beuther et al., eds.), pp. 715–738. Univ. of Arizona, Tucson.

Foreman-Mackey D., Hogg D. W., and Morton T. D. (2014) Exoplanet population inference and the abundance of Earth analogs from noisy, incomplete catalogs. *Astrophys. J., 795(1)*, 64.

Forgan D. (2014) Assessing circumbinary habitable zones using latitudinal energy balance modelling. *Mon. Not. R. Astron. Soc., 437(2)*, 1352–1361.

Forgan D. (2016) Milankovitch cycles of terrestrial planets in binary star systems. *Mon. Not. R. Astron. Soc., 463(3)*, 2768–2780.

Forgan D. H. (2017) On the feasibility of exo-moon detection via exoplanet phase curve spectral contrast. *Mon. Not. R. Astron. Soc., 470(1)*, 416–426.

Forgan D. and Dobos V. (2016) Exomoon climate models with the carbonate-silicate cycle and viscoelastic tidal heating. *Mon. Not. R. Astron. Soc., 457(2)*, 1233–1241.

Forgan D. and Kipping D. (2013) Dynamical effects on the habitable zone for Earth-like exomoons. *Mon. Not. R. Astron. Soc., 432(4)*, 2994–3004.

Fortney J. J., Mordasini C., Nettelmann N., Kempton E. M. R., Greene T. P., and Zahnle K. (2013) A framework for characterizing the atmospheres of low-mass low-density transiting planets. *Astrophys. J., 775(1)*, 80.

Frost D. J. and McCammon C. A. (2008) The redox state of Earth's mantle. *Annu. Rev. Earth Planet. Sci., 36*, 389–420.

Fujii Y., Del Genio A., and Amundsen D. (2017) NIR-driven moist upper atmospheres of synchronously rotating temperate terrestrial exoplanets. *Astrophys. J., 848*, 13.

Fujii Y., Angerhausen D., Deitrick R., Domagal-Goldman S., Grenfell J. L., Hori Y., Palle E., Siegler N., Stapelfeldt K., and Rauer H. (2018) Exoplanet biosignatures: Observational prospects. *Astrobiology, 18(6)*, 739–778.

Fulton B. J. and Petigura E. A. (2018) The California-Kepler Survey. VII. Precise planet radii leveraging Gaia DR2 reveal the stellar mass dependence of the planet radius gap. *Astron. J., 156(6)*, 264.

Fulton B. J., Petigura E. A., Howard A. W., Isaacson H., Marcy G. W., Cargile P. A., Hebb L., et al. (2017) The California-Kepler Survey. III. A gap in the radius distribution of small planets. *Astron. J., 154(3)*.

Gaidos E., Mann A. W., Kraus A. L., and Ireland M. (2016) They are small worlds after all: Revised properties of Kepler M dwarf stars and their planets. *Mon. Not. R. Astron. Soc., 457(3)*, 2877–2899.

Garcia-Sage K., Glocer A., Drake J., Gronoff G., and Cohen O. (2017) On the magnetic protection of the atmosphere of Proxima Centauri b. *Astrophys. J. Lett., 844(1)*, L13.

Gehrels N., Laird C. M., Jackman C. H., Cannizzo J. K., Mattson B. J., and Chen W. (2003) Ozone depletion from nearby supernovae. *Astrophys. J., 585*, 1169–1176.

Genda H. and Abe Y. (2005) Enhanced atmospheric loss on protoplanets at the giant impact phase in the presence of oceans. *Nature, 433(7028)*, 842–844.

Gillon M., Jehin E., Lederer S. M., Delrez L., de Wit J., Burdanov A., Van Grootel V., et al. (2016) Temperate Earth-sized planets transiting a nearby ultracool dwarf star. *Nature, 533(7602)*, 221–224.

Gillon M., Triaud A. H., Demory B.-O., Jehin E., Agol E., Deck K. M., Lederer S. M., et al. (2017) Seven temperate terrestrial planets around the nearby ultracool dwarf star TRAPPIST-1. *Nature, 542(7642)*, 456–460.

Ginzburg S., Schlichting H. E., and Sari R. (2018) Core-powered mass-loss and the radius distribution of small exo-planets. *Mon. Not. R. Astron. Soc., 476(1)*, 759–765.

Gizis J. E., Reid I. N., and Hawley S. L. (2002) The Palomar/MSU nearby star spectroscopic survey. III. Chromospheric activity, M dwarf ages, and the local star formation history. *Astron. J., 123(6)*, 3356.

Godolt M., Grenfell J. L., Hamann-Reinus A., Kitzmann D., Kunze M., Langematz U., von Paris P., Patzer A. B. C., Rauer H., and Stracke B. (2015) 3D climate modeling of Earth-like extrasolar planets orbiting different types of host stars. *Planet. Space Sci., 111*, 62–76.

Gold T. and Soter S. (1969) Atmospheric tides and the resonant rotation of Venus. *Icarus, 11*, 356–366.

Goldblatt C., Matthews A. J., Claire M., Lenton T. M., Watson A. J., and Zahnle K. J. (2009) There was probably more nitrogen in the Archean atmosphere and this would have helped resolve the faint young Sun paradox. *Geochim. Cosmochim. Acta, Suppl., 73*, 446.

Goldblatt C., Robinson T. D., Zahnle K. J., and Crisp D. (2013) Low simulated radiation limit for runaway greenhouse climates. *Nature Geosci., 6(8)*, 661–667.

Goldreich P. (1966) Final spin states of planets and satellites. *Astron. J., 71*, 1.

Gowanlock M. G. (2016) Astrobiological effects of gamma-ray bursts in the Milky Way galaxy. *Astrophys. J., 832(1)*, 38.

Green J., Huber M., Waltham D., Buzan J., and Wells M. (2017) Explicitly modelled deep-time tidal dissipation and its implication for lunar history. *Earth Planet. Sci. Lett., 461*, 46–53.

Grimm S. L., Demory B.-O., Gillon M., Dorn C., Agol E., Burdanov A., Delrez L., et al. (2018) The nature of the TRAPPIST-1 exoplanets. *Astron. Astrophys., 613*, A68.

Grinspoon D. H. (1993) Implications of the high D/H ratio for the sources of water in Venus' atmosphere. *Nature, 363*, 428–431.

Güdel M. (2004) X-ray astronomy of stellar coronae. *Astron. Astrophys. Rev., 12(2–3)*, 71–237.

Haghighipour N. and Kaltenegger L. (2013) Calculating the habitable zone of binary star systems. II. P-type binaries. *Astrophys. J., 777(2)*, 166.

Hamano K., Abe Y., and Genda H. (2013) Emergence of two types of terrestrial planet on solidification of magma ocean. *Nature, 497*, 607–610.

Haqq-Misra J. and Heller R. (2018) Exploring exomoon atmospheres with an idealized general circulation model. *Mon. Not. R. Astron. Soc., 479(3)*, 3477–3489.

Haqq-Misra J., Domagal-Goldman S. D., Kasting P. J., and Kasting J. F. (2008) A revised, hazy methane greenhouse for the Archean Earth. *Astrobiology, 8*, 1127–1137.

Haqq-Misra J., Kopparapu R. K., Batalha N. E., Harman C. E., and Kasting J. F. (2016) Limit cycles can reduce the width of the habitable zone. *Astrophys. J., 827(2)*, 120.

Haqq-Misra J., Wolf E. T., Joshi M., Zhang X., and Kumar R. (2018) Demarcating circulation regimes of synchronously rotating terrestrial planets within the habitable zone. *Astrophys. J., 852(67)*, 16.

Harman C. E., Schwieterman E. W., Schottelkotte J. C., and Kasting J. F. (2015) Abiotic O$_2$ levels on planets around F, G, K, and M stars: Possible false positives for life? *Astrophys. J., 812(2)*, 137.

Hart M. H. (1978) The evolution of the atmosphere of the Earth. *Icarus, 33(1)*, 23–39.

Hart M. H. (1979) Habitable zones about main sequence stars. *Icarus, 37(1)*, 351–357.

Hawley S. L., Reid I. N., and Tourtellot J. (2000) Properties of M dwarfs in clusters and the field. In *Very Low-Mass Stars and Brown Dwarfs* (R. Rebolo and M. R. Zapatero-Osorio, eds.), pp. 109–118. Cambridge Univ., Cambridge.

Heller R. (2012) Exomoon habitability constrained by energy flux and orbital stability. *Astron. Astrophys., 545*, L8.

Heller R. and Barnes R. (2013) Exomoon habitability constrained by illumination and tidal heating. *Astrobiology, 13(1)*, 18–46.

Heller R. and Barnes R. (2015) Runaway greenhouse effect on exomoons due to irradiation from hot, young giant planets. *Intl. J. Astrobiol., 14(2)*, 335–343.

Heller R., Leconte J., and Barnes R. (2011) Tidal obliquity evolution of potentially habitable planets. *Astron. Astrophys., 528*, A27.

Henning W. G. and Hurford T. (2014) Tidal heating in multi-layered terrestrial exoplanets. *Astrophys. J., 789*, 30.

Hess S. and Zarka P. (2011) Modeling the radio signature of the orbital parameters, rotation, and magnetic field of exoplanets. *Astron. Astrophys., 531*, A29.

Hill M. L., Kane S. R., Seperuelo Duarte E., Kopparapu R. K., Gelino D. M., and Wittenmyer R. A. (2018) Exploring Kepler giant planets in the habitable zone. *Astrophys. J., 860(1)*, 67.

Hinkel N. R. and Kane S. R. (2013) Habitability of exomoons at the hill or tidal locking radius. *Astrophys. J., 774(1)*, 27.

Holman M. J. and Murray N. W. (2005) The use of transit timing to detect terrestrial-mass extrasolar planets. *Science, 307(5713)*, 1288–1291.

Hu Y. and Yang J. (2014) Role of ocean heat transport in climates of tidally locked exoplanets around M dwarf stars. *Proc. Natl. Acad. Sci. U.S.A., 111(2)*, 629–634.

Huang S.-S. (1959) The problem of life in the universe and the mode of star formation. *Publ. Astron. Soc. Pac., 71*, 421, DOI: 10.1086/127417.

Hunten D. and Donahue T. M. (1976) Hydrogen loss from the terrestrial planets. *Annu. Rev. Earth Planet. Sci., 4(1)*, 265–292.

Hut P. (1981) Tidal evolution in close binary systems. *Astron. Astrophys., 99*, 126–140.

Ingersoll A. P. (1969) The runaway greenhouse: A history of water on Venus. *J. Atmos. Sci., 26(6)*, 1191–1198.

Jackson B., Barnes R., and Greenberg R. (2008) Tidal heating of terrestrial extrasolar planets and implications for their habitability. *Mon. Not. R. Astron. Soc., 391*, 237–245.

Jakosky B. M., Brain D., Chaffin M., Curry S., Deighan J., Grebowsky J., Halekas J., et al. (2011) Loss of the martian atmosphere to space: Present-day loss rates determined from MAVEN observations and integrated loss through time. *Icarus, 315*, 146–157.

Jenkins G. (2000) Global climate model high-obliquity solutions to the ancient climate puzzles of the faint-young Sun paradox and low-latitude Proterozoic glaciation. *J. Geophys. Res., 105(D6)*, 7357–7370.

Jin S., Mordasini C., Parmentier V., van Boekel R., Henning T., and Ji J. (2014) Planetary population synthesis coupled with atmospheric escape: A statistical view of evaporation. *Astrophys. J., 795(1)*, 65.

Joshi M. (2003) Climate model studies of synchronously rotating planets. *Astrobiology, 3(2)*, 415–427.

Joshi M., Haberle R., and Reynolds R. (1997) Simulations of the atmospheres of synchronously rotating terrestrial planets orbiting M dwarfs: Conditions for atmospheric collapse and the implications for habitability. *Icarus, 129(2)*, 450–465.

Kadoya S. and Tajika E. (2014) Conditions for oceans on Earth-like planets orbiting within the habitable zone: Importance of volcanic CO_2 degassing. *Astrophys. J., 790(2)*, 107.

Kadoya S. and Tajika E. (2015) Evolutionary climate tracks of Earth-like planets. *Astrophys. J. Lett., 815(1)*, L7.

Kaib N. A. (2018) Galactic effects on habitability. In *Handbook of Exoplanets* (H. Deeg and J. Belmonte, eds.), pp. 3091–3109. Springer, Cham.

Kaib N. A., Quinn T., et al. (2011) Sedna and the Oort cloud around a migrating Sun. *Icarus, 215(2)*, 491–507.

Kaib N. A., Raymond S. N., and Duncan M. (2013) Planetary system disruption by galactic perturbations to wide binary stars. *Nature, 493(7432)*, 381-384.

Kaltenegger L. and Haghighipour N. (2013) Calculating the habitable zone of binary star systems. I. S-type binaries. *Astrophys. J., 777(2)*, 165.

Kaltenegger L. and Sasselov D. (2009) Detecting planetary geo-chemical cycles on exoplanets: Atmospheric signatures and the case of SO_2. *Astrophys. J., 708(2)*, 1162.

Kaltenegger L., Henning W., and Sasselov D. (2010) Detecting volcanism on extrasolar planets. *Astron. J., 140(5)*, 1370.

Kane S. R. and Hinkel N. R. (2013) On the habitable zones of circumbinary planetary systems. *Astrophys. J., 762(1)*, 7.

Kasting J. F. (1988) Runaway and moist greenhouse atmospheres and the evolution of Earth and Venus. *Icarus, 74(3)*, 472–494.

Kasting J. F. and Pollack J. B. (1983) Loss of water from Venus. I. Hydrodynamic escape of hydrogen. *Icarus, 53*, 479–508.

Kasting J. F., Whitmire D. P., and Reynolds R. T. (1993) Habitable zones around main sequence stars. *Icarus, 101*, 108–128.

Katz R. F., Spiegelman M., and Langmuir C. H. (2003) A new parameterization of hydrous mantle melting. *Geochem. Geophys. Geosyst., 4*, 1073.

Kawahara H. and Fujii Y. (2010) Global mapping of Earth-like exoplanets from scattered light curves. *Astrophys. J., 720(2)*, 1333.

Kilic C., Raible C., and Stocker T. (2017) Multiple climate states of habitable exoplanets: The role of obliquity and irradiance. *Astrophys. J., 844(2)*, 13.

Kipping D. M. (2011) Transit timing effects due to an exomoon. In *The Transits of Extrasolar Planets with Moons*, pp. 127–164. Ph.D. thesis, University College London, Springer-Verlag, Berlin.

Kipping D. M., Bakos G. Á., Buchhave L., Nesvorný D., and Schmitt A. (2012) The hunt for exomoons with Kepler (HEK) I. Description of a new observational project. *Astrophys. J., 750(2)*, 115.

Kipping D. M., Forgan D., Hartman J., Nesvorný D., Bakos G. Á., Schmitt A., and Buchhave L. (2013) The hunt for exomoons with Kepler (HEK) III. The first search for an exomoon around a habitable-zone planet. *Astrophys. J., 777(2)*, 134.

Kipping D. M., Nesvorný D., Buchhave L. A., Hartman J., Bakos

G. Á., and Schmitt A. R. (2014) The hunt for exomoons with Kepler (HEK) IV. A search for moons around eight M dwarfs. *Astrophys. J., 784(1)*, 28.

Kipping D. M., Schmitt A. R., Huang X., Torres G., Nesvorný D., Buchhave L. A., Hartman J., and Bakos G. Á. (2015) The hunt for exomoons with Kepler (HEK) V. A survey of 41 planetary candidates for exomoons. *Astrophys. J., 813(1)*, 14.

Kivelson M., Khurana K., Coroniti F., Joy S., Russell C., Walker R., Warnecke J., Bennett L., and Polanskey C. (1997) The magnetic field and magnetosphere of Ganymede. *Geophys. Res. Lett., 24(17)*, 2155–2158.

Kley W. and Haghighipour N. (2014) Modeling circumbinary planets: The case of Kepler-38. *Astron. Astrophys., 564*, A72.

Kodama T., Genda H., Abe Y., and Zahnle K. J. (2015) Rapid water loss can extend the lifetime of planetary habitability. *Astrophys. J., 812(2)*, 165.

Koll D. D. and Abbot D. S. (2016) Temperature structure and atmospheric circulation of dry tidally locked rocky exoplanets. *Astrophys. J., 825(2)*, 99.

Koll D. D. B. and Cronin T. W. (2019) Hot hydrogen climates near the inner edge of the habitable zone. *Astrophys. J., 881(2)*, 120.

Komacek T. D. and Abbot D. S. (2016) Effect of surface-mantle water exchange parameterizations on exoplanet ocean depths. *Astrophys. J., 832(1)*, 54.

Komacek T. D. and Abbot D. S. (2019) The atmospheric circulation and climate of terrestrial planets orbiting Sun-like and M dwarf stars over a broad range of planetary parameters. *Astrophys. J., 871(2)*, 245.

Kopparapu R. K., Ramirez R., Kasting J. F., Eymet V., Robinson T. D., Mahadevan S., Terrien R. C., Domagal-Goldman S., Meadows V., and Deshpande R. (2013) Habitable zones around main-sequence stars: New estimates. *Astrophys. J., 765*, 16.

Kopparapu R. K., Ramirez R. M., SchottelKotte J., Kasting J. F., Domagal-Goldman S., and Eymet V. (2014) Habitable zones around main-sequence stars: Dependence on planetary mass. *Astrophys. J. Lett., 787(2)*, L29.

Kopparapu R. K., Wolf E. T., Haqq-Misra J., Yang J., Kasting J. F., Meadows V., Terrien R., and Mahadevan S. (2016) The inner edge of the habitable zone for synchronously rotating planets around low-mass stars using general circulation models. *Astrophys. J., 819(1)*, 84.

Kopparapu R. K., Wolf E. T., Arney G., Batalha N. E., Haqq-Misra J., Grimm S. L., and Heng K. (2017) Habitable moist atmospheres on terrestrial planets near the inner edge of the habitable zone around M dwarfs. *Astrophys. J., 845(5)*, DOI: 10.3847/1538-4357/aa7cf9.

Kopparapu R. K., Hébrard E., Belikov R., Batalha N. M., Mulders G. D., Stark C., Teal D., Domagal-Goldman S., and Mandell A. (2018) Exoplanet classification and yield estimates for direct imaging missions. *Astrophys. J., 856*, 122.

Korenaga J. (2008) Urey ratio and the structure and evolution of Earth's mantle. *Rev. Geophys., 46*, RG2007, DOI: 10.1029/2007RG000241.

Kostov V. B., McCullough P. R., Hinse T. C., Tsvetanov Z. I., Hébrard G., Díaz R. F., Deleuil M., and Valenti J. A. (2013) A gas giant circumbinary planet transiting the F star primary of the eclipsing binary star KIC 4862625 and the independent discovery and characterization of the two transiting planets in the Kepler-47 system. *Astrophys. J., 770(1)*, 52.

Kostov V. B., Orosz J. A., Welsh W. F., Doyle L. R., Fabrycky D. C., Haghighipour N., Quarles B., et al. (2016) Kepler-1647b: The largest and longest-period Kepler transiting circumbinary planet. *Astrophys. J., 827(1)*, 86.

Kraus S., Kreplin A., Fukugawa M., Muto T., Sitko M. L., Young A. K., Bate M. R., et al. (2017) Dust-trapping vortices and a potentially planet-triggered spiral wake in the pre-transitional disk of v1247 Orionis. *Astrophys. J. Lett., 848(1)*, L11.

Kreidberg L., Line M. R., Bean J. L., Stevenson K. B., Désert J.-M., Madhusudhan N., Fortney J. J., et al. (2015) A detection of water in the transmission spectrum of the hot Jupiter WASP-12b and implications for its atmospheric composition. *Astrophys. J., 814*, 66.

Kreidberg L., Koll D. D. B., Morley C., Hu R., Schaefer L., Deming D., Stevenson K. B., et al. (2019) Absence of a thick atmosphere on the terrestrial exoplanet LHS 3844b. *Nature, 573*, 87–90.

Kulikov Y. N., Lammer H., Lichtenegger H. I. M., Terada N., Ribas I., Kolb C., Langmayr D., et al. (2006) Atmospheric and water loss from early Venus. *Planet. Space Sci., 54*, 1425–1444.

Kump L. R. (2008) The rise of atmospheric oxygen. *Nature, 451(7176)*,

277–278.

Lammer H. (2012) *Origin and Evolution of Planetary Atmospheres: Implications for Habitability.* Springer-Verlag, Berlin. 98 pp.

Lammer H., Kasting J. F., Chassefiere E., Johnson R. E., Kulikov Y. N., and Tian F. (2008) Atmospheric escape and evolution of terrestrial planets and satellites. *Space Sci. Rev., 139(1-4),* 399–436.

Lammer H., Schiefer S.-C., Juvan I., Odert P., Erkaev N. V., Weber C., Kislyakova K. G., Güdel M., Kirchengast G., and Hanslmeier A. (2014) Origin and stability of exomoon atmospheres: Implications for habitability. *Origins Life Evol. Biosph., 44(3),* 239–260.

Lammer H., Zerkle A. L., Gebauer S., Tosi N., Noack L., Scherf M., Pilat-Lohinger E., et al.(2018) Origin and evolution of the atmospheres of early Venus, Earth, and Mars. *Astron. Astrophys. Rev., 26(1),* 2.

Laughlin G. and Adams F. C. (1998) The modification of planetary orbits in dense open clusters. *Astrophys. J. Lett., 508,* L171–L174.

Leconte J., Forget F., Charnay B., Wordsworth R., Selsis F., Millour E., and Spiga A. (2013a) 3D climate modeling of close-in land planets: Circulation patterns, climate moist bistability, and habitability. *Astron. Astrophys., 544,* A69.

Leconte J., Forget F., Charnay B., Wordsworth R., and Pottier A. (2013b) Increased insolation threshold for runaway greenhouse processes on Earth-like planets. *Nature, 504(7479),* 268–271.

Leconte J., Wu H., Menou K., and Murray N. (2015) Asynchronous rotation of Earth-mass planets in the habitable zone of lower-mass stars. *Science, 347,* 632–635.

Lenardic A. and Crowley J. W. (2012) On the notion of well-defined tectonic regimes for terrestrial planets in this solar system and others. *Astrophys. J., 755(2),* 132.

Lenardic A., Jellinek A., Foley B., O'Neill C., and Moore W. (2016) Climate-tectonic coupling: Variations in the mean, variations about the mean, and variations in mode. *J. Geophys. Res.–Planets, 121(10),* 1831–1864.

Lenton T. M., Crouch M., Johnson M., Pires N., and Dolan L. (2012) First plants cooled the Ordovician. *Nature Geosci., 5(2),* 86–89.

Lincowski A. P., Meadows V. S., Crisp D., Robinson T., and Luger R. (2018) Evolved climates and observational discriminants for the TRAPPIST-1 planetary system. *Astrophys. J., 867,* 34.

Lincowski A. P., Lustig-Yaeger J., and Meadows V. S. (2019) Observing isotopologue bands in terrestrial exoplanet atmospheres with the James Webb Space Telescope: Implications for identifying past atmospheric and ocean loss. *Astron. J., 158(1),* 26.

Linsenmeier M., Pascale S., and Lucarini V. (2015) Climate of Earth-like planets with high obliquity and eccentric orbits: Implications for habitability conditions. *Planet. Space Sci., 105,* 43–59.

Lissauer J. J., Fabrycky D. C., Ford E. B., Borucki W. J., Fressin F., Marcy G. W., Orosz J. A., et al. (2011) A closely packed system of low-mass, low-density planets transiting Kepler-11. *Nature, 470(7332),* 53–58.

Loftus K., Wordsworth R. D., and Morley C. V. (2019) Sulfate aerosol hazes and SO_2 gas as constraints on rocky exoplanets' surface liquid water. *Astrophys. J., 887(231),* DOI: 10.3847/1538-4357/ab58cc.

López-Morales M., Ben-Ami S., Gonzalez-Abad G., García-Mejía J., Dietrich J., and Szentgyorgyi A. (2019) Optimizing ground-based observations of O_2 in Earth analogs. *Astron. J., 158(1),* 24.

Lovelock J. E. and Margulis L. (1974) Homeostatic tendencies of the Earth's atmosphere. *Orig. Life, 5(1-2),* 93–103.

Lovis C., Snellen I., Mouillet D., Pepe F., Wildi F., Astudillo-Defru N., Beuzit J. L., et al. (2017) Atmospheric characterization of Proxima b by coupling the SPHERE high-contrast imager to the ESPRESSO spectrograph. *Astron. Astrophys., 599,* A16.

Luger R. and Barnes R. (2015) Extreme water loss and abiotic O_2 buildup on planets throughout the habitable zones of M dwarfs. *Astrobiology, 15,* 119–143.

Luger R., Barnes R., Lopez E., Fortney J. J., Jackson B., and Meadows V. (2015) Habitable evaporated cores: Transforming mini-Neptunes into super-Earths in the habitable zones of M dwarfs. *Astrobiology, 15(1),* 57.

Luger R., Sestovic M., Kruse E., Grimm S. L., Demory B.-O., Agol E., Bolmont E., et al. (2017a) A seven-planet resonant chain in TRAPPIST-1. *Nature Astron., 1,* 0129.

Luger R., Kruse E., Foreman-Mackey D., Agol E., and Saunders N. (2017b) An update to the EVEREST K2 pipeline: Short cadence, saturated stars, and Kepler-like photometry down to Kp = 15. *Astron. J, 156,* 99.

Luger R., Lustig-Yaeger J., Fleming D. P., Tilley M. A., Agol E.,

Meadows V. S., Deitrick R., and Barnes R. (2017c) The pale green dot: A method to characterize Proxima Centauri b using exo-aurorae. *Astrophys. J., 837(1),* 63.

Lundin R., Lammer H., and Ribas I. (2007) Planetary magnetic fields and solar forcing: Implications for atmospheric evolution. *Space Sci. Rev., 129(1-3),* 245–278.

Lustig-Yaeger J., Tovar G., Schwieterman E., Fujii Y., and Meadows V. (2017) Detecting oceans on exoplanets using phase-dependent mapping with next-generation coronagraph-equipped telescopes. In *Habitable Worlds 2017: A System Science Workshop,* Abstract #4110, LPI Contribution No. 2042, Lunar and Planetary Institute, Houston.

Lustig-Yaeger J., Meadows V. S., Tovar Mendoza G., Schwieterman E. W., Fujii Y., Luger R., and Robinson T. D. (2018) Detecting ocean glint on exoplanets using multiphase mapping. *Astron. J., 156(6),* 301.

Lustig-Yaeger J., Meadows V. S., and Lincowski A. P. (2019) The detectability and characterization of the TRAPPIST-1 exoplanet atmospheres with JWST. *Astron. J., 158(1),* 27.

Lyons T. W., Reinhard C. T., and Planavsky N. J. (2014) The rise of oxygen in Earth's early ocean and atmosphere. *Nature, 506(7488),* 307–315.

Marois C., Macintosh B., Barman T., Zuckerman B., Song I., Patience J., Lafrenière D., and Doyon R. (2008) Direct imaging of multiple planets orbiting the star HR 8799. *Science, 322(5906),* 1348.

Marois C., Zuckerman B., Konopacky Q. M., Macintosh B., and Barman T. (2010) Images of a fourth planet orbiting HR 8799. *Nature, 468(7327),* 1080–1083.

Mayor M. and Queloz D. (1995) A Jupiter-mass companion to a solar-type star. *Nature, 378(6555),* 355–359.

Mayor M. and Queloz D. (2012) From 51 Peg to Earth-type planets. *New Astron. Rev., 56(1),* 19–24.

McKay C. P., Lorenz R. D., and Lunine J. I. (1999) Analytic solutions for the antigreenhouse effect: Titan and the early Earth. *Icarus, 137(1),* 56–61.

Meadows V. S. (2017) Reflections on O_2 as a biosignature in exoplanetary atmospheres. *Astrobiology, 17(10),* 1022–1052.

Meadows V. S. and Barnes R. K. (2018) Factors affecting exoplanet habitability. In *Handbook of Exoplanets* (H. J. Deeg and J. A. Belmonte, eds.), pp. 2771-2794. Springer, Cham.

Meadows V. S., Arney G. N., Schwieterman E. W., Lustig-Yaeger J., Lincowski A. P., Robinson T., Domagal-Goldman S. D., et al. (2018) The habitability of Proxima Centauri b: Environmental states and observational discriminants. *Astrobiology, 18(2),* 133–189.

Melott A. L., Lieberman B. S., Laird C. M., Martin L. D., Medvedev M. V., Thomas B. C., Cannizzo J. K., Gehrels N., and Jackman C. H. (2004) Did a gamma-ray burst initiate the late Ordovician mass extinction? *Intl. J. Astrobiol., 3,* 55–61.

Menou K. (2015) Climate stability of habitable Earth-like planets. *Earth Planet. Sci. Lett., 429,* 20–24.

Misra A., Krissansen-Totton J., Koehler M. C., and Sholes S. (2015) Transient sulfate aerosols as a signature of exoplanet volcanism. *Astrobiology, 15(6),* 462–77.

Mojzsis S. J., Brasser R., Kelly N. M., Abramov O., and Werner S. C. (2019) Onset of giant planet migration before 4480 million years ago. *Astrophys. J., 881(44).*

Moore W. B. and Webb A. A. G. (2013) Heat-pipe Earth. *Nature, 501,* 501–505.

Moore W. B., Simon J. I., and Webb A. A. G. (2017) Heat-pipe planets. *Earth Planet. Sci. Lett., 474,* 13–19.

Moran S. E., Hörst S. M., Batalha N. E., Lewis N. K., and Wakeford H. R. (2018) Limits on clouds and hazes for the TRAPPIST-1 planets. *Astron. J., 156(6),* 252.

Morley C. V., Fortney J. J., Marley M. S., Visscher C., Saumon D., and Leggett S. K. (2012) Neglected clouds in T and Y dwarf atmospheres. *Astrophys. J., 756(2),* 172.

Morley C. V., Kreidberg L., Rustamkulov Z., Robinson T., and Fortney J. J. (2017) Observing the atmospheres of known temperate Earth-sized planets with JWST. *Astron. J., 850,* 18.

Morowitz H. and Sagan C. (1967) Life in the clouds of Venus? *Nature, 215,* 1259–1260.

Mulders G. D., Pascucci I., Apai D., and Ciesla F. J. (2018) The exoplanet population observation simulator. I. The inner edges of planetary systems. *Astron. J., 156(1),* 24.

Ness N. F. (1978) Mercury: Magnetic field and interior. *Space Sci. Rev., 21(5),* 527–553.

Nimmo F. (2002) Why does Venus lack a magnetic field? *Geology,* *30(11),* 987–990.

Nisbet E., Zahnle K., Gerasimov M. V., Helbert J., Jaumann R., Hofmann B. A., Benzerara K., and Westall F. (2007) Creating habitable zones, at all scales, from planets to mud micro-habitats, on Earth and on Mars. *Space Sci. Rev., 129(1–3),* 79–121.

Noda S., Ishiwatari M., Nakajima K., Takahashi Y. O., Takehiro S., Onishi M., Hashimoto G. L., Kuramoto K., and Hayashi Y. Y. (2017) The circulation pattern and day-night heat transport in the atmosphere of a synchronously rotating aquaplanet: Dependence on planetary rotation rate. *Icarus, 282,* 1–18.

Olson P. and Christensen U. R. (2006) Dipole moment scaling for convection-driven planetary dynamos. *Earth Planet. Sci. Lett., 250,* 561–571.

Orgel L. E. (2004) Prebiotic chemistry and the origin of the RNA world. *Crit. Rev. Biochem. Mol. Biol., 39,* 99–123.

Orosz J. A., Welsh W. F., Carter J. A., Brugamyer E., Buchhave L. A., Cochran W. D., Endl M., et al. (2012a) The Neptune-sized circumbinary planet Kepler-38b. *Astrophys. J., 758(2),* 87.

Orosz J. A., Welsh W. F., Carter J. A., Fabrycky D. C., Cochran W. D., Endl M., Ford E. B., et al. (2012b) Kepler-47: A transiting circumbinary multiplanet system. *Science, 337(6101),* 1511.

Owen J. E. and Mohanty S. (2016) Habitability of terrestrial-mass planets in the HZ of M dwarfs — I. H/He-dominated atmospheres. *Mon. Not. R. Astron. Soc., 459(4),* 4088–4108.

Owen J. E. and Wu Y. (2013) Kepler planets: A tale of evaporation. *Astrophys. J., 775(2),* 105.

Peale S. J. (1977) Rotation histories of the natural satellites. In *IAU Colloq. 28: Planetary Satellites* (J. A. Burns, ed.), p. 87. Proceedings of IAU Colloq. 28, Ithaca, New York, August 1974, Univ. of Arizona, Tucson.

Pepin R. O. (2006) Atmospheres on the terrestrial planets: Clues to origin and evolution. *Earth Planet. Sci. Lett., 252(1-2),* 1–14.

Petigura E. A., Howard A. W., and Marcy G. W. (2013) Prevalence of Earth-size planets orbiting Sun-like stars. *Proc. Natl. Acad. Sci. U.S.A., 110(48),* 19273–19278.

Pierrehumbert R. T. and Ding F. (2016) Dynamics of atmospheres with a non-dilute condensible component. *Proc. R. Soc. London, Ser. A, 472,* 20160107.

Pierrehumbert R. and Gaidos E. (2011) Hydrogen greenhouse planets beyond the habitable zone. *Astrophys. J. Lett., 734(1),* L13.

Pierrehumbert R., Abbot D., Voigt A., and Koll D. (2011) Climate of the Neoproterozoic. *Annu. Rev. Earth Planet. Sci., 39,* 417–460.

Piran T. and Jimenez R. (2014) Possible role of gamma ray bursts on life extinction in the universe. *Phys. Rev. Lett., 113(23),* 231102.

Popp M. and Eggl S. (2017) Climate variations on Earth-like circumbinary planets. *Nature Commun., 8,* 14957.

Quanz S. P., Crossfield I., Meyer M. R., Schmalzl E., and Held J. (2015) Direct detection of exoplanets in the 3–10 μm range with E-ELT/METIS. *Intl. J. Astrobiol., 14(02),* 279–289.

Quintana E. V., Barclay T., Borucki W. J., Rowe J. F., and Chambers J. E. (2016) The frequency of giant impacts on Earth-like worlds. *Astrophys. J., 821(2),* 126.

Rajan A., Rameau J., De Rosa R. J., Marley M. S., Graham J. R., Macintosh B., Marois C., et al. (2017) Characterizing 51 Eri b from 1 to 5 μm: A partly cloudy exoplanet. *Astron. J., 154(1),* 10.

Ramirez R. M. and Kaltenegger L. (2014) The habitable zones of pre-main-sequence stars. *Astrophys. J. Lett., 797(2),* L25.

Ramirez R. M. and Kaltenegger L. (2017) A volcanic hydrogen habitable zone. *Astrophys. J. Lett., 837(1),* L4.

Ramirez R. and Kaltenegger L. (2018) A methane extension to the classical habitable zone. *Astrophys. J., 858(72),* 10.

Ramircz R. M., Kopparapu R. K., Lindner V., and Kasting J. F. (2014) Can increased atmospheric CO_2 levels trigger a run-away greenhouse? *Astrobiology, 14(8),* 714–731.

Rasio F. A., Tout C. A., Lubow S. H., and Livio M. (1996) Tidal decay of close planetary orbits. *Astrophys. J., 470,* 1187.

Rauer H., Catala C., Aerts C., Appourchaux T., Benz W., Brandeker A., Christensen-Dalsgaard J., Deleuil M., Gizon L., Goupil M.-J., et al. (2014) The Plato 2.0 mission. *Exp. Astron., 38(1–2),* 249–330.

Raymond S. N. and Izidoro A. (2017) Origin of water in the inner solar system: Planetesimals scattered inward during Jupiter and Saturn's rapid gas accretion. *Icarus, 297,* 134–148.

Raymond S. N., Quinn T., and Lunine J. I. (2004) Making other Earths: Dynamical simulations of terrestrial planet formation and water delivery. *Icarus, 168(1),* 1–17.

Raymond S. N., Scalo J., and Meadows V. S. (2007) A decreased probability of habitable planet formation around low-mass stars. *Astrophys. J., 669(1),* 606–614.

Raymond S. N., Barnes R., and Mandell A. M. (2008) Observable consequences of planet formation models in systems with close-in terrestrial planets. *Mon. Not. R. Astron. Soc., 384(2),* 663–674.

Raymond S. N., Armitage P. J., Moro-Martin A., Booth M., Wyatt M. C., Armstrong J. C., Mandell A. M., Selsis F., and West A. A. (2012) Debris disks as signposts of terrestrial planet formation. II. Dependence of exoplanet architectures on giant planet and disk properties. *Astron. Astrophys., 541,* A11.

Read P., Barstow J., Charnay B., Chelvaniththilan S., Irwin P., Knight S., Lebonnois S., Lewis S., Mendonça J., and Montabone L. (2015) Global energy budgets and Trenberth diagrams for the climates of terrestrial and gas giant planets. *Q. J. R. Meteor. Soc., 142(695),* 703–720.

Ribas I., Bolmont E., Selsis F., Reiners A., Leconte J., Raymond S. N., Engle S. G., et al. (2016) The habitability of Proxima Centauri b. I. Irradiation, rotation, and volatile inventory from formation to the present. *Astron. Astrophys., 596,* A111.

Robinson T. D., Meadows V. S., and Crisp D. (2010) Detecting oceans on extrasolar planets using the glint effect. *Astrophys. J. Lett., 721(1),* L67–L71.

Rodenbeck K., Heller R., Hippke M., and Gizon L. (2018) Revisiting the exomoon candidate signal around Kepler-1625 b. *Astron. Astrophys., 617,* A49.

Rodler F. and López-Morales M. (2014) Feasibility studies for the detection of O_2 in an Earth-like exoplanet. *Astrophys. J., 781(1),* 54.

Rodríguez A., Callegari N., Michtchenko T. A., and Hussmann H. (2012) Spin-orbit coupling for tidally evolving super-Earths. *Mon. Not. R. Astron. Soc., 427,* 2239–2250.

Rogers L. A. (2015) Most 1.6 Earth-radius planets are not rocky. *Astrophys. J., 801,* 41.

Rose B. E., Cronin T. W., and Bitz C. M. (2017) Ice caps and ice belts: The effects of obliquity on ice-albedo feedback. *Astrophys. J., 846,* 28, DOI: 10.3847/1538-4357/aa8306.

Roškar R., Debattista V., Loebman S., Ivezić, Ž., and Quinn T. (2011) Implications of radial migration for stellar population studies. *ArXiv e-prints,* arXiv:1101.1202.

Roškar R., Debattista V. P., Quinn T. R., and Wadsley J. (2012) Radial migration in disc galaxies. I. Transient spiral structure and dynamics. *Mon. Not. R. Astron. Soc., 426(3),* 2089–2106.

Rugheimer S., Kaltenegger L., Segura A., Linsky J., and Mohanty S. (2015) Effect of UV radiation on the spectral fingerprints of Earth-like planets orbiting M stars. *Astrophys. J., 809(1),* 57.

Sagan C. and Chyba C. (1997) The early faint Sun paradox: Organic shielding of ultraviolet-labile greenhouse gases. *Science, 276(5316),* 1217–1221.

Sandu C., Lenardic A., and McGovern P. (2011) The effects of deep water cycling on planetary thermal evolution. *J. Geophys. Res.–Solid Earth, 116(B12).*

Schaefer L. and Sasselov D. (2015) The persistence of oceans on Earth-like planets: Insights from the deep-water cycle. *Astrophys. J., 801(1),* 40.

Schaefer L., Wordsworth R. D., Berta-Thompson Z., and Sasselov D. (2016) Predictions of the atmospheric composition of GJ 1132b. *Astrophys. J., 829(2),* 63.

Schoeberl M. and Hartmann D. L. (1991) The dynamics of the stratospheric polar vortex and its relation to springtime ozone depletions. *Science, 251(4989),* 46–52.

Schubert G. and Soderlund K. (2011) Planetary magnetic fields: Observations and models. *Phys. Earth Planet. Inter., 187(3-4),* 92–108.

Schwieterman E., Binder B., Tremmel M., Garofali K., Agol E., and Meadows V. (2015) Promoting diversity in STEM through active recruiting and mentoring: The pre-major in astronomy program (Pre-MAP) at the University of Washington. *AAS/Division for Planetary Sciences Meeting Abstracts, 47,* #202.08.

Schwieterman E. W., Reinhard C. T., Olson S. L., Harman C. E., and Lyons T. W. (2019) A limited habitable zone for complex life. *Astrophys. J., 878(1),* 19.

Seager S. (2014) The future of spectroscopic life detection on exoplanets. *Proc. Natl. Acad. Sci. U.S.A., 111(35),* 12634–12640.

Segura A., Krelove K., Kasting J. F., Sommerlatt D., Meadows V., Crisp D., Cohen M., and Mlawer E. (2003) Ozone concentrations and ultraviolet fluxes on Earth-like planets around other stars.

Astrobiology, 3(4), 689–708.

Segura A., Kasting J. F., Meadows V., Cohen M., Scalo J., Crisp D., Butler R. A. H., and Tinetti G. (2005) Biosignatures from Earth-like planets around M dwarfs. *Astrobiology, 5(6)*, 706–725.

Segura A., Walkowicz L. M., Meadows V., Kasting J., and Hawley S. (2010) The effect of a strong stellar flare on the atmospheric chemistry of an Earth-like planet orbiting an M dwarf. *Astrobiology, 10*, 751–771.

Sekiya M., Nakazawa K., and Hayashi C. (1980) Dissipation of the rare gases contained in the primordial Earth's atmosphere. *Earth Planet. Sci. Lett., 50(1)*, 197–201.

Sellwood J. A. and Binney J. J. (2002) Radial mixing in galactic discs. *Mon. Not. R. Astron. Soc., 336*, 785–796.

Selsis F., Kasting J. F., Levrard B., Paillet J., Ribas I., and Delfosse X. (2007) Habitable planets around the star Gliese 581? *Astron. Astrophys., 476(3)*, 1373–1387.

Sherwood S. C. and Huber M. M. (2010) An adaptability limit to climate change due to heat stress. *Proc. Natl. Acad. Sci., 107(21)*, 9552–9555.

Shields A. L., Meadows V. S., Bitz C. M., Pierrehumbert R. T., Joshi M. M., and Robinson T. D. (2013) The effect of host star spectral energy distribution and ice-albedo feedback on the climate of extrasolar planets. *Astrobiology, 13*, 715–739.

Shields A. L., Barnes R., Agol E., Charnay B., Bitz C., and Meadows V. S. (2016) The effect of orbital configuration on the possible climates and habitability of Kepler-62f. *Astrobiology, 16(6)*.

Simon A. E., Szabó G. M., Kiss L. L., Fortier A., and Benz W. (2015) CHEOPS performance for exomoons: The detectability of exomoons by using optimal decision algorithm. *Publ. Astron. Soc. Pac., 127(956)*, 1084.

Sleep N. H., Bird D. K., and Pope E. (2012) Paleontology of Earth's mantle. *Annu. Rev. Earth Planet. Sci., 40*, 277–300.

Snellen I., de Kok R., de Mooij E., and Albrecht S. (2010) The orbital motion, absolute mass, and high-altitude winds of exoplanet HD 209458b. *Nature, 465*, 1049–1051, DOI: 10.1038/nature09111.

Snellen I., de Kok, R., Birkby J. L., Brandl B., Brogi M., Keller C., Kenworthy M., Schwarz H., and Stuik R. (2015) Combining high-dispersion spectroscopy with high contrast imaging: Probing rocky planets around our nearest neighbors. *Astron. Astrophys., 576*, A59.

Solomatov V. S. and Moresi L.-N. (2000) Scaling of time-dependent stagnant lid convection: Application to small-scale convection on Earth and other terrestrial planets. *J. Geophys. Res., 105*, 21795–21818.

Spiegel D. S., Menou K., and Scharf C. A. (2009) Habitable climates: The influence of obliquity. *Astrophys. J., 691(1)*, 596.

Spiegel D. S., Raymond S. N., Dressing C. D., Scharf C. A., and Mitchell J. L. (2010) Generalized Milankovitch cycles and long-term climatic habitability. *Astrophys. J., 721(2)*, 1308.

Spurzem R., Giersz M., Heggie D. C., and Lin D. N.C. (2009) Dynamics of planetary systems in star clusters. *Astrophys. J., 697*, 458–482.

Stanley S. and Glatzmaier G. A. (2010) Dynamo models for planets other than Earth. *Space Sci. Rev., 152(1-4)*, 617–649.

Stark C. C., Belikov R., Bolcar M. R., Cady E., Crill B. P., Ertel S., Groff T., et al. (2019) Exo Earth yield landscape for future direct imaging space telescopes. *J. Astron. Telesc. Instrum. Syst., 5*, 024009.

Stevenson D.J. (2010) Planetary magnetic fields: Achievements and prospects. *Space Sci. Rev., 152(1–4)*, 651–664.

Stevenson D. J. (2018) Commentary: The habitability mantra: Hunting the snark. *Phys. Today, 71(11)*, 10–12.

Szabó R., Szabó G. M., Dálya G., Simon A. E., Hodosán G., and Kiss L. L. (2013) Multiple planets or exomoons in Kepler hot Jupiter systems with transit timing variations? *Astron. Astrophys., 553*, A17.

Tasker E., Tan J., Heng K., Kane S., Spiegel D., Brasser R., Casey A., et al. (2017) The language of exoplanet ranking metrics needs to change. *Nature Astron., 1*, 0042.

Teachey A. and Kipping D. M. (2018) Evidence for a large exomoon orbiting Kepler-1625b. *Sci. Adv., 4(10)*, eaav1784.

Thomas B. C., Jackman C. H., Melott A. L., Laird C. M., Stolarski R. S., Gehrels N., Cannizzo J. K., and Hogan D. P. (2005) Terrestrial ozone depletion due to a Milky Way gamma-ray burst. *Astrophys. J. Lett., 622(2)*, L153–L156.

Tian F. and Ida S. (2015) Water contents of Earth-mass planets around M dwarfs. *Nature Geosci., 8*, 177–180.

Tilley M. A., Segura A., Meadows V., Hawley S., and Davenport J.

(2017) Modeling repeated M dwarf flaring at an Earth-like planet in the habitable zone: I. Atmospheric effects for an unmagnetized planet. *Astrobiology, 19(1)*, DOI: 10.1089/ast.2017.1794.

Trenberth K. E., Fasullo J. T., and Kiehl J. (2009) Earth's global energy budget. *Bull. Am. Meteor. Soc., 90*, 311–324.

Turbet M., Leconte J., Selsis F., Bolmont E., Forget F., Ribas I., Raymond S. N., and Anglada-Escudé G. (2016) The habitability of Proxima Centauri b. II. Possible climates and observability. *Astron. Astrophys., 596*, A112.

Turbet M., Forget F., Head J. W., and Wordsworth R. (2017a) 3D modelling of the climatic impact of outflow channel formation events on early Mars. *Icarus, 288*, 10–36.

Turbet M., Forget F., Leconte J., Charnay B., and Tobie G. (2017b) CO_2 condensation is a serious limit to the deglaciation of Earth-like planets. *Earth Planet. Sci. Lett., 476*, 11–21.

Turbet M., Ehrenreich D., Lovis C., Bolmont E., and Fauchez T. (2019) The runaway greenhouse radius inflation effect. An observational diagnostic to probe water on Earth-sized planets and test the habitable zone concept. *Astron. Astrophys., 628*, A12.

Turner G. (1989) The outgassing history of the Earth's atmosphere. *J. Geol. Soc. London, 146(1)*, 147–154.

Turnpenney S., Nichols J. D., Wynn G. A., and Burleigh M. R. (2018) Exoplanet-induced radio emission from M dwarfs. *Astrophys. J., 854*, 72.

Tziperman E., Halevy I., Johnston D. T., Knoll A. H., and Schrag D. P. (2011) Biologically induced initiation of Neoproterozoic snowball-Earth events. *Proc. Natl. Acad. Sci. U.S.A., 108(37)*, 15091–15096.

Unterborn C. T., Dismukes E. E., and Panero W. R. (2016) Scaling the Earth: A sensitivity analysis of terrestrial exoplanetary interior models. *Astrophys. J., 819(1)*, 32.

Unterborn C. T., Desch S. J., Hinkel N. R., and Lorenzo A. (2018) Inward migration of the TRAPPIST-1 planets as inferred from their water-rich compositions. *Nature Astron., 2*, 297–302.

Vidotto A., Jardine M., Morin J., Donati J.-F., Lang P., and Russell A. (2013) Effects of M dwarf magnetic fields on potentially habitable planets. *Astron. Astrophys., 557*, A67.

Walker J. C., Hays P., and Kasting J. F. (1981) A negative feedback mechanism for the long-term stabilization of Earth's surface temperature. *J. Geophys. Res.–Oceans, 86(C10)*, 9776–9782.

Wang Z. and Cuntz M. (2019) S-type and P-type habitability in stellar binary systems: A comprehensive approach. III. Results for Mars, Earth, and super-Earth planets. *Astrophys. J., 873(2)*, 113.

Watson A. J., Donahue T. M., and Walker J. C. G. (1981) The dynamics of a rapidly escaping atmosphere — Applications to the evolution of Earth and Venus. *Icarus, 48*, 150–166.

Way M. J. and Georgakarakos N. (2017) Effects of variable eccentricity on the climate of an Earth-like world. *Astrophys. J. Lett., 835*, L1.

Way M. J., Del Genio A. D., Kelley M., Aleinov I., and Clune T. (2015) Exploring the inner edge of the habitable zone with fully coupled oceans. *ArXiv e-prints*, arXiv:1511.07283.

Way M. J., Del Genio A. D., Kiang N. Y., Sohl L. E., Grinspoon D. H., Aleinov I., Kelley M., and Clune T. (2016) Was Venus the first habitable world of our solar system? *Geophys. Res. Lett., 43*, 8376–8383.

Weiss L. M. and Marcy G. W. (2014) The mass-radius relation for 65 exoplanets smaller than 4 Earth radii. *Astrophys. J. Lett., 783*, L6.

Weiss M. C., Sousa F. L., Mrnjavac N., Neukirchen S., Roettger M., Nelson-Sathi S., and Martin W. F. (2016) The physiology and habitat of the last universal common ancestor. *Nature Microbiol., 1*, 16116.

Welsh W. F., Orosz J. A., Short D. R., Cochran W. D., Endl M., Brugamyer E., Haghighipour N., et al. (2015) Kepler 453 b — The 10th Kepler transiting circumbinary planet. *Astrophys. J., 809(1)*, 26.

West A. A. and Basri G. (2009) A first look at rotation in inactive late-type M dwarfs. *Astrophys. J., 693(2)*, 1283.

West A. A., Hawley S. L., Bochanski J. J., Covey K. R., Reid I. N., Dhital S., Hilton E. J., and Masuda M. (2008) Constraining the age-activity relation for cool stars: The Sloan Digital Sky Survey Data Release 5 low-mass star spectroscopic sample. *Astron. J., 135(3)*, 785.

Wheatley D. F., Chan M. A., and Okubo C. H. (2019) Clastic pipes and mud volcanism across Mars: Terrestrial analog evidence of past martian groundwater and subsurface fluid mobilization. *Icarus, 328*, 141–151.

White R. J., Gabor J. M., and Hillenbrand L. A. (2007) High-dispersion optical spectra of nearby stars younger than the Sun. *Astron. J., 133(6)*, 2524.

Wielen R., Fuchs B., and Dettbarn C. (1996) On the birth-place of the Sun and the places of formation of other nearby stars. *Astron. Astrophys., 314,* 438.

Williams D. M. and Pollard D. (2002) Earth-like worlds on eccentric orbits: Excursions beyond the habitable zone. *Intl. J. Astrobiol., 1(1),* 61–69.

Williams D. M., Kasting J. F., and Wade R. A. (1997) Habitable moons around extrasolar giant planets. *Nature, 385(6613),* 234.

Windemuth D., Agol E., Ali A., and Kiefer F. (2019) Modelling Kepler eclipsing binaries: Homogeneous inference of orbital and stellar properties. *Mon. Not. R. Astron. Soc., 489(2),* 1644–1666.

Winn J. N. (2010) Exoplanet transits and occultations. In *Exoplanets* (S. Seager, ed.), pp. 55–77. Univ. of Arizona, Tucson.

Winn J. N. (2018) Planet occurrence: Doppler and transit surveys. In *Handbook of Exoplanets* (H. Deeg and J. Belmonte, eds.), pp. 1949–1966. Springer, Cham

Wolf E. T. (2017) Assessing the habitability of the TRAPPIST-1 system using a 3D climate model. *Astrophys. J. Lett., 839,* L1.

Wolf E. and Toon O. (2010) Fractal organic hazes provided an ultraviolet shield for early Earth. *Science, 328(5983),* 1266–1268.

Wolf E. T. and Toon O. B. (2015) The evolution of habitable climates under the brightening Sun. *J. Geophys. Res.–Atmos., 120,* 5775–5794.

Wolf E. T., Kopparapu R. K., and Haqq-Misra J. (2019) Simulated phase-dependent spectra of terrestrial aquaplanets in M dwarf systems. *Astrophys. J., 877(35),* 18.

Wolfgang A., Rogers L. A., and Ford E. B. (2016) Probabilistic mass-radius relationship for sub-Neptune-sized planets. *Astrophys. J., 825(1),* 19.

Wordsworth R. (2015) Atmospheric heat redistribution and collapse on tidally locked rocky planets. *Astrophys. J., 806(2),* 180.

Wordsworth R. and Pierrehumbert R. (2013a) Hydrogen-nitrogen greenhouse warming in Earth's early atmosphere. *Science, 339(6115),* 64–67.

Wordsworth R. D. and Pierrehumbert R. T. (2013b) Water loss from terrestrial planets with CO_2-rich atmospheres. *Astrophys. J., 778,* 154.

Wordsworth R. and Pierrehumbert R. (2014) Abiotic oxygen-dominated atmospheres on terrestrial habitable zone planets. *Astrophys. J. Lett., 785(2),* L20.

Wordsworth R., Forget F., Selsis F., Madeleine J.-B., Millour E., and Eymet V. (2010) Is Gliese 581d habitable? Some constraints from radiative-convective climate modeling. *Astron. Astrophys., 522,* A22.

Wordsworth R. D., Schaefer L. K., and Fischer R. A. (2018) Redox evolution via gravitational differentiation on low-mass planets: Implications for abiotic oxygen, water loss, and habitability. *Astron. J., 155(5),* 195.

Wright J. T. (2018) Radial velocities as an exoplanet discovery method. In *Handbook of Exoplanets* (H. Deeg and J. Belmonte, eds.), pp. 619–631. Springer, Cham.

Wright J., Marcy G., Howard A., Johnson J. A., Morton T., and Fischer D. (2012) The frequency of hot Jupiters orbiting nearby solar-type stars. *Astrophys. J., 753(2),* 160.

Yang J. and Abbot D. S. (2014) A low-order model of water vapor, clouds, and thermal emission for tidally locked terrestrial planets. *Astrophys. J., 784(2),* 155.

Yang J., Cowan N. B., and Abbot D. S. (2013) Stabilizing cloud feedback dramatically expands the habitable zone of tidally locked planets. *Astrophys. J. Lett., 771,* L45.

Yang J., Boué G., Fabrycky D. C., and Abbot D. S. (2014) Strong dependence of the inner edge of the habitable zone on planetary rotation rate. *Astrophys. J. Lett., 787(1),* L2.

Yang J., Abbot D. S., Koll D. B. D., Hu Y., and Showman A. P. (2019) Ocean dynamics and the inner edge of the habitable zone for tidally locked terrestrial planets. *Astrophys. J., 871(29),* 17.

Zahnle K. J. and Catling D. C. (2017) The cosmic shoreline: The evidence that escape determines which planets have atmospheres, and what this may mean for Proxima Centauri b. *Astrophys. J., 843(2),* 122.

Zarka P. (2007) Plasma interactions of exoplanets with their parent star and associated radio emissions. *Planet. Space Sci., 55(5),* 598–617.

Zsom A., Seager S., de Wit J., and Stamenković V. (2013) Toward the minimum inner edge distance of the habitable zone. *Astrophys. J., 778(2),* 109.

Walker S. I., Cronin L., Drew A., Domagal-Goldman S., Fisher T., Line M., and Millsaps C. (2020) Probabilistic biosignature frameworks. In *Planetary Astrobiology* (V. Meadows et al., eds.), pp. 477–503. Univ. of Arizona, Tucson, DOI: 10.2458/azu_uapress_9780816540068-ch018.

Probabilistic Biosignature Frameworks

Sara Imari Walker
Arizona State University

Leroy Cronin
University of Glasgow

Alexa Drew
Arizona State University

Shawn Domagal-Goldman
NASA Goddard Space Flight Center

Theresa Fisher
Arizona State University

Michael Line
Arizona State University

Camerian Millsaps
Arizona State University

The preponderance of false positives for many candidate biosignatures has led to an emerging area of probabilistic approaches to life detection. Thinking statistically about biosignatures allows the advantage of assigning confidence intervals to the discovery of alien life, through explicit treatment of observational uncertainties and uncertainties about the measurable properties of life. Here we review a number of approaches that could be incorporated into a probabilistic framework, enabling new, more robust biosignatures that are less susceptible to false positives. We discuss how such approaches could unify solar system and extrasolar searches, as well as the search for bio- and technosignatures within a common framework.

"The laws of physics and chemistry are statistical throughout." – E. Schrödinger (1943)

1. INTRODUCTION

Our capacity to search for life on other worlds is rapidly advancing, both within our solar system and beyond. In the last few decades robotic missions have explored every major target of astrobiological interest in our solar system, collecting data relevant to assessing habitability across diverse environments, from rocky planets to icy satellites. Outside our solar system, we have now identified thousands of worlds orbiting other stars, many of which may be habitable, suggesting there are many places beyond our solar system where life could, at least in principle, emerge and persist. The next challenge in advancing the search for alien life is to determine whether any of these worlds are not just habitable, but actually inhabited.

Determining whether or not an environment is inhabited requires capabilities to unambiguously identify signs of life, either biosignatures or technosignatures, above the abiotic background. A key challenge is that currently any universal properties that could define living systems remain uncertain, as do their resultant signals. To add to this uncertainty, observational signals for most proposed biosignatures and technosignatures have mimics that can be generated in environments entirely devoid of life. These *false positives* arise when a putative "biosignature" is produced by abiotic processes. The hallmark example in exoplanetary settings is abundant atmospheric O_2, which was once thought plausible only for planets that support photosynthesizing life on their surfaces. We now know that atmospheric O_2 can be readily produced and accumulated under abiotic conditions: For example, O_2 is produced on water-rich worlds due to photolysis *in the absence of life*. Thus in some contexts, abundant O_2 can be a biosignature, but in other contexts it is not, potentially fooling us into falsely assuming it is produced by life (*Meadows et al.,* 2018). Other *prima facie* signs of life and likely false positives include the light fluctuations of KIC 8462852 (Boyajian's Star) (*Boyajian et al.,* 2016) and seasonal methane cycling on Mars (*Webster et al.,* 2018). Both examples raise speculation of technological or biological sources, respectively, but neither is believed to actually be a definitive sign of life because there exist *more likely* abiotic explanations.

Just as observables for uninhabited worlds can provide mimics of inhabited ones, inhabited worlds can also mimic the observational signals of uninhabited ones. This occurs with "cryptic biospheres," where production levels of atmospheric gases associated with life are below detectable levels, leading to the possibility of "false negatives" — where life is present but not observable. This is likely the case for exoplanet biospheres that are similar to the Archean Earth, where production of atmospheric O_2 by life is thought to not have been remotely detectable by current technology (*Reinhard et al.,* 2017) (see also the chapter by Robinson and Reinhard in this volume). It could also be the case on worlds with subsurface biospheres as might be the case for Mars if it is inhabited (*Michalski et al.,* 2013). Avoiding both "false positives" and "false negatives" requires the capability to disentangle biological from abiotic signals, and/or focusing our search on worlds and in environments where the likelihood of biological signals dwarfs that of their abiotic counterparts.

The above considerations highlight the most significant challenge facing the astrobiology community if we seek to claim definitive detection of alien life in the coming decades: Living and non-living physical and chemical processes can — *and in many cases do* — produce the same observational signals. To date, the abiotic explanation has been the more likely one for *all in situ* or remote observations of other worlds. However, we will soon be positioned to take more numerous and nuanced measurements to test for life, and we will need to determine with a high degree of confidence whether or not a biological (or technological) explanation is more likely than a strictly abiotic one. In this chapter, we discuss motivations for why astrobiologists must adopt probabilistic approaches for next-generation life detection efforts. This approach will not only unify a broad cross-section of astrobiology research efforts by providing a multivariate framework for life detection, but also enable astrobiology to leverage a suite of *agnostic biosignatures*, i.e., those not dependent on the particular chemistry underlying living processes (see also the chapter by Johnson et al. in this volume). This enables the possibility for astrobiology to emerge as a quantitative scientific discipline where we

can assign a community-consensus statistical likelihood and confidence intervals to any discovery announcing we have found alien life, whether in the solar system on an exoplanet or even in the lab.

2. A MULTIVARIATE PROBABILISTIC APPROACH TO LIFE DETECTION

At present, no scientific theory exists that can fundamentally explain what life is. While many definitions for life exist, none are widely accepted, and nearly all are descriptive rather than quantitative. This introduces an obvious challenge for astrobiologists seeking to identify life on other worlds. Without a theory describing universal properties all living matter should be expected to share, we are left in the dark with respect to predicting its properties and signatures in alien chemistries.

The physicist Erwin Schrödinger was among the first scientists to significantly influence the debate about life's fundamental properties, addressing the question *"What is life?"* in a series of lectures delivered in 1943 at the Dublin Institute for Advanced Studies. These were published as a highly influential book in the subsequent year (*Schrödinger,* 1943). Most widely noted is his influence on Watson and Crick to search for the structure of genetic material, which Schrödinger correctly predicted as an "aperiodic crystal": what we now know as deoxyribonucleic acid (DNA). Schrödinger's prediction was based solely on logical reasoning about how heredity must work to be consistent with the known laws of physics. He also explored an apparent incompatibility between life and the second law of thermodynamics, addressing how it is that life increases order when the order in the universe should always decrease (as necessitated by the second law). Schrödinger's answer was that life must feed on what he called negative entropy (free energy), often referred to as "negentropy," i.e., food (or sunlight).

The idea that life might be explained by non-equilibrium physics remains an appealing one among some astrobiologists, where disequilibria is often discussed as a potential biosignature (e.g., *Krissansen,* 2016). However, there are fundamental differences between the modes of disequilibria found on the surfaces of inhabited and uninhabited worlds. Both Jupiter's Great Red Spot and the nighttime glow of Earth's massive cities are examples of far-from-equilibrium dissipative structures, but biology's role in the latter may not be apparent from a single disequilibrium parameter. Additional tools are necessary to distinguish or deconvolve the disequilibria characteristic of life from other kinds of disequilibria.

This raises an important issue in identifying candidate biosignatures distinctive to living matter: Life does not violate any of the known laws of physics or chemistry (nor should we expect it to). What life does do is control physical and chemical processes. It is the conjecture of some of the authors that this occurs via life's informational properties (see, e.g., *Davies and Walker,* 2016), which means living systems can turn events and objects that have exceedingly low probability to occur in the absence of life into having high probability via information control; examples include building and launching satellites to space [a biotechnological process that requires evolution to produce intelligent civilizations capable of understanding some of the laws of physics (see, e.g., *Walker,* 2016, for discussion)], or generating very high complexity molecules that require biological or technological processes for their production (*Marshall et al.,* 2017) (see section 3.2). If a NASA rover discovered a screwdriver-like object on Mars, we would be fairly confident it was a biosignature because the probability of such an object spontaneously forming in the absence of an evolutionary system capable of producing it is miniscule if not precisely zero.

Philosophy aside, for the practical purposes of life detection, we need to know whether or not the predictions for an observable (e.g., screwdrivers, atmospheric composition, or apparent disequilibrium therein) should differ in the case where life is present as opposed to where life is not present. Ultimately this will depend on what physics underlie life (i.e., the answer to Schrödinger's question), but in the absence of an explanation for what life is, astrobiologists can at least proceed on the knowledge of what life does on Earth to construct testable hypotheses about the relevant likelihoods of biological and abiotic processes. If an observable is distinguishable between these hypotheses in some environment, the signal is a good candidate biosignature or technosignature in that environment.

2.1. Bayesian Approaches to Biosignatures

The boundary between abiotic and biological signals is for many practical cases rarely a hard one, especially for signals amenable to remote observational study. Many researchers are now shifting toward statistical approaches, including applying Bayesian methods (*Walker et al.,* 2018; *Catling et al.,* 2018) to deal with this problem. In particular, the Bayesian approach permits systematically determining the likelihood that a signal is due to biological or technological processes, against its abiotic explanation, allowing one to assess quantitatively the confidence in biological origin. While other statistical frameworks can and should be explored by the community, we focus on Bayesian approaches herein. The aim of a Bayesian approach is to determine the *posterior* probability of life, i.e., the likelihood of life being the correct explanation for a given observation. This is calculated using Bayes theorem

$$P\left(\text{life|obs}\right) = \frac{P\left(\text{obs|life}\right)P(\text{life})}{P(\text{obs})} \quad (1)$$

where *obs* indicates *any* observable(s) that could indicate the presence of life on a solar system body or exoplanet or even in an experiment in a lab here on Earth. Bayes theorem relies on the concept of conditional probabilities, which occur in probability theory as a measure of the likelihood of an event occurring given that another event has occurred. P (life|obs) is therefore the conditional probability that life is detected, given observation of the observable(s): This is what we want to know. Bayes theorem allows calculating this based on the conditional probabilities of the inverse observation(s), i.e., on how likely it is to obtain the observable if life is present [P (obs|life)]. It also requires estimates of P (life), the prior probability of life occurring in the environment where the observation was made, and P (obs), the prior probability of the observable itself.

Observables here can refer to a wide variety of observations, including statistics from planetary surveys, stellar properties, remote observations of a specific planet, *in situ* measurements on a planetary surface, mass spec measurements, etc. Specific examples include the abundance of atmospheric O_2 in a terrestrial exoplanet atmosphere, or the distribution of amino acids in a sample collected by a robotic mission to Enceladus. It is important to stress that the general framework here is sufficiently general to include any putative biosignature, so long as the likelihood of it being produced in the presence and absence of life can be estimated, and a prior for life in the proper environmental context can also be estimated (these values turn out to be highly nontrivial to estimate, especially the latter; more on this is discussed below).

The denominator in equation (1) is the total prior probability for the observable(s) of interest. Since we are interested in isolating whether or not life has produced the observation, this can be expanded to isolate contributions from living sources from everything else

$$P\left(\text{life|obs}\right) = \frac{P\left(\text{obs|life}\right)P(\text{life})}{P\left(\text{obs|life}\right)P(\text{life})+P\left(\text{obs|abiotic}\right)P(\text{abiotic})} \quad (2)$$

Here, "abiotic" is meant to denote all sources that are *not* life, which can encompass both abiotic sources that could produce the same signal or sources of noise in the observation [see, e.g., *Walker et al.* (2018) for a treatment of noise in this framework, which we do not explicitly include here]. We denote these combined signals as abiotic as we aim to isolate contributions from living sources, and because the expectation is we will identify signatures where the expected noise is small (this is a challenging task in its own right, but as we are focused on theory in this chapter we leave observational challenges to other discussions).

To estimate and constrain the conditional probabilities P (obs|life) and P (obs|abiotic), simulations and experiments are necessary. *Walker et al.* (2018) provides an extensive review of possible paths forward for estimating both conditional probabilities for exoplanets. This includes simulations of potential exoplanet atmospheres and extraterrestrial ecosystems, and there exists a large community effort to simulate the chemistry of potential exoplanet atmospheres on both inhabited and uninhabited worlds (see, e.g., review by *Schwieterman et al.,* 2018). Initially, analysis was limited to modeling of exoplanet atmospheres: These simulations indicated that biosignature gases once thought to be "smoking gun" signatures of life such as O_2 and CH_4 could be produced abiotically (*Domagal-Goldman et al.,* 2014). The co-existence of O_2 and simple organics in an atmosphere was once thought to be a potentially definitive biosignature (*Seager et al.,* 2013), because these gases co-exist only in a state of chemical disequilibrium driven by life. Furthermore, O_2 is considered a relatively robust biosignature, although its status as such has been challenged more recently, particularly regarding planets orbiting stars with high FUV flux (such as young M dwarfs), where photochemical reactions involving CO_2 and H_2O could give rise to O_2 abiotically (*Meadows,* 2017). However, further experimental work is often required to help constrain these scenarios (*Fortney et al.,* 2016), and adding a probabilistic assessment of the relevant likelihoods of abiotic and biotic production across different planetary contexts will help enable identifying the best targets for detecting life with high confidence [see, e.g., *Walker et al.* (2018) for an example of a Bayesian approach to assessing O_2 as a biosignature within the context of other observables].

At the ecosystem level, the study of analog environments on Earth (such as subglacial lakes for Europa or Enceladus, and the Antarctic Dry Valleys for Mars) can also help constrain priors on biological production rates (*Amils et al.,* 2007; *Marion et al.,* 2003; *Fisher and Schulze-Makuch,* 2013), necessary for estimating P (obs|life) in analogous environments within the solar system and on exoplanets. These can be aided and accompanied by *in vitro* analyses of extremophile growth rates and productivity to better understand the boundaries of life's ability to produce detectable levels of biosignature gases (*Cavicchioli,* 2002).

For most observables, the signals plausibly attributable to abiotic sources and noise will be considerable. Even our best instruments and most rigorous inverse models will be limited in their ability to minimize uncertainty and noise, frustrating our attempts to distinguish biogenic signals from background. However, a statistical framework could help us develop strategies to control and allocate our uncertainties across observations, based on which parameters seem most consequential for calculating P (life|obs) and and P (obs|abiotic). It could help us to refine potential targets, select appropriate observables, and design instruments and tools around those observables, based on the anticipated *relative* contributions of biotic and abiotic processes to an observable within a given planetary context. It could even help us to constrain the prior probability of life's emergence [P (life)] in specific environments, by steering experimental origins-of-life studies away from strictly historical reconstruction and toward predictive frameworks applicable (and testable) in diverse planetary environments (*Scharf and Cronin,* 2016) (see also the chapter by Hoehler et al. in this volume).

Specific research programs to address individual terms in equation (1) were discussed in detail in *Walker et al.* (2018). Currently, our most well-studied candidates for remotely detectable or *in situ* biosignatures are small molecules that are commonly used or produced by life, such as methane, oxygen, and amino acids. Interpretation of these molecules as biological in origin is confounded by abiotic production mechanisms. The Bayesian approach is most useful in cases where abiotic and biotic processes can produce the same signatures, such that the answer to whether a given observation is produced by biology or not is not a simple "yes" or "no." In such cases, we must assign a statistical likelihood, as captured by the conditional probabilities P (obs|life) and P (obs|abiotic). False-positives are likely when P (obs|abiotic) $\geq$ P (obs|life), i.e., when the likelihood of a signal being produced by an abiotic process is comparable to or greater than the likelihood from a biological one. Thus to minimize the possibility of being confounded by false positives, searches for life should focus efforts on observables that maximize the odds ratio in favor of life being the cause of the signal

$$D = \frac{P\left(obs|life\right)}{P\left(obs|abiotic\right)} \qquad (3)$$

such that $D \gg 1$. D can be roughly considered the *detectability* of a given biosignature (see, e.g., *Walker et al.,* 2018), but it should be noted that detectability also depends on our ability to measure the particular observable in the first place, not just on the comparison of its likelihood to be produced by living vs. abiotic processes.

Many of our current biosignature candidates do not necessarily meet this criteria for detectability. While the challenges of detectability in the face of false positives have been most prominently discussed in the exoplanet literature, this is true even for observations within our own solar system. Like Earth, Mars appears to exhibit seasonal variations in methane abundance (more recently observations indicate O_2 variation as well). However, this is likely a false positive observation because martian methane cycling may be plausibly attributed to adsorption and diffusion by its surface materials (*Moores et al.,* 2019), with baseline methane levels potentially sourced from a variety of abiotic processes (*Oehler and Etiope,* 2017), rather than being biologically driven as on Earth. NASA's Curiosity mission has also identified complex organic molecules whose origins, although ambiguous, are plausibly abiotic (*Krissansen-Totton et al.,* 2018). A host of other complex organics, as well as amino acids, have been found on the Murchison meteorite, prompting the identification of abiotic molecules for biomolecule production in meteoritic and hydrothermal settings (*Schmitt-Kopplin et al.,* 2010; *Summons et al.,* 2008). In these examples, P (obs|abiotic) > P (obs|life) or at least P (obs|abiotic) ~ P (obs|life), meaning it is impossible to attribute these "biosignatures" to processes that are biological in origin without additional context such as other observables or more detailed knowledge of abiotic or biotic production mechanisms.

Beyond the challenges posed by false positives, exoplanetary settings provide additional challenges. Present data takes the form of sparse surveys. Current and future planet-specific observations will be subject to low spectral and limited to no spatial resolution, requiring extensive integration times, advanced deconvolution techniques, and/or major investments in instrumental design and stability. Current modeling and inversion techniques will be subject to further uncertainty as we depart from the dynamical, chemical, and radiation regimes for which we have solar-system analogs. Observations of giant planets in particular have motivated a suite of forward and inverse models by which astronomers can constrain the (abiotic) process couplings compatible with acquired data. Detailed observations of smaller planets will require more advanced telescopes, but will likely leverage these models and resulting insights, incorporating couplings uniquely relevant to habitability and the evolution of rocky planets (*Fujii et al.,* 2018; *Foley and Driscoll,* 2016).

The astrobiology community will need to more tightly constrain the likelihoods of biological and abiotic production mechanisms for a given single or composite signal, to better infer the presence or absence of life from the limited data we can acquire. The exoplanet community is acutely aware of this for atmospheric biosignatures such as oxygen and methane, and is doubling efforts on understanding the role of planetary context in producing these signals abiotically. Similar approaches can and should be adopted in the search for life within the solar system, where the likelihoods of both biological and abiotic production mechanisms are weighted against one another conditioned on the appropriate environmental context and constrained by multivariate observations. We next consider alternative statistical approaches, and then for the remainder of the chapter explore systems biosignatures that take advantage of a probabilistic framework for life detection in that the biosignatures themselves are statistically derived by considering the properties of ensembles of living chemistries and comparing to non-living ones.

2.2. Other Statistical Approaches

A Bayesian approach to life detection, by logically establishing a false-positive/false-negative formalism and providing a mathematical framework for their treatment, encourages us to question, refine, and expand our repertoire of potential biosignatures. Given the effort we expect the community to put in to making the future detection of candidate biosignatures as unambiguous as possible with respect to their biogenecity, it is tempting to assume that our data itself will be relatively unambiguous, i.e., that the primary uncertainty could be our data's origins in biotic or abiotic processes. However, even the data from which we source these inferences will be inherently uncertain, even if our inferences (about living and abiotic processes) are ultimately good ones (a big "if" at this stage). Recognizing and propagating the inherent uncertainties in our measurements, particularly for exoplanets, will be necessary to accurately calculate our resulting probabilities and confidence intervals.

Fortunately, a variety of statistical methods have been developed and widely employed to quantify and address diverse forms of uncertainty and noise in astrophysical settings. Some such methods are themselves built on probabilistic frameworks, and a few could even be adapted for a top-level life detection inference framework. For example, Bayesian nonparametric methods (*Hjort et al.,* 2010) have been applied to the prediction of sparse but interrelated planetary parameters (*Ning et al.,* 2018) and to inferring the properties of living systems (e.g., *Hjort et al.,* 1990). Other statistical methods could form the backbone of subsystems embedded in some term of equation (1), with implications and linkages subsequently cascading up our assessment framework. A detailed treatment of such approaches is beyond the scope of this chapter, but a brief overview of their demonstrated use cases may be worthwhile, keeping the above agnostic biosignatures in mind. For example, Bayesian tools have been designed to spectrally and spatially resolve exoplanetary surfaces (*Farr et al.,* 2018), leveraging slight fluctuations in albedo over very few extremely coarse-grained observations. Similarly enhanced spatial resolution was achieved for planetary thermal emission spectra using an analytical (i.e., less explicitly probabilistic) framework with a theoretical precedent and expected near-future refinement in giant-planet applications (*Luger et al.,* 2019). Multiple Bayesian-based approaches have been developed for retrieving planetary spectra by taking advantage of forward modeling to increase the confidence of priors used (*Line et al.,* 2013).

The forgoing provides just a few examples, and the community should explore more ways of leveraging statistical methods for building better quantitative tools for dealing with limited data inherent to life detection efforts on other worlds. Additionally, by focusing on agnostic biosignatures (discussed in subsequent sections) and processes manifesting across a wide range of signals, we minimize our risk of both false positives and false negatives, reducing our reliance on models specific to the properties of life on Earth in the process. For the remainder of this chapter we discuss some of the examples of new kinds of biosignature frameworks opened by adopting a statistical approach to life detection.

3. EXAMPLES OF PROBABILISTIC BIOSIGNATURES IN CHEMICAL SPACE

The advantage of adopting a Bayesian or any other probabilistic method for life detection is that it opens up possibilities for new classes of biosignatures that are not candidates as signatures for life outside of statistical approaches. So far, the majority of biosignature frameworks have focused on the presence of molecules implicated in living processes, e.g., amino acids in solar system searches for life or O_2 on extrasolar planets. A statistical approach enables shifting this focus from the particular chemical composition of life on Earth to characterizing those statistical properties common across life on Earth, opening the possibility of more agnostic biosignatures that might not only apply to life as we know it, but also life as we don't know it. In what follows, we illustrate the approach with a few toy examples. The examples, while based on real data, are necessarily contrived, as properly working out each would be a research project in its own right. Our aim here is to merely illustrate how a probabilistic approach can work in practice and what kinds of new biosignature frameworks are opened up by thinking statistically, with the ultimate goal being to inspire the reader to consider possibilities for probabilistic biosignatures in their own research program.

3.1. Molecular Weight and Diversity

A key concern of astrobiology is not only that we identify signatures that are strongly indicative of the presence of life, but also that these must be observable. Nearly every mission within our solar system is fitted with a mass spectrometer, and calculations of molecular mass can be derived from atmospheric spectra in cases where we can identify specific molecules that give rise to specific spectral features. Thus, both within and beyond the solar system molecular weight is readily accessible observationally, fulfilling a key requirement of a good biosignature. But, is it discriminatory of life? There are some arguments that high-complexity, high-molecular-weight molecules are most likely to be produced by life, but we also know of examples of abiotic processes that can produce high-molecular-weight compounds; these are typically less "complex" with repeating units as found in polycyclic aromatic hydrocarbons (PAHs), which are abundant abiotically, found even in interstellar space (*Sloan et al.,* 1999; *Szczepanski and Vala,* 1993). Thus, molecular weight alone is not necessarily discriminatory of life, unless we can probabilistically anchor the distributions for the suites of molecules that we expect living and non-living processes to produce, and subsequently verify they are statistically distinguishable.

In what follows, we consider a toy model of a Bayesian approach to biosignatures based on a contrived example considering molecular weight as a biosignature. This not meant to be quantitatively exact, but is instead intended to be qualitatively illustrative.

One reason the examples are contrived is that we do not know the expected distribution of molecular weights in abiotic vs. biological samples. A major challenge is sorting out which molecules are produceable by abiotic processes and which have a high likelihood of being produced only by living processes. This is the problem of determining where in chemical space lies the boundary between non-living and living chemistry. Chemical space is defined by the set of all possible molecules that could exist. This set is so vast there are not enough resources in the entire universe to ever explore all of it; not all molecules that could exist will ever exist. To make this concrete, consider the comparatively small subspace of chemical space defined by all possible "biological" proteins that are 200 amino acids long and composed of the 20 most commonly coded amino acids; the size of this subspace is 20^{200} (*Walker,* 2019). This is

infinitesimally smaller than the size of chemical space, but nonetheless so exponentially large that there are insufficient resources in the universe to produce just one copy of each protein in this space. [It is a philosophical question as to whether or not the entire set considered should be termed "biological" since not all of these proteins will ever exist and the majority would in any case be non-functional, so although Earth life could in principle produce any member of this set of proteins the vast majority either will (1) never be produced by a life-form or (2) if produced, be non-functional.] For a molecule as large and presumably complex as a length 200 protein, we might expect only biology to be capable of producing it because it is exceeding unlikely such a molecule could be produced from a rare statistical fluctuation in a random abiotic chemical mixture. [It should be noted that some origin of life arguments rely on such a rare statistical fluke producing the first "living" molecule (*Monod,* 1971), but that is not a view shared by the authors here.] However, there are many other examples of molecules that are readily made by both living and non-living processes. Take for example the experimental enterprise of prebiotic chemistry, which aims to produce biologically important molecules, such as nucleobases and amino acids, through purely abiotic mechanisms. Prebiotic chemistry succeeds because some molecules produceable by life are also produceable abiotically and vice versa. This suggests that while there are regions of chemical space inaccessible to abiotic chemistry (e.g., exploration of protein sequence space), which might be thresholded by setting a lower bound on where biological production is the only reasonable explanation (see section 3.2), the precise boundary between non-life and life in chemical space is not a hard one. It is in fact quite fuzzy and environmentally driven since which molecules are produced depend on environmental conditions such as pH, temperature, salinity, etc. This lack of a hard boundary in chemical space between non-living and living chemistries necessitates a probabilistic approach so that we can assign a likelihood that a given molecule or system of molecules is produced by life in the given environment.

We focus here on developing an approach treating molecular weight as a potential probabilistic biosignature by first considering how molecular diversity, quantified in terms of the number of distinct molecular compositions found in a given molecular weight range, might distinguish biological samples from abiotic ones. *Kauffman et al.* (2019) used the PubChem database to study how many different molecular compositions share the same molecular weight, with the aim of determining how chemical evolution has explored composition space from the early universe to the present day. Based on our foregoing discussion, we should already be ready to anticipate the number of possible compositions should increase exponentially with molecular weight. For example, a biological protein of length L has 20^L possible compositions (where we considered L = 200 above). We should expect that the length L scales roughly with molecular weight and thus that the number of compositions does as well.

If instead of considering permuting over sequences of amino acids we considered the number of all possible molecular compositions of equivalent molecular weight, the number of possibilities would be exponentially higher still; we would now be referring to combinatorics over atoms instead of amino acids (which are already a collection of atoms with specific composition). Kauffman et al. estimated an upper bound for the number of hypothetically possible compositions of a given molecular weight, providing an approximation on the total size of chemical space at a given mass scale

$$N(M) \sim \frac{1}{\Gamma(d+1)(M/M_0)^d} \qquad (4)$$

where M is molecular weight, M_0 is the geometric mean of the masses of d different atoms and $\Gamma(d + 1)$ is the complete gamma function. In practice, d ends up being an effective dimensionality, capturing the number of frequently used high-molecular weight elements, and M_0 the mean of their masses (see description of equation (5) below). Their hypothetical upper bound is plotted in Fig. 1 (gray dashed line), which clearly exhibits the requisite exponential growth in the number of compositions as a function of molecular weight.

Kauffman et al. compare this hypothetical upper bound on what is possible (assuming no resource limitations) to the number of distinct compositions chemical evolution has actually been able to explore in the universe's history, leveraging data from PubChem on known molecular weights and compositions across 3.5-M molecules. In their work, the distribution of the number of compositions vs. molecular weight derived from the PubChem data was fit to two regimes: a diffusive regime describing low-molecular-weight compounds, and a preferential attachment regime describing high-molecular-weight compounds. At low molecular weights the number of possible compositions for a given molecular weight is fairly small and it is very likely most molecules in the space will be accessible by transformations (reactions) of members already in the space. The model they used (see *Kauffman et al., 2019*) assumes an ensemble of molecules described in terms of a time-dependent density in composition space, which is initialized with only atoms having non-zero abundance. The subsequent time-dependent dynamics lead to an expansion into composition space toward larger compounds involving more elements, much like a diffusion process. The model they derived to describe the distributions generated by this diffusive process is given by

$$n(M) = \frac{1}{\Gamma(d)M_0} \left(\frac{M}{M_0} \right)^{d-1} \frac{\Gamma(d/2, M^2/2\sigma^2)}{\Gamma(d/2)} \qquad (5)$$

where σ is fit to data, and M_0 is the geometric mean of the masses of d different atoms and $\Gamma(d + 1)$ is the complete gamma function as in equation (4). The parameter d can be interpreted as the effective number of elements in the different compositions, and takes the value of d = 3.5 for the fits in *Kauffman et al.* (2019), consistent with most compounds being dominated by just three heavy atoms (C, O, N) with lighter H playing an ancillary role with respect to molecular weight. The fits for the PubChem data in Kauffman et al. also yield values $\sigma_{PC} = 263.3 \pm 0.7$ and $M_0 = 3.81$ (these values are used in Fig. 1).

Notably, the fit to the number of compositions shown in Fig. 1 quickly diverges from the number of hypothetically possible compositions. This is expected as the hypothetical size of chemical space is much larger than what the physical universe can instantiate. The divergence is higher for large molecular weights, where only an exponentially vanishing fraction of this hypothetical space is ever explored by the physical universe (recall our length L = 200 protein example). The high-molecular-weight regime (M > 660 Da) from the PubChem database has what is called a "heavy-tail," i.e., the distribution is more weighted there than one might *a priori* expect based on predictions of the diffusive model. In Kauffman et al. a preferential attachment model was used to explain this. The preferential attachment model assumes higher-molecular-weight compounds are more likely to increase in size more quickly (preferentially), such as occurs in polymerization. For example, polypeptides or PAHs are not built via random accumulation of atoms (as the diffusive process would describe), but instead mostly from the accumulation of large repeating blocks such as

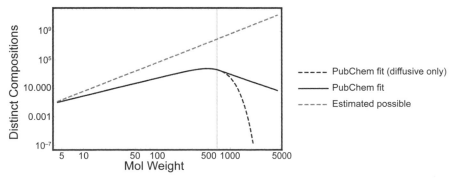

Fig. 1. The estimated number of molecular compositions as estimated in *Kauffman et al.* (2019), plotted against a fit for the number observed from data-mining the PubChem database.

amino acids and aromatic rings. This yields more observed heavy molecules than one might expect by the naïve diffusive model, and results in a heavy-tailed, power-law distribution at high molecular weight

$$n(M) = CM^{3.5} \qquad (6)$$

where we fit $C = 3.12 \times 10^{13}$ for their model, assuming that equation (5) and equation (6) match values at M = 660 Da, where the model of Kauffman et al. transitions from the diffusive to the preferential attachment model (see *Kauffman et al., 2019,* for discussion). The model fit including both the low-molecular-weight (diffusive) and high-molecular-weight (preferential attachment) regimes is shown in Fig. 1 (black solid line).

PubChem, like all major databases inventorying chemical compounds, includes molecules produced abiotically, those produced by life, and molecules produced by technology (e.g., pharmaceutical drugs). A research program aiming to determine the statistical signatures of living chemistries would necessarily need to deal with determining what compounds in databases such as PubChem are exclusively abiotic (assuming such compounds exist), exclusively biological, or are produceable across living and non-living systems. To do so will require constructing probability distributions directly from experimental and observational data on non-living and living chemistries. In particular, we expect the PubChem database is dominated by biologically derived compounds at high molecular weight and therefore for purposes of discussion herein, view the diffusive approximation as a better estimate of the likely abiotic distribution, which we might expect to find in non-living chemistries. However, Kauffman et al. demonstrate that the full model (including the heavy-tailed power law) is a good approximation to the observed distribution of compounds found in the Murchison meteorite. Since here we are interested primarily in illustrating the Bayesian approach, and not developing a full statistical framework for molecular weight in abiotic and biological samples (a subject of future work), we do not address these issues. Instead, we take either the diffusive model or the diffusive model with the power-law tail as the assumed abiotic distribution, acknowledging the caveat that the PubChem data is not strictly abiotic, and much more work needs to be done to develop accurate distributions across non-living chemistries under different conditions.

To apply the Bayesian framework presented in section 2 requires knowing the likelihoods for both the abiotic *and* the biological distributions. For our toy example, we consider the number of compositions vs. molecular weight for biochemical compounds cataloged in the Kyoto Encyclopedia of Genes and Genomes (KEGG), which, like PubChem, also yields a distribution with a heavy tail, as shown in Fig. 2. The distributions in both Fig. 1 and Fig. 2 count the number of distinct compositions for a given molecular weight, which in our toy example we consider as the abiotic and biological distributions respectively. The distribution appears "quantized" due to binning compounds together with similar weight; the bin size shown is 1 Da. Unlike with the PubChem data, here we are not showing a fit to the distribution, but the actual binned values from the raw data.

To turn the binned data into probability distributions we have to make a few assumptions (on top of the first assumption of choosing a bin size). For simplicity we

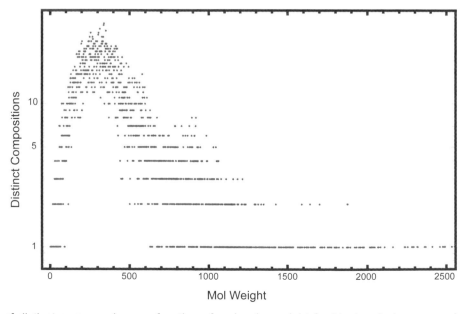

Fig. 2. Number of distinct compounds as a function of molecular weight for biochemical compounds cataloged in the Kyoto Encyclopedia of Genes and Genomes. The discrete pattern arises due to binning compounds with similar molecular weight; here the bin size is 1 Da.

assume a uniform distribution over observed abundances across molecular compositions. While it is unlikely that high-molecular-weight compounds are as abundant as low molecular weight ones in real samples, we expect the resultant distribution to roughly track the real abundance distribution of molecular weights. There exist many more distinct compositions at low molecular weights for both the toy abiotic and biological distributions, despite the combinatorial explosion of possible molecular formulas at high molecular weights (see Fig. 1), meaning the "abundance" distribution generated by this estimate will favor low-molecular-weight compounds as we might expect the real distributions to do. Our second assumption is that frequency-sampling the molecular weight abundance distribution (approximated by the number of distinct compositions) provides sufficient statistics to accurately approximate the true distributions for the abiotic and biological distributions (i.e., we are using frequentist statistics; in reality our view is that the real distributions should be inferred by a combination of experimental observation, statistical inference, and supporting theory). The resulting probability distributions are shown in Fig. 3.

In a Bayesian framework, the goal is to identify the likelihood that a molecule of a given weight is produced by a living process; i.e., we want to know the following likelihoods (with associated confidence intervals, which we do not address herein)

$$P\left(\text{life}|M_w\right) = \frac{P\left(M_w \mid \text{life}\right)P(\text{life})}{P\left(M_w \mid \text{life}\right)P(\text{life}) + P\left(M_w \mid \text{abiotic}\right)(1-P(\text{life}))} \quad (7)$$

or conversely

$$P\left(\text{abiotic}|M_w\right) = \frac{P\left(M_w \mid \text{abiotic}\right)(1-P(\text{life}))}{P\left(M_w \mid \text{life}\right)P(\text{life}) + P\left(M_w \mid \text{abiotic}\right)(1-P(\text{life}))} \quad (8)$$

We use the probability distributions in Fig. 3 to approximations $P(M_w|\text{life})$ and $P(M_w|\text{abiotic})$, as generated from KEGG and PubChem respectively (for the latter we have two alternatives we explore: the diffusive model or the full PubChem fit with diffusive and preferential attachment regimes).

A challenge is determining $P(\text{life})$ in a specific environment (note we are not explicitly treating the role of environment here, since distributions in KEGG and PubChem are sampled across many environments). Currently $P(\text{life})$ is unconstrained on the interval $(0,1]$, i.e., we do not know the probability of life to emerge in *any* environment. We here assume a value of $P(\text{life}) = 0.1$; i.e., for the inventory of chemical systems we are analyzing we expect *a priori* that 10% will contain living matter. This number is a complete guess, as we do not currently have any constraints on $P(\text{life})$ or even the environments we should consider calculating our prior over. We emphasize this point to illustrate the magnitude of the problem of an unconstrained prior for life. It is our view that a major effort in the community should be to direct research efforts in life detection to constraining $P(\text{life})$ (see *Walker et al., 2018*, for discussion).

The resulting likelihoods are shown in Figs. 4 and 5 for the diffusive model and the diffusive model with a power-law tail as the assumed abiotic distribution, respectively. If the diffusive approximation for the abiotic molecular weight distribution is the more accurate one, high-molecular-weight compounds with $M_w > \sim 1100$ Da are a reasonably good biosignature as $P(M_w|\text{life}) > P(M_w|\text{abiotic})$. Under the assumption of a diffusive abiotic model with a power-law tail, this threshold shifts to $M_w > \sim 2500$ Da, demonstrating how the ability to infer whether a given "biosignature" is indeed indicative of life depends on both the assumed abiotic and biological distributions expected for that observable. The values where $P(M_w|\text{life}) > P(M_w|\text{abiotic})$ will also depend on the prior probability of life (as well as the priors that we do not explore here); if the prior for life is lower than assumed here $P(\text{life}) < 0.1$), these thresholds would shift

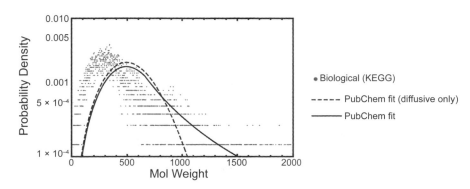

Fig. 3. Estimated probability density of molecular weights for biology (generated from the KEGG database, binning data in 1-Da intervals; see Fig. 2) as compared to estimated distribution for molecules inventoried in the PubChem database in the diffusive (low molecular weight regime) and the full PubChem distribution (see Fig. 1), generated from fits in *Kauffman et al.* (2019).

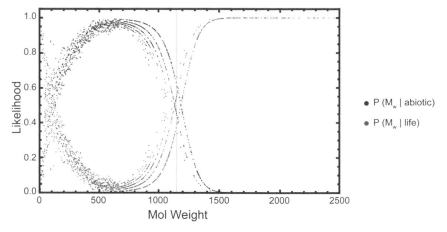

Fig. 4. Likelihood of an abiotic vs. biological explanation as a function of molecular weight, assuming a prior P_{life} = 0.1, with the abiotic distribution described by the diffusive regime only (from Fig. 3).

to higher values, as life would be a less likely explanation for the data (and vice versa if we assumed a higher value for the prior probability of life P (life) > 0.1). Thus, the threshold for life detection changes with our assumptions about the expected abiotic distribution in a given environment, and also the expectation on the prior probability for life in that environment. Even in the absence of knowing the prior for life, we can at least assess, within the Bayesian framework, if a biosignature is a good one by considering the ratio in equation (3).

It may be the case that only very high-molecular-weight molecules, which are also very rare, are indeed biosignatures with respect to their molecular weight. For more moderate molecular weights, which are both more amenable to observation and more likely to be observed, it will likely be difficult to disentangle a biological vs. nonbiological explanation based on molecular weight alone. For example, for a molecule of M_W = 500 Da, assuming the diffusive model, there is only a 6% likelihood it is produced by life under the contrived assumptions of our example. Additional measures beyond molecular weight may therefore likely be necessary,

demonstrating how a multivariate suite of metrics may be necessary to infer the presence of life (see, e.g., discussion in *Walker et al.,* 2018). An alternative approach is to not just compare an observed molecule of a given molecular weight, but the entire observed distribution of molecules in the sample as the biosignature, which should be more discriminatory depending on how close abiotic and biological molecular weight distributions are in real samples.

3.2. Thresholding the Products of Living Processes in Chemical Space with Molecular Assembly

To build probabilistic frameworks for life detection requires identifying quantitative properties of life that we expect to statistically distinguish living systems from nonliving ones. As we showed in the last section, basic observables of living chemistries such as the distribution of molecular weights could provide probabilistic biosignatures, but may not be discriminatory enough. It is likely that additional metrics that better distinguish living matter will be required, opening a frontier for development of new probabilistic

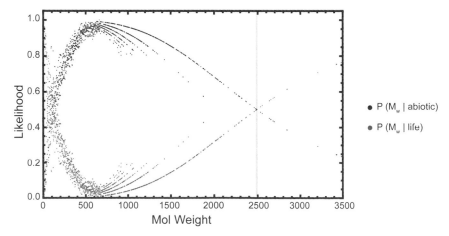

Fig. 5. Likelihood of an abiotic vs. biological explanation as a function of molecular weight, assuming a prior P_{life} = 0.1, with the abiotic distribution derived from the fit to the PubChem distribution including both the diffusive and power-law tail regimes (from Fig. 3).

biosignatures, whose individual components could not be considered signatures of life outside of a statistically significant ensemble.

One property that discriminates living things from inorganic matter is life's capacity to generate non-random or complex structures in large abundance. This is obvious for artifacts such as satellites, cities, and other non-random objects generated by technological civilizations. Importantly, it is also true for molecules: The space of all possible molecules is immense, but biology (*Pace,* 2001) can only produce a nonrandom subset of these (see, e.g., Fig. 2). As we saw in section 3.1 the number of possible compositions scales exponentially with molecular weight (see Fig. 1). Importantly, the increase in number of possible compositions can also correspond to an increase in the complexity of molecular structures. While complex molecules are not themselves living, they cannot randomly form in any abundance and hence are good candidates for either biological or technological signatures. Consider again our example of a length L = 200 protein; such a molecule with a precise sequence is exceedingly unlikely to be produced in the absence of biology as there is no information to specify its production. Even if we did find an abiotic system that could produce L = 200 proteins (a big "if"), it would likely do so without any specificity, so we would never expect to find a precise sequence in high abundance (as it is just 1 out of the 20^{200} possibilities). To understand how to identify high-complexity molecules such as proteins or other biological products, as distinct from abiotic structures such as PAHs, an approach that aims to evaluate complex molecules as possible biosignatures could be useful to explore the cosmos for new life forms.

One promising method developed to achieve this involves an information measure called molecular assembly (MA) (*Marshall et al.,* 2017). Molecular assembly is a metric to quantify the number of steps required to assemble an object based on a hypothetical history of an object's formation. These steps do not necessarily correspond to a reaction sequence, but instead consider the intrinsic complexity of an object and the number of repeating units necessary to assemble it. Specifically, by partitioning an object into its irreducible parts and counting the steps by which the object can be assembled from those parts, and considering the probabilities of such steps, the total probability that the object could form in the absence of biological or technologically-driven processes can be estimated. This allows thresholding the abiotic-biotic divide by identifying the threshold above which a biological process was necessary to construct a given object (see Fig. 6 for an example from language computing the steps to construct the word "banana"). The idea of a biological threshold does not presume objects below the threshold are not made by life, but only that anything *above* the threshold must necessarily have been made by life in order for it to be found in any abundance (i.e., to not be an accident of a random statistical fluke, but made by a reliable process that repeatedly occurs with a specific outcome). Since the attributes measured (number

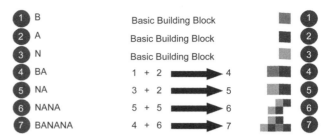

Word Assembly No. = No. Joining Operations = 4

Fig. 6. Object assembly for a text string and a simple block shape giving the assembly to construct the word "banana" from its basic building blocks value of 4.

of steps) are generic and applicable across a wide range of complex biological and technological systems, MA provides a fairly robust agnostic biosignature applicable across *both* exoplanetary and solar system settings (e.g., *Walker et al.,* 2018) and in laboratory efforts to experimentally search for new life forms (*Cronin and Walker,* 2016).

Marshall et. al. implemented MA as an algorithm and validated MA [referred to as "molecular complexity" in *Marshall et al.* (2017)] for molecules. A preliminary approach using molecular weight, MA, and relative abundance has shown that molecules derived from living systems have a distinct signature. By constructing a workflow that can separate an environmental sample (solid, liquid, gas, or mixture of these) using chromatography followed by mass spectrometry (MS) and MS/MS fragmentation of the molecular ions, it is possible to plot frequency of occurrence against MA (see Fig. 7). Thus a mass spectrometer can be used to determine for a given sample or mixture (1) the molecular ions present. (2) the fragmentation pattern associated with the molecular ion, and (3) the number of fragmentations associated with a given molecular ion. These can be used to approximate MA. Molecular assembly does not use the MS data to obtain the structure of the molecule and hence calculate the assembly number. The MA value is calculated independently based on the known structure of the molecule, and correlates with the fragmentation in the MS data. Some examples of biologically derived molecules with similar molecular weights but different MA values are shown in Fig. 8.

Molecular assembly provides a probabilistically-rooted biosignature by thresholding the abiotic-biotic transition in molecular complexity space. It also provides another candidate measure for a Bayesian approach to biosignatures if we consider the expected MA in biological and non-biological samples. We illustrate this with a second toy model. Molecular data such as that in Fig. 7 suggest that the average value of MA scales differently for biological and abiotic samples as a function of molecular weight. This scaling behavior remains to be rigorously confirmed experimentally, but for purposes of our toy models exploring a probabilistic approach to biosignatures, we assume the following scaling behavior for the average MA of biochemical compounds derived from living systems

$$\langle MA(M_W) \rangle_{life} = C_1 \cdot M_W^{1/2} + C_2 \cdot M_W \qquad (9)$$

and from non-living systems

$$\langle MA(M_W) \rangle_{abiotic} = C_3 \cdot M_W^{1/2} \qquad (10)$$

In what follows, we assume $C_1 = -5 \times 10^{-17}$ and $C_2 = 0.05$, which gives a reasonable trend consistent with that of Fig. 7. We also assume $C_3 \sim 0.2$, which is consistent with the currently estimated upper bound of $\langle MA(M_W) \rangle_{abiotic} \sim 15$ observed empirically in preliminary laboratory studies (in progress; see *Marshall et al., 2017*). Both scaling trends are plotted in Fig. 9. To generate a probability density over average MA, we make the same assumptions as in section 3.1 and consider a toy case where all distinct compositions are uniformly abundant and our sampling is sufficient to generate a reasonable frequency distribution. Noting that we can generate an abundance distribution as a function of molecular weight for biological and abiotic examples with these assumptions (see section 3.1, e.g., Fig. 3), we likewise generate a probability distribution for MA values using the relations plotted in Fig. 9 to determine the frequency of observing a specific MA value based on the expected molecular weight for that value. The resulting probability distributions are shown in Fig. 10. Consistent with expectation (since we constructed the distributions to meet this expectation for our toy system) the range for the threshold delineating biological production as the only explanation occurs around MA = 15.

With these distributions we might, for example, be interested in calculating the likelihood that a molecule observed on Mars or Titan is produced by living processes. To do so we need to determine the posterior likelihoods

$$P(life|MA) = \frac{P(MA|life)\,P(life)}{P(MA|life)\,P(life) + P(MA|abiotic)(1 - P(life))} \qquad (11)$$

or conversely

$$P(abiotic|MA) = \frac{P(MA|abiotic)(1 - P(life))}{P(MA|life)\,P(life) + P(MA|abiotic)(1 - P(life))} \qquad (12)$$

These are plotted in Fig. 11 assuming a uniform prior probability for life P (life) = 0.1, as in section 3.1 (corresponding to an expectation that there is a 10% prior probability for life in the sample). We use the probability distributions from Fig. 10 for P (life|MA) and P (abiotic|MA). In section 3.1 we considered the likelihood of an observed molecule with $M_W = \sim 500$ Da being produced by life,

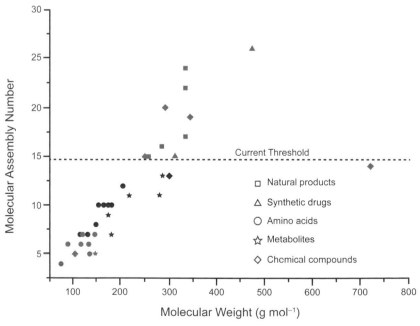

Fig. 7. Molecular assembly (MA) plotted against molecular weight for a range of molecules including natural products, synthetic drugs, amino acids, metabolites, and chemical compounds. MA roughly captures the number and probability of each step to assemble a molecule from its component parts. The dotted line indicates the current threshold for biological production, on the basis of preliminary data from the Cronin lab. Note, biological molecules are produceable below the threshold, but by definition the threshold will fall at the observed MA value where there are there no abiotic molecules are found with an MA value above the threshold (so one can be confident biology is the production mechanism).

which was unlikely given that abiotic processes can produce molecules with similar molecular weight. If we now consider MA we find that there is the possibility of discriminating these molecules as biologically-sourced, i.e., if trends such as those in Fig. 9 hold up. For the molecules in Fig. 8, we have the following likelihoods based on molecular weight (assuming the diffusive model): P (life M_W = 524 Da) = 0.023, P (life$|M_W$ = 531 Da) = 0.029, P (life$|M_W$ = 524 Da) = 0.023, and P (life$|M_W$ = 401 Da) =

0.083; i.e., we cannot have a high confidence that any are consistent with life based on molecular weight alone. If we instead consider likelihoods based on MA, we have P (life$|$MA = 21) = 1, P (life$|$MA = 27) = 1, P (life$|$MA = 14) = 0.96, and P (life$|$MA = 12) = 0.85. In other words, within our constructed model there is a high likelihood the molecules are biologically derived due to their high MA values, despite the fact their molecular weight alone is not a distinguishing factor.

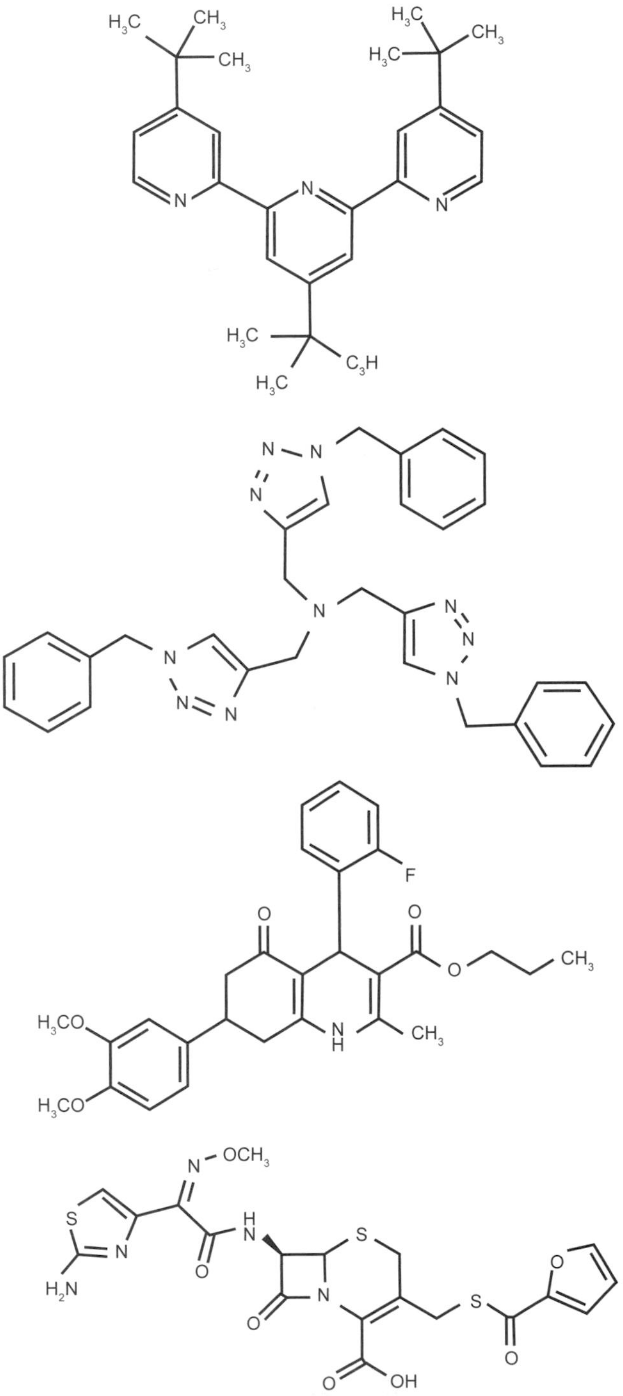

(a) $C_{27}H_{35}N_3$, M_w = 401.587 Da, MA = 12

(b) $C_{30}H_{30}N_{10}$, M_w = 531.27 Da, MA = 14

(c) $C_{28}H_{30}FNO_5$, M_w = 480.24Da, MA = 21

(d) $C_{19}H_{17}N_5O_7S_3$, M_w = 524.53Da, MA = 27

Fig. 8. Some examples of molecules produced by biology with similar molecular weights but varied molecular assembly ranging from values below the threshold (top two examples with MA = 12 and 14 respectively) to above the threshold (bottom two examples with MA = 21 and 27 respectively).

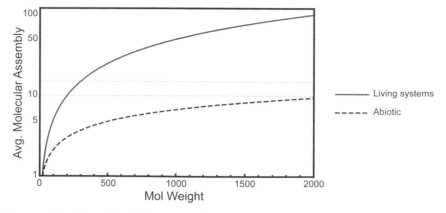

Fig. 9. Toy model example of hypothetical scaling of average molecular assembly index over compositions for a given molecular weight for living and abiotic systems. Trends are consistent with current observations (see Fig. 7), but are not confirmed rigorously against empirical results. Horizontal lines indicate the current estimate range for the threshold for biological production based on preliminary data, which is expected to lie somewhere between MA = 10 and 20.

Whether or not MA or other measures are adopted by the community as probabilistic biosignatures for life detection, the key feature is that the likelihood distributions for the signature to be produced by living vs. non-living processes should be as divergent as possible for a useful biosignature. Our two examples illustrate this quite well, as the distributions for molecular weight in abiotic and biological cases were similar and overlapping but for MA were quite different. We assumed a reasonably high prior probability for life, P (life) = 0.1, but if life is rare, the more discriminating probabilistic biosignatures will be not just useful, but necessary.

4. STATISTICALLY DISTINGUISHING LIVING CHEMICAL NETWORKS

In the foregoing examples we considered distributions over observables such as molecular weight and PA as possible probabilistic biosignatures that characterize properties across living systems and can be analyzed

within a Bayesian framework. We next turn to discussing statistical properties characterizing life itself as an emergent property of a complex web of interacting molecules. Many biological and technological systems share a similar web-like structure, being comprised of many components that are heterogeneously interconnected. Over the past few decades new statistical approaches have been developed to describe such systems. These go beyond the nineteenth century statistical physics of idealized non-interacting particles to include the topology of *interactions* among system components and their resultant dynamics (*Albert and Barabási,* 2002). These interacting systems are described mathematically with network theory, which projects the complex web of interactions in real systems onto an abstract representation as a graphical object (*Barabási et al.,* 2016). Networks are important mathematical descriptors in cases where the structure of interactions matters more than counting individual component parts, e.g., what we think happens as non-living matter transitions to life. Due to its utility in concisely describing complex, interacting systems,

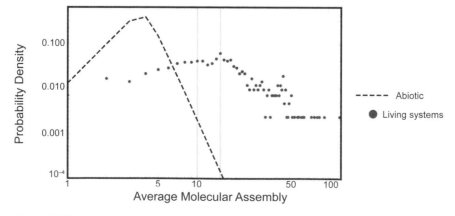

Fig. 10. Estimated probability density of molecular assembly values assuming the toy model in Fig. 9 and that the number of compositions for a given molecular weight is a good proxy for the frequency of observing molecules with that weight. Shown are abiotic and biological distributions. Gray vertical lines indicate the expectation that the threshold should be in the range MA ~ 10–15 from Fig. 7. Note the statistical assessment places the threshold slightly lower because of our low assumed prior probability for life.

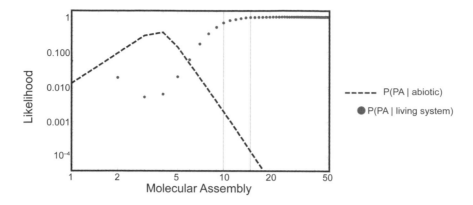

Fig. 11. Estimated likelihoods of abiotic and biological explanations generated from probability distributions in Fig. 10, calculated for a molecule as determined by its molecular assembly assuming a uniform prior probability of life $P_{life} = 0.1$.

network theory has been applied to an increasing number of systems in fields ranging from biology (e.g., *Barabasi and Oltvai,* 2004) to the social sciences (e.g., *Wasserman and Faust,* 1994). Adopting a network theory approach to characterizing living chemistries could therefore provide new insights into the statistical patterns that distinguish living chemical networks from non-living ones. This is because no individual molecules or reactions are alive, but taken together the *patterns* in how molecules interact through biochemical reactions may lead precisely to the emergent property we call life.

Mathematically, networks are studied using the tools of graph theory, where entities are represented by nodes (also called vertices) and their interactions by edges (also called links). Familiar examples include social networks, such as Twitter, where the nearly 126 million daily users could be mathematically represented by nodes and their relationships "following" one another by edges. In practice, it's computationally difficult to construct and analyze networks this large, but many networks of interest are smaller than Twitter, or subnetworks corresponding to specific communities or spatial locations can be studied. In a graph-theoretic representation of Twitter, an individual would be connected to every individual they "follow," and network dynamics might include studying how information propagates as new headlines or memes become viral.

For the purposes of applying network theory to life detection, astrobiologists are most likely not concerned with social networks (unless we are looking for intelligent and social aliens), but instead the properties of *chemical reaction networks* (CRNs), where nodes represent individual molecular species and edges represent the reactions between them. This is because chemistry is our first point of contact for future and planned life detection efforts. If network theory can be developed and validated as a tool that accurately captures features of the chemistry of living matter that are statistically distinguishable from those of non-living matter it could provide novel quantitative tools for inferring the presence of life.

One of the subtleties in network analysis is choosing the proper representation to accurately capture the relevant statistical properties of the systems of study. There are many different ways to graphically represent chemical systems, each permitting quantitative analysis of different aspects of their structure and organization. Chemical species reacting with one another can be represented by networks where nodes represent molecular species or reactions, and edges represent connections of molecular species to the reactions they participate in. Two commonly implemented network representations for chemical systems are shown in Fig. 12. Shown is a *bipartite*, reaction-substrate graph, where substrates (reactants and products) and their reactions are nodes, and edges connect substrates to their relevant reactions. Bipartite networks are so-called because there exist two distinct types of nodes in the network: Here, molecular species represent one type of node (circles) while chemical reactions represent the other (squares). Also shown is a *unipartite*, substrate-substrate graph, representation of the same chemical reaction network where reactions are abstracted away and reactants and products are directly connected by edges. In this representation, an edge between a pair of nodes can be thought of as a group of processes that convert some molecular species to others (*Jolley and Douglas,* 2012). A further refinement is these networks can be weighted by the relative strength of interactions, e.g., by adding weighting to the edges corresponding to the rates of reactions and modeling the network with what is called a *weighted network* representation (*Newman,* 2004), where edges have a strength of weight associated with them. There are many other types of network descriptors including multilayer networks, which contain different types of edges representing different connections (*Kivelä et al.,* 2014) that could be implemented for use in astrobiological searches for life. A review of many of the different representations of biochemical networks and their utility and shortcomings as applied to different scientific questions are discussed in *Montanez et al.* (2010).

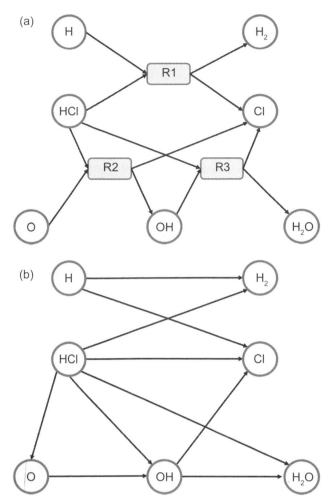

Fig. 12. Two different network representations of the same atmospheric reaction system, with reactions H + HCl → H₂ + Cl, HCl + O → Cl + OH, and HCl + OH → Cl + H₂O. **(a)** Bipartite substrate-reaction. **(b)** Unipartite substrate-substrate graph.

Once a representation is established, the topology and properties of a network can be inferred using a suite of mathematical measures. A common statistical property studied in graph theory is node degree, k, which counts the number of edges connected to the node. The *mean degree*, averaged over an entire network, provides a statistical measure of global topology and is mathematically defined as

$$\langle k \rangle = \frac{1}{N} \sum_{\forall i} k_i \qquad (13)$$

for a network with N nodes. In the context of chemical reaction networks, mean degree is equivalent to the average number of reactions any given species in the network participates in. The degree distribution, or degree sequence, is the probability distribution of node degree taken over an entire network, described by P (k), which is the frequency (probability) distribution of a node in the network having degree k. In the simple substrate-substrate graph of

Fig. 12, the degrees are 2, 2, 5, 3, 2, 4, 2 for compounds H, H₂, HCl, Cl, O, OH, and H₂O respectively, yielding a degree distribution of P (k) = 4/7, 1/7, 1/7, 1/7 for k = 2, 3, 4, 5 respectively. This distribution has a mean degree ⟨k⟩ = 2.86, and there is a tail of outlier nodes with a higher degree than the others. In this respect, the network is heterogeneous (this network is of course too small to make statistically meaningful statements, but it serves for illustrative purposes).

For decades it was thought most networks were homogeneous, but in the late 1990s and early 2000s as more data became available on real-world systems, it was discovered that most biological and technological networks are in fact very heterogeneous, with heavy-tailed degree distributions consistent with power-law or log-normal fits (see, e.g., *Barabási,* 2009). In many real-world networks, most nodes have very few connections, but a few nodes called "hubs" have many connections, and link less connected nodes together. A classic example of a heterogenous network would be the "hub and spoke" arrangement of many airlines, where there are only a few airports with a very high number of connections to other airports (hubs) where flights tend to be routed through due to their high connectivity. These types of networks are thought to have some measure of built-in redundancy, as the failure of any randomly selected node is unlikely to knock out large portions of the system (*Barabási et al.,* 2016). In a highly influential paper, *Jeong et al.* (2000) reported that metabolic networks of 43 distinct organisms sampled across all three domains of life are "scale-free." In scale-free networks, the degree distribution counting all the instances of nodes with specific k is heavy-tailed, and roughly follows a power law

$$P(k) \sim k^{-\alpha} \qquad (14)$$

An example of a scale-free network is shown in Figs. 13b,d, with an exemplary power-law "heavy-tailed" degree distribution.

As with social networks, metabolic networks also contain hubs, which include highly utilized molecules in biochemistry such as H₂O and ATP (*Wagner,* 2002). These molecules participate in hundreds of reactions, with a comparably high node degree, whereas the mean degree of metabolic networks globally is in the range of just α = 2–5 connections (*Jeong et al.,* 2000; *Kim et al.,* 2019). Projecting metabolism onto a substrate-substrate network representation yields fits for the degree sequence that in general follow a power-law fit, indicative of scale-free structure. However, rigorously confirming a power-law fit for a given degree distribution is a challenging technical problem and an active area of research in the network inference community. In recent years, new tools have been developed for reliably determining cases of scaling consistent with a true power-law behavior, as opposed to other heavy-tailed degree distributions, such as log-normal, revealing that scale-free networks may be rarer than previously supposed (see, e.g., *Broido and*

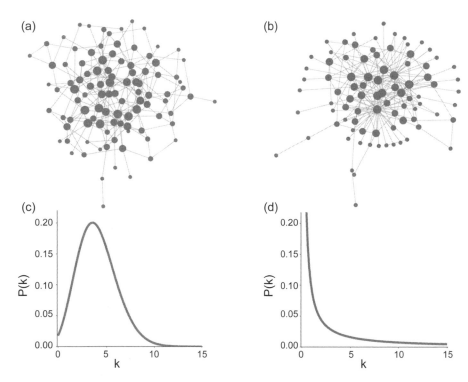

Fig. 13. Comparison of the topology of scale-free and Erdős-Renyii networks. Adopted from *Barabási et al.* (2016).

Clauset, 2019). For example, our recent analysis applying these tools to a dataset of >28,000 biochemical networks constructed from genomic and metagenomic data reveals that only a subset of biochemical networks can plausibly be fit to true power-law scaling (*Smith et al.,* 2019).

Network structure can depend on the network projection (e.g., its representation as unipartite or bipartite) (*Montanez et al.,* 2010), leading to complications in interpreting results of fits to degree distributions without reference to other properties of the network or randomized controls. Nonetheless, important structural differences between networks can often be seen directly from the degree distribution and other topological measures. In many cases, these are indicative of properties that seem to be distinctive to living networks. For example, random networks, such as Erdős-Rényi (ER) networks (*Erdős and Rényi,* 1960), are characterized by degree distributions that are Poisson distributed, such that most nodes share roughly the same number of edges and the probability of finding nodes with very high degree is exponentially small

$$P(k) \sim e^{-k} \text{ for } k \gg 1 \tag{15}$$

Random networks are described as homogenous in their distribution of edges among nodes since most nodes share similar degree. An example of a homogenous network is shown Figs. 13a,c with an exemplary Poisson degree distribution describing the degree sequence. As one can see in Fig. 13, network structure and degree distribution are visually very different for homogeneous networks when compared to heavy-tailed or scale-free networks. Due to these structural differences, the systematic observations of heavy-tailed networks (independent of precise fit) across biological and technological provides a window into their statistical properties that distinguishes living networks from "random" ones (see section 4.1).

There are other network measures that could aid in efforts to apply network theory to the astrobiological problem of distinguishing statistical properties of living and non-living systems. One such measure is *average neighbor degree*, $\langle k_{nn} \rangle$, which is (as the name suggests) the average degree of the nodes in the neighborhood of a given node, defined as

$$\langle k_{nn} \rangle \frac{1}{|N(i)|} \sum_{j \in N(i)} k_j \tag{16}$$

This can be useful in determining how densely connected a given node is to the rest of the network. Another measure, *average shortest path length*, $\langle \ell \rangle$, defined as

$$\langle \ell \rangle = \frac{1}{N(N-1)} \sum_{\forall i,j} d(i,j) \tag{17}$$

represents the average minimum distance between any nodes i, j. Chemically, it represents the average smallest number of intermediate reactions between any two species. Another important statistic is *betweenness centrality*, which measures how often a particular node is on the shortest path between all other nodes in the network. Nodes with high betweenness centrality can sometimes be low degree, but nonetheless essential to dynamics and function since they play a key structural role by connecting many otherwise disconnected or distant nodes. High betweenness centrality is often correlated with high degree (hubs). In biochemical networks, molecules like H_2O and ATP tend to have both very high degrees and high centrality, due to their fundamental roles in aqueous organic chemistry and in metabolism, respectively. *Node betweenness centrality*, $\langle g(v) \rangle$, of node v is mathematically defined as

$$\langle g(v) \rangle = \sum_{s \neq v \neq t} \frac{\sigma_{st}(v)}{\sigma_{st}} \qquad (18)$$

here σ_{st} is the total number of shortest paths from node s to node t, and $\sigma_{st}(v)$ is the number of those paths that pass through v. In a chemical network, this measure quantifies how important a given molecule is in mediating multi-step reactions occurring across the network (i.e., how many shortest reaction sequences between two other molecules the given molecule participates in). Many of the measures discussed so far track statistical properties of individual nodes, or paths between two nodes, but there are also many measures for higher-order properties of networks. For example, clustering coefficient tracks how many tightly knit communities exist within a given network, typically measured by counting the number of complete triangles connecting three nodes. The *average clustering coefficient* $\langle C \rangle$ is defined as

$$\langle C \rangle = \frac{1}{N} \sum_{\forall i} \frac{2e_i}{E_i (E_i - 1)} \qquad (19)$$

and describes how densely linked the communities in the immediate neighborhood of a node are. In chemical reaction networks, this can be thought of as how likely it is that species that participate in a reaction together will participate in other reactions together. Networks with high clustering coefficients have many clusters of nodes with above average connections between them (relative to the rest of the network). Complete triangles represent one example of a network motif. Network motifs are subgraphs that have specific connection patterns, and that are overrepresented in biological systems with respect to randomized networks (*Milo et al.,* 2002). They were first uncovered in networks as diverse as biochemistry, ecology, neurobiology, and engineering, and have been proposed as a means to uncover the building blocks of functional networks (*Alon,* 2003). From this perspective, network motifs are an important concept for biosignature development; identifying the network motifs that readily form under abiotic conditions vs. those that combine to form more complex, "lifelike" systems would advance our understanding of key structural properties needed for identifying distinctive properties of living networks.

4.1. Statistical Approaches to Characterizing Biochemical Reaction Networks

A major challenge for any claims about universal properties of life is the common ancestry of all life on Earth (*Pace,* 2001; *Woese,* 1998). Discovery of alien life would obviously enable us to identify any universalities, if they exist. However, in the absence of such a discovery, can astrobiologists confidently make claims of universality? This question is among the most important in astrobiology: In order to search for alien life we must know what to look for. This implies we cannot wait until the discovery of a second sample of life to understand life's universal properties if we are to most effectively conduct our search for alien life. This necessitates developing better quantitative methods for inferring universal properties from the examples of life we already have on Earth (see also that discussion in the chapter by Hoehler et al. in this volume).

A key advantage of a network theoretic approach is that it can be used as a quantitative method to extrapolate away from known biochemistry because it deals with the statistical properties of interacting molecules and reactions rather than their precise properties. Shifting thinking from the specific molecules of life as we know it to the statistical properties of living chemistries can be accomplished by studying probabilistically the properties of biochemical networks across different examples of life on Earth. In fact, the entire biosphere can be studied as a chemical reaction network (see Fig. 14), allowing a new window into studying statistical regularities that occur across all of known biochemistry. In particular, we have proposed in prior work a network approach to quantifying universal properties of life that permits comparison of living systems across multiple scales including individuals and ecosystems for common features (*Kim et al.,* 2019; *Smith et al.,* 2019). If common patterns are found in how living matter organizes across scales within the biosphere, it increases our confidence that those patterns are derivative of universal laws, rather than being merely a product of shared common ancestry (see Fig. 15).

We recently took a "big data" approach to characterize the properties of biochemical networks across multiple levels of organization in the biosphere ranging from the chemical reaction networks within cells, to ecosystems, to the biosphere as a whole (*Kim et al.,* 2019). Biochemical reaction networks were constructed using annotated genomic data (see, e.g., methods in *Jeong et al.,* 2000), collected from the Joint Genome Institute (JGI) and Pathosystems Resource Integration Center (PATRIC) (*Wattam et al.,* 2016) databases, including annotated genomes from 21,637 bacteria taxa, 845 archaea taxa, 77 eukaryotic taxa, and 5587 annotated metagenomes. Annotation provides information on the enzyme commission (EC) numbers associated with a given gene, i.e., it provides a map from genomic information to the likely enzymes encoded in that genome. Known enzymes are cataloged by their EC number in the KEGG database (*Kanehisa and Goto,* 2000) with detailed information on the reactions they catalyze. Therefore for annotated data

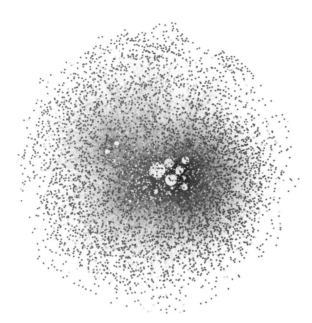

Fig. 14. Earth's biosphere, comprising all known catalyzed biochemical reactions cataloged in the Kyoto Encyclopedia of Genes and Genomes (KEGG), represented as substrate-substrate network where nodes (substrates) are connected by an edge if they participate in a common reaction. Node size corresponds to relative degree, such that the largest nodes (shown in white) are have the highest degree, e.g., are "hubs" (these are molecules such as H_2O and ATP that participate in many biochemical reactions). The biosphere visually demonstrates a key characteristic of many heterogenous networks with heavy-tail degree distributions; a few highly connected hubs are connected to many more lower-degree nodes.

we were able to construct the chemical reaction networks catalyzed by the organism(s) as encoded in their genome or metagenome [see *Kim et al.* (2019) for a discussion of limitations of annotated data and the robustness of network models to those limitations]. The biosphere-level network was constructed by constructing a network including *all* enzymatically catalyzed reactions cataloged in the KEGG database (*Kanehisa and Goto,* 2000) (see Fig. 14).

Scaling laws, relating properties such as metabolic rate to body size, represent one of the most consistent known mathematical regularities across all life on Earth and are often cited as a candidate for universal biology as they unify trends across different biological organisms and scales of organization (*West et al.,* 2002; *Gisiger,* 2001).

Familiar examples of scaling behavior from physics include critical phenomena near-phase transitions, where physical properties such as heat capacity, correlation length, and susceptibility all follow power-law behavior. Scaling relations, due to their ability to predict the values of system parameters based on other measured quantities, represent one of the closest approaches so far to a predictive theoretical (astro)biology, akin to theoretical physics [see, e.g., *Walker et al.* (2018) for a discussion of future

applications of scaling laws to exoplanet biosignatures]. Analyzing scaling of topological properties of biochemical networks as a function of their size (number of compounds) reveals universal structural properties across biochemical networks on Earth.

Randomly sampling reactions from known biochemistry to construct networks of similar size to organismal and ecosystem-level biochemical networks does not reproduce the scaling observed for living biochemical networks (see Fig. 16). This suggests it is possible to *statistically* distinguish chemical networks drawn from living (evolved) systems from chemical networks that are random (unevolved), and indeed in *Kim et al.* (2019) it was shown that average topological measures and network size (number of compounds) was alone enough to predict with high accuracy the evolutionary domain (archaea, bacteria, or eukarya) of a given biochemical network, validating that aggregate network measures can statistically distinguish different examples of life. The differences between biochemical and random networks indicate it is the particular manner in which reactions are organized in living matter, and not the compounds or set of reactions alone, which yield the distinctive properties of living systems. Using the observation that cells and organisms are constrained in their growth by resource distribution networks, predictive models can be generated that accurately provide values for the scaling exponents observed in a number of diverse biological systems (*West et al.,* 1997). Similar predictive models should be generated for biochemical network scaling (a work in progress). These would provide insights into universal constraints on biochemical architecture that could be used as a statistical signature of living chemistries and adapted within Bayesian or alternative statistical frameworks.

4.2. Statistical Approaches to Characterizing Atmospheric Chemical Reaction Networks

Exoplanet science has rapidly matured from an era of detection to an era of characterization, with atmospheric characterization leading the current frontier. Studying exoplanetary atmospheres enables determining fundamental properties of planets, including composition, climate, formation avenues, whether or not they can host life (their habitability), and (ultimately) whether or not they do host life. Even with the limited data we have obtained so far, a surprise has been the diversity of exoplanet atmosphere properties, many of which have no solar system analogs. In the next decade, we are poised to see significant advances in our understanding of the atmospheres of ultra-irradiated hot to temperate terrestrial worlds with missions including the James Webb Space Telescope, ESA's Atmospheric Remote-sensing Infrared Exoplanet Large-survey (ARIEL) and groundbased high-resolution spectroscopy with the 30-m-class telescopes, and other proposed missions such as the Origins Space Telescope, the Large UV-Optical Infrared Telescope, and the Habitable Worlds Explorer (*Cowan et al.,* 2015).

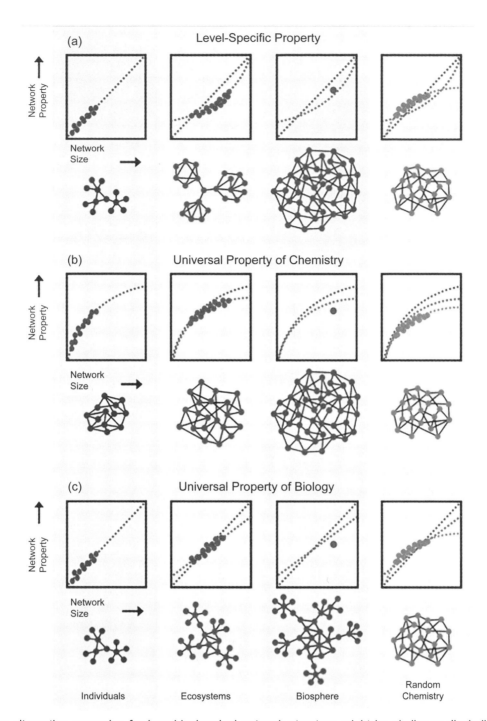

Fig. 15. Three alternative scenarios for how biochemical network structure might be similar or dissimilar from random chemistries: **(a)** Biochemistry does not exhibit common network structure across levels; **(b)** biochemistry has a common network structure across all levels, but this structure is also shared by random chemical networks (chemical network structure is universal); and **(c)** biochemistry has shared structure across all levels, which is different from that of random chemical networks (biochemical network structure is universal but distinguishable from random chemistry). From *Kim et al.* (2019).

While the broad diversity in planetary properties observed so far is daunting, it also presents a tremendous opportunity to develop new statistical frameworks for future life detection efforts. One such approach is to apply network theory as a statistical tool to characterize atmospheric properties. In other words, one can apply networks as a statistical tool to characterize the atmospheric chemistry of non-living and living worlds in direct analogy to how

networks are applied to characterize non-living and living chemistries as described in section 4. Prior work by *Solé and Muntaneu* (2004) has demonstrated that the chemical composition of planetary atmospheric chemical networks has a quantifiable impact on their topology, specifically highlighting how Earth's atmospheric chemistry exhibits properties differing from other atmospheres in our solar system (*Solé and Munteanu,* 2004). Their analysis revealed that Earth's atmospheric network shares similar heavy-tailed heterogenous structural features to that of biological networks (e.g., like Fig. 13d), in agreement with an earlier report (*Gleiss et al.,* 2001). In their study, other planetary atmospheres — including those of Venus, Mars, Titan, and the jovian planets — were shown to have a different, more homogeneous network structure more like an ER network.

Complex networks are difficult to visualize in a manner that readily reveals statistical features while minimizing

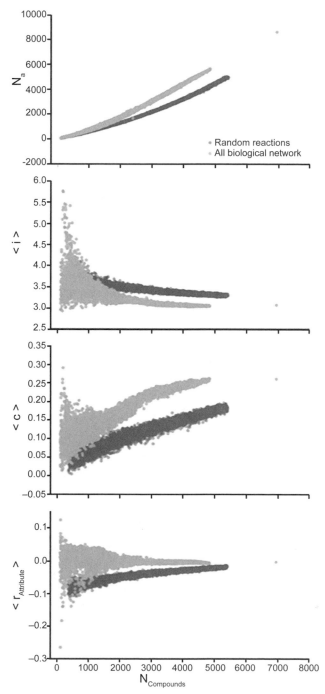

Fig. 16. Scaling of network attributes with network size. Biological networks scale differently from random collections of biochemical reactions. Adapted from *Kim et al.* (2019).

visual artifacts (for example, the network representation of the biosphere in Fig. 14, while visually appealing, can mask important features such as clusters of highly connected nodes or other important structural features). Hive plots provide an alternative visualization method for drawing networks, where the nodes are mapped to and positioned along radially distributed linear axes, clearing showing statistical properties of the network such as the structure of the degree distribution. Hive plots presenting a visualization of the degree distribution for the atmospheric networks of Mars and Earth from *Solé and Munteanu* (2004) are shown in Fig. 17, where each axis includes nodes with values for their degree k in a specific range, and connections between the axes indicate when nodes are connected that are in different degree ranges. Visually this shows how statistical patterns among the interactions of molecules in the two atmospheres differ.

Presumably the distinctive organization of Earth's atmosphere as compared to Mars and other atmospheres in our solar system arises due to the nonlinear coupling between Earth's atmosphere and biosphere, which generates structural properties in Earth's atmospheric chemical reaction network that are similar to those observed across other biological networks. However, the results of *Solé and Munteanu* (2004), while suggestive, are not conclusive. Modeling of planetary atmospheres is sensitive to the included reaction list, as well as properties of the chemical and physical environment such as temperature, pressure, UV environment, etc. To rigorously determine whether the atmospheres of living planets are indeed statistically distinct in terms of their reaction network topology from non-living planets, ensemble approaches are necessary. Since we do not have an ensemble of living worlds (or even non-living worlds) to do such analyses, the path forward is to couple complex systems approaches to ongoing modeling efforts of the exoplanet community to study the properties of planets with and without biology across a variety of planetary and stellar contexts to determine if we can rigorously adopt network biosignatures as a statistical signature of an inhabited planet.

If it can indeed be found that life produces an indelible imprint on the topology of planetary atmospheric chemical reaction networks, a further challenge will be mapping the relevant biosignature network properties (which could be measures such as those discussed in section 4 or other statistical properties of the networks) to remotely detectable observables. There is some cause for optimism that remotely detectable observables, such as abundant atmospheric O_2, will allow inferring properties of the underlying atmospheric network. As discussed in the introduction, O_2 has long been considered the prime candidate biosignature gas for exoplanets. More recently, it was recognized that additional context beyond just the presence of O_2 is, however, necessary to determine biogenecity (*Meadows,* 2017; *Meadows et al.,* 2018). Currently the community is considering planetary and stellar environment as additional contextual information, including detecting the presence of other gases like CO or CO_2, or lack of other gases such as CH_4 or N_2, which depending on the context could indicate the O_2 is produced

(a)
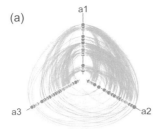

(b)
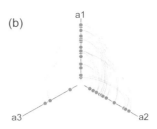

Fig. 17. Hive plots comparing properties of atmospheric chemical reaction networks generated from models of **(a)** Earth's and **(b)** Mars' atmospheres [models derived from *Solé and Munteanu* (2004)]. Nodes are organized along three axes, where each axis includes nodes whose degree k is within a specified range: a1 includes nodes with k < 2, a2 includes nodes with $2 \leq k \leq 5$, and a3 includes nodes with k > 5. Of note, Earth's atmosphere contains both more high-degree nodes and more connectivity of these nodes to nodes of varying degree, consistent with the heterogenous structure reported in Solé and Munteanu. However, it is difficult to directly compare the two networks given their different sizes and differences in the confidence of our models (for example, the reaction list is much more constrained and detailed for Earth than it is for Mars). Statistical ensembles of planetary atmospheres modeled for inhabited and uninhabited planets must be developed to make a rigorous comparison.

by life rather than an abiotic mechanism (see *Meadows et al.*, 2018). Network properties could also provide additional context. It is already known that the presence of O_2 has a dramatic impact on the network structure of the biosphere (Fig. 14). Studying the impact of O_2 on the evolution of complex life, *Raymond and Segrè* (2006) demonstrated removal of O_2 from the biosphere-level network generated from KEGG, and showed that large fractions of the biosphere-level network — even compounds that do not contain O atoms — are inaccessible without molecular O_2. The effect of oxygen was to lead to a massive expansion in possible reactions, not all of which directly involve O_2 as a substrate, which in turn leads to an expansion in the size of the biosphere network. The scaling relations reported by *Kim et al.* (2019) for biochemical network topology suggest that both aerobic and anaerobic organisms share the same overall average topology, scaled by network size (with aerobes having larger networks). It remains to be verified if (1) topological properties of biochemistries that have never been in contact with O_2 statistically differ from those that have; and (2) whether we could distinguish the topology of a O_2 network based on the sources of molecular oxygen, e.g., whether biological or produced by photolysis. Presumably these two mechanisms will have different effects on the kinds of reactions taking place that would leave an imprint on the global topology. More work remains to be done to determine the potential of network biosignatures for planetary atmospheres.

5. CONSTRAINING PRIORS

We are continually improving our understanding of abiotic planetary contexts and expanding our repertoire of candidate biosignatures, as detailed above. However, no matter how refined our models for determining the likelihood of biological and abiotic sources for a given candidate biosignature become, we will continue to be constrained in our confidence in detection of life unless we can better understand the environments and contexts where life is most likely to emerge and evolve. This can be done by constraining P (life) across different environments and contexts.

While we remain uncertain about what life *is* at a fundamental level and how it emerged, an origin event can be considered a minimal instantiation of life, and the phenomena and states we associate with it (*Scharf et al.*, 2015, *Mathis et al.*, 2017). If we can identify the minimum necessary requirements for even a single such origin event, we can plausibly begin to constrain the likelihood of life's presence; i.e., P (life) in equation (1) from that of its emergence (*Chen and Kipping*, 2018; *Scharf and Cronin*, 2016). For example, if the development of an autocatalytic set, or the production of lipid vesicles, were *necessary* precursor events for life's emergence, then theoretically or experimentally bounding the likelihood of these events over a given region of geochemical or cosmochemical space would be informative.

Much effort has been expended in investigating proposed minimal requirements for life's origins and constraining the independent and joint likelihoods of such requirements from physical priors informed by early Earth settings. Emphases range from mechanisms promoting the assembly and concentration of component structures (*Chen and Walde*, 2010; *Hazen and Sverjensky*, 2010), to the potential for component processes or their predecessors to evolve and proliferate before coupling (*Vasas et al.*, 2009; *Hordijk et al.*, 2018), to the structure and dynamics of protocellular systems (*Adamala and Luisi*, 2011). Models of artificial life have also enabled *in silico* investigations of proposed origin sequences (*CG et al.*, 2017), although this has historically proven challenging (*Bedau et al.*, 2000).

Within each of the above approaches is embedded a slightly different interpretation of what constitutes a *minimally* lifelike system or life-relevant process (*Scharf et al.*, 2015). There is no single quantitative consensus about the marginal gain in "lifelike" character for any given attribute of life, and one discipline's proposed minimal satisfactory representation of "life" might fail to capture what another discipline considers a fundamental aspect of living systems or aspect of life's origins. Depending on how "life" and life's origins are framed, a clear understanding about the process by which life originated on Earth may not suggest an obvious, self-evident definition of "life" (*Szostak,* 2012). Which types of factors, for example, can be considered irreducibly or uniquely fundamental to living systems: component parts, individual-or-population-scale processes, individual or collective energetics, or network or informational character? At what stage in life's trajectory toward "life as we know it" did a complex dynamical system transition to a living state occur? Are viruses alive? What about synthetic cells? Would an "RNA world" be a living world? Origins hypotheses mandating a specific *order* of process innovation, e.g., strict interpretations of "genetics-first" or "metabolism-first" hypotheses (*Joyce and Orgel,* 2006), add aspects of probabilistic conditionality to what would otherwise be independent minimal requirement likelihoods.

The most obvious trigger of consensus about these *actual* minimal requirements would be an emergence of life *de novo* in a laboratory setting. This has proved elusive. We may not have coupled the right requirements yet, or if we have, we may have captured the resulting outcome at uninformative scales using the methods of a single discipline. It may be that we have been overly conservative in the scope of our research (e.g., focusing too much on "life as we know it"), such that efforts to ensure experimental settings are "prebiotically plausible" have disincentivized, or diminished insights from, alternative complex systems. It is worth noting that the metric of "prebiotic plausibility" has shifted with our understanding (*Benner,* 2018) (much like notions of habitability) and will continue to shift as we learn about plausible prebiotic settings in extrasolar environments.

However, we need not wait for a successful independent-origin event to start exploring the effects of possible successful outcomes. The laws of physics and chemistry are continuous across disciplines, allowing for frameworks that unify insights from many disciplines across many scales (*Smith and Morowitz,* 2016). At this stage in origins research, fruitful insights may emerge from different disciplines characterizing the same systems at *different* levels of coarse-graining, at *different* scales, and with *different* frameworks for understanding systems and codifying understanding. If life represents a fundamental physical phenomena or an expression of other fundamental phenomena, its outcomes might be sufficiently approximated from many angles. Or they might be captured precisely within a multivariate framework with several equivalent expressions. The goal is to arrive at a multivariate framework for constraining P (life), as a function of the independent and joint probabilities of each of life's actual required attributes. A truly minimal representation might involve parts of multiple proposed representations, or it might reflect more general underlying principles, of which each discipline's formalism is a special case.

Whatever the case may be, future work seeking to constrain the basic requirements for an origin event would do well to frame these requirements in the context of their broad (i.e., cosmic) likelihoods. We may find that several hypotheses, if eventually shown to be correct, would suggest a similar cosmic likelihood, despite being informed by vastly different methodologies. Alternatively, we may find patterns or clusters in the likelihoods projected, suggesting potential correlations or methodological biases. Either outcome would offer a starting point for constraining P (life) while we work toward more comprehensive fundamental understanding.

Another approach to bounding the likelihood of life, and one that is abstracted from life's minimum requirements, would involve bounding the likelihood of "lifelike" states or system-level properties, whether or not they relate to hypotheses about life's emergence. If, for example, life's *present* use of chemistry is fundamentally an informational phenomena, even at a biosphere level, then biochemistry's bulk statistical properties (e.g., see section 4.1) could be used to estimate the frequency or uniqueness of comparable states. Such a frequency bound might not provide *robust* limits in itself, but it would allow us to place life in a context with potentially analogous complex systems and structures.

A subset of this method involves monitoring the "lifelike" statistical character of mixtures emerging from unconstrained ensemble chemical reactions in a "messy chemistry" approach motivated by the emergence of living systems in a complex, stochastic planetary environment (*Guttenberg et al.,* 2017). In the messy chemistry framework, chemical mixtures are permitted to evolve in minimal analog planetary environments comprised of water, dissolved salts, organic compounds, and specified mineral substrates on which prebiotic assembly might take place. Unlike directed prebiotic synthesis experiments, reaction space is not constrained along specific moleculars, or even monitored for traversals through *specific* regions of chemical space. Instead, product *distributions* are monitored, more readily allowing for statistical comparisons to messy biological datasets.

This emphasis on ensemble properties over specific products and moleculars also enables a wider exploration of chemical space than that considered "prebiotically plausible" for early Earth. Complex product ensembles embed statistically distinguishable signatures of their environmental history (*Surman et al.,* 2019). Comparisons between product distributions with slightly different known histories could offer insights into how various mechanistic controls identified in less-coarse-grained experimental settings might actually translate to the ensemble properties of prebiotic environments.

As stated earlier, no single coherent metric is universally acknowledged to impart upon any system the quality of being "lifelike"; some attributes of living systems may be difficult to assess in messy chemistry settings. However, many *generic* attributes of messy and assuredly biochemical mixtures have been identified (*Summons et al.,* 2008).

Drawing on these mixture-level attributes, and on information-based metrics, may permit more robust parallels to be drawn from messy chemistry to biochemistry.

To summarize, the following are a few lines of inquiry relevant to constraining the likelihood of life's emergence, or of the presence of complex mixtures sharing qualities with life.

Top-down, specifically: What fraction of chemical space, and what portion of geochemical and cosmochemical environments [e.g., *Shock et al.* (2019) for early Earth], are themselves amenable to:

- Specific chemicals (e.g., liquid water) — if we have reason to believe these offer uniquely advantageous system-level properties
- Specific types of chemicals (e.g., amino acids) — if we believe these are uniquely effective at imparting properties fundamental to living systems
- Specific molecules, directions, or types of reactions (e.g., carboxylation and decarboxylation)
- Specific reaction networks or system-level structures [e.g., production of metabolic precursors (*Muchowska et al.*, 2019)]

Bottom-up, generically: What can we infer about the fraction of chemistry amenable to life, from *generic* properties of the chemical space it has explored (*Summons et al.*, 2008)?

- Generic chemicals (e.g., C, Si, metal cofactors) — What, besides availability, fundamentally limits or guides life's use of chemistry?
- Generic chemical patterns (e.g., C number) — How non-random is biochemistry in an abstract sense, and how does this non-randomness differ from non-random systematics in nonliving systems and processes?
- Generic mixture demographics (e.g., portion of straight-chain or branched-chain polymers)
- Informational attributes (e.g., asymmetry, genetic inheritance, molecular recognition) — What range of information content/density/character has life chosen to explore? How non-unique are the kinds of information architectures and networks life leverages?

Addressing these questions will require a concerted effort by the origins community to constrain P (life) across diverse planetary environments, thus providing a critical set of quantitative bounds on the likelihood for life to emerge in different planetary contexts.

6. UNIFYING THE SEARCH FOR LIFE ACROSS DIVERSE PLANETARY ENVIRONMENTS

Astrobiologists have ambitious plans to search for signs of life in myriad environments near and far from our home planet. In this chapter, we have reviewed how a quantitative, probabilistic approach to biosignature science provided many benefits to a broad cross-section of astrobiology, from exoplanet and solar system biosignatures to the search for the origin of life in the lab. There remains one additional benefit, which we now discuss — the ability to unify these searches under a common quantitative framework.

Despite the common goal between the search for life in the solar system and the search for life beyond, many reject any comparative work across the two endeavors due to their stark differences. The search for life in the solar system will primarily rely on *in situ* techniques that measure the composition of a soil or water sample in tremendous detail, so that individual or small assemblages of microbes, and/or their complex molecular constituents or byproducts, can be identified. Compare this with the search for life on exoplanets, which will primarily rely on remote detection of the major constituents of a planet's atmosphere and the dominant reflection features from the planet's surface, so that global-scale biospheres and their impacts on the planet's atmosphere and/or surface can be identified. These approaches differ in the spatial scale we must probe of the respective target environments, the spatial scale over which they require biology to be active to produce a detectable signal, the details to which they assess that environment, and even in their ability to detect extant vs. extinct life. As a result, they call for different kinds of measurements, lead to the development of different instruments, and create differing mission strategies for searching for life. The search for life in the lab has so far hardly made contact with these efforts as most origin of life efforts to date have focused on anthropocentric searches for producing one or a few molecules important to life on Earth, synthesized under conditions plausible on the prebiotic Earth. However, a focus on such dissimilarities ignores the commonalities between these endeavors, and the degree to which all of them benefit from each other.

There seems to be general agreement that these searches for life are at the very least linked theoretically. A widely accepted discovery of life on any one world or in the lab makes the detection of life on another world less sensational and more likely to be accepted. Similarly, the search for life in the varied environments expressed in the solar system and expected beyond it offer the potential to understand what life is and how it manifests as a function of its environment. Finally, an improvement of our knowledge of non-biological planetary processes that act in one environment can inform our ability to understand that same process in the other environment.

Probabilistic frameworks for life detection have the ability to simultaneously acknowledge the differences between these approaches, while also highlighting how they are related. Such approaches allow for comparisons between otherwise disparate biosignature observations, instruments, standards, and techniques. They also provide a pathway for the results from the search for life on one target to inform the assessment of the search for life elsewhere. For example, the likelihoods of biological or abiotic production are general frameworks that can be assessed across a wide variety of environments, both in the solar system and on exoplanets, once they are developed for a given candidate biosignature. Additionally, statistical frameworks like molecular composition, MA, and network measures as explored here can all be applied both *in situ* within the solar system, and remotely for exoplanet searches. The generality of these agnostic approaches is their strength; the same statistical measures can be used for chemical soups or planetary atmospheres, with their capacity to infer biogenecity being intimately tied to the context of the measurements being taken, other data available, and relevant priors. This generality allows unification of our biosignatures and what they tell us about what life is and the properties of inhabited worlds.

If we can successfully create an origin of life event *de novo* in the lab, it would provide incredible information about the processes underlying life's emergence in our search for other such events. Conversely, constraints from searches for life on other worlds, whether successful or not, will provide important constraints on emergence of life experiments by merging statistical approaches to life detection with statistical approaches to exploring its origins. An improved understanding of lifeless environments in either setting would also improve our ability to model abiotic processes and their likelihood to generate data observed elsewhere. Finally, by using a unified approach, the advantages and disadvantages of the approach itself will be shared and commonly advanced across these efforts.

Probabilistic frameworks ultimately provide a quantitative means of not only locating life, but also being capable of adapting to changing experimental, theoretical, and observational variables over time with the improvement of new instruments and better techniques. Unlike more traditional approaches, statistical methods allow also bounding the probability of life *not* inhabiting certain environments (giving us information even if no discovery is made), as well as assessing the likelihood of life having existed in an environment in the past. Both are equally important when trying to locate life and understand its fundamental properties. The possibilities for statistical approaches to astrobiological questions are quite limitless, as their application, when directed properly, can be used for planning missions, developing new tools and techniques, and avoiding false positives, and most importantly, they can be applied universally across the board regardless of what is being analyzed. Thus, a statistical framework provides the first quantitative steps toward unifying astrobiology as a discipline, uniting our search for new examples of life at home, close by, and light years away from us. In a best-case scenario, this would usher in a new era of *comparative* astrobiology that could yield observation-driven theories on habitability and life in a planetary context. Astrobiology is poised to enter a new frontier to actually solve the problem of understanding what life is, how it originates, and how widespread it is in the universe by recognizing that these are not separate problems but the same problem all astrobiologists share an interest in solving.

Acknowledgments. S.I.W. was supported by NASA Exobiology grant NNX15AL24G. This work benefited from collaborations and/or information exchange within NASA's Nexus for Exoplanet System Science (NExSS) research coordination network, sponsored by NASA's Science Mission Directorate.

REFERENCES

Adamala K. and Luisi P. L. (2011) Experimental systems to explore life origin: Perspectives for understanding primitive mechanisms of cell division. In *Cell Cycle in Development* (J. Z. Kubiak, ed.), pp. 1–9. Springer, Berlin.

Albert R. and Barabási A.-L. (2002) Statistical mechanics of complex networks. *Rev. Mod. Phys., 74(1),* 47.

Alon U. (2003) Biological networks: The tinkerer as an engineer. *Science, 301(5641),* 1866–1867.

Amils R., González-Toril E., Fernández-Remolar D., Gómez F., Aguilera A., Rodríguez, N., Malki M., et al. (2007) Extreme environments as Mars terrestrial analogs: The Rio Tinto case. *Planet. Space Sci., 55(3),* 370–381.

Barabási A.-L. (2009) Scale-free networks: A decade and beyond. *Science, 325(5939),* 412–413.

Barabási A.-L. and Oltvai Z. N. (2004) Network biology: Understanding the cell's functional organization. *Nature Rev. Genet., 5(2),* 101.

Barabási A.-L. et al. (2016) *Network Science.* Cambridge Univ., Cambridge. 456 pp.

Bedau M. A., McCaskill J. S., Packard N. H., Rasmussen S., Adami C., Green D. G., Ikegami T., Kaneko K., and Ray T. S. (2000) Open problems in artificial life. *Artif. Life, 6(4),* 363–376.

Benner S. A. (2018) Prebiotic plausibility and networks of paradox-resolving independent models. *Nature Commun., 9,* 5173.

Boyajian T., LaCourse D., Rappaport S., Fabrycky D., Fischer D., Gandolfi D., Kennedy G. M., et al. (2016) Planet hunters IX. KIC 8462852 — Where's the flux? *Mon. Not. R. Astron. Soc., 457(4),* 3988–4004.

Broido A. D. and Clauset A. (2019) Scale-free networks are rare. *Nature Commun., 10(1),* 1017.

Catling D. C., Krissansen-Totton J., Kiang N. Y., Crisp D., Robinson T. D., DasSarma S., Rushby A. J., Del Genio A., Bains W., and Domagal-Goldman S. (2018) Exoplanet biosignatures: A framework for their assessment. *Astrobiology, 18(6),* 709–738.

Cavicchioli R. (2002) Extremophiles and the search for extraterrestrial life. *Astrobiology, 2(3),* 281–292.

CG N., LaBar T., Hintze A., and Adami C. (2017) Origin of life in a digital microcosm. *Philos. Trans. R. Soc., A, 375(2109),* 20160350.

Chen I. A. and Walde P. (2010) From self-assembled vesicles to protocells. *Cold Spring Harbor Perspect. Biol., 2(7).*

Chen J. and Kipping D. (2018) On the rate of abiogenesis from a Bayesian informatics perspective. *Astrobiology, 18(12),* 1574–1584.

Cowan N., Greene T., Angerhausen D., Batalha N., Clampin M., Colón K., Crossfield I., et al. (2015) Characterizing transiting planet atmospheres through 2025. *Publ. Astron. Soc. Pac., 127(949),* 311.

Cronin L. and Walker S. I. (2016) Beyond prebiotic chemistry. *Science, 352(6290),* 1174–1175.

Davies P. C. and Walker S. I. (2016) The hidden simplicity of biology. *Rept. Prog. Phys., 79(10),* 102601.

Domagal-Goldman S. D., Segura A., Claire M. W., Robinson T. D., and Meadows V. S. (2014) Abiotic ozone and oxygen in atmospheres similar to prebiotic Earth. *Astrophys. J., 792(2),* 90.

Erdős P. and Rényi A. (1960) On the evolution of random graphs. *Publ. Math. Inst. Hung. Acad. Sci., 5(1),* 17–60.

Farr B., Farr W. M., Cowan N. B., Haggard H. M., and Robinson T. (2018) Exocartographer: A Bayesian framework for mapping exoplanets in reflected light. *Astron. J., 156(4),* 146.

Fisher T. M. and Schulze-Makuch D. (2013) Nutrient and population dynamics in a subglacial reservoir: A simulation case study of the Blood Falls ecosystem with implications for astrobiology. *Intl. J. Astrobiol., 12(04),* 304–311.

Foley B. J. and Driscoll P. E. (2016) Whole planet coupling between climate, mantle, and core: Implications for rocky planet evolution. *Geochem., Geophys., Geosyst., 17(5),* 1885–1914.

Fortney J. J., Robinson T. D., Domagal-Goldman S., Amundsen D. S., Brogi M., Claire M., Crisp D., et al. (2016) The need for laboratory work to aid in the understanding of exoplanetary atmospheres. *ArXiv e-print,* arXiv:1602.06305.

Fujii Y., Angerhausen D., Deitrick R., Domagal-Goldman S., Grenfell

J. L., Hori Y., Kane S. R., et al. (2018) Exoplanet biosignatures: Observational prospects. *Astrobiology, 18(6),* 739–778.

Gisiger T. (2001) Scale invariance in biology: Coincidence or footprint of a universal mechanism? *Biol. Rev. Cambridge Philos. Soc., 76(2),* 161–209.

Gleiss P. M., Stadler P. F., Wagner A., and Fell D. A. (2001) Relevant cycles in chemical reaction networks. *Adv. Complex Syst., 4(02n03),* 207–226.

Guttenberg N., Virgo N., Chandru K., Scharf C., and Mamajanov I. (2017) Bulk measurements of messy chemistries are needed for a theory of the origins of life. *Philos. Trans. R. Soc., A, 375(2109),* 20160347.

Hazen R. M. and Sverjensky D. A. (2010) Mineral surfaces, geochemical complexities, and the origins of life. *Cold Spring Harbor Perspect. Biol., 2(5).*

Hjort N. L. et al. (1990) Nonparametric Bayes estimators based on beta processes in models for life history data. *Annu. Stat., 18(3),* 1259–1294.

Hjort N. L., Holmes C., Müller P., and Walker S. G., eds. (2010) *Bayesian Nonparametrics.* Cambridge Series in Statistical and Probabilistic Mathematics, Vol. 28, Cambridge Univ., Cambridge. 299 pp.

Hordijk W., Naylor J., Krasnogor N., and Fellermann H. (2018) Population dynamics of autocatalytic sets in a compartmentalized spatial world. *Life, 8(3),* 33.

Jeong H., Tombor B., Albert R., Oltvai Z. N., and Barabási A.-L. (2000) The large-scale organization of metabolic networks. *Nature, 407(6804),* 651.

Jolley C. and Douglas T. (2012) Topological biosignatures: Large-scale structure of chemical networks from biology and astrochemistry. *Astrobiology, 12(1),* 29–39.

Joyce G. F. and Orgel L. E. (2006) 2 progress toward understanding the origin of the RNA world. *Cold Spring Harbor M. Arch., 43,* 23–56.

Kanehisa M. and Goto S. (2000) KEGG: Kyoto encyclopedia of genes and genomes. *Nucleic Acids Res., 28(1),* 27–30.

Kauffman S. A., Jelenfi D. P., and Vattay G. (2019) Theory of chemical evolution of molecule compositions in the universe, in the Miller-Urey experiment and the mass distribution of interstellar and intergalactic molecules. *J. Theor. Biol., 486,* 110097.

Kim H., Smith H. B., Mathis C., Raymond J., and Walker S. I. (2019) Universal scaling across biochemical networks on Earth. *Sci. Adv., 5(1),* eaau0149.

Kivelä M., Arenas A., Barthelemy M., Gleeson J. P., Moreno Y., and Porter M. A. (2014) Multilayer networks. *J. Complex Netw., 2(3),* 203–271.

Krissansen-Totton, J., Olson S., and Catling D. C. (2018) Disequilibrium biosignatures over Earth history and implications for detecting exoplanet life. *Sci. Adv., 4(1),* eaao5747.

Line M. R., Wolf A. S., Zhang X., Knutson H., Kammer J. A., Ellison E., Deroo P., Crisp D., and Yung Y. L. (2013) A systematic retrieval analysis of secondary eclipse spectra. I. A comparison of atmospheric retrieval techniques. *Astrophys. J., 775(2),* 137.

Luger R., Agol E., Foreman-Mackey D., Fleming D. P., Lustig-Yaeger J., and Deitrick R. (2019) Starry: Analytic occultation light curves. *Astron. J., 157(2),* 64.

Marion G. M., Fritsen C. H., Eicken H., and Payne M. C. (2003) The search for life on Europa: Limiting environmental factors, potential habitats, and Earth analogues. *Astrobiology, 3(4),* 785–811.

Marshall S. M., Murray A. R., and Cronin L. (2017) A probabilistic framework for identifying biosignatures using pathway complexity. *Philos. Trans. R. Soc., A, 375(2109),* 20160342.

Mathis C., Bhattacharya T., and Walker S. I. (2017) The emergence of life as a first-order phase transition. *Astrobiology, 17(3),* 266–276.

Meadows V. S. (2017) Reflections on O_2 as a biosignature in exoplanetary atmospheres. *Astrobiology, 17(10),* 1022–1052.

Meadows V. S., Reinhard C. T., Arney G. N., Parenteau M. N., Schwieterman E. W., Domagal-Goldman S. D., Lincowski A. P., et al. (2018) Exoplanet biosignatures: Understanding oxygen as a biosignature in the context of its environment. *Astrobiology, 18(6),* 630–662.

Michalski J. R., Cuadros J., Niles P. B., Parnell J., Rogers A. D., and Wright S. P. (2013) Groundwater activity on Mars and implications for a deep biosphere. *Nature Geosci., 6(2),* 133.

Milo R., Shen-Orr S., Itzkovitz S., Kashtan N., Chklovskii D., and Alon U. (2002) Network motifs: Simple building blocks of complex networks. *Science, 298(5594),* 824–827.

Monod J. (1971) *Chance and Necessity: An Essay on the Natural Philosophy of Modern Biology.* Knopf, New York. 198 pp.

Montanez R., Medina M. A., Sole R. V., and Rodríguez-Caso C. (2010) When metabolism meets topology: Reconciling metabolite and reaction networks. *BioEssays, 32(3),* 246–256.

Moores J. E., Gough R. V., Martinez G. M., Meslin P.-Y., Smith C. L., Atreya S. K., Mahaffy P. R., Newman C. E., and Webster C. R. (2019) Methane seasonal cycle at gale crater on Mars consistent with regolith adsorption and diffusion. *Nature Geosci., 12,* 321–325.

Muchowska K. B., Varma S. J., and Moran J. (2019) Synthesis and breakdown of universal metabolic precursors promoted by iron. *Nature, 569(7754),* 104.

Newman M. E. (2004) Analysis of weighted networks. *Phys. Rev. E, 70(5),* 056131.

Ning B., Wolfgang A., and Ghosh S. (2018) Predicting exoplanet masses and radii: A nonparametric approach. *Astrophys. J., 869(1),* 5.

Oehler D. Z. and Etiope G. (2017) Methane seepage on Mars: Where to look and why. *Astrobiology, 17(12),* 1233–1264.

Pace N. R. (2001) The universal nature of biochemistry. *Proc. Natl. Acad. Sci. U.S.A, 98(3),* 805–808.

Raymond J. and Segrè D. (2006) The effect of oxygen on biochemical networks and the evolution of complex life. *Science, 311(5768),* 1764–1767.

Reinhard C. T., Olson S. L., Schwieterman E. W., and Lyons T. W. (2017) False negatives for remote life detection on ocean-bearing planets: Lessons from the early Earth. *Astrobiology, 17(4),* 287–297.

Scharf C. and Cronin L. (2016) Quantifying the origins of life on a planetary scale. *Proc. Natl. Acad. Sci. U.S.A, 113(29),* 8127–8132.

Scharf C., Virgo N., Cleaves H. J., Aono M., Aubert-Kato N., Aydinoglu A., Barahona A., et al. (2015) A strategy for origins of life research. *Astrobiology, 15(12),* 1031–1042.

Schmitt-Kopplin P., Gabelica Z., Gougeon R. D., Fekete A., Kanawati B., Harir M., Gebefugi I., Eckel G., and Hertkorn N. (2010) High molecular diversity of extraterrestrial organic matter in Murchison meteorite revealed 40 years after its fall. *Proc. Natl. Acad. Sci. U.S.A, 107(7),* 2763–2768.

Schrödinger E. (1943) What is life? The physical aspect of the living cell. Lecture, Trinity College, Dublin, February, 1943.

Schwieterman E. W., Kiang N. Y., Parenteau M. N., Harman C. E., DasSarma S., Fisher T. M., Arney G. N., et al. (2018) Exoplanet biosignatures: A review of remotely detectable signs of life. *Astrobiology, 18(6),* 663–708.

Seager S., Bains W., and Hu R. (2013) Biosignature gases in H_2-dominated atmospheres on rocky exoplanets. *Astrophys. J., 777(2),* 95.

Shock E., Bockisch C., Estrada C., Fecteau K., Gould I. R., Hartnett H., Johnson K., et al. (2019) Earth as organic chemist. In *Deep Carbon: Past to Present* (B. N. Orcutt et al., eds.), pp. 415–446. Cambridge Univ., Cambridge.

Sloan G., Hayward T., Allamandola L., Bregman J., DeVito B., and Hudgins D. (1999) Direct spectroscopic evidence for ionized polycyclic aromatic hydrocarbons in the interstellar medium. *Astrophys. J. Lett., 513(1),* L65.

Smith E. and Morowitz H. J. (2016) *The Origin and Nature of Life on Earth: The Emergence of the Fourth Geosphere.* Cambridge Univ., Cambridge. 677 pp.

Smith H. B., Kim H., and Walker S. I. (2019) Biochemical networks display universal structure across projections and levels of organization. *Artificial Life Conference Proceedings 2019, 31,* 282–283, DOI: 10.1162/isal_a_00176.

Solé R. V. and Munteanu A. (2004) The large-scale organization of chemical reaction networks in astrophysics. *Europhys. Lett., 68(2),* 170–176.

Summons R. E., Albrecht P., McDonald G., and Moldowan J. M. (2008) Molecular biosignatures. *Space Sci. Rev., 135(1),* 133–159.

Surman A. J., Rodriguez-Garcia M., Abul-Haija Y. M., Cooper G. J., Gromski P. S., Turk-MacLeod R., Mullin M., Mathis C., Walker S. I., and Cronin L. (2019) Environmental control programs the emergence of distinct functional ensembles from unconstrained chemical reactions. *Proc. Natl. Acad. Sci. U.S.A, 116(12),* 5387–5392.

Szczepanski J. and Vala M. (1993) Laboratory evidence for ionized polycyclic aromatic hydrocarbons in the interstellar medium. *Nature, 363(6431),* 699.

Szostak J. W. (2012) Attempts to define life do not help to understand the origin of life. *J. Biomol. Struct. Dyn., 29(4),* 599–600.

Vasas V., Szathmáry E., and Santos M. (2009) Lack of evolvability in self-sustaining autocatalytic networks: A constraint on the metabolism-first path to the origin of life. *Proc. Natl. Acad. Sci. U.S.A., 107(4),* 1470-1475.

Wagner A. (2002) The large-scale structure of metabolic networks: A glimpse at life's origin? *Complexity, 8(1),* 15–19.

Walker S. I. (2016) The descent of math. In *Trick or Truth? The Mysterious Connection Between Physics and Mathematics* (A. Aguirre et al., eds.), pp. 183–192. Springer, Switzerland.

Walker S. (2019) The new physics needed to probe the origins of life. *Nature, 569(7754),* 36–38.

Walker S. I., Bains W., Cronin L., DasSarma S., Danielache S., Domagal-Goldman S., Kacar B., et al. (2018) Exoplanet biosignatures: Future directions. *Astrobiology, 18(6),* 779–824.

Wasserman S. and Faust K. (1994) *Social Network Analysis: Methods and Applications, Vol. 8.* Cambridge Univ., Cambridge.

Wattam A. R., Davis J. J., Assaf R., Boisvert S., Brettin T., Bun C., Conrad N., et al. (2016) Improvements to PATRIC, the all-bacterial bioinformatics database and analysis resource center. *Nucleic Acids Res., 45(D1),* D535–D542.

Webster C. R., Mahaffy P. R., Atreya S. K. A., Moores J. E., Flesch G. J., Malespin C., McKay C. P., et al. (2018) Background levels of methane in Mars' atmosphere show strong seasonal variations. *Science, 360(6393),* 1093–1096.

West G. B., Brown J. H., and Enquist B. J. (1997) A general model for the origin of allometric scaling laws in biology. *Science, 276(5309),* 122–126.

West G. B., Woodruff W. H., and Brown J. H. (2002) Allometric scaling of metabolic rate from molecules and mitochondria to cells and mammals. *Proc. Natl. Acad. Sci. U.S.A, 99(1),* 2473–2478.

Woese C. (1998) The universal ancestor. *Proc. Natl. Acad. Sci. U.S.A, 95(12),* 6854–6859.

Meadows V. S., Arney G. N., Des Marais D. J., and Schmidt B. E. (2020) Unifying themes and future work in planetary astrobiology. In *Planetary Astrobiology* (V. Meadows et al., eds.), pp. 505–516. Univ. of Arizona, Tucson, DOI: 10.2458/azu_uapress_9780816540068-ch019.

Unifying Themes and Future Work in Planetary Astrobiology

Victoria S. Meadows
University of Washington

Giada N. Arney
NASA Goddard Space Flight Center

David J. Des Marais
NASA Ames Research Center

Britney E. Schmidt
Georgia Institute of Technology

In this concluding chapter we identify and outline current major cross-cutting themes in astrobiology that hold significant promise for future research, and point the reader to chapters of the book and other resources that discuss and elaborate on those themes. Since its inception in the late 1990s, the field of astrobiology has advanced in both complexity and rigor. The field's required interdisciplinarity has matured into a systems-level approach to planetary habitability that appreciates the potential habitability of both surface and subsurface environments, and recognizes that life itself is a *planetary process*. Similarly, new frameworks are being developed to search for life that emphasize the importance of multiple lines of evidence in interpreting biosignatures. These frameworks combine potential signs of life and the environmental context to arrive at a probabilistic assessment of life's presence. These promising new approaches present a path toward a robust, consensus assessment of habitability and biosignatures, but require further work by the community to develop and adopt these frameworks. There is also still considerable room for astrobiology to continue to grow and to combine needed research from many disciplines to answer questions that are too large for a single discipline to answer. Robotic exploration of the solar system continues apace, revealing a diversity of terrestrial chemistry and ocean worlds under ice shells. Meanwhile, exoplanet scientists attempt the very first measurements to constrain the nature of likely terrestrial exoplanet atmospheres. To identify the best targets on which to search for life, and to understand how best to conduct the search, stronger coordination and exchange of expertise between solar system planetary and exoplanetary scientists, as well as incorporation of additional expertise from stellar astronomy, heliophysics, and Earth science could lead to powerful outcomes. Ultimately, astrobiology and its focus on interdisciplinarity have the potential to forge a planetary science community that integrates the detail available in the solar system with the context available from multiple planetary systems to improve our understanding of all planets — near and far.

I believe that the extraordinary should certainly be pursued. But extraordinary claims require extraordinary evidence." — Carl Sagan

1. INTRODUCTION

The development of astrobiology, the study of life in the universe, is intimately tied to the development of spaceflight in the mid to late twentieth century, which enabled the discovery and exploration of non-Earth environments that might harbor life. In 1976, the Viking missions to Mars supported a core goal to search for signs of life beyond Earth (*Klein et al.,* 1972). However, the Viking life detection experiments provided inconclusive evidence of organics and life (e.g.,

Klein, 1979; *Biemann,* 1979; *Quinn et al.,* 2013), a result that put a damper on both Mars exploration and exobiology (as astrobiology was known then). Meanwhile, researchers continued to explore and understand the limits of life on this planet, and in the late 1980s and 1990s, humanity returned to exploration of Mars' radiation environment, magnetic field, atmosphere, and surface from orbit. In the mid-1990s, astrobiology was revived by the near-simultaneous claims of putative life in the martian meteorite Allan Hills (ALH) 84001 (*McKay et al.,* 1996), a result that remains controversial, and the — ultimately Nobel-Prize-winning — discovery of the first exoplanet orbiting a main-sequence star (*Mayor and Queloz,* 1995). These discoveries sparked interest in the search for life in our solar system and beyond, and led to the formation in 1998 of the NASA Astrobiology Institute (*Blumberg,* 2003), with the overarching goal to provide the scientific framework needed to support the robotic and telescopic exploration for habitable environments and life in the solar system and on exoplanets.

As a scientific field, astrobiology has many characteristics that make it strongly attractive, including its compelling and fundamental importance to understanding our cosmic context and its emphasis on pioneering massively interdisciplinary science to answer truly big scientific questions. By addressing questions of life's origin, evolution, and distribution in the universe, astrobiology tackles questions that are essential to understanding the cosmic context of Earth and humanity. The field encompasses the origin and evolution of habitable planets, life's origins, and the co-evolution of life with its environment. It is also central to current and future searches for past or present habitable environments and life on solar system worlds and exoplanets [*Des Marais et al.,* 2003, 2008; *Hays et al.,* 2015; *National Academies of Sciences, Engineering, and Medicine (NASEM),* 2018a,b]. Astrobiology is a "systems science" that studies interactions between physical, chemical, biological, geological, planetary, and astrophysical systems. Over the last two decades, astrobiology has grown in rigor and complexity, and has provided key topics and avenues of discovery that thrive on, and indeed require, interdisciplinary collaboration. In particular, understanding whether an exoplanet has the potential to be habitable requires that we understand how terrestrial planets form, evolve, and interact with their parent star and planetary system, thereby combining fields that address planet formation and volatile delivery, interior and surface evolution, atmospheric photochemistry and climate, aeronomy, and stellar astrophysics.

Important major themes in current and future astrobiology research require this level of interdisciplinarity. These include understanding the role of Earth's environment in the origin of life, life's coevolution with Earth's environment throughout 4 G.y. of our planet's history, studying planetary habitability using a systems science approach (*Hays et al.,* 2015; *NASEM,* 2018a), and understanding and interpreting biosignatures in the context of their environments. This last point is critical for any attempts at life detection, because no biosignature can be understood or interpreted independently of its environment, whether that environment is the surface or subsurface of Mars, the oceans of Europa or Enceladus, or on an exoplanet. This integrated approach has enabled the field to develop beyond initial, simplistic frameworks like the habitable zone (HZ) concept, and the search for abundant O_2. Astrobiology now embraces a more complex and rigorous approach that requires characterization of processes in addition to observable characteristics, one that calls for multiple lines of evidence and a more probabilistic approach to life detection. Below we expand upon several of the key themes in astrobiology research that hold the most promise for future advances in our understanding of life. These themes will help improve how to search for life on other worlds and thereby improve our confidence in its detection. Throughout we refer the reader to chapters in this book, or other resources, that expand upon these themes. We conclude with an outline of areas in which solar system and exoplanet scientists could benefit from collaborative research as we work toward a broader, unified field of planetary science.

2. LIFE IS A PLANETARY PROCESS

Life has substantially modified Earth's atmosphere, surface, and even interior over time, and investigations of environments on other habitable worlds must consider any life there as a potentially significant planetary process (see the chapter in this volume by Stüeken et al., which reviews the state of our understanding on life's coevolution with Earth environments). Life's current prominence on Earth arose from a presumably favorable trajectory of habitable environments that unfolded over billions of years. Life likely began sometime within the first few hundred million years of Earth history, although this is inferred from the scant evidence that survived. This evidence includes mineralogical records of planetary change (*Pearce et al.,* 2018; *Ohtomo et al.,* 2014), isotopic clues (ca. 3.95 Ga) suggesting metabolic activity (e.g., *Tashiro et al.,* 2017), and ancient fossils (ca. 3.5+ Ga) whose preservation was sufficient to have retained biogenic evidence (e.g., *Dodd et al.,* 2017). Prior to the advent of photosynthesis, Earth's anaerobic biosphere depended critically upon volcanism, hydrothermal activity, and atmospheric photochemical reactions as sources of redox energy and chemical reducing power for biosynthesis. Subsequently, oxygenic photosynthesis increased global biological productivity by orders of magnitude, transforming life into a globally-significant phenomenon (see the chapters by Stüeken et al. and Robinson and Reinhard for discussions on the environmental impact of photosynthesis on the early Earth). Long-term trends toward enhanced crustal weathering and oxidation, as well as atmospheric oxygenation, probably required feedbacks between global tectonic transformations and biological innovations.

Recognizing that other planetary environments probably differ from those on Earth, our search for habitable environments elsewhere requires an improved understanding of the environmental needs and fundamental chemical and physical mechanisms that underpin life's functionality. As stated in their chapter, Hoehler et al. stress that "The life-hosting potential of environments on Earth — not just whether they support life, but how abundantly and robustly they could support it — is expected to be as diverse as the conditions and processes that prevail there." Life as we know it requires energy, chemical building blocks to construct cellular components, a solvent that can facilitate biochemical processes, and environmental conditions that help to maintain the diversity of essential biomolecules and biochemical transformations. Improving our understanding of life's fundamental requirements would allow us to determine which particular environments might sustain life in greater abundances, making it potentially easier to detect. It also would enhance efforts to recognize alternative biochemistries and their potential environmental impacts. Analog research at the extremes of Earth's environments can help us better understand the limits of life and delineate the potential range of life-hosting environments (see the chapter by Schmidt for a description of Earth analog environments for icy worlds). These studies can also reveal the strategies that organisms evolved to utilize energy in these limiting environments. These data are important for understanding the environmental underpinnings of ecosystem architecture, and thus help provide a guide for life-hosting requirements to be sought in diverse planetary environments beyond Earth. Within our solar system, promising environments for life and prebiotic chemistry are being sought and studied on Mars (see the chapters by Davila et al. and Amador and Ehlman), Titan (chapter by Lunine et al.), Enceladus (chapter by Cable et al.), and the jovian moons (chapter by Schmidt).

While origins of life research has often focused on laboratory experiments that explore the chemistry that may have led to the initiation of life, there is a growing awareness of the key role that Earth's early environments — including interactions within and between those environments (*Stüeken et al., 2016*) — could have played in driving or inhibiting life's origins. A host of planetary factors could create conditions favorable to life's origins, from endogenic hydrothermal systems on the surface, to deep ocean vents, to impacts (see the chapters by Baross et al. and Zahnle and Carlson for a discussion of Earth's earliest environments and planetary processes that are relevant for life's origins). Other factors such as climatic events, changes in planetary heat loss, and intense periods of bombardment could impede life's origins. If life can be shown to have emerged as a consequence or extension of a planetary environmental process, this would have important implications for the search for life. Correspondingly, whether diverse environments could promote the origin of life is a rich area of inquiry, and attempting to constrain the nature and processes that were prevalent on Earth at the time of the origin of life is now a key, growing component of origins of life research (see the chapter by Baross et al.) (*NASEM, 2018a*).

Studying the coevolution of life with Earth's environment over its history is a strongly interdisciplinary area of research that encourages an improved understanding of the nature of Earth's earliest environments, including evidence of when particular metabolisms likely evolved and thrived, and how the organisms that used them impacted biogeochemical cycles or otherwise directly influenced many key aspects of the planet's environment (see the chapter by Stüeken et al. for a discussion of these impacts). In essence, the early Earth presents a series of habitable, and inhabited, environments that are alien when compared to our modern Earth, but for which we have geological and biological constraints that allow us to understand life and its environmental impact. Research into the use of early Earth environments as analogs

for alien habitable environments will likely continue to be an expanding and fruitful area of research (see the chapter by Robinson and Reinhard for more detail on this concept).

3. A SYSTEMS LEVEL APPROACH TO HABITABILITY

In the past decades, our understanding of the generation and maintenance of habitable environments has evolved from the focused but somewhat simplistic "follow the water" approach to encompass a more multifactorial process-based definition of habitability. An environment may transition from uninhabitable to habitable (or vice versa) over different spatial and temporal scales as a function of stellar, planetary, and environmental evolution; the presence of life; and the feedbacks between related complex physical, chemical, and biological parameters and processes (*NASEM*, 2018a) (see the chapters by Kopparapu et al. and Del Genio et al. for a broader overview and more details on processes and interactions that impact planetary habitability). Studies of Earth's environmental metamorphoses over its history show us how the Earth system interacts to produce alternative examples for habitable environments with different degrees of habitability, and that support different organisms (see the chapter by Robinson and Reinhard). Significantly, the environmental conditions required to maintain life once it has a foothold may not necessarily be the same as those that were essential for the origins of life.

In our solar system, studying key astrobiological targets such as Mars, Europa, Enceladus, and Titan with a systems science approach to evaluating their potential habitability has encouraged an exploration of the current state and past evolution of their chemical composition and physical processes. Systems science for potentially habitable worlds focuses on the *function* of a planetary process or energy source, which feeds directly into understanding how life might emerge from or be sustained by the resulting environments. For Mars, this includes understanding climatic evolution and surface/subsurface interactions (see the chapters by Amador and Ehlmann and Davila et al. for more description of these processes). Surface-subsurface interactions are also highly relevant for the icy moons (see the chapters by Schmidt, Cable et al., and Lunine et al.). The energy budgets of the icy moons are also strongly impacted by the gravitational influences of their host planets, and this context is required to understand the formation and persistence of energetic reservoirs as these planetary bodies evolve [see the chapters by Schmidt (jovian moons), Cable et al. (Enceladus), and Lunine et al. (Titan) for discussions on the roles of the host planet in controlling the interior structure and energy availability of the icy satellites of Jupiter and Saturn].

For exoplanets, surface liquid water remains the key characteristic of a habitable environment that could potentially support a surface biosphere, which increases the probability that the habitable planet can be identified and searched for life. The concept of the HZ, defined as the range of distances around a star of a given spectral type in which an Earth-like planet is more likely to be able to maintain liquid water on its surface (*Kasting et al.*, 1993; *Kopparapu et al.*, 2013), is still a useful construct for zeroth-order assessment of potential habitability for a newly discovered exoplanet. A focus on planets orbiting in this "light on water" zone (see the chapter by Zahnle and Carlson for an introduction to the concept of the HZ, and the chapter by Kopparapu et al. for a more detailed discussion) not only enhances the possibility of detecting a surface ocean, but also increases the chances that the biosphere is on the planetary surface. A surface biosphere can utilize photosynthesis, which increases the probability that life can interact with the surface and atmosphere in ways that are telescopically detectable.

However, for a more detailed assessment of a planet's potential habitability, the HZ is being supplanted by an expanding, systems-level understanding of the importance of interactions between the planetary environment, star, and planetary system that allow a planet to acquire and maintain liquid water on its surface over a significant period of time [see the chapter by Kopparapu et al. for an overview of these environmental parameters and interactions that support habitability (*Meadows and Barnes*, 2018)]. These processes include those that allow terrestrial planets to form (see the chapter by Raymond et al.) and that govern volatile delivery and accretion (see the chapters by Zahnle and Carlson and by Meech and Raymond for a more detailed discussion of these processes). Processes that govern whether planets with different interior compositions and outgassing can produce habitable surface environments is also a promising new area of research. Key research will also be needed to understand under what conditions a planet can maintain surface liquid water in the face of atmospheric and ocean loss processes over geological timescales. To support this latter research, improved understanding of star-planet interactions, including knowledge to be gained from heliophysics, and a deeper understanding of the impact of stellar luminosity evolution, winds, particles, and magnetic fields on planetary atmospheric lifetimes and compositions are a rich area for future study. These studies have the potential to revolutionize our understanding of the maintenance and loss of planetary habitability over time (see the chapters by Arney and Kane and by Amador and Ehlmann, which describe Venus' and Mars' role in improving our understanding of what makes a planet uninhabitable, and see the chapter by Del Genio et al. for an overview of multiple processes that can cause planets to lose atmospheres and oceans). However, while the HZ concept and an improved theoretical understanding of processes, interactions, and planetary characteristics that are more likely to result in a habitable world are potentially useful for identifying high-priority targets for observations, these theoretical studies alone cannot prove that a given planet is or is not habitable. Accurate habitability assessments will require astronomical observations of the planetary target to search for signs of habitability.

In the near-term, an exciting frontier will be the first acquisition of observational data on terrestrial exoplanets that may illuminate key processes that inform our understanding of planetary habitability and allow us to search for evidence of habitable environments. To date, our understanding of exoplanet habitability has advanced via computer modeling and comparisons with processes observed on Venus, Earth, and Mars. Recent observational constraints on the properties of small exoplanets suggest that they may indeed be terrestrial in nature (*Rogers*, 2015; *Fulton et al.*, 2017). Preliminary observations of some of the most promising targets orbiting nearby M dwarfs suggest near-terrestrial, if volatile-rich, bulk densities (*Grimm et al.*, 2018; *Luger et al.*, 2017), and have ruled out hydrogen-dominated atmospheres on several high-priority targets (*de Wit et al.*, 2016, 2018; *Moran et al.*, 2018), which suggests that these planets either have high molecular weight, terrestrial-like atmospheres (e.g., N_2-, O_2-, or CO_2-dominated), or no atmospheres at all. The TRAPPIST-1 seven-planet system (*Gillon et al.*, 2016, 2017) is a high-priority target for the James Webb Space Telescope (JWST), which is slated for launch in 2021. JWST transmission spectroscopy of these planets may reveal that they lack atmospheres — which will inform our understanding of whether M-dwarf terrestrials can maintain atmospheres and oceans — or may confirm the presence of terrestrial-type post-ocean-loss, Venus-like, or even habitable atmospheres (*Morley et al.*, 2017; *Lincowski et al.*, 2018, 2019, *Wunderlich et al.*, 2019; *Lustig-Yaeger et al.*, 2019a,b).

However, JWST will not be able to probe the surface and near-surface environments of these worlds for signs of habitability, and we will need to rely on future direct imaging missions to produce more robust assessments of planetary habitability. JWST will be limited to transmission spectroscopy for HZ planets, as these more temperate worlds are too cold to allow the use of emission spectroscopy (*Lustig-Yaeger et al.*, 2019a). However, due to geometry, refraction (*García-Muñoz, et al.*, 2012), the intrinsic density of a planetary atmosphere, and the likely presence of clouds (*Komacek et al.*, 2019), transmission spectroscopy cannot probe the surface and near-surface environments of a terrestrial planet, and must rely on attempts to measure water vapor in higher atmospheric layers to assess habitability. Ironically, for habitable planets there will be far less water in the upper troposphere and stratosphere due to the presence of the cold trap, which keeps water in the lower troposphere (*Meadows et al.*, 2018a), and so detection of water vapor on a habitable planet will be quite challenging with transmission (*Lincowski et al.*, 2018; *Lustig-Yaeger et al.*, 2019a), especially if clouds are also present (*Komacek et al.*, 2019). However, specific signs that the planet has lost an ocean, or no longer hosts one, may be gleaned by JWST from O_2-O_2 collisionally-induced absorption due to a massive, post-ocean-loss O_2-atmosphere, or an enhanced atmospheric D/H ratio (*Lincowski et al.*, 2019), both of which are potentially more detectable than an Earth-like water vapor profile (*Lustig-Yaeger et al.*, 2019a). In the longer term, space-based direct-imaging missions could provide more robust planetary habitability assessments by probing through the atmospheric column to the planetary surface. In addition to providing a higher likelihood of success in detecting near-surface water vapor, direct imaging measurements can also be used to map the planetary surface and search for glint from a planetary ocean (see the chapter by Robinson and Reinhard for an overview of exoplanet characterization strategies) (*Robinson et al.*, 2010, 2014; *Lustig-Yaeger et al.*, 2018).

To support the interpretation of these upcoming observations, it is important that the community develop and adopt metrics and standards for habitability assessment. These standards should include observational environmental and evolutionary markers that can be sought and combined in a probabilistic fashion to provide assessments of habitability that have varying degrees of robustness. In addition to assessing the habitability of newly observed exoplanet environments, this process has been identified by the astrobiology community as a key step in the development of a probabilistic assessment framework for potential exoplanet biosignatures, as it helps to establish environmental context for biosignature interpretation (*Catling et al.*, 2018; *Meadows et al.*, 2018b; *Walker et al.*, 2018).

4. THE FRONTIERS OF BIOSIGNATURE SCIENCE

One of the principal frontiers in biosignature science is the development of novel techniques to identify new biosignatures that have the highest likelihood of successful detection and correct interpretation. Biosignatures record the measurable impact of life on its environment, and may be preserved in the rock record or subsurface, on the surface, or in the liquids or atmosphere of a planet [see the chapters by Johnson et al. and Walker et al., as well as *Des Marais et al.* (2008) and *Schwieterman et al.* (2018) for biosignature definitions and examples of established and new biosignatures]. The search for biosignatures can be done via multiple techniques. *In situ* measurement techniques require visiting solar system worlds. Remote-sensing techniques involve orbiting or flyby spacecraft for solar system worlds — or astronomical telescopes for exoplanets.

While a limited number of existing potential global biosignatures that could be searched for via remote sensing on exoplanets have long been known — such as abundant O_2 and the O_2/CH_4 and O_2/N_2O disequilibria (*Hitchcock and Lovelock*, 1967) — more recent work has started to explore biosignatures for novel metabolisms in different environmental contexts. This focus also builds on initial explorations of possible biosignatures for early metabolisms (*Pilcher*, 2003) that were likely present for long periods before the rise of O_2 in our planet's atmosphere. To evaluate the robustness and utility of any novel biosignature, careful consideration must be taken to understand the reliability of the potential biosignature gas being produced by biological processes, its survivability against destruction by planetary processes such as geological activity or photochemistry, and its ultimate detectability (*Meadows*, 2017). Examples of new remote-sensing biosignatures identified using this process include the production of C_2H_6 (*Domagal-Goldman et al.*, 2011) or

hydrocarbon haze (*Arney et al.*, 2018) by sulfur biospheres in anoxic environments orbiting M dwarfs. Another example is high (and therefore more likely to be biogenic) surface fluxes of CH_4 that could result in detectable CO_2/CH_4 disequilibrium pairs in Archean-Earth-like planets orbiting M dwarfs (*Krissansen-Totton et al.*, 2018a,b). Possible biosignatures for more reducing, but still terrestrial, environments—such as NH_3 (*Seager et al.*, 2013) and PH_3 (*Bains et al.*, 2019; *Sousa-Silva et al.*, 2019a) — are also being explored. In an alternative approach, researchers are also cataloging a large number of potentially volatile molecules that might be detected in a planetary atmosphere (*Seager et al.*, 2016). Models are also being developed to simulate approximate molecular spectra based on functional groups for a very large number of these molecules (*Sousa-Silva et al.*, 2019b) until the more accurate laboratory spectroscopic opacities needed can be obtained for key species (*Fortney et al.*, 2016). To determine which of these diverse volatile gases are good candidate biosignatures, future work would be needed, potentially combining theoretical, laboratory, and field work to study the interaction of these biosignatures in a variety of planetary contexts. This will help us understand which of these numerous species can be produced by life and which survive to produce detectable spectral signatures that might be sought with future observations.

Identifying the best *in situ* biosignatures to search for on solar system planets faces similar challenges, although organic compounds are often considered prime targets. Biosignatures that can be measured using *in situ* techniques include objects, substances, and/or patterns whose origins specifically require biological processes (*Des Marais et al.*, 2008). Categories of biosignatures for *in situ* search include organic biomolecules, other biogenic chemical and stable isotopic patterns, certain minerals, macroscale and microscale fabrics and structures (e.g., stromatolites and microfossils), and active biological processes (see the chapter by Johnson et al.). Organic compounds are prime exploration targets because they harbor potentially abundant information that can be diagnostic of the conditions and processes that created them. Biology uses enzyme-driven reactions to generate organic entities that are biochemically useful, while non-biological reactions are rarely as specific in the generation of their products (see the chapter by Hoehler et al. for discussion of rules governing the production and use of biomolecules, and the chapter by Walker et al. for a description of how patterns in biomolecule distributions could be sought as a biosignature). For example, our biosphere is dominated by membrane fatty acids having 16 and 18 carbon atoms because these particular fatty acids impart favorable properties to cellular membranes. Analogous membrane lipids on other planets may reveal the specific influence of life. Detailed analyses of organic molecular structures can also reveal information concerning the taxonomic affinities of ancient biota on Earth (with varying degrees of specificity), their physiologies, the environmental conditions that prevailed in the depositional environment, and the degradation of the host organic matter. Stable carbon isotopic patterns within and between specific molecules can be diagnostic of biosynthesis. Accordingly, organic matter should be characterized regarding its abundance, molecular structures, stable isotopic patterns, and molecular weight distributions.

Other types of biosignatures also deserve attention as potential *in situ* biosignatures. Some carbon-bearing minerals probably occur exclusively as a result of biological activity (*Hazen et al.,* 2008). Because microbial biofilms can alter the chemistry and physical properties of sediments, microscale or macroscale rock or mineral fabrics and structures could indicate past life. Biological activity can alter the chemical composition of its surroundings as it acquires energy *via* redox reactions or chemical building blocks by altering inorganic phases. Such patterns can indicate conditions of formation and alteration and also might provide evidence of chemical equilibria or disequilibria that is inconsistent with abiotic processes and thus might indicate biological activity. Finally, as is the case for remote-sensing biosignatures, which rely on the survivability of a molecule in its environment, any *in situ* biosignature must also be preserved in order to be useful. An improved understanding is needed of which environmental processes can alter or destroy biosignatures, and which processes can preserve them, to help identify those biosignatures most likely to persist (see the chapter by Johnson et al. for a discussion of biosignature preservation).

Perhaps the most innovative approach to date for potential biosignature identification has been the pursuit of agnostic biosignatures, i.e., patterns in a planetary environment or sample of sufficiently high complexity to be unlikely to be produced by planetary processes (*Cronin and Walker,* 2016; *Cabrol,* 2016; *Johnson et al.,* 2018). This type of biosignature may be useful for both *in situ* and remote-sensing searches. Agnostic biosignatures may manifest as complex molecules that require multiple formation steps that are unlikely without biochemistry, suites of molecules that have unlikely distributions, or atmospheric chemical networks that are significantly more complex than abiotic examples (see the chapters by Johnson et al. and Walker et al. for discussions of agnostic biosignatures and their identification). Compared to the more targeted searches for specific molecules known to be produced by Earth metabolisms, agnostic biosignatures can be more generalized, providing the distinct advantage of allowing us to search a planetary environment for unknown forms of life (life as we do not know it), without assuming a particular biochemistry. However, while the detection of complex chemical systems as potential evidence of life is exciting, it requires a sufficiently comprehensive assay of the planetary environment to confirm that, for example, the patterns of molecules or reaction networks are indeed anomalous.

The second major frontier in biosignatures science is the recently initiated development of comprehensive, probabilistic frameworks for the assessment of observed biosignatures [see the chapter in this volume by Walker et al., as well as *Walker et al.* (2018), *Catling et al.* (2018),

Meadows (2017), *Meadows et al.* (2018b), and *Neveu et al.* (2018)]. This work is driven by the realization that simply detecting a potential biosignature is likely insufficient, as false positives and environmental context must also be identified, and that further assessment will be required to interpret the observed phenomenon as being more likely to be due to life than abiotic processes. These frameworks are being developed for both *in situ* and remote-sensing life detection.

For remote-sensing biosignatures, the field has moved beyond the more simplistic idea that (for example) abundant O_2 is always a rigorous biosignature into an understanding that the planetary environment and its interaction with the host star can work to mask, destroy, enhance, or even mimic a biosignature gas. This understanding is the result of a synthesis of research and perspectives from multiple fields. Our improving understanding of the rise of O_2 over the age of our planet, and its likely environmental suppression for at least hundreds of millions of years between the evolution of oxygenic photosynthesis and its abundant presence in the atmosphere, highlights the possibility of false negatives for life [see the chapter in this volume by Robinson and Reinhard as well as *Reinhard et al.* (2017) and *Meadows et al.* (2018b) for discussions of biosignature false negatives]. Planetary processes that are most likely to result in a false negative detection of life are important to identify to vet potential targets for biosignature searches. Similarly, modeling over the past five years has developed an improved understanding of star-planet interactions that may generate abiotic O_2 in the atmospheres of terrestrial planets orbiting M dwarfs. These false-positive mechanisms for abiotic O_2 generation include those that lead to atmospheric and ocean loss (*Luger and Barnes,* 2015), and photolysis of atmospheric H_2O and CO_2 (*Domagal-Goldman et al.,* 2014; *Wordsworth and Pierrehumbert,* 2014; *Tian et al.,* 2014; *Harman et al.,* 2015, 2018; *Tian et al.,* 2015; *Gao et al.,* 2015) [see the chapters in this volume by Robinson and Reinhard and Walker et al. for a discussion of false positives; see *Meadows* (2017) for a review of O_2 false positives]. Consequently, a conservative assumption is that any observed potential biosignature gas may have an abiotic origin, and a biosignature assessment framework must use complementary observations and models to determine how likely it is that the observed biosignature is in fact due to life, rather than to an abiotic planetary process (see the chapter in this volume by Walker et al. for a discussion of probabilistic biosignature assessment frameworks).

The challenge for the field going forward is to identify the multiple lines of evidence that strengthen the conclusion that potential biosignatures indeed arise from life, and to continue to develop and adopt a community-consensus, comprehensive, probabilistic framework that incorporates this additional information to assess observed biosignatures. There will likely be a significant scientific and societal impact of finding extraterrestrial life, and these extraordinary claims will indeed require extraordinary evidence. Consequently, it is particularly important to develop rigorous, uniform community standards and processes for biosignature detection, assessment, and verification that can be adopted by those reporting discoveries of habitable planets or possible signs of life. A large effort toward this goal was initiated by the exoplanet astrobiology community at the 2016 NExSS Exoplanet Biosignatures workshop, which resulted in six review papers that address the current state of the field (*Kiang et al.,* 2018; *Schwieterman et al.,* 2018), our current understanding of false positives and negatives, the initial concepts for the probabilistic framework to assess biosignatures (*Meadows et al.,* 2018b; *Catling et al.,* 2018; *Walker et al.,* 2018), and the upcoming observing capabilities that might make those assessments (*Fujii et al.,* 2018). These broad community-authored papers outline the critical information still needed for biosignature assessment, including environmental context, such as stellar UV spectra and activity levels, and/or the observed presence of water, both of which affect atmospheric chemistry. Additional information required to strengthen biosignature interpretation includes observations of multiple spectral features that can help to "rule in" life and "rule out" abiotic processes. These features include corroborating biosignatures: For the case of O_2, this includes the presence of CH_4, N_2O, and/or a spectral reflectance signature from plants, or signs of O_2 or CO_2 seasonality. Additionally, we must search for the presence of false positive discriminants that indicate abiotic processes are unlikely to be at work, or the absence of indicators of abiotic processes. While progress is being made in these areas, there is still significant work needed to develop and test the statistical framework for biosignature assessment that uses these parameters as input, and to develop and adopt community standards for assessing and reporting life detection. This will be a major area for community development of the field over the next decade.

For *in situ* life detection the development of a similar comprehensive framework for biosignature assessment has been initiated. Referred to as the "ladder of life detection," this framework proposes a small put potentially expandable set of rules that can be used to assess and integrate sets of measurements to provide a stepwise assessment of putative biosignatures informed by their environmental context (*Neveu et al.,* 2018). This is a living document, and the latest version can be found at *https://astrobiology.nasa. gov/research/life-detection/ladder*. Although the Viking lander life detection experiments were not conclusive, they provided key lessons on the importance of understanding environmental context (both the likelihood of habitability and possible abiotic processes that could mimic biological outcomes) and the potential for terrestrial contamination. Consequently the ladder of life detection outlines the following key criteria for assessing life-detection measurements: that the measurements be sufficiently *sensitive*, *contamination-free*, and *repeatable* (see the chapter by Rummel for a description of efforts to reduce contamination when exploring other planetary environments), and that one or more features be sufficiently *preserved*, *detectable*, measurably *different from expected abiotic signals*, and

compatible with life processes as we know them. Perhaps most importantly, the biological interpretation must be *the last-resort hypothesis* (*Neveu et al.,* 2018).

There are four key considerations regarding *in situ* exploration for life on other planets that feed into this ladder of life detection assessment framework, to increase the probability that we will detect and correctly interpret biosignatures:

• It is necessary to characterize the environmental attributes, both past and present, of potential sites of exploration to evaluate their potential for biosignature preservation.

• Assessments should be made of the likelihood that non-biological features could mimic biosignatures, via characterization of the environment to the extent necessary to mitigate the risk of false positives, and to thereby strengthen the interpretation that a potential biosignature is more likely to be due to life.

• The biosignature interpretation can be significantly strengthened by searching for multiple categories of biosignatures in a given sample/location, including organic molecules, other chemical substances and patterns, minerals, stable isotopic patterns, and sedimentary macrostructures and microstructures.

• In support of the previous consideration, it is important to develop flight instrument packages for astrobiology missions that include arrays of measurement types (e.g., imaging, spectroscopy, chemistry, mineralogy) that collectively could identify several different categories of biosignatures.

The ladder of life detection is intended as a starting point to stimulate discussion, debate, and further research on life's characteristics, what constitutes a definitive biosignature, and the optimal means of measuring them. It still requires significant community input. As is the case for exoplanets, there is a great deal of work to be done over the next 5–10 years to refine this assessment framework for searches for life in our solar system, and to work toward widespread adoption of life detection criteria that would convince a majority of the scientific community.

In the next 5–10 years, there are several exciting, although limited, opportunities to search for biosignatures on exoplanets and in our solar system. For exoplanets, transmission spectroscopy with JWST of planets orbiting in the HZ of the TRAPPIST-1 M dwarf star may be capable of detecting the CO_2/CH_4 biosignature for anoxic environments (*Krissansen-Totton et al.,* 2018b). However, detection of O_2 or O_3 is likely to be extremely challenging due to the wavelength range covered by JWST and the anticipated strengths of O_2 and O_3 features for inhabited worlds (*Wunderlich et al.,* 2019; *Lustig-Yaeger et al.,* 2019a). On the ground, high-resolution spectrometers planned for upcoming extremely large telescopes, scheduled to become operational in the mid-2020s, will be able to search for O_2 in the atmospheres of a handful of terrestrial planets orbiting M dwarfs using transmission (*López-Morales et al.,* 2019) or reflected light spectroscopy (*Lovis et al.,* 2017).

In our solar system, the search for *in situ* evidence for life has already begun on Mars, with sophistication increasing from spectroscopic tools capable of assessing environmental attributes (i.e., the Mars Exploration Rovers) to pyrolytic measurements (i.e., the Mars Science Laboratory) and potentially onward to further *in situ* and returned sample analysis (i.e., Mars 2020; see the chapters by Amador and Ehlmann and Davila et al.). The international campaign to select and return samples from Mars is a highly significant near-term opportunity (*iMOST*, 2019). The broadest, most rigorous investigations of potential biosignatures can only be achieved in Earth-based laboratories. State-of-the-art laboratory instrumentation typically requires sophisticated sample preparation techniques that in turn require initial laboratory-based observations that guide the selection and adaptation of these techniques. A Curiosity-class rover launched in 2020 (*Mustard et al.*, 2013) would explore and acquire samples from Jezero Crater, where a delta and deposits of phyllosilicates, carbonates, and silica have been identified. NASA and the European Space Agency (ESA) are collaborating to develop subsequent orbiter and landed "fetch rover" missions to return the samples.

In the longer term, missions to the outer solar system to search for life are being developed, and these have benefited from an already holistic picture of biosignature science. Two orbiting spacecraft will explore the moons of the jovian system, focusing on the icy moons Europa and Ganymede (see the chapter by Schmidt). NASA's Europa Clipper will launch ca. 2025 and ESA's JUpiter ICy moons Explorer (JUICE) will launch in 2022. Europa Clipper will explore Europa through a diverse set of interdisciplinary investigations with a focus on assessing its potential for habitability (*Phillips and Pappalardo*, 2014). Europa Clipper will also undertake the next generation of measurements started by Cassini at Enceladus (see the chapter by Cable et al.). In particular, it will employ higher mass resolution and a wider mass range to further biosignature science (see chapter by Schmidt). Beyond Europa Clipper, concepts for a landed mission to Europa and plume science missions to Enceladus are being planned. These are the first two studied missions to explicitly include coordinated sets of measurements that build toward understanding higher-order signatures of life, and to have the capability to begin to measure agnostic signatures such as chemical complexity and energy (e.g., *Lunine et al.*, 2015; *Reh et al.*, 2016; *Hand et al.*, 2016). Missions through and below the ice, while potentially a few decades away, are approaching technical readiness for flight (see the chapter by Schmidt). Dragonfly, NASA's New Frontiers 4 mission to Saturn's moon Titan, is an eight-bladed rotorcraft lander designed to investigate sites on the dunes and elsewhere, including localities where aqueous chemistry might have occurred (*Lorenz et al.*, 2018) (see the chapter by Lunine et al.). Dragonfly will make the second (after the Huygens probe) and most sophisticated *in situ* measurements of Titan's organic chemistry. Dragonfly will search for prebiotic processes (see the chapter by Lunine et al.)

while also performing meteorological, seismic, and other measurements at each landing site, and profiling Titan's atmosphere while in flight.

In the longer term, NASA is developing a suite of space-based telescope mission concepts that will have the capability to detect and characterize Earth-like HZ exoplanets, to search for signs of habitability and life. These telescope concepts include large, spacebased direct imaging telescopes, which, if selected, will launch in the 2030s and 2040s. These next-generation observatories (*Gaudi et al.*, 2019; *Roberge et al.*, 2019) would have the capability to search hundreds of stars for Earth-like HZ exoplanets, and explore the most promising targets for signs of habitability and life throughout the UV to near-infrared wavelength range. These telescopes could image and obtain time-resolved photometry and spectroscopy of terrestrial HZ planets orbiting Sun-like stars (FGK spectral type) inaccessible to near future facilities like JWST, as well as larger nearby M dwarfs. In addition to allowing direct detection of glint from exoplanet oceans (*Robinson et al.*, 2010, 2014; *Lustig-Yaeger et al.*, 2018), these missions will also enable the search for a comprehensive suite of biosignature gases and surface features, including seasonal variability (*Schwieterman et al.*, 2018; *Olson et al.*, 2018). Most importantly, these missions may obtain a large sample size of potentially habitable and even inhabited worlds, allowing us to consider these planets, and our own, in a broad comparative context. These telescopes would also have the capability to obtain exquisite, flyby-quality images of ocean worlds in the outer solar system over long time baselines. These telescopes could monitor dynamical activity due to plumes and interior oceans — thereby expanding the range of ocean worlds that can be monitored remotely.

5. EXOPLANET AND SOLAR SYSTEM SYNERGIES

As we push toward a system- and process-based approach to understanding planetary habitability and biosignatures, it has become increasingly clear that many of the most exciting and fruitful areas of research lie at the intersection of the traditionally-separated domains of exoplanet and solar system research. Just as comparative planetology within our solar system has empowered a deeper understanding of general planetary phenomena on a wider scale than studies of a single planet had achieved, comparative planetology within and across other planetary systems will teach us about these same, and other, processes on a much grander scale.

In the solar system, we have access to a diverse suite of worlds in our own cosmic backyard that can be examined in up-close, exquisite detail, providing a rich trove of data that illuminates how planets appear and operate on an intimate level. In contrast, the field of exoplanet characterization will remain in a comparatively data-scarce regime for years to come, and ground-truth, *in situ* measurements are likely decades to centuries away. Consequently, one of the most valuable synergies between solar system and exoplanet science is the availability of solar system *in situ* and remote-sensing observations that can be used to illuminate planetary processes that may be universal, and to validate models of exoplanet atmospheres and surfaces. Although solar system observations provide a much more rigorous means of validating exoplanet models than the more commonly employed comparison of a new model to an existing one, it can be challenging for members of the exoplanet community to identify and obtain relevant solar system observations. Conversely, the critical importance to exoplanet scientists of some of the relatively simple observations of solar system planets (e.g., phase-dependent albedos, occultation spectra) that would mimic exoplanet observations is not widely understood in the solar system community. Future work to enhance discussions between these two communities could result in rapid gains in understanding of the significance and usefulness of each other's datasets and results, and lead to improvements in rigor for exoplanet models and interpretation — which could in turn support key astrobiological goals. In addition, by viewing solar system planets through the lens of their relevance as exoplanet analogs, it may also be possible to enhance the overall science return from planetary science missions by including exoplanet-relevant objectives and measurements. This may be most effectively achieved by engaging representatives from the exoplanet community in planetary science mission development, similar to the way that planetary scientists have been included as team members for recent astrophysics mission concept development efforts.

Earth itself will always be our best studied planet; Earth and its nearest neighbors can also illuminate key processes for exoplanet astrobiologists. Earth has shown us how habitability can arise (see the chapters by Zahnle and Carlson and by Meech and Raymond) and has taught us everything we know about how life originates and can co-evolve with its environment over time (see the chapters by Stüeken et al. and by Robinson and Reinhard). By examining Earth's biosphere, we have begun to understand the limits and requirements for life's origin and survivability, a crucial starting point in our searches for life elsewhere (see the chapter by Baross et al. for environments that support life's origins, and the chapter by Hoehler et al. for life's environmental requirements). Observations of Earth, from simple photometric light curves (e.g., *Cowan et al.,* 2009; *Robinson et al.,* 2010, 2011; *Livengood et al.,* 2011) to exquisitely detailed occultation observations (*MacDonald and Cowan,* 2019), can inform future attempts to study habitable exoplanets (see the chapter by Robinson and Reinhard). Looking beyond Earth to our nearest planetary neighbors, the histories of Venus (see the chapter by Arney and Kane) and Mars (see the chapter by Amador and Ehlmann), when combined with our understanding of heliophysical processes, have taught us about how planetary habitability can evolve — and even be lost — over time (see the chapter by Del Genio et al.). Mars, in particular, is currently the most visited and arguably best-studied body other than Earth in the solar system. The processes that have shaped its atmosphere and surface, and its current environmental state, have stimulated considerable astrobiological interest in this target (see the chapters by

Davila et al. and by Amador and Ehlmann). In particular, understanding the star-planet interactions that result in Mars' atmospheric loss processes may be relevant to planets orbiting active M dwarfs (*Jakosky,* 2019; *Lillis et al.,* 2015; *Garcia-Sage et al.,* 2017). Venus, while not currently habitable, has a mysterious past that may include more clement conditions. It serves as a nearby representative of the vast numbers of likely Venus-like exoplanets that represent the most observable terrestrial-sized planets for JWST (*Kane et al.,* 2014), and it may even allow us to test ideas about O_2 false-positive photochemistry in CO_2-rich atmospheres. Farther out in our solar system, the icy moons of the outer solar system show us a different kind of possible habitability: subsurface oceans. Although these kinds of deep biospheres would be unobservable across interstellar distances, these icy moons nevertheless will allow us to generalize and test our ideas of what life requires, helping guide which observable exoplanets we do target [see the chapters by Schmidt (Europa/jovian moons), Cable et al. (Enceladus), and Lunine et al. (Titan) for discussions of potential habitable environments for these bodies], and contextualizing different types of planets and moons as they become observable.

While exoplanets cannot be studied in as much detail as solar system bodies, they exist in vast numbers and in different configurations, so they enrich solar system science by opening the door to understanding statistics of planetary processes and properties. By comparing the properties of our solar system with exoplanetary systems, we have already learned that our solar system's architecture is "weird" compared to most exoplanetary systems (see the chapter by Raymond et al.). The outcomes and processes of planetary formation are complex (see the chapters by Raymond et al., Meech and Raymond, and Zahnle and Carlson), and we did not fully appreciate how incomplete our understanding of the solar system's formation was until the discovery of the first exoplanets. These discoveries revealed examples of planetary systems sculpted by processes like migration, which were not widely considered for our solar system using the old frameworks, and which have significant impacts on volatile delivery to forming planets.

Exoplanets also allow us to study types of planets that do not occur in our solar system. Although we currently know little about most exoplanets beyond their basic size and/or mass and orbital properties, we have already discovered unanticipated key classes of planets that do not exist locally: hot Jupiters, sub-Neptunes, and super-Earths. As we begin to better understand the atmospheric properties of these worlds, we will almost certainly continue to encounter surprises. Future exoplanet observations of potentially habitable worlds will allow us to empirically test our ideas of where habitable planets can and cannot be found (see the chapter by Kopparapu et al.) by searching exoplanets for signs of past ocean loss. Exoplanets will also be observed at different ages and stages of evolution, embedded in different planetary architectures, with different host stars. This diversity of examples will allow us to understand general principles of how planets evolve over time, and how a different star and planetary system alter the planetary processes we think we understand within our solar system. If we can characterize these distant worlds, we may be able to provide snapshots of planets akin to past states of solar system bodies (e.g., young, habitable Venus-like exoplanets might be observed, or early steam atmospheres, informing our understanding of venusian history). Ultimately, exoplanet and solar system science must support each other, and move together toward a grander, more unified understanding of planets, both in our solar system and beyond.

REFERENCES

Arney G., Domagal-Goldman S. D., and Meadows V. S. (2018) Organic haze as a biosignature in anoxic Earth-like atmospheres. *Astrobiology, 18(3),* 311–329.

Bains W., Petkowski J. J., Sousa-Silva C., and Seager S. (2019) New environmental model for thermodynamic ecology of biological phosphine production. *Sci. Total Environ., 658,* 521–536.

Biemann K. (1979) The implications and limitations of the findings of the Viking Organic Analysis Experiment. *J. Mol. Evol., 14,* 65–70.

Blumberg B. S. (2003) The NASA Astrobiology Institute: Early history and organization. *Astrobiology, 3(3),* 463–470.

Cabrol N. A. (2016) Alien mindscapes — A perspective on the search for extraterrestrial intelligence. *Astrobiology, 16(9),* 661–676.

Catling D. C., Krissansen-Totton J., Kiang N. Y., Crisp D., Robinson T. D., DasSarma S., Rushby A. J., Del Genio A., Bains W., and Domagal-Goldman S. (2018) Exoplanet biosignatures: A framework for their assessment. *Astrobiology, 8(6),* 709–738.

Cowan N. B., Agol E., Meadows V. S., Robinson T., Livengood T. A., Deming D., Lisse C. M., et al. (2009) Alien maps of an ocean-bearing world. *Astrophys. J., 700(2),* 915.

Cronin L. and Walker S. I. (2016) Beyond prebiotic chemistry. *Science, 352(6290),* 1174–1175.

Des Marais D. J., Allamandola L. J., Benner S. A., Boss A. P., Deamer D., Falkowski P. G., Farmer J. D., et al. (2003) The NASA Astrobiology Roadmap. *Astrobiology, 3(2),* 219–235.

Des Marais D. J., Nuth J. A. III, Allamandola L. J., Boss A. P., Farmer J. D., Hoehler T. M., Jakosky B. M., et al. (2008) The NASA Astrobiology Roadmap. *Astrobiology, 8(4),* 715–730.

de Wit J., Wakeford H. R., Gillon M., Lewis N. K., Valenti J. A., Demory B. O., Burgasser A. J., et al. (2016) A combined transmission spectrum of the Earth-sized exoplanets TRAPPIST-1 b and c. *Nature, 537(7618),* 69.

de Wit J., Wakeford H. R., Lewis N. K., Delrez L., Gillon M., Selsis F., Leconte J., et al. (2018) Atmospheric reconnaissance of the habitable-zone Earth-sized planets orbiting TRAPPIST-1. *Nature Astron., 2(3),* 214.

Dodd M. S., Papineau D., Grenne T., Slack J. F., Rittner M., Pirajno F., O'Neil J., and Little C. T. (2017) Evidence for early life in Earth's oldest hydrothermal vent precipitates. *Nature, 543(7643),* 60.

Domagal-Goldman S. D., Meadows V. S., Claire M. W., and Kasting J. F. (2011) Using biogenic sulfur gases as remotely detectable biosignatures on anoxic planets. *Astrobiology, 11(5),* 419–441.

Domagal-Goldman S. D., Segura A., Claire M. W., Robinson T. D., and Meadows V. S. (2014) Abiotic ozone and oxygen in atmospheres similar to prebiotic Earth. *Astrophys. J., 792(2).*

Fortney J. J., Robinson T. D., Domagal-Goldman S., Amundsen D. S., Brogi M., Claire M., Crisp D., et al. (2016) The need for laboratory work to aid in the understanding of exoplanetary atmospheres. *ArXiv e-prints, arXiv:1602.06305.*

Fujii Y., Angerhausen D., Deitrick R., Domagal-Goldman S., Grenfell J. L., Hori Y., Kane S. R., et al. (2018) Exoplanet biosignatures: Observational prospects. *Astrobiology, 18(6),* 739–778.

Fulton B. J., Petigura E. A., Howard A. W., Isaacson H., Marcy G. W., Cargile P. A., Hebb L., et al. (2017) The California-Kepler Survey. III. A gap in the radius distribution of small planets. *Astron. J., 154(3),* 109.

Gao P., Hu R., Robinson T. D., Li C., and Yung Y. L. (2015) Stabilization of CO_2 atmospheres on exoplanets around M dwarf stars. *Astrophys. J., 806,* 249–261.

García-Muñoz A., Zapatero Osorio M. R., Barrena R., Montañes-Rodríguez P., Martín E. L., and Pallé E. (2012) Glancing views of the Earth: From a lunar eclipse to an exoplanetary transit. *Astrophys. J., 755(2),* 103.

Garcia-Sage K., Glocer A., Drake J. J., Gronoff G., and Cohen O. (2017) On the magnetic protection of the atmosphere of Proxima Centauri b. *Astrophys. J. Lett., 844(1),* L13.

Gaudi S., Seager S., Kiessling A., Mennesson B., and Warfield K. (2019) The Habitable Exoplanet Observatory (HabEx). Astro2020: Decadal Survey on Astronomy and Astrophysics, APC white papers. *Bull. Am. Astron. Soc., 51(7),* 89.

Gillon M., Jehin E., Lederer S. M., Delrez L., de Wit J., Burdanov A., Van Grootel V., et al. (2016) Temperate Earth-sized planets transiting a nearby ultracool dwarf star. *Nature, 533(7602),* 221.

Gillon M., Triaud A. H., Demory B. O., Jehin E., Agol E., Deck K. M., Lederer S. M., et al. (2017) Seven temperate terrestrial planets around the nearby ultracool dwarf star TRAPPIST-1. *Nature, 542(7642),* 456.

Grimm S. L., Demory B. O., Gillon M., Dorn C., Agol E., Burdanov A., Delrez L., et al. (2018) The nature of the TRAPPIST-1 exoplanets. *Astron. Astrophys., 613,* A68.

Hand K. P., Murray A. E., Garvin J. B., Brinkerhoff W. B., Edgett K. S., Ehlmann B. L., German C. R., et al. (2016) *Report of the Europa Lander Science Definition Team.* JPL D-97667, NASA, Washington, DC.

Harman C. E., Schwieterman E. W., Schottelkotte J. C., and Kasting J. F. (2015) Abiotic O_2 levels on planets around F, G, K, and M stars: Possible false positives for life? *Astrophys. J., 812(2),* 137.

Harman C. E., Felton R., Hu R., Domagal-Goldman S. D., Segura A., Tian F., and Kasting J. F. (2018) Abiotic O_2 levels on planets around F, G, K, and M stars: Effects of lightning-produced catalysts in eliminating oxygen false positives. *Astrophys. J., 866(1),* 56.

Hays L., Achenbach L., Bailey J., Barnes R., Baross J., Bertka C., Boston P., et al. (2015) *NASA Astrobiology Strategy 2015.* Available online at *https://nai.nasa.gov/media/medialibrary/2016/04/NASA_Astrobiology_Strategy_2015_FINAL_041216.pdf.*

Hazen R. M., Papineau D., Bleeker W., Downs R. T., Ferry J. M., McCoy T. J., and Yang H. (2008) Mineral evolution. *Am. Mineral., 93(11–12),* 1693–1720.

Hitchcock D. R. and Lovelock J. E. (1967) Life detection by atmospheric analysis. *Icarus, 7,* 149–159.

iMOST (2019) The potential science and engineering value of samples delivered to Earth by Mars sample return. *Meteoritics & Planet. Sci., 54(3),* 667–671, DOI: 10.1111/maps.13232.

Jakosky B. M. (2019) The CO_2 inventory on Mars. *Planet. Space Sci., 175,* 52–59.

Johnson S. S., Anslyn E. V., Graham H. V., Mahaffy P. R., and Ellington A. D. (2018) Fingerprinting non-terran biosignatures. *Astrobiology, 18(7),* 915–922.

Kane S. R., Kopparapu R. K., and Domagal-Goldman S. D. (2014) On the frequency of potential Venus analogs from Kepler data. *Astrophys. J. Lett., 794(1),* L5.

Kasting J. F., Whitmire D. P., and Reynolds R. T. (1993) Habitable zones around main sequence stars. *Icarus, 101(1),* 108–128.

Klein H. P. (1979) The Viking mission and the search for life on Mars. *Rev. Geophys., 17,* 1655.

Klein H. P., Lederberg J., and Rich A. (1972) Biological experiments: The Viking Mars Lander. *Icarus, 16,* 139–146.

Kiang N. Y., Domagal-Goldman S., Parenteau M. N., Catling D. C., Fujii Y., Meadows V. S., Schwieterman E. W., and Walker S. I. (2018) Exoplanet biosignatures: At the dawn of a new era of planetary observations. *Astrobiology, 18(6),* 619–629.

Komacek T. D., Fauchez T. J., Wolf E. T., and Abbot D. S. (2019) Clouds will likely prevent the detection of water vapor in JWST transmission spectra of terrestrial exoplanets. *Astrophys. J. Lett., 888(2).*

Krissansen-Totton J., Garland R., Irwin P. and Catling D. C. (2018a) Detectability of biosignatures in anoxic atmospheres with the James Webb Space Telescope: A TRAPPIST-1 e case study. *Astron. J., 156(3),* 114.

Krissansen-Totton J., Olson S., and Catling D. C. (2018b) Disequilibrium biosignatures over Earth history and implications for detecting exoplanet life. *Sci. Adv., 4(1),* eaao5747.

Kopparapu R. K., Ramirez R., Kasting J. F., Eymet V., Robinson T. D., Mahadevan S., Terrian R. C., Domagal-Goldman S., Meadows V. S., and Deshpande R. (2013) Habitable zones around main-sequence stars: New estimates. *Astrophys. J., 765(2)*, DOI: 10.1088/0004-637X/765/2/131.

Lillis R. J., Brain D. A., Bougher S. W., Leblanc F., Luhmann J. G., Jakosky B. M., Modolo R., et al. (2015) Characterizing atmospheric escape from Mars today and through time, with MAVEN. *Space Sci. Rev., 195(1–4)*, 357–422.

Lincowski A. P., Meadows V. S., Crisp D., Robinson T. D., Luger R., Lustig-Yaeger J., and Arney G. N. (2018) Evolved climates and observational discriminants for the TRAPPIST-1 planetary system. *Astrophys. J., 867(1)*, 76.

Lincowski A. P., Lustig-Yaeger J., and Meadows V. S. (2019) Observing isotopologue bands in terrestrial exoplanet atmospheres with the James Webb Space Telescope: Implications for identifying past atmospheric and ocean loss. *Astron. J., 158(1)*, 26.

Livengood T. A., Deming L. D., A'Hearn M. F., Charbonneau D., Hewagama T., Lisse C. M., McFadden L. A., et al. (2011) Properties of an Earth-like planet orbiting a Sun-like star: Earth observed by the EPOXI mission. *Astrobiology, 11(9)*, 907–930.

López-Morales M., Ben-Ami S., Gonzalez-Abad G., Garcia-Mejia J., Dietrich J., and Szentgyorgyi A. (2019) Optimizing ground-based observations of O$_2$ in Earth analogs. *Astron. J., 158(1)*, 24.

Lorenz R. D., Turtle E. P., Barnes J. W., Trainer M. G., Adams D. S., Hibbard K. E., Sheldon C. Z., et al. (2018) Dragonfly: A rotorcraft lander concept for scientific exploration at Titan. *Johns Hopkins APL Tech. Dig., 34(3)*.

Lovis C., Snellen I., Mouillet D., Pepe F., Wildi F., Astudillo-Defru N., Beuzit J.-L., et al. (2017) Atmospheric characterization of Proxima b by coupling the SPHERE high-contrast imager to the ESPRESSO spectrograph. *Astron. Astrophys., 599*, A16.

Luger R. and Barnes R. (2015) Extreme water loss and abiotic O$_2$ buildup on planets throughout the habitable zones of M dwarfs. *Astrobiology, 15(2)*, 119–143, DOI: 10.1089/ast.2014.1231.

Luger R., Sestovic M., Kruse E., Grimm S. L., Demory B. O., Agol E., Bolmont E., et al. (2017) A seven-planet resonant chain in TRAPPIST-1. *Nature Astron., 1(6)*, 0129.

Lustig-Yaeger J., Meadows V. S., Mendoza G. T., Schwieterman E. W., Fujii Y., Luger R., and Robinson T. D. (2018) Detecting ocean glint on exoplanets using multiphase mapping. *Astron. J., 156(6)*, 301.

Lustig-Yaeger J., Meadows V. S., and Lincowski A. P. (2019a) The detectability and characterization of the TRAPPIST-1 exoplanet atmospheres with JWST. *Astron. J., 158(1)*, 27.

Lustig-Yaeger J., Meadows V. S., and Lincowski A. P. (2019b) A mirage of the cosmic shoreline: Venus-like clouds as a statistical false positive for exoplanet atmospheric erosion. *Astrophys. J. Lett., 887(1)*, L11.

Lunine J., Waite H., Postberg F., Spilker L., and Clark K. (2015) Enceladus life finder: The search for life in a habitable moon. *EGU General Assembly Abstracts, 17*, 14923.

Macdonald E. J. and Cowan N. B. (2019) An empirical infrared transit spectrum of Earth: Opacity windows and biosignatures. *Mon. Not. R. Astron. Soc., 489(1)*, 196–204.

Mayor M. and Queloz D. (1995) A Jupiter-mass companion to a solar-type star. *Nature, 378(6555)*, 355.

McKay D. S., Gibson E. K. Jr., et al. (1996) Search for past life on Mars: Possible relic biogenic activity in martian meteorite ALH84001. *Science, 273(5277)*, 924–930.

Meadows V. S. (2017) Reflections on O$_2$ as a biosignature in exoplanetary atmospheres. *Astrobiology, 17(10)*, 1022–1052.

Meadows V. S. and Barnes R. K. (2018) Factors affecting exoplanet habitability. In *Handbook of Exoplanets* (H. J. Deeg and J. A. Belmonte, eds.), pp. 2771–2794. Springer, Cham.

Meadows V. S., Arney G. N., Schwieterman E. W., Lustig-Yaeger J., Lincowski A. P., Robinson T., Domagal-Goldman S. D., et al. (2018a) The habitability of Proxima Centauri b: Environmental states and observational discriminants. *Astrobiology, 18(2)*, 133–189.

Meadows V. S., Reinhard C. T., Arney G. N., Parenteau M. N., Schwieterman E. W., Domagal-Goldman S. D., Lincowski A. P., et al. (2018b) Exoplanet biosignatures: Understanding oxygen as a biosignature in the context of its environment. *Astrobiology, 18(6)*, 630–662.

Moran S. E., Hörst S. M., Batalha N. E., Lewis N. K., and Wakeford H. R. (2018) Limits on clouds and hazes for the TRAPPIST-1 planets. *Astron. J., 156(6)*, 252.

Morley C. V., Kreidberg L., Rustamkulov Z., Robinson T., and Fortney J. J. (2017) Observing the atmospheres of known temperate Earth-sized planets with JWST. *Astrophys. J., 850*, 121.

Mustard J. F., Adler M., Allwood A., Bass D., Beaty D., Bell J. F., et al. (2013) *Report of the Mars 2020 Science Definition Team*. Available online at https://mepag.jpl.nasa.gov/reports/MEP/Mars_2020_SDT_Report_Final.pdf.

National Academies of Sciences, Engineering, and Medicine (2018a) *An Astrobiology Science Strategy for the Search for Life in the Universe*. National Academies, Washington, DC. 188 pp.

National Academies of Sciences, Engineering, and Medicine (2018b) *Exoplanet Science Strategy*. National Academies, Washington, DC. 186 pp.

Neveu M., Hays L. E., Voytek M. A., New M. H., and Schulte M. D. (2018) The ladder of life detection. *Astrobiology, 18(11)*, 1375–1402.

Ohtomo Y., Kakegawa T., Ishida A., Nagase T., and Rosing M. T. (2014) Evidence for biogenic graphite in early Archaean Isua metasedimentary rocks. *Nature Geosci., 7(1)*, 25.

Olson S. L., Schwieterman E. W., Reinhard C. T., Ridgwell A., Kane S. R., Meadows V. S., and Lyons T. W. (2018) Atmospheric seasonality as an exoplanet biosignature. *Astrophys. J. Lett., 858(2)*, L14.

Pearce B. K., Tupper A. S., Pudritz R. E., and Higgs P. G. (2018) Constraining the time interval for the origin of life on Earth. *Astrobiology, 18(3)*, 343–364.

Phillips C. B. and Pappalardo R. T. (2014) Europa Clipper mission concept: Exploring Jupiter's ocean moon. *Eos Trans. AGU, 95(20)*, 165–167.

Pilcher C. B. (2003) Biosignatures of early Earths. *Astrobiology, 3(3)*, 471–486.

Quinn R. C., Martucci H. F. H., Miller S. R., et al. (2013) Perchlorate radiolysis on Mars and the origin of martian soil reactivity. *Astrobiology, 13*, 515–520.

Reh K., Spilker L., Lunine J. I., Waite J. H., Cable M. L., Postberg F., and Clark K. (2016) Enceladus Life Finder: The search for life in a habitable moon. *2016 IEEE Aerospace Conference*, 1–8. Institute of Electrical and Electronics Engineers, Piscataway, New Jersey.

Reinhard C. T., Olson S. L., Schwieterman E. W., and Lyons T. W. (2017) False negatives for remote life detection on ocean-bearing planets: Lessons from the early Earth. *Astrobiology, 17(4)*, 287–297.

Roberge A., Fischer D., Peterson B., et al. (2019) The Large UV/Optical/Infrared Surveyor (LUVOIR): Telling the story of life in the universe. Astro2020: Decadal Survey on Astronomy and Astrophysics, APC white papers. *Bull. Am. Astron. Soc., 51(7)*, 199.

Robinson T. D., Meadows V. S., and Crisp D. (2010) Detecting oceans on extrasolar planets using the glint effect. *Astrophys. J. Lett., 721(1)*, L67.

Robinson T. D., Meadows V. S., Crisp D., Deming D., A'Hearn M. F., Charbonneau D., Livengood T. A., et al. (2011) Earth as an extrasolar planet: Earth model validation using EPOXI Earth observations. *Astrobiology, 11(5)*, 393–408.

Robinson T. D., Ennico K., Meadows V. S., Sparks W., Bussey D. B. J., Schwieterman E. W., and Breiner J. (2014) Detection of ocean glint and ozone absorption using LCROSS Earth observations. *Astrophys. J., 787(2)*, 171.

Rogers L. A. (2015) Most 1.6 Earth-radius planets are not rocky. *Astrophys. J., 801(1)*, 41.

Schwieterman E. W., Kiang N. Y., Parenteau M. N., Harman C. E., DasSarma S., Fisher T. M., Arney G. N., et al. (2018) Exoplanet biosignatures: A review of remotely detectable signs of life. *Astrobiology, 18(6)*, 663–708.

Seager S., Bains W., and Hu R. (2013) Biosignature gases in H$_2$-dominated atmospheres on rocky exoplanets. *Astrophys. J., 777(2)*, 95.

Seager S., Bains W., and Petkowski J. J. (2016) Toward a list of molecules as potential biosignature gases for the search for life on exoplanets and applications to terrestrial biochemistry. *Astrobiology, 16(6)*, 465–485.

Sousa-Silva C., Seager S., Ranjan S., Petkowski J. J., Zhan Z., Hu R., and Bains W. (2019a) Phosphine as a biosignature gas in exoplanet atmospheres. *Astrobiology, 20(2)*, 235–268, DOI: 10.1089/ast.2018.1954.

Sousa-Silva C., Petkowski J. J., and Seager S. (2019b) Molecular simulations for the spectroscopic detection of atmospheric gases. *Phys. Chem. Chem. Phys.*, *21(35)*, 18970–18987.

Stüeken E. E. (2016) A test of the nitrogen-limitation hypothesis for retarded eukaryote radiation: Nitrogen isotopes across a Mesoproterozoic basinal profile. *Geochim. Cosmochim. Acta, 120*, 121–139.

Tashiro T., Ishida A., Hori M., Igisu M., Koike M., Méjean P., Takahata N., Sano Y., and Komiya T. (2017) Early trace of life from 3.95 Ga sedimentary rocks in Labrador, Canada. *Nature, 549(7673),* 516.

Tian F. (2015) History of water loss and atmospheric O_2 buildup on rocky exoplanets near M dwarfs. *Earth Planet. Sci. Lett., 432,* 126–132.

Tian F., France K., Linsky J. L., Mauas P. J. D., and Vieytes M. C. (2014) High stellar FUV/NUV ratio and oxygen contents in the atmospheres of potentially habitable planets. *Earth Planet. Sci. Lett.,* *385*, 22–27, DOI: 10.1016/j.epsl.2013.10.024.

Walker S. I., Bains W., Cronin L., DasSarma S., Danielache S., Domagal-Goldman S., Kacar B., et al. (2018) Exoplanet biosignatures: Future directions. *Astrobiology, 18(6),* 779–824.

Wordsworth R. and Pierrehumbert R. (2014) Abiotic oxygen-dominated atmospheres on terrestrial habitable zone planets. *Astrophys. J. Lett., 785(2),* L20.

Wunderlich F., Godolt M., Grenfell J. L., Städt S., Smith A. M., Gebauer S., Schreier F., Hedelt P., and Rauer H. (2019) Detectability of atmospheric features of Earth-like planets in the habitable zone around M dwarfs. *Astron. Astrophys., 624,* A49.

Index

A

Abiotic nutrient cycles   79
Acetogenesis   388
Acetogens   75, 83
Adenosine triphosphate   42, 55, 93, 103, 233, 492, 494
Aerobic respiration   94
Akatsuki mission   357, 359
Albedo
    apparent   399
    Bond   433ff
Algae   127, 132, 135
    chlorarachniophytes   110
    green   110
    red   110
Alkalithermophilic bacteria   231
Alkaphilic organisms   352
Allochthonous burial   140
Aluminum, radioactive isotope   7, 8, 16, 16, 296
Amide bonds   42
Amino acids   72, 122, 125, 126, 129, 131, 140, 141, 220, 236, 480ff, 500
    in meteorites   236
    in space   142
    on Enceladus   236ff
Amino acids
    synthesis of   73, 77, 78, 81, 85
Ammonia
    as a greenhouse gas   387
Andromeda Strain   268
Antarctic Dry Valleys   480
Antarctica   122, 177, 203
    as an analog for Europa   204
Anthropic principle   4
Anthropogenic greenhouse warming   438
Apex basalt   134
Apex cherts   100
Apollo program   162, 278, 279
    lunar quarantine activity   273
Aqua satellite   400
Archaea   74, 75, 101, 121, 122, 128, 237
    Asgard superphylum   75
Archaeoglobules   101
Archaeplastida   110
Archean Earth   71, 74, 94, 478
    atmosphere   97ff, 384ff, 434ff

atmospheric pressure   386
    climate   381, 388, 434ff
    crust   107
    electrochemical energy   83
    environment of   94ff
    hydrothermal processes   100
    hydrothermal vents   77, 99, 100
    impacts   23
    mantle properties   423, 425
    microfossil record   134, 135
    ocean chemistry   77
    plume volcanism   80
    stromatolites   133
    sulfur   99
    tectonism   95, 106
    temperature of   97
    volcanism   94ff, 106, 111
    xenon   28
Aromatic molecules   41
Asteroid belt   300, 303, 331, 344, 463
    challenges in formation models   290, 304
    formation of   304ff
    low-mass model   339, 342
Asteroids
    active   345
    as a source of Earth water   304
    carbonaceous chondrites
        as a source of Earth water   29
    C-type   300, 304, 306, 307, 331, 338
    delivery of organics   19
    planetary protection of   274
    S-type   300, 304, 306, 307, 331, 338
    Themis family   345
Asteroids, named
    90 Antiope   345
    Bennu   314
    Ceres   345
    433 Eros
        planetary protection of   273
    25143 Itokawa   277
    Ryugu   314
    24 Themis   345
Astrobiology
    characteristics of   506
    history of   505ff
    major themes   506
Astrometry   451, 458
Atacama Desert   63, 177ff
Atacama Large Millimeter Array (ALMA)   304, 333, 340

Page numbers refer to specific pages on which an index term or concept is discussed. "ff" indicates that the term is also discussed on the following pages.

N

Y

Z

Color Section

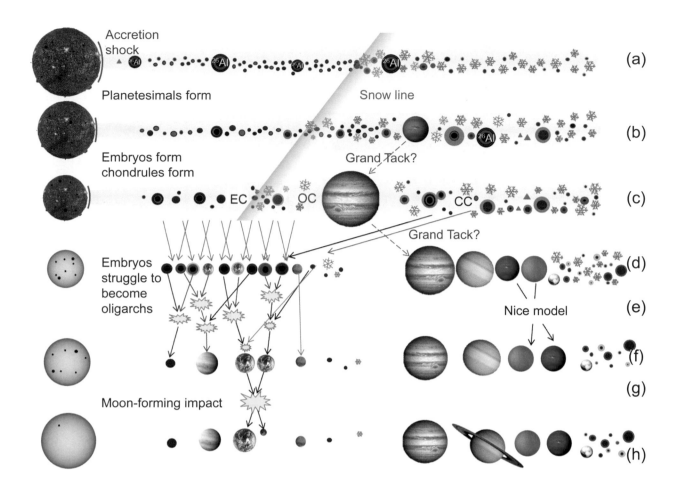

Plate 1. A pictorial synopsis of standard solar system formation. **(a)** 0–1 m.y.: 100-km-sized NC planetesimals quickly assemble and melt from heating by ^{26}Al. **(b)** 0–2 m.y.: Snowline planetesimals/boulders/pebbles assemble into a proto-Jupiter the size of Neptune, which balloons into something more Jupiter-like with the gravitational capture of nebular gas. CC planetesimals assemble and melt from heating by ^{26}Al. Jupiter separates NC (inside Jupiter's orbit) from CC planetesimals (outside); the two isotopic reservoirs remain separate while the nebula endures. **(c)** 0–4 m.y.: Planetesimals and embryos merge to make proto-planets. The Grand Tack is optional. The snowline moves inward as the nebula thins and the Sun fades. **(d),(e)** 4–60 m.y.: Embryos and proto-planets mix and merge to create planets. Mars survives as a fossil protoplanet or embryo by practicing nonintervention. **(e),(f)** Hypothesized Nice model rearrangement (timing uncertain). **(g)** 60–150 m.y.: The Moon and Earth form as products of the collision of two planets. **(h)** Later. Saturn acquires rings. Throughout, the size of the Sun suggests how bright it is (*D'Antona and Mazzitelli*, 1994), and the indications of activity (spots, prominences) on it suggest its EUV luminosity.

Accompanies chapter by Zahnle et al. (pp. 3–36).

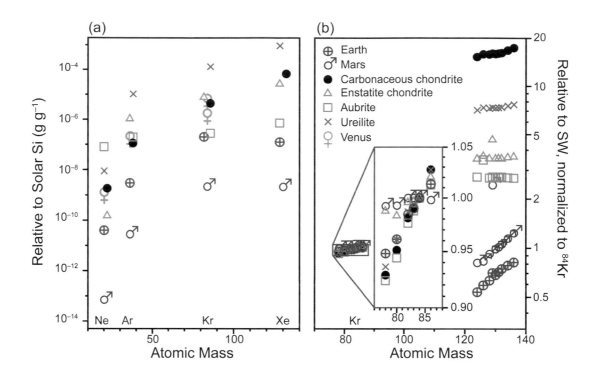

Plate 2. Noble gases. **(a)** Elemental abundances normalized by mass with respect to solar abundances. The planetary pattern — monotonically increasing relative abundances of the heavier elements — is evident in the meteorites and the planets, although offset by orders of magnitude. Note that Venus' bulk Ne and Ar abundances are similar to those in carbonaceous chondrites and enstatite chondrites. Both reported measurements of Kr on Venus are shown. Earth is depleted in all noble gases compared to Venus and the chondrites, and Mars is extremely depleted. **(b)** Comparison of isotopic structures of Xe and Kr; Kr has been magnified to make the differences between chondritic Kr and solar Kr visible. Xenon in Earth and Mars is depleted and very strongly mass fractionated, observations that together imply that Xe has escaped from both planets. No data from Venus are plotted. The ureilite is Havero, the aubrite is Pesyanoe, the enstatite chondrite is South Oman.

Accompanies chapter by Zahnle et al. (pp. 3–36).

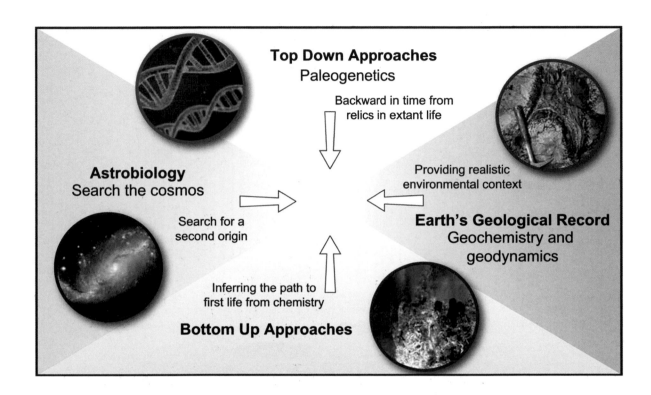

Plate 3. Approaches to understanding the origin of life can be broadly categorized into "top-down" and "bottom-up" approaches, with top-down approaches focusing on bioinformatics methods that can reveal clues about the nature of our common ancestors by examining the genomes of extant organisms, whereas bottom-up approaches focus on principles of synthetic and experimental chemistry to recreate the most likely steps leading to early life. Earth's geological record, by providing information about the Hadean Earth, contextualizes the results from both top-down and bottom-up approaches in a realistic environment and can validate results from both methods. Astrobiology could yield further information by identifying a second origin of life.

Accompanies chapter by Baross et al. (pp. 71–92).

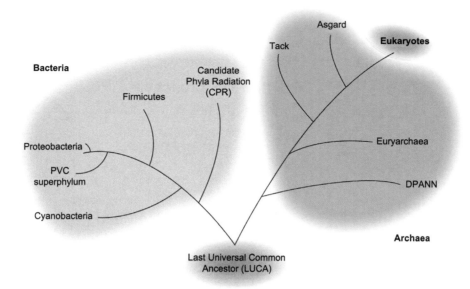

Plate 4. Schematic representation of our current understanding of the tree of life. Discovery of novel lineages through genomes assembled from environmental metagenomes has revealed archaeal superphyla that may provide the closest link yet discovered between the archaeal domain and the eukaryotic ancestors. While debate remains ongoing, if true these results would indicate that the tree of life is a two-domain tree of life rather than a three-domain tree of life, with the eukaryotes falling within the archaeal clade. The last universal common ancestor (LUCA) is at the root of the tree, representing the ancestor of all extant cellular life.

Accompanies chapter by Baross et al. (pp. 71–92).

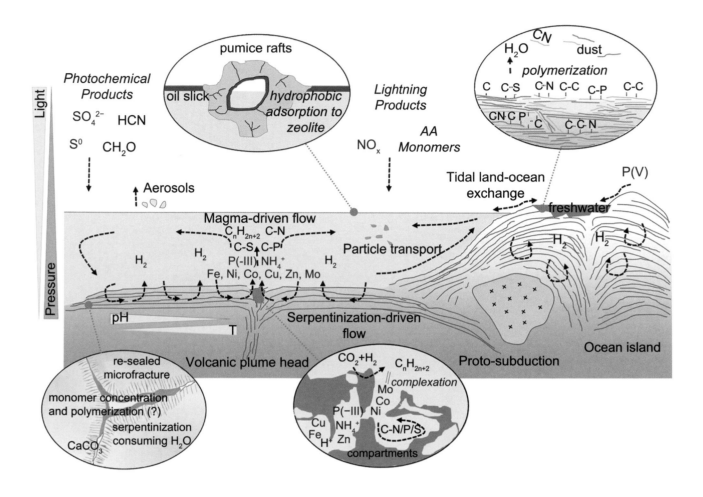

Plate 5. Schematic cross section of Hadean surface environments. Extensive plume volcanism as inferred from the early rock record (*O'Neill and Debaille,* 2014) would have led to widespread hydrothermal circulation in the crust (black arrows). Magma-driven hydrothermal vents may have formed proximal to volcanic fissures whereas distal circulation may have been driven by serpentinization of fresh (ultra-)mafic lava flows. Also, land masses were likely more mafic in composition and supporting H_2 production through serpentinization (*Smit and Mezger,* 2017). Major localities for the production of simple organic molecules were probably lightning and photochemistry in the atmosphere (*Tian et al.,* 2011; *Miller and Urey,* 1959) and Fischer-Tropsch-type reactions in high-temperature hydrothermal vents (*McCollom and Seewald,* 2007). The latter would also have released reactive forms of nitrogen, phosphorus, and transition metals. Polymerization into larger more complex molecules likely occurred in lower-energy environments such as evaporitic lakes and perhaps in dehydrated fractures in serpentinized crust. It is unlikely that monomers and polymers were produced in the same environmental setting, but the respective settings were likely linked by numerous transport processes, including adsorption onto suspended minerals with diverse surface properties.

Accompanies chapter by Baross et al. (pp. 71–92).

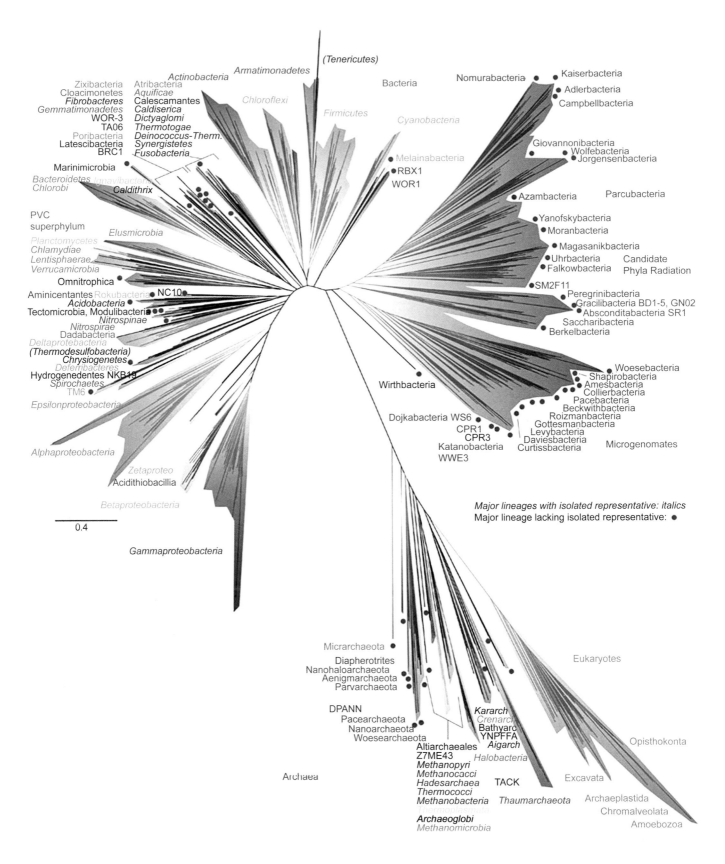

Plate 6. Using new genomic data from over 1000 uncultivated and little-known organisms, together with published sequences, *Hug et al.* (2016) recently published this expanded version of the tree of life.

Accompanies chapter by Johnson et al. (pp. 121–150).

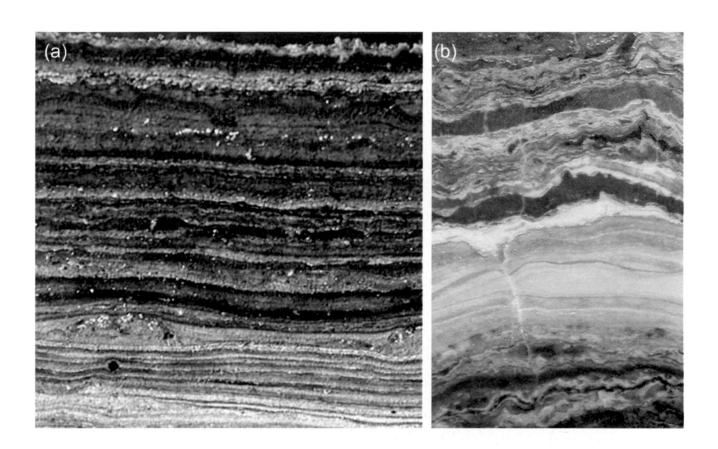

Plate 7. **(a)** Vertical (3 cm) section of a hypersaline mat from Guerrero Negro, Mexico. The crenulated fabric at the surface is lost and the mat layers become thinner with depth due to the degradation of the organic matter. **(b)** Vertical (3 cm) stromatolite section from an evaporite deposit, 1.35-Ga Sibley Formation, Ontario, Canada. The crenulated mat fabric was retained at depth, indicating that degradation rates were attenuated likely because mineralization had rapidly entombed the organic matter of the mat.

Accompanies chapter by Johnson et al. (pp. 121–150).

Plate 8. Vertical cross-section of a omal stromatolite (16 cm wide), Morrison Formation, Triassic, New Mexico, USA, as an example of morphologies.

Accompanies chapter by Johnson et al. (pp. 121–150).

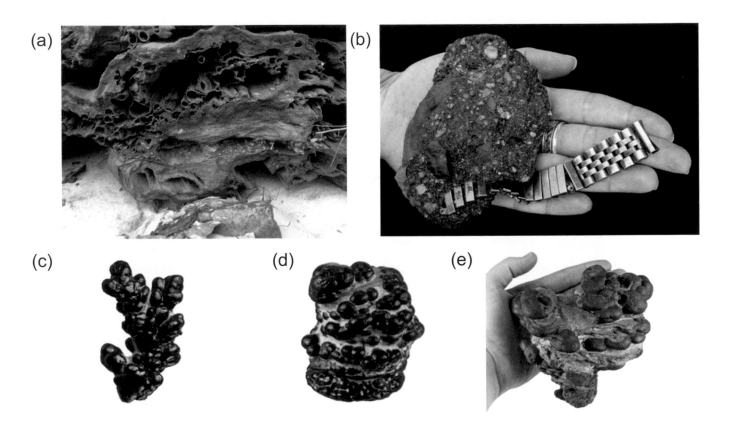

Plate 9. Iron and manganese biominerals. The distinctive morphologies of many oxide-hydroxide iron and manganese mineral deposits reflect their nanocrystalline constituents, which are rapidly precipitated at numerous nucleation sites through microbial redox reactions. Massive ferric iron oxide-hydroxide deposits [principally goethite, FeO(OH)] form with sufficient rapidity to preserve localized fluid flow patterns [**(a)** field of view ~1 m; photo by Robert Hazen] and to encase recently lost objects such as this watchband [**(b)** photo courtesy of Stephen Godfrey]. Nodules and rosettes of nanophase Mn^{3+}-Mn^{4+} minerals, including birnessite [$(Na,Ca,K)_{0.6}(Mn^{4+},Mn^{3+})_2O_4 \cdot 1.5H_2O$], hollandite [$Ba(Mn_6^{4+}Mn_2^{3+})O_{16}$], romanechite [$(Ba,H_2O)_2(Mn^{4+},Mn^{3+})_5O_{10}$], and todorokite [$(Na,Ca,K,Ba,Sr)_{1-x}(Mn,Mg,Al)_6O_{12} \cdot 3-4H_2O$], may form at the end of this process after most of the iron has been precipitated [**(c)** (height ~3 cm), **(d)** height ~3 cm), and **(e)** courtesy of Rob Lavinsky]. All specimens from Driftwood Beach, Calvert County, Maryland.

Accompanies chapter by Johnson et al. (pp. 121–150).

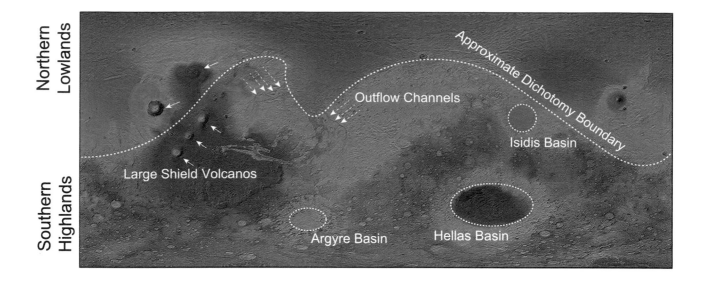

Plate 10. MOLA colorized topography of Mars with major landforms referenced. Dashed white arrows mark the large outflow channels, solid white arrows indicate large shield volcanos, dashed circles indicate large impact basins, and dashed white line indicates the approximate boundary of the hemispheric dichotomy (*Andrews-Hanna et al.,* 2008).

Accompanies chapter by Amador and Ehlmann (pp. 153–167).

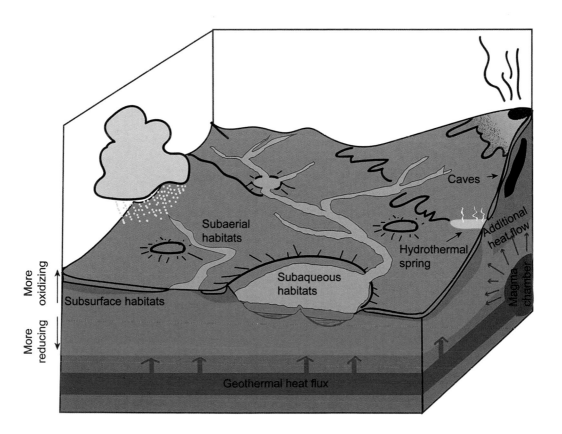

Plate 11. Schematic of early martian habitats.

Accompanies chapter by Amador and Ehlmann (pp. 153–167).

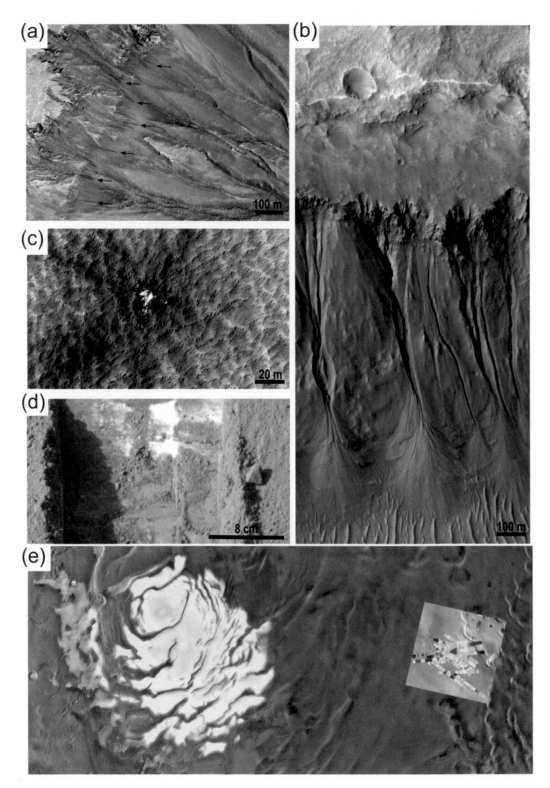

Plate 12. (a) Recurring slope lineae on the equator-facing wall of Palikir Crater. Credit: NASA/JPL-Caltech/Univ. of Arizona. (b) Gullies with clearly defined alcoves, channels, and terminal aprons (HiRISE image ESP_011727_1490). Credit: NASA/JPL/University of Arizona. (c) Ground ice (white) uncovered by NASA's Phoenix Lander. Credit: NASA/JPL-Caltech/University of Arizona/Texas A&M University. (d) Newly formed impact crater that has excavated down to buried ice (HiRISE image ESP_016954_2245). Credit: NASA/JPL/University of Arizona. (e) Mosaic image of the south polar region on Mars where a body of liquid water might exist 1.5 km below the surface (blue area in the center of the color tracks). Adapted from USGS Astrogeology Science Center, Arizona State University, INAF.

Accompanies chapter by Davila et al. (pp. 169–184).

Plate 13. **(a)** The upper reaches of the McMurdo Dry Valleys in Antarctica. This is the largest ice-free region in Antarctica. Here, primary productivity almost exclusively occurs a few millimeters under the surface of sandstone cliffs and boulders (inset). **(b)** Salt-encrusted playa in the hyperarid core of the Atacama Desert in northern Chile. Here, primary productivity almost exclusively occurs inside hygroscopic halite nodules (inset).

Accompanies chapter by Davila et al. (pp. 169–184).

Plate 14. Europa's icy surface is characterized by a young geological age (having few impact craters) and both tectonic and endogenic geologic processes. Generally, young materials are dark in color, and age to brighter white with time. Both implanted materials from particles swept up by Europa (leading hemisphere) and from bombardment by ions trapped in Jupiter's rapidly rotating magnetosphere (trailing, mid-latitudes) can cause discoloration in addition to the emplacement of salt-rich materials from below. Credit: NASA/JPL-Caltech/SETI Institute PIA 19048.

Accompanies chapter by Schmidt et al. (pp. 185–216).

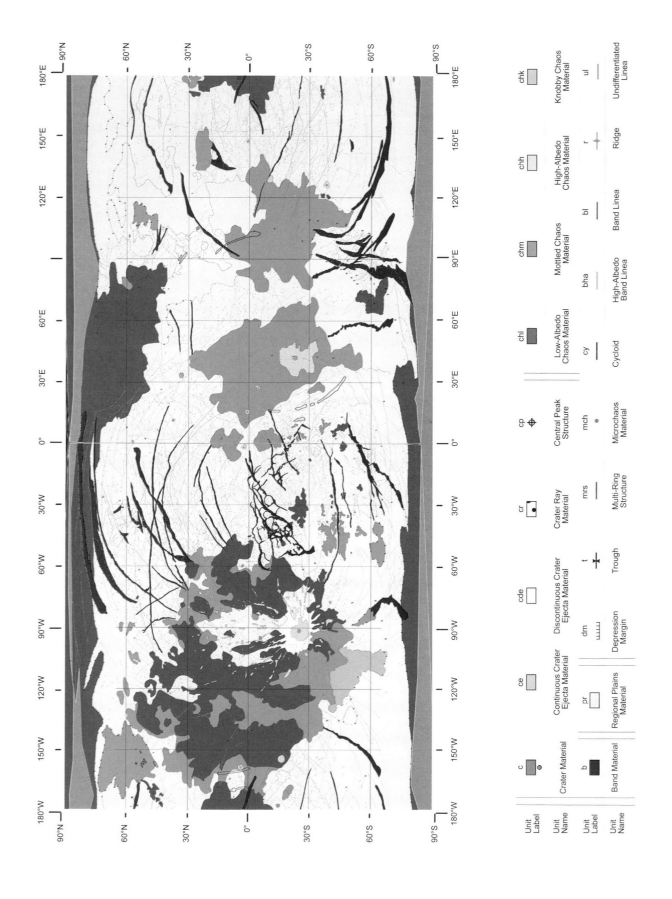

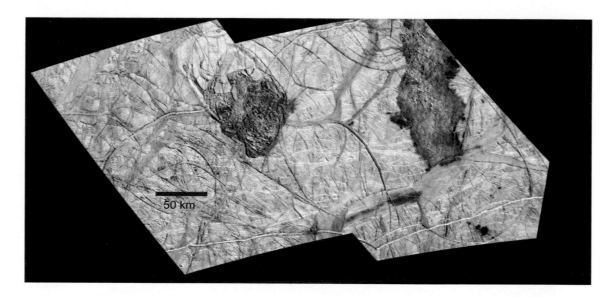

Plate 16. Thera Macula (left) and Thrace Macula (right) are two of Europa's largest chaos terrains. Multiple lines of evidence exist for liquid water and brines being involved in the formation of these features, including (1) their reddish color, which is associated with young age and high salt content; (2) the appearance of brines "soaking" from the features into surrounding terrain; and (3) their topography. Initially thought to be surface flows in Voyager spacecraft data, this image and locally-controlled topography reveal that Thera Macula is generally a large, circular depression with inward-facing scarps. The morphology suggests that the southern tip was the first area to rupture, and that the disruption of the ice may be proceeding northward and opening fractures creating new icebergs. Comparisons to Conamara Chaos fueled the formation of chaos evolution models (*Schmidt et al., 2011*). Credit: NASA/JPL/University of Arizona PIA02099.

Accompanies chapter by Schmidt et al. (pp. 185–216).

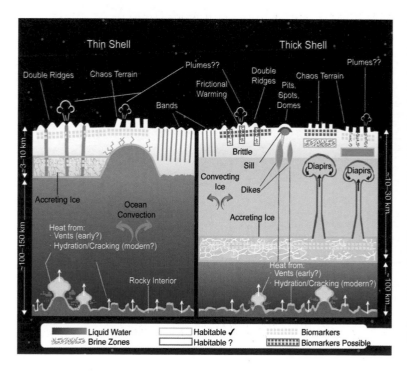

Plate 17. Schematic of habitability, biomarker production, and preservation on Europa as a function of geological processes for both a conductive (thin) ice shell and a convecting (thick) ice shell. All the colors and textures are chosen based on assessing hypothesized geological and physical processes for the corresponding feature. Locations within the system that have conditions hypothesized to be within the range of habitable conditions in ice on Earth are denoted as "Habitable" whereas regions that could once have had habitable conditions or for which the conditions are less favorable are marked "Habitable?" Regions where biomarkers could be produced and/or survive for long periods of time are marked "Biomarkers," and areas where the lifetime might be shorter or conditions relatively less favorable for producing biomarkers are marked "Biomarkers Possible."

Accompanies chapter by Schmidt et al. (pp. 185–216).

Plate 15. (facing page) Global geologic map of Europa showing the distribution and juxtaposition of major terrains (*Leonard et al., 2018*). The map was built by compiling images from Galileo and Voyager missions, and used the highest-resolution data to clarify local-scale crosscutting relationships. This is a draft of the global map; the final version will be Leonard et al. (in preparation) Geologic Map of Europa, U.S. Geological Survey Scientific Investigations Map, scale 1:15,000,000. Credit: NASA/Erin Leonard, Alex Patthoff, and Dave Senske, building on work by Ron Greeley, Thomas Doggett, and Melissa Bunte.

Accompanies chapter by Schmidt et al. (pp. 185–216).

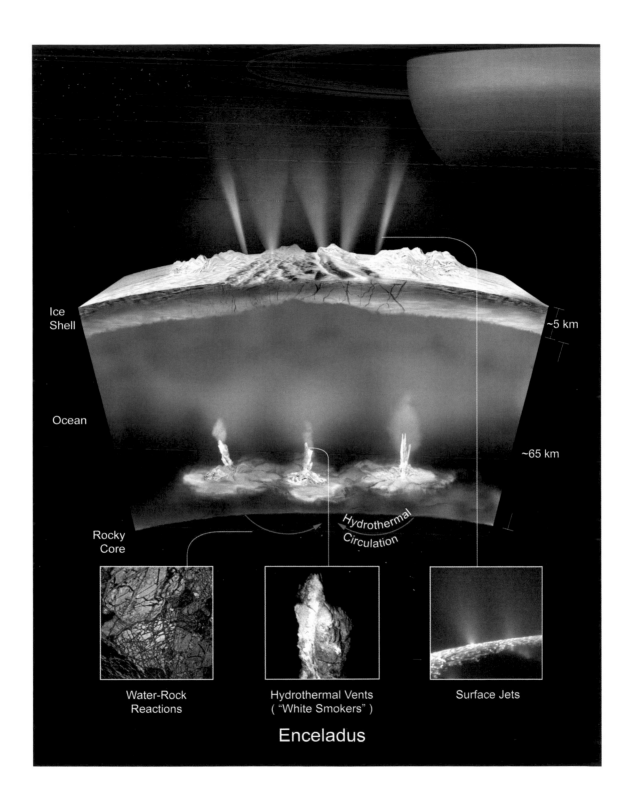

Ice
Shell

Ocean

Rocky
Core

~5 km

~65 km

Hydrothermal
Circulation

Water-Rock
Reactions

Hydrothermal Vents
("White Smokers")

Surface Jets

Enceladus

Plate 18. Enceladus' plume is sourced from a subsurface ocean in contact with the rocky core below at temperatures indicating hydrothermal activity. Credit: NASA/JPL-Caltech/SwRI.

Accompanies chapter by Cable et al. (pp. 217–246).

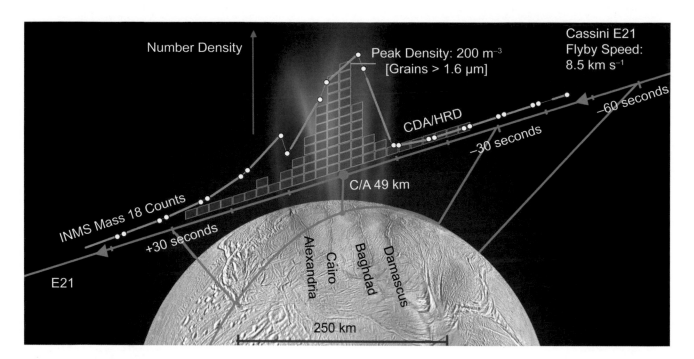

Plate 19. Plume ice grain and neutral gas profiles for the Cassini E21 fly-through of the Enceladus plume measured by the CDA High Rate Detector (HRD) (green boxes) and INMS (orange trace). The neutral gas profile shown is of water vapor (H_2O has a mass of 18 u). Closest approach (C/A) for this fly-through is indicated at 49 km altitude (this was the last and one of the closest plume crossings), with the time-delineated groundtrack outlined on Enceladus' surface in blue. In this image, the south pole of Enceladus is oriented up.

Accompanies chapter by Cable et al. (pp. 217–246).

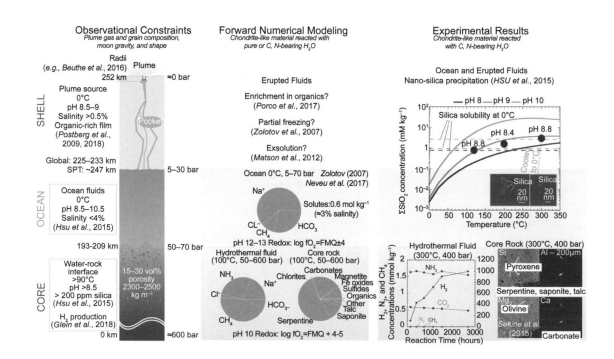

Plate 20. Physical and chemical connections between the plume, the ocean, and the core as evidenced by converging observational (section 3), modeling (section 4), and experimental results. SPT = south polar terrain.

Accompanies chapter by Cable et al. (pp. 217–246).

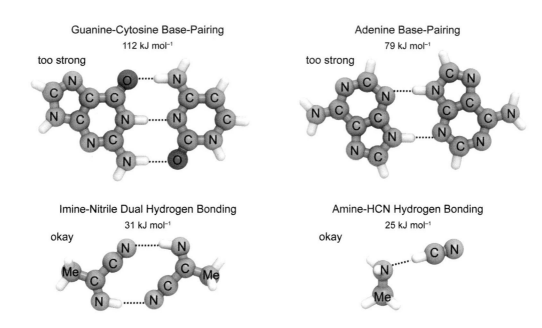

Guanine-Cytosine Base-Pairing
112 kJ mol⁻¹
too strong

Adenine Base-Pairing
79 kJ mol⁻¹
too strong

Imine-Nitrile Dual Hydrogen Bonding
31 kJ mol⁻¹
okay

Amine-HCN Hydrogen Bonding
25 kJ mol⁻¹
okay

Plate 21. Examples of structure-directing hydrogen-bonds with varying interaction strength. Stacked guanine-cytosine base pairs are sufficiently strong to maintain the DNA double helix on Earth, but much too strong to allow for dynamic chemistry on Titan. On Titan, weaker C/N-H—N hydrogen bonds could be the primary linkages holding together biostructures. The interactions between adenine molecules, an HCN-pentamer that might form on Titan, exceeds the estimated 17–35 kJ mol⁻¹ range in which sufficiently rapid dynamics is possible. Interactions in a more favorable range is possible, for example, between amines, imines, nitriles, and HCN. Binding energies have been calculated at the DLPNO-CCSD(T)/ aug-cc-pVTZ//wB97X-D3/Def2-SVPD level of theory, and are corrected for zero-point energy vibrations using the ORCA code (*Neese,* 2012).

Accompanies chapter by Lunine et al. (pp. 247–266).

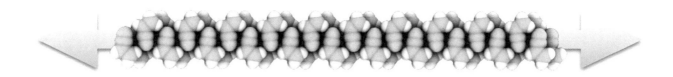

Plate 22. Proposed "azotosome" cell-membrane made from acrylonitrile (*Stevenson et al.,* 2015a).

Accompanies chapter by Lunine et al. (pp. 247–266).

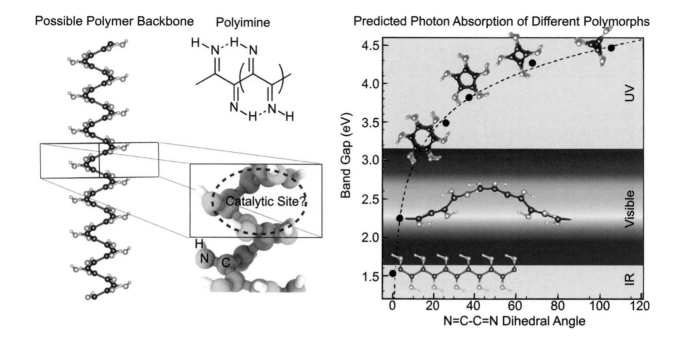

Possible Polymer Backbone

Polyimine

Predicted Photon Absorption of Different Polymorphs

Plate 23. Polyimine, one possible product of HCN polymerization on Titan, might utilize inter- and intramolecular =N-H···N hydrogen bonds to govern the formation of different partially ordered structures, some of which may synergize with photon absorption and act catalytically. The band gaps of possible polyimine polymorphs span a range across the entire visible spectrum and into the ultraviolet. From *Rahm et al.* (2016); used with permission.

Accompanies chapter by Lunine et al. (pp. 247–266).

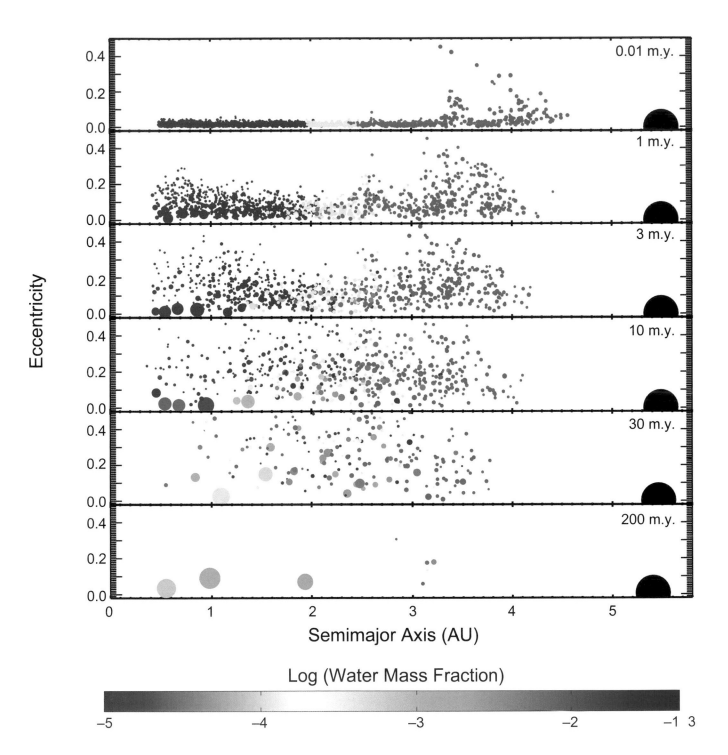

Plate 24. A simulation of the classical model of terrestrial planet formation (adapted from *Raymond et al.*, 2006b). The simulation started from 1886 self-gravitating planetary embryos, represented as dots with size proportional to its mass$^{1/3}$. Jupiter is fully formed (large black dot) on a near-circular orbit at the start of the simulation. The color of each embryo represents its water content (see color bar at the bottom), with red objects being dry and the darkest blue containing 5% water by mass (see *Raymond et al.*, 2004). This simulation produced quite good Earth and Venus analogs, a very poor Mars analog, and a plausible, albeit far too massive, asteroid belt. Note that time zero for this simulation corresponds to the dissipation of the gaseous disk, a few million years after CAIs. A movie of this simulation can be viewed at *https://youtu.be/ m7hNlg9Gxvo*.

Accompanies chapter by Raymond et al. (pp. 287–324).

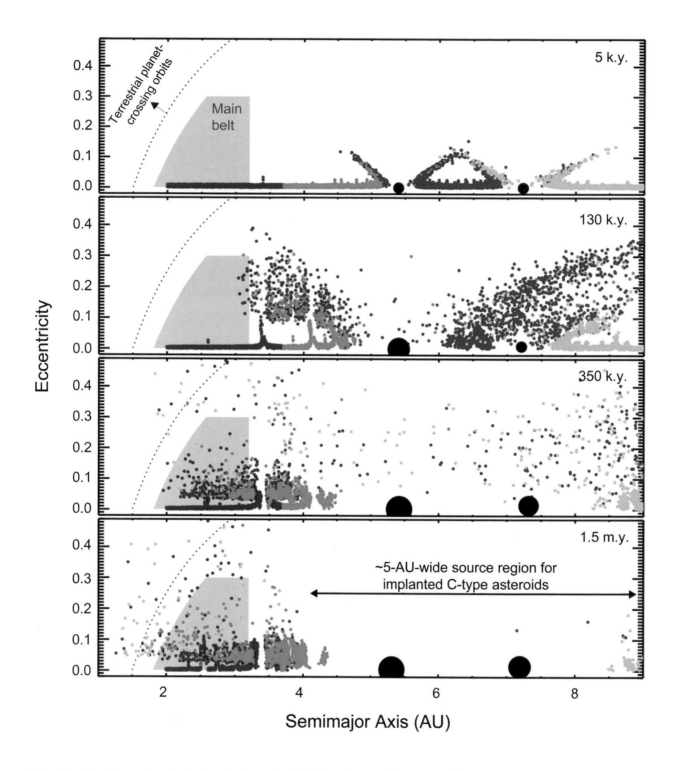

Plate 25. Injection of planetesimals into the asteroid belt and toward the terrestrial planet region as a consequence of giant planet formation (from *Raymond and Izidoro, 2017a*). This simulation starts with Jupiter and Saturn's 3 $M_\oplus$ cores embedded in a realistic gas disk [disk model from *Morbidelli and Crida* (2007), adapted to include gaps carved by the planets] including a population of 10^4 planetesimals assumed to be 100 km in diameter for the gas drag calculation. Jupiter grew (and carved a gap in the disk) from 100 to 200 k.y. and Saturn from 300 to 400 k.y., and the disk dissipated on a 200-k.y. exponential timescale (uniformly in radius). The colors of planetesimals represent their starting location. This case is the least dynamic possible scenario as it neglects migration of the gas giants and the formation and migration of the ice giants. Including those factors the source region of planetesimals implanted into the main belt extends out to 20 AU (*Raymond and Izidoro, 2017a*). An animation of this simulation can be viewed at *https://youtu.be/Ji5ZC7CP5to*.

Accompanies chapter by Raymond et al. (pp. 287–324).

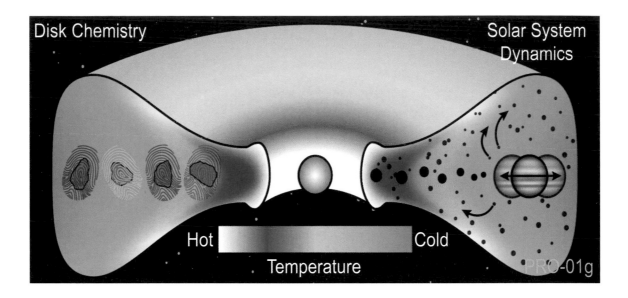

Plate 26. Protoplanetary disk chemical signatures are implanted on planetesimals as ices freeze near the midplane. This fingerprint is a sensitive measure of the planetesimal formation location. Subsequent planetesimal scattering as the giant planets grew will scramble this signature. The key to understanding where inner solar system volatiles originated is to measure the isotopes in a primitive reservoir of material that records the dynamical history during formation.

Accompanies chapter by Meech and Raymond (pp. 325–353).

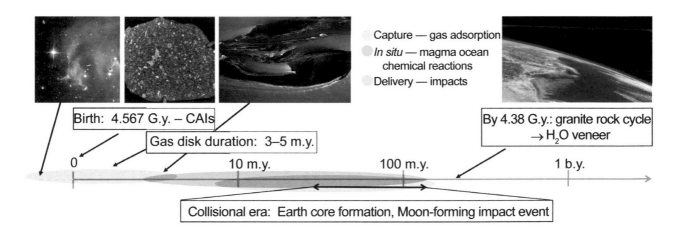

Plate 27. Earth received its water early; by 4.38 Ga Earth had oceans. Credit: ESA and the Hubble Heritage team (STScI/auRA)-ESA/Hubble Collaboration; AMNH-Creative commons license, 2.0 (*https://creativecommons.org/licenses/by/2.0/deed.en*); NASA.

Accompanies chapter by Meech and Raymond (pp. 325–353).

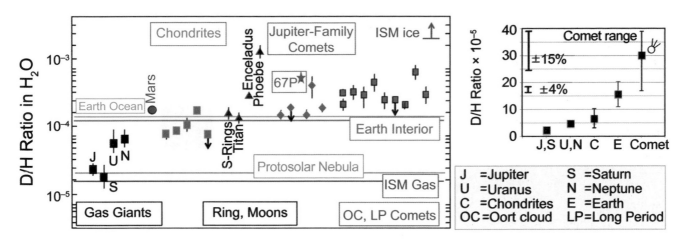

Plate 28. D/H measurements from a variety of solar system reservoirs. The values and references are shown in Table 1. While high-precision measurements of D/H can discriminate between source reservoirs, some reservoirs have similar D/H values. D/H alone cannot identify the source of inner solar system water.

Accompanies chapter by Meech and Raymond (pp. 325–353).

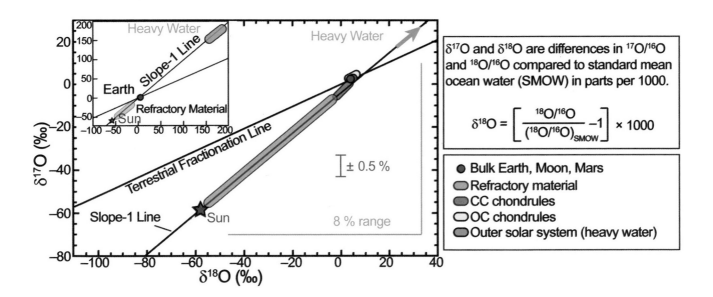

Plate 29. The range of oxygen isotopic variation in the solar system is small, so distinguishing reservoirs requires very-high-precision measurements.

Accompanies chapter by Meech and Raymond (pp. 325–353).

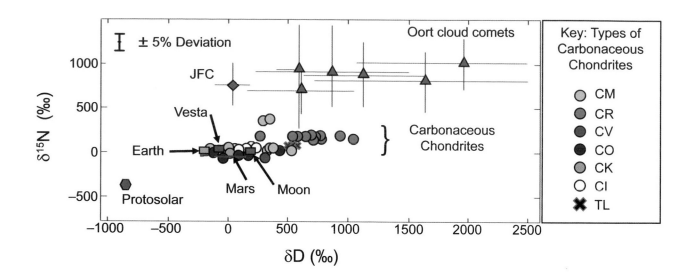

Plate 30. Combining nitrogen and hydrogen isotope ratios helps discriminate between reservoirs. The N-isotope difference between the group defined by Earth, Vesta, the Moon, Mars, chondrites, the Oort cloud comets, and the protosolar value is large. The difference between the CR chondrites and the other chondrites is 15%. The cometary values are measured in a variety of different molecules including water, HCN, and NH₃. It is challenging to infer the bulk isotopic compositions.

Accompanies chapter by Meech and Raymond (pp. 325–353).

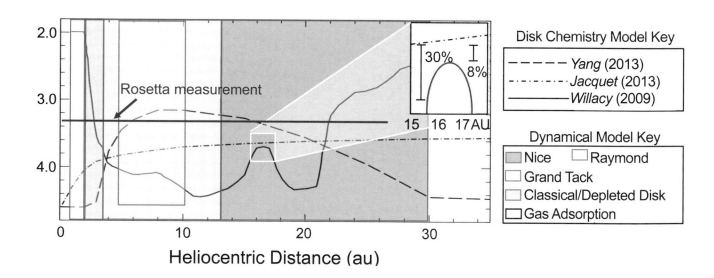

Plate 31. Measuring only D/H cannot uniquely determine an origin location. The Rosetta measurement from Comet 67P/ Churyumov-Gerasimenko is consistent with at least two disk chemical models, predicting formation at several heliocentric distances compatible with several dynamical models. Furthermore, disk chemical models can have similar D/H at different distances, and different dynamical models can scatter from overlapping regions.

Accompanies chapter by Meech and Raymond (pp. 325–353).

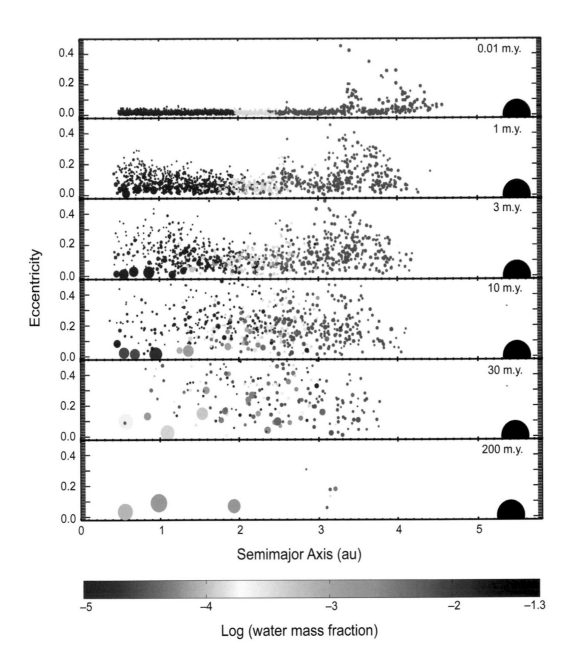

Plate 32. Snapshots from a simulation of the classical model of terrestrial planet formation. Jupiter was included from the start, on a low-eccentricity orbit at 5.5 au (large black circle). Almost 2000 planetary embryos are represented by their relative sizes (proportional to their masses$^{1/3}$) and their water contents, which were imposed at the start of the simulation (red = dry, black = 5% water by mass; see color bar). Three terrestrial planets formed, and each acquired material from beyond 2.5 au that delivered its water. Adapted from *Raymond et al.* (2006b).

Accompanies chapter by Meech and Raymond (pp. 325–353).

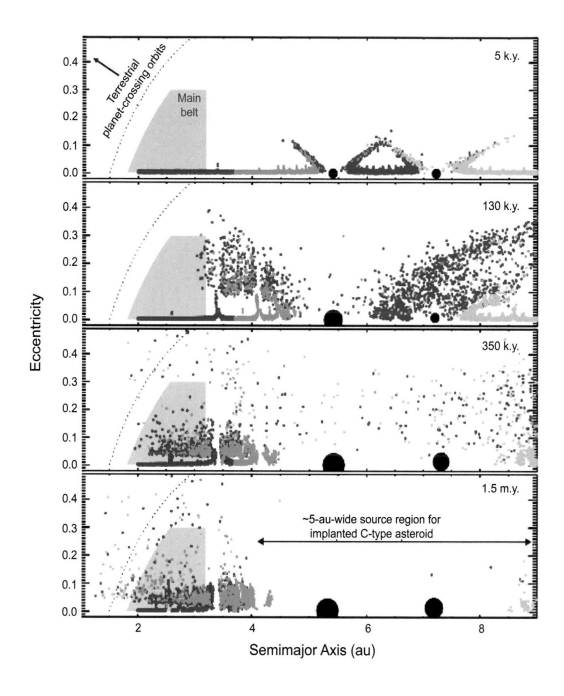

Plate 33. Dynamical injection of planetesimals into the inner solar system during Jupiter and Saturn's growth. This figure shows snapshots of a simulation in which 100-km planetesimals interact gravitationally with the growing gas giants and by gas drag with the disk. Jupiter's mass was increased from a core to its final mass from 100 to 200 k.y. and Saturn from 300 to 400 k.y. The colors of planetesimals serve to indicate their starting location. A large number of planetesimals were captured on stable orbits in the outer asteroid belt, providing a good match to the C-type asteroids. Many planetesimals were scattered onto high-eccentricity orbits that cross the growing terrestrial planets' and may have delivered water to Earth. The source region for implanted planetesimals was from 4 to 9 au in this example, but can extend out to 20 au when migration and the ice giants are considered. From *Raymond and Izidoro* (2017a).

Accompanies chapter by Meech and Raymond (pp. 325–353).

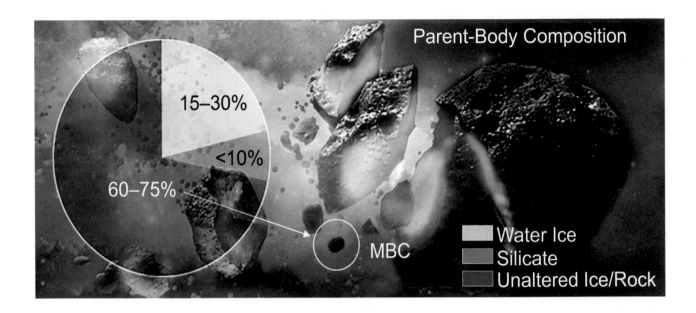

Plate 34. Thermal models show that MBC parent bodies preserve unaltered materials throughout 60–75% of the body for their history (*Castillo-Rogez and Schmidt,* 2010). MBCs are from this unaltered fraction (*Marsset et al.,* 2016).

Accompanies chapter by Meech and Raymond (pp. 325–353).

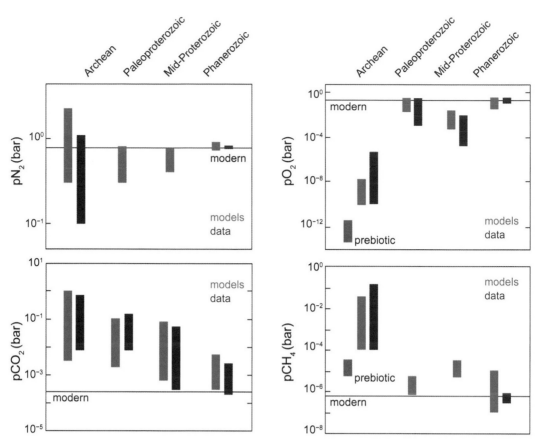

Plate 35. Summary of theoretical and empirical constraints on the abundances of N_2, O_2, CO_2, and CH_4 in Earth's atmosphere for the four major time periods discussed in the text (Archean, Paleoproterozoic, mid-Proterozoic, and Phanerozoic). Blue bars show reconstructions from models, while red bars show inferences based on empirical data. Also shown for the Archean are model-based estimates of prebiotic O_2 and CH_4 levels (gray bars). The ranges are meant to be inclusive, and some of the variability in a given time period should be considered to arise from time-dependent variability rather than uncertainty (e.g., *Olson et al.,* 2018b). Constraints are as described in Table 1.

Accompanies chapter by Robinson et al. (pp. 379–416).

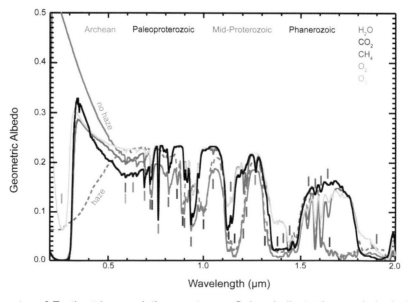

Plate 36. Simulated spectra of Earth at key evolutionary stages. Colors indicate time period: Archean (orange), Paleoproterozoic (dark gray), mid-Proterozoic (light gray), and Phanerozoic (black). Both hazy and haze-free Archean models are shown, and all models include fractional water cloud coverage. Key absorption features are indicated. Original sources for spectra are *Arney et al.* (2016) (Archean), *Robinson et al.* (2011) (Phanerozoic), and E. Schwieterman (personal communication) (Paleoproterozoic and mid-Proterozoic).

Accompanies chapter by Robinson et al. (pp. 379–416).

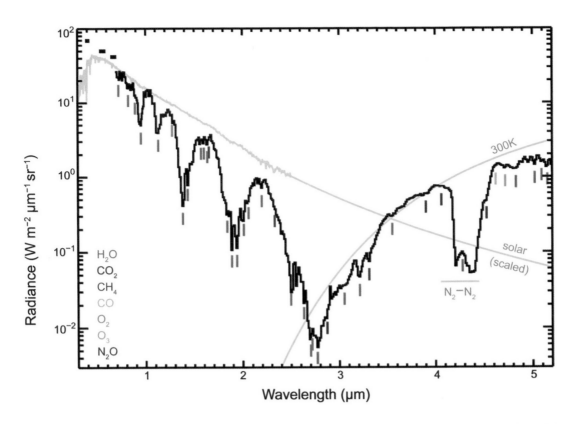

Plate 37. Spatially-resolved spectrum of Earth from the second Galileo flyby, acquired for a small region of the planet in the Indian Ocean and near the terminator at the time of observation. Visible SSI photometry (in the violet, green, and red filters) and NIMS spectroscopy are both shown, and key absorption bands are indicated. Refected light contributes signifcantly below about 3 μm, while thermal emission dominates at longer wavelengths. Light gray curves indicate a scaled solar spectrum and a 300 K blackbody.

Accompanies chapter by Robinson et al. (pp. 379–416).

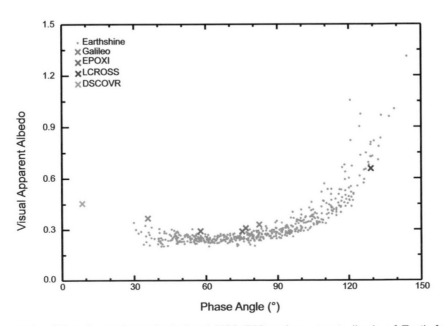

Plate 38. Measurements of the phase-dependent visual (400–700 nm) apparent albedo of Earth from Earthshine data spanning several years (from *Pallé et al., 2003*), and from several spacecraft missions. Apparent albedo values larger than unity indicate stronger directional scattering than can be produced by a sphere whose surface reflects light isotropically (i.e., a Lambert sphere). The DSCOVR datapoint is derived from the available four narrowband channels that span the visible range since an integrated 400–700-nm observation cannot be produced from DSCOVR data.

Accompanies chapter by Robinson et al. (pp. 379–416).

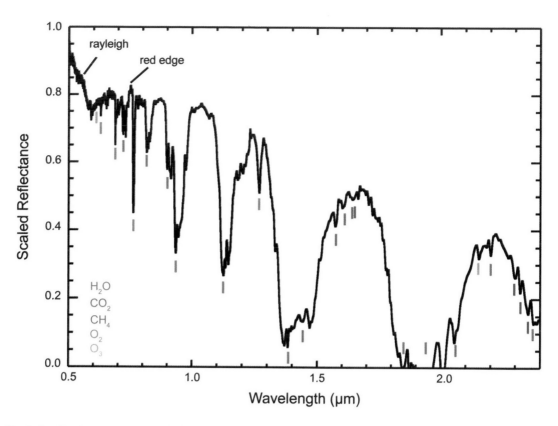

Plate 39. Scaled reflectance spectrum of Earth at visible and near-infrared wavelengths measured from Earthshine. Key absorption and refection features are indicated. Data courtesy of M. Turnbull from *Turnbull et al.* (2006).

Accompanies chapter by Robinson et al. (pp. 379–416).

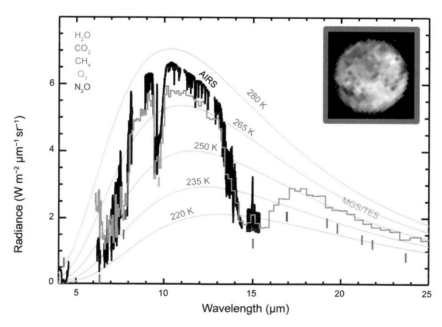

Plate 40. Disk-integrated thermal infrared spectra of Earth from the AIRS instrument (*Hearty et al.,* 2009) and from the Mars Global Surveyor Thermal Emission Spectrometer (MGS/TES) (*Christensen and Pearl,* 1997), where differences are due to the combined effects of seasons, climate, clouds, and viewing geometry. Key features are labeled and blackbody spectra at different emitting temperatures are shown. Inset is a broadband thermal infrared (6–10 μm) image of Earth from the LCROSS mission (*Robinson et al.,* 2014).

Accompanies chapter by Robinson et al. (pp. 379–416).

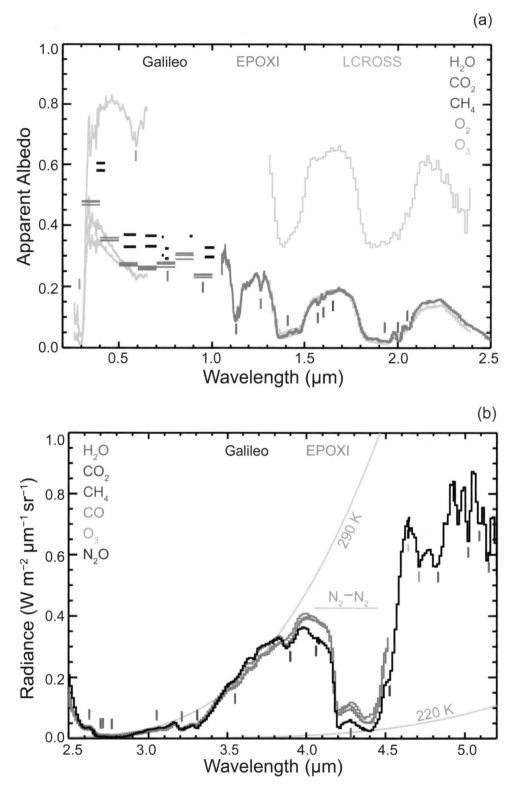

Plate 41. Summary of published observations of the distant Earth at ultraviolet, visible, and near-infrared wavelengths. The first figure presents spectra of Earth's apparent albedo from Galileo (*Sagan et al.*, 1993), EPOXI (*Livengood et al.*, 2011), and LCROSS (*Robinson et al.*, 2014). A crescent-phase observation from LCROSS is marked by large apparent albedo, which was driven primarily by forward scattering from a glint spot. The second figure presents near-infrared emission observations from Galileo and EPOXI, and blackbody spectra are provided. Key absorption features are indicated in both figures.

Accompanies chapter by Robinson et al. (pp. 379–416).

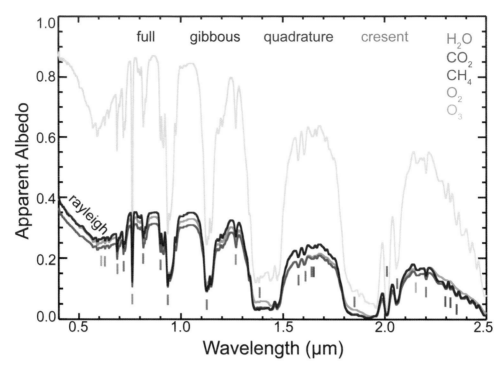

Plate 42. Simulations of Earth's phase- and wavelength-dependent apparent albedo (from *Robinson et al., 2010*). Models are averaged over a full rotation at each phase, and the angles of the given phases are 0° (full), 45° (gibbous), 90° (quadrature), and 135° (crescent). Large apparent albedos at crescent phase are primarily due to ocean glint and cloud forward scattering. The similarity in apparent albedo scales for the quadrature, gibbous, and full spectra indicate that Earth largely scatters like a Lambert sphere across these phase angles. A slight enhancement in apparent albedo at full phase is due to cloud back scattering.

Accompanies chapter by Robinson et al. (pp. 379–416).

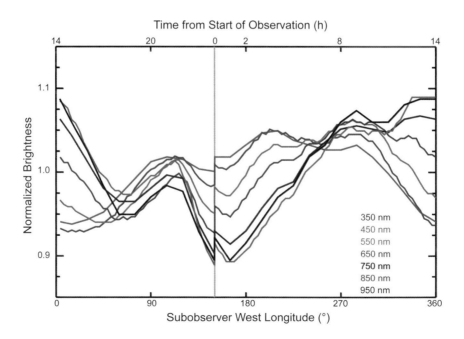

Plate 43. Full-rotation lightcurves for Earth from the March 2008 EPOXI dataset in all shortwave filters. Filter bandpasses are 100 nm wide, and filter center wavelengths are indicated. Overall brightness is driven largely by clouds at shorter wavelengths and by both clouds and continents at longer wavelengths (*Robinson et al., 2011*). Variability is primarily due to Earth's rotation; however, differences in brightness after a full rotation are due to longer-term evolution of cloud patterns.

Accompanies chapter by Robinson et al. (pp. 379–416).

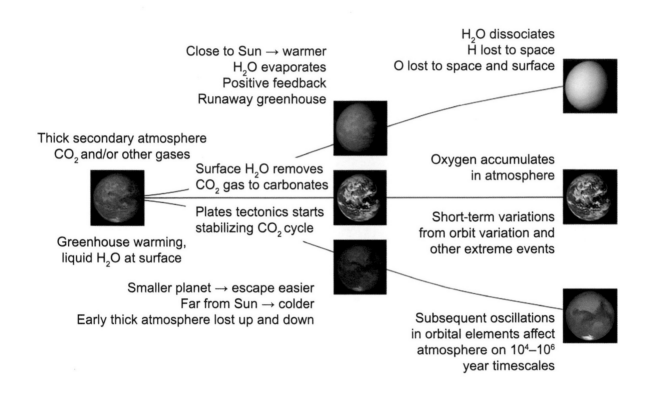

Plate 44. Venus, Earth, and Mars may have formed with similar inventories of chemical elements and could conceivably have had similar surface conditions early in solar system history. Differences in their sizes and/or distances from the Sun, as well as the emergence of life on Earth (see the chapter by Baross et al. in this volume), are the most likely causes of their subsequent divergent evolutions and explain their disparate present climates.

Accompanies chapter by Del Genio et al. (pp. 419–447).

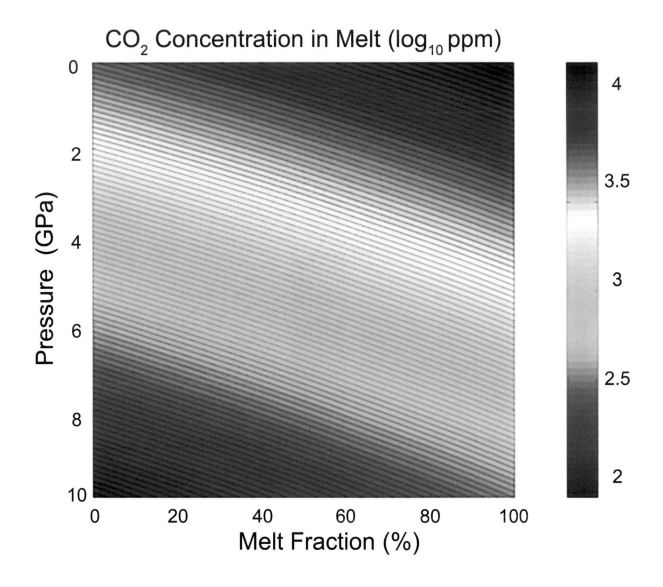

Plate 45. Amount of CO_2 that can be dissolved in graphite-saturated, tholeiitic melt for an oxygen fugacity at the iron-wustite buffer, based on *Holloway et al.* (1992) and *Grott et al.* (2011).

Accompanies chapter by Del Genio et al. (pp. 419–447).

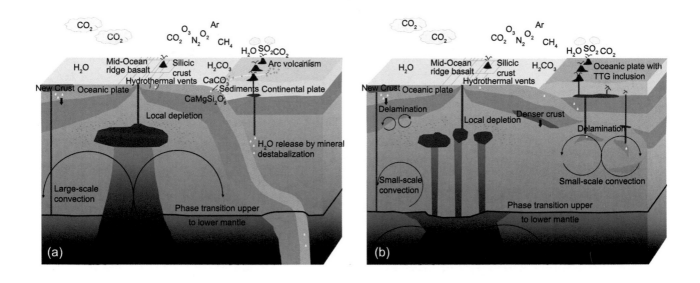

Plate 46. Sketch of **(a)** Earth's present-day volatile cycles, including outgassing of greenhouse and trace gases, recycling of carbonates and water, and deep water release due to destabilization of serpentinites, in comparison to **(b)** less-efficient material recycling mechanisms that have been suggested for early Earth in the absence of (present-day-like) plate tectonics and with a possible layering of convection between upper and lower mantle (as discussed by *Faccenda and DalZilio*, 2017). The depth profile of the sketch is logarithmic.

Accompanies chapter by Del Genio et al. (pp. 419–447).

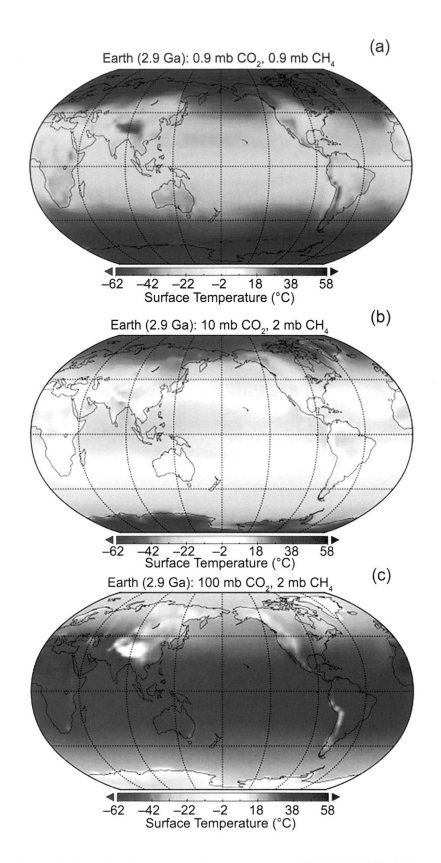

Plate 47. 2.9-Ga Archean Earth surface temperature simulated by the ROCKE-3D GCM (Way et al., 2017) assuming atmospheric compositions for cases **(a)** A, **(b)** B, and **(c)** C of *Charnay et al.* (2013) and the solar spectrum from *Claire et al.* (2012). Uncertainties due to the unknown land-ocean distribution are a few degrees. Simulations courtesy of Michael Way.

Accompanies chapter by Del Genio et al. (pp. 419–447).

Plate 48. Classification of exoplanets into different categories (*Kopparapu et al., 2018*). The boundaries of the boxes represent the regions where different chemical species are condensing in the atmosphere of that particular-sized planet at that stellar flux, according to equilibrium chemistry calculations. The radius division is from *Fulton et al.* (2017) for super-Earths and sub-Neptunes, and from *Chen and Kipping* (2017) for the upper limit on jovians..

Accompanies chapter by Kopparapu et al. (pp. 449–476).

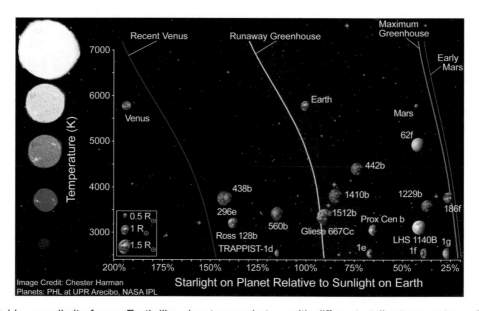

Plate 49. Habitable zone limits for an Earth-like planet around stars with different stellar temperatures (vertical axis) in terms of incident stellar flux (horizontal axis) on the planet, from a one-dimensional climate model. The "conservative HZ" is the region between the runaway greenhouse and maximum greenhouse limits. The "optimistic HZ" is the region between recent Venus and early Mars limits. See text for HZ definitions. Currently confirmed terrestrial exoplanets, along with the solar system ones, are also shown.

Accompanies chapter by Kopparapu et al. (pp. 449–476).

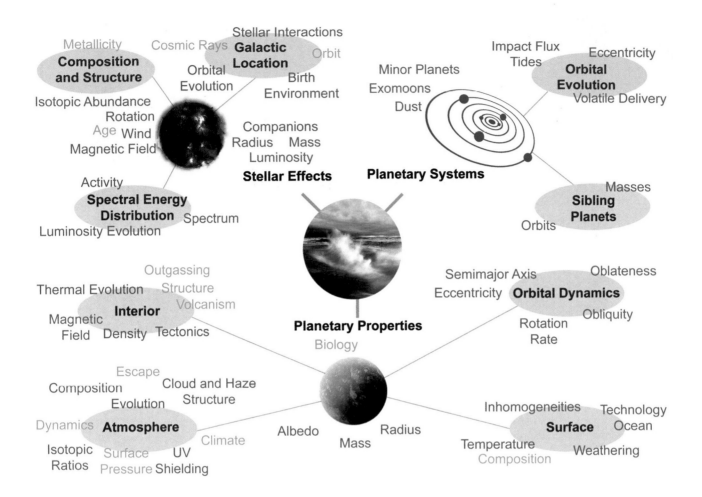

Plate 50. Currently understood planetary, stellar, and planetary system properties that may impact planetary habitability. The larger the number of these factors that can be determined for a given HZ candidate, the more robust our assessment of habitability will be. Colors of type denotes characteristics that could be observed directly with sufficiently powerful telescopes (blue); those that require modeling interpretation, possibly constrained by observations (green); and the properties or processes that are accessible primarily through theoretical modeling (orange). From *Meadows and Barnes* (2018).

Accompanies chapter by Kopparapu et al. (pp. 449–476).